WITHDRAWN

Nazca Plate:
Crustal Formation and Andean Convergence

The Geological Society of America, Inc.
Memoir 154

Nazca Plate:
Crustal Formation and Andean Convergence

A Volume Dedicated to George P. Woollard

Editors

LaVerne D. Kulm
School of Oceanography
Oregon State University
Corvallis, Oregon 97331

Jack Dymond
School of Oceanography
Oregon State University
Corvallis, Oregon 97331

E. Julius Dasch
Department of Geology
Oregon State University
Corvallis, Oregon 97331

Donald M. Hussong
Hawaii Institute of Geophysics
2525 Correa Road
Honolulu, Hawaii 96822

Associate Editor

Roxanne Roderick
School of Oceanography
Oregon State University
Corvallis, Oregon 97331

Copyright © 1981 by The Geological Society of America, Inc.

Copyright is not claimed on any material prepared by Government employees within the scope of their employment.

All material subject to this copyright and included in this volume may be photocopied for the noncommercial purpose of scientific or educational advancement.

Library of Congress Cataloging in Publication Data
Main entry under title:

Nazca plate.

    (Memoir / Geological Society of America ; 154)
    Includes bibliographies.
    1. Plate tectonics—Addresses, essays, lectures.
2. Geology—Pacific Ocean—Addresses, essays, lectures.
3. Geology—Andes—Addresses, essays, lectures.
4. Woollard, George Prior, 1908-   . 5. Geologists—United States—Biography. I. Woollard, George Prior, 1908-   II. Kulm, LaVerne D., 1936-   .
III. Series: Memoir (Geological Society of America) ; 154.
QE511.4.N39   551.1'36   81-7167
ISBN 0-8137-1154-1   AACR2

Contents

Preface ix

INTRODUCTION

History of the Nazca Plate Project 3
   George P. Woollard and LaVerne D. Kulm

DIVERGENT BOUNDARY

Tectonics of the Nazca-Pacific divergent plate boundary 27
   David K. Rea
Structure and evolution of the Easter plate 63
   D. W. Handschumacher, R. H. Pilger, Jr., J. A. Foreman, and J. F. Campbell
Petrogenesis and secondary alteration of upper layer 2 basalts of the Nazca plate 77
   K. F. Scheidegger and J. B. Corliss
Temporal variations in secondary minerals from Nazca plate basalts, diabases, and microgabbros 109
   Debra S. Stakes and K. F. Scheidegger

METALLIFEROUS SEDIMENTS

Geochemistry of Nazca plate surface sediments: An evaluation of hydrothermal, biogenic, detrital, and hydrogenous sources 133
   Jack Dymond
Metalliferous-sediment deposition in time and space: East Pacific Rise and Bauer Basin, northern Nazca plate 175
   G. Ross Heath and Jack Dymond
Lead isotopic composition of metalliferous sediments from the Nazca plate 199
   E. Julius Dasch
Sediment accumulation rate patterns on the northwest Nazca plate 211
   G. M. McMurtry, H. H. Veeh, and C. Moser
Uranium and thorium isotopic investigations in metalliferous sediments of the northwestern Nazca plate 251
   H. Herbert Veeh
Formation and growth of ferromanganese oxides on the Nazca plate 269
   Mitchell Lyle
Sediment and associated structure of the northern Nazca plate 295
   D. L. Erlandson, D. M. Hussong, and J. F. Campbell

Economic appraisal of Nazca plate metalliferous sediments 315
   Cyrus W. Field, Dennis G. Wetherell, and E. Julius Dasch

CONTINENTAL MARGIN AND TRENCH

Tectonics, structure, and sedimentary framework of the Peru-Chile Trench............. 323
   W. J. Schweller, L. D. Kulm, and R. A. Prince

Coastal structure of the continental margin, northwest Peru and southwest
Ecuador.. 351
   Glenn L. Shepherd and Ralph Moberly

Sedimentary basins of the Peru continental margin: Structure, stratigraphy,
and Cenozoic tectonics from 6°S to 16°S latitude................................ 393
   T. Thornburg and L. D. Kulm

Crustal structures of the Peru continental margin and adjacent Nazca plate,
9°S latitude.. 423
   Paul R. Jones III

Crustal structure and tectonics of the central Peru continental margin
and trench... 445
   L. D. Kulm, R. A. Prince, W. French, S. Johnson, and A. Masias

Late Cenozoic carbonates on the Peru continental margin: Lithostratigraphy,
biostratigraphy, and tectonic history 469
   LaVerne D. Kulm, Hans Schrader, Johanna M. Resig, Todd M. Thornburg,
   Antonio Masias, and Leonard Johnson

Vertical movement and tectonic erosion of the continental wall of the
Peru-Chile Trench near 11°30'S latitude 509
   Donald M. Hussong and Larry K. Wipperman

Shallow structures of the Peru Margin 12°S - 18°S 525
   S. H. Johnson and G. E. Ness

Clay mineralogy of the Peru continental margin and adjacent Nazca plate:
Implications for provenance, sea level changes, and continental accretion 545
   Victor J. Rosato and LaVerne D. Kulm

Structures of the Nazca Ridge and continental shelf and slope of
southern Peru.. 569
   Richard Couch and Robert M. Whitsett

Tectonics of the Nazca plate and the continental margin of western South
America, 18° to 23°S 587
   William T. Coulbourn

Biogeography of benthic foraminifera of the northern Nazca plate and
adjacent continental margin... 619
   Johanna M. Resig

Estimation of depth to magnetic source using maximum entropy power spectra,
with application to the Peru-Chile Trench 667
   Richard J. Blakely and Siamak Hassanzadeh

An active spreading center collides with a subduction zone: A geophysical
survey of the Chile Margin triple junction 683
   E. M. Herron, S. C. Cande, and B. R. Hall

Structures of the continental margin of Peru and Chile 703
   Richard Couch, Robert Whitsett, Bruce Huehn, and Luis Briceno-Guarupe

ANDEAN CONVERGENCE ZONE

Volcanic gaps and the consumption of aseismic ridges in South America 729
   Amos Nur and Zvi Ben-Avraham

Geological and geophysical variations along the western margin of Chile
near lat 33° to 36°S and their reaction to Nazca plate subduction 741
   Allen Lowrie and Richard Hey

Chile Margin near lat 38°S: Evidence for a genetic relationship between
continental and marine geologic features or a case of curious coincidences? 755
   E. M. Herron

Convergence and mineralization — Is there a relation? 761
   C. Wayne Burnham

Role of subducted continental material in the genesis of calc-alkaline
volcanics of the central Andes 769
   David E. James

Isotopic composition of Pb in Central Andean ore deposits 791
   George R. Tilton, Robert J. Pollak, Alan H. Clark, and Ronald C. R. Robertson

Epilogue: Geostill reconsidered 817
   Cyrus W. Field and E. Julius Dasch

PREFACE

This Memoir is dedicated to George P. Woollard who conceived the Nazca plate study and who served as director of the Nazca Plate Project. Dr. Woollard died in 1979 during the late stages of preparation of manuscripts for this publication. He wrote the bulk of the paper "History of the Nazca Plate Project" in this volume, but unfortunately his additional papers on the gravity of the Nazca plate were not sufficiently completed to be included herein.

As early as twelve years ago, George Woollard believed that an integrated investigation of a relatively small lithospheric plate could yield basic information for the developing theory of plate tectonics. He selected the Nazca plate as the primary area for such a study. During a period of several decades, Dr. Woollard had initiated a number of fundamental gravity studies on the South American continent. He formed an enduring relationship with his South American colleagues, on a personal as well as a scientific basis. This close working relationship lead to the endorsement of the Nazca Plate Project by his South American colleagues and produced an administrative and scientific structure that allowed for effective cooperative studies and dissemination of the resulting scientific information.

George Woollard led a productive and creative life. His formal education culminated with B.S. and M.S. degrees in engineering (1932, civil; 1934, geologic) from the Georgia Institute of Technology and M.A. and Ph.D. degrees (1935, mineral deposits; 1937, structural geology) from Princeton University. He held several postdoctoral and honorary fellowships as well as honorary degrees from Uppsala University, Sweden, and the University of Wisconsin. Dr. Woollard developed two centers for geophysical research: The Geophysical and Polar Research Center at the University of Wisconsin and the Hawaii Institute of Geophysics at the University of Hawaii. He served as President of the American Geophysical Union and received the William Bowie Medal of that society. He held positions on the Council of the Geological Society of America and on the Board of Directors of the American Geological Institute. He was a member of many standing and special committees at the university, national, and international levels.

His scientific contributions, fundamental and numerous, are primarily in the field of geophysics and the geological interpretation of geophysical data. Although his major works were in gravity measurements and their standardization and interpretation, he also contributed in many ways to the disciplines of magnetics, seismic refraction, and marine acoustics. Nearly forty students worked under his supervision; many now hold important positions in science, administration, and industry. An earlier scientific Memoir, *The Geophysics of the Pacific Ocean Basin and Its Margin* (American Geophysical Union Geophysical Monograph 19), published in 1976, was dedicated to George Woollard. In addition to the excellent papers contained therein, the reader is referred to that volume for further details on Dr. Woollard's life. One should not miss the monograph's epilogue which consists of affectionate doggerel by G. G. Shor and by Woollard himself. Finally, a comment should be included about the man himself: Everyone liked and respected him! One measure of his life is simply the fact that he and his wife, Eleanore, reared nine children, five of whom they adopted; supported eighteen foster children; and were devoted participants in children placement programs.

Sixty-eight papers based on information collected as part of the Nazca Plate Project have been published, not including those contained in this Memoir. The thirty-five papers herein include twenty-eight authored by participants in the project and seven by investigators independent of the project. A complete listing of these papers is given in "History of the Nazca Plate Project" in this volume.

Two sets of maps published by the Geological Society of America are supplemental to the materials in this Memoir. They were funded by IDOE/NSF through the Nazca Plate project and are:

Bathymetry of the Southeast Pacific. J. Mammerickx and S. M. Smith. 1978. GSA Map and Chart Series MC-26.

Part I: Bathymetry of the Peru-Chile Continental Margin and Trench. R. A. Prince, W. J. Schweller, W. T. Coulbourn, G. L. Shepherd, G. E. Ness, A. Masias. *Part II. Chemical Composition of Nazca Plate Surface Sediments.* Jack Dymond, John B. Corliss. GSA Map and Chart Series MC-34.

This Memoir is organized into five sections. The history of the Nazca Plate Project—project initiation, objectives, participants, cruise activities, and publications—is described in chapter one. Subsequent chapters are arranged in the natural order of the evolution of the oceanic crust at the spreading East Pacific Rise (DIVERGENT BOUNDARY); the sedimentation on the evolving crust as it is transported across the Nazca plate (METALLIFEROUS SEDIMENTS); the disruption and destruction of the subducting plate along the Peru-Chile Trench and its influence on the evolution of the Andean continental margin (CONTINENTAL MARGIN AND TRENCH); and the effect of subduction on the volcanism, mountain building, and ore deposits of the Andes (ANDEAN CONVERGENCE ZONE).

We thank our many South American colleagues who participated in the Nazca Plate Project in so many different ways. This project would not have been possible without their gracious assistance; see "History of the Nazca Plate Project" for special thanks to individuals.

The editors gratefully acknowledge the assistance of Mary Jo Armbrust in typing the majority of the tables appearing in the Memoir and for reading and correcting many of the galley proofs. We especially thank the Associate Editor, Roxanne Roderick, for her invaluable assistance: for typing several of the manuscripts and all of the reference material in GSA format, for coordinating the review of the papers, for communicating with the authors, and for reading the galley and page proofs.

We thank the more than 70 colleagues external to the Nazca Plate Project for their review of the papers in this Memoir. Their time and effort are greatly appreciated.

Funding for the research activities conducted under the auspices of the Nazca Plate Project was provided by the International Decade of Ocean Exploration (IDOE) of the U.S. National Science Foundation (NSF). Supplemental funds were also provided by IDOE/NSF for the publication of this Memoir.

LAVERNE D. KULM
JACK DYMOND
E. JULIUS DASCH
DONALD M. HUSSONG

INTRODUCTION

History of the Nazca Plate Project

GEORGE P. WOOLLARD*
LAVERNE D. KULM
School of Oceanography
Oregon State University
Corvallis, Oregon 97331

INTRODUCTION

The Nazca Plate Project was the outgrowth of an invited paper presented by the senior author at the symposium "Results of Upper Mantle Investigations with Emphasis on Latin America" held in conjunction with the 1970 Buenos Aires International Union of Geodesy and Geophysics (IUGG) Conference on Solid Earth Problems. The paper (unpublished) dealt with the evidence for active plate motion in the Pacific Ocean as exemplified by geological and geophysical relations both offshore and onshore in Mexico, Central America, and South America. Among other things, it pointed to the need for studying in some detail an oceanic plate from its locus of origin on an oceanic rise to its ultimate destruction by subduction beneath a continental margin with resulting offshore-onshore geological and geophysical consequences. Because the Nazca plate off South America (Fig. 1) appeared ideal for such a study, and because there was an ongoing land program for the geological and geophysical study of the Andes that would complement the investigation visualized, informal presentations for study of the Nazca plate were made in 1970 to both the Pan American Institute of Geography and History (PAIGH) and the National Science Foundation (NSF). The program proposed was endorsed by PAIGH under Resolution 27 adopted at their Mexico City meeting in 1970 and endorsed in principle by the NSF as a program that might well qualify under the newly organized program for the International Decade of Ocean Exploration (IDOE). However, it was pointed out by IDOE that any official proposal submitted should include metallogenesis on the sea floor as a pertinent part of the investigation so that the proposal would fully qualify for funding under IDOE program Category 3 (Seabed Assessment). This new addition to the crustal structure program originally envisioned for the Nazca plate and the large size of the area proposed for study dictated that the project be a cooperative investigation involving more than one institution.

Because the oceanographic group at Oregon State University (OSU) had been carrying out a program of study in the Gulf of Panama similar to that envisioned for the Nazca plate study, and because the OSU group included a number of geochemists with strong interests in metallogenesis on the sea floor, this group was approached as a working partner with the Hawaii Institute of Geophysics (HIG) in studying the Nazca plate. The only problem visualized by the OSU in joining with HIG on the proposed study of the Nazca plate was their ship schedule which only permitted participation in the field program on an alternate year basis. This was not regarded as a serious problem; in fact, it later

---
*Deceased, formerly of Hawaii Institute of Geophysics, 2525 Correa Road, Honolulu, Hawaii 96822.

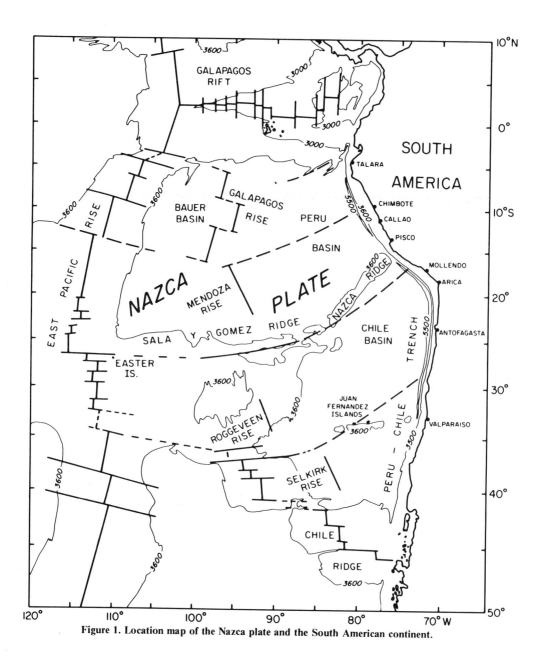

Figure 1. Location map of the Nazca plate and the South American continent.

proved to be a blessing in disguise since it allowed one group adequate time to digest the large mass of data obtained before being inundated with new material.

Accordingly, in 1971 a joint proposal to study the Nazca plate was prepared and submitted by the HIG and the School of Oceanography of OSU to the IDOE Office of the NSF. The proposal preamble is of interest in that it recognized at the outset the complementary strengths of the participating institutions and the roles that each would play in carrying out the study. As much of the success of the program—indeed the survival of the project beyond its first funding period—can be attributed to this recognition of the need for complementary strengths, the preamble is cited herein:

The following proposal represents a collaborative effort under the IDOE by the Hawaii Institute of Geophysics, University of Hawaii, and the School of Oceanography, Oregon State University. The Hawaii Institute of Geophysics has the scientific manpower and the experience to conduct large-scale geophysical-geological studies of the lithosphere. The Geological Oceanography group of Oregon State University, on the other hand, has developed its strengths in the study of smaller scale specific problems related particularly to sedimentological, geochemical, and geophysical processes on the continental margin and oceanic rise areas. The two groups thus complement each other in interests and capabilities with sufficient overlap to assure both institutional cooperation and collaboration and an effective overall program.

PROGRAM RATIONALE AND OBJECTIVES

The proposed program emphasized the following: (1) that despite the general acceptance of the concept of global tectonics, there was only limited knowledge regarding the complete tectonic cycle of oceanic rift-plate-trench systems; (2) that although there was increasing knowledge of the oceanic rift systems, there was a lack of knowledge of subsequent geophysical and geological events during transport of the newly formed lithospheric plate by crustal spreading, and a similar lack of knowledge concerning the processes and events occurring in the zone of convergence where an oceanic plate is subducted beneath a continental boundary; (3) that the convergence zone is not only of scientific interest but also has sociological and economic importance because it is the site of intense seismicity, recurring volcanism, intrusion, formation of ore deposits, crustal uplift, and mountain building; and (4) that to relate phenomena including metallogenesis, known to be associated with new crust formed at the oceanic rise rift zones, to those deposits associated with the convergence zone of subducted old-age oceanic crust, it would be necessary to carry out geophysical, gological, and geochemical studies not only at the rise crests and the convergence zone but also across the entire crustal plate if the processes related to crustal aging and transport were to be understood.

FIELD PROGRAM PROPOSED

Fourteen traverses were envisioned in the initial field program for the East Pacific Rise to the South American continental boundary between 1°N and 47°S latitudes; auxilliary short traverses were concentrated along the East Pacific Rise crest area of new crust formation and along the Peru-Chile Trench and continental margin of crustal subduction. During the first year (1972), we proposed to concentrate on the northern part of the Nazca plate area. Under the first phase, four parallel East-West trans-plate multidisciplinary traverses would be established between 10° and 13°S latitude. One traverse was planned for a two-ship operation using reversed profile seismic refraction measurements for determining a nearly continuous sectioned crustal and upper mantle structure across the Nazca plate. Concentration of these regional traverses over a restricted range of latitutde also would help establish a magnetic lineation pattern across the plate. Gravity, heat flow, and other observations, would extend our knowledge gained about the lithosphere from the two-ship traverse to the other traverses. A second phase of work proposed for the first year was the establishment of three additional multidisciplinary traverses between 4° and 6°S latitude, one of which would involve two-ship seismic refraction measurements in the northwest corner of the plate on and adjacent to the East Pacific Rise. A third phase was to be concentrated on the Peru-Chile trench and adjacent continental shelf in two areas: one lying between 4° and 6°S latitude, and the other between 10° and 17°S latitude. A fourth phase of the work proposed involved detailed multidisciplinary studies of two areas of about 0.5° by 0.5° size on the crest of the East Pacific Rise. One of these areas, located at about 5°S latitude, occurs where the rise is offset on a transform fault; the other area, at about 11°S latitude, was where metalliferous sediments were known to be abundant from previous studies. Generally, dredging, piston coring, heat flow, and sonobuoy ASPER measurements would be done on a routine basis and in areas of special interest, as indicated by the underway measurements defining bathymetry, sediment character and thickness, gravity, and magnetic field. Auxiliary work would involve occasional seismic velocity anisotrophy measurements.

The described field program, it was thought, would provide the data needed to satisfy the scientific objectives which were divided to form three subprograms as follows:

I. Regional crustal and mantle structure of the Nazca plate as a function of distance and crustal age from the crest of the East Pacific Rise

II. Tectonic, stratigraphic, and structural studies of the Peru-Chile Trench and adjacent South American continental margin area of crustal convergence and oceanic plate subduction

III. Structural, volcanic, and sedimentary studies on the crest of the East Pacific Rise with emphasis on the processes leading to metal enrichment

This program, in terms of the scientific objectives and field operations proposed, was approved by the NSF-IDOE office and funded on a 2-yr basis with 18 weeks of field work scheduled for 1972 to start in January after a 6-month period for both institutional groups to gear up for the program and to assemble and study existing data on the Nazca plate and adjacent areas.

MODE OF OPERATION

It was agreed that where two-ship operations were required for the crustal seismic refraction measurements each institutional group would operate independently with its own ships (HIG, R/V *Kana Keoki;* OSU R/V *Yaquina* and in 1977 R/V *Wecoma*) in jointly agreed upon areas and along traverses specified in advance so that the field program would have coherence at all times in meeting the objectives of the three subprograms. Arrangements were also made for an exchange of the data obtained and for an interchange of personnel between the institutions as well as for the inclusion of Latin American scientific colleagues in carrying out the field program and in analyzing the data. Another significant arrangement that proved extremely valuable was to have group meetings of all the scientists from both institutions periodically to review progress and to discuss results. Through these meetings, to be held alternately at OSU and HIG, everyone involved would be kept aware of results as well as problems. In brief, a mechanism was established to assure close collaboration on every aspect of the program—planning, field operation, data analysis, international cooperation, and publication of results.

1972 FIELD PROGRAM

1972 Cruise Tracks and Data

The 1972 cruise tracks for R/V *Kana Keoki* and R/V *Yaquina* are shown in Figure 2. The approximate number of nautical miles logged by each ship over the area of investigation on which continuous underway measurement of bathymetry, gravity, and magnetic and seismic reflection measurements of the bottom sediments were made were: R/V *Kana Keoki,* 12,800 nautical miles; R/V *Yaquina,* 10,200 nautical miles.

Geophysical Data

The locations of reversed profiles of seismic refraction crustal measurements and the two measurements for mantle seismic velocity anisotropy established jointly as two-ship operations in 1972 are identified by number in Figure 3 (profiles 1-30), which is a composite plot for all such measurements made during the life of the project. It is to be noted that, except on the East Pacific Rise, most of the seismic refraction profiles were oriented East-West in the direction of crustal spreading and potential maximum velocity anisotropy. In addition to the seismic velocity anisotropy measurements, several of the profiles were shot normal to the direction of crustal spreading in order to obtain additional information on the degree to which mantle velocity anisotropy is important across the Nazca plate. These deep crustal measurements were supplemented by single-ended sonobuoy

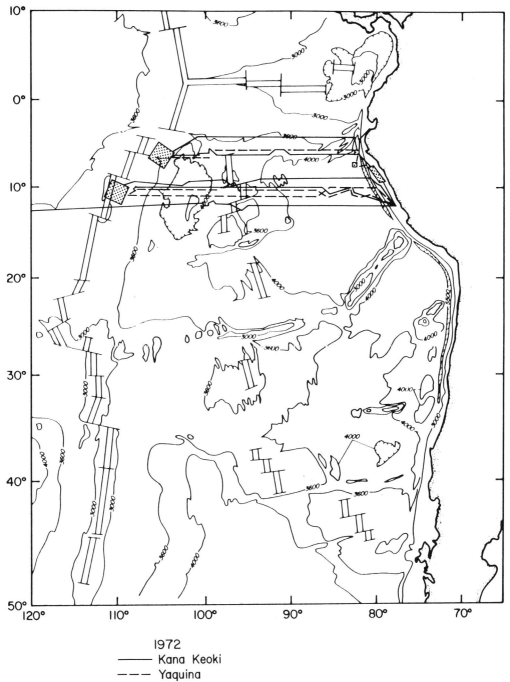

Figure 2. Tracklines for 1972 cruises on the Nazca plate and the Peru continental margin and trench. Dotted boxes are regions of detailed studies.

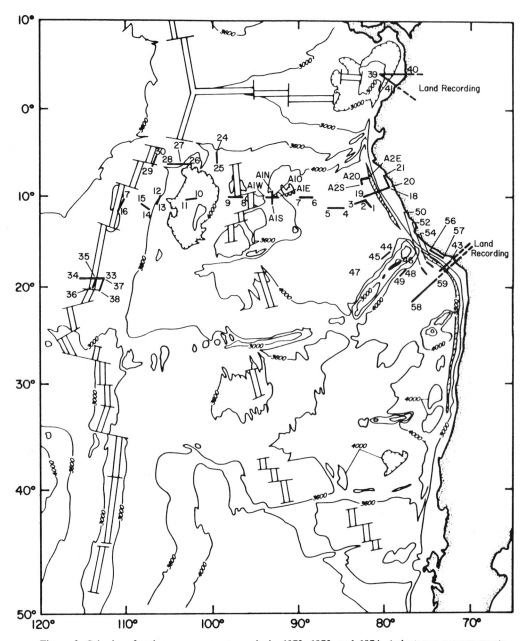

Figure 3. Seismic refraction measurements made in 1972, 1973, and 1974. Anisotropy measurements indicated by the letter A (A1N, A2S).

ASPER profiles whose locations are indicated in Figure 4, the composite figure for all ASPER lines. The 1972 sonobuoy ASPER measurements made with the 300-cu in. airgun on the R/V *Kana Keoki* defined the mantle at nine locations. These measurements provided valuable additional information on crustal thickness and velocity structure on the Nazca plate. An additional supplement to the seismic program was the recordings taken with a four-component, short-period seismograph that had been installed on Easter Island by the HIG. These data in combination with land seismograph station

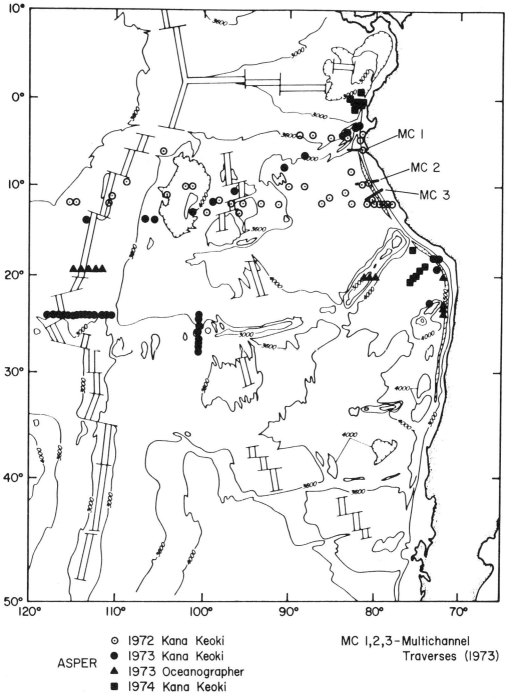

Figure 4. Location of Airgun Sonobuoy Precision Echo Recorder (ASPER) stations on Nazca plate and adjacent continental margin. The 24-channel seismic reflection lines (MC) are shown on the Peru margin.

records made it possible to determine time relations between energy release on the East Pacific Rise and energy release in the Peru-Chile Trench.

Geologic Sampling and Heat-Flow Measurements

Geologic sampling of the ocean floor was carried out by the two ships. The R/V *Yaquina* collected 115 piston and gravity cores, grab samples, and dredge samples in 1972 (Fig. 5) and 7 additional cores from heat-flow measurements (Fig. 6). The R/V *Kana Keoki* collected 156 core and dredge samples of which 23 were piston cores, 104 free-fall cores, 18 gravity cores, 6 large-diameter cores, 4 dredge samples, and 2 rock cores (Fig. 5). Cores were obtained from 44 additional heat-flow measurements (Fig. 6).

New Information

The objectives of the three subprograms were closely adhered to by both groups, and a wealth of new data were obtained in 1972 for the northern part of the Nazca plate from the East Pacific Rise to the South American borderland area of crustal subduction. Important new information concerning the Nazca plate that was realized in 1972 was the confirmed presence of a former (fossil) crustal spreading center in the mid-plate area. This spreading center, which correlated in location with that of the Galapagos Rise separating the Bauer Basin from the North Peru Basin, was determined by Ellen Herron (1972) of the Lamont-Doherty Geological Observatory on the basis of magnetic data obtained earlier. Herron's analysis showed that the Galapagos Rise had been an active crustal spreading center prior to the time of magnetic anomaly 5 (10 m.y. B.P.), at which time the locus of crustal spreading jumped to its present East Pacific Rise location. The courtesy of Ellen Herron in making this information available just before the start of the 1972 field season and 6 months prior to publication in June permitted some modifications in the field program that resulted in more significant results than otherwise might have been obtained.

In addition, extensive metalliferous sediment deposits were outlined and sampled in 1972. These deposits were known to occur on the East Pacific Rise (Bostrom and Peterson, 1966), and they were shown, based on our surveys, to be quite extensive over the northern East Pacific Rise and in the Bauer Basin to the east of the rise.

Another independent study carried out in 1972 on the Nazca plate was a cruise of the USSR R/V *Dimitri Mendelev*. On this cruise some 45 locations were sampled for metalliferous sediments in the area of the Bauer Basin and over the East Pacific Rise.

Along the Peruvian convergence zone, large crustal rupture features were discovered in the Peru Trench. These basaltic ridges rise up to 900 m above the trench floor. Seismic refraction studies also confirmed the existence of an outer continental shelf structural high suspected from geologic studies of coastal strata.

1973 FIELD PROGRAM

Although the preliminary results obtained for the first year's work (1972) were reviewed favorably at a meeting at IDOE headquarters in October 1972, this marked the end of the program as originally planned. Subsequent work in 1973 and 1974 followed a somewhat different plan in which Subprogram I, the study of the Nazca lithospheric plate between the East Pacific Rise and the Peru-Chile Trench, was de-emphasized. Because IDOE decided to focus its attention on metallogenesis, our portion of their Seabed Assessment Program, the Nazca Plate Project, was required to change its emphasis to those facets that would produce new information on the processes relating to the formation of metalliferous deposits along the divergent boundary of the Nazca plate, including the Bauer Basin, and on those processes responsible for volcanism and the formation of ore deposits above the Andean

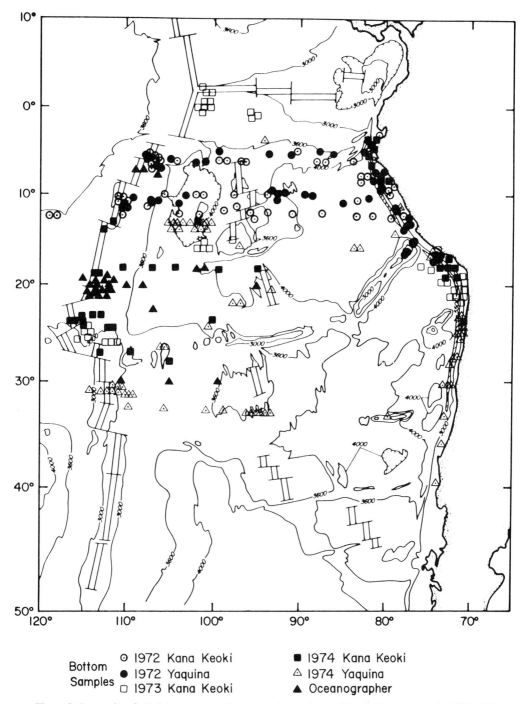

Figure 5. Composite of all piston core, gravity core grab samples, and dredge locations for the 1972, 1973, and 1974 cruises.

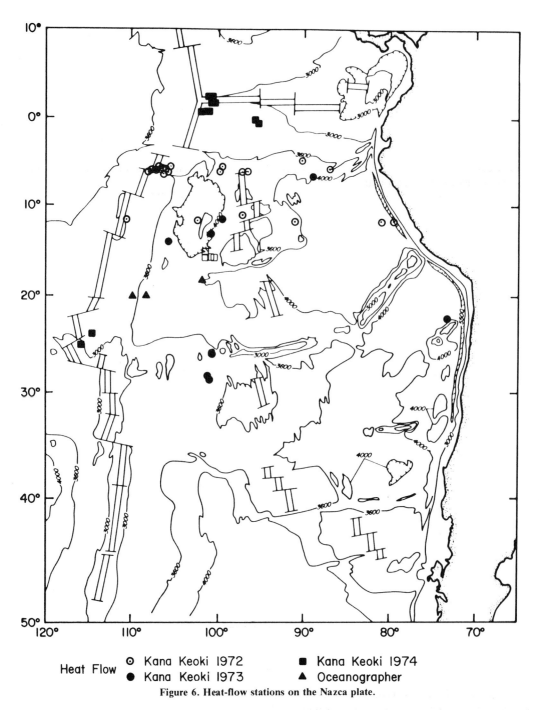

Figure 6. Heat-flow stations on the Nazca plate.

convergence zone. The new objectives for the project are discussed in a later section on Reorientation of the Nazca Plate Project.

Because the OSU group could not schedule the R/V *Yaquina* for operations on the Nazca plate in 1973, the HIG would either have to work alone or find an alternate working partner if additional seismic refraction traverses of the crust and upper mantle across the Nazca plate were to be made that year. The search for another partner was regarded as highly desirable since the first crustal and upper mantle traverse made with OSU at about 10°S latitude had shown that the crust was highly variable in thickness and velocity structure and was underlain by highly variable upper mantle material of rather high velocity.

The NOAA Pacific Oceanographic Laboratory at Seattle, which independently had been making underway measurements of bathymetry, gravity, and magnetism across the Nazca plate, was invited by HIG and OSU to join the program. This group accepted the invitation and was able to schedule OSS *Oceanographer* to work with the HIG vessel R/V *Kana Keoki* for 4 months in 1973. On a continuing basis, the combiantion of the R/V *Kana Keoki*, R/V *Yaquina*, and OSS *Oceanographer* in 1974 considerably sped up the program, which at that time still envisioned 14 seismic refraction traverses across the Nazca plate.

Unfortunately, the inclusion of the Pacific Oceanographic Laboratory (POL) in the program did not result in any additional seismic refraction traverses across the Nazca plate in 1973. Although seismic recording equipment was built for the OSS *Oceanographer* and persons were trained for the work at HIG, permission could not be obtained from NOAA headquarters for the *Oceanographer* to carry and fire explosives. This meant that the *Oceanographer* could only be used as a recording ship. The leap-frog method of alternately shooting and recording that had proved so effective in establishing the seismic refraction traverses with the R/V *Kana Keoki* and the R/V *Yaquina* in 1972, therefore, could not be used. Each ship, therefore, operated essentially independently in 1973 except in the area of the East Pacific Rise crest near Easter Island and over the Sala y Gomez Ridge where two-ship seismic refraction measurements were made with the R/V *Kana Keoki* firing and the OSS *Oceanographer* recording. Although the R/V *Kana Keoki* established only two new East-West traverses across the Nazca plate in 1973 and spent most of its time studying the East Pacific Rise, Bauer Basin, and parts of the trench and margin, the OSS *Oceanographer* traveling at 17 to 20 knots (twice the speed of R/V *Kana Keoki*) established six such traverses of underway measurements across the plate. In addition, the OSS *Oceanographer* made detailed surveys involving bottom samples, heat flow, seismic reflection profiles, and sonobuoy ASPER measurements in two areas on the East Pacific Rise and over the Peru-Chile Trench area in the vicinity of Valparaiso and between Antofagasta and Arica, Chile.

In view of the above considerable contribution to the Nazca Plate Project by the POL, it was with considerable regret that the participation of this partner had to end at the completion of the 1973 field season. This was a result of a reorganization in NOAA which eliminated the marine geological-geophysical program at the POL.

1973 Cruise Tracks and Data

The 1973 cruise tracklines for the OSS *Oceanographer* and the R/V *Kana Keoki* are shown in Figure 7. Both ships made underway bathymetric, gravity, and magnetic measurements, and the R/V *Kana Keoki* also made seismic reflection measurements on a routine basis. Several areas were studied in detail as shown in Figure 7. The R/V *Kana Keoki* logged approximately 20,940 nautical miles on and adjacent to the Nazca plate, and the OSS *Oceanographer* logged 25,940 miles.

Seismic Data Obtained

As previously indicated, two-ship seismic refraction measurements were limited; four reversed profiles were shot on the East Pacific Rise near Easter Island and one reversed profile on the Sala y Gomez Ridge. The locations (33–38) of these measurements are indicated as numbered line segments in Figure 3. Forty-two single-ended sonobuoy ASPER measurements were made by R/V *Kana Keoki* and fifteen made by the OSS *Oceanographer*. The locations of these measurements are shown in Figure 4.

Also shown in Figure 3 are the locations of two lines (39-40 and 39-41) of large explosive charges fired by R/V *Kana Keoki* off Buenaventura and Tumaco, Colombia, for onshore recording across the

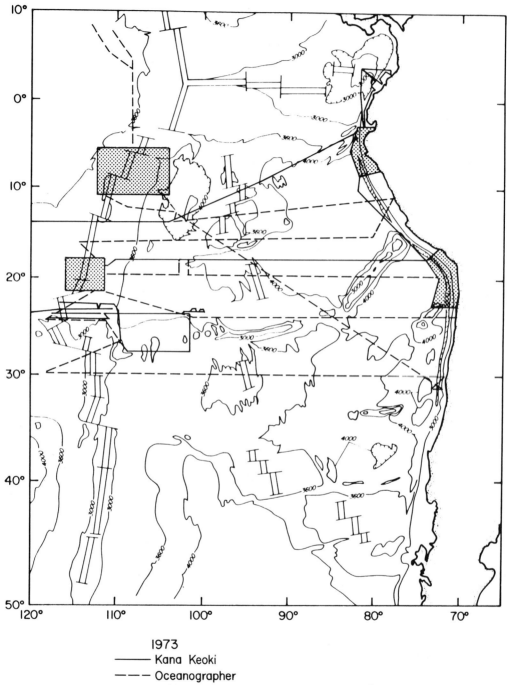

Figure 7. Tracklines for the 1973 cruises. Dotted boxes are areas of detailed studies.

continental shelf and up to the crest of the Andes. These constituted part of Project Nariño, a seismic refraction study of the nature of the crustal transition between the ocean and continent in an area where geologic and gravitational data suggest oceanic crust.

In the area of the bight (20° S) in the South American coast line between Mollendo, Peru, and Arica,

Chile, large explosive charges were fired by the R/V *Kana Keoki* from both outside and inside the Peru-Chile Trench for onland recording. These represented a preliminary test for a planned investigation of crustal transition where there appeared to be normal subduction of the oceanic lithosphere. Not only were these explosive charges recorded at the receiving station located at Arequipa, Peru, in the Andean region, but also signals were recorded from the R/V *Kana Keoki*'s 300-cu in. airguns.

Geologic Sampling and Heat-Flow Measurements

A total of 12 piston cores, 37 free-fall cores, and 12 dredge hauls augmented with bottom photography at 21 sites and heat-flow measurements were taken by R/V *Kana Keoki* in 1973. The OSS *Oceanographer* collected 32 cores with auxiliary heat-flow measurements at 5 sites. The locations of the sites for the respective types of data are shown in the composite plot for bottom samples (Fig. 5) and heat flow (Fig. 6).

REORIENTATION OF THE NAZCA PLATE PROJECT

The year 1973 marked the end of the initial Nazca Plate Project. In that year a decision was reached by IDOE to curtail future work over the central plate area and to emphasize the relation between plate tectonics and metallogenesis on the East Pacific Rise and in the Peru-Chile margin and trench area of crustal convergence. The program was redesigned to concentrate on the plate margins and to minimize both regional geological sampling and regional geophysical studies. Three of the principal objectives under subprogram I (changes with crustal age in sediment character and thickness, heat flow, and oceanic plate crust and upper mantle characteristics) were deleted from the Nazca plate study. The only regional investigations retained under subprogram I were the traverses required to cross the plate to conduct the detailed studies of the plate boundaries.

The new objectives for the Nazca Plate Project were defined as follows:

Divergent Boundary

1. Identification of areas, nature, and origin of primary mineralization along the spreading East Pacific Rise and within the anomalous Bauer Basin
2. Relate changes in mineralization to changes in spreading rate, magmatic differentiation and/or primary magma source, heat flow, and crustal structure

Convergent Boundary

1. Determine thickness and composition of oceanic crust and sediments just prior to subduction
2. Determine degree and composition of oceanic sediments accreted to the continent
3. Identify regions of accretion and consumption by their structural styles
4. Relate items 1–3 to crustal structure, seismicity patterns, volcanism, and ore formation on the continent

Continental Metallogenesis

1. Summarize the pertinent land, isotopic, and geochemical data on volcanic rocks and ore deposits and attempt to relate these to similar data on the oceanic Nazca plate rocks and sediments to test for an oceanic origin and concomitant areal changes
2. Relate convergent boundary information to temporal and spatial distribution of continental metallogenic and volcanic provinces.

The defined objectives clearly emphasized the importance of studying metallogenesis. In this regard, two of the more significant results obtained under the project during 1972 and further verified in 1973 were that one of the most highly mineralized oceanic areas known occurs in the Bauer Basin and that the degree of metallogenesis correlates with high heat flow. The Bauer Basin occurs between the East Pacific Rise and the mid-plate fossil Galapagos crustal spreading center defined by Herron (1972). The Bauer Basin was therefore included as part of the metallogenesis study, and more sampling was done in the Bauer Basin and on the former Galapagos spreading center. However, coring, heat flow, seismic refraction, and sonobuoy ASPER measurements in all other parts of the plate, except on its boundaries, were excluded from the 1974 field program.

1974 FIELD PROGRAM

The field program for the OSU vessel R/V *Yaquina* on the Nazca Plate Project in 1974 actually started in October 1973 and continued through the latter part of March 1974. The HIG vessel R/V *Kana Keoki* started in early January 1974 and continued through mid-June 1974. Figure 8 shows the locations of the regional traverses established and areas studied in detail by both R/V *Yaquina* and R/V *Kana Keoki*. The R/V *Yaquina* logged approximately 19,500 nautical miles, and the R/V *Kana Keoki* 18,800 nautical miles. For the most part, the two ships operated independently; about 3 weeks, however, was spent in making two-ship refraction crustal measurements over the eastern end of the Nazca Ridge and adjacent trench and continental shelf area. The five essentially East-West trans-plate traverses along which underway measurements were made were located adjacent to traverses established earlier (Fig. 9). The proximity of data sets made it possible to establish, with some assurance, a regional pattern of bathymetry, sediment character and thickness, gravity magnetic field and crustal age across the plate. Bottom sampling, heat flow, and sonobuoy ASPER measurements were confined to the areas of detailed study, except for a series of closely spaced bottom samples taken by R/V *Yaquina* across the postulated location of the former Galapagos spreading center defined by Herron (1972) at 33°S latitude.

Data Obtained in 1974

Figure 3 shows the locations of the two-ship seismic refraction lines established in conjunction with study of the eastern end of the Nazca Ridge. The southernmost long seismic line (58-59) was the location of a series of large explosive charges fired by R/V *Kana Keoki* and recorded offshore by R/V *Yaquina* and onshore by the Instituto Geofisica del Peru and the Carnegie Institution of Washington, Department of Terrestrial Magnetism.

In addition, the R/V *Kana Keoki* carried out 19 sonobuoy ASPER measurements whose lcoations are shown in Figure 3. These and the passive monitoring of earthquake activity using both sonobuoys and an on-bottom seismometer on the Easter Island miniplate (33-38; Handschumacher and others, this volume) represent the extent of the seismic work done in 1974.

Geologic Sampling and Other Data

The R/V *Yaquina* collected 81 bottom samples; 16 were in the Bauer Basin, 20 on and adjacent to the East Pacific Rise, 27 in the southern extension of the Bauer Basin and on the fossil Galapagos spreading center, and 28 in and adjacent to the Peru-Chile Trench (Figure 5). The R/V *Kana Keoki*, despite problems with its cable winch, collected 58 bottom samples; 20 were in the area of the East Pacific Rise–Galapagos triple junction, 7 in the Bauer Basin, 7 on the Easter Island miniplate, and 14 in the area of the Peru-Chile Trench and adjacent continental slope and shelf. The locations of the areas sampled by both R/V *Yaquina* and R/V *Kana Keoki* are shown in Figure 5.

Although no heat-flow measurements were made by R/V *Yaquina*, the R/V *Kana Keoki* made 12 successful heat-flow measurements in the area of the East Pacific Rise–Galapagos triple junction, 7 measurements in the area of the Bauer Basin, and 2 measurements on the Easter Island miniplate. The

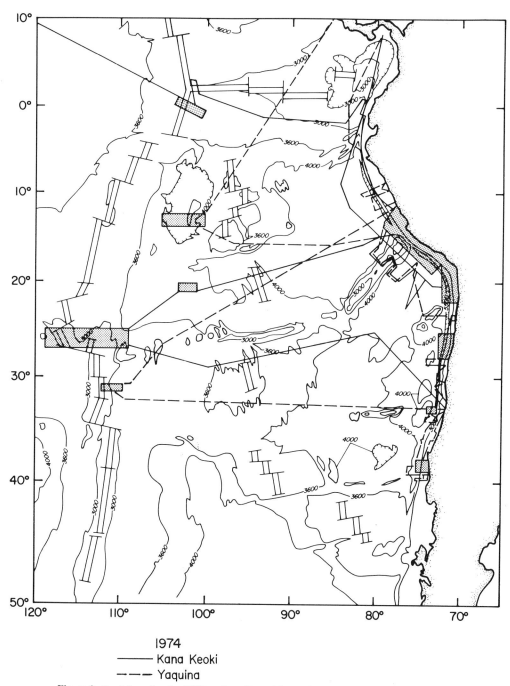

Figure 8. Tracklines for the 1974 cruises. Dotted boxes indicate regions of detailed studies.

locations of these measurements are shown in the composite plot for heat-flow measurements (Fig. 6).

Bottom samples and magnetic data also were obtained through participation of scientific staff from OSU and HIG on the 1974–75 cruise of the USSR R/V *Dimitri Mendelev* over the Nazca plate. While this cruise primarily obtained physical oceanographic information plus bottom samples in areas of

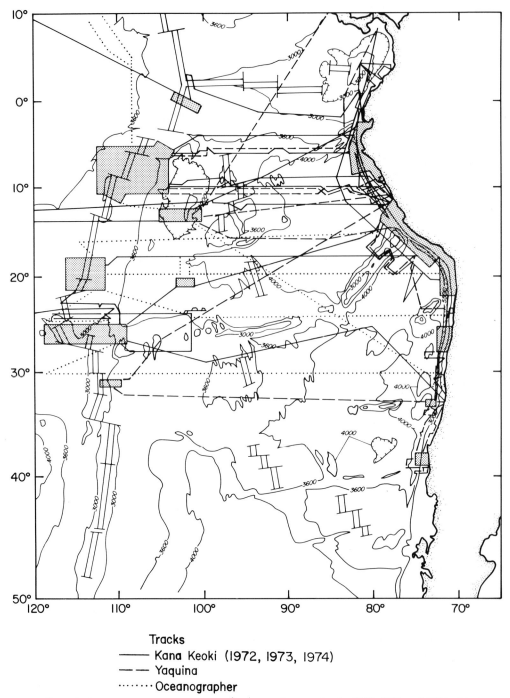

Figure 9. Composite of all tracklines and detailed study areas for the 1972, 1973, and 1974 cruises.

probable metalliferous sediments, it was possible for HIG to install a proton precession magnetometer on the R/V *Dimitri Mendelev* when she made a port call in Samoa so that magnetic data could be obtained on the cruise legs over the Nazca plate.

A second independent program that contributed significantly to knowledge concerning the distribution of metalliferous sediments with depth on the Nazca plate was the JOIDES Deep Sea Drilling Project. Four holes to basement were drilled by the *Glomar Challenger* in the northern part of the plate in January 1974. Two of these holes, sites 319 and 319A, were adjacent to each other in the central part of the Bauer Basin, and the other two sites, 320 and 321, were just seaward of the Peru-Chile Trench on the eastern border of the North Peru Basin. The locations of these drilling sites are shown in Figure 1.

A third independent program that contributed to the project was the U.S. Navy Project Magnet. The release of Project Magnet data, taken along certain ship's tracklines with an airplane, permitted the differentiation of ship-defined magnetic anomalies having actual geological age significance from apparent anomalies having similar amplitude and wavelength related to diurnal variation when operating near the magnetic equator.

1975 FIELD PROGRAM

OSU conducted a brief field program along the Peru-Chile Trench from 33°S to 9°S latitude to collect 8 piston cores on sediments covering structual features within the trench axis and to make 10 dredges over the basaltic ridges exposed within the trench. This work was done on the FDRAKE cruise of the R/V *Melville* of the Scripps Institution of Oceanography by OSU investigators under sponsorship of NSF/IDOE during April to May 1975. The sample locations and results of this cruise are described by Scheidegger and others (1978).

1977 FIELD PROGRAM

The final cruise of the Nazca Plate Project was conducted during June 1977 aboard the R/V *Wecoma* of OSU in conjunction with an Office of Naval Research cruise to the Peru-Chile region. Seven dredges were taken on the continental slope and outer continental shelf off Peru (12° to 3.5°S) and on the continental slope off Chile (22° to 23°S). The locations of the dredge samples are given in Kulm and others (this volume).

SOUTH AMERICAN PARTICIPATION

This project would not have been possible without the participation of numerous colleagues from the South American countries of Colombia, Ecuador, Peru, Bolivia, and Chile. These scientists assisted in the Nazca Plate Project in many different ways during the 9-yr period from 1970 to 1979. Giving recognition where recognition and thanks are due, we identify each scientist who contributed to the success of this project.

Bolivia
Instituto Panamericano de Geografia e Historia: Reynaldo Salgueiro Pabón
Observatorio San Calixto: Ramón Cabre R., S.J.

Chile
Departamento de Geofisica y Geodesia: Alfredo Eisenberg; Peter Welkner
Instituto Hidrografico de la Armada: Raúl Herrera

Colombia
Instituto Geofisico de los Andes Colombianos: Jesús Emilio Ramírez, S.J.

Ecuador
Servicio Hidrografico y Oceanografico, Armada del Ecuador: Raúl Canizares; Samuel Franco
Universidad del Litoral: Hugo Reinoso; José Santoro
Universidad Nacional de Guayaquil: Héctor Ayón; Flor de María Valverde

Peru
Armada del Peru: César Vargas
Direccion General de Hidrocarburos: Manuél Naranjo A.
Instituto Geofisico del Peru: Matéo Casaverde; Feliciano Huacache; Alberto A. Giésecke M.; Leonidas C. Ocola
Petroleos del Peru: Juan Antonio Masías E.; Manuél Paredes; Fernando Zúñiga
Servicio de Geologia y Mineria: Hugo Jaén; Fernando Perales; Luís Reyes; Humberto Salazar
Servicio Nacional de Meteologia e Hidrologia: Jorge Valdería
Sociedad Geologico del Peru: Enrico Fuentes

SYNTHESIS OF INFORMATION

The final synthesis of Nazca plate data was started in 1978 and was completed by mid-1980 with the preparation of this Memoir. Additional papers published or in press as part of the Nazca Plate Project are given in Appendix 1. Approximately 100 papers have been or will be published, including those in this Memoir.

Several papers were contributed to this Nazca plate Memoir by investigators who conducted their research independently of the Nazca Plate Project. We welcome their contributions which are cited in Appendix 2. Because the Nazca Plate Project was not funded to conduct a research program in the Andes Mountains adjacent to the Nazca plate, a synthesis of the Andean convergence zone would not have been as complete without the addition of these papers.

REFERENCES CITED

Bostrom, K., and Peterson, M.N.A., 1966, Precipitates from hydrothermal exhalations on the East Pacific Rise: Economic Geology, v. 61, p. 1258–1265.

Handschumacher, D., and others, 1981, Structure and evolution of the Easter plate, in Kulm, L. D., and others, eds., Nazca plate: Crustal formation and Andean convergence: Geological Society of America Memoir 154 (this volume).

Herron, E., 1972, Sea-floor spreading and the Cenozoic history of the East-Central Pacific: Geological Society of America Bulletin, v. 83, p. 1671–1692.

Kulm, L. D., and others, 1981, Late Cenozoic carbonates on the Peru continental margin: Lithostratigraphy, biostratigraphy, and tectonic history, in Kulm, L. D., and others, eds., Nazca plate: Crustal structure and Andean convergence: Geological Society of America Memoir 154 (this volume).

Scheidegger, K. F., and others, 1978, Fractionation and mantle heterogeneity in basalts from the Peru-Chile Trench: Earth and Planetary Science Letters, v. 37, p. 409–420.

MANUSCRIPT RECEIVED BY THE SOCIETY NOVEMBER 12, 1980
MANUSCRIPT ACCEPTED DECEMBER 30, 1980
CONTRIBUTION 1098, HAWAII INSTITUTE OF GEOPHYSICS, UNIVERSITY OF HAWAII

APPENDIX 1. ADDITIONAL PAPERS PUBLISHED AS PART OF THE NAZCA PLATE PROJECT

Berg, E., Sutton, G. H., and Walker, D. A., 1977, Dynamic interaction of seismic activity along rising and sinking edges of plate boundaries: Tectonophysics, v. 39, p. 559–578.

Blakely, R. J., 1976, An age-dependent, two-layer model for marine magnetic anomalies, *in* Sutton, G. H., and others, eds., The geophysics of the Pacific ocean basin and its margin: American Geophysical Union Monograph 19, p. 227–235.

Blakely, R. J., and Cox, A., 1975, Comment on "Stacking marine magnetic anomalies: A critique" by Robert L. Parker: Geophysical Research Letters, v. 2, p. 185–187.

Blakely, R. J., and Lynne, W. S., 1977, Reverse transition width in oceanic crust as a function of spreading rate: Earth and Planetary Science Letters, v. 33, p. 321–330.

Blakely, R. J., Klitgord, K. D., and Mudie, J. D., 1975, Analysis of marine magnetic data: Reviews of Geophysics and Space Physics, v. 13, p. 182–185 and p. 221–223.

Burnett, W. C., 1977, Geochemistry and origin of phosphorite deposits from off Peru and Chile: Geological Society of America Bulletin, v. 88, p. 813–823.

Burnett, W. C., and Veeh, H. H., 1977, Uranium-series disequilibrium in phosphorite nodules from the west coast of South America: Geochimica et Cosmochimica Acta, v. 41, p. 755–764.

Corliss, J. B., 1974, The sea as alchemist: Oceanus, v. 17, p. 38–43.

Corliss, J. B., Dymond, J., and Lopez, C., 1976, Elemental abundance patterns in Leg 34 rocks, *in* Yeats, R. S., Hart, S. R., and others, eds., Initial reports of the Deep Sea Drilling Project, Volume 34: Washington, D.C., U.S. Government Printing Office, p. 293–299.

Coulbourn, W. T., 1980, Relationship between the distribution of foraminifera and geologic structures of the Arica Bight, South America: Journal of Paleontology, v. 54, p. 676–718.

Coulbourn, W. T., and Moberly, R., 1976, Structural evidence of the evolution of fore-arc basins: Canadian Journal of Earth Sciences, v. 14, p. 102–116.

Coulbourn, W. T., and Resig, J. M., 1979, Middle Eocene pelagic microfossils from the Nazca plate: Geological Society of America Bulletin, v. 90, p. 643–650.

Cox, M. E., and McMurtry, G. M., 1981, Vertical distribution of mercury in sediments from the East Pacific Rise: Nature, v. 289, p. 789–792.

Dasch, E. J., 1974, Metallogenesis in the southeastern Pacific: A progress report on the IDOE Nazca Plate Project: Physics of the Earth and Planetary Interiors, v. 9, p. 249–258.

—— 1976, Nazca Plate studied: The Geostill Conference: Geotimes, v. 21, p. 24–25.

Dasch, E. J., Dymond, J. R., and Heath, G. R., 1971, Isotopic analysis of metalliferous sediment from the East Pacific Rise: Earth and Planetary Science Letters, v. 13, p. 175–180.

Dasch, E. J., Hedge, C. E., and Dymond, J. R., 1973, Effect of sea-water interaction on strontium isotope composition of deep-sea basalt: Earth and Planetary Science Letters, v. 19, p. 177–183.

Dymond, J., and Eklund, W., 1978, A microprobe study of metalliferous sediment components: Earth and Planetary Science Letters, v. 40, p. 243–251.

Dymond, J., and Veeh, H. H., 1975, Metal accumulation rates in the southeast Pacific and the origin of metalliferous sediments: Earth and Planetary Science Letters, v. 28, p. 13–22.

Dymond, J., and others, 1973, Origin of metalliferous sediments from the Pacific Ocean: Geological Society of America Bulletin, v. 84, p. 3355–3372.

Dymond, J., Corliss, J. B., and Stillinger, R., 1976, Chemical composition and metal accumulation rates of metalliferous sediments from Sites 319, 320B and 321, *in* Yeats, R. S., Hart, S. R., and others, eds., Initial reports of the Deep Sea Drilling Project, Volume 34: Washington, D.C., U.S. Government Printing Office, p. 575–588.

Dymond, J. R., Corliss, J. B., and Heath, G. R., 1977, History of metalliferous sedimentation at Deep Sea Drilling Site 319, in the Southeastern Pacific: Geochimica et Cosmochimica Acta, v. 21, p. 741–753.

Field, C. W., and others, 1976, Metallogenesis in the southeast Pacific Ocean: The Nazca Plate Project: American Association of Petroleum Geologists Memoir 25, p. 539–550.

Field, C. W., and others, 1976, Sulfur isotope reconnaissance of epigenetic pyrite in ocean-flow basalts, Leg 34 and elsewhere, *in* Yeats, R. S., Hart, S. R., and others, eds., Initial reports of the Deep Sea Drilling Project, Volume 34: Washington, D.C., U.S. Government Printing Office, p. 381–384.

Handschumacher, D. W., 1976, Post-Eocene plate tectonics of the Eastern Pacific, *in* Sutton, G. H., and others, eds., The geophysics of the Pacific Ocean basin and its margin (Woollard volume): American Geophysical Union Geophysical Monograph 19, p. 177–202.

Hussong, D. M., Odegard, M. E., and Wipperman, L. K., 1975, Compressional faulting of the oceanic crust prior to subduction in the Peru-Chile Trench: Geology, v. 3, p. 601-604.

Hussong, D. M., Campbell, J. F., and Sutton, G. H., 1976, Crustal structure of the Peru-Chile Trench, 8°-13°S latitude, *in* Sutton, G. H., and others, eds., The geophysics of the Pacific Ocean basin and its margin (Woollard volume): American Geophysical Union Geophysical Monograph 19, p. 71-85.

Johnson, S. H., 1976, Interpretation of split-spread refraction data in terms of plane dipping layers: Geophysics, v. 41, p. 418-424.

Kulm, L. D., and others, 1973, Tholeiitic basalt ridge in the Peru Trench: Geology, v. 1, p. 11-14.

Kulm, L. D., and others, 1974, Transfer of Nazca Ridge pelagic sediments to the Peru continental margin: Geological Society of America Bulletin, v. 85, p. 769-780.

Kulm, L. D., and others, 1976, Lithologic evidence for convergence of the Nazca plate with the South American Continent, *in* Yeats, R. S., and Hart, S. R., and others, eds., Initial reports of the Deep Sea Drilling Project, Volume 34: Washington, D.C., U.S. Government Printing Office, p. 795-801.

Kulm, L. D., Schweller, W. J., and Masias, A., 1977, A preliminary analysis of the geotectonic processes of the Andean continental margin, 6° to 45°S, *in* Talwani, M., and Pitman, W. C. III, eds., Island arcs, deep sea trenches, and back-arc basins: American Geophysical Union, Maurice Ewing Series, v. 1, p. 285-301.

Kulm, L. D., and others, 1981, Cenozoic structure, stratigraphy and tectonics of the central Peru forearc, *in* Leggett, J. K., ed., Trench and forearc sedimentation and tectonics in modern and ancient subduction zones: London, Geological Society of London (in press).

Kureth, C. L., and Rea, D. W., 1981, Large-scale oblique features in an active transform fault, the Wilkes fracture zone near 9°S on the East Pacific Rise: Marine Geophysical Research (in press).

Lyle, M. W., 1976, Estimation of hydrothermal manganese to the oceans: Geology, v. 4, p. 733-736.

Lyle, M., and Dymond, J., 1976, Metal accumulation rates in the southeast Pacific—errors introduced from assumed bulk densities: Earth and Planetary Science Letters, v. 30, p. 164-168.

Lyle, M. W., Dymond, J., and Heath, G. R., 1977, Copper-nickel-enriched ferromanganese nodules and associated crusts from the Bauer Basin, northwest Nazca plate: Earth and Planetary Science Letters, v. 35, p. 55-64.

McMurtry, G. M., 1981, Metallogenesis on oceanic plates—the East Pacific Rise and Bauer Basin, *in* Energy resources of the Pacific region: American Association of Petroleum Geologists (in press).

McMurtry, G. M., and Burnett, W. C., 1975, Hydrothermal metallogenesis in the Bauer Deep of the South-eastern Pacific: Nature, v. 254, p. 42-44.

McMurtry, G. M., and Yeh, H-W., 1981, Hydrothermal clay mineral formation of East Pacific Rise and Bauer Basin sediments: Chemical Geology, v. 32 (in press).

Meyer, R. P., and others, 1976, Project Narino III: Refraction observation across a leading edge, Malpelo Island to the Colombian Cordillera Occidental, *in* Sutton, G. H., and others, eds., The geophysics of the Pacific Ocean basin and its margin (Woollard volume): American Geophysical Union Geophysical Monograph 19, p. 105-132.

Moberly, R., Shepherd, G. L., and Coulbourn, W. T., 1981, Forearc and other basins, continental margin of northern and southern Peru and adjacent Ecuador and Chile, *in* Leggett, J. K., ed., Trench and forearc sedimentation and tectonics in modern and ancient subduction zones: London, Geological Society of London (in press).

Nelson, D. O., and Dasch, E. J., 1976, Disequilibrium of strontium isotopes between mineral phases of parental rocks during magma genesis: A discussion: Journal of Volcanology, v. 1, p. 183-191.

Ness, G. E., and others, 1976, Geología estructural del margen continental en la región del sur de Peru adyacente a la cordillera Nazca [Structural geology of the continental margin of southern Peru adjacent to the Nazca Ridge], *in* Perez-Rodriguez, R., and Suarez-Zozaya, M. R., eds., I Reunión Latinoamericana sobre Ciencia y Tecnología de los Oceanos, Volume 2: Mexico, D.F., Secretaria de Marina, p. 162-179.

Prince, R. A., and Kulm, L. D., 1975, Crustal rupture and the initiation of imbricate thrusting in the Peru-Chile Trench: Geological Society of America Bulletin, v. 86, p. 1639-1653.

Prince, R. A., and Schweller, W. J., 1978, Dates, rates and angles of faulting in the Peru-Chile Trench: Nature, v. 271, p. 743-745.

Prince, R. A., and others, 1974, Uplifted turbidite basins on the seaward wall of the Peru Trench: Geology, v. 2, p. 607-611.

Rea, D. K., 1975, Model for the formation of topographic features of the East Pacific Rise crest: Geology, v. 3, p. 77-80.

——1976, Changes in the axial configuration of the East Pacific Rise near 6°S during the past 2 m.y.: Journal of Geophysical Research, v. 81, p. 1495-1504.

——1976, Analysis of a fast-spreading rise crest: The East Pacific Rise, 9° to 12°S: Marine Geophysical Research, v. 2, p. 291-313.

——1977, Local axial migration and spreading rate

variations, East Pacific Rise, 31°S: Earth and Planetary Science Letters, v. 34, p. 78–84.
——1978, Asymmetric sea-floor spreading and a nontransform axis offset: The East Pacific Rise 20°S survey area: Geological Society of America Bulletin, v. 89, p. 836–844.
Rea, D. K., and Blakely, R. J., 1975, Short-wavelength magnetic anomalies in a region of rapid seafloor spreading: Nature, v. 255, p. 126–128.
Rea, D. K., and Malfait, B. T., 1974, Geologic evolution of the northern Nazca Plate: Geology, v. 2, p. 317–320.
Rea, D. K., and others, 1973, New estimates of rapid sea-floor spreading rates and the identification of young magnetic anomalies on the East Pacific Rise, 6° and 11°S: Earth and Planetary Science Letters, v. 19, p. 225–229.
Resig, J. M., 1976, Benthic foraminiferal stratigraphy, eastern margin, Nazca Plate, *in* Yeats, R. S., and Hart, S. R., and others, eds., Initial reports of the Deep Sea Drilling Project, Volume 34: Washington, D.C., U.S. Government Printing Office, p. 743–759.
——1981, *Nodellum moniliforme, Ammomarginulina hadalensis,* and *Favocassidulina subfavus,* three new species of Recent deepwater benthic foraminifera: Journal of Paleontology (in press).
Resig, J. M., and others, 1980, An extant opaline foraminifer: Test ultrastructure, mineralogy, and taxonomy, *in* Memorial to Orville L. Bandy: Cushman Foundation Special Publication 18, p. 205–214.
Rosato, V. J., Kulm, L D., and Derks, P. S., 1975, Surface sediments of the Nazca plate: Pacific Science, v. 29, p. 117–130.
Scheidegger, K. F., and Stakes, D. S., 1977, Mineralogy, chemistry and crystallization sequence of clay minerals in altered tholeiitic basalts from the Peru Trench: Earth and Planetary Science Letters, v. 36, p. 413–422.
Scheidegger, K. F., and others, 1978, Fractionation and mantle heterogeneity in basalts from the Peru-Chile Trench: Earth and Planetary Science Letters, v. 37, p. 409–420.
Schweller, W. J., and Kulm, L. D., 1978, Depositional patterns and channelized sedimentation in active Eastern Pacific trench, *in* Stanley, D. J., and Kelling, G., eds., Sedimentation in submarine canyons, fans and trenches: Stroudsburg: Dowden, Hutchinson, and Ross, p. 311–324.
——1978, Extensional rupture of oceanic crust in the Chile Trench: Marine Geology, v. 28, p. 271–291.
Senechal, R. G., and Dasch, E. J., 1974, The Oregon State University solid-source mass spectrometer: A new instrument for research and instruction. Part I: Description and system evaluation: Proceedings of the Oregon Academy of Sciences, v. 10, p. 15–46.
Woollard, G. P., 1975, The interrelationships of crustal and upper mantle parameter values in the Pacific: Review of Geophysics and Space Physics, v. 13, p. 87–137.
——1977, The geological and geophysical setting of Latin America, *in* Tanner, J. G., and Dence, M. R., eds., Geophysics in the Americas, Volume 46: Ottawa, Department of Energy, Mines and Resources, Earth Physics Branch, p. 17–36.
——1981, Mineral deposits and plate tectonics, *in* Energy resources of the Pacific region: American Association of Petroleum Geologists (in press).
Woollard, G. P., and McMurtry, G. M., 1981, Mineralization in relation to the dynamic processes along the Peru-Chile Trench, *in* Proceedings, Latin American Geological Congress, 4th, Trinidad & Tobago, W. I. (in press).
Woollard, G. P., and Ocola, L. C., 1973, The tectonic pattern of the Nazca plate: Geofisica Panamericana, v. 2, p. 125–149.

APPENDIX 2. CONTRIBUTIONS TO THIS MEMOIR BY INVESTIGATORS INDEPENDENT OF THE NAZCA PLATE PROJECT

Burnham, C. W., 1981, Convergence and mineralization: Is there a relation?, *in* Kulm, L. D., and others, eds., Nazca plate: Crustal formation and Andean convergence: Geological Society of America Memoir 154 (this volume).
Herron, E. M., 1981, Chile margin near 38°S: Evidence for a genetic relationship between continental and marine geologic features—or a case of curious coincidences?, *in* Kulm, L. D., and others, eds., Nazca plate: Crustal formation and Andean convergence: Geological Society of America Memoir 154 (this volume).
Herron, E. M., Cande, S. C., and Hall, B. R., 1981, An active spreading center collides with a subduction zone: A geophysical survey of the Chile margin triple junction, *in* Kulm, L. D., and others, eds., Nazca plate: Crustal formation and Andean convergence: Geological Society of America Memoir 154 (this volume).
James, D. E., 1981, Role of subducted continental

material in the genesis of calc-alkaline volcanics of the central Andes, *in* Kulm, L. D., and others, eds., Nazca plate: Crustal formation and Andean convergence: Geological Society of America Memoir 154 (this volume).

Lowrie, A., and Hey, R., 1981, Geological and geophysical discontinuities in Chile near 33°–36°S and their relationship to Nazca plate subduction, *in* Kulm, L. D., and others, eds., Nazca plate: Crustal formation and Andean convergence: Geological Society of America Memoir 154 (this volume).

Nur, A., and Ben-Avraham, Z., 1981, Volcanic gaps and the consumption of aseismic ridges in South America, *in* Kulm, L. D., and others, eds., Nazca plate: Crustal formation and Andean convergence: Geological Society of America Memoir 154 (this volume).

Tilton, G. R., and others, 1981, Isotopic composition of lead in central Andean ore deposits, *in* Kulm, L. D., and others, eds., Nazca plate: Crustal formation and Andean convergence: Geological Society of America Memoir 154 (this volume).

DIVERGENT BOUNDARY

# Tectonics of the Nazca-Pacific divergent plate boundary

DAVID K. REA
Oceanography Program
Department of Atmospheric and Oceanic Science
The University of Michigan
Ann Arbor, Michigan 48109

ABSTRACT

Five regions along the Nazca-Pacific plate boundary are the sites of detailed surveys completed during the Nazca Plate Project, four in relatively typical axial regions and one in the area of the tectonically complex Easter plate. Deep tow surveys have also been conducted on the spreading center and in a fracture zone. In addition to the survey data, many tracklines of a reconnaissance nature cross the spreading center. Sea-floor spreading occurs at about 160 mm/yr (whole rate) along this part of the East Pacific Rise. South of the Garret Fracture Zone at 13.5°S spreading is asymmetrical, being faster to the east than to the west. The lithosphere near fast-spreading rises is quite thin and therefore more susceptible to deformation than along slow-spreading ridges. The effects of a weak lithosphere take several forms: a long oblique ridge in the active portion of the Wilkes Fracture Zone at 9°S; twinned spreading centers surrounding the small Easter plate; and small, 10 to 15 km, offsets of the spreading center that can be formed or healed rapidly. These offsets are all formed within a time span of 0.5 m.y. or less and may involve segments of the rise axis up to 200 km long. Apparently they are the result of small, discrete axis jumps facilitated by the unusually thin lithosphere. Reconnaissance data between the 13.5°S and 4.5°S fracture zones are adequate to decipher the history of formation of the East Pacific Rise in this region. Spreading activity shifted 600 to 850 km westward from the old Galapagos Rise, now in the center of the Nazca plate, to its present location by three large jumps. Each of these jumps resulted in the formation of a fracture zone–bound section of the new rise; the jumping process covered about 2.5 m.y., from 8.2 to 5.7 m.y. ago.

INTRODUCTION

The spreading center forming the western edge of the Nazca lithospheric plate is that part of the East Pacific Rise (EPR) extending about 4,000 km south from the Galapagos triple junction at 2°N to the triple junction with the Chile Ridge at about 35°S (Fig. 1). This section of the EPR is a geologically young feature that has been created since late Oligocene time by a series of westward spreading-center jumps spanning hundreds of kilometres from the now extinct Galapagos Rise (Heron, 1972a). Present crustal accretion rates measured along the Nazca-Pacific boundary range from 151 to 162 mm/yr, increasing to the south. The Nazca plate boundary, therefore, occupies the fast-spreading end of the spectrum of the world rift system, and because of the rapid accretion rate has a tectonic style very different from slower, Atlantic-type spreading centers.

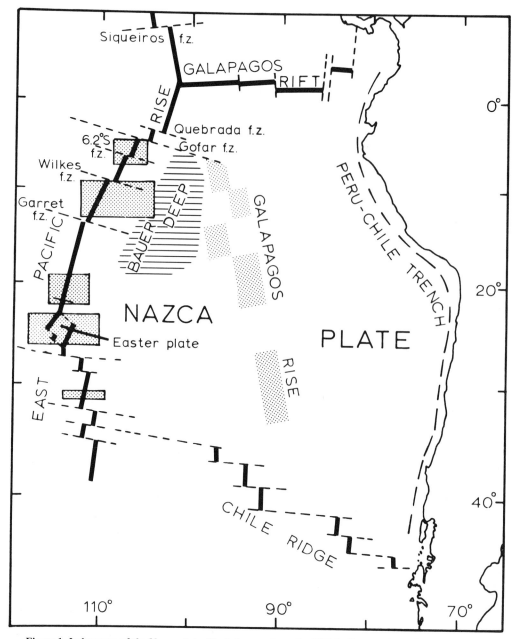

Figure 1. Index map of the Nazca plate. Shaded areas along the EPR indicate regions of detailed surveys centered at 6°S, 10°S, 20°S, 25°S, and 31°S. Locations of fracture zones discussed in text are shown.

Scope of This Report:

This paper is intended to accomplish two goals. The first is to summarize the available geophysical data bearing upon the present tectonic regime of the EPR. This effort will concentrate upon the bathymetric and magnetic anomaly data; other papers in this volume treat petrology and geochemistry of crustal rocks (Scheidegger and Corliss, 1981). Data sets from the Nazca-Pacific boundary regions will be discussed in order of increasing geological complexity.

The second objective of this paper is to interpret from the data the structural and dynamic nature of the Nazca-Pacific plate boundary concentrating upon aspects of this topic applicable to a region broader than a single survey area. These topics pertain to spatial and temporal variations observed in the spreading regime: the along-strike changes in EPR crustal elevation; the nature of asymmetrical sea-floor spreading; the origin of the several small, 10 to 15 km, axial offsets; and the history of the large axis jumps from the extinct Galapagos Rise to the EPR that occurred north of the fracture zone at 13.5°S.

History of Exploration

The EPR was discovered by the H.M.S.*Challenger* which crossed it at about 39°S in October of 1875 (Murray, 1895). The first marine geological expedition to reach the vicinity of the Nazca-Pacific plate boundary was lead by Alexander Aggasiz aboard the *Albatross* during 1904-1905 (Agassiz, 1906). Apparently the next ship to study the area was the *Carnegie*, which in 1928-1929 sounded and named the Carnegie Ridge and the Bauer Deep (Ault, 1946).

Essentially no additional knowledge was gained about this region until during and after World War II when the increasing interest of the U.S. Navy in Antarctica resulted in many ships traversing and sounding the southeast Pacific. A series of long research cruises to the South Pacific initiated by the Scrips Institution of Oceanography in the 1950's and by Lamont-Doherty Geological Observatory in the 1960s contributed much new information and helped refine the older data.

Apparently the EPR was never named as such. The term first appears in the early to mid-1950s in articles written by marine geologists at Scripps, although the term "Albatross Plateau" was still being used by the continental geologists (compare Gilluly, 1955, with Revelle and others, 1955). At the same time, workers at Lamont referred to the EPR as the Easter Island Ridge (Ewing and Heezen, 1956).

Menard (1960) outlined the entire EPR system from the Gulf of California to south of New Zealand and later (Menard and others, 1964) pointed out the existence of the Galapagos Rise in the center of the Nazca plate. In 1972 Herron published a compilation of much of the marine geophysical data available for the Nazca plate and attempted to decipher the Cenozoic history of the area (Herron, 1972a). Handschumacher (1976) has addressed the post-Eocene history of the same area utilizing the large amount of data collected during the Nazca Plate Project in addition to the prior information. Lonsdale, in a series of three papers, has reported on a detailed study of the EPR crestal region near 3°25'S (Lonsdale, 1977a), the nearby Quebrada transform fault (Lonsdale, 1978), and presented a regional study of EPR morphology and tectonics covering that portion of the rise between the Siqueiros Fracture Zone near 8°N and about 5°S (Lonsdale, 1977b).

Most of the data collection along the EPR under the auspices of the Nazca Plate Project was accomplished during 1972, 1973, and 1974 by the R/V *Yaquina* of Oregon State University, the R/V *Kana Keoki* of Hawaii Institute of Geophysics, and the NOAA ship *Oceanographer* from the Pacific Marine Environmental Laboratory. Among them, these ships completed underway geophysical surveys, seismic refraction, coring, and dredging of five regions along the EPR crest; the regions span approximately 5° to 7°S, 9° to 12°S, 18.7° to 22°S, 23° to 27°S, and 30.5° to 31.3°S, a total of more than 35% of the entire Nazca-Pacific plate boundary. Data from the underway surveys, predominantly bathymetry and magnetics, have been published for four of these regions (Rea, 1976a, 1976b, 1977, 1978a). The reader is referred to these and other publications, given below in the discussion of the individual areas, for the various details of data processing methods. Mammerickx and her co-workers

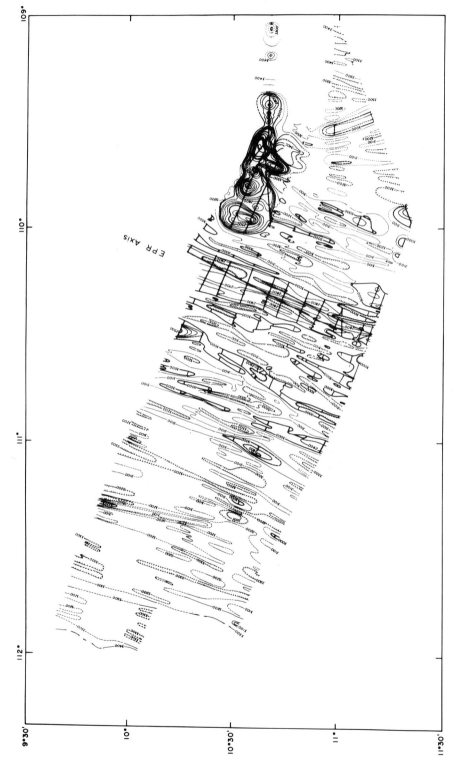

Figure 2. Bathymetric map of the 10.5° S survey area. Depths are in corrected metres; contour interval is 100 m. Regions less than 3,000 m deep are shaded.

have provided the best bathymetric maps of the Nazca plate and adjacent regions (Mammerickx and others, 1975; Mammerickx and Smith, 1978).

DATA FROM THE RISE CREST

9° to 12°S Region

The development of the EPR axis in the region between about 9° and 12°S is the simplest of the areas surveyed. Data in this region (Rea, 1976a) consist of a detailed survey area centered on the rise crest at about 10.5°S and several reconnaissance lines that cross the east flank and crest of the rise. A bathymetric contour map of the 10.5°S survey area (Fig. 2) depicts the typical features of the EPR crest. The precise axis of the EPR is a low ridge about 300 m high and 15 to 20 km wide (Figs. 2, 3). The shallowest depth of this feature is 2,600 to 2,700 m, and the 3,000-m contour line (Fig. 2) generally defines its lateral extent. The axis is aligned along a trend of 018° in the northern two-thirds of the area, but changes trend at 11°S to 012° in the southern part of the survey area. Low ridges commonly occur along the flanks of the axial ridge and where well developed (Fig. 3, lines 4 and 5) divide the axial block into three peaks of subequal width; the middle peak in all cases is the highest one. Such an axial ridge characterizes the Nazca-Pacific plate boundary throughout its length, except at the Galapagos triple junction where there is rift-valley type topography (Lonsdale, 1977b).

Away from the axial ridge, the topography of the upper flank of the EPR is characterized by abyssal hills 3 to 5 km wide and 100 to 200 m in relief. These hills and the intervening valleys are lineated parallel to the rise axis, and some of them may be continuous across the entire area surveyed. More commonly, these features are 20 to 30 km long. A large seamount of about 1,500-m relief occurs in the northeast portion of the survey area (Fig. 2).

Magnetic anomalies are somewhat difficult to interpret at 10.5°S because the crest of the EPR is within a few degrees of the equator and the trend of the axis is not far from north, a geometric condition that results in low amplitude sea-floor spreading magnetic anomalies (Schouten, 1971). Magnetic anomalies in the 10.5°S survey area are irregular, and no obvious sea-floor spreading anomaly stripes extend across this area (Rea, 1976a).

Linear magnetic anomalies are more readily observed on magnetic profiles than on the magnetic anomaly map. Correlation of magnetic anomalies in the entire 9° to 12°S region of the EPR has been done elsewhere (Rea, 1975a, 1976a), and the results of this correlation are presented on Figure 4, a map of the significant bathymetric and magnetic features of the region. Figure 4 shows two small axial offsets at 10.0°S and 11.5°S which divide the EPR axial region into three subregions. In the northern two subregions the axis position is approximately centered within the magnetic anomalies on either flank. The central subportion is offset about 8 to 10 km left-laterally from the northern portion at approximately 10.0°S. This small offset evidently has existed for at least 0.9 m.y., as both the Jaramillo and Brunhes-Matuyama anomalies are symmetrical about the axis segments and are offset by an amount similar to the axis. Anomaly 2 to the east apparently is unaffected by this offset, implying that the offset originated sometime between about 1.7 and 1.0 m.y. ago. Definition of anomaly 2 west of the small offset, however, is ambiguous.

Another small axial offset occurs at about 11.5°S and is right lateral in sense. Unlike the one at 10.0°S, the axial offset in the southern part of the region is not reflected in the magnetic anomalies, and at 12°S the EPR axis lies 10 km west of a median line between the flanking Brunhes-Matuyama anomalies (Fig. 4). Thus, this offset formed sometime during the past 0.7 m.y., since the time of the Brunhes-Matuyama reversal. Both offsets, therefore, were achieved within a maximum time span of 0.6 to 0.7 m.y.

Sea-floor spreading near 10.5°S on the EPR has been essentially symmetrical since the time of anomaly 2 and presently occurs at a whole rate of about 158 mm/yr (Rea, 1976a).

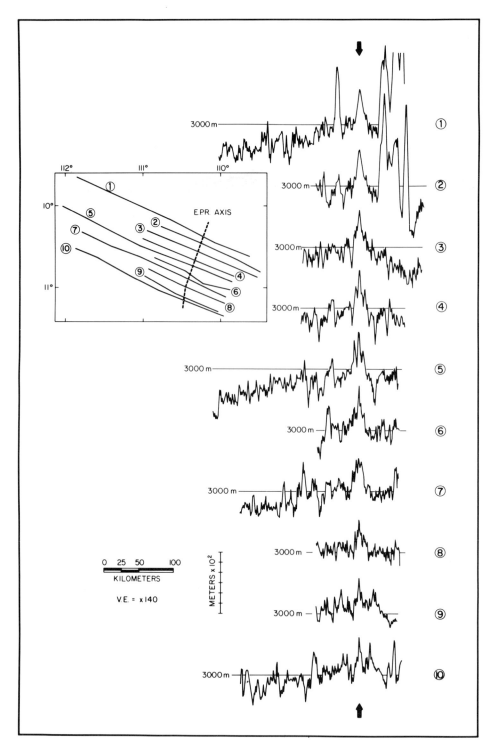

Figure 3. Bathymetric profiles from the 10.5° S survey area. Trackline locations given on inset. West is to the left. Profiles are aligned on the EPR axial ridge (arrows).

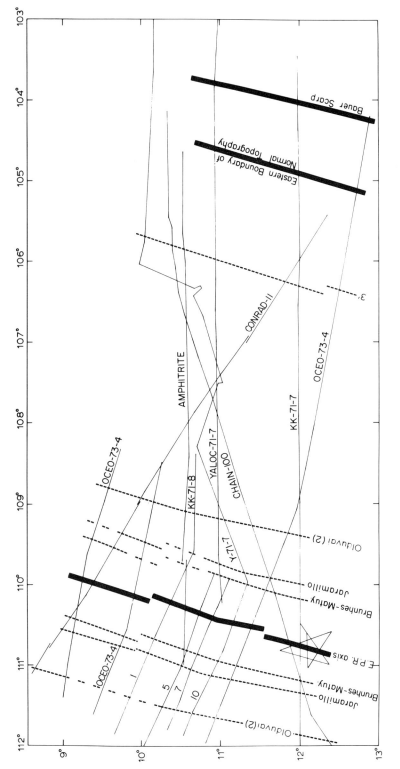

Figure 4. Map of the EPR showing magnetic anomaly (dashed lines) and bathymetric (solid lines) correlations.

Deep-Tow Survey Near 3.5°S

A 90-km long traverse of the EPR axis at 3°23'S by the Scripps deep-tow vehicle shows the topography of the axial region, discussed above in a general fashion, in detail (Lonsdale, 1977a). The EPR axial ridge at 3.5°S is about 15 km wide, 300 m high, and trends 008°. It is characterized by a central shield volcano 2 km wide that is flanked by an area of open fissures and low-relief, fault-bound ridges. The sea floor away from the axial ridge displays fault-bound abyssal hills 2 to 4 km wide and 100 to 300 m high (Lonsdale, 1977a).

The deep-tow survey was centered on the axial shield volcano, a constructional feature characterized by a summit graben approximately 35 m deep and a few hundred metres wide. Nearly vertical fault scarps bound the graben, and fresh pillow lavas form long parallel ridges along its floor. There is also evidence for the ponding of lava flows within the summit graben (Lonsdale, 1977a).

18.7° to 22.°S Region

The EPR axial region between 18.7° and 22°S has the least ambiguous magnetic anomaly data of the several survey areas. Unlike the regions to the north, the linear magnetic anomalies are well developed here and present a readily decipherable record of spreading activity during the past 2.4 m.y. (Rea and Blakely, 1975; Rea, 1978a).

Topographic features of the EPR crestal area near 20°S are shown on the bathymetric map (Fig. 5). The spreading center is denoted by a low ridge about 15 km wide and 300 to 400 m high. This axial ridge trends 013° near 113.5°W in the northern part of the area; near 20.7°S it is offset to the west where it occurs at 114.2°W along the southern edge of the survey area. The axial offset is approximately 15 km in extent and right lateral in nature. It occurs across a rather broad, ill-defined zone extending 30 to 40 km along the general trend of the axis, which is neither a well-defined small transform fault nor just a bend in the axial ridge (Rea, 1978a). Flanking the axial ridge are abyssal hills that commonly are 100 to 200 m high and lineated subparallel to the axis. Larger hills occur in the extreme northwest corner of the area surveyed, and a small seamount lies near 20.8°S, 114.5°W. The overall depth of the 20°S survey area is asymmetric about the rise axis; the west flank averages about 50 m shallower than the east flank near the axial ridge and 100 m shallower farther away (Fig. 5; Rea, 1978a).

Linear magnetic anomalies aligned parallel to the axial ridge are well developed at 20°S. Their unusually clear and unambiguous nature permits ready correlation of the anomalies produced by the Matuyama-Gauss reversal boundary, anomaly 2', dated at 2.41 m.y. old; the Olduvai event, anomaly 2, centered at 1.73 m.y. ago; the Jaramillo event centered at 0.92 m.y. ago; and the Brunhes-Matuyama reversal boundary 0.70 m.y. old (Fig. 6; anomaly ages from Klitgord and others, 1975). In addition to these more obvious features, the magnetic data from this survey area contain evidence of shorter wavelength magnetic anomalies probably produced by brief reversals of the Earth's field (Rea and Blakely, 1975).

Mapping the magnetic anomalies in the 20°S survey area (Rea, 1978a; Fig. 7) reveals several interesting aspects of the spreading regime. All the anomalies younger than the Matuyama-Gauss anomaly are offset similarly to the topographic axis, 15 km right laterally, indicating that this small offset originated some time between about 2.4 and 1.8 m.y. ago. The bathymetric and magnetic trace left by this small axis offset does not follow the Pacific-Nazca transform direction, 285°–105°, but rather forms a broad, south-pointing, V (Fig. 7). The creation of the broad V-pattern suggests that the axial offset is not a transform fault governed by the geometries of rigid plate motions but rather is nontransform in nature (Rea, 1978a). And finally, the spacing of the anomalies on either side of the rise axis record ongoing asymmetrical spreading averaging 70 mm/yr to the west and 92 mm/yr to the east during the past 2.41 m.y. Spreading rates between each anomaly pair determined along each profile and shown on Figure 7, reveal that spreading rates have not been constant during the past 2.4 m.y. but have varied over about 10% of their value from a high of 170 mm/yr whole rate during the Matuyama-Gauss to Olduvai interval to a low value of 156 mm/yr during the Olduvai to Jaramillo interval (Rea and Scheidegger, 1979).

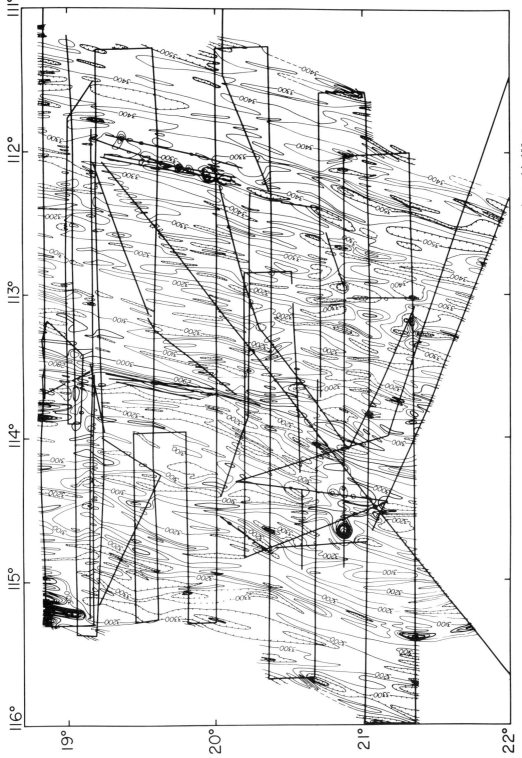

Figure 5. Bathymetric map of the 20°S survey area. Depths are in corrected metres; contour interval is 100 m.

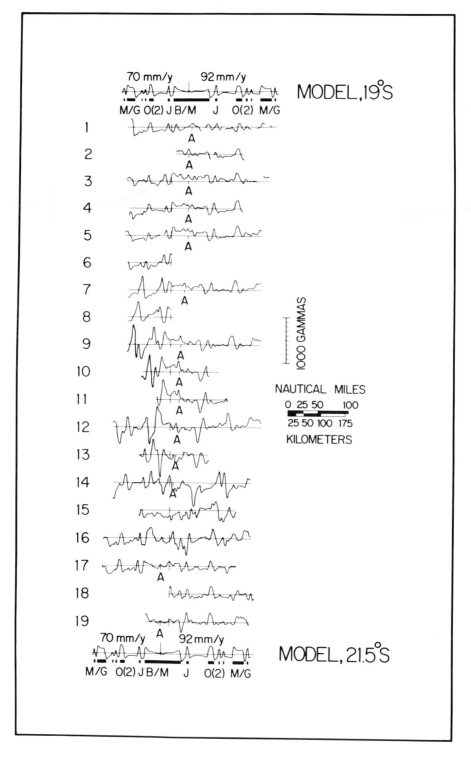

Figure 6. Magnetic anomaly profiles of the EPR axial region near 20°S. M/G is the Matuyama-Gauss anomaly (2'); O, the Olduvai anomaly (2); J., the Jaramillo anomaly; and B/M, the Brunhes-Matuyama anomaly. "A" denotes location of topographic axis of the EPR. Vertical dashes mark the location where profiles cross 114°W.

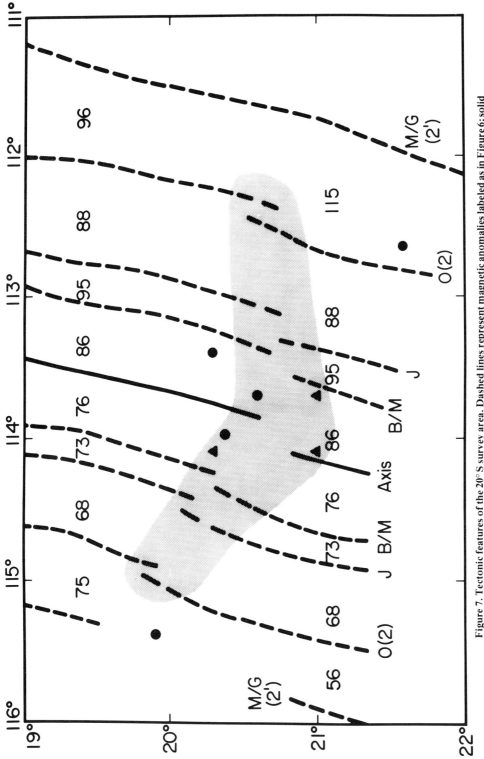

Figure 7. Tectonic features of the 20°S survey area. Dashed lines represent magnetic anomalies labeled as in Figure 6; solid line is EPR axis. Numbers give spreading rates in mm/yr for each anomaly-bound interval. Dots denote better-located earthquake epicenters (1961-1975); triangles denote less well located epicenters (all 1960). Shading outlines the broad V-shaped region of magnetic and bathymetric disturbance.

30.5° to 31.3°S Region

The region near 31°S on the EPR is the southernmost extent of relatively uncomplicated sea-floor spreading along the Nazca-Pacific plate boundary (Rea, 1977). South of about 32°S lies the geologically complex region that may either reflect the tectonic influence of the EPR–Chile Ridge triple junction (Forsyth, 1972; Stover, 1973) or be the location of a small "mini-plate" (Herron, 1972b, Anderson and others, 1974). The data set at 31°S is more restricted than in the other regions studied, consisting of only four tracklines across the axial region.

The bathymetric features of the EPR axial region at 31°S are generally similar to those observed elsewhere along the Nazca-Pacific plate boundary. The spreading axis is marked by a low ridge 450 to 500 m high, about 20 km wide at the base, and trending 012° along about 111.9°W (Fig. 8). Flanks of the axial ridge are fairly smooth, and abyssal hills 100 to 200 m high occur on either side of the axis. In other EPR crest survey areas (Figs. 2, 5; Lonsdale, 1977a) these hills are strongly lineated parallel to the axis, but such a lineation is not as apparent in this area (Fig. 8), and bathymetric contours, except for those along the axial ridge, have not been extended between the more widely spaced tracklines. Several isolated peaks occur in the survey area, the largest of which is a 2,050-m-high seamount occurring at about 31°S, 113°W; other smaller peaks occur throughout the area.

Linear magnetic anomalies within the region surveyed are 400 to 600 γ in amplitude and are easily identified (Fig. 9; Rea, 1977). Examination of the magnetic anomaly map (Fig. 9) reveals that the oldest anomaly shown, 2', is straight, anomaly 2 is curved, and the Jaramillo anomaly is offset about 10 km right laterally. This offset apparently was soon healed as the Brunhes-Matuyama anomaly is continuous, although curved. At present, the spreading center is linear (Fig. 8). Additional variations in the spreading regime are revealed along the southern part of the small survey area (Fig. 9) where the contoured Jaramillo anomalies on either flank of the rise are not parallel to each other between the southern two lines. This is in conflict with the basic concept of linear magnetic anomaly generation at a single axis and suggests that additional complications in the spreading process occurred about the time of the Jaramillo event.

Calculation of crustal accretion rates at 31°S show a total rate of 163 mm/yr since the time of the Matuyama-Gauss reversal. Spreading half-rates, however, have not been equal during the past 2.41 m.y., as crustal accretion has averaged about 77 mm/yr to the west and 86 mm/yr to the east. During the past 0.7 m.y., however, spreading has been symmetrical. Whole rates of spreading have also varied, ranging from a maximum of 176 mm/yr during the interval between anomalies 3.1 and 2' to a minimum of 145 mm/yr during Olduvai to Jaramillo time. Presently, crustal accretion is occurring at about 162 mm/yr at 31°S (Rea, 1977; Rea and Scheidegger, 1979).

5° to 7°S Region

The section of the EPR near 6°S is fairly complex; fortunately the trackline coverage here is very good (Fig. 10). A 55 km, right-lateral offset of the spreading center occurs in the southern half of the survey area. Magnetic anomalies are poorly developed and difficult to measure in this region because of its proximity to the geographic equator and position directly on the geomagnetic equator (Rea, 1976b).

The rise axis is represented in the northern half of the survey area by a ridge about 300 m high and 15 to 20 km wide (Figs. 10, 11). The trend of the axis is 014° in the central part of the survey area, and changes to 022° in the northern part. A scarp occurs near 107°W, about 20 km west of and subparallel to the axial ridge. Across this scarp the ocean floor is stepped down to the west 100 to 200 m (Figs. 10, 11). South of 5°S, the axial peak is lower and merges with other hills into topographic block 70 to 75 km wide and elevated 200 m above the surrounding area. This wider block is bounded by the scarp noted above on the west and by a similar, although less extensive, feature on the east (Fig. 11, lines 6 and 7). The highest of the several small peaks on this 75-km-wide block lies on trend with the EPR axis to the north. The southern edge of this feature occurs at the fracture zone at about 6.4°S. South of the fracture zone the EPR axis is again represented by an axial ridge about 300 m high, 10 to 15 km wide,

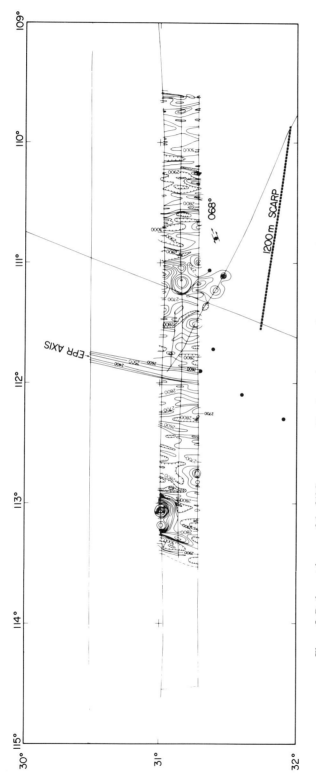

Figure 8. Bathymetric map of the 31° S survey area. Depths are in corrected metres; contour interval is 100 m. Dots denote earthquake epicenters.

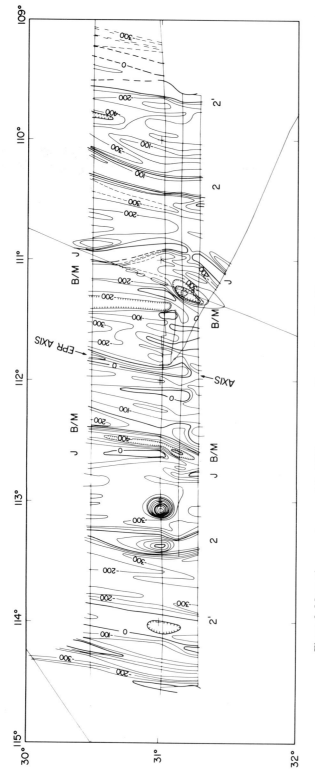

Figure 9. Magnetic anomaly map of the 31°S survey area. Contour interval is 100 gammas. Nomenclature as in Figure 6.

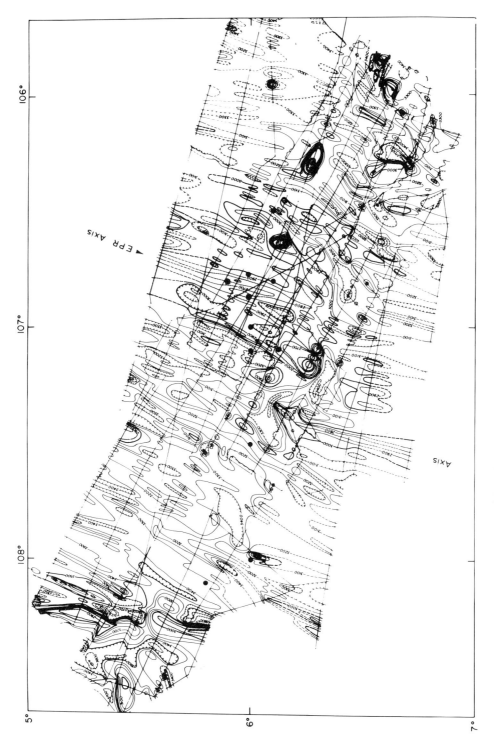

Figure 10. Bathymetric map of the 6° S survey area. Depths are in corrected metres; contour interval is 100 m. Regions less than 3,000 m deep are shaded. Dots denote earthquake epicenters.

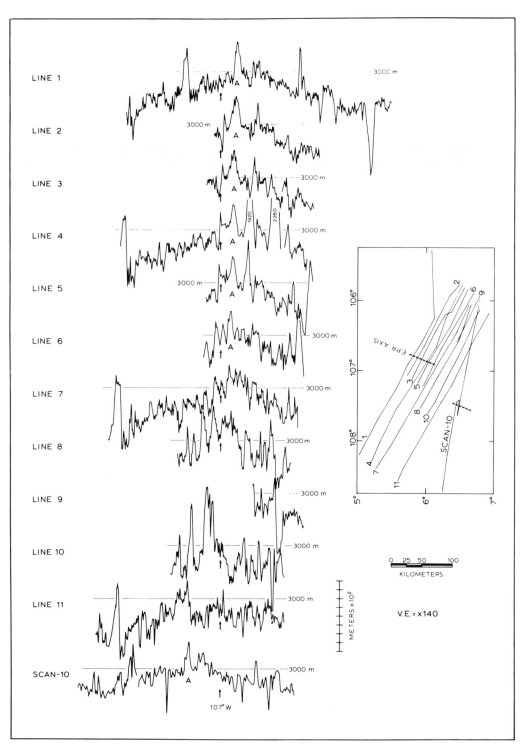

Figure 11. Bathymetric profiles from the 6°S survey area. Trackline locations given on inset. West is to the left. "A" denotes EPR axial ridge.

and trending 015°. Abyssal hills in the survey area are similar in form to those found elsewhere on the EPR, having a relief of 100 to 150 m and a wavelength of 3 to 5 km. The hills are generally elongate subparallel to the rise axis.

Two large topographic blocks lie in the western portion of the survey area. They range from 6 to 10 km wide, having about 1,000 m of relief, and trend slightly east of north along about 108.2°W. Deeps paralleling the blocks on the east are 200 to 300 m below the regional depth level of about 3,400 m. A similar but slightly smaller block occurs at 107.4°W along the northern edge of the survey area (Figs. 10, 11). All blocks have their steeper slopes facing east and their gentler slopes to the west (Fig. 11, lines 1, 4, 7, and 11). The morphology of these features suggests that they are large fault blocks, with the fault scarp facing the axis, similar to many smaller bathymetric features (Rea, 1975b).

An 800-m-deep depression 4 to 9 km wide and about 30 km long occurs near 6.6°S, 106.3°W in the southeastern part of the survey area. This and another depression of similar magnitude found on line 1 northeast of the survey area (Fig. 11) may also be fault-formed features (Rea, 1976b).

Because of the low-amplitude magnetic anomalies at 6°S and the large-amplitude diurnal variation, the electroject effect (Rea, 1976b), it was not possible to construct a coherent magnetic anomaly contour map of the region. However, a few of the individual profiles do contain usable magnetic information after data taken during the time of the electroject effect, about 2 or 3 h on either side of the local noon, were discarded. These data show several things. Anomaly 2 trends 042°, 20° east of the present regional axis trend. This more easterly trend apparently records an episode of oblique spreading as there is no evidence elsewhere for a large change in the direction of Nazca-Pacific opening. The trends of the Jaramillo and Brunhes-Matuyama anomalies cannot be reliably established; the rise axis is not centered between these anomalies but lies 8 to 10 km west of the medial position. This implies that during the past 0.7 m.y. there has been asymmetric spreading faster to the east or a 10 km westward jump of the axis. Spreading rates for the 6°S area are presently about 151 mm/yr (Rea, 1976b).

Earthquake locations in the vicinity of the 6.2°S fracture zone do not coincide with the axial offset but lie northward about 40 km (Fig. 10). Although the epicentral determinations may be biased by irregularities in seismic travel times, such errors should be much less than 40 km. Also the epicenters associated with larger fracture zones at 4.5°S and 9°S are located within the active portion of those transform faults. Thus, most of the earthquakes are probably correctly located and represent seismic activity away from the fracture zone as defined on the basis of the axial offset.

The eight epicenters that lie north of the fracture zone appear to be associated with the unusually broad development of the EPR axial ridge and the apparent pedestal that it sets on. The 6.2°S fracture zone does not generate recordable earthquakes and must be well lubricated. All the earthquakes and the unusual axial bathymetry occur near the part of the EPR axis that trends 014°. Between the northern part of the 6°S survey area and 4.5°S fracture zone the EPR axial block trends 022° (Fig. 13) and has neither a pedestal nor recorded earthquakes.

22° to 27°S Region

Herron (1972b) first noticed the unusual pattern of seismic activity west of Easter Island in the region between about 22° and 27°S on the EPR where epicenter locations form a rough circle 4° to 5° in diameter. From an examination of bathymetric and magnetic anomaly information, she postulated that in this region the spreading center was doubled and that a small, growing plate lay between the two axes (Herron, 1972b). Forsyth (1972) showed that the first motions of two strike-slip earthquakes near 22.2°S diverged widely from the normal Nazca-Pacific transform direction and briefly mentioned some possible tectonic complications, including Herron's small-plate suggestion.

Anderson and others (1974) combined all the available bathymetric, magnetic, and focal-mechanism data from the 22° to 27°S region to define more accurately the small Easter plate. They envision the Easter plate as being bound by transform faults on the northwest and northeast which join the EPR at about 22.2°S in a ridge-fault-fault triple junction. The southwestern and southeastern edges of the plate are dominated by divergent boundaries that join the EPR at about 27°S in a ridge-

ridge-ridge triple junction (Anderson and others, 1974). The plate so described is an equidimensional figure roughly 500 km across (Fig. 12a).

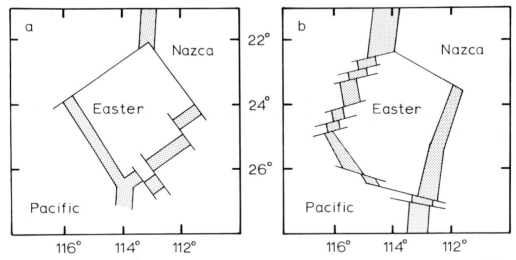

Figure 12. Configuration of the Easter plate as envisioned by (a) Anderson and others (1974), and (b) Handschumacher and others (this volume). After Handschumacher and others.

During the course of the Nazca Plate Project, the *Kana Keoki* and *Oceanographer* systematically collected a large amount of new data in the region of the Easter plate (Handschumacher and others, 1975). Handschumacher and others (this volume) have made a comprehensive study of all the data and find the boundaries of the small plate to be somewhat similar to that described by Anderson and others (1974) (Fig. 12b). They suggest that the eastern divergent edge of the small plate is becoming dominant and eventually will be the site of all spreading activity. When extinct, the western spreading center will join the Pacific plate and migrate westward with the former Easter plate (Handschumacher and others, this volume).

Fracture Zones

Five large fracture zones, ranging in offset from 55 to 200 km, cross the EPR axis between the Galapagos triple junction and the Easter plate (Table 1). The northern two fracture zones have a combined right-lateral offset of about 400 km that occurs in roughly equal amounts across the Quebrada Fracture Zone at about 3.8°S and the Gofar Fracture Zone at 4.5°S (names from Lonsdale, 1977b; Mammerickx and Smith, 1978). A deep-tow study conducted in the eastern part of the Quebrada transform revealed that the active portion of that fault is characterized by a shallow, steep-sided trough 100 to 200 m wide, perhaps 30 m deep, and trending 098° (Lonsdale, 1978). The fracture valley shoals at the axis intersection, the opposite effect from that observed in Atlantic-type fracture zones (Sleep and Beihler, 1970). Lonsdale's (1978) deep-tow survey also revealed the presence of several 1-, to 2-km-long scarps trending 050°, about 50° to the main transform direction. These features are the result of normal faulting and apparently are oblique tensional fractures occasioned by the horizontal shear couple (Lonsdale, 1978). The Quebrada Fracture Zone does not extend west beyond 107°W or 108°W. It does not occur on the *Oceanographer* tracklines along 108.5°W, but has a clear bathymetric and magnetic expression along the YALOC-71-9 track at 106°W (Figs. 13, 14).

The Gofar Fracture Zone at 4.5°S is the most extensive fracture zone between the Galapagos triple junction and the Easter plate. It is about 100 km wide, has an internal relief of 1.5 km, and a basement drop 500 m down to the north where it is crossed by the two *Oceanographer* tracklines (Rea and Malfait, 1974, their Fig. 3c). Reconnaissance data shown on Figures 13 and 14 and reported by Rea

TABLE 1. CHARACTERISTICS OF LARGER FRACTURE ZONES ALONG THE NAZCA-PACIFIC PLATE BOUNDARY, NORTH OF THE EASTER PLATE

Latitude (°S)	Name*	Offset right lateral (km)	Trend Bathymetric (°)	Trend Magnetic (°)	Trend Seismic (°)
3.8	Quebrada	400†	098†		104§
4.5	Gofar				103.5§
					100#
6.2	Unnamed	55**		108**	
9.0	Wilkes	200††		097±2††	104§
13.5	Garret	130§§	115±10§§	105±5§§	102##

*Mammerickx and Smith, 1978.
†Lonsdale, 1978.
§Anderson and Sclater, 1972.
#Forsyth, 1972.
**Rea, 1976b.
††Rea and Kureth, 1979.
§§Anderson and others, 1978.
##Anderson and others, 1974.

and Malfait (1974) show that the extinct Galapagos Rise [as mapped by Herron (1972a) and by Mammerickx and Smith (1978)] in the center of the Nazca plate (Fig. 1) does not extend north of about 7°S. Both the old rise and the Bauer Deep between it and the modern EPR are apparently bounded on the north by the southeastern extension of the Gofar Fracture Zone. Furthermore, of the four fracture zones shown on Figure 14, only one at 4.5°S extends very far westward (Rea and Malfait, 1974). North of about 3°S, the fossil rise system in the east-central Pacific lies to the west of the present spreading center (Sclater and others, 1971; Anderson and Davis, 1973; van Andel and others, 1975). Prior to approximately 10 m.y. ago, the southern end of this northern system, now at 3°S, 115°W, must have been joined to the northern end of the Galapagos Rise, now at about 7°S, 97°W, by a long transform plate boundary with left-lateral offset. The 4.5°S fracture zone is the only feature extensive enough to have been this plate boundary·[see Fig. 7 of van Andel and others (1975) for present configuration of these features]. Allowing for the amount of sea floor created since the spreading-center jumps to the present configuration, this old offset between the northern end of the extinct Galapagos Rise and the southern end of the fossil Clipperton-Mathematicians Ridge must have been about 750 km long (Rea, 1978b).

Plots of epicenters from earthquakes associated with the Gofar Fracture Zone reveal an easterly (Stover, 1973) or northeasterly (Forsyth, 1972) trend, but earthquake focal mechanisms from this region show strike-slip motion in the direction about 103° (Forsyth, 1972; Anderson and Sclater, 1972). This discrepancy between epicenter trends and first-motion directions is best explained by presuming the existance of a series of small, active en echelon faults (Forsyth, 1972).

The 6.2°S fracture zone offsets the EPR axis 55 km right laterally (Fig. 10). The fracture zone is not well expressed bathymetrically, having neither a continuous ridge or trough along it, nor even an obvious system of discontinuous ridges and troughs. The deepest part of the fracture valley lies between the offset axes rather than at the axial intersections (Fig. 10), as is true for the Quebrada Fracture Zone (Lonsdale, 1978). Outside the offset axes, the bathymetric expression of the fracture zone is markedly reduced. In the southeastern portion of the survey area the abyssal hills change trend, but not character, along the probable southeastern extension of the fracture zone. To the west of the southwestern portion of the EPR axis, the only apparent bathymetric expression of the fracture zone is a slight southerly rise in the sea floor between 107.5° and 108°W (Fig. 10).

A high-amplitude magnetic anomaly characterizes the 6.2°S fracture zone. This dipole anomaly has an amplitude of up to 1,100 γ and trends 108° (Rea, 1976b). Magnetic anomalies associated with the 6.2°S fracture zone are, like the bathymetric expression, much more pronounced in the active portion of the fracture zone and extend only a few tens of kilometres beyond the offset ends of the axis. Identification of sea-floor spreading magnetic anomalies south of the fracture zone (Rea, 1975a)

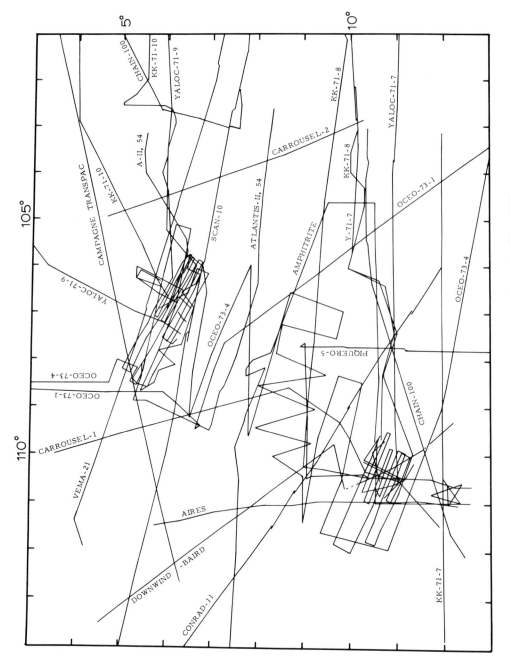

Figure 13. Tracklines across EPR axial region, 3° to 13° S.

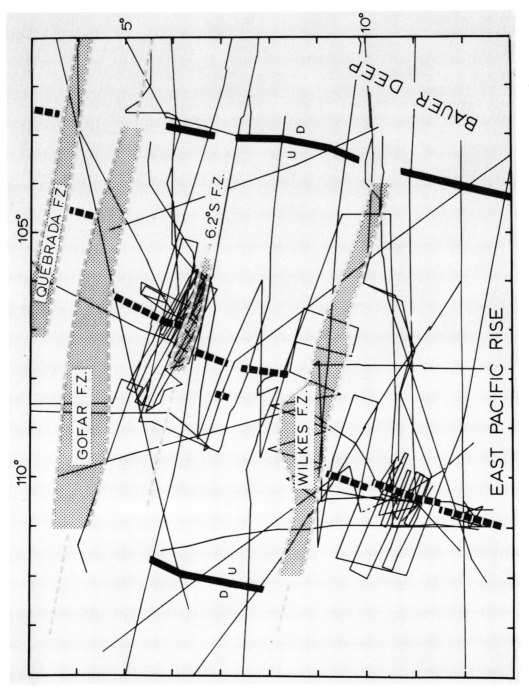

Figure 14. Regional structure of the EPR, 3° to 13°S. Black line is the rise axis. Gray lines represent the Bauer scarp on the east edge of the modern EPR and the equivalent feature on the west flank. Dashed lines outline fracture zones.

suggests that the EPR has been offset at this location for at least 1.7 m.y. The rather abrupt changes in the geophysical expression of the 6.2°S fracture zone occurring just beyond the offset ends of the axis, then, are interpreted as reflecting some change in the process of formation of the fracture zone that occurred a few hundred thousand years ago (Rea, 1976b).

The Wilkes Fracture Zone at 9°S offsets the EPR axis a distance of about 200 km right laterally. The topographic expression of the 9°S fracture zone is very different along the active and inactive portions. Outside the offset rise axis, the 9°S fracture zone is represented by a simple topographic step or slope downthrown in the proper direction. About 75 km from the offset rise axes, however, the fracture zone becomes much broader and is 100 km wide in the active portion. The southern margin of this broad zone is characterized by a large continuous magnetic anomaly up to 1,200 γ in amplitude that trends 097°. A low ridge a few hundred metres high and perhaps 190 km long and an associated magnetic anomaly trend 070° across the wide portion of the fracture zone between the ends of the offset axes. This feature, oriented about 25° to the main transform trend as defined by the large magnetic anomaly may be a shear related feature as are the much smaller oblique ridges in the Quebrada Fracture Zone (Lonsdale, 1978; Rea and Kureth, 1979).

Anderson and others (1978) gave a brief description of the Garret Fracture Zone that offsets the EPR axis 130 km right laterally at 13.5°S. The Garret Fracture Zone is characterized by a 60-km-wide region of complicated topography with an overall trend of 115±10°. The fracture zone is shallower at the axis intersections and deeper midway between them. Magnetic anomaly data are much more straightforward; a single large anomaly with amplitudes up to 1,250 γ trends 105±5° along the fracture zone (Anderson and others, 1978).

These five fractures zones (Table 1) have several common characteristics. Their active zones are about half as wide as the offset is long and are characterized by apparently discontinuous or oblique topographic features. The simple trough-and-wall topography typical of Atlantic fracture zones and those on the Pacific-Cocos plate boundary does not occur along the Pacific-Nazca plate boundary. Magnetic anomalies along the EPR fracture zones are of large amplitude and are presumed to give a more reliable indication of the the direction of transform motion than does the bathymetry (Rea, 1976b; Anderson and others, 1978). The trends of the large anomalies associated with the 6.2°S, Wilkes, and Garret Fracture Zones are slightly divergent, however, and results of earthquake first-motion studies may give a better indication of the direction of relative plate motion, 102±2° along this section of the rise (Table 1; Anderson and Sclater, 1972; Forsyth, 1972; Anderson and others, 1974).

In addition to the five transform faults described above, there are some smaller axial offsets. Two of these at 10°S and 11.5°S, discussed above and by Rea (1976a), span about 10 km each and appear to be similar to other small axial offsets recorded along the Nazca-Pacific boundary. Reconnaissance data between 6.5°S and 8.5°S reveal what may be another axis offset. The 1973 *Oceanographer* crossing of the axial region (Oceo-73-1) showed no axial ridge at the expected location near 7.4°S, but rather an axis-like ridge lying 65 km to the west (Rea, 1976c). Two other transform faults offset the rise axis 25 km each at 1°35′N and 2°50′S; the northern one separates the rift-valley topography at the triple junction from the usual axial ridge to the south (Lonsdale, 1977b).

Reconnaissance data

In addition to data from the rise-crest survey areas and from the region of the Easter Island mini-plate, there exists a considerable amount of reconnaissance data from the EPR. Most of the data are in the form of individual profiles across the crestal region and were completed by Lamont-Doherty Geological Observatory (L-DGO), Scripps Institution of Oceanography (SIO), Woods Hole Oceanographic Institution (WHOI), or the Pacific Oceanographic Laboratory (now Pacific Marine Environmental Laboratory) of NOAA. Many of these tracklines are shown on the bathymetric maps published by Mammerickx and others (1975) and Mammerickx and Smith (1978). Herron (1972a) has given bathymetric and magnetic profiles from the L-DGO data in the area. Handschumacher and others (1975) published a data report containing bathymetric and magnetic profiles from the southeast Pacific but did not include either SIO or WHOI data.

The reconnaissance data support the general view of the EPR being a broad region of low topographic relief with an axial ridge a few hundred metres high and between 15 and 25 km wide. There are two regions, however, where these data can contribute more than routine information to the study of EPR tectonics. The first is a survey of the EPR axis between about 17° and 18.5°S completed in 1970 by L-DGO cruise *Conrad*-13. Information from this survey has not been published as such, but Handschumacher and others (1975) showed some of the bathymetric and magnetic anomaly profiles obtained. The region of the *Conrad*-13 survey is, as might be expected, very similar to the northern part of the 20°S survey area discussed above (Figs. 5, 6). The EPR axis is a low ridge trending slightly east of north. Magnetic anomaly profiles indicate sea-floor spreading rates of about 160 mm/yr since anomaly 2 time and show that the spreading at 17° to 18.5°S is asymmetrical, faster to the east. Asymmetrical sea-floor spreading, faster to the east, occurs in all regions studied along the EPR between the Garret Fracture Zone and the 35°S triple junction.

The region between 3° and 13°S along the EPR is the location of, in addition to the areas at 6° and 10.5°S surveyed in detail, 12 cruises of a reconnaissance nature (Fig. 13). These tracklines reveal the major structural features of the region (Fig. 14). Herron (1972a) and Anderson and Sclater (1972) have demonstrated independently and from different data that the EPR in this region originated, probably during the late Miocene, by one or more westward jumps of spreading activity. When a new spreading center suddenly begins to create new material in old lithosphere, one result is the formation of a topographic form peculiar to this situation. This form is shaped like a broad building with steep sides and a "roof" sloping gently up to the axis. Vertical offset on the boundary-scarp "sides" is proportional to the age difference between the old and new lithosphere. Similarly, the slope of the "roof" decreases with increasing spreading rate (Sclater and others, 1971). The present topography of the EPR at 6° to 7°S may be the best example of this general form; the Scan-10 profile (Fig. 15) shows the steep boundary scarps and gently sloping upper surface of the rise. Other crossings of the steep scarps bounding this portion of the EPR have been shown in profile form by Anderson and Sclater (1972), Handschumacher and others (1975) and Rea (1976a). Locations of these scarps are mapped on Figure 14. The scarp along the east edge of the EPR between 5° and about 13°S and separating it from the Bauer Deep to the east is the Bauer Scarp (Rea, 1976a; Mammerickx and Smith, 1978). Dating this feature would reveal the time of origin of the modern EPR.

Seismicity

EPR seismicity was first described by Sykes (1963) who noticed less seismic activity between 9° and 20°S than on parts of the rise to the north and south. In more recent studies of the distribution of earthquakes along the EPR, both Northrop and others (1970) and Stover (1973) pointed out that essentially all the teleseismic activity is associated with fracture zones, a characteristic of fast-spreading rises. Berg and Sutton (1974) studied the cumulative strain release along the western edge of the Nazca plate and found that periods of high and low strain release correlated with strain release fluctuations along the Peru-Chile subduction zone.

Several investigations have published focal-mechanism studies from earthquakes along the Nazca-

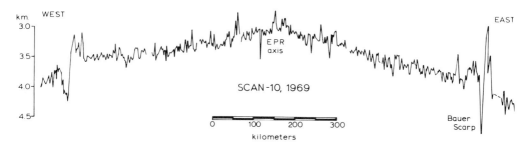

Figure 15. Scan-10 profile across the modern EPR. Profile location on Figure 13.

Pacific boundary: Stauder and Bollinger (1964, 1966), Sykes (1967), Anderson and Sclater (1972), Forsyth (1972), and Anderson and others (1974). The last two papers are the most inclusive. The 22 mechanisms published, apparently all from fracture zones, record the general transform motion between the Nazca and Pacific plates, approximately 102° (Table 1), and reveal the more complex dynamics in the vicinity of the small Easter plate and near the EPR–Chile Ridge triple junction at 32° to 34°S.

Heat Flow

The first few heat-flow data in the vicinity of the Nazca-Pacific plate boundary were published by von Herzen (1959). Since then more extensive lists of heat-flow data have been given by von Herzen and Uyeda (1963), Lee and Uyeda (1965), von Herzen and Anderson (1972) and Anderson and others (1976, 1978). These authors and others who have discussed the available data along the EPR (von Herzen and Lee, 1969; Le Pichon and Langseth, 1969; Langseth and von Herzen, 1971) all report a broad band of high and variable heat flow along the crest of the EPR.

DISCUSSION: REGIONAL TECTONICS

Tectonic aspects of the several individual study areas along the Nazca-Pacific boundary have been adequately discussed in the publications pertaining to those regions, Rea, (1976a, 1976b, 1977, and 1978a) and Lonsdale (1977a, 1978) for the rise crest survey areas, and Anderson and others (1974) and Handschumacher and others (this volume) for the Easter plate. The formation of the EPR axial ridge, its shoulders, and the flanking tilted abyssal hills has also been discussed in detail (Lonsdale, 1977a; Rea, 1975b; Anderson and Noltimier, 1973). changes in the rate of sea-floor spreading both with distance along the axis and with time over the past 4 m.y. have been detailed by Rea and Scheidegger (1979). It is more appropriate in this presentation to discuss those aspects of EPR tectonics that entail structural and tectonic information from more than one location along the Nazca-Pacific divergent boundary. Four such aspects of the regional tectonic scheme will be discussed as outlined in the introduction: along-axis variations in crustal topography; asymmetrical sea-floor spreading; the five 10- to 15-km axial offsets observed; and the history of spreading-center jumps that created the modern EPR between the fracture zones at 4.5°S and 13°S.

Topographic Profile Along the EPR Crest

If magma plumes rising through the mantle (Morgan, 1971, 1972) contribute additional material, beyond that normally obtained from the asthenosphere (Schilling, 1973; Schilling and others, 1976), to some location beneath the spreading center, then a topographic expression of this phenomenon should be expected. The plume crest should underlie a volcanic center that may rise kilometres above the regional depth along a spreading ridge. Ridge crestal depths will increase away from the volcanic center, and plume-generated magma may flow down this gradient along the conduit formed by the subaxial magma chamber (Vogt and Johnson, 1975; Vogt, 1976). Significant offsets of this conduit at transform faults may act as dams and impede the flow, resulting in mass accumulations (shallow axial depths) on the upflow side of the offsets (Vogt and Johnson, 1975). These concepts have been utilized to explain topographic gradients and offsets along the spreading center away from Iceland, the Azores, the central Juan de Fuca Ridge (Vogt and Johnson, 1975), and the Galapagos Islands (Vogt, 1976; Johnson and others, 1976).

Axial topography north of 18°S along the Nazca-Pacific plate boundary has been shown in a general manner by Lonsdale (1976) who also presented a detailed profile along the EPR axis extending from 10°N to 6°S (Londsale, 1977b). A detailed profile depicting the axial topography along the EPR from 10°N to 40°S is shown in Figure 16.

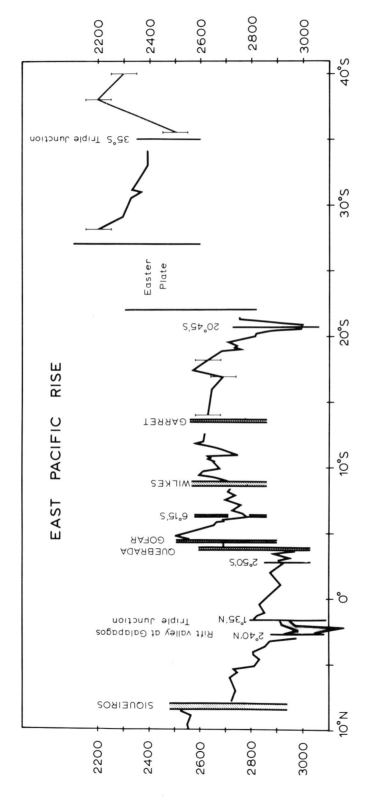

Figure 16. Topographic profile along EPR crest, 10° N to 40° S. Error bars ± 50 m on points picked from profiles, other points plotted at observed depth in corrected metres. Location of fracture zones indicated. Data north of 5° S from Lonsdale (1977b).

The Nazca-Pacific spreading center does not display a simple, to the first order, crestal profile as does the northern Mid-Atlantic Ridge (Vogt, 1976). Rather, there are three distinct topographic levels, 1,000 to 2,000 km long, that ascend to the south: the regional depth is 2,850±50 m between 8°N to 4°S; from about 5°S to 22°S it is 2,650±50 m; and south of the Easter plate the axial depth is 2,350±50 m. Several shorter wavelength, approximately 100 km, topographic features are superimposed upon the three broad steps (Fig. 16).

In the deepest region, north of the Quebrada and Gofar Fracture Zones, the sea floor has an overall southerly slope broken by the deep section along the rift-valley section of the EPR near 2°N. The floor of the small rift valley may reach 3,150 m, about 300 m below normal axial depths in this region (Lonsdale, 1977b; Fig. 16).

Across the Quebrada-Gofar fault system, the sea floor steps up to the south over 400 m to a relatively shallow peak near 5°S. The ocean floor then slopes down almost 300 m to a relative low at the 6.2°S fracture zone. Three minor peaks between the Wilkes and Garret Fracture Zones correspond to the three axial segments mapped between 9°S and 13°S and the two intervening lows to the small axial offsets near 10.0°S and 11.5°S (Figs. 4, 16).

South of the Garret Fracture Zone there is another relatively high area near 18°S. Near 20.5°S occurs a section of the axis that deepens to 3,000 m, 350 m below the regional axial depth (Fig. 16). This depression, which is relatively deeper than that at the Galapagos triple junction, occurs in conjunction with the small, probably nontransform, axial offset at 20.7°S (Figs. 5, 7; Rea, 1978a) and serves to emphasize the unusual nature of that offset.

Just south of the Easter plate the rise axis stands about 550 m higher than immediately north of it and slopes downhill from 28°S to the triple junction near 35°S. The rise crest appears to have a saddle centered at this triple junction with elevations rising again to the south (Fig. 16).

Two general observations arise from examination of this EPR axial profile. First, there is almost always a deepening of the axial ridge at intersections with axis offsets. Only the Garret Fracture Zone does not display this characteristic. Even the small axial offsets at 10°S, 11.5°S, and 20.7°S display moderate to large amounts of axial deepening. There is no apparent correlation between the distance of axis offset and the magnitude of crestal depression. A deepening of the spreading center at transform fault intersections is a common occurrence along spreading centers (van Andel and others, 1971; Ramberg and van Andel, 1977; Lonsdale, 1977b), and such along-axis relief often approaches 500 m on slow-spreading ridges (Needham and Francheteau, 1974).

The second general observation is that the EPR axial profile of Figure 16 is very difficult to reconcile with Vogt's plume and subaxial flow model (Vogt, 1976); alternatively, there may not be any plumes along the EPR. Reversals in the regional slope occur every few hundred kilometres along the EPR axis negating the possibility of any long-distance subaxial magma flux. Furthermore, Schilling and Bonatti (1975) have concluded on the basis of geochemical evidence that the olivine tholeiite basalts generated along the EPR between 2°S and 19°S are normal mid-ocean ridge basalts and show no influence of plume-related compositions.

Farther south, however, the rise crest reaches its maximum elevation, about 2,200 m, just south of the Easter plate and near Easter Island (Fig. 16). It is possible that the Easter plate serves to dam any northward magma flux from a presumed Easter volcanic center, but the southerly slope extends only to the ridge-fault-fault triple junction at 35°S before it begins to rise again.

Asymmetrical Sea-floor spreading

Asymmetrical sea-floor spreading, faster to the east, has been documented by magnetic anomaly data along the Nazca-Pacific plate boundary from 17°S on south. In the 20°S survey area there is a strong depth asymmetry, deeper to the east, associated with the spreading asymmetry. The detailed map of the southeastern Pacific Ocean published by Mammerickx and Smith (1978) shows that this depth asymmetry continues north to the 13.5°S fracture zone (see also the discussion by Lonsdale, 1977b). Farther north, depths and spreading rates are symmetrical. Mapped depth asymmetries, therefore, suggest that asymmetrical accretion along the EPR extends north to the south side of the Garret Fracture Zone at 13.5°S. A change from asymmetrical to symmetrical spreading across the Garret Fracture Zone may account for the more complex nature of this feature (Mammerickx and others, 1975; Anderson and others, 1978).

Two aspects of the regional asymmetrical spreading process require consideration, the continuity of the process and the causes of asymmetrical accretion. More crust of a given age span may occur on one side of a spreading center as a result of either discontinuous, probably fault-controlled, processes such as axis jumps or continuous volcanic-thermal processes such as true asymmetrical spreading. However, there may be a complete spectrum of smaller and smaller axis jumps as one progresses from an obviously discontinuous process to apparently continuous process. Sea-floor spreading seems continuous when discrete axis jumps can no longer be discerned. The definition of continuous, then, is empirical and depends upon the nature of the data available.

Magnetic anomaly data in the 20°S area where the spreading asymmetry is best displayed were obtained by the *Oceanographer*. The digitization interval of these data is about 2.25 km (5 min). An analysis of anomaly widths on the faster and slower spreading flanks at 20°S have shown that, within the limits of these types of data, asymmetrical spreading can be considered to be a continuous process along this part of the Nazca-Pacific plate boundary (Rea, 1978a).

Causes of true regional asymmetric spreading have been mentioned by several authors but discussed in detail by only a few (Hayes, 1976, Weissel and Hayes, 1971, 1974). Separating oceanic plates, in the most simple case, should split at their weakest part, which is presummed to be the hot center of the dike most recently emplaced along the spreading axis. Regardless of the motion of the individual plates, this process should result in symmetrical accretion to each (Morgan, 1971). The only way asymmetrical accretion can occur appears to be if the newly emplaced dike regularly cools in an asymmetrical fashion and therefore does not split down the middle as plate separation continues. This requires that one side of the dike emplacement zone be hotter than the other; more material will adhere to the cooler side.

An important corollary of the ocean-floor age-depth relationship (Sclater and others, 1971) is that the faster spreading flank of a rise should be shallower than the slower spreading flank at the same distance from the axis while age-depth curves of both flanks are the same. Neither of these holds true for the 20°S EPR survey area where the slower spreading west flank of the rise is 100 m or more shallower than the faster spreading east flank (Fig. 5). Theoretical depths for any age of ocean floor less than about 80 m.y. old can be calculated if an appropriate age-depth relationship is known (Parker and Oldenburg, 1973; Davis and Lister, 1974). Such a relationship has been determined for the symmetrically spreading part of the EPR near 11°S (Rea, 1976a) and is

$$h(\text{depth, km}) = 0.40 t(\text{age})^{1/2} + 2.82 \ .$$

Application of this relationship to the EPR at 20°S shows that the west flank is unusually shallow and the east flank is about at the expected depth (Rea, 1978a). This depth asymmetry can be explained by postulating an asymmetrical supply of heat beneath the west flank of the EPR which retards the cooling and subsidence of the lithosphere.

An additional indication of the suggested asymmetrical heat supply can be found in available heat-flow data. Measurements have been taken along profiles crossing the EPR at 17°S (*Vema* -19; Lee and Uyeda, 1965) and at 18.5°S (Amphitrite; von Herzen and Anderson, 1972). Values from 22 stations within 300 km of the axis along these traverses show an average heat flow of $2.5 \pm 1.8\ \mu\text{cal cm}^{-2}\ \text{s}^{-1}$ on the west flank and $1.7 \pm 0.8\ \mu\text{cal cm}^{-2}\ \text{s}^{-1}$ east of the axis.

At 20°S on the EPR the asymmetries in spreading rate, depth, and heat flow probably result from a heat source lying west of the axis. Origins of such a heat source must remain speculative. Hayes (1976) suggests two types of sources, one consisting of a higher than normal concentration of radioactive minerals and the other being convective mantle plumes or hot spots. In either case the source must lie beneath the lithosphere because the heat source remains relatively stationary as the lithosphere migrates away from the axis. The supply of heat from this source must be less than that available at the spreading center or the axis would somehow migrate to that hottest magma source and then stay there. Since asymmetrical accretion often continues for millions of years, hottest temperatures and weakest lithosphere always occur at the original spreading center, and the asymmetrical heat source is secondary, only acting to retard the lithospheric cooling process on one flank of the rise.

This line of reasoning suggests that magmatic processes may be somehow different beneath the two portions of the rise. Scheidegger and Corliss (this volume) have demonstrated that a third and equally fundamental characteristic of EPR geology also changes from north to south at 13.5°S where the morphological and rate symmetries change. They completed a factor analysis of 256 Nazca plate basalt samples which divided that population into five distinct subgroups. Two of these groups lie along the EPR, and the boundary between them appears to be at the Garret Fracture Zone. Basalts from north of 13.5°S are more primitive than those occurring to the south. Since these geochemical changes are the result of subtle changes in the process of shallow-level fractional crystallization (Scheidegger and Corliss, this volume), it does seem that magmatic processes are different beneath the symmetrically and asymmetrically spreading portions of the EPR.

Origin of Small Axial Offsets

Five axial offsets of 10 to 15 km are recorded in the data from the rise crest. Three of these are present axis offsets, at 10°S, 11.5°S, and 20.7°S (Figs. 4, 5), and two are recorded in the magnetic anomaly patterns, at 6°S and 31°S (Rea, 1976b, 1977). All of these offsets have happened within roughly 0.5 m.y. and have affected a part of the axis only 100 to 200 km long. Similar sudden small axis shifts have occurred along the Galapagos spreading center during the past few million years. These shifts involve axial segments only tens of kilometres long and span 5 to 30 km (Hey and Vogt, 1977, Hey, 1979).

Two possible causes exist for the formation of these small axial offsets. The first involves a brief episode of extremely asymmetrical sea-floor spreading that results in a relative axis migration of about 10 km. Such an episode must be confined both temporally to less than 0.6 m.y. and spatially to short lengths of the rise crest. If asymmetrical spreading is the effect of an asymmetrical heat supply, then this heat source must also be rather rigidly confined in both space and time.

There is no question that this process occurs to a moderate degree on a regional basis and over periods extending for millions of years. At the sites of the axis offsets, this process would have to have an asymmetry value (fast rate/slow rate) of 1.5 to 2.0 to accomplish the axis shifts, values that are much higher than the average of about 1.25 from regions of known asymmetrical spreading along moderate and fast-spreading rise crests (Rea, 1978a). Bathymetric data from the 20°S area tend to support the idea of extreme asymmetry in spreading because the greatest bathymetric asymmetry coincides geographically with the area of the largest spreading rate asymmetry (compare Figs. 5 and 7, the deepest part of the survey area in the southeast corner coincides with the maximum distance between anomalies 2' and 2). The most severe criticism of highly asymmetrical spreading as a cause of the axis shifts occurs when the heat source presumably responsible is considered. It is difficult to imagine a heat source that can be turned on, function, and then turned off all in 0.5 m.y. or less. Furthermore, the lateral boundaries of the region presumably affected by the heat asymmetry are unusually sharp, probably only a few kilometres wide. Hayes (1976), in discussing the boundaries between spreading regimes, invoked some critical threshold of temperature or petrologic properties to account for the abrupt spatial changes he observed. In the case of the small axis shifts along the EPR, any thermal event must be rigidly bound in time as well as space.

The other process that could result in the 10- to 15-km axial offsets, discrete jumps of the spreading center, has also been documented on large and medium scales. The northern part of the present EPR was created during the Pliocene and earlier by eastward jumps of spreading activity north of the present 4.5°S fracture zone (Sclater and others, 1971; Anderson and Davis, 1973; van Andel and others, 1975) and the more southerly portion by westward jumps from the fossil Galapagos Rise (Herron, 1972a). These jumps ranged in distance from several hundred to a thousand kilometres. Shih and Molnar (1975) have postulated a series of lesser axis jumps, spanning 40 to 70 km, to account for the changing magnetic anomaly patterns in the northeast Pacific near the Surveyor Fracture Zone. The lengths of axis involved in those jumps, 100 to 200 km, are similar to those involved in the small axis shifts along the Nazca-Pacific plate boundary, but longer than those involved with the small shifts along the Galapagos rift zone (Hey and Vogt, 1977).

These small jumps may be the result of a fault-controlled process. Abyssal-hill topography is created by downfaulting the flanks of the axial ridge along faults that extend parallel to the ridge. These faults may dip toward the central zone of intrusion at the axis (Rea, 1975b) or may be more nearly vertical (Lonsdale, 1977a). It is possible that one of them could tap the underlying magma chamber, found at very shallow depths, perhaps less than 2 km below the sea floor (Orcutt and others, 1976) on fast-spreading rises and become the preferred conduit for extrusion. The pre-existing axial ridge would then be broken into parallel ridges and down-faulted to merge with the adjacent abyssal hills. This mechanism would result in an axial shift spanning roughly the half width of the axial ridge, would leave essentially no trace in the topography, and could occur fairly quickly, all of which is in accordance with observations along the EPR.

As in the case of the asymmetrical sea-floor spreading processes, these two mechanisms may be end-members of a continuum. However, axial jumps seem much more likely to occur on a time scale of 0.5 m.y. or less than does axial migration by asymmetrical spreading. At 20°S the southern axial segment, if the jump hypothesis is correct, would have jumped toward the hotter side of the pre-existing asymmetrically spreading axis, a more likely direction than to the colder side away from the magma (and heat) supply. In addition to their geological rapidity, one more bit of evidence supports the axis jump hypothesis. This is the observation that all five observed axial shifts cover nearly the same distance, approximately 10 km. It seems unlikely that a brief spurt of highly asymmetrical spreading would always result in a 10-km axial offset, whereas a fault-controlled axis jump might be expected to cover a similar distance each time, assuming regular positioning of the faults (Rea, 1975b). The concept of fault-controlled axis jumps as the cause of the small axial offsets along the EPR is preferred to that of highly asymmetrical sea-floor spreading as the cause of the 10- to 15-km shifts of the spreading center. Small axis jumps observed along the Galapagos rift zone also appear to be tectonically controlled rather than the result of asymmetrical spreading (Hey and Vogt, 1977; Hey, 1979). The evidence is not conclusive on this subject, but the fault-controlled procedures seems more likely than the short-lived, spatially-bound asymmetrical heat source required for extraordinary asymmetrical spreading.

Geologic History of the EPR, 4.5° to 13°S

The reconnaissance data available between the fracture zones at 4.5° and 13°S, discussed above, are sufficient to allow a reconstruction of the axis jumps that resulted in the formation of the modern EPR in this region. A map of the active spreading centers during the later part of the middle Miocene time can be created from information in the literature. North of 4°S, spreading was occurring from the southern extension of the Mathematicians-Clipperton Ridge now at 115°W (van Andel and others, 1975). South of about 13°S, spreading activity had begun about 20 m.y. ago along the portion of the EPR extending south from the 13.5°S fracture zone (Herron, 1972a; Handschumacher, 1976). Between about 7°S and 13°S, the active spreading center was the Galapagos Rise, now located near 97°W (Herron, 1972a, Handschumacher, 1976). This part of the Galapagos Rise was then the last active segment of the old ridge system on the Nazca plate. It was connected to the Mathematicians-Clipperton spreading system by the predecessor of the present Gofar Fracture Zone which then had a 750-km, sinistral offset. A dextral offset of about 1,050 km must have connected the southern Galapagos Rise with the northern end of the middle Miocene EPR (Fig. 17a).

Spreading activity along the southern extension of the Mathematicians-Clipperton Ridge ceased 10 to 12 m.y. ago (van Andel and others, 1975) and presumably shifted east at that time to the present EPR north of the Gofar Fracture Zone. Between 5° and 13°S along the EPR, the Bauer scarp (Fig. 14) marks the inception of spreading so the problem of historical reconstruction reduces to obtaining a plausible age for this feature. In determining the age of the Bauer scarp, three assumptions must be made, all of which seem sound. First, sea-floor spreading has occurred symmetrically from the EPR axis, and therefore the east-flank rate is half of the whole spreading rate. Second, the present spreading

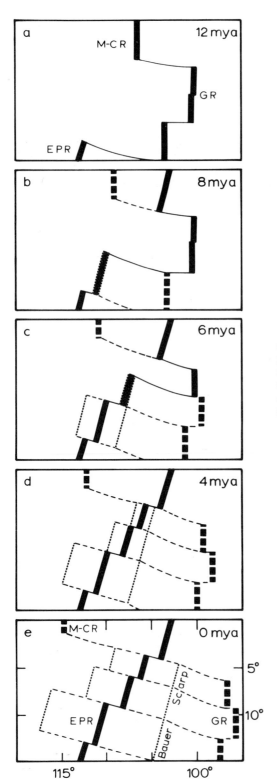

Figure 17. Schematic diagram of changing spreading configurations along the Nazca-Pacific plate boundary during the past 12 m.y. Wide lines represent active (solid) and extinct (dashed) spreading centers. Thin lines represent active (solid) and inactive (dashed) transform faults. Dotted lines represent EPR boundary scarp.

rates closely approximate spreading rates since the EPR began. The third assumption is that the age of anomaly 3' which can be correlated across the east flank of the EPR between the 9°S and 13.5°S fracture zone (fig. 5) is 5.4 m.y. (see discussion in Rea, 1976a). The first assumption is supported by the medial position of the EPR axis between the bounding scarps (Fig. 14; Rea, 1978b) and the mapping of younger magnetic anomalies (Fig. 4, Rea, 1975a, 1976a, 1976b). The assumption of uniform spreading rates is only adequate, but the present rates seem to be a reasonable, probably within 5%, estimate of overall post-Miocene rates (Rea and Scheidegger, 1979).

A region of blocky, high-relief topography, often more than 50 km wide and 500 to 1,200 m in relief, commonly occurs along the young side of the boundary scarps of newly formed spreading centers. This feature can be seen on the Scan-10 profile (Fig. 15) where it crosses either boundary scarp and on almost all other crossings of these features (Sclater and others, 1971; Rea, 1976a). This feature can be correlated among the several tracklines crossing the Bauer scarp; its western extent is shown on Figure 4 and by Rea (1978b). This topographic feature may mark the place where a newly created rise crest must heat, expand, and crack the initially cold lithosphere before it can function normally.

Inception of spreading between the Wilkes and Garret Fracture Zones can be determined by extrapolating eastward from anomaly 3'. The distance from anomaly 3' to the Bauer Scarp is about 230 km and to the western extent of the high-relief region is about 140 km (Fig. 4). The spreading rate in this region is assumed to be 80 mm/yr (Rea, 1976a). From this information, the time of initiation of the modern EPR between the 9°S and 13.5°S fracture zones is estimated to be 8.2 m.y. ago and the time when it began to function smoothly to be about 7.1 m.y. ago.

North of the Wilkes Fracture Zone, there are not any magnetic anomalies that can be correlated across the lower part of the rise flank. The age of the Bauer scarp in this case is estimated by extrapolating spreading rates over the axis-to-scarp distance. In the region between the 6.2°S and Wilkes Fracture Zones the distance between the two boundary scarps is almost exactly 1,000 km (Fig. 14), and the whole spreading rate is about 156 mm/yr (Rea, 1975a, 1976c). These figures result in an age of 6.4 m.y. for the initiation of activity in this region. Anderson and Sclater (1972) previously determined an age of 6.5 m.y., on the basis of bathymetric evidence alone, for the Galapagos Rise–EPR ridge jump in this general region; most of their data came from north of the Wilkes Fracture Zone.

Between the fracture zones at 6.2°S and 4.5°S the distance from the axis to the Bauer scarp is about 430 km (Fig. 14) and the spreading rate is approximately 151 mm/yr. Thus, the Bauer scarp in this region is about 5.7 m.y. old, just slightly younger than to the south of the 6.2°S fracture zone.

The process of shifting spreading activity that results in the final extinction of the Galapagos Rise and the completion of the EPR apparently occurred in three separate jumps over about 2.5 m.y. The distance spanned by these jumps was about 750 km at 9° to 13°S, 850 km at 6.5° to 9°S, and 650 km at 4.5° to 6.5°S (Rea, 1978b). The nature of these axial jumps and the evolving plate boundaries shown in Figure 17 have implications for the tectonics of transform faults and of large axis jumps.

Different fracture zone bound blocks of the EPR were formed at different times by the three westward jumps in spreading activity. Specifically, the part of the EPR south of the Wilkes Fracture Zone originated about 8.2 m.y. ago and the part to the north about 6.4 m.y. ago. This implies that for a period of approximately 1.8 m.y. the Wilkes Fracture Zone was a 900-km-long, active transform fault that connected the new EPR with the old Galapagos Rise (Fig. 17b). Similarly, the 6.2°S fracture zone may have been about the same length (Fig. 17) for a period of less than 1 m.y. That such faults could propagate in considerably less than 1 m.y. supports the suggestion that the lithosphere offers relatively little resistance to shearing (Lachenbruch and Thompson, 1972; Lachenbruch, 1973; Froidevaux, 1973).

One of the more striking tectonic features of the rise is the remarkable linearity of the Bauer scarp (Fig. 14; Mammerickx and Smith, 1978). Sea-floor spreading began at different times in different fracture zone bound segments of the rise. One would expect a newly forming ridge to align with the existing one because of the proximity to heat source. However, the new ridge between 6.5° and 9°S began in the same relative position, now seen as the Bauer scarp, as did the rise between 9° and about 13.5°S. The situation across the fracture zone extension at 6.5°S is similar to that at 9°S (Fig. 17). This preference for the segments of the new rise to begin along the same trend at different times suggests the pre-existence of some linear zone of the incipient melting in the asthenosphere, since any lithospheric zone of weakness should move westward with ongoing spreading from the old Galapagos Rise.

Tectonic Style of Fast-Spreading Rises

The major topographic and tectonic differences between fast-spreading rises and rifted, Atlantic-type ridges result from the fundamental dynamic difference, the difference in spreading rates. These rates range from 20 to 40 km/m.y. for the Mid-Atlantic Ridge (MAR) (Talwani and others, 1971; Macdonald, 1977; van Andel and Moore, 1970) and are roughly 160 km/m.y. for the Nazca-Pacific section of the EPR.

The major topographic features of the rifted rises, the rift valleys, and the high, steep valley walls are the result of vertical uplift caused by still-buoyant material cooling and adhering to the walls of the magma conduit (Osmaston, 1971). Eventually the accumulated buoyancy is relieved by up-faulting the walls of the rift valley. On fast-spreading rises, the walls of the conduit do not have time to cool appreciably before they are removed from the magma chamber and therefore do not hinder magma upwelling. Lava is allowed to reach the ocean floor where it is extruded to form the broad low axial ridge (Rea, 1975b; Lonsdale, 1977a). Flanks of this ridge are then down-faulted as lateral motion removes them from atop the supporting magma column. A reasonably complete continuum both topographically and mechanically exists between the depressed axis and elevated flanks of slow-spreading rises and the elevated axis with down-faulted flanks of the EPR (Lachenbruch, 1973; Lonsdale, 1977a; Sleep and Rosendahl, 1979).

The thickness of oceanic lithosphere appears to be proportional to the square root of its age (Parker and Oldenburg, 1973). Therefore, at any given distance from the axis, the lithosphere along fast-spreading rises is much thinner than along slow-spreading rises. Lithosphere at 20°S on the EPR would be $\sqrt{8}$ or a factor of 2.8 thinner than that in the FAMOUS area at 37°N on the MAR. The thin lithosphere of fast-spreading rises is much more susceptible to deformation than the thicker lithosphere of the MAR-type spreading centers. For this reason, the variety of tectonic complications in the axial region of the EPR, such as the Easter plate, the complications associated with the EPR–Chile Ridge triple junction, the nontransform axis offset at 20.7°S, and the small sudden axis jumps should not be unexpected.

The combination of thin lithosphere and unhindered upwelling of magma found on fast-spreading rises results in the formation of a very different type of fracture zone than occurs across slow-spreading ridges. The deep, narrow troughs so typical of MAR fracture zones are not found on the western edge of the Nazca plate, probably for two reasons. First, magma upwelling is not greatly hindered by either walls of the axial conduit or the additional fracture zone wall, precluding development of the deep troughs found on the MAR (Sleep and Biehler, 1970). EPR fracture zones shoal at axis intersections in contrast to the MAR where they deepen, but axial depths do become greater toward offsets along the EPR, presumably for the same reason as along the MAR (Sleep and Biehler, 1970). The second reason concerns plate thickness, and therefore strength, along the transform offset. For an axial offset of 100 km, plate thickness on the MAR would be 21 km at the midpoint, but only 8.6 km on the EPR [calculated from relation of Parker and Oldenburg (1973)]. If a component of closing occurs across these features as a result of a small change in spreading direction, the tectonic response will probably be quite different. The Atlantic-type fracture zones bounded by thick plates may merely change trend slightly or, in the extreme case, close, which occasions either an axis jump or the creation of a new transform fault. In contrast to this, what must have been a rather small change in the spreading regime resulted in a very large change in both the topographic and magnetic character of the 6.2°S fracture zone. The change in morphology of the Wilkes Fracture Zone from a simple, 10- to 20-km-wide topographic step to a 100-km-wide zone with a large oblique ridge may have resulted from large-scale rupture of thin lithosphere in response to the shear couple across this feature (Rea and Kureth, 1979). The detailed nature of fracture-zone tectonics on fast-spreading rises is not well understood and remains a fruitful area for further study.

ACKNOWLEDGMENTS

The dilligent efforts of the many scientists and crew members aboard the *Yaquina*, *Kana Keoki*, and *Oceanographer* were essential in making the Nazca Plate Project a success. Data from other institutions were kindly provided by W. C. Pitman III and D. E. Hayes of Lamont-Doherty Geological Observatory, T. Chase of the Scripps Institution of Oceanography, and R. P. von Herzen of Woods Hole Oceanographic Institute. Dave Handschumacher kindly provided a preprint of his manuscript concerning the Easter plate. Ken Macdonald and an anonymous reviewer read this manuscript and provided many useful suggestions.

Support for this project has been provided by the National Science Foundation through the IDOE Nazca Plate Project.

REFERENCES CITED

Agassiz, A., 1906, General report of the ('Albatross') Expedition: Cambridge, Massachusetts, Harvard College, Museum of Comparative Zoology, Memoir, v. 33, p. 1-75.

Anderson, R. N., and Davis, E. E., 1973, A topographic interpretation of the Mathematician Ridge, Clipperton Ridge, East Pacific Rise system: Nature, v. 241, p. 191-193.

Anderson, R. N., and Noltimier, H. C., 1973, A model for the horst and graben structure of midocean ridge crests based upon spreading velocity and basalt delivery to the oceanic crust: Royal Astronomical Society Geophysical Journal, v. 34, p. 137-147.

Anderson, R. N., and Sclater, J. G., 1972, Topography and evolution of the East Pacific Rise between 5° and 20°S: Earth and Planetary Science Letters, v. 14, p. 433-441.

Anderson, R. N., and others, 1974, Fault plane solutions of earthquakes on the Nazca Plate boundaries and the Easter Plate: Earth and Planetary Science Letters, v. 24, p. 188-202.

Anderson, R. N., and others, 1976, New terrestrial heat flow measurements on the Nazca plate: Earth and Planetary Science Letters, v. 29, p. 243-254.

Anderson, R. N., and others, 1978, Geophysical surveys on the East Pacific Rise–Galapagos Rise system: Royal Astronomical Society Geophysical Journal, v. 54, p. 141-166.

Ault, J. P., 1946, The captain's report, *in* The work of the *Carnegie* and suggestions for future scientific cruises, oceanography - IV: Washington, D.C., Carnegie Institution of Washington Publication 571, p. 1-28.

Berg, E., and Sutton, G. H., 1974, Dynamic interaction of seismic and volcanic activity of the Nazca plate edges: Physics of the Earth and Planetary Interiors, v. 9, p. 175-182.

Davis, E. E., and Lister, C.R.B., 1974, Fundamentals of ridge crest topography: Earth and Planetary Science Letters, v. 21, p. 405-413.

Ewing, M., and Heezen, B. C., 1956, Some problems of Antarctic submarine geology, *in* Crary, A. P., and others, eds., Antarctica in the International Geophysical Year: Washington, D.C., American Geophysical Union, Geophysical Monograph 1, p. 75-81.

Forsyth, D. W., 1972, Mechanisms of earthquakes and plate motions in the east Pacific: Earth and Planetary Science Letters, v. 17, p. 189-193.

Froidevaux, D., 1973, Energy dissipation and geometric structure at spreading plate boundaries: Earth and Planetary Science Letters, v. 20, p. 419-424.

Gilluly, J., 1955, Geologic contrasts between continents and ocean basins, *in* Poldervaart, A., ed., Crust of the Earth: Geological Society of America Special Paper 62, p. 7-18.

Handschumacher, D. W., 1976, Post-Eocene plate tectonics of the eastern Pacific, *in* Sutton, G. H., and others, eds., The geophysics of the Pacific Ocean basin and its margin, a volume in honor of George P. Woollard: Washington, D.C., American Geophysical Union, Geophysical Monograph 19, p. 177-202.

Handschumacher, D. W., and others, 1975, Magnetic and bathymetric profiles from the central and southeastern Pacific: 10°N-45°S, 70°W-150°W: Honolulu, Hawaii, University, Institute of Geophysics, Data Report 29.

Handschamacher, D. W., and others, 1981, Structure and evolution of the Easter plate, *in* Kulm, L. D., and others, eds., Nazca plate: Crustal formation and Andean convergence: Geological Society of America Memoir 154 (this volume).

Hayes, D. E., 1976, Nature and implications of asymmetric sea-floor spreading—"Different rates for different plates": Geological Society of America Bulletin, v. 87, p. 994-1002.

Herron, E. M., 1972a, Sea-floor spreading and the

Cenozoic history of the East-Central Pacific: Geological Society of America Bulletin, v. 83, p. 1671–1692.

——1972b, Two small crustal plates in the South Pacific near Easter Island: Nature; Physical Science, v. 240, p. 35–37.

Hey, R., 1979, Evidence for spreading-center jumps from fine-scale bathymetry and magnetic anomalies near the Galapagos Islands: Geology, v. 7, p. 504–506.

Hey, R., and Vogt, P., 1977, Spreading center jumps and sub-axial asthenosphere flow near the Galapagos hotspot: Tectonophysics, v. 37, p. 41–52.

Johnson, G. L., and others, 1976, Morphology and structure of the Galapagos Rise: Marine Geology, v. 21, p. 81–120.

Klitgord, K. D., and others, 1975, An analysis of near-bottom magnetic anomalies: Sea-floor spreading and the magnetized layer: Royal Astronomical Society Geophysical Journal, v. 43, p. 387–424.

Lachenbruch, A. H., 1973, A simple mechanical model for oceanic spreading centers: Journal of Geophysical Research, v. 78, p. 3395–3417.

Lachenbruch, A. H., and Thompson, G. A., 1972, Oceanic ridges and transform faults: Their intersection angles and resistance to plate motion: Earth and Planetary Science Letters, v. 15, p. 116–122.

Langseth, M. G., Jr., and von Herzen, R. P., 1971, Heat flow through the floor of the world's oceans, *in* Maxwell, A. E., ed., The sea, Volume 4, Part 1: San Francisco, Wiley-Interscience, p. 299–353.

Lee, W.H.K., and Uyeda, S., 1965, Review of heat flow data, *in* Lee, W.H.K., ed., Terrestrial heat flow, Geophysical Monograph 8: Washington, D.C., American Geophysical Union, p. 87–190.

Le Pichon, X., and Langseth, M. G., Jr., 1969, Heat flow from the mid-ocean ridges and sea-floor spreading: Tectonophysics, v. 8, p. 319–344.

Lonsdale, P., 1976, Abyssal circulation of the southeastern Pacific and some geological implications: Journal of Geophysical Research, v. 81, p. 1163–1176.

——1977a, Structural geomorphology of a fast-spreading rise crest: The East Pacific Rise near 3°25'S: Marine Geophysical Researches, v. 3, p. 251–293.

——1977b, Regional shape and tectonics of the Equatorial East Pacific Rise: Marine Geophysical Researches, v. 3, p. 295–315.

——1978, Near-bottom reconnaissance of a fast-slipping transform fault zone at the Pacific-Nazca plate boundary: Journal of Geology, v. 86, p. 451–472.

Macdonald, K. C., 1977, Near-bottom magnetic anomalies, asymmetric spreading, oblique spreading, and tectonics of the Mid-Atlantic Ridge near lat 37° N: Geological Society of America Bulletin, v. 88, p. 541–555.

Mammerickx, J., and Smith, S. M., 1978, Bathymetry of the Southeast Pacific: Geological Society of America Map and Chart Series, MC-26, scale 1:6,442,194.

Mammerickx, J., and others, 1975, Morphology and tectonic evolution of the east-central Pacific: Geological Society of America Bulletin, v. 86, p. 111–118.

Menard, H. W., 1960, The East Pacific Rise: Science, v. 132, p. 1737–1746.

Menard, H. W., Chase, T. E., and Smith, S. M., 1964, Galapagos Rise in the southeastern Pacific: Deep-Sea Research, v. 11, p. 233–242.

Morgan, W. J., 1971, Convection plumes in the lower mantle: Nature, v. 230, p. 42–43.

——1972, Plate motions and deep mantle convection, *in* Shagam, R., and others, eds., Studies in earth and space sciences, a Memoir in honor of Harry Hammond Hess: Geological Society of America Memoir 132, p. 7–22.

Muray, J., 1895, A summary of the scientific results, *in* Report on the scientific results of the voyage of H.M.S. *Challenger* during the years 1873-76: London, Her Majesty's Government, 1608 p.

Needham, H. D., and Francheteau, J., 1974, Some characteristics of the rift valley in the Atlantic Ocean near 36°48' North: Earth and Planetary Science Letters, v. 22, p. 29–43.

Northrop, J., Morrison, M. F., and Duennebier, F. K., 1970, Seismic slip rate versus sea-floor spreeading rate on the Eastern Pacific Rise and Pacific Antarctic Ridge: Journal of Geophysical Research, v. 75, p. 3285–3290.

Orcutt, J. A., Kenneth, B.L.M., and Dorman, L. M., 1976, Structure of the East Pacific Rise from an ocean bottom seismometer survey: Royal Astronomical Society Geophysical Journal, v. 45, p. 305–320.

Osmaston, M. F., 1971, Genesis of ocean ridge median valleys and continental rift valleys: Tectonophysics, v. 11, p. 387–405.

Parker, R. L., and Oldenburg, D. W., 1973, Thermal model of ocean ridges: Nature, Physical Science, v. 242, p. 137–139.

Ramberg, I. B., and van Andel, Tj. H., 1977, Morphology and tectonic evolution of the rift valley at lat 36°30'N, Mid-Atlantic Ridge: Geological Society of America Bulletin, v. 88, p. 577–586.

Rea, D. K., 1975a, Tectonics of the East Pacific Rise, 5° to 12°S [Ph.D. thesis]: Corvallis, Oregon State University, 139 p.

——1975b, Model for the formation of the topographic features of the East Pacific Rise crest: Geology, v.

3, p. 77–80.

———1976a, Analysis of a fast-spreading rise crest: The East Pacific Rise, 9° to 12° South: Marine Geophysical Researches, v. 2, p. 291–313.

———1976b, Changes in the axial configuration of the East Pacific Rise near 6°S during the past 2 m.y.: Journal of Geophysical Research, v. 81, p. 1495–1504.

———1976c, Possible spreading-center jumps on the East Pacific Rise at 6.5° and 7°S: American Geophysical Union, Transactions, v. 57, p. 752.

———1977, Local axial migration and spreading rate variations, East Pacific Rise, 31°S: Earth and Planetary Science Letters, v. 34, p. 78–84.

———1978a, Asymmetric sea-floor spreading and a non-transform axis offset: The East Pacific Rise 20°S survey area: Geological Society of America Bulletin, v. 89, p. 836–844.

———1978b, Evolution of the East Pacific Rise between 3°S and 13°S since the middle Miocene: Geophysical Research Letters, v. 5, p. 561–564.

Rea, D. K., and Blakely, R. J., 1975, Short-wavelength magnetic anomalies in a region of rapid seafloor spreading: Nature, v. 255, p. 126–128.

Rea, D. K., and Kureth, C. L., Jr., 1979, Structural geology of the East Pacific Rise 9°S (Wilkes) fracture zone: American Geophysical Union Transactions, v. 60, p. 377.

Rea, D. K., and Malfait, B. T., 1974, Geologic evolution of the northern Nazca plate: Geology, v. 2, p. 317–320.

Rea, D. K., and Scheidegger, K. F., 1979, Eastern Pacific spreading rate fluctuation and its relation to Pacific area volcanic episodes: Journal of Volcanology and Geothermal Research, v. 5, p. 135–148.

Revelle, R., and others, 1955, Pelagic sediments of the Pacific, *in* Poldervaart, A., ed., Crust of the Earth: Geological Society of America Special Paper 62, p. 221–235.

Scheidegger, K. F., and Corliss, J., 1981, Petrogenesis and secondary alteration of upper layer 2 basalts of the Nazca plate, *in* Kulm, L. D., and others, eds., Nazca plate: Crustal formation and Andean convergence: Geological Society of America Memoir 154 (this volume).

Schilling, J. G., 1973, Iceland mantle plume: Geochemical study of Reykjanes Ridge: Nature, v. 242, p. 565–571.

Schilling, J. G., and Bonatti, E., 1975, East Pacific Ridge (2°S–19°S) versus Nazca intraplate volcanism: Rare-earth evidence: Earth and Planetary Science Letters, v. 25, p. 93–102.

Schilling, J. G., Anderson, R. N., and Vogt, P., 1976, Rare earth, Fe and Ti variations along the Galapagos spreading centre, and their relationship to the Galapagos Mantle plume: Nature, v. 261, p. 108–113.

Schouten, J. A., 1971, A fundamental analysis of magnetic anomalies over oceanic ridges: Marine Geophysical Researches, v. 1, p. 111–144.

Sclater, J. G., Anderson, R. N., and Bell, M. L., 1971, The elevation of ridges and the evolution of the central eastern Pacific: Journal of Geophysical Research, v. 76, p. 7888–7915.

Shih, J., and Molnar, P., 1975, Analysis and implications of the sequence of ridge jumps that eliminated the Surveyor transform fault: Journal of Geophysical Research, v. 80, p. 4815–4822.

Sleep, N. H., and Biehler, S., 1970, Topography and tectonics at the intersections of fracture zones with central rifts: Journal of Geophysical Research, v. 75, p. 2748–2752.

Sleep, N. H., and Rosendahl, B. R., 1979, Topography and tectonics of mid-oceanic ridge axes: Journal of Geophysical Research, v. 84, p. 6831–6839.

Stauder, W., and Bollinger, G. A., 1964, The S-wave project for focal mechansim studies, earthquakes of 1962: Seismological Society of America Bulletin, v. 54, p. 2199–2208.

———1966, The S-wave project for focal mechanism studies, earthquakes of 1963: Seismological Society of America Bulletin, v. 56, p. 1363–1371.

Stover, C. W., 1973, Seismicity and tectonics of the East Pacific Ocean: Journal of Geophysical Research, v. 78, p. 5209–5220.

Sykes, L. R., 1963, Seismicity of the South Pacific Ocean: Journal of Geophysical Research, v. 68, p. 5999–6006.

———1967, Mechanism of earthquakes and nature of faulting on the mid-oceanic ridges: Journal of Geophysical Research, v. 72, p. 2131–2153.

Talwani, M., Windisch, C. C., and Langseth, M. G., Jr., 1971, Reykjanes Ridge crest: A detailed geophysical study: Journal of Geophysical Research, v. 76, p. 473–517.

van Andel, Tj. H., and Moore, T. C., Jr., 1970, Magnetic anomalies and seafloor spreading rates in the northern South Atlantic: Nature, v. 226, p. 328–330.

van Andel, Tj. H., von Herzen, R. P., and Phillips, J. D., 1971, The Vema fracture zone and the tectonics of transverse shear zones in oceanic crustal plates: Marine Geophysical Researches, v. 1, p. 261–283.

van Andel, Tj. H., Heath, G. R., and Moore, T. C., Jr., 1975, Cenozoic history and paleoceanography of the central Equatorial Pacific Ocean: Geological Society of America Memoir 143, 134 p.

Vogt, P. R., 1976, Plumes, subaxial pipe flow, and topography along the mid-oceanic ridge: Earth and Planetary Science Letters, v. 29, p. 309–325.

Vogt, P. R., and Johnson, G. L., 1975, Transform faults and longitudinal flow below the midoceanic

ridge: Journal of Geophysical Research, v. 80, p. 1399–1428.
von Herzen, R., 1959, Heat-flow values from the southeastern Pacific: Nature, v. 183, p. 882–883.
von Herzen, R. P., and Anderson, R. N., 1972, Implications of heat flow and bottom water temperature in the eastern equatorial Pacific: Royal Astronomical Society Geophysical Journal, v. 26, p. 427–458.
von Herzen, R. P., and Lee, W.H.R., 1969, Heat flow in oceanic regions, *in* Hart, P. J., ed., The earth's crust and upper mantle: Washington, D.C., American Geophysical Union Geophysical Monograph 13, p. 88–95.
von Herzen, R. P., and Uyeda, S., 1963, Heat flow through the eastern Pacific Ocean floor: Journal of Geophysical Research, v. 68, p. 4219–4250.
Weissel, J. K., and Hayes, D. W., 1971, Asymmetric seafloor spreading south of Australia: Nature, v. 231, p. 518–522.
——1974, The Australian-Antarctic discordance: New results and implications: Journal of Geophysical Research, v. 79, p. 2579–2587.

MANUSCRIPT RECEIVED BY THE SOCIETY NOVEMBER 12, 1980
MANUSCRIPT ACCEPTED DECEMBER 30, 1980

Printed in U.S.A.

Geological Society of America
Memoir 154
1981

Structure and evolution of the Easter plate

D. W. Handschumacher
Sea Floor Division, Code 360, Naval Ocean Research and Development Activity, NSTL Station, Mississippi 39529

R. H. Pilger, Jr.
Department of Geology, Louisiana State University, Baton Rouge, Louisiana 70803 and Sea Floor Division, Code 360, Naval Ocean Research and Development Activity, NSTL Station, Mississippi 39529

J. A. Foreman
Conservation Division, U.S. Geological Survey, 1304 W. 6th Street, Los Angeles, California 90017

J. F. Campbell
Hawaii Institute of Geophysics, University of Hawaii, Honolulu, Hawaii 96822

ABSTRACT

Seismicity, magnetic anomaly, and bathymetric data are used to interpret the tectonic structure and history of the Easter plate. The Easter plate is a small oceanic plate in an area of locally anomalous sea-floor spreading activity that internally disrupts the tectonic continuity of the Nazca-Pacific divergent plate boundary west of Easter Island. Active sea-floor spreading centers, mapped as the plate's eastern and western boundaries, are positioned along the crests of two northwardly-trending topographic rises, herein referred to as the Este and Oeste Rises. Active fault zones characterized by rugged topography are mapped as the plate's northern and southern boundaries. The fault zones are inferred by constraints of plate tectonics theory to presently connect the Este and Oeste spreading centers to segments of the Nazca-Pacific spreading center north and south of the Easter plate through ridge-fault-fault or ridge-ridge-fault triple junctions. Magnetic anomaly and bathymetric lineations, identified on the Easter plate and adjacent portions of the Nazca and Pacific plates, provide constraints used for reconstructing the evolution of the Easter plate. Our reconstruction shows the Easter plate to be 3.2 m.y. old, having been progressively generated by and between the Este and Oeste spreading centers since that time. We speculate that the Este spreading center is assuming a more dominant kinematic role relative to Nazca-Pacific plate divergence and that it will eventually be incorporated as a segment of the Nazca-Pacific spreading center. At that time, spreading activity on the Oeste Rise would have ceased, leaving the Eater plate as a captured crustal component of the Pacific plate. Furthermore, we speculate on possible genetic relationships between the Easter plate and the Easter-Sala y Gomez island-seamount chain, a major structural lineament transecting the Nazca plate eastward of the Easter plate.

INTRODUCTION

Forsyth (1972) and Herron (1972a) first proposed the existence of a small oceanic plate on the East Pacific Rise west of Easter Island. They located the plate, herein referred to as the Easter plate, in an area of locally anomalous seismicity that is seen (Fig. 1) to internally disrupt the tectonic continuity of the Nazca-Pacific divergent plate boundary between 22°S and 26°S. Focal mechanism solutions derived for earthquakes occurring in this area are inconsistent with theoretical solutions that would be associated with simple Nazca-Pacific transform faulting (Forsyth, 1972; Anderson and others, 1974). Forsyth (1972) suggested that this inconsistency could be explained if the anomalous earthquake activity was occurring on the margins of a small crustal plate temporarily trapped between the Nazca and Pacific plates. Herron (1972a) used magnetic anomaly and bathymetric data to support her contention that the epicenter distribution in the anomalous area forms a "seismic ring" outlining the present boundaries of such a plate. With these data, active sea-floor spreading centers were located on the eastern and western limits of the epicenter pattern. Identification of magnetic anomalies out to anomaly 2 about the eastern spreading center and resultant spreading rate calculations led Herron to conclude that the small plate may have existed for the past 3 m.y. Anderson and others (1974) provided additional documentation of the existence of this plate by using focal mechanism solutions and magnetic anomaly and bathymetric data as criteria to map its present boundary configuration. The resultant map showed the plate, therein named the Easter plate, to be diamond-shaped, bounded on the north by transform faults and on the south by spreading centers.

The Easter plate provides a unique opportunity for the study of microplate behavior because its boundary kinematics are constrained by rapid, Pliocene-Pleistocene Nazca-Pacific plate divergence that is realtively well known from earlier regional studies (for example, Herron, 1972b;

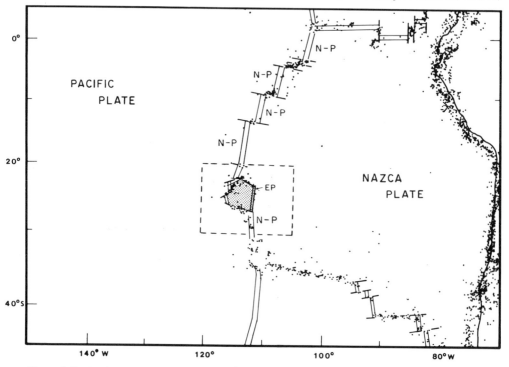

Figure 1. Regional tectonic map locating area of anomalous seismicity (box) on the Nazca-Pacific sea-floor spreading center (N-P) that contains the Easter plate (EP, stipple pattern). Earthquake epicenter locations (solid dots) from NOAA sources. Tectonic elements from regional studies of Herron (1972b), Handschumacher and others (1975), Mammerickx and others (1975), and Handschumacher (1976).

Handschumacher, 1976). During participation in the IDOE Nazca Plate Project, ships from the Hawaii Institute of Geophysics (HIG) and the Pacific Oceanographic Laboratories (POL) of the National Oceanic and Atmospheric Administration (NOAA) conducted a detailed survey of the Easter plate. This paper uses magnetic anomaly and bathymetric data from this survey, together with the data previously reported by Forsyth (1972), Herron (1972a), and Anderson and others (1974), to reinterpret the tectonic structure and evolution of the Easter plate.

MAGNETIC ANOMALY DATA

Residual magnetic anomaly profiles from the HIG-POL survey of the anomalous area are plotted in Figure 2; the International Geomagnetic Reference Field epoch 1965.0 (Fabiano and Peddie, 1967) has been removed. Symmetrical anomaly patterns to 2' are identified on the profiles across the eastern and western margins of the postulated Easter plate. The eastern pattern is best defined by the magnetic data, and is mapped by identification of the axial anomaly (1) to be composed of an active spreading center trending N18°E, herein referred to as the Este spreading center. On profiles C through G, the axial anomaly is recognizable as a broad positive anomaly; of some interest is its apparent narrowing from south to north. This narrowing trend, originally noted by Herron (1972a), is much more apparent in older anomaly identifications of the pattern, from the Jaramillo (1a) to the 2' anomaly.

From the magnetic data, the nature of the postulated western margin of the plate, also described as being a spreading center (Herron 1972a), is more difficult to define than the eastern margin. The difficulty appears to arise from the complex character of this margin, which is believed to result from disruption of the magnetic signature by several fault zones offsetting the spreading center. The axial zone of only one segment of the western spreading center, herein referred to as the Oeste spreading center, could be mapped from the magnetic data. For the segment, identification anomalies out to 2' are again made about axial anomaly identifications (profiles A and B).

BATHYMETRIC DATA

Each of the axial anomaly identifications described above is seen in Figure 3 to coincide with a topographic rise crest shoaling at less than 3,000 m on corresponding bathymetric profiles. On the basis of bathymetry alone, the axial zone of the Este spreading center is inferred to continue northward beyond the limits of an identifiable axial anomaly by northward location of the eastern rise crest through profile A. Confidence in extension of the spreading center in the absence of a magnetic signature is strengthened by the consistent correlation of the axial anomaly with similar bathymetric expression of the rise crests. Bathymetric traverses of the eastern rise, herein referred to as the Este Rise, also show a gradual northward increase in the slope of its flanks that is recognizable from profiles G through C. The phenomenon corresponds to the northward narrowing of the magnetic anomaly pattern described earlier. Tectonic boundaries of the Este Rise, indicated by dotted lines in Figure 4, have been positioned on topographic scarps or slope reversals tenuously identified in the outer edges of the magnetic anomaly pattern. These boundaries provide important constraints for reconstructions of past spreading activity presented later in the paper. The axial anomaly identifications made for the Oeste spreading center also coincide with bathymetric expression of a topographic rise cresting at less than 3,000 (profiles A and B). Another spreading segment of the Oeste spreading center is located from similar patterns of topographic relief to exist to the south (profiles C, D, E and F). Although it is poorly located by the bathymetric data, this spreading segment may have an almost southwest-northeast trend in contrast to the N7°W trend indicated for the northern segment.

Fault zones shown to connect the Este and Oeste spreading centers to active spreading segments of the Nazca-Pacific boundary north and south of the anomalous zone (Fig. 4) are characterized by extremely rugged topography. The northernmost fault zones (FZ) shown to effectively form the northern boundary of the Easter plate are traversed by profile I in Figure 3. Topographic expression of

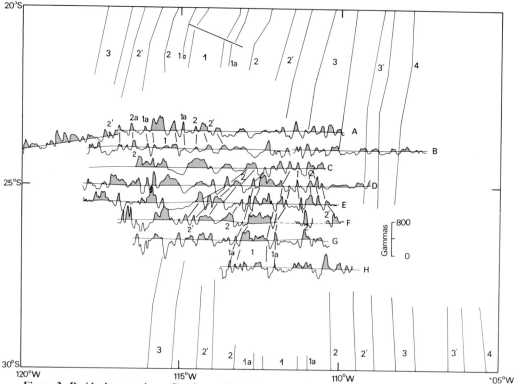

Figure 2. Residual magnetic profiles from HIG-POL survey of anomalous area indicated in Figure 1. Numbering of sea-floor spreading anomalies; axial anomaly as 1, Jaramillo anomaly as 1a, others as shown. Anomaly lineations outside anomalous area shown on this and subsequent figures are from Herron (1972a), Handschumacher and others (1975), and Handschumacher (1976). Refer to Figures 3, 4, and 5 for location of Este and Oeste spreading centers.

the Este and Oeste Rises terminates to the south of this profile. Rugged topographic relief consisting of, from west to east, an irregular ridge cresting to 2,000 and a series of narrow trough-ridge structures seen on the profile is interpreted to reflect a low-angle crossing of one or more fault zones. The southernmost fault zone is located primarily by seismicity data and inferred tectonic constraints. The fault zone is located along a band of concentrated earthquake activity (Fig. 5) and tectonically connects the Oeste spreading center to the Nazca-Pacific boundary. On the bathymetric profiles, the western end of this fault zone is positioned by rugged topography seen on the western end of profiles F and G (Fig. 3). Because of the low-angle crossing of survey tracks, the fault zone offsetting the Oeste spreading center is not detectable on the bathymetric profiles. As will be discussed later, each of the described fault zones probably contains a series of closely spaced faults offsetting small segments of Este and Oeste spreading centers that cannot be mapped with the existing data base.

INTERPRETATION

Present Boundaries

The mapped magnetic anomaly and bathymetric trends deduced from the HIG-POL survey data (Fig. 2, 3) partially define the present tectonic boundaries of the Easter plate. Active sea-floor spreading centers on the eastern and western margins of the plate are located by identifications of the

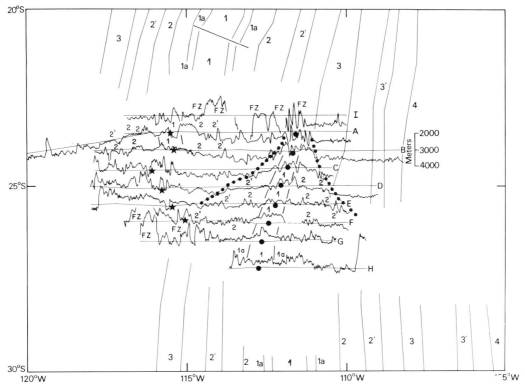

Figure 3. Bathymetric profile from HIG-POL survey corresponding to magnetic profiles in Figure 2 with the addition of profile I. Ship tracks represent baseline depth of 3,000 m for profile plots. Large dots and stars are inferred rise crests where sea-floor spreading activity is occurring on the Este and Oeste spreading centers. Small dots outline inferred boundaries of crust generated by the Este spreading center (see text). Suggested fault zone (FZ) crossings indicated. Magnetic anomaly lineations and numbers from Figure 2.

axial magnetic anomaly, or on topographic rises cresting at less than 3,000 m. Active fault zones are located where one or more of the following criteria exist: (1) distorted magnetic signature, (2) irregular topography, and (3) earthquake activity. As noted earlier, the seismotectonic studies of Forsyth (1972) and Andrson and others (1974) indicate that earthquake activity within the anomalous area does not conform to that expected for simple transform faulting. For this reason, the use of transform fault nomenclature in the descriptive phase of this study has been intentionally avoided.

The above data interpretations were found to be consistent with magnetic anomaly data reported by Herron (1972a) and Anderson and others (1974). These identifications and inter-trackline correlations were utilized, together with inferences required by plate tectonics theory, to construct the map of the configuration of the present boundaries of the Easter plate shown in Figure 5. On this map, the eastern and western boundaries of the Easter plate are northward-trending spreading centers, the Este and Oeste spreading centers. The northern and southern boundaries of the plate are indicated to be closely spaced fault zones that tectonically connect the Este and Oeste spreading centers to the Nazca-Pacific spreading center through ridge-fault-fault triple junctions. Earthquake epicenters are generally concentrated on or near these fault zones.

Anderson and others (1974) arrived at a significantly different interpretation for present boundary configuration of the Easter plate, as illustrated by comparison of the interpretations made in Figure 6. Superimposed on both interpretations are the earthquake epicenter locations and selected magnetic profile segments utilized by Anderson and others to constrain their interpretation. The seismic activity

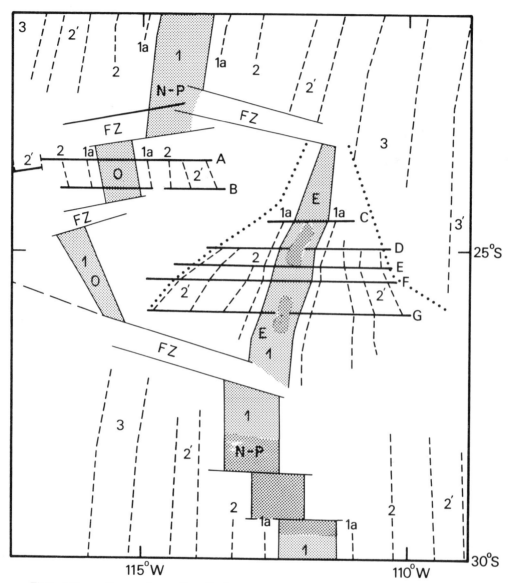

Figure 4. Composite map of magnetic and bathymetric constraints defined by magnetic anomaly data (Fig. 2) and bathymetric data (Fig. 3). Este (E), Oeste (O), and Nazca-Pacific (N-P) spreading centers indicated.

is better explained by the interpretation presented in this study, in which earthquake epicenters appear to be more typically located on or near fault zones. Axial anomaly identifications also appear to be recognizable over segments of the Este and Oeste spreading centers and show minimal skewness, suggesting a more northerly trend of the spreading axis than indicated by Anderson and others. Because the focal mechanism solutions for earthquakes in the anomalous zone contain both normal and thrust components (Forsyth, 1972), it is questionable whether resulting determinations should have been used by Anderson and others to orient transform faults. At best, the mechanism suggests a transition in the evolution of the plate, or even non-rigid behavior in this region, as will be discussed.

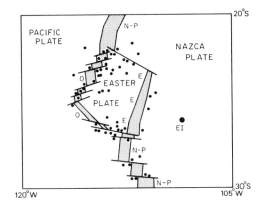

Figure 5. Present boundary map for the Easter plate constructed from data constraints shown in Figure 4. Small dots are earthquake epicenter locations. Large dot shows present site of Easter Island (EI).

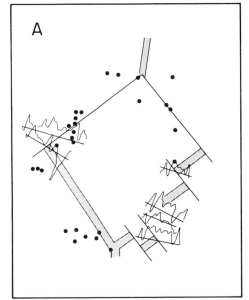 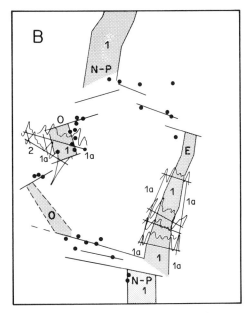

Figure 6. (A) Present boundary map for the Easter plate by Anderson and others (1974) with epicenter locations and selected magnetic anomaly profiles presented as constraints for this map. (B) Present boundary map for the Easter plate from this study with same constraints. Note good correlation of the axial anomaly (1) and Jaramillo (1a) anomaly identifications with the Este (E) and Oeste (O) spreading centers in this figure. Also note that earthquake epicenters are located on or near inferred zones on this map.

PLIOCENE-PLEISTOCENE TECTONIC TRENDS

This section describes the identification of fossil magnetic and bathymetric trends associated with, and generated by, the tectonic regimes just described (Fig. 4). The findings of this study suggest that the present tectonic configuration of the anomalous zone evolved in the past 3.2 m.yy. an age close to that originally proposed by Herron (1972b).

Roughly symmetric magnetic anomaly patterns out to anomaly 2' (3.2 m.y.B.P.) exist about portions of both Este and Oeste spreading centers (Fig. 4). On the Este Rise, anomalies to 2' are identified on profiles C through F on its western flank, and on profiles C through G on the eastern flank. A degree of asymmetry is noted, especially in the location of anomaly 2, which is interpreted to reflect a minor westward shift of the spreading ridge as this anomaly was generated. Again, the northward narrowing of older anomaly lineations on the rise flanks is more pronounced than that suggested for the axial and 1a anomalies.

The northward narrowing of the older anomaly trends coincides with interpreted bathymetric boundaries of the Este Rise regime (Fig. 3). The boundaries are approximately symmetrically located immediately outside of anomaly 2' on the southern part of the rise and on bathymetric scarps on the northern part of the rise. Topographically, the inferred boundary lines change from south to north; the western boundary is mapped as a topographic low or slope reversal on profiles D and C, and is mapped on profile B as a bathymetric scarp, shoaling to the east. South of profile E, the eastern boundary cannot be mapped because its location lies at or beyond the limits of our detailed data coverage. North of this profile, it is mapped as a bathymetric scarp, shoaling to the west, on profiles D through A.

Identification of the older magnetic patterns generated by the Oeste spreading center is limited to those located about the northern spreading segment where anomalies out to 2' have been identified on profiles A and B (Fig. 2). Failure to identify older anomalies generated elsewhere by the Oeste spreading system was attributed to its segmented character. Although the flanks of the Oeste Rise are more irregular than those of the Este Rise, they are defined by regional slopes away from the mapped ridge axis. The eastern boundary of the Oeste Rise is interpreted to be the previously described western boundary of the Este Rise. The western boundary, if it exists, would lie outside data coverage and thus remains undefined.

North and south of the Easter plate, roughly symmetric patterns of Pliocene-Pleistocene anomalies exist about axial anomaly identification on segments of the Nazca-Pacific spreading center (Figs. 2, 4). However, the axial anomaly is asymmetrically shifted to the west relative to older anomaly identification on segments of this boundary just south of the Easter plate. Similar asymmetry appears just north of the plate at a bend in axial anomaly lineation at 21°S. In both cases, the asymmetry indicates recent westward jumps of the Nazca-Pacific spreading center in these areas.

EVOLUTION

This section describes an attempt to reconstruct the Pliocent-Pleistocene tectonic history of the anomalous zone area and the evolution of the Easter plate. The reconstruction, presented in Figure 7, is a schematic interpretion of the configuration of operative sea-floor spreading activity that generated the Easter plate at 5 m.y.B.P., 3 m.y.B.P., and present. Because of limitations in the data constraints described in the previous section, the interpretation is somewhat speculative but reflects the simplest model, which would conform with both the available data and constraints inferred by plate tectonics theory. For convenience, the Nazca plate has been arbitrarily fixed throughout the reconstructions in Figure 7. Descriptions are qualitative in nature because past or present, quantitative kinematic description in anlysis of the anomalous area containing the Easter plate is not possible because of the youth and complexity of the region as well as the uncertainties arising in interpretation of the data.

The first reconstruction at 5 m.y.B.P. (anomaly 3; Fig. 7A) predates the Easter plate and shows a continuous Nazca-Pacific boundary with sinistral offset on a series of closely spaced transform faults in the future location of the Easter plate. The offsets are schematic but confined to the region indicated in Figure 7A by the anomaly 3 lineations mapped north and south of the anomalous zone. No evidence exists for locating double spreading centers bounding a small plate on the boundary at this time in either magnetic or bathymetric data.

At 3 m.y.B.P. (Fig. 7B), double spreading centers have developed and started to generate the Easter plate as these necessarily migrated away from each other. The plate consequently expanded in an east-west direction roughly parallel to that of Nazca-Pacific plate divergence. Magnetic evidence for the early evolution of the eastern Este spreading center is the symmetric lineations of anomaly 2' mapped on the flanks of the Este Rise. The topographic low or slope reversal identified earlier is interpreted to

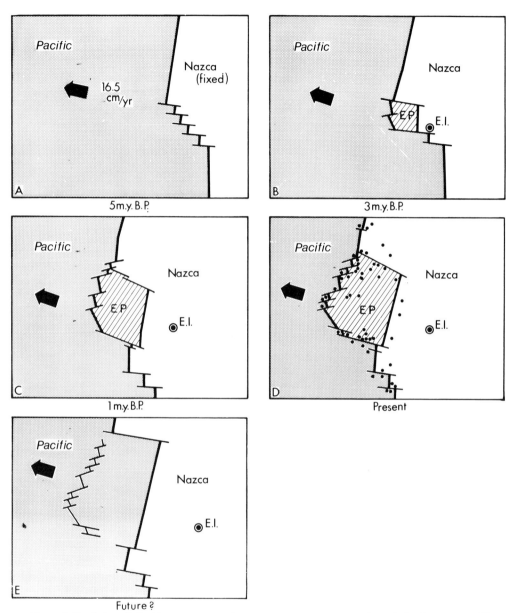

Figure 7. Evolution of the Easter plate. (A) 5 m.y.B.P., predates plate (B) 3 m.y.B.P., Easter plate (EP) appears with development of double spreading centers on a section of Nazca-Pacific plate boundary. Easter Island (EI) forms at this time. (C) 1 m.y.B.P., Easter plate grows larger by east-west spreading on the Este and Oeste spreading centers and by the northward extension of these spreading centers. (D) Present. (E) Future, Este spreading center becomes component of Nazca-Pacific spreading center as the Oeste spreading center becomes extinct, making the former Easter plate a crustal component of the Pacific plate.

be the fossil boundary separating crust generated by the two spreading centers. Such a boundary would be formed by the development of a double spreading center regime. Based on depth versus age relationships of oceanic crust (e.g., Sclater and others, 1971), shoaling of the crust to the east and west of the topographic low indicates subsequent spreading of the Este and Oeste spreading centers away from the boundary. This is schematically illustrated in a hypothetical model reconstructing the bathymetric evolution of two spreading centers shown in Figure 8.

The initial development of the Easter plate is approximately coeval with the early formation of Easter Island (Baker and othrs, 1974). Currently located on the Nazca plate to the east of the anomalous zone, this island is located by the reconstruction near the southern triple junction of the dual spreading system at 3 m.y.B.P., during its earliest formative stages. The possible genetic implications of the relationship between the initiation of the dual spreading activity and the formation of Easter Island will be discussed later.

The next reconstruction, Figure 7C, shows that the dual ridge system had expanded northward by 1 m.y.B.P. Principal evidence for this expansion is found in the magnetic and bathymetric trends associated with the Este Rise. Across the northern portion of the rise, magnetic identifications of anomaly 1a are found adjacent to bathymetric scarps interpreted to be the fossil tectonic boundaries of crust generated by this spreading center. The scarps suggest that the northern part of the Este Ridge developed in older crust of the Nazca plate, and is, therefore, elevated relative to the older, bordering crust. As theoretically modeled in Figure 8, the middle profile stack at 1 m.y.B.P. shows the elevated crust and bathymetric scarps produced by the northern extension of the Este spreading center. The only difference between this topographic model and that presented for a ridge jump in Sclater and others (1971) is that the western (Oeste) spreading center remains active instead of becoming extinct. However, the degree of segmentation of the Oeste spreading regime, which is attributed to the development of the Este spreading center, may foreshadow the eventual extinction of the Oeste spreading center. The northward propagation of the Este spreading center resembles the propagating

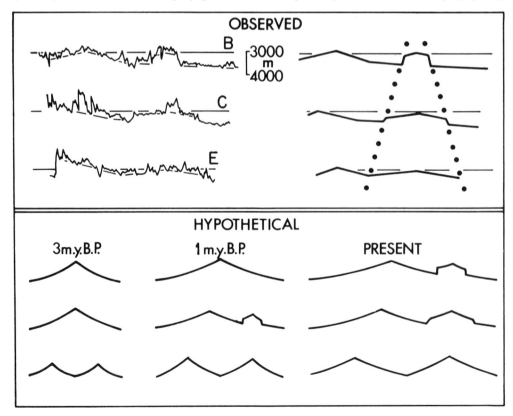

Figure 8. Topographic model used in reconstruction of the Eater plate as discussed in the text. Refer to Figure 3 for location of profiles B, C, and E.

ridge hypothesis of Hey (1977) for some other spreading systems. However, the small size of the Easter plate may introduce complicating elements into the simple model of a propagating rift developed by Hey. By 1 m.y.B.P., the model shows the westward jumping of the Nazca-Pacific boundary just north and south of the Easter plate as previously discussed.

Figure 7 shows the present tectonic configuration of the anomalous zone as mapped in this study. Further northward lengthening of the Este Ridge and the dual spreading system is accompanied by increased segmentation of the former Nazca-Pacific Ridge sections as they are assimilated by the Oeste spreading center. The continued northward extension of the Este Ridge is interpreted from topography where elevated crust bordered by bathymetric scarps is apparent. Again, Figure 8 shows the correspondence of smoothed observed bathymetric profiles with those illustrated in the topographic model for the evolution of the dual ridge system.

The "fanning" of anomaly identifications across the Este Rise is similar to that inferred by Talwani and Eldholm (1977) in the Norwegian-Greenland Sea for magnetic anomalies 20 to 23. Talwani and Eldholm, however, were unable to find what would be an extinct counterpart to the Oeste Ridge in the Norwegian-Greenland Sea, although rigid plate kinematics would require it be present.

The lessening in the degree of northward convergence of the Pliocene-Pleistocene anomaly lineations by anomaly 1a time on the Este Rise suggests that spreading rates on the Este spreading center are becoming progressively more uniform. Also, while the width of the axial anomaly still displays a gradual narrowing to the north, its width adjacent to the southern triple junction is approximately equal to that of the Nazca-Pacific boundary south of the triple junction. The present full spreading rate at the southern end of the Este Ridge is, therefore, approaching that of the Nazca-Pacific boundary, which is approximately 16.5 cm/yr in this region (Handschumacher, 1976). Since the axial trend of the Este spreading center is compatible with the trend of the Nazca-Pacific boundary, this spreading center may be assuming an increasingly dominant role in accommodating the required Nazca-Pacific plate divergence on this portion of the Nazca-Pacific boundary. At the same time, the Oeste spreading center has apparently deteriorated as a component of the Nazca-Pacific boundary. Such a deterioration may be indicated by both the segmented nature of this ridge and the inferred clockwise rotation of the axial trends of its component segments from north to south.

Construction of a future tectonic configuration for the region is shown in Figure 7E based on the above observation. In this totally speculative interpretation, the Oeste spreading center has become extinct and the Este spreading center has been assimilated as a component of the Nazca-Pacific spreading center. With the extinction of the Oeste spreading center, the Easter plate has become a crustal component of the Pacific plate. Prior to that event, additional northward growth of the dual ridge system is predicted. The minor fault zone seen developing on the Nazca-Pacific boundary north of the Easter plate at 21°S (Fig. 7D) may mark the next position of the northward migrating triple plate junction.

If this interpretation is correct, the Easter plate is a transitory feature resulting from a localized eastward shift of the Nazca-Pacific boundary. The shifting of the portion of the Nazca-Pacific boundary is occurring through a progressive transformation of a complex tectonic regime over a period of time in excess of 3 m.y. During the transformation process, it is possible that some degree of nonrigid plate behavior takes place, which would account for the anomalous focal mechanism solutions obtained for earthquakes in this region (Forsythe, 1972; Anderson and others, 1974). It is further suggested that the difficulty encountered in resolving anomaly patterns about extinct spreading centers may be partially attributed to the complex structure acquired by a ridge as it becomes extinct.

GENETIC AND REGIONAL CONSIDERATIONS

Up to this point the discussion has been limited to defaning the Pliocent-Pleistocene tectonic evolution of the Easter plate without regard to its genetic or regional significance. As indicated in Figure 9, the Easter plate is located at the intersection of the Nazca-Pacific plate boundary with a

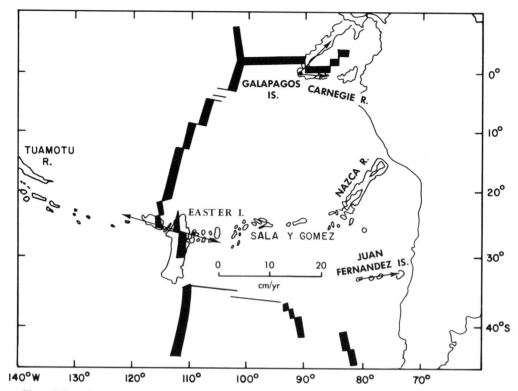

Figure 9. Location of the Easter plate in relation to the Tuamotu and Nazca Ridges and Easter–Sala y Gomez island-seamount chain.

major structural trend of the southeastern Pacific. From east to west, this trend is composed of the Nazca Ridge, the Sala y Gomez–Easter island-seamount chain, and the Tuamotu Ridge. In addition, the anomalous zone is also located at or near the equator of spreading for the Nazca-Pacific boundary where the highest spreading rate (16.5 cm/yr) for any active spreading ridge is found (Herron, 1972a; Handschumacher, 1976). These factors strongly suggest that a genetic relationship may exist between the anomalous activity and the regional tectonic fabric and evolution of the southeastern Pacific.

Such a relationship has been considered in light of the hypotehsis that attributes the formation of linear oceanic ridges and seamount chains to the existence of hot spots located in the Earth's mantle (Wilson, 1963; Morgan, 1971). A hot spot of this type has been proposed by Morgan (1971) to exist beneath the Nazca-Pacific boundary near Easter Island and to have generated the Line-Tuamotu Ridges on the Pacific plate and the irregular Easter Sala y Gomez island-seamount chain on the Nazca plate. Reconstructions by Pilger (1978), based on magnetic anomaly identifications of Handschumacher (1976) and Herron (1972b) and bathymetric data of Mammerickx and others (1975), result in juxtaposition of the Tumaotu and Nazca Ridges prior to anomaly 11 time (Pilger and Handschumacher, 1981), and support the hypothesis that the two ridges have a common origin due to a melting anomaly centered beneath the fossil Pacific-Farallon spreading center (Fig. 10). The hypothesis requires that the Pacific and Farallon plate had a northwesterly and northeasterly motion, respectively, relative to the melting anomaly. At anomaly 11 time, the melting anomaly would have encountered a fracture zone and passed beneath the Nazca plate, thereby terminating production of the Tuamotu Ridge, an event supported by the apparent termination of the Tuamotu Ridge at anomaly 11 and the absence of a seamount chain on the Pacific plate corresponding with the Easter–Sala y Gomez chain on the Nazca plate. At anomaly 7 time, a major reorganization of eastern Pacific sea-floor spreading patterns occurred, and the Nazca plate, formed by fragmentation of the Farallon

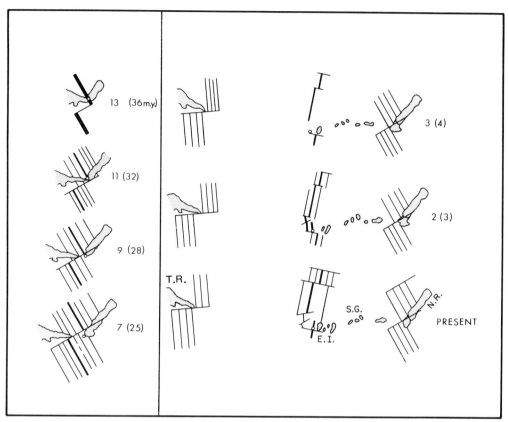

Figure 10. "Hot-spot" model for origin of the Easter–Sala y Gomez chain and Tuamotu and Nazca Ridges (from Pilger and Handschumacher, 1981). Small open circle indicates hot-spot. Nazca Ridge (N.R.), Sala y Gomez (S.G.), Tuamotu Ridge (T.R.), and Easter Island (E.I.) shown on present configuration of these features. Magnetic anomaly numbers (ages in parenthesis) identify time stages for the model.

plate by this reorientation, began moving more nearly east with respect to the melting anomaly (Handschumacher, 1976).

Given the age of the oldest volcanic rocks on Easter Island, 3 m.y.B.P. (Baker and others, 1974), instantaneous plate motion models suggest a present location of the melting anomaly (hot spot) responsible for the Tuamotu and Nazca Ridges and the Easter–Sala y Gomez chain could be beneath the Easter plate (Pilger and Handschumacher, 1981). It is conceivable that the melting anomaly is related to the existence of the Easter plate, as also noted by Anderson and others (1974). The melting anomaly could produce thinning of the overlying lithosphere, as has been inferred for intraplate island-seamount chains (Detrick and Crough, 1973). As a consequence dual spreading centers might be more favored in the vicinity of a melting anomaly, particularly along fast spreading ridges.

Understanding the origin and nature of such a melting anomaly beneath the Easter plate is clearly critical to interpreting the origin of the small plate. It is conceivable that both the melting anomaly and the anomalous spreading represent effects of another process. Pilger and Handschumacher (1981) elsewhere demonstrate that a melting anomaly beneath the Easter plate cannot be fixed relative to a Hawaiian hot spot and still produce the Easter–Sala y Gomez island-seamount chain and the Tuamotu and Nazca Ridges unless either the age of the Hawaiian–Emperor bend is significantly older than dated bottom samples or accepted geomagnetic time scales are in significant error. The existence of young volcanism over the length of the Easter-Sala y Gomez chain (Clark and Dymond, 1977; Bonatti and others, 1977) further argues against a fixed hot-spot origin for the chain.

Perhaps a more attractive model for the origin of island-seamount chains might involve intraplate stress due to asthenospheric drag (Solomon and Sleep, 1974) and cooling and contraction of oceanic plates (Turcotte and Oxburgh, 1973). In the anomalous plate region, both processes might be expected to result in lithospheric stresses of the appropriate orientation to produce the Easter-Sala y Gomez chain and the Easter plate.

REFERENCES CITED

Anderson, R. N., and others, 1974, Fault plane solutions of earthquakes on the Nazca plate boundaries and the Easter plate: Earth and Planetary Science Letters, v. 24, p. 188–202.

Baker, P. E., Buckley, F., and Holland, J. G., 1974, Petrology and geochemistry of Easter Island: Contributions to Mineralogy and Petrology, v. 44, p. 85–100.

Bonatti, E., and others, 1977, Easter volcanic chain (southeast Pacific): A mantle hot line: Journal of Geophysical Research, v. 82, p. 2457–2478.

Clark, J. G., and Dymond, J., 1977, Geochronology and petrochemistry of Easter and Sala y Gomez Islands: Implications for the origin of the Sala y Gomez Ridge: Journal of Volcanology and Geothermal Research, v. 2, p. 29–48.

Detrick, R. S., and Crough, S. T., 1978, Island subsidence, hot spots, and lithospheric thinning: Journal of Geophysical Research, v. 83, p. 1236–1244.

Fabiano, E. B., and Peddie, N. W., 1967, Grid values of total magnetic intensity IGRF-1965: ESSA Technical Report C and G538, United States Department of Commerce, 55 p.

Forsyth, D. W., 1972, Mechanisms of earthquakes and plate motions in the East Pacific: Earth and Planetary Science Letters, v. 17, p. 189–193.

Handschumacher, D. W., 1976, Post-Eocene plate tectonics of the Eastern Pacific, *in* Sutton, G. H., and others, eds., The geophysics of the Pacific Ocean basin and its margin, a volume in honor of George P. Woollard: American Geophysical Union, Geophysical Monograph 19, p. 117–202.

Handschumacher, D. W., Okamura, S. T., and Wong, P. K., 1975, Magnetic and bathymetric profiles from the central and southeastern Pacific: 10°N-45°S, 70°W-150°W: Honolulu, Hawaii Institute of Geophysics Report HIG-75-18-1975.

Herron, E. M., 1972a, Two small crustal plates in the South Pacific near Easter Island: Nature, v. 240, p. 35–37.

——1972b, Sea-floor spreading and the Cenozoic hitory of the East Central Pacific: Geological Society of America Bulletin, v. 83, p. 1671–1692.

Hey, R. N., 1977, A new class of pseudofaults and their bearing on plate tectonics: A propagating rift model: Earth and Planetary Science Letters, v. 37, p. 321–325.

Mammerickx, J., and others, 1975, Morphology and tectonic evolution of the East Central Pacific: Geological Society of America Bulletin, v. 86, p. 111–118.

Morgan, W. J., 1971, Convection plumes in the lower mantle: Nature, v. 230, p. 42.

Pilger, R. H., Jr., 1978, A method for finite plate reconstructions, with applications to Pacific-Nazca plate evoltuion: Geophysical Research Letters, v. 5, p. 469–472.

Pilger, R. H., Jr., and Handschumacher, D. W., 1981, The fixed hotspot hypothesis and origin of the Easter-Sala y Gomez-Nazca trace: Geological Society of America Bulletin (in press).

Sclater, J. G., Anderson, R. N., and Bell, M. I., 1971, Elevation of ridges and evolution of the central eastern Pacific: Journal of Geophysical Research, v. 76, p. 7888–7915.

Solomon, S. C., and Sleep, N. H., 1974, Some simple physical models for absolute plate motions: Journal of Geophysical Research, v. 79, p. 2557–2567.

Talwani, M., and Eldholm, O., 1977, Evolution of the Norwegian-Greenland Sea: Geological Society of America Bulletin, v. 88, p. 969–999.

Turcotte, D. L., and Oxburgh, E. R., 1973, Mid-plate tectonics: Nature, v. 244, p. 337–339.

Wilson, J. T., 1963, Evidence from islands on the spreading of ocean floors: Nature, v. 198, p. 925–929.

MANUSCRIPT RECEIVED BY THE SOCIETY NOVEMBER 12, 1980
MANUSCRIPT ACCEPTED DECEMBER 30, 1980
CONTRIBUTION 1099, HAWAII INSTITUTE OF GEOPHYSICS, UNIVERSITY OF HAWAII

Printed in U.S.A.

Petrogenesis and secondary alteration of upper layer 2 basalts of the Nazca plate

K. F. Scheidegger
J. B. Corliss
School of Oceanography
Oregon State University
Corvallis, Oregon 97331

ABSTRACT

This paper examines the factors controlling the petrogenesis and chemical alteration of oceanic layer 2 basalts recovered from the margins and interior of the Nazca plate in the southeast Pacific. Specifically, the extent and nature of fractional crystallization, spreading-rate variations, proximity to mantle plumes and prominent fracture zones, mantle source-rock heterogeneities, and secondary alteration are evaluated as some of the more salient factors influencing the composition of upper oceanic lithosphere of the plate. A total of 274 analyses from 88 locations are available for this purpose, and many analyses are of crust generated at the fastest spreading part of the world's mid-ocean ridge system, the East Pacific Rise.

We find that oxidative, low-temperature sea-water alteration, extensive shallow-level fractional crystallization, and mantle source-rock inhomogeneities are the most important factors. The most pronounced effects of oxidative alteration (high Fe_2O_3/FeO ratios and K_2O contents; low MgO abundances; preponderance of a celadonite–iron oxide secondary mineral assemblage) are observed in older basalts (<10 m.y.) dredged from topographic highs on the plate interior, whereas basalts dredged from the crestal portions of the East Pacific Rise and the Galapagos spreading center show only minimal chemical effects of alteration. Only where drilling or recent tectonism has made possible the recovery of basalts from lower in the crustal section are the effects of nonoxidative (hydrothermal) alteration preserved (low Fe_2O_3/FeO ratios and K_2O contents; high MgO abundances; smectite-dominated secondary mineral assemblage). Extensive shallow-level fractional crystallization, involving plagioclase, clinopyroxene, and olivine, is clearly the most important process controlling the range in composition of basalts observed in many areas; Fe-Ti basalts are the end products of this process and are common on the plate. Clinopyroxene fractionation is required for the formation of many of these highly evolved basalts, and this is indicated by observed phenocryst assemblages, normative mineralogy, standard variation diagrams and petrogenetic modeling. First-order regional differences in the composition of mantle source rocks are required to explain the pronounced light-rare-earth-element enrichment and high $^{87}Sr/^{86}Sr$ ratios of basalts associated with suspected mantle-plume activity of Easter Island and the Galapagos Isalnds. Less pronounced mantle source-rock heterogeneity may be responsible for the occurrence of mid-ocean ridge basalts with similar major-element abundances but markedly different rare-earth-element patterns. Except for fracture-zone offsets near 85° and 95°W longitude along the Galapagos spreading center, we find little evidence supporting the concept that such prominent offsets may be compositional interfaces between opposing rise-crest segments of the East Pacific Rise. Continuous compositional variation exists along the East Pacific Rise, although basalts tend to become somewhat more primitive as the Nazca-Pacific-Cocos triple junction is approached.

Compared to basalts from the slow-spreading Mid-Atlantic Ridge, basalts analyzed from the fast-spreading East Pacific Rise and the Galapagos Rise are significantly more evolved (higher FeO^*/MgO ratios and TiO_2 contents; lower CaO and Al_2O_3 abundances). Such systematic differences in major-element abundances appear related to a near order-of-magnitude difference in spreading rates of the divergent plate margins and to the size and continuity of subaxial magma chambers that can be physically maintained beneath them. It is proposed that an enhanced thermal regime beneath a fast-spreading center favors the existence of a large, steady-state magma chamber. Primitive magma entering such a reservoir will mix with much larger volumes of highly evolved magma, thereby reducing the probability that primitive magma can be erupted on the sea floor. The observed extensive fractionation required to account for the highly evolved character of Nazca plate basalts would require a thicker layer of oceanic layer 3 cumulate; available seismic refraction data support the existence of such a layer for older crust of the Nazca plate.

INTRODUCTION

The primary goals of most studies of oceanic crust have concerned defining and explaining the chemical diversity of basalts recovered by dredging, diving, and drilling from the world's ocean basins. A large variety of explanations have been invoked to account for observed, first-order compositional differences between areas, including differences in spreading rates and depths and degrees of partial melting (Bass, 1971; Scheidegger, 1973; Melson and others, 1976a), mantle source-rock inhomogeneities (Corliss, 1970; Tarney and others, 1979), proximity to mantle plumes and hot spots (Schilling, 1973), fracture-zone offsets of mid-ocean ridges (Melson and others, 1976a; Byerly and others, 1976), and the nature and degree of secondary alteration (Hart, 1970; Hekinian, 1971). Similarly, shallow-level fractional crystallization, fractional fusion, batch partial melting, zone refining, magma mixing, and dynamic melting are other processes that have been proposed to account for the compositional variability of ocean-floor basalts (Kay and others, 1970; Bryan and Moore, 1977; Langmuir and others, 1977). With such a large number of potentially interrelated processes and factors, it is not surprising that petrologists have commonly developed nonunique petrogenetic models to account for the formation of basement rocks of the ocean basins.

Progress toward more fully understanding the petrogenesis of oceanic crust has been hindered by insufficient sampling programs and incomplete sample analysis. Despite repeated attempts to drill into oceanic layer 2 during the DSDP and IPOD programs, only the upper few hundred metres have been sampled in a very limited number of locations (Bryan and others, 1979). Fault scarps (Scheidegger and others, 1978) and walls of fracture zones (Bonatti and Honnorez, 1976) have provided natural "windows" into the middle and lower crust, but these remain to be extensively sampled by high-resolution, acoustically navigated dredging, or by submersible. Most of our existing knowledge about ocean crust has come from conventional dredging, and more recently submersibles have played a very significant role in detailed studies of small areas of the Mid-Atlantic Ridge (FAMOUS area), Galapagos spreading center, and the East Pacific Rise (21°N). Regardless of the sampling method, only a very small proportion of the spreading centers in the world's oceans have been systematically sampled, and most work has focused on the slow-spreading Mid-Atlantic Ridge in the North Atlantic where most of our models concerning petrogenesis of oceanic crust have been developed. Equally troublesome has been the general lack of integrated analysis of the recovered samples. Evaluation of the processes responsible for the formation of oceanic basalts require isotopic analysis, major- and trace-element analyses, mineral analysis, and petrographic data on the *same* samples; rarely has this been accomplished.

A major objective of the present study has concerned an evaluation of existing data on basalts recovered from the margins and interior of the Nazca plate, a relatively small oceanic plate in the southeast Pacific. The major focus of these efforts has concentrated on an examination of possible first-order controls on the petrogenesis of upper layer 2 crustal rocks in an area characterized by extremely rapid sea-floor spreading. The data set compiled for this purpose is one of the more comprehensive ones available, although it is still subject to the same limitations noted previously of insufficient sampling in some areas and incomplete analysis. It includes 94 previously unpublished and 180 published major-element analyses of basalts from 88 different locations. We have concentrated on these data, mainly because petrographic descriptions and trace-element and isotopic data are much less complete. This paper represents a synthesis of the available data. It is hoped that this will provide a useful focal point for future investigations of oceanic crust in this area as well as for those interested in future comparative studies of oceanic crust found in other parts of the world's ocean basins.

PHYSIOGRAPHIC FEATURES OF THE NAZCA PLATE

A number of studies (Herron, 1972a; Mammerickx and others, 1975; Handschumacher, 1976; Rea, 1976; Bonatti and others, 1977) have considered the origin and evolution of the prominent topographic features of the plate. We present here a necessary synopsis of this work and refer the reader to the appropriate papers for more detailed discussions of particular areas.

Three spreading centers (Galapagos, East Pacific, and Chile), transform faults, and the Peru-Chile Trench are the present plate boundaries of the Nazca plate (Fig. 1), although some 10 m.y. ago the north-south–trending Galapagos Rise was an active spreading center and plate boundary between the Nazca and Pacific plates (Herron, 1972a). Studies of magnetic anomalies on the plate by Herron (1972a) and Handschumacher (1976) indicated that all crust is less than about anomaly 21, or 50 m.y. old. Much of the older crust (10 to 50 m.y.) of the plate was formed at the fast-spreading Galapagos Rise (total spreading rate ~14 cm/yr; Mammerickx and others, 1975), whereas the younger crust on the western part was formed at the East Pacific Rise, the world's most rapid spreading center. Present total sea-floor spreading rates vary latitudinally on the East Pacific Rise: 6°S (15.1 cm/yr); 10.5°S (15.8 cm/yr); 20°S (16.2 cm/yr); and 31°S (16.2 cm/yr) (Rea and Scheidegger, 1979). Slow (4.0 to 6.0 cm/yr; Herron and Hayes, 1969) and moderate (6.0 to 7.0 cm/yr; Anderson and others, 1975) total sea-floor spreading rates have characterized the Chile Rise and the Galapagos spreading center, respectively.

Mantle-plume and related hot-spot activity have been postulated for the Galapagos Islands and Easter Island (Morgan, 1972). The aseismic Cocos, Carnegie, and Nazca ridges represent traces of such activity according to such a model. However, a more recent evaluation of the topographic, geochemical, and age relationships of features associated with Easter Island has lead Bonatti and others (1977) to conclude that the east-west–trending Easter Fracture Zone (Fig. 1) may be a mantle hot line, a locus of perhaps several active volcanic centers. If this feature continues into the Chile Trench near 27°30'S (Bonatti and others, 1977), then the origin of the Nazca Ridge remains uncertain.

Other features of interest include the numerous, prominent east-west–trending fracture zones and transform faults, and a very small oceanic plate, the Easter plate, on the East Pacific Rise near 25°S (Herron, 1972b; Fig. 1).

PREVIOUS WORK

Numerous studies have been conducted on crustal rocks of the Nazca plate and its plate boundaries, but nearly all of them have concentrated on small geographic areas. Most early work concentrated on the islands, including Easter (Lacroix, 1928; Bandy, 1937), Galapagos (McBirney and Aoki, 1966; McBirney and Williams, 1969), Sala y Gomez (Fisher and Norris, 1960; Engel and Engel, 1964), Juan Fernandez (Lacroix, 1928; McBirney and Williams, 1969), and San Felix (Willis and Washington,

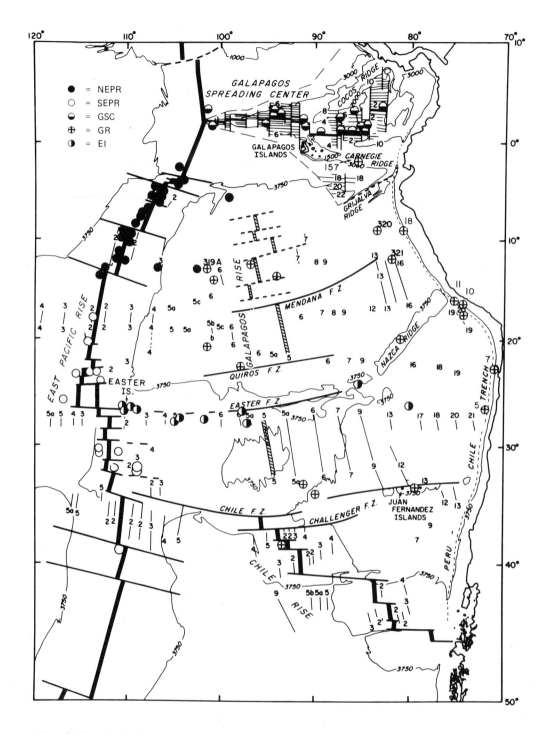

Figure 1. Map showing locations of sample sites (all round symbols) relative to principal tectonic features and identified magnetic anomalies (smaller numbers) of the Nazca plate. Larger numbers correspond to DSDP sites and dredge locations in the Peru-Chile Trench. Five different symbols (see legend on map) are used to show locations corresponding to populations of samples discussed in text.

1924). Subsequent studies have concentrated on Easter Island (Baker and others, 1974; Clark and Dymond, 1977; Bonatti and others, 1977) and the Galapagos Islands (Schilling, 1971) because of suspected mantle-plume activity. McBirney and Gass (1967) summarized much of the earlier work on the islands of the eastern Pacific and noted that the most siliceous rocks are found on Easter Island (sodic rhyolites) and the Galapagos Islands (trachytes and quartz syenites) with much larger volumes of tholeiitic basalts; more silica-undersaturated alkalic basalts were characteristic of the other islands. Recent evaluation of these and other data (Schilling and others, 1976; Bonatti and others, 1977) indicated that the light-rare-earth-element (LREE) enrichment, high FeO^*/MgO ratios, high TiO_2 and K_2O concentrations, and wide variation in silica contents are consistent with geochemical anomalies in the mantle.

Seamounts of the Nazca plate have received little attention. Funkhouser and others (1968) presented age data for four seamounts, and Bonatti and Fisher (1971) have compared the ages and compositions of seamount basalts with data from the adjacent East Pacific Rise. Bonatti and Fisher (1971) concluded that the basalts are anomalously young and indicate midplate volcanism, and that the compositions of the basalts are comparable to those dredged from the East Pacific Rise.

Sampling of vertical successions of flow units from oceanic crustal layer 2 of the plate has not been very successful. During DSDP Leg 16, site 157 was drilled near the Carnegie Ridge; a few metres of altered tholeiitic basalt were recovered (Yeats and others, 1973). Similarly, during DSDP Leg 34, the first leg devoted to drilling into oceanic basement, a few tens of metres were recovered from DSDP sites 319A, 320, and 321 (Fig. 1); reasonably fresh tholeiitic basalts were recovered from several different flow units (see Yeats and others, 1976; Mazzullo and Bence, 1976). Dredging of prominent (~1 km) fault scarps in the Peru-Chile Trench (Fig. 1; Scheidegger and others, 1978) provided access to surprisingly unaltered middle and lower layer 2 basalts. Fresh glasses from each fault scarp exhibited considerable compositional variations, an observation consistent with the sampling of several flow units in each area. Shallow-level fractional crystallization was called upon by Scheidegger and others (1978) to explain the chemical variation at each fault scarp.

Parts of the East Pacific Rise and the Galapagos spreading center have been sampled and studied by several different investigators. Analyses of East Pacific Rise basalts have been presented by Engel and Engel (1964), Kay and others (1970), Hekinian (1971), Bonatti and Fisher (1971), Scheidegger (1973), Schilling (1975), Schilling and Bonatti (1975), and Melson and others (1976a); analyses of basalts from the Galapagos spreading center are given by Campsie and others (1973), Anderson and others (1975), Schilling and others (1976), Byerly and others (1976), and Byerly (1980). These studies have indicated that basalts from the East Pacific Rise have typical mid-ocean ridge basalt (MORB) compositions, whereas segments of the Galapagos spreading center have basalts with more MORB and Fe-Ti–enriched compositions. Many additional samples from the Galapagos spreading center are being studied by Clague, Corliss, Sinton, and Schilling (1979, personal commun.). By comparison, the Chile Rise to the south has only been studied in a preliminary fashion (Hekinian, 1971).

ADDITIONAL SAMPLING, ANALYSIS, AND SYNTHESIS

Our compilation of published data on Nazca plate basalts resulted in 180 analyses from 52 locations. Most studies from which these data were taken were focused on reasonably small geographic areas; no attempt was made by previous investigators to systematically sample particular topographic features, and as a result, many gaps existed.

During the past decade, several oceanographic cruises to the southeast Pacific collected samples from the Nazca plate, but many of these samples had not been described in the literature. By obtaining representative samples from these collections we were able to add an additional 94 analyses from 36 locations to our data set. Thus, we had available 274 analyses from 88 locations (Fig. 1). In Table 1 (microfiche) we identify the locations, water depths, source of all samples, and available geochemical data discussed in this paper.

A combination of atomic absorption spectrophotometric, colorimetric, and instrumental neutron activation techniques were used in our laboratories to analyze the samples we obtained for as many as 28 elements. Major elements and Co, Cu, Zn, and Mn were determined by atomic absorption following the procedures of Omang (1969); rare earth elements (REE) and other selected trace elements were determined by instrumental neutron activation analysis (Gordon and others, 1968); and ferrous iron abundance was determined by titration using the metavanadate–ferrous ammonium sulfate method. Repeated analysis of U.S. Geological Survey standard rock BCR-1 during the course of the analyses indicated that both the accuracy and reproducibility of the results are satisfactory (see Scheidegger and others, 1978).

Although samples from many different locations on the Nazca plate are discussed in this report, the completeness of the sampling is less than ideal. Large sections of the East Pacific Rsie and the Chile Rise still remain to be sampled (Fig. 1); basement rocks exposed along transform faults, fracture zones, and adjacent crest of spreading centers should be investigated; sampling of oceanic crust at different levels in crustal sections, at different ages, and at different locations on the plate relative to mantle flow lines remain to be performed in a more comprehensive way.

One of the initial problems we faced was developing criteria for distinguishing between fresh and altered samples. Following the work of Frey and others (1974) and Scheidegger and others (1978), we have adopted an Fe_2O_3/FeO weight percent ratio of 0.5 as one arbitrary criterion; samples with a ratio of 0.5 or greater were categorized as altered, and less oxidized samples as fresh. Many samples in our data set were studied by electron microprobe, and ferrous and ferric iron abundance were not presented. Because these samples were glasses, a low Fe_2O_3/FeO ratio was assumed, and the samples were considered as being unaltered. Total iron was reported for many holocrystalline basalts but not Fe_2O_3 and FeO abundances. For many of these samples we used total water contents less than or greater than 0.7 wt % (Frey and others, 1974) to classify them as being fresh or altered, respectively. If iron and water data were not available, we relied on petrographic descriptions. In the next section of this report we discuss how secondary alteration has influenced the compositions of Nazca plate basalts.

CHEMICAL CONSEQUENCES OF LOW-TEMPERATURE ALTERATION

With few exceptions basement rocks considered here were dredged from the uppermost, surficial outcroppings of oceanic layer 2. Two dissimilar alteration processes can control the stability of secondary minerals and the resulting elemental abundance patterns of such rocks. If little or no free oxygen is available in the environment of alteration, nonoxidative alteration would be expected to be prevalent. This is favored by hydrothermally induced circulation through the piles of flow units during the first few million years following their formation (Hart and Staudigel, 1978). Fe-rich saponites, talc, and sulfides are common alteration products (Bass, 1976; Scheidegger and Stakes, 1977), and oxygen isotopic data (Seyfried and others, 1978) indicate that such minerals form at low temperature ($>20°C$). By comparison, oxidative alteration can occur at any time if crustal rocks are exposed for prolonged periods of time to cold, oxygenated sea water. Iron oxides and hydroxides and celadonite are the characteristic stable secondary minerals of oxidative alteration (Bass, 1976).

Compositional effects of both types of alteration can be most easily observed from chemical analysis of fresh glass–altered holocrystalline basalt pairs recovered from the Peru-Chile Trench (Stakes and Scheidegger, 1981). In this tectonic setting, nonoxidatively altered pillow basalts have been tectonically exposed to sea water for varying, but short periods of time (<0.5 m.y.; Scheidegger and Stakes, 1977), and they exhibit varying chemical and mineral effects associated with this more oxygenated environment of alteration. Altered holocrystalline interiors of pillow basalts with $Fe_2O_3/FeO \leq 1.0$ are dominated by a nonoxidative secondary mineral assemblage which presumably developed several tens of millions of years ago. Such rocks show slight enrichments in the alkali metals Ti and Mg, and depletions in Fe relative to coexisting fresh glasses (Fig. 2; Stakes and Scheidegger, 1981). These compositional changes largely reflect the effects of additions of Mg-rich secondary minerals (smectites) in fractures and other available pore spaces in hydrothermally altered rocks. With increased oxidation, such smectites are destroyed and celadonite becomes the dominant phyllosilicate (Bass, 1976; Stakes and Scheidegger, 1981); this mineral change is accompanied by the expected pronounced enrichment in K and Fe, and by a depletion in Mg (Fig. 2). Clearly, oxidative sea-water alteration can bring about very significant changes in major-element abundances. Ludden and Thompson (1979) have also described LREE enrichments in highly oxidized basalts relative to their fresher, nonoxidatively altered counterparts. These trends are consistent with the work of Hart (1970) who shows that enrichments in potassium, iron, titanium, and sodium, and depletions in magnesium accompany hydration and increased oxidation.

Both the nature and degree of alteration of basalts recovered from the East Pacific Rise (0 to 10 m.y.) and from oceanic crust formed at the Galapagos Rise (10 to 50 m.y.) can be evaluated qualitatively on similar AFM diagrams (Fig. 3). Using the Fe_2O_3/FeO ratio as an index of alteration, we find that for the East Pacific Rise "altered" basalts ($Fe_2O_3/FeO \geq 0.5$) define surprisingly the same compositional field as the less oxidized samples. Three explanations can be given for the apparent lack of significant chemical alteration: (1) Most of the "altered" East Pacific Rise samples are not highly oxidized (Fe_2O_3/FeO ratios range from 0.5 to 1.0); (2) nearly all of the East Pacific Rise samples were dredged from the crest of the rise and are quite young; and (3) most of the samples show little evidence of minerals characteristic of nonoxidative alteration. This suggests that because of thin surficial exposures the samples may have never experienced much hydrothermally induced alteration. From these and other data described below it is clear that all samples collected from the crestal portions of the East Pacific Rise have experienced minimal chemical alteration.

By comparison, the more oxidized basalts dredged or cored from Nazca plate crust formed at the Galapagos Rise exhibit typically pronounced alkali enrichment relative to the less oxidized samples (Fig. 3). Their much greater age (10 to 50 m.y.) is believed to be responsible for their more complete oxidation and chemical alteration. We do not see limited alkali and magnesium enrichment expected for nonoxidative, hydrothermally induced alteration (Fig. 2). We conclude that low-temperature, oxidative sea-water alteration is the dominant factor responsible for the chemical alteration of the uppermost exposures of upper layer 2 crustal rocks of the Nazca plate.

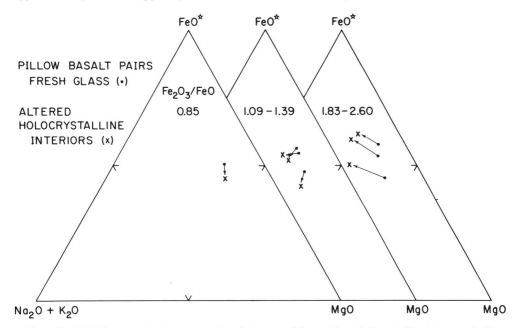

Figure 2. AFM diagrams showing compositional changes of the oxidized, holocrystalline interiors of pillow basalts relative to fresh glass from the margins of each pillow. Fe_2O_3/FeO ratios noted on each portion of figure correspond to altered, holocrystalline interiors. Note iron enrichment that accompanies increased oxidation.

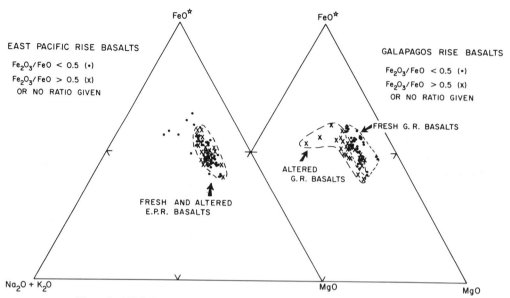

Figure 3. AFM diagrams for fresh and altered basalts from the EPR and the GR.

GEOCHEMICAL POPULATIONS OF NAZCA PLATE BASALTS

Before discussing the factors that exert first-order influence on the initial compositions of Nazca plate basalts, we summarize the nature of the compositional variability of basalts associated with specific geographical regions. Our goal here is to provide a brief synthesis of both published and unpublished data so that the reader can more easily follow subsequent discussions on the petrogenesis of basalts recovered from different parts of the Nazca plate.

From a combination of multivariate statistical tests (factor analysis) and standard geochemical variation diagrams of major- and trace-element data, five major populations of samples can be recognized (Fig. 1): (1) North East Pacific Rise (NEPR; 3°N to 17°S lat); (2) South East Pacific Rise (SEPR; 17° to 33°S); (3) Galapagos spreading center (GSC); (4) Galapagos Rise (GR); and (5) Easter Island (EI) and associated islands and seamounts of the Easter "Hot Line" (Bonatti and others, 1977). Samples from NEPR and SEPR consist of young (<10 m.y.) glassy to microcrystalline fragments of pillow basalts dredged from what Schilling (1975) has considered to be "normal" segments of the spreading center of the world's ocean basins. The GSC has both "normal" segments (west of 95°W and east of 85°W), but the intervening portion north of the Galapagos Islands is suspected of being influenced by the Galapagos hot spot (Schilling and others, 1976; Byerly and others, 1976); equally fresh pillow basalts have been dredged from these areas. Nearly all samples in the GR population are believed to have been produced 10 to 50 m.y. ago along normal segments of the GR, whereas samples in the EI population are believed to have been influenced by mantle-plume activity (Bonatti and others, 1977). As noted previously (Fig. 3) and as described by Bonatti and others (1977), many of the older samples from the GR and EI groups have experienced significant chemical alteration.

Major-element characteristics of fresh samples ($Fe_2O_3/FeO < 0.5$) from each group are summarized on AFM diagrams (Fig. 4), and Table 2 presents average major-element compositions of basalts from each population. Several interesting observations can be made from these data:

1. Except for the EI group, characteristic trends of iron enrichment are apparent for each population. This can be quite pronounced (GSC), or it can be moderate (SEPR and GR).

2. Many of the groups appear compositionally distinct. As examples, the EI population is very different from the others, and the GR is somewhat less enriched in the alkalis as compared to the EPR populations.

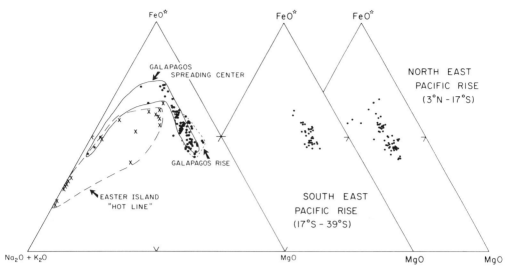

Figure 4. AFM diagrams showing compositional fields of EI, GSC, GR, SEPR, and NEPR populations. Note the great compositional variability of NEPR and SEPR groups and the distinctiveness of the EI and GSC populations.

TABLE 2. AVERAGE COMPOSITIONS OF FIVE POPULATIONS OF BASALTS FROM THE NAZCA PLATE

	SF1(NEPR)	SF2(SEPR)	SF3(GR)	SF4(GSC)	SF5(EI)	MAR†	MORB§	MORB#
SiO_2	50.26	50.29	50.39	50.25	49.52	50.68	49.61	49.84
Al_2O_3	14.75	14.68	14.35	14.00	15.66	15.60	16.01	17.25
FeO*	10.15	10.39	11.03	12.70	11.13	9.85	11.49	8.71
MgO	7.35	7.08	7.72	6.16	4.70	7.69	7.84	7.28
CaO	11.29	11.39	11.75	10.99	8.28	11.44	11.32	11.68
Na_2O	2.89	2.95	2.37	2.57	3.93	2.66	2.76	2.76
K_2O	0.18	0.18	0.13	0.15	1.20	0.17	0.22	0.16
TiO_2	1.67	1.69	1.61	2.08	3.15	1.49	1.43	1.51

Note: Only fresh basalts (Fe_2O_3/FeO<0.5) or basaltic glasses were used in the average. For comparison, an average composition of basaltic glasses from the MAR and MORB are given. FeO* = total iron calculated as FeO.

†Melson and others, 1976a.

§Cann, 1971.

#Engel and others, 1965.

3. Except for the NEPR data, which show some evidence of biomodality, continuous variations exist within each group.

The significance of these differences among groups of Nazca plate volcanic rocks will be developed in subsequent sections of this report.

Further insight into the major-element characteristics of each population of samples was obtained from factor analysis. An *R*-mode analysis was performed on 256 samples with eight variables (SiO_2, Al_2O_3, FeO*, MgO, CaO, Na_2O, K_2O, and TiO_2) per sample. All data for each variable were normalized to equal means prior to analysis, thereby ensuring that each variable was given equal weight. Three factors accounted for 99.2% of the total variance in the data set. In Figure 5 we plot on a ternary diagram the factor loadings of each sample with Fe_2O_3/FeO < 0.5. Elements listed by each apex of the triangle correspond to those most important in characterizing the three factors. Factor 1 (Mg, Ca, Al, and Si) accounts for most of the variance (65.0%), and samples with high factor 1 loadings are the least evolved (high MgO, CaO, and Al_2O_3 contents). Factor 2 (Fe and Ti; 23.8% variance), and factor 3 (K; 10.4% variance) were of lesser importance, but corresponded to samples exhibiting Fe-enrichment effects of fractional crystallization (Figs. 4, 5) and pronounced K enrichment associated with magma generation. For EI, high factor loadings in factor 3 may be also associated with mantle-plume activity.

Many of the same compositional distinctions noted above on the AFM plots (Fig. 4) are also apparent from the factor analysis data (Fig. 5). We see the obvious differences between the GSC and EI populations, the interesting relationships between the NEPR and SEPR groups, and the considerable overlap of compositional fields, particularly for samples with high factor 1 loadings.

Chondrite-normalized REE patterns provide additional insight into the general nature of basalt compositions present in the Nazca plate data set. With the exception of strong LREE-enriched patterns of basalts from the EI populations and the Galapagos Islands, most basalts exhibit the characteristic LREE-depleted patterns of MORB (Fig. 6). However, there are a number of exceptions. We find that flat and LREE-enriched patterns are common for both the EPR and GR populations (Fig. 6). Some of these samples have $(La/Sm)_{ef} > 1$, yet are not associated with topographic features such as seamounts where such LREE enrichment is common (compare Figs. 1 and 7). Clearly, the compositional diversity of basalts found on the Nazca plate or its constructive plate boundaries is as great as that associated with well-studied areas of the Mid-Atlantic Ridge (for example, FAMOUS area; Bryan and Moore, 1977).

In summary, both major- and trace-element data support the hypothesis that low-potassium, LREE-depleted tholeiitic basalt is the most common rock type present on the Nazca plate. Most of these basalts exhibit moderate to strong iron enrichment and expected variability in related major and trace elements. Highly evolved (siliceous) volcanic products, on the basis of present sampling, appear to be associated with EI (Bonatti and othes, 1977), Galapagos Islands (McBirney and Gass, 1967), GSC near 95.24°W (Byerly and others, 1976), and one location on the NEPR at 12.87°S where a basaltic andesite (56.82% SiO_2) has been recovered (Table 1, microfiche). Several samples from SEPR, NEPR, and GR populations have flat to LREE-enriched patterns, but have more typical MORB major-element abundances with similar degrees of silica saturation. LREE-enriched, silica-undersaturated basalts are typically found on the Galapagos Islands and seamounts and islands of the EI population. Considerable compositional variability exists in each population. We must now

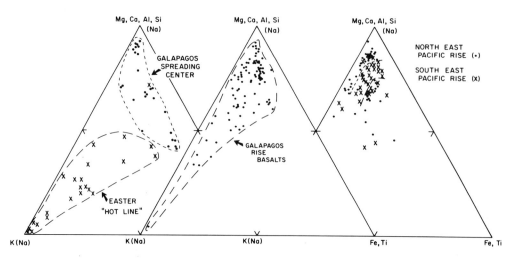

Figure 5. Ternary plots of factor loadings resulting from a factor analysis of five populations of Nazca plate basalts.

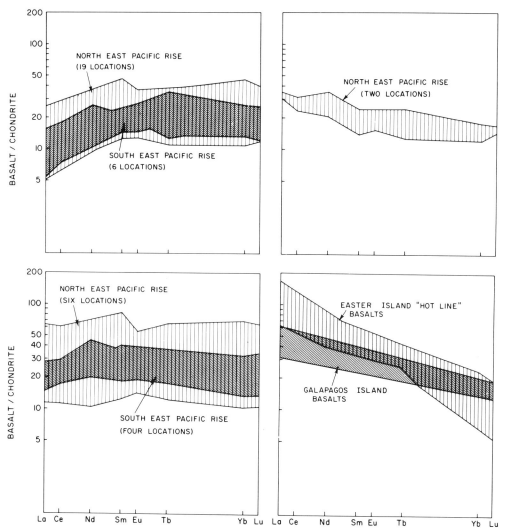

Figure 6. Summary diagram of the variability in REE abundances for basalts recovered from different portions of the Nazca plate. Although LREE-depleted patterns are by far the most common, varying degrees of LREE-enrichment are observed for many samples.

examine more closely some of the factors that we believe can account for such primary compositional diversity and for the differences among populations of Nazca plate basalts.

PRIMARY CONTROLS ON COMPOSITIONS OF BASALTS FROM THE NAZCA PLATE

Fractional Crystallization

It has been commonly acknowledged by most petrologists that primary basaltic liquids cannot ascend to the crestal portions of mid-ocean ridges without experiencing significant shallow-level fractional crystallization (O'Hara, 1968). Kay and others (1970) suggested that separation of variable proportions of magnesia olivine and calcic plagioclase phenocrysts from such magmas can explain much of the observed compositional diversity of MORB-type basalts, although others (Clague and

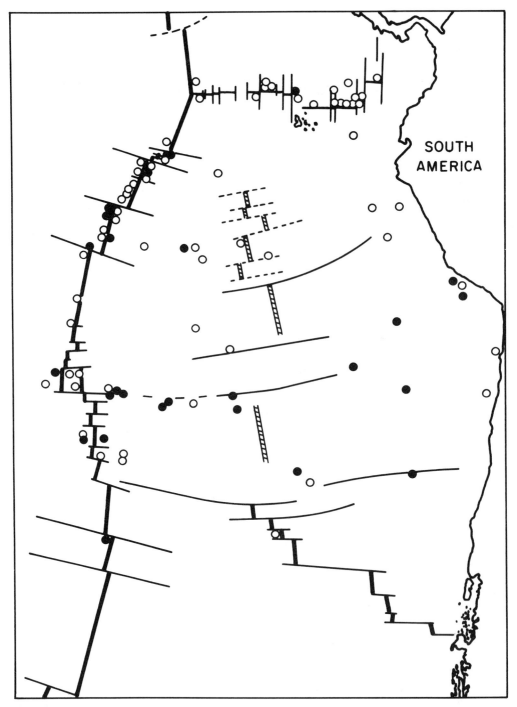

Figure 7. Map showing locations of samples with either flat or LREE-enriched patterns (●). Note common occurrence of such samples along the EPR, a presumed normal portion of the mid-ocean ridge system of the world's oceans.

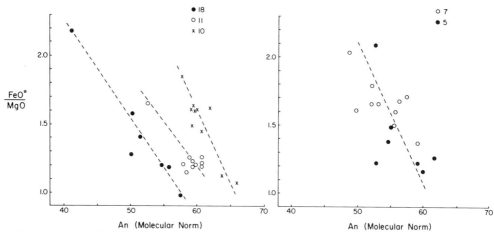

Figure 8. A plot of FeO*/MgO versus An (molecular norm) for basalts recovered from five separate ~1-km fault scarps in the Peru-Chile Trench (see Fig. 1; Scheidegger and others, 1978). Note range in compositions and distinctiveness of groups of samples from dredge areas 10, 11, and 18.

Bunch, 1976; Rhodes and Dungan, 1979) noted that subcalcic augite may also be an important phenocryst phase. In this section we assess the evidence for the nature and extent of fractional crystallization for different areas of the Nazca plate.

With the exception of the alkali-rich population of samples from E1, familiar trends of iron enrichment noted previously characterize groups of samples recovered from specific areas of the Nazca plate (Fig. 4). Such trends have been recognized for many populations of basalts and gabbros recovered from the ocean basins (Miyashiro and others, 1970). Low-pressure fractional crystallization under low oxygen fugacity is required to account for such iron enrichment.

Similar trends are also apparent for basalts dredged from individual fault-scarp exposures of oceanic crust in the Peru-Chile Trench (Fig. 8; Scheidegger and others, 1978). Two significant observations can be made about these data, and they may be crucial to our understanding of the petrogenesis of basalts produced at fast-spreading centers: (1) Glassy fragments of basalts recovered from each fault scarp define surprising coherent trends on variation diagrams. (2) Fractionation of distinctive magmas are suggested for some crustal sections by the lack of overlap of apparent trends (Fig. 8). Scheidegger and others (1978) have called upon shallow-level fractional crystallization of compositionally distinct magma batches to account for these data. For each exposure, as much as 1 km of vertical section was dredged, but there was no way of knowing from where the different flow units were sampled.

We have found many elemental covariances for basalts from specific areas or regions which suggest that plagioclase and olivine are the important phases in the initial fractionation of the primary liquids, followed by clinopyroxene fractionation in the more evolved basaltic liquids. Plagioclase fractionation is required to explain the correlation between increasing FeO*/MgO ratios and increasingly negative Eu anomalies in chondrite-normalized REE patterns, both for particular dredge hauls (AMPHITRITE 3, 12°52'S on EPR; Fig. 9a) and for specific regions (GSC; Fig. 9b). Plagioclase fractionation is not indicated by the REE data for relative concentrations less than about 20 times chondrites although Eu partition coefficients for calcic plagioclase liquid may not be greater than one. Negative correlations between Al_2O_3 and CaO when plotted against FeO*/MgO, an index of differentiation (Bryan and others, 1979), are consistent with the removal of plagioclase during fractionation (Fig. 10). Olivine fractionation is suggested by many of the same geochemical data, particularly by regular changes in FeO*/MgO ratios with increased differentiation (Figs. 4, 8, 10) and by the parallelism of chondrite-normalized REE patterns at both low and high REE concentrations (Fig. 9a, 9b). If clinopyroxene were fractionated early in the evolution of a given magma, some enrichment of the LREE relative to the HREE (heavy-rare-earth-elements) would be expected (Schilling, 1975).

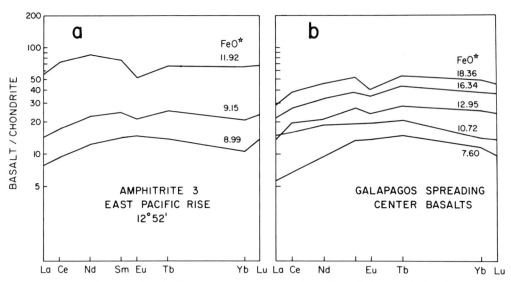

Figure 9. Chondrite-normalized REE patterns for basalts from the GSC population and the Ampitrite 3 dredge haul on the EPR. Eu anomalies occur in basalts with REE concentrations of about 20 times chondrites.

Plots of CaO/Al_2O_3 versus $Mg/Mg + Fe^{2+}$ (Fig. 11; Bence and others, 1979) and Sc versus MgO (Fig. 12) point to the importance of clinopyroxene fractionation in the middle to late stages of fractionation. For the EPR we find that as the $Mg/Mg + Fe^{2+}$ ratio decreases the CaO/Al_2O_3 ratio first increases and then decreases. Bence and others (1979) have shown that an increase in the CaO/Al_2O_3 ratio is consistent with fractionation involving olivine and plagioclase (Fig. 11), whereas a decrease in CaO/Al_2O_3 with decreasing $Mg/Mg + Fe^{2+}$ accompanies clinopyroxene fractionation. The data in Figure 11 thus indicate early fractionation involving olivine and plagioclase, followed by clinopyroxene. A similar conclusion can be drawn from a plot of Sc versus MgO (Fig. 12). With decreasing MgO (increasing fractionation), most basalts show a slight increase in Sc and then a leveling off or slight decrease in Sc. Because partition coefficients for Sc in olivine and plagioclase are low and in clinopyroxene they are high, such trends would be anticipated if clinopyroxene were a late crystallizing phase. The presence of clinopyroxene phenocrysts, particularly for the more evolved Fe- and Ti-enriched basalts, and petrogenentic modeling calculations also add additional support that clinopyroxene is important in the later stages of fractional crystallization (Mazzullo and Bence, 1976; Clague and Frey, 1979; Fisk and Bence, 1979; Byerly, 1980).

Further evidence for the importance of plagioclase-olivine-clinopyroxene fractionation in the petrogenesis of Nazca plate basalts comes from a ternary plot of these normative minerals (Fig. 13). We find that unaltered samples ($Fe_2O_3/FeO < 0.5$) from the GSC, NEPR, and SEPR define linear trends on such a diagram, approximating the plagioclase-olivine cotectic described by Miyashiro and others (1970) for Mid-Atlantic Ridge basalts and gabbros. In addition, many of the Nazca plate samples cluster on the plagioclase-clinopyroxene join, and this also suggests that multiphase fractionation is important (Bence and others, 1979). These data suggest that clinopyroxene would be expected to be an important phenocryst for many of the more evolved Nazca plate samples. As an example, Byerly (1980) has shown that formation of Fe-Ti basalts of the GSC require 50% to 75% crystallization of five parts plagioclase, three parts augite, and one part olivine from a parental, MORB-type tholeiitic magma. Clinopyroxene was observed as a common phenocryst in such fractionated basalts.

Fe-Ti basalts ($FeO^* > 12.0\%$, $TiO_2 > 2.0\%$; Byerly and others, 1976) are regarded as the most common end products of shallow-level fractional crystallization of mid-ocean ridge magmas. In Figure 14 we note locations on the Nazca plate where Fe-Ti basalts occur. Clearly Fe-Ti basalts have a

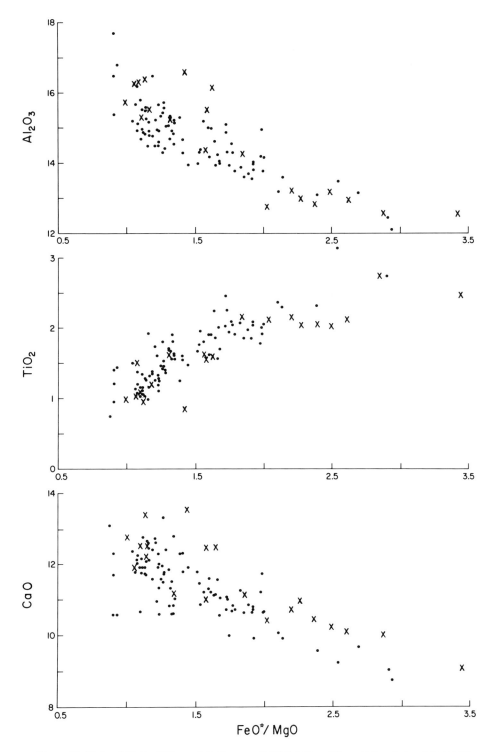

Figure 10. Plots of Al_2O_3, TiO_2, and CaO versus FeO*/MgO, a differentiation index, for samples from the EPR (•) and GSC (×).

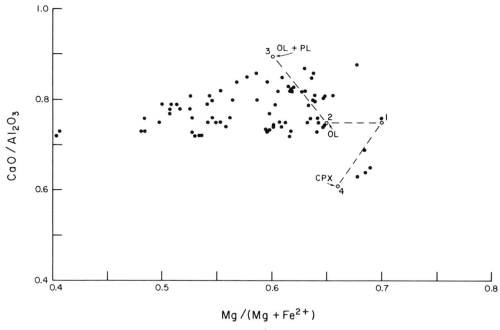

Figure 11. A plot of CaO/Al_2O_3 versus $Mg/(Mg + Fe^2)$ for basalts from the EPR. The initial increase in CaO/Al_2O_3 ratio with decreasing $Mg/(Mg + Fe^{2+})$ ratio (0.70 to 0.60) suggests the importance of initial olivine and plagioclase fractionation, whereas the decreasing CaO/Al_2O_3 ratio for samples with lower $Mg/(Mg + Fe^{2+})$ ratios (<0.60) indicates that clinopyroxene fractionation is also required to explain the compositional diversity of Nazca plate basalts. Numbers along least-squares fractionation paths of Bence and others (1979) correspond to (1) starting composition FAMOUS basalt 527-1-1; (2) cotectic composition following 6% olivine fractionation; (3) residual liquid after 30% crystallization with olivine and plagioclase fractionated in molar proportions 1:2; and (4) clinopyroxene fractionation of (1) under more elevated pressures.

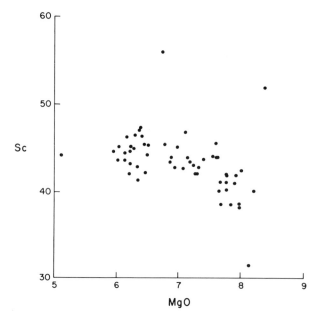

Figure 12. A plot of Sc versus MgO for EPR samples. Note tendency of Sc abundance to increase, then level off or decrease with decreasing MgO. Such data are consistent with clinopyroxene being involved in the fractionation for samples with MgO abundances less than about 7.5%.

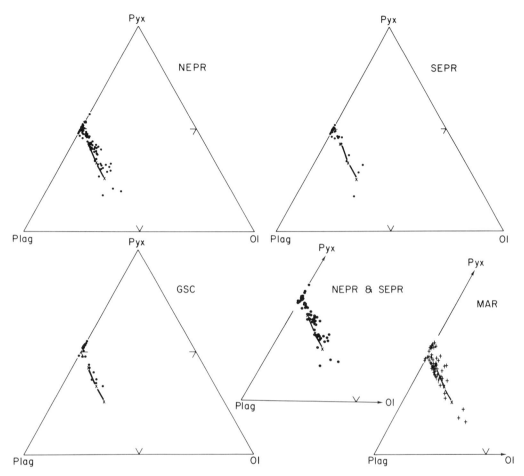

Figure 13. Ternary plots of molecular normative plagioclase, pyroxene, and olivine for NEPR, SEPR, GSC, and MAR samples discussed in text. Note how samples from the NEPR, SEPR, and GSC groups plot either on or to the right of the presumed olivine-plagioclase cotectic described by Miyashiro and others (1970) for MAR basalts and gabbros. Many samples from these three groups also plot on the plagioclase-pyroxene join. From the diagrams in the lower right, note how the MAR samples are generally richer in normative plagioclase than their Nazca plate counterparts.

random distribution. This suggests that pronounced fractional crystallization has occurred beneath all present and past constructional plate boundaries of the Nazca plate. With the exception of basalts from two dredges on the GSC near 85.50° and 95.24°W which have >15% FeO*, Fe-Ti basalts from the Nazca plate commonly have 12% to 15% FeO*.

Fracture Zone Offsets of Spreading Centers

Prominent fracture zones commonly offset segments of the spreading centers of the world's oceans, and across many such features mid-ocean ridge basalts may exhibit strikingly different compositions. Along the Mid-Atlantic Ridge major-element data have been used to demonstrate compositional differences of opposing segments of the spreading center (Melson and others, 1976a; Melson and O'Hearn, 1979). Similar differences have been reported for the Gorda–Juan de Fuca Ridge in the northeast Pacific (Wakeham, 1978) and for the GSC (Clague and Bunch, 1976; Byerly and others, 1976; Schilling and others, 1976). The major physical factor involved in such compositional contrasts

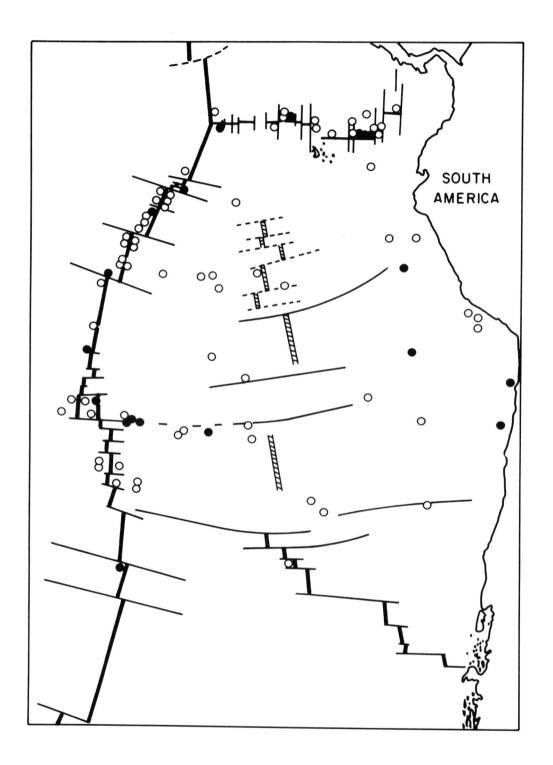

Figure 14. Map showing locations of highly fractionated (Fe-Ti) basalts on the Nazca plate (•). Note rather random occurrences of such samples.

may be that fracture zones either dam subaxial asthenospheric flow or block the overlying axial continuity of shallow magma reservoirs (Vogt and Johnson, 1975). Recent geophysical studies (Rosendahl, 1976; Rosendahl and others, 1976) and thermal modeling considerations of mid-ocean ridges (Sleep, 1975; Sleep and Rosendahl, 1979) indicate that either consequence may be likely. It follows that if compositionally distinct magmas are generated beneath opposing segments of a mid-ocean ridge, they may remain distinct because their lateral flow and consequent mixing will be impeded by crustal and subcrustal "dams." In addition, it is also possible that the abutment of a spreading center against older and cooler lithosphere of a fracture zone may influence the degree of magma evolution. Questions of interest to us are: (1) Are compositional contrasts characteristic of prominent fracture-zone offsets of the constructive plate boundaries of the Nazca plate? (2) Is the intersection of a transform fault with a spreading center a preferred setting for pronounced hydrothermal circulation, enhanced cooling of shallow magma reservoirs, and the occurrence of highly fractionated magmas?

The first question has been addressed by others for the GSC along the northern boundary of the Nazca plate (for example, Byerly and others, 1976; Schilling and others, 1976). They found that two prominent fracture zones at 85°W and 95°W offset the GSC by 120 km and 25 km, respectively. Basalts recovered between the fracture zones were commonly Fe, Ti, and K enriched, and some had flat REE patterns, particularly those immediately north of the Galapagos Islands. Highly siliceous volcanic products (andesites and rhyodacites) were also reported for this area near 95.24°W. Beyond the fracture zones at 85°W and 95°W, basalts with typical MORB compositions were found. However, MORB-type basalts were also common between the fracture zones. As Schilling and others (1976) noted, the unusual character of the GSC segment between 85°W and 95°W may be associated with the crustal or subcrustal mixing of magma derived from the Galapagos hot spot with magma derived from a more typical, depleted mantle source beneath the GSC. According to such a hypothesis, fracture zones near 85°W and 95°W may block the lateral flow of magma and/or mantle associated with a mantle plume ascending beneath the Galapagos Islands.

Along the EPR from 33°S to near the equator (Fig. 1), several prominent fracture-zone offsets are indicated by bathymetric surveys. However, we do not find evidence for obvious compositional contrasts across the fracture zone even though at 9°S and 12°S lateral offsets are about 150 km. In Figure 15 we plot FeO*, Yb, FeO*/MgO, and La/Sm of fresh basalts ($Fe_2O_3^*/FeO < 0.5$) as a function of latitude along the EPR. We find that (1) considerable compositional variation exists within and among individual dredge hauls from particular segments of the EPR, a finding consistent with the influence of fractional crystallization decribed previously; (2) continuous, not discontinuous, compositional variability exists among these spreading-center segments, and we see no obvious compositional contrast across fracture zones; (3) basalts found between 17°S and 33°S latitude (SEPR population) are less evolved (intermediate FeO*/MgO ratios) than those north of 17°S (NEPR population; see also Fig. 5), and it appears that basalts gradually become more primitive (low FeO*/MgO ratios) toward the equator; and (4) basalts with flat to LREE-enriched chondrite-normalized REE patterns (Figs. 6, 7) are characteristically found all along the EPR, a presumed "normal" portion of the mid-ocean ridge system (Schilling, 1975), and are not confined to any particular segment of the EPR.

The main conclusion to be drawn from these data is that fracture-zone offsets of the EPR are not compositional interfaces. Available major- and trace-element data for the EPR (Fig. 15) are consistent with the hypothesis that a rather uniform mantle source may exist all along the EPR. Thus, even though fracture zones may block axial flow of magma beneath the EPR, the magmas erupted at the crest of the EPR are so similar that compositional contrasts cannot be observed.

Assessing the degree of magma evolution along the EPR in relationship to proximity to transform faults is made difficult by the general lack of required systematic sampling and analysis (Fig. 1). In Figure 16 we plot the FeO*/MgO ratios of basalts versus distance to fracture zones of the GSC and EPR. There is some suggestion that the degree of Fe enrichment (magma evolution) increases as fracture zones of the GSC are approached, but this relationship is not convincing for the EPR. As noted above, the GSC may be a special case where normal segments of the spreading center (west of

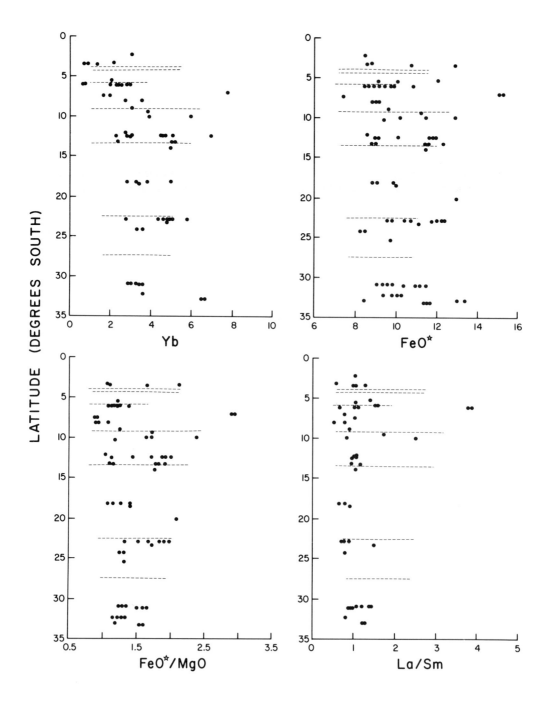

Figure 15. Plots of Yb, FeO*, FeO*/MgO, and La/Sm versus latitude along the EPR. Horizontal dashed lines indicate locations of prominent fracture-zone offsets of the EPR. Note considerable compositional variation of basalts of particular rise-crest segments and the rather continuous compositional variation along the EPR. The lower FeO*/MgO ratios and Yb abundances north of about 10°S indicate that some of the least differentiated (primitive) samples are found in this area.

95°W and east of 85°W) are separated from an intervening, transitional segment by north-south–trending transform faults (Schilling, 1975). For the EPR, some basalt found within about 100 km of ridge-ridge transform faults are slightly more evolved, but this trend may be of questional significance. These data lead us to suggest that proximity to fracture zones of the EPR does not control the degree of fractional crystallization of magmas. For the northern Mid-Atlantic Ridge, the situation may be different (Hekinian and Thompson, 1976).

Spreading Rate Variations

Bass (1971), Scheidegger (1973) and Melson and others (1976a) have suggested that the rate of sea-floor spreading at a mid-ocean ridge or rise may be an important factor in influencing the compositions of sea-floor basalts. Specifically, these studies have suggested that basalts produced at a fast spreading center may be enriched in FeO* and TiO_2, depleted in MgO, and more silica saturated than their slow-spreading-center counterparts. Shallower depth of magma segregation beneath fast spreading centers was suggested as a possible explanation of the observed differences.

The general validity of this factor has not received wide acceptance, mainly because of the following considerations: (1) Basalts recovered from small, well-studied parts of the Mid-Atlantic Ridge (for example, FAMOUS area, Bryan and Moore, 1977; DSDP Leg 37 results; Langmuir and others, 1977) exhibit considerable compositional diversity in small areas; (2) basalts from mantle-plume or transitional ridge segments (Schilling, 1975) may have been included with those from "normal" segments in the earlier surveys; and (3) excluding samples from the Juan de Fuca Ridge and the GSC where mantle-plume activity has been documented, there has never been a large number of analyses

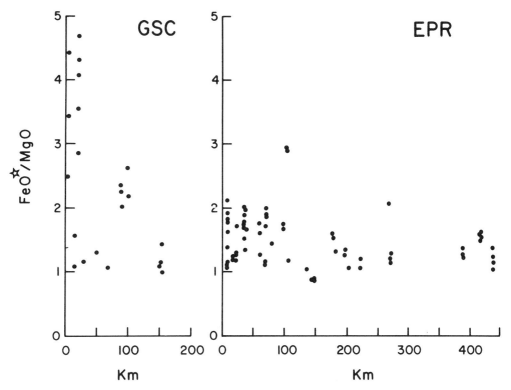

Figure 16. A plot of FeO*/MgO versus distance of samples relative to prominent (>25 km) fracture-zone offsets of the GSC and EPR. For the GSC, samples become more fractionated as the fracture zones are approached, but for the EPR, no trend is observed.

available from the fastest spreading part of the mid-ocean ridge system, the EPR (total spreading rate: ~15 cm/yr), for comparative purposes.

In this section we use the larger data base now available from normal segments of the EPR (3°N to 33°S; Schilling, 1975) and representative, high quality analyses of glasses from the Mid-Atlantic Ridge (70°N to 31°S; Melson and others, 1976b) to re-examine this important hypothesis. We have not included much additional data available from the Mid-Atlantic Ridge (MAR) (Bence and others, 1979) because they were obtained by many investigators on glassy to holocrystalline and altered materials, and the results are not always internally consistent (Byerly and Wright, 1978). The basic question that we are addressing here is: Are populations of basalts obtained from the fastest and slowest spreading portions of the world's mid-ocean ridge system fundamentally different in their major-element chemistries?

Because of the large amounts of data involved, our approach to this problem again concentrated on factor analysis, and it was applied to the major-element data (eight elements) for the NEPR and SEPR populations described previously (94 analyses from 42 locations) and to a representative sampling of data from the MAR (Melson and others, 1976b; 93 analyses from 53 locations). In Figure 17 we present histograms of some of the data used in the analysis, and in Figure 18 the results of a three-factor analysis are illustrated. In the analysis, data for each major element were again column-normalized to equal means. Three factors accounted for 99.6% of the total variance in the data set. The factors had positive coefficients of variation for MgO, Al_2O_3, CaO, and SiO_2 (factor 1), for FeO* and TiO_2 (factor 2), and for K_2O (factor 3); Na_2O had similar coefficients of variation for all factors. Again, we have found that samples with high factor 1 loadings are the most primitive samples (low FeO*/MgO, high Al_2O_3, less silica saturated), whereas those with high factor 2 and factor 3 loadings are more evolved.

The results of this statistical analysis confirm that although there is considerable overlap between the EPR and MAR populations, the two exhibit interesting differences (see Figs. 17, 18). Specifically, the MAR samples are, on the average, more primitive (higher Al_2O_3 and MgO; somewhat lower FeO*

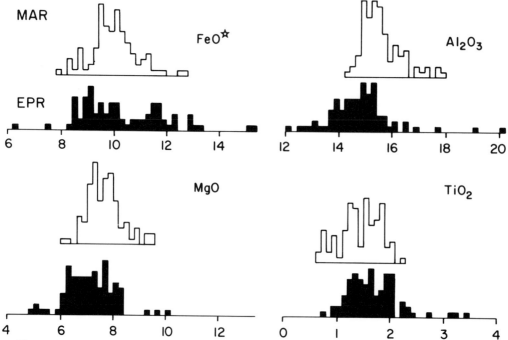

Figure 17. Histograms of major-element data of basalts from the MAR and the EPR. Note systematic differences in compositions of basalt populations.

and TiO_2) than the EPR samples. In this regard, none of the 279 analyses presented by Melson and others (1976b) for MAR would be classified as Fe-Ti basalts (FeO* > 12.0%, TiO_2 > 2.0%; Byerly and others, 1976), yet approximately 10% of the EPR samples are so classified. The data in Figure 17 also show that it is very common for EPR basalts to have greater than 11.0% FeO*. These important systematic differences in major-element chemistry are consistent with the earlier work of Bass (1971) and Scheidegger (1973), and point to fundamental differences in the factors that control the composition of oceanic crust. In the Discussion section of this report we examine the petrological significance of these data.

Evaluation of possible correlations between basalt compositions and spreading rates of present and past constructive plate boundaries of the Nazca plate can also be considered, but many populations of samples are not large. The GSC population described previously is an example. Although GSC is characterized by a moderate total rate of spreading (6.0 to 7.0 cm/yr; Anderson and others, 1975), both normal and mantle-plume–influenced segments have been recognized (Schilling, 1975; Schilling and others, 1976), and few samples are available from the normal segments west of 95° W and east of 85° W latitude. Thus, we cannot directly attribute possible differences between the GSC and EPR (Fig. 5) to spreading-rate differences. Along the EPR, spreading rates vary by less than 10% (total spreading rates: 15.1 cm/yr at 6° S increasing to 16.2 cm/yr at 20° to 31° S; Rea and Scheidegger, 1979). As noted above, some chemical variation exists along the EPR (Fig. 15), but it cannot be easily related to spreading-rate differences. For the GR population (total spreading rate of GR ~14.0 cm/yr; Mammerickx and others, 1975), we find that it is quite similar to the EPR populations described previously (Fig. 5); this result, including the rather ubiquitous occurrence of Fe-Ti basalts (Fig. 14) would be anticipated on the basis of the comparable rapid spreading rates of the EPR and the GR.

Mantle Source Heterogeneities

Basalts associated with EI and the Galapagos Islands, two suspected locations of mantle-plume activity (Morgan, 1972), demonstrate the possible influence of mantle source heterogeneities on the composition of crust of the Nazca plate. The basalts are strongly LREE enriched (Fig. 6), and those from the EI typically have high $^{87}Sr/^{86}Sr$ ratios of 0.7030 to 0.7033 (for example, Bonatti and others, 1977), values much higher than MORB ratios of about 0.7026 (Hart, 1971). In our opinion, a less depleted mantle source rock, perhaps one ascending from the middle or lower mantle beneath the

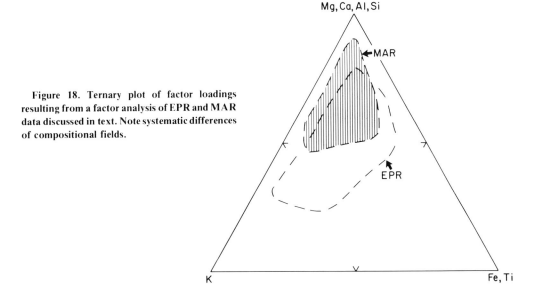

Figure 18. Ternary plot of factor loadings resulting from a factor analysis of EPR and MAR data discussed in text. Note systematic differences of compositional fields.

islands, is needed to explain such first-order compositional differences, although some workers (for example, Langmuir and others, 1977) have suggested other models.

Influence of suspected Galapagos mantle-plume activity on the neighboring GSC has been described previously in this report and in more detail by other investigators (Schilling and others, 1976; Byerly and others, 1976; Clague and Frey, 1979; Fisk and Bence, 1979). They suggested that basalts with flat REE patterns on the GSC north of the Galapagos Islands may be attributed to the mixing of magmas from two sources (Schilling and others, 1976), and that the high FeO^*, TiO_2, and K_2O abundances of basalts between 85° and 95°W may be related to a similar process. Although data on basalts from the neighboring Carnegie and Cocos aseismic ridges are not available, they would be expected to show affinities with those present on the Galapagos Islands.

Mantle-plume activity of the southern Nazca plate is associated not only with EI but also with Sala y Gomez and other islands and seamounts of the east-west-trending Easter "Hot Line" (Fig. 1; Bonatti and others, 1977). These workers envision a zone of mantle upwelling perhaps 100 km or more wide, extending from EI to the Chile Trench. Insufficient data are available from basalts of the northeast-trending Nazca Ridge (Fig. 1) to discuss their petrogenesis. However, of the three analyses available from a single dredge haul (Fig. 1), one is a Fe-Ti basalt, and all show evidence of LREE enrichment. A possible extension of mantle-plume activity to the west of EI was postulated by Bonatti and others (1977), but data on basalts recovered near 25°S on the EPR make it very unlikely that such activity strongly influences basalt compositions; basalts recovered from this area are typically LREE-depleted, tholeiitic basalts (Fig. 7).

Seamount basalts of the Nazca plate appear to require mantle source rocks similar to those involved in the petrogenesis of basalts dredged from neighboring segments of the EPR (Schilling and Bonatti, 1975). Although K-Ar dating indicates the possibility that such products were associated with midplate volcanism (Funkhouser and others, 1968), older ages may be possible for these samples if excess Ar and/or alteration were significant problems. Batiza's (1979) study of similar types of seamounts north of the GSC indicated that they may form rapidly near the crestal portions of the EPR.

For the EPR and the GR a depleted mantle source associated with the asthenosphere is required to account for the typical MORB character of most basalts recovered from the Nazca plate. However, available data indicate that this source cannot be compositionally invariant. Some of the best data that support the presence of compositional variability in a depleted mantle source come from analyses of basalts recovered from different oceanic crustal sections in the Peru-Chile Trench (Scheidegger and others, 1978). Basalts with nearly identical major-element compositions had strikingly dissimilar trace-element abundances, some showing LREE depletion and others LREE enrichment. Scheidegger and others (1978) noted that basalts dredged from individual fault scarps maintained either LREE enriched or depleted patterns over a large range in FeO^*/MgO ratios, indicating that shallow-level fractional crystallization was not responsible for the initial differences in trace-element abundance patterns. Similar arguments can be made for tholeiitic basalts dredged from the NEPR and SEPR; some have flat or LREE-enriched REE patterns (Fig. 7), but major-element abundances are typical of olivine tholeiites requiring similar degrees of melting of mantle source rocks. Although we do not have trace-element and isotopic data needed to set constraints on mantle sources and partial melting and fractionation models, our preliminary interpretations require mantle source-rock heterogeneities in explaining some of the regional compositional diversity of Nazca plate basalts.

A puzzling observation to us is the common occurrence of tholeiitic basalts found along the EPR which show moderate degrees of LREE enrichment (Figs. 6, 7). Two of the samples with the most pronounced degree of LREE enrichment (Fig. 6) were dredged at 9° and 10°S from the central uplifted axial block of the EPR away from seamounts or fracture zones. Samples are unavailable from the areas between these sampling localities. Nevertheless, the possibility exists that a rather large segment of the EPR may have been derived by partial melting of an "anomalous" mantle source rock. Observation such as these bring into question the validity of using REE patterns to distinguish between "normal" and "mantle-plume-influenced" segments of mid-ocean ridge segments (Schilling, 1975), because this area is far from such activity.

PETROGENESIS OF OCEANIC CRUST AT FAST SPREADING CENTERS

With few exceptions (GSC along the northern Nazca plate; EI "Hot Line," and Chile Rise to the south), nearly all oceanic crust of the Nazca plate was generated at the fast-spreading GR or more recently the EPR (Fig. 1). Our examination of data on basalts from the latter two areas reveals that several observations must be satisfied in order to explain the composition of basalts erupted at a fast-spreading mid-ocean ridge. These include: (1) near absence of primitive (MgO > 9.0%) tholeiitic basalts relative to the MAR; (2) dominance of moderately to highly fractionated basalts, including common Fe-Ti basalts; (3) apparent absence of compositional discontinuities across fracture-zone offsets of the EPR; (4) occurrence of LREE-enriched tholeiitic basalts along "normal" segments of the EPR and in older crustal sections associated with the GR; (5) coherent and distinctive fractionation trends associated with crustal sections for which samples of several flow units in a restricted area are available; (6) magma temperatures for basalts from fast spreading centers which may be on the order of 70°C cooler than those from slow spreading centers (Scheidegger, 1973); and (7) oceanic crust that is somewhat thinner than other oceanic crustal sections in other ocean basins and that has a thick, high-velocity (7.2 km/s) basal crustal layer (Hussong and others, 1976).

Unlike earlier conclusions of Bass (1971) and Scheidegger (1973), we now believe many of these observations can be best reconciled with fundamental differences in the size, continuity, and internal conditions within subaxial magma chambers beneath fast and slow spreading centers, and not with differences in depth at which magmas segregate from mantle source rocks. In Figuer 19, we present schematic models of subaxial magma chambers for the EPR (Rosendahl and others, 1976; Rosendahl, 1976) and the MAR (Bryan and Moore, 1977). The model for the EPR was developed largely on the basis of seismic refraction studies using sonobuoys for the 9°N area of the EPR, but it was also consistent with models based on the Samail ophiolite (Hopson and Pallister, 1980). For the purposes of discussion, we have liberally interpreted Rosendahl's (1976) model for the EPR by suggesting that even greater than 30% melt may be possible within the axial magma chamber. The MAR model was based upon petrological considerations. Both models are consistent with thermal modeling considerations of fast and slow spreading centers (Sleep, 1975; Sleep and Rosendahl, 1979), although the width and axial continuity of the magma chamber beneath the MAR can be best described as tentative (Nisbet and Fowler, 1978). Geophysical studies across the MAR have not detected evidence for subaxial magma chambers, and some petrologists (Langmuir and others, 1977) thought that the compositions of basalts in the FAMOUS area of the MAR can be best interpreted in terms of evolution of small, compositionally distinct magma batches, and not of one large magma reservoir.

Such considerations lead us to suggest that it is the fundamental physical difference between a continuous, steady-state magma chamber beneath the EPR, and smaller, less continuous in space and time, magma chambers beneath the MAR, which accounts for the observed differences for EPR and MAR, MORB-type basalts (Figs. 17, 18). Specifically, we suggest that if a batch of primitive magma ascends from the upper mantle and encounters a small magma chamber or no magma chamber at crustal depths, it may pass rapidly through the crust without experiencing much shallow-level fractional crystallization. Higher MgO and Al_2O_3 values would be expected (Figs. 17, 18). Considerable compositional diversity could be produced by incomplete mixing of different magma batches, or by fractionation in short-lived, crustal-depth magma reservoirs. By comparison, a similar, primitive batch of magma ascending beneath a fast spreading center should encounter a much larger magma reservoir at crustal levels (Fig. 19). Magma entering such a reservoir would be expected to mix with a much larger reservoir of more evolved magma, thereby causing the recently injected magma to lose its primitive character (Dugan and Rhodes, 1978). Consequences of such magma mixing would be coherent trends of magma evolution (Fig. 8), lack of MgO-rich basalts and a preponderance of moderately to highly evolved basalts (Fig. 18), and a tendency for basalts to plot either on the olivine-plagioclase cotectic or on the plagioclase-olivine-clinopyroxene pertectic (Fig. 13). Additionally, a large, steady-state magma chamber would favor extensive shallow-level fractionation with associated lower magma temperatures, and this could lead to the formation of a thicker layer of olivine-clinopyroxene cumulates at the base of recently generated oceanic crust (Fig. 19). It is also apparent

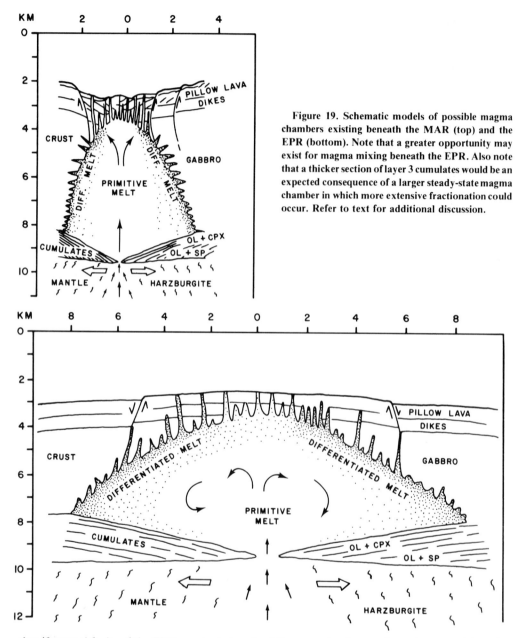

Figure 19. Schematic models of possible magma chambers existing beneath the MAR (top) and the EPR (bottom). Note that a greater opportunity may exist for magma mixing beneath the EPR. Also note that a thicker section of layer 3 cumulates would be an expected consequence of a larger steady-state magma chamber in which more extensive fractionation could occur. Refer to text for additional discussion.

that if normal faults of the EPR penetrate to a few kilometres depth away from the crest of the axial block (Fig. 19), fractionated magma from the margins of the magma reservoir may be tapped.

A difficulty for either model is the occurrence of LREE-enriched basalts along normal segments of the mid-ocean ridge system (Fig. 7). For the Nazca plate it appears that such occurrences are not confined to particular crustal sections or to specific areas of the EPR. The most dramatic examples for the EPR concern the recovery of only LREE basalts from the crest of the axial block of the EPR at 9° and 10°S (Fig. 7); north and south of this area more typical LREE-depleted basalts have been recovered. Such data are not inconsistent with there being large reservoirs of magmas in some areas that may have unusual compositions, perhaps inherited from an inhomogenous upper-mantle source.

The same data may also be interpreted in terms of more complex melting and fractionation models (for example, Langmuir and others, 1977).

Although the general model outlined above remains to be tested by additional field and laboratory studies, there can be little remaining doubt that differences in thermal regimes between fast and slow spreading centers are first-order features that can influence the evolution of magma and the formation of oceanic crust.

CONCLUSIONS

Our synthesis of published and previously unpublished data on basalts recovered from the Nazca plate allows us to make the following conclusions:

1. Except for basalts dredged from the crestal portions of the EPR and the GSC, most surficial outcroppings of basalt show evidence of low-temperature oxidative alteration, including pronounced enrichments in alkali metals, TiO_2 and FeO^*, and depletions in MgO relative to original basalt compositions. Effects of nonoxidative diagenesis on basalt compositions appear confined to samples recovered by drilling on the plate, or to those basalts that have been recently tectonically re-exposed to sea water.

2. Most basalts recovered from the Nazca plate are LREE depleted, MORB-type tholeiitic basalts, exhibiting moderate to pronounced iron enrichment, although LREE-enriched basalts are also common.

3. Mantle-plume activity characterizes the Galapagos Islands and EI. For EI, the effects of such activity may extend as a mantle "hot line" as far east as 27°30'S in the Chile Trench, but not to the EPR on the west; for the Galapagos Islands, mantle-plume activity may influence the compositions of basalts on the GSC immediately north of the islands as well as between 85° and 95°W.

4. Except for the prominent fracture-zone offsets of the GSC at 85° and 95°W, similar offsets do not influence the compositions of basalts recovered from opposing rise-crest segments of the EPR. Reasonably continuous compositional variations characterize the composition of basalts recovered from the crestal portions of the EPR.

5. Basalts with flat and LREE-enriched patterns are found along normal segments of spreading centers, far from suspected mantle-plume activity. Partial melting or heterogeneous asthenospheric mantle appears necessary to explain the major- and trace-element abundances of these basalts.

6. Seamounts not related to the EI "hot line" consistently have LREE-depleted, MORB-type compositions, suggesting an origin near the axial portions of spreading centers. Uncertainties about published age data do not preclude the possibility of midplate volcanism.

7. Oceanic crust generated at fast-spreading EPR is generally more fractionated (higher FeO^*/MgO, lower Al_2O_3, higher TiO_2, more silica-saturated) than crust generated along the slow-spreading MAR.

8. Magma chambers beneath the EPR are large, steady-state reservoirs, which allow primitive magmas to enter but not to pass through without first mixing with a much larger volume of more evolved magma; absence of primitive magmas and presence of ubiquitous fractionated magmas from the crest of fast spreading centers is thus ensured. The absence of large, steady-state magma reservoirs beneath the MAR may facilitate the ascent to the surface of less fractionated magma or the fractionation of compositionally distinct magma batches in short-lived crustal magma chambers.

ACKNOWLEDGMENTS

This study was made possible by the assistance and support of many individuals. We wish to specifically thank Enrico Bonatti, Charles Fein, and Stephen Johnson for making available samples or unpublished data on basalts from the East Pacific Rise; Mark Hower and Roberta Conard for performing many of the geochemical analyses; Chi Muratli and Sarah Hoffman for assisting with data reduction, and Bill Melson and Rodey Batiza for providing helpful comments on an earlier draft of this manuscript. In addition, samples were also obtained directly from the Scripps dredge collection from the southeast Pacific. Financial support for this study was received from the National Science Foundation through grants IDOE 71-04208 A07 and OCE 76-05903 A02.

REFERENCES CITED

Anderson, R. N., and others, 1975, Magnetic and petrologic variations along the Galapagos spreading center and their relation to the Galapagos melting anomaly: Geological Society of America Bulletin, v. 86, p. 683-694.

Baker, P. E., Buckley, F., and Holland, J. G., 1974, Petrology and geochemistry of Easter Island: Contributions to Mineralogy and Petrology, v. 44, p. 85-100.

Bandy, M. C., 1937, Geology and petrology of Easter Island: Geological Society of America Bulletin, v. 48, p. 1589-1610.

Bass, M. N., 1971, Variable abyssal basalt populations and their relation to sea-floor spreading rates: Earth and Planetary Science Letters, v. 11, p. 18-22.

———1976, Secondary minerals in oceanic basalts with special reference to Leg 34, Deep Sea Drilling Project, *in* Hollister, C. D., and others, eds., Initial reports of the Deep Sea Drilling Project, Volume 34: Washington, D.C., U.S. Government Printing Office, p. 393-432.

Batiza, R., 1979, Geologic evolution of small oceanic volcanoes: Preliminary results [abs.]: Geological Society of America Abstracts with Programs, Annual Meeting, v. 11, p. 385.

Bence, A. E., and others, 1979, Controls on the major and minor element chemistry of mid-ocean ridge basalts and glasses, *in* Talwani, M., and others, eds., Deep drilling results in the Atlantic Ocean: Ocean crust, Maurice Ewing Series 2: Washington, D.C., American Geophysical Union, p. 331-341.

Bonatti, E., and Fisher, D. E., 1971, Oceanic basalts: Chemistry versus distance from oceanic ridges: Earth and Planetary Scinece Letters, v. 11, p. 307-311.

Bonatti, E., and Honnorez, J., 1976, Sections of the Earth's crust in the equatorial Atlantic: Journal of Geophysical Research, v. 81. p. 4104-4116.

Bonatti, E., and others, 1977, Easter volcanic chain (southeast Pacific): A mantle hot line: Journal of Geophysical Research, v. 82, p. 2457-2478.

Bryan, W. B., and Moore, J. G., 1977, Compositional variations of young basalts in the Mid-Atlantic Ridge rift valley near 36°49'N: Geological Society of America Bulletin, v. 88, p. 556-570.

Bryan, W., Thompson, G., and Frederick, F., 1979, Petrologic character of the Atlantic crust from DSDP and IPOD drill sites, *in* Talwani, M., and others, eds., Deep drilling results in the Atlantic Ocean: Ocean crust, Maurice Ewing Series 2: Washington, D.C., American Geophysical Union, p. 273-284.

Byerly, G. R., 1980, The nature of differentiation trends in some volcanic rocks from the Galapagos spreading center: Journal of Geophysical Research, v. 85, p. 3797-3810.

Byerly, G. R., and Wright, T. L., 1978, Origin of major element chemical trends in DSDP Leg 37 basalts, Mid-Atlantic Ridge: Journal of Volcanology and Geothermal Research, v. 3, p. 229-279.

Byerly, G. R., Melson, W. G., and Vogt, P. R., 1976, Rhyodacites, andesites, ferrobasalts and ocean tholeiites from the Galapagos spreading center: Earth and Planetary Science Letters, v. 30, p. 215-221.

Campsie, J., Bailey, J. C., and Rasmussen, M., 1973, Chemistry of tholeiites from the Galapagos Islands and adjacent ridges: Nature; Physical Science, v. 245, p. 122-137.

Cann, J. R., 1971, Major element variations in ocean-floor basalts: Royal Society of London, Philosophical Transactions, Series A, Mathematical and Physical Sciences, v. 268, p. 495-505.

Clague, D. A., and Bunch, T. E., 1976, Formation of ferrobasalt at east Pacific midocean spreading centers: Journal of Geophysical Research, v. 81, p. 4247-4256.

Clague, D. A., and Frey, F. A., 1979, Trace element constraints on the origin of Galapagos spreading center FeTi basalt, andesite and rhyodacite [abs.]: Hilo, Hawaii symposium on intraplate volcanism and submarine volcanism, p. 57.

Clark, J. G., and Dymond, J., 1977, Geochronology and petrochemistry of Easter and Sala y Gomez Islands: Implications for the origin of the Sala y

Gomez ridge: Journal of Volcanology and Geothermal Research, v. 2, p. 29–48.

Corliss, J. B., 1970, Mid-ocean ridge basalts [Ph.D. thesis]: La Jolla, California, Scripps Institution of Oceanography, 147 p.

Corliss, J. B., Dymond, J., and Lopez, C., 1976, Elemental abundance patterns in Leg 34 rocks, *in* Hollister, C. D., and others, eds., Initial reports of the Deep Sea Drilling Project: Washington, D.C., U.S. Government Printing Office, v. 34, p. 393–432.

Dugan, W. A., and Rhodes, J. M., 1978, Residual glasses and melt inclusions in basalts from DSDP Legs 45 and 46: Evidence for magma mixing: Contributions to Mineralogy and Petrology, v. 67, p. 417–431.

Engel, A.E.J., and Engle, C. G., 1964, Igneous rocks of the East Pacific Rise: Science, v. 146, p. 477–485.

Engel, A.E.J., Engle, C. G., and Havens, R. G., 1965, Chemical characteristics of oceanic basalts and the upper mantle: Geological Society of America Bulletin, v. 76, p. 719–734.

Fisher, R. L., and Norris, R. M., 1960, Bathymetry and geology of Sala y Gomez, southeast Pacific: Geological Society of America Bulletin, v. 71, p. 497–502.

Fisk, M. R., and Bence, A. E., 1979, Fractionation trends of basaltic glasses from the Galapagos spreading center [abs.]: Geological Society of America Abstracts with Programs, Annual Meeting, v. 11, p. 426.

Frey, F. A., Bryan, W. B., and Thompson, G., 1974, Atlantic Ocean floor: Geochemistry and petrology of basalts from Legs 2 and 3 of the Deep-Sea Drilling Project: Journal of Geophysical Research, v. 79, p. 5507–5527.

Funkhouser, J., Fisher, D. E., and Bonatti, E., 1968, Excess argon in deep sea rocks: Earth and Planetary Science Letters, v. 5, p. 95–100.

Gordon, G. E., and others, 1968, Instrumental activation analysis of standard rocks with high-resolution x-ray detectors: Geochimica et Cosmochimica Acta, v. 32, p. 364–396.

Handschumacher, D. W., 1976, Post-Eocene plate tectonics of the eastern Pacific, *in* Woollard, G. P., and others, eds., The geophysics of the Pacific Ocean and its margin, Geophysical Monograph Series, Volume 19: Washington, D.C., American Geophysical Union, p. 177–202.

Hart, R., 1970, Chemical exchange between seawater and deep ocean basalts: Earth and Planetary Science Letters, v. 9, p. 269–279.

Hart, S. R., 1971, K, Rb, Ca, Sr and Ba contents and Sr isotope ratios of ocean floor basalts: Royal Society of London, Philosophical Transactions, Series A, Mathematical and Physical Sciences, v. 268, p. 573–587.

Hart, S. R., and Staudigel, H., 1978, Oceanic crust: Age of hydrothermal alteration: Geophysical Research Letters, v. 5, p. 1009–1012.

Hekinian, R., 1971, Chemical and mineralogical differences between abyssal hill basalts and ridge tholeiites in the eastern Pacific Ocean: Marine Geology, v. 11, p. 77–91.

Hekinian, R., and Thompson, G., 1976, Comparative geochemistry of volcanics from rift valleys, transform faults and aseismic ridges: Contributions to Mineralogy and Petrology, v. 57, p. 145–162.

Herron, E. M., 1972a, Seafloor spreading and Cenozoic history of the east central Pacific: Geological Society of America Bulletin, v. 83, p. 1671–1691.

———1972b, Two small crustal plates in the South Pacific near Easter Island: Nature; Physical Science, v. 240, p. 35–37.

Herron, E. M., and Hayes, D. E., 1969, A geophysical study of the Chile ridge: Earth and Planetary Science Letters, v. 6, p. 77–83.

Hopson, C. A., and Pallister, J. S., 1980, Samail ophiolite plutonic suite: field relations, phase variation, and layering: A model of a spreading ridge magma chamber: Journal of Geophysical Research (in press).

Hussong, D. M., and others, 1976, Crustal structure of the Peru-Chile Trench: 8°–12°S latitude, *in* Sutton, G. H., and others, eds., The geophysics of the Pacific Ocean basin and its margin: Washington, D.C., American Geophysical Union Monograph, v. 19, p. 71–85.

Kay, R., Hubbard, N. J., and Gast, P. W., 1970, Chemical characteristics and origin of oceanic ridge volcanic rocks: Journal of Geophysical Research, v. 75, p. 1585–1613.

Kulm, L. D., and others, 1973, Tholeiitic basalt ridge in the Peru Trench: Geology, v. 1, p. 11–14.

Lacroix, A., 1928, Composition chemique des laves de l'Ile de Paque: Academie des Sciences, Comptes Rendus Hebdomadaires des Seances, v. 202, p. 601–605.

Langmuir, C. H., and others, 1977, Petrogenesis of basalt from the FAMOUS area, Mid-Atlantic Ridge: Earth and Planetary Science Letters, v. 36, p. 133–156.

Ludden, J. N., and Thompson, G., 1979, An evaluation of the behavior of the rare earth elements during the weathering of seafloor basalt: Earth and Planetary Science Letters, v. 43, p. 85–92.

Mammerickx, J., and others, 1975, Morphology and tectonic evolution of the east central Pacific: Geological Society of America Bulletin, v. 86, p. 111–118.

Mazzullo, L. J., and Bence, A. E., 1976, Abyssal tholeiites from DSDP Leg 34: The Nazca plate:

Journal of Geophysical Research, v. 81, p. 4327–4351.

McBirney, A. R., and Aoki, K., 1966, The Galapagos: Berkeley, University of California Press, 256 p.

McBirney, A. R., and Gass, I. G., 1967, Relations of oceanic volcanic rocks to mid-ocean rises and heat flow: Earth and Planetary Science Letters, v. 2, p. 265–276.

McBirney, A. R., and Williams, H., 1969, Geology and petrology of the Galapagos Islands: Geological Society of America Memoir 118, 197 p.

Melson, W. G., and O'Hearn, T., 1979, Basaltic glass erupted along the Mid-Atlantic ridge between 0-37°N: Relationships between composition and latitude, in Talwani, M., and others, eds., Deep drilling results in the Atlantic Ocean: Ocean crust, Maurice Ewing Series 2: Washington, D.C., American Geophysical Union, p. 249–261.

Melson, W. G., and others, 1976a, Chemical diversity of abyssal volcanic glasses erupted along Pacific, Atlantic, and Indian Ocean seafloor spreading centers, in Sutton, G. H., and others, eds., The geophysics of the Pacific Ocean Basin and its margin: Washington, D.C., American Geophysical Union Monograph, v. 19, p. 351–367.

Melson, W. G., and others, 1976b, A catalog of the major element chemistry of abyssal volcanic glasses: Washington, D.C., Smithsonian Institution Internal Document, p. 31–60.

Miyashiro, A., Shido, F., and Ewing, M., 1970, Crystallization and differentiation in abyssal tholeiites and gabbros from mid-ocean ridges: Earth and Planetary Science Letters, v. 7, p. 361–365.

Morgan, W. J., 1972, Deep mantle convection plumes and plate motions: American Association of Petroleum Geologists Bulletin, v. 56, p. 203–213.

Nisbet, E. G., and Fowler, C.M.R., 1978, The Mid-Atlantic Ridge at 37° and 45°N: Some geophysical and petrological constraints: Geophysical Journal of Royal Astronomical Society, v. 54, p. 631–660.

O'Hara, M. J., 1968, The bearing of phase equilibrium studies in synthetic and natural systems on the origin and evolution of basic and ultrabasic rocks: Earth-Science Reviews, v. 4, p. 69–133.

Omang, S. H., 1969, A rapid fusion method for decomposition and comprehensive analysis of silicates by atomic absorption spectrophotometry: Analytica Chimica Acta, v. 46, p. 225–230.

Rea, D. K., 1976, Analysis of a fast-spreading rise crest: The East Pacific Rise, 9° to 12° South: Marine Geophysical Researches, v. 2, p. 291–313.

Rea, D. K., and Scheidegger, K. F., 1979, Eastern Pacific spreading rate fluctuation and its relation to Pacific area volcanic episodes: Journal of Volcanology and Geothermal Research, v. 5, p. 135–148.

Rhodes, J. M., and Dungan, M. A., 1979, The evolution of ocean-floor basaltic magmas, in Talwani, M., and others, eds., Deep drilling results in the Atlantic Ocean: Ocean crust, Maurice Ewing Series 2: Washington, D.C., American Geophysical Union, p. 262–272.

Rosendahl, B. R., 1976, Evolution of oceanic crust 2: Constraints, implications and inferences: Journal of Geophysical Research, v. 81, p. 5305–5314.

Rosendahl, B. R., and others, 1976, Evolution of oceanic crust 1: A physical model of the East Pacific Rise crest derived from seismic refraction data: Journal of Geophysical Research, v. 81, p. 5294–5304.

Scheidegger, K. F., 1973, Temperatures and compositions of magmas ascending along midocean ridges: Journal of Geophysical Research, v. 78, p. 3340–3355.

Scheidegger, K. F., and Stakes, D. S., 1977, Mineralogy, chemistry and crystallization sequence of clay minerals in altered tholeiitic basalts from the Peru Trench: Earth and Planetary Science Letters, v. 36, p. 413–422.

Scheidegger, K. F., and others, 1978, Fractionation and mantle heterogeneity in basalts from the Peru-Chile Trench: Earth and Planetary Science Letters, v. 37, p. 409–420.

Schilling, J.-G., 1971, Sea-floor evolution: Rare earth evidence: Royal Society of London, Proceedings, Series A, v. 268, p. 663–706.

——1973, Iceland mantle plume: Geochemical study of Reykjanes ridge: Nature, v. 242, p. 565–573.

——1975, Rare-earth variations across 'normal segments' of the Reykjanes ridge, 60°-53°N, Mid-Atlantic Ridge, 29°S, and East Pacific Rise, 2°-19°S, and evidence on the composition of the underlying low-velocity layer: Journal of Geophysical Research, v. 80, p. 1459–1473.

Schilling, J.-G., and Bonatti, E., 1975, East Pacific Ridge (2°S-19°S) versus Nazca intraplate volcanism: Rare earth evidence: Earth and Planetary Science Letters, v. 25, p. 93–102.

Schilling, J.-G., Anderson, R. N., and Vogt, P. R., 1976 Rare earth, Fe and Ti variations along the Galapagos spreading center, and the relationships to the Galapagos mantle plume: Nature, v. 261, p. 108–113.

Seyfried, W. F., Jr., Shanks, W. C., and Dibble, W. E., Jr., 1978, Clay mineral formation in DSDP Leg 34 basalt: Earth and Planetary Science Letters, v. 41, p. 265–276.

Sleep, N. H., 1975, Formation of oceanic crust: Some thermal constraints: Journal of Geophysical Research, v. 80, p. 4037–4042.

Sleep, N. H., and Rosendahl, B. R., 1979, Topography and tectonics of mid-oceanic ridge axes: Journal of Geophysical Research, v. 84, p. 6831–6839.

Stakes, D. S., 1979, Submarine hydrothermal systems: Variations in mineralogy, chemistry, temperatures and the alteration of oceanic layer II [PhD thesis]: Corvallis, Oregon State University, 189 p.

Stakes, D., and Scheidegger, K. F., 1981, Temporal variations in secondary minerals from Nazca plate basalts, in Kulm, L D., and others, eds., Nazca plate: Crustal formation and Andean convergence: Geological Society of America Memoir 154 (this volume).

Tarney, J., and others, 1979, Nature of mantle heterogeneity in the North Atlantic: Evidence from Leg 49 basalts, in Talwani, M., and others, eds., Deep drilling results in the Atlantic Ocean: Ocean crust, Maurice Ewing Series 2: Washington, D.C., American Geophysical Union, p. 285–301.

Vogt, P. R., and Johnson, G. L., 1975, Transform faults and longitudinal flow below the midoceanic ridge: Journal of Geophysical Research, v. 80, p. 1399–1428.

Wakeham, S., 1978, Petrochemical patterns in young pillow basalts dredged from Juan de Fuca and Gorda ridges [Master's thesis]: Corvallis, Oregon State University.

Willis, B., and Washington, H. S., 1924, San Felix and San Amrosio: Their geology and petrology: Geological Society of America Bulletin, v. 35, p. 365–384.

Yeats, R. S., and others, 1973, Petrology and geochemistry of DSDP Leg 16 basalts, eastern equatorial Pacific, in Kaneps, A. G., ed., Initial reports of the Deep-Sea Drilling Project, Volume 16: Washington, D.C., U.S. Government Printing Office, p. 617–640.

——— 1976, Initial reports of the Deep Sea Drilling Project, Volume 34: Washington, D.C., U.S. Government Printing Office, 814 p.

MANUSCRIPT RECEIVED BY THE SOCIETY NOVEMBER 12, 1980
MANUSCRIPT ACCEPTED DECEMBER 30, 1980

Printed in U.S.A.

… Geological Society of America
Memoir 154
1981 …

Temporal variations in secondary minerals from Nazca plate basalts, diabases, and microgabbros

DEBRA S. STAKES
K. F. SCHEIDEGGER
School of Oceanography
Oregon State University
Corvallis, Oregon 97331

ABSTRACT

The mineralogy and chemistry of secondary phases observed in altered basalts from the Nazca plate provide a record of temporal variations in the processes that accompany the aging of oceanic crust. The earliest formed phase in vein and vesicle fillings is Fe- and Mg-rich, Al-poor saponite; it is the most abundant alteration mineral in younger rocks dredged from near the East Pacific Rise, and it lines many fractures and vesicles in much older pillow basalts recovered from the plate. Saponite formation is enhanced by low pH values, high water-rock ratios, and Mg- and Fe-rich solutions, conditions that persist until fractures and other pore spaces become closed and the hydrothermally induced circulation becomes negligible. During this phase of alteration, constituents derived from high-temperature leaching of underlying holocrystalline diabasic and gabbroic rocks are redistributed, both into secondary minerals found in overlying pillow basalts and into sea water. Holocrystalline, upper layer 2 basalts containing abundant smectite are richer in Mg, Na, K, and Ti and poorer in Ca relative to unaltered fresh glasses. During the waning of the hydrothermal system, lower temperatures and slightly alkaline and oxidizing solutions are indicated by the formation of calcite, ferric oxides and hydroxides, and celadonite. Calcite commonly fills all remaining void spaces in hydrothermally altered rocks. If such hydrothermally altered crust is re-exposed to oxygenated sea water by tectonic uplift or by other means, intense low-temperature oxidation can dramatically affect its bulk chemical composition and secondary mineralogy. Earlier formed saponites and ferrosaponites are destroyed and replaced by celadonite and ferric oxides, and bulk-rock compositions show pronounced losses in Mg and enrichments in Fe and K relative to fresh glasses. It is argued that secondary mineral assemblages in altered layer 2 crustal rocks are very sensitive indicators of conditions of alteration. With few exceptions, evidence of nonoxidative, hydrothermal alteration is ubiquitous in upper layer 2 crustal rocks recovered from many different locations on the Nazca plate.

INTRODUCTION

The secondary mineralogy of altered layer 2 rocks recovered from the ocean floor is a potential clue to the processes associated with the formation and evolution of oceanic crust. Such secondary mineral products may provide information on changes in pore-water chemistry and water-rock ratios. For the northern Nazca plate, Bass (1976) has previously classified a sequence of alteration processes and their associated mineral assemblages found in DSDP sites 319, 320, and 321. These include (1) late magmatic deuteric alteration, (2) sea-water alteration (preburial), (3) nonoxidative diagenesis (postburial, limited oxidation), and (4) oxidative diagenesis (postburial, extensive oxidation). This classification restricts sea-water alteration to the formation of "palagonite" and replacement of olivine by smectite (2) or to the destruction of smectites during highly oxidative sea-water alteration (4). All other minerals are depicted as products of in situ reactions between the rocks and either late-stage magmatic liquids or modified pore fluids. Possible remobilization and large-scale migration of elements during hydrothermally induced circulation and alteration are not specifically addressed by Bass.

Smectites (saponite and nontronite) and celadonite are the most abundant secondary minerals identified in the altered rocks from all areas of the Nazca plate. They are reasonably uniform in chemistry and structure, but are markedly different in composition from the host basalt (Seyfried and others, 1978; Scheidegger and Stakes, 1977). The smectites from the Nazca plate are typically Fe- and Mg-rich and Al-poor, and much of the difference in bulk composition of concentrations of these secondary minerals in vesicle and vein fillings is a result of varying proportions of the different species of phyllosilicates. Similar Mg-rich smectites have been identified as replacement products of basaltic glass during hydrothermal alteration (Bischoff and Dickson, 1975; Hajash, 1975; Mottl, 1976) and as precipitates from mixtures of such derived hydrothermal solutions and normal sea water (Seyfried and Bischoff, 1977). Widespread mobility of several important major elements may thus be common during hydrothermal alteration and may be more important than localized diagenetic reactions in producing the smectite that commonly lines or fills prominent fractures and vesicles in pillow basalts recovered from the Nazca plate.

A first step toward understanding the processes responsible for the alteration of oceanic crust requires identification of phyllosilicates and other associated secondary minerals, a task made more difficult by variations in chemistry, structure, optical and physical properties, and degree of crystallinity of the minerals. In this report we first characterize the most common secondary minerals separated from altered Nazca plate crustal rocks. Our goal is to establish criteria for their recognition for comparison with analogous materials from other areas. Second, we suggest that such alteration products provide important clues about the nature of the alteration processes and, in some cases, about temporal variations in the chemistry of secondary fluids that have permeated through oceanic crustal rocks (Scheidegger and Stakes, 1977). Finally, we argue that with the exception of the late-stage intense oxidative diagenesis of earlier formed mineral phases, many of the secondary minerals described by Bass (1976) can be interpreted as products of hydrothermally induced circulation and alteration initiated at the East Pacific Rise or the Galapagos Rise.

SAMPLE SELECTION AND ANALYTICAL METHODS

Samples selected for this study represent a broad spectrum of primary texture and mineralogy as well as nature and extent of alteration. Seven dredge hauls, each from a different prominent (~1 km) fault scarp, provided representative samples of altered oceanic crust of different ages from a variety of geographical locations on the Nazca plate (Fig. 1). The dredged areas include (1) the eastern flank of the East Pacific Rise (EPR) at about 32°S (dredges 52 and 1011) in crust 3 m.y. old (Rea, 1977); (2) the Quiros Fracture Zone (dredge 28) in crust estimated to be about 20 m.y. old; and (3) fault blocks of oceanic crust in the Peru-Chile Trench (dredges 7, 10, 11, 18) believed to be about 50 m.y. old (Scheidegger and Stakes, 1977; Scheidegger and others, 1978). Dredge haul 52 from the EPR contains over 300 samples spanning almost the entire range of textures and alteration effects; it thus provides an opportunity to evaluate how these factors vary in a single section. In addition, mineralogical and chemical data from altered rocks recovered from DSDP site 319 (15.2 ± 3.6 m.y.) and sites 320 and

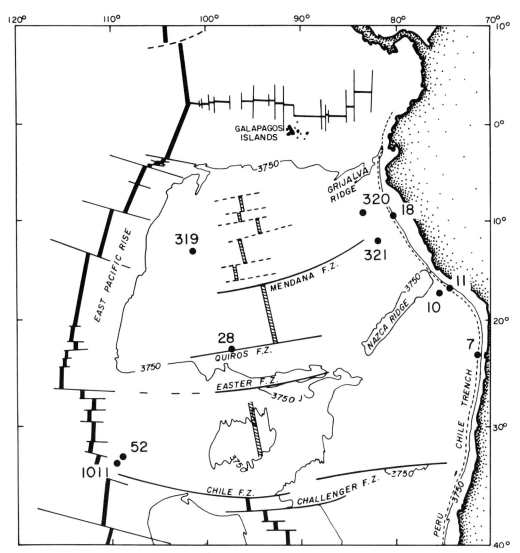

Figure 1. Map showing locations of dredges (52, 1011, 7, 10, 11, 18, and 28) and DSDP sites (319, 320, and 321) on the Nazca plate.

321 (43.7 ± 1.2 m.y.) are available for comparison (Hogan and Dymond, 1976; Yeats and others, 1976).

A variety of techniques was used to identify and characterize the alteration minerals. A combination of petrographic descriptions, chemical analysis, and X-ray diffraction study proved to be a powerful method of discriminating between various clay minerals. Bulk samples were ground under butanol for 1 h, and the randomly oriented powder mounts were scanned from 3° to 70° 2θ for total mineralogy. Clay mineral separates were prepared either by scraping large clay-filled veins or vesicles or by separating the <2-μm fraction from the bulk pulverized rock by repeated settling and decantation. Mineral separates thus obtained were Mg-saturated, vacuum suctioned onto pressed silver planchets, solvated with either ethylene glycol or glycerol, and scanned from 3° to 30° 2θ. Both bulk samples and oriented clay separates were step-scanned in 0.02° 2θ increments at 4-s count times. Selected clay separates were also K-saturated and rapidly scanned after heating to 105 °C, 300 °C, and 550 °C to discern the temperatures of collapse of the various clay species (Harward and others, 1962). Finally, the random mounts were scanned slowly (0.02° 2θ steps; 10- or 20-s count times) between 59.0° and 62.5° 2θ to study (060) reflections.

Major-element analyses of bulk samples and pure mineral separates were obtained by atomic absorption spectroscopy (AA) following dissolution in aqua regia and hydrofluoric acid (Fukui, 1976) or by electron microprobe. A metavanadate—ferrous ammonium sulfate titration was used to determine ferrous iron, and water content was determined with a Dupont 26-231 solids moisture analyzer. Altered glasses and large clay aggregates were analyzed by microprobe using a defocused beam (Melson and Thompson, 1973). Chemical analyses of the same clay species by AA and microprobe provided consistent results.

SAMPLE PETROGRAPHY AND MINERALOGY

The entire suite of altered rocks recovered from various parts of the Nazca plate can be divided into four main groups: (1) fresh pillow basalts (FPB), (2) altered pillow basalts and massive basalts (AB), (3) saponite-rich pillow breccias (SPB), and (4) coarsely crystalline diabases (DB) and microgabbros (GB). Gabbros containing hydrous minerals are denoted by HGB. The primary mineralogy and common alteration phases characterizing each group are described here, and Table 1 gives a more complete listing of these observations. Later sections discuss in detail the important alteration phases and the nature of the alteration processes responsible for their formation.

The FPB commonly contain plagioclase, olivine, and clinopyroxene with typical quench morphologies (Bryan, 1972). The outer glassy margin is completely isotropic in this section and is representative of the original erupted liquid (Melson and others, 1976). The only observed alteration is yellow or orange superficial discolorations on cooling surfaces. Even though these samples appear completely fresh, X-ray analyses of bulk clay separations still show traces of poorly crystalline smectite.

AB are characterized by the presence of smectites (saponite and nontronite), celadonite, Fe-oxides, and calcite filling veins and vesicles. Olivine phenocrysts are partially replaced either by

TABLE 1. SUMMARY OF SAMPLE MINERALOGY

Classification	Secondary mineralogy	Abundance	Mode of occurrence	Degree of oxidation
Fresh pillow basalts (FPB)	Sm-Cel RML Fe oxide	T	Along cooling fractures and in vesicles	Limited
Altered basalts (AB)				
EPR	Sm-Cel RML	P	Replaces olivine and fills vesicles	Limited
Site 319	Sm-Cel Talc Palagonite Phillipsite	A P	Fills veins and vesicles, replaces olivine and clinopyroxene and replaces glass	Decreasing downcore
Trench	Sm-Cel RML Sm + Cel Palagonite Calcite	A P A	Fills veins and vesicles, replaces olivine and clinopyroxene and replaces glass	Pervasive, increases with length of exposure to sea water
Saponite-rich pillow breccias (SPB)	Pure saponite	A	Fills veins	Limited
	Palagonite Sm-Cel RML Calcite	A P	Replaces glass and found in veins	Limited
Holocrystalline rocks (EPR) (DB + GB + MGB)	Saponite Talc Chlorite Apatite Actinolite Calcite Vermiculite } Celadonite	A A P P P T P	Replaces olivine and clinopyroxene, or present interstitially	Limited

Note: Abbreviations: P, present; A, abundant; T, trace; Sm-Cel RML, smectite-celadonite random mixed layer.

smectites or by calcite and Fe oxide. Near the crest of the EPR (dredges 52 and 1011) alteration is minimal with only a few vesicles and intergrain boundaries lined with poorly crystalline smectite and Fe oxides and hydroxides. The older basalts from the DSDP sites and the Peru-Chile Trench contain well-crystallized saponite, nontronite, celadonite, talc, calcite, aragonite, Fe oxide, and pyrite. Deposition of such secondary minerals in veins and veiscles greatly decreases the porosity of the rocks (Bass, 1976; Seyfried and others, 1978; Scott and Swanson, 1976; Scheidegger and Stakes; 1977).

In contrast to the AB, in which secondary minerals appear confined to veins, vesicles, and fractures due to precipitation from circulating fluids, large amounts of glass in the SPB are replaced by saponite and celadonite. For such pillows the once vitreous outer margins are altered to a bluish-green color and cross cut by thick veins of blackish-green saponite and sporadic calcite, giving the rock a brecciated appearance. In thin section (see SEM photograph, Fig. 2A), these margins appear as islands of nonisotropic, cryptocrystalline saponite surrounded by numerous microfractures lined with highly birefringent saponite or bright green celadonite. The resulting brecciated appearance is thought to be the result of original cooling fractures along which solutions of sea-water origin permeated to alter the edges of the glass. The "brecciated" texture ends abruptly at a more crystalline variolitic zone, thus preserving an original transition zone to the slightly more crystalline interior (Fig. 2B).

DB and GB were recovered from the two dredge hauls near the EPR and from the lower portions of DSDP site 319A. Maximum grain size in these samples varies from 1 to 33 mm, and they characteristically have subophitic to ophitic intergrowths of calcic plagioclase (An_{75}-An_{60}) and augite with minor rounded grains of olivine. All of the diabases from the EPR contain talc or smectite pseudomorphs after olivine and clinopyroxene, respectively, and interstices are filled with smectite, chlorite, and celadonite. Some of the finer-grained samples of diabase contain tiny laths of actinolite fringing the plagioclase-pyroxene grain boundaries.

Altered gabbroic rocks contain a large variety of primary and secondary minerals. Interstitial patches around ophitic intergrowths of plagioclase and augite contain hydrous minerals in a fine-grained mesostasis, including smectite, chlorite, phlogopite, talc, hornblende, apatite, and actinolite, in addition to the anhydrous minerals pigeonite and magnetite (Table 1) (M. Bass, 1979 personal commun.; Yeats and others, 1976). Unusual patches containing granophyric intergrowths of quartz, albite, and K-feldspar also occur interstitially in sample HGB 1515 from dredge 1011.

ALTERATION PRODUCTS IN CRUSTAL ROCKS OF THE NAZCA PLATE

A brief catalogue of major alteration minerals reported in Nazca plate samples, including their optical properties and X-ray diffraction characteristics, is included in this section, both as a summary of available data and as an aid in identifying clay mineral species. Bass (1976) provided a very comprehensive description of the textural relationships of secondary minerals, and only a brief synopsis of this work is given here.

Smectite

The smectite clay minerals ferrosaponite and saponite are the most abundant alteration phases identified, filling fractures, vesicles, and intergrain boundaries and replacing primary olivine, clinopyroxene, or glass. The dioctahedral species nontronite has been reported only as vesicle fillings in samples from the Peru-Chile Trench and as interlayers in a saponite from a vein from site 319 (Seyfried and others, 1978; Scheidegger and Stakes, 1977). The position of the (060) peak is a sensitive indicator of chemical substitution in the octahedral sheet of smectites. Characteristic (060) spacings are 1.492 to 1.522 Å for dioctahedral species and 1.520 to 1.536 Å for trioctahedral species, with substitution of Fe^{2+} producing the larger d-spacings in these two intervals (MacEwan, 1961). Observed (060) spacings for the saponites in this study typically vary from 1.535 to 1.529 Å, with the smaller values found in the more oxidized species. The nontronite vesicle fillings from the older trench samples have (060) peaks near 1.522 Å (Scheidegger and Stakes, 1977). Nontronitic smectite may be a

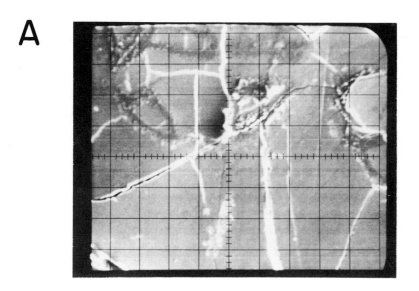

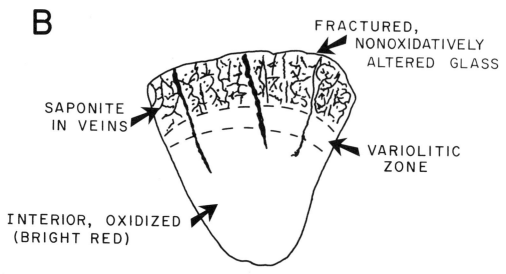

Figure 2. (A) SEM photograph of microfractures in saponite-rich pillow breccia. (B) Relationship of nonoxidatively altered glass (shown in A) to variolitic and holocrystalline interior of pillow basalt.

ferrosaponite that acquired dioctrahedral character during low temperature oxidation. It is possible that as the Fe^{+2} was oxidated to Fe^{+3} in the octahedral sites, resulting excess Fe could have been expelled as observed intermixed Fe-oxides. In Figure 3 representative diffraction patterns of smectites are given.

Heat treatments of saponites separated from EPR rocks show that they resist collapse at 300 °C (Fig. 4), indicating partial $Fe(OH_2)$ or $Mg(OH)_2$ interlayers (Harward and others, 1962). Increase in the Fe^{+2} content increases relief, color intensity, and pleochroism of these clays (Scheidegger and Stakes, 1977). Ferrosaponties in the SPB are strongly pleochroic, highly birefringent, coarsely crystalline, and have high relief in thin section. Smectites in vesicles in the AB tend to be only slightly pleochroic and finely crystalline.

Chemical composition of the saponite minerals from different locations on the Nazca plate are characteristically high in Fe and Mg and low in Al relative to the compositions of the host basalts (Table 2). Textural relationships of vesicle and fracture fillings in basalts from the Peru-Chile Trench (dredge 11; Fig. 1) indicate that the more Fe-rich smectites tended to form first, followed by the more Mg-rich smectites and calcite (Scheidegger and Stakes, 1977). This trend was interpreted to reflect temporal changes in pore-solution chemistry.

Results of several studies in which basaltic glass and sea water were reacted under various conditions demonstrate that smectite is the first and most abundant phase to form and is usually a replacement of basaltic glass (Mottl, 1976; Bischoff and Dickson, 1975; Hajash, 1975; Seyfried and Bischoff, 1977). Smectite continues to form until the available Mg is exhausted. During the formation of smectite, the pH of the solution declines as hydroxyl groups are taken up by clays, and this decline may be an important factor in remobilizing further metals into solution (Seyfried and Mottl, 1977). High water-rock (>50) ratios enhance smectite formation by providing more Mg (Seyfried and Mottl, 1977). Under these conditions, a smectite-chlorite, mixed-layer phase forms which is structurally similar to the saponite with hydroxy interlayers previously identified in the SPB. At lower water-rock ratios, the Mg supply is exhausted and other secondary minerals begin to form (Seyfried and Mottl, 1977).

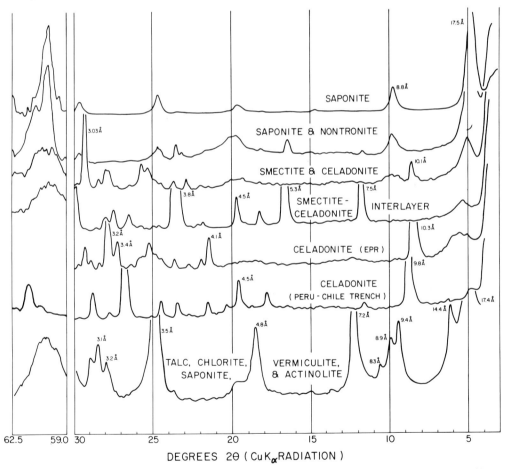

Figure 3. Representative diffractograms of major secondary minerals found in altered basalts from the Nazca plate. Oriented slides were prepared from Mg-saturated, glycerol-solvated samples. Note break in 2θ scale at left and characteristic (060) reflections.

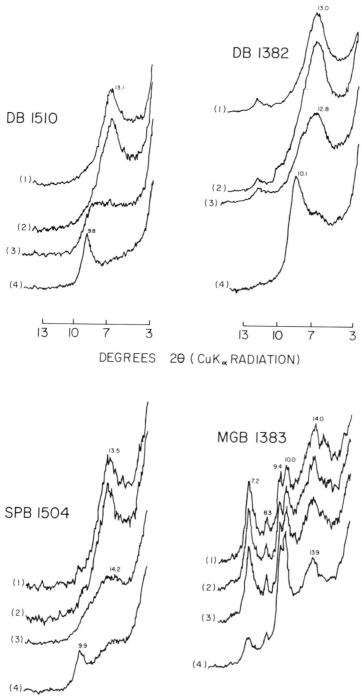

Figure 4. X-ray diffraction data on four K-saturated clay mineral separates from altered rocks dredged from area 52 on the EPR. Conditions were (1) dried at 110 °C and equilibrated at 54% relative humidity; (2) dried at 110 °C and kept in dry air; (3) heated to 300 °C and kept in dry air; and (4) heated to 550 °C and kept in dry air.

Chlorite

Chlorite was found in the centers of some veins in the DSDP Leg 34 basalts (Bass, 1976), in trace amounts mixed with saponite in the SPB, and in the micrograbbros associated with interstitial patches of talc, vermiculite, and actinolite. The addition of Fe- or Mg-hydroxy interlayers to a primary smectite can convert the clay structure to that of a nonexpandable chlorite (Carstea and others, 1970). The appearance of chlorite in veins and in the SPB can probably be attributed to such a diagenetic origin, because some incomplete interlaying was apparent during the heat treatments (Fig. 4). Bass (1976) concluded also that chlorite in veins in DSDP cores is primarily diagenetic.

The chlorite associated with actinolite and talc in the micrograbbros probably formed at temperatures above 200 °C, but is still unlike high-temperature chlorites. Using X-ray diffraction data for (060) and (001) d-spacings and a formula devised by Hayes (1970), Stakes (1978) calculated the following structural formula for the chlorite from HGB 1383:

$$(Mg_{4.62}Fe^{+2}_{0.81}Al_{0.57})(Si_{3.43}Al_{0.57})O_{10}(OH)_8 .$$

This calculated composition is unusually high in Mg and low in Al, and is not comparable to the compositions of regional metamorphic chlorites compiled by Hayes (1970) or even the compositions of chlorites from greenschist facies metamorphosed basalts from the Mid-Atlantic Ridge (Stakes, 1978). Further, metamorphic chlorites are usually IIb polytypes, as distinguished by characteristic (h01) peaks in the 1.6- to 3.0-å region of the diffractogram (Hayes, 1970). The peaks for this polytype are present in the diffractogram of the HGB chlorite, but are very weak. The conspicuously Mg-rich and Al-poor nature of the chlorite from HGB 1383 may suggest that it is transitional to the chemically similar ferrosaponite (Table 2).

Talc

Talc was identified in all the holocrystalline rocks and in the few glassy rocks that contained olivine phenocrysts. This association suggests that it is most common as a replacement of primary olivine and

TABLE 2. CHEMICAL COMPOSITIONS OF REPRESENTATIVE SMECTITES AND CELADONITE IN ALTERED BASALTS FROM THE NAZCA PLATE

	(1)	(2)	(3)	(4)	(5)	(6)	(7)
SiO_2	43.40	44.77	43.25	43.78	54.77	45.12	42.33
Al_2O_3	7.32	3.12	5.79	5.80	3.30	5.13	6.03
FeO	5.01	19.43	6.78	1.72	4.41	4.77	4.62
Fe_2O_3	8.97	*	9.73	10.03	15.03	11.14	9.15
MgO	17.56	18.22	16.62	18.25	7.18	17.06	17.11
CaO	1.26	1.01	0.65	1.74	0.36	0.21	0.00
Na_2O	2.41	0.26	1.95	2.48	0.27	2.68	1.18
K_2O	0.45	0.30	1.14	0.91	9.64	0.85	0.78
TiO_2	0.35	0.10	0.18	0.39	0.19	0.23	1.04
H_2O	13.63	13.60	13.59	15.28	5.40	13.60	17.53
Total	100.36	100.81	99.68	100.38	100.55	100.79	99.77

Note: Column (1), saponite vein in saponite-rich pillow breccia, EPR breccia, EPR dredge 52 (this study). Column (2), pyroxene pseudomorph in saponite-rich breccia, EPR dredge 52 (this study). Column (3), green smectite vesicle filling, dredge 11 in Peru-Chile Trench (Scheidegger and Stakes, 1977). Column (4), brown smectite vesicle filling, dredge 11 in Peru-Chile Trench (Scheidegger and Stakes, 1977). Column (5), celadonite vesicle filling, dredge 11 in Peru-Chile Trench (Scheidegger and Stakes, 1977). Column (6), smectite vein, DSDP site 321 (Seyfried and others, 1978). Column (7), green smectite, DSDP site 321 (Bass, 1976).

*Total Fe as FeO.

indicates locally high Si and Mg activities (Bass, 1976). Talc was identified by a peak at 9.4 å (Fig. 3), which suggests that it is Fe-rich like the smectites and chlorites. Colorless in this section, it has high birefringence and relief.

Talc is typically intergrown with chlorite, actinolite, and saponite in the HGB, suggesting a genetic relationship between these phases. Experimental sea-water–basalt reactions at temperatures above 400 °C produced similar mineral assemblages containing actinolite, smectite, talc, and albite (Mottl, 1976). This association may thus be indicative of reactions between basalt and a hydrothermal or deuteric fluid at comparable temperatures.

Mica and Vermiculite

There may be two different micas present in the alteration assemblages, and each is associated with a different mode of formation. A green phlogopitic mica is present in small quantitites associated with actinolite and pigeonite in some of the GB (Bass, 1979, personal commun.), suggesting a high-temperature origin. This mica exists as hexagonal plates with birds-eye mottled extinction, strong birefringence, and distinctive pleochroism from bright green to pale yellow (or brown in oxidized rocks). It was identified optically and has a 2V of approximately 15°. Vermiculite was identifed in HGB 1383 by the nonreversible collapse of the (001) peak from 14 å to 10 å, following K-saturation and drying at 110 °C (Fig. 4). It otherwise resembles the phlogopite and may result from alteration of this phase.

Celadonite is the most common mica. It is associated with smectite in veins and vesicles in DSDP Leg 34 basalts (Bass, 1976; Seyfried and others, 1978), in vesicles in rocks recovered from dredge area 11 in the Peru-Chile Trench (Scheidegger and Stakes, 1977), and in bulk clay separates from the SPB and the holocrystalline basalts from the EPR. In thin section it is often slightly pelochroic in shades of green or yellow, and can be microcrystalline with very low relief. However, Scheidegger and Stakes (1977) noted that it can also be strongly pleochroic, ranging from colorless to vivid turquoise blue. Celadonite is a dioctahedral mica, rich in ferric iron and low in Al, and is distinguished from glauconite only by its mode of occurrence (Wise and Eugster, 1964). Both celadonite and phlogopite have (110) peaks near 10.2 å, very weak (002) reflections, and strong (003) reflections (Seyfried and others, 1978). Celadonite breaks down at high temperatures to a phlogopitic mica (Wise and Eugster, 1964), suggesting that temperatures of formation may control the stability of the two minerals. Under strongly oxidizing conditions, celadonite is rich in Fe^{+3} compared to Fe^{+2}, and its basal reflections shift to smaller d-spacings (Wise and Eugster, 1964). The celadonite described by Scheidegger and Stakes (1977) has a basal spacing of only 9.8 å, which may reflect formation during late-stage, highly oxidative conditions. Celadonite commonly has a (060) spacing of 1.507 å (Scheidegger and Stakes, 1980). Except for the common occurrence of K in interlayer positions in smectites, Fe-rich, Al-poor mica is the only highly K-rich mineral phase observed, and its distribution must be controlled by the availability of K and requisite oxidizing conditions.

Spectite-Celadonite Interlayer

A mixed-layer, smectite-celadonite species was identified most often in strongly oxidized samples. This interlayer species is characterized by a weak (001) peak at 15 å, and very strong peaks at 7.5 å, 5.3 å, 4.8 å, and 3.04 å (Fig. 3). The mixed-layer peaks usually appear only in X-ray patterns for samples solvated with glycol; whereas samples solvated with glycerol yield peaks expected for a mixture of smectite and mica. Ethylene glycol is a more polar solvent than glycerol, and this evidently produces a coherent partial expansion of the mixed-layer species resulting in the stronger peaks of a monomineralic species. The association of the interlayer with strongly oxidized samples also suggests that such a mixed-layer mineral could be produced by partial oxidation of an Fe-rich saponite to celadonite.

Variations in celadonite compositions may also suggest a gradual transition from ferrosaponite to celadonite. Chemical analyses of celadonite from oxidation zones and vesicle fillings in DSDP Leg 37 basalts showed decreases in the Al/Si ratio accompanied by increases in K + Na + Ca content with the resulting charge balance maintained by variable substitution of Mg^{+2} in the octahedral layer (Andrews, 1977). This increased substitution of Fe^{+3} for Mg^{+2} or Fe^{+2} in the octahedral layer of a saponite with concomitant increases in the K content would change a ferrosaponite to a celadonite. Thus, increased oxidation accompanied by loss of Mg and increase in Fe^{+3} could cause a transition from smectite clays to micas with nontronite as an intermediate composition.

Nonoxidativley Altered Glass

The outmost glassy margins of pillow basalts are broken up by microfractures along which movement of water and cations would be much more rapid than bulk diffusion through the glass (Fig. 2). Water and cations diffuse in or out along these microfractures during initial alteration and turn the normally isotropic black sideromelane to a translucent pale brown or yellow color. More extensive alteration produces devitrification with tiny fibers of birefringent smectite replacing the glass. This fibropalagonite texture has been described by Bass (1976), and its formation has been characterized as a low-temperature, diffusion-controlled process occurring under oxidizing conditions (Moore, 1966). Palagonite formation has been used to estimate the ages of basalt flows erupted onto the sea floor (Bryan and Moore, 1977).

Although pale brown or yellow palagonite is a very common alteration product of Nazca plate pillow basalts, altered glass in the bluish outer margin of the SPB (Fig. 2) shows many of the same petrographic characteristics as the palagonite and fibropalagonite described by Bass (1976) for DSDP Leg 34 basalts. However, this much darker colored, altered glass formed under reducing, not oxidizing, conditions. This is suggested by the coexistence of abundant smectite and pyrite in the veins that cut the altered glass. Such material does not fit the strict definition of palagonite suggested by Bass (1976) because of the differing conditions of formation. In the following discussion, Doremus's (1973) general model of glass alteration is used as an analog of palagonitization in the deep sea, and then possible chemical and mineralogical differences of glass alteration resulting from oxidative and nonoxidative conditions are examined.

Doremus (1973) considered the reaction between water and glass as a two-step process: (1) inward diffusion of water into the glass and (2) ion exchange between cations in the glass (for example, Na or Ca) and H^+ in the water. Although the first step is a slow diffusion-controlled process ($r^2 \alpha$ time), the second step greatly enhances the reaction rate by creating a more open network. The silicate glass structure breaks down and devitrification occurs as the glass alteration proceeds. Diffusion rates associated with the replacement of a silicate framework and loss of Si from the glass (step 2 above) may be several orders of magnitude faster than the rates associated with step 1 above (Doremus, 1973). High temperatures and a low pH would be expected to accelerate glass alteration. Oxygen isotope studies of saponite separated from veins in the SPB samples (Fig. 2) indicate temperatures of formation exceeding 100 °C (Stakes and O'Neil, 1977), and the appearance of pyrite noted above in the same veins indicates reducing conditions.

Chemical analyses of altered glass found in the SPB from dredge 52 compared to an average analysis of fresh glass recovered from dredge 52 provide insight into the bulk compositional changes accompanying glass alteration under nonoxidative conditions (Table 3). Even fresh-appearing residual islands of glass in the SPB show an extensive loss of Ca and gain in Si (Fig. 5), observations which are consistent with Doremus's (1973) hydration and cation exchange model and the breakdown of the glass structure. Compared to the composition of tholeiitic glass, the alteration of glass under nonoxidative conditions entails increases in Mg, Fe, and Ti and decreases in Ca and Si (Table 3). The high Al content of the cryptocrystalline smectite which replaces the glass (Table 3) is much higher than that associated with smectite in veins and vesicles (Table 2).

The formation of palagonite under low-temperature, oxidizing conditions results in quite different chemical exchanges. These have been well described for DSDP Leg 37 basalts, and include large increases in K and Fe^{3+}, small increases in Ti, and losses in Ca, Mg, and Na (Andrews, 1977; Baragar and others, 1977). The formation of palagonite under oxidizing conditions produces much larger

TABLE 3. REPRESENTATIVE ANALYSES OF SAMPLES OBTAINED FROM SAPONITE-RICH PILLOW BRECCIAS (SPB) IN DREDGE 52 FROM THE EPR

	(1)	(2)	(3)	(4)
SiO_2	56.20	44.17	44.64	51.29
Al_2O_3	14.44	14.16	13.90	14.97
FeO^*	9.96	19.52	18.53	10.28
MgO	7.85	14.41	16.54	7.78
CaO	3.29	1.02	2.03	10.99
Na_2O	5.26	3.42	0.70	2.81
K_2O	0.49	0.78	0.92	0.10
TiO_2	2.49	2.52	2.74	1.76

Note: Samples include (1) fresh-appearing, nonoxidatively altered glass; (2) dark bluish-green altered glass with texture of palagonite; and (3) fibropalagonite. Average composition of fresh tholeiitic glass (4) from the same dredge is shown for comparision. All analyses have been calculated on a water-free basis.

*Total Fe as FeO.

increases in K than observed under more reducing conditions. In addition, Mg is strongly depleted in comparison with the gain in Mg observed in the SPB (Table 3). Celadonite is the predominant clay mineral in the Leg 37 basalts, and its occurrence may be a consequence of more oxidizing conditions associated with palagonite formation.

Calcite

Bass (1976) provided an extensive discussion of calcite and aragonite distributions in altered basalts. Calcite is always a late-forming mineral, found only in the larger pore spaces that were not completely filled by the earlier clay minerals. Apparently, the H^+ associated with smectite formation lowers the pH in the modified pore fluids enough to preclude the early formation of calcite, even though these pore fluids are probably enriched in calcium. Progressive vein and vesicle fillings follow a predictable sequence from saponite, as the early formed phase, to Fe oxide, and then calcite as the later formed phases. Thus, the precipitation of calcite and Fe oxide signals a return to oxidative, slightly alkaline conditions, perhaps due to the influx of normal bottom water. Calcite is ubiquitous in altered basalts recovered from the Nazca plate, and it appears to be the major sink for Ca cations mobilized during glass or plagioclase alteration.

ALTERATION PROCESSES OF NAZCA PLATE CRUSTAL ROCKS

The chemical composition and secondary mineralogy of altered layer 2 crustal rocks are controlled

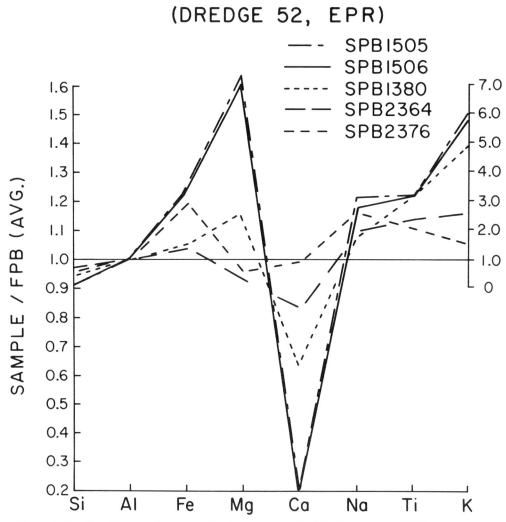

Figure 5. Relative differences in composition between saponite-rich pillow breccias (SPB) and an average analysis of glasses from fresh pillow basalts (FPB) present in dredge 52. Samples were normalized to constant Al. Scale at right pertains to K only.

by several different processes. These processes include the following: (1) Short-lived, high temperature "leaching" reactions that remove some elements from pillow basalts and deeper-seated diabases and gabbros. Elements mobilized by such reaction can be redistributed throughout upper layer 2 rocks and perhaps even into overlying sediments by the hydrothermally induced circulation (Stakes, 1979). (2) Longer-lived, hydrothermally induced reactions brought about by the circulation of such chemically modified sea water through cooling oceanic crustal rock masses, a process that ultimately declines as geothermal gradients diminish and as secondary minerals fill available voids. Rb-Sr isochron age dating of secondary minerals in oceanic crust from the Atlantic Ocean indicates that precipitation of secondary minerals may persist 5 to 10 m.y. (Hart and Staudigel, 1978), and thermal modeling considerations suggest that thermally driven circulation may persist up to 15 m.y. (Sleep, 1975). Little or no free oxygen is available for these reactions. (3) Oxidative alteration involving a reequilibration of a nonoxidative (hydrothermal) secondary mineral assemblages to intrusions of cold oxygenated sea water. This type of alteration would be expected to occur after hydrothermally

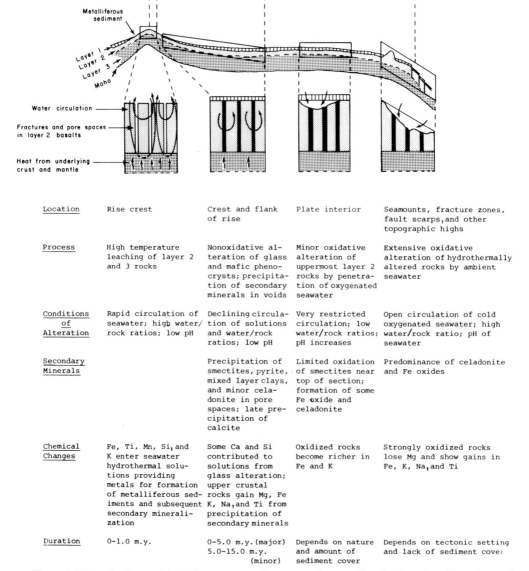

Location	Rise crest	Crest and flank of rise	Plate interior	Seamounts, fracture zones, fault scarps, and other topographic highs
Process	High temperature leaching of layer 2 and 3 rocks	Nonoxidative alteration of glass and mafic phenocrysts; precipitation of secondary minerals in voids	Minor oxidative alteration of uppermost layer 2 rocks by penetration of oxygenated seawater	Extensive oxidative alteration of hydrothermally altered rocks by ambient seawater
Conditions of Alteration	Rapid circulation of seawater; high water/rock ratios; low pH	Declining circulation of solutions and water/rock ratios; low pH	Very restricted circulation; low water/rock ratios; pH increases	Open circulation of cold oxygenated seawater; high water/rock ratio; pH of seawater
Secondary Minerals		Precipitation of smectites, pyrite, mixed layer clays, and minor celadonite in pore spaces; late precipitation of calcite	Limited oxidation of smectites near top of section; formation of some Fe oxide and celadonite	Predominance of celadonite and Fe oxides
Chemical Changes	Fe, Ti, Mn, Si, and K enter seawater hydrothermal solutions providing metals for formation of metalliferous sediments and subsequent secondary mineralization	Some Ca and Si contributed to solutions from glass alteration; upper crustal rocks gain Mg, Fe K, Na, and Ti from precipitation of secondary minerals	Oxidized rocks become richer in Fe and K	Strongly oxidized rocks lose Mg and show gains in Fe, K, Na, and Ti
Duration	0-1.0 m.y.	0-5.0 m.y. (major) 5.0-15.0 m.y. (minor)	Depends on nature and amount of sediment cover	Depends on tectonic setting and lack of sediment cover

Figure 6. Schematic diagram illustrating processes of importance in controlling the alteration of layer 2 crustal rocks of the Nazca plate.

induced circulation ceases, either by intrusion of oxygenated sea water through overlying sediments and rock masses (Bass, 1976) or by tectonic uplift and exposure of altered crustal rocks to ambient sea water (Scheidegger and Stakes, 1977). Each process listed above is associated with its own pore-water chemistry, and it would be expected to bring about the formation of characteristic secondary mineral assemblages. We suggest that the observed changes in the bulk chemistry of the altered crustal rocks from the Nazca plate can be explained by variations in the nature and extent of these reactions and by corresponding changes in secondary mineral assemblages. In Figure 6 we illustrate schematically the probable mineral and chemical consequences of each process and the locations on the Nazca plate where such types of alteration may occur.

Processes Associated with Cooling of Oceanic Crust

The loss of Fe, Ti, Mn, K, and rare-earth elements from crystalline pillow interiors and holocrystalline diabases has been attributed to leaching by sea water at very high temperatures (Corliss, 1971; Scott and Hajash, 1976). Holocrystalline interiors of pillow basalts from the Northern Mid-Atlantic Ridge (Corliss, 1971) and from the East Pacific Rise (Stakes, 1979) have been found to be consistently depleted in these metals relative to glasses from the pillow margins. Unfortunately, too few data are presently available to adequately document these high-temperature leaching reactions of pillow basalts, and because of their short-lived character, little time was available to develop diagnostic secondary mineral products. For more coarsely crystalline crystal rocks (gabbros) recovered from dredges 52 and 1011 on the Nazca plate, Stakes (1979) has suggested that interstitial concentrations of late-stage fluids existed and cooled rapidly, yielding phlogopite, pigeonite, red-brown amphibole, and abundant magnetite with quench morphology. Equant, subparallel laths of albite and apatite also project into these finer grained interstitial regions. Thompson and others (1976) have documented similar occurrences in diabases recovered from DSDP site 319. The late-stage fluids required for the formation of these minerals must be rich in Fe, Ti, K, and P, and they would be expected to form during the final stages of crystallization of tholeiitic magmas. Stakes (1978) proposed that if sea water comes in contact with such fluids, these metals could be easily extracted and thereby could contribute to the composition of circulating hydrothermal solutions.

Fe-rich hydrothermal solutions derived from such high-temperature sea water–basalt interactions can react with rock masses as they exit under variable conditions of temperature and water-rock ratios. Reaction with diabasic or gabbroic rocks at temperatures near 300 °C can result in the formation of greeschist facies minerals (for example, albite, chlorite, and actinolite; Stakes, 1979), but these reactions are characteristically incomplete, which suggests rapid cooling by circulating sea water. The interaction of the exiting solutions with shallower glassy rocks at temperatures of 100 °C to 200 °C (Stakes, 1978) may lead to the nonoxidative alteration of glass in the SPB described previously. As a result of these interactions, some Mg, Fe, K, Na, Ti, and H_2O would be lost from solution and some Si and Ca would enter the solutions (Fig. 5), perhaps contributing to the eventual formation of calcite and phillipsite.

Under conditions of declining temperatures and dilution of hydrothermal solutions with normal sea water, Al-deficient, Fe-rich smectites derived from sea-water–hydrothermal solutions precipitate in vesicles and fractures with little interaction with the host basalt. Where K and Si concentrations are high enough and where sufficient Fe^{+3} is present, celadonite may also form. These minerals steadily deplete the solution in Fe, Mg, Si, and K. The Fe derived from high-temperature leaching reactions may be depleted most rapidly as recorded in the transition from the earliest formed Fe-enriched smectite to the later formed Mg-enriched smectites (Scheidegger and Stakes, 1977). Smectites form until the supply of Mg is exhausted or until the circulation of new sea water and the influx of Mg is impeded by the clogging of veins and fractures. When Mg is no longer available for smectite formation, the pH of the solution begins to increase (Seyfried and Mottl, 1977), and other minerals will begin to form. Thus, the few remaining void spaces are generally filled with phillipsite and calcite, which precipitate from slightly alkaline Ca-, Al-, and Si-bearing solutions. The initiation of plagioclase alteration (Scheidegger and Stakes, 1977) may also provide some Al during late-stage hydrothermal alteration. The association of pervasive plagioclase alteration and formation of Al-rich secondary minerals found in altered Cretaceous crustal rocks from DSDP Leg 51, site 417A (Scheidegger and Stakes, 1980) also substantiates Al mobility during more intensive alteration.

In summary, we interpret the nature and distribution of many secondary minerals observed in altered Nazca plate crustal rocks as evidence of the generation and aging of hydrothermal solutions, perhaps lasting a few million years (Hart and Staudigel, 1978). Many of the metals involved in such secondary mineral formation may have been carried by the circulation of hydrothermal fluids derived from high-temperature reactions occurring in underlying crustal rocks. It follows that not all secondary mineral formation need involve only localized diagenetic reaction. It is perhaps more likely that many of the metals in secondary minerals of altered Nazca plate rocks have precipitated in rock masses far removed from where they were derived by high-temperature reactions.

Oxidative Alteration at Low Temperature

Oxidation of pre-existing, hydrothermally formed Fe-rich smectites by low-temperature oxygenated sea water is expected to introduce nontronite layers by oxidizing the ferrous iron to ferric iron. However, if the oxidation is intense, the smectites may be destroyed completely (Bass, 1976). During such low-temperature oxidation, Mg-rich smectite replacements of olivine or glass would themselves be replaced by amorphous ferric hydroxides and/or celadonite, and calcic pyroxene would be replaced by Fe oxide and calcite (Mueller, 1973), resulting in a loss of Mg and Si and a gain of K and Fe in the bulk rock. Thus, under highly oxidative conditions, many of the Fe- and Mg-rich smectites formed under nonoxidative conditions would be destroyed, and celadonite and Fe oxides would be the stable secondary phases (Bass, 1976; Scheidegger and Stakes, 1980). If primary plagioclase feldspar is involved in alteration, Ca- and Al-rich secondary minerals (phillipsite, analcite, potassium feldspar, and calcite) would also be expected (Scheidegger and Stakes, 1980).

Examples of such compositional and mineral changes accompanying oxidation of old hydrothermally altered basalts can be best documented for pillow basalts dredged from fault scarps within the Peru-Chile Trench (Scheidegger and Stakes, 1977; Scheidegger and others, 1978). The prominent fault scarps (~1 km vertical offsets) are believed to have developed during the past 0.5 m.y. as oceanic crust about 50 m.y. old entered the trench (Scheidegger and others, 1978). Basalts exposed along the individual scarps have experienced highly oxidative conditions since the faulting occurred. Some of the dredged pillow basalts have some fresh glass preserved in the rims with variably oxidized and altered crystalline interiors. Presumably, soon after formation 50 m.y. ago these pillow basalts were exposed to hydrothermally derived solutions that produced secondary minerals (smectites) characteristic of nonoxidative alteration. In Table 4 and Figure 7 we present the results of the study of the bulk chemistry of such fresh-glass, altered-basalt pairs from four separate fault scarps in the Peru-

TABLE 4. CHEMICAL ANALYSES OF GLASS (G) FROM THE GLASSY RINDS OF PILLOW BASALTS COMPARED WITH ALTERED (A) BASALT FROM THE MORE HOLOCRYSTALLINE INTERIORS OF THE SAME PILLOWS

	7-72(G)	7-72(A)	7-166(G)	7-166(A)	7-183(G)	7-183(A)	10-77(G)
SiO_2	49.35	49.54	50.03	50.54	49.72	50.71	49.96
TiO_2	1.50	1.67	1.76	2.03	1.72	2.11	1.62
Al_2O_3	14.77	15.81	13.98	14.98	14.20	14.90	13.24
Fe_2O_3	0.42	9.21	3.40	6.13	1.04	5.68	2.91
FeO	11.35	3.54	8.71	4.94	11.52	5.22	11.19
MgO	7.72	4.81	7.26	5.86	7.06	6.03	7.88
CaO	12.37	12.12	12.33	12.19	12.18	12.15	11.14
Na_2O	2.40	2.65	2.40	2.80	2.43	2.71	2.05
K_2O	0.12	0.65	0.13	0.54	0.13	0.48	0.10

	10-77(A)	11-3(G)	11-3(A)	11-47(G)	11-47(A)	18-34(G)	18-34(A)
SiO_2	46.60	50.42	51.91	51.14	50.37	50.30	50.15
TiO_2	1.87	1.48	1.59	1.47	1.75	1.21	1.38
Al_2O_3	14.88	15.39	15.05	15.32	15.96	15.77	16.99
Fe_2O_3	10.43	1.19	4.46		5.29	0.09	6.40
FeO	5.71	9.19	5.25	9.46	3.81	9.12	3.02
MgO	6.13	7.41	7.85	7.80	7.99	8.42	4.83
CaO	11.11	12.30	10.89	12.21	11.54	12.46	13.47
Na_2O	2.45	2.38	2.64	2.40	2.83	2.58	3.23
K_2O	0.73	0.14	0.88	0.21	0.47	0.05	0.51

Note: Insufficient glass was available from sample 11-47(G) to perform titration. Total Fe for this sample calculated as FeO. Analyses have been calculated on a water-free basis.

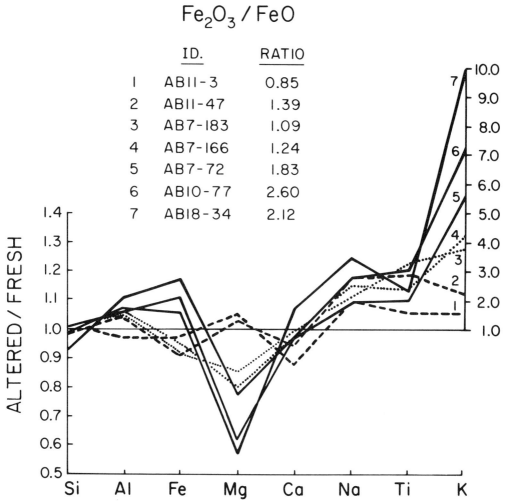

Figure 7. Chemical changes in altered holocrystalline interiors of pillow basalts from dredge areas 7, 10, 11, and 18 in the Peru-Chile Trench relative to the composition of fresh glass preserved on the glassy margins of each pillow. Fe_2O_3/FeO ratios noted by identification numbers apply only to the altered holocrystalline basalt; glasses had low Fe_2O_3/FeO ratios (<0.39; see Table 4). Scale at right pertains to K only.

Chile Trench. The altered-basalt samples were obtained from the pillows approximately 10 cm in and normal to the glassy margins. Data on each altered basalt were compared to the composition of the corresponding fresh glass ($Fe_2O_3/FeO \leq 0.39$) to show relative changes in composition.

We find that as the degree of oxidation of the altered holocrystalline basalt increases, Mg decreases and Al, Fe, and K increase (Fig. 7). For Si, Ca, Na, and Ti the chemical changes accompanying oxidation are not as clear and consistent, although it appears that the most oxidized samples are richest in Ca, Na, and Ti and have somewhat less Si. The chemical trends noted in Figure 7 can be interpreted in terms of the action of oxygenated sea water on oceanic crust that has been previously altered by hydrothermal solutions. Two of the least oxidized samples (samples 11-3 and 11-47) appear megascopically to be unaltered, but show an increase in Mg, a result consistent with the presence of dark green ferrosaponite filling available pore spaces (Fig. 8). In contrast, the three most highly oxidized smaples (7-72, 10-77, and 18-34) are reddish in color and are characterized by brown smectite, celadonite, iron oxides, and open void spaces. For these samples we find sharper and stronger 060 reflections of celadonite (1.507 Å) and weaker more diffuse 060 reflections of trioctahedral smectite (1.530 to 1.535 Å; Fig. 8). The high Al, Fe, and K, and low Mg contents of these samples are consistent with the oxidation of the pre-existing ferrosaponite. As such minerals are oxidized, much of the Mg is remobilized, Fe is retained as iron oxides or celadonite, and K increases by removal from sea water during celadonite formation. In addition, the observed increase in Al is consistent with the tendency for more Al-rich smectite to form in oxidized samples (Melson and Thompson, 1973; Scheidegger and Stakes, 1980).

Many of the changes that accompany variable degrees of oxidation can also be observed in individual hand specimens (Scheidegger, 1979). In Figure 9 we have one example of (060) reflections of phyllosilicates separated from the reddish, oxidized exterior and the light gray interior of a

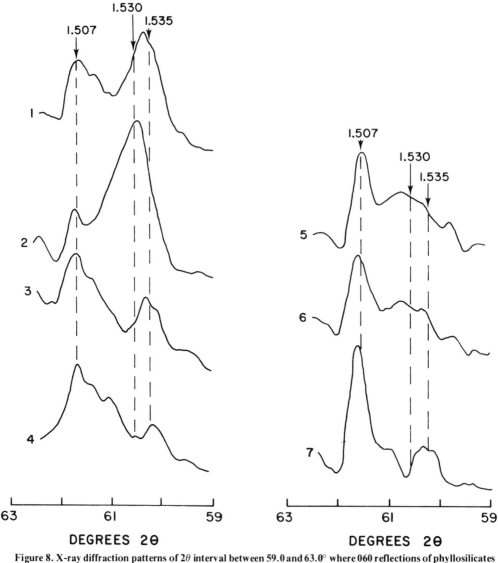

Figure 8. X-ray diffraction patterns of 2θ interval between 59.0 and 63.0° where 060 reflections of phyllosilicates occur (1.507 Å : celadonite; 1.530-1.535 Å : trioctahedral smectite). The diffractograms are of <2 μm material separated from altered basalts oxidized to varying degrees. Numbers 1 to 7 refer to samples identified in Figure 7. Note decrease in smectite and increase in celadonite peaks with increased Fe and K of altered basalt.

holocrystalline basalt fragment dredged from 17°S in the Peru Trench. The most oxidized portion is clearly dominated by celadonite, whereas the interior still retains smectite, presumably related to an earlier episode of hydrothermal alteration under nonoxidative conditions.

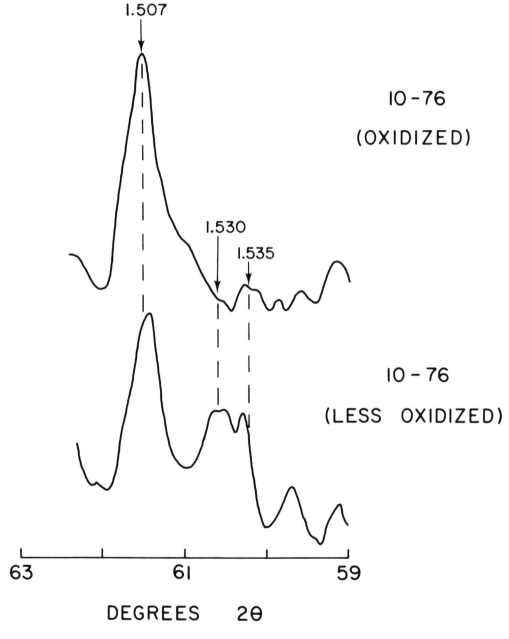

Figure 9. Diffraction patterns similar to those illustrated in Figure 8, but these were obtained on one holocrystalline sample recovered from dredge area 10. Diffractogram at top was from the reddish oxidized exterior, and the one at the bottom was from dark gray interior. Note dominance of celadonite (1.507 å peak) in the more oxidated material.

SUMMARY

The interaction of sea water and cooling holocrystalline basalts results in oxygen-poor, heated solutions rich in Fe, Mg, Ti, K, and Na. These solutions may react with basaltic glass to produce dark-colored, nonoxidatively altered glass which is texturally similar to palagonite but chemically quite different. Such hydrothermally altered glass is enriched in Fe, Mg, K, Na, and Ti, and impoverished in Ca and Si. Smectite formation is the main consequence of hydrothermal alteration, and it proceeds until the Mg supply in the circulating fluids is exhausted; during active smectite crystallization, the pH is kept low, thus enhancing transport of metals to overlying basal sediments and preventing the precipitation of calcite. Calcic plagioclase is unaltered under these conditions, and Al appears relatively immobile. Nonoxidative, hydrothermal alteration leads to the widespread formation of clay minerals rich in Fe, Mg, and K and poor in Al. Subsequently when Fe and Mg have disappeared from the solution and available pore spaces become nearly filled with secondary minerals, clay mineral formation ceases, alkaline conditions ensue, and calcite may fill all remaining voids. This phase of alteration is controlled by hydrothermal circulation and may last only a few million years. Oxidation of nonoxidative secondary mineral assemblages can occur as a result of the tectonic exposure of basement rocks to ambient sea water or to the percolation of oxygenated sea water through overlying rock masses. If the oxidation is sufficiently intense, smectites are destroyed and Mg is remobilized. In addition, the K and Fe contents of the rock can increase as residual Fe oxide and celadonite become the stable secondary phases. The increase in total Fe is only an apparent one as Fe is retained as an oxide or in celadonite after the destruction of the precursor smectites. Under highly oxidizing, low-temperature conditions, primary plagioclase may alter, thereby contributing Al to solutions and leading to more Al-rich secondary minerals. The secondary minerals and bulk compositions of crustal rocks from the Nazca plate provide evidence of the pervasive effects of hydrothermal circulation initiated at the East Pacific Rise and the Galapagos Rise and of low-temperature sea-water alteration in redistributing metals throughout layer 2 rocks.

ACKNOWLEDGMENTS

We thank Mark Hower for assisting with much of the analytical work required by this project and Dan Weill (Volcanology Center, University of Oregon) for making his electron microprobe available to us. We also commend Roger Hart, Manuel Bass, and William Seyfried for their thoughtful and thorough reviews of an earlier draft of this paper. This study was made possible by financial support from the National Science Foundation through grants IDOE 71 04208 A07 and OCE 76-05903 A02.

REFERENCES CITED

Andrews, A. J., 1977, Low temperature fluid alteration of oceanic layer 2 basalts, DSDP Leg 37: Canadian Journal of Earth Sciences, v. 14, p. 911–926.

Baragar, W.R.A., and others, 1977, Petrology and alteration of selected units of Mid-Atlantic Ridge basalts sampled from sites 332 and 335, DSDP: Canadian Journal of Earth Sciences, v. 14, p. 837–874.

Bass, M. N., 1976, Secondary minerals in oceanic basalt, with special reference to Leg 34, Deep Sea Drilling Project, *in* Yeats, R. S., Hart, S. R., and others, eds., Initial reports of the Deep Sea Drilling Project, Volume 34: Washington, D.C., U.S. Government Printing Office, p. 393.

Bischoff, J. L., and Dickson, F. W., 1975, Seawater-basalt interaction at 200 °C and 500 bars: Implications for origin of seafloor heavy-metal deposits and regulation of seawater chemistry: Earth and Planetary Science Letters, v. 25, p. 385–397.

Bryan, W. B., 1972, Morphology of quench crystals: Journal of Geophysical Research, v. 77, p. 5812–5819.

Bryan, W. B., and Moore, J. G., 1977, Compositional variations of young basalts in the Mid-Atlantic Ridge rift valley near 36°49′N: Geological Society of America Bulletin, v. 88, p. 556–570.

Carstea, D. D., Howard, M. E., and Knox, E. G., 1970,

Comparison of iron and aluminum hydroxy interlayers in montmorillonite and vermiculite: I. Formation; II. Dissolution: Soil Science of America, Proceedings, v. 34, p. 517–520.

Corliss, J. B., 1971, The origin of metal-bearing submarine hydrothermal solutions: Journal of Geophysical Research, v. 76, p. 8128–8138.

Doremus, R. H., 1973, Glass science: New York, Wiley, 349 p.

Fukui, S., 1976, Laboratory techniques used for atomic absorption spectrophotometric analysis of geologic samples: Oregon State University, Reference 76-10.

Hajash, A., 1975, Hydrothermal processes along mid-ocean ridges: An experimental investigation: Contributions to Mineralogy and Petrology, v. 53, p. 205–226.

Hart, S. R., and Staudigel, H., 1978, Oceanic crust: Age of hydrothermal alteration: Geophysical Research Letters, v. 5, p. 1009–1012.

Harward, M. E., Theisin, A. A., and Evans, D. D., 1962, Effect of iron removal and dispersion methods on clay mineral identification by X-ray diffraction: Soil Science of America, Proceedings, v. 26, p. 535–540.

Hayes, J. B., 1970, Polytypism of chlorite in sedimentary rocks: Clays and Clay Minerals, v. 18, p. 285–306.

Hogan, L., and Dymond, J., 1976, K-Ar and ^{40}Ar-^{36}Ar dating of sites 319 and 321 basalts, in Yeats, R. S., Hart, S. R., and others, eds., Initial reports of the Deep Sea Drilling Project, Volume 34: Washington, D.C., U.S. Government Printing Office, p. 439.

MacEwan, D.M.C., 1961, The montmorillonite minerals (montmorillonoids), in Brown, G., ed., The X-ray identification and structure of clay minerals: London, Mineralogical Society of Great Britain, p. 86.

Melson, W. G., and Thompson, G., 1973, Glassy abyssal basalts, Atlantic sea floor near St. Paul's Rocks: Petrography and composition of secondary clay minerals: Geological Society of America Bulletin, v. 84, p. 703.

Melson, W. G., and others, 1976, Chemical diversity of abyssal glass erupted along Pacific Atlantic and Indian Ocean sea-floor spreading centers, in Sutton, G. H., and others, eds., The geophysics of the Pacific Ocean Basin and its margin: A volume in honor of G. P. Woollard: American Geophysical Union Geophysical Monograph, v. 19, p. 351–367.

Moore, J. G., 1966, Rate of palagonitization of submarine basalt adjacent to Hawaii: U.S. Geological Survey Professional Paper 550-D, p. D163–D171.

Mottl, M. J., 1976, Chemical exchange between seawater and basalt during hydrothermal alteration of the ocean crust [Ph.D. dissert.]: Cambridge, Harvard University.

Mueller, R. F., 1973, System $CaO-MgO-FeO-SiO_2-C-H_2-O_2$: Some correlations from nature and experiment: American Journal of Science, v. 273, p. 152–170.

Rea, D. K., 1977, Local axial migration and spreading rate variations, East Pacific Rise, 31°S: Earth and Planetary Science Letters, v. 34, p. 78–84.

Scheidegger, K. F., 1979, An oxidative diagenetic overprint on hydrothermally altered tholeiitic basalts: Chemical and mineral consequences [abs.]: Geological Society of America Abstracts with Programs, Annual Meeting, v. 11, p. 511.

Scheidegger, K. F., and Stakes, D. S., 1977, Mineralogy, chemistry and crystallization sequence of clay minerals in altered tholeiitic basalts from the Peru Trench: Earth and Planetary Science Letters, v. 36, p. 413–422.

——1980, X-ray diffraction and chemical study of secondary minerals from Deep Sea Drilling Project Leg 51, holes 417A and 417D, in Donnelly, T., and others, eds., Initial reports of the Deep Sea Drilling Porject, Volumes 51, 52, 53: Washington, D.C., U.S. Government Printing Office, p. 1253–1263.

Scheidegger, K. F., and others, 1978, Fractionation and mantle heterogeneity in basalts from the Peru-Chile Trench: Earth and Planetary Science Letters, v. 37, p. 409–420.

Scott, R. B., and Hajash, A., Jr., 1976, Initial submarine alteration of basaltic pillow lavas: A microprobe study: American Journal of Science, v. 276, p. 480–501.

Scott, R. B., and Swanson, S. B., 1976, Mineralogy and chemistry of hydrothermal veins and basaltic host rocks at hole 319A and site 321, in Yeats, R. S., Hart, S. R., and others, eds., Initial reports of the Deep Sea Drilling Project, Volume 34: Washington, D.C., U.S. Government Printing Office, p. 377–380.

Seyfried, W. E., and Bischoff, J. L., 1977, Hydrothermal transport of heavy metals by seawater: The role of seawater/basalt ratio: Earth and Planetary Science Letters, v. 34, p. 71–77.

Seyfried, W. E., and Mottl, M. J., 1977, Origin of submarine metal-rich hydrothermal solutions: Experimental basalt-seawater interaction in a seawater-dominated system at 300 °C, 500 bars, in Paquet, H., and Tardy, Y., eds., Proceedings of the Second International Symposium on Water-Rock Interaction: Strasbourg, France, p. 173–180.

Seyfried, W. E., Shanks, W. C., and Bischoff, J. L., 1978, Clay mineral formation in DSDP Leg 34 basalt: Earth and Planetary Science Letters, v. 41, p. 265–276.

Sleep, N. H., 1975, Formation of ocean crust: Some thermal constraints: Journal of Geophysical Research, v. 80, p. 4037–4042.

Stakes, D. S., 1978, Mineralogy and geochemistry of the Michigan Deep Hole metagabbro compared to seawater hydrothermal alteration: Journal of Geophysical Research, v. 83, p. 5820–5324.

——1979, Submarine hydrothermal systems: Variations in mineralogy, chemistry, temperatures and the alteration of oceanic layer 2 [Ph.D. thesis]: Corvallis, Oregon State University, 189 p.

Stakes, D. S., and O'Neil, J. R., 1977, Stable isotope compositions of hydrothermal minerals from altered rocks from the East Pacific Rise and the Mid-Atlantic Ridge [abs.]: EOS (American Geophysical Union Transactions), v. 58, p. 1151.

Thompson, G., and others, 1976, Petrology and geochemistry of basalts from DSDP Leg 34, Nazca palte, *in* Yeats, R. S., Hart, S. R., and others, eds., Initial reports of the Deep Sea Drilling Project, Volume 34: Washington, D.C., U.S. Government Printing Office, p. 215–226.

Wise, W. S., and Eugster, H. P., 1964, Celadonite: Synthesis, thermal stability and occurrences: American Mineralogist, v. 49, p. 1031–1083.

Yeats, R. S., Hart, S. R., and others, 1976, Initial reports of the Deep Sea Drilling Project, Volume 34: Washington, D.C., U.S. Government Printing Office, 814 p.

MANUSCRIPT RECEIVED BY THE SOCIETY NOVEMBER 12, 1980
MANUSCRIPT ACCEPTED DECEMBER 30, 1980

METALLIFEROUS SEDIMENTS

Geological Society of America
Memoir 154
1981

Geochemistry of Nazca plate surface sediments: An evaluation of hydrothermal, biogenic, detrital, and hydrogenous sources

JACK DYMOND
School of Oceanopgraphy
Oregon State University
Corvallis, Oregon 97331

ABSTRACT

Strong compositional gradients in Nazca plate sediments result from variability in the sedimentary sources in different geographic regions. At the western margin of the plate, precipitates from hydrothermal sources associated with the East Pacific Rise dominate the sediment budget. In the northern part of the plate, biogenic tests that reflect the high productivity along the equator are important. Near the continent the sediments are largely detrital aluminosilicates. The basins, which lie in the central portions of the plate, have strong enrichments in the transition metals whose source is either hydrothermal precipitates or authigenic ferromanganese deposition from normal sea water.

This study develops and applies a normative composition model that converts bulk chemical composition data into quantitative estimates of the weight percent of five distinct components: (1) hydrothermal precipitates, (2) biogenic tests, (3) detrital aluminosilicates, (4) hydrogenous ferromanganese precipitates, and (5) the insoluble residue of organisms. The model assumes that each component has a constant composition anywhere on the plate and uses linear programming to arrive at a mixture of components that best accounts for the measured composition of any sample. The chemical compositions of 425 samples were determined and analyzed using the normative model.

Samples from within 100 km of the rise crest generally have >80% hydrothermal components. The geographic distribution of the hydrothermal components indicates that north of approximately 30°S bottom currents transport hydrothermal precipitates to the west, whereas south of 30°S bottom currents carry the hydrothermal precipitates eastward into the Roggeveen Basin. The biogenic contents of the sediments decrease latitudinally from approximately 80% near the equator to values less than 10% south of 20°S latitude. The detrital abundance is generally >80% in regions within 1,000 km of the South American Continent; however, between 20° and 25°S unusually high concentrations of the detrital component extend away from the continent and into the Yupanqui Basin. These data suggest the importance of bottom nepheloid-layer transport in this part of the plate. The deep basins lying east of the East Pacific Rise have the highest contents of hydrogenous component, presumably caused by a constant deposition of ferromanganese material into a region with small contributions from other sources. Elements remaining after the dissolution of biogenic matter are termed "dissolution residue" in the model. This component reaches a maximum abundance in the basins on the southern edge of the equatorial high productivity zone, a pattern caused by decreased preservation of biogenic matter in regions of lower biogenic deposition.

Published sedimentation rates have been used to calculate rates of the accumulation of the 5 components. Although far fewer in number of samples, these rate data support the conclusions based solely on the relative abundances of the sources.

Except for an obvious dilution by biogenic opal in the more northern samples, compositional range of rise-crest sediments is small. This observation is remarkable, considering the broad range in compositions previously reported for hydrothermal precipitates recovered from the oceans, and confirms the validity of the assumed constant composition of the hydrothermal source. These data suggest that either the Nazca plate metalliferous sediments are the result of large-scale homogenization of precipitates from a broad range of hydrothermal vent types or the Nazca plate portion of the East Pacific Rise has a magmatic and tectonic regime which produces a consistent hydrothermal fluid composition and precipitation conditions.

INTRODUCTION

Studies of the composition of deep-sea sediments have clarified the sources, processes of distribution, and chemical changes that are important for their formation. Studies of Nazca plate

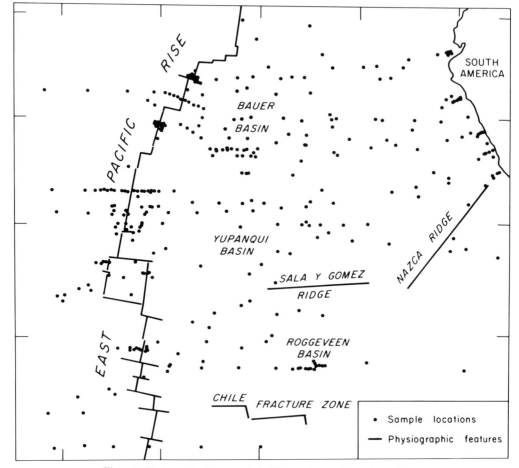

Figure 1. Bathymetric features of the Nazca plate and locations of samples.

sediments have focused primarily on the metalliferous deposits, following the observation that sediments near the East Pacific Rise (EPR) are strongly enriched in Fe, Mn, and other transition elements relative to normal pelagic sediments (Bostrom and Peterson, 1966; Piper, 1973). More recent studies have confirmed the anomalous chemical composition of these rise-crest sediments (Bostrom, 1973) and have suggested that sea-water hydrothermal systems are the mechanism for leaching metals from the crust and depositing them on the sea floor (Corliss, 1971). Isotopic studies support a mantle source for a significant part of the Pb, but indicate that other elements, such as U and Sr, have a sea-water origin (Bender and others, 1971; Dasch and others, 1971; Dymond and others, 1973).

The importance of nonhydrothermal sources of metals in Nazca plate sediments is reflected in the observed chemical variability. For example, the Bauer Basin, a deep basin lying a few hundred kilometres to the east of the crest of the EPR, is accumulating sediments with significantly higher Ni-Fe values than sediments on the rise crest (Dymond and others, 1973). The addition of hydrogenous ferromanganese phases of a composition similar to nodules and crusts recovered in the area (Lyle and others, 1977) appears to be important in this basin.

Metal accumulation rates of Nazca plate sediments have provided important data confirming the anomalous accumulation of Fe and Mn on the EPR and for determining the dispersal of hydrothermal precipitates by bottom currents (Bender and others, 1971; Dymond and Veeh, 1975; Dymond and others, 1977; McMurtry and others, this volume). Ultimately, accumulation rate data can provide a measure of the hydrothermal flux of elements into the oceans; for example, see Lyle (1976).

To determine the influence of sea-water hydrothermal systems on the composition of deep-sea sediments, it is necessary to identify and quantify the other sources and processes that modify the sediment composition. For the Nazca plate there have been previous efforts to quantify the contribution of detrital, biogenic, and hydrogenous sources relative to the hydrothermal sources using different normative sediment analysis schemes (Bostrom, 1973; Heath and Dymnd, 1977). In this paper I will present an extension of previous normative sediment approaches and attempt to show that the influence of different sources varies strongly over the Nazca plate. Because no single source dominates all areas of the plate, the sources can be identified and quantified. As a result of this analysis, the nature and composition of Nazca plate sea-water hydrothermal systems can be clarified.

METHODOLOGY

Samples

Samples were selected primarily from cores recovered during the cruises conducted by Oregon State University and Hawaii Institute of Geophysics as part of the Nazca Plate Project; additional samples were obtained from the core collections of Lamont-Doherty Geological Observatory and the Scripps Institution of Oceanography. Nearly all samples were taken from the 5 to 10 cm level of gravity cores to avoid the disturbed upper portions of the cores, although where this material was unavailable, a few samples from deeper levels and piston cores were used. The sample locations for 425 analyzed samples are shown on Figure 1 and listed in Table 1 (microfiche).

Analytical Procedures

Wet samples from cores were freeze-dried without previous treatment and dry-split into sizes appropriate for analyses. Salt content of the dry samples was computed from the water content determined by weighing before and after freezing and assuming $35^0/_{00}$ salts in the pore waters. Water content and salt content from dried Lamont-Doherty Geological Observatory cores were computed from the empirical relationship between carbonate and salt content (Lyle and Dymond, 1976). Since the salt content of dried samples varies between 1.5% and 19%, all compositional data have been recomputed to a salt-free basis.

Also, because of the large variations in CaCO₃ content observed in sediments recovered from different depths on the plate (0% to 97% $CaCO_3$) (Fig. 2), all compositional data used in this paper have been corrected to a carbonate-free basis. The $CaCO_3$ content has been determined by measuring the total Ca content of the samples and computing $CaCO_3$ with corrections for Ca in the sea salts and the non carbonate fraction (Dymond and others, 1976). This approach has been compared extensively to other direct $CaCO_3$ measurements, such as CO_2 volume released by acid dissolution and CO_2 analysis with a Leco carbon analyzer. My approach is subject to error for samples with low carbonate abundances because of uncertainties in the Ca content of the noncarbonate fraction. For high carbonate samples, where correction for carbonate abundance is very sensitive, however, the total Ca measurement with computation for $CaCO_3$ content is unequalled in precision and accuracy by these other techniques.

Four hundred and twenty-five samples were analyzed for a standard set of elements consisting of Al, Si, Ca, Mn, Fe, Ni, Cu, Zn, and Ba by atomic absorption spectrophotmetry. Each sample was dried overnight at 100 °C, and approximately 400 mg were dissolved in a teflon-lined steel bomb with 2 ml of Aqua Regia and 6 ml of hydrofluoric acid heated to 100 °C for 2 to 3 h. After additions of boric acid and $CsCl_2$, all 9 elements from various dilutions of a single dissolution were analyzed by atomic absorption spectrophotometer (AAS) (Fukui, 1976). The AAS data used in this paper are the results of duplicate analyses of each sample. In cases where two determinations disagreed significantly, a third analysis was made to resolve discrepancies. The precision and accuracy of the AAS data are indicated by Table 2 which compares the composition of U.S. Geological Survey standard rock BCR-1 determined during the period the samples were analyzed with the published data of other workers. My data and the published results agree within the precision estimates and suggest that there are no systematic errors in the data.

The data for many of the samples near the South American Continent were supplied by Ken Scheidegger of Oregon State University. A total of 83 samples were analyzed by Scheidegger by AAS using a slightly different dissolution and analytical scheme (Omang, 1969).

RESULTS

Chemical Composition of Nazca Plate Sediments

Figures 2 to 6 exhibit the $CaCO_3$ and abundances of selected elements in Nazca plate surface sediments. Color contour maps of elemental abundance data are part of a Geological Society of

TABLE 2. ANALYSIS OF UNITED STATES GEOLOGICAL SURVEY STANDARD ROCKS

	W-1		BCR-1		GSP-1	
	OSU*	Published†	OSU*	Published†	OSU*	Published†
Al (%)	7.89 ± .08	7.94	7.22 ± .12	7.22	7.82	7.99
Si (%)	24.47 ± .43	24.61	25.90 ± .33	25.47	31.76	31.43
Ca (%)	7.65 ± .10	7.83	4.81 ± .08	4.97	1.40	1.45
Mn (ppm)	1,380 ± 59	1,320	1,430 ± 18	1,317	296	310
Fe (%)	7.74 ± .10	7.76	9.42 ± .07	9.45	3.05	3.03
Ni (ppm)	80 ± 13	76	10 ± 5	15	10	11
Cu (ppm)	118 ± 4	110	18 ± 2	22	32	35
Zn (ppm)	88 ± 4	86	134 ± .5	132	102	143
Ba (ppm)	171 ± 14	160	742 ± 25	790	1,340	1,360
Number of analyses	20		5		2	

*All data determined by atomic absorption spectrophotometry; see text. Values represent the means and standard deviations of the analyses.

†Flanagan, 1973.

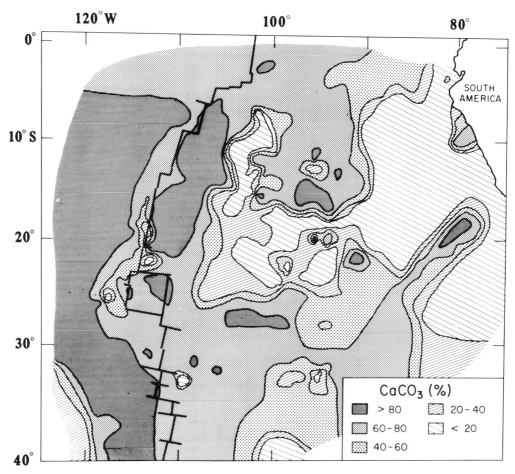

Figure 2. CaCO₃ content of Nazca plate sediments.

America map series (Dymond and Corliss, 1980). A qualitative interpretation for the obvious patterning in these compositional data is relatively easy for some elements. The strong enrichment in Mn (Fig. 3) in the western plate is undoubtedly the result of hydrothermal activity along the EPR. Fe, Cu, and Zn exhibit similar enrichments at the rise crest. The enrichment of these transition elements is in contrast to a depletion in Al over the EPR, and reflects the very low Al abundance in precipitates from hydrothermal solutions (Bostrom and Peterson, 1966). Al is enriched most strongly in the eastern margin of the Nazca plate (Fig. 4). This supports the commonly held concept that Al is associated with detrital phases. The influence of detritus from the South American Continent on the plate is undoubtedly reduced by the existence of the trench; however, the trend toward higher Al values in proximity to South America is clear. A tongue of higher Al seems to follow the 4,000-m contour at 15° to 20°S, which would suggest that some detrital transport into the Yupanqui Basin is in the form of near-bottom transport. The abundance of Si in the sediments follows a latitudinal banding with the highest values in the northern plate, near the equator (Fig. 5). The first-order pattern of Si abundance in the sediment thus appears to be controlled by the productivity of surface waters which results in the rain and partial preservation of biogenic opal. A tongue of high Si at 15° to 20°S extends away from the South American Continent, however, which indicates the importance of detrital sources for Si as well. The minor elements Ni (Fig. 6), Cu, Zn, and Ba appear to have the greatest abundance in sediments recovered from the Bauer, Yupanqui, and Roggaveen Basins. These are

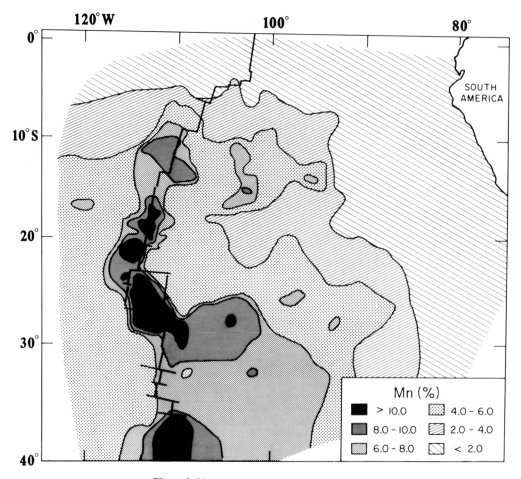

Figure 3. Mn content of Nazca plate sediments.

regions on the plate of low sediment accumulation, and in these regions it is likely that a hydrogenous source of elements, that is, direct precipitation of elements from sea water, would be most important (Kirshnaswami, 1976). Thus, it is reasonable to suggest that the distribution of these minor elements is controlled by incorporation of hydrogenous Fe and Mn phases into the sediments.

Normative Sediment Analysis

To analyze the sediment compositions more quantitatively, I have used a normative sediment analysis model designed to extract source information from a given sediment chemical composition. With this analysis the composition of a model sediment and the measured composition can be compared and qualitative ideas on the important sources of particular elements can be tested. Normative sediment analyses of Nazca plate sediments have been done previously by Bostrom (1973) and by Heath and Dymond (1977). These studies, as in this approach, have attempted to define the compositions of pure end-member sources and to determine what mixture of sources can best account for the bulk sediment composition. The current approach differs in some of the assumptions and in the use of linear programming to solve a series of linear equations that relate the bulk composition to the composition of individual sources.

Five components of sediment are evaluated by the model: (1) detrital, (2) hydrothermal,

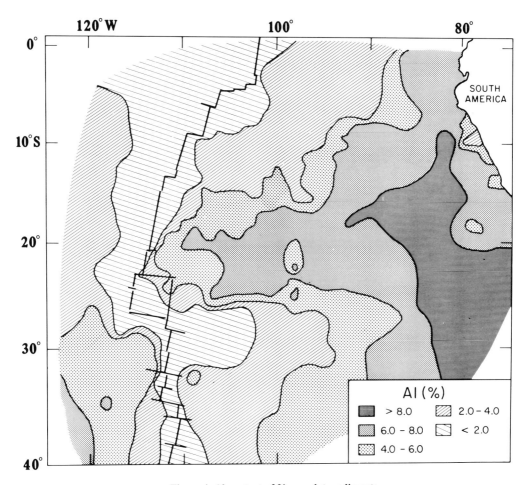

Figure 4. Al content of Nazca plate sediments.

(3) biogenic, (4) hydrogenous, and (5) dissolution residue. The model is based on the assumption that each of these components has a constant composition over the whole plate.

The composition of the detrital source has been taken from summary analyses of igneous and sedimentary rock (Poldervaart, 1955; Wedepohl, 1969). The values used in the model are similar to those for intermediate igneous rocks.

The hydrothermal source term includes both primary elements carried by the hydrothermal solutions and elements that are coprecipitated from sea water by formation of hydrothermal phases. While a distinction between the elements that are deposited directly from hydrothermal solutions and those that coprecipitate on hydrothermal ferromanganese hydroxide is important for geochemical budget considerations, this is only unequivocally possible for a few elements such as Pb, Nd, and Sr where isotopic distinctions can be made. The composition of this source has been evaluated by examination of the elemental ratios of samples close to the crest of the EPR.

The biogenic source term is meant to include elements that reach the sea floor by way of the hard and soft parts of marine organisms. Since marine organisms undergo oxidation and dissolution reactions during their fall through the water column as well as after reaching the bottom, the elemental proportions of biogenic material reaching the sea floor are fractionated from those observed in plankton. Because the degree of fractionation will vary with such variables as depth and sedimentation rate, I have chosen the biogenic composition to be similar to marine plankton (Martin and Knauer,

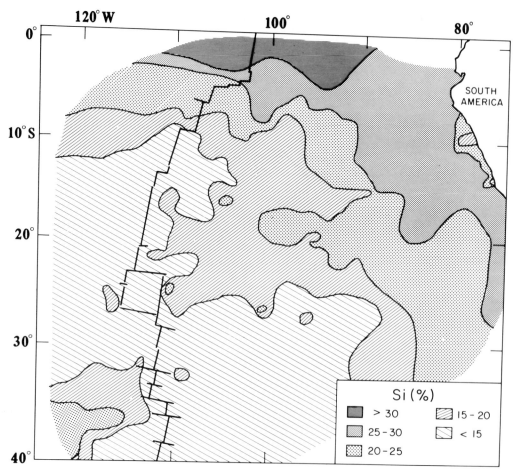

Figure 5. Si content of Nazca plate sediments.

1973) rather than attempt to estimate an average fractionation for all locations. Thus, the biogenic source term is for essentially unfractionated biogenic debris, and fractionation of biogenic matter will be accounted for by another term discussed below. Also, because the elemental data are corrected for carbonate content, the biogenic source is composed predominantly of biogenic opal and refractory organic matter.

Composition of the hydrogenous source term in the model is taken from ferromanganese nodule and crust data with the assumption that these phases represent the most important direct precipitation from sea water. Because ferromanganese nodules may form by a variety of processes, some of which require relatively unknown reactions in the sediment, determining the best value for the composition of a phase precipitating directly from sea water is difficult. Fortunately, several dredge hauls from the Nazca plate have recovered ferromanganese crusts that have compositions distinct from nodules which in some cases were recovered in the same dredge haul (Lyle and others, 1977). I have selected a hydrogenous composition by an evaluation of these crust data. The supposition is that ferromanganese crusts are removed from the sediment influence and that the dissolved elements in sea water provide the only source. The hydrogenous elemental ratios used in the model were further modified by evaluation of analyses of that portion of Nazca plate nodules and crusts that was soluble in an ammonium oxalate-oxalic acid leach, a dissolution procedure that does not dissolve silicate phases in nodules (Lyle, 1979). It is possible that the bottom waters in the Nazca plate have a minor-

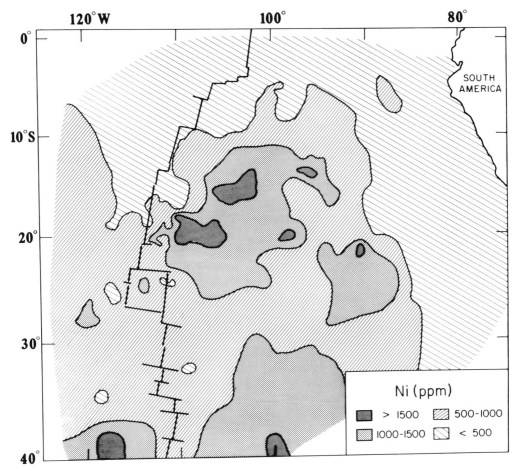

Figure 6. Ni content of Nazca plate sediments.

element content that has been modified by the extensive hydrothermal activity on the EPR. Consequently, the Nazca plate ferromanganese crusts may show a hydrothermal influence and not truly represent precipitation from "normal" sea water. Comparisons between analyses of crusts recovered from the Nazca plate and from other parts of the oceans could evaluate this question.

The dissolution residue is composed of relatively insoluble elements that do not return to the water column upon dissolution and oxidation of the hard and soft parts of carbonate and siliceous organisms. This component is of biogenic origin; however, it has a highly fractionated composition composed of the nonlabile elements carried by organisms. The model uses Ba as the index element for dissolution residue. This choice is based on the common beliefs that Ba has strong biogenic associations (Goldberg and Arrhenius, 1958; Gurvich and others, 1978; Suess, in prep.) and that it is relatively refractory, since barite is saturated in pore waters (Church and Wolgemuth, 1972). The composition of the model dissolution residue has been obtained from an evaluation of element-to-Ba ratios measured in plankton. For example, the Al/Si and Ba/Si values are both 0.002 in our model biogenic matter (Table 3). Consequently, biogenic matter falling to the sea floor should have an Al/Ba value of 1.0. If Al and Ba were equally refractory during complete dissolution of biogenic opal and carbonate, the dissolution residue would have an Al/Ba value of 1.0 as well. However, based on evidence that some Al is easily lost from biogenic phases reaching the sea floor (Mackenzie and others, 1978), I have assigned an Al-Ba ratio of 0.5 for our dissolution residue source. Thus,

TABLE 3. ELEMENTAL RATIO COEFFICIENTS OF THE FIVE COMPONENTS USED IN THE NORMATIVE ANALYSIS

	$(E/Al)_D$	$(E/Fe)_H$	$(E/Si)_B$	$(E/Ni)_A$	$(E/Ba)_R$
Al	1.0000	0.0060	0.002000	1.5	0.5000
Si	3.0000	0.1300	1.000000	4.5	0.0000
Mn	0.0160	0.2900	0.000023	30.0	0.0070
Fe	0.7000	1.0000	0.001000	15.0	0.3500
Ni	0.0015	0.0009	0.000040	1.0	0.0065
Cu	0.0012	0.0042	0.000050	0.5	0.0160
Zn	0.0014	0.0019	0.000080	0.1	0.0040
Ba	0.0120	0.0050	0.002000	0.2	1.0000

Note: $(E/Al)_D$, $(E/Fe)_H$, $(E/Si)_B$, $(E/Ni)_A$, and $(E/Ba)_{DR}$ are the elemental ratios in pure detrital, hydrothermal, biogenic, authigenic, and dissolution residue components, respectively

$$(\text{Element}/\text{Ba})_{\text{Dissolution Residue}} = (\text{Element}/\text{Ba})_{\text{Biogenic}} \times \alpha \quad (1)$$

where α is a measure of how labile an element is compared to Ba.

Evaluation of α, the index of relative refractivity, is one of the most uncertain aspects in the model and in marine geochemistry as well. Transition element versus depth profiles in the ocean (Bruland and others, 1978; Boyle, 1976) provide some evidence for the relative ease at which Ni, Cu, Zn, and Ba are released from decaying and dissolving biogenic matter as it sinks through the water column or lies exposed on the bottom. At the present I justify the selected α values only as an intuitive estimate based on a consideration of available water column compositional data.

The detrital, hydrothermal, biogenic, hydrogenous, and dissolution residue sources to the sediments are solved using data for 8 elements: Al, Si, Mn, Fe, Ni, Cu, Zn, and Ba. The bulk (carbonate and salt corrected) content of any element in a given sample, E_T, is given by the equation:

$$E_T = (E/Al)_D D_{Al} + (E/Fe)_H H_{Fe} + (E/Si)_B B_{Si} + (E/Ni)_A A_{Ni} + (E/Ba)_R R_{Ba} \quad (2)$$

where $(E/Al)_D$ is the relevant element-to-Al ratio in detrital matter; $(E/Fe)_H$ is the element-to-Fe ratio in hydrothermal material; $(E/Si)_B$ is the element-to-Si ratio in biogenic debris; $(E/Ni)_A$ is the element-to-Ni ratio in the authigenic (hydrogenous) source; $(E/Ba)_R$ is the element-to-Ba ratio in the dissolution residue; and the variables D_{Al}, H_{Fe}, B_{Si}, A_{Ni}, and R_{Ba} are the quantities of detrital Al, hydrothermal Fe, biogenic Si, hydrogenous Ni, and dissolution residue Ba, respectively, in the sample. Thus, using the bulk analysis of these 8 elements, I have set up 8 equations with 5 unknowns (D, H, B, A, and R) and 40 fixed elemental ratio coefficients (Table 3) for every sample.

For any element the product of each of the variables and its elemental ratio coefficient gives the concentration of that element from each of the 5 sources. Thus, $(Si/Al)_D \times D_{Al}$ = concentration of detrital Si in the sample. The sum of the right-hand side of equation 2 is the model concentration of that element, which for a perfect fit of the model to the data would equal the measured concentration of the element in question (that is, the left-hand side of the equation).

Since there are 5 unknowns and 8 equations, this is an over-determined system of linear equations. An optimum solution to the 5 unknowns is obtained for each sample using linear programming (Simmons, 1972; Narula and Wellington, 1977). Linear programming differs from a least-squares solution of an over-determined set of linear equations by enabling one to set certain restrictions to the solution. Here I have restricted the solutions to the unknowns, D_{Al}, H_{Fe}, B_{Si}, A_{Ni}, and R_{Ba}, to positive or zero values, since negative values do not have physical significance. For each sample the optimum solution to the linear equations is one in which the sum of the differences between the computed concentrations and the measured concentrations is a minimum:

$$E^i_T - (E^i/Al_D)D_{Al} - (E^i/Fe_H)H_{Fe} - (E^i/Si_B)B_{Si} - (E^i/Ni_A)A_{Ni} \qquad (3)$$
$$- (E^i/Ba_R)R_{Ba} = B^i$$

Thus, an optimum solution is one where $\Sigma B^{i=1\ to\ 8}$ is a minimum. To obtain a solution that was influenced by the minor elements, Cu, Ni, and Zn, as well as the major elements, the minor-element equations were weighted by multiplying these equations by 20. The factor 20 was chosen by repeated runs using different scaling factors from 1 to 100, and it was found that between 1 and 20 the optimal solutions had very low B-sums; however, the B-values for Cu, Ni, and Zn were relatively large compared to the total abundance E^i_T. Weighting factors between 15 and 50 gave the same solutions, and although these solutions commonly had higher B-sums, the fit for the minor elements was better. For minor-element weighting factors >50, the B-sums became unacceptably high for many samples. Minor-element weighting between 15 and 50 gave these elements some influence in the solution without letting these elements dominate the solution.

With this formulation, the optimum solution for the 5 unknowns, D_{Al}, H_{Fe}, B_{Si}, A_{Ni}, and R_{Ba}, can be computed for each of the 425 samples. The product of the unknown and the appropriate coefficients (Table 3) is the concentration of detrital, hydrothermal, biogenic, hydrogenous, or dissolution residue contributions of each element in that sample. Further, by assuming a value for (1) the concentration of Al in pure detritus (84,000 ppm), (2) the concentration of Fe in pure hydrothermal material (348,000 ppm), (3) the concentration of Si in pure biogenic opal (360,000 ppm), (4) the concentration of Ni in hydrogenous material (10,000 ppm), and (5) the concentration of Ba in the dissolution residue (270,000 ppm), the weight fractions of detritus, hydrothermal material, biogenic matter, hydrogenous material, and dissolution residue can be computed for each sample. These absolute concentrations were taken from the same literature sources that were used to obtain the elemental ratios found in Table 3.

The mathematic formulation of this normative analysis resembles petrographic mixing equations (Bryan and others, 1969), particularly those which use linear programming rather than least squares for a solution (Wright and Doherty, 1970). The value of this type of analysis is that it provides quantitative estimates of the hypothesized source contributions for each sample. The pattern of these inputs over the whole plate can clarify the geologic and oceanographic processes that operate in this part of the ocean. In addition, this analysis can further our understanding of the geochemical behavior of individual elements. Samples and/or elements that cannot be satisfactorily explained by the model indicate areas of incomplete understanding.

The use of the normative analysis as a quantitative estimate of sources requires that the basic assumption of constant source compositions be valid, at least to a first approximation. Variations in hydrothermal solution compositions might be expected to result from differences in the water-rock ratio or temperature of basalt–sea-water interaction at different loctions on the plate (Bischoff and Dickson, 1975; Seyfried and Bischoff, 1977; Mottl and Holland, 1978). In addition, detritus at the ridge crest is most likely to result from weathering and erosion of tholeiitic basalt and differs strongly from our chosen detritus which has a more intermediate composition (Table 3). The success of previous attempts at explaining Nazca plate sediment compositions by mixing equations (Heath and Dymond, 1977; Bostrom, 1973), however, suggests that the approach is not unreasonable.

The relatively constant Fe/Mn, Fe/Cu, and Fe/Zn values of the EPR sediments (Fig. 7) supports a

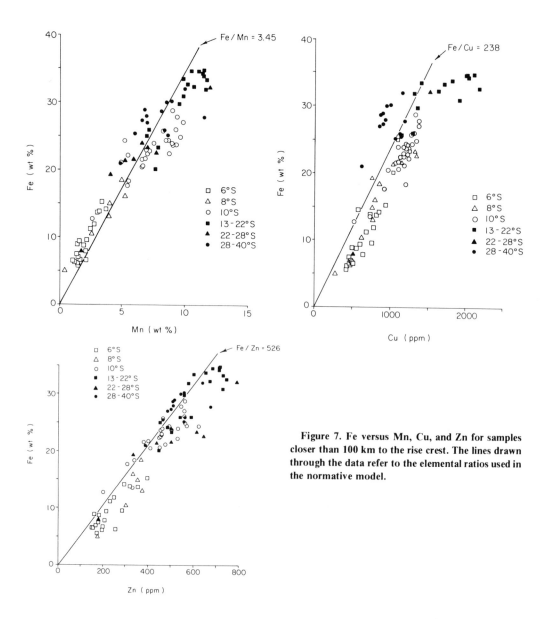

Figure 7. Fe versus Mn, Cu, and Zn for samples closer than 100 km to the rise crest. The lines drawn through the data refer to the elemental ratios used in the normative model.

constant hydrothermal composition. Nonetheless, there are regional variations in these predominantly hydrothermal elements along the EPR which are exhibited most clearly in Figure 8, a plot of Fe/Mn values for rise-crest samples versus latitude. Rise-crest sediments from the 6°S area have a Fe/Mn value generally greater than the value of 3.4 used in our model. Other rise-crest areas have an Fe/Mn value near the model value; however, there is more scatter in this value in some areas. I have made numerous runs on the model using different coefficients to evaluate the effects of a breakdown in the constant composition assumption. It appears that the weight fraction contribution from the 5 sources, which is computed by the model, is not greatly sensitive to changes in the element ratios. To illustrate, in Table 4 I list the results of 3 samples from the 6°S rise-crest area that have high Fe-Mn ratios and consequently large negative Mn residuals using the model coefficients (Table 3). Rerunning the model with lower hydrothermal Mn/Fe values and higher hydrothermal

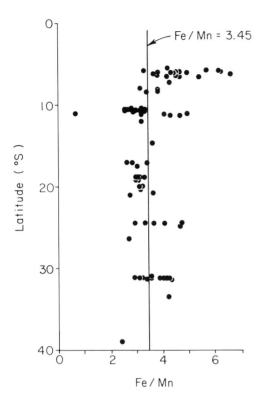

Figure 8. Fe/Mn versus latitude from samples less than 100 km from the rise crest.

Fe/Ni, Fe/Cu, and Fe/Zn values (values which appear to be more appropriate for the hydrothermal contribution in this area), lowers the Mn residual greatly, but changes the weight percent contribution from the 5 sources only slightly (Table 4).

Figure 9 is a comparison between the computed values obtained by summing the contributions from the 5 sources and the measured values of the 8 elements. The model accounts for the Si, Fe, and Ba contributions to the sediment nearly perfectly. This probably results both from the consistent geochemical behavior of these elements and the fact that Si, Fe, and Ba are strongly linked to biogenic, hydrothermal, and dissolution residue sources, respectively.

Al abundances in most samples are accounted for very well by the model, although a number of samples have higher measured Al content than can be accounted for by the Al contribution from the 5 hypothesized sources (positive residuals). Most of the samples with positive Al residuals are from rise-crest locations, which suggests that these samples have basaltic detritus rather than intermediate detritus with the elemental ratios used in the model.

Samples from the Roggeveen Basin systematically have positive Al residuals and negative Mn residuals which are not easily explained. The Roggeveen Basin sediments have Si/Al values that lie below the detrital ratio of 3.0 used in the normative model (Table 3). Only the hydrogenous and the dissolution residue sources have Si-Al ratios less than 3.0 (2.7 and 0, respectively). The Ba content in Roggeveen Basin sediments is not sufficiently high to account for the excess Al by the dissolution residue source. Consequently, the normative model accounts for most of the Al by the detrital source. As a result, sufficient Fe is modeled as being of detrital origin that the hydrothermal source no longer accounts for all the Mn. Thus, the Roggeveen Basin residuals in Mn and Al suggest that either the Si/Al values for detritus are too low for this region or the Al/Ba values for dissolution residue are too low. X-ray diffraction of Roggeveen Basin samples indicate abundant goethite, plagioclase, smectite, fish debris, and minor amounts of quartz. The abundance of plagioclase, a silicate with low Si/Al and low Fe abundances, favors the low Si/Al detritus explanation. The effect of this perturbation to the

TABLE 4. COMPARISON BETWEEN NORMATIVE ANALYSIS WHICH USE DIFFERENT ELEMENTAL RATIOS FOR THE HYDROTHERMAL SOURCE

	MS454	MS487	MS489
Weight % of sources			
Detritus (1)	0	1.5	3.1
Detritus (2)	0	1.5	3.1
Hydrothermal (1)	19.1	23.7	29.3
Hydrothermal (2)	19.3	23.8	29.3
Biogenic (1)	75.3	71.8	63.4
Biogenic (2)	75.4	71.9	63.4
Hydrogenous (1)	.8	.4	0
Hydrogenous (2)	.4	.1	0
Dissolution residual (1)	4.8	3.6	4.2
Dissolution residual (2)	4.9	2.8	4.2
Residuals			
Al (1)	-641	0	0
Al (2)	-597	0	0
Si (1)	0	0	0
Si (2)	0	0	0
Mn (1)	-7,242	-8,294	-10,569
Mn (2)	0	122	-809
Fe (1)	0	0	0
Fe (2)	0	0	0
Ni (1)	4	0	-31
Ni (2)	18	0	-58
Cu (1)	0	30	58
Cu (2)	-91	-89	-102
Zn (1)	77	-3	-11
Zn (2)	56	-30	-46
Ba (1)	0	0	0
Ba (2)	0	0	0

Note: (1) refers to a normative analysis which uses Table 3 coefficients. (2) refers to a normative analysis with the following model coefficients: (Mn/Fe) = 0.18, (Ni/Fe) = 0.0012, $(Cu/Fe)_H$ = 0.006, $(Zn/Fe)_H$ = 0.0023.

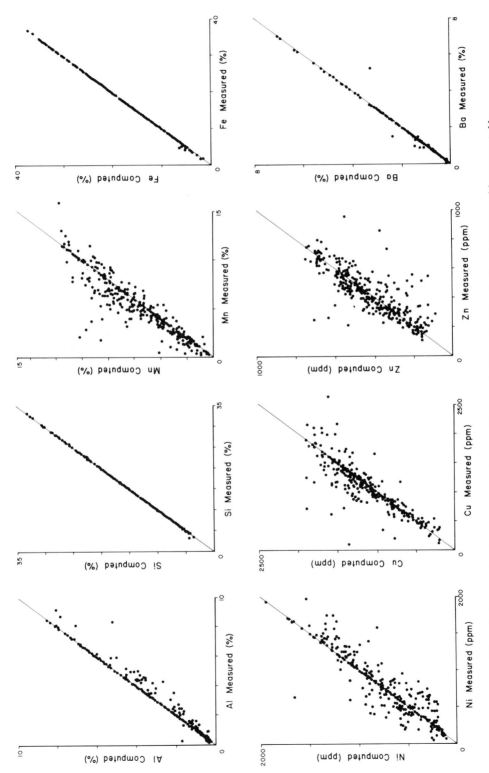

Figure 9. A comparison between measured composition on a CaCO₃ and salt-free basis and the composition computed by summing the 5 components in the normative model.

model is to underestimate the hydrothermal source slightly, and possibly overestimate the detrital source, depending on the actual Fe/Al in the Roggeveen Basin detritus.

Mn, Ni, Cu, and Zn are accounted for with less precision by the model. The Mn and Ni residuals indicated by the model for some samples may be caused by a diagenetic mobilization within the sediment, which can cause loss of Mn from the sediment or enrichment in an oxic surface layer. These effects are not considered by the model. The inability of the model to account precisely for the observed Cu and Ni abundances in the sediment probably reflects a breakdown in the constant composition assumption for these minor elements.

Distribution of the Normative Sediment Components

I have used the normative analysis of the entire surface-sediment data to construct contour maps of the contributions from the 5 hypothesized sources in the Nazca plate (Fig. 10–14). Although these maps have provided a quantitative estimate of the relative importance of various processes that introduce sediments to different parts of the plate, without sedimentation rate data they do not give the absolute variations in the 5 source contributions. Nonetheless, these maps are instructive, and further implications from available accumulation rate data will be discussed later in this paper.

Detrital Component. The detrital contribution increases strongly as a function of proximity to the

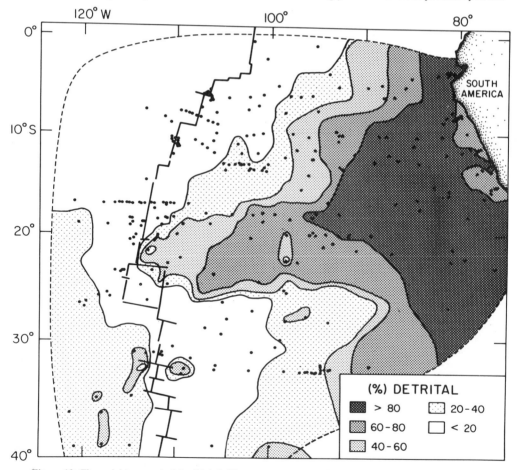

Figure 10. The weight percent of the "detrital" component on a carbonate and salt-free basis in the Nazca plate sediments.

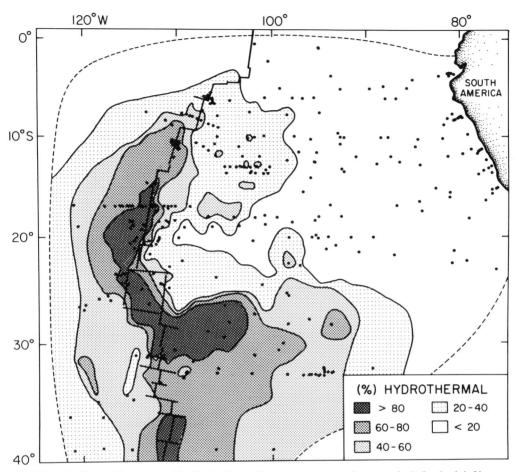

Figure 11. The weight percent of the "hydrothermal" component on a carbonate and salt-free basis in Nazca plate sediments.

South American Continent (Fig. 10). Except for the coastal upwelling region where biogenic dilution occurs, sediments from areas closer than 1,000 to 1,500 km are composed of greater than 80% detritus. A prominent feature of the detrital abundance pattern is the high detrital lobe that extends westward from South America and is centered between 20° and 25°S latitude. This lobe is in part a result of the north-south gradient in biogenic contribution which dilutes the detrital input to the north (Fig. 16); however, this explanation does not account for the intensity of the detrital lobe or the relatively low detrital input south of 25°S in the central Nazca plate.

An examination of the bathymetry of the plate (Mammerickx and Smith, 1978) reveals that the detrital lobe follows closely a 4,000-m contour that swings seaward and defines the Yupanqui Basin. In the region of 15° to 20°S, the extinct spreading center, the Galapagos Rise, is only weakly developed and provides less of a topographic barrier to near-bottom-sediment movement than it would to the north and south. To the south the Nazca Ridge and the Sala y Gomez Ridge are additional topographic barriers that may guide the seaward movement of detritus from the continent. The southern edge of the high detrital lobe falls very close to the northern edge of the Sala y Gomez Ridge. These observations suggest that the predominant mode of detrital input to the eastern Nazca plate is by way of near-bottom nepheloid-layer transport. Eolian input cannot be ruled out, however, because the high detrital lobe lies at the southern edge of the trades, and detrital phases could be transported from the South American deserts by the generally easterly winds.

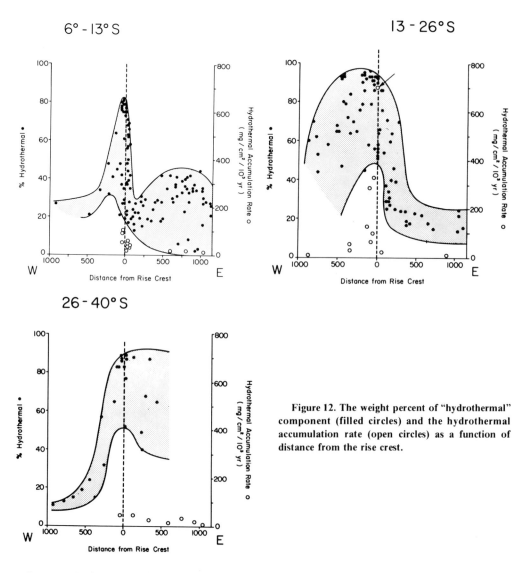

Figure 12. The weight percent of "hydrothermal" component (filled circles) and the hydrothermal accumulation rate (open circles) as a function of distance from the rise crest.

In general, rise-crest samples have detritus concentrations that are less than 10 wt%; however, scattered samples of high detrital content are found along the EPR. These samples represent local contribution of the basaltic basement exposed on the rise crest and fracture zones.

Hydrothermal Component. As expected, sediments with the highest hydrothermal contribution are found near the rise crest (Fig. 11). The pattern of the hydrothermal component around the rise crest reflects processes that disperse hydrothermal precipitates as well as possible hydrothermal activity at some distance from the rise crest. The center of the band of sediments with the highest hydrothermal concentration lies to the west of the rise crest between 6° and 26°S latitude (Fig. 12). This contrasts with the lobe of hydrothermally enriched sediments that extends to the east of the rise crest and into the Roggeveen Basin between 26° and 40°S latitude (Figs. 11, 12).

Distinguishing between hydrothermal enrichments that are caused by near-bottom-current dispersal and those that could result from off-rise hydrothermal activity has been discussed previously relative to Bauer Basin sediments which are composed of 20% to 40% hydrothermal contribution (Dymond and Veeh, 1975; Anderson and Halunen, 1974) and with regard to the sediments from the

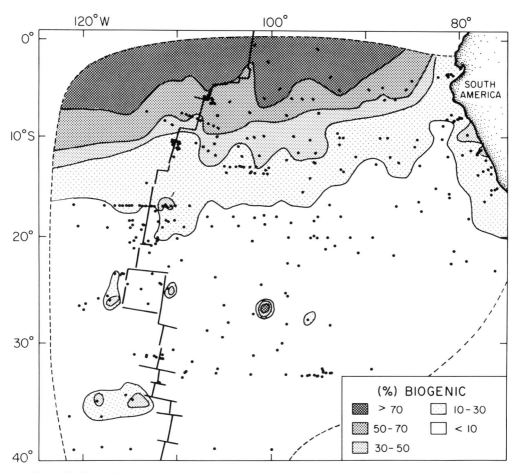

Figure 13. The weight percent of the "biogenic" component on a carbonate and salt-free basis in Nazca plate sediments.

EPR at 11°S (Heath and Dymond, this volume). For the Bauer Basin, downslope wafting of hydrothermal precipitates has been suggested to account for the metal-rich nature of these slowly accumulating sediments. This suggestion appears to be compatible with Lonsdale's (1976) suggestion that bottom water enters the Bauer Basin through the deeper transform faults on the EPR. Detailed sampling at 11°S on the rise crest, however, indicates a strong shift in the hydrothermal sediment to the west of the rise crest, which suggests westward-flowing bottom waters in this region. Bottom-water flow to the west is further supported by the westward shift in a "plume" of bottom water containing enrichments of ^{3}He in this area of the rise crest (Lupton and Craig, 1978). The scale of the westward displacement of hydrothermally rich sediments at 5° to 13°S and the eastward displacement at 25° to 32°S suggests that dispersal by water flowing over the rise crest and through fracture zones is the cause. On the basis of the hydrothermal sediment distribution alone, bottom water appears to be entering the Nazca plate to the south at the large fracture zones on the EPR between 30° and 35°S and probably along the Chile Rise, as suggested by Lonsdale (1976). This influx of bottom water carries hydrothermal precipitates eastward into the Roggeveen Basin. Water must exit from the Nazca plate westward at least in part along the 5° to 13°S portion of the rise crest, either through large fracture zones such as the Barret (13°S) or Wilkes (9°S) or by flow over the crest of the rise, which in the 5° to 13°S region has a very narrow region shallower than 3,000 m.

Biogenic Component. The biogenic content of the sediments follows basically a latitudinal pattern

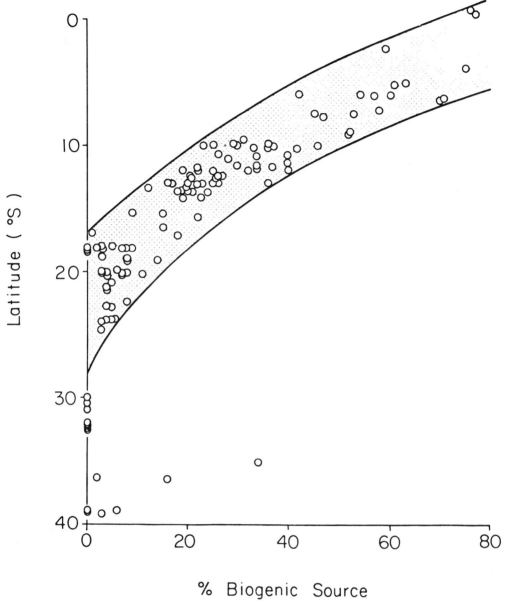

Figure 14. Weight percent of the biogenic component versus latitude for samples from the Bauer, Yupanqui, Roggeveen, and Southern Basins.

(Figs. 13, 14). Samples north of 5°S are composed of >70% biogenic material, and there is a systematic decrease in this component southward such that <10% of the sediments south of 10° to 15°S are of biogenic origin. The simple latitudinal pattern is disturbed slightly at the rise crest where the hydrothermal contribution is important and in the eastern Nazca plate where detrital sources and high coastal productivity occur. The generally latitudinal pattern breaks down in the region closer than ~1,500 km from the continent. The pattern here reflects dilution by detrital sources and enhanced biological inputs in response to high productivity in the coastal current and upwelling systems.

Hydrogenous Component. The hydrogenous content of Nazca plate sediments is greatest in the

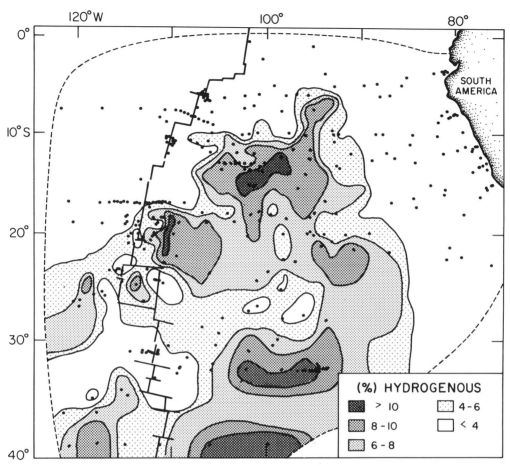

Figure 15. The weight percent of the "hydrogenous" component on a carbonate and salt-free basis in Nazca plate sediments.

basins that lie east of the rise crest (Fig. 15). It is hypothesized that this component is being deposited at a constant rate at any location on the sea floor (Krishnaswami, 1976). Consequently, its abundance in sediments would be controlled by the rate of accumulation of the other sources and would be inversely proportional to the total sedimentation rate. Although this concept cannot be rigorously tested without sedimentation rate data, the observed distribution pattern is consistent with a constant accumulation rate of hydrogenous material. Both the northern plate with high biogenic accumulations and the eastern plate in proximity to detrital sources have sediments that contain low abundances of the hydrogenous component. The various basins, however, have sediments composed of greater than 10% hydrogenous component. Measured sedimentation rates in the Bauer Basin are less than 1mm/1,000 yr (Dymond and Veeh, 1975; McMurtry and others, this volume). Sedimentation rate data do not exist for the western Yupanqui Basin. Because of its location east of the rise crest, in the central gyre, and a great distance from the continent (Fig. 1), it is probable that the noncarbonate accumulation rates are among the lowest in the oceans. The locations of the other basins with high content of hydrogenous sediments are consistent with low bulk sedimentation rates.

Dissolution Residue. Nazca plate sediments having the highest content of dissolution residue component occur between 10° and 22°S (Figs. 16, 17). The rate of accumulation of this component, which is viewed as the remains of dissolution and decomposition of tests and organic debris, should be high in areas of high productivity. The more rapid burial of biogenic particles in areas of high productivity decreases the time biogenic material is in contact with oxidative and corrosive bottom waters. Consequently, a smaller fraction of the biogenic particles reaching the bottom in productive areas would undergo dissolution, and the proportion of the dissolution residue component relative to the biogenic component would be lower near the equator. The 10° to 22°S region lies on the southern edge of the high productivity region, and because of the slower deposition rates, a larger proportion of biogenic particles are dissolved and/or decomposed forming dissolution residue. This effect, along with dilution by detritus, can account in a qualitative way for the maximum in the dissolution residue between 10° and 22°S observed in Fig. 17.

Estimates of the fraction of opal that has undergone dissolution can be made from the fraction of dissolution residue computed by the normative model. The biogenic source is assumed to have a Ba/Si value of 0.002 (Table 3); therefore, the biogenic Si that has been lost to dissolution, Si_{DR}, is given by the equation

$$Si_{DR} = Ba_{DR}/0.002$$

where Ba_{DR} is the dissolution residue Ba computed from the model. And

$$\text{Fraction of opal preserved} = \frac{Si_B}{Si_B + Si_{DR}}$$

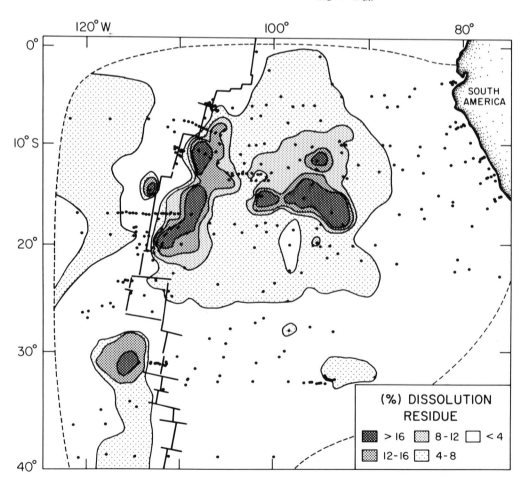

Figure 16. The weight percent of the "dissolution residue" component on a carbonate and salt-free basis in Nazca plate sediments.

where Si_B is the Si from the biogenic source computed by the model. Assuming that the Si and Ba proportions in biogenic material are constant, this calculation is valid regardless of what the specific carriers of Ba are. Although the good correlation of Ba and SiO_2 in Geosecs profiles (Bacon and Edmond, 1972) suggest that Ba is being carried by opaline tests, recent studies suggest it is associated with refractory organic material (Gurvich and others, 1978; Suess, in prep.). Dehairs and others (1980) observed discrete barite particles of biogenic origin in suspended matter. If these barite particles fall to the sea floor free of associated opal, the above opal dissolution calculation does not refer strictly to dissolution on the sea floor. Also, if significant Ba dissolves along with the opal upon reaching the sea floor, the computed fraction of opal preserved will be too large.

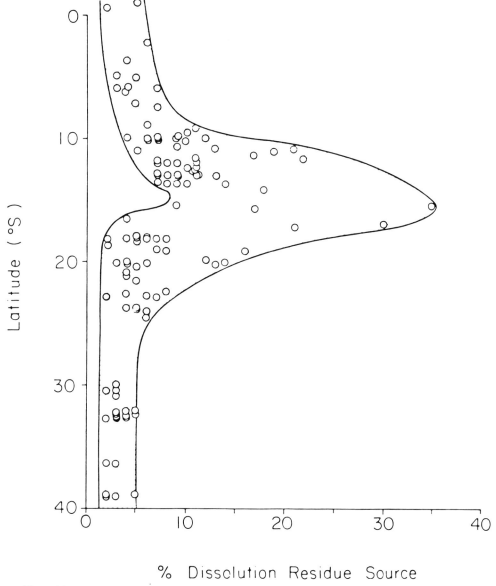

Figure 17. Weight percent of the dissolution residue component versus latitude for samples from the Bauer, Yupanqui, Roggeveen, and Southern Basins.

In Figure 18 I have plotted the percentage of opal preserved for all Nazca plate basin samples as a function of latitude. This plot shows that greater than 90% of the biogenic opal dissolves even in samples underlying the equatorial high productivity zone. Consequently, even for this region, the predominant imprint of biological productivity on the elemental abundance patterns of sediments will be made by the dissolution residue and not by the preserved biogenic debris. The smooth decrease in the fraction preserved away from the equator is caused by more rapid burial of biogenic debris in regions of high biogenic sedimentation near the equator.

Accumulation Rates of Sediment Sources

Sedimentation rates for a limited number of cores from the Nazca plate have been measured using a variety of techniques, and the existing data have been summarized in McMurtry and others (this volume), Heath and Dymond (this volume), and Moser (1980). Rate data are available for 39 of the samples analyzed in this study, and these data have been converted to accumulation rates of the 5 sediment sources (Table 5). These data, although fewer than the compositional data, provide additional insight to the geochemical processes operating on the Nazca plate.

Detrital Rates. Figure 19 summarizes the detrital rates according to their locations on the rise crest (within 100 km) or basins and flanks of the EPR. The rise crest has detrital accumulation rates that are

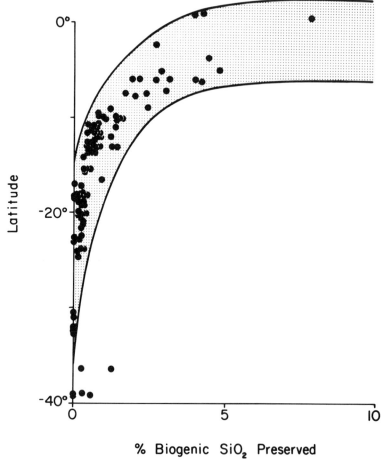

Figure 18. Weight percent of biogenic SiO$_2$ which is preserved as a function of latitude.

TABLE 5. ACCUMULATION RATE DATA

	Lat (S)	Long (W)	Sedimentation rate (cm/10³yr)	Dry density (g/cm³)	Component rates (mg·cm⁻²·10⁻³yr⁻¹)				
					Det	Hyd	Bio	Auth	DR
Rise crest, N									
KK71-PC86*	5°48'	107°01'	1.8	0.49	2.1	51	150	0.6	6.4
KK71-PC88*	6°06'	106°43'	1.2	0.54	2.7	19	65	0.4	3.7
KK71-PC90*	6°11'	106°36'	1.6	0.59	1.0	42	47	0	5.7
KK71-GC14*	10°37'	110°30'	0.62	0.71	0	97	20	3.7	1.2
471-44†	10°42'	110°01'	1.23	0.66	35	54	0	2.8	2.8
471-52P†	10°43'	110°42'	0.71	0.58	0	91	22	6.1	1.2
471-53MG†	10°53	110°44	1.3	0.58	0	185	33	14	2.3
Y71-45P§	11°05'	110°06'	0.93	0.71	0	30	11	15	4.2
Y71-46MG†	11°10'	110°05	0.82	0.72	0.1	35	11	1.5	2.5
Y71-47MG†	11°18'	110°06'	0.76	0.73	1.7	27	9.5	0.82	2.1
Y71-48P†	11°20'	110°11'	1.2	0.69	0	29	14	0	3.2
KK71-FFC109#	12°05'	110°37'	1.1	0.67	2.4	94	21	3.7	2.4
Rise crest, S									
V19-53**	17°01'	113°31'	2.1	0.46	0	329	15	22	3.7
V19-54††	17°02'	113°54'	1.5	0.52	0	290	8.9	1.5	1.1
OC73-20MG**	19°15'	113°34'	3.4	0.41	7.6	700	15	46	0.6
OC73-25MG**	20°00'	114°31'	0.39	0.69	1.0	65	0.6	2.8	0.6
OC73-16P**	20°19'	113°13'	0.40	0.86	2.9	22	0	2.6	1.4
OC73-27MG**	21°01'	114°44'	0.61	0.51	1.1	98	0.5	7.5	0.6
Y73-40MG**	31°07'	112°26'	0.29	0.79	2.4	43	0	1.4	1.0
Basin and flanks, N									
KK71-GC10§	9°59'	106°02'	0.43	0.86	1.4	4.4	6.9	0.6	1.8
Y71-36MG§	10°08'	102°51'	0.14	0.22	4.7	10	8.1	0.1	1.5
KK71-FFC132#	11°33'	97°25'	0.20	0.49	6.9	4.6	7.8	1.2	2.8
KK71-FFC115#	11°58'	103°21'	0.21	0.23	8.8	14	9.6	3.0	3.1
Y73-13MG§§	12°59'	101°33'	0.11	0.22	4.1	5.5	3.6	1.3	1.2
Y73-21K*	13°37'	102°32'	0.13	0.28	7.7	10	5.6	3.9	2.1
V19-41**	14°06'	96°12'	1.67	0.82	11	4.3	7.5	3.9	5.9
Basin and flanks, S									
V19-64††	16°56'	121°12'	0.45	0.72	2.7	10	1.0	1.2	1.7
V19-61††	16°57'	116°18'	0.70	0.72	1.9	30	14	0.1	1.4
V19-55**	17°00'	114°11'	1.11	0.54	0	130	46	0	1.8
OC73-8MG**	18°00'	102°01'	0.16	0.77	47	8.7	2.0	6.1	4.0
OC73-6P**	20°03'	95°18'	0.21	0.20	29	2.9	1.1	2.1	1.5
RC9-99**	24°36'	115°27'	0.70	0.70	3.3	55	0	3.9	3.3
RC8-94**	27°17'	102°05'	0.18	0.78	0.8	6.4	0	0.3	0.3
RC8-93**	29°22'	105°14'	0.35	0.67	1.5	34	0	1.9	0.8
OC73-30MG**	30°30'	110°29'	0.28	0.70	2.4	43	0	2.5	1.0
Y73-34P**	31°09'	113°29'	0.39	0.97	280	24	0	0	0
RC8-92**	31°33'	108°30'	0.29	0.69	1.4	25	0	1.1	1.0
Y73-55MG**	32°38'	105°55'	0.15	0.73	2.7	14	0	1.1	0.8
Y73-86**	32°39'	102°16'	0.15	0.58	4.5	16	0	2.6	0.7
Y73-64**	32°46'	94°42'	0.43	0.43	37	54	0	10	3.1

Note: Column headings under component rates are as follows: Det = detrital, Hyd = hydrothermal, Bio = biogenic, Auth = hydrogenous, DR = dissolution residues.
*Source: McMurtry and others, this volume.
†Source: Heath and Dymond, this volume.
§Source: Dymond and Veeh, 1975.
#Source: McMurtry and Burnett, 1975.
**Source: Moser, 1980.
††Source: Bender and others, 1971.
§§Source: Dymond and others, 1977.

less than 1 mg · cm⁻² · 10⁻³ yr⁻¹; however, 2 samples have higher values which probably reflect local basaltic sources. Basin and flank samples have generally greater rates of detrital accumulation. Two of the highest rates (29 and 47 mg · cm⁻² · 10⁻³ yr⁻¹) come from the Yupanqui Basin, confirming that the high detrital abundances in this region are not due to low inputs by the other sources. For comparison,

4 samples from the Bauer Basin, approximately the same longitude as the Yupanqui Basin samples, have an average detrital accumulation rate of 6.3 ± 3.3 mg $\cdot$ cm^{-2} $\cdot$ 10^{-3} yr^{-1}.

Hydrothermal Rates. The data for hydrothermal rates (Fig. 19) have been separated into those samples north or south of 13°S latitude to examine any variations in the hydrothermal output from these two segments of the rise crest. The highest hydrothermal rates are found on the southern rise crest where values greater than 100 mg $\cdot$ cm^{-2} $\cdot$ 10^{-3} yr^{-1} are common, and one sample, OC73-20Mg, has a hydrothermal rate of approximately 700 mg $\cdot$ cm^{-2} $\cdot$ 10^{-3} yr^{-1}. An exponential increase in hydrothermal rates toward 20°S has been suggested by McMurtry and others (this volume) who found the highest Mn and Fe accumulation rates in this region of the rise crest.

It should be noted, however, that the accumulation rates on both the northern and southern parts of the spreading center are highly variable (Table 5; Figs. 12, 19). Consequently, there is no significant statistical difference in rates between the northern and southern parts of the rise crest. A "run test" (Tate and Clelland, 1957) indicated that the two populations are not different at the 0.2 significance level. Thus, there may not be a difference in the rate of hydrothermal input to the two areas, and the high values noted in the southern rise crest may reflect sampling that, by chance, was close to a hydrothermal vent. It should also be recognized that the existing sampling is very coarse and broad-scale relative to the probable locations of hydrothermal vents. The locations of the hot springs studied on the Galapagos spreading center and 21°N on the EPR suggest that active hydrothermal vents are at the very crest of the spreading center, the locus for the most recent volcanism (Corliss and others, 1979). Because the sediment cover is too thin for conventional coring techniques, very few of our samples are located closer than 10 km to the spreading center. While the distance factor may have resulted in homogenizing the hydrothermal precipitates from the vents, possibly accounting for their compositional uniformity, it also biases sampling toward depressions where ponding effects are important. Thus, the strong variability in hydrothermal rates may reflect variations in proximity to vents, topographic control, and errors in the sedimentation rate measurement. Uncertainty in the latter should not be underestimated. For example, sedimentation rate for core OC73-20P, which has the highest hydrothermal rate recorded, has been measured by ^{230}Th dating as 1.94 cm/10^3 yr (McMurtry and others, this volume) and by thickness of sediment and basement age as 3.35 cm/10^3 yr (Moser, 1980).

The hydrothermal rate data also support the concept that hydrothermal precipitates are accumulating more rapidly to the west of the rise crest in the northern part of the spreading center as the decrease in hydrothermal rate is much more gradual to the west than it is to the east (Fig. 12). There is a suggestion that hydrothermal products are being transported to the east in the 26° to 40°S region; however, the lack of accumulation rate data to the west of the crest prevents an unequivocal test of this question.

Hydrothermal rates are generally low for the basins except in the southern plate where hydrothermal material appears to be transported to the west (Fig. 19).

Biogenic Accumulation Rates. The biogenic accumulation rates decrease sharply with increasing latitude so that essentially no biogenic material is preserved south of 20° latitude (Fig. 20). In addition, for a given latitude, the biogenic accumulation rate is greater on the rise crest than it is in the flanks and basins (Figs. 19, 20). This is a preservation effect caused by the greater sedimentation rate on the rise crest. Because both CaCO$_3$ and hydrothermal precipitates are accumulating relatively rapidly on the shallow and volcanically active spreading centers, biogenic silica is buried more quickly, decreasing the fraction of biogenic matter which is dissolved.

Hydrogenous Accumulation Rates. Hydrogenous accumulation rates are small compared to the inputs from detrital, biogenic, and hydrothermal sources (Fig. 19). Also, the hydrogenous rate is relatively uniform except for a few anomalous samples that have very high rates. I suspect these high rates are not valid and result from a breakdown in the constant composition assumption in the normative model, which causes unrealistically high hydrogenous abundance estimates in the sediment. Most of these unusually high hydrogenous accumulation rates occur on the rise crest and probably reflect small variations in the elemental ratios in the hydrothermal precipitates at these locations. Variations in the hydrothermal Fe/Mn ratio, for example, have little effect on the fraction

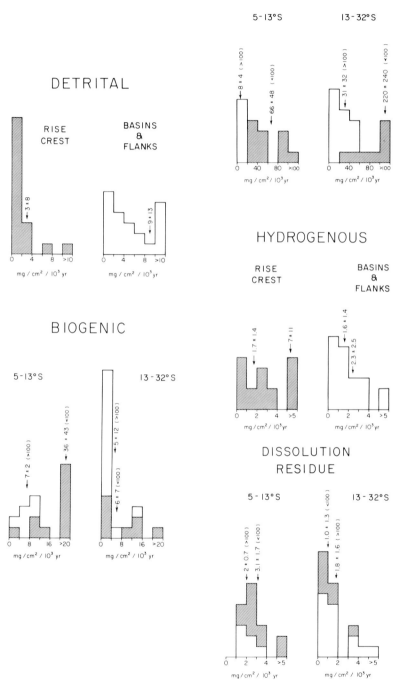

Figure 19. Histograms of accumulation rates of the 5 components. Shaded areas are samples less than 100 km from the rise crest. Means and standard deviations are indicated by arrows and adjacent numbers. The two values indicated for the hydrogenous rates are those omitting and including the anomalously high values.

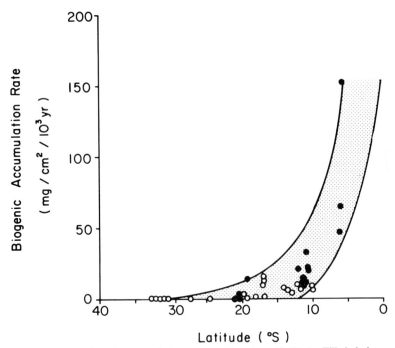

Figure 20. Biogenic accumulation rate as a function of latitude. Filled circles refer to samples within 100 km of the rise crest.

of hydrothermal source computed but can produce a 2 to 4 times variation in the hydrogenous content (Table 4).

Ignoring these few anomalous values, the rise-crest and basin hydrogenous accumulation rates are nearly the same (1.7 ± 1.4 and 1.6 ± 1.4, respectively). These rates are equivalent to a deposition of approximately 0.5 mg · cm^{-2} · 10^{-3} yr^{-1} of hydrogenous Mn and 0.25 mg · cm^{-2} · 10^{-3} yr^{-1} of hydrogenous Fe. These values fall within the range of hydrogenous Mn and Fe deposition rates suggested by Krishnaswami (1976) for the deep sea and provide futher support for the normative model.

Dissolution Residue Rates. The rate of accumulation of the dissolution residue source decreases with increasing latitude in response to the decrease in productivity away from the equator (Fig. 19). However, because the rate of accumulation of this source does not decrease as rapidly to the south as does the biogenic rate and becomes relatively constant south of 20° latitude (Figs. 20, 21), the maximum in dissolution residue content between 15°S and 25°S noted in Figures 16 and 17 and can be accounted for.

Elemental Distribution Patterns

To evaluate the elemental distribution patterns on the Nazca plate, I have computed the proportion of each element that is introduced via the 5 sources of the normative model. For this purpose all samples have been assigned to four regions of the Nazca plate: (1) the rise crest, ±100 km from the spreading center; (2) the northern Nazca plate, non-rise crest samples north of 10°S latitude; (3) the eastern Nazca plate, samples east of 90°W longitude; and (4) the basin samples, all other samples. For each area the contribution of the elements Al, Si, Mn, Fe, Ni, Cu, Zn, and Ba from the 5 hypothesized sources were averaged and the results plotted in Figure 22.

Aluminum. The Al distribution in deep-sea sediments has generally been considered to be controlled by detrital sources. My data support this general concept for the Nazca plate, since the highest Al

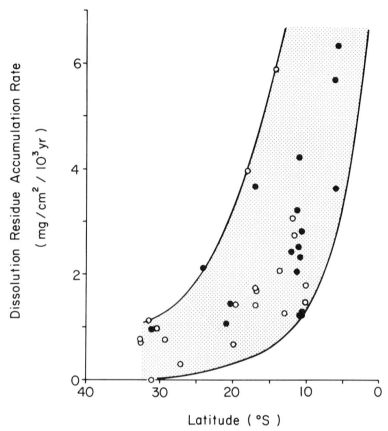

Figure 21. The accumulation rate of the dissolution residue versus latitude. Filled circles refer to samples within 100 km of the rise crest.

abundances are found near the South American Continent (Fig. 4) and detrital Al is the largest source in the eastern Nazca plate (Fig. 22). Within individual areas, however, the dissolution residue source is significant, accounting for 35% of the Al in the northern Nazca plate, 22% of the Al in the basins, and 39% of the Al at the rise crest. At the rise crest the hydrothermal source of Al is significant as well. Hydrothermal Al has been previously suggested on the basis of the enhanced accumulation of Al at spreading centers (Dymond and others, 1977). Also, corundum has been observed as a precipitate around the hydrothermal vents in the 21°N area of the EPR (Haymon and others, 1979). The likelihood that hydrothermal systems are a source for Al is further supplied by recent laboratory hydrothermal experiments (Seyfried and Bischoff, 1977) in which approximately 0.2 ppm of Al was measured in solution. It is of interest that this experiment with a high water-to-rock ratio and at 260 °C has very nearly the same Fe/Al value (~20) as that used in the normative analysis (Fe/Al = 16.7; Table 3).

Silicon. Si distribution in Nazca plate sediments is controlled primarily by biogenic, detrital, and hydrothermal sources (Fig. 22). As expected, the biogenic source is most important in the northern Nazca plate where the normative model suggests that 62% of the Si is in the form of biogenic opal. Detrital silica sources predominate in the basins and eastern Nazca plate. On the rise crest biogenic, detrital, and hydrothermal sources are all significant contributors to the observed silica content. The high Si/Al values for rise-crest samples compared to basin samples from the same latitude (Fig. 23) result both from preferential preservation of biogenic opal and the abundant hydrothermal input with a Si/Al value of approximately 22 (Table 3).

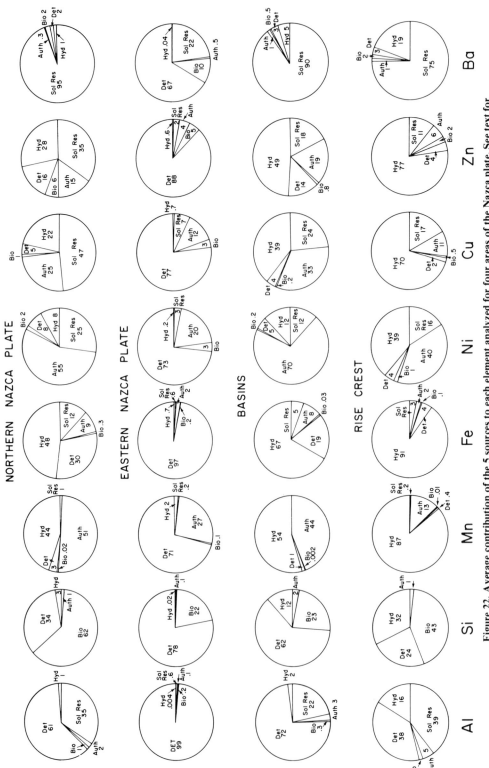

Figure 22. Average contribution of the 5 sources to each element analyzed for four areas of the Nazca plate. See text for location of the areas.

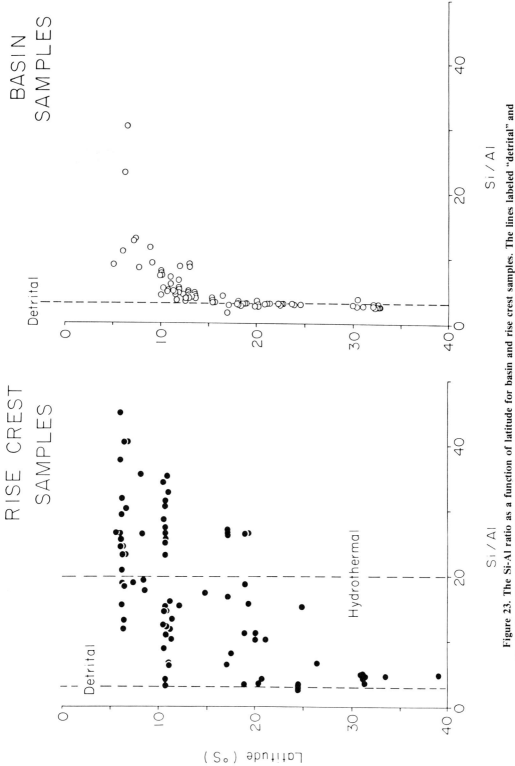

Figure 23. The Si-Al ratio as a function of latitude for basin and rise crest samples. The lines labeled "detrital" and "hydrothermal" refer to the Si-Al values of these components used in the normative model.

Manganese. Except for the eastern Nazca plate where detrital inputs so greatly predominate, the Mn distribution is largely controlled by hydrothermal and hydrogenous contributions (Fig. 22). The importance of transport and dispersal of hydrothermal precipitates is indicated by the fact that even in the basins, which are hundreds of kilometres from a rise-crest hydrothermal source, hydrothermal Mn predominates hydrogenous Mn (54% versus 44%, respectively).

Iron. The Fe abundance in the eastern Nazca plate can be explained almost exclusively as detrital contributions. Elsewhere on the plate, however, hydrothermal sources are predominant. The normative model suggests that in the northern plate and basins there are important contributions of detrital Fe as well as significant inputs of Fe by dissolution residue and hydrogenous sources.

Nickel. As seen in Figure 5, the greatest Ni abundance occurs in the basins, particularly the Bauer and western Yupanqui Basins. Because these areas accumulate sediments at very low rates, it is reasonable to suspect the high Ni concentrations result from the relatively greater contribution of hydrogenous Ni bearing ferromanganese micronodules and coatings in these basins. The normative model suggests that hydrogenous Ni is the most important source in all areas except in the eastern Nazca plate. In the basins hydrogenous Ni accounts for 70% of the Ni flux to the sediment (Fig. 22). Hydrothermal Ni is important only for rise-crest samples and basin samples, where this source contributes an average of 39% and 12%, respectively, of the observed Ni concentration in the sediment.

Copper. The copper distribution shows a complex pattern with the highest values occurring along the rise crest and the basins to the east and west. This pattern reflects relatively high concentrations of this element from hydrothermal sources (1,460 ppm), hydrogenous sources (8,350 ppm), and the dissolution residue source (4,320 ppm). Except for the eastern plate where the detrital contribution is predominant, these 3 sources provide most of the copper preserved in the sediment. In the northern plate the dissolution residue source provides the largest share of Cu (47%). This is consistent with the high productivity and probable increased rate of biogenic matter dissolution in this region. In rise-crest sediment and basins, however, the hydrothermal source predominates, providing 70% and 39% of the Cu, respectively.

Zinc. Zn and Cu distributions in the sediments have similar patterns, reflecting similar chemistries that result in enrichments in the hydrothermal (660 ppm), hydrogenous (1,000 ppm), and dissolution residue sources (1,080 ppm). Again, the dissolution residue source predominates in the northern plate (35%), while the hydrothermal sources predominate in the rise-crest samples (77%) and basin samples (49%).

Barium. The distribution of Ba on the Nazca plate indicates the greatest abundances are found on the flanks of the EPR and into the Bauer Basin, between 10° and 20°S latitude. Previous studies have suggested that Ba is introduced into the sediments either by biogenic particles (Goldberg and Arrhenius, 1958; Church, 1970) or by hydrothermal processes (Arrhenius and Bonatti, 1965). The biogenic association of Ba is supported by the 25× greater BaO/TiO_2 in sediments at the equatorial high productive zone compared to higher latitude samples and the relatively high Ba contents in some marine organisms (Arrhenius, 1963; Thompson and Bowen, 1969). The hydrothermal origin of Ba in sediments is indicated by high Ba content in EPR sediments (Arrhenius and Bonatti, 1965) and by the Ba-rich (up to 6%) Mn oxide deposits in the Afar Rift (Bonatti and others, 1972). In addition, recent laboratory hydrothermal experiments show that Ba can readily be leached from basalts by heated sea water (Seyfried and Bischoff, 1977; Mottl and Holland, 1978). The Galapagos thermal springs also have strong enrichments of Ba (Edmond and others, 1979).

The Nazca plate sediment data do not support a predominantly hydrothermal origin for Ba. On the basis of the Ba concentration of the Fe-rich rise-crest sediments, we have used a Ba concentration in the normative model of 1,740 ppm for the hydrothermal end member. Although Arrhenius and Bonatti (1965) suggested the hydrothermal origin for the Ba on the basis of Ba-rich sediments along the EPR, a close inspection of their data indicates that the high-Ba sediments are mostly on the eastern flanks of the EPR, which is very similar to my observations. Church (1970) also found that in a traverse across the EPR at 15°S, the highest Ba abundances on a carbonate-free basis occurred well away from the rise crest.

The normative model indicates that even in rise-crest samples 75% of the Ba is introduced by way of the dissolution residue, and 90% and 95% of the Ba in the northern plate and basin samples, respectively, come from this source. Because of the relatively low abundances of Ba in biogenic matter, detritus, and ferromanganese phases, these sources can account for only a small fraction of the observed Ba in all samples except the very detrital-rich samples from the eastern Nazca plate.

HYDROTHERMAL SOURCE AND NATURE OF SEA-WATER HYDROTHERMAL SYSTEMS

With regard to the origin of hydrothermal sediments on the Nazca plate, it is important to understand the degree to which hydrothermal deposits reflect initial variations in hydrothermal solution composition or varying conditions of mixing between hydrothermal fluids and ambient sea water. Laboratory experimental studies demonstrate that primary hydrothermal solution compositions can vary widely in response to variations in the conditions of water and rock interaction. Edmond and others (1979a, 1979b) and Corliss and others (1979) have emphasized, however, that mixing and precipitation processes can produce a broad range of solution and solid-phase compositions from a single, unmixed hydrothermal solution composition.

Analyses of deposits forming from sea-water hydrothermal systems suggest that their chemical and mineralogical composition can vary greatly. Examples of the type of deposits believed to form from sea-water hydrothermal systems are as follows:

1. Oxyhydroxides of Fe and Mn, which vary from predominantly Fe-rich as in "goethite amorphous" underlying the Red Sea brines (Bischoff, 1969), the precipitates from shallow-water Mediterranean hot springs (Bonatti and others, 1971), the amorphous Fe crusts associated with the Galapagos "mounds" (Corliss and others, 1978), the umbers overlying the basaltic crust in the Cyprus ophiolite (Robertson and Hudson, 1973), and the Fe crusts dredged from seamounts (Bonatti and Joensuu, 1966; Piper and others, 1975) to predominantly Mn-rich, as in the Trans-Atlantic Geotraverse (TAG) deposits from the Mid-Atlantic Ridge (Scott and others, 1974), the Galapagos "mounds" crusts (Corliss and others, 1978), and the Mn crusts from near the Galapagos spreading center–EPR triple junction (Moore and Vogt, 1975);

2. Nontronite or iron-montmorillonite as found in the Red Sea brine deposits (Bischoff, 1969), the "mounds" deposits near the Galapagos spreading center (Corliss and others, 1978), and in the Gulf of Aden (Cann and others, 1977). These deposits have very low abundances of Mn and other transition elements, being predominantly low-Al iron silicates;

3. Sulfide deposits, which are predominantly iron sulfides as found in Red Sea brine deposits (Bischoff, 1969), and the 21°N region of the EPR (Haymon and others, 1979). In addition, the massive sulfides observed within the basaltic crust of ophiolites have an unquestionable sea-water hydrothermal origin.

The interaction of sea water with hot basalt results in a solution that is reducing and acidic as a result of the utilization of the primary oxidants in sea water, O_2 and $SO_4^=$, and the removal of Mg^{++} and OH^- during the formation of Mg-rich secondary minerals (Bischoff and Dickson, 1975). The ability of sea-water hydrothermal solutions to leach elements from the oceanic crust is related to the temperature, pH, and reducing nature of the solutions. Laboratory experimental studies (Seyfried and Bischoff, 1977, 1981; Hajash and Archer, 1980) have shown that the water-rock ratio is important in determining the pH of hydrothermal solutions. Under sea-water–dominated conditions the continual formation of Mg-bearing secondary minerals is possible since Mg is never totally depleted. Consequently, a low pH is maintained. Under rock-dominated conditions, Mg in the fluid may be quantitatively removed, and the pH will rise as a result of hydrolysis reactions, reducing the ability of the solution to carry metals. Also, for a given water-rock ratio, the laboratory studies suggest that temperature may be an important variable in determining the relative proportion of transition metals. For example, the Fe/Mn ratio changes from approximately 1:1 at 350 °C to 5:1 at 500 °C (Bischoff and Dickson, 1975; Seyfried and Bischoff, 1977, and Mottl and Holland, 1978).

Thus, there may be primary differences in the composition of hydrothermal fluid that reflect the age of the hydrothermal system, the temperature of rock-fluid interaction, and the permeability of the oceanic crust. These primary differences in fluid composition are modified by mineral precipitation which can take place through both subsea-floor mixing between hydrothermal fluid and cooler oxygenated sea water and venting of hydrothermal fluids into the bottom waters. At one end of the spectrum, hot unmixed solutions may vent directly into the bottom water and result in the formation of sulfides on the sea floor like those observed at 21°N on the EPR (Hekinian and others, 1980; Haymon and others, 1979). At the other end of the spectrum, extensive subsea-floor mixing of hydrothermal fluid results in the precipitation of Fe as sulfides, nontronite, and hydroxyoxides prior to exiting. The diluted warm fluids venting into sea water may carry only Mn as the predominant transition metal, which precipitates in the form of the very pure todorokite and birnessite crusts observed in areas like the Galapagos "mounds" and the TAG area.

Nontronite or the amorphous iron hydroxyoxides may represent an intermediate state of mixing between sea water and hydrothermal fluid. Experimental studies of Harder (1976) suggested that for nontronite formation, slightly acid, mildly reducing conditions with moderate concentrations of Fe and Si in solution are required (4 to 7.5 ppm of Fe, 20 ppm of SiO_2, and $Eh = -0.1$ to -0.8). With higher concentrations of Fe and Si and more oxidizing conditions, Harder found goethite and quartz precipitates instead of nontronite. The Galapagos mounds deposits are perhaps one example of the range of deposits produced by variation in subsea-floor mixing. A common sequence observed during *Alvin* dives to the mounds is a surficial layer of Mn-rich crust underlain by an amorphous, iron-rich layer with significant silica content, underlain by Al-poor nontronite (Dymond and others, 1980; Williams and others, 1980). The diffuse hydrothermal flow through the sediment establishes a steady-state redox-gradient that precipitates nontronite at depth in the sediment, Fe hydroxides at shallow depths, and Mn phases near the sediment water interface (Corliss and others, 1978).

The variety of factors discussed above and the observed variations in other hydrothermal deposits suggest that the most surprising aspect of the Nazca plate deposits is their uniformity in composition. In Figure 24 I have plotted the relative contents of Fe, Mn, and Si of deposits from a number of hypothesized sea-water hydrothermal systems. The great variability in composition reflects the well-known ease of separation of Fe and Mn by sedimentary processes (Krauskopf, 1957) as well as the factors discussed above which cause formation of nontronite as opposed to Fe oxyhydroxides. The range in composition exhibited by Nazca plate sediments is largely explained by variations in the abundance of biogenic opal, as there is a systematic decrease in Si content away from the equatorial high productivity zone. Note also that basal metalliferous sediments recovered by deep-drilling in the Pacific have Fe-Mn-Si abundances similar to their modern analogs. This suggests that similar processes of formation occurred 30 to 60 m.y. ago (Fig. 28; Dymond and others, 1973, 1977). Significant variations in the Fe/Mn values for sediments from different sections of the rise crest are apparent only for samples from 5° to 6°S (Fig. 8), and these variations are small compared to the total range observed in sea-water hydrothermal deposits. Fe/Mn variations in the sediments as a function of distance from the rise crest are not apparent either (Fig. 25). Consequently, existing data on Nazca plate metalliferous sediments indicate that compositional variability through time and space is surprisingly small, considering the processes that could affect the initial composition of the hydrothermal fluid and the variety of fractionating precipitating reactions that can occur both beneath the sea floor and upon venting into bottom waters.

Moreover, the observed mineralization and water compositions of the two deep-sea hydrothermal systems that have been directly studied by submersible, the Galapagos spreading center vents (Corliss and others, 1979) and the 21°N EPR vents (Edmond and others, 1979a, 1979b: Haymon and others, 1979), suggest that these systems may not be representative of the hydrothermal system that produces the metalliferous sediments distributed over the Nazca plate. Sulfide and nontronite precipitation prior to venting is indicated for the Galapagos area by the presence of pyrite and saponite particles in the venting waters (Corliss and others, 1979), the Cu, Ni, and Cd content less than ambient sea water, and the relatively low and highly variable Fe/Mn values (~0.01 to 0.7, Edmond and others, 1979b). Only the waters from the Clambake vent of the Galapagos hydrothermal area carry significant Fe (Edmond and others, 1979b). Preferential removal of Fe in the immediate vicinity of the Clambake

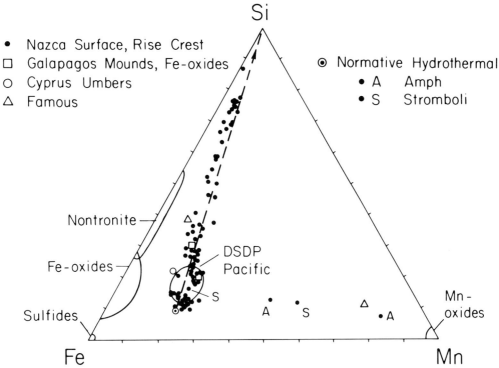

Figure 24. Si-Mn-Fe abundances of hydrothermal deposits. Data from Dymond and others (1973), Corliss and others (1978), Toth (1980), Robertson and Hudson (1973), Bonatti and others (1971), Bischoff (1969), Cann and others (1977), Piper and others (1975), and Scott and others (1974).

vent could produce a precipitate with an Fe-Mn ratio of 3 even though the venting solution had a ratio of 0.5 to 0.7. However, precipitation would have to occur very near the vent because Fe values in the hydrothermal water drop to near or less than ambient sea-water values within a few metres of the vent. Because Mn from these vents behaves as a more or less conservative element with mixing (Edmond and others, 1979b; Klinkhammer and others, 1977), the Fe/Mn values of any precipitates would be expected to decrease very sharply away from the hydrothermal vents. This is not observed in Nazca plate sediments (Fig. 25) and has not been tested in the Galapagos speading center area.

Although complete compositional data for the hot springs at 21°N are not available, reported end-member compositions have Fe-Mn and Fe-Cu ratios similar to Nazca plate metalliferous sediments (Edmond, 1980). Much of the Fe, however, appears to precipitate as sulfide chimneys (Haymon and others, 1979; Hekinian and others, 1980). Edmond and others (1979b) suggested that this system represents a high temperature end member in which the fluids exit with high velocity, undergoing little mixing before venting and producing a heterogeneous deposit of sulfides, native sulfur, and opal near the vent. Although sulfides are not observed in the oxidized metalliferous sediments of the Nazca plate, oxidation of dispersed sulfide particles and absorption of Mn on the surfaces could possibly result in a deposit with the composition of metalliferous sediments. Analyses of alteration products of the sulfides ("gossan") at the 21°N area, however, have no detectable Mn (Hekinian and others, 1980). Alternatively, ferromanganese oxyhydroxides with approximately 3:1 Fe-Mn ratios may precipitate at some distance from the sulfides at the vents. Unfortunately, analyses of sediments and precipitates away from the vents are not available to test this possibility.

Consequently, it appears that the Galapagos hydrothermal system is near the heavily mixed, low-flow end member and is not producing significant quantities of an Fe-Mn oxyhydroxide on the sea floor. The 21°N area is perhaps at the other extreme, where much of the Fe precipitates immediately at the vent as sulfide. Despite the strong differences in these two observed systems, it appears that precipitation from neither system can explain the relatively uniform composition of Nazca plate metalliferous sediments.

Two explanations are offered for consideration: (1) The metalliferous sediments on the Nazca plate represent a physical homogenization of the full range of precipitation processes that may take place at spreading centers. (2) The Nazca plate metalliferous sediments form from hydrothermal processes that are unique to the rapid spreading portion of the EPR.

The first explanation would allow for strong fractionation of Mn and Fe by the precipitation processes discussed above; however, the dispersal of fine precipitates over distance of tens and hundreds of kilometres is hypothesized to result in a relatively uniform hydrothermal component. Because of the low accumulation rates over most of the Nazca plate, it is difficult to sample close to the spreading center. Consequently, the samples analyzed in all studies of Nazca plate metalliferous sediments may be biased toward locations that receive a hydrothermal contribution from some relatively distant hydrothermal vent. This would provide the time and distance necessary for physical homogenization of fractionating precipitation reactions that may occur immediately around a vent.

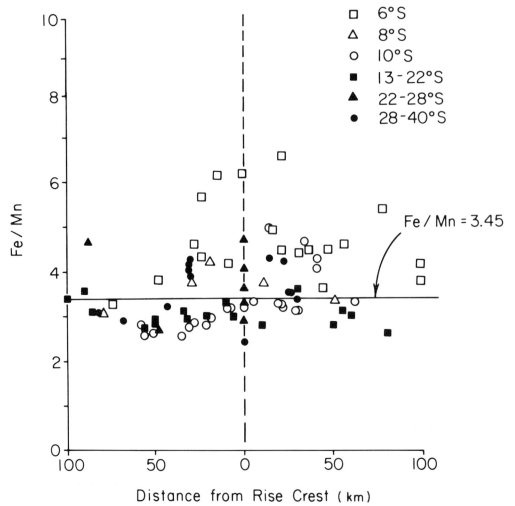

Figure 25. Fe/Mn versus distance from rise crest. Fe/Mn = 3.45 refers to the hydrothermal ratio in the normative model.

The second explanation was first suggested by Stakes (1979) who observed distinct alteration mineralogies in EPR samples compared to Mid-Atlantic Ridge (MAR) samples. Furthermore, stable isotope studies by Stakes suggested that the greenstones, which are the predominant high temperature alteration product of the MAR, formed between 200 °C and 350 °C with water-rock ratios of less than 10:1. Higher water-rock ratios (>50:1) and alteration temperatures of <200 °C characterize EPR alteration products. She suggested that the presence of deep faults characteristic of slow spreading ridges produce a rock-dominated hydrothermal system, and that rapid circulation through pillow basalts may produce a sea-water–dominated system on the EPR. In contrast, Edmond and others (1979b) emphasized that the Fe-sulfide ratio in the unmixed hydrothermal fluid is the important variable in determining whether sulfides will precipitate much of the Fe before or immediately upon exiting from a vent. With high Fe-sulfide ratios most of the Fe will precipitate in the water column and scavenge Mn, Cu, Zn, and "soluble" elements such as Si and Ba. Edmond and others suggested that the more intense tectonic activity of the MAR would increase permeability and sea-water circulation, leading to low exit temperatures.

With existing data it is difficult to evaluate the importance of precipitate homogenization compared to primary fluid compositional differences as an explanation for EPR metalliferous sediment compositions. Examination of data of the Troodos umbers provides some support for the concept that physical mixing of Mn-rich and Fe-rich precipitates is important in establishing the metalliferous sediment composition. Although the average Fe-Mn ratio of these umbers falls very close to the field defined by Nazca plate metalliferous sediments (Fig. 24), Robertson and Hudson (1973) reported Fe/Mn values that range from 2.3 to 28. It should be recognized, however, that hydroxyoxides with compositions similar to Nazca plate metalliferous sediments can precipitate directly from hydrothermal solutions and that physical mixing of an Fe hydroxyoxide and a Mn oxide is not required to form these deposits. Evidence for direct precipitation is found in the amorphous Fe crusts at the Galapagos mounds (Corliss and others, 1978) (Fig. 24) and in crusts forming from a sea-water hot spring in the Solomon Islands (Dymond, unpub. data), both of which have compositions very similar to the Nazca plate metalliferous sediments. Clearly, more studies of precipitation from various sea-water hydrothermal solutions are needed if we are to make knowledgeable inferences of the nature of hydrothermal solutions by study of their deposits.

SUMMARY

The compositional variability of Nazca plate sediments can be accounted for by mixtures of 5 components: (1) detrital—solid weathering products, forming primarily from erosion of the South American Continent; (2) hydrothermal—precipitates from solutions produced by interaction of sea water and the recently emplaced crust of the EPR; (3) biogenic—remains of planktonic organisms that fall to the sea floor; (4) hydrogenous—direct ferromanganese precipitates from sea water, which are hypothesized to accumulate everywhere in the oceans at very slow rates; (5) dissolution residue—Ba-rich remains from the dissolution of planktonic organisms.

The detrital source dominates sediments near the South American Continent, but the Yupanqui Basin in the central Nazca plate has particularly strong detrital inputs. The apparent topographic control on the movement of detritus with increasing distance from the continent suggests the importance of near-bottom movement of detritus away from South America. For much of the Nazca plate, detrital inputs are so low that Al is the only element whose distribution is dominated by this source.

The hydrothermal component, although strongly associated with the rise crest, appears to have the greatest abundance west of the spreading axis in the northern part of the plate and to the east in the southern plate. The most reasonable explanation for this distribution pattern is that bottom waters are entering the Nazca plate through the large fracture zones located 30° to 40°S and moving out of the plate over the rise crest to the north. Fe, Mn, Cu, and Zn are the elements analyzed that most strongly reflect the hydrothermal source. The hydrothermal precipitates appear to have a relatively narrow compositional range. However, rare samples show strong fractionations between Mn and Fe, and samples from 5° to 6°S on the rise crest are relatively enriched in Fe compared to more southern locations.

The biogenic source is of greatest importance in the biologically productive equatorial region of the northern plate and some areas near the South American continental margin. Enhanced biogenic opal preservation is apparent on the EPR and some shallow seamount samples due to the accumulation of biogenic carbonate. Because of the low abundances of Al, Ba, and transition metals in biogenic opal, Si is the only important biogenic element.

The hydrogenous source is accumulating at a realtively constant rate over the plate. Therefore, its abundance in the sediments is greatest in the basins that are distal from other sources. In these regions important fractions of the Ni, Mn, and Cu contents are from the hydrogenous source.

The dissolution residue component reaches the maximum abundance in the southern Bauer Basin and western Yupanqui Basins. This mid-latitude maximum can be explained by a decrease in the fraction of biogenic debris which is preserved with increasing distance from the equatorial high productivity region. Because 95% to 100% of the biogenic opal dissolves upon reaching the sea floor, the dissolution residue source is the major means for introducing biogenically associated elements into the sediments. Important contributions of Ba, Zn, Cu, Ni, and Al are introduced to the sediments in the northern Nazca plate via the dissolution residue component, and this source appears to be the predominant means for incorporating Ba into the sediment in all areas of the plate.

The small compositional range of rise-crest sediments and the generally good fit of the 5-source model suggest the hydrothermal precipitates that reach the sediments have a uniform composition. This observation is surprising considering the broad range in compositions that has been reported for hydrothermal precipitates from other regions. Moreover, it seems unlikely that the hydrothermal vents which have been sampled by submersible at 21°N EPR and the Galapagos spreading center have the appropriate composition to form the Nazca plate sediments of hydrothermal origin. Either the Nazca plate hydrothermal products are the result of large-scale homogenization of precipitates from a broad range of hydrothermal vent types or the Nazca plate portion of the EPR has a magmatic and tectonic regime that consistently produces hydrothermal fluids that are uniform and distinct from other locations. The solution to this question may not be possible until direct sampling of hydrothermal vents on the Nazca plate is carried out.

ACKNOWLEDGMENTS

I thank the following persons whose efforts and analytical skills provided the basic data set: Christine Chou, Chi Muratli, and Ron Stillinger. Special thanks go to Billy Chou who provided the expertise in linear programming needed in the normative model. Chi Muratli did all the data processing and patiently carried out many revisions in the normative model. Numerous discussions with colleagues were very helpful, and special acknowledgment should go to G. Ross Heath, Mitch Lyle, Erwin Suess, and Julius Dasch. Ken Scheidegger graciously provided unpublished compositional data for 83 samples near South America. The collection of these data was supported by a grant from the Office of Naval Research. This paper and the data collecting were supported by the NSF/IDOE Nazca Plate Project.

REFERENCES CITED

Anderson, R. N., and Halunen, A. J., Jr., 1974, The implications of heat flow for metallogenesis in the Bauer Deep: Nature, v. 251, p. 473-475.

Arrhenius, G., 1963, Pelagic sediments, in Hill, M., ed., The sea: New York, Interscience Publishers, Inc., v. 3, p. 655-727.

Arrhenius, G., and Bonatti, E., 1965, Neptunism and volcanism in the ocean, in Sears, M., ed., Progress in oceanography: Oxford, Pergamon Press, Ltd., p. 7-22.

Bacon, M. P., and Edmond, J. M., 1972, Barium at Geosecs III in the southwestern Pacific: Earth and Planetary Science Letters, v. 16, p. 66-74.

Bender, M., and others, 1971, Geochemistry of three cores from the East Pacific Rise: Earth and Planetary Science Letters, v. 12, p. 425-433.

Bischoff, J. L., 1969, Red Sea geothermal brine deposits: Their mineralogy, chemistry, and genesis, in Degans, E. T., and Ross, D. A., eds., Hot brines and recent heavy metal deposits in the Red Sea: New York, Springer-Verlag, p. 368.

Bischoff, J. L., and Dickson, F. W., 1975, Seawater-basalt interactions at 200°C and 500 bars: Implications for origin of heavy metal deposits and regulation of seawater chemistry: Earth and Planetary Science Letters, v. 25, p. 385-397.

Bonatti, E., and Joensuu, O., 1966, Deep sea iron deposits from the South Pacific: Science, v. 154, p. 643-645.

Bonatti, E., and others, 1971, Submarine iron deposits from the Mediterranean Sea, in Stanley, D. J., ed., The Mediterranean Sea: A natural sedimentation laboratory: Stroudsburg, Pennsylvania, Dowden, Hutchinson and Ross, p. 701-710.

Bonatti, E., and others, 1972, Iron-manganese-barium deposits from the northern Afar Rift (Ethiopia): Economic Geology, v. 67, p. 717-730.

Bostrom, K., 1973, The origin and fate of ferromanganoan active ridge sediments: Stockholm Contributions in Geology, v. 27, p. 149-241.

Bostrom, K., and Peterson, M.N.A., 1966, Precipitates from hydrothermal exhalations on the East Pacific Rise: Economic Geology, v. 61, p. 1258-1265.

Boyle, E. A., 1976, The marine geochemistry of trace elements [Ph.D. dissert.]: Massachusetts Institute of Technology and Woods Hole Oceanographic Institution, 156 p.

Bruland, K. W., Knauer, G. A., and Maring, J. H., 1978, Zinc in northeast Pacific water: Nature, v. 271, p. 741-743.

Bryan, W. B., Finger, L. W., and Chayes, F., 1969, Estimating proportions in petrographic mixing equations by least squares approximation: Science, v. 163, p. 926-927.

Cann, J. R., Winter, C. K., and Pritchard, R. G., 1977, A hydrothermal deposit from the floor of the Gulf of Aden: Mineralogical Magazine, v. 41, p. 193-199.

Church, T. M., 1970, Marine barite [Ph.D. dissert.]: San Diego, University of California, 100 p.

Church, T. M., and Wolgemuth, K., 1972, Marine barite saturation: Earth and Planetary Science Letters, v. 15, p. 35-44.

Corliss, J. B., 1971, The origin of metal-bearing hydrothermal solutions: Journal of Geophysical Research, v. 76, p. 8128-8138.

Corliss, J. B., and others, 1978, The chemistry of hydrothermal mounds near the Galapagos Rift: Earth and Planetary Science Letters, v. 40, p. 12-24.

Corliss, J. B., Edmond, J. M., and Gordon, L I., 1979, Some implications of heat/mass ratios in Galapagos Rift hydrothermal fluids for models of seawater-rock interaction and the formation of oceanic crust, in Talwani, M., and others, eds., Deep drilling results in the Atlantic Ocean: Continental margins and paleoenvironment: Washington, D.C., American Geophysical Union, v. 3, p. 391-402.

Dasch, E. J., Dymond, J., and Heath, G. R., 1971, Isotopic analysis of metalliferous sediments from the East Pacific Rise: Earth and Planetary Science Letters, v. 13, p. 175-180.

Dehairs, F., Chesselet, R., Jedwab, J., 1980, Discrete suspended particles of barite and the barium cycle in the open ocean: Earth and Planetary Science Letters v. 49, p. 528-550.

Dymond, J., and Corliss, J. B., 1980, Chemical composition of Nazca plate sediments: Geological Society of America Map and Chart Series, MC-34, lat 0°-40°S.

Dymond, J., and Veeh, H. H., 1975, Metal accumulation rates in the southeast Pacific and the origin of metalliferous sediments: Earth and Planetary Science Letters, v. 28, p. 13-22.

Dymond, J., and others, 1973, Origin of metalliferous sediments from the Pacific Ocean: Geological Society of America Bulletin, v. 84, p. 3355-3372.

Dymond, J., Corliss, J. B., and Stillinger, R., 1976, Chemical composition and metal accumulates rates of metalliferous sediments from sites 319, 320B and 321, in Yeats, R. S., Hart, S. R., and others, eds., Initial reports of the Deep Sea Drilling Project, Volume 34: Washington, D.C., U.S. Government Printing Office, p. 575-588.

Dymond, J., Corliss, J. B., and Heath, G. R., 1977, History of metalliferous sedimentation at Deep Sea Drilling Site 319, in the South Eastern Pacific: Geochemica et Cosmochimica Acta, v. 41, p. 741-753.

Dymond, J., and others, 1980, Composition and origin of sediments recovered by deep drilling of sediment mounds, Galapago spreading center, in Rosendahl, B. R., Hekinian, R., and others, eds., Initial reports of the Deep Sea Drilling Project, Volume 54: Washington, U.S. Government Printing Office, p. 377–385.

Edmond, J. M., 1980, The chemistry of the 350°C hot springs at 21°N on the East Pacific Rise [abs.]: EOS (American Geophysical Union Transactions). v. 61, p. 992.

Edmond, J. M., and others, 1979a, Ridgecrest hydrothermal activity and the balances of the major and minor elements in the ocean: The Galapagos data: Earth and Planetary Science Letters, v. 46, p. 1–19.

Edmond, J., and others, 1979b, On the formation of metal-rich deposits at ridge crests: Earth and Planetary Science Letters, v. 46, p. 19–30.

Flanagan, F. J., 1973, 1972 values for international geochemical reference samples: Geochimica et Cosmochimica Acta, v. 37, p. 1189–1200.

Fukui, S., 1976, Laboratory techniques used for atomic absorption spectrophotometric analysis of geologic samples: Oregon State University, School of Oceanography Report 76-10, 126 p.

Goldberg, E. D., and Arrhenius, G.O.S., 1958, Chemistry of Pacific pelagic sediments: Geochimica et Cosmochimica Acta, v. 13, p. 153–212.

Gurvich, Ye. G., Bogdanov, Yu. A., and Lisitsin, A. P., 1978, Behavior of barium in recent sedimentation in the Pacific: Translation from Geokhimiya (by Scripta Publishing Co.) v. 3, p. 359–374.

Hajash, A., and Archer, P., 1980, Experimental seawater/basalt interactions: Effects of cooling: Contributions to Mineralogy and Petrology, v. 75, p. 1–13.

Harder, H. 1976, Nontronite synthesis at low temperatures: Chemical Geology, v. 18, p. 169–180.

Haymon, R., and others, 1979, Mineralogy and chemistry of hydrothermal sulfide deposits and sediments at EPR 21°N: EOS (American Geophysical Union Transactions), v. 60, p. 864.

Heath, G. R., and Dymond, J. R., 1977, Genesis and diagenesis of metalliferous sediments from the East Pacific Rise, Bauer Deep, and Central Basin, northwest Nazca plate: Geological Society of America Bulletin, v. 88, p. 723–733.

—— 1981, Metalliferous sediment deposition in time and space: East Pacific Rise and Bauer Basin, northern Nazca plate, in Kulm, L. D., and others, eds., Nazca plate: Crustal formation and Andean convergence: Geological Society of America Memoir 154 (this volume).

Hekinian, R., and others, 1980, Sulfide deposits from the East Pacific Rise near 21°N: Science, v. 207, . 1433–1444.

Klinkhammer, G., Bender, M. L., and Weiss, R. F., 1977, Hydrothermal manganese in the Galapagos Rift: Nature, v. 269, p. 319–320.

Krauskopf, K. B., 1957, Separation of manganese from iron in sedimentary processes: Geochimica et Cosmochimica Acta, v. 12, p. 61–84.

Krishnaswami, S., 1976, Authigenic transition elements in Pacific pelagic clays: Geochimica et Cosmochimica Acta, v. 40, p. 425–434.

Lonsdale, P., 1976, Abyssal circulation of the southeast Pacific and some geological implications: Journal of Geophysical Research, v. 81, p. 1163–1176.

Lupton, J. E., and Craig, H., 1978, ^{3}He in the Pacific Ocean: Injection at active spreading centers and applications to deep circulation studies: EOS (American Geophysical Union Transactions), v. 59, p. 1105–1106.

Lyle, M., 1976, Estimation of hydrothermal manganese input to the oceans: Geology, v. 4, p. 733–736.

—— 1979, The formation and growth of ferromanganese oxides on the Nazca plate [Ph.D. dissert.]: Corvalis, Oregon State University, 172 p.

Lyle, M. W., and Dymond, J., 1976, Metal accumulation rates in the south-east Pacific—errors introduced from assumed bulk densities: Earth and Planetary Science Letters, v. 30, p. 164.

Lyle, M., Dymond, J. R., and Heath, G. R., 1977, Copper-nickel enriched ferromanganese nodules and associated crusts from the Bauer Deep, northwest Nazca plate: Earth and Planetary Science Letters, v. 35, p. 55–64.

Mackenzie, F. T., Stoffyn, M., and Wollast, R., 1978, Aluminium in seawater: Control by biological activity: Science, v. 199, p. 680–682.

Mammerickx, J., and Smith, S. M., 1978, Bathymetry of the Southeast Pacific: Geological Society of America Map and Chart Series, MC-26, scale 1:6,442,194.

Martin, J. H., and Knauer, G. A., 1973, The elemental composition of plankton: Geochimica et Cosmochimica Acta, v. 37, p. 1639–1654.

McMurtry, G. M., and Burnett, W. C., 1975, Hydrothermal metallogenesis in the Bauer Deep of the southeastern Pacific: Nature, v. 254, p. 42–44.

McMurtry, G. M., Veeh, H. H., and Moser, C., 1981, Sediment accumulation rate patterns on the northwet Nazca plate, in Kulm, L. D., and others, Nazca plate: Crustal formation and Andean convergence: Geological Society of America Memoir 154 (this volume).

Moore, W. S., and Vogt, P. R., 1975, Hydrothermal manganese crusts from two sites near the Galapagos spreading axis: Earth and Planetary Science Letters, v. 29, p. 349–356.

Moser, J. C., 1980, Sedimentation and accumulation rates of Nazca plate metaliferous sediments by high resolution Ge (Li) gamma ray sepctrometry of uranium series isotopes [M.S. thesis]: Corvallis, Oregon State University, 65 p.

Mottl, M. J., and Holland, H. D., 1978, Chemical exchange during hydrothermal alteration of basalt by seawater - I. Experimental results for major and minor components of seawater: Geochimica et Cosmochimica Acta, v. 42, p. 1103–1115.

Narula, S. C., and Wellington, J. F., 1977, Multiple linear regression with minimum sum of absolute errors: Applied Statistics, v. 26, p. 106–111.

Omang, S. H., 1969, A rapid fusion method for decomposition and comprehensive analysis of silicates by atomic absorption spectrophotometry: Analytica Chimica Acta, v. 46, p. 225–230.

Piper, D. Z., 1973, Origin of metalliferous sediments from the East Pacific Rise: Earth and Planetary Science Letters, v. 19, p. 75–82.

Piper, D. Z., and others, 1975, An iron-rich deposit from the Northeast Pacific: Earth and Planetary Science Letters, v. 26, p. 114–120.

Poldervaart, A., 1955, Chemistry of the Earth's crust: Geological Society of America Special Paper 62, p. 119–144.

Poldervaart, A., 1955, Chemistry of the Earth's crust: Geological Society of America Special Paper 62, p. 119–144.

Robertson, A.H.F., and Hudson, J. D., 1973, Cyprus umbers: Chemical precipitates on a Tethyan ocean ridge: Earth and Planetary Science Letters, v. 18, p. 93–101.

Scott, M. R., and others, 1974, Rapidly accumulating manganese deposit from the median valley at the Mid-Atlantic Ridge: Geophysical Research Letters, v. 1, p. 355–358.

Seyfried, W. E., Jr. and Bischoff, J. L., 1977, Hydrothermal transport of heavy metals by seawater: The role of seawater/basalt ratio: Earth and Planetary Science Letters, v. 34, p. 71–77.

——1981, Experimental seawater-basalt interaction at 300°C, 500 bars, chemical exchange, secondary mineral formation and implications for transport of heavy metals: Geochimica et Cosmochimica Acta, v. 45, p. 135–147.

Simmons, D. M., 1972, Linear programming for operations research: San Francisco, Holden-Day, Inc., 288 p.

Stakes, D., 1979, Submarine hydrothermal systems: Variations in mineralogy, chemistry, temperatures and the alteration of oceanic layer II [Ph.D. dissert.]: Corvallis, Oregon State University, 188 p.

Tate, M. W., and Clelland, R. C., 1957, Nonparametric and shortcut statistics: Danville, Illinois, Interstate Printers and Publishers, Inc., 171 p.

Thompson, G., and Bowen, V., 1969, Analyses of coccolith ooze from the deep tropical Atlantic: Journal of Marine Research, v. 27, p. 32–38.

Toth, J. R., 1980, Deposition of submarine crusts rich in manganese and iron: Geological Society of America Bulletin, v. 91, p. 44–54.

Wedepohl, K. H., 1969, Handbook of geochemistry: New York, Springer-Verlag, 5 v.

Williams, D., and others, 1980, The hydrothermal mounds of the Galapagos Rift: observations with DSRV *Alvin* and detailed heat flow studies: Journal of Geophysical Research, v. 84, p. 7467–7484.

Wright, T. L., and Doherty, P. C., 1970, A linear programming and least squares computer method for solving petrologic mixing problems: Geological Society of America Bulletin, v. 81, p. 1995–2008.

MANUSCRIPT RECEIVED BY THE SOCIETY NOVEMBER 12, 1980
MANUSCRIPT ACCEPTED DECEMBER 30, 1980

Metalliferous-sediment deposition in time and space: East Pacific Rise and Bauer Basin, northern Nazca plate

G. Ross Heath
Jack Dymond
School of Oceanography
Oregon State University
Corvallis, Oregon 97331

ABSTRACT

Comparison of surface and down-core samples from northern Nazca plate sediment cores shows that for most elements the temporal variability over 10^5 to 10^6 yr is comparable to the modern spatial variability over 10^3 to 10^4 km^2. Even mildly reducing conditions in the sediments lead to depletion of Mn down core, however. Systematic enrichments of Fe, Cu, and Zn in surface versus down-core samples west of the crest of the East Pacific Rise and depletion of the same elements in surface versus down-core samples from the Bauer Basin imply a recent change of near-bottom circulation that now tends to sweep hydrothermal debris west from the rise crest.

East-west profiles of the accumulation of the five sediment components, detritus, hydrothermal material, biogenic debris, authigenic precipitates, and solution residue from dissolved opal, show a westerly decrease in detrital input, except for a maximum of presumed volcanic origin on the rise crest. Hydrothermal material accumulates at more than 200 mg·cm^{-2}·10^{-3} yr^{-1} on the East Pacific Rise, but decreases rapidly within a few hundred kilometres from the rise crest, and is essentially zero 2,500 km to the east. The other components accumulate fairly uniformly across the region as expected from their sources in the relatively monotonous surface and bottom waters of the central southeast Pacific Ocean.

INTRODUCTION

Studies of the properties of core-top sediments have done much to relate presently active oceanographic and geologic processes to the geochemistry and distribution of deep-sea deposits. Studies of the Nazca plate by Dymond (this volume), Heath and Dymond (1977), Sayles and Bischoff (1973), Dymond and others (1973), Bostrom (1973), Bender and others (1971), and Bostrom and Peterson (1966), to name but a few, have shown that precipitates derived from rise-crest hydrothermal

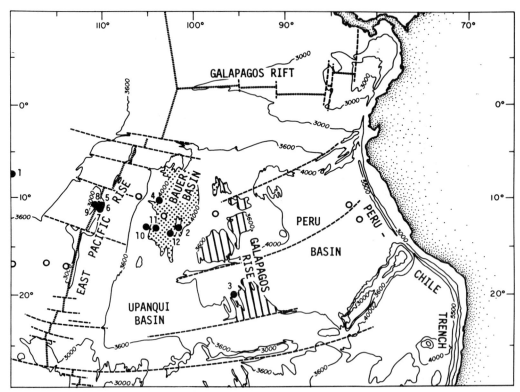

Figure 1. Sample locations relative to major physiographic units of the northern Nazca plate. Filled circles = cores listed in Table 1. Open circles = additional dated cores used to construct east-west profiles (Table 18). Isobaths in metres.

systems, authigenic precipitates from sea water, terrigenous debris, and the remains of planktonic marine organisms all contribute major fractions of one or more elements to sea-floor sediments.

Studies of the properties of subbottom sediments yield somewhat different results. Such studies allow us to reconstruct the history of sedimentation at a particular site (Dymond and others, 1977, for example), to determine the *rates* at which the various components are supplied to the sea floor (Dymond and Veeh, 1975, for example), to estimate the temporal variability of depositional processes and compare them to the modern spatial variability and to determine the extent to which diagenesis modified primary depositional patterns.

Details of the cores analyzed during the present study are summarized in Table 1 (microfiche). Their locations are shown in Figure 1. In addition, we have drawn on the published analyses of Bender and others (1971), Dymond and others (1977), McMurtry and Burnett (1975), and Dymond and Veeh (1975). Our cores are concentrated on the crest of the East Pacific Rise (EPR) at 10° to 11°S lat (Rea, 1976) and in the Bauer Basin between the EPR and the now-extinct Galapagos Rise. Most of the rise-crest cores penetrated the full sedimentary section. In contrast, our Bauer Basin cores penetrated only the uppermost sediments. Fortunately, we have the continuously cored Deep Sea Drilling Project site 319, which penetrated the full sedimentary section in the Basin, for control (Hart and others, 1976; Dymond and others, 1977).

METHODOLOGY

Samples were analyzed by atomic absorption spectrophotometry for the elements Si, Al, Fe, Mn, Cu, Ni, Zn, Ca, and Ba. The sample-treatment and analytical procedures are described in Dymond and others (1976) and will not be repeated here. Because the depths of the cores straddle the calcite compensation depth, absolute elemental concentrations in the samples vary widely because of variable dilution by biogenic calcite. To reduce this variability, we expressed all analyses as fractions of carbonate-free sediment, using the method of Dymond and others (1976). We also corrected for the presence of salt derived from pore waters collected with the original samples, using the method of Dymond and others (1976). A "correction factor" tabulated with the elemental analyses shows the ratio of the salt- and carbonate-corrected concentrations to the original uncorrected bulk-sediment concentration (that is, division of the elemental concentrations in Table 3 (microfiche) by the appropriate correction factors yields the measured bulk compositions).

The analytical data have been manipulated in several ways in an attempt to extract the underlying patterns of elemental supply and redistribution.

Normative Partitioning of Elements

The original version of this technique, described by Heath and Dymond (1977), assumed that for each element the ratios of the detrital component of that element to detrital Al, of the hydrothermal component to hydrothermal Fe, and of the authigenic component to authigenic Ni are constant throughout the region. Furthermore, it assumed that all the Fe and Al are either detrital or hydrothermal and that all the Ni is either detrital, hydrothermal, or authigenic. The remainder of other elements was attributed to a biogenic component.

This model has been very successful in providing plausible explanations for the distribution of elements in northern Nazca plate surface sediments (Heath and Dymond, 1977), in the sequence of sediments deposited at DSDP site 319 in the Bauer Basin (Dymond and others, 1977), and in DOMES sediments from the central and eastern Pacific (Bischoff and others, 1979a). For this and a companion study (Dymond, this volume), however, we have modified our assumptions to allow for the presence of authigenic and biogenic Al and for biogenic Ni. Analyses of pure biogenic material and of authigenic ferromanganese deposits prompted the addition of a fifth component, which is Ba rich. We believe that this component is the residue after dissolution of biogenic opal (Dymond, this volume).

The new model carries out the partitioning of elements between the five possible sources simultaneously rather than sequentially, as in the original simpler model. In order that the contributions from each source to each element be non-negative, we have solved the set of simultaneous equations by linear programming (Geary and McCarthy, 1964; Dymond and Eklund, 1978; Heath and Pisias, 1979; Dymond, this volume). The interelemental ratios specified for the sources for this study are listed in Table 2 (microfiche).

The elemental ratios to aluminum of the detrital source (Table 2) have been derived from published data for rocks of intermediate composition (Wedepohl, 1969) with some consideration of data available for shales and other clayey sedimentary rocks (see Dymond, this volume, for details). The elemental ratios to iron in the hydrothermal source are based on the average composition of sediments recovered from the crest of the EPR (Dymond, this volume), particularly those areas south of $15°S$ where the biogenic inputs are low. We have used published analyses of plankton (Martin and Knauer, 1973; Bostrom and others, 1978) for the elemental ratios to Si in the biogenic source term of the model. The ratios to Ni for hydrogenous components have been taken from data for Nazca plate ferromanganese crusts (Lyle, 1979). The elemental ratios to Ba for the dissolution residue have been estimated from the biogenic sources and from available information on the relative ease with which different elements are incorporated into the sediments rather than being returned in solution to sea water. The relative immobility of different elements in the deep sea is poorly known, but studies of elemental distributions in the water column (Boyle and others, 1976; Sclater and others, 1976; Bruland and others, 1978) and of sediment-trap debris (Cobler and Dymond, 1977) have helped us choose the dissolution-residue elemental ratios.

Q-mode Factor Analysis

This widely used technique (Imbrie and van Andel, 1964; Heath and Dymond, 1977) groups correlated elements into a small number of independent artificial variables that can be thought of as ideal end-member sediments. Although the end members are sometimes difficult to interpret and can vary according to the exact form in which data are entered into the analysis, the technique is a powerful method of extracting underlying patterns from a complex data set.

Rates of Accumulation

If the rate of sedimentation (cm/1,000 yr) and bulk density of the sediments are known, the elemental analyses of the bulk sediment can be converted to mass accumulation rates. Such rates enhance the value of the preceding models by allowing not only the relative contribution but also the flux of each element from each source to be estimated. We have used sedimentation rates from Bender and others (1971), Dymond and Veeh (1975), McMurtry and Burnett (1975), McMurtry and others (this volume), Moser (1980), and Sayles and Bischoff (1973). For many cores, measured bulk densities are available. For dried cores or cores with only published sedimentation rates and elemental analyses, however, we have estimated bulk densities using the technique of Lyle and Dymond (1976).

RESULTS

Rise-Crest Samples

The analytical data for 5 cores from the crest of the EPR at 10° to 11°S are summarized in Table 3 (microfiche). In addition, the table includes averaged analyses from Dymond (this volume) for 13 surface-sediment samples collected west of the spreading axis and 13 collected east of the axis. Figure 2 shows vertical profiles of the elements analyzed in a typical rise-crest core, Y71-7-45P.

Variability. The spatial and temporal variability of rise-crest sediments were assessed by applying t tests to the two populations of surface samples and to the sets of down-core samples. For the surface sets, only Mn and Ni were not different at the $p = 0.1$ level (Table 4). Given the authigenic character of these two elements, their uniform deposition over an area of a few thousand square kilometres is not surprising. The strong fractionation of the other elements, with the eastern samples markedly enriched in Al, Si, and Ba but depleted in Fe, Cu, and Zn, was unexpected, however, and points to spatially variable hydrothermal acitivity, to east-to-west dispersal of fine hydrothermal material by near-bottom currents, or to enrichment of the eastern samples by a nonhydrothermal sediment component.

In general, eastern down-core samples strongly resemble the eastern surface samples; only core Y71-7-44P, which is enriched in Al and depleted in Si, is anomalous. The spectacular color variations down this core suggest proximity to a volcanic source, perhaps resulting in abnormally high input of local detritus and the high down-core variability which appears to mask additional compositional differences from the surface sample set.

Manganese is depleted in down-core samples relative to surface samples both east and west of the rise crest. We attribute this to diagenetic remobilization of manganese, which has been measured in similar brown (that is, visibly oxidized) sediments from the eastern North Pacific (M. Bender, 1978, oral commun.). Such mobilization also can explain the erratic behavior of Mn in Heath and Dymond's (1977) model.

Barium behaves erratically, being enriched in some cores and depleted in others relative to surface samples. Whether this behavior reflects remobilization or temporal variability in supply (due to climatically induced variations in biologic productivity during the Quaternary) is unclear.

Down-core samples from both western cores differ markedly from the surface suite. In fact, the down-core enrichments of Si, Ba, and to a lesser extent, Al, and depletions of Fe, Cu, Mn, and Zn make these samples resemble the eastern surface rise suite more than the surrounding western suite. This similarity suggests that the transition metal enrichment observed west of the rise crest today is a very recent phenomenon, perhaps due to an unusual Holocene pattern of bottom circulation (leading to less detrital-biogenic deposition) or of hydrothermal activity.

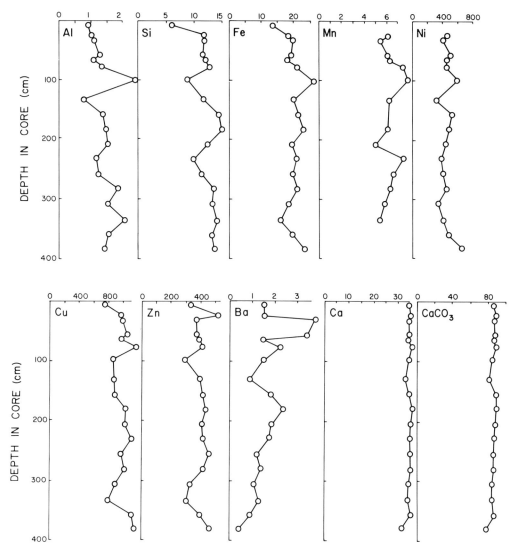

Figure 2. Vertical profiles of elemental abundances in a typical East Pacific Rise core, Y71-7-45P (Tables 1, 3).

Overall, the rise-crest data suggest that single core samples adequately represent a large surrounding area for Al, Fe, Ni, Cu, and Zn. Core samples are of marginal adequacy for Si and are unreliable for Ba and Mn, reflecting the distorting effects of diagenesis and, perhaps, of temporally variable deposition.

The variability of analyses for each element within each data set is summarized by the table of coefficients of variation (Table 5). The major features are the great homogeneity of the western relative to the eastern suite and the low variability of Si, Fe, Cu, and Zn, elements that Heath and Dymond (1977) inferred were readily incorporated in an iron-rich smectite.

Factor Analysis. Because the preceding statistical analysis demonstrated that surface sediments west of the spreading axis are richer in metalliferous components than those to the east, we factored the two groups of analyses separately as well as together. In each analysis, 3 factors accounted for more than 96% of the variance of the data set. One factor, which accounts for 48% of the variance, is dominated by Mn, Fe, Ni, Cu, and Zn, and clearly reflects hydrothermal input. A second factor (27% of the variance) is dominated by Al (as well as Si for eastern samples), and reflects detrital input. The

TABLE 4. SUMMARY OF T TESTS OF POPULATIONS OF ELEMENTAL ANALYSES OF EAST PACIFIC RISE (EPR) AND BAUER BASIN SURFACE AND DOWN-CORE SAMPLES

	Surface EPR West versus EPR East	Analyses of surface samples compared to cores							
		EPR East			EPR West		Bauer Basin		
		Y71-7-44P	Y71-7-45P	Y71-7-46P	Y71-7-52P	Y71-7-53P	Y73-3-20P	Y71-7-36P	DSDP-319
Al	++	++	·	·	·	+	·	--	·
Si	++	--	·	·	++	+	·	·	·
Mn	·	-	·	-	--	--	·	·	·
Fe	--	·	·	·	-	-	++	++	+
Ni	·	·	·	·	·	-	·	++	·
Cu	--	·	·	·	--	--	+	++	·
Zn	--	·	·	·	--	--	+	·	·
Ba	--	-	·	++	++	+	-	--	·

Note: Period (·) = populations indistinguishable at 90% confidence level. Single plus sign (+) or single minus sign (-) = core values higher (or lower) at 90% but not 98% confidence level. Double plus sign (++) or double minus sign (--) = core values higher (or lower) at 98% confidence level. Data from Tables 2 and 6.

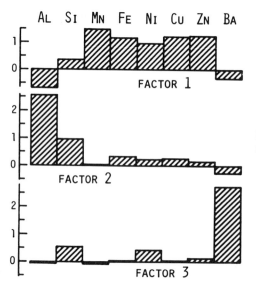

Figure 3. Factor scores for EPR analyses. Factor 1, hydrothermal; factor 2, detrital; factor 3, solution residue.

third factor (22% of the variance) is primarily Ba, with strong Si west of the axis, and reflects biogenic input. Figure 3 summarizes the scaled varimax factor scores for the lumped rise-crest data.

Figure 4, derived from a Q-mode factor analysis of all the rise-crest samples, shows that there are systematic differences between the eastern and western sample sets. The western samples clearly are depleted in the biogenic factor and are enriched in the metalliferous factor. The latter difference could be attributed to bottom currents preferentially carrying hydrothermal precipitates to the west, a conclusion supported by Lupton and Craig's (1978) He-3 data for deep waters in this region. We are not, however, aware of any oceanographic factor that can account for the fractionation of biogenic components across the spreading axis.

Figure 4 also emphasizes that surface samples west of the spreading axis are more metal-rich than most down-core samples. No such difference is evident in the eastern sample suite. The visually heterogeneous core Y71-7-44P appears to contain a highly variable detrital component superimposed on a more average metalliferous background. The biogenic component is most prominent in cores 45 (Fig. 5) and 46. Even in these cores, however, this factor does not change monotonically with depth. Such a pattern is suggestive of glacial-interglacial changes in surface biologic productivity (Pisias, 1976) which our sampling interval cannot fully resolve.

Partitioning of Elements. The proportions of sediment derived from the five major sources are summarized in Table 6 (microfiche). All the rise-crest samples are dominated by hydrothermal material, which ranges from 48% in Y71-7-44P to 78% in western surface deposits (Fig. 6). As

TABLE 5. COEFFICIENTS OF VARIATION (STANDARD DEVIATION/MEAN) FOR INDIVIDUAL ELEMENTS IN EPR AND BAUER BASIN SURFACE AND DOWN-CORE SAMPLE SUITES

	Al	Si	Mn	Fe	Ni	Cu	Zn	Ba
EPR East								
Surface	64	24	61	19	58	21	24	49
Y71-7-44P	67	26	38	28	45	45	31	78
-45P	30	18	47	14	59	12	14	53
-46P	24	9	24	30	43	29	47	22
EPR West								
Surface	43	10	12	7	17	8	10	20
Y71-7-52P	70	18	36	12	37	28	47	18
-53P	44	45	35	18	41	28	27	35
Mean EPR	49	21	36	18	42	24	28	39
Bauer Basin								
Surface	24	9	19	12	26	17	13	19
Y73-3-20P	26	13	21	10	35	13	11	21
Y71-7-36P	13	6	12	8	10	5	7	24
DSDP-319	9	1	16	3	22	8	6	11
Mean Bauer	18	7	17	8	23	11	9	19
Y73-3-7P	20	15	21	17	25	27	25	25
Y73-3-9P	17	9	18	10	51	7	18	11
Y73-3-18P	61	20	17	16	63	17	29	26
OC73-3-6P	21	19	22	15	24	20	13	29

Note: Data from Tables 2 and 7. Values in percent.

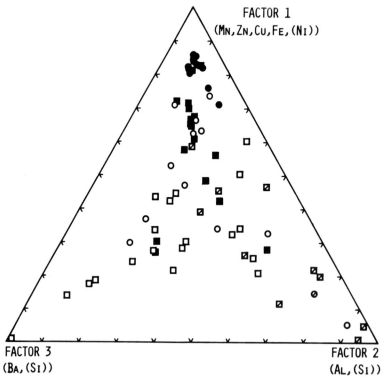

Figure 4. Relative factor loadings of EPR samples. Open symbols, samples collected east of the EPR crest; filled symbols, samples west of the crest; circles, surface samples; squares, down-core samples; diagonal bars, Y71-7-44P samples.

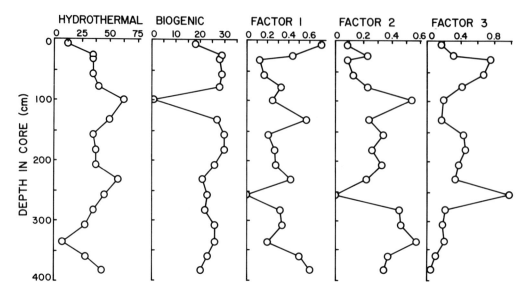

Figure 5. Vertical profiles of component abundances (from linear programming model) and Q-mode factor scores for samples from EPR core Y71-7-45P.

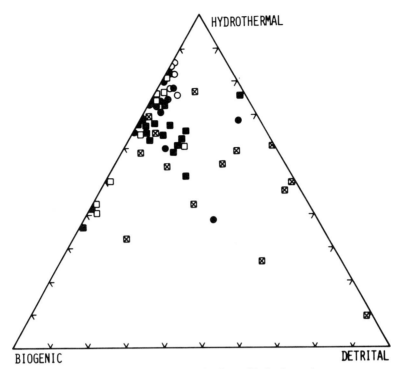

Figure 6. Relative contributions of hydrothermal, detrital, and biogenic components to EPR samples. Open symbols, eastern samples; filled symbols, western samples; circles, surface samples; squares, down-core samples; symbols with crosses, Y71-7-44P.

mentioned previously, western surface rise-crest sediments are enriched in hydrothermal materials and depleted in biogenic debris relative to all the down-core samples and to the eastern surface deposits. Y71-7-44P, which stands out in the factor analysis, contains 29% detrital material—more than three times as much as any other sample set. The isolation of the rise-crest from terrigenous sources and the relatively rapid accumulation rate of rise-crest deposits are reflected in the low-detrital, authigenic, and solution-residue contributions.

Even though the bulk sediment is dominated by hydrothermal and biogenic debris, the individual elements can show markedly different provenances depending on their concentrations in the source materials and on the contributions from the various sources. Tables 7 to 14 (microfiche) show the provenances of the elements studied. Not surprisingly, rise-crest iron is predominantly hydrothermal, whereas silicon is largely biogenic and hydrothermal. Nickel, on the other hand, is largely authigenic and hydrothermal, with an important solution-residue contribution despite the minimal contribution of authigenic and solution-residue material to the bulk sediment. Similarly, barium is dominated by the solution-residue input, even though this component makes up an average of only 4% of rise-crest sediments.

The component breakdowns show essentially the same agreement between core and surface suites of samples as do the raw elemental analyses (Tables 4 and 5). Within the resolution of the sample spacing and the statistical uncertainties of the data, the temporal variability at a single location over 10^5 to 10^6 yr is comparable to the spatial variabilty over 10^3 to 10^4 km². As with the bulk data, barium is very erratic, with marked differences between surface and down-core samples, as well as between east and west sample suites, in the detrital, hydrothermal, and solution-residue contributions. Similarly, surface samples are enriched relative to down-core samples in manganese from both hydrothermal and authigenic sources, with the difference particularly striking in the case of authentic manganese west of the rise-crest.

For the rise-crest samples, factor analysis (reflecting correlated variations in elemental abundances) and model partitioning (in which the analytical data are forced to fit a five-component system) yield very comparable pictures. For example, Figure 7 shows the correlation between the hydrothermal fractions of rise-crest sediments and the loading of factor 1 on the same samples. With the exception of five samples with unusually low iron contents (and, therefore, low hydrothermal contents), the correlations of the two measures ($r^2 = 0.76, p < 0.001$) is striking. This suggests that the partitioning model is realistic, and that in situ physical and chemical processes that act on the deposits at the sea floor have not perturbed the primary components.

Sedimentation Rates. The elemental accumulation rates are discussed in more detail in a subsequent section. Because the spreading history of the EPR at 10° to 11°S has been studied in detail (Rea, 1976), and because most of our cores penetrated the full sedimentary section, we have an unusually good opportunity to determine the variability of sedimentation rates over a rise-crest area of a few thousand square kilometres (Fig. 8). The relevant data are summarized in Table 15. The mean sedimentation rate for the five cores is 9.7 m/m.y. (0.97 cm/1,000 yr) with a coefficient of variation of less than 30%. The agreement of this rate with Dymond and Veeh's (1975) radiometric value of 0.93 cm/1,000 yr for core Y71-7-45P probably is fortuitous, but the similarity of their value and ours for this core (Table 15) suggests that our method is valid.

Bauer Basin Samples

This section focuses on cores from the deepest parts of the Bauer Basin and its southern neighbor, the Upanqui Basin ("Central Basin" of Heath and Dymond, 1977). Typical profiles of elemental abundances are shown in Figure 9. The Basin cores (Tables 1, 16, 17 microfiche) lie below the present calcite compensation depth of about 4,300 m. All cores bottom in carbonate-rich sediments. This

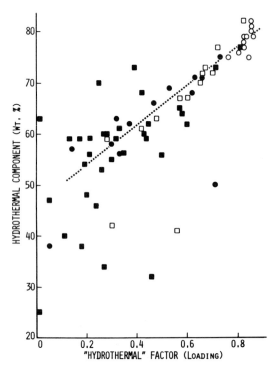

Figure 7. Proportion of hydrothermal component (from linear programming model) versus "hydrothermal" factor 1 (from Q-mode factor analysis) for EPR samples. Open symbols, eastern samples; filled symbols, western samples; circles, surface samples; squares, down-core samples. Points far below the best-fit line have unusually low Fe contents.

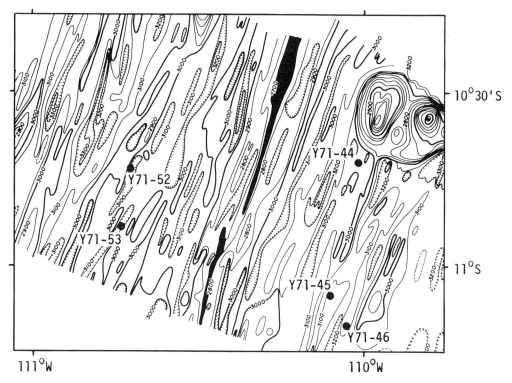

Figure 8. Location of EPR-crest piston cores at 10° to 11° S. Bathymetry after Rea (1976). Isobaths in metres. Crestal block (axis of rifting) shaded.

pattern of carbonate-poor over carbonate-rich deposits has been investigated in detail by Kendrick (1974) and Dymond and others (1976, 1977).

The change is a consequence of the subsidence of the oceanic crust as it ages and cools (Sclater and others, 1971). The actual date of the lithologic change is a function both of the depth of the calcite compensation level, which has varied over geologic time (van Andel and others, 1975; van Andel, 1975; Heath and others, 1977), and the age of the crust. In the Bauer Basin, the transition occurred about 4 m.y. ago, during early Pliocene time (Kendrick, 1974). In order to assess the temporal variability of the present depositional regime, we have calculated means and standard deviations of elemental abundances for the upper (noncarbonate) sections as well as for the entire cores from the Bauer Basin (Table 16).

Variability. The striking feature of the carbonate-free Bauer Basin samples, particularly those from the vicinity of DSDP site 319, is their great uniformity relative to rise-crest samples. For virtually all elements, the coefficients of variation for Bauer Basin sample suites are one-half to less than one-third of the values for rise-crest suites (Table 5).

This uniformity is consistent with thorough mixing of the sediment components during lateral transport from the rise crest and other distant source areas (Heath and Dymond, 1977), but it is difficult to reconcile with the local hydrothermal activity in the Bauer Basin proposed by Anderson and Halunen (1974) and McMurtry and Burnett (1975). As with the rise-crest samples, the inferred "smectite-affiliated" elements, Si, Fe, Cu, and Zn, show remarkably low coefficients of variation (in the 10% range; Table 5). Such uniformity is consistent with the concept of the "maturation" of metalliferous sediments (Lyle and others, 1977; Bischoff and others, 1979b), where the great thermodynamic stability of iron-manganese–rich smectite in sea water favors the reaction of primary hydrothermal ferric hydroxide with silica (either biogenic opal or, perhaps, even dissolved silica) to form this silicate.

TABLE 15. VARIABILITY OF SEDIMENTATION RATES AT THE CREST OF THE EAST PACIFIC RISE AT 11°S

Core Y71-7	Km from axis	Basement age (m.y.)	Core length (cm)	Hit basalt?	Sed. rate (m/m.y.)
44P	30.6E	0.38	464	Yes	12.3
45P	38.9E	0.48	385	Yes	8.0 (9.3*)
46P	48.0E	0.59	484	?	>8.2
52P	39.8W	0.50	355	Yes	7.1
53P	33.9W	0.43	561	Yes	13.1

Note: See text for details.

*Radiometric rate determination (Th^{230}_{excess} decay) by Dymond and Veeh (1975).

The degree to which down-core samples are representative of the surrounding surface sediments varies from element to element (Table 4). Al, Si, Mn, Ni, and Zn are well behaved, whereas Cu tends to be enriched and Ba depleted down core. Fe, surprisingly, is significantly enriched relative to surface samples in all the cores from the vicinity of DSDP site 319. We have no explanation for this pattern, although it is consistent with the very recent westward shift in transition metal deposition recorded by the rise-crest cores discussed previously. The small variability of the iron analyses for each sample suite undoubtedly accentuates the statistical significance of the differences recorded in Table 4.

One striking difference between the rise-crest and Bauer Basin data sets is in the behavior of Mn. All but one of the EPR crest cores are depleted in Mn relative to surrounding surface samples. In contrast, there is no statistical difference between the Mn content of Bauer Basin surface and down-core samples. This distinction suggests widespread near-surface reduction of Mn at the rise crest, but strongly oxidizing conditions throughout the Bauer Basin sections. Such a pattern is consistent with the relatively rapid sedimentation rates and high biogenic content of EPR crest sediments, contrasted with the extremely slow sedimentation rates of the opal- and carbonate-free Bauer Basin deposits.

Factor analysis. Q-mode factor analysis reveals that more than 99% of the variance in the Bauer Basin analyses can be accounted for by 3 factors. The scaled varimax factor scores are summarized in Figure 10. The first factor, which is dominated by Al, Ni, and Ba, accounts for 48% of the variance and reflects detrital and solution-residue inputs. The reason for the grouping of these two components is unclear. Heath and Dymond (1977) suggested that the deposition of elements like Ni and Ba is governed by the available surface area of detritus. There is no independent evidence for such a hypothesis, however. Inasmuch as high detrital loadings in this area imply minimal input of other components, such locations would be likely sites for severe degradation of biogenic debris, with a resultant accumulation of solution residues.

The second factor which accounts for 42% of the variance is dominated by Si, Fe, and Zn. This association, unlike most of the others, does not reflect a common provenance, but rather the occurrence of authigenic or diagenetic iron-rich smectite. This smectitie is inferred to form by the reaction of hydrothermal ferric oxyhydroxide (containing minor Zn, Ni, and Cu) with sea-water Mg and either biogenic opal (Heath and Dymond, 1977; Lyle and others, 1977) or even silica dissolved in the bottom water of the ocean (Bischoff and others, 1979b).

The third factor, which is less important (accounting for only 9.6% of the total variance) is dominated by Mn and Ni with lesser Cu and Fe. It reflects an authigenic oxyhydroxide contribution to the sediments. This factor is a "background" of all the Nazca plate sediments, becoming prominent

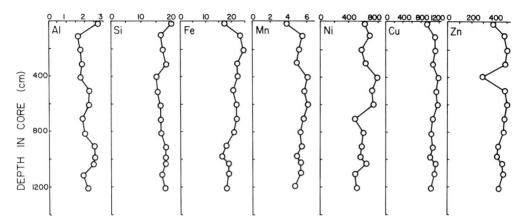

Figure 9. Vertical profiles of elemental abundances in a typical Bauer Basin core, Y71-7-36P (Kendrick, 1974).

TABLE 16. ANALYTICAL DATA FOR SEDIMENT CORES FROM THE BAUER BASIN

Sample ID no.	Depth in core (cm)	Al (%)	Si (%)	Mn (%)	Fe (%)	Ni (ppm)	Cu (ppm)	Zn (ppm)	Ca (%)	Ba (%)	CaCO$_3$ (%)	Correction factor
CORE Y71-7-36P												
497	7	2.73	20.22	3.81	17.75	625	980	386	..	1.90	1.2	1.25
384	807	2.02	16.88	5.29	21.11	598	1038	467	..	0.86	0.7	1.20
388	907	2.52	18.03	5.43	18.48	577	1089	426	..	1.32	0.6	1.20
393	982	2.53	17.91	4.81	16.35	553	1012	410	..	1.48	0.5	1.20
395	1032	2.45	18.06	5.34	18.77	626	1123	455	..	1.23	0.5	1.20
398	1107	1.89	16.97	5.33	19.28	501	1097	464	..	1.18	0.4	1.21
402	1207	2.15	17.99	4.47	18.07	524	1036	429	..	1.50	0.4	1.20
$\bar{x}$		2.33	18.01	4.93	18.54	572	1054	433		1.35	0.6	
SD		0.31	1.10	0.6	1.46	49	51	30		0.32	0.3	
435	1502	1.50	18.47	2.48	12.83	318	1076	381	32.94	1.37	82.0	6.97
437	1563	1.45	23.25	2.80	15.99	366	1116	421	12.80	0.25	30.6	1.75
472	1625	1.39	22.40	2.52	14.50	375	1090	424	20.16	0.24	49.3	2.39
478	1787	1.68	22.57	3.00	14.77	308	1269	417	22.78	0.29	56.0	2.83
508	1882	1.51	21.69	2.80	14.42	353	1199	402	22.34	0.27	54.9	2.77
$\bar{x}$		1.99	19.54	4.01	16.86	477	1094	423	22.20	0.99	23.1	
SD		0.48	2.35	1.23	2.44	124	80	28	7.22	0.59	29.9	
CORE Y73-3-20P												
1366	7	3.71	16.79	6.85	16.66	1561	1385	469	9.04	2.29	21.0	1.48
2101	35	3.51	15.87	6.44	16.55	1493	1452	464	5.23	2.37	11.3	1.35
2102	70	3.66	16.32	6.00	16.59	1315	1401	443	2.01	2.42	3.1	1.23
2103	170	3.26	14.49	8.14	18.54	1594	1653	506	2.32	2.10	3.2	1.19
2104	270	2.88	13.77	7.43	21.41	1465	1481	607	2.59	1.45	4.6	1.19
2105	370	3.26	17.87	5.74	17.70	971	1213	471	3.21	1.99	6.2	1.24
2106	450	1.35	19.89	3.91	18.41	386	1090	493	4.57	1.44	9.6	1.36
$\bar{x}$		3.09	16.43	6.36	9.62	1255	1382	493	4.14	2.01	8.53	
SD		0.82	2.06	1.36	6.51	438	184	54	2.47	0.41	6.27	
2107	520	1.28	18.38	3.63	16.99	315	1050	440	18.95	1.39	46.3	2.30
2108	660	1.14	19.13	3.63	15.92	288	1060	432	20.48	1.24	50.1	2.46
2109	690	1.32	14.90	3.49	17.29	377	1291	411	33.61	1.45	84.3	8.24
2110	770	1.17	18.61	4.05	17.35	292	1141	398	23.45	1.16	57.7	2.83
1356	902	1.42	17.61	3.56	18.71	329	1193	442	22.96	1.33	56.5	2.72
1357	1002	1.74	17.39	3.64	18.14	314	1206	440	22.79	1.21	56.0	2.69
$\bar{x}$		2.28	17.00	5.12	17.71	823	1278	463	13.17	1.60	31.6	
SD		1.08	1.86	1.70	1.41	576	185	53	10.83	0.58	27.7	

TABLE 16. (Continued)

Sample ID no.	Depth in core (cm)	Al (%)	Si (%)	Mn (%)	Fe (%)	Ni (ppm)	Cu (ppm)	Zn (ppm)	Ca (%)	Ba (%)	CaCO$_3$ (%)	Correction factor
CORE DSDP-319												
1289	8	3.65	17.51	4.94	16.45	1259	1360	472	9.31	2.36	21.7	1.53
1290	28	3.15	17.36	6.05	16.68	1161	1167	503	3.32	2.34	6.4	1.28
1133	211	3.35	17.33	5.39	16.25	874	1172	442	2.70	2.27	4.9	1.23
1134	381	2.94	17.67	4.14	17.50	797	1184	450	2.51	2.19	4.4	1.21
x̄		3.27	17.47	5.13	16.72	773	1221	467	4.46	3.05	9.4	
SD		0.30	0.16	0.80	0.55	455	93	27	3.25	0.08	8.3	
1135	774	2.80	17.60	4.15	18.20	690	1190	461	6.39	2.24	14.3	1.35
1389	1030	1.42	17.73	4.18	17.85	396	1102	492	25.06	1.49	61.8	3.32
1390	1211	1.00	20.70	3.44	15.33	314	1088	442	27.42	1.55	67.8	3.68
1391	1321	1.31	19.22	3.78	18.77	280	1198	462	24.50	1.54	60.4	2.98
1136	1441	1.14	19.33	3.46	16.86	261	1044	418	23.42	1.62	57.7	2.78
1392	2121	1.34	19.40	3.63	18.68	287	1173	480	23.46	1.46	57.7	2.86
x̄		2.21	18.39	4.32	15.26	532	1168	462	14.81	1.91	35.7	
SD		1.05	1.18	0.87	4.69	356	85	26	10.74	0.40	27.4	
1341	7	4.11	16.16	6.88	16.40	1745	1571	497	11.74	2.50	27.9	1.59
1342	102	1.51	9.16	5.86	22.93	600	1656	805	35.87	2.04	90.0	13.16
1343	202	1.68	15.24	5.83	24.23	880	1802	1010	37.52	2.75	93.0	23.09
1345	402	1.16	13.32	4.56	18.71	492	1275	663	35.93	1.52	90.0	13.25
1346	502	1.47	15.03	4.83	18.68	493	1201	572	32.63	1.61	81.1	6.55
x̄		1.99	13.78	5.59	20.19	842	1501	709	30.74	2.08	76.4	
SD		1.20	2.78	0.93	3.27	529	255	204	10.77	0.54	27.5	
Surface suite of 22 Bauer Basin samples												
x̄		3.20	17.36	5.38	15.63	1124	1224	435		2.23		
SD		0.78	1.48	1.02	1.93	296	213	55		0.43		

Note: Values are given on a carbonate- and salt-free basis. Division of the tabulated values by the "correction factors" yields bulk-sediment compositions. Averaged surface-sediment data from Dymond (this volume).

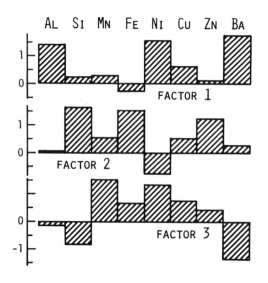

Figure 10. Factor scores for Bauer Basin analyses. Factor 1, detrital-authigenic-solution residue; factor 2, smectite; factor 3, oxyhydroxide.

only where it is not diluted by other components (that is, where the bulk accumulation rate is very slow).

Figure 11 is a triangular plot of the factor loadings of Bauer Basin surface and down-core samples. The two sets show considerable overlap, but it is clear that the down-core samples are markedly enriched relative to the surface samples in the smectite factor. This points to diagenesis in situ. The pore-water data of Bischoff and Sayles (1972), however, show no clear evidence of subsurface uptake of dissolved silica. This suggests that if smectite does form diagenetically, the reaction is slow relative to molecular diffusion and to the dissolution reactions that supply soluble silica to the interstitial waters.

Partitioning of Elements. The major differences between the provenance of EPR and Bauer Basin sediments (Table 6) are the greater relative importance of detrital, and to a lesser extent, authigenic and solution-residue inputs to the Basin, versus the dominance of hydrothermal material at the EPR crest. A secondary feature is the striking uniformity of the Basin deposits, indicative of thorough mixing of all components during their transport from distant sources, as well as remarkably little variation in provenance for the past several million years. As with the bulk sediment data (Table 5), the partitioned data show few differences between surface and down-core samples (Tables 7 to 14). For Fe, which is enriched down core (Table 4), the enrichment seems to be restricted to the hydrothermal components (Table 4). This supports the suggestion that the recent westward deflection of hydrothermal deposition on the EPR crest has partially starved the Bauer Basin to the east. The temporal patterns of deposition of the hydrothermal components of the other transition metals show similar patterns.

For the Bauer Basin, factor analysis and elemental partitioning yield very different views of the sediments. This contrasts strikingly with the results for the rise crest. Such a difference is consistent

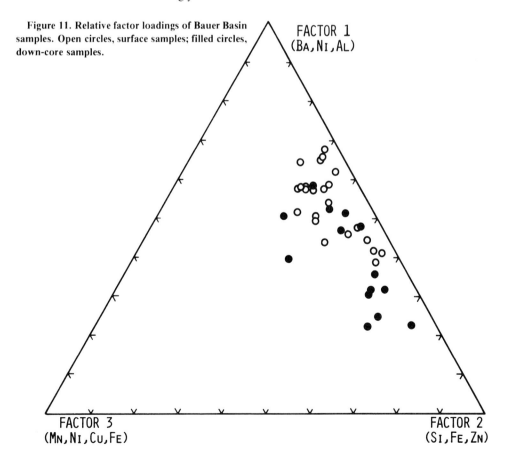

Figure 11. Relative factor loadings of Bauer Basin samples. Open circles, surface samples; filled circles, down-core samples.

TABLE 18. CORES USED FOR ACCUMULATION-RATE SECTIONS

	Lat (S)	Long (W)	Depth (m)	Distance to rise crest (km)	Sediment rate determination (references)*
V19-64	16°56'	121°12'	3540	860W	1
V19-61	16°57'	116°18'	3407	340	1
V19-54	17°02'	113°54'	2830	80	1
Y71-52	10°43'	110°42'	3198	40	8
Y71-53	10°53'	110°44'	3180	34	8
Y71-44	10°42'	110°01'	3167	31E	8
Y71-45	11°05'	110°06'	3096	39	2
Y71-46	11°10'	110°03'	3180	48	8
K71-10	9°59'	106°02'	3447	315	2
Y71-36	10°08'	103°51'	4541	585	2
Y73-7	13°00'	105°14'	3832	600	6
K71-115	11°58'	103°21'	4465	740	3
DSDP-319	13°01'	101°31'	4290	950	4
K71-132	11°33'	97°27'	3996	1180	3
OC73-6	20°03'	95°18'	4309	1845	7
K71-7	10°33'	83°02'	4514	2380	3
DSDP-321	12°01'	81°54'	4827	2550	5

*1. Bender and others (1971).
 2. Dymond and Veeh (1975).
 3. McMurtry and Burnett (1975).
 4. Dymond and others (1977).
 5. Hart and others (1976).
 6. McMurtry and others (this volume).
 7. Moser (1980).
 8. This paper.

with the dominance of rise-crest sediments by relatively unaltered primary components, but with the dominance of Basin deposits by secondary minerals (smectite and oxyhydroxides) formed from the hydrothermal, authigenic, and biogenic components. Slight variations in the relative abundances of smectite and oxyhydroxide from sample to sample lead the factor analysis to select phases rather than components as end members. At the same time, the provenance-controlled elemental associations have been insufficiently perturbed to prevent the partitioning model from diagnosing the contributions of the five selected components.

Rates of Accumulation

Seventeen cores (Table 18) have been used to estimate elemental accumulation rates for an east-west profile across the northern Nazca plate. The bulk accumulation rates vary from almost 1,000 mg $\cdot$ cm^{-2} $\cdot$ 10^{-3} yr^{-1} on the rise crest, to a few tens of mg $\cdot$ cm^{-2} $\cdot$ 10^{-3} yr^{-1} in the Bauer Basin (Table 19). Much of the variation as well as the spikiness of the profile (Fig. 12) is due to variations in biogenic calcite accumulation, depending on whether the site lies above or below the local calcite compensation level. The carbonate- and salt-free accumulation rates (Table 19, Fig. 12) show a much simpler pattern, with very low values everywhere except on the crest of the East Pacific Rise and approaching South America to the east. The "baseline" values, in the 20 mg $\cdot$ cm^{-2} $\cdot$ 10^{-3} yr^{-1} range, are among the lowest to be found anywhere in the oceans. Such low values reflect the isolation of the region from sources of terrigenous debris, as well as its location beneath surface waters of very low biological productivity.

The contributions of the five sediment components to the cores listed in Table 18 show a fairly simple east-west pattern (Fig. 13). The importance of detrital material decreases almost monotonically from the Peru Basin adjacent to South America to the crest of the EPR. The relative increase in detritus west of the EPR could reflect influx from the Pacific basin or reduced hydrothermal dilution.

The importance of the hydrothermal component increases from essentially zero some 2,400 km east of the EPR to a maximum of 96% in V19-54, 80 km west of the EPR crest (Fig. 13). The westward

TABLE 19. ACCUMULATION RATE (LINEAR, TOTAL MASS, CARBONATE-FREE MASS, CARBONATE- AND SALT-FREE MASS), WATER CONTENT, BULK DENSITY, AND CARBONATE CONTENT OF CORES USED IN ACCUMULATION-RATE CROSS SECTIONS

Core No.	Sedimentation rate (mm/1,000 yr)	Bulk density (g/ml)	Water content (wt %)	Accumulation rate (mg/cm^2/1,000 yr)	Mean CaCO$_3$ (wt %)	CaCO$_3$-free accumulation rate (mg/cm^2/1,000 yr)	CaCO$_3$-free and salt-free accumulation rate (mg/cm^2/1,000 yr)
V19-64	4.5	1.47	51	324	90	32	22
V19-61	7.0	1.48	51	508	90	51	35
V19-54	15.0	1.34	61	784	65	274	238
Y71-52	7.1	1.38	58	412	70.6	121	104
Y71-53	13.1	1.37	58	754	68.7	236	205
Y71-44	12.3	1.43	54	809	80.9	155	126
Y71-45	9.3	1.40	50	651	86.6	87	68
Y71-46	8.2	1.47	51	591	87.6	73	55
K71-10	4.3	1.57	45	371	89.5	39	30
Y71-36	1.4	1.11	80	31	0.6	31	27
Y73-7	3.7	1.48	51	268	90.0	27	18
K71-115	2.1	1.15	80	48	7.0	45	39
DSDP-319	1.1	1.16	79	29	9.4	24	21
K71-132	2.0	1.40	65	98	75.0	24	19
OC73-6	2.1	1.14	82	43	2.1	42	36
K71-7	2.3	1.37	70	94	2.0	93	86
DSDP-321	10.0	1.28	66	435	1.0	431	406

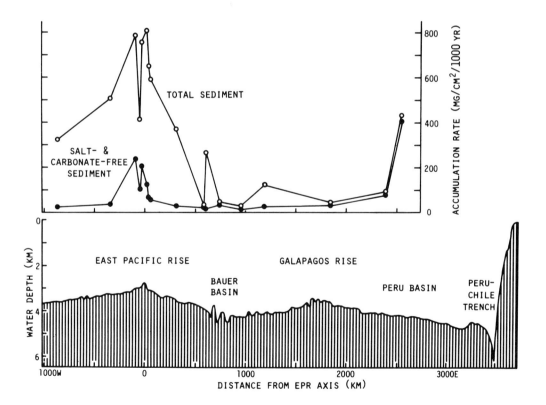

Figure 12. Sediment accumulation rates (total and carbonate/salt-free) in an east-west profile of cores across the northern Nazca plate. Bathymetric profile from 12° S.

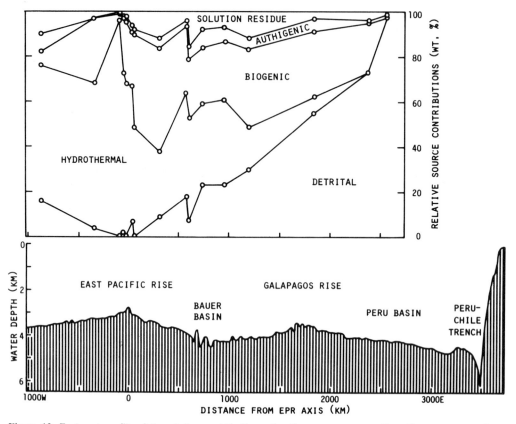

Figure 13. East-west profile of the relative contributions of sediment components (from linear programming model) to northern Nazca plate cores. Bathymetric profile from 12° S.

displacement of the most hydrothermal sediments evident in the factor analysis of the EPR samples clearly persists more than 1,000 km to the west of the rise crest.

The proportions of the remaining components do not vary systematically across the profile. This is not surprising, as all the cores lie beneath relatively oligotrophic surface waters (hence the biogenic and solution-residue inputs should be low but fairly uniform), and as authigenic sedimentation should be insensitive to location. For all three components, however, the influence of dilution by hydrothermal sediments on the EPR and by detritus toward South America is very obvious.

The core data of Tables 6 and 19 can be combined to yield accumulation rates of the individual components across the northern Nazca plate (Fig. 14). These profiles show the dominance of detritus toward South America and of hydrothermal material on the crest and west flank of the EPR. In addition, though, Figure 14 suggests that the rapid deposition rates on the rise crest enhance the preservation of biogenic silica. The minor components, authigenic and solution-residue debris, accumulate at rates that are almost uniform across the profile. The amount of variability (about a factor of 3) in these components is no greater than the core-to-core differences in sedimentation rates found over distances of a few kilometres in pelagic environments (Moore and Heath, 1966; Prince and others, 1980). Such variability probably reflects local deposition patterns governed by bottom-steered tidal circulation. The rate of authigenic sedimentation varies little from the EPR crest to the Bauer Basin, despite an order-of-magnitude reduction in total sedimentation rate (Table 19). This finding argues against major sediment loss or gain by either environment, and should serve as a warning against using profiles like Figure 13 to infer rates of sedimentary processes.

The accumulation rates of individual elements as a function of provenance resemble Figure 14.

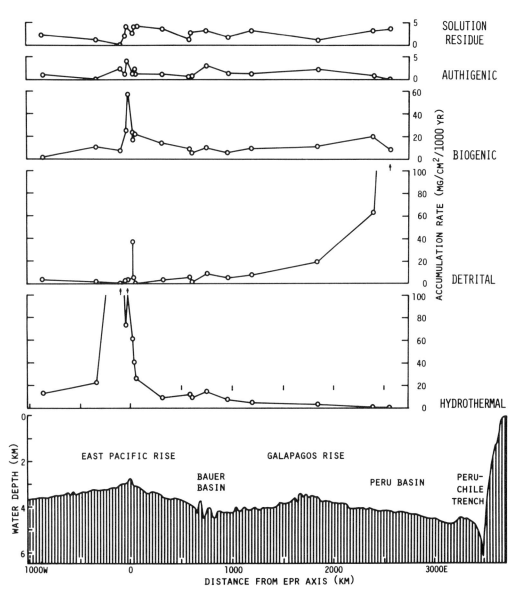

Figure 14. East-west profile of the rates of accumulation of sediment components (from linear programming model) in northern Nazca plate cores (Tables 6, 19). Bathymetric profile from 12°S.

However, the relative importance of the various components can be quite variable. Thus, in the case of Ni (Fig. 15), the authigenic component is dominant everywhere except on the EPR crest and in the eastern Peru Basin. This fact simply reflects the abundance of Ni in authigenic material. An alternative way of expressing this concept is that the shapes of the five curves of Figure 14 are the same for all elements; only the relative ordinate values for each profile change. Thus the *sum* of all the profiles (the total accumulation rate for each element) can be very different from element to element.

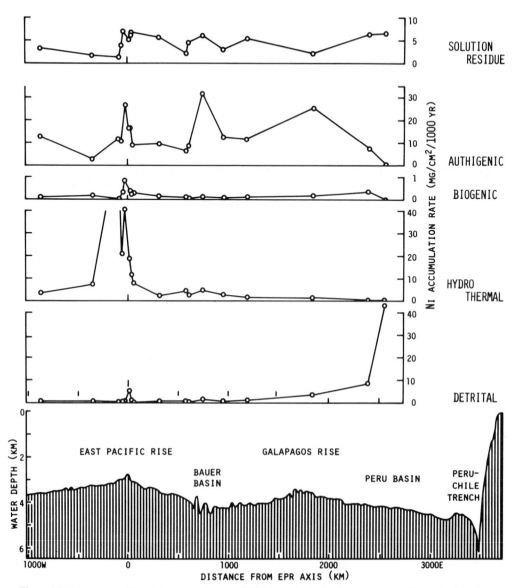

Figure 15. East-west profiles of the rates of input of Ni from principal sediment sources (determined by linear programming model) to cores of the northern Nazca plate (Tables 9, 19). Bathymetric profile from 12°S.

CONCLUSIONS

1. For most elements, the temporal variability in abundance down core over 10^5 to 10^6 yr is comparable to the present spatial variability over areas of 10^3 to 10^4 km^2.

2. Even very mildly reducing conditions down core (as inferred at the EPR crest, for example) leads to the loss of Mn from older sediments. Only in highly oxidized deposits like those of the Bauer Basin are surface and down-core Mn values statistically indistinguishable.

3. Surface sediments west of the EPR crest are significantly enriched in Fe, Mn, Ni, Cu, and Zn and depleted in Ba, Si, and Al relative to down-core samples. This pattern suggests a recent shift of hydrothermal activity to the west or a change of bottom currents so that they now carry a major

fraction of hydrothermal precipitates west from the EPR crest. The latter explanation is supported by a deficiency of Fe, Cu, and, to a lesser extent, Ni and Zn in Bauer Basin surface relative to down-core sediments. This suggests a recent reduction in the supply of the hydrothermal component rich in these elements from the EPR crust.

4. Factor analysis shows that virtually all the variance in either EPR or Bauer Basin samples can be explained by 3 factors. The EPR factors are dominated by transition metals (hydrothermal), Al with lesser Si (detrital), and Ba with lesser Si (biogenic). In contrast, the Bauer Basin factors are dominated by Al, Ni, and Ba (detrital and solution residue), Si, Fe, and Zn (authigenic or diagenetic smectite), and in a less important third factor, by Mn and Ni, with lesser Cu and Fe (authigenic oxyhydroxide).

5. Partitioning of elements according to provenance by linear programming suggests that variable inputs of three major components (detrital, hydrothermal, and biogenic) and two minor components (authigenic and solution residue) can account for the compositional variability of northern Nazca plate sediments). The detrital component generally decreases in importance away from South America, whereas the hydrothermal component is undetectable 2,500 km east of the EPR, but peaks within 1,000 km of the rise crest. The other three components do not vary systematically across the plate.

6. An east-west profile of carbonate- and salt-free sedimentation rates show maxima toward South America and on the EPR crest, with extremely low values (~ 20 mg $\cdot$ cm$^{-2} \cdot 10^{-3}$ yr^{-1}) at intermediate locations and west of the rise crest.

7. Accumulation-rate profiles of the five components show an almost monotonic east-to-west decrease in detrital input, broken by a small peak of presumed volcanic origin at the EPR crest. Hydrothermal input is sharply peaked within a few hundred kilometres of the spreading axis. Biogenic input is highest to the east and on the EPR crest, which suggests enhanced preservation in areas of higher sedimentation. The authigenic and solution-residue components are being deposited fairly uniformly across the region, with no more variability than expected from local lateral reworking of sediments at the sea floor.

8. The combination of factor analysis with partitioning by linear programming can allow elemental correlations due to provenance to be distinguished from those due to authigenic or diagenetic reaction at the sea floor. Such a distinction, particularly if combined with sedimentation-rate data, offers the prospect of a comprehensive understanding of the geochemical budget of a number of elements in the deep sea, as well as a means of establishing the genesis of pelagic sediments thrugh space and time.

ACKNOWLEDGMENTS

We are grateful to Charlotte Meredith Muratli for many of the analyses and for data presentation. Christin Chou performed the bulk of the analyses. The critical reviews of J. Bischoff and M. Leinen helped focus our ideas. This research was supported by the NSF/IDOE Nazca Plate Program.

REFERENCES CITED

Anderson, R. N., and Halunen, A. J., 1974, The implications of heat flow for metallogenesis in the Bauer Deep: Nature, v. 251, p. 473-475.

Bender, M., and others, 1971, Geochemistry of three cores from the East Pacific Rise: Earth and Planetary Science Letters, v. 12, p. 425-533.

Bischoff, J. L., and Sayles, F. L., 1972, Pore fluid and mineralogical studies of recent marine sediments: Bauer Depression region of the East Pacific Rise: Journal of Sedimentary Petrology, v. 42, p. 711-724.

Bischoff, J. L., Heath, G. R., and Leinen, J., 1979a, Geochemistry of deep-sea sediments from the Pacific manganese nodule province: DOMES sites A, B, and C, in Bischoff, J. L., and Piper, D. Z., eds., Marine geology and oceanography of the Pacific Manganese Nodule Province: New York, Plenum, p. 397-436.

Bischoff, J. L., Piper, D. Z., and Quinterno, P., 1979b, Nature and origin of metalliferous sediment in DOMES site C, Pacific Manganese Nodule Province, in Lalou, C., ed., International Colloquium on the Genesis of Manganese Nodules: Paris, Colloques International du C.N.R.S., p. 119-137.

Bostrom, K., 1973, The origin and fate of ferromanganoan active ridge sediments: Stockholm Contributions in Geology, v. 27, p. 149-243.

Bostrom, K., and Peterson, M.N.A., 1966, Precipitates from hydrothermal exhalations on the East Pacific Rise: Economic Geology, v. 61, p. 1258-1265.

Bostrom, K., Lysen, L., and Moore, C., 1978, Biological matter as a source of authigenic matter in pelagic sediments: Chemical Geology, v. 23, p. 11-20.

Boyle, E. A., Sclater, F., and Edmond, J. M., 1976, On the marine geochemistry of cadmium: Nature, v. 263, p. 43-44.

Bruland, K. W., Knauer, G. A., and Maring, J. H., 1978, Zinc in Northeast Pacific water: Nature, v. 271, p. 741-743.

Cobler, R., and Dymond, J., 1977, Sediment trap experiment at the Galapagos spreading center: Preliminary results: EOS (American Geophysical Union Transactions), v. 58, p. 1172.

Dymond, J., 1981, Geochemistry of Nazca plate surface sediments: An evaluation of hydrothermal, biogenic, detrital, and hydrogenous sources, in Kulm, L. D., and others, eds., Nazca plate: Crustal formation and Andean convergence: Geological Society of America Memoir 154 (this volume).

Dymond, J., and Eklund, W., 1978, A microprobe study of metalliferous sediment components: Earth and Planetary Science Letters, v. 40, p. 243-251.

Dymond, J., and Veeh, H. H., 1975, Metal accumulation rates in the southeast Pacific and the origin of metalliferous sediments: Earth and Planetary Science Letters, v. 28, p. 13-22.

Dymond, J., and others, 1973, Origin of metalliferous sediments from the Pacific Ocean: Geological Society of America Bulletin, v. 84, p. 3355-3372.

Dymond, J., Corliss, H. B., and Stillinger, R., 1976, Chemical composition and metal accumulation rates of metalliferous sediments from sites 319, 320B, and 321, in Initial reports of the Deep Sea Drilling Project: Washington, D.C., U.S. Government Printing Office, v. 34, p. 575-588.

Geary, R. C., and McCarthy, M. D., 1964, Elements of linear programming, with economic applications: New York, Hafner, 126 p.

Hart, S. R., and others, 1976, Initial reports of the Deep Sea Drilling Project: Washington, D.C., U.S. Government Printing Office, v. 34, 814 p.

Heath, G. R., and Dymond, J., 1977, Genesis and transformation of metalliferous sediments from the East Pacific Rise, Bauer Deep, and Central Basin, northwest Nazca plate: Geological Society of America Bulletin, v. 88, p. 723-733.

Heath, G. R., and Pisias, N. J., 1979, A method for the quantitative estimation of clay minerals in North Pacific deep-sea sediments: Clays and Clay Minerals, v. 27, p. 175-184.

Heath, G. R., Moore, T. C., Jr., and van Andel, Tj. H., 1977, Carbonate accumulation and dissolution in the equatorial Pacific during the past 45 million years, in Anderson, N. R., and Malahoff, A., eds., The fate of fossil fuel CO_2 in the oceans: New York Plenum, p. 627-639.

Imbrie, J., and van Andel, Tj. H., 1964, Vector analysis of heavy mineral data: Geological Society of America Bulletin, v. 75, p. 1131-1156.

Kendrick, J. W., 1974, Trace element studies of metalliferous sediments in cores from the East Pacific Rise and Bauer Deep, 10°S [M.S. thesis]: Corvallis, Oregon State University, 117 p.

Lupton, J. E., and Craig, H., 1978, ^{3}He in the Pacific Ocean: Injection at active spreading centers and applications to deep circulation studies: EOS (American Geophysical Union Transactions), v. 59, p. 1105-1106.

Lyle, M., 1979, The formation and growth of ferromanganese oxides on the Nazca plate [Ph.D. dissert.]: Corvallis, Oregon State University, 172 p.

Lyle, M. W., and Dymond, J., 1976, Metal accumulation rates in the southeast Pacific—Errors introduced from assumed bulk densities: Earth and Planetary Science Letters, v. 30, p. 164-168.

Lyle, M., Dymond, J., and Heath, G. R., 1977, Copper-nickel-enriched ferromanganese nodules and associated crusts from the Bauer Basin, northwest Nazca plate: Earth and Planetary Science Letters, v. 35, p. 55-64.

Martin, J. H., and Knauer, G. A., 1973, The elemental composition of plankton: Geochimica et Cosmochimica Acta, v. 37, p. 1639-1653.

McMurtry, G. M., and Burnett, W. C., 1975, Hydrothermal metallogenesis in the Bauer Deep of the southeastern Pacific: Nature, v. 254, p. 42-44.

McMurtry, G. M., Veeh, H. H., and Moser, C., 1981, Sediment accumulation rate patterns on the northwest Nazca plate, in Kulm, L. D., and others, eds., Nazca plate: Crustal formation and Andean convergence: Geological Society of America Memoir 154 (this volume).

Moore, T. C., Jr., and Heath, G. R., 1966, Manganese nodules, topography, and thickness of Quaternary sediments in the central Pacific: Nature, v. 212, p. 983-985.

Moser, J. C., 1980, Sedimentation and accumulation rates of Nazca plate metalliferous sediments by high resolution Ge (Li) gamma ray spectrometry of uranium series isotopes [M.S. thesis]: Corvallis, Oregon State University, 65 p.

Pisias, N. G., 1976, Late Quaternary sediment of the Panama Basin: Sedimentation rates, periodicities, and controls of carbonate and opal accumulation: Geological Society of America Memoir 145, p. 375-391.

Prince, R. A., Heath, G. R., and Kominz, M., 1980, Paleomagnetic studies of central North Pacific sediment cores: Stratigraphy, sedimentation rates, and the origin of magnetic instability: Geological Society of America Bulletin, v. 91, p. 447-449.

Rea, D. K., 1976, Analysis of a fast-spreading rise crest: The East Pacific Rise, 9° to 12° south: Marine Geophysical Researches, v. 2, p. 291-313.

Sayles, F. L., and Bischoff, J. L., 1973, Ferromanganoan sediments in the equatorial east Pacific: Earth and Planetary Science Letters, v. 19, p. 330-336.

Sclater, F. R., Boyle, E., and Edmond, J. M., 1976, On the marine geochemistry of nickel: Earth and Planetary Science Letters, v. 31, p. 119-128.

Sclater, J. G., Anderson, R. N., and Bell, M. L., 1971, Elevation of ridges and evolution of the central eastern Pacific: Journal of Geophysical Research, v. 76, p. 7888-7915.

van Andel, Tj. H., 1975, Mesozoic/Cenozoic calcite compensation depth and the global distribution of calcareous sediments: Earth and Planetary Science Letters, v. 26, p. 187-194.

van Andel, Tj. H., Heath, G. R., and Moore, T. C., Jr., 1975, Cenozoic history and paleoceanography of the central equatorial Pacific Ocean: Geological Society of America Memoir 143, 134 p.

Wedepohl, K. H., ed., 1969, Handbook of geochemistry, Volume II: New York, Springer-Verlag, 92 unnumbered sections.

MANUSCRIPT RECEIVED BY THE SOCIETY NOVEMBER 12, 1980
MANUSCRIPT ACCEPTED DECEMBER 30, 1980

Printed in U.S.A.

Lead isotopic composition of metalliferous sediments from the Nazca plate

E. JULIUS DASCH
Department of Geology
Oregon State University
Corvallis, Oregon 97331

ABSTRACT

Metal-rich sediments recently deposited over the northern part of the Nazca plate have lead isotopic compositional ranges as follows: $^{206}Pb/^{204}Pb = 18.123$ to 18.734; $^{207}Pb/^{204}Pb = 15.472$ to 15.640; $^{208}Pb/^{204}Pb = 37.796$ to 38.729. On $^{207}Pb/^{204}Pb$ versus $^{206}Pb/^{204}Pb$ and $^{208}Pb/^{204}Pb$ versus $^{206}Pb/^{204}Pb$ graphs, data for these sediments plot both within and between the otherwise separate fields on the diagrams that represent Pb isotopic ranges from mid-ocean ridge basalts (MORB; mantle-derived Pb) and ferromanganese nodules (MNOD; marine Pb). Least radiogenic (MORB) Pb is derived from the most volcanogenic sediments—along the East Pacific Rise and within the Bauer Deep. These leads exhibit extensive isotopic ranges that may in part reflect isotopic variations in mantle source rocks. Scatter plots of Pb isotopic composition of the samples versus a variety of elemental abundances, ratios, and combinations of elemental and other data show highest (positive) correlation coefficients with parameters that can best be described as "detrital."

Pb isotopic data for the sediments may be viewed graphically as mixing bands between mantle and marine Pb. Least radiogenic (but isotopically variable) mantle Pb is incorporated within the more volcanogenic sediments by coprecipitation with and absorption on Fe and Mn oxides and hydroxyoxides. Pb abundance in the sediments is increased by the adsorption and precipitation of Pb from sea water and pore waters. The marine Pb is more radiogenic and apparently has been of rather uniform isotopic composition in this region of the Pacific during the time of deposition and diagenesis of the sediment. These most radiogenic leads are isotopically indistinquishable from lead derived from deep ocean manganese nodules.

INTRODUCTION

One of the goals of the Nazca Plate Project has been to understand better the distribution, mineralogical and chemical makeup, origin, and postdeposition modifications of the potentially economic metalliferous sediments found on the plate. Discussions dealing with the overall physical and chemical nature of these sediments are collected herein (for example, Dymond, this volume; Heath and Dymond, this volume). The purpose of this article is to provide a description and interpretation of the lead isotopic composition of selected sediment samples for which pertinent supporting data are available. Samples were chosen mainly with respect to their geologic and geographic positioning on the plate (Fig. 1) and to their chemical compositions, which can be termed "hydrothermal," "detrital," "authigenic," biogenic," and "dissolution residue" percentages of the bulk sediment (Dymond, this volume).

Papers describing these metalliferous sediments have been published during the project (for example, Dasch and others, 1971; Dymond and others, 1973; Field and others, 1976), and these works are updated in this Memoir. Although these metal-rich sediments were noted 100 yr ago (Murray and Renard, 1891), it has been only within the past 15 yr (Bostrom and Peterson, 1966) that significant interest has been expressed in their geochemical and economic importance.

Present knowledge of these sediments is briefly summarized in the following sentences as a basis for discussion of the Pb isotopic data. Long thought to be essentially amorphous, the sediments are known to have a complex mineralogy consisting of geoethite, Fe-rich smectite, and Mn hydroxyoxide, along with lesser amounts of biogenic and detrital components. The dominant phases form

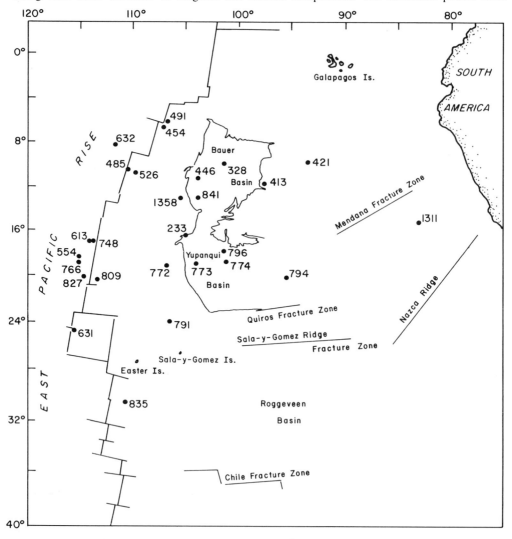

Figure 1. Location of samples analyzed for Pb isotopic composition.

aluminosilicate agregates and also Fe-Mn micronodules that appear to be diagenetic products resulting from reactions between Fe-Mn hydroxides and biogenic silica. Major element chemistry of these unique pelagic sediments shows that their main features are high transition metal contents and low aluminum abundances. Trace element and isotopic studies indicate that data for each element must be interpreted individually, with regard to ultimate and more immediate provenance of that element. Elemental and isotopic concentrations within the various physical compounds have been modified by extensive diagenetic transformations. Elements contained in the sediments appear to originate from four "original" sources: hydrothermal, involving primarily the leaching of cooling basalt by sea water (along the East Pacific Rise, perhaps within the Bauer Deep, and near the Galapagos Rift) and subsequent coprecipitation with Fe and Mn oxides and hydroxyoxides; hydrogenous (authigenic), or precipitation or absorption from sea water; biogenous, or mainly the contributions of silica, calcite, and carbonate apatite; and detrital, or principally aluminosilicate debris from the land and from alteration of sea-floor basalt.

The toxic metal lead has very low concentrations in sea water (<80 ng/kg; Patterson, 1974) but can be markedly enriched (34 to 130 ppm) in slowly accumulated deep ocean metalliferous and more typical pelagic sediments (Chow and Patterson, 1962; Senechal and others, 1975) and in ferromanganese modules (21 to 2,500 ppm; Chow and Patterson, 1959; Reynolds and Dasch, 1971). Contrastingly, rapidly accumulated material, for example some biogenic oozes (Chow and Patterson, 1962) and hydrothermal crusts (R. B. Scott, personal commun.), contain extremely low lead concentrations. Because three of the four stable isotopes of Pb are radiogenic, its isotopic composition is useful as an indicator of geologic provenance. The Pb isotopic composition of worldwide pelagic sediments was shown by Chow and Patterson (1962) to cover a wide range and to have a geographic patterning that could be identified with weathering provenance. Owing to partial correction of an earlier problem of mass fractionization during analysis, Reynolds and Dasch (1971) showed that the isotopic range for marine Pb of ferromanganese nodules was much more restricted than previously thought, and that there was an indication that $^{207}Pb/^{204}Pb$ ratios from nodules with a clearly "continental" aspect (as ascertained from the detrital mineralogy of adjacent sediments and their strontium isotopic composition; Dasch, 1969) were distinct from nodules recovered from sea-floor sediments with a more "volcanogenic" aspect.

Lead in marine materials has isotopic characteristics that may be separable into four "end-members" (Fig. 2).

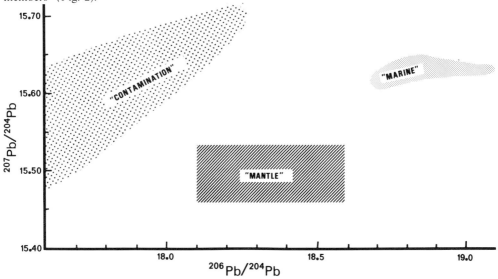

Figure 2. $^{207}Pb/^{204}Pb$ versus $^{206}Pb/^{204}Pb$ for possible ranges of Pb provenance. "Marine" is equivalent to MNOD, and "Mantle" is equivalent to MORB in Figures 3 and 4.

"Marine" Pb

The isotopic composition of "marine" Pb is that of the mixture of trace amounts of Pb dissolved in sea water and ancient sea water from all natural sources, particularly from continental runoff and from marine volcanic and hydrothermal processes. For this discussion I assume that the isotopic ranges for marine Pb are approximated by Pb from deep ocean ferromanganese nodules, or "MNOD" (Reynolds and Dasch, 1971; Figs. 2-4). Modern surface sea water apparently is contaminated by Pb primarily of industrial origin (Chow and others, 1973); some geochemists believe that this contamination may also have reached the deepest layers of the ocean (B.L.K. Somayajulu, personal commun.) especially by way of rapid transport in fecal pellets.

"Mantle" Pb

"Mantle" Pb refers to the comparatively less radiogenic trace Pb found in modern, primarily theoleiitic basalts at diverging plate boundaries (MORB), and unaltered basalts recovered from other parts of the sea floor. Analyses defining the Pb isotopic characteristics of these rocks, as shown in Figures 2-4, were made by Tatsumoto (1966), Unruh and Tatsumoto (1976), and Sun (1973). Although MORB Pb has restricted isotopic ranges (Figs. 2-4), measurable variation exists between samples. These data have been interpreted as indicating that mantle source rocks, which yield flow rocks by partial melting, are isotopically heterogenous (Tatsumoto, 1966); (Brooks and others, 1976) have also suggested that the data reflect meaningful ages in the source rocks.

"Continental" Pb

Waters draining continental rocks, and aluminosilicate detritus from the continents, contain very small amounts of Pb, although some minerals and rock fragments may contain higher concentrations (Chow and Patterson, 1962). This "continental" Pb, supplied to the seas by rivers (in solution and by bottom transport) and by wind, however, can form a significant percentage of material that has accumulated very slowly, such as pelagic "red" clay and deep ocean ferromanganese nodules. Unlike mantle Pb, continental Pb can have extremely varied isotopic composition, comprising the known range for Pb isotopic variation (thus not shown in Fig. 2). Ocean processes may, in places, effectively homogenize the isotopic composition by mixing. Continental Pb with an isotopic composition that happens to fall within the ranges of other types of Pb cannot be distinguished from them; only isotopic compositions clearly outside the other varietal ranges can indicate continental Pb components.

"Contamination" Pb

Surface waters and perhaps even the benthic boundary layer of the oceans may contain trace amounts of Pb of anthropogenic origin. "Contamination" Pb results mainly from atmospheric and stream washout of combustible Pb from automobile alkyls and smelter exhaust. Extensive global contaimination of Pb alkyls began with the widespread introduction of Pb as a gasoline additive in the early 1920s (Patterson, 1971). Chow and others (1973) reported that such Pb accounts for half or more of the total Pb recently supplied to sediments in marine basins near heavily populated and industrialized southern California. Although the isotopic composition of contamination Pb varies according to its geologic provenance, most of it has come from a relatively few large ore deposits. The isotopic ratios of Pb added to gasoline (Chow, 1971) differ from the ranges for marine Pb and for mantle Pb (Fig. 2).

ANALYTICAL PROCEDURES

Samples for Pb isotopic analysis (total dissolution) were chosen from the large group of surface samples collected over the northern part of the Nazca plate primarily on the basis of existing chemical information. An effort was made to include a variety of sediments having disparate provenance. Sample preparation and mass spectrometry procedures were similar to those reported in Barnes and others (1973). Sample weights varied from about 0.1 to 0.5 g. Fractionation in the Oregon State Univerity spectrometer (Senechal and Dasch, 1974) varied from 0.1% to 0.2% per mass unit. Reproducibility for the ratios is better than 0.1%.

RESULTS AND DISCUSSION

Locations of the 27 samples of sediment analyzed for Pb isotopic composition are shown in Figure 1. Lead isotopic and support data are listed in Table 1, and Pb isotopic compositional data are plotted in Figures 3, 4, and 5.

Lead isotopic ranges for the sediments are as follows: $^{206}Pb/^{204}Pb = 18.123$ to 18.734; $^{207}Pb/^{204}Pb = 15.472$ to 15.640; $^{208}Pb/^{204}Pb = 37.796$ to 38.729. There is considerably more variation in the $^{206}Pb/^{204}Pb$ range (3.4%) than in the $^{207}Pb/^{204}Pb$ range (1.2%), or in the intermediate $^{208}Pb/^{204}Pb$ range. If the conventional Pb isotopic plots of Figures 3 and 4 included primary growth curves for the Earth, Pb for all the sediments would be seen to plot as negative (future) model ages, a point made by Chow and Patterson (1962) for sediment Pb from the world oceans. Negative model Pb ages are believed to reflect open-system behavior of the upper mantle through geologic time (provided that the immediate source of Pb is from the upper mantle), with concomitant increases in U/Pb and Th/Pb ratios in these reservoir rocks. Slight but apparently systematic differences in Pb isotopic compositions (and other elemental, isotopic, and mineralogic variations) have been determined for a size-fractionated, surface sample of metalliferous sediment from the Bauer Deep (Senechal and others, 1975). Although the Pb isotopic compositions of the different size fractions show systematic trends when compared with other data, the ratios are indistinguishable at the 95% confidence level. Thus, the Pb distributed in this sample, at least to a first approximation, may be characterized as isotopically homogeneous. Lead isotopic compositions of top and bottom material from several carefully collected ferromanganese nodules that appear to have remained in their natural position for long periods of time also have uniform isotopic composition, although the sample pairs have very different Pb concentrations, transition metal abundances, and Si/Al ratios (Senechal and Heath, in prep.). These data suggest, at least on the scale of the samples studied, either that the source of Pb is

TABLE 1. LEAD ISOTOPIC COMPOSITION OF SEDIMENTS FROM THE NAZCA PLATE

Sample no.	Latitude (°S)	Longitude (°W)	Water depth (m)	$\frac{^{206}Pb}{^{204}Pb}$	$\frac{^{207}Pb}{^{204}Pb}$	$\frac{^{208}Pb}{^{204}Pb}$
MS 233	16.35	104.80	4157	18.573	15.578	38.410
MS 328	9.92	101.15	4520	18.499	15.567	38.307
MS 413	11.55	97.45	3850	18.589	15.580	38.459
MS 421	9.60	93.33	4622	18.713	15.620	38.671
MS 446	11.00	103.75	5343	18.472	15.523	38.213
MS 454	6.60	106.80	3319	18.291	15.472	37.863
MS 485	10.55	110.60	2970	18.376	15.513	38.002
MS 491	6.18	106.60	3020	18.500	15.555	38.186
MS 526	10.75	109.68	3440	18.432	15.537	38.137
MS 554	18.40	114.95	3075	18.123	15.490	37.796
MS 613	17.03	113.90	2964	18.471	15.484	38.016
MS 631	24.60	115.45	2625	18.684	15.553	38.308
MS 632	8.35	111.52	3627	18.441	15.513	38.096
MS 748	17.02	113.52	3058	18.487	15.498	38.009
MS 766	18.80	114.90	3286	18.497	15.537	38.140
MS 772	19.00	106.55	3935	18.696	15.626	38.661
MS 773	18.88	103.65	4122	18.661	15.628	38.608
MS 774	18.72	100.92	4162	18.712	15.638	38.643
MS 791	23.87	106.15	3836	18.710	15.615	38.654
MS 794	20.05	95.30	4309	18.597	15.611	38.520
MS 796	17.87	101.07	4192	18.685	15.640	38.671
MS 809	20.32	113.22	3313	18.510	15.549	38.177
MS 827	20.00	114.52	3247	18.395	15.502	37.942
MS 835	30.50	110.48	3003	18.734	15.563	38.328
MS 841	13.00	103.50	4320	18.457	15.566	38.308
MS 1311	15.15	82.80	4870	18.716	15.632	38.729
MS 1358	13.00	105.23	3832	18.596	15.583	38.463

204 DASCH, E. J.

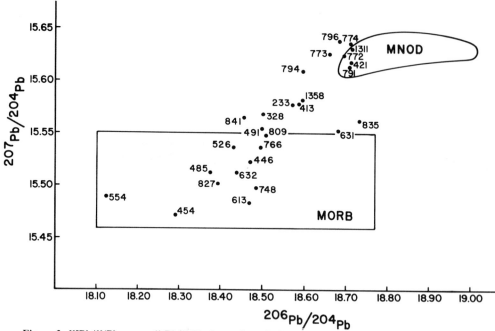

Figure 3. $^{207}Pb/^{204}Pb$ versus $^{206}Pb/^{204}Pb$ for surface (bulk) samples of sediment from the Nazca plate. "MNOD" is equivalent to Marine and "MORB" is equivalent to Mantle in Figure 1.

isotopically uniform or that the Pb has been effectively redistributed by diagenetic alteration. The latter explanation is preferable in that extensive redistribution of many elements and diagenetic transformation have been well documented for Nazca plate sediments (Heath and Dymond, 1977; Dymond and Eklund, 1978; Lyle and others, 1977).

Thus, the main Pb isotopic variability found in metalliferous sediments of the Nazca plate occurs on a scale larger than that of the size of the sample. Lead isotopic compositions of bulk samples can be plotted as a function of many chemical analyses (carbonate-free basis) and other factors that are available from the Nazca plate study (Dymond, this volume) as a help in delimiting the sources of isotopic variability. Correlation coefficients (r) for $^{206}Pb/^{204}Pb$, $^{207}Pb/^{204}Pb$, and $^{208}Pb/^{204}Pb$ versus chemical and other parameters that appear to have the greatest genetic significance (major and trace element concentrations, elemental ratios, geographic locations) have been calculated. Elsewhere in this collection of papers (Dymond, this volume) the genetic significance of these parameters is codified as hydrothermal, biogenic, detrital, authigenic (hydrogenous), and dissolution residue. Lead isotopic ratios show significant correlation coefficients ($r = 0.7$ and above) only for parameters that are characterized as detrital or hydrothermal (Table 2). Highest (positive) coefficients exist only with

TABLE 2. SELECTED CORRELATION COEFFICIENTS (r) FOR LEAD ISOTOPIC COMPOSITION OF NAZCA PLATE SEDIMENTS VERSUS ELEMENTAL AND OTHER PARAMETERS

	Al	Si	Mn	Fe	Fe/Ni	Th	Sc	Distance from EPR	Detrital* (%)	Hydrothermal* (%)	Authigenic* (%)	Biogenic* (%)	Solution Residue* (%)
$\frac{^{206}Pb}{^{204}Pb}$	0.70	0.37	-0.39	-0.44	-0.57	0.56	0.64	0.56	0.68	-0.49	0.43	-0.31	+0.33
$\frac{^{207}Pb}{^{204}Pb}$	0.92	0.61	-0.62	-0.66	-0.79	0.94	0.88	0.79	0.89	-0.74	0.53	-0.26	+0.45
$\frac{^{208}Pb}{^{204}Pb}$	0.92	0.63	-0.63	-0.68	-0.79	0.90	0.90	0.79	0.89	-0.76	0.53	-0.23	+0.47

*For definition of terms, see Dymond (this volume).

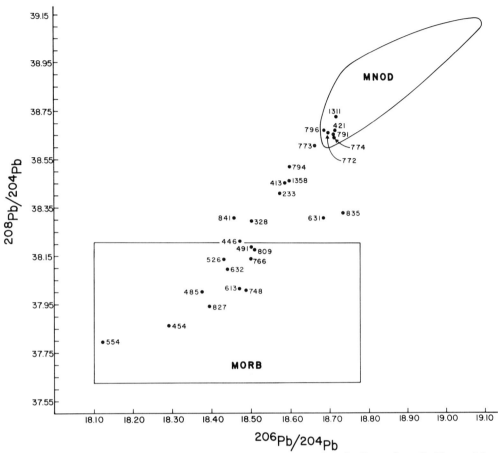

Figure 4. $^{208}Pb/^{204}Pb$ versus $^{206}Pb/^{204}Pb$ for surface (bulk) samples of sediment from the Nazca plate. "MNOD" is equivalent to Marine and "MORB" is equivalent to Mantle in Figure 1.

parameters believed to represent detrital influence (see also Figure 5). Authigenic processes may account for some of the isotopic variability ($r \sim +0.5$, Table 2). Biogenic influence on the Pb isotopic compositions is not obvious (rs ~ 0.2 to -0.3). Of perhaps more significance than biogenic is the dissolution residue term described by Dymond (this volume); this interpretative elemental parameter apparently reflects the dissolution of 90% to 100% of biogenic material, and thus may concentrate markedly the trace elements, perhaps including Pb, released from organic matter. Although r values for Pb isotopic composition versus dissolution residue are higher ($\sim +0.3$ to $+0.5$) than those for the biogenic parameter, they do not show the significant increases that might be expectable in view of the large amounts of dissolution that have been suggested.

Possible provenance of Pb for these sediments is shown graphically in Figure 2. The earliest analysis (from the Bauer Deep; Dasch and others, 1971) plotted within the field of Pb from mid-ocean ridge basalts (MANTLE). We concluded that the Pb in this sample was volcanogenic, either resulting from hydrothermal processes at the East Pacific Rise (EPR) spreading center, or released by the weathering of ocean-floor basalts. As Figures 3 and 4 show, however, the sediments have Pb isotopic compositions that plot not only within the MORB field but also within the ferromanganese nodule (MNOD) field, and scattered between these fields. The points fall essentially in a band formed by connecting the extremities of the MORB and MNOD fields, on both the $^{207}Pb/^{204}Pb$ versus $^{206}Pb/^{204}Pb$ and $^{208}Pb/^{204}Pb$ versus $^{206}Pb/^{204}Pb$ graphs. The rough linearity of points on these diagrams could be interpreted as having chronologic significance. I believe, however, that the

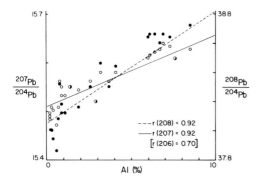
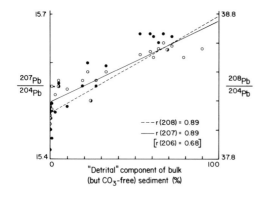
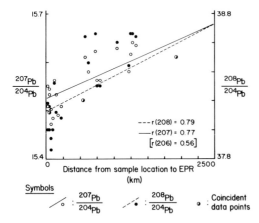

Figure 5. $^{207}Pb/^{204}Pb$ and $^{208}Pb/^{204}Pb$ versus Al(%), detrital component of bulk sediment (%), and distance from sample location to EPR (km).

distribution of data is more logically explained by mixing of two principal kinds of Pb—volcanogenic and detrital. This interpretation is consistent with the higher correlation coefficients for Pb isotopic composition versus detrital and hydrothermal factors.

The youngest sediments—those collected from along or very near the crest of the EPR—contain the least radiogenic Pb (plotting within or very near the MORB field) and exhibit the widest isotopic variability (as contrasted with sediments most distant from the EPR, many of which have Pb isotopic compositions that fall within or very near the MNOD field). Much of the sediment Pb from the highly metalliferous Bauer Deep also plots in the MORB field. Provided that postdepositional alteration and contamination has been minimal, Pb contained in the youngest, most hydrothermal sediments along the EPR may reflect the isotopic composition of Pb in the underlying source rocks, maintaining isotopic identity through the intermediate processes of partial melting, hydrothermal activity, and rapid precipitation on the sea floor. Viewed this way, Pb in these young sediments indicates heterogeneity and some apparently systematic variability in the mantle underlying the EPR from about 6° to 30.5°S (Table 1). In general, Pb is more radiogenic in sediments from the southern part of the EPR that was sampled, but the composition-latitude covariance is not exact. Four of the northernmost samples closest to the EPR (samples 491, 454, 485, 526) have Pb isotopic compositions that are linear on the isotopic plots (Figs. 3, 4), and thus could result from temporal (or mixing) characteristics of source rock within this limited region. As a group these youngest sediments show the greatest amount of the total Pb isotopic variability determined (100% of the ^{206}Pb/^{204}Pb variability, and about 50% of the ^{207}Pb/^{204}Pb and ^{208}Pb/^{204}Pb variabilities); the much higher variability for ^{206}Pb may result from greater variations in source rock concentrations of its more abundant, longer lived parent, ^{238}U. This primary variability may also cause the uniformly lower correlation coefficients involving ^{206}Pb, relative to those involving ^{207}Pb and ^{208}Pb (Table 2).

In contrast to the less radiogenic Pb from metalliferous sediments along the EPR, the most radiogenic Pb is found within sediments that show the strongest detrital influences, such as highest concentrations of Al, Th, Sc, and distance from the EPR (Table 2; Fig. 5). These data form a much tighter cluster on the isotopic composition graphs (samples 421, 772, 773, 774, 791, 796, and 1311; Figs. 3, 4) and plot within or very close to the MNOD field. In samples closer to South America, and depending on the extent of sediment mixing, one might expect wider ranges in isotopic variability (greater than the entire field of Figure 2, for ^{206}Pb and ^{207}Pb, for example). Although sediment Pb is more radiogenic as a function of increasing distance from the EPR, its isotopic variability is greatly reduced rather than enhanced. Although Pb isotopic composition in typical river-and-air-transported detritus may be quite variable, Pb concentrations are very low (Chow and Patterson, 1962). Therefore, I believe that much of the Pb in the sediment results from sea-water precipitation, a general conclusion of Chow and Patterson.

If a significant amount of Pb in sediments away from the EPR is derived from Pb dissolved in sea water, this source apparently is characterized by a very restricted isotopic range. Such a restricted range for the "end-member," most radiogenic Pb must result from a realtively restricted or homogeneous Pb isotopic marine reservoir in this section of the Pacific. The residence time for Pb in sea water is short ($\leq$400 yr—Goldberg and others, 1971; perhaps 50 yr—Somayajula and Craig, 1976) relative to the mixing rates for the oceans, and Pb in pelagic sediments of the world oceans is isotopically variable. (Extensive regions of approximate isotopic homogeneity, however, can be seen in the ^{206}Pb/^{207}Pb plots of Chow and Patterson, 1962). Consequently, modern (precontamination) marine Pb from the northern Nazca plate region of the Pacific may have been reasonably homogeneous, with an isotopic composition approximated by the cluster of points centering around the lower left portion of MNOD on Figures 3 and 4. If Pb in these sediments becomes homogenized (and more radiogenic) as a function of contact with sea water or pore waters, the isotopic data, when plotted for example against distance from the EPR (Fig. 5), may wedge toward, or point to, the approximate Pb isotopic composition of sea water for this region of the Pacific. This graphical analysis, which is strongly controlled by the most radiogenic samples, gives the following composition for northern Nazca plate sea water Pb over the time period of sedimentation and diagenesis: ^{206}Pb/^{204}Pb ~ 18.73; ^{207}Pb/^{204}Pb ~ 15.63; ^{208}Pb/^{204}Pb ~ 38.73.

ACKNOWLEDGMENTS

The IDOE-NSF Nazca Plate Project funded the purchase and installation of our National Bureau of Standards, Shields-type mass spectrometer, its associated chemical laboratory, and most of the maintenance and recent improvements to the spectrometer and laboratory. The NBS staff, especially W. R. Shields, I. L. Barnes, W. A., Bowman, E. L. Garner, and L. J. Moore, were generous in their help with the early operation of the spectrometer. We thank L. D. Kulm for his early and continuing support of this research, through his direction of the Project.

I have benefited from discussions with Ronald G. Senechal, Jack Dymond, John B. Corliss, and G. Ross Heath. Ronald G. Senechal performed the analytical work. Charlotte M. Murati was very helpful with data reduction. Jack Dymond and Ronald G. Senechal reviewed and improved the manuscript.

REFERENCES CITED

Barnes, I. L., and others, 1973, Lead separation by anodic deposition and isotope ratio mass spectrometry of microgram and smaller samples: Analytica Chimica Acta, v. 45, p. 1881–1884.

Bostrom, K., and Peterson, M.N.A., 1966, The origin of aluminum-poor ferromanganous sediments in areas of high heat flow on the East Pacific Rise: Marine Geology, v. 7, p. 427–447.

Chow, T. J., and Patterson, C. C., 1959, Lead isotopes in manganese nodules: Geochimica et Cosmochimica Acta, v. 17, p. 21–31.

——1962, The occurrence and significance of lead isotopes in pelagic sediments: Geochimica et Cosmochimica Acta, v. 26, p. 263–308.

Chow, T. J., and others, 1973, Lead pollution: Records in southern California coastal sediments: Science, v. 181, p. 551–552.

Dasch, E. J., 1969, Strontium isotopes in weathering profiles, deep-sea sediments, and sedimentary rocks: Geochimica et Cosmochimica Acta, v. 33, p. 1521–1552.

Dasch, E. J., Dymond, J., and Heath, G. R., 1971, Isotopic analysis of metalliferous sediment from the East Pacific Rise: Earth and Planetary Science Letters, v. 13, p. 175–180.

Dymond, J., 1981, The geochemistry of Nazca plate surface sediments: An evaluation of hydrothermal, biogenic, detrital, and hydrogenous sources, *in* Kulm, L. D., and others, eds., Nazca Plate: Crustal formation and Andean convergence: Geological Society of America Memoir 154 (this volume).

Dymond, J., and others, 1973, Origin of metalliferous sediments from the Pacific Ocean: Geological Society of America Bulletin, v. 84, p. 3355–3372.

Dymond, J., and Eklund, W., 1978, A microprobe study of metalliferous sediment components: Earth and Planetary Science Letters, v. 40, p. 243–251.

Field, C. W., and others, 1976, Metallogenesis in the southeast Pacific Ocean: the Nazca Plate Project: American Association of Petroleum Geologists Memoir, no. 25, p. 539–550.

Goldberg, E. D., and others, 1971, Radioactivity in the marine environment, *in* Marine chemistry: Washington, National Academy of Sciences, p. 137–146.

Heath, G. R., and Dymond, J., 1977, Genesis and diagenesis of metalliferous sediments from the East Pacific Rise, Bauer Deep, and Central Basin, northwest Nazca plate: Geological Society of America Bulletin, v. 88, p. 723–733.

——1981, Distribution of metalliferous sediments on the Nazca plate through time and space, *in* Kulm, L. D., and others, eds., Nazca Plate: Crustal formation and Andean convergence: Geological Society of America Memoir 154 (this volume).

Lyle, M., Dymond, J., and Heath, G. R., 1977, Copper-nickel-enriched ferromanganese nodules and associated crust from the Bauer Basin, northwest Nazca plate: Earth and Planetary Science Letters, v. 35, p. 55–64.

Murray, J., and Renard, A. F., 1891, Deep sea deposits: Report of the scientific results of the H.M.S. *Challenger*, 1873–1876: Her Majesty's Government Printing Office.

Patterson, C. C., 1971, Lead, *in* Hood, D. W., ed., Impingement of man on the oceans: New York, John Wiley & Sons, p. 245–258.

——1974, Lead in seawater: Science v. 183, p. 553–554.

Reynolds, P. H., and Dasch, E. J., 1971, Lead isotopes in marine manganese nodules and the ore-lead growth curve: Journal of Geophysical Research, v. 76, p. 5124–5129.

Senechal, R. G., and Dasch, E. J., 1974, The Oregon State University solid-source mass spectrometer: A new instrument for research and instruction. Part I: Description and system evaluation: Proceedings of the Oregon Academy of Sciences, v. 10, p. 14–46.

Senechal, R. G., and others, 1975, Nazca plate metalliferous sediments: III. Isotopic, elemental, and mineralogic distributions in size fractions of a Bauer Deep sample: EOS (Transactions, American Geophysical Union), v. 56, p. 446.

Somayajulu, B.L.K., and Craig, H., 1976, Particulate and soluble ^{210}Pb activities in the deep Sea: Earth and Planetary Science Letters, v. 32, p. 268–276.

Sun, S. S., 1973, Lead isotope studies of young volcanic rocks from oceanic islands, mid-ocean rises, and island arcs [Ph.D. dissert.]: New York, Columbia University, 139 p.

Tatsumoto, M., 1966, Genetic relations of oceanic basalts as indicated by lead isotopes: Science, v. 153, p. 1094–1101.

Unruh, D. M., and Tatsumoto, M., 1976, Lead isotopic composition and uranium, thorium and lead concentrations in sediments and basalts from the Nazca plate, *in* Yeats, R. S., and Hart, S. R., and others, eds., Initial report of the Deep Sea Drilling Project, Volume 34: Washington, D.C., U.S. Government Printing Office, p. 341–347.

Manuscript Received by the Society November 12, 1980
Manuscript Accepted December 30, 1980

Geological Society of America
Memoir 154
1981

Sediment accumulation rate patterns on the northwest Nazca plate

G. M. McMurtry
Department of Oceanography
University of Hawaii
Honolulu, Hawaii 96822

H. H. Veeh
School of Earth Sciences
The Flinders University of South Australia
Bedford Park, S. A. (Australia)

C. Moser
School of Oceanography
Oregon State University
Corvallis, Oregon 97331

ABSTRACT

More than 50 sediment cores from the Nazca plate have now been analyzed for sedimentation rates and bulk chemical composition. Sedimentation rates determined by direct assessment of unsupported ionium (^{230}Th) and microfossil zonation have been augmented by gamma-ray spectrometry techniques; a comparison of sedimentation rates determined by both direct (alpha) and indirect (gamma) ^{230}Th spectrometric techniques on 6 cores shows a general agreement between the two methods. Measured and calculated dry bulk densities and bulk chemical composition were used to calculate regional patterns of metal accumulation from the sedimentation rates. The regional pattern of Fe accumulation parallels those for Mn and U, whereas those of Cu, biogenic Si, Al, Ti, and Th generally display an inverse relationship. Multivariate factor analysis was employed to include correlation of accumulation rates for Na, K, Mg, Ca, Ba, Ni, Zn, and P, and to distinguish hydrothermal from detrital, hydrogenous, biogenous, and diagenetic associations. A ^{230}Th inventory, defined as the ratio of the ^{230}Th accumulation rate to the ^{230}Th sea-water production rate, was employed as an index of sediment loss or enhanced sediment accumulation. The ^{230}Th inventory is greatest on the East Pacific Rise, which suggests that some sediment ponding occurs there; it is generally lowest in the Bauer Basin, which suggests that sediment has been lost there by either coring disturbance, $CaCO_3$ dissolution, or bottom-current winnowing.

The regional pattern of Fe and Mn accumulation indicates that their rates of accumulation on the crest of the East Pacific Rise (EPR) vary by more than an order of magnitude; values near the equator (6°S) correspond to normal authigenic accumulation and exponentially increase toward 20°S latitude. The lower Fe and Mn accumulation rates in the northern EPR area suggest that (1) there is preferential loss of the metalliferous component to the north because of increased bottom-current activity, or (2) hydrothermal circulation is more intense to the south because of a lack of regional sediment cover or more extensive fracturing, or (3) mantle heterogeneity exists along the rise crest. Bottom-current winnowing is the explanation most consistent with the existing data on the distribution of conductive heat flow and crustal production along the EPR, and can explain both the variation in metal accumulation along the EPR crest and the appearance of a second area of metal enrichment in the Bauer Basin. Global budget calculations indicate that roughly 50% or more of the Mn produced at ridge crests may be lost to the adjacent sea floor by this mechanism, which has implications for the magnitude of hydrothermal input to the ocean and the origin of Mn nodules.

INTRODUCTION

Studies of the distribution of transition metals and other metals in sediments of the southeast Pacific have revealed the regional association of metal enrichment to the EPR spreading center (Bostrom and Peterson, 1966, 1969; Bostrom and others, 1969; Bostrom, 1973; Piper, 1973). This association led early workers to the conclusion that the cause of enrichment was hydrothermal deposition in association with local volcanism. Although additional circumstantial evidence, including the sampling of emanating submarine hot springs (Corliss and others, 1979; Spiess and others, 1980), now indicates that these deposits do indeed form by precipitation from submarine hydrothermal solutions, the direct application of the various geochemical parameters used to investigate the origins of these unusual deposits has generally led to ambiguous and inconclusive results. For example, $^{234}U/^{238}U$, $^{34}S/^{32}S$, $^{18}O/^{16}O$, and $^{87}Sr/^{86}Sr$ ratios and rare-earth-element patterns indicate a sea-water origin for those elements, whereas $^{207}Pb/^{206}Pb$ ratios indicate that this element derives from basalt (Bender and others, 1971; Dasch and others, 1971; Dymond and others, 1973). Further, the $^{207}Pb/^{206}Pb$ data are ambiguous in that these ratios do not allow separation of Pb derived from hydrothermal leaching from that derived from weathering or deuteric alteration (Dymond and others, 1973). The only supposedly unambiguous test for a hydrothermal origin has been to compare rates of metal accumulation. Bostrom (1970) and Bender and others (1971) first compared metal accumulation rates from the EPR to other areas on the sea floor and found that Fe and Mn were accumulating at about 25 to 30 times the normal authigenic rate. These workers concluded that at least Fe, Mn, and Pb were of local volcanic origin.

McMurtry and Burnett (1975) studied the rates of metal accumulation on a profile across the Nazca plate and found rates of Fe and Mn accumulation in a core from the Bauer Basin that are several times the normal authigenic values, whereas Fe and Mn are accumulating at close to these values in cores from the adjacent Galapagos Rise and EPR (Fig. 1). Dymond and Veeh (1975) studied metal accumulation rates on a profile close to that of McMurtry and Burnett (1975). These workers concluded that although the existing data ruled out normal authigenic precipitation of Fe and Mn in the EPR and Bauer Basin areas, they could not unequivocally distinguish between the "transport" and "localized" hypotheses of origin for the Bauer Basin deposits. Dymond and Veeh (1975) also applied a ^{230}Th inventory[1] to the radiometrically dated cores in these areas and found that ^{230}Th accumulation exceeds its sea-water production on the EPR, whereas ^{230}Th accumulation is less than its sea-water production in the Bauer Basin. This situation is opposite to observations in the southwest Pacific (Cochran and Osmond, 1976) that topographic highs exhibit lower ^{230}Th inventories because of bottom-current winnowing and transport, whereas basins generally display higher inventories because of enhanced sedimentation in topographic lows.

Dymond and Veeh (1975) attributed this reverse pattern of ^{230}Th accumulation to different removal processes in the EPR and Bauer Basin areas. An alternative explanation is that the EPR sediments are ponded, whereas the Bauer Basin sediments are winnowed by bottom currents. Ponding of rise-crest

---

1. The ^{230}Th inventory = excess ^{230}Th accumulation rate/excess ^{230}Th sea-water production rate. Ideally, ^{230}Th inventory = 1.0.

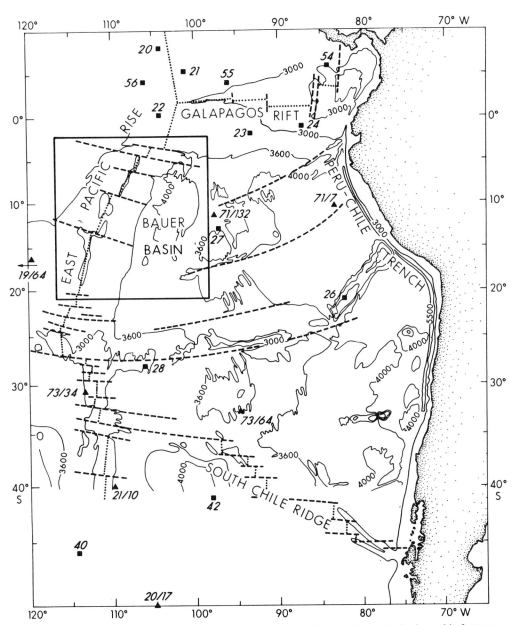

Figure 1. Index map of the Nazca plate, showing station locations and general physiographic features. Bathymetry after Mammerickx and others (1975) with contours in metres. Dotted lines indicate approximate positions of axial rift zones; dashed lines indicate approximate positions of major transform faults. On this and all subsequent figures, except where otherwise noted, triangles denote cores analyzed by alpha spectrometry, circles by gamma spectrometry and squares by microfossil zonation. Insert is area shown in Figure 2.

sediments would raise questions about the validity of the rapid metal accumulation rates calculated for the EPR. In addition, the lower metal accumulation rates found in an EPR core by McMurtry and Burnett (1975) would imply that the metal flux along the crest of the EPR is highly variable. Dymond and Veeh (1975) suggested that this core was anomalous. An alternative explanation would suggest that the early models of hydrothermal deposition are oversimplified.

These observations have raised fundamental questions to be addressed in this study regarding (1) the validity of metal accumulation rates as an indication of hydrothermal origin, (2) the representativeness of the rapidly accumulating EPR sediments, and (3) the extent and cause of the variability in metal accumulation on the EPR. Although 15 cores had been dated in the EPR and Bauer Basin areas, the coverage was sparse over this immense region of the sea floor. This study has employed rapid, indirect ^{230}Th assessment by gamma-ray spectrometry to more than double sedimentation rates determined by the direct assessment of ^{230}Th and microfossil zonation. Because gamma spectrometry is a relatively new technique, this study includes a comparison of the sedimentation rates derived by both the direct and indirect ^{230}Th techniques on 6 cores. To test the possibility of a variable metal flux on the crest of the EPR, 7 cores were investigated within a detailed survey area at 6°S latitude, and additional cores were investigated at latitudes ranging from 10°N to 50°S (Figs., 1, 2). Additional cores were also investigated in the Bauer Basin in an attempt to resolve the origin of these deposits. Multivariate factor analysis has been applied to the metal and other element accumulation rates to distinguish hydrothermal from biogenic, detrital, and hydrogenous

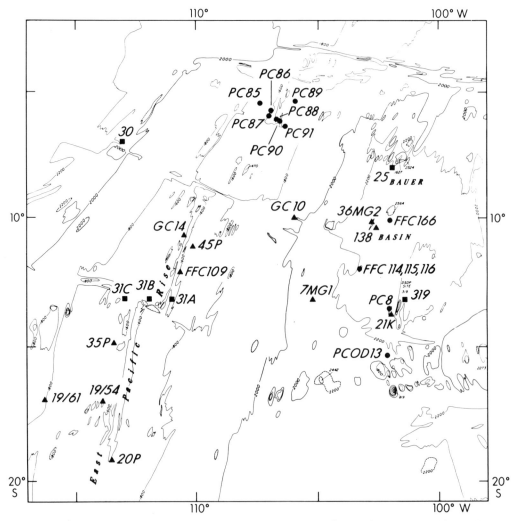

Figure 2. Station locations in the East Pacific Rise and Bauer Basin areas of the northwest Nazca plate. Contours in fathoms.

associations. Finally, a detailed assessment of the ^{230}Th inventory in the EPR and Bauer Basin region is presented in order to gain insight into the various sedimentological controls of accumulation rates.

RADIOCHEMICAL DATING

Activity versus depth plots for 18 of the cores investigated are presented in Figures 3, 4, and 5. Most of the cores have been dated at the University of Hawaii by a nondestructive, scanning NaI (Tl) gamma-ray spectrometer. Several of the cores have been dated at Oregon State University by a Ge(Li) gamma-ray spectrometer (cores Y73-3-21K, Y73-4-64K, OC73-3-20MG and P, and Y73-3-34MG and PC; see Fig. 5). Excess ^{214}Bi and ^{214}Pb was determined as the difference of the total activity and the ^{238}U-supported activity, which was either measured by alpha spectrometry or calculated by the differential method of Cochran and Osmond (1974). Excess ^{230}Th has similarly been determined by alpha spectrometry on 10 of the cores (McMurtry, 1975; Veeh, this volume), of which 6 serve as direct comparisons of the two methods (core KK71-FFC109, McMurtry, 1979; cores KK71-PC87, KK71-PC88, KK71-FFC115, Y73-3-21K and Y73-4-64K; Figs. 3, 4, 5) and two serve as comparisons of the two methods in cores taken at the same sampling site (cores OC73-3-20MG and P, and Y73-4-34MG and PC; Fig. 5).

The experimental procedure used for direct ^{230}Th measurement in this study is a modification of that of Ku (1965, 1966). The separation and purification of U and Th in the samples involved total dissolution in mixtures of HF, HClO$_4$, HCl and HNO$_3$, except for those cores analyzed by the leaching method of Goldberg and Koide (1962): cores KK71-FFC109, KK71-FFC115, KK71-FFC132, and KK71-GC07 (McMurtry, 1975). All samples were spiked with gravimetrically calibrated ^{232}U and ^{234}Th tracers, and then passed through a series of hydroxide precipitations, ion exchanges, and solvent extraction steps. The samples were finally electroplated onto stainless steel counting discs; surface-barrier detectors connected to a 1024-channel pulse-height analyzer were used to count the samples. Further details of these procedures are found in Veeh (this volume) and McMurtry (1979).

The scanning gamma-ray spectrometer developed for this study is a modification of that described by Cochran and Osmond (1974). The spectrometer was calibrated with ^{238}U (^{214}Bi) and ^{232}Th (^{208}Tl) counting standards certified to be in secular equilibrium. The standards were cast into the geometry of a core half to eliminate the necessity of absolute calibration via the estimation of detection volume. Archive core halves were sequentially counted under a lead-shielded NaI(Tl) detector connected to a 1024-channel pulse-height analyzer. Energy spectra were analyzed by the stripping technique (Rybach, 1971; Fankhauser, 1976).

The gamma-ray technique assumes that secular equilibrium has been established in the decay series between the parent nuclides (^{238}U, ^{230}Th) and the gamma-ray emitting daughter nuclides (^{214}Bi, ^{214}Pb). Disequilibrium has been observed in some cases because of the postdepositional migration of ^{226}Ra (Kröll, 1955; Koczy and Bourret, 1958; Osmond and Pollard, 1967). However, the magnitude of the calculated diffusion coefficient, 10^{-8} to 10^{-9} cm^{2}s^{-1} or smaller, indicates that ^{226}Ra disequilibria will be confined to within the topmost metre of the sediment and will not seriously affect sedimentation rate determinations for sediments accumulating at a few cm/10^3 yr or greater, or for sediments possessing uniformly high adsorption capacities (Koczy and Bourret, 1958; Goldberg and Koide, 1963; Cochran and Osmond, 1974; McMurtry, 1979). The most serious disequilibrium problem encountered with the scanning gamma-ray spectrometer was the gaseous mobility, or emanation, of ^{222}Rn. The ^{222}Rn disequilibria affects the determination of the absolute ^{230}Th activity, but does not affect the apparent sedimentation rate of cores scanned as an open system. The absolute ^{230}Th activity of samples that are individually sealed prior to analysis, such as those determined by the Ge(Li) gamma-ray spectrometer, are not affected by ^{222}Rn emanation. Corrections for ^{222}Rn emanation based on individually sealed samples have been applied to the absolute ^{230}Th activities of the scanner-analyzed cores in Table 3. Further details of these procedures are discussed elsewhere (McMurtry, 1979; McMurtry and others, in prep.).

With the exception of cores KK71-PC87 and KK72-PC008, all the activity versus depth plots have

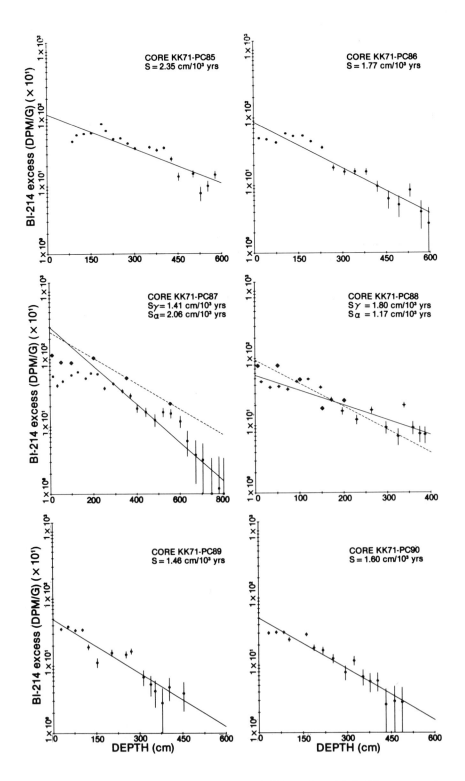

Figure 3. Plots of excess ^{214}Bi activity versus depth. Diamonds and dashed lines are excess ^{230}Th activities and regression fits, respectively. The comparatively lower excess ^{214}Bi activities of cores KK71-PC87 and KK71-PC88 are largely attributed to ^{222}Rn emanation (see Radiochemical Dating section).

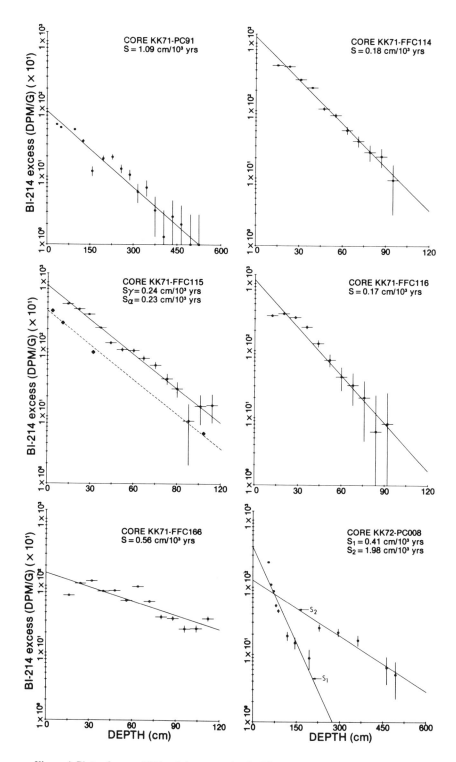

Figure 4. Plots of excess ^{214}Bi activity versus depth. Diamonds and dashed lines are excess ^{230}Th activites and regression fits, respectively. The comparatively lower excess ^{230}Th activities of core KK71-FFC115 are attributed to loss during sample preparation procedures. The core was analyzed by the leaching method of Goldberg and Koide (1962), which does not ensure total sample dissolution.

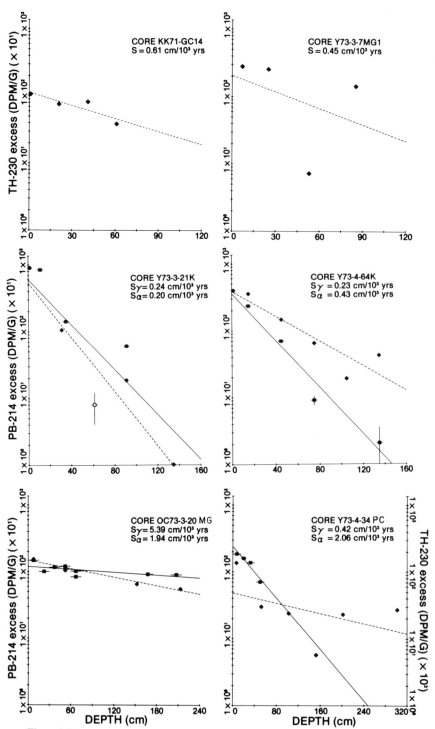

Figure 5. Plots of excess ^{214}Pb activity versus depth. Diamonds and dashed lines are excess ^{230}Th activities and regression fits, respectively. For sites OC73-3-20 and Y73-4-34; squares and solid lines represent data for gravity cores (MG) and diamonds and dashed lines represent data for piston cores (PC).

been fitted by unweighted linear regression according to the philosophy of Ku and others (1968). This treatment disregards the fine structure in the activity versus depth profiles (Goldberg, 1968), but effectively averages scatter due to small-scale physical mixing or changes in the sedimentation rate or deposition of ^{230}Th. Changes in structure on the order of 1 m or more in length are difficult to disregard, however. Analysis by both gamma and alpha spectrometry of core KK71-PC87 indicates that the top 2 m are disturbed (Fig. 3), so sedimentation rate determinations were made only on data below this depth. Similarly disturbed sections may also exist within the topmost metre of the other cores shown in Figure 3, but the cores show no visual evidence for slumping or coring disturbance. The apparent flattening of the trend of data points in the uppermost parts of these cores may be the result of ^{226}Ra migration out of the sediment (Osmond and Pollard, 1967). In any event, the inclusion of the uppermost data points should not affect the average sedimentation rates derived as greatly as that of KK71-PC87. Core KK72-PC008 displays a break in slope at approximately 225 cm, which may be due to fivefold decrease in the rate of sedimentation accompanied by about the same decrease in the deposition of ^{230}Th (Fig. 4). Similar profiles have been observed on the Mid-Atlantic Ridge (Goldberg and Griffin, 1964). However, a decrease in ^{230}Th deposition is opposite to expectations based on ^{230}Th scavenging by slowly depositing particles (Ku and others, 1968). The second apparent sedimentation rate was therefore disregarded in the following accumulation rate calculations pending further verification by alpha spectrometry.

Inspection of the gamma-alpha comparison cores in Figures 3 to 5 shows a general agreement of sedimentation rates to within 5%, 20%, and 30%, respectively, for cores KK71-FFC115, Y73-3-21K, and KK71-PC87, and to within 50% for cores KK71-PC88 and Y73-4-64K. Figure 6 presents a comparison of the sedimentation rates derived by the two methods for the 6 cores investigated in this study and for 4 cores investigated by Cochran and Osmond (1974). The error of each sedimentation rate has been estimated from the standard deviation of the slope of best fit as determined by linear regression analysis. The general trend of the cores around the 45° line of perfect fit (Fig. 6) suggests that as the sedimentation rate increases, either the agreement or the certainty of the derived sedimentation rate decreases. This can be ascribed to an increase in the probability of a violation of the uniform deposition assumption in the more rapidly accumulating cores. For example, cores KK71-PC87 and KK71-PC88 contain several ash layers and other changes in lithology (Andrews and others, 1975) that violate the assumption of uniform deposition. It is therefore likely that most of the disagreement in the sedimentation rates of these 2 cores derives from the lack of comparable coverage in the alpha and gamma methods (Fig. 3). When a closer coverage exists for both methods, the agreement is improved, as shown for those cores investigated by Cochran and Osmond (1974). This illustrates the importance of carrying out determinations at exactly the same intervals when comparing methods on cores of nonuniform lithology. It also illustrates the danger in relying on only a

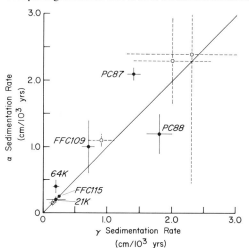

Figure 6. Comparison of sedimentation rates determined by alpha and gamma spectrometry. Errors estimated from standard deviation of slope of best fit. Open circles denote data from Cochran and Osmond (1974).

few determinations, regardless of method, for a reliable sedimentation rate in cores of nonuniform lithology. For cores such as Y73-4-64K, where more comparable sampling coverage exists, the disagreement in sedimentation rates is most likely due to a lack of uniform uranium support, as this nonuniformity violates the differential method of calculating the level of uranium support in gamma analysis (Cochran and Osmond, 1974).

There is little comparison of the results of the two methods in cores taken at the same sampling site (cores OC73-3-20P and MG, Y73-4-34PC and MG; Fig. 5), although large differences in the sedimentation rates of closely spaced cores have been observed elsewhere (Goldberg and Koide, 1963; Cochran and Osmond, 1976). In choosing a particular sedimentation rate on those cores where there is more than one method applied, the basis has been to rely on the method with the largest data coverage. Exception has been made to core KK71-PC88 because the fit of the gamma data to theoretical ^{226}Ra diffusion curves (McMurtry, 1979) indicated that the apparent alpha-derived rate was more nearly correct. In choosing a sedimentation rate for the OC73-3-20P site (Fig. 5), the alpha-derived rate was preferred because it fell more in the range with the other EPR sedimentation rates. For site Y73-4-34, the gamma-derived rate for the trigger gravity core was preferred because of its lack of scatter and more reasonable ^{230}Th inventory. Note that although there is an order of magnitude difference in activity between these cores, the topmost alpha activity in core Y73-4-34PC follows the same trend as the gamma activity of core Y73-4-34MG (Fig. 5), which suggests that the top section of the piston core has been lost in the coring operation (see ^{230}Th budget discussion below).

ELEMENT ACCUMULATION RATES

Element accumulation rates were calculated for each radiometrically dated core using the data presented in Table 1 and chemical analyses from each core (McMurtry, 1979; Dymond, unpub. data). Included are all the available rate data from the literature for the Nazca plate region. The element accumulation rates, expressed in either mg or μg · cm^{-2}10^{-3} yr^{-1}, are presented in Table 2. These rates were calculated as follows:

$$A_e = C_e \rho S, \qquad (1)$$

where A_e = the accumulation rate of element e, C_e = the average bulk concentration of element e in either mg or μg/g, ρ = the average dry bulk density in g/cm^3, and S = the average sedimentation rate in cm/10^3 yr. Figure 7 presents the average sedimentation rates used in the calculation of the accumulation rate pattern in the EPR and Bauer Basin areas.

The dry bulk densities used in these calculations have been a problem to previous work because they were assumed to be constant at 0.70 to 0.75 g/cm^3 (Bender and others, 1971; Bostrom, 1973; McMurtry and Burnett, 1975), whereas in fact they can range by as much as 0.2 to 1.0 g/cm^3 in these sediments depending mainly on lithology. To avoid this problem, measured dry bulk densities have been used whenever possible. The precision of the measured densities is generally within 10% of the value reported (Table 1). Dry bulk densities were calculated where these measurements could not be made. The calculated dry bulk densities were obtained from empirical relations to carbonate content for southeast Pacific sediments developed by Dymond and Veeh (1975) and Lyle and Dymond (1976). These relations are

$$1/\rho_b = 0.88 - 0.20 X_c - 0.03 X_c^2 \qquad (2)$$

and

$$w = 0.83 - 0.36 X_c, \qquad (3)$$

where ρ_b = the wet bulk density in g/cm^3, w = the weight fraction of water, and X_c = the weight

TABLE 1. CORE LOCATIONS AND ACCUMULATION RATES, NAZCA PLATE

Core description	Lat (°)	Long (°W)	Water depth (m)	Average density (g/cm^3)	Sedimentation rate (cm/10^3 yr)	Bulk accumulation rate (g/cm^{2}10^3 yr)	Average carbonate (%)	Noncarbonate accumulation rate (g/cm^{2}10^3 yr)
KK71-PC85	5-29S	107-23	3357	0.46 ± 0.04	2.35 ± 0.27	1.08 ± 0.16	83	0.18 ± 0.03
KK71-PC86	5-48S	107-00	3157	0.49 ± 0.04	1.77 ± 0.14	0.87 ± 0.10	81	0.16 ± 0.02
KK71-PC87	5-55S	107-03	3170	0.49 ± 0.05	1.41 ± 0.12	0.69 ± 0.09	81	0.13 ± 0.02
KK71-PC88	6-06S	106-43	3073	0.54 ± 0.05	1.17 ± 0.30	0.63 ± 0.17	81	0.12 ± 0.03
KK71-PC89	5-23S	106-08	3160	0.61 ± 0.08	1.46 ± 0.16	0.89 ± 0.15	90	0.09 ± 0.02
KK71-PC90	6-11S	106-36	2893	0.59 ± 0.07	1.60 ± 0.11	0.94 ± 0.13	89	0.10 ± 0.01
KK71-PC91	6-24S	106-19	3285	0.63 ± 0.07	1.09 ± 0.08	0.69 ± 0.09	90	0.07 ± 0.01
KK71-FFC109*	12-05S	110-37	3079	0.56 ± 0.04	1.08 ± 0.45	0.60 ± 0.25	83	0.10 ± 0.04
KK71-FFC114	11-58S	103-22	4489	0.25 ± 0.01	0.18 ± 0.01	0.05 ± 0.01	2.1	0.04 ± 0.01
KK71-FFC115*	11-58S	103-21	4465	0.21 ± 0.01	0.24 ± 0.02	0.05 ± 0.01	1.7	0.05 ± 0.01
KK71-FFC116	11-58S	103-21	4453	0.23 ± 0.02	0.17 ± 0.01	0.04 ± 0.01	3.1	0.04 ± 0.01
KK71-FFC132*	11-33S	97-25	4094	0.34 ± 0.02	0.20 ± 0.02	0.07 ± 0.01	56	0.03 ± 0.01
KK71-FFC166	10-04S	102-09	4460	0.33 ± 0.10	0.56 ± 0.10	0.18 ± 0.06	9.7	0.17 ± 0.05
KK72-PC8	13-19S	102-06	4458	0.28 ± 0.03	0.41 ± 0.06	0.11 ± 0.02	1.4	0.11 ± 0.02
KK74-PCOD13	15-13S	102-10	4309	0.30 ± 0.04	0.31 ± 0.09	0.09 ± 0.03	1.4	0.09 ± 0.03
KK71-GC7*	10-31S	83-04	4520	0.41 ± 0.07	0.23 ± 0.01	0.09 ± 0.02	0	0.09 ± 0.02
KK71-GC10+	9-59S	106-02	3447	0.74 ± 0.07	0.43 ± 0.05	0.32 ± 0.05	90	0.03 ± 0.01
KK71-GC14	10-37S	110-31	3095	0.71 ± 0.07	0.61 ± 0.23	0.43 ± 0.17	54	0.20 ± 0.08
OC73-3-20P	19-15S	113-35	3081	0.38§± 0.10	1.94 ± 0.34	0.74 ± 0.23	40	0.44 ± 0.14
Y73-4-34PC	31-09S	113-21	3064	0.70§± 0.18	0.42 ± 0.10	0.29 ± 0.10	73	0.08 ± 0.03
Y73-3-7MG1	13-00S	105-14	3832	0.62§± 0.16	0.49 ± 0.82	0.30 ± 0.51	75	0.08 ± 0.13
Y73-3-21K	13-37S	102-32	4410	0.28 ± 0.03	0.20 ± 0.06	0.06 ± 0.02	5.1	0.06 ± 0.02
Y73-4-64K	32-46S	94-42	3871	0.43 ± 0.04	0.43 ± 0.09	0.18 ± 0.04	10.2	0.16 ± 0.04
Y71-7-36MG2+	10-08S	103-51	4541	0.22 ± 0.02	0.14 ± 0.03	0.03 ± 0.01	1.4	0.03 ± 0.01
Y71-7-45P+	11-05S	110-06	3096	0.70 ± 0.07	0.93 ± 0.11	0.65 ± 0.10	89	0.07 ± 0.01
V19-54#	17-02S	113-54	2830	0.52§± 0.13	1.50 ± 0.19	0.78 ± 0.22	65	0.27 ± 0.08
V19-61#	16-57S	116-18	3407	0.73§± 0.18	0.70 ± 0.07	0.51 ± 0.14	90	0.05 ± 0.01
V19-64#	16-56S	121-12	3540	0.73§± 0.18	0.45 ± 0.04	0.33 ± 0.09	90	0.03 ± 0.01
GS-7202-35P**	14-48S	113-30	3044	0.59 ± 0.06	0.74 ± 0.20	0.44 ± 0.13	68	0.14 ± 0.04
E21-10††	40-00S	109-51	3146	0.72§± 0.18	0.89 ± 0.19	0.64 ± 0.21	89	0.07 ± 0.02
E20-17††	50-51S	104-56	3895	0.73§± 0.18	2.79 ± 0.59	2.04 ± 0.66	90	0.20 ± 0.07
138§§	10-22S	102-38	4267	0.25 ± 0.03	0.25 ± 0.02	0.06 ± 0.01	1.0	0.06 ± 0.01
DSDP Site 319+##	13-01S	101-31	4290	0.24§	0.11	0.03	11	0.03
20##	8-00N	104-00	..	0.19§	1.20	0.23	0	0.23
21##	5-00N	101-00	..	0.19§	1.20	0.23	0	0.23
22##	0-00N	104-00	..	0.19§	0.70	0.13	0	0.13
23##	2-00N	93-00	..	0.19§	1.20	0.23	0	0.23
24##	1-00N	87-00	..	0.19§	1.10	0.21	0	0.21
25##	8-00N	102-00	..	0.19§	0.73	0.14	0	0.14
26##	21-00S	82-00	..	0.19§	0.04	0.01	0	0.01
27##	13-00S	97-00	..	0.19§	0.15	0.03	0	0.03
28##	28-00S	106-00	..	0.19§	0.08	0.02	0	0.02
30##	7-00S	113-00	..	0.19§	0.16	0.03	0	0.03
31A##	13-00S	111-00	..	0.19§	1.40	0.27	0	0.27
31B##	13-00S	112-00	..	0.19§	0.52	0.10	0	0.10
31C##	13-00S	113-00	..	0.19§	0.40	0.08	0	0.08
40##	46-00S	114-00	..	0.19§	0.06	0.01	0	0.01
42##	41-00S	98-00	..	0.19§	0.06	0.01	0	0.01
54##	6-00N	84-00	..	0.19§	2.90	0.55	0	0.55
55##	4-00N	96-00	..	0.19§	0.53	0.10	0	0.10
56##	4-00N	106-00	..	0.19§	0.65	0.12	0	0.12
61##	13-00S	113-00	..	0.19§	0.38	0.07	0	0.07

*Data from McMurtry, 1975.

†Data from Dymond and Veeh, 1975.

§Calculated dry bulk densities using the empirical relationships of Dymond and Veeh, 1975; Lyle and Dymond, 1976.

#Data from Bender and others, 1971.

**Data from Rydell and others, 1974.

††Data from Kraemer, 1971.

§§Data from Sayles and others, 1975.

##Sedimentation rate from microfaunal determinations; noncarbonate sedimentation rates from Bostrom, 1973.

TABLE 2. ELEMENT ACCUMULATION RATES, NAZCA PLATE

Core Description	Fe	Mn	Si	Si*	Al	Ti	Na	K	Mg	Ca+	P	Ba	Ni	Cu	Zn	U	Th
	(mg/cm²/10³ yrs)												(μg/cm²/10³ yrs)				
KK71-PC85	15	1.4	64	61	1.4	0.65	1.4	1.2	6.0	3.1	0.85	...	...	290	39	0.84	1.1
KK71-PC86	16	1.1	50	41	3.4	0.37	1.4	1.4	5.5	4.2	0.80	...	...	240	34	0.24	0.7
KK71-PC87	14	2.3	35	30	1.7	0.21	1.4	1.3	4.5	5.9	1.0	...	...	190	24	0.25	0.4
KK71-PC88	7.4	0.68	41	26	5.7	0.53	1.5	0.68	4.7	4.1	0.22	...	...	140	19	0.13	0.3
KK71-PC89	6.9	0.68	41	30	3.9	0.65	1.5	0.44	3.7	3.3	0.31	...	...	190	26	0.16	0.6
KK71-PC90	8.1	1.8	21	8.6	4.6	0.62	1.2	1.1	4.5	11	0.53	...	...	190	22	0.20	0.6
KK71-PC91	4.1	0.86	22	20	0.91	0.45	0.87	0.46	1.7	1.9	0.36	...	...	170	24	0.14	0.5
KK71-FFC109	9.5	2.7	14	3.5	4.0	0.00	16s	1.8	3.5	4.2	0.92	...	66	90	...	1.4	0.1
KK71-FFC114	7.0	2.7	6.9	1.6	2.0	0.23	0.39	0.46	1.0	0.72	0.33	...	...	49	5	0.12	0.3
KK71-FFC115	7.9	3.3	6.5	2.7	1.4	0.08	2.8s	0.69	0.99	0.99	0.29	...	66	46	...	0.14	0.3
KK71-FFC116	6.6	2.3	5.4	1.3	1.6	0.19	0.58	0.43	0.82	0.52	0.29	...	...	41	4	0.12	0.3
KK71-FFC132	3.2	1.5	8.6	3.3	2.0	0.05	2.7s	0.70	1.1	2.8	0.51	...	45	33	...	0.11	0.4
KK71-FFC166	25	4.4	36	16	7.4	0.89	2.6	1.6	4.2	1.2	0.75	...	...	140	22	0.23	0.6
KK72-PC8	22	6.5	17	3.3	5.0	0.58	1.6	1.3	2.7	2.0	1.1	...	...	120	13	0.42	0.6
KK74-PCOD13	9.6	2.8	8.4	1.4	2.6	0.30	0.92	0.77	1.4	2.4	0.96	...	...	70	7	0.31	0.4
KK71-GC7	4.4	0.22	25	7.9	6.5	0.31	2.9s	2.3	1.5	0.77	0.08	...	23	15	...	0.52	0.4
KK71-GC10	1.3	0.45	2.3	1.5	0.28	...	...	...	...	...	...	0.62	4	14	4	0.05	0.1
KK71-GC14	34	12	12	11	0.42	...	...	...	...	...	...	0.58	...	170	69	2.8	0.2
OC73-3-20P	120	36	28	18	4.1	...	...	...	...	...	...	0.50	...	590	250	4.7	0.2
Y73-4-34PC	21	6.8	4.4	1.6	1.0	...	...	...	...	...	...	2.1	...	80	42	0.36	0.4
Y73-3-7MG1	8.0	3.2	9.3	3.3	2.2	...	...	...	...	...	...	0.96	...	72	29	0.12	0.2
Y73-3-21K	8.7	3.1	7.9	3.7	1.6	...	...	...	...	...	...	1.3	...	65	24	0.21	0.2
Y73-4-64K	37	10	17	0.0	6.7	...	...	...	...	...	...	...	...	240	79	0.65	1.1
Y71-7-36MG2	4.3	0.99	4.8	3.2	0.63	...	...	...	...	...	...	...	16	25	11	0.05	0.1
Y71-7-45P	11	5.8	4.8	2.7	0.78	...	...	...	...	...	...	...	56	51	21	0.32	0.2
V19-54	82	28	19	15	1.2	...	...	...	...	...	...	...	160	...	...	4.0	0.2
V19-61	10	2.0	6.2	1.9	1.6	...	...	...	...	...	...	...	27	...	...	0.27	0.1
V16-64	3.8	0.63	3.1	1.7	0.52	...	...	...	...	...	...	...	17	...	...	0.11	0.1

Sample																	
GS-7202-35P	40	15		9.3	8.2	0.45	0.05	..	..	..	2.6	0.61	97	210	89	6.5	..
E21-10	27	11		..	..	..	..	..	..	..	..	..	15	89	61	0.38	0.6
E20-17	11	3.2		..	..	..	..	..	..	..	..	..	20	160	44	0.17	2.2
138	12	3.4	11	6.7	1.4	..	0.44	0.58	1.5	0.77	..	0.88	61	81	41	0.18	..
DSDP Site 319#	3.5	1.0	3.6	1.7	0.64	..	..	..	..	..	..	0.05	19	24	9	..	..
20	16	4.9	27	6.8	7.5	0.36	..	..	..	..	..	1.5	51	72	..	..	..
21	14	3.0	34	6.3		0.52	..	..	..	..	..	1.1	82	52	..	..	..
22	2.9	0.33			1.3		..	..	..	..	..	0.40	11	35	..	..	..
23	8.0	3.2	67	56	4.1	0.18	..	..	..	..	..				..	..	..
24	5.9	0.56	43	20	8.8	0.40	..	..	..	..	..	0.69	27	30	..	..	..
25	18	5.7	16		4.4	0.21	..	..	..	..	..	2.1	140	180	..	..	..
26	0.51	0.05	1.1	0.14	0.35	0.03	..	..	..	..	..	0.09	1	2	..	..	..
27	2.8	0.88	3.2	0.25	1.1	0.04	..	..	..	..	..	0.77	25	25	..	..	..
28	2.3	0.58	1.4	0.48	0.33	0.06	..	..	..	..	..	0.10	12	20	..	..	..
30	2.6	0.82	5.1	4.5	0.22	0.01	..	..	..	..	..	0.18	5	12	..	..	..
31A	56	18	16	14	0.67	0.07	..	..	..	..	..	0.40	110	200	..	..	..
31B	23	8.7	6.0	4.7	0.49	0.02	..	..	..	..	..	0.25	51	85	..	..	..
31C	17	6.0	3.9	2.3	0.61	0.02	..	..	..	..	..	0.19	36	69	..	..	..
40	0.89	0.29	0.85	0.19	0.25	0.01	..	..	..	..	..	0.10	2	5	..	..	..
42	1.60	0.54	1.7	0.00	0.65	0.04	..	..	..	..	..	0.11	8	9	..	..	..
54	33	0.28	140	19	46		..	..	..	..	..		46	24	..	..	..
55	5.9	2.7	20	5.7	5.3	0.20	..	..	..	..	..	0.58	40	24	..	..	..
56	5.9	3.6	12	2.4	3.5	0.16	..	..	..	..	..	0.73	42	41	..	..	..
61	20	7.7	4.7	4.1	0.23	0.02	..	..	..	..	..	0.30	49	110	..	..	..

*Biogenic Si calculated from excess over average crustal ratio of Si/Al = 2.65.

+Non-carbonate Ca calculated from difference of Ca determined by bulk analysis and gas-volume $CaCO_3$ analysis.

§Samples not washed to remove sea salts.

#Additional analyses from Dymond and others, 1977.

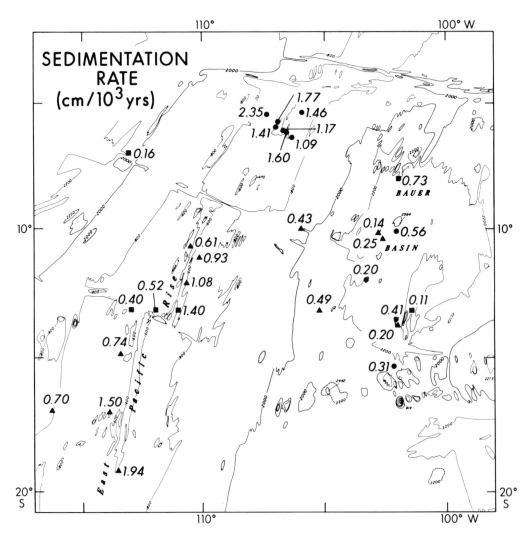

Figure 7. Sedimentation rate pattern on the northwest Nazca plate. Units of cm/10^3 yr. Contours in fathoms in this figure and in following Figures 8 to 16.

fraction of carbonate. The dry bulk density, ρ, is then obtained by

$$\rho = \rho_b(1.00 - w) . \qquad (4)$$

The precision of the calculated densities is about 25% of the reported value (Lyle and Dymond, 1976). Figure 8 presents the average density of the cores in the EPR and Bauer Basin areas.

Chemical analyses were made by X-ray fluorescence spectrometry at the University of Hawaii and by atomic absorption spectrophotometry and neutron activation analysis at Oregon State University. Instrument precision for the ARL model 72000 X-ray fluorescence quantometer used in this study is within 6% for all the major elements presented in Table 2, with Na displaying by far the largest variability (5.8%). Precision for the trace elements is generally lower, with values ranging from within 3% for U, 15% for Cu and Ni, to within 50% for Th and Zn. U was determined by alpha spectrometry, whereas Th was determined by both alpha and gamma spectrometry. For more detailed information

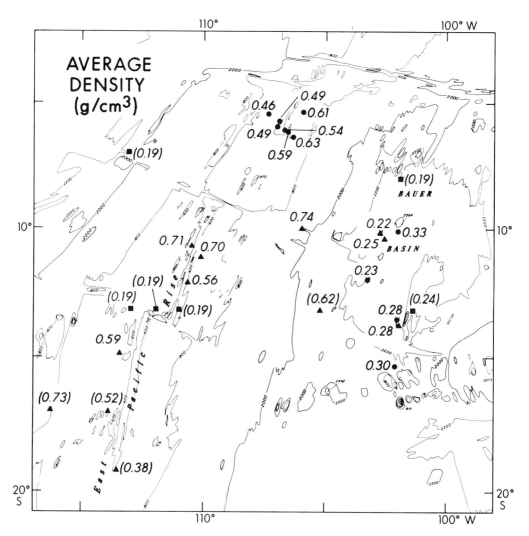

Figure 8. Average measured densities of cores on the northwest Nazca plate. Values in parentheses are calculated from empirical relation to $CaCO_3$.

on analytical procedures, see Dymond and others (1976, 1977), McMurtry (1975, 1979) and Veeh (this volume).

An inspection of the errors estimated for the sedimentation rates (Table 1) shows a range from less than 10% to greater than 50% depending on the scatter and on the data coverage in each core. Most of the sedimentation rates display a precision within 20% of the reported value, in good agreement with the findings of Ku and others (1968). The accuracy of these values can be estimated from the comparison of the alpha and gamma methods on the same core; most cores display an agreement of sedimentation rates to within 50% (Radiochemical Dating section). Inspection of the errors propagated from the bulk accumulation rates (Table 1), which are used in the calculation of the element accumulation rates in Table 2, indicates that these values are generally within 30% of the reported value. Values displaying the greatest uncertainty in Table 2 are accordingly those for Th and Zn and all values for cores displaying bulk accumulation rate errors in excess of 50% in Table 1. Unfortunately, no estimation of error is possible for those cores taken from Bostrom (1973).

ACCUMULATION RATE PATTERNS

Figures 9 through 16 present the accumulation pattern for Fe, Mn, U, biogenic Si[2], Cu, Al, Ti, and Th, respectively, in the East Pacific Rise and Bauer Basin areas of the northwestern Nazca plate. The patterns of accumulation for Fe, Mn, and U correlate strongly in these areas, the regression coefficients for Mn to Fe and U equaling 0.98 and 0.83, respectively, and there is a systematic increase in their accumulation along the crest of the EPR toward 20°S latitude (Figs. 9, 10, 11). Here, Mn is accumulating at 28 to 36 times the average authigenic rate established by Bender and others (1970). This indicates that this part of the EPR may be dominated by hydrothermal deposition. The accumulation pattern in the adjacent Bauer Basin displays no systematic areal trend for these elements, but does indicate that Mn is accumulating in this basin at 3 to 6 times the average authigenic

---

2. Determined indirectly as the excess amount over that required to match the crustal ratio of $Si/Al = 2.65$. This procedure, which corrects for detrital silica, has been criticized because it generally underestimates the amount of structural silica (Leinen, 1977). However, most of the nondetrital structural silica in these sediments is thought to be ultimately of biogenic origin (Heath and Dymond, 1977).

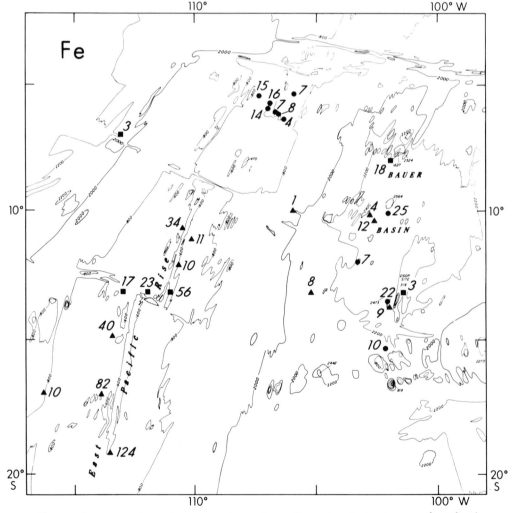

Figure 9. Fe accumulation rate pattern on the northwest Nazca plate. Units of $mg \cdot cm^{-2} \cdot 10^{-3} \, yr^{-1}$.

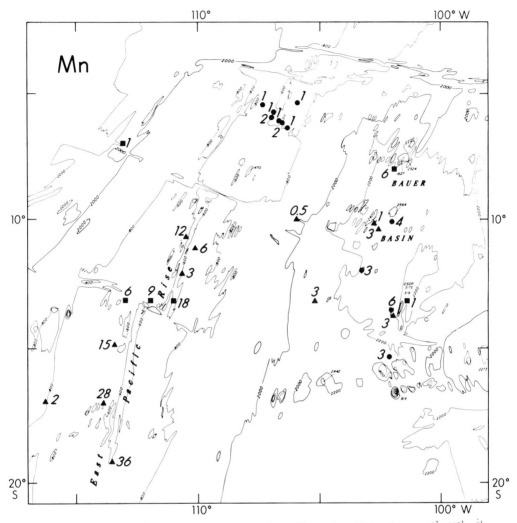

Figure 10. Mn accumulation rate pattern on the northwest Nazca plate. Units of mg · cm^{-2} · 10^{-3} yr^{-1}.

rate. Values for cores on the EPR flanks generally display the lowest rates for Fe, Mn, and U accumulation.

The large U accumulation rate values on the East Pacific Rise crest reflect enrichments for this element of approximately an order of magnitude over typical pelagic values of 1 to 3 ppm. Rydell et al. (1974) have found evidence for possible post-depositional injections of U-rich hydrothermal solutions in at least one core from this region (GS-7202-35P, Table 1), and such injections apparently enrich the sediment in U that is out of equilibrium with ^{230}Th, causing a negative ^{230}Th excess in some sections of the core. This source of enrichment, if prevalent, would have serious effects on the excess ^{230}Th dating method, and may account in part for the anomalous scatter frequently observed. However, the strong correlation of U to Fe and Mn together with the ^{234}U/^{238}U isotopic ratios of these sediments (Veeh, this volume) suggest that most of the U is absorbed onto ferromanganese phases from seawater.

Accumulation rates for Cu correlate less strongly with those for Fe, Mn, and U ($r = 0.72$ for Mn and Cu) but are highest near 20°S, where Fe, Mn, and U are also highest. There is some correlation of Cu accumulation to that of biogenic Si ($r = 0.58$), which increases northward toward the equatorial zone of high productivity where $r = 0.88$ in the 6°S EPR sediments (Figs. 12, 13). The accumulation rate patterns for Al, Ti, and Th generally parallel the trend for biogenic Si. As these elements reflect detrital inputs (Goldberg and Arrhenius, 1958; Ku, 1976), their accumulation patterns indicate an increased detrital flux to the northern sediments (Figs. 14, 15, 16). Visual inspection and chemical analysis of the northern EPR cores indicate that this detrital flux originates largely from basaltic ash inputs, with distinct layers observed in some of the cores. The regional Al accumulation is highly variable and does not correlate as closely to the regional Ti and Th accumulations ($r = 0.72$ and 0.32 for Al to Ti and Th, respectively). This may be reflecting multiple Al inputs; possibly a combination of basaltic detritus in the north and a hydrothermal input, as suggested by Dymond and Veeh (1975), which increases toward the south.

MULTIVARIATE FACTOR ANALYSIS

The accumulation rate patterns presented in Figures 9 through 16 are limited to those elements that have sufficient data coverage to produce regional patterns. To expand this coverage to include correlations to the other elements presented in Table 2, the element accumulation rate data have been

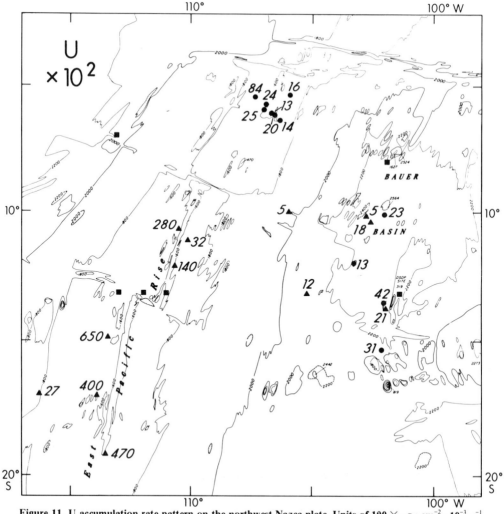

Figure 11. U accumulation rate pattern on the northwest Nazca plate. Units of $100 \times \mu g \cdot cm^{-2} \cdot 10^{-3} yr^{-1}$.

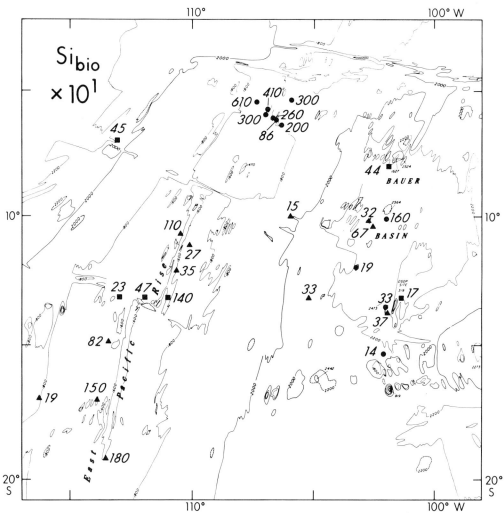

Figure 12. Biogenic Si accumulation rate pattern on the northwest Nazca plate. Units of $10 \times mg \cdot cm^{-2} \cdot 10^{-3} \, yr^{-1}$.

subjected to multivariate factor analysis. Figure 17 presents the three principal factors that explain at least 2σ (67%) of the total variance in the element accumulation rates for the entire Nazca plate, the EPR from 6° to 40°S, the Bauer Basin, and the EPR at 6°S. All the elements listed in Table 2 are represented as loadings on each varimax rotated factor, with the exception of the 6°S EPR where there are no data available for Ba and Ni. These factor loadings may be considered as standardized regression coefficients of each variable (element) to that factor (Smith, 1971), and as such, readily display interelement correlations.

Thus, factor analysis is a statistical method of showing the degree of association among variables, in this case the elements contained in these deep-sea sediments. The associations can be used as guides for the interpretation of the genesis of the sediment. That is to say, elements commonly associated together may be in the same phase, or scavenged by a common phase, or concentrated by a common process. The characterization of a factor is based on previous observations of sediment mineralogy and the location of "end-member" sediment types. A detrital sediment is characterized by the abundance of aluminosilicate phases, and a biogenous sediment ($CaCO_3$-free) is characterized by the abundance of opal. A "hydrothermal" sediment contains abundant Fe and Mn phases and occurs on

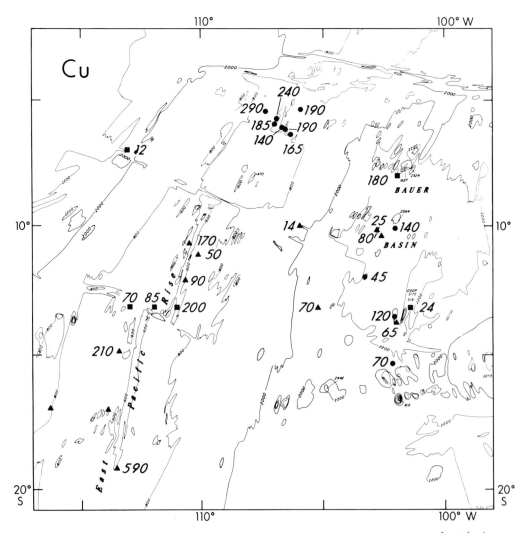

Figure 13. Cu accumulation rate pattern on the northwest Nazca plate. Units of $\mu g \cdot cm^{-2} \cdot 10^{-3} yr^{-1}$.

the EPR. Hydrogenous sediments contain authigenic phases such as barite, whereas diagenetic sediments contain authigenic phases such as smectite.

Heath and Dymond (1977) and Dymond and others (1977) have applied factor analysis to their compositional data to check their conclusions regarding hypothesized sources for the EPR and Bauer Basin sediments. Heath and Dymond (1977) found that four factors, which they suggested to represent detrital-hydrogenous, hydrothermal, biogenous, and diagenetic (smectite) sources, together accounted for 98% of the variance of scaled composition data. These genetic designations were based on the following factor loadings: factor 1 (detrital-hydrogenous)—high loadings on Al, leachable Ni, Ba, and residual Fe; factor 2 (hydrothermal)—high loadings on leachable Fe, Zn, Cu, and Mn; factor 3 (biogenous)—high loadings on Si and Ba; and factor 4 (diagenetic)—high loadings on residual Cu, Fe, and Zn, with a strong negative association to Al. Dymond and others (1977) found that only three factors accounted for most of the variance in the Bauer Basin. These factors were designated biogenous (Si, with minor Fe, Cu, Zn, and Ba), hydrothermal (Mn, Fe, Cu, and Zn), and combined detrital-hydrogenous (Al, Ni, and Ba). The appearance of the same element in more than one factor

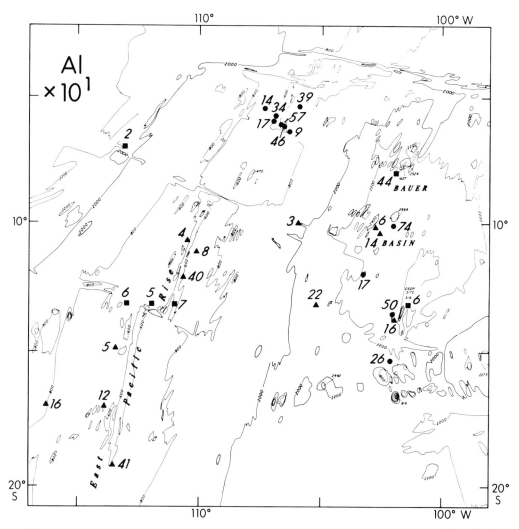

Figure 14. Al accumulation rate pattern on the northwest Nazca plate. Units of $10 \times mg \cdot cm^{-2} \cdot 10^{-3} yr^{-1}$.

suggests its partitioning among the various sediment sources. The appearance of a combined factor suggests either a spatial or temporal association of sources.

In this study, for the element accumulation rate data on the Nazca plate, factor 1, which accounts for 33% of the total variance, is dominated by Fe, Mn, Zn, P, U, Cu, and Ni (Fig. 17a). This factor can therefore be assigned as hydrothermal, and its loadings display the extent of correlation among all the elements. Unfortunately, these correlations cannot distinguish between elements that are truly of hydrothermal origin and those that are subsequently adsorbed onto surface-active Fe and Mn hydroxides. P and U are probably adsorbed from sea water as PO_4^{3-} and UO_2^{2+} radicals (Berner, 1973; Veeh, this volume). Adsorption from sea water may also account for a large part of the Cu, Ni, and Zn. Factor 2 (19% of total variance) is dominated by biogenic Si, total Si, Mg, Cu, and Ti, and minor Al and Th contributions, and appears to be a combined biogenic and diagenetic (smectite) factor. A correlation of Cu to biogenic Si has also been reported by Heath and Dymond (1977). A biogenic source for the Mg and Ti is less certain. A small fraction of the Mg may be associated with the carbonate material (Chave, 1954), and Ti has been suggested to be in small part biogenic (Goldberg

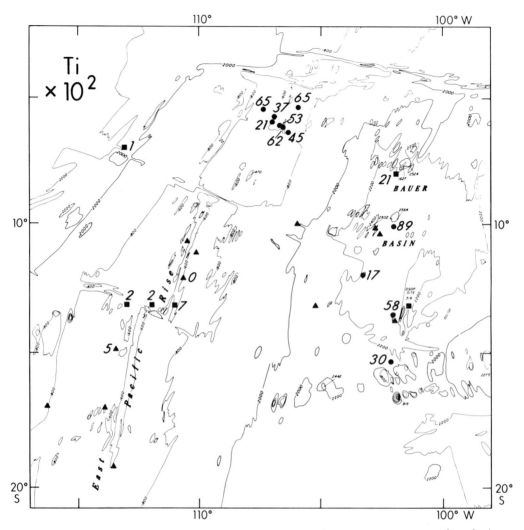

Figure 15. Ti accumulation rate pattern on the northwest Nazca plate. Units of $100 \times mg \cdot cm^{-2} \cdot 10^{-3} yr^{-1}$.

and Arrhenius, 1958). The loadings on Al and Th suggest, however, the Mg and Ti are more likely to be incorporated into a diagenetically formed aluminosilicate phase. X-ray mineralogy studies (Sayles and Bischoff, 1973; Heath and Dymond, 1977; McMurtry, unpub. data) indicate that the major mineral in the noncarbonate fraction of metalliferous sediments is an authigenic smectite. Heath and Dymond (1977) considered the smectite to be an Fe-rich and Al-poor nontronite. The factor loadings presented here suggest a more Mg-rich smectite, such as saponite.

Factor 3 (15% of total variance) is dominated by Ba, Al, Ti, total Si, Th, and Ni, and is apparently another combined factor representing hydrogenous and detrital inputs. The remaining factors in Figure 17 are based on subsets represented by fewer samples and are therefore less reliable. These factors also become increasingly difficult to interpret. The 6° to 49°S EPR subset (Fig. 17b) shows essentially the same factor loadings as the entire Nazca plate, which indicates that the entire region is affected by these EPR factors. However, when analyzed separately, the Bauer Basin displays significantly different factor loadings (Fig. 17c). Here, factor 1 (38% of total variance) is dominated by Mg, K, Ti, Al, total and biogenic Si, Th, and Fe, and may be a combined biogenic-diagenetic (smectite)

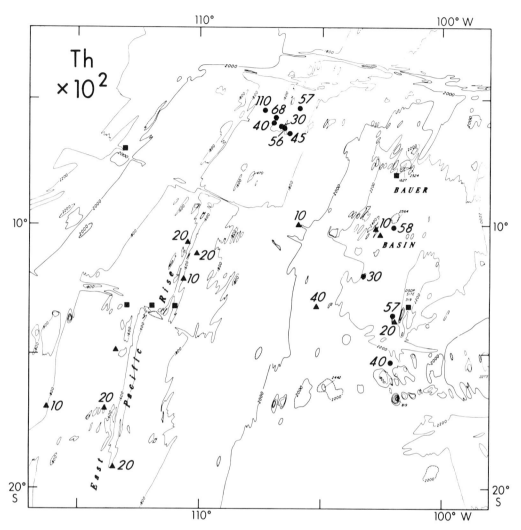

Figure 16. Th accumulation rate pattern on the northwest Nazca plate. Units of $100 \times \mu g \cdot cm^{-2} \cdot 10^{-3} yr^{-1}$.

factor. Factor 2 is high in Fe, Mn, Cu, Ni, Zn, U, total Si, Al, and Ba, possibly reflecting hydrothermal, hydrogenous, and detrital inputs. Factor 3 is dominated by Ca, P, U, and Th. This is apparently another biogenic factor representing fish-bone apatite. The association of U with fish-bone apatite in the Bauer Basin has been directly confirmed by fission-track studies (Veeh, this volume). For the 6°S EPR (Fig. 17d), there is another significant difference in the factor loadings, and in general this area's factors resemble those of the Bauer Basin more than those of the EPR. Here, factor 1 (40% of total variance) is characterized by high loadings on Cu, Zn, U, Th, total and biogenic Si, and lesser loadings on Fe, Mg, K, and Ti. This factor is similar to the Bauer Basin combined biogenic-smectite factor. Factor 2 has high loadings on Fe, Mn, P, and K, and would appear to be a hydrothermal factor. Factor 3 is most likely a detrital factor, and the high loadings on Na, Mg, Al, and Si may be reflecting the basaltic ash inputs noted above for this region of the EPR.

The results of the element accumulation rate factor analysis compares favorably with the statistical analyses of Heath and Dymond (1977) and Dymond and others (1977). Their detrital-hydrogenous factor compares well with factor 3 in Figures 17a and 17b. Factor 1 in Figures 17a and 17b corresponds

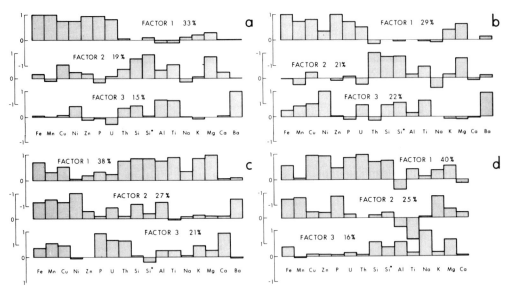

Figure 17. Contributions of the elements listed in Table 2 to the three principal varimax rotated factors (factor loadings) for (a) all Nazca plate stations; (b) 6° to 40° S EPR; (c) Bauer Basin; (d) 6° S EPR. Si = total silica, Si* = biogenic silica.

to their hydrothermal factor. Factor 2 in Figures 17a and 17b would correspond to a combination of their biogenous and diagenetic (smectite) factors. These last two factors were found to account for most of the variance in the Bauer Basin and northern EPR sediments, which is also in agreement with the results presented in Figures 17c and 17d. The extent of agreement between this and previous work is remarkable when the different approach and data sets used are considered. This study has employed element accumulation rates as opposed to scaled composition data and has analyzed the additional elements Na, K, Ca, Mg, Ti, P, U, and Th. The only major difference is the suggestion of Mg-rich, as opposed to an Fe-rich, smectite in the biogenic-diagenetic association.

SEDIMENTOLOGICAL CONTROLS—THE EXCESS ^{230}Th BUDGET

No interpretation of element accumulation rates in sediments should be made without some consideration of the regional features and oceanographic factors that can affect rates of sediment accumulation. Previous work has been limited by a lack of consideration for the effects of topography and bottom currents on sediment accumulation rates (Bostrom, 1970, 1973; Bender and others, 1971; McMurtry and Burnett, 1975; Dymond and Veeh, 1975). In order to gain some insight into these parameters, bulk and noncarbonate sediment accumulation rates have been plotted versus water depth (Fig. 18). Ideally, under conditions of uniform surface productivity, zero bottom-current movement and decreasing water depth, the bulk accumulation rate should increase with the degree of carbonate preservation. This relationship was first observed on the EPR by Broecker and Broecker (1974), who suggested that the decreased accumulation with depth of water was due to carbonate dissolution below an apparent compensation depth of about 3,950 m. The systematic relation observed in Figure 18 substantiates their work, with two exceptions. First, the considerable scatter observed at about 3,100-m water depth may be due to bottom-current winnowing (low points) and sediment buildup, or ponding (high points). Second, a core taken off the rise near the Antarctic convergence (E20-17, Fig. 1) displays anomalously high bulk (carbonate) accumulation for its water depth (Fig. 18). Cochran and Osmond (1976) found similar high carbonate accumulation off ridges in the Tasman Basin and concluded that the area was gaining sediment from bottom-current winnowing

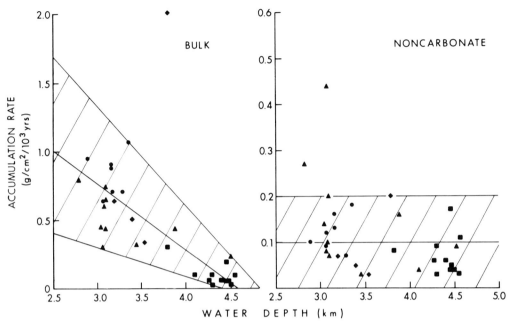

Figure 18. Bulk and noncarbonate sediment accumulation rates versus water depth. Circles, 6° EPR stations; squares, Bauer Basin stations; triangles, all other Nazca plate stations; diamonds, stations off Nazca plate.

on the South Tasman Rise, also near the Antarctic convergence. Noncarbonate accumulation averages over the Nazca plate region at 0.1 ± 0.1 g $\cdot$ cm^{-2} $\cdot$ 10^{-3}yr^{-1}, with the exception of 2 cores on the EPR crest near 20°S. These large noncarbonate accumulations may reflect increased hydrothermal deposition in this area, but could also be due to sediment ponding in rift valleys.

To more definitively test the possibility of sediment ponding on the EPR, the budget of excess ^{230}Th has been employed as an index of enhanced or depleted sediment accumulation, depending on deviations from the ideal situation where ^{230}Th accumulation in the sediment matches its production from U in the sea-water column above the sediment (Dymond and Veeh, 1975; Cochran and Osmond, 1976). When a sediment has been winnowed or eroded by bottom currents, it should display a ^{230}Th accumulation less than its sea-water production. When the sediment has gained sediment from other areas, it should display a ^{230}Th accumulation in excess of the sea-water production at that site. This excess ^{230}Th accumulation can be due to (1) the accumulation of resuspended ^{230}Th-enriched particles, or (2) the accumulation of particles that have scavenged additional ^{230}Th in transit, or (3) both of these processes. The method is based on the uniform distribution of U in sea water at 3.30 ± 0.16 µg/L (Turekian and Chan, 1971; Ku and others, 1974), which allows the ^{230}Th production rate to be calculated as:

$$F_p = DCR\lambda(^{234}U/^{238}U)_w, \qquad (5)$$

where F_p = the ^{230}Th production rate in dpm $\cdot$ cm^{-2}10^{-3}yr^{-1}, D = the water depth in centimetres, C = the U concentration in sea water (3.30 $\pm$ 0.16 µg/L), R = the U mass-activity conversion factor (7.42 $\times$ 10^5 dpm/g ^{238}U), λ = the ^{230}Th decay constant (9.2 $\times$ 10^{-6}yr^{-1}), and $(^{234}U/^{238}U)_w$ = the U activity ratio in sea water (1.14 $\pm$ 0.01) (Koide and Goldberg, 1965; Ku and others, 1974). The excess ^{230}Th accumulation rate, F_a, is calculated by substitution of the core-top extrapolated ^{230}Th activity into the C_e term in equation (1). These values and their inventory, or ratio of F_a/F_p, are presented in Table 3.

Figure 19 presents a plot of excess ^{230}Th accumulation versus water depth for the Nazca plate region. The ^{230}Th sea-water production rate curve is also shown, about which the ^{230}Th accumulations should

TABLE 3. EXCESS ^{230}THORIUM BUDGET, NAZCA PLATE

Core description	^{230}Th$_{ex}$ at surface (dpm/g)	^{230}Th$_{ex}$ accumulation (dpm/cm^{2}10^3 yrs)	^{230}Th$_{ex}$ production (dpm/cm^{2}10^3 yrs)	Inventory (accumulation/ production)
KK71-PC85	19 ± 3*	20 ± 5	8.6 ± 0.3	2.3 ± 0.5
KK71-PC86	14 ± 2*	12 ± 2	8.1 ± 0.3	1.5 ± 0.3
KK71-PC87	15 ± 5*	10 ± 4	8.2 ± 0.3	1.2 ± 0.5
KK71-PC88	10 ± 3	6 ± 2	7.9 ± 0.3	0.8 ± 0.3
KK71-PC89	8 ± 1*	7 ± 2	8.2 ± 0.3	0.8 ± 0.2
KK71-PC90	8 ± 1*	8 ± 1	7.5 ± 0.2	1.0 ± 0.2
KK71-PC91	16 ± 4*	11 ± 3	8.5 ± 0.3	1.3 ± 0.3
KK71-FFC109	12 ± 3	7 ± 3	7.9 ± 0.3	0.9 ± 0.4
KK71-FFC114	141 ± 21*	7 ± 1	11.6 ± 0.4	0.6 ± 0.1
KK71-FFC115	138 ± 25*	7 ± 1	11.5 ± 0.4	0.6 ± 0.1
KK71-FFC116	176 ± 42*	7 ± 2	11.5 ± 0.4	0.6 ± 0.2
KK71-FFC132	48 ± 5	3.4 ± 0.6	10.6 ± 0.3	0.3 ± 0.05
KK71-FFC166	65 ± 13*	12 ± 5	11.5 ± 0.4	1.0 ± 0.4
KK72-PC8	182 ± 71*	20 ± 9	11.8 ± 0.4	1.7 ± 0.7
KK74-PCOD13	11 ± 7*	1.0 ± 0.7	11.1 ± 0.4	0.1 ± 0.1
KK71-GC7	27 ± 2	2.4 ± 0.6	11.7 ± 0.4	0.2 ± 0.05
KK71-GC10	19 ± 4	6 ± 2	8.9 ± 0.3	0.7 ± 0.2
KK71-GC14	11 ± 3	5 ± 2	8.0 ± 0.3	0.6 ± 0.3
OC73-3-20P	15 ± 2	11 ± 4	8.0 ± 0.3	1.4 ± 0.5
Y73-4-34PC	23 ± 4	7 ± 3	7.9 ± 0.3	0.9 ± 0.4
Y73-3-7MG1	21 ± 56	6 ± 20	9.9 ± 0.3	0.7 ± 2.2
Y73-3-21K	40 ± 50	2 ± 3	11.4 ± 0.4	0.2 ± 0.3
Y73-4-64K	39 ± 15	7 ± 3	10.0 ± 0.3	0.7 ± 0.3
Y71-7-36MG2	183 ± 130	6 ± 4	11.7 ± 0.4	0.5 ± 0.4
Y71-7-45P	12 ± 3	8 ± 3	8.0 ± 0.3	1.0 ± 0.3
V19-54	18 ± 4	14 ± 5	7.3 ± 0.2	2.0 ± 0.7
V19-61	35 ± 11	18 ± 7	8.8 ± 0.3	2.0 ± 0.8
V19-64	30 ± 9	10 ± 4	9.1 ± 0.3	1.1 ± 0.5
GS-7202-35P	22 ± 11	10 ± 6	7.9 ± 0.3	1.2 ± 0.7
E21-10	0.93 ± 0.42	0.6 ± 0.3	8.1 ± 0.3	0.1 ± 0.05
E20-17	10 ± 2	21 ± 8	10.0 ± 0.3	2.1 ± 0.8
138	15 ± 6	0.9 ± 0.4	11.0 ± 0.4	0.1 ± 0.04

*Scanning gamma-ray spectrometer values were corrected for ^{222}Rn emanation by comparison of ^{214}Bi activities to those determined at the same interval on individually sealed samples in a well-type gamma-ray spectrometer and by comparison of ^{214}Bi activities to alpha spectrometry data at similar intervals. The corrections for cores KK71-PC85, 86, 89, 90 and 91 are based on measurements of activity differences in two companion cores, KK71-PC87 and KK71-PC88. The average activity difference determined for these cores is 38 ± 5%. The rest of the corrections are as follows: KK71-FFC114, 44 ± 10%; KK71-FFC115, 33 ± 8%; KK71-FFC116, 45 ± 10%; KK71-FFC166, 85 ± 1%; KK72-PC8, 78 ± 3%; KK74-PCOD13, 74 ± 4% (see "Radiochemical Dating" section).

cluster if the sediments are in equilibrium with the overlying sea-water column. However, most of the sediments in the Nazca plate region are not in equilibrium. Instead, two trends are apparent. First, with the exception of the anomalously accumulating core near the Antarctic convergence and 1 core in the Bauer Basin, all cores displaying ^{230}Th accumulations greater than their respective production rates are located on the EPR. Second, again with the exception of these 2 cores, there appears to be a general depletion of ^{230}Th accumulation with increasing depth of water, which is opposite to expectations based on observations of sediment ponding in deep-water basins (Cochran and Osmond, 1976). The implications of these trends for the rapid metal accumulation rates on the EPR and for the origin of the Bauer Basin sediments are discussed below.

Figure 20 presents the geographic locations of the ^{230}Th inventory, or ratio of F_a/F_p, for the EPR and Bauer Basin areas. Also presented is the bottom-current pattern for this region deduced primarily from hydrocast data (potential temperature, salinity, and oxygen measurements) and supported by a few direct-current measurements (Lonsdale, 1976). Lonsdale (1976) has suggested that most of the bottom-current scour in this region should be confined to the transform fault troughs across the EPR crest, where bottom-current flow would be strongest. This conclusion is supported by the ^{230}Th inventories on the EPR; 9 of the 13 rise-crest cores in this region display ^{230}Th inventories greater than or equal to 1.0. Further, Lonsdale (1976) and Heath and Dymond (1977) have suggested that the Bauer Basin deposits receive their hydrothermal component from these EPR fracture zones by bottom-current winnowing and transport. This conclusion is not supported by the ^{230}Th inventories in the Bauer Basin; 7 out of 9 cores in this basin display ^{230}Th inventories that are less than 1.0. The two major bottom-water masses in this region converge near 17° S (Lonsdale, 1976), and divergence in this area could cause a mass outlet to the south (dashed arrow, Fig. 20) and possibly carry sediment from the Bauer Basin into the Central (Yupanqui) Basin and adjacent southern basins. However, there are other controls on the ^{230}Th inventory which can more reasonably account for its geographic distribution.

Dymond and Veeh (1975) suggested that different removal processes in the EPR and Bauer Basin could account for the observed ^{230}Th accumulation pattern. One possibility is that freshly precipitated colloidal Fe and Mn hydroxides, which are more abundant in EPR sediments (Heath and Dymond, 1977), are more effectively scavenging ^{230}Th from sea water. The ^{230}Th removal by or in association with Fe and Mn colloids has been suggested to operate elsewhere on the world rift system (Scott and others, 1972). A plot of ^{230}Th versus Fe accumulation (Fig. 21A) indicates that this mechanism is not operating as effectively here. This plot and a comparison of Figures 9 and 10 to Figure 20 also indicate that there is no systematic relation of Fe and Mn accumulation to the ^{230}Th inventory. This implies that although there may be significant ponding of sediment on the EPR, this effect has little impact on those cores displaying the more rapid accumulations of Fe and Mn.

There is some evidence that ^{230}Th is brought down by the skeletal remains of organisms, particularly CaCO$_3$-secreting organisms such as foraminifera (Goldberg and Griffin, 1964; Goldberg, 1968). This relationship is also suggested by plots of ^{230}Th accumulation versus that of CaCO$_3$ and biogenic Si

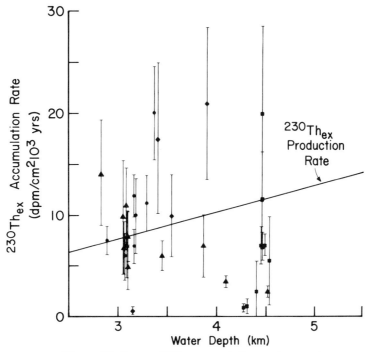

Figure 19. Excess ^{230}Th accumulation and sea-water production rates verus water depth. Circles, 6° EPR stations; squares, Bauer Basin stations; triangles, all other Nazca plate stations; diamonds, stations off Nazca plate.

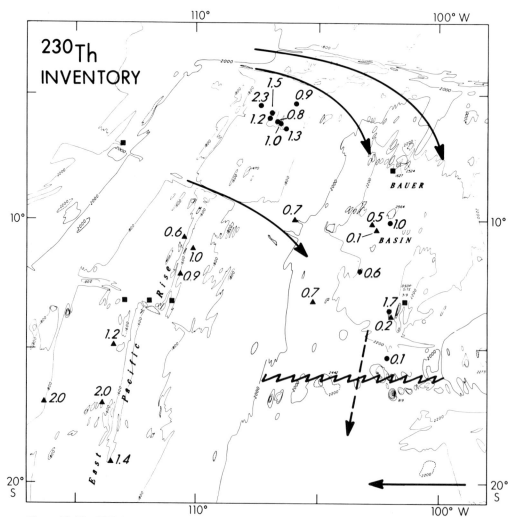

Figure 20. The ^{230}Th inventory on the northwest Nazca plate. Bottom current directions (arrows) modified after Lonsdale (1976). The zig-zag line denotes the convergence of the two major bottom water masses in this area.

(Figs. 21B, 21C). The general decrease of ^{230}Th accumulation with increasing depth of water (Fig. 19) could be explained if ^{230}Th were lost to the overlying water column when $CaCO_3$ dissolves at the sediment-water interface, as most of the ^{230}Th-depleted sediments lie near or below the carbonate compensation depth of 3,950 m. Such a remobilization during biological and chemical reworking of freshly deposited material has been suggested to explain dissolved Cu-depth profiles in the Pacific (Boyle and others, 1977). However, this mechanism implies a dualistic nature for ^{230}Th, which is hard to reconcile on the basis of existing data.

Two other controls may explain the ^{230}Th depletions in the Bauer Basin. One is that there have been systematic losses of ^{230}Th during sample preparation. Although care was taken to ensure total sample dissolution, some Th loss could have occurred in the alpha spectrometry procedures. Systematic losses of ^{226}Ra and ^{222}Rn could lower the apparent ^{230}Th activity of the cores analyzed by gamma spectrometry, but ^{226}Ra diffusion is apparently not a significant problem (see Radiochemical Dating), while corrections have been applied for ^{222}Rn loss (Table 3). The second possible control is illustrated

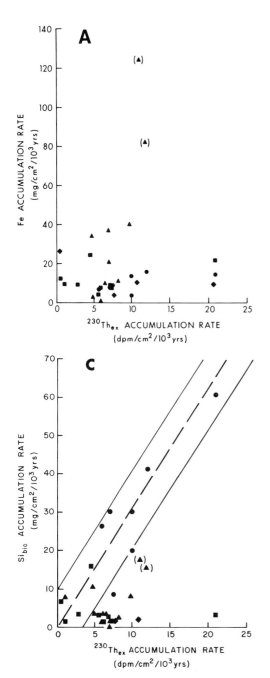
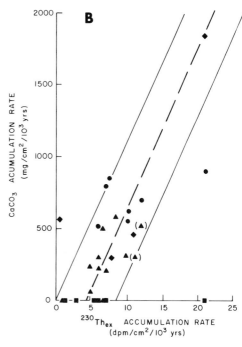

Figure 21A. Plot of Fe versus excess ^{230}Th accumulation. Circles, 6° EPR stations; squares, Bauer Basin stations; triangles, all other Nazca plate stations; diamonds, stations off Nazca plate. Bracketed triangles indicate the same cores that exhibit large noncarbonate accumulations in Fig. 18.

Figure 21B. Plot of CaCO$_3$ versus excess ^{230}Th accumulation. Circles, 6° EPR stations; squares, Bauer Basin stations; triangles, all other Nazca plate stations; diamonds, stations off Nazca plate.

Figure 21C. Plot of biogenic silica versus excess ^{230}Th accumulation. Circles, 6° EPR stations; squares, Bauer Basin stations; triangles, all other Nazca plate stations; diamonds, stations off Nazca plate.

in Figure 22, where the average sedimentation rates of the ^{230}Th-deficient Bauer Basin cores have been projected to levels of $F_a/F_p = 1.0$ in order to estimate the amount of core material these ^{230}Th deficiencies represent. The largest projected sediment losses are displayed by 2 piston cores (PCOD13 and 138), whereas the smallest projected losses are displayed by the free-fall and gravity cores (Fig. 22). This suggests that sediment has been lost during the coring procedure. It is especially likely that some sediment has been lost or disturbed during coring procedures for the longer piston cores that require

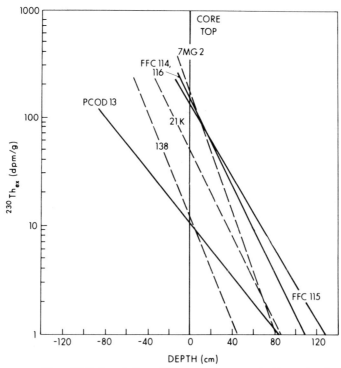

Figure 22. Extrapolations of the regression curves in Figures 3 to 5 to $F_a/F_p = 1.0$ (ideal ^{230}Th inventory) for those cores from the Bauer Basin that display $F_a/F_p < 1.0$. Solid lines, gamma spectrometry curves; dashed lines, alpha spectrometry curves.

triggering, leveling, and sectioning (D. Byrne, 1978, personal commun.). This loss would probably be enhanced for cores with high water contents (low dry bulk densities), such as those from the Bauer Basin (Table 1; Fig. 8). Sediment losses could also occur during free-fall, gravity, and box-core (21K) attempts due to a "pressure barrier" effect, although losses exceeding about 10 cm would be unusual (A. Soutar, 1979, personal commun.).

The existing data suggest that core-top losses can explain much of the apparent ^{230}Th depletions observed in the Bauer Basin. For cores such as 21K, where at least 40 cm of sediment loss is indicated (Fig. 22), loss of ^{230}Th in the alpha spectrometry procedures would seem more reasonable. The ^{230}Th loss by either or both processes has implications for the higher ^{230}Th inventories displayed on the EPR: these values could be even higher. However, the more rapidly accumulating sediments on the EPR will be less affected by a comparable core-top loss because as the accumulation rate increases, the ^{230}Th concentration gradient in the sediment decreases. Clearly, more work is needed on monitoring the distribution of ^{230}Th in the various phases of these sediments, similar to that described by Arrhenius and others (1957), and on monitoring the behavior of the coring process, similar to that described by Seyb and others (1977), before ^{230}Th inventories can become more conclusive.

OCEANOGRAPHIC VERSUS GEOLOGIC CONTROLS ON METALLOGENESIS

Bostrom (1973) developed an empirical relation between crustal spreading rates and the ratio (Fe + Mn)/Al in pelagic sediments which suggested that the metal enrichment process is associated with increased volcanic activity at fast-spreading centers. Figure 23 presents a profile of Mn accumulation along the EPR crest from 50°S to 20°N latitude based on data from Figures 1 and 10. Also shown are the mean and range of normal authigenic accumulation (Bender and others, 1970), the averaged Pliocene-Pleistocene spreading rates in the area, and estimates of regional sediment thickness. Mn (and Fe) accumulations vary exponentially along this section of the EPR from normal authigenic values near the equatorial zone of high productivity and the Antarctic convergence to extremely high values near 20°S latitude, whereas spreading rates vary slightly. Clearly, Bostrom's model cannot explain this variability.

It is appreciated that the lack of large Mn accumulations at 6°S on the EPR crest could be caused by a biased sampling of normal pelagic sediments in this area, and, conversely, that the high accumulations of Mn found in cores to the south of this area is caused by a chance sampling of scattered areas of anomalously intense hydrothermal activity. That this is not the case, however, is suggested by the behavior of hydrothermal Mn in the deep sea (Weiss, 1977). Using a one-dimensional scavenging model, Weiss (1977) calculated a residence time for hydrothermal Mn in the water column overlying the Galapagos Rift of 51 years and a horizontal propagation distance on the order of 1000 km. Even assuming a narrow hydrothermal vent field that is confined to the EPR axial rift valley, as is apparent on the Galapagos Rift (Corliss et al., 1979) and the EPR at 21°N (Spiess et al., 1980), the seven cores investigated at 6°S are distributed within approximately 50 km on either side of the axial rift zone defined by paleomagnetic measurements at this latitude (Rea, 1976), a distance that is well within the zone of horizontal Mn propagation. On the other hand, that the EPR toward 20°S latitude is truly an area of high Mn (and Fe) accumulation is corroborated by the recent discovery of a major

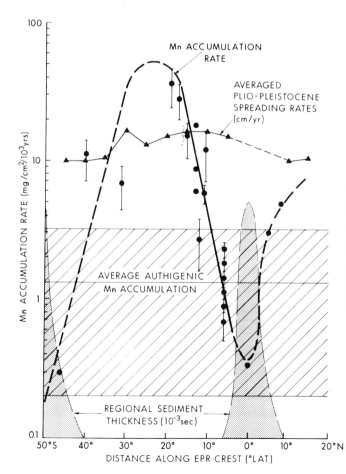

Figure 23. Mn accumulation rate versus distance along EPR crest. Averaged Pliocene-Pleistocene spreading rates versus distance along EPR crest from Handschumacher (1976). Authigenic Mn accumulation from Bender and others (1970). Regional sediment thickness from D. Erlandson (1978, personal commun.) and Lisitzin (1972). Note same logarithmic scale used throughout.

³He source in this vicinity as compared to water column ³He measurements at 6°S, 21°N and the Galapagos Rift (Lupton and Craig, in prep.).

Recent evidence suggests that sea-water circulation leaches cooling basalt intrusives at depth and subsequently deposits the leached metals on the sea floor (McMurtry, in press). The circulating sea water lowers the expected conductive heat flow near the rise crests, and the difference between the theoretical and observed conductive heat flow has been used to estimate the amount of hydrothermal heat release and sea-water circulation (Williams and others, 1974; Wolery and Sleep, 1976). Analysis of the observed conductive heat flow along the EPR from 6° to 40°S has revealed no significant changes in the heat flux that could explain the exponential variation in Fe and Mn accumulations by variations in hydrothermal circulation. This conclusion is based on the observations of nearly constant crustal production over this region (Fig. 23) and on six profiles of conductive heat flow across the EPR (Fig. 24). Although there is some indication that the mean conductive heat flow increases toward the north, the means of each profile are not statistically different at the 67% confidence level (Fig. 24). Shallow-focus seismicity, an index of volcanic activity, also is generally uniform along the EPR crest (Barazangi and Dorman, 1969), as is basalt chemistry (C. D. Fein, 1978, personal commun.). Mantle heterogeneity could account for the accumulation rate variability if these

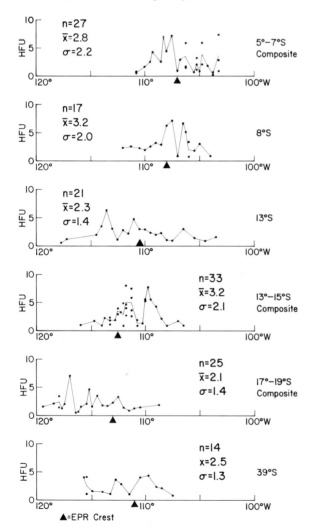

Figure 24. Conductive heat flow on six profiles across the East Pacific Rise. Data from Anderson and Hobart (1976).

sediments reflect a deep-seated source (Bostrom, 1973), but oceanographic controls offer a simple explanation.

Bottom-current distributions suggest sediment loss in the northern EPR area but could explain the Mn and Fe variability only if there is a preferential loss of the metalliferous component. This would require a selective winnowing process that depletes these sediments in Fe and Mn relative to ^{230}Th because the maximum variation in the ^{230}Th accumulation is a factor of 2 as opposed to 18 and 28 for Fe and Mn accumulations, respectively, from 6° to 20°S latitudes. Figure 21A indicates that a decoupling of the Fe and ^{230}Th accumulations does exist here, and thus a selective winnowing process could operate. If Fe and Mn are precipitated from debouching hydrothermal solutions and ^{230}Th is predominantly scavenged from larger, settling biogenic particulates, currents could winnow these sediments of colloidal Fe and Mn without seriously affecting the ^{230}Th inventory. The same situation may hold for the EPR near the Antarctic convergence, as bottom currents have been suggested to affect the sediment distribution in that region (Ewing and others, 1969).

There also appears to be an inverse relationship between Mn (and Fe) accumulation and regional sediment thickness (Fig. 23). Sediment thickness has been suggested as a control on hydrothermal circulation (Lister, 1972) and may be regulating the degree of metallogenesis under the equatorial zone of high productivity and under the zone of high productivity near the Antarctic convergence. However, if sediment thickness is the control, then variations in hydrothermal circulation exist that are not resolvable by existing heat-flow measurements. A variation in the extent of crustal fracturing would be another possible control on the degree of metallogenesis via hydrothermal circulation, but this is difficult to verify.

GLOBAL HYDROTHERMAL CONTRIBUTION TO THE OCEANS

The extreme variability exhibited by the Mn and Fe accumulations along the crest of the EPR has implications for global budget calculations. Lyle (1976) and Froelich and others (1977) have used the extremely high Mn and P flux observed near 20°S in their budget calculations, which could certainly overestimate the global hydrothermal contribution to the oceans. Lyle (1976) calculated that the annual world production of Mn due to hydrothermal sources was about 9×10^{11} g, based on graphical integration of an accumulation rate profile across the EPR at 10° to 17°S latitude and extrapolation to the world crustal production of 3.94 km^2/yr^{-1} estimated by Williams and von Herzen (1974). As this value is about 3 times greater than the dissolved Mn load of rivers (2.5×10^{11} g/yr), Lyle has suggested that hydrothermal sources are an important input of oceanic Mn. Inspection of the Mn accumulation rate profile along the EPR crest (Fig. 23) shows how extrapolation of the high values near 20°S latitude will overestimate the hydrothermal Mn flux. Graphical integration of this profile along the 20°N to 50°S part of the EPR crest gives an average Mn accumulation rate of about 12 mg · cm^{-2} · 10^{-3} yr^{-1}, less than half the values used by Lyle (28 mg · cm^{-2} · 10^{-3} yr^{-1}). Applying this new crestal Mn value to Lyle's calculations gives an EPR hydrothermal production of about 3.5×10^7 g yr^{-1}km^{-1} of rise crest and an annual world hydrothermal Mn production of 6.4×10^{11}g, both of which are 30% lower than Lyle's estimates (Lyle, 1976). Low accumulations have been observed elsewhere on the world rift system (Scott and others, 1972; McArthur and Elderfield, 1977) and Lyle (1976) suggested that the slower spreading mid-Atlantic Ridge may lower his estimate by about 20%. Therefore, even a 50% lower annual world production of 4.5×10^{11} g could be an upper limit to the hydrothermal Mn contribution to the ocean.

Table 4 presents a summary of various estimates of the global oceanographic flux for Mn. Wolery and Sleep (1976) have suggested that the release of residual magmatic fluids (deuteric alteration), high-temperature basalt alteration, and submarine weathering all contribute roughly equal amounts to the submarine Mn flux, which compares well with the total excess Mn flux from the indirect balance estimates of Mackenzie and Wollast (1977) (Table 4). However, there are discrepancies with the total submarine Mn flux estimated from sediment accumulation rate data, especially if the lower estimate of 4.5×10^{11} g/yr is considered an upper limit. There appear to be two paths of reconciliation. First, there

TABLE 4. GLOBAL OCEANOGRAPHIC BUDGET ESTIMATES FOR MANGANESE

Source	Rate (g/yr)	Remarks	Reference
Stream input	$0.25 - 5.72 \times 10^{12}$ (1.55×10^{13})	Dissolved plus suspended loads, Pre-man. Present-day flux in parentheses	Garrels and Mackenzie (1971); Mackenzie and Wollast (1977)
Atmospheric input	4.6×10^{11} (6.7×10^{11})	Pre-man dust. Present-day flux in parentheses	Mackenzie and Wollast (1977)
Hydrothermal flux	$0.5 - 4.5 \times 10^{12}$ (1.4×10^{12})	Calculated from estimated flow rate of $1.3 - 9 \times 10^{17}$ g/yr and gain in conc. from 2 ppb to 4-5 ppm. Best estimate assuming 200°C exit temperature in parentheses	Wolery and Sleep (1976)
	$0.4 - 1 \times 10^{13}$	Estimated from Mn/^{3}He injection relationships	Weiss (1977); Corliss and others (1979)
	$4.5 - 9 \times 10^{11}$	Calculated from Mn accumulation rates in EPR sediments and global crust production	Lyle (1976); this study
Volcanic residual liquids (deuteric alteration)	1.1×10^{12}	Estimated for 120 ppm Mn lost from flow interiors to depth of 1 km (pervasive)	Corliss (1971)
Submarine weathering	0.79×10^{12}	Estimated for 0.66 km of pervasive low-temperature basalt alteration	Hart (1973)
Total submarine volcanism and mafic alteration	$1.82 - 2.83 \times 10^{12}$	Indirect balance estimate. Range from pre-man to present-day flux	Mackenzie and Wollast (1977)
Total excess Mn flux to oceans	$1 - 7 \times 10^{12}$	Estimated by subtracting dissolved stream input (0.2×10^{12} g/yr) from various balance models	Elderfield (1976)

is the possiblity that the calculated flux values are real and the difference seen is due to Mn transport away from the ridge crests. If we adopt the best estimate value of Wolery and Sleep (1976) as representative of the combined fluxes of residual magmatic fluids, hydrothermal alteration, and submarine weathering—which seems reasonable because the estimated discharge flux could be the result of all three alteration sequences—then an approximate 50% loss from the ridge crests leaves a residual accumulation that falls within the range calculated near the crests. This would be in agreement with the loss estimated by Elderfield (1976) for Mn precipitation on ridge crests at an average of about 1 order of magnitude more rapidly than elsewhere on the sea floor. Second, there is the possibility that most of the more indirect estimates are simply too high. Weiss (1977) allowed that his estimates of ~1 $\times 10^{13}$ g/yr may be too high because of the questionable assumption that Mn and He behave similarly during basalt alteration. The range of Mn fluxes estimated from the Galapagos thermal springs data by Corliss et al. (1979) are of similar magnitude, but are based on data that may not be representative of all ridge crest hydrothermal systems. The fluxes for residual magmatic fluids and submarine weathering are probably overestimated because of the assumption of pervasive alteration. As alteration is apparently fault controlled (Spooner and Fyfe, 1973; Crane and Normark, 1977), these fluxes will depend on the fracture permeability, an unknown but probably large constraint. Finally, it should be noted that Elderfield's (1976) large range of estimates for excess Mn does not include corrections for suspended stream load and atmospheric input to the oceans (Table 4).

Of the possible explanations for the variable metal flux on the EPR, Fe and Mn loss by bottom-current winnowing and transport appears to be the most consistant with the existing data. Crustal fracturing and regional sediment thickness should play important roles in regulating hydrothermal circulation and metallogenesis, the latter especially in areas of great surface productivity and sedimentation, but these constraints are not resolvable by the existing heat-flow data. On the other hand, the distribution of bottom currents is poorly defined, and few direct measurements have been made (Lonsdale, 1976). The winnowing mechanism proposed here to explain the lower metal accumulation rates on the northern EPR crest requires that Fe and Mn and ^{230}Th occupy separate phases in the sediment. This separation is tenuous, as correlations of Fe, Mn, and **230**Th are indicated elsewhere on the world rift system (Scott and others, 1972). Further, Bender and others (1977) have suggested that since no dissolved Mn anomalies of mid-oceanic ridge depth are observed, submarine volcanism cannot be the dominant source of Mn in pelagic sediments away from ridge crests. However, Klinkhammer and others (1977) have reported the injection of dissolved Mn into bottom waters on the Galapagos spreading center, and Betzer and others (1974) have reported Fe- and Mn-rich suspended particulate matter in bottom water near the crest of the Mid-Atlantic Ridge. Thus, bottom-current transport of dissolved and particulate Mn, possibly in nepheloid layers, could circumvent Bender and others' (1977) conclusions based on observations of total dissolved Mn at mid-ocean ridge depths. This conclusion is in agreement with the suggestions of extensive transport from ridges made by Goodell and others (1971), Elderfield (1976), Wolery and Sleep (1976), Lonsdale (1976), and Heath and Dymond (1977), and has important implications for the magnitude of hydrothermal input to the ocean, the origin of the Bauer Basin deposits, and the origin of manganese nodules.

CONCLUSIONS

Value of Gamma-Ray Spectrometry

Gamma spectrometry has proved to be a successful means of determining rates of sediment accumulation on the Nazca plate. The sedimentation rates determined by both alpha and gamma spectrometry are in general agreement; most cores display an agreement of sedimentation rates to within 50%. When the two methods are in poor agreement, the source of disagreement is either the lack of an equivalent number of alpha measurements in cores of nonuniform lithology or the lack of uniform uranium support in the gamma-analyzed cores. Cores displaying the largest errors in sedimentation rate are generally those with the lowest number of data points, regardless of technique. The net advantage of gamma techniques over alpha spectrometry is their rapidity, both in sample preparation and in counting, thereby allowing a better sampling coverage. Gamma spectrometry, however, cannot match the alpha technique in analytical precision, but is useful in regional sedimentation studies where time and cost are important factors.

Patterns of Element Accumulation

The expanded coverage of cores analyzed by gamma spectrometry, combined with elemental analysis by X-ray fluorescence spectrometry, has shown that rapid rates of metal accumulation are not typical in sediments of the East Pacific Rise. Further, high rates of metal accumulation may be atypical throughout the world rift system. Rates of Fe and Mn accumulation vary from values corresponding to normal authigenic accumulation on the northern EPR to values exceeding 30 times normal authigenic accumulation near 20°S latitude. The Bauer Basin exhibits rates of Fe and Mn accumulation that range from normal values to 6 times the authigenic rate. This variation indicates that sedimentary processes other than normal authigenic ones exist in this part of the Eastern Pacific Ocean.

Sources of Element Accumulation

Multivariate factor analysis was used to suggest possible sources of the components in the sediment. Its results indicate that three associations characterize most of the variance in element accumulation rates on the Nazca plate. These associations are interpreted as being the result of (1) hydrothermal, (2) biogenic-diagenetic, and (3) detrital-hydrogenous processes. Most of the EPR is dominated by a hydrothermal association, whereas the Bauer Basin and the northern EPR sediments are dominated by a biogenic-diagenetic association. Compared with previous work, these results are based on different samples and on more elements and are approached through rates of accumulation rather than relative abundances of elements, but they are in general agreement with the previous work. Further, the factor analysis suggests the association of the northern EPR and Bauer Basin sediments.

Comparison of ^{230}Th Budget with Rates of Fe and Mn Accumulation

The ^{230}Th inventory indicates that sediment ponding does occur on the EPR, but ponding does not seriously affect the rapid rates of metal accumulation observed on the EPR near 20°S latitude. This conclusion is based on the lack of correspondence between the ^{230}Th inventory and the accumulation rates of Fe and Mn. The lower ^{230}Th inventory displayed by most cores in the Bauer Basin can be better attributed to core-top loss and ^{230}Th losses in sample preparation for alpha spectrometry than to ^{230}Th losses by $CaCO_3$ dissolution or bottom-current winnowing, or to different ^{230}Th removal processes than those on the EPR. Future work should concentrate on the distribution of ^{230}Th in these sediments and on monitoring the behavior of the coring process before interpretation of ^{230}Th budgets can become more conclusive. Knowledge of the ^{230}Th distribution will also test the hypothesized selective winnowing of the northern EPR sediments.

Control of Metallogenesis in Nazca Plate Sediments

Bottom-current winnowing and transport, sediment thickness, and the extent of crustal fracturing are more reasonable controls on the distribution of metallogenesis than mantle heterogeneity. Of these possible controls, bottom currents offer the most attractive, albeit unproved, explanation for the existing patterns and have the broadest implications for the role of hydrothermal input to the ocean: 50% or more of the metals introduced to the ocean on and near the crests of mid-ocean ridges could be distributed to adjacent areas of the sea floor such as the Bauer Basin and in manganese nodules.

Directions of Future Studies

Future work should be aimed at direct verification of bottom-current transportation by current measurements and the analysis of bottom waters for dissolved and particulate Fe and Mn. The coverage of heat-flow and metal-accumulation measurements on the EPR will have to be increased, especially in areas of rapid sedimentation, and will have to be made in conjunction with the bottom-current measurements and nephelometry before a separation of these controls can be carried out adequately.

ACKNOWLEDGMENTS

We thank Kwan Tak and Karen Young for their assistance in sample preparation, Barry Fankhauser for his advice and assistance in the radiochemical laboratory, R. C. Jones for his help with the X-ray fluorescence quantometer, and J. Dymond for the use of his unpublished chemical analyses. We also thank R. Moberly, R. W. Buddemeier, D. C. Hurd, D. M. Hussong, and K. A. Pankiwskyj for helpful comments on the manuscript and T.-L. Ku and M. R. Scott for their helpful reviews. Support for this study was granted by the National Science Foundation, Office of International Decade of Ocean Exploration, under the Nazca Plate Project.

REFERENCES CITED

Anderson, R. N., and Hobart, M. A., 1976, The relation between heat flow, sediment thickness, and age in the eastern Pacific: Journal of Geophysical Research, v. 81, p. 2968–2969.

Andrews, J. E., Foreman, J. A., and scientific staff, 1975, Sediment core descriptions: R/V *Kana Keoki* 1971 cruise, eastern and western Pacific Ocean: Honolulu, University of Hawaii, Hawaii Institute of Geophysics Report no. 28, 314 p.

Arrhenius, G., Bramlette, M. N., and Picciotto, E., 1957, Localization of radioactive and stable heavy nuclides in ocean sediments, Nature, v. 180, p. 85–86.

Barazangi, M., and Dorman, J., 1969, World seismicity maps compiled from ESSA, Coast and Geodetic Survey, epicenter data, 1961–1967: Seismological Society of America Bulletin, v. 59, p. 369–380.

Bender, M., Ku, T.-L., and Broecker, W. S., 1970, Accumulation rates of manganese in pelagic sediments and nodules: Earth and Planetary Science Letters, v. 8, p. 143–148.

Bender, M., and others, 1971, Geochemistry of three cores from the East Pacific Rise: Earth and Planetary Science Letters, v. 12, p. 425–433.

Bender, M. L., Klinkhammer, G. P., and Spencer, D. W., 1977, Manganese in seawater and the marine manganese balance: Deep-Sea Research, v. 24, p. 799–812.

Berner, R. A., 1973, Phosphate removal from seawater by adsorption on volcanogenic ferric oxides: Earth and Planetary Science Letters, v. 18, p. 77–86.

Betzer, P. R., and others, 1974, The Mid-Atlantic Ridge and its effect on the composition of particulate matter in the deep ocean: American Geophysical Union, Transactions, v. 55, p. 293.

Bostrom, K., 1970, Submarine volcanism as a source for iron: Earth and Planetary Science Letters, v. 9, p. 348–354.

—— 1973, The origin and fate of ferromanganoan active ridge sediments: Stockholm Contributions in Geology, v. 27, p. 149–243.

Bostrom, K., and Peterson, M.N.A., 1966, Precipitates from hydrothermal exhalations on the East Pacific Rise: Economic Geology, v. 61, p. 1258–1265.

—— 1969, The origin of aluminum-poor ferromanganoan sediments in areas of high heat on the East Pacific Rise: Marine Geology, v. 7, p. 427–447.

Bostrom, K., and others, 1969, Aluminum-poor ferromanganoan sediments on active oceanic ridges: Journal of Geophysical Research, v. 74, p. 3261–3270.

Boyle, E. A., Sclater, F. R., and Edmond, J. M., 1977, The distribution of dissolved copper in the Pacific: Earth and Planetary Science Letters, v. 37, p. 38–54.

Broecker, W. S., and Broecker, S., 1974, Carbonate dissolution on the western flank of the East Pacific Rise, *in* Hay, W. W., ed., Studies in paleo-oceanography: Society of Economic Paleontologists and Mineralogists Special Publication 20, p. 44–57.

Chave, K. E., 1954, Aspects of the biogeochemistry of magnesium I. Calcareous marine organisms: Journal of Geology, v. 62, p. 266–283.

Cochran, J. K., and Osmond, J. K., 1974, Gamma spectrometry of deep-sea cores and sediment accumulation rates: Deep-Sea Research, v. 21, p. 721–737.

—— 1976, Sedimentation patterns and accumulation rates in the Tasman Basin: Deep-Sea Research, v. 23, p. 193–210.

Corliss, J. B., 1971, The origin of metal-bearing submarine hydrothermal solutions: Journal of Geophysical Research, v. 76, p. 8128–8138.

Corliss, J. B., and others, 1979, Submarine thermal springs on the Galapagos rift: Science, v. 203, p. 1073–1083.

Crane, K., and Normark, W. R., 1977, Hydrothermal activity and crustal structure of the East Pacific Rise at 21°N: Journal of Geophysical Research, v. 82, p. 5336–5348.

Dasch, E. J., Dymond, J., and Heath, G. R., 1971, Isotopic analysis of metalliferous sediments from the East Pacific Rise: Earth and Planetary Science Letters, v. 13, p. 175–180.

Dymond, J., and Veeh, H. H., 1975, Metal accumulation rates in the southeast Pacific and the origin of metalliferous sediments: Earth and Planetary Science Letters, v. 28, p. 13–22.

Dymond, J., Corliss, J. B., and Stillinger, R., 1976, Chemical composition and metal accumulation rates of metalliferous sediments from sites 319, 320B, and 321, *in* Yeats, R. S., and Hart, S. R., eds., Initial Reports of the Deep-Sea Drilling Project, Volume 34: Washington, D.C., U.S. Government Printing Office, p. 575–588.

Dymond, J., Corliss, J. B., and Heath, G. R., 1977, History of metalliferous sedimentation at Deep-Sea Driling Site 319, in the South Eastern Pacific: Geochimica et Cosmochimica Acta, v. 41, p. 741–753.

Dymond, J., and others, 1973, Origin of metalliferous sediments from the Pacific Ocean: Geological Society of America Bulletin, v. 84, p. 3355–3372.

Elderfield, H., 1976, Manganese fluxes to the oceans: Marine Chemistry, v. 4, p. 103–132.

Ewing, M., Houtz, R., and Ewing, J., 1969, South Pacific sediment distribution: Journal of Geophysical Research, v. 74, p. 2477–2493.

Fankhauser, B. L., 1976, Thermoluminescence dating

applied to Hawaiian basalts [M.S. thesis]: Honolulu, University of Hawaii, 188 p.

Froelich, P. N., Bender, M. L., and Heath, G. R., 1977, Phosphorous accumulation rates in metalliferous sediments on the East Pacific Rise: Earth and Planetary Science Letters, v. 34, p. 351–359.

Garrels, R. M., and Mackenzie, F. T., 1971, Evolution of sedimentary rocks: New York, W.W. Norton and Co., 397 p.

Goldberg, E. D., 1968, Ionium/thorium geochronologies: Earth and Planetary Science Letters, v. 4, p. 17–21.

Goldberg, E. D., and Arrhenius, G.O.S., 1958, Chemstry of Pacific pelagic sediments: Geochimica et Cosmochimica Acta, v. 13, p. 153–212.

Goldberg, E. D., and Koide, M., 1962, Geochronological studies of deep-sea sediments by the ionium/thorium method: Geochimica et Cosmochimica Acta, v. 26, p. 417–450.

——1963, Rates of sediment accumulation in the Indian Ocean, in Geiss, J., and Goldberg, E. D., eds., Earth science and meteoritics: New York, John Wiley & Sons, p. 90–102.

Goldberg, E. D., and Griffin, J. J., 1964, Sedimentation rates and mineralogy in the South Atlantic: Journal of Geophysical Research, v. 69, p. 4293–4309.

Goodell, H. M., Meylan, M. A., and Grant, B., 1971, Ferromanganese deposits of the South Pacific Ocean, Drake Passage, and Scotia Sea, in Reid, J. L., ed., Antarctic oceanology I: Antarctic Research Series, v. 15, p. 27–92.

Handschumacher, D. W., 1976, Post-Eocene plate tectonics of the Eastern Pacific, in Sutton, G. H., and others, eds., The geophysics of the Pacific Ocean basin and its margin (Woollard Volume): Washington, D.C., American Geophysical Union Geophysical Monograph 19, p. 177–202.

Hart, R. A., 1973, A model for chemical exchange in the basalt-seawater system of oceanic layer II: Canadian Journal of Earth Sciences, v. 10, p. 799–816.

Heath, G. R., and Dymond, J., 1977, Genesis and transformation of metalliferous sediments from the East Pacific Rise, Bauer Deep, and Central Basin, northwest Nazca plate: Geological Society of America Bulletin, v. 88, p. 723–733.

Klinkhammer, G., Bender, M., and Weiss, R. F., 1977, Hydrothermal manganese in the Galapagos Rift: Nature, v. 269, p. 319–320.

Koczy, F. F., and Bourret, R., 1958, Diffusion of radioactive nuclides in a moving and turbulent medium, progress report: University of Miami, Marine Laboratory, NSF Grant A-3995.

Koide, M., and Goldberg, E. D., 1965, Uranium-234/uranium-238 ratios in sea water, in Sears, M., ed., Progress in oceanography, Volume 3: New York, Pergamon Press, p. 173–177.

Kraemer, T. F., 1971, Rates of accumulation of iron, manganese, and certain trace elements on the East Pacific Rise [M.S. thesis]: Tallahassee, Florida State University, 94 p.

Kröll, V., 1955, The distribution of radium in deep-sea cores, in Pettersson, H., ed., Reports of the Swedish deep-sea expedition: Göteborg, Elanders Boktryckeri Aktiebolag, v. 10, p. 3–32.

Ku, T.-L., 1965, An evaluation of the $^{234}U/^{238}U$ method as a tool for dating pelagic sediments: Journal of Geophysical Research, v. 70, p. 3457–3474.

——1966, Uranium series disequilibrium in deep-sea sediments [Ph.D. dissert.]: New York, Columbia University, 157 p.

——1976, The uranium series methods of age determination: Earth and Planetary Sciences Annual Review, v. 4, p. 347–379.

Ku, T.-L., Broecker, W. S., and Opdyke, N., 1968, Comparison of sedimentation rates measured by paleomagnetic and the ionium methods of age determination: Earth and Planetary Science Letters, v. 4, p. 1–16.

Ku, T.-L., Knauss, K. G., and Mathieu, G. G., 1974, Uranium in open ocean: Concentration and isotopic composition: American Geophysical Union Transactions, v. 55, p. 314.

Leinen, M., 1977, A normative calculation technique for determining opal in deep-sea sediments: Geochimica et Cosmochimica Acta, v. 41, p. 671–676.

Lisitzin, A. P., 1972, Sedimentation in the world ocean: K. S. Rodolfo, ed., Society of Economic Paleontologists and Mineralogists Special Publication 17, Tulsa, Oklahoma, 218 p.

Lister, C.R.B., 1972, On the thermal balance of a mid-ocean ridge: Geophysical Journal of the Royal Astronomical Society, v. 26, p. 515–535.

Lonsdale, P., 1976, Abyssal circulation of the southeastern Pacific and some geological implications: Journal of Geophysical Research, v. 81, p. 1163–1176.

Lyle, M., 1976, Estimation of hydrothermal manganese input to the oceans: Geology, v. 4, p. 733–736.

Lyle, M. W., and Dymond, J., 1976, Metal accumulation rates in the southeast Pacific - errors introduced from assumed bulk densities: Earth and Planetary Science Letters, v. 30, p. 164–168.

Mackenzie, F. T., and Wollast, R., 1977, Sedimentary cycling models of global processes, in Goldberg, E. D., ed., The sea, Volume 6: New York, John Wiley & Sons, p. 739–785.

Mammerickx, J., and others, 1975, Morphology and tectonic evolution of the east-central Pacific: Geological Society of America Bulletin, v. 86, p. 111–118.

McArthur, J. M., and Elderfield, H., 1977, Metal accumulation rates in sediments from mid-Indian Ocean Ridge and Marie Celeste Fracture Zone: Nature, v. 266, p. 437–439.

McMurtry, G. M., 1975, Geochemical investigations of sediments across the Nazca plate at 12°S, in Contributions to the geochemistry of Nazca plate sediments, southeast Pacific: Hawaii Institute of Geophysics Report no. 75-14, 40 p.

——1979, Rates of sediment accumulation and their bearing on metallogenesis on the Nazca plate, southeast Pacific [Ph.D. dissert.]: Honolulu, University of Hawaii, 232 p.

——1981, Metallogenesis on oceanic plates: The East Pacific Rise and Bauer Basin, in Circum-Pacific energy and mineral resources, Volume 2: American Association of Petroleum Geologists Memoir (in press).

McMurtry, G. M., and Burnett, W. C., 1975, Hydrothermal metallogenesis in the Bauer Deep of the southeastern Pacific: Nature, v. 254, p. 42–44.

Osmond, J. K., and Pollard, L. D., 1967, Sedimentation rate determinations in deep-sea cores by gamma-ray spectrometry: Earth and Planetary Science Letters, v. 3, p. 476–480.

Piper, D. Z., 1973, Origin of metalliferous sediments from the East Pacific Rise: Earth and Planetary Science Letters, v. 19, p. 75–82.

Rea, D. K., 1976, Changes in the axial configuration of the East Pacific Rise near 6°S during the past 2 m.y.: Journal of Geophysical Research, v. 81, p. 1495–1504.

Rybach, L., 1971, Gamma-ray spectrometry for simultaneous U, Th, and K determinations, in Wainerdi, R. E., and Uken, E. A., eds., Modern methods of geochemical analysis: New York, Plenum Press, 397 p.

Rydell, H., and others, 1974, Postdepositional injections of uranium-rich solutions into East Pacific Rise sediments: Marine Geology, v. 17, p. 151–164.

Sayles, F. L., and Bischoff, J. L., 1973, Ferromanganoan sediments in the equatorial East Pacific: Earth and Planetary Science Letters, v. 19, p. 330–336.

Sayles, F. L., Ku, T. -L., and Bowker, P. C., 1975, Chemistry of ferromanganoan sediment of the Bauer Deep: Geological Society of America Bulletin, v. 86, p. 1423–1431.

Scott, M. R., Osmond, J. K., and Cochran, J. K., 1972, Sedimentation rates and sediment chemistry in the South Indian Basin, in Antarctic oceanology II: Antarctic Research Series, v. 19, p. 317–334.

Seyb, S. M., Hammond, S. R., and Gilliard, T., 1977, A new device for recording the behavior of a piston core: Deep-Sea Research, v. 24, p. 943–950.

Smith, S. V., 1971, Factor analysis: A tool for environmental studies: Marine Technology Society Journal, v. 5, p. 15–19.

Speiss, F. N., and others, 1980, Hot springs and geophysical experiments on the East Pacific Rise: Science, v. 207, p. 1421–1432.

Spooner, E.T.C., and Fyfe, W. S., 1973, Sub-seafloor metamorphism, heat and mass transfer: Contributions to Mineralogy and Petrology, v. 42, p. 287–304.

Turekian, K. K., and Chan, L. H., 1971, The marine geochemistry of the uranium isotopes, Th-230 and Pa-231, in Brunfelt, A. O., and Steinnes, E., eds., Activation analysis in geochemistry and cosmochemistry: Universitetsforlaget, p. 311–320.

Veeh, H. H., 1981, Uranium and thorium isotopic investigations in metalliferous sediments of the Northwestern Nazca plate, in Kulm, L. D., and others, eds., Nazca plate: Crustal formation and Andean convergence: Geological Society of America Memoir 154 (this volume).

Weiss, R. F., 1977, Hydrothermal manganese in the deep sea: Scavenging residence time and $Mn/^3He$ relationships: Earth and Planetary Science Letters, v. 37, p. 257–262.

Williams, D. L., and others, 1974, The Galapagos spreading center: Lithospheric cooling and hydrothermal circulation: Geophysical Journal of the Royal Astronomical Society, v. 38, p. 587–608.

Williams, D. L., and von Herzen, R. P., 1974, Heat loss from the Earth: New estimate: Geology, v. 2, p. 327–328.

Wolery, T. J., and Sleep, N. D., 1976, Hydrothermal circulation and geochemical flux at mid-ocean ridges: Journal of Geology, v. 84, p. 249–275.

MANUSCRIPT RECEIVED BY THE SOCIETY NOVEMBER 12, 1980
MANUSCRIPT ACCEPTED DECEMBER 30, 1980
HAWAII INSTITUTE OF GEOPHYSICS CONTRIBUTION NO. 1100

Printed in U.S.A.

ns of America
Memoir 154
1981

Uranium and thorium isotopic investigations in metalliferous sediments of the northwestern Nazca plate

H. HERBERT VEEH
School of Earth Sciences
Flinders University of South Australia
Bedford Park, S.A. 5042
Australia

ABSTRACT

Metalliferous sediments from the East Pacific Rise and the Bauer Deep on the northwestern Nazca plate, as well as size-separated sediment components, have been analyzed for uranium (U) and thorium (Th) isotopes by alpha spectrometry in an attempt to determine the origin and mode of emplacement of uranium in the sediments. In addition, the partition of uranium between different sediment components has been investigated, using fission-track techniques.

The results indicate that the uranium in metalliferous sediments has been derived from sea water and detrital sources in various proportions, depending on time and location. Along the crest of the East Pacific Rise, the $^{234}U/^{238}U$ activity ratio is indistinguishable from that in modern sea water, indicating that the uranium on the rise crest is dominated by the uranium of hydrogenous origin, whereas lower $^{234}U/^{238}U$ and U/Th ratios in one area on the east flank of the East Pacific Rise and everywhere in the Bauer Deep may reflect relatively more input of uranium from detrital sources, coupled with a reduced hydrogenous input.

Isotopic analysis of individual sediment components, coupled with fission-track studies, reveals that the hydrogenous uranium in the Bauer Deep resided predominantly in fish-bone apatite, with smaller amounts in manganese micronodules, whereas the uranium of detrital origin appears to be restricted largely to the clay fraction. By contrast, all size fractions on the East Pacific Rise are dominated by hydrogenous uranium. Moreover, the distribution of uranium in metalliferous sediments is quite heterogeneous. This suggests that locally reducing conditions, perhaps associated with fish debris, fecal pellets, or decaying organic matter inside foraminiferal shell cavities, may enhance the fixation of uranium in the sediments.

On the basis of these data, coprecipitation of uranium from sea water with volcanogenic iron oxides, followed by fixation in the sediments by local reducing conditions, is considered to be the predominant mechanism for the enrichment of uranium in metalliferous sediments on the East Pacific Rise.

The mechanism of hydrogenous uranium input in the Bauer Deep, on the other hand, probably involves redeposition of laterally transported rise-crest sediments, followed by diagenetic release of uranium and its fixation in the sediment by fish-bone apatite and manganese micronodules. It is estimated that a maximum of 70% of the total uranium content in the Bauer Deep may have originated in this way, the rest having been supplied from detrital sources.

The data presented in this study do not provide any support for significant contributions of uranium from hydrothermal solutions on the Nazca plate.

INTRODUCTION

Unusually high concentrations of uranium have been found in surface sediments on mid-ocean ridges (Fisher and Boström, 1969; Bender and others 1971; Turekian and Bertine, 1971; Scott and others, 1972; Rydell and others, 1974), in manganese crusts immediately overlying basaltic rocks near hydrothermally active spreading centers (Scott and others, 1974; Moore and Vogt, 1976), in many iron-rich, submarine, hot-spring deposits (Rydell and Bonatti, 1973; Bonatti and others, 1972), and in the iron-rich sediments of the Red Sea geothermal area (Ku, 1969). Despite intensified research during the past few years, there is still no general agreement on the origin of the uranium, nor on the mode of its emplacement in the metalliferous sediments. According to one model, the uranium, along with iron and other transition metals, was deposited in metalliferous sediments by "hydrothermal solutions," derived by degassing from the mantle beneath spreading ridges (Fisher and Boström, 1969; Boström, 1973). Alternative models for the origin of the uranium in metalliferous sediments include coprecipitation of uranium from sea water with iron, presumably of hydrothermal origin (Ku; 1969; Veeh and Boström, 1971; Bender and others, 1971; Scott and others, 1974), or precipitation of uranium under anaerobic conditions near the sea floor, caused by the formation of ephemeral restricted basins and the high rate of supply of organic matter to the relatively shallow ridge crests

The question of the origin of uranium and its mode of emplacement is important for at least two reasons. First, in the geochemical balance of uranium, the sources and sinks of uranium in the ocean are not yet adequately defined. In particular, the effect of seawater penetration deep into the ocean crust near mid-ocean ridges on the marine budget of uranium requires further evaluation (Aumento, 1971; Macdougall, 1977; Edmond and others, 1979; Bloch, 1980). Although the amount of uranium deposited annually in sediments on the sea floor is more than accounted for by river input (Veeh, 1967; Turekian and Chan, 1971; Sackett and others, 1973), this may not necessarily be valid for the uranium isotope ^{234}U for which additional sources have been postulated (Ku, 1965). Second, the process or processes by which uranium becomes fixed in the metalliferous sediments, aside from being of general interest to the marine geochemist, may provide some clues on the pathways and enrichment processes of other metals as well.

The purpose of this study is to provide some insight into the relative importance of several possible sources and depositional processes for the uranium as outlined above. The Nazca plate (Fig. 1) is ideally suited for a study of this kind, as all of the likely inputs for a given element in the ocean, namely, hydrothermal, terrigenous, hydrogenous, and biogenous, are represented there, with one or the other dominating depending on location (Heath and Dymond, 1977). Thus, the East Pacific Rise as the site of submarine volcanic activity of variable intensity along its crest may be dominated by hydrothermally derived metals, whereas sediments in the Bauer Deep should receive relatively more hydrogenous inputs. The influence of terrigenous components, which normally dominate in pelagic clays, is subdued on the northwestern Nazca plate owing to the effective sediment trap of the Peru-Chile Trench and the topographic barrier of the Galapagos Rise. Inasmuch as organic productivity in surface water is higher along the equatorial divergence, any biogenous input would increase toward the northern boundary of the Nazca plate. An additional variable is provided by water depth, as it controls the proportion of biogenic material (that is, foraminiferal shells and organic matter) surviving vertical transport from shallow water to the deep sea floor. As can be seen in Figure 1, the water depth in the area of study varies from less than 3,000 m over the East Pacific Rise to more than 4,000 m in the Bauer Deep, with the result that the Bauer Deep lies beneath the carbonate compensation depth.

The large number of sediment cores taken in the course of the Nazca Plate Project provides an adequate sample coverage for the evaluation of regional patterns in the distribution of uranium on the

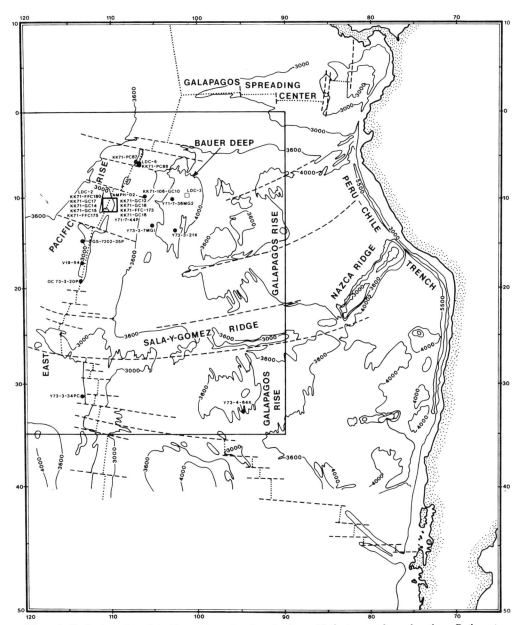

Figure 1. Geologic setting of the Nazca plate, showing physiographic features and core locations. Bathymetry after Mammerickx and others (1975). Depth contours are in metres. The large and small inserts refer to Figures 3 and 4, respectively.

northwestern Nazca plate, while the detailed survey area at 10°S allows some insight into the relationship between the uranium isotopic composition of ridge-crest sediments and bathymetric features such as seamounts or fracture zones. Down-core variations in uranium concentrations, combined with radioactive disequilibrium relationships among the various members of the uranium decay series, should reveal any changes of uranium input with time. Metal accumulation rates, derived from $^{230}Th_{ex}$ measured in these cores will be discussed in a companion paper (McMurty and others, this volume).

PROCEDURES

Unless otherwise indicated, the data in this study consist of bulk analyses, on a dry-weight basis. Sediment samples were either oven dried at 110 °C, or freeze dried.

The methods for the chemcial separation of uranium and thorium prior to their radiometric assay were similar to procedures described by Ku (1965). Samples were completely dissolved in mixtures of HF, $HClO_4$, HCl, and HNO_3, and the tracers ^{234}Th and ^{232}U were added. Any insoluble residue, usually barite, remaining after repeated acid treatment was dissolved at elevated temperature and pressure in the presence of cation exchange resin, as described by Church (1970).

After purification by ion exchange and solvent extraction methods, uranium and thorium isotopes were electroplated onto stainless steel discs and their abundances measured by alpha spectrometry, using a Nuclaer Data ND 2400 pulse-height analyzer with ORTEC surface-barrier detectors. The beta activity of the ^{234}Th yield tracer was determined under an end-window Geiger counter. The overall efficiency of the surface-barrier detector used for thorium analysis was repeatedly checked with a gravimetric thorium standard of known activity.

In an attempt to assess the partition of uranium between different sediment components, size-separated sediment fractions of representative cores from the East Pacific Rise and the Bauer Deep were analyzed individually by the same methods as the bulk sediments. An alternative method to determine the partition of uranium between coexisting sediment components was provided by the study of thermal neutron-induced fission tracks registered in Lexan plastic films (Fleischer and others, 1975), using either open-faced thin sections of impregnated, undisturbed sediments, or special slides prepared from epoxy-mounted individual sediment components of the coarse (<63 μm) fraction. After irradiation by slow neutrons in a nuclear reactor, the Lexan track detectors were etched in 6 N NaOH at 68 °C for 10 min, and the fission-track distribution pattern was observed under an optical microscope.

RESULTS AND DISCUSSION

The geological setting of the Nazca plate, with pertinent bathymetric features and core locations, is shown in Figure 1. Figures 2 and 6 compare the uranium and thorium isotopic data (Tables 1, 2) with those from sediment "end members" of known origins and with well-defined isotopic compositions.

A U-Th diagram (Fig. 2) shows the uranium and thorium concentrations, on a carbonate-free basis, in relation to those from sediments and deposits with predominantly hydrothermal (for example, submarine, hot-spring deposits), hydrogenous (for example, manganese nodules), or terrigenous (for example, normal pelagic clays) origins. It should be noted that the term "hydrothermal sediment" in this context is descriptive rather than genetic, that is, it does not necessarily imply a truly juvenile origin for all of its components.

Figures 3 to 7 display the sediments or sediment-size fractions in terms of their ^{234}U/^{238}U activity ratios and U/Th weight ratios. Inasmuch as ^{234}U/^{238}U in sea water has a constant value of 1.14 ±0.014 (Koide and Goldberg, 1965), this parameter serves as a convenient tracer for uranium of marine (that is, hydrogenous or biogenous) origin, easily distinguishable from uranium of terrigenous origin, as in river muds (Scott, 1968), and normal pelagic clays (Ku, 1965; Heye, 1969), where the ^{234}U/^{238}U ratio tends to be somewhat less than 1.00, presumably because of selective leaching of ^{234}U under oxidizing conditions (Chalov and Merkulova, 1966) in the course of weathering and transport.

The U/Th weight ratio, on the other hand, has a well-defined value of 0.25 in primary igneous rocks (Clark and others, 1966) and ranges between 0.1 and 0.4 in terrigenous sediment components (Scott, 1968; Ku, 1965; Aller and Cochran, 1976), so any additional, nondetrital input of uranium can be easily recognised. Because of the extremely low concentration of thorium in sea water, the U/Th ratio in a hydrogenous deposit also serves as a measure of accumulation rate (Bonatti, 1975). This is easily seen in Figure 2, where the thorium content in slowly accumulating manganese nodules and crusts is shown to be several orders of magnitude higher than in the more rapidly accumulating manganese

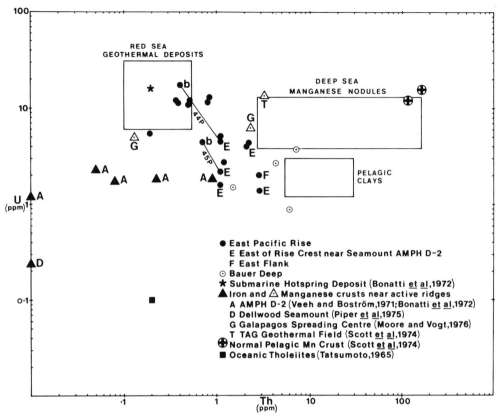

Figure 2. Uranium and thorium concentrations of sediment surface samples, recalculated to a carbonate-free basis, in relation to those from sediment "end numbers" of known origin. The bottom values, marked "b" of 2 cores (Y71-7-44P and Y71-7-45P) located east of the rise crest at 10°S are also shown. Data from Ku (1965, 1969); Ku and Broecker (1969); Heye (1969), and as shown in the legend.

crusts near active spreading centers, although their uranium content is not much different. Some iron crusts on seamounts contain virtually no thorium at all.

Although not immediately obvious from the data in Tables 1 and 2, the uranium concentrations in the metalliferous sediments on the Nazca plate, when expressed on a carbonate-free basis, are significantly higher than those in normal pelagic clays from other areas (Fig. 2). Greatest enrichment in uranium is shown for sediments from the crust of the East Pacific Rise, whereas sediments from the Bauer Deep show only marginally higher uranium content than typical pelagic clays and sediment components of terrigenous origin. Even among sediments from the East Pacific Rise there is considerable variation in uranium and thorium contents, with a strong regional dependence. Thus, sediments from the southern section of the East Pacific Rise between 10° and 20°S resemble "hydrothermal" deposits, whereas sediments from the east flank of the rise at 10°S and on the rise crest farther to the north are almost indistinguishable from sediments in the Bauer Deep.

This regional pattern is also borne out by the $^{234}U/^{238}U$ and U/Th data (Figs. 3, 4) which indicate predominantly nondetrital input of uranium along the southern section of the rise crest, but suggest a mixed input of both detrital and nondetrital uranium for the remaining areas of the rise crest as well as for the Bauer Deep.

The sharp contrast between the west flank and the east flank of the East Pacific Rise in the detailed survey area at 10°S is puzzling. Judging from the change of U/Th and $^{234}U/^{238}U$ with depth in the two available cores from the east flank (Y71-7-44P, Y71-7-45P), there is some indication that in the past

TABLE 1. URANIUM AND THORIUM ISOTOPIC DATA IN SEDIMENT CORES AND CORE TOPS FROM THE NORTHWESTERN NAZCA PLATE

Core no. depth (cm)	Location	Water depth (m)	CaCO$_3$ (%)	U (ppm)	Th (ppm)	U/Th	$^{234}U/^{238}U$	$^{230}Th_{ex}$ (d/m/g)
KK71-PC87	5°55'S 107°03'W	3170						
6-10			74	0.71	0.36	2.0	1.12±.03	11.6
50-55			89	0.24	0.19	1.3	1.12±.03	8.93
100-105			82	0.35	0.23	1.5	1.07±.04	8.68
200-205			82	0.39	0.52	0.8	1.09±.06	10.4
350-355			83	0.27	0.17	1.6	1.08±.03	5.12
550-555				0.21	0.19	1.1	1.14±.04	2.23
KK71-PC88	6°06'S 106°43'W	3073						
5-10			95	0.21	0.11	1.9	1.18±.04	7.73
50-55			90	0.18	0.19	0.9	1.15±.04	7.92
100-105			88	0.23	0.17	1.4	1.20±.06	4.91
150-155			92	0.19	0.15	1.3	1.13±.05	1.87
200-205			59	0.23	0.28	0.8	1.22±.06	2.37
KK71-GC14	10°37'S 110°30'W	3095						
0-2			69	4.05	0.26	16	1.15±.02	10.7
20.22			43	5.84	0.20	29	1.13±.02	7.55
40-42			40	5.14	0.43	12	1.13±.02	8.19
60-62			48	10.76	0.20	54	1.13±.02	3.85
KK71-GC15	10°33'S 110°36'W	2293						
0-5			59	4.98	0.21	24	1.13±.01	8.49
KK71-GV17	10°34'S 110°23'W	3172						
0-5			54	5.47	0.17	32	1.13±.01	9.81
KK71-FFC175	10°31'S 110°49'W	3248						
0-3			60	4.74	0.32	15	1.13±.02	11.5
KK71-FFC180	10°23'S 110°27'W	3132						
0-3			60	4.63	0.15	31	1.14±.04	8.96
Y71-7-44P	10°42'S 110°01'W	3167						
0-5			86	0.66	0.16	4.1	1.09±.07	12.9
88-93			87	0.60	0.14	4.3	1.23±.06	6.68
210-215			86	0.54	0.56	1.0	1.02±.10	1.52
226-231			78	0.62	0.07	8.9	1.09±.07	0.62
290-295			83	0.81	0.14	5.8	1.23±.04	4.54
310-315			53	0.70	0.15	5.8	1.08±.05	1.51
330-335			88	1.00	0.08	12	1.05±.04	1.87
371-376			81	0.92	0.11	8.4	0.97±.05	1.80
442-447			78	3.40	0.11	31	1.15±.03	0.05
Y71-7-45P*	11°05'S 110°06'W	3096						
10-15			87	0.29	0.14	2.1	1.10±.03	6.97
50-55			89	0.32	0.25	1.3	1.06±.03	7.75
92-97			91	0.49	0.12	4.1	1.09±.03	5.11
142-147			86	0.39	0.20	2.0	1.09±.03	3.33
192-197			91	0.35	0.53	0.7	1.09±.03	3.08
260-265			90	0.69	0.20	3.4	1.16±.03	1.16
300-305			89	0.71	0.28	2.5	1.13±.03	0.35
343-348			85	0.66	0.11	6.0	1.08±.03	0.36
KK71-GC12	10°32'S 110°03'W	3116						
3-8			84	0.23	0.47	0.5	1.04±.05	10.8
KK71-GC16	10°49'S 110°08'W	3315						
2-5			78	0.36	0.25	1.4	1.14±.03	14.4
KK71-GC18	10°33'S 110°13'W	2933						
0-4			63	1.66	0.42	4.0	1.14±.02	8.24

TABLE 1. - Continued

Core no. depth (cm)	Location	Water depth (m)	CaCO$_3$ (%)	U (ppm)	Th (ppm)	U/Th	^{234}U/^{238}U	^{230}Th$_{ex}$ (d/m/g)
KK71-FFC173	10°45'S 109°41'W	3410						
10-13			87	0.52	0.27	1.9	1.10±.03	18.4
OC73-3-20P	19°15'S 113°35'W	3081						
5-12			21	4.29	0.15	29	1.19±.03	15.8
50-55				7.01	0.35	20	1.13±.01	10.4
150-155				7.21	0.38	19	1.12±.02	6.52
215-220				6.65	0.23	29	1.14±.02	5.61
Y73-3-34PC	31°09'S 113°21'W	3064						
5-12			17	0.25	0.37	0.7	1.00±.06	1.36
50-55				1.67	0.74	2.2	1.04±.05	0.30
100-105				1.91	0.95	2.0	1.07±.03	0.24
150-155				1.88	0.74	2.5	1.03±.04	0.06
200-205				1.55	0.46	3.4	1.04±.04	0.23
300-305				1.19	0.50	2.4	1.00±.04	0.27
KK71-GC10*	9°59'S 106°02'W	3447						
10-15			91	0.18	0.25	0.7	1.10±.04	12.4
42-47			86	0.17	0.25	0.7	1.03±.03	8.44
72-77			90	0.19	0.38	0.5	0.94±.04	4.46
112-116			91	0.15	0.31	0.5	1.00±.04	1.46
Y71-7-36MG2*	10°08'S 102°51'W	4541						
0-5			1.2	1.48	1.48	1.0	1.04±.02	101
15-20				1.53	4.57	0.3	1.04±.02	76.9
35-40			1.6	1.45	3.57	0.4	1.02±.02	23.2
50-55			1.3	2.14	2.95	0.7	1.00±.02	2.80
Y73-3-7MG1	13°00'S 105°14'W	3822						
5-10			74	0.23	1.57	0.1	1.11±.05	28.0
20-25				0.50	1.64	0.3	1.04±.04	25.5
50-55				0.32	0.70	·0.5	0.98±.07	0.73
80-85				0.48	1.81	0.3	1.01±.04	14.5
Y73-3-21K	13°37'S 102°32'W	4410						
0-2			5.1	2.60	4.07	0.6	1.08±.04	89.4
30-32					4.76			10.7+
60-62				4.45	2.67	1.7	1.08±.04	0.82
90-92				4.25	2.38	1.8	1.06±.02	1.83
134-136				3.73	2.33	1.6	1.06±.02	-0.38
Y73-4-64K	32°46'S 94°42'W	3871						
0-2			30.1	2.56	4.93	0.5	1.07±.04	39.7
14-16				2.81	4.98	0.6	1.05±.04	35.8
44-46				3.71	5.67	0.7	1.04±.05	14.7
74-76				4.08	6.19	0.7	1.04±.05	6.42
104.106				4.32	6.98	0.6	1.06±.04	1.91
134-136				4.18	7.04	0.6	1.04±.04	4.22

Note: All values on total dry weight basis. ^{234}U, ^{238}U, ^{230}Th$_{ex}$ in disintegrations min^{-1}g^{-1}(d/m/g). Errors quoted are based on counting statistics (±1σ). Other errors are U(∼3%); Th(5-20%); ^{230}Th$_{ex}$(∼3%). ^{230}Th$_{ex}$ = unsupported ^{230}Th.

*Data from Dymond and Veeh, 1975.

+Correction for ^{234}U supported ^{230}Th based on average ^{234}U content of adjoining core sections.

the sediments there more closely resembled the present surface sediments on the west flank (Fig. 2). The change in conditions is particularly evident in core Y71-7-44P (Fig. 5). It would appear that the large seamount in the northeast corner of the detailed survey area has something to do with this anomaly, perhaps as a source of detrital material (volcanic ash) which would lower both the $^{234}U/^{238}U$ and U/Th ratios in the sediments near its base. However, the thorium accumulation rates on either side of the rise crest are similar, whereas the uranium accumulation rate is almost one order of magnitude higher on the west flank (McMurtry and others, this volume), suggesting instead that nonterrigenous input of uranium on the rise crest has been variable, both in space and time. A striking example for down-core variation in uranium content has been cited by Rydell and others (1974) for a core from the southern East Pacific Rise and interpreted as evidence of postdeposition injection of uranium-rich solutions, presumably of hydrothermal origin.

In terms of $^{234}U/^{238}U$ and U/Th, the sediments from the southern rise crest are quite similar to the hydrothermal sediments from the Red Sea geothermal area and the Trans-Atlantic Geotraverse (TAG) geothermal field, but they have lower U/Th ratios than iron and manganese deposits directly associated with submarine volcanoes (Fig. 6), perhaps because of slower accumulation rates than the latter. In most cases, the $^{234}U/^{238}U$ value of modern sea water is approached, but not exceeded. It would appear, therefore, that the excess uranium found in metalliferous sediments was derived entirely from sea water, presumably by coprecipitation with volcanogenic iron, as suggested previously (Ku, 1969; Scott and others, 1972; Bender and others, 1971). Unfortunately, the evidence provided by $^{234}U/^{238}U$ is ambiguous (Rydell and Bonatti, 1973). If hydrothermal solutions are metal enriched, recycled sea-water solutions, emanating along active ridges, the $^{234}U/^{238}U$ ratio in these solutions may be indistinguishable from normal sea water. Only where extensive alpha-recoil induced leaching of ^{234}U along the path of these circulating sea-water solutions has taken place would a measurable increase of $^{234}U/^{238}U$ above that of original sea water be expected. Considering the low

TABLE 2. URANIUM AND THORIUM ISOTOPIC DATA IN SIZE-SEPARATED SEDIMENT COMPONENTS

Sample no. Size fraction (μm)	Location	Water depth (m)	CaCO$_3$ (%)	U (ppm)	Th (ppm)	U/Th	$^{234}U/^{238}U$	$^{230}Th_{ex}$ (d/m/g)
KK71-LDC6	6°06'S 106°43'W	3073						
< 2			84	0.62	0.28	2.2	1.17±.05	16.7
>61			99	0.075	0.07	1.1	1.05±.08	3.03
KK71-LDC2	10°38'S 110°32'W	3145						
< 2			66	1.72	0.60	2.9	1.18±.03	7.69
2-61			78	1.57	0.13	12	1.16±.02	4.37
>61			95	1.04	0.10	10	1.16±.03	1.10
KK71-LDC3	9°55'S 101°09'W	4493						
< 2				1.90	4.8	0.4	0.99±.04	139
2-5				1.85	4.8	0.4	1.07±.03	98
5-10				1.58	3.4	0.5	1.05±.03	82
10-63				1.67	2.3	0.7	1.07±.02	63
>63			3	1.57	2.4	0.7	1.09±.02	68
Bulk Sediment (untreated)				1.25	3.3	0.4	1.03±.04	79

Note: All values on total dry weight basis. ^{234}U, ^{238}U, $^{230}Th_{ex}$ in disintegrations min^{-1}g^{-1}(d/m/g). Errors quoted are based on counting statistics (±1σ). Other errors are U(∼3%); Th(5-20%); $^{230}Th_{ex}$(∼3%).

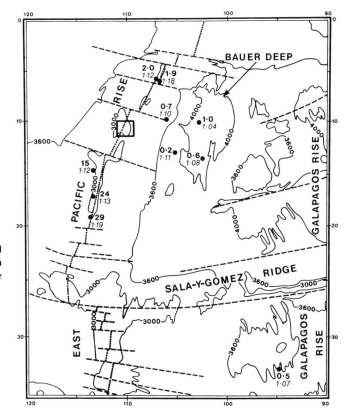

Figure 3. U/Th (bold face) and $^{234}U/^{238}U$ (italics) in core tops from the northwestern Nazca plate. The small insert refers to Figure 4.

uranium content in unaltered submarine basalts (Tatsumoto and others, 1965) and the relatively young age of the basaltic crust beneath active ridges, it is not very likely that sufficient ^{234}U has been generated by in situ decay of ^{238}U to measurably affect the $^{234}U/^{238}U$ in the hydrothermal solutions, unless the uranium content in these solutions happens to be abnormally low.

There is some evidence that in the course of sea-water reaction with submarine basalt, uranium actually enters the rocks rather than the reverse (Aumento, 1971; MacDougall, 1977); therefore, any sea water solution re-emerging on the sea floor could be depleted in uranium. Such a process, presumably caused by the reduction of soluble hexavalent uranium in sea water to insoluble tetravalent uranium, was first proposed by Rydell and Bonatti (1973) and recently has been confirmed for hot springs at the Galapagos spreading center (Edmond and others, 1979). Because of the lowered uranium content, the $^{234}U/^{238}U$ activity ratio in such hot-spring water would be more susceptible to change by additional input of ^{234}U, the magnitude of change depending on the length of time the water spends in contact with the rock.

In this connection, it may be significant that the only submarine deposits with $^{234}U/^{238}U$ ratios above that of sea water are iron or manganese encrustations with relatively low uranium content but high U/Th ratios on seamounts or submarine ridges at some distance from currently active spreading centers (Fig. 6). For instance, the seamount from which core AMPH D-2 was dredged has a basement age of at least 500,000 yr, allowing sufficient for in situ production of mobile ^{234}U. Even here, this process does not invariably lead to anomalous $^{234}U/^{238}U$ ratios, and 4 out of 5 different samples of core AMPH D-2 contained uranium which is indistinguishable from that in normal water (Fig. 6). Unless complete mixing of hydrothermal solutions with normal sea water has been prevented by rapid deposition of the uranium, any uranium with anomalous isotopic composition would be masked by uranium of hydrogenous origin (Veeh and Boström, 1971; Rydell and Bonatti, 1973; Piper and others, 1975).

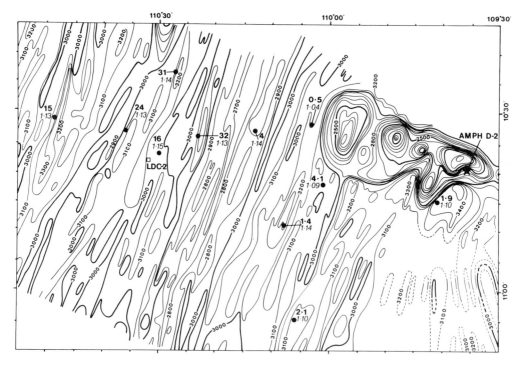

Figure 4. U/Th (bold face) and ^{234}U/^{238}U (italics) in core tops from detailed survey area across rise crest near 10° S. Bathymetry after David K. Rea, Oregon State University. Note contrast in U/Th and ^{234}U/^{238}U between the areas east and west of the rise crest.

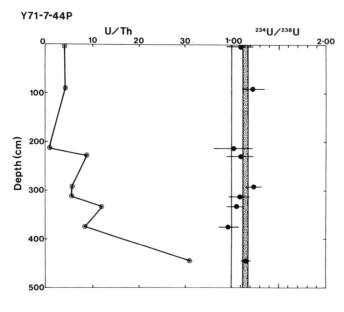

Figure 5. Down-core variation in U/Th and ^{234}U/^{238}U in core from east flank of East Pacific Rise near 10° S. Errors bars for ^{234}U/^{238}U indicate ±2σ of counting statistics. Stippled area shows ^{234}U/^{238}U in modern sea water.

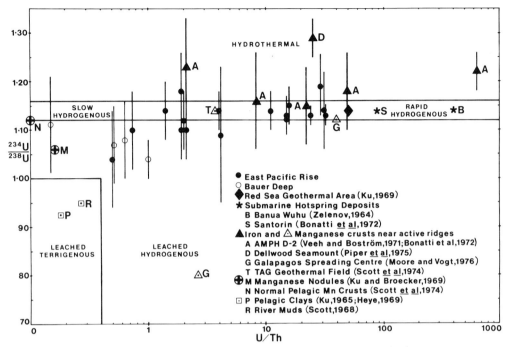

Figure 6. $^{234}U/^{238}U$ versus U/Th in surface sediments from the northwestern Nazca plate in relation to other sediments and deposits of known origins. Error bars for $^{234}U/^{283}U$ indicate $\pm 2\sigma$ of counting statistics. The low $^{234}U/^{238}U$ value for average manganese nodules shown in this figure probably reflects inclusion of some material from old nodule interiors in several of the surface values.

That selective leaching of ^{234}U does take place in the marine environment and in deposits associated with submarine hot-spring activity is shown by $^{234}U/^{238}U$ ratios which are highly depleted in ^{234}U beneath the outer surface of a manganese crust occurring on a submarine ridge with an estimated basement age of about 2 m.y. near the Galapagos spreading center (Moore and Vogt, 1976).

In an attempt to discover the pathways of uranium and the possible mechanism of its fixation in the metalliferous sediments, several size-separated sediment compounds of representative cores from the East Pacific Rise and the Bauer Deep were analyzed individually. This approach was not entirely successful, evidently because of aggregation of clay-size material into larger particles and trapping of fine material inside empty foraminiferal shells.

Nevertheless, a slight but systematic increase in $^{234}U/^{238}U$ and U/Th toward the coarse fraction is apparent in core LDC-3 from the Bauer Deep (Fig. 7). In particular, the < 2 μm fraction appears to contain predominantly uranium of terrigenous or detrital origin, while the remaining size fractions show variable mixtures of detrital and hydrogenous uranium. The coarse fraction (>63 μm) from another core in the Bauer Deep (Y73-3-21K) was further subdivided by hand picking into fish-bone apatite, manganese micronodules, phillipsite, and foraminiferal shells, and subjected to fission-track analysis. Although a determination of absolute uranium content was not feasible by this method, mainly because the actual density of individual minerals could not be determined accurately, the relative uranium concentrations were found to be highest in fish-bone apatite, followed by manganese micronodules, and phillipsite, with negligible uranium content in foraminiferal shells. From this it would appear that hydrogenous uranium in the Bauer Deep resides predominantly in fish-bone apatite and perhaps manganese micronodules, while detrital uranium is confined mainly to the clay-size fraction. An association of uranium with fish debris in deep-sea sediments has previously been reported by Arrhenius and others (1957). It is interesting to note that apatite has also been singled out

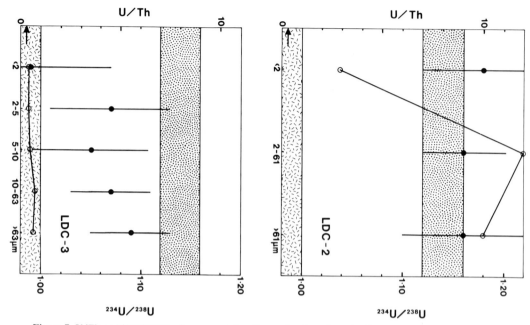

Figure 7. U/Th and ^{234}U/^{238}U in size-separated sediment fractions from the East Pacific Rise at 10°S (LDC-2) and the Bauer Deep (LDC-3). Error bars indicate ±2σ of counting statistics. Stippled area and cross-marked area refer to ^{234}U/^{238}U values in modern sea water, and normal pelagic clays, respectively. Arrow indicates U/Th in primary igneous rocks.

as the most likely host mineral for uranium in the Bauer Deep, on the basis of multivariate factor analysis of bulk sediments (McMurtry and others, this volume).

On the rise crest, all size fractions including the <2 μm fraction appear to be dominated by hydrogenous uranium. Since the course fraction consists almost entirely of foraminiferal shells, it would appear that some of the uranium is carried into the sediments associated with biogenic carbonate. Fission-track studies revealed, however, that the foraminiferal shells are virtually free of uranium. Rather, several of the conspicuous fission-track clusters of the type shown in Plate 1 can be traced to ferruginous, fine material trapped inside foraminiferal shells. This suggests that locally reducing conditions in the shell cavities may aid in the fixation of uranium by reduction. Precipitation of uranium under reducing conditions, presumably within ephemeral stagnant basins receiving a high influx of organic matter, has been proposed as the main mechanism for uranium enrichment on mid-ocean ridges (Turekian and Bertine, 1971). However, an entirely biological pathway for uranium on the East Pacific Rise is unlikely in view of the geographic distribution pattern of ^{234}U/^{238}U and U/Th (Figs. 3, 4), the accumulation rate pattern of the uranium, and the results of multivariate factor analysis (McMurtry and others, this volume). For instance, the highest influx of uranium is observed along the southern rise crest and not toward the north where biogenous inputs predominate. Multivariate factor analysis shows clearly that uranium is not associated with biogeneous silica, but with Fe, Mn, P, Cu, Zn, and Ni and that highest factor loading is in sediments from the southern rise crest. While Fe and Mn are probably of hydrothermal origin (Dymond and others, 1973; Dymond and Veeh, 1975; Heath and Dymond, 1977), the covariance of P with Fe can best be explained by absorption of phosphate ion from sea water onto volcanogenic ferric oxides (Berner, 1972). The fact that coprecipitation or adsorption of uranium from sea water, using freshly prepared ferric hydroxide, is routinely carried out in the laboratory or in situ at sea (Somayajulu and Goldberg, 1966), indicates that the removal of uranium from sea water can take place under oxidizing conditions in the presence of a suitable collecting agent.

A hydrogenous origin for most of the uranium in sediments on the East Pacific Rise is certainly not in conflict with any of the data presented here. At a uranium concentration of 3 μg/l in sea water (Turekian and Chan, 1971) and a renewal rate of 10^3 yr for bottom water in the Pacific Ocean, the amount of uranium required for the maximum input rate of 6 μg · cm^{-2} · 10^{-3}yr^{-1} measured on the East Pacific Rise (McMurtry and others, this volume) can easily be supplied by a 20-m sea-water column above 1 cm² of the rise crest, hence there is no need to invoke additional sources for the uranium. But how was hydrogenous uranium supplied to sediments in the Bauer Deep? Heath and Dymond (1977) have proposed a model for the origin of metalliferous sediments in the Bauer Deep according to which most of the iron and several other "hydrothermal" metals were transported laterally from the northern section of the East Pacific Rise into the Bauer Deep by bottom currents. It is possible that much of the hydrogenous uranium was carried into the Bauer Deep together with volcanogenic iron oxides generated on the rise crest, and was subsequently fixed in the sediments by incorporation into fish-bone apatite and manganese micronodules.

It would be most interesting to estimate the proportion of hydrogenous uranium input into the Bauer Deep, based on the uranium isotopic data. If we let

$$R_T = R_H + R_D$$

and

$$R_T A_T = R_H A_H + R_D A_D ,$$

then

$$R_D = \frac{R_T(A_H - A_T)}{A_H - A_D} ,$$

where R_T, R_H, and R_D are the accumulation rates of total uranium, hydrogenous uranium, and detrital uranium, respectively, in the Bauer Deep, and A_T, A_H, and A_D are the $^{234}U/^{238}U$ activity ratios of the total uranium, hydrogenous uranium, and detrital uranium, respectively, then R_D and R_H can be calculated from the remaining data as follows: $A_T = 1.08$ (average $^{234}U/^{238}U$ in surface sediments of the Bauer Deep; see Fig. 3). $A_H = 1.14$ ($^{234}U/^{238}U$ in sea water, and hence in hydrogenous uranium by definition). $A_D = 0.93$ (average $^{234}U/^{238}U$ in normal pelagic clay, taken here as representative of detrital uranium; see Fig. 5). $R_T = 0.13$ μg · cm^{-2} · 10^{-3}yr^{-1} (average uranium accumulation rate in the Bauer Deep, based on those cores for which surface $^{234}U/^{238}U$ values have been measured; McMurtry and others, this volume). Using the values, we get $R_D = 0.04$ μg · cm^{-2} · 10^{-3} yr^{-1}, and $R_H = 0.09$ μg · cm^{-2} · 10^{-3} yr^{-1}. This suggests that up to 70% of the total uranium input in the Bauer Deep is hydrogenous and could have been supplied by lateral transport from the East Pacific Rise, the rest being of detrital origin.

CONCLUSIONS

Uranium isotopic, U/Th, and fission-track data indicate that the uranium in metalliferous sediments of the Nazca plate has been derived from sea water and terrigenous or local detrital sources in variable proportions, depending on time and location.

Along the crest of the East Pacific Rise, the uranium in the sediment is dominated by uranium of hydrogenous origin, and the $^{234}U/^{238}U$ ratio is indistinguishable from that of modern sea water. In the Bauer Deep, the input of hydrogenous uranium is relatively less important, so uranium of terrigenous or local detrital origin makes up a significant proportion of the total uranium content.

The regional pattern of $^{234}U/^{238}U$ and U/Th and down-core variations of these two parameters in a number of cores on the East Pacific Rise indicate that the input of hydrogenous uranium has varied in time and in space. While this could be interpreted as evidence for a hydrothermal origin of the uranium, it can be equally well explained by a mechanism of uranium removal from sea water which is dependent on a supply of volcanogenic iron oxide phases, so the hydrogenous supply of uranium varies with the hydrothermal supply of iron.

Coprecipitation of uranium from sea water with volcanogenic iron oxides is considered to be quantitatively the most important mechanism for the observed enrichment of uranium in metalliferous sediments on the East Pacific Rise. Locally reducing conditions within the sediments, perhaps associated with decaying organic matter in fecal pellets, fish debris, or inside foraminiferal shell cavities, may aid in the fixation of uranium in the sediments.

The mechanism of hydrogenous uranium input in the Bauer Deep is less clear, but it may involve redeposition of rise-crest sediments by bottom currents (Heath and Dymond, 1977; McMurtry and others, this volume) followed by diagenetic release of the uranium and its fixation in the sediment by fish-bone apatite and manganese micronodules.

Although any hydrothermal contribution of uranium to sediments on the Nazca plate, as suggested by occasional observations of anomalous $^{234}U/^{238}U$ ratios in iron encrustations on seamounts, cannot be dismissed, it would be quantitatively insignificant, and certainly negligible in the global geochemical balance of uranium.

AKNOWLEDGMENTS

The sediment samples used in this study were supplied by the Hawaii Institute of Geophysics and Oregon State University. I thank J. Dymond and C. Lopez of Oregon State University, and G. McMurtry of the Hawaii Institute of Geophysics for several of the carbonate analyses and the size-separated sediment fractions.

Financial assistance was provided by the International Decade of Ocean Exploration Office of the National Science Foundation, the Australian Research Grants Committee, and the Australian Institute of Nuclear Science and Engineering. Continued interest in this study has been sustained by several short sabbaticals at the Hawaii Institute of Geophsyics and Oregon State University.

Plate 1 (facing page). Fission-track distribution maps of impregnated sediment section of core KK71-FFC-180 from East Pacific Rise, showing enrichment pattern of uranium in some of the coarse-fraction material. Although association of fission-track clusters with foraminiferal shell (A) or with materials trapped inside foraminiferal shell chambers (B) is evident, most of the foraminifera in the same section are virtually free of uranium. This suggests that the uranium is not fixed by the shell carbonate, but rather by locally reducing conditions caused by decaying organic matter associated with some of the foraminifera. Width of photographs is approximately 500 μm.

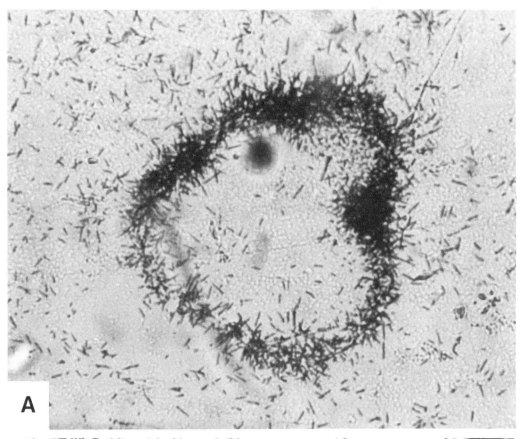

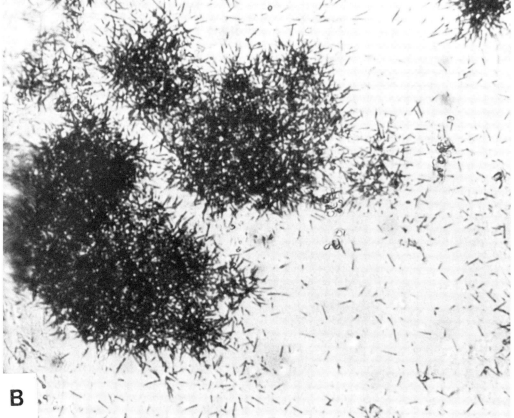

REFERENCES CITED

Aller, R. C., and Cochran, J. K., 1976, ^{234}Th/^{238}U disequilibrium in nearshore sediment: Particle reworking and diagenetic time scales: Earth and Planetary Science Letters, v. 29, p. 37–50.

Arrhenius, G., Bramlette, M. N., and Piciotto, E., 1957, Localization of radioactive and stable heavy nuclides in ocean sediments: Nature, v. 180, p. 85–86.

Aumento, F., 1971, Uranium content of mid-oceanic basalts: Earth and Planetary Science Letters, v. 11, p. 90–94.

Bender, M., and others, 1971, Geochemistry of three cores from the East Pacific Rise: Earth and Planetary Science Letters, v. 12, p. 425–433.

Berner, R. A., 1972, Phosphate removal from sea water by adsorption on volcanogenic ferric oxides: Earth and Planetary Science Letters, v. 18, p. 77–84.

Bloch, S., 1980, Some factors controlling the concentration of uranium in the world ocean: Geochimica et Cosmochimica Acta, v. 44, p. 373–377.

Bonatti, E., 1975, Metallogenesis at oceanic spreading centers: Earth and Planetary Science Letters, Annual Review, v. 3, p. 401–431.

Bonatti, E., and others, 1972, Submarine iron deposits from the Mediterranean Sea, *in* Stanley, D. J., ed., The Mediterranean Sea: Stroudsburg, Pennsylvania, Dowden, Hutchinson and Ross, Inc., p. 701–710.

Boström, K., 1973, The origin and fate of ferromanganoan active ridge sediments: Stockholm Contributions in Geology, v. 27, p. 149–243.

Chalov, P. I., and Merkulova, K. I., 1966, Comparative rate of oxidation of ^{234}U and ^{238}U atoms in certain minerals: Academiya Nauk SSSR, Doklady, Proceedings Earth Science Section, v. 167, p. 146–148.

Church, T. M., 1970, Marine barite [Ph.D. thesis]: San Diego, University of California.

Clark, S. P., Peterman, Z. E., and Heier, K. S., 1966, Abundances of uranium, thorium, and potassium, *in* Clark, S. P., ed., Handbook of physical constants: Geological Society of America Memoir 97, p. 521–541.

Dymond, J., and others, 1973, Origin of metalliferous sediments from the Pacific Ocean: Geological Society of America Bulletin, v. 84, p. 3355–3372.

Dymond, J. and Veeh, H. H., 1975, Metal accumulation rates in the southeast Pacific and the origin of metalliferous sediments: Earth and Planetary Science Letters, v. 28, p. 13–22.

Edmond, J. M., and others, 1979, On the formation of metal-rich deposits at ridge crests: Earth and Planetary Science Letters, v. 46, p. 19–30.

Fisher, D. E., and Boström, K., 1969, Uranium rich sediments on the East Pacific Rise: Nature, v. 224, p. 64–65.

Fleischer, R. L., Price, P. B., and Walker, R. M., 1975, Nuclear tracks in solids: Berkeley, University of California Press, 605 p.

Heath, G. R., and Dymond, J., 1977, Genesis and diagenesis of metalliferous sediments from the East Pacific Rise, Bauer Deep and Central Basin, northwest Nazca plate: Geological Society of America Bulletin, v. 88, p. 723–733.

Heye, D., 1969, Uranium, thorium and radium in ocean water and deep-sea sediment: Earth and Planetary Science Letters, v. 6, p. 112–116.

Koide, M., and Goldberg, E. D., 1965, ^{234}U/^{238}U ratios in sea water: Progress in Oceanography, v. 3, p. 173–177.

Ku, T. L., 1965, An evaluation of the ^{234}U/^{238}U method as a tool for dating pelagic sediments: Journal of Geophysical Research, v. 70, p. 3457–3474.

———1969, Uranium series isotopes in sediments from the Red Sea hot-brine area, *in* Degens, E. T., and Ross, D. A., eds., Hot brines and recent heavy metal deposits in the Red Sea: New York, Springer-Verlag, p. 521–524.

Ku, T. L., and Broecker, W. S., 1969, Radiochemical studies on manganese nodules of deep-sea origin: Deep Sea Research, v. 16, p. 625–637.

Macdougall, D., 1977, Uranium in marine basalts: Concentration, distribution and implications: Earth and Planetary Science Letters, v. 35, p. 65–70.

Mammerickx, J., and others, 1975, Morphology and tectonic evolution of the East-Central Pacific: Geological Society of America Bulletin, v. 86, p. 111–118.

McMurtry, G. M., Veeh, H. H., and Moser, C., 1981, Sediment accumulation rate patterns on the northwest Nazca plate, *in* Kulm, L. D., and others, eds., Nazca plate: Crustal formation and Andean convergence: Geological Society of America Memoir 154 (this volume).

Moore, W. S., and Vogt, P. R., 1976, Hydrothermal manganese crusts from two sites near the Galapagos Spreading Axis: Earth and Planetary Science Letters, v. 29, p. 349–356.

Piper, D. W., and others, 1975, A marine iron deposit from the northeast Pacific: Earth and Planetary Science Letters, p. 114–120.

Rydell, H., and others, 1974, Postdepositional injections of uranium-rich solutions into East Pacific Rise sediments: Marine Geology, v. 17, p. 151–164.

Rydell, H. S., and Bonatti, E., 1973, Uranium in submarine metalliferous deposits: Geochimica et Cosmochimica Acta, v. 37, p. 2557–2565.

Sackett, W. M., and others, 1973, A re-evaluation of the marine geochemistry of uranium: International Atomic Energy Agency, Vienna, IAFA-SM-158/51, p. 757–769.

Scott, M. R., 1968, Thorium and uranium concentrations and isotope ratios in river sediments: Earth and Planetary Science Letters, v. 4, p. 245–252.

Scott, M. R., Osmond, J. K., and Cochran, J. K., 1972, Sedimentation rates and sediment chemistry in the South Indian Basin, in Hayes, D. E., ed., Antarctic oceanology II: The Australian–New Zealand sector: Antarctic Research Series, v. 19, p. 317–334.

Scott, M. R., and others, 1974, Rapidly accumulating manganese deposits from the median valley of the Mid-Atlantic Ridge: Geophysical Research Letters, v. 1, p. 355–358.

Somayajulu, B.L.K., and Goldberg, E. D., 1966, Thorium and uranium isotopes in sea water and sediments: Earth and Planetary Science Letters, v. 1, p. 102–106.

Tatsumoto, M., Hedge, C. E., and Engel, A.E.F., 1965, Potassium, rubidium, strontium, thorium, uranium, and the ratio Sr-87/Sr-86 in oceanic tholeiitic basalts: Science, v. 150, p. 886–888.

Turekian, K. K., and Bertine, K. K., 1971, Deposition of molybdenum and uranium along the major ocean ridge systems: Nature, v. 229, p. 250–251.

Turekian, K. K., and Chan, L. H., 1971, The marine geochemistry of the uranium isotopes, ^{230}Th and ^{231}Pa, in Brunfelt and Steinnes, E., eds., Activation analysis in geochemistry and cosmochemistry: Oslo: Universitetsforlaget, p. 311–320.

Veeh, H. H., 1967, Deposition of uranium from the ocean: Earth and Planetary Science Letters, v. 3, p. 145–150.

Veeh, H. H., and Boström, K., 1971, Anomalous ^{234}U/^{238}U on the East Pacific Rise: Earth and Planetary Science Letters, v. 10, p. 372–374.

Zelenov, K. K., 1964, Iron and manganese in exhalations of the submarine Banu Wahu Volcano (Indonesia): Academiya Nauk SSSR, Doklady, v. 155, p. 1317–1320.

MANUSCRIPT RECEIVED BY THE SOCIETY NOVEMBER 12, 1980
MANUSCRIPT ACCEPTED DECEMBER 30, 1980

Formation and growth of ferromanganese oxides on the Nazca plate

MITCHELL LYLE
School of Oceanography
Oregon State University
Corvallis, Oregon 97331

ABSTRACT

Marine ferromanganese oxides form four major types of deposits: hydrothermal crust, ferromanganese coatings on basalt, ferromanganese nodules, and a mixture of micronodules and other dispersed oxyhydroxides within sediments.

Hydrothermal crusts grow only near active marine hydrothermal systems that cool newly emplaced basaltic crust. The crusts are characterized by rapid growth rates, extreme fractionation of Mn from Fe, low accessory-element concentrations, and well-crystallized Mn minerals.

Ferromanganese coatings on basalt can receive an Fe-rich component from a hydrothermal source, but are mainly composed of ferromanganese oxides grown by direct preceipitation from sea water (hydrogenous formation). They have δ-MnO_2 mineralogy, have almost equal Mn and Fe abundances, are relatively enriched in the rare-earth elements, are highly enriched in Ce and Co, and have relatively low Cu, Ni, and Zn abundances. Coatings with a large hydrothermal component are more enriched in Fe and generally have lower trace-element contents.

Nodule and micronodule compositional variations resulting from different sources of transition metals may be further modified by diagenetic reactions within the sediments. The most extreme type of diagenesis is preferential reductive mobilization of Mn within the sediment column by oxidation of organic carbon and subsequent Mn diffusion to the sea-water–sediment interface. Ferromanganese oxides formed or altered by this process lie beneath highly productive equatorial waters and near the South American continent, because supply of organic carbon to marine sediments comes primarily from biological productivity in the surface waters. Ferromanganese oxides modified in this manner are characterized by relatively pure and well-crystallized Mn oxides, geneally the 7 Å mineral birnessite, and by rapid growth rates. Mn/Fe weight ratios are 5 or greater, and accessory-element contents are low.

The fourth type of ferromanganese deposit is enriched in Mn relative to hydrogenous preceipitates and is found in the Bauer Deep and other areas along the fringes of the most highly productive regions. There sediments are not reducing enough to remobilize Mn. Oxic diagenetic reactions, such as Fe-smectite formation from biogenic opal, fractionate Fe from ferromanganese hydroxyoxides. The released Mn precipitates as nodules and micronodules.

Fe and Mn that consitute by far the dominant components of the ferromanganese deposits have only two ultimate sources: runoff from the continents and hydrothermal interactions between sea water and basalts formed on the mid-ocean ridges. Sea water acts as a reservoir and mixing medium for these two sources as currents carry the introduced metals far from their input point. Hydrothermal input of Mn introduces most of this element to the Nazca plate, although terrigenous input probably dominates worldwide.

INTRODUCTION

The ferromanganese oxides form four major types of marine deposits: nodules, coatings on rocks, hydrothermal crusts, and micronodules. Most spectacular are the huge deposits of ferromanganese nodules found throughout the oceans in regions of slowly accumulating sediments (Mero, 1965). The nodules are at least 1 cm in diameter, and some grow to be greater than 10 cm across. They have shapes that range from flat discs to almost spherical concretions and are primarily composed of Fe and Mn oxides, along with lesser amounts of the other transition metal oxides.

Ferromanganese oxides also occur as coatings on basalt or other hard substrates. Normally these basalt coatings are on the order of 1 mm thick on fresh basalts, but may reach several centimetres in thickness on older rocks. Their chemistry to a certain extent resembles that of the nodules (Cronan, 1975), although they tend to be more Fe rich and have systematic variations in the other transition metals (Lyle and others, 1977; Toth, 1980).

The most recently documented ferromanganese deposits occur near mid-ocean ridges and are associated with cooling of newly emplaced crust by sea water (Corliss and others, 1978; Moore and Vogt, 1976; Cann and others, 1977; Lalou and others, 1977; Scott and others, 1974; Toth, 1980). They occur as crusts coating basalt or sediment and can be distinguished from other basalt coatings by the strong fractionation of Fe and Mn from each other. Typical deposits consist of nearly pure Mn oxides, or conversely, of nearly pure Fe oxide or Fe silicate material. New discoveries at 21°N show that sulfides, barite, and opal can also form (Haymon and others, 1979).

The fourth type of ferromanganese oxides are micronodules and dispersed oxides within sediments. Micronodules are small concretions less than 1 mm in diameter that grow as discrete bodies within the sediment or attached to other sedimentary grains. Much of the dispersed oxide fraction, as distinct from the micronodules, occurs as coatings on other sediment components, although a significant fraction may also be contained in the extremely fine fraction of the sediment. It appears from the little data available that micronodules have similar mineralogy and chemical compositions as do the larger concretions (Dymond and Eklund, 1978: Lopez, 1978; Friedrich, 1976). The dispersed oxide fraction is also dominated by Fe and Mn oxides. Fe/Mn ratios of this easily leachable fraction correlate significantly with the Fe/Mn ratios in coexisting nodules, although they tend to be richer in iron (Volkov, 1977).

Although the ferromanganese oxides occur in a range of morphologies, form diverse mineral phase, and vary in chemical composition, this diversity may be accounted for by relatively few processes. In Figure 1, I have illustrated the pathways through which ferromanganese oxides enter the oceans and then become redistributed. As I will show in later sections, the ultimate source of Fe, Mn, and other elements in ferromanganese deposits can only rarely be distinguished. Hydrothermally influenced deposits can be recognized near the axis of the mid-ocean ridge; terrigenous deposits are generally so reorganized by diagenesis that only the characteristics of the diagenetic process can be recognized. The majority of elements that eventually find their way into the various types of concretions have been relatively well mixed after being added to the oceans. Precipitation from the oceanic reservoir forms what others have called hydrogenous or authigenic deposits (Goldberg, 1961; Krishnaswami, 1976).

Diagenesis due to interactions between newly precipitated ferromanganese hydroxyoxides and other sedimentary components dominates the regional geochemistry of the ferromanganese deposits. In regions of relatively low surface productivity or in deposits that grow only in contact with sea water, little or no diagenesis occurs. These deposits have elemental abundances and a mineralogy that can be considered hydrogenous (authigenic).

Interactions of ferromanganese oxides with biogenic silica occur where productivity of the surface waters is of intermediate magnitude. Formation of Fe-rich smectites may fractionate Fe from Mn;

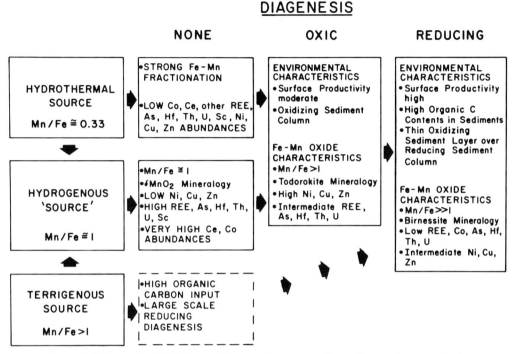

Figure 1. Pathways by which Mn and Fe enter the oceans and are subsequently reorganized.

more Mn-rich oxides can thus be formed (Heath and Dymond, 1977; Lyle and others, 1977). Ni, Cu, and Zn absorbed or incorporated in the microfossil tests and released by dissolution can also be added to the growing concretions (Calvert and Prince, 1977; Greenslate and others, 1973; Piper and Williamson, 1977).

Under regions of high productivity, large amounts of organic carbon are added to the sediments. Reduction of Mn^{IV} to Mn^{II} coupled to the oxidation of organic matter remobilizes Mn within the sediment column; it may diffuse to and deposit in an oxidized surface layer. Some of the remobilized Mn can be added to surface concretions, but because Fe and other transition metals are not mobilized concurrently, oxides rich in Mn are precipitated (Froelich and others, 1979).

This paper will explore sources of Fe and Mn on the Nazca plate and the effects that source and diagenesis have on the major ferromanganese oxide deposits. I will first describe the various ferromanganese oxide deposits that occur on the plate and discuss the geochemical evidence for the processes that formed them. I will then estimate the magnitude of the Mn and Fe sources to the Nazca plate and suggest possible dispersal paths for these two elements.

SAMPLE LOCATIONS

Locations for all samples analyzed for this paper are shown in Figure 2 and listed in Table 1 (microfiche). Ferromanganese coatings that comprise the data set were recovered from dredgings of rock outcrops in the Peru-Chile Trench at 17°S (FDR75-3-10-45), from the Nazca Ridge at 21°S (DM1028), from the Sala y Gomez Ridge at 25°S (DM1016), and from the East Pacific Rise at 23° and 33°S (KK72-33-2 and DM1011, respectively); also, two ferromanganese coatings were recovered from the Galapagos Rise (DM1 and DM2; reported in Corliss and others, 1978) and one from the Bauer Basin at 14°S (Y73-3-22 DC; reported in Toth, 1980; Lyle and others, 1977).

Nodule samples were chosen to cover the widest possible geographic range. Three of the samples reported in this study are not from the Nazca plate. DM981 is from the southern central gyre at 161°W

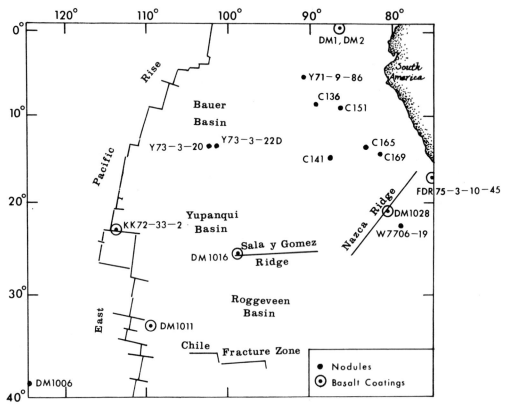

Figure 2. Sample locations. All nodule stations have accompanying sediment analyses. Locations of DM 981 and DM 984 are not shown.

and was chosen because of the likelihood that it would be formed primarily by hydrogenous precipitation. DM1006 was recovered southwest of the Nazca plate, and DM994 was recovered to the south.

Most of the nodules analyzed for this study came from the Peru Basin near the South American coast (C136, C141, C151, C165, and C169). The Basin sediments are primarily continentally derived, but near the coast increased productivity due to upwelling adds a significant biogenic fraction (see Dymond, this volume).

The nodule recovered in Y71-9-86MGI at 5°S, 91°W, formed on pelagic carbonate ooze, and was added to the study to provide an end-member formed under highly productive surface waters. Two nodules are also included from the metalliferous sediments of the Bauer Deep (Y73-3-20 and Y73-3-22DN; analyzed by Toth, 1980; Lyle and others, 1977), as well as one that was recovered from an extensive nodule field just south of the Nazca Ridge (W7706-19) where sedimentation is similar to low-productivity central gyre red clays.

PROCEDURES

Ferromanganese Nodules and Coatings

Either one-half or one-quarter of each ferromanganese nodule in the study was coarsely ground (~30 mesh) in an agate mortar. In addition, portions were scraped from the tops and bottoms of each of three nodules (C141, C165, C169) that had distinquishable top and bottom features, as described by Raab (1972); they were subjected to separate analysis. A split was taken for x-ray diffraction analysis and ground fine (<325 mesh) under butanol in an autogrinder. Random powder mounts were x-rayed at 500 seconds per degree 2Θ. A second split was dissolved under heat and pressure in hydrofluoric acid and aqua regia and analyzed for Mg, Al, Si, Ca, Mn, Fe, Co, Ni, Cu, and Zn by atomic absorption spectrophotometry (see Dymond and others, 1973, for more complete description of the technique). The precision of analyses was better than 8% for Al, Si, and Ca and better than 5% for the other elements. Accuracy determined by in-house standards was at the same level. A third split was leached in oxalic acid buffered at pH 3 by ammonium oxalate (Heath and Dymond, 1977) to separate the oxide fraction of the nodule from silicate and other refractory components. the solution was filtered through prewashed, preweighed 0.45 um filters to determine the percentage of non-oxide components within the nodule and to allow calculation of the composition of the oxide fraction and of the residue. Residues collected on the filters were also analyzed by x-ray diffraction. The leachate was analyzed by atomic absorption spectrophotometry for Mg, Al, Si, Mn, Fe, Co, Ni, Cu, and Zn. Ca was not analyzed because it forms an insoluble oxalate precipitate during this treatment. Precision of the analysis was determined by repeated leaching of a Bauer Deep sediment standard. The analyses indicate a precision of better than 15% for Si, better than 10% for Co, and better than 5% for the other elements.

Two other splits of the ground nodule material were taken for instrumental neutron activation analysis (Gordon and others, 1968). One was irradiated for the elements Sc, Co, Ag, As, Sb, Ba, La, Ce, Nd, Sm, Yb, Tb, Lu, Hf, Th, and U. The second was leached as described above. The residue was collected and weighed, and a portion irradiated to analyze for the elements listed above. There was not enough material scraped from the tops and bottoms of the nodules for the residue study. Among the basalt coatings, only one sample, FDR75, was sufficiently thick for the complete range of sampling and analytical procedures. The other samples were analyzed partially depending on the quantities of material recovered. The data from all these analyses are presented in Tables 2 (microfiche) and 3 (microfiche). Also included in the tables are four samples from other recent studies located on the Nazca plate (Corliss and others, 1978; Toth, 1980).

Micronodule and Sediment Analysis

Surface-sediment samples were taken from the same stations as the nodules in this study, except for Y73-3-22D, in the Bauer Deep. The samples were first split for micronodule analysis and a bulk sediment study. The split for bulk sediment analysis was further subdivided for bulk chemical analysis, leach chemical analysis, and for x-ray diffractometry. Bulk chemical analysis was accomplished by a combination of atomic absorption spectrophotometry and neutron activation analysis, as had been done for the nodule study.

Leach chemical analysis was performed to separate the micronodule and dispersed oxide fraction from the other sediment components by the same technique as for the nodules. The oxalic acid leach solution was filtered through prewashed, preweighed 0.45 um filters, and the leachate was analyzed for Mg, Al, Si, Mn, Fe, Co, Cu, Ni, and Zn by atomic absorption spectrophotometry.

A third part of the bulk sediment was examined by x-ray diffractometry. The subsample was first leached with acetic acid buffered at pH 5 with sodium acetate to remove any calcium carbonate. A preliminary study revealed that this treatment did not chemically attack micronodule material to any significant extent. The acetic acid leached sample was ground fine to less than 325 mesh under butanol, and random powder mounts of the material were X-rayed using Cu-K alpha radiation at a scan rate of 500 seconds per degree 2θ over a range of $5°$ to $70°$ 2θ.

Micronodules were separated from the second split of the sediment by a combined physicochemical technique. The sediment samples were first sieved wet at 44 um (325 mesh) to remove the clay fraction. Clay aggregates were dispersed and removed by gently rubbing the sample through the sieve under flowing water. Calcium carbonate debris was removed by leaching with acetic acid buffered at pH 5 with sodium acetate, which has the additional effect of removing any absorbed or surface-coating ferromanganese oxides.

The final separation of micronodule material from the coarse fraction of the sediment was achieved by the buffered oxalic acid leach technique. The leach was filtered again at 0.45 um, and the difference in weight between the residue and original sample was used to determine the weight of micronodules analyzed. Accuracy of the micronodule weight is between 1 and 2 mg for typical residue recovery.

Atomic absorption analysis was perfomed for Mg, Al, Si, Mn, Fe, Co, Ni, Cu, and Zn. Precision determined by replicate analyses is better than 10% for all elements. Absolute composition seems good for samples where greater than 5 mg of micronodule material were leached. Relative abundances for those samples where less micronodule material was leached still seem good, except for sample Y71-9-86 MG1. So little micronodule material was leached from this sample that the data was ignored. Sediment and micronodule analyses are reported in Tables 3 and 4 (microfiche).

HYDROTHERMAL FERROMANGANESE DEPOSITS ON THE NAZCA PLATE

Hydrothermal ferromanganese deposits are one type of ferromanganese deposits where source has a demonstrable effect on the composition of the deposit. Marine hydrothermal systems debouch fluids relatively rich in both Mn and Fe, but carry only trace amounts of the other elements commonly associated with ferromanganese deposits. High concentrations of H_2S, such as found in the Galapagos hydrothermal fluids (Corliss and others, 1979), suggest that Fe and some other trace metals could be retained in the rocks as sulfides.

The huge influx of Fe and Mn causes the deposits to grow quite rapidly, as has been reported by Scott and others (1974), Moore and Vogt (1976), and has been inferred by Corliss and others (1978). The rapid growth rate precludes incorporation of large abundances of trace metals from sea water (Toth, 1980; Corliss and others, 1978) and also prevents large-scale preferential enrichment of elements such as Ce and Co from sea water, as may be seen in more slowly growing deposits.

Because of the different geochemical behavior of Fe and Mn, fractionation of the two elements is common in hydrothermal-type deposits. Fe may precipitate as sulfides before the hydrothermal fluid leaves the vent (Corliss and others, 1979). It may also precipitate as silicates (Stakes, 1978) or as relatively pure Fe oxides before Mn will precipitate (Bonatti and others, 1972; Krauskopf, 1957). Relatively pure Mn oxides and Fe silicates, sulfides, or oxides instead of mixed phases are thus normally found.

Documented hydrothermal deposits are all local features near vents (Lalou and others, 1977; Scott and others, 1974; Cann and others, 1977; Corliss and others, 1978). The influence of hydrothermal activity extends further, however, as is indicated by metalliferous sediments near active spreading centers (Bostrom and Peterson, 1969; Dymond and others, 1973; Dymond, this volume), and by compositions of ferromanganese coatings as discussed by Toth (1980) and the next section.

FERROMANGANESE COATINGS ON BASALT

Toth (1980) has suggested that ferromanganese coatings on basalt are formed by hydrogenous precipitation of ferromanganese oxides and associated elements from sea water, diluted near the rise crest by hydrothermal Fe and Si. Findings in this study agree with his conclusions. Hydrogenous formation of ferromanganese oxides is assumed to occur because sea water is supersaturated with respect to Fe and Mn oxides. The excess will precipitate in sediments, in nodules, and on basalt and will coprecipitate other elements from sea water. Calvert and Price (1977) have assumed that ferromanganese nodules that have the same Mn/Fe weight ratio as the oxide fraction of the associated sediment have not been diagenetically altered and thus are formed solely by authigenic precipitation. By use of this criterion they suggest that hydrogenous ferromanganese oxides are distinguished by the presence of δ-MnO_2, a Mn/Fe weight ratio near unity, relatively high concentrations of Ce, Co, Pb, and Ti, and relatively low concentrations of Cu, Ni, Zn, and Mo.

On the other hand, hydrothermal precipitates are distinguishable by very low trace-element concentrations and by rare-earth-element abundance patterns similar to sea water (Toth, 1980; Corliss and others, 1978). The Ce content is low in these precipitates, presumably because of the low trace-element content of the hydrothermal fluid and the relatively rapid growth which precludes large-scale cerium adsorption from sea water. Toth (1980) also suggested that hydrothermal precipitates incorporated into ferromanganese coatings will be Fe and Si rich.

Figure 3 illustrates the regional distribution of Mn, Fe, Ce, and Co in ferromanganese coating from this study. Although Fe is not always most highly enriched in coatings recovered near the rise crest, trace-element contents, illustrated by Co and Ce, are lowest there as would be expected if ferromanganese coatings growing near the rise crest have a hydrothermal component. It is unlikely that the high Ce contents near the continent is caused by terrigenous input, since enrichments as high or higher can be found in ferromanganese nodules recovered far from the continents (for example nodule DM981, Ce = 1,770 ppm, cersus coating FDR75, Ce = 769 ppm).

Additional support for a hydrothermal component in rise-crest ferromanganese coatings is illustrated by Figures 4 and 5. Rare-earth abundance patterns of ferromanganese coatings (Figure 4) show that rise-crest coatings typically have lower absolute abundances of the rare earths and patterns more similar to sea water than coatings from off the rise crest. Figure 5 demonstrates that the La/Ce ratio of coatings from the rise crest is similar to sea water, whereas coatings from off the rise crest are much more highly enriched in Ce.

FERROMANGANESE NODULES

While ferromanganese oxide coatings on basalt derive their distinct chemical compositions through an interplay between hydrothermal and hydrogenous precipitation, nodule chemistry is further complicated by interactions with the sediment. Diagenetic reactions in sediments have been considered as a possible formation mechanism for Mn nodules since the time that they were first discovered (Murray and Irvine, 1894). One of the original hypotheses was that oxidation of organic carbon in the sediment column would reduce Mn from Mn^{IV} to the more soluble Mn^{II} form. Mn^{II} then diffuses back to the sediment surface to be deposited as nodules.

This type of mobilization and deposition of Mn can only occur under certain special sedimentation conditions. There must be a reservoir of a reduced species to be oxidized during Mn reduction, most probably organic carbon, since it is by far the largest sedimentary electron reservoir (Stumm and Morgan, 1970) and biogenic catalysis of its oxidation is quite common. The amount of organic carbon that will accumulate in the sediment is a function of both sedimentation rate and biological productivity in the surface waters. Muller and Suess (1979) have empirically determined that organic carbon content in the sediment is directly proportional to organic carbon fixation in surface waters, provided that sedimentation rate is constant. They also have shown that a tenfold increase in sedimentation rate will double the organic carbon preservation in the sediment, assuming that all other factors are constant. Slowly accumulating sediments under low productivity surface waters will thus have the least amount of organic carbon contents.

For Mn to be reduced, higher Eh oxidants must first be removed from the pore waters. They must be consumed by the oxidation of organic carbon faster than they can diffuse into the sediment from bottom water. Since the rate that oxidants are consumed in a sediment should be a function of the organic carbon content, there is some critical combination of surface productivity and sedimentation rate that will drive a sediment anoxic. This should occur under high productivity regions and on continental shelves.

Even in sediments that go strongly reducing, a surface layer where oxidants from sea water are being consumed remains more oxidizing. There is commonly a brown to gray-green color change in the tops of cores from reducing environments that marks the Fe^{III}-Fe^{II} oxidation boundary (Lynn and Bonatti, 1965; Hartmann and others, 1976; Bender and others, 1978). Since this boundary marks the level at which the diffusion of oxidants into the sediment from sea water is no longer sufficient to

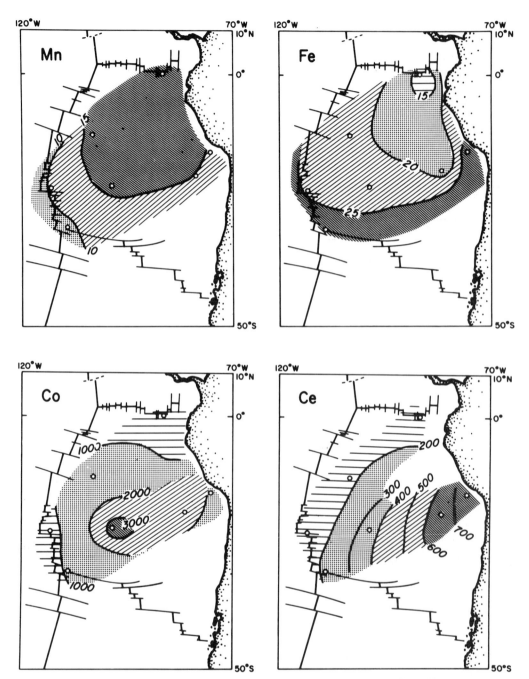

Figure 3. Regional distribution of Mn, Fe, Co, and Ce in ferromanganese coatings, this study.

prevent the reduction of Fe^{III} (Eh of $\sim$-250 mV for sea water; see Fig. 6), the depth to this boundary provides a simple semiquantitative measure of the intensity of reducing conditions within the sediment. Figure 7 maps the depth to this boundary measured in 161 cores taken by Oregon State University in the eastern Pacific and combined with the data set of Lynn and Bonatti (1965). Near the equator, the maximum depth to this boundary is between 20 and 30 cm, but farther south the

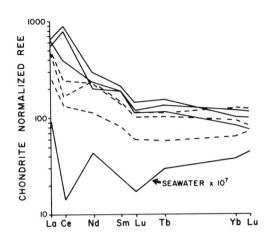

Figure 4. Rare-earth abundance patterns for ferromanganese coatings normalized to chondrites. Dashed patterns are from coatings on the East Pacific and Galapagos Rise. Solid lines are for off-crest coatings.

maximum depth is only about 15 cm. Cores slightly more oxidizing show no Fe^{II}-Fe^{III} boundary at all but stay more oxidizing than this Eh couple for their entire length. As would be expected, Figure 7 shows that the most intensely reducing environments occur along the continental shelf and slope, underneath the highly productive coastal upwelling near the coast of Peru, and underneath the equatorial upwelling regions.

In order for reduced Mn to be incorporated into a nodule, it must first pass through the oxidized surface layer. Boudreau and Scott (1978) have estimated that an oxidized surface layer 40 cm thick and probably as thin as 20 cm will block the upward diffusion of Mn to the surface sediments. Such a surface layer would block Fe even more effectively. The oxidation of Fe to its +3 valence occurs at a much lower Eh, and Fe^{II} minerals are less soluble than the Mn^{II} counterparts (Krauskopf, 1957).

Nodules in this study that were recovered from within the region where sediments become anoxic at depth (Y71-9-86, C151, C165, C169) all have distinctive chemical compositions and mineralogy. Birnessite (following the usage of Burns and Burns, 1977) is generally the dominant Mn mineral present, although one sample (C151) was composed of a well-crystallized todorokite. Mn/Fe ratios are extremely high, in all cases greater than 5 for the bulk nodule compositions (see Table 2). Other transition metal and trace-element contents are relatively low. For example, all samples contain less than 1% Ni, and the rare-earth concentrations are about 10% to 20% of hydrogenous ferromanganese coatings (for example, FDR75 3-10-45 or Y73-3-22D coating).

Sb and Ag are both enriched in the nodules formed by reducing diagenesis. The good correlation between Sb and Mn (Fig. 8) at higher Mn contents, and the Similarity between Mn^{IV} and Sb ionic radii (Whittaker and Muntus, 1970) suggest that it may replace Mn^{IV} in the Mn lattice. Ag is very actively sorbed by Mn oxides (Anderson and others, 1973). Any available silver, perhaps from the continents, may enter these nodules.

Oxic diagenesis

The classic hypothesis for diagenetic enrichment of Mn and other transition metals in Mn nodules requires that the sediment become reducing enough to remobilize Mn at depth. As the last section has illustrated, conditions that would enable reducing diagenesis to form Mn nodules only occur in a small area of the world ocean. At the fringes of high productivity regions are large zones of sediment dominantly composed of biogenic debris, but which remain oxic throughout the sediment column. Despite the lack of a reducing sedimentary environment, nodules from these regions are enriched in Mn above hydrogenous end-member nodules and have the highest quantities of the economically interesting elements Ni and Cu (Calvert and Price, 1977; Piper and Williamson, 1977).

Lyle and others (1977) have shown that the enrichment of Mn, Ni, and Cu in these nodules must be due to some type of sediment interaction. Ferromanganese nodules recovered in the Bauer Deep on the Nazca plate that grow on the sediment are consistently enriched in Mn, Ni, Cu, and Zn above ferromanganese basalt coatings recovered in the same dredge haul that grew in contact with sea water alone.

Ferromanganese nodules themselves have a similar pattern of metal distribution. Many have morphologically distinguishable tops and bottoms, the tops being exposed to sea water and the bottoms resting in the sediment. Raab (1972) discovered that the bottoms of nodules are generally enriched in Mn, Ni, and Cu with respect to the top, and depleted in Fe, Co, and Pb. Moore and others (1981) have measured growth rates and elemental fluxes to an equatorial Pacific nodule from the oxic diagenetic regime. Fluxes of Mn, Ni, and Cu are at least a factor of 4 higher to the bottom of the nodule than to the top. The data support the hypothesis that an additional component rich in Mn, Ni, and Cu can be added to nodules from the sediments, even where the sediments do not become anoxic.

Three nodules from this study had morphologically distinguishable tops and bottoms. Of these nodules, one (C141) came from a region of oxic diagenesis and two (C165, C169) came from a region of

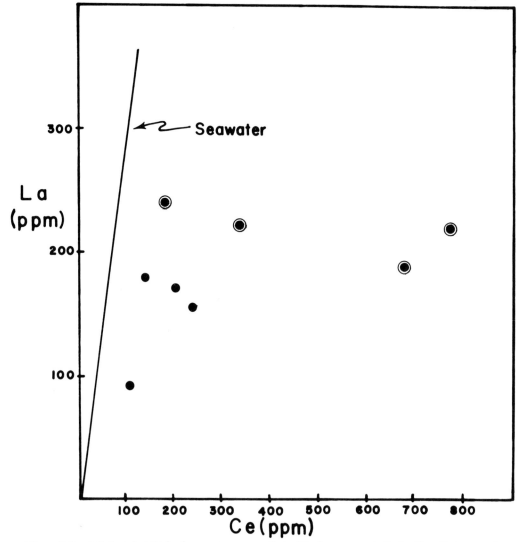

Figure 5. La plotted against Ce for ferromanganese coatings. The line represents the La/Ce ratio in sea water. Solid circles are coatings from the rise crest.

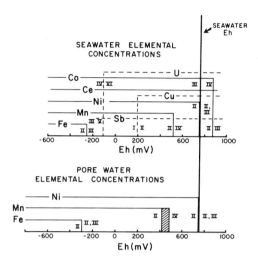

Figure 6. Redox equilibria at pH 8 for elemental concentrations as in sea water (Brewer, 1975) and in pore waters (Hartmann and Muller, 1981). Lines go to the baseline at the Eh that the elemental oxide precipitates (data from Pourbaix, 1974).

suboxic or anoxic diagenesis. C141, from the region of oxic diagenesis, has a similar pattern of enrichment as reported by Raab (1972); the bottom is richer in Mn, Ni, and Cu. C165 and C169 have different enrichment patterns though. Cu and Ni are still enriched in the nodule bottoms, but additionally Fe and Co are enriched and Mn is slightly depleted there. The enrichment of Mn in the tops of these nodules may be due to the flushing of Mn out of highly reducing sediments nearer to the continent and its subsequent precipitation on nodules farther offshore. This possiblity will be explored more fully in a later section.

The data presently available suggest that some type of diagenetic reaction(s) occurs in oxic sediments that releases Mn, Ni, and Cu to be later incorporated into nodules growing at the sediment surface. Lyle and others (1977) noted that oxic diagenetic enrichments of nodules seem only to occur in slowly accumulating sediments rich in biogenic silica that are found at the fringes of high productivity regions. Because of this observation, they suggested that a diagenetic reaction between colloidal-size ferromanganese hydroxyoxides dispersed in the sediment and biogenic silica could form an Fe-rich smectite and release Mn. The Mn so released could be later precipitated on a ferromanganee nodule. Cu and Ni contained in the biogenic silica or contained in a labile fraction and sedimented with biogenic silica could be released by dissolution and be coprecipitated on the nodule with the Mn. Thermodynamic modeling by Bischoff and others (1979) suggested that formation of smectite is favorable enough ($\log k = 2.57$) that amorphous ferric hydroxide might actually even extract silica from bottom waters. Oxygen isotope data on smectites from the DOMES Mn nodule study areas in the central Pacific show that the smectite in the sediment has been formed at sea-floor temperatures, and confirm that the reaction actually proceeds (Hein and others, 1980).

An alternative way to remobilize Mn is for reducing diagenesis to occur on a microscale in oxic sediments. Wilson (1978) has presented evidence that nitrates are being reduced to nitrogen in the uppermost sediments of an oxic pelagic sequence in the North Atlantic. He suggested that interiors of fecal pellets deposited in the oxic sediments may become anoxic, and within this microenvironment denitrification will occur. If this is true, any Mn also incorporated in the fecal pellet could be reduced and remobilized to the oxic sediments surrounding it.

Oxic diagenetic reactions that are most important for nodule growth probably occur in the uppermost layer of the sediment. Sediment trap studies (for example, Cobler and Dymond, 1980) have shown that particulate matter falling through the water column immediately above the sea bottom contains much more silica, organic carbon, and calcium carbonate than surface sediments. Presumably this "excess" material is consumed in the upper few centimetres of the sediments and could drive small-scale diagenetic reactions. Hartman and Muller (1981) reported that Mn, Cu, Ni, and Zn are all enriched in pore waters from oxic sediments from the north equatorial Pacific, up to 28 times higher than deep ocean water. They also showed that the pore-water concentrations of these metals decrease with depth. Both of these observations are consistent with remobilization of metals in the upper few centimetres of oxic sediments (oxic diagenesis). Some of these metals could be added to a Mn nodule growing at the sediment surface.

Nodules from the oxic diagenetic fringes of high productivity regions (C136, C141, Y73-3-22D, and Y73-3-20P in this study) have Mn/Fe weight ratios of 2 or greater. They are also characterized by high

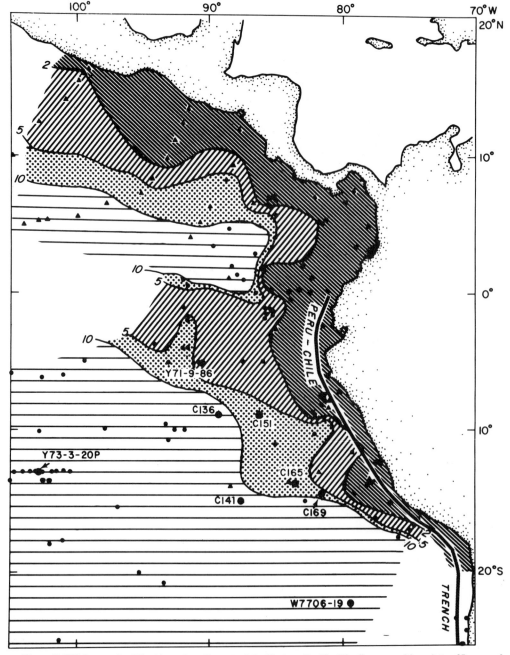

Figure 7. Thickness of oxidized layer in sediments of the Eastern Pacific. The figure combines data of Lynn and Bonatti, 1965, (triangles) with that from Oregon State University (circles).

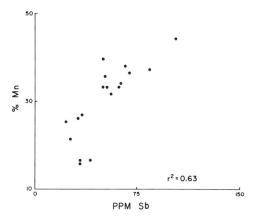

Figure 8. Scattergram of Mn against Sb. Note the good correlation in nodules with greater than 20% Mn.

Ni and Cu contents, moderate to high rare-earth and other "hydrogenous" elements (see next section), and a mineral composition dominated by todorokite.

Piper and Williamson (1977) pointed out that nodule compositions and their measured growth rates are related. Lyle (1978) enlarged upon this observation in order to estimate nodule growth rates and to determine elemental accumulation rates. I present a further refinement here, which will also be published elsewhere in more detail. The 15 nodules that have been dated by either ^{230}Th or ^{10}Be and which also have published chemical compositions (Krishnaswami and Lal, 1972; Somayajulu and others, 1971; Sharma and Somayajulu, 1979; Moore and others, 1981) can be examined for relationships among accumulation rates of different elements. One of the most prominent of these is that Mn accumulation in this set of nodules is proportional to the square of Fe accumulation. By substituting the definition of accumulation rate into the above proportionality, one can derive that the growth rate of Mn nodules should be proportional to $E_{Mn}/(E_{Fe})^2$, where E_{Mn} is concentration of Mn in the nodule. The best-fit regression line from the measured set of nodules is defined by the following equation:

$$\text{Growth rate} = 16.0[E_{Mn}/(E_{Fe})^2] + 0.448$$

Figure 9 combines this growth rate estimate with elemental composition data for a suite of southeast Pacific nodules to show that nodule accumulation rates of Mn, Ni, and Cu all increase underneath waters of high primary productivity. Due to its preferential reductive remobilization, however, Mn accumulation rates to nodules increase much more rapidly near shore, where suboxic and anoxic diagenesis occurs, than do the rates for Ni and Cu. The parallel increase of Ni and Cu accumulation rates with productivity supports the hypothesis that much of the reactive Ni and Cu is transported to the sediment by the rain of biogenic debris (Greenslate and others, 1973). Ni and Cu concentrations are highest in the regions of oxic diagenesis, however, due to Mn dilution in more productive regions.

The dilution of Cu and Ni by Mn is similar to dilution of metalliferous sediments by calcium carbonate on the East Pacific Rise. The highest rates of accumulation of Mn and Fe in sediments occur at the ridge axis of the East Pacific Rise (Dymond and Veeh, 1975), but the highest concentrations of these two elements are found in the sediments of deep basins to the east and west of the rise. A large flux of calcite to the ridge axis dilutes the ferromanganese precipitate that falls to the sediment there; the deep basins are below the carbonate compensation depth and receive little or no carbonate.

Hydrogenous Nodules

Nodules from low productivity regions apparently receive metals only through direct precipitation from the water column. The amount of unstable reactive phases added to the sediment by biogenic processes and needed to drive oxic or suboxic diagenetic reactions are negligible. As Calvert and Price (1977) have shown, ferromanganese nodules from the poorly productive central gyre regions have compositions similar to the oxide fraction of the associated sediments. From this observation they

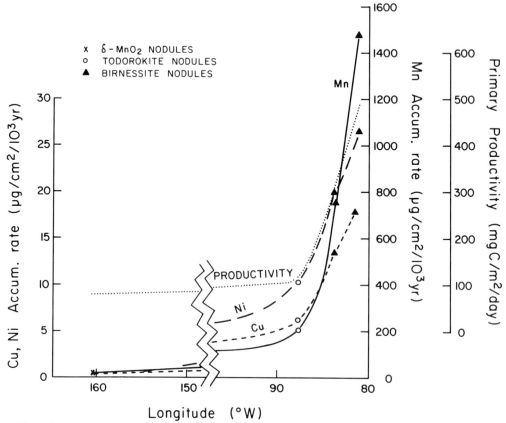

Figure 9. Mn, Ni, and Cu accumulation rates in nodules from a transect extending from the central gyre to highly productive waters near the west coast of South America, assuming a bulk density of 2.5 g/cm^3. The nodules that form the transect are from west to east, DM 981, C141, C165, and C 169. Note that the accumulation rate for all elements is highest under the most highly productive regions, but the extreme flux of Mn to nodules then produces nodules with low Cu and Ni concentrations.

have concluded that the oxide precipitate from sea water consists of the disordered δMnO_2 phase and has a relatively low Mn/Fe weight ratio and low minor-element contents.

Nodules from low productivity regions in this study (DM981, DM994, DM1006, and W7706-19) appear to be formed primarily by such hydrogenous precipitation. They contain only δ-MnO$_2$, have low Mn/Fe weight ratios, high Ce and Co contents, and relatively low Cu, Ni, and Zn abundances. The nodules from this study also exhibit high rare-earth elements, Sc, As, Th, and U abundances (Table 2). Nodules considered hydrogenous by Calvert and Price (1977) and those from low productivity regions in this study show little compositional difference with respect to the oxide fraction of the sediments, defined here to be the fraction leachable by oxalic acid.

Hydrogenous precipitation most probably adds Mn, Fe, and co-precipitated trace elements in relatively constant proportions, and will continue to add elements even to those nodules highly modified by diagenesis. Good correlations should thus exist for all elements whose prime source is from hydrogenous precipitation alone. Good correlations are found between Fe, As, and rare earths, Hf, Th, and Sc in all the nodules of this study, as illustrated in Figure 10. Fe hydroxides are used analytically to strip rare earths from sea water (Hogdahl and others, 1968) and to strip Th and U from solution during separation for isotope studies (Ku, 1966). Fe hydroxides are also known to quantitatively coprecipitate As (Onishi, 1969). These elements should coprecipitate with Fe as it precipitates from sea water and help support the hydrogenous nodule concept.

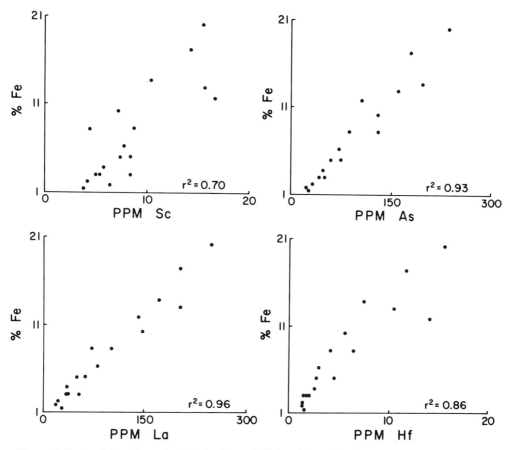

Figure 10. Scatterplots of Fe against Sc, As, La, and Hf from Mn nodules in this study to demonstrate the coherent behavior of these elements. They probably are all added to nodules by hydrogenous precipitation.

Nodules from low productivity regions can be strongly enriched in Ce with respect to other rare earths (Figure 11A), but this may be due to oxidation to Ce^{IV} in highly oxidizing marine environments (Piper, 1974). Co also can be oxidized from Co^{II} to Co^{III} at about the same Eh (Fig. 6). Both are most highly enriched in nodules from what appear to be the most oxidizing environments, and each is enriched in constant proportion to the other (Fig. 11B). Therefore, high Ce and Co contents in nodules can be considered additional evidence for hydrogenous precipitation.

Micronodules and the Dispersed Oxide Fraction of the Sediment

Micronodules and the dispersed oxide fraction of sediments are probably the least understood of the ferromanganese deposits. Glover (1977) reported the presence of either birnessite or a birnessite-todorokite mixture in micronodules from a Caribbean core, while Dymond and Eklund (1978) found todorokite in micronodules from the Bauer Deep. Glover reported chemical compositions of micronodules that are essentially pure Mn oxides, while Dymond and Eklund found that micronodules had similar compositions to nodules, except for slightly lower Ni and Cu contents. Friedrich (1976) reported micronodule compositions similar to those of nodules recovered from the same location in the central Pacific.

Absolute elemental abundances for the micronodule analyses in this study (Table 5, microfiche) may not be highly accurate, since the weight of micronodules analyzed was low and was determined

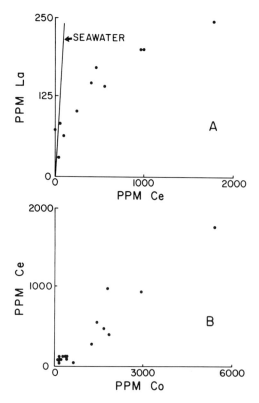

Figure 11. Scatterplots of (A) La against Ce and (B) Ce against Co. They show that Ce can be highly enriched with respect to the other rare earths, but covaries with Co.

from the difference between the original sample weight and the residue collected on a 0.45-μm filter. Although I estimate the weight of the micronodule sample determined by this technique to be accurate to within 1 to 2 mg, micronodule samples at stations C141, C136, C169, and C151 were less than 5 mg in weight. Significant errors in absolute abundance for these stations could therefore occur. In order to eliminate the weighing error, I will compare the chemical compositions of these elements on a ternary diagram of Fe, Mn, and (Ni + Cu + Co $\times$ 10), as in Bonatti and others (1972) (see Fig. 13).

Nodule and Fe-Mn coating data for the oxide fraction only (fraction leachable with oxalic acid) as well as top and bottom data from the nodule are included in the figure to enable comparison with the micronodule data. Also included in the figure is an approximate field for hydrogenous precipitates, similar to hydrogenous nodules and ferromanganese coatings. Finally, the effects of dilution with a hydrothermal source and the effects of diagenetic changes on the ferromanganese oxides are indicated by arrows. Oxic diagenesis will not only add Mn but also will add transition metals from biogenic debris. The addition of remobilized Mn during reducing diagenesis will dilute Fe and trace-metal contents.

Micronodule and nodule compositions lie in similar field on this diagram. All the micronodules are relatively enriched in Cu, Ni, and Co, and except for DM981, the station from the central Pacific, all are more enriched in Mn than indicated by the hydrogenous "source" material.

The composition of the sediment fraction leachable by oxalic acid is also given in Figure 12. This fraction is composed of both micronodules and the dispersed oxide fraction of the sediment, and its composition is similar to the hypothesized hydrogenous precipitate, except for the samples C169 and Y71-9-86. The former is enriched in Ni, Cu, and Co; the latter is depleted in Mn. Because Y71-9-86 probably experiences the most severe reducing diagenesis, I speculate that the Mn depletion represents actual loss of Mn from the sediment either to sea water or to the nodule at the surface of the core. The enrichment of C169 in trace metals may also demonstrate reducing diagenesis of the sedimentary oxides to a less severe degree.

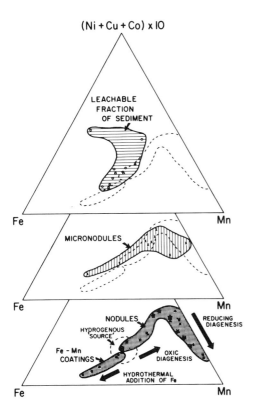

Figure 12. Composition of micronodule and leachable fraction of sediments plotted with ferromanganese nodule and Fe-Mn coating oxide fraction on a ternary diagram of Mn, Fe, and (Ni + Cu + Co) × 10. Stars indicate nodules for which micronodule compositions are reported.

Figure 13, which shows the distribution of the Mn/Fe weight ratio in the leachable fraction of the sediment, demonstrates that there are systematic variations in its composition. Except for Y71-9-86, which may have lost Mn from the sediment, the highest Mn/Fe weight ratios occur where reducing conditions are most intense (see fig. 7), which confirms that a component of Mn is added to the surface sediments by remobilization.

The variations of the chemical compositions of nodules, micronodules, and the leachable component of the sediment form a pattern that can be explained by the following sedimentary processes. Hydrogenous precipitation provides ferromanganese hydroxyoxides to sediments and nodules. In regions of oxic diagenesis, the ferromanganese hydroxyoxides within the surface sediments reconstitute to make more Mn-rich micronodules and a more Fe-rich residue of dispersed oxides or cryptocrystalline smectite. A portion of the released Mn is added to the nodule to make it relatively Mn rich. Dissolution of biogenic silica (which could also drive the oxic diagenesis) or another biogenic component will release Cu, Ni, and Zn to be incorporated in nodules and the sediment oxide fraction.

In regions of reducing diagenesis, more dramatic effects occur. An additional component of Mn cycles back to the surface sediments from below and is added both to nodules and to micronodules, although there may be preferential uptake by the nodules.

Nodules continue to grow and micronodules do not, most probably because micronodules stay in their stratigraphic position in sediment. After a period of time, all available ferromanganese hydroxyoxides at a stratigraphic level will have altered diagenetically, or all readily soluble biogenic opal will have been used and small-scale diagenesis will stop. Growth rates of nodules demand that they be continually moving to the benthic boundary layer, either by benthos pushing them upward or by some other mechanism. They are thus continually surrounded by reactive components and claim their share of newly precipitated ferromanganese oxides.

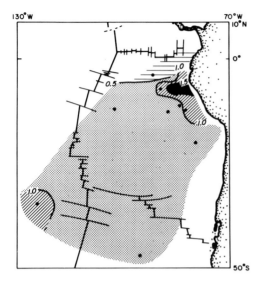

Figure 13. Distribution of Mn/Fe ratios in the leachable fraction of the sediment from the Nazca plate.

SOURCES OF Mn AND Fe ON THE NAZCA PLATE AND THEIR DISPERSAL AGENTS

Ferromanganese hydroxyoxides are the only major components of all oceanic ferromanganese deposits. There are only two ultimate sources of Mn and Fe in sea water and marine sediments—the continents and the mantle. Erosion strips sediments and dissolved constituents from the land and cycles them into the oceans. Emplacement of basaltic crust derived from the mantle and the high temperature interaction between sea water and rock will leach Mn, Fe, and other constituents and deliver them to the oceans.

The relative proportions of Mn and Fe that enter the oceans at rise crests are probably best estimated from the composition of sedimentary deposits at the East Pacific Rise (Dymond and others, 1973; Heath and Dymond, 1977). Although local conditions probably produce major variations in the composition of hydrothermal solutions (Stakes, 1978), sediments most probably collect precipitates from large numbers of hydrothermal fields in all stages of development, thereby averaging local variations. The Fe/Mn weight ratio of East Pacific Rise sediments is essentially constant at 3.5 all along the rise crest (Dymond, this volume; Heath and Dymond, 1977). I will use this as the ratio of Fe to Mn from hydrothermally derived material. I earlier estimated the hydrothermally derived Mn input to the oceans to be 9×10^{11} g of Fe/yr are released by the same process (Lyle, 1976).

Mn and Fe are removed from the continents through the erosion of highlands and transport of dissolved and particulate erosion products to the sea by rivers. Although Mn and Fe are dissolved components in river water, most are associted with the particulate fraction carried by the river (Boyle and others, 1977; Turekian, 1969). As the Mn and Fe pass into the marine environment, the dissolved fraction converts to particulate form (Boyle and others, 1977; Graham and others, 1976) and either settles out of the water column within estuaries or passes into the open ocean to be deposited on the continental shelf along with the suspended fine aluminosilicate fraction (Yeats and others, 1979).

Sediments on the Peru Margin are strongly reducing throughout the entire sediment column because of the high productivity associated with the coastal upwelling and the burial and preservation of organic carbon by the high rates of sedimentation typical of active continental margins (Muller and Suess, 1979). At the Peru Margin much of the Mn^{IV} in particulate oxides or sorbed to other grains will be remobilized and cycled back to the sea. Consequently, much of the Mn carried in both dissolved and particulate form by rivers should escape from the margin sediments and cycle to the open ocean. The loss of Mn by shelf and slope sediments is shown in Figure 14. The average Mn/Al ratio of three

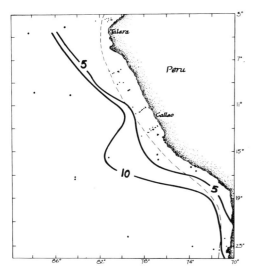

Figure 14. Distribution of Mn/Al ratio in sediments from the Peru coast. Average South America river Mn/Al is 6.4×10^{-3}, while worldwide average is 10.8×10^{-3}. Note that all the shelf and slope sediments have lost Mn. Data from K. Scheidegger (unpub.).

South American rivers is 6.4×10^{-3} (Martin and Meybeck, 1979) and 9.8×10^{-3} worldwide. The average Mn/Al ratio for the Peru shelf is 4.2×10^{-3} (K. Scheidegger, unpub. data).

Fe that is not contained in silicate phases is also reduced in shelf sediments, but I will assume that much of this Fe is fixed by formation of diagenetic minerals such as sulfides, clays, and phosphates (Krauskopf, 1957). The continentally derived Mn and Fe, remobilized from the margin sediments, is carried away by surface currents and mixed with normal ocean waters.

Continental Mn and Fe Inputs to the Nazca Plate

The continental Mn and Fe contribution from South America to the Nazca plate can be estimated from runoff data for the Pacific coast of South America. Since little runoff data for rivers on the western side of the continent are available, however, I have used maps of average rainfall for Chile and Colombia (Prohaska, 1976; Snow, 1976) and contoured the sparse data for the Pacific coast of Ecuador presented in Ferdon (1950) to estimate average rainfall to these countries. Holland (1978) has compiled the water balance for the continents. From the works he has gathered, it appears that the average runoff/rainfall ratio for South America is between 0.35 and 0.41. I will use 0.40 for runoff/rainfall to calculate runoff to the Nazca plate, and estimate that 1.2×10^{14} L/yr flows from Colombia, while 8×10^{13} L/yr and 4×10^{14} L/yr flow from Ecuador and Chile, respectively. Average total runoff from Peru is 4×10^{14} L/yr, primarily from the north coast (Zuta and Guillen, 1970). The total for the entire coast is 1.0×10^{15} L/yr.

The total mass of Mn and Fe that reaches the shelf and continental slope will be equivalent to the dissolved plus particulate load of the rivers. The amount of continental material added to the continental shelf can be estimated by assuming that the particulate load of the river is proportional to runoff. On the Pacific coast of South America, this does not seem to be a bad assumption; Scholl and others (1970) have demonstrated that the volume of Cenozoic sediments on the Chile Margin is proportional to the rainfall on land. Scholl and others' (1970) maximum denudation rate suggested that Chile supplies approximately 10^{14} g of sediment/yr to the shelf, and implies an average suspended load of 250 mg/L. Turekian (1969) has estimated average suspended load for the world's rivers to be 330 mg/L, which I will use for a maximum suspended load. The rivers of South America thus supply a suspended sediment load of 2.5 to 3.3×10^{14} g/yr to the shelf and slope of the continental margin. Multiplying the annual fluvial sediment supply to the shelf by the average Al content of South American river particulates (11.1%, Martin and Meybeck, 1979) and by the average Mn to Al ratio in South American river sediments (6.4×10^{-3}, Martin and Meybeck, 1979) yields the annual Mn input to the shelf, 1.8 to 2.3×10^{11} g/yr. Multiplying the annual fluvial sediment supply by the average Al of river sediments and the average Mn/Al ratio of shelf-slope sediments (4.2×10^{-3}, Scheidegger, unpub. data) yields the annual flux of Mn stored on the shelf, 1.2 to 1.6×10^{11} g/yr. The difference between the two represents the Mn lost by reductive remobilization.

Accordingly, 6 to 7×10^{10} g of Mn would be remobilized from particulates each year on the continental shelf of South America based on these assumptions. If the same fraction of Mn is remobilized from the suspended load discharged by the world's rivers, continental Mn flux should be 2.1 to 3.6×10^{12} g/yr. (This estimate uses Turekian's [1969] estimate of 3.6×10^{16} L/yr and Kozoun and others' [1974] estimate of 4.7×10^{16} L/yr as the range for worldwide continental runoff.) Elderfield (1976) calculated that precipitation of Mn in deep-sea sediments requires a yearly flux of 1 to 7×10^{12} g of Mn in excess of the dissolved stream supply. Elderfield's value is for all other sources including hydrothermal sources; nonetheless, the similarity of the numbers provides support for my estimate. Total Mn removed from the continent will also include the dissolved Mn, which will have a worldwide flux of 1×10^{10} g/yr (based on 7 ppb Mn in average river water; Turekian, 1969), and 2.5×10^{9} g/yr from South America alone. The total Mn flux from South America should therefore be about 6 to 7×10^{10} g/yr. Table 6 lists the fluxes of Mn and Fe from terrigenous and hydrothermal sources around the Nazca plate.

DISTRIBUTION OF TERRIGENOUS AND HYDROTHERMAL Fe AND Mn

Fe and Mn entering at mid-ocean ridges will be dispersed in a fashion different from that derived from the South American Ccontinent. Hydrothermal Mn and Fe is injected in dissolved form into the bottom waters, so bottom flow will govern the hydrothermal dispersal. Terrigenous Mn and Fe will be added to near-surface waters passing over the continental shelf, so that surface currents or mid-water flow will govern the dispersal in this case. Reducing abyssal sediments near the continents may also cause remobilization of terrigenous Mn and addition into bottom waters, however.

Figure 15 illustrates the generalized surface circulation for the Nazca plate and shows where maximum river runoff occurs. As shown in this figure, most continentally derived elements will travel along the coast and be carried offshore between 15°S and the equator.

As terrigenous Mn and Fe are being carried by the near-surface currents, they will be continually

TABLE 6. MANGANESE AND IRON FLUXES TO THE NAZCA PLATE (x 10^{11} G/YR)

	Fe	MN
Nazca plate		
Terrigenous		
Dissolved	?	0.025
Remobilized from particulates	?	0.6 - 0.7
Total terrigenous	?	0.6 - 0.7
Hydrothermal	11	3
Total	>11	3.6 - 3.7
World ocean		
Terrigenous	?	21 - 36
Hydrothermal	27	9

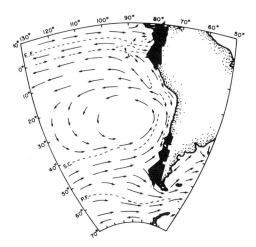

Figure 15. Generalized surface circulation of the Eastern Pacific Ocean after Molina-Cruz (1978).

stripped from the waters by adsorption onto biogenic tests or by incorporation into the planktonic organisms. The highest productivity occurs along the continental margin (Zuta and Guillen, 1970); Mn and Fe may thus cycle many times between waters and sediments before escaping the margin. The surface waters of the eastern equatorial Pacific are also highly productive (Love and Allen, 1975), so the cycling of continental Mn and Fe to the sediments near the equator should be very effective.

Circulation of abyssal waters that controls the distribution of hydrothermal Mn and Fe is less well understood then surface circulation. Lonsdale (1976) has developed general circulation patterns for the southeast Pacific, however. Flow over the East Pacific Rise most probably passes through fracture zones in the north and is forced south by topographic confinement of the fossil Galapagos Rise (Figure 16). Hydrothermal components for this highly active section of the mid-ocean ridge are thus largely confined in the Bauer Deep. A second flow of abyssal waters enters through fracture zones on the southern boundary of the Nazca plate. There is much less crustal production on this segment of the

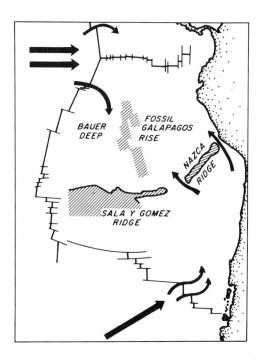

Figure 16. Abyssal circulation in the Eastern Pacific Ocean, after Lonsdale (1976). Lined pattern represents elevated topographic features.

mid-ocean ridge, however, so hydrothermal influence is probably not as great as in the northwest corner of the plate.

Most Mn and Fe from either hydrothermal or terrigenous sources get stripped from sea water near where they enter the oceans. Biogenic debris strips much of the terrigenous Fe and Mn and deposits it in sediments along the margin and in the northeast corner of the plate. Hydrothermal Fe and Mn precipitate rapidly to the sediment, as indicated by accumulation rates (Dymond and Veeh, 1975) and modeling studies (Weiss, 1977). The remains mix together to form the hydrogenous "source."

CONCLUSIONS

The geochemistry of ferromanganese oxide deposits is governed by source and by diagenetic process (Figure 1). Near mid-ocean ridges they can be demonstrably influenced by a hydrothermal source; deposits near the continents are so diagenetically altered that source cannot truly be distinguished. Most of the ferromanganese deposits appear to have received ferromanganese hydroxyoxides from a well-mixed reservoir of terrigenous and hydrothermal source material that is commonly known as the hydrogenous or authigenic "source."

The hydrothermal source adds about 4 to 5 times as much Mn as the terrigenous source to the ocean around the rim of the Nazca plate. The Nazca plate is atypical of the world ocean, however; it has a low amount of terrigenous input while a large fraction of new oceanic crust is produced along its borders. Mn input is dominated worldwide by the terrigenous source. Total Mn input to the oceans is about 3.0 to 4.5×10^{12} g/yr. This estimate is two orders of magnitude higher than that based on the dissolved load of rivers alone. The residence time of Mn in the oceans is therefore much more similar to that of Fe (40 to 160 yr) than previously estimated (7,000 yr; Goldberg and Arrhenius, 1958).

Diagenesis can profoundly modify the chemical composition and mineralogy of the ferromanganese oxides. Both oxic and anoxic diagenesis are most probably strongly influenced by surface productivity. Dissolution of biogenic silica and formation of authigenic smectites is a possible fractionation process that fixes Fe from mixed Fe-Mn oxyhydroxides; this process I have referred to as oxic diagenesis. The released Mn can precipitate as oxides and coprecipitate Cu, Ni, and Zn also released by the dissolution of biogenic silica or other biogenic remains.

High productivity regions receive a large organic carbon flux along with high fluxes of other biogenic remains. Oxidation of the organic carbon lowers redox conditions sufficiently for both Fe and Mn to be reduced. Lower solubilities of Fe minerals bind Fe in the solid phase. Mn, however, will diffuse back to the surface sediments and nodules that grow there. The added component of Mn that diffuses from below causes these nodules to grow at a fast rate and to form nodules of relatively pure Mn oxides lower in Cu, Ni, and Zu than oxic diagenetic nodules.

Because productivity of surface waters controls the type and intensity of diagenesis, regional distributions of ferromanganese oxide compositions and mineralogy will mirror surface productivity. Hydrogenous-type nodules will be found in the central gyres of the oceans, while nodules enriched in Ni, Cu, and Zn will occur at the edges of high productivity regions, as is reported by Piper and Williamson (1977). Highest Mn concentrations will be found underneath the most productive regions. However, ferromanganese oxides grown only in contact with sea water will have hydrogenous compositions no matter where they are found; they are not influenced by diagenetic reactions in the sediments.

ACKNOWLEDGMENTS

I thank Jack Dymond and Erwin Suess for the many fruitful discussions we have had about this paper. I also thank Ken Scheidegger for providing me with unpublished sediment analyses from the Peru Margin. The paper has also benefited from comments by Jack Corliss, E. Julius Dasch, G. Ross Heath, and Caroll DeKock.

The study has been funded by the NSF/IDOE Nazca Plate Project, as well as an International Nickel Company fellowship and by the NSF/IDOE Manganese Nodule Project.

REFERENCES CITED

Anderson, B. J., Jenne, E. A., and Chao, T. T., 1973, The sorption of silver by poorly crystalline manganese oxides: Geochimica et Cosmochimica Acta, v. 37, p. 611–622.

Bender, M., and others, 1978, The significance of color banding in eastern equatorial Pacific sediments from MANOP study sites: EOS (American Geophysical Union Transactions), v. 59, p. 1117.

Bischoff, J. L., Piper, D. Z., and Quinterno, P., 1979, Nature and origin of metalliferous sediment in DOMES site C, Pacific manganese nodule province, in Lalou, C., ed., International colloquium on the genesis of manganese nodules: Paris, Colloques Internationaux du C.N.R.S.

Bonatti, E., Kramer, T., and Rydell, H., 1972, Classification and genesis of submarine iron-manganese deposits, in Horn, D. R., ed., Papers from a Conference on Ferromanganese Deposits on the Ocean Floor: National Science Foundation/International Decade of Ocean Exploration Publication, p. 149–166.

Bostrom, K., and Peterson, M.N.A., 1969, The origin of aluminum-poor ferromanganoan sediments in areas of high heat flow on the East Pacific Rise: Marine Geology, v. 7, p. 427–447.

Boudreau, B. P., and Scott, M. R., 1978, A model for the diffusion-controlled growth of deep-sea manganese nodules: American Journal of Science, v. 278, p. 903–929.

Boyle, E. A., Edmond, J. M., and Sholkovitz, E. R., 1977, The mechanism of iron removal in estuaries: Geochimica et Cosmochimica Acta, v. 41, p. 1313–1324.

Brewer, P. G., 1975, Minor elements in sea water, in Riley, J. P., and Skirrow, G., eds., Chemical oceanography, Volume 1 (second edition): New York, Academic Press, p. 415–497.

Burns, R. G., and Burns, V. M., 1977, Mineralogy, in Glasby, G. P., ed., Marine manganese deposits: Amsterdam, Elsevier, p. 185–249.

Calvert, S. E., and Price, N. B., 1977, Geochemical variation in ferromanganese nodules and associated sediments from the Pacific Ocean: Marine Chemistry, v. 5, p. 43–74.

Cann, J. R., Winter, C. K., and Pritchard, R. G., 1977, A hydrothermal deposit from the floor of the Gulf of Aden: Mineralogical Magazine, v. 41, p. 193–199.

Cobler, R., and Dymond, J., 1980, Sediment trap experiment on the Galapagos Spreading Center, equatorial Pacific: Science, v. 209, p. 801–803.

Corliss, J. B., and others, 1978, The chemistry of hydrothermal mounds near the Galapagos Rift: Earth and Planetary Science Letters, v. 40, p. 801–803.

Corliss, J. B., and others, 1979, Submarine thermal springs on the Galapagos Rift: Science, v. 203, p. 1073–1083.

Cronan, D. S., 1975, Manganese nodules and other ferromanganese oxide deposits from the Atlantic Ocean: Journal of Geophysical Research, v. 80, p. 3831–3837.

Dymond, J., 1981, The geochemistry of Nazca plate surface sediments: An evaluation of hydrothermal, biogenic, detrital, and hydrogenous sources, in Kulm, L. D., and others, eds., Nazca plate: Crustal formation and Andean convergence: Geological Society of America Memoir 154 (this volume).

Dymond, J., and Eklund, W., 1978, A microprobe study of metalliferous sediment components: Earth and Planetary Science Letters, v. 40, p. 243–251.

Dymond, J., and Veeh, H. H., 1975, Metal accumulation rates in the southeast Pacific and the origin of metalliferous sediments: Earth and Planetary Science Letters, v. 28, p. 13–22.

Dymond, J., and others, 1973, Origin of metalliferous sediments from the Pacific Ocean: Geological Society of America Bulletin, v. 84, p. 3355–3372.

Elderfield, H., 1976, Manganese fluxes to the oceans: Marine Chemistry, v. 4, p. 103–132.

Ferdon, E. N., Jr., 1950, Studies in Ecuadorian geography: School of American Research Monograph 15, 86 p.

Friedrich, G. H., 1976, Manganese micronodules and their relation to manganese nodule fields, in Glasby, G. P., and Katz, H. R., eds., United Nations Economic and Social Commission for Asia and the Pacific CCOP/SOPAC Technical Bulletin 2, p. 39–53.

Froelich, P. H., and others, 1979, Early oxidation of organic matter in pelagic sediments of the eastern Equatorial Atlantic: Suboxic diagenesis: Geochimica et Cosmochimica Acta, v. 43, p. 1075–1090.

Glover, E. D., 1977, Characterization of a marine birnessite: American Mineralogist, v. 62, p. 278–285.

Goldberg, E. D., 1961, Chemical and mineralogical aspects of deep-sea sediments, in Physics and chemistry of the Earth, Volume 4: Oxford, Pergamon Press, p. 281–301.

Goldberg, E. D., and Arrhenius, G., 1958, Chemistry of Pacific pelagic sediments: Geochimica et Cosmochimica Acta, v. 13, p. 153–212.

Gordon, G. E., and others, 1968, Instrumental neutron activation analysis of standard rocks with high resolution γ-ray detectors: Geochimica et Cosmochimica Acta, v. 32, p. 665–673.

Graham, W. F., Bender, M. L., and Klinkhammer, G.

P., 1976, Manganese in Narragansett Bay: Limnology and Oceanography, v. 21, p. 665–673.

Greenslate, J. L., Frazer, J. Z., and Arrhenius, G., 1973, Origin and deposition of selected transition elements in the seabed, in Morgenstein, M., ed., Papers on the origin and distribution of manganese nodules in the Pacific and prospects for exploration: Hawaii Institute of Geophysics, p. 45–70.

Hartmann, M., and Muller, P. J., 1981, Trace metals in interstitial waters from Central Pacific Ocean sediments, in Manheim, F., and Fanning, K., eds., The dynamic environment of the ocean floor: Lexington, Massachusetts, D. C. Heath and Co. (in press).

Hartmann, M., and others, 1976, Chemistry of late Quaternary sediments and their interstitial waters from the NW African continental margin: "Meteor" Forschungsergebnisse, Reihe C, no. 24, p. 1–67.

Haymon, R., and others, 1979, Mineralogy and chemistry of hydrothermal sulfide deposits and sediments at EPR 21°N: EOS (American Geophysical Union Transactions), v. 60, p. 84.

Heath, G. R., and Dymond, J., 1977, Genesis and transformation of metalliferous sediments from the East Pacific Rise, Bauer Deep, and Central Basin, northwest Nazca plate: Geological Society of America Bulletin, v. 88, p. 723–733.

Hein, J. R., Yeh, H.-W., and Alexander, E. R., 1980, Origin of iron-rich montmorillonite from the manganese nodule belt of the north equatorial Pacific: Clays and Clay Minerals, v. 27, p. 185–193.

Hogdahl, O. T., Melsom, S., and Bowen, V. T., 1968, Neutron activation analysis of lanthanide elements in sea water: American Chemical Society, Advances in Chemistry no. 73, p. 308–325.

Holland, H. D., 1978, The chemistry of the atmosphere and oceans: New York, John Wiley & Sons, 351 p.

Kozoun, V. I., and others, 1974, World water balance and water resources of the Earth: USSR National Committee for the International Hydrological Decade.

Krauskopf, K. B., 1957, Separation of manganese from iron in sedimentary processes: Geochimica et Cosmochimica Acta, v. 12, p. 61–84.

Krishnaswami, S., 1976, Authigenic transition elements in Pacific pelagic clays: Geochimica et Cosmochimica Acta, v. 40, p. 425–434.

Krishnaswami, S., and Lal, D., 1972, Manganese nodules and the budget of trace solubles in oceans, in Dyrssen, D., and Jagner, D., eds., The changing chemistry of the oceans, Nobel Symposium 20: New York, Wiley Interscience, p. 307–321.

Ku, T. L., 1966, Uranium series disequilibrium in deep sea sediments [Ph.D. thesis]: New York City, Columbia University, 157 p.

Lalou, C., and others, 1977, Radiochemical scanning electron microscope (SEM) and X-ray dispersive energy (EDAX) studies of a FAMOUS hydrothermal deposit: Marine Geology, v. 24, p. 245–258.

Lonsdale, P., 1976, Abyssal circulation of the southeastern Pacific and some geological implications: Journal of Geophysical Research, v. 81, p. 1163–1176.

Lopez, C., 1978, Elemental distributions in the components of metalliferous sediments from the Bauer and Roggeveen Basins—Nazca plate [M.S. thesis]: Corvallis, Oregon State University, 153 p.

Love, C. M., and Allen, R. M., 1975, EASTROPAC Atlas, Biological and nutrient chemistry data from principal participating ships third survey cruise, February–March 1968: Washington, D.C., U.S. Department of Commerce.

Lyle, M., 1976, Estimation of hydrothermal manganese input to the oceans: Geology, v. 4, p. 733–736.

——1978, The formation and growth of ferromanganese oxides on the Nazca plate [Ph.D. thesis]: Corvallis, Oregon State University, 172 p.

Lyle, M., Dymond, J., and Heath, G. R., 1977, Copper-nickel-enriched ferromanganese nodules and associated crusts from the Bauer Basin, northwest Nazca plate: Earth and Planetary Science Letters, v. 35, p. 55–64.

Lynn, D. C., and Bonatti, E., 1965, Mobility of manganese in the diagenesis of deep-sea sediments: Marine Geology, v. 3, p. 457–474.

Martin, J. M., and Meybeck, M., 1979, Elemental mass balance of material carried by major world rivers: Marine Chemistry, v. 7, p. 173–206.

Mero, J. L., 1965, The mineral resources of the sea: Amsterdam, Elsevier, 312 p.

Minster, J. B., and others, 1974, Numerical modeling of instantaneous plate tectonics: Geophysical Journal of the Royal Astronomical Society, v. 36, p. 541–576.

Molina-Cruz, A., 1978, Late Quaternary oceanic circulation along the Pacific coast of South America [Ph.D. thesis]: Corvallis, Oregon State University, 246 p.

Moore, W. S., and Vogt, P. R., 1976, Hydrothermal manganese crusts from two sites near the Galapagos spreading axis: Earth and Planetary Science Letters, v. 29, p. 349–356.

Moore, W. S., and others, 1981, Fluxes of metals to a manganese nodule: Radiochemical, chemical, structural, and mineralogical studies: Earth and Planetary Science Letters, v. 52, p. 151–171.

Muller, P. J., and Suess, E., 1979, Productivity, sedimentation rate, and sedimentary organic matter in the oceans. I. Organic carbon preservation: Deep Sea Research, v. 26A, p. 1347–1362.

Murray, J., and Irvine, R., 1894, On the manganese oxides and manganese nodules in marine deposits: Royal Society of Edinburgh, Transactions, v. 37, p. 721-742.

Onishi, H., 1969, Arsenic, in Wedepohl, K. H., ed., Handbook of geochemistry, Volume II: Berlin, Springer-Verlag.

Piper, D. Z., 1974, Rare earth elements in ferromanganese nodules and other marine phases: Geochimica et Cosmochimica Acta, v. 38, p. 1007-1022.

Piper, D. Z., and Williamson, M. E., 1977, Composition of Pacific Ocean ferromanganese nodules: Marine Geology, v. 23, p. 285-303.

Pourbaix, M., 1974, Atlas of electrochemical equilibria in aqueous solutions: Houston, National Association of Corrosion Engineers, 644 p.

Prohaska, F., 1976, The climate of Argentina, Paraguay, and Uruguay, in Schwerdtfeger, W., ed., World survey of climatology, Volume 12—climates of Central and South America: Amsterdam, Elsevier, p. 13-113.

Raab, W., 1972, Physical and chemical features of Pacific deep sea manganese nodules and their implications to the genesis of nodules, in Horn, D. R., ed., Papers from a conference on ferromanganese deposits on the ocean floor: Washington, D.C., National Science Foundation/ International Decade of Ocean Exploration Publication, p. 31-50.

Scholl, D. W., and others, 1970, Peru-Chile Trench sediments and sea-floor spreading: Geological Society of America Bulletin, v. 81, p. 1339-1360.

Scott, R. B., and others, 1974, The TAG hydrothermal field: Nature, v. 251, p. 301-302.

Sharma, P., and Somayajulu, B.L.K., 1979, Growth rates and composition of two ferromanganese nodules from the central north Pacific, in Gènese des nodules de manganese [Colloque international sur la gènese des nodules manganese]: Paris, France, Centre National de la Recherche Scientifique, no. 289, p. 281-288.

Snow, J. W., 1976, The climate of northern South America, in Schwerdtfeger, W., ed., World survey of climatology, Volume 12—climates of Central and South America: Amsterdam, Elsevier, p. 295-405.

Somayajulu, B.L.K., and others, 1971, Rates of accumulation of manganese nodules and associated sediment from the equatorial Pacific: Geochimica et Cosmochimica Acta, v. 35, p. 621-624.

Stakes, D. S., 1978, Submarine hydrothermal systems: Variations in mineralogy, chemistry, temperature, and the alteration of oceanic layer II [Ph.D. thesis]: Corvallis, Oregon State University, 188 p.

Stumm, W., and Morgan, J. J., 1970, Aquatic chemistry: New York, Wiley-Interscience, 583 p.

Toth, J. R., 1980, Deposition of submarine hydrothermal manganese and iron, and evidence for hydrothermal input of volatile elements to the ocean: Geological Society of America Bulletin, v. 91, p. 44-54.

Turekian, K. K., 1969, The oceans, streams, and atmosphere, in Wedepohl, K. H., ed., Handbook of geochemistry, Volume 1: Berlin, Springer-Verlag, p. 227-249.

Volkov, I. I., 1977, The mode of formation of Fe-Mn nodules in recent sediments: Geochemistry International, v. 6, p. 150-156.

Weiss, R. F., 1977, Hydrothermal manganese in the deep sea: Scavenging residence time, and $Mn/^3He$ relationships: Earth and Planetary Science Letters, v. 37, p. 257-262.

Whittaker, E.J.W., and Muntus, R., 1970, Ionic radii for use in geochemistry: Geochimica et Cosmochimica Acta, v. 34, p. 945-956.

Wilson, T.R.S., 1978, Evidence for denitrification in aerobic pelagic sediments: Nature, v. 274, p. 354-356.

Yeats, P.A., Sundby, B., and Bewers, J. M., 1979, Manganese recycling in coastal waters: Marine Chemistry, v. 8, p. 43-55.

Zuta, S., and Guillen, O., 1970, Oceanografia de las aguas costeras del Peru: Instituto del Mar del Peru Boletin, v. 2, p. 161-323.

MANUSCRIPT RECEIVED BY THE SOCIETY NOVEMBER 12, 1980
MANUSCRIPT ACCEPTED DECEMBER 30, 1980

Geological Society of America
Memoir 154
1981

Sediment and associated structure of the northern Nazca plate

D. L. ERLANDSON
D. M. HUSSONG
J. F. CAMPBELL
Hawaii Institute of Geophysics
University of Hawaii
Honolulu, Hawaii 96822

ABSTRACT

An isopach map of total sediment thickness of the Nazca plate north of 30°S latitude shows deviations from the the obvious sediment trends that are governed by crustal age, sea-floor depth, and surface biologic productivity. Sediment on the eastern flank of the East Pacific Rise has accumulated at half the rate of that on the western rise flank; this difference is attributed in part to asymmetric spreading. Broad areas of thin sediment occur over the crest of the fossil Galapagos Rise and in a 300-km-wide zone on the eastern plate margin parallel to the Peru-Chile Trench. Whereas the thin sediment over the fossil rise crest is the natural result of relatively young crust, the thin sediment paralleling the trench is characterized by postdepositional volcanic rocks and is related to, but cannot be totally explained by, the rupturing of the upper oceanic plate prior to subduction.

Seismic reflection records reveal that volcanism and basement structure are responsible for the initiation of submarine valley erosion near the Carnegie Ridge and the confining of sediments on the axis of the Nazca Ridge. Reflection records also indicate that two periods of tectonic activity have occurred in the Bauer Basin: the first when the basin crust was generated at the Galapagos Rise and left seamounts that are now covered with sediment; the second when spreading activity migrated westward to the East Pacific Rise and left younger, sediment-free seamounts.

INTRODUCTION

The physiographically and structurally complex northern Nazca plate is composed of crust generated by three spreading centers, the active East Pacific Rise (EPR), the active Cocos-Nazca Rise, and the fossil Galapagos Rise (Fig. 1). Since Menard and others (1964) first noted the fossil rise, many investigators have considered a variety of geophysical and geological relationships to help explain the structure and evolution of the plate. The distribution of sediments and their relationship to basement topography, however, have received little attention. This paper describes the thickness and distribution of sediments on the northern part of the Nazca plate and some tectonically significant variations in the sediment thickness trends.

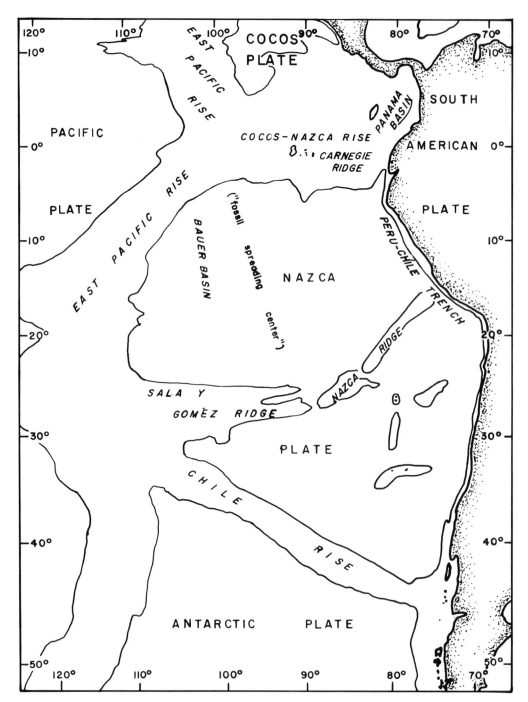

Figure 1. Physiographic features of the Nazca plate.

The amount and composition of the mid-ocean sea-floor sediment are controlled by the carbonate compensation depth (CCD), the surface productivity, the crustal age, and the bottom conditions that cause dissolution and erosion. These effects vary geographically and temporally and are complexly interrelated. We will discuss the effect of these parameters on the broad sedimentation patterns.

Certain well-known influences on sedimentation trends were anticipated prior to the preparation of the sediment isopach map (Fig. 2). The lysocline and the CCD in the equatorial eastern Pacific were determined by Berger and Winterer (1974) to be 3,700 m and <5,000 m, respectively, becoming significantly shallower southward on the Nazca plate. Biologic productivity at the sea surface is also greatest at the equator in the equatorial zoogeographic zone (Fisher, 1958) and decreases southward, and is also great near the coast of South America where there is upwelling of nutrient-rich bottom waters. Sediments accumulate on new crust along spreading ridges. Initially this rate is high because the near-axial zones of spreading ridges are shallower than the CCD. However, as the age of the crust increases, the sea-floor depth also increases (Sclater and others, 1971). As the sea floor approaches the CCD, the sediment accumulation rate decreases markedly (often at crustal ages between 15 and 32 m.y.B.P.).

On the Nazca plate, the areas of thickest sediments are the areas of highest productivity (northern boundary), oldest crust (Bauer Basin and the eastern plate boundary), and high elevations (Nazca Ridge). The areas of thin sediments are over young crust (EPR and the fossil Galapagos Rise) and low surface productivity. This paper will describe some relationships between the physiography, topography, and sediment cover for each of these areas, and point out some geologically significant variations from these relationships.

DATA REDUCTION

The total-sediment-thickness isopach chart (Fig. 2) was prepared from seismic reflection profiles acquired in 1972, 1973, and 1974 on board the R/V *Kana Keoki* of the Hawaii Institute of Geophysics (HIG) and the R/V *Yaquina* of Oregon State University (OSU). The track lines are shown in Figure 3. Although various combinations of sound sources and receiver filter settings were used by different investigators on the cruises, in most cases air-gun sources were used, and reflections within the frequency range of 30 to 300 Hz provided good definition of acoustic basement. Navigation was satellite controlled.

The seismic reflection profiles were widely spaced, so it was not meaningful to try to compile sediment-thickness data in detail except in areas of special interest with high track density (such as the Bauer Basin). Representative sediment thickness was determined irregularly, from intervals as small as 2 km in areas of large variation and greater track density, to intervals as great as 10 km in regions of low variability. The data were picked and digitized directly from reflection records and merged with ship's navigation for plotting purposes. A computer program then averaged sediment thickness over a 28-km grid and generated a contour chart, which was subsequently adjusted by hand to produce the isopach map. Averaging the data tends to filter out the small-scale sediment thickness variations associated with features such as small seamounts and the relatively narrow sediment-free axis of the EPR north of 4°S latitude.

EAST PACIFIC RISE

The EPR exhibits features typical of a fast spreading mid-ocean ridge (Rea, this volume). Generally, the EPR rises to axial depths of less than 3 km and sometimes has an axial crestal block about 15 km wide that rises an additional 300 to 350 m above the regional bathymetric trends. The flanks of the EPR are covered with small abyssal hills (typically <200 m in relief) that are elongated parallel to the spreading axis.

The axial sediment-free zone on the EPR (Fig. 2) widens southward of 5°S, primarily in response to

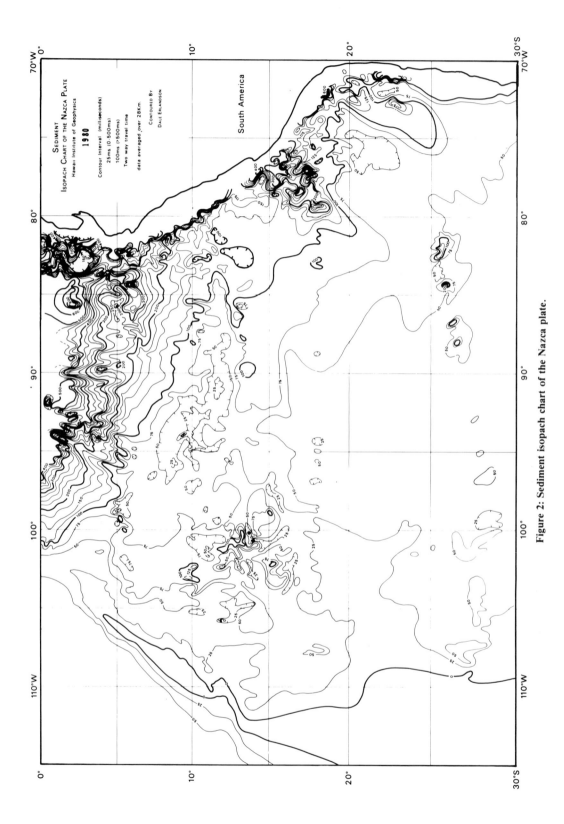

Figure 2: Sediment isopach chart of the Nazca plate.

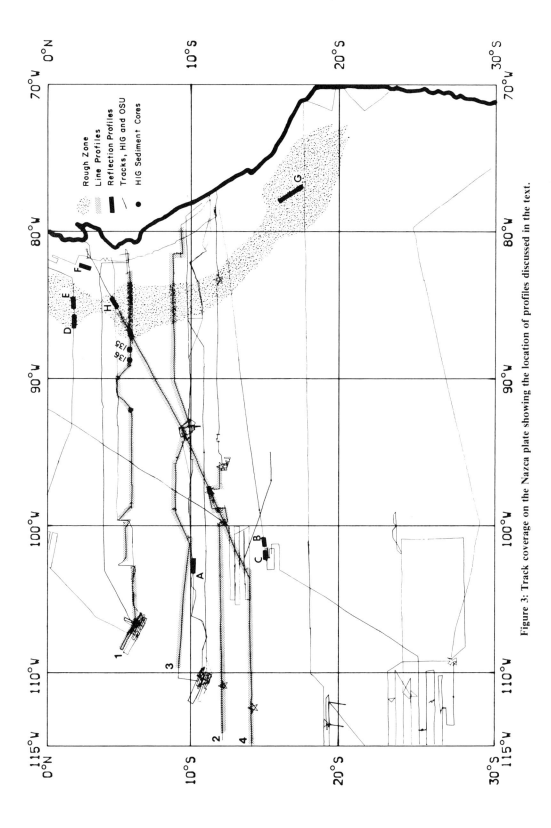

Figure 3: Track coverage on the Nazca plate showing the location of profiles discussed in the text.

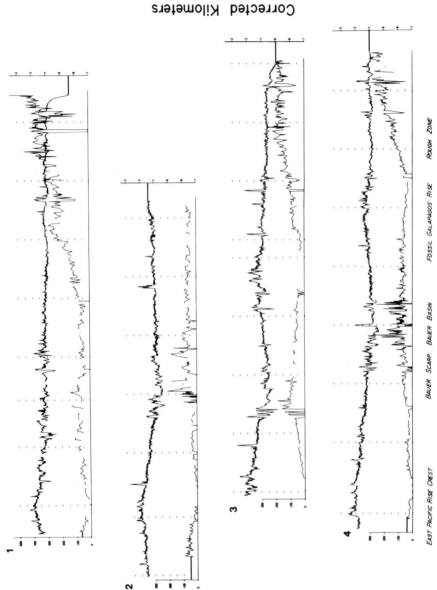

Figure 4. Line profiles across the Nazca plate showing bottom topography (upper trace) and sediment thickness (lower trace). Note the change in the character of the bottom and sediment thickness over the various regions. Profile 1 is located at the northern limit of the Bauer Basin, so the phenomena associated with the scarp and basin are subdued. The profile locations are indicated on Figure 3. Bathymetry corrected according to the variations of sound in sea water (Matthews, 1939).

declining surface productivity. The increasing crustal spreading rate and shoaling of the CCD cause a decrease in the amount of sediment on the rise crest, but these factors appear less important than surface productivity. On the scale of our isopach chart, the EPR north of 4°S appears totally covered by sediment, although on a finer scale some localized topographic highs and the narrow axial zone of near-zero-age crust are free of sediment, even near the equator.

The isopach chart and sediment-thickness profiles (Fig. 4) suggest an asymmetric rate of crustal accretion on the EPR north of 20°S as sediments thicken more rapidly on the western side than on the eastern side of the crest. At equal distances from the rise crest, the sediment on the western flank of the EPR is twice as thick as the sediment on the eastern flank. This difference is likely to result from a combination of biologic, oceanographic, and geophysical effects.

Lonsdale (1976) has described bottom currents on the EPR as running eastward, restricted largely to the topographic lows of the fracture zones. Since the majority of our reflection data were gathered on east-west traverses, the effects of such bottom currents were not readily discernible. The accumulation of sediment on the flanks of the rise appears to have been continuous. No unconformities or other evidence of scour were seen on the reflection records.

South of 20°S on the EPR, marked asymmetry has been noted in previous descriptions of magnetics and bathymetry (Herron, 1972; Molnar and others, 1975; Rea, 1978 and this volume). This asymmetry is characterized by faster spreading on the eastern (Nazca plate) side. Rea (this volume) further suggests that short and erratic episodes of asymmetric spreading may be deduced from small axial offsets in magnetics data near 6°S, 10°S, and 11.5°S. North of 20°S, however, the character of the magnetic anomalies are so degraded because of the proximity to the magnetic equator that they provide little evidence relative to spreading symmetry.

Anderson and Sclater (1972) have noted asymmetry in bathymetric profiles across the rise. The Nazca plate is considerably deeper than the Pacific plate at equal distances from the crest. This depth asymmetry is thoroughly documented by Mammerickx and others (1975) and Mammerickx and Smith (1978), who suggest that the Nazca plate is accreting slower than the Pacific plate, which is in the opposite sense from the asymmetry deduced from the previously mentioned magnetics data. In addition, the distances from the rise crest to the scarps associated with the jump in the spreading center are inconsistent with the observed depth asymmetry. The distance to the eastern (Bauer) scarp is consistently greater than the distance to the western scarp (Anderson and Sclater, 1972; Ade-Hall, 1976). Both scarp features were formed at approximately the same time, suggesting faster spreading to the east.

We conclude that the sediment thicknesses over the flanks of the EPR are not great enough to alter the basic bathymetric asymmetry noted above. At distances of 100 to 200 km from the rise crest, for instance, the Nazca plate is about 100 m deeper than the Pacific plate. At these distances the sediment on the Pacific plate is about 50 m thick, whereas on the Nazca plate it is only about 25 m thick. Therefore, after correcting for isostatic loading, the sediment blanket will make a difference of only about 15 m between depths on either side of the plate. It is possible that the thicker sediment on the Pacific plate proportionally impedes hydrothermal circulation to the point where plate cooling, and therefore contraction and crustal subsidence, is slowed relative to this process on the Nazca plate. It is not clear how such thin sediment cover could have such a drastic effect on crustal cooling.

On the eastern edge of the EPR at 10° to 13°S, about 50 m of sediment has accumulated on crust generated just prior to the Bauer scarp. The age of this crust should be 6 to 9 m.y. on the basis of the time of initiation of EPR-spreading calculated by Herron (1972), and 7 to 8 m.y. on the basis of the recent spreading rate at 10°S calcualted by Rea (1976). Thus, the sediment accumulation rate averaged 5 to 9 m/m.y. in this region.

CENTRAL NAZCA PLATE

The north-central Nazca plate is dominated by the fossil Galapagos Rise. Anderson and Sclater (1972) and Herron (1972) suggested that the Galapagos Rise was active from prior to ~6 to 9 m.y.B.P., when the center of spreading jumped about 900 km west to the EPR. Our sediment-thickness profiles (Fig. 4) support this model. Bathymetric profiles across the Galapagos Rise show a typically elevated rise crest with correspondingly thinner sediments. The older flanks of the fossil rise are deeper and have appropriately thicker sediments.

The bathymetry of the fossil Galapagos Rise has been described and contoured by Menard and others (1964), Anderson and Sclater (1972), Herron (1972), Mammerickx and others (1975) and Mammerickx and Smith (1978). The western slope of the rise is steeper than the eastern slope; depths on the eastern flank are ~250 km farther from the fossil rise crest than the same depths on the western side. Assuming that there was no vertical tectonic movement of the Galapagos Rise during the spreading center jump, this bathymetric asymmetry suggests that the spreading rate on the fossil rise was asymmetric in the same sense as that indicated by sediment thickness, magnetic anomaly patterns, and bathymetry across the EPR (depth asymmetry between the Pacific and Nazca plates excepted).

The northern limit of the fossil rise has been difficult to delineate from bathymetric or magnetic data. A depth-to-basement map (Fig. 5) traces the fossil rise to 5°S. This map suggests that the asymmetry described in the bathymetry is controlled by the basement.

The crust generated by the Galapagos Rise is separated from the younger EPR crust by the Bauer scarp, a topographic step that strikes parallel to the EPR from ~6°S to 15°S. The area of the scarp is composed of a series of linear ridges and deep intervening valleys with relief of over 600 m striking parallel to the EPR (Fig. 6). These high relief features, which are unique on the Nazca plate, may be remnants of the chaotic rifting and subsidence that accompanied the initiation of spreading on the present EPR.

The sea-floor depth in the central Bauer Basin is 4,000 to 4,400 m (Fig. 6). The basin is deepest in the west, adjacent to the Bauer scarp. Topography in the center of the basin is irregular, without the lineation apparent near the scarp. Most of this area is covered by abyssal hills and depressions with relief of 500 to 800 m and a wavelength of 8 to 10 km. The relief has no apparent trend, although the hills are more numerous around the edge of the basin. The larger and better-charted seamounts are about 20 km in diameter and demonstrate no preferred orientation relative to each other or to the major tectonic features on the Nazca plate.

Only thin sediment blankets most of the central part of the Nazca plate, and more sediment is accumulated in the older, deeper regions than in the younger, shallower regions. The thickest accumulations of sediment in the central Nazca plate are in the deep valleys of the Bauer scarp, although reflection records show that these sediments are not significantly thicker than in the immediately adjacent Bauer Basin. Sediments that have collected in these topographic lows are largely derived from natural oceanic sedimentation and slumps from topographic highs in the immediate area (Erlandson, 1975). Reflection records show no evidence of mass movement from the relatively undisturbed sediments on the flank of the EPR into the deeps. This suggests that bottom transport must be infrequent, not from distant sources, and a fairly noncatastrophic process.

Other evidence for local sources of the Bauer sediments lies in the fact that in at least the upper 100 cm of the cores collected in water deeper than 4,400 m the sediments appear homogeneous (Andrews, Foreman and others, 1975, 1976; Theyer, Mato, and others, 1977). No carbonates have survived transportation into the deeps. Lonsdale (1976) also found that bottom currents in the Bauer Basin were weak and unable to cause erosion.

Some sedimentary and basement features in the Bauer Basin suggest that a younger, postdepositional period of tectonic activity has occurred. Normal pelagic sedimentation has covered existing small topographic features and caused a general flattening of the bottom. Some faults and seamounts that formed early in the spreading history of the fossil rise are buried (Yeats and Heath, 1976), whereas others nearby are not (Fig. 7). A postdepositional period of volcanism might explain the recent faulting and sediment-free seamounts. It may also be reflected in the high heat-flow measurements in the Bauer Basin, which Anderson and Halunen (1974) associated with the movement of the spreading center beneath the basin during the jump.

Sediments on the east flank of the fossil ridge increase only slightly in thickness toward the South American Continent. Surface productivity is low almost to the trench axis.

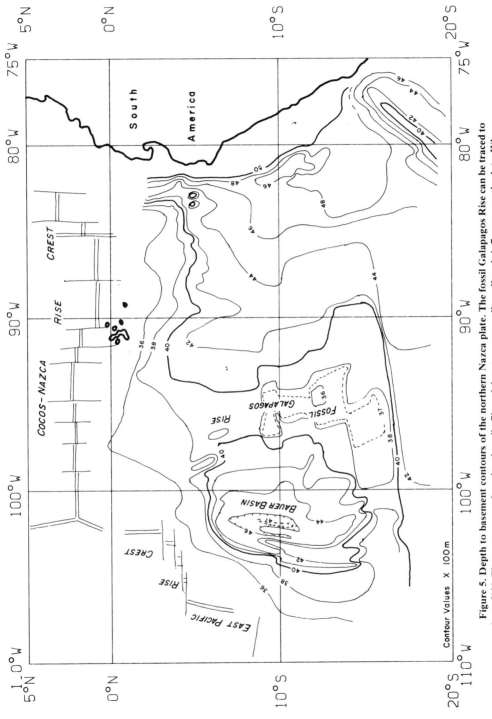

Figure 5. Depth to basement contours of the northern Nazca plate. The fossil Galapagos Rise can be traced to about 5°N. The contours are based on heavily filtered data to remove all small-scale influences on the data. Where data were lacking, the contours follow the bathymetric trends.

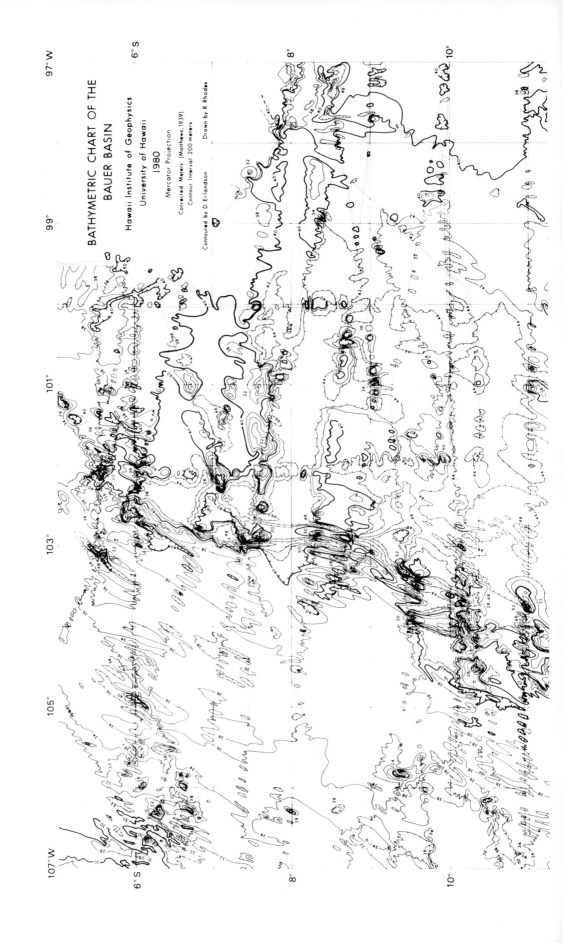

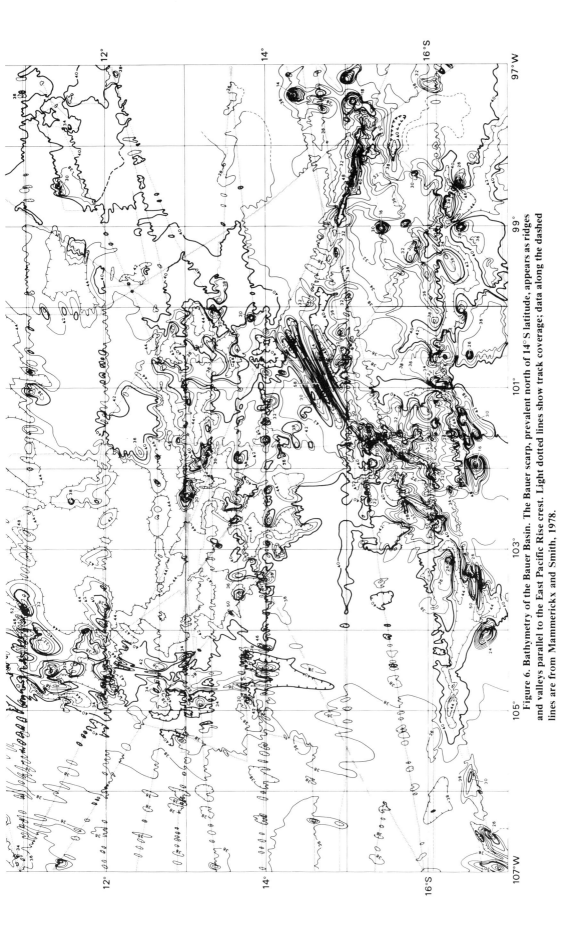

Figure 6. Bathymetry of the Bauer Basin. The Bauer scarp, prevalent north of 14°S latitude, appears as ridges and valleys parallel to the East Pacific Rise crest. Light dotted lines show track coverage; data along the dashed lines are from Mammerickx and Smith, 1978.

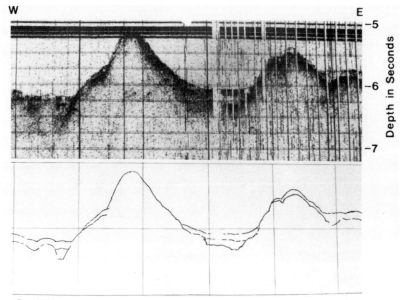

Profile A

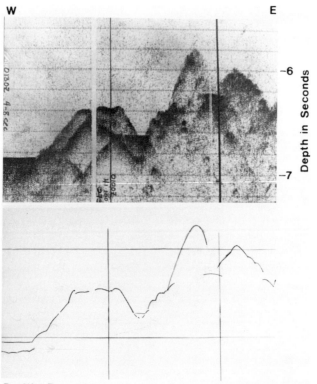

Profile B

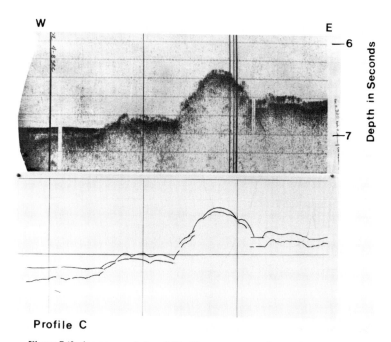

Profile C

Figure 7 (facing page and above). Profiles suggest two phases of tectonic activity in the Bauer Basin by showing sediment-covered and uncovered features. We interpret the barren features as having been formed when active spreading jumped, 6 to 9 m.y.B.P. Profile locations indicated on Figure 3; vertical time lines on tracing are ~5 nautical miles. Vertical exaggeration ≈12:1.

NORTHERN NAZCA PLATE, COCOS-NAZCA RISE, AND CARNEGIE RIDGE

Except for the Galapagos Islands, the thickly sedimented southern flank of the Cocos-Nazca Rise has less extreme small-scale topography than either the Bauer Basin or the fossil Galapagos Rise. Very few large seamounts disrupt the level ocean floor (Mammerickx and Smith, 1978).

The regional bathymetric trend in the northern Bauer Basin, south of the rough-smooth boundary, unexpectedly shoals perpendicular to the direction of age progression. The crust was generated in an east-west direction but shoals northward toward the rough-smooth boundary. This phenomonon is primarily a function of sediment thickness. The sediments increase in thickness from 100 milliseconds (ms) (two-way travel time, that corresponds to approximately 100 m of sediment) at 7°S to over 400 to 500 ms (400 to 500 m) at the equator. The sea floor rises ~600 m over this distance. After correction for sediment loading, the top of the crust, if there had been no sedimentation, would only rise about 300 m from 7°S to the equator. The increase in sediment thickness can be accounted for by the previously mentioned high productivity in the equatorial zoogeographic zone.

Submarine erosion in the area of the equator has resulted in incised valleys and a regional thinning of the sediment cover. The incised valleys are numerous in the saddle area between the Carnegie Ridge and the coast of Ecuador, where bottom currents are as great as 33 cm/s (Lonsdale, 1976). Some valleys cut through sequences of sediment with up to 500 ms of reflection time (Fig. 9, Lonsdale, 1976; Heath and van Andel, 1973). In many areas near the Carnegie Ridge there appears to be a relation

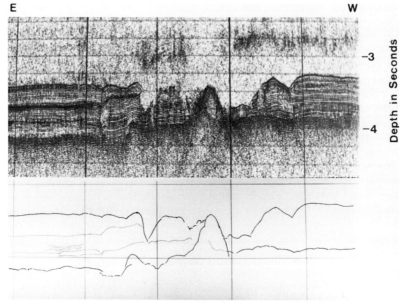

Profile D

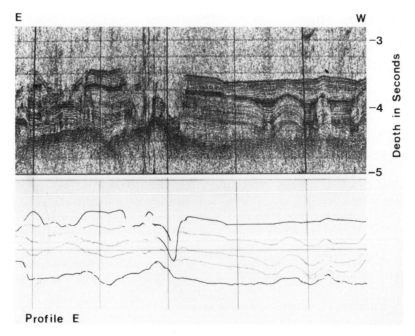

Profile E

Figure 8. The association between incised valleys and volcanic piercements. Profiles D and E show that valleys cut the entire sequence of sediments where secondary volcanism has occurred. Profile locations indicated on Figure 3; vertical time lines on tracing are ~5 nautical miles. Vertical exaggeration ≈ 12:1.

between the location of a major valley and a postdepositional volcanic event. Figure 8 is illustrative of HIG reflection records in this area that, along with data from Lonsdale (1976) and Heath and van Andel (1973), commonly show volcanic piercements or faults in close association with the incised valley. This coincidence, coupled with the bottom currents and potential gravitational instabilities which do not appear adequate to initiate valley formation, point to a tectonic event as the origin of the erosional valley.

The north-striking "channels" on the sedimentary isopach chart (Fig. 2) south of the equator may be artifacts of the contouring process. On the reflection records they appear not as channels but as piercements and extrusive volcanic rocks. Whether they can be reliably correlated on the scale of the chart is speculative.

The large area of thin and missing sediments (Fig. 2; 3°S, 83'W) corresponds to the crestal area of the Carnegie Ridge. Heath and van Andel (1973) suggested that the sediments had been stripped from the top of the ridge rather than that they never had been deposited there. Our reflection records also show adjacent areas with surface sediments removed by erosion and other areas with distinct disconformities and sediment wedges, all attesting to the presence of strong bottom currents in the general area.

A large portion of the north-moving bottom current is likely to have been diverted westward by the Carnegie Ridge and may have aided in the enlarging of the channels and incised valleys on the southern side of the ridge. The sediments are surprisingly thin (Fig. 2) on the oldest Cocos-Nazca generated crust (22 to 23 m.y. old, according to Hey, 1977). This area is slightly south of the ridge, within 6° of the equator, in the equatorial zoogeographic zone. Sediment scouring has been found in the Ecuador trench (Lonsdale, 1976) and on our reflection records (Fig. 9). The presence of mid-Miocene and Pliocene fossils at the surface of two nearby HIG cores (stations 135 and 136, respectively; Andrews, Foreman, and others, 1975) also suggests that sediment scouring may be an explanation for the general thinning of Nazca plate sediments.

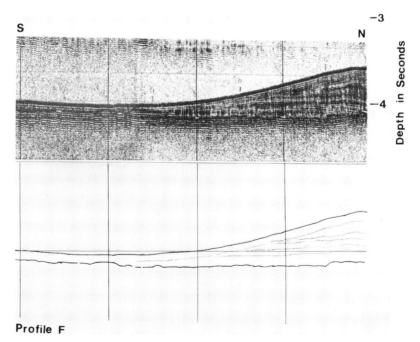

Figure 9. Regional erosion in the area southeast of the Carnegie Ridge. Profile location indicated on Figure 3; vertical time lines are ~5 nautical miles. Vertical exaggeration ≈ 12:1.

EASTERN MARGIN OF THE NAZCA PLATE

The Nazca Ridge is a 1,100-km-long, northeast-striking, aseismic linear ridge that intersects the Peru-Chile Trench (Fig. 1). The average depth of the ridge is between 2,000 and 3,000 m, and the southwestern end is generally shallower than the northeastern end. Bathymetric profiles show that the ridge is also asymmetric—the southern slope is steeper than the northern (Fisher, 1958; our data).

Cutler (1977) proposed that the Nazca Ridge formed at the spreading center and is consequently the same age as the surrounding sea floor, or 26 to 36 m.y. (Handschumacher and others, this volume).

The crest of the Nazca Ridge is almost totally above the lysocline and the CCD. This allows sediments to accumulate to sequences greater than 300 ms in places. The thickest sediments along the crest appear to be confined between volcanic and basement structural highs (Fig. 10). The axial structure of the ridge is controlled by apparent large basement relief, suggesting block faulting that is mostly masked by the intervening thick sedimentary sequences.

Sediment cover on the surrounding Nazca plate is thin (Fig. 2), less than 125 ms. However, the sediment cover on the northern side and flank of the Nazca Ridge is thicker than on the southern side, which may in part account for the gentler northern slope.

From our reflection records, erosion does not appear to have significantly changed the sedimentary patterns on the ridge. Antarctic bottom currents apparently move around either end of the ridge without affecting the crestal sedimentary sequences.

A region of relatively rough basement and thinner sediments parallels the Peru-Chile Trench. The geographical extent of the region is shown in Figure 3. It begins over 300 km west of the trench axis and extends through the entire area mapped in Figure 2. The thinning of the sediments in the rough region is shown in Figure 4. These profiles are typical of those obtained throughout the region on ocean crust of varying thickness and age and on regions with and without an outer gravity high (as defined by Getts, 1975). The rough zone is not a function of known fracture zones or bathymetric discontinuities and is observed on all eleven of the plate traverses we have studied so far. Inspection of the seismic reflection records shows that the relative roughness is due to extensive faulting and upper-crustal deformation as well as intrusive and extrusive volcanism. We hypothesize that the rough zone that covers the same area of the plate as the outer gravity high of Getts (1975) is related to the movement of the Nazca plate into the widespread stress field of the Peru-Chile subduction zone. The rough zone extends farther seaward than the horst and graben faulting caused by extension of the top of the lithospheric slab as it bends into the subduction zone. The apparent sediment thinning that persists in the region may result from secondary (off-ridge) volcanism, producing sills and interlayering of basalt flows that form an acoustic basement shallower than the deepest sediments. These intrusive and extrusive volcanic rocks have been observed on our reflection records (Fig. 11) and are perhaps also responsible for initiating the valley erosion mentioned above.

Tholeiitic basalt sampled from a ridge in the Peru-Chile Trench axis was dated by Kulm and others (1973) as having a *minimum* age of 8.7 m.y., which is much younger than that of the adjacent Nazca plate crust. Basalts sampled in DSDP holes 320 and 31 (within the rough zone) were initially described as being surprisingly unweathered, suggestive of a young age. However, the samples were subsequently radiometrically dated, and results match the magnetic anomaly age (Hogan and Dymond, 1976; Mitchell and Aumento, 1976).

The age distribution of the topmost part of layer 2 and possible secondary volcanism on the eastern margin of the Nazca plate are poorly determined. With existing data we can only say that sediment thinning seems related to, but not directly caused by, upper crustal deformation prior to subduction. No widespread scouring or bottom erosion is observed in the area. On the basis of the inconsistent basement-age data in the trench region, it is tempting to speculate that off-ridge volcanism has produced sills and apparent sediment thinning. Unfortunately, DSDP age dates are consistent with basement formation at the ridge crest, and no other direct evidence of mid-plate volcanism has been observed in this region.

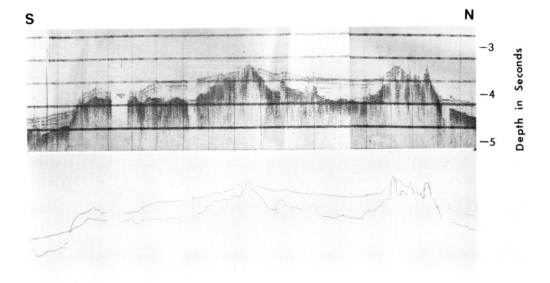

Profile G

Figure 10. Profile of sedimentary and basement character across the Nazca Ridge. Profile location indicated on Figure 3; vertical time lines on profile are ~5 nautical miles. Vertical exaggeration ≈ 19:1.

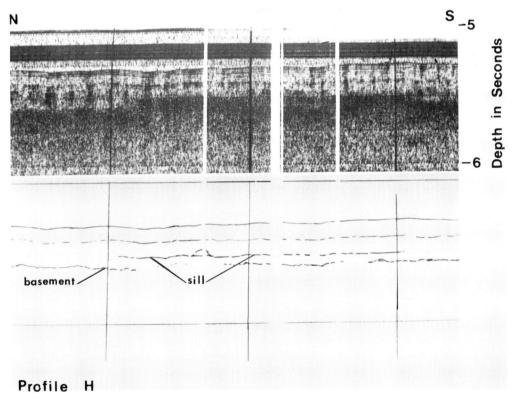

Profile H

Figure 11. Profile H, intrusive volcanic rocks in the "rough zone" of the Nazca plate. The sill, most evident in the line tracing, shows posterosional volcanism. The profile location is indicated on Figure 3; vertical time lines on tracing are ~5 nautical miles. Vertical exaggeration ≈ 12:1.

SUMMARY AND CONCLUSIONS

Generalized patterns of sediment accumulation on the northern Nazca plate fit trends that can be readily explained by crustal age, depth of the sea floor relative to the lysocline and the CCD, and surface productivity. Only on the Cocos-Nazca Rise and Carnegie Ridge do we find any evidence of persistent scouring and erosion of sediments by bottom currents.

Sediment distribution across the EPR, however, shows a consistent asymmetry; western flank sediment is twice as thick as on the east. Although there is strong evidence of crustal spreading asymmetry on the EPR, the 2:1 asymmetry in the sediment is too great and must, in part, result from other causes.

The orderly sedimentation pattern versus age in the center of the Nazca plate helps to delineate the crest of the fossil Galapagos Rise, and enables us to trace this feature farther north beneath the thicker equatorial sediments that obscure it.

A zone of marked sediment thinning and rough bathymetry on the eastern margin of the plate parallels the Peru-Chile Trench. Although these effects are likely caused by the fracturing of the top of the oceanic plate as it bends in response to entering the subduction zone, this faulting of the top of layer 2 does not explain the loss of sediments.

There is no evidence to support this huge area having been affected by sea-floor erosion or nondeposition. We speculate that the region may be subjected to off-ridge volcanism producing sills that elevate the acoustic basement.

The relationship between the postdepositional volcanism in the region of rough bottom and the tectonic events responsible for the incised valleys south of the Carnegie Ridge is not known. It may be speculated, however, that the stress field associated with the Peru-Chile Trench is responsible for both.

ACKNOWLEDGMENTS

We thank the crews of the University of Hawaii and Oregon State University research ships for their hours of taking and annotating seismic reflection records. We also thank Paul Grim from The National Geophysical Solar-Terrestrial Data Center, J. Mammerickx, and Barbara Long from Scripps Institute of Oceanography, and Steven Johnson from Oregon State University for supplying bathymetric data. In addition, we thank Chris Mato from the HIG core lab for help in dating cores from stations 135 and 136, George Woollard for providing us with the 1972 Mendeleev data, and the HIG Geophysical Data Analysis Lab for acquiring, preparing, and plotting the geophysical data.

Research was funded and supported by National Science Foundation grant OCE 76-24177 IDOE Nazca Plate Project.

REFERENCES CITED

Ade-Hall, J. M., 1976, Underway surveys, leg 34, *in* Yeats, R. S., Hart, S. R., and others, Initial reports of the Deep Sea Drilling Project, Volume 34: Washington, D.C., U.S. Government Printing Office, p. 163-181.

Anderson, R. N., and Halunen, J., Jr., 1974, Implications of heat flow for metallogenesis in the Bauer Deep: Nature, v. 251, no. 5475, p. 474-475.

Anderson, R. N., and Sclater, J. G., 1972, Topography and evolution of the East Pacific Rise between 5°S and 20°S: Earth and Planetary Science Letters, v. 14, p. 433-441.

Andrews, J. E., Foreman, J. A., and others, 1975, Sediment core description: R/V *Kana Keoki* 1971 cruise, Eastern and Western Pacific Ocean: Hawaii Institute of Geophysics Technical Report HIG-75-15, 14 p. and Appendix, 314 p.

—— 1976, Sediment core descriptions, R/V *Kana Keoki* 1972 cruise, Eastern and Western Pacific Ocean: Hawaii Institute of Geophysics Technical Report HIG-76-13, 10 p. and Appendix, 112 p.

Berger, W. H., and Winterer, F. L., 1974, Plate stratigraphy and the fluctuating carbonate line, *in* Hsu, K. J., and others, eds., Pelagic sedimentation on land and under the sea: International Association of Sedimentologists, Special Publication Number 1, p. 11.

Cutler, S. T., 1977, Geophysical investigation of the Nazca ridge [M.S. thesis]: Honolulu, University of Hawaii, 83 p.

Erlandson, D. L., 1975, Sediment accumulation in structural traps in the southern Bauer Basin, southeast Pacific [abs.]: EOS (American Geophysical Union Transactions), v. 56, p. 445.

Fisher, R. L., 1958, Downwind investigations of the Nazca ridge; Preliminary report on Expedition Downwind, *in* Fisher, R. L., ed., IGY General Report Series No. 2: Washington, D.C., IGY World Data Center A, p. 58.

Getts, T. R., 1975, Gravity and tectonics of the Peru-Chile Trench and eastern Nazca plate 0°-33°30'S [M.S. thesis]: Honolulu, University of Hawaii, p. 103.

Handschumacher, D. W., Pilger, R. H., Jr., and Campbell, J. F., 1981, Structure and evolution of the Easter plate, *in* Kulm, L. D., and others, eds., Nazca plate: Crustal formation and Andean convergence: Geological Society of America Memoir 154 (this volume).

Heath, G. R., and van Andel, Tj. H., 1973, Tectonics and sedimentation in the Panama basin: Geological results of leg 16, *in* van Andel, Tj. H., and Heath, G. R., eds., Initial reports of the Deep Sea Drilling Project, Volume 16: Washington, D.C., U.S. Government Printing Office, p. 899-913.

Herron, E. M., 1972, Sea-floor spreading and the Cenozoic history of the east-central Pacific: Geological Society of America Bulletin, v. 83, p. 1671-1692.

Hey, R., 1977, Tectonic evolution of the Cocos-Nazca spreading center: Geological Society of America Bulletin, v. 88, p. 1404-1420.

Hogan, L., and Dymond, J., 1976, K-Ar and ^{40}Ar-^{39}Ar dating of site 319 and 321 basalts, *in* Yeats, R. S., Hart, S. R., and others, Initial reports of the Deep Sea Drilling Project, Volume 34: Washington, D.C., U.S. Government Printing Office, p. 439-442.

Kulm, L. D. and others, 1973, Tholeiitic basalt ridges in the Peru Trench: Geology, v. 1, no. 1, p. 11-14.

Lonsdale, P., 1976, Abyssal circulation of the southeastern Pacific and some geological implications: Journal of Geophysical Research, v. 81, no. 6, p. 1163-1176.

Mammerickx, J., and Smith, S. M., 1978, Bathymetry of the southeast Pacific: Chart published by Geological Society of America for Scripps Institution of Oceanography, scale 1:6,442,194.

Mammerickx, J., and others, 1975, Morphology and tectonic evolution of the east-central Pacific: Geological Society of American Bulletin, v. 86, p. 111-118.

Matthews, D. S., 1939, Tables of the velocity of sound in pure water and sea water for use in echo-sounding and echo-ranging: London, Admiralty Hydrographic Department, 52 p.

Menard, H. W., Chase, T. E., and Smith, S. M., 1964, Galapagos Rise in the southeastern Pacific: Deep Sea Research, v. 11, p. 233.

Mitchell, W. S., and Aumento, F., 1976, Fission track chronology and uranium content of basalts from DSDP Leg 34, *in* Yeats, R. S., Hart, S. R., and others, Initial reports of the Deep Sea Drilling Project, Volume 34: Washington, D.C., U.S. Government Printing Office, p. 452-453.

Molnar, P., and others, 1975, Magnetic anomalies, bathymetry and the tectonic evolution of the South Pacific since the Late Cretaceous: Geophysical Journal of the Royal Astronomical Society, v. 40, p. 383-420.

Rea, D. K., 1976, Analysis of a fast-spreading rise crest: the East Pacific Rise, 9° to 12° South: Marine Geophysical Researches, v. 2, p. 291-313.

—— 1978, Asymmetric sea-floor spreading and a nontransform axis offset: The East Pacific Rise 20°S survey area: Geological Society of America Bulletin, v. 89, p. 836-844.

—— 1981, Tectonics of the Nazca-Pacific divergent plate boundary, *in* Kulm, L. D., and others, eds., Nazca plate: Crustal formation and Andean convergence: Geological Society of America Memoir 154 (this volume).

Sclater, J. G., Anderson, R. N., and Bell, M. L., 1971, Elevation of ridges and evolution of the central eastern Pacific: Journal of Geophysical Research, v. 76, no. 32, p. 7888.

Theyer, F., Mato, C., and others, 1977, Sediment core descriptions: R/V *Kana Keoki* 1973 North Central Pacific Cruise, 1974 Southeastern Pacific Cruise, and a 1974 Mid-Atlantic Ridge IPOD Site Survey: Hawaii Institute of Geophysics Technical Report HIG-77-9, 13 p. and Appendix, 114 p.

Yeats, R. S., and Heath, G. R., 1976, Bathymetry and structure of the Bauer Deep around DSDP site 319, *in* Yeats, R. S. Hart, S. R., and others, Initial reports of the Deep Sea Drilling Project, Volume 34: Washington, D.C., U.S. Government Printing Office, p. 157–162.

MANUSCRIPT RECEIVED BY THE SOCIETY NOVEMBER 12, 1980
MANUSCRIPT ACCEPTED DECEMBER 30, 1980
CONTRIBUTION 1101, HAWAII INSTITUTE OF GEOPHYSICS

Economic appraisal of Nazca plate metalliferous sediments

Cyrus W. Field
Department of Geology
Oregon State University
Corvallis, Oregon 97331

Dennis G. Wetherell
C. C. Hawley & Associates, Inc.
Box 78-D, Star Route A
Anchorage, Alaska 99507

E. Julius Dasch
Department of Geology
Oregon State University
Corvallis, Oregon 97331

ABSTRACT

Metalliferous sediments over the Nazca plate constitute an enormous potential resource because of their widespread surficial distribution. The total value of contained metals is nearly $67.50 per metric ton based on current prices for iron, manganese, cobalt, copper, molybdenum, nickel, zinc, and silver. However, economic comparisons are unfavorable with alternative or existing sources of metals such as the manganese nodules and ores of the Cuajone and Cerro de Pasco mines. The principal obstacles are the relatively low concentrations for all metals and the high proportion of total value in the metallurgically complex association of iron and manganese. Other largely negative uncertainties notwithstanding, we conclude that these metalliferous sediments will remain an untapped resource into the foreseeable future.

INTRODUCTION

Regardless of speculations concerning the role of metalliferous sediments in cordilleran metallogeny (Field and Dasch, this volume, and references cited), these sediments constitute a geologically unique and geochemically anomalous product of dominantly submarine hydrothermal activity (Dymond, this volume, and references cited). Because of the serious efforts and substantial expenditures that have been directed to the eventual mining of ocean resources for the past decade or more, it is of interest to make a preliminary evaluation of the Nazca plate metalliferous sediments as a possible future source of metals. We do so by comparison to the manganese nodules, another potential resource of the ocean basins, and to the well-known ore deposits of Cuajone and Cerro de Pasco in Peru. The latter are representative of porphyry and fissure-replacement types of hydrothermal mineralization, respectively. Although the following analysis of necessity involves many speculative assumptions, it clearly demonstrates that the metalliferous sediments are not an economically viable source of metals under present conditions of supply and demand and the existing state of mining and extractive technology.

ASSUMPTIONS AND UNCERTAINTIES

Data pertaining to the content (kilograms per metric ton) and value (dollars per metric ton) of metal contained in the metalliferous sediments, manganese nodules, and ores of Cuajone and Cerro de Pasco are listed in Table 1. The content of metal (kg/mt) was obtained by multiplying the percentage concentration of the metal by 10. Concentrations of metals (not given in Table 1; simply divide kg/mt by 10) in the resources and ores were derived from sources as follows: Those for the metalliferous sediments represent the approximate average for the more metal-rich samples collected from the Bauer Deep (Dymond and Field, unpub. absolute analyses; see Dymond, this volume, for corrected, carbonate-free analyses). Metal concentrations for the manganese nodules are the mid-points of ranges cited by Agarwal and others (1979). Those for ore at the Cuajone mine were provided by personnel of the Southern Peru Copper Corporation in 1978, and those for Cerro de Pasco represent a composite average of ore mined during the interval from 1906 to 1976 (see Einaudi, 1977, p. 895). The individual and total dollar values of metal contained in these resources and ore materials ($/mt) as given in Table 1 were determined from content and price. Prices of the metals ($/kg) according to The Engineering and Mining Journal (1980) were as follows: iron ($0.06); manganese ($0.64) as ferromanganese; cobalt ($55.12); copper ($2.25); lead ($0.79); molybdenum ($19.84); nickel ($7.60); zinc ($0.79), and silver ($516.32). Although the prices for most of these metals are fairly representative of long-term stability, those for cobalt, molybdenum, and silver have increased as much as fivefold from 1977 to 1980. Thus, potential fluctuations in metal prices contribute uncertainties to the conclusions of these and similar evaluations.

The mines of Cuajone and Cerro de Pasco are economically profitable operations under normal conditions. Accordingly, comparisons of the metal content and dollar value of these ores to those of the ocean resources provide a useful, but inexact, measure of apparent relative worth. At least two problems relate to figures given for the total value of ores and resources (Table 1). First, metallurgical efficiencies for the extraction of metals from ores rarely exceed 90%. Thus, the gross, or recoverable, values (90% of total value) listed in Table 1 provide a more realistic dollar estimate of the ore and resource material. Second, the similar chemical properties of iron and manganese commonly preclude their recovery from materials in which they are mutually associated. Because both are major

TABLE 1. CONTENT OF METALS AND TOTAL AND GROSS DOLLAR VALUES PER METRIC TON IN METALLIFEROUS SEDIMENTS, MANGANESE NODULES, AND ORES OF THE CUAJONE AND CERRO DE PASCO MINES, PERU

Metal	Metalliferous sediments		Manganese nodules		Cuajone Mine		Cerro de Pasco mines	
	(kg/mt)	($/mt)	(kg/mt)	($/mt)	(kg/mt)	($/mt)	(kg/mt)	($/mt)
Iron	150.0	9.00	60.0	3.60				
Manganese	50.0	32.00	275.0	176.00				
Cobalt	0.2	11.02	3.0	165.36				
Copper	1.1	2.48	15.0	33.75	10.0	22.50	14.3	32.18
Lead							15.1	11.93
Molybdenum	0.15	2.98	0.75	14.88	0.14	2.78		
Nickel	1.0	7.60	15.0	114.00				
Zinc	0.4	0.32					38.5	30.42
Silver	0.004	2.07					0.112	57.83
Total value		67.47		507.59		25.28		132.36
excl. Fe-Mn		26.47		327.99				
Gross value		60.72		456.83		22.75		119.12
excl. Fe-Mn		23.82		295.19				

constituents of the metalliferous sediments and manganese nodules, and yet may be of little economic importance if not recovered, the total and gross values of the ocean resources are listed both inclusive and exclusive of contributions from these metals in Table 1. The gross values exhibit a remarkable range among and between the ores ($22.75 to $119.12) and resources ($60.72 to $467.83) on a per ton basis. High values for the manganese nodules relate to large concentrations of numerous metals. In contrast, the lower values for the metalliferous sediments are derived from a larger number of metals, but from metals whose individual concentrations are lower and that range from one-half to one-tenth the value of those contained in primary ores of these metals. Thus, the value of metalliferous sediment resides in the combined input of eight different metals, all at relatively low concentrations, and of which iron and manganese constitute 61% of the total. The commercial viability of the Cuajone mine, whose ore contains fewer metals at lower concentrations than those of Cerro de Pasco, is attributable to operational economies of a large, high-volume, openpit mining operation that are characteristic of porphyry-type deposits.

Other statistics concerning the physical characteristics of the mineralization, rates of production, and financial estimates of capitalization, costs, and revenues that are useful or critical to the comparative evaluation of these deposits are listed in Table 2. Some are factual or at least are reasonable estimates, such as the shapes, attitudes, dimensions, densities, volumes, and tonnages of the metal-bearing materials. However, the Cerro de Pasco orebody is a composite of convenience that

TABLE 2. SUMMARY OF PHYSICAL CHARACTERISTICS AND ESTIMATES OF PRODUCTION, CAPITALIZATION, COSTS, AND REVENUES FOR THE MINING OF METALLIFEROUS SEDIMENTS, MANGANESE NODULES, AND ORES OF THE CUAJONE AND CERRO DE PASCO MINES, PERU

	Metalliferous sediments	Manganese nodules	Cuajone Mine	Cerro de Pasco mines
Shape of orebodies	Stratiform	Stratiform	Cylinder	Tabular
Attitude	Horizontal	Horizontal	Vertical	Dipping
Dimensions (m)				
Length	15,100	69,400	680	1,000
Width	15,100	69,400	(Diameter)	25
Depth	1	0.005	470	590
Surface area (km^2)	227	4,810	0.4	0.03
Density of ore (g/cm^3)				
Wet	1.11	2.00		
Dry	0.22	1.40	2.55	3.25
Volume of ore (m^3/mt)				
Wet	0.90	0.50		
Dry	4.54	0.71	0.39	0.31
Tons of ore (mt)				
Wet	252,300,000	71,400,000		
Dry	50,000,000	50,000,000	430,000,000	48,000,000
Production (mt/d)				
Wet	28,800	8,200		
Dry	5,700	5,700	50,000	5,500
Production (mt/yr)				
Wet	10,100,000	2,900,000		
Dry	2,000,000	2,000,000	17,500,000	1,925,000
Longevity of mine (yr)	25	25	25	25
Capitalization ($)	360,000,000	390,000,000	1,100,000,000	175,000,000
excl. Fe-Mn	310,000,000	340,000,000		
Capitalization ($/mt ore mined)	1.43	5.46	2.56	3.65
excl. Fe-Mn	1.23	4.76		
Gross Revenue ($/mt)	60.72	456.83	22.75	119.12
excl. Fe-Mn	23.82	295.19		
Costs ($/mt)	57.50	57.50	13.50	30.00
excl. Fe-Mn	27.50	27.50		
Net revenue ($/mt)	3.22	399.33	9.25	89.12
excl. Fe-Mn	-3.68	267.69		
Net revenue ($/yr)	6,440,000	798,660,000	161,880,000	171,560,000
excl. Fe-Mn	-7,360,000	535,380,000		
Payout period (yr)	55.9	0.5	6.8	1.0
excl. Fe-Mn	Never	0.6		

represents one openpit and several smaller underground mines. Other statistics are derivatives of specific assumptions. For example, tonnages of the ocean resources, although essentially infinite, are estimated from assumed rates of production and longevity, whereas all three of these statistics are reasonably factual for the two operating mines. Internally consistent operational longevities of 25 yr fortuitously result from current reserves and production rates at the mines, and from economically optimum constraints assumed for production of the ocean resources. The truly enormous sizes of the ocean mining "claims" would appear to be unrealistic as compared to normally smaller areas on the continents for which rights to mineral property are secured. These large areas are a function of the thin, horizontal, and stratiform distribution of the ocean resources, and of the necessity to have sufficient reserves to justify a costly mining operation (see Table 2). The assumed thickness of the metalliferous sediment (1 m) may be increased by more than one order of magnitude provided the thickness is recoverable by the method of mining, whereas that of the manganese nodules (0.005 m) is essentially fixed by an assumed mineable concentration of 10 kg/m^2 of nodules (Li and Tinsley, 1975; Tinsley, 1977). However, to the extent that the efficiency of the mining method is less than 100%, the sizes of the mining "claims" will have to be increased proportionately to obtain the necessary tonnages of "ore." Nonetheless, the sizes of these "claims" are far less than 1% of the total area considered favorable for metal-rich sediments and nodules in the Bauer Deep and equatorial Pacific southeast of Hawaii, respectively.

The largest uncertainties and potential errors relate to our estimates of metal recovery, capitalization, and costs, which in turn directly affect the final derivative conclusions of profitability (net revenue) and return of capital (payout period). Because of the relatively lower concentrations of all metals, except iron, in the sediments, our assumed 90% level of metallurgical recovery may be unduly optimistic as the efficiency of the hydrometallurigcal extractive process may be proportionately diminished. The extent to which this may be true, and the likelihood that iron cannot be recovered at all, will thereby reduce both the dollar value of the metalliferous sediment and the revenues generated during a mining operation. Estimates of capital requirements are stated in terms of 1980 dollars, and those for Cuajone and Cerro de Pasco are not likely to be in error by more than 10% and 20%, respectively. Estimates for the ocean resources are based entirely on data provided by Li and Tinsley (1975), National Academy of Sciences (1975), Sisselman (1975), Tinsley (1977), Agarwal and others (1979), McKelvey (1980), and other references cited therein. Although the efforts of these authors were almost exclusively concerned with manganese nodules, we have assumed that technology, costs, and capital requirements will be analogous and nearly identical to those required for the mining of metalliferous sediments. The slightly lower estimate for capitalization reflects an assumption of spin-off benefits in mining and extractive technology gained from prior efforts with manganese nodules. However, it is possible that our estimates of capitalization and costs for the mining of ocean resources may be understated by as much as 100%. For example, we assume identical costs for the resources, yet the mining of sediments requires the handling of nearly four times more material (wet) than with the mining of nodules to obtain equivalent tonnages of the dry product. This large differential, however, may be partly balanced by economies associated with the method of mining (fluid slurry versus bucket line) and the geologic occurrence (larger thickness and smaller areal distribution) that favor the sediments. In summary, the data and computations listed in Table 2 contain potentially large errors that could affect profitability because of assumptions related to the (1) price of metals, (2) efficiency of the mining method, (3) recovery of metals, (4) capitalization, and (5) costs of mining and processing the ores. Because of these and other uncertainties, including expenditures for complying with environmental regulations and the normally unfavorable "surprises" that may accompany new mining endeavors, the summation that follows is necessarily preliminary at best.

CONCLUSIONS

With the foregoing discussion intended to be a clear warning of the potential errors inherent to this evaluation, inspection of the data confined to the bottom three rows of Table 2 permits a tentative comparison of the relative financial merits of these four deposits. Net revenues given per ton and per year provide a crude measure of relative profitability, but they do not take into account other probable expenses such as taxes, royalties, and interest on funds expended for capitalization. Nonetheless, on the basis of all prior assumptions, the mines of Cuajone and Cerro de Pasco, and especially the operation postulated for the mining of manganese nodules, will collectively generate much larger revenues than the operation postulated for the mining of metalliferous sediments. Moreover, the latter incurs a deficit if iron and manganese are not extracted from the sediments, whereas the effect of this loss in the recovery of metals is less dramatic and not critical to the economic viability of mining nodules. However, considerations of profitability based entirely on revenues neglect the fundamentally important concept of capital return. In essence, this factor represents the length of time needed to recover the cost of capitalization, or funds expended in exploration, research, and development of a mining property to the state of production. It may be approximated by the payout period (see Table 2), obtained by dividing capitalization by net annual revenues, which gives the time (in years) required to recover the original capital investment. Because of risks normally inherent to the mining and petroleum industries, proposed ventures having payout periods of more than 10 to 15 yr are considered to be economically unattractive. In contrast to the long payout period (60 yr) for the mining of metalliferous sediments, the payout periods for the mining of nodules and the polymetallic ores of Cerro de Pasco are extraordinarily short (1 yr or less), whereas the payout period for the Cuajone mine (7 yr) is more consistent with the industry-wide norm. Again, the exclusion of recoverable iron and manganese eliminates the mining of metalliferous sediments from economic consideration, but has little effect on the payout period for the mining of nodules. Values listed in Table 2 are crude approximations at best because they do not take into account the prolonging effect on capital recovery of imposed taxes, royalties, and interest. For example, if a 10% rate of interest is assumed on the funds required for capitalization, the payout periods for the mining of manganese nodules and the Cerro de Pasco mine are little affected because of the rapidity of capital return. In contrast, that for the Cuajone mine is extended to nearly 12 yr, and that for the mining of metalliferous sediments is changed to "never" because interest charges exceed net revenues annually.

Thus, the results of this comparative evaluation clearly suggest the profitable recovery of metals from manganese nodules. They are consistent with, but excessively more favorable than, those obtained by previous investigators. An important question is why, in spite of apparently favorable economics and the expenditure of several hundred millions of dollars to date, has not a full-scale ocean-mining venture been undertaken? The answer is undoubtedly less related to economics and technology than it is to current problems of sea-floor ownership, taxation, royalties, and rights to proprietary technology that are now of international concern (see McKelvey, 1980). Regardless of outcome, the mining of metalliferous sediments as we have described for the Nazca plate will not become a commercial reality in the foreseeable future.

ACKNOWLEDGMENTS

We acknowledge the thoughtful suggestions provided by Alexander G. Jones of the Hanna Mining Company, Werner J. Raab of the Anaconda Company, and Alan R. Wallace of the U.S. Geological Survey who reviewed an earlier draft of this manuscript, but accept responsibility for the premises and conclusions stated herein.

REFERENCES CITED

Agarwal, J. C., and others, 1979, Kennecott process for recovery of copper, nickel, cobalt, and molybdenum from ocean nodules: Mining Engineering, v. 31, no. 12 (December), p. 1704–1707.

Dymond, J., 1981, The geochemistry of Nazca plate surface sediments: An evaluation of hydrothermal, biogenic, detrital, and hydrogenous sources, *in* Kulm, L. D., and others, eds., Nazca plate: Crustal

formation and Andean convergence: Geological Society of America Memoir 154 (this volume).
Einaudi, M. T., 1977, Environment of ore deposition at Cerro de Pasco, Peru: Economic Geology, v. 72, p. 893–924.
Engineering and Mining Journal, 1980, Markets: Engineering and Mining Journal, v. 181, no. 8 (August), p. 19–25.
Field, C. W., and Dasch, E. J., 1981, Epilogue: Geostill reconsidered, *in* Kulm, L. D., and others, eds., Nazca plate: Crustal formation and Andean convergence: Geological Society of America Memoir 154 (this volume).
Li, T. M., and Tinsley, C. R., 1975, Meeting the challenge of material demands from the oceans: Mining Engineering, v. 27, no. 4 (April), p. 28–55.
McKelvey, V. E., 1980, Seabed minerals and the law of the sea: Science, v. 209, p. 464–472.
National Academy of Sciences, 1975, Mining in the outer continental shelf and in the deep oceans: Washington, D.C., National Academy of Sciences, 119 p.
Sisselman, R., 1975, Ocean miners take soundings on legal problems, development alternatives: Engineering and Mining Journal, v. 176, no. 4 (April), p. 75–86.
Tinsley, C. R., 1977, Nodule miners ready for prototype testing: Engineering and Mining Journal, v. 178, no. 1 (January), p. 80–81 and 101.

MANUSCRIPT RECEIVED BY THE SOCIETY NOVEMBER 12, 1980
MANUSCRIPT ACCEPTED DECEMBER 30, 1980

Printed in U.S.A.

CONTINENTAL MARGIN AND TRENCH

Tectonics, structure, and sedimentary framework of the Peru-Chile Trench

W. J. Schweller
L. D. Kulm
R. A. Prince
School of Oceanography
Oregon State University
Corvallis, Oregon 97331

ABSTRACT

A comprehensive data set of more than 200 profiles across the Peru-Chile Trench between 4° and 45°S is used to describe the morphology and shallow structure of the trench axis and the downbending oceanic plate just prior to subduction. Five morphotectonic provinces (4°–12°, 12°–17°, 17°–28°, 28°–45°S) show distinct changes in trench depth, axial sediment thickness, oceanic plate fault structures, and dip of the seaward trench slope. In general, the northern and southern regions are characterized by relatively shallow axial depths, moderate to thick trench axis turbidites, and a gently dipping seaward trench slope that exhibits minor normal faults. The deeper central area is almost barren of axial sediments and bends downward more steeply prior to subduction; bending has developed an extensive network of major faults with up to 1,000 m vertical offset on the seaward slope.

Two systems of faulting occur in conjunction with subduction. Bending of the oceanic plate causes extensional stress and brittle failure of the upper oceanic crust, resulting in step faults, grabens, and tilted fault blocks on the seaward trench slope. Extensional faulting begins near the outer edge of the trench and develops progressively toward the trench axis. Basaltic ridges and tilted, uplifted trench fill at several locales along the trench can both be explained by thrust faulting. Compressional stress due to plate convergence occasionally can be transmitted seaward from beneath the continental margin through the oceanic plate, emerging as thrust faults within the oceanic crust near the trench axis. Axial turbidites are commonly tilted landward as they are uplifted, probably as a result of downward curving of the underlying thrust fault. Faulting of the oceanic crust prior to and during subduction may have important implications for evolution of convergent continental margins.

INTRODUCTION

Deep-sea trenches are as vital to the concept of plate tectonics as mid-ocean ridges, yet far less detailed structural analysis has been focused on trenches. Part of our current lack of knowledge can be attributed to the difficulty of obtaining good quality data from an extremely deep, narrow feature that is often buried beneath a thick cover of sediment. In addition, structural features created on the

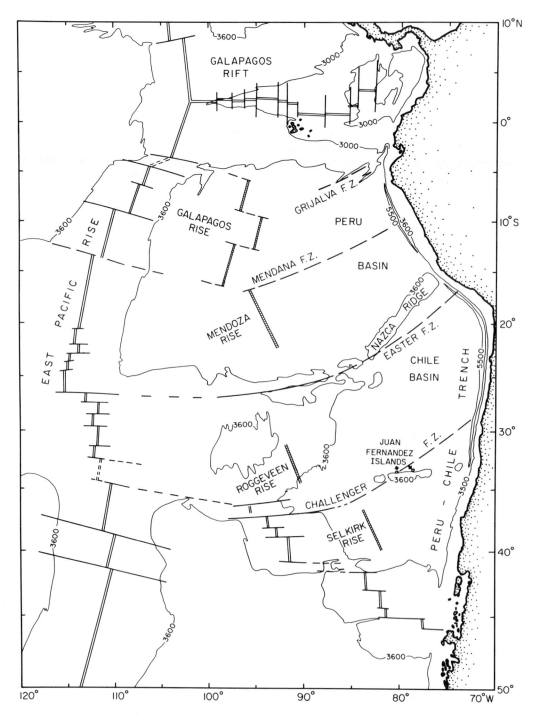

Figure 1. Regional features of the Nazca plate, modified after Mammerickx and others (1980).

oceanic crust by subduction-related processes are only visible for a short time before they disappear beneath the continental margin. Despite these difficulties, an understanding of the mechanisms of subduction of oceanic crust is crucial to deciphering the evolution of active continental margins and perhaps the generation of earthquakes, arc volcanism, and metal ore deposits associated with convergent plate margins.

A major aim of the Nazca Plate Project, conducted jointly by Oregon State University (OSU) and the Hawaii Institute of Geophysics (HIG), has been the integrated study of the structures and sediment of the subducting Nazca plate along the Peru-Chile Trench (Fig. 1). This trench is a classic example of oceanic crust that is underthrusting a continental block in an uncomplicated plate tectonic framework. The convergence rate along the 4,000 km length of the trench is a relatively rapid 9 to 11 cm per year, and the convergence angle is nearly perpendicular to most of the trench (Minster and others, 1974). A well-studied dipping seismic zone extends the description of the underthrusting system beneath the Andes (Stauder, 1973, 1975; Barazangi and Isacks, 1976).

Several models for trench axis faulting have been proposed on the basis of a series of detailed structural studies of limited parts of the Peru-Chile Trench. Prince and Kulm (1975) suggested that normal faults within the descending oceanic plate reversed to thrust faults within the trench axis, resulting in axial ridges between 6° and 10°S. Coulbourn and Moberly (1977) described the oceanic plate seaward of the trench between 17° and 23°S as a composite of faulted blocks with seaward rotation of some blocks near the trench axis. The adjacent area to the south was characterized by Schweller and Kulm (1978b) as having predominantly extensional faulting on the oceanic plate, with pervading horst-graben development. Lister (1971) used an acoustic reflection profile at 33°S as the

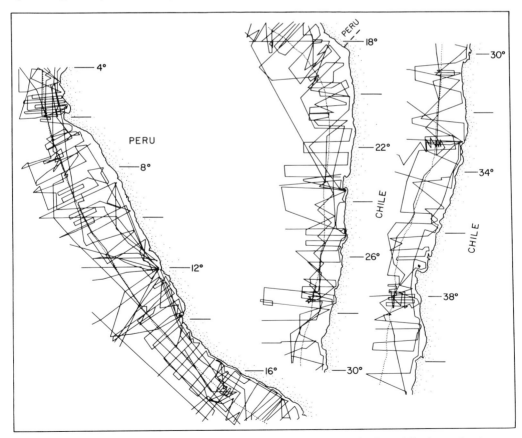

Figure 2. Tracklines along the Peru-Chile Trench and continental margin. Dotted line is trench axis.

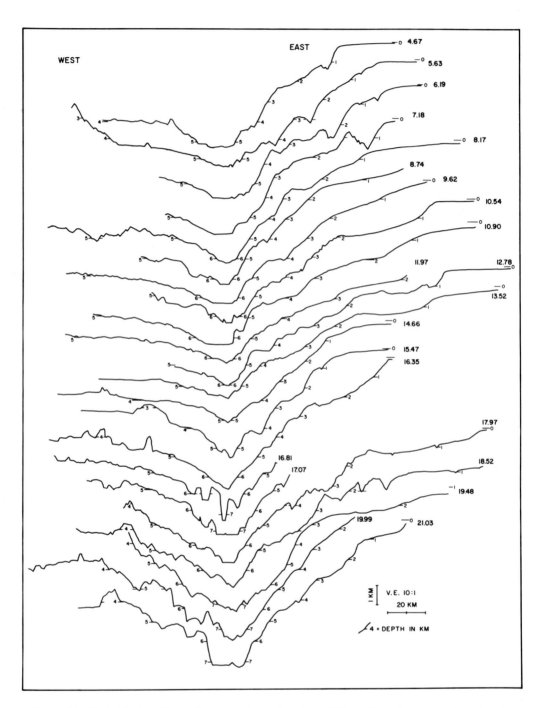

Figure 3A. Selected bathymetric profiles across the northern Peru-Chile continental margin and trench. Index numbers give the latitude of the trench axis.

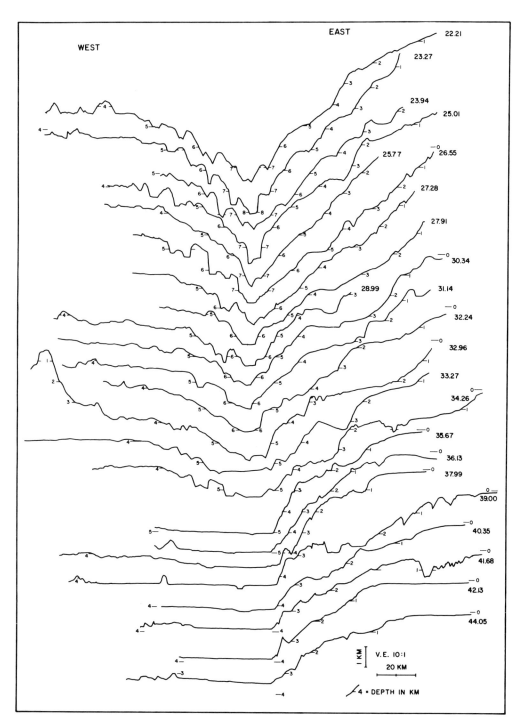

Figure 3B. Selected bathymetric profiles across the northern Peru-Chile continental margin and trench.

basis for a model of vertical shearing of the subducting lithosphere. Scholl and others (1968, 1970) found no evidence of compressional deformation within the trench axis sediments between 23° and 44°S.

The diversity of these models gives the impression of complex and sometimes contradictory mechanisms of faulting along an otherwise continuous feature. In this study, we combine the previously reported evidence and models with a more extensive set of unpublished data to evaluate whether a consistent model can be formulated to explain various types of faulting along the entire Peru-Chile Trench. We then discuss the implications of this faulting on continental margin development.

More than 200 bathymetric profiles across the Peru-Chile Trench, which were positioned by satellite navigation, have been collected in conjunction with the Nazca Plate Project during the past six years (1971–1977, Fig. 2). More than one-half of these profiles include single-channel seismic reflection data. Three multichannel seismic reflection profiles were taken across the Peru Trench at 3°, 9°, and 12°S. An additional set of profiles from the 1967 *Davis* cruise, first described by Scholl and others (1970), was used to augment the trackline coverage in the Chile Trench. This data set provides unprecedented data coverage of the entire length of a major deep-sea trench. Selected profiles from the above data set were computer processed to uniform vertical exaggerations to show the general trends of structure and morphology (Figs. 3A,B). The bathymetric data were used to construct a new set of bathymetric maps of the trench and continental margin from 3°S to 40°S (Prince and others, 1980).

TECTONIC ELEMENTS OF THE PERU-CHILE TRENCH

Like most oceanic trenches, the Peru-Chile Trench resembles a giant check mark ($\swarrow$) in cross section, with the steeper side representing the continental or overthrust block. Four basic morphologic units can be used to describe the trench: the trench axis, the downbending oceanic plate, here termed the seaward trench slope, the outer bathymetric high, and the continental slope.

Trench Axis

The Peru-Chile Trench is 2 to 3 km shallower than most western Pacific island arc trenches (Fisher and Hess, 1963, their Table 1). Peru-Chile Trench axis depths are shown in Figure 4, which was compiled from about 250 complete crossings of the trench axis. Overall, the trench is relatively shallow off Peru and southern Chile, with axial depths of 6.5 km or less. The deeper central province off northern Chile, with axis depths of 7 to 8 km, is bounded to the north by the Nazca Ridge and to the south by a 1 km scarp near 28°S.

Where the aseismic Nazca Ridge intersects the trench at 15°S (Figs. 1,4), the trench axis shoals 2 km. This shoaling is about equal to the relief of the Nazca Ridge away from the trench. However, the shoaling in the trench is a broader feature than the Nazca Ridge further seaward on the Nazca plate (Fig. 4). Unlike many trench-aseismic ridge junctures in the western Pacific, the intersection of the Nazca Ridge with South America does not perceptibly alter the arcuate planimetric curvature of the Peru Trench (Vogt and others, 1976). However, there is a marked steepening of the continental slope opposite the ridge (Thornburg and Kulm, this volume).

Seaward Trench Slope

The seaward trench slope is steepest where the trench depths are maximum. Off Peru and central Chile south of 28°S, the seaward slope dips between 2° and 3° near the trench axis, while off northern Chile between 18° and 28°S, dip angles of 5° to 8° are typical (Fig. 3). The width of the seaward slope is about 50 to 80 km in most areas. Surprisingly, this width does not significantly increase as the trench axis depths increase. Instead, the apparent flexure of the oceanic plate becomes sharper along the deeper sections of the trench. Most of the bending seems to occur along a narrow zone near the outer or seaward edge of the trench, and below 5.5 km there is little increase in the dip of the oceanic plate (Fig. 3, 18°–27°S).

In order to qualitatively evaluate whether significant differences occur in the curvature of the downbending oceanic plate, all available bathymetric profiles were grouped in sections at intervals of two degrees of latitude (approximately 220 km wide). Each two-degree set of profiles was visually averaged and smoothed to a single curve in order to reduce the effects of local topographic anomalies such as seamounts. These smoothed curves were then compared for similarity. Four distinct areas of similar curvature were defined by this analysis (Fig. 5). Although the method is subjective, the differences in curvature between areas are larger than the uncertainties in the curve-smoothing technique. These variations in curvature generally coincide with major and abrupt changes in the depth of the trench (compare Fig. 5 with Fig. 4).

Outer Bathymetric High

The outer bathymetric high or outer rise commonly associated with oceanic trenches (Parsons and Molnar, 1976; Caldwell and others, 1976) is poorly developed along most of the Peru-Chile Trench. In general, long profiles across the trench and the adjacent ocean floor show a much lower outer rise than is present along most island arc trenches in the western Pacific (Caldwell and others, 1976). Off Peru, the elevation above the regional depth averages 200 to 300 m, whereas off Chile, there is often no significant rise adjacent to the trench (Fig. 4). One notable exception is the area of the Chile bight near 20°S, where the trench has a sharp concave-seaward curvature (Fig. 1). At this locale, a rise of nearly 1,000 m occurs just seaward of the trench. The associated outer gravity high of +80 mgal (Getts, 1975) is considerably greater than the maximum anomalies of 50 to 60 mgal found along other Pacific trenches by Watts and Talwani (1974). The large outer rise in the Chile bight may be related to stresses developed in conjunction with subduction along this unusual concave trench curvature.

Continental Slope

The continental slope is consistently steeper than the seaward trench slope in most areas, ranging

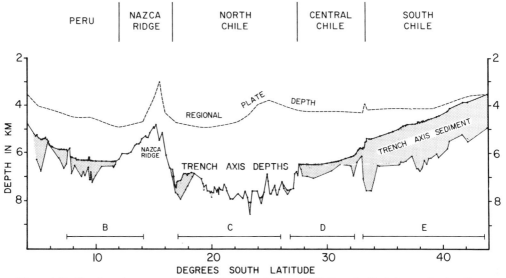

Figure 4. Profile of maximum trench axis depths along the Peru-Chile Trench. Shaded area of axis sediments does not include oceanic plate sediments. Regional plate depth at 300 km from trench axis, from Mammerickx and others (1978). Names at top refer to trench morphotectonic provinces. Bars (B-E) at bottom refer to seismic zone segments of Barazangi and Isacks (1976).

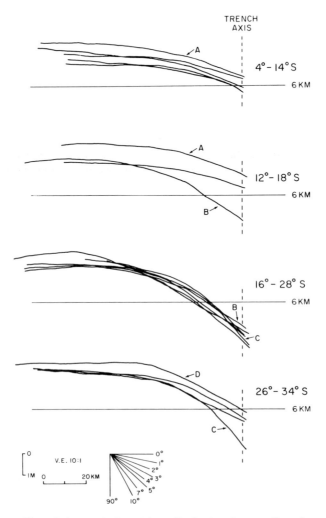

Figure 5. Average bathymetric profiles for two-degree sections along the Peru-Chile Trench, grouped into regions of similar bending curvature. Lettered profiles are transitional sections between regions: A, 12–14°S; B, 16–18°S; C, 26–28°S; D, 32–34°S.

between 5° and 10° on the lower 2 to 3 km above the trench axis (Fig. 3). Numerous structural benches occur along the lower slope at various levels and extend laterally 10 to 30 km. Single-channel seismic reflection records reveal virtually no internal structures within the highly diffracting subsurface of the lower slope. Interpretations of structures resolved by the three multichannel seismic reflection lines are discussed in other parts of this volume (Kulm and others, this volume; Shepherd and Moberly, this volume).

The slope is dissected by a number of submarine canyons north of 7° and south of 33°S adjacent to regions of high rainfall on the continent (GSA map series; Prince and others, 1980; Galli-Olivier, 1969). Offshore of the coastal deserts between about 10° and 30°S there are neither canyons nor thick accumulations of sediments on the lower slope, although upper slope sedimentary basins are present in some areas (Kulm and others, this volume; Coulbourn and Moberly, 1977; Schweller and Kulm, 1978b). This coincides with a general paucity of trench sediment adjacent to the arid regions.

RELATION OF STRUCTURAL PROVINCES TO TRENCH SEDIMENT DISTRIBUTION

The features and structural trends outlined above can be used to subdivide the trench into five major morphotectonic provinces (Fig. 4): Peru (4°–12°S), Nazca Ridge (12°–17°S), North Chile (17°–28°S), Central Chile (28°–33°S), and South Chile (33°–44°S). The southern one third of Chile (44°–55°S) does not border on the Nazca plate and is not included in this classification. The boundaries between these provinces vary in their widths, and the divisions are somewhat arbitrary, because no single criterion defines all of the transitions. Some boundaries, such as the Peru-Nazca Ridge transition, are gradual structural or morphologic changes that extend over a degree or more of latitude. Others, such as the North Chile–Central Chile break near 27°30′S (Fig. 4), are sharply defined over a few tens of kilometres by abrupt jumps in the trench axis depth and the curvature of the seaward trench slope.

These structural changes strongly influence the patterns of sediment distribution along the trench. Where possible, maximum sediment thicknesses in the trench were measured to the nearest one tenth of a second of two-way travel time from seismic reflection profiles. Travel times were converted to thicknesses in metres using sonobuoy refraction velocity data from the Aleutian Trench (Hamilton and others, 1974), because no sediment velocity data are available for the Peru-Chile Trench. Because both trenches contain rapidly deposited, land-derived turbidites, the sediment interval velocities should be roughly comparable.

The wide range in axial sediment thickness along the trench reflects the variability of rainfall on the adjacent landmass, with some modification by transport processes along the trench axis (Fig. 4). Extremely wet climates near the equator and south of 40°S provide great amounts of sediment to the continental shelf and slope and eventually to the trench via turbidity flows and hemipelagic sedimentation (Figs. 4,6). In contrast, the coastal regions of northern Chile are one of the world's most arid deserts, with no permanent streams reaching the shoreline between 18° and 30°S. In addition, much of the sediment from erosion of the Andes is trapped by interior drainage patterns; hence little material reaches the shoreline, and even less is supplied to the trench (Galli-Olivier, 1969).

North of the Nazca Ridge in the Peru province, trench axis turbidites supplied via submarine canyons on the continental slope (Prince and others, 1974) form a series of terraces or level basins ponded behind basement ridges (Fig. 4). This distribution suggests a source area to the north. In seismic reflection profiles, these sediments form horizontally layered, asymmetric basins 2 to 10 km wide and generally a few hundred metres thick (Fig. 4). Piston cores recovered from these basins contain predominantly silt and sand turbidites interlayered with mud.

No ponded axial sediments are observed anywhere in the Nazca Ridge province of the trench (12°–17°S, Fig. 4). The trench axis slopes northward relatively steeply, blocking further southward transport in the axis from the Peru province. Climatic conditions onshore are more arid than to the north so that little sediment is available for downslope transport. In addition, there is minimal structural relief on the basement flooring the trench to provide traps for any turbidity flows coming off the adjacent margin.

In the North Chile province (17°–28°S), the trench consists of a series of saddles and basins with 200 to 800 metres of structural relief on the oceanic basement (Figs. 4,6). Most of the small basins contain up to a few hundred metres of horizontally layered turbidites; the saddle areas lack any ponded sediment and are mantled by a thin covering of oceanic plate sediments, generally only about 100 m thick.

A single Kasten core in the axial basin at 23° recovered a series of silt and silty sand turbidites overlain by 50 cm of hemipelagic mud, suggesting an intermittent supply of sediment from the continental margin. None of the other basins have been sampled to verify the origin of their deposits.

Between 28° and 33°S, the trench is partially filled by a long, continuous basin underlain by a few hundred metres of sediment that abruptly thicken to considerably more than 1 km thick south of 33°S (Fig. 4). The basin in the Central Chile trench segment is comparable in size to the Peru trench segment basins and contains horizontally layered sediment less than 500 m thick and a few kilometres wide (Fig. 6). Unlike the level sediment ponds of the Peru province, the trench sediments between 28° and

44°S have a uniform northward gradient of about 1:650 except for a steep incline at about 33°S (Fig. 4). A deep-sea channel, first described by von Huene (1974) and studied in detail by Schweller and Kulm (1978a), begins south of 40°S and extends northward into the Central Chile province. Channel levees can be seen on the axis basin between 33° and 30°S but are absent north of 30°S (Fig. 3B). The distribution of the axial sediment can be attributed largely to northward transport of turbidites along the axis from the high sediment input region south of 40°S (Schweller and Kulm, 1978a).

Lister (1971) described the areas at 33°S as a transition between a sediment-filled trench to the south and an empty trench to the north. Figures 4 and 6 demonstrate that the axis sediments do not terminate at 33°S but are much reduced in thickness and width north of this point to 28°S. A topographic constriction of the trench axis, perhaps related to the intersection of the Juan Fernadez Island lineation with the trench (Fig. 1), apparently restricts the northward transport of sediment via turbidity current flows. The axial deep-sea channel becomes deeply incised into the trench wedge within the area of steepened gradient around 33°, suggesting erosion within the constricted area (Lister, 1971; Schweller and Kulm, 1978a). The channel resumes its normal depositional character northward to 28°S (Fig. 3B).

A large, wedge-shaped deposit of sediment up to 2 km thick and 75 km wide nearly fills the trench south of 33° (Figs. 3,4,6,7,8). Seismic reflection profiles show horizontal layering at the surface and an increasing landward dip at depth within this wedge, as well as scattered faults and channel structures (Fig. 7). Under the present convergence rates, 9–11 cm/yr (Minster and others, 1974), all of the axis

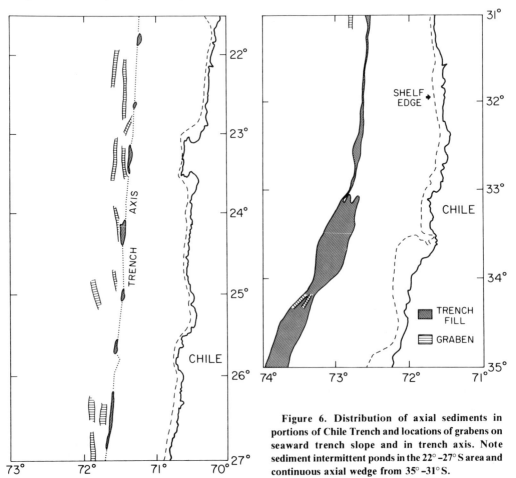

Figure 6. Distribution of axial sediments in portions of Chile Trench and locations of grabens on seaward trench slope and in trench axis. Note sediment intermittent ponds in the 22°–27°S area and continuous axial wedge from 35°–31°S.

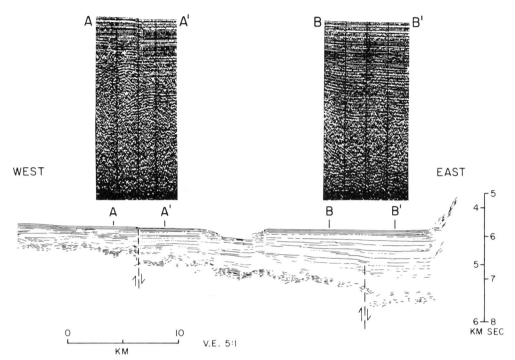

Figure 7. Seismic reflection profile across the Chile Trench at 39°S, demonstrating the progressive downward increase in offset of sediment layers above step faults in the oceanic basement. Section A-A' shows a recent 10 m offset in the surface layer, while section B-B' is apparently inactive. A large deep-sea channel cuts the turbidites in the center of the profile.

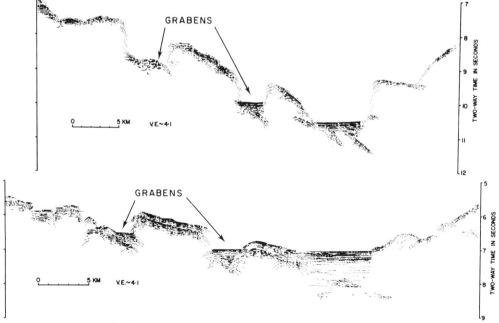

Figure 8. Line tracings of seismic reflection profiles across the Chile Trench at 23.27° (top) and 34.26°S (bottom). Note the similarity of graben development despite large differences in trench depth and axial sediment thickness.

sediments now in the trench wedge would have been deposited within the past one million years (mid–late Pleistocene). Dividing this age into the maximum axial sediment thicknesses of 1.5–2 km in the South Chile province (Fig. 4) yields average sedimentation rates of 1,500 to 2,000 m/m.y. These rates are comparable to turbidite sedimentation rates given for the eastern Aleutian Trench (Kulm and others, 1973a).

Pelagic and hemipelagic sediments on the oceanic crust seaward of the trench are generally thin, ranging from about 200 m off Peru and central Chile to only about 100 m off northern Chile. The 40- to 50-m.y. age of the oceanic crust near the trench (Mammerickx and others, 1980) indicates that sedimentation rates are low—about 2-5 m/m.y. In constrast, the Nazca Ridge is blanketed with a thicker pelagic sediment cover due to its elevation above the regional calcite compensation depth (Kulm and others, 1974; Rosato and others, 1975). The oceanic pelagic sediment cover also thickens near the equator because of increased input from the continent and higher biologic productivity along the eastern equatorial divergence zone. The thicker plate sediment cover south of 40°S is probably due to increased hemipelagic sedimentation rates and some continentally derived turbidites that spill over the filled trench onto the adjacent ocean floor.

EVIDENCE FOR FAULTING WITHIN THE TRENCH

Extensional Fault Examples

The seaward slope of the Peru-Chile Trench exhibits much more structural relief than the adjacent flat ocean floor, suggesting that the slope has been extensively faulted. Several types of structures described in previous studies of the seaward slope and axis of the Peru-Chile Trench have been attributed to faulting within the oceanic crust just prior to subduction. For example, Prince and Kulm (1975) show numerous steplike offsets in the oceanic basement and overlying sediment between 6° and 10°S (Fig. 9), which they interpret as normal faults due to flexure of the oceanic crust. Ridges in the trench axis (Fig 9, profiles 9.01, 9.16, and 9.47) are attributed to thrust faulting from compression near the trench axis. Coulbourn and Moberly (1977) also found a network of step faults on the seaward trench slope in the North Chile province between 18° and 23°S. In addition, they noted a seaward rotation of some large blocks in the Arica bight area (19°–20°S, Fig. 10) and a gradual transition to horst-graben structures toward the south. Schweller and Kulm (1978b) recognized predominant horst-graben faulting in the trench between 23° and 34°S and noted that step faults are slightly more common south of 28°S.

A review of seismic reflection data from other parts of the Peru-Chile Trench shows that step faults are found in almost all sections, including the relatively shallow area in the Peru province north of 6°S (Shepherd and Moberly, this volume) and the sediment-filled South Chile province from 33° to 44°S (Figs. 7,11). Step faulting on the seaward trench slope is easily recognized in the Nazca Ridge province (12°–17°S), where the thick cover of pelagic sediment clearly records recent offsets (Fig. 12). The offset on a single step fault ranges up to a kilometre in the North Chile province, but elsewhere, the average offset is a hundred metres or less.

Grabens are another prominent structural element on the seaward trench slope and consist of a downdropped block between a pair of inward-facing normal faults. Examples of grabens within oceanic trenches were first recognized by Ludwig and others (1966) as being in the Japan Trench and were described by Schweller and Kulm (1978b) as occurring in the Chile Trench. Grabens are not as ubiquitous as step faults along the Peru-Chile Trench but are limited mainly to the North Chile province (17°–28°S) where the oceanic plate bends downward sharply (Fig. 6). The seaward slope in this region is commonly broken by one or two major grabens, and vertical fault offsets often exceed 500 m (Figs. 3,13). Unlike step faults, some of the major grabens can be correlated between adjacent profiles with apparent continuities of several tens of kilometres subparallel to the trench axis (Fig. 6).

Only a few examples of grabens can be found in other provinces along the trench. Notable among these are a shallow graben in a region of predominant step faulting (13.52°, Fig. 13) and a pair of large grabens partially covered by axial turbidites at 34.26°S (Fig. 8). The latter occurrence demonstrates

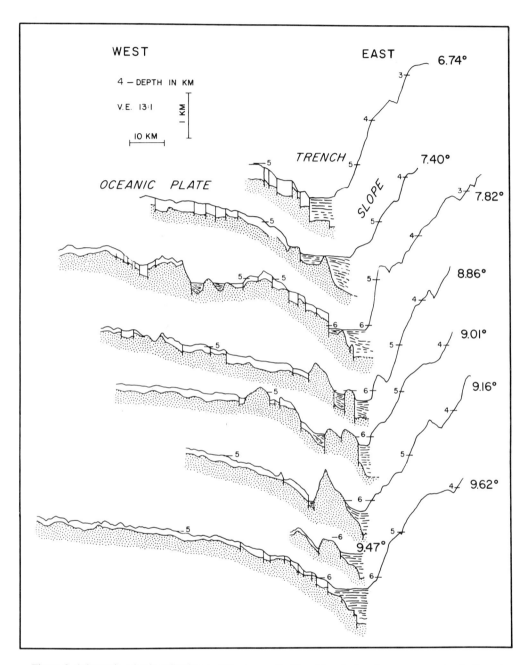

Figure 9. Schematic seismic reflection profiles across the Peru Trench showing numerous step faults and a basement ridge at 9.16° S (after Prince and Kulm, 1975, their Fig. 4).

that the processes which generate grabens are independent of the presence or absence of axial sediment. A comparison of the profile at 34°S with a profile at 23°S, 1,200 km to the north in the deepest part of the trench, shows a remarkable similarity of graben features, despite radical differences in the depth of the trench and the amount of axial sediment (Fig. 8).

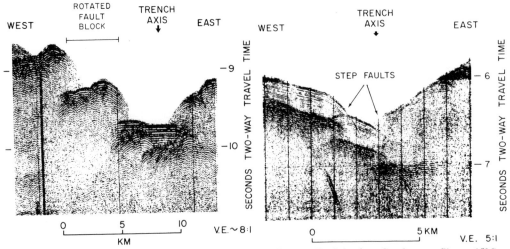

Figure 10. Seismic reflection profile across the trench axis near 19°S showing rotated fault blocks on the seaward trench slope and normal faulting of trench axis sediments.

Figure 12. Seismic reflection profile at 15°S showing two step faults on the oceanic plate near the trench axis. The apparent decrease in dip of the oceanic plate beneath the continental slope is an artifact of the increased sound velocity in the overlying edge of the continental margin.

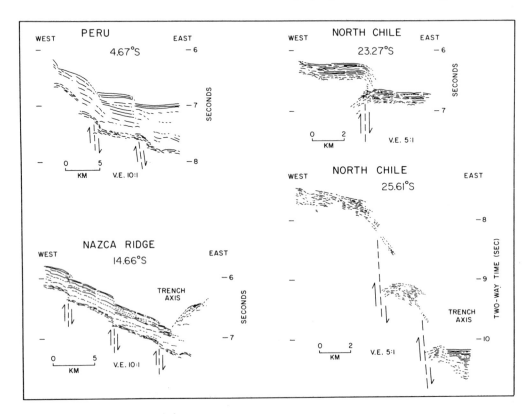

Figure 11. Examples of step faulting at various sites on the seaward slope of the Peru-Chile Trench.

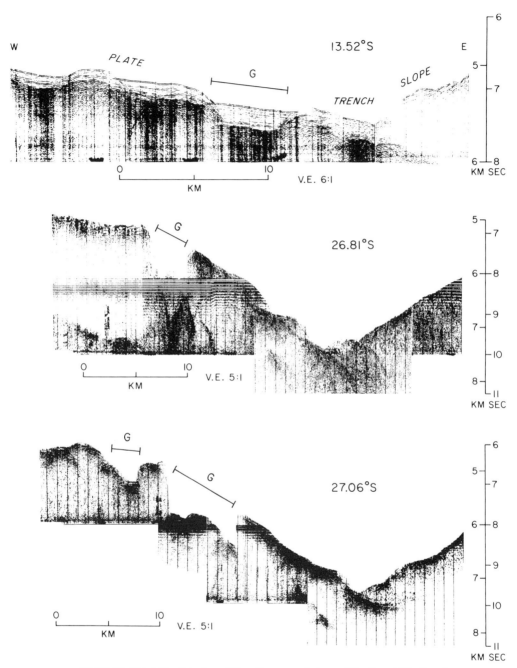

Figure 13. Examples of grabens (G) on the seaward slope of the Peru-Chile Trench.

Thrust Fault and Sediment Uplift Examples

The convergence of plates at trenches causes compression and thrust faulting along the zone of contact between the top of the oceanic plate and the base of the continental block (Isacks and others, 1968). Low-angle thrusting beneath the South American continental margin is abundantly demonstrated by earthquake focal mechanism solutions (Stauder, 1973; 1975). Most shallow thrust mechanisms occur at depths of 30 km or more beneath the South American continental margin, although the complex velocity structure of this margin increases the uncertainty of the locations.

If extensional faulting dominates on the seaward trench slope, as proposed in this study, and if the interior portion of the overriding edge of the continental margin is under compression, then the trench axis becomes a middle area where either type of deformation might be expected. Although step faults persist into the axis in some areas (for example, Fig. 12), evidence from other locales along the Peru-Chile Trench suggests that thrust faults occur slightly seaward of the base of the continental slope. Features that indicate compressional deformation of oceanic crust and sediment in and near the trench axis include asymmetrical basalt ridges, faulted and tilted trench-wedge turbidites, and uplifted areas of former trench deposits.

The strongest evidence for thrusting in the Peru-Chile Trench includes a 900-m-high basaltic ridge in the trench axis near 9°S (Fig. 9, 9.16°), initially described by Kulm and others (1973b) and discussed in more detail by Prince and Kulm (1975). Identification of this ridge as a thrust feature is based on the presence of Pleistocene trench axis turbidites on the ridge crest, far above the present trench axis depth. Subsequent radiocarbon dating of turbidites from the top of the ridge indicates at least 700 m of vertical displacement within the past 3,155 years (Prince and Kulm, 1975).

To improve the resolution of this feature, a multichannel seismic reflection profile was run across the ridge and the adjacent continental margin (Kulm and others, this volume). Although the turbidites cored on the ridge crest are not acoustically resolvable (Fig. 14B), landward-tilted turbidites in the trench extend up the eastern flank of the ridge (Figs. 14B,C,D). The longitudinal axis profile (Fig. 4) shows a small area of trench fill next to the ridge that is raised over 100 m above the level of the axis to the north and south. Prior to uplift, these axis sediments may have been connected to the turbidites on the ridge crest and to the small sediment pond west of the ridge (Fig. 14A) as part of a former continuous trench wedge. We infer that faulting since 3,155 yr B.P. has uplifted and tilted some sections of this basin and destroyed the acoustic continuity between the resolvable sections.

A strong reflector can be seen dipping relatively steeply beneath the seaward edge of the basalt ridge (Fig. 14E). Time migration of this basement reflector confirms that there is a significant extension of the reflector beneath the ridge (Kulm and others, this volume). This reflector is assumed to be the seaward end of a thrust fault along which uplift occurred.

Several smaller but analogous ridges associated with apparent thrust faults have been identified on single-channel seismic reflection profiles from the North Chile province (Fig. 15). The slightly different ridge configurations present in these separate locales may represent different stages of thrust ridge formation and incorporation into the continental slope. We cannot prove that the observed seismic reflection horizons are thrust faults rather than strong hyperbolic tails on these single-channel records. Our interpretations are based on the asymmetric surface forms of the ridges and on

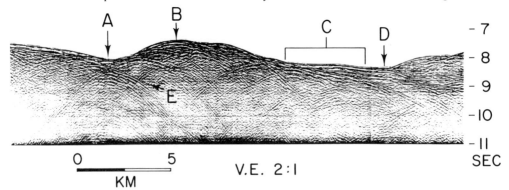

Figure 14. Multichannel seismic reflection profile across the thrust ridge near 9°S (time section, after Kulm and others, this volume). Important features include a small sediment pond seaward of the ridge (A), uplifted turbidites cored on the ridge crest (B), a landward-tilted section of axis turbidites (C), the main trench axis (D), and a strong reflector presumed to be the main fault surface (E).

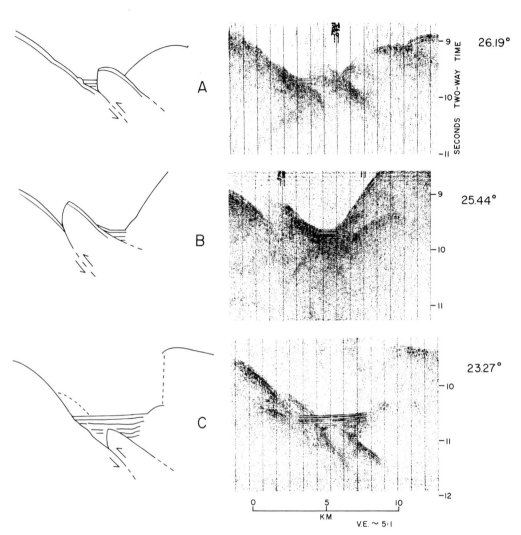

Figure 15. Examples of postulated thrust ridges in the axis of the northern Chile Trench. The continental slope is to the right on all profiles.

comparisons with the multichannel profile described above (Fig. 14). The lack of resolution of the hypothesized thrust plane on these records is not surprising, since a basalt-to-basalt fault contact would not have enough acoustic contrast to produce a strong reflecting horizon. Little of the history of faulting and uplift of these minor ridges can be deduced because of the lack of axial sediments in this part of the trench.

Recent relative uplift of former trench axis turbidites now seaward of and above the present axis has been demonstrated at several sites along the Peru-Chile Trench (Fig. 16). Uplift rates calculated from radiocarbon ages of the latest turbidite range from 3–22 cm/yr for uplifts of as much as 990 m (Prince and Schweller, 1978). Steeply dipping reverse faults (up to 66°) are required to produce these rapid vertical rates under conditions of steady-state convergence. Unfortunately, none of these sites have enough lateral continuity of the sediment layers to provide a complete structural record of the mechanisms of uplift.

The trench in the North Chile province contains a relatively thick section of horizontally layered turbidites with a slight seaward tilt (Fig. 17). A seismic reflection profile at 17.67°S (Fig. 17B) shows an

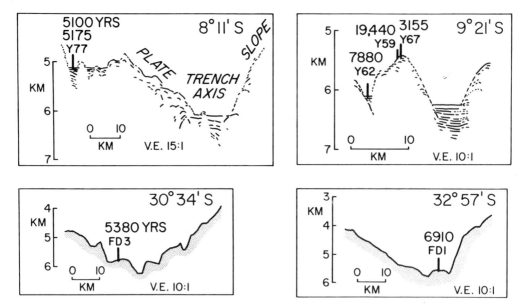

Figure 16. Sites of dated uplift of trench axis turbidites along the Peru-Chile Trench (from Prince and Schweller, 1978). The age given (upper bold number) is for the last turbidite deposited prior to uplift. The smaller letters and vertical bar indicate piston core position. The continental slope is to the right on all profiles.

abrupt 120 m offset of the sediment surface over a basement ridge in the middle of the trench. The central section of the trench fill appears to have been uplifted and tilted landward, while both the seaward and landward edges of the trench axis have remained essentially undisturbed.

An adjacent profile (Fig. 17A) 7 km to the north reveals a slightly different pattern of deformation, including a 60 m offset in the center of the axis but not tilting of the near-surface sedimentary layers landward of the fault. A third profile 11 km to the south of profile B shows no offsets in the axial basin, although there appears to be a basement ridge beneath the axial strata (Fig. 17C). The axial depth across this southern profile (C) is about equal to the depths of the seaward (downthrown) edges of profiles A and B, suggesting that the landward and central sections of the two northern profiles are locally uplifted. Although the age of this faulting is unknown, dates of other uplifted trench sediments suggest that the faulting is probably no older than a few thousand years.

MODELS OF CRUSTAL DEFORMATION IN TRENCHES

Extensional Faulting Model

The types and orientation of faulting at an active trench can be modeled by applying a simple, predictable stress system to a non-homogeneous ocean crust as it is subducted. Tectonic analysis of the East Pacific Rise (EPR) has shown that the newly formed oceanic crust is not a single integral slab but is broken by normal faults on several scales (Rea, 1975). This faulting is caused by extensional stress that predominates parallel to the direction of spreading. Although the offsets of these faults decrease away from the ridge crest, the oceanic crust probably remains a mosaic of competent blocks separated by structurally weak zone (old ridge crest faults) throughout its history.

As the ocenic plate enters the trench, the plate bends approximately as an elastic beam with a load applied at one end (Caldwell and others, 1976). The convex upper surface of the elastic layer of the bending lithosphere is subjected to extensional or tensile stress, while below some depth there is a complimentary zone of compressional stress (Fig. 18). The transition zone between these two regions of opposite stress is termed the neutral surface. Le Pichon and others (1973, p. 224) estimate that for an

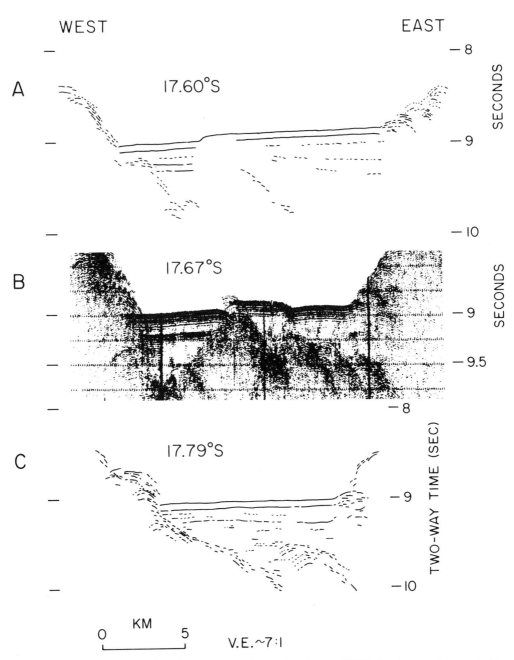

Figure 17. Closely spaced triad of profiles across the trench axis near 17°40'S showing recent reverse faulting (upthrown trench floor to the east) of trench axis turbidites (A,B) adjacent to an unfaulted section of the same axis basin (C). Note the landward tilting of the central uplifted section of profile B.

elastic plate 50 km thick, the strain on the upper surface of a typically bending oceanic plate would range between 1 and 10 parts in 100, corresponding to tensional stress of tens of kilobars. The lesser elastic thicknesses of 20 to 30 km predicted by Calwell and others (1976) would result in somewhat smaller amounts of strain. However, even for the minimum elastic plate thickness, the tensile stress

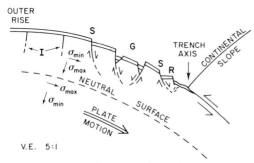

Figure 18. Schematic diagram of various types of extensional faulting related to bending and subduction of an oceanic plate. S = step fault; G = graben; I = inactive fault formed at the spreading center; R = rotated fault block.

produced by bending would far exceed the strength of the oceanic crust. Unfractured basalt fails under tension at less than 1 kb (Handin, 1966), and the fractured pillows, sills, and interlayered breccias of the oceanic crust would probably fail under considerably less tensile stress.

The pattern of focal mechanisms predicted by the bending plate model has been verified by a recent detailed study of earthquake events in the oceanic plate on the seaward side of several trenches in the Pacific and Indian oceans (Chapple and Forsyth, 1979). Normal faulting (extensional mechanisms) predominates down to 25 to 30 km below the ocean floor, while thrust or reverse faulting (compressional mechanisms) occurs at depths of 40 to 50 km. These data were fit to a two-layer bending plate model with a weak upper layer, a strong lower layer, and a total plate thickness of 50 km (Chapple and Forsyth, 1979).

Theoretical orientations of fracturing in response to a range of geologic conditions were predicted by Hafner (1951). For a system of extensional stress similar to that of a bending plate, predicted rupture patterns are normal faults with nearly vertical dips at the surface but gradually curving at depth (Fig. 18). The strike of the theoretical fault system is perpendicular to the axis of least principal stress, which is parallel to the bending axis. Along much of the Peru-Chile Trench, the original ridge crest structures (for example, the magnetic anomaly lineations from the fossil Galapagos Rise [GR]) in the oceanic plate adjacent to the trench are nearly parallel to the axis of bending (Fig. 1; Mammerickx and others, 1980). The favorable orientation of old ridge crest faults to the bending stress field may result in some control of trench faulting by reactivation of these faults, as suggested by Coulbourn and Moberly (1977). However, bending of the seaward trench slope is sufficient to rupture the oceanic crust irregardless of preexisting fault zones.

The evidence presented above shows that normal faulting on the seaward trench slope produces two primary types of structures: step faults and grabens (Fig. 18). Step faults are simple offsets with downward displacements on the landward side of the fault, implying that the fault dips steeply landward. Grabens require a pair of inward-dipping normal faults that may flatten and intersect at depth. These patterns are analogous to well-studies extensional fault systems on land (for example, de Sitter, 1964) and agree with the theoretical studies of Hafner (1951).

In general, both the frequency and the offset of step faults increase along a given profile as the dip of the oceanic plate steepens. This distribution of structures, together with the averaged oceanic plate profiles (Fig. 5), suggests that faulting initiates near the outer edge of the trench and develops as the plate bends into the trench axis. Evidence supporting this trend is contained in the thick sedimentary wedge of the South Chile province, where the downward increase in the offset of sedimentary layers over a basement step fault indicates continued growth of the fault during turbidite deposition (Fig. 7).

Grabens are large and numerous in the North Chile province where the oceanic plate bends sharply downward (Figs. 5,6,8,13). Elsewhere, step faults far outnumber grabens, especially where the plate dip is less than 3°. This distribution suggests that grabens may result from larger total strains and/or more rapid strain rates than are required for step faults. Data from other trenches support this trend. Numerous large grabens dissect the steeply dipping seaward trench slope in the southern part of the Japan Trench, while deformation on the less steep northern portion consists mainly of small offsets (Ludwig and others, 1966).

Two types of secondary faulting features associated with these major extensional fault structures in the Peru-Chile Trench can also be explained by downward-curving faults. Rotated step fault blocks, first described by Coulbourn and Moberly (1977), are best developed in the Chile Trench near 20°S, where the entire seaward trench slope is extensively disrupted (Fig. 10). Rotation of fault blocks requires a downward flattening of the fault and is enhanced by incompetent materials at the base of the blocks (de Sitter, 1964; Fig. 18, position marked R). The floors of some large grabens in the Central Chile province have an arched character that appears to have resulted from collapse at the base of the graben walls (Fig. 13, 26.81S, 27.06S). This collapse can be attributed to antithetic normal faulting within the hanging wall, which allows downward adjustment into the void space created by vertical movement along a curving main fault (Fig. 18, position marked G).

Both antithetic faulting and rotating blocks are common accessory features in major extensional faulting systems on land and reinforce the interpretation of prevalent extensional faulting on the seaward trench slope. De Sitter (1964, p. 127) concludes that "the forming of antithetic blocks and of tilted step blocks are alternative solutions of the same problem of tectonic forces: both indicate a flattening of the hade of the main fault with increasing depth."

Thrust Faulting and Uplift Model

The compressive stress that arises from plate convergence is normally relieved along thrust faults beneath the continental margin either as continuous creep or as sudden rupture events (Isacks and others, 1968) (Fig. 19A). In order to produce thrust faulting seaward of the base of the continental slope, there must be some means of transmitting this compressional stress seaward through the oceanic crust into the trench axis region. Although both aseismic creep and sudden failure events may be active simultaneously, the periodic recurrence intervals of major earthquakes along the South American margin are strong evidence that periodic locking of sections of the main thrust zone is a fairly common phenomenon (Kelleher, 1972). During these periods of locking, a buildup of compressive stress occurs in the vicinity of the slip zone, including the upper oceanic plate. When the increasing stress exceeds the strength of the weakest zone within the stress field, the crust ruptures along this zone.

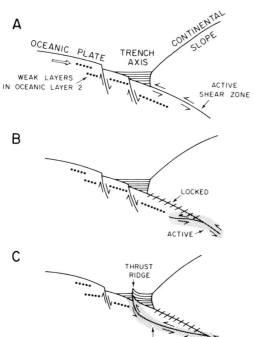

Figure 19. Schematic diagram of the formation of a thrust ridge within the trench axis due to locking of a portion of the main underthrusting surface. Diagram illustrates uplift and landward tilting of axial turbidites as a result of upward curving of the leading edge of a thrust fault in the oceanic basement.

In most cases, the previous thrust fault surface is already weakened and will be the most favorable zone of renewed motion. However, there may be occasions when the main plate thrust surface is not the weakest plane within the stressed zone. Recent drilling of upper oceanic crust by the Deep-Sea Drilling Project (Aumento and others, 1977) has revealed that the upper 500 m of oceanic crust are not as internally competent as previously thought. Rather, this segment of oceanic crust has numerous structurally weak layers of basalt breccia and/or sediment interlayered between basalt flows. Dredging of several fault scarps 500 m to 1,100 m high within the Peru-Chile Trench yielded only pillow basalts and volcanic breccia with no coarse-grained rocks (Scheidegger and others, 1978), indicating a relatively weak crust composed of many separate flow units. Such lithologies are characteristic of the upper few hundred metres of crustal layer 2. If there is less resistance to renewed motion along one of the internal weak layers than along the locked primary thrust plane, the active thrust surface may branch off the preexisting fault zone and shift downward to a more easily fractured horizon (Fig. 19B). The upper section of oceanic crust above the new fault surface would become detached from the rest of the oceanic plate and fixed with respect to the continental margin.

After the initial branching, the new thrust propagates seaward through the oceanic crust and eventually surfaces through a vertically oriented weak zone (Fig. 19C). If the thrust fault intersets the surface seaward of the base of the continental slope along a preexisting normal fault, the motion along this fault will reverse, creating a basaltic ridge or offsetting axial sediments.

The detailed pattern of sediment uplift and tilting due to reverse faulting can be used to infer some of the geometry of the underlying fault surface. As noted by Prince and Schweller (1978, their Fig. 3), the rate of uplift on the hanging wall of a reverse fault under condititions of uniform horizontal convergence is proportional to the dip of the underlying fault surface. If the dip of the fault changes laterally, the overlying areas can be uplifted in different amounts by the same horizontal motion. The landward tilting of uplifted axis turbidites shown in Figure 14 indicates that the underlying fault dips more steeply near the seaward edge. This pattern, illustrated in Figure 19C, is similar to the geometry of thrust sheets on land (Hills, 1963, p. 197). Uplift of the landward edge of the trench fill would be negligible because of the nearly horizontal dip of the thrust surface away from the leading edge (Fig. 19C). This mechanism enables the central part of the trench fill to be elevated as an isolated ridge above the regional level of the rest of the trench fill.

DISCUSSION

Causes of Changes in Oceanic Plate Bending

The regional trends in the depth of oceanic basement beneath the trench axis appear to correlate with the degree of bending of the oceanic plate on the seaward trench slope (Fig. 3). The north-south changes in the bending curvature are fairly distinct and are confined to transition zones 100–200 km wide at trench province boundaries; the plate curvature is relatively constant within each province (Fig. 5). These transition zones correspond well with seaward extensions of the changes in dip of the inclined seismic zone as described by Barazangi and Isacks (1976) (Fig. 4).

The north-south changes in bending curvature of the seaward trench slope do not follow the age relationship proposed by Caldwell and Turcotte (1978) for western Pacific trenches. In their model, the effective elastic thickness of the oceanic plate increases with age, making the plate more resistant to bending. The sharpest bending along the Peru-Chile Trench is off northern Chile, where the Nazca plate is oldest and should be most resistant to bending.

Part of this paradox may be inherent in the degree of faulting on the seaward trench slope. As noted by Le Pichon and others (1973, p. 224), fracturing of the upper part of an elastic slab will decrease the effective elastic thickness of the slab and allow the slab to flex more sharply. The pervasive faulting of the upper several kilometres of oceanic crust in the North Chile province could decrease the elastic thickness of the lithosphere and might explain why the oldest sections of the plate bend more sharply than the younger regions.

However, extensive faulting will not occur unless an initial force causes the downbending. A more fundamental control on the bending of the plate and the depth of the trench may be the thickness of the Andean continental margin. Several studies have shown that a very deep continental root, up to 300 km, may exist beneath northern Chile, with decreasing thicknesses to the north and south (James, 1971, 1978; Sacks and Snoke, 1977). While the implications of the geophysical data are presently a matter of controversy, there is reason to believe that the shape of the overriding plate may partially control the geometry of the subducting plate.

Depth of Extensional Faulting

Extensional faulting should not extend completely down to the neutral surface (Fig. 18) but rather to some depth above it where the material strength of the lithosphere is sufficient to accommodate the tensile bending stresses by elastic deformation rather than rupture. Caldwell and others (1976) calculated an effective elastic thickness for bending oceanic lithosphere of 20–30 km. Assuming that the neutral surface is midway within this slab, the extensional zone would extend between 10 and 15 km down into the oceanic crust, and the depth of faulting would be somewhat less. Recent focal mechanism studies in the Middle Americas Arc (Dean and Drake, 1978) show normal fault solutions near the trench axis at depths of 18–20 km below sea level, which correspond to the upper 10–15 km of the oceanic crust. A slightly shallower estimate for the depth of faulting of 4–7 km was derived from structural considerations of probable fault angles and the consistent widths of downdropped graben floors in the Chile Trench (Schweller and Kulm, 1978B). This depth approximately coincides with the 5–7 km crustal thickness to the mantle on the eastern Nazca plate as determined by the refraction study of Hussong and others (1976). Thus, it seems reasonable that extensional faulting may penetrate most or all of the thickness of the oceanic crust within the Peru-Chile Trench. The data of Chapple and Forsyth (1979) from several trenches in the Pacific and Indian oceans indicate normal faulting down as far as 25–30 km below the ocean floor. However, some of these deeper events may not be linked directly to surface faulting features.

The normal faults that we have discussed are surface effects of bending of the oceanic plate as opposed to normal faults that have been postulated as completely penetrating the oceanic lithosphere. Lister (1971) proposed the latter type as the primary mechanism for subduction based on observations of faulting in the Chile Trench. Not only is it mechanically difficult to produce such deep, closely spaced vertical faults through tens of kilometres of lithosphere, but this vertical fault model also fails to account for the commonly observed landward tilting of faulted blocks of oceanic crust (Figs. 11, 13). In contrast, fracturing of a few kilometres of relatively weak crust in response to bending requires far less force and readily explains both the faults and the general curvature of the subducting crust. In addition, the vertical distribution of earthquake focal mechanisms reported by Chapple and Forsyth (1979) precludes normal faults below 30 km within the bending oceanic lithosphere.

Thickness of Thrust Sheets

The thickness of the thrust sheets near the trench axis has not been observed, nor can it be calculated directly from the data in this study. However, basalt pillow lavas were commonly dredged from the axial ridges in the trench (Scheidegger and others, 1978), which suggests a relatively shallow depth of faulting and consequently the occurrence of relatively thin thrust sheets. The model proposed here (Fig. 19) requires weak layers known to occur within oceanic layer 2 but not thought to be present in layer 3. A minimum thickness of at least several tens of metres is probably required for lateral strength and continuity of a few kilometres, while the thickness of oceanic layer 2 is roughly 2 km (Hussong and others, 1976). Thus, indirect estimates of thrust sheet thickness range from about 100 m to 1 or 2 km.

Hussong and others (1975) proposed a deep crustal thrust fault on the basis of a large apparent offset in velocity layers of the Nazca plate 300 km west of the Peru Trench at 12°S. Because of the lack of any surface expression of faulting above the velocity layer offset, we favor the alternate hypothesis that the offset is a relict plate structure unrelated to the subduction process. In any case, the type of deep crustal thrust fault postulated by Hussong and others (1975) does not fit the model for thrust faults within the upper oceanic crust as outlined in this study.

Influence of the Subducting Plate on the Andean Margin

Faulting on the oceanic plate may influence the evolution of the continental margin, particularly if trench fill is absent. The relatively low structural relief of unfaulted oceanic crust can be smoothed over by a fairly thin cover of pelagic and trench axis sediment, allowing accretion of excess sediment on to the leading edge of the continental margin (Fig. 20A). However, if the relief of the oceanic crust is greatly increased by large fault offsets, this rough surface will cause more resistance during subduction. In areas with thick trench fill, large volumes of sediment will be trapped within the structural lows and will be subducted along with the oceanic plate, leaving less material available for accretion (Fig. 20B). In areas without thick sediments, the major scarps may grind directly against the underside of the continental margin, possibly resulting in the breakup and tectonic erosion of material from the underside of the margin (Fig. 20C). Partially consolidated sediment in the accretionary prism would be more susceptible to tectonic erosion than crystalline continental basement at the leading edge of the continental margin.

Shallow focus thrust-mechanism earthquakes along the Chile Trench are generally closer to the trench axis in the highly faulted, sediment-poor regions north of 28°S than in the less-faulted areas with thicker axial sediments south of 30°S (Stauder, 1973). Although more mechanisms, particularly

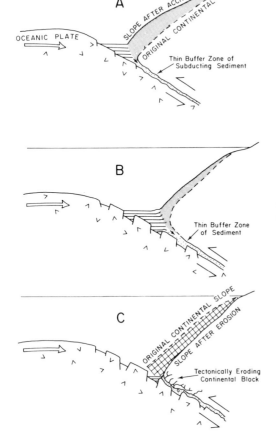

Figure 20. Influence of trench axis sediment and faulting of the oceanic plate on the evolution of the continental margin. See text for discussion.

south of 34°S, are needed to confirm this trend, it would seem that large amounts of trench axis sediment act as a cushion or buffer zone to minimize friction along the upper seismic zone just landward of the trench axis. The increased resistance to underthrusting in highly faulted areas without sediments enhances the possibility of internal shearing of the oceanic plate and the development of axial thrust ridges with an accompanying high level of seismic activity just landward of the trench axis.

Large thrust ridges such as the one at 9°S do not seem to be common features of present trench axes, although evidence for reverse faulting and sediment uplift on a smaller scale is widespread in the Peru-Chile Trench. Either large ridges form very rarely, or they have a fairly short life span before they disappear from the trench axis. Because thrust sheets are generally thin slabs, they would probably be susceptible to breakup and piecemeal removal by continued convergence and under-thrusting. Thrust ridges apparently migrate gradually landward beneath the continental margin over time. Some of the prominent lower slope benches (Fig. 3) may be the elevated and partly eroded remnants of older thrust ridges. If a portion of the thrust sheet remains attached to the underside of the continental margin, subsequent accretion of sediments may permanently incorporate a basalt slab into a mélange similar to the Franciscan complex.

CONCLUSIONS

More than 200 trench crossings were used to construct a detailed longitudinal profile along 4,000 km of the Peru-Chile Trench (Fig. 4). Five morphotectonic provinces 4°–12°, 12°–17°, 17°–28°, 28°–33°, 33°–45°) are defined by major changes in trench depth, axis sediment thickness, and structural features. The trench is generally shallow and partially sediment-filled at the north and south ends and is deeper and generally empty in the central region.

Sedimentary fill in the trench varies from small isolated ponds separated by barren saddles to a thick wedge up to 2 km thick (Fig. 4). The thickness and surface gradients of these turbidites suggest transport of continental sediment southward along the trench from near the equator to about 12°S and northward from 40°S to 27°S.

The depth of oceanic basement in the trench axis varies from 5 to 8 km along the trench. Major changes in depth roughly correlate with boundaries between segments of the inclined seismic zone and with changes in bending of the oceanic plate. The seaward trench slope bends more sharply downward adjacent to the deepest province in the trench.

The seaward trench slope undergoes normal faulting in response to extensional stress created by downbending of the oceanic plate. Two prominent types of features, step faults and grabens, offset the oceanic crust as much as a kilometre. Grabens have floors generally 4–7 km wide and are best developed in the North Chile province (17°–28°S). Step faults are distributed more uniformly along the trench and progressively increase in offset as the oceanic plate descends into the trench.

Compressional stress from plate convergence causes thrust faulting (demonstrated by focal mechanism solutions) beneath the continental slope. Faulted and tilted trench axis turbidites, asymmetrical basalt ridges, and uplifted trench fill indicate that reverse or thrust faulting extends into the trench axis in some locales. Locking of the main interplate thrust zone allows compressional rupture to propagate seaward through weak zones in the upper oceanic crust and to emerge in the trench axis. Downdropped graben floors, rotated fault blocks, and tilted trench axis turbidites all suggest that both normal and reverse faults tend to flatten with depth.

All of the evidence presented here indicates shallow faulting with the Peru-Chile Trench at depths less than 20 km and probably less than 10 km. Normal faults that extend all the way through the oceanic lithosphere (Lister, 1971) or thrust faults that extend through the crust into the mantle (Hussong and others, 1975) are inconsistent with the structural data and with the models presented in this paper for the trench region between the outer bathymetric high and the base of the continental slope.

Extensive faulting of the oceanic plate can influence the evolution of a convergent plate boundary. Large offsets in the oceanic crust entrap more sediment with the subducting plate, leaving less to be accreted. If axial sediments are lacking, the direct abrasion of oceanic crust against the continental slope may gradually erode the edge of the continental margin.

ACKNOWLEDGEMENTS

We thank Roland von Huene and Tom Chase of the U.S. Geological Survey for their generous assistance in providing bathymetric and seismic reflection data from the *Davis* cruise off Chile. W. M. Chapple and D. W. Forsyth provided a preprint of their focal mechanism study. This research was supported by the National Science Foundation, Office of the International Decade of Ocean Exploration Grant Nos. GX 28675, IDOE 71-04208, and OCE 76-05903.

REFERENCES CITED

Aumento, F., and others, 1977, Initial reports of the Deep Sea Drilling Project, Volume 37: Washington, D.C., U.S. Government Printing Office, 1008 p.

Barazangi, M., and Isacks, B. L., 1976, Spatial distribution of earthquakes and subduction of the Nazca plate beneath South America: Geology, v. 4, p. 686–692.

Caldwell, J. G., and others, 1976, On the applicability of a universal elastic trench profile: Earth and Planetary Science Letters, v. 31, p. 239-246.

Caldwell, J. G., and Turcotte, D. L., 1978, Lithosphere elastic thickness versus lithosphere age [abs.]: EOS (American Geophysical Union, Transactions), v. 59, p. 372.

Chapple, W. M., and Forsyth, D. W., 1979, Earthquakes and bending of plates at trenches: Journal of Geophysical Research, v. 84, p. 6729-6749.

Coulbourn, W. T., and Moberly, R., 1977, Structural evidence of the evolution of fore-arc basins off South America: Canadian Journal of Earth Sciences, V. 14, p. 102–116.

Dean, B. W., and Drake, C. L., 1978, Focal mechanism solutions and tectonics of the Middle America Arc: Journal of Geology, v. 86, p. 111-128.

de Sitter, L. U., 1964, Structural geology: New York, McGraw-Hill, 551 p.

Fisher, R. L., and Hess, H. H., 1963, Trenches, *in* Hill, M. N., ed., The sea: New York, Wiley-Interscience, v. 3, p. 411-436.

Galli-Olivier, C., 1969, Climate—a primary control of sedimentation in the Peru-Chile Trench: Geological Society of America Bulletin, v. 80, p. 1849–1852.

Getts, T. R., 1975, Gravity and tectonics of the Peru-Chile Trench and eastern Nazca plate, 0°–33°S [M.S. thesis]: Honolulu, University of Hawaii, 103 p.

Hafner, W., 1951, Stress distributions and faulting: Geological Society of America Bulletin, v. 62, p. 373–398.

Hamilton, E. L., and others, 1974, Sediment velocities from sonobuoys: Bay of Bengal, Bering Sea, Japan Sea, and North Pacific: Journal of Geophysical Research, v. 79, p. 2653–2668.

Handin, J., 1966, Strength and ductility, *in* Clark, S. P., Jr., ed. Handbook of physical constants: Geological Society of America Memoir 97, p. 223–289.

Hills, E. S., 1963, Elements of structural geology: New York, John Wiley & Sons, 483 p.

Hussong, D. M., Odegard, M. E., and Wipperman, L. K., 1975, Compressional faulting of the oceanic crust prior to subduction in the Peru-Chile Trench: Geology, v. 3, p. 601–604.

Hussong, D. M., and others, 1976, Crustal structure of the Peru-Chile Trench: 8°S–12°S latitude, *in* Sutton, G. H., and others, eds., The geophysics of the Pacific Ocean basin and its margins: American Geophysical Union, Geophysical Monograph 19, p. 71–85.

Isacks, B., Oliver, J., and Sykes, L. R., 1968, Seismology and the new global tectonics: Journal of Geophysical Research, v. 73, p. 5855–5899.

James, D. E., 1971, Andean crustal and upper mantle structure: Journal of Geophysical Research, v. 76, p. 3246–3271.

—— 1978, Subduction of the Nazca plate beneath central Peru: Geology, v. 6, p. 174–178.

Kelleher, J. A., 1972, Rupture zones of large South American earthquakes and some predictions: Journal of Geophysical Research, v. 77, p. 2087–2103.

Kulm, L. D., and others, 1973a, Initial reports of the Deep Sea Drilling Project, Volume 18: Washington, D.C., U.S. Government Printing Office, 930 p.

Kulm, L. D., and others, 1973b, Tholeiitic basalt ridge in the Peru Trench: Geology, v. 1, p. 11–14.

Kulm, L. D., and others, 1974, Transfer of Nazca Ridge pelagic sediments to the Peru continental margin: Geological Society of America Bulletin, v. 85, p. 769–780.

Kulm, L. D., and others, 1981, Crustal structure and tectonics of the central Peru continental margin and trench, in Kulm, L. D., and others, eds., Nazca plate: Crustal formation and Andean convergence: Geological Society of America Memoir (this volume).

Le Pichon, X., Francheteau, J., and Bonnin, J., 1973, Plate tectonics: Amsterdam, Elsevier Scientific Publishing Company, 311 p.

Lister, C. R. B., 1971, Tectonic movement in the Chile Trench: Science, v. 173, p. 719–722.

Ludwig, W. J., and others, 1966, Sediments and structure of the Japan Trench: Journal of Geophysical Research, v. 71, p. 2121–2137.

Mammerickx, J., Herron, E., and Dorman, L., 1980, Evidence for two fossil spreading ridges in the southeast Pacific: Geological Society of America Bulletin, v. 91, p. 263–271.

Minster, J. B., and others, 1974, Numerical modeling of instantaneous plate tectonics: Royal Astronomy Society Geophysical Journal, v. 36, p. 541–576.

Parsons, B., and Molnar, P., 1976, The origin of outer topographic rises associated with trenches: Royal Astronomy Society Geophysical Journal, v. 45, p. 707–712.

Prince, R. A., and others, 1974, Uplifted turbidite basins on the seaward wall of the Peru Trench: Geology, v. 2, p. 607–611.

Prince, R. A., and Kulm, L. D., 1975, Crustal rupture and the initiation of imbricate thrusting in the Peru-Chile Trench: Geological Society of America Bulletin, v. 86, p. 1639–1653.

Prince, R. A., and Schweller, W. J., 1978, Dates, rates, and angles of faulting in the Peru-Chile Trench: Nature, v. 271, p. 743–745.

Prince, R. A., and others, 1980, Bathymetry of the Peru-Chile continental margin and trench: Geological Society of America Map and Chart Series, MC-34, scale 1: 1,095,706 at equator.

Rea, D. K., 1975, Model for the formation of topographic features of the East Pacific Rise crest: Geology, v. 3, p. 77–80.

Rosato, V. J., Kulm, L. D., and Derks, P. S., 1975, Surface sediments of the Nazca plate: Pacific Science, v. 29, p. 117–130.

Sacks, I. S., and Snoke, J. A., 1977, The use of converted phases to infer the depth of the lithosphere-asthenosphere boundary beneath South America: Journal of Geophysical Research, v. 82, p. 2011–2017.

Scheidegger, K. F., and others, 1978, Fractionation and mantle heterogeneity in basalts from the Peru-Chile Trench: Earth and Planetary Science Letters, v. 37, p. 409–420.

Scholl, D. W., and others, 1970, Peru-Chile Trench sediments and sea-floor spreading: Geological Society of America Bulletin, v. 81, p. 1339–1360.

Scholl, D. W., von Huene, R., and Ridlon, J. B., 1968, Spreading of the ocean floor—undeformed sediments in the Peru-Chile Trench: Science, v. 159, p. 869–871.

Schweller, W. J., and Kulm, L. D., 1978a, Depositional patterns and channelized sedimentation in active eastern Pacific trenches, in Stanley, D. J., and Kelling, G., eds., Sedimentation in submarine canyons, fans, and trenches: Stroudsburg, Pennsylvania, Dowden, Hutchinson & Ross, p. 311–324.

—— 1978b, Extensional rupture of oceanic crust in the Chile Trench: Marine Geology, v. 28, p. 271–291.

Shepherd, G., and Moberly, R., 1981, Coastal structure of the continental margin, northwest Peru and Southwest Ecuador, in Kulm, L. D., and others, eds., Nazca plate: Crustal formation and Andean convergence: Geological Society of America Memoir 154 (this volume).

Stauder, W., 1973, Mechanism and spatial distribution of Chilean earthquakes with relation to subduction of the oceanic plate: Journal of Geophysical Research, v. 78, p. 5033–5061.

—— 1975, Subduction of the Nazca plate under Peru as evidenced by focal mechanisms and by seismicity: Journal of Geophysical Research, v. 80, p. 1053–1064.

Thornburg, T., and Kulm, L. D., 1981, Sedimentary basins of the Peru continental margin: Structure, stratigraphy, and Cenozoic tectonics from 6°S to 16°S latitude, in Kulm, L. D., and others, eds., Nazca plate: Crustal formation and Andean convergence: Geological Society of America Memoir 154 (this volume).

Vogt, P. R., and others, 1976, Subduction of aseismic oceanic ridges: Effects on shape, seismicity, and other characteristics of consuming plate boundaries: Geological Society of America Special Paper 172, 59 p.

von Huene, R., 1974, Modern trench sediments, in Burk, C. A., and Drake, C. L., eds., The geology of continental margins: New York, Springer-Verlag, p. 207–211.

Watts, A. G., and Talwani, M., 1974, Gravity anomalies seaward of deep-sea trenches and their tectonic implications: Royal Astronomy Society Geophysical Journal, v. 36, p. 57–90.

MANUSCRIPT RECEIVED BY THE SOCIETY NOVEMBER 12, 1980
MANUSCRIPT ACCEPTED DECEMBER 30, 1980

Printed in U.S.A.

Geological Society of America
Memoir 154
1981

Coastal structure of the continental margin, northwest Peru and southwest Ecuador

GLENN L. SHEPHERD*
RALPH MOBERLY
Hawaii Institute of Geophysics
University of Hawaii
Honolulu, Hawaii 96822

ABSTRACT

With the exception of the lower continental slope, or inner wall of the Peru Trench, normal faulting characteristic of extensional tectonics predominates in the shallow structure from the outer rise on the Nazca plate, across the trench, the contenental margin, the Coastal province, and into the High Cordillera province of the Andes of northern Peru. Marine seismic reflection and gravity traverses are used to trace the major basins, horsts, and large faults seaward from where they have been mapped on land and in the subsurface. Although these large-scale structures can be traced geophysically, neither single-channel nor multichannel seismic records reveal the wealth of detail of block faults and olistostrome-like low-angle slides known from subsurface studies of the onshore and offshore oil fields in the fore-arc basin along the coast at Talara, Peru.

The subduction process has been efficient during the Cenozoic at the Peruvian margin. The accretionary wedge under the inner wall of the trench is small compared to the volume of sediment that entered the trench. Moreover, plutons near the coast and the trends that strike seaward in the structures of the igneous and metamorphic basement there indicate that a substantial width of the lower outer edge of the continental crust has been removed. The evidence strongly suggests that the crust has been stoped away by subduction.

Whereas coastal Peru is underlain by a metamorphic, plutonic, and volcanic basement that is termed continental, coastal Ecuador has a belt of ophiolitic rocks regarded as former oceanic crust. The Dolores–Guayaquil Megashear separates the two terranes. It is probably a transform-fault boundary setting aside the ophiolites of coastal Ecuador and Colombia as a mini-plate. The Progreso Basin lies beneath the Gulf of Guayaquil and contains hydrocarbons. It was formed as a pull-apart basin by dextral movements along the Dolores-Guayaquil Megashear where that fault strikes westward into the trench. Banco Peru may be a horst of mafic or ultramafic basement left as a slice when the Progreso Basin was formed.

INTRODUCTION

As part of the Nazca Plate Project, this study attempts to describe the structure of the South

---
*Present address: Cities Service East Asia Inc., Suite 604, Cathay Bldg., Mount Sophia, Singapore 0922.

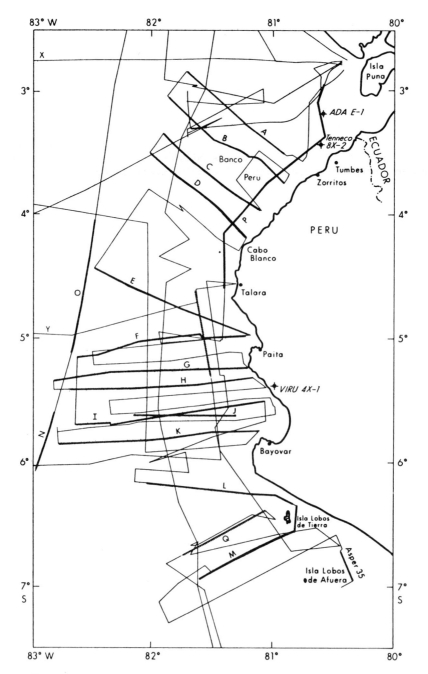

Figure 1. Area of study showing tracks of cruises that provided data. All are HIG lines except J (Seis. Comm. Delta, 1974), X (NOAA, 1970), and Y (OSU, 1972). Seismic reflection profiles of track sections indicated by heavy lines and letters are described or illustrated.

American continental margin between 2°30′ and 7°00′S. We use information provided by seismic reflection and other marine surveys and existing knowledge of surface and subsurface geology, and draw upon analogous studies of other trench margins.

Northwesternmost Peru is a terrane of igneous and metamorphic basement overlain by Cretaceous and Cenozoic sedimentary rocks that are exposed as close as 70 km to the axis of the Peru-Chile Trench. The Nazca plate, of oceanic lithosphere, is being subducted below the South American plate, of continental lithosphere, along the boundary marked by this trench.

To the north, that is, north of the Gulf of Guayaquil, western Ecuador is a terrane of ophiolitic and associated rocks in marked contrast to the Peruvian coast. We also report on the geologic boundary between the margins of the two countries, because oil fields at the coasts and off the shores near this boundary are among the few commercial fields on active trench margins.

METHODS OF STUDY

Data for this study were obtained mainly on the R/V *Kana Keoki* (Hawaii Institute of Geophysics [HIG]) in March and April 1972, February and March 1973, and February and April 1974 and were supplemented by lesser amounts of data collected by R/V *Yaquina* (Oregon State University) and R/V *Oceanographer* (NOAA). Seiscom Delta Inc. recorded a multichannel seismic reflection profile at 5°30′S.

About 7,400 km of single-channel continuous seismic reflection profiles and 8,100 km of bathymetric profiles and seaborne gravity furnish the principal data used (Fig. 1). The reflection system included airgun plus sparker, hydrophone streamers, and recorders. After filtering, generally these three frequency ranges were recorded: 40–80 Hz, 50–200 Hz, and a mixture of 40–80 plus 150–300 Hz. A 3.5 kHz system recorded bathymetry, and gravity data were obtained with a LaCoste-Romberg stable-platform gravimeter. ASPER (Airgun-Sonobuoy Precision Echosounder), piston cores, free-fall cores, and dredging were used occasionally. Magnetic data were collected but not analyzed due to our proximity to the magnetic equator and the electrojet.

In addition to having the geological literature of Peru and Ecuador, we were fortunate that petroleum companies and geologists made available to us unpublished details about the geology.

GEOLOGY OF COASTAL NORTHERN PERU AND ADJACENT SOUTH AMERICA

The geology of the Andes varies from place to place along their length, and published opinions of the geology along the Andean Range also vary. This discussion, however, will focus on northwestern Peru and southern Ecuador, adjacent to the marine survey (Fig. 2). The spatial distribution of rock types and their ages and structural relationships in some areas are poorly known; some areas have yet to be studied even on a reconnaissance basis, while others are known in extreme detail, such as in the oil fields of northwestern Peru and the mining districts of the Cordillera.

The Andean Range separates Peru into three morphotectonic and physiographic regions (Ham and Herrera, 1963): the Coastal (including the submerged continental shelf), the Cordilleran, and the Subandean.

The dry northern coast region is recognized as a massif of low discontinuous hills trending in an arcuate pattern from the Cerros de Amotape, Silla de Paita, and Cerros de Illescas to the offshore islands of Lobos de Tierra and Lobos de Afuera (Fig. 3). This massif exposes the oldest known rocks in northwest Peru. They are composed of more than 4,500 m of predominantly detrital marine Devonian through Pennsylvanian, and possibly Permian, sedimentary and metasedimentary rocks (J. Paredes, 1966, unpub. data; Martinez, 1970; Paz, 1974) and contain turbidite sequences. They lap onto the Precambrian Brazilian shield east of the Andes (Hosmer, 1959) and are unconformably overlain by Triassic-Jurassic and Cretaceous sedimentary rocks, as seen in outcrops and as known from wildcat

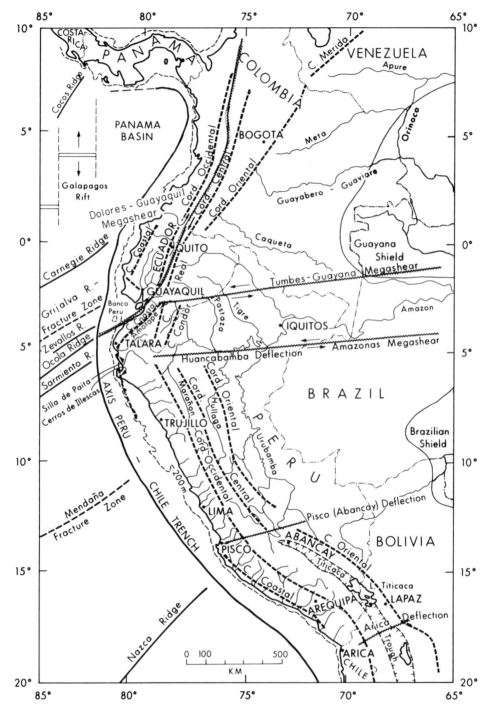

Figure 2. Morphotectonic map of northwestern South America and eastern part of Nazca plate.

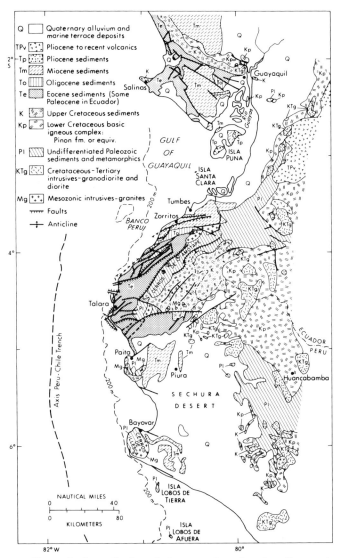

Figure 3. Generalized geologic map of northwestern Peru and southwestern Ecuador. Data from Bellido, E., 1969; Servicio Nacional de Geología y Mineria, 1969, Mapa Geologico do la Republica del Ecuador; Zevallos, O., 1969, Mapa Estructural Generalizado del Noroeste Peru, Primer Congreso Latinoamericano de Geología, Tomo III, November 1970; and unpublished sources.

oil wells. The Paleozoic rocks crop out on uplifts or hills of the northern Coastal Cordillera from southwestern Ecuador, across northwest Peru, and through the islands to the south and were encountered in wells drilled on the continental shelf at 8° S. They show diverse metamorphism ranging from contact to regional. The metamorphic processes have resulted predominantly in phyllites but the rocks also include slates, quartzites, hornfels, and mica schists. Some, however, are only lightly metamorphosed. They are intruded by Jurassic granitic rocks (Bellido, 1969; A. Aleman, pers. comm. 1974) and Late Cretaceous (?) (O. Zevallos, pers. commun., 1974) dioritic rocks that may represent deeper exposures of a former volcanic arc in that region (Fischer, 1956; Morris and Aleman, 1975). The volcanic rocks were a source of thick volcaniclastic flysch sequences east of the coastal massif.

It is inferred that the Paleozoic rocks of northwestern Peru were deposited on Precambrian crust (Harrington, 1975), although there are no exposures within the coastal zone showing such a relationship. About 300 km southeast of Talara along the Rio Marañon in the Cordillera Central, young Precambrian metamorphic and intrusive rocks are unconformably overlain by lower Paleozoic slates (Hosmer, 1959; Bellido, 1969). Supportive evidence for the inference that Precambrian crust underlies coastal northwestern Peru comes from both the south and the north along the structural grain. In the southern Coastal Cordillera, Paleozoic lies on Precambrian, a relationship that presumably is linked as a linear structural element to the massifs of the northern Coastal Cordillera. A similar association is reported in the Amotape-Chanchan Range of southern Ecuador (Campbell, 1975).

Northwestern Peru was a region of thick sedimentation at least to mid-Permian time, when diastrophism, represented by both epeirogenic and orogenic movements, affected the whole of western South America (Fischer, 1956; Paredes, M., 1958; Harrington, 1962; Megard, 1973). Movements culminated in the Middle to Late Jurassic, and the region of the massifs emerged as volcanic island arcs. In central and northern Peru, coeval intrusive diorites and extrusive andesites (Helwig, 1972; Megard, 1973) suggest that active subduction began in mid-Permian time and possibly as early as Late Silurian if granitic intrusives and orogenic events reported by Megard and others (1971) in the central Andes presage subduction.

The discontinuous arcuate low hills composing the massif of coastal northwestern Peru are horsts bounded by faults that trend orthogonally to the coast. They form the cores of the capes and seaward prominences. On the coastal plain and continental shelf, some 10,000 m of marine Cenozoic sediments, deposited in four cycles (Paz, 1974), overlie Late Cretaceous marine flysch that is thin on the massif. The flysch may aggregate 15,000 m (Ham and Herrera, 1963), however, where it filled the grabens flanking the massif. It is notable that the Andean coastal batholith formed largely in the Paleocene and Eocene (Giletti and Day, 1968) while deposition occurred in northwestern Peru.

Thick Triassic-Jurassic volcaniclastic sediments, flows, tuffs, pillow lavas with minor marine limestone, and shale are known east of the coastal massifs, but none is known to the west (Fischer, 1956; Zuñiga and Travis, 1975). Likewise, extremely thick Late Cretaceous marine volcaniclastic flysch occurs east of the massif, but it is absent to the west.

The coastal region underwent a massive taphrogenic breakdown contemporaneously with deposition, affecting Cretaceous and Cenozoic detrital paralic to marine sediments. Faulting began in Late Cretaceous time and has persisted energetically through the Cenozoic to the Quaternary (M. Paredes, 1958). The structure is characterized by exclusively normal block faulting in which folding is absent and in which even minor warping is rare (Travis, 1953), as indicated by subsurface data from more than 9,000 wells in the coastal zone. The master faults cut into basement with as much as 10 km of displacement, as, for example, in the Lagunitos trough (A. Fischer, pers. commun., 1975). Involved with the taphrogeny are gravity slides inclined about 7° to bedding planes (Baldry, 1938; Brown, 1938) and extending more than 10 km. They crop out near the coast and trend offshore. Cenozoic sediments also exhibit rubble zones from mudflows and graded deposits from turbidity currents (Dorreen, 1951).

The taphrogeny also affected the north side of the Gulf of Guayaquil, where normal faults show that the continental margin of that part of South America has been uplifted in an extensional stress-field since at least the Late Cretaceous. The Gulf of Guayaquil, formed as the result of extension of older crustal rocks, is characterized by subsidence and exclusive normal faulting (growth faults). This extension provided the framework for the accumulation of at least 6,000 m of essentially Middle to Late Cenozoic sediments that compose the east-west-trending Progreso Basin and are gently folded within it. The genesis of this region is discussed more fully in the section on the Progreso Basin and the Dolores-Guayaquil Megashear.

The next physiographic province east of the Coastal province is the High Cordillera, and east of that is the Subandean province (Fig. 2). It is not possible to describe the details of their geology in this paper, although any overall analysis of the effects of subduction along western South America must include them, for two reasons. The first is self-evident; they are a part of the arch-typical example of orogeny between oceanic and continental lithosphere ("Andean type"). The highly elevated western part of the High Cordillera province is a Mesozoic foldbelt whose rocks probably represent the remains of an island arc (Hosmer, 1959; Megard, 1973). Its geosynclinal rocks are overlain by Eocene molasse sediments strongly deformed under a cover of Oligocene and Miocene-Pliocene silicic volcanic rocks that are slightly folded and faulted. These terranes are intruded by the great Andean composite batholith, which ranges in age from Late Cretaceous to Late Cenozoic. Summaries of the geology are in Fischer (1956), Hosmer (1959), Cossio and Jaen (1967), Bellido (1969), J. Paredes (1972), Myers (1974), and Shepherd (1979).

A second reason is that any overall analysis must consider whether or not plate convergence here included the accretion of exotic terranes or microcontinents. The concept that plate movement juxtaposes terranes that formed far apart is being thoroughly explored with reference to western North America (for example, summaries in Howell and McDougall, 1978) and Europe (for example, Dewey and others, 1973) and has been used to explain anomalous sedimentary sequences in the Andes of Bolivia (Helwig, 1972) and Chile (Bruhn and Dalziel, 1977). A major conclusion of this present paper is that scant evidence exists along the inner wall of the Peru Trench for any substantial accretion of sediment that may have been scraped off the top of the subducting plate during the major part of the Cenozoic. One possibility is that subduction is an efficient process. An alternative, however, is that one or more small plates consisting of coastal Peru and part of the Andes have collided with South America and that the trench formed in its present position only recently. Detailed studies in the Andes of paleomagnetism, sedimentary facies, and faunal provinces, and searches for sutures are needed to test this alternative.

The area of southern Ecuador and northern Peru is the locus of several regional geologic trends (Fig. 2). There, the Andean ranges rather abruptly change trend from southeasterly to northeasterly. This change is named the Chicama Trend (Jenks, 1956) or the Huancabamba Deflection (Ham and Herrera, 1963). The deflection parallels east-west folds whose axial planes dip southward and align with the Amazonas Megashear (C. Ham, pers. commun., 1974). The deflection also trends along the south side of the Amazon Basin and separates contrasting basement known from wells drilled in that region. The western massifs of the northern Coastal Cordillera of northwestern Peru bend in line with the Huancabamba Deflection. As yet, however, there is no strong evidence of a continuation of the deflection west of the Cordillera Occidental (F. Zuñiga, pers. commun., 1974). The tripartite cordilleras south of the Huancabamba Deflection terminate against it; two other ranges continue north of the Tumbes-Guayana Megashear (C. Ham, pers. commun.), also known as the Amotape Trend (J. Baldock, pers. commun.), into Ecuador. The elevations of the Andes are about 1,000 m lower between the Huancabamba Deflection and the Tumbes-Tuyana Megashear than to the north or south. The Tumbes-Guayana Megashear trends out of the Precambrian Guayana Shield toward the south side of the Gulf of Guayaquil and aligns with mineralization along old east-west fault zones, a major east-west magnetic disruption in the southern part of the Ecuadorean Andes (Goosens, 1972), and with "relics of early easterly structural grain" (Campbell, 1974).

Carey (1958), Krause (1965), De Loczy (1970), and Campbell (1974) discuss major intracontinental structural lineaments that coincide with the north and south sides of the Amazon Basin, whose trends align with the Huancabamba-Amazonas and Tumbes-Guayana megashears. It can be reasoned that they are old reactivated lines of stress in which the depressed Marañon Portal (Campbell, 1974) between them moved westward. Gregory (1929), quoted in Du Toit (1937), remarked about "the westerly urge of the continent" in northern Peru. The megashears may have been the boundaries of an aborted aulacogen that became a reentrant for marine deposition that persisted from the Paleozoic to as late as the Oligocene (Weeks, 1947; Hosmer, 1959; Harrington, 1962). The marine rocks were deposited prior to the principal Andean orogeny. It is noteworthy that prior to the Andean orogeny, the Subandean region drained westward into the Pacific.

Calc-alkaline igneous rocks so dominant in outcrops of the coastal batholith to the south become fewer in the Marañon Portal. They may be more extensive at depth, however, and not yet unroofed.

The Dolores-Guayaquil Megashear (Case and others, 1971) aligns with the Cordillera Occidental in

Ecuador and Colombia and separates continental crust on the east from former oceanic crust on the west (Case and others, 1971; Goosens, 1973; Campbell, 1974; for another view, see Henderson and Evans, 1980). It bends westerly into the Gulf of Guayaquil and the east-west–trending Cenozoic Progreso Basin, which separates a basement of granitic and metamorphic Paleozoic rocks on the south from a basement of mafic to ultramafic rocks with oceanic or ophiolitic affinities (Goosens and Rose, 1973; Lonsdale, 1978) on the north.

SUBMARINE MORPHOLOGY

The bathymetry of the northwestern Peruvian margin and the Gulf of Guayaquil is displayed in the Geological Society of America Map Series MC-34 (Sheet 1).

The submarine topography is characterized by (1) a continental shelf of varying width; (2) a steeply declining continental slope and inner trench wall crossed by submarine canyons; (3) a narrow, flat trench floor; and (4) a gently declining outer trench slope with ridges striking obliquely into the trench.

Continental Shelf

The sharp change in gradient that marks the shelf edge ranges between 100 and 300 m in depth. The shelf is more than 60 km wide in the south, where it is underlain by the seaward extension of the Sechura Basin (Masias, 1976). Between Isla Lobos de Tierra and Cabo Blanco, the shelf is as narrow as 1 or 2 km off capes and submarine canyons, but it is 50 km wide in the bay north of Bayovar (Fig. 3). North of Cabo Blanco, the shelf again widens to accommodate the Progreso Basin centered about the Gulf of Guayaquil, where abundant detrital sediment is delivered by the Rio Guayas system.

Continental Slope and Trench-Slope Break

The upper slope off Peru varies from 1° to 10° and averages 3.5°. As elsewhere off Peru and northern Chile (Kulm and others, 1977; Coulbourn and Moberly, 1977), one or more breaks may occur in the slope. For reasons to be discussed later, the trench-slope breaks that are about 20 to 30 km from the base of the inner wall of the trench locate the main structural change. They are at steps and benches of a fairly consistent 2,500 m depth from north to south, but their vertical elevation above the trench floor is about 1,000 m greater in the south due to a general deepening of the trench in that direction (MC-34). Below that trench-slope break, the lower slope averages 6.5° but may exceed 15° in steepness. Trends of most ridges below the break strike northwest, but some strike in other directions, implying a diverse pattern to the underlying structure. Near the trench, a few linear ridges on the lowest slope are parallel to the trench.

The slopes are crossed by several "v"-shaped submarine canyons. Their heads generally coincide with the 100 m depth contour or the edge of the shelf, and most canyons continue to the trench. A few begin at the shelf edge and end at mid-slope benches, and others begin at a trench-slope break and end at the trench. This morpholoy contrasts with the observations of Prince (1974) south of 75°S, where canyons terminate at the trench-slope break and benches occur below the break.

Trench

The Peru-Chile Trench is the dominant morphotectonic feature of the area studied. The trench is fairly uniform along its length, but it does have important geological and geophysical differences along certain sectors (Hayes, 1966; Scholl and others, 1968; Kulm and others, 1973; Prince, 1974; Barazangi and Isacks, 1976; Hussong and others, 1976; Coulbourn and Moberly, 1977; Schweller and others, this volume). The northern Peruvian part of the trench is partially filled with a wedge of sediments of varying thickness believed to include contributions from each of: (1) the oceanic side, as pelagic and hemipelagic sediments carried down by the descending Nazca plate or redeposited by turbidity flows;

(2) the landward side, including detrital sediments carried by turbidity flows through canyons; mass movement and redeposition of hemipelagic surficial sediment down the slope; rocks eroded from the shelf, upper slope, and lower slope carried through canyons; and massive slumps from the lower inner slopes; and (3) a modest contribution of hemipelagic sediment directly into the trench.

The trench floor is generally smooth and flat and varies in width from 2 to 17 km. Constrictions occur where large ridges from the oceanic plate intercept the trench or where immense slumps from the lower continental slope cover part of the floor of the trench.

Outer Trench Slope

The slope of the oceanic side of the trench is more gentle than the slope on the continental side. The outer slope is about 50 km wide from the outer edge of the trench floor to the crest of an outer ridge. Normal faults that are downdropped toward the trench characteristically offset both oceanic basement and the overlying sediments of the outer trench slope. The outer ridge and the normal faults are believed to be consequences of the flexing and tension in the upper lithosphere prior to subduction.

En echelon ridges trending obliquely to the trench are as much as 1,600 m above their surrounding sea floor. They are essentially free of sediment. Mammerickx and others (1975) described them as fracture zones. Hey (1975, 1977) and Handschumacher (1976) implied that the ridges conform with fracture zones or transform faults created by changes of spreading regimes as determined from paleomagnetic constructions (Herron, 1972). They may be analogous to similar features in the Atlantic that von Andel and others (1969) called "leaky transforms."

A further discussion of morphotectonics, submarine canyons, sedimentary processes, and the en echelon ridges is in Shepherd (1979).

STRUCTURAL INTERPRETATION OF SEISMIC REFLECTION PROFILES

Seismic reflection profiles of the continental margin, shown here as line drawings, were constructed by examining and tracing the mixed-frequency records (usually 40–80 and 150–300 Hz). In addition to this base, records of other frequencies were composited wherever needed for clarification. Maximum penetration observed on the single-channel reflection profiles in this area is generally less than 1 sec of two-way travel time. Velocity measurements from ASPER record number 35 suggest that this penetration is less than 1 km.

Profiles A through M and Q (Fig. 1) transect the trench and continental margin orthogonally from north to south and demonstrate considerable variation in morphology and underlying structure, especially on the continental margin. Gross similarities in morphology prevail, but profiles separated by less than 5 km may show wide variances. This implies that morphology and structure may be more complicated, especially on the inner wall of the trench, than can be deciphered under the constraint of the density of the data acquired. Profile J is a multichannel seismic profile obtained by Seiscom Delta. Profiles are horizontally scaled in kilometres from zero on the landward end. Profiles N and O are short north-south segments on the oceanic plate.

Profile A

Oceanic basement shows normal, down-to-the-trench faults offsetting the overlying pelagic and hemipelagic sediments (Fig. 4). Here the flat trench floor is obliterated by materials that apparently slumped from the inner wall, as indicated by the hummocky topography and discontinuous and incoherent reflectors.

A trench-slope break is prominent at km 65 and, as for all these crossings, is based on the concurrence of topographic and gravity profile inflections at the middle part of the inner trench wall. Incoherent internal reflectors are noted between the trench-slope break and the trench. Landward of the trench-slope break, shallow reflectors gradually become more continuous and traceable, as seen on

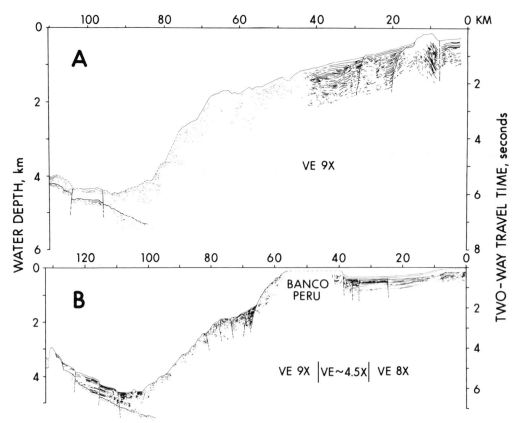

Figure 4. *Profile A* extends northwest from near Zorritos, Peru, showing south end of Banco Peru (km 10), faulted fore-arc basin (km 20–40), trench-slope break (km 65), and slumped sediment in axis of Peru Trench (km 90). Positions of this and other profiles are shown on Figure 1. *Profile B* crosses Banco Peru, which lacks internal reflectors, and a deep sedimentary basin between it and the coastline. Faults are interpreted from offset beds on opposite sides of Banco Peru. Those on the seaward side are associated with complexly oriented reflectors, suggesting considerable deformation. Normal faults on the seaward wall of the trench offset basement on overlying sediments. Oceanic basement reflectors extend under the landward wall of the trench. The small bump in the middle of the trench is believed to be a sediment diapir. Variable vertical exaggerations are due to changes in ship speed.

other profiles with similar settings. Between km 40 and 30, angular discordance between families of reflectors suggests unconformities, and between km 30 and 0, three prominent normal faults offset beds. The topographic high at km 10 is an extension of Banco Peru, discussed in Profile B.

Profile B

The oceanic side of the trench shows normal faults that are interrupted by a sharp prominence at km 130 (Fig. 4). This prominence may be an igneous intrusion emplaced along a recent fault zone that was created by the bending of the oceanic plate prior to its entry into the subduction zone.

From the trench to Banco Peru, only incoherent, short reflectors are seen. Faulting from km 80 to 65, shown by discontinuous reflectors, is interpreted to coincide with topographic notches.

Flat-topped Banco Peru lacks internal reflectors except on its sides. On one crossing, siltstone with abundant burrows was dredged from 142 m depth of water. It may be only a thin capping over an unknown but very dense rock type, as indicated by a high free-air gravity anomaly. The relationship of Banco Peru to its surroundings is discussed more fully in the section on free-air gravity.

Between Banco Peru and the eastern end of the profile, continuous reflectors outline a sedimentary basin with a few well-defined faults. Subsurface studies by Zevallos (1970) show the basin to have sediments greater than 5,500 m thick. A large negative free-air gravity anomaly substantiates this viewpoint. Between km 10 and 0, dips of the reflectors suggest progradation of sediments.

Profiles C and D

Well-defined offsets in the oceanic basement and continuous sediment reflectors of the oceanic plate and wedge of sediments in the trench characterize Profile D, whereas Profile C shows the trench wedge prograding over nondisrupted oceanic-plate deposits (Fig. 5).

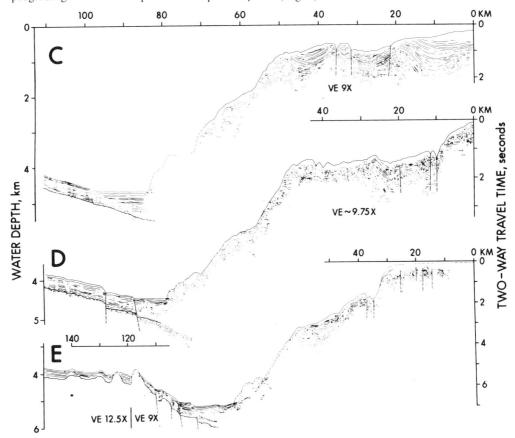

Figure 5. *Profile C*, oriented perpendicular to the coast, shows the outer wall of the trench unaffected by faulting. Large-amplitude folds with converging reflectors suggest unconformities or penecontemporaneous folding and deposition. Faults between km 40 and km 30 are parts of the same trend of faulitng found on the landward side of Banco Peru, as seen in Profile B (Fig. 4). The sea floor of the lower inner trench wall could not be deciphered with accuracy, probably because of slopes and rugose topography. *Profile D* off Cabo Blanco shows a broad, uneven bench between km 50 and km 10 underlain by a fore-arc basin whose complex structure is defined by short, discontinuous reflectors. A trench-slope break is selected at km 50 and is coincident with an inflection in the gravity and topography. Landward inclination of reflectors at km 29 suggests landward migration of the axis of sedimentation. *Profile E* transects the trench obliquely between Talara and Paita. Sediments of the oceanic plate drape over basement highs at kms 140 and 130, whereas igneous piercements at kms 123 and 118 have little to no sediment cover. The trench shows an undulating floor and a mid-trench fold. Below the trench-slope break at km 60, the reflectors are few and incoherent. At km 35, the profile crosses a submarine canyon showing deeper sedimentary beds that could crop out.

A 40-km-wide bench is seen on Profiles C and D; it is deeper (1 km depth) on D than on C. Its uneven topography and rather poorly defined, discontinuous reflectors suggest that this fore-arc basin is structurally more complicated than single-channel reflection profiles can detect. Landward inclination of reflectors at km 28 on D suggests migration of the axis of the sedimentation, similarly noted by Coulbourn and Moberly (1977) off northern Chile.

Faults at km 40 on D are continuations of those that skirt the southeast side of Banco Peru.

Profile E

On the oceanic plate, sediments are draped over basement elevations at kms 140 and 130 (Fig. 5). Apparently they are piercements of igneous rocks through the oceanic sediments. Normal faults are seen near the trench on the oceanic plate. The piercements at kms 123 and 118 have little to no sediment cover, which indicates either that they are young or that bottom currents are strong. Both the piercements and the sediment drapes are part of the Ocola Ridge (Figs. 2 and 12).

The undulating trench floor is quite uncommon in this part of the Peru-Chile Trench. Whether it was deformed tectonically or by other processes is not known.

Below the trench-slope break at km 60, reflectors are few and incoherent, but reflectors become more distinct landward from km 60 to 40. The topographic low at km 35 is a submarine canyon that crosses obliquely to the profiles. Reflectors indicate that deeper sedimentary beds may crop out on the canyon wall, but dredging to sample these "outcrops" was unsuccessful in similar settings elsewhere.

Landward of the continental shelf breaks, offsets of families of reflectors suggest faulting and folding. The appearance of folding is deceiving, because folding does not exist. The subsurface geology is well known from more than 9,000 oil wells and detailed geophysical data in the onshore and offshore oil fields of northwestern Peru south of about 4°S. Figure 6 is a simplified cross section, with well control projected as necessary into the line of section, through the Restin-Cabo Blanco oil field and typifies the structural style believed to exist landward of the continental shelf break. The degree of stratigraphic and structural control is, by oil field standards, the best possible. The entire length of the cross section, about 11 km, can be located between about km 20 and 10 of Profile E (Fig. 5). Comparison between the two profiles shows clearly that these single-channel seismic reflection profiles cannot discern the complex style of structural deformation characteristic of this part of the continental margin. Of significance is the lack of folding and even minor warping in any of the drilled areas, offshore and onshore, at the La Brea-Parinas-Talara oil fields. There is a general concurrence, however, between the seismic profiles and cross section in that stratigraphic units dip landward, shoreward of km 25. No reversal of this trend has been found from drilling to date, which may suggest that exploration for additional hydrocarbon traps should continue seaward.

Profile G

This traverse crosses the trenchward end of the Sarmiento Ridge, but with no noticeable response of the ridge in the free-air gravity anomaly curve plotted above this seismic reflection profile (Fig. 7). Evenly bedded and supratenuously folded sediments are apparent on the oceanic plate, as demonstrated by continuous reflectors. Little or no sediment can be seen on the ridge, attesting to recent volcanic activity or removal of sediment by bottom currents.

A reduced exaggeration (~2.5×) profile below tracings of the original reflection record graphically illustrates the advantages and disadvantages in visualizing the structural and geomorphic setting. The slope between the trench floor and the trench-slope break at km 50 is about 6°, and almost no subbottom reflectors were recorded there.

A submarine landslide probably extends between kms 58 and 28. This interpretation is based on discontinuities of a structure and faults across the presumed slide plane, and on relatively lumpy topography.

The slide plane closely corresponds with dark shading or bottom simulating reflections (BSR), which cannot be shown on this line tracing from the original record. The BSR is believed to be the

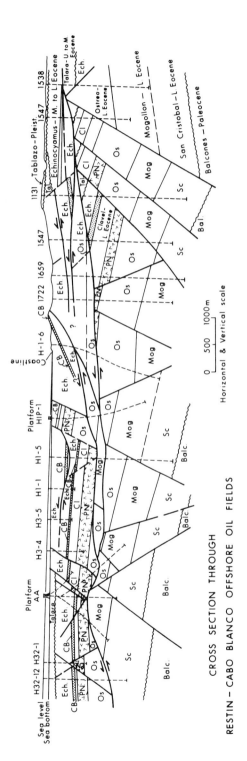

Figure 6. A simplified cross section through one of the offshore-onshore Restin–Cabo Blanco oil fields south of Cabo Blanco, northwestern Peru. It shows the complexity of subsurface structure believed to exist elsewhere landward of the continental shelf break. Note the exclusively normal block faulting cut by gravity-slide planes.

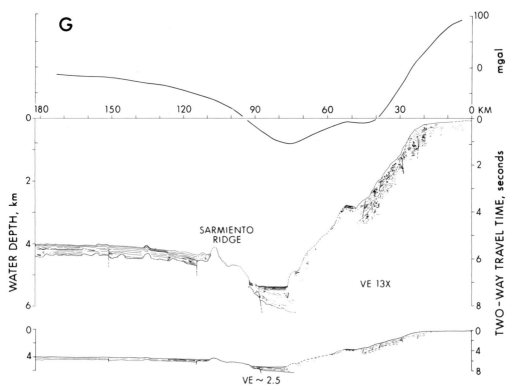

Figure 7. *Profile G* passes over the landward end of the Sarmiento Ridge. The gravity profile plotted above the reflection tracing shows the landward offset of the gravity minimum with respect to the trench bathymetric minimum. The offset probably is due to the mass of sediments, less dense than the oceanic crust, in the accreted wedge below a trench-slope break at km 50. A reduced exaggeration profile of about 2.5× below the tracings of the original reflection record illustrates the advantages and disadvantages in visualizing exaggeration of the structural and geomorphic setting. A gravity slide is interpreted to lie between kms 59 and 28.

result of gas-hydrates or clathrates that accumulate in shallow sedimentary horizons. Summerhayes and others (1979) discuss the existence of sediment slides on the Southwest African continental slopes as being associated with clathrate horizons that may act as a triggering mechanism for the slides. Triggering by simple overloading of sediments or earthquakes, however, cannot be discounted in this setting. Submarine slumps of this scale may be modern-day example of olistostromes (Maxwell, 1959).

The gravity minimum of the trench is skewed landward of the bathymetric minimum, probably in response to the mass of less dense sediments of an accreted wedge noted elsewhere in the Peru-Chile Trench by Hayes (1966). The inflection of the gravity profile at km 50, seen similarly on other profiles in mid-slope regions, corresponds to the trench-slope break.

Profiles H and I

Irregular topography on the slopes of Sarmiento Ridge (Fig. 8) suggests that hemipelagic sediment has slumped there. A piston core taken along Profile I at km 150 and below the carbonate compensation depth contained calcareous foraminifera that J. Resig (pers. commun., 1976) believes came from higher on the Sarmiento Ridge by downslope movements of sediment. Continuous reflectors and well-defined acoustic basement show bending of the oceanic plate toward the trench. The oceanic-plate sediment reflectors rapidly lose their continuity approaching the trench, where they

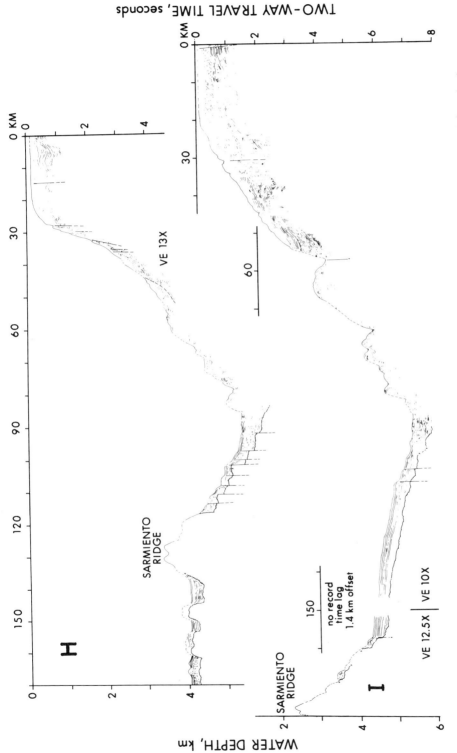

Figure 8. *Profile H* shows the apparently sediment-free Sarmiento Ridge and closely spaced faults of the oceanic plate. The trench floor that is uneven suggests deformation as do diffuse, chaotic reflectors and hyperbolae on the lower part of the inner wall of the trench. Slumping is suspected between kms 60 and 40, and reflectors under km 30 suggest that landward-dipping sediments may crop out in that area. *Profile I.* The Sarmiento Ridge appears free of sediment on its summit, but piles of sediment indicated by chaotic reflectors occur on its flank. Continuous reflectors of oceanic sediments and well-defined acoustic basement show bending of the oceanic plate. Irregular topography, few and short reflectors below the trench-slope break at km 65, and lack of flat-lying reflectors in the trench suggest slumping. Internal reflectors at km 55 intersect the bottom on the wall of a submarine canyon.

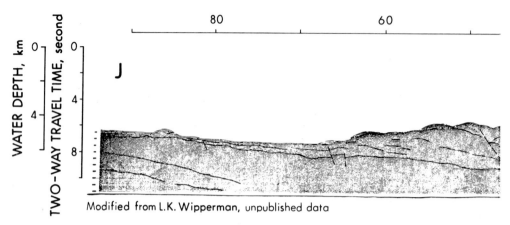

Figure 9. *Profile J* (Seiscom Delta Line). A multichannel CDP seismic reflection profile that extends across the trench and margin from the oceanic plate to the upper slope. Approximate ages of formations shown on the right-hand side are projected from a wildcat well drilled onshore about 40 km from the landward end of the profile. Reflector labeled M on the oceanic plate is believed to be the Moho discontinuity. Horizon B is the oceanic

become completely chaotic, suggesting deformation. The deepest part of the trench on these traverses is devoid of the usual continuous flat-lying reflectors from turbidites. The diffraction-like events and irregular topography suggest extreme slumping or tectonic complexities or both between the trench and the trench-slope break at km 65. As is seen on other profiles, there are few definable internal reflectors below the inner trench wall. At km 60, the ship's track crossed a portion of a lateral ridge that trends into a canyon so that the topographic prominence of the trench-slope break is accentuated. Note that the landward sediments intersect the bottom at km 55 on the wall of a submarine canyon. Under the continental shelf, the appearance of the record suggests that folded sediments are present, but the structure may actually be blocks of beds tilted by faulting, as shown on the oil field section (Fig. 6).

Seiscom Delta Line, Profile J

In order to improve resolution of structural details on the northern Peru margin, Seiscom Delta Inc. was contracted to record and process a 24-channel CDP (common depth point) digital unmigrated seismic reflection profile over the sector in question as an extension of work being done by them for oil exploration on the continental shelf. Velocity analysis and processing to produce deconvolved and 1200%-stacked time sections were done by Seiscom Delta Inc. in cooperation with Exxon Production Research Co. L. K. Wipperman (1974, unpub.) used velocity spectra to calculate the internal velocity-depth functions at appropriate parts of the profile. These spectra aided in correlation of layers with a velocity-age function, allowing considerable resolution. A nonmigrated depth section was used to modify the interpretations on the time section. Velocities selected by an interpreter may either create or eliminate reflections and structural configurations. The time and depth sections used are in general compatible with major structural geometries but may vary widely with respect to minor ones. Only those segments along the depth section which made geologic sense or were compatible with the time section were used for interpretation.

The velocity-formational function for the upper slope part of the profile was obtained from a velocity survey in a wildcat well (VIRU 4-X-1) drilled onshore about 40 km from the landward end of the line. Additional and confirming velocity-formational data came from an offshore well drilled on the continental shelf at 8°S and from a seismic refraction line crossing the shelf and trench at about 8°30'S (Shepherd and others, 1973; Hussong and others, 1976). All the velocity-formational determinations show good compatibility.

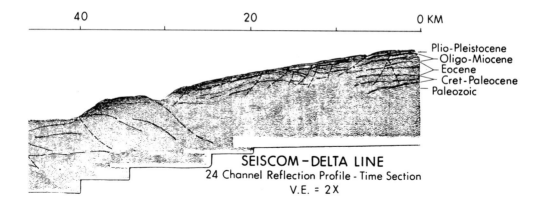

acoustic basement, which can be traced conservatively 23 km under the landward wall of the trench and possibly as much as 32 km. Incoherent reflectors and a mass of diffractions between the trench-slope break at km 35 and the trench represent accreted sediments.

Prominent coherent reflections are emphasized by solid dark lines on the profile, whereas faults are dashed. Subsurface formations with their ages are shown on the right-hand side of Profile J (Fig. 9). About 10 km seaward from the landward end of the line, interpretations of the ages and formational contacts become increasingly unreliable. This uncertainty may be due to low velocity contrasts across the sediment interfaces caused by facies changes (at least below the base of the Oligocene-Miocene), but it is more probably caused by lack of coherent reflections as a result of intense faulting of the Paleocene to Paleozoic section. It seems reasonable to project the Cretaceous to Pliocene-Pleistocene section, presumably lying on Paleozoic basement, to km 30, where the more coherent reflectors dissipate within a faulted area. The internal constitution of the prominent fault-bounded trench-slope break between km 30 and 40 is conjectural. Well-determined reflectors are lacking. Those that are traced have geometries contrasting to those farther landward but are similar to those farther seaward.

On the ocean side of the profile at 9.3 sec two-way travel time, a reflector labeled M is believed to be the Moho discontinuity and correlates with its determination from a refraction line adjacent to another CDP profile at 11°40′S (Hussong and others, 1976). The reflector at 8 sec is an undetermined horizon within the oceanic crust. Horizon B, which separates oceanic layer 1 (sediments) from oceanic layer 2 (igneous), can be conservatively traced under the landward slope of the trench as a continuous reflector 23 km, and possibly as far as 32 km, from the intersection of the slope and trench floor. From the slope-trench intersection to km 37, the B horizon appears nearly flat and even rises landward. This record, however, is a time section, and the actual configuration of this horizon cannot be determined without constucting a migrated depth section. The landward increase in thickness of higher velocity, overlying sediments decreases the travel time to horizon B. The dip of oceanic layer 2 is actually about 9° east and therefore is compatible with that determined by Hussong and others (1976) on a similarly oriented 24-channel CDP reflection profile at 11°40′S. It is interesting to note that about 15 km of crust of unknown type remains between the identified landward projection of oceanic crust under the inner trench wall and the western or seaward extent of continental elements, if our interpretations are valid. Although the vertical relationships are less clear, the depth section suggests that about 6,800 m of tectonized sediments or slices of oceanic crustal material may be present as a mélange.

The zone from the trench-slope break to the base of the landward slope, and above horizon B, lacks coherent reflectors and is represented by a mass of diffraction patterns seen similarly on single-channel reflection profiles. Seely and others (1974) for the Middle America Trench, Beck and Lehner (1974) for the Java Trench, and Kulm and Fowler (1974) off Oregon found similar reflection signatures that were interpreted as intensely folded sediments or randomly tectonized sequences separated by faults which provide the point sources for the mass of the diffractions beginning abruptly at the trench-landward slope interface. It is believed that the bulk of the accreted sediments in this zone are pervasively sheared and that the horizons marked by dark lines may be acoustic artifacts or bedding plane faults.

Sediments in the trench have a landward dip of 1.5° to 2°, indicating that turbidite sedimentation may not be keeping pace with the rate of oceanic crustal subduction. A small hump at km 85 is interpreted as a tangential crossing near a topographic prominence out of the line of the profile. A gravity slide is interpreted to lie between km 20 and 30, and a thin turbidite channel deposit is in the prominent notch of a submarine canyon at km 30. Significantly, the trench-slope break separates intense discrete block faults on the upper slope from abundant hyperbolic diffraction patterns, probably associated with abundant anastomosing faults, in the lower slope.

Profile K

Trenchward from the spur of the Sarmiento Ridge, a rather uneven acoustic basement prevails under some supratenuous folds (Fig. 10). The usual flat trench bottom is completely obliterated by a massive slump from the inner wall. The average slope from km 97 to 77 is about 1.5°, but its irregularity indicates slump topography. The position at which the break-away for the slump occurs is difficult to identify, but the length of the slump is suggested to be in the range of 20 to 30 km.

Profiles L and M

Profile L enters the north side of a broad embayment in the continental shelf and passes 8 km north of Isla Lobos de Tierra, which exposes metamorphosed Paleozoic rocks (Fig. 1). The island is on a trend with similar rocks that crop out on Silla de Paita at 6°S, 81°W (Fig. 3). No deeper reflectors or other seismic indication of these rocks can be seen on the record (Fig. 11). Therefore, the graben among the faults that cut the arcuate massif from the Amotape Mountains southward on the continental margin contain thick sections of sediments younger than the Paleozoic basement.

Profile M is almost a continuation of the landward end of Profile L. Before the change in course at km 0, the ship's track passed Isla Lobos de Tierra 5 km abeam. Paleozoic metamorphic rocks that crop out on the island are belived to be discernible as deeper reflectors of one character that are abutted by reflectors of younger sediments of another character (Fig. 11). This does not occur in Profile L. Shelf reflectors intersect the walls of a submarine canyon at km 15 on the seaward leg of the ship's track. The canyon bottom appears to be devoid of younger sediment fill.

Profiles N and O

A line drawing of a short segment of traverse across the Sarmiento Ridge shows no sediment cover over the ridge (Fig. 12). Chaotic, lumpy masses can, however, be seen on the flanks. Perhaps they are slumped sediments from higher on the ridge. If the ridge has a sediment cover, it is transparent or too thick to detect. The latter alternative is favored, as Kulm and others (1973) piston-cored the top ridge adjacent to the trench at 9°20'S and recovered 5.5 m of sediment and tholeiitic basalt at the base of the core. The seismic reflection record of that ridge also showed little or no sediment.

The contact relationship between the igneous rocks of the ridge and the sediments of the oceanic plate is unclear. Certainly the sediment reflectors appear to abut the ridge, and the igneous rocks of the ridge appear to pierce the sediments. If, however, multiple volcanic flows cover oceanic sediments within the flanks of the ridge, or if the two rock types interdigitate, they would not be seen on a seismic reflection profile.

The contrasting nature of the Ocola and Zevallos ridges is shown on Profile O (Fig. 12). Ocola Ridge is very much like the Sarmiento Ridge, whereas the subdued topography, multiple topographic highs, and rough acoustic basement of the Zevallos Ridge have discernible sediment cover. The apparent piercement to the south of Ocola Ridge shows no sediment cover.

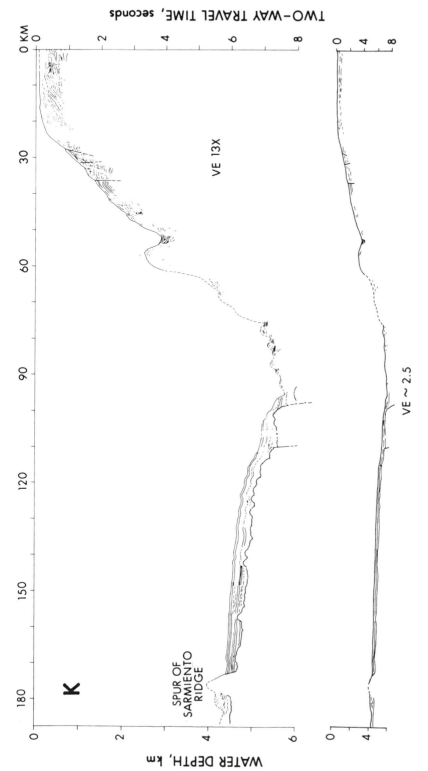

Figure 10. *Profile K* is an east-west profile off Bayovar. An uneven acoustic basement on the oceanic plate is unfaulted except near and in the trench. The trench is completely covered by a slump. A trench-slope break at km 60 is accentuated as the profile obliquely crosses a submarine canyon that may have some sediment fill at km 55. A reduced vertical exaggeration (~2.5×) is shown below the 13× exaggeration for a different perspective.

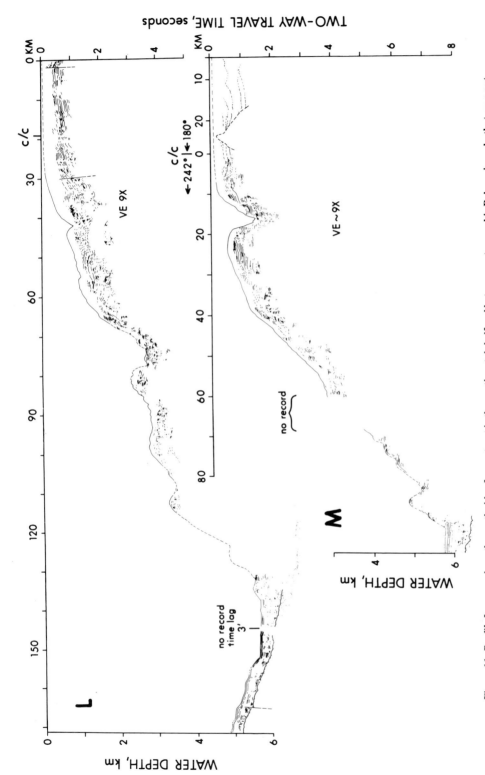

Figure 11. *Profile L* passes along the north side of a reentrant in the continental shelf and between metamorphic Paleozoic rocks that crop out onshore as well as on Isla Lobos de Tierra. Deeper reflectors representing those rocks cannot be seen, indicating thick post-Paleozoic rocks deposited in a graben. A trench-slope break is located at km 100. An anticlinal fold is suggested at the trench-lower slope interface. Neither bottom nor coherent internal reflectors could be deciphered between kms 118 and 133. Along *Profile M*, younger Cenozoic sediments of the Sechura Basin appear to lap onto Paleozoic metamorphic rocks that crop out on Isla Lobos de Tierra about 5 km abeam. Shelf sediment reflectors appear to crop out in a submarine canyon at km 15. An anticlinal fold is expressed as a ridge that parallels the trench on the lower inner wall.

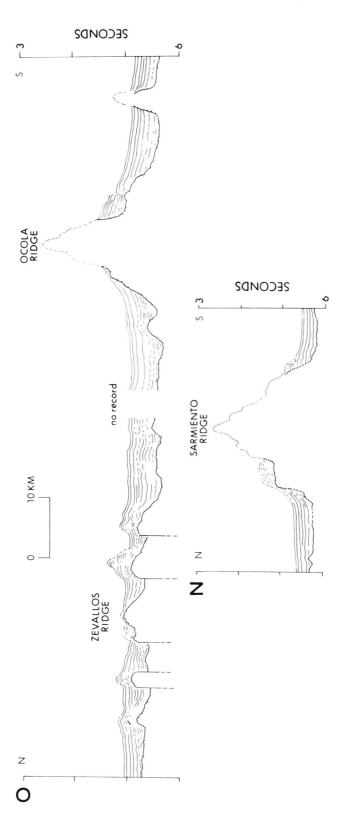

Figure 12. *Profile N* (lower) over the Sarmiento Ridge shows apparent lack of sedimentation on the summit of the ridge and possible slumped sediments on its lower flanks. *Profile O* (upper) is across the Ocola and Zevallos ridges. The Ocola Ridge appears free of sediment on its summit, whereas the multiple Zevallos Ridge shows sediment cover, rough acoustic basement, and possible intrusions.

Some topographically higher parts of the Zevallos Ridge allow supratenuous folds of the sediment cover, and faults may be present on their flanks, but the appearance of faults may simply be due to very steep sides of an intrusion. The topographic high on the central part of the Zevallos Ridge has thin but chaotically defined reflectors, and it may be an igneous intrusion.

Profile P

This line drawing is of more than 200 km of reflection profile recorded at a 2 sec sweep (Fig. 13). The track parallels the coastline at about 20 km distance and crosses submarine canyons as well as the Progreso Basin, which contains thick Cenozoic sediments. Subparallel reflectors intersect the canyons, attesting to their erosional origin. There is only a slight amount of sediment fill in Talara Canyon, whereas the deepest part of Cabo Blanco Canyon shows more fill. Shoreward of this track, the bathymetry shows closed contours in the Cabo Blanco Canyon, indicating a possible trap of recent sediments.

Several significant faults known onshore and offshore between Talara and Zorritos (Fig. 3) trend into the line of profile but are not clearly recognized on the profiles. One is inferred to be at km 70 where a break in slope and apparent offsets of reflectors coincide. Disharmonic folds are suggested by reflectors within the confines of Cabo Blanco Canyon.

A fault zone in the axial part of a large anticline at km 130 is believed to be the same zone that borders the landward side of Banco Peru. The angular relationship of reflectors on the flanks of this fold at km 125 and 140 implies an unconformity. Oblique progradation is shown between km 11 and 125, which is indicative of deposition in an environment of high energy from waves commonly seen aligned with some deltas (Sangree and Widmier, 1978). Anticlines are aligned with the two wells drilled a short distance from the trackline. An oil field has been developed in Peruvian waters from Tertiary sediments on the anticline on which Tenneco-Union 8X-2 was drilled. The ADA E-1, in Ecuadorean waters, is a shut-in gas well on a rather broad anticline with several faults. The well, with commercial productive rates, is shut in pending contractual agreements with the government. Intensity of folding and structural complexity increase with depth, at least within the field that includes the Tenneco-Union well (M. Naranjo, pers. commun., 1974).

ASPER Run 35

This sonobuoy run, analyzed by L. Wipperman (Fig. 14), is located on a wide part of the continental shelf (Fig. 1) within the offshore part of the north-south-trending Sechura Basin. The axis of the basin is east of the Paleozoic massif composed of Cerro de Illesca and the islands of Lobos de Tierra and Lobos de Afuera (Masias, 1976).

Zuñiga and Travis (1975) report sediments as thick as 6,000 m overlying Paleozoic basement; therefore, the sonobuoy-determined thickness of about 5,000 m is in good agreement.

FREE-AIR GRAVITY

The most significant free-air gravity anomaly is the strong negative belt of −110 to −150 mgal, associated with the Peru-Chile Trench (Fig. 15). The belt is found over the entire length of the trench (Hayes, 1966; Getts, 1975). Off Peru, the axis of minimum gravity is 2.5 to 8 km landward of the trench axis due to the mass effect of the continental margin. This relationship can also be seen on the profiles of Hayes (1966) over most of the Peru-Chile Trench.

Seaward of the trench, the free-air gravity anomaly increases gradually to an average of 0 mgal, which is the outer gravity high commonly seen elsewhere along the Peru-Chile Trench (Getts, 1975) and outside other trenches of the world (Watts and Talwani, 1975). The Grijalva, Sarmiento, and Ocola ridges are strongly positive (+20 to +40 mgal) in response to the increased mass and topographic relief of their igneous rocks emplaced along fracture zones.

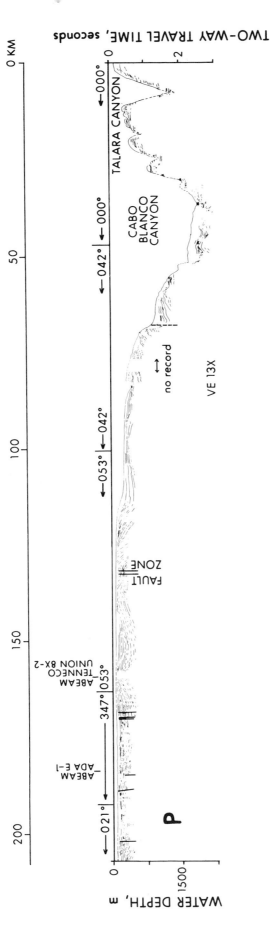

Figure 13. *Profile P* passes parallel to and about 30 km from the coastline from Talara to Tumbes, and thence into the Progreso Basin underlying the Gulf of Guayaquil. The ship's track passes across the Talara and Cabo Blanco submarine canyons. A fault zone in the axial part of a large anticline at km 130 continues past the landward side of Banco Peru and terminates at the trench-slope break of the Peru-Chile Trench. Closely abeam of the track are two hydrocarbon-discovery wells located on anticlines. North is on the left side of the profile.

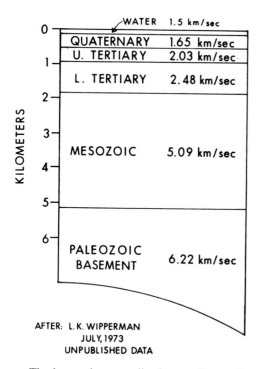

Figure 14. ASPER Run No. 35 is located in the offshore part of the Sechura Basin, between Isla Lobos de Afuera and the coast (Fig. 3). The plot depicts thicknesses and compressional velocities of underlying rocks. Length of the interpreted ASPER is about 12 km.

The lower slope usually shows a flattened gradient that can be attributed to a mass deficiency provided largely by a wedge of tectonized sediments accumulated by subduction processes below the trench-slope break. There is also some effect of the draping of less dense hemipelagic younger sediments there. An additional flattening effect may be seen on some profiles near and parallel to submarine canyons whose topography provides mass deficiency.

Landward, the free-air anomaly increases to more than +100 mgal with a rather abrupt inflection in the upper-slope region. On most profiles, this inflection coincides with the main topographic break, and if there is more than one topographic break, we identify the one at the gravity inflection to be the principal trench-slope break. The trench-slope break that we identify on the basis of the reflection profiling and gravity interpretation is not the same as the ridge at and near the Peruvian coastline that Lonsdale (1978) termed the tench-slope break. Lonsdale's information came mainly from his work at the Ecuador trench and slope, which he apparently projected south of the Gulf of Guayaquil.

The inflection continues from Cabo Blanco (Fig. 15) to the southern boundary of the study area but is not seen north of Cabo Blanco or opposite the Gulf of Guayaquil. The strong gradient reappears off southern Ecuador at about 2°15'S and continues northward to 0°30'S (Getts, 1975), where basement is presumed to be the mafic and ultramafic Piñon Formation (Hector Ayon, pers. commun., 1975) with oceanic crustal affinities (Goosens and Rose, 1973). Such relatively dense rocks could logically provide the strong gravity gradients in that sector. From Cabo Blanco southward, however, the strong gradient may reflect the edge of continental crust against accreted sediments.

The continental shelf is characterized by free-air anomalies of greater than +50 mgal that may locally exceed +100 mgal. Along the coast, the capes are horst blocks composed of metamorphosed Paleozoic sediments and Jurassic(?) instrusive rocks. The capes are characterized by free-air anomalies with steep gradients. Intervening graben are mainly filled with early Cenozoic sediments that lap onto the horsts. Closed free-air gravity contours surround Isla Lobos de Tierra and Isla Lobos de Afuera where metamorphosed Paleozoic sediments crop out. Gravity gradients there also suggest

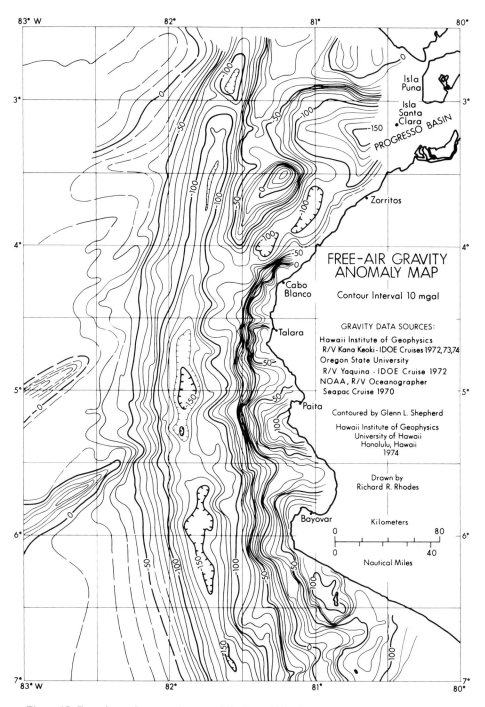

Figure 15. Free-air gravity anomaly map of the Peru-Chile Trench and environs from 2°30′ to 7°S latitude. Axis of gravity minima is 2.5–8 km landward of the bathymetric axis. Steep gradients surrounding the capes accentuate the denser older rocks composing them. A contrasting gravity high underlies Banco Peru at 3°30′S and 81°15′W. The low gravity minima in the Gulf of Guayaquil represents more than 6,000 m of young sediments of the Progreso Basin.

segmentation by faults, a situation similar to the capes immediately to the north. The steep gradients of the capes facing the trench suggest that Paleozoic rocks are 40 to 50 km from the axis of the trench.

Near Zorritos, oil wells drilled to basement revealed granitic rocks of undetermined age but that are believed to be correlative to Mesozoic granitic rocks that crop out about 22 km farther east, on the west side of Cerros de Amotape (O. Zevallos, pers. commun., 1974).

North of Cabo Blanco, complex gravity anomalies indicate an altogether different shallow crustal structure than to the south. Eastward along $3°30'S$, values of -120 mgal in the trench rise to -50 mgal, then again decrease to -150 mgal on the shelf areas of the Gulf of Guayaquil. That gravity minimum coincides with the axis of the Progreso Basin. The northern part of the Gulf of Guayaquil has a strong gravity gradient, believed to be the locus of maximum faulting for the northern margin of the Progreso Basin. It is worth noting that basement rocks on the north side of the Gulf of Guayaquil are mafic ones that Goosens and Rose (1975) believed originated as oceanic crust, whereas those on the south side are continental granitic rocks. The strong gravity gradients may represent the margins of a pull-apart basin (Crowell, 1974), which very likely is the structural origin of the widening and subsiding Progreso Basin, filled with more than 6,000 m of Cenozoic sediments (Zevallos, 1970).

The strong positive (+30 mgal) gravity anomaly over Banco Peru at $3°30'$ and $81°15'W$ indicates an excess mass of high-density rocks. T. Getts (pers. commun., 1976) determined density contrasts with respect to the thick sedimentary sections on each side of Banco Peru. Several simple geometric configurations were used, all of which yielded density contrasts of 0.3 to 0.6 gm/cc. On the basis of empirical mean densities with depth of sediments (Woollard, 1962, unpub. data), the rocks underlying Banco Peru have a minimum density of at least 2.7 gm/cc and a positive maximum density of 3.0 gm/cc. These densities fall within the range typical of mafic rocks or of mixtures of ultramafic and less dense rocks. Two dredges consisting of dense dolomicrite were recovered from the western and eastern sides of the bank (Kulm and others, this volume). If the dolomite section is thick, it may account for a large part of the density contrast.

Between Banco Peru and the coast, the negative values that prevail coincide with a basin containing in excess of 5,500 m of Cenozoic sediments. Along the coast in this region, gravity contours parallel the basin and align with basinward-dipping antithetic growth faults (H. Mason, pers. commun., 1975) on the northwest side of a shallow granitic basement high known from wells drilled around Zorritos (Zevallos, 1970).

SUBDUCTION TECTONICS

In many of the best-studied zones of convergence of oceanic and continental or island-arc-bounded plates, the process of convergence seems to add material above the subducting oceanic plate. In some zones, however, the evidence suggests that little accretion takes place. Considerable efforts have been made to discover: (1) the way in which an accreted wedge forms in the convergent margin and the composition of the wedge; (2) the relative roles of extension and compression across the margin and of uplift and sinking across the orogenic belt; and (3) the relative importance among these: accretion to the margin by abduction, tectonic erosion of the margin by the subducting oceanic plate, and tectonic removal of the margin by lateral faulting. The following discussion shows our interpretation of the evidence off northwest Peru that bears on these problems.

Cross sections of trench turbidites and of the oceanic-plate sediments lying under them were measured on the 8 seismic reflection records where these sediments were most clearly defined. Calculations indicate that the turbidite content ranges from 26% to 51% and averages 38%; the rest is pelagic sediment. Thus, in the recent past, about three-fifths of the sediment in the trench came from the oceanic plate.

According to Larson and Pitman (1972), subduction has occurred here at least since Cretaceous time. An interesting comparison is between the amount of sediment that could have been accreted to the Peruvian margin since that time and the size of the existing sediment prism defined by seismic profiling. From the multichannel seismic profile (Fig. 9, Profile J), the distance of the inner wall and

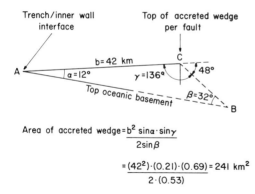

Figure 16. Trigonometric solution for cross-sectional area of the accreted wedge along Profile J (Fig. 9). A-B distance (42 km) between the trench-inner wall interface and a fault at km 31. The fault is believed to be the upper limit of the accreted wedge. Angle α is the sum of the 9° dip of the subducting oceanic basement plus the 3° slope of the sea floor at the inner wall. The dip of the fault (48°) at km 31 along segment B-C of the oblique triangle is determined from a depth section at two points on the fault. Dashed lines projected to B close the triangle. Area of the triangular cross section is 241 km². See text for discussion.

slope can be readily determined as can the dip of the acoustic basement of the oceanic plate under the wedge. This information is used to construct two sides of an oblique triangle (Fig. 16) whose other side must be approximated. The upper limit of the wedge is assumed to coincide with an interpreted fault at km 31 on Profile J. The dip of the fault is acoustically difficult to resolve but, using L. K. Wipperman's depth section, is estimated to be about 48°. The triangle is closed, and its area of 241 km² is assumed to represent the total cross-sectional area of accreted sediments within that traverse of the convergent margin.

For comparison, to calculate the area that could have accreted, the assumptions are:

1. Subduction of 5,000 km since Cretaceous time (Larson and Pitman, 1972) has been at the present trench; no mini-plates have collided to change the locus of subduction.
2. All oceanic plate sediment is scraped off into the wedge because there are soft sediments on top of the basalt throughout the Nazca plate section at DSDP sites 320 and 321 (Yeats and Hart, 1976). This situation is unlike that proposed by Moore (1975), that the scrape-off should occur above lithified sediment.
3. Because there has been little change in latitude of the Nazca plate (Yeats and Hart, 1976), an average thickness of oceanic sediment of about 250 m in this part of the trench has probably prevailed since Cretaceous time.
4. Accreted oceanic-plate sediments are assumed to have but 62% of their original volume due to compaction and tectonic dewatering. This assumption is based on average porosity reductions from DSDP sites 298 (Ingle and others, 1975) and 320 (Yeats and Hart, 1976).
5. The hemipelagic component is considered to have settled equally on all areas within the trench and margin and is not considered in the calculations, nor is the recycling or cannibalization of the materials slumped from the inner trench wall considered.

Therefore, 5,000 km $\times$ 0.25 km $\times$ 62% = 775 km² of cross-sectional area of pelagic sediment, which is about three-fifths of total trench sediment. This means that 1,292 km² entered the trench and could have accreted, compared to 241 km² now occupying the prism. Apparently, about four-fifths of the sediment is being removed from the convergent margin by subduction. Hussong and others (1976) came to the same conclusion from data gathered between 8° and 12°S, where oceanic-plate sediments are only about 100 m thick off that part of Peru.

For the following discussion of the broader aspects of subduction, refer to Figure 17, which schematically depicts major geologic features that are along a transect at 5°S that extends from the oceanic plate eastward about 720 km to the Iquitos Arch on the Brazilian craton. This section implies that subduction tectonics is not limited to the trench but affects the entire Andean orogen.

The Nazca plate is shown as a flexed slab outside the trench. On the basis of seismicity, this subducting slab dips gently, about 10°, under this part of the Andes (James, 1971; Ocola and Meyer, 1973; Stauder, 1975). At 5°S, the principal morphotectonic interpretation of compression is restricted to the rather minuscule folds at the base of the inner trench wall. There is no information to indicate

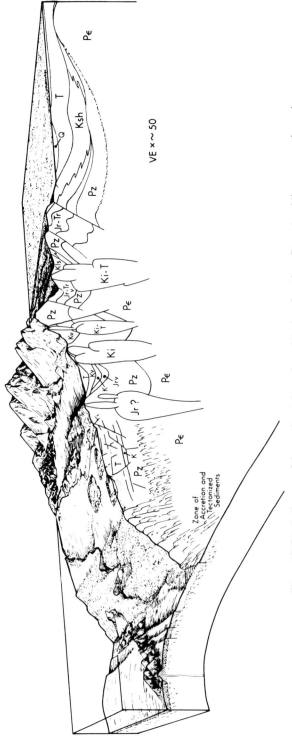

Figure 17. Block diagram of the continental margin and Andes of northern Peru. An oblique perspective and schematic cross section that extends 720 km along 5° S from the oceanic plate to the Amazon Basin. Vertical exaggeration varies but averages 50×. The oceanic part of the cross section is about 3× larger than the landward part in order to accentuate the features determined from this study. Drawn by W. T. Coulbourn.

compressional thrusting within the outer ridge, as has been suggested at 12°S (Hussong and others, 1976), and only a few of the seismic reflection records off northwest Peru show the folding in fore-arc basins and their landward migration from compressional uplift of the trench-slope break that exists off southern Peru and northern Chile (Coulbourn and Moberly, 1977). In fact, on land and at least as far seaward as there is good structural control (for example, Fig. 6 of oil field), the fault blocks indicate that there has been a tensional regime in the late Cenozoic.

Processes causing vertical tectonics and deformation in the Andes are undoubtedly deep-seated ones. Previous Andean geological studies were not viewed within the context of plate tectonics. Recent work by French geologists (Megard and others, 1971; Megard, 1973; Auboin and others, 1973; Audebaud and others, 1973; Julivert, 1973; Faucher and Savoyat, 1973) and the Peruvian geologist J. Paredes (1972) have provided new insights on the chronology and tectonics within the context of plate tectonics and have pointed out some contradictions and imprecisions in the theory.

One of the major conclusions of most of these authors is that the main Andean cordilleras have structures dictating an extensional stress field concomitant with 3,000-4,000 m of uplift since Pliocene time (Quechua phase). Woollard and Ocola (1973) in a study of shallow- and intermediate-depth earthquakes in Peru conclude that tensional fractures dipping toward the Andes from the east as well as from the west are evidence that the Andes are in an extensional stress field. The Andean orogeny with four phases began in Late Cretaceous (Santonian) (Megard, 1971; 1973) or mid-Cretaceous (Cenomanian) (Rutland, 1971) followed by Oligocene-Eocene, mid-Miocene, and the Quechua phase, which still continues with measurable uplift (G. P. Woollard, pers. commun., 1976).

Several mechanisms have been proposed for vertical uplift in the Andes, such as thermal expansion of magma and compositional contamination (Gough, 1973) or hydration along a subducting slab (Fyfe and McBirney, 1975), both of which would create essentially irreversible volume changes corresponding to uplift and tensional stress fields. If the Fyfe and McBirney model is applied to the cross section (Fig. 17), we postulate that hydration along the subducting slab underlies known igneous intrusives and volcanism, and depressions relate to dehydration. It is implicit that considerably more intrusives lie at depth below any surface manifestations for this area. As pointed out, depression with volcanism is inconspicuous, because the volume of volcanic rocks keeps pace with subsidence as magma is removed from below the volcanic axis. One degree farther south of the cross section, thousands of metres of the Cenozoic and Quaternary Calipuy Volcanics (Bellido, 1969), with no exposed base, are cut by stocks related to the composite Andean batholith. The model of Fyfe and McBirney in this case would have to be modified, because it does not consider the lateral variations in the geometry (Stauder, 1975) or segmentation (Rodriguez and others, 1976) of the subducting slab. These variations that occur along the Andean margin were determined from seismicity studies. Neither does the model consider probable changes through time in the dip of the subducting slab, which might allow a better fit to the known geology.

Although ages of igneous rocks in northern Peru are determined mostly by cross-cutting relationships rather than by radiometric methods, they appear to be younger to the east. The Jurassic age assigned to the intrusion near the coast is extrapolated by A. Aleman (pers. commun., 1974) from a radiometrically dated pluton in the Coastal Cordillera to the northeast. From cross-cutting relationships, O. Zevallos (pers. commun., 1974) can date the westernmost intrusives of northwest Peru only as pre–Late Cretaceous. They may be the root of a Mesozoic volcanic island arc that has since eroded, and they seem anomalously located with respect to the present subducting slab and nearness to the trench. It does not seem possible that temperatures could be high enough at 20 km depth and 75 km from the trench to mobilize magma from the subducting slab for the intrusions. It seems that (1) the Mesozoic subducting slab had to be dipping steeper than at present to attain melting temperatures, or (2) melting occurred with excess water (Fyfe and McBirney, 1975) for shallow generation, or (3) tectonic erosion has removed some of the leading edge of continental crust. The removal may have been by subduction, by a mechanism of lateral translation (Karig, 1974) of continental crust, as suggested by Megard (1973) in studies of oblique subduction in Peru, or by combinations thereof. No conclusions can be made from existing data, although the truncation of trends of crystalline basement at the coast suggests (3) tectonic erosion. Gough (1973), however, presumed from the presence of electrical conductivity zones supported by abnormal delay and absorption of seismic waves that a zone of partial melting could occur at a depth of 20 km, but this zone is under the Cordillera Oriental and above a slab dipping 25°.

An alternative application of the Fyfe and McBirney hypothesis might be that hydration events successively stepped eastward with time in correspondence to a decreasing dip of the subducting slab. Today, however, the slab may be shallower than the geotherm necessary for melting, which may account for the lack of recent volcanism in this sector of the Andes chain. Barazangi and Isacks (1976) remarked on the correlation of flat segments of the subducting Nazca plate with lack of recent volcanism here and in parts of Chile.

On the basis of eugeosynclinal-miogeosynclinal paired sedimentary facies, the Cretaceous instrusives (Ki on Fig. 17) are thought to be the root of a volcanic arc that persisted in the Cretaceous (Fischer, 1956; Hosmer, 1959; Megard, 1973). The intrusives may be composite and have extrusive counterparts that crop out to the northeast and continue into Ecuador, where several thousands of metres of andesites (Kennerly, 1973) grade westward into Late Cretaceous marine volcaniclastic formations. The Late Cretaceous formations of the Lancones synclinorium east of the Cerros de Amotape have many facies variations but are essentially flysch successions deposited as thick as 3,700 m in a rapidly subsiding basin (Morris and Aleman, 1975). Work (A. Aleman, pers. commun., 1974) south of the Lancones synclinorium has encountered dunites and peridotites associated with volcanic rocks which may have been partial sources for the marine volcanisclastic sediments. The significance and relationships of these rocks are not yet determined. A. Aleman provisionally suggested "[back-arc] rifting caused by a thermal anomaly forming a trough in which a series of basic volcanic rocks were emplaced" (free translation from Spanish).

In addition, the ages of some of the intrusives south and east of the Cerros de Amotape are in doubt. After they are dated, evaluation of their ages may materially modify this part of the schematic cross section and portrayal of subduction tectonics. Ongoing work by the Institute of Geological Sciences of Great Britain is clarifying some of the poorly understood geology of Ecuador (J. Baldock, pers. commun., 1975; Henderson and Evans, 1980). One of the significant findings is that two volcanic arcs are associated with the Cordillera Occidental of Ecuador: an older one to the west partly covered by a younger one to the east. Institute geologists have also better delineated the Dolores-Guayaquil Megashear north of Quito.

PROGRESO BASIN AND THE DOLORES-GUAYAQUIL MEGASHEAR

The Gulf of Guayaquil is an unusual embayment in the generally straight west coast of South America (Fig. 3). It is also structurally unusual in that it is underlain by the Progreso Basin, which has an east-west trend that is contrary to other Cenozoic basins that parallel the gross north-south structural grain of the Andes (Fig. 2). Marchant (1961), Malfait and Dinkelman (1972), Goosens (1973), Faucher and Savoyat (1973), and Campbell (1974, 1975) have suggested that the origin of the gulf and basin lies with right-lateral slip on the Dolores-Guayaquil Megashear (Case and others, 1971).

The Dolores-Guayaquil Megashear, a major structural discontinuity in northwestern South America, trends about 1,600 km south-southwest from northwestern Colombia into the Gulf of Guayaquil (Fig. 18). From the mapped limit of the fault zone on land, Shepherd and Moberly (1975), on the basis of these marine geophysical investigations, extended the Dolores-Guayaquil Megashear system southwestward across the continental margin into the landward wall of the Peru-Chile Trench. They also proposed a model for the origin of the Progreso Basin based on the right-lateral slip movement along the megashear.

Earthquake epicenters align with the southern part of the megashear (Goosens, 1973; Campbell, 1974), and on its trace in Ecuador, Stauder (1975) determined a right-lateral focal mechanism. The region encompassing coastal Ecuador and Colombia thus appears to be a separate mini-plate. Paucity of earthquakes along the megashear, however, indicates that the megashear is not particularly active at

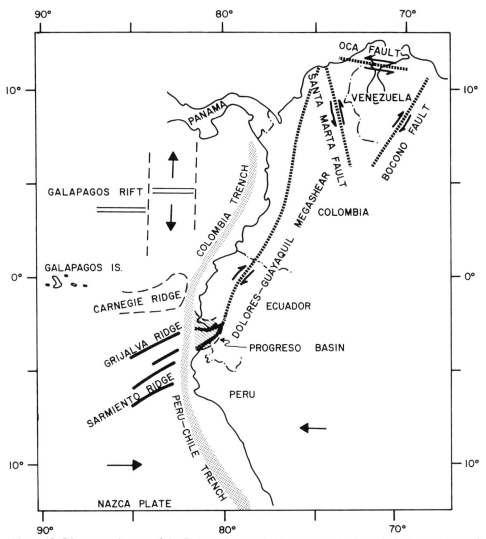

Figure 18. Diagrammatic map of the Dolores-Guayaquil Megashear and other principal morphotectonic features. Contemporary relative movements of plates and blocks are shown by arrows.

present. Indeed, Lonsdale (1978) has concluded that at the latitutde of Ecuador, the Ecuador Trench is the only significant boundary today between the Nazca and South American plates.

In Colombia and Ecuador, the megashear separates rocks on the west, interpreted as having been Mesozoic oceanic crust, from Precambrian and Paleozoic continental crust on the east (Case and others, 1971; Case and others, 1973; Gansser, 1973; Campbell, 1974; Irving, 1975). In Colombia and Ecuador, the former oceanic crust is overlain by Mesozoic and Cenozoic open-ocean, deep-water sandstones, cherts, tuffs, and siliceous shales associated with diabases and pillowed basalts, and intruded by granitic to dioritic plutons (Case and others, 1971; Malfait and Dinkelman, 1972). In Ecuador, the tholeiitic, former oceanic crust is called Piñon Formation, Piñon Complex, or Basic Igneous Complex (Goosens and Rose, 1973; Goosens and others, 1977) and Diabase Formation (Sauer, 1971). There are petrologically similar terranes in Colombia, but apparently with fewer known structural complications. Contributors to the study of the Piñon Formation and its correlatives include Sauer (1971), Goosens and Rose (1973, 1975), Goosens and others (1977), Faucher and

Savoyat (1973), Kennerly (1973), Ruegg (1967), Thalmann (1943, 1946), and Henderson (1979). Most of these persons point out that the petrologic nature of the Piñon Formation is similar wherever it crops out in the coastal lowlands and on the western flank of the Cordillera Occidental of Ecuador, Colombia, and northernmost Peru. They also affirm that the Piñon Formation is found in widely varying structural settings and diachronous ages. Although Henderson (1979) believes that the Piñon Formation is an old arc terrane rather than allochthonous oceanic crust emplaced landward of the present Peru-Chile Trench, we favor the interpretation of many of the other investigators, namely, that at the present day, younger oceanic crust is subducting under older oceanic crust (Piñon Formation), whose boundary with continental crust is rather sharply delineated along the Dolores-Guayaquil Megashear and coincides with the Cordillera Occidental of Ecuador and Colombia. Further, Faucher and Savoyat (1973) suggested that the eastern boundary of the Piñon Formation was a subduction zone in the Jurassic that would have had to be aborted or become inactive since the Late Cretaceous, as constrained by absolute dates of 110–135 m.y. for the Piñon Formation. A new subduction zone would then have had to be stepped westward possibly by mid-oceanic-plate subduction in the Late Cretaceous (Faucher and Savoyat, 1973) or Paleocene (Malfait and Dinkelman, 1972). Perhaps the Early Cenozoic subduction was oblique (Karig, 1974).

The occurrence of former oceanic crust on the north side of the Gulf of Guayaquil and continental crust on the south has been a tectonic problem comparable to the juxtaposition of the two crusts at the Dolores-Guayaquil Megashear. Campbell (1975) suggested that a "great E-W transverse line of faulting" appears to align with the Tumbes-Guayana Megashear (Fig. 2) on the north side of the Amazon Basin. It also aligns with old metallogenic transverse faults in the high Andes (Goosens, 1972), earthquake focal centers (Goosens, 1973; Campbell, 1975) in that region, and a cluster of focal centers on the south side of Guayaquil. Campbell (1975) further alluded that the Gulf of Guayaquil is due to the interaction between the Tumbes-Guayana and Dolores-Guayaquil megashears.

Thus, within our area of investigation, the Dolores-Guayaquil Megashear is a first-order tectonic boundary, comparable in scale with the Peru-Chile Trench. The scale of movement along the megashear and the history of its interaction with the present trench deserve comment. In Colombia, strain on the Dolores-Guayaquil Megashear is distributed over several faults in a broad zone. Feininger (1970) and Irving (1975) have mapped dextral offsets of intersecting faults and plutons that may represent only the latest displacements of about 30 km. In the Gulf of Guayaquil, if the steep free-air gravity-anomaly gradients on opposite sides of the gulf (Fig. 15) are the deep-seated margins of a rhombochasm-like basin, then the dextral displacement is about 80 to 100 km.

We believe that the curved trace of the Dolores-Guayaquil Megashear into the Gulf of Guayaquil and across the continental margin and its right-lateral movement (Case and others, 1971; Campbell, 1975) formed a pull-apart basin (Crowell, 1974), namely, the Progreso Basin (Fig. 18), with more than 6,000 m of sediment fill. As movement along the megashear progressed, sediments from high-standing lands poured into a continuously widening hole. With widening, younger stratigraphic units lap basinward, as is known from outcrop (Fig. 3) and subsurface data (Zevallos, 1970).

Between Isla Puna (Fig. 3) and the coastline and extending southeasterly toward the Andes lies the Jambeli Graben, the extension of the Progreso Basin. It follows the curvature of the converging margins of the basin, where 9,000 m of post-Cretaceous sediments are estimated to be present (Faucher and Savoyat, 1973). In the same area, H. Mason (pers. commun., 1975) calculated at least 5,200 m of throw on the southern margin of the graben.

Also in the Gulf of Guayaquil is Banco Peru. Its position on the continental margin, apparent isolation, and presumed rock constitution need explanation. It has been determined from gravity calculations that Banco Peru is underlain by rocks having densities in the range of mafic to utlramafic rocks, such as those of the Piñon Complex. It is surrounded by strong gravity gradients with faults seen on seismic reflection profiles on the northwest and southwest sides; the latter follows a bathymetric notch to the trench-slope break about 1,300 m abvoe the trench floor (MC-34).

Figure 19 illustrates one possible explanation for subduction of younger oceanic crust under older oceanic crust of the Piñon Formation south of the Carnegie Ridge, the curve of present-day Dolores-Guayaquil Megashear, and the origin of Progreso Basin and Banco Peru. No specific times for stages 2

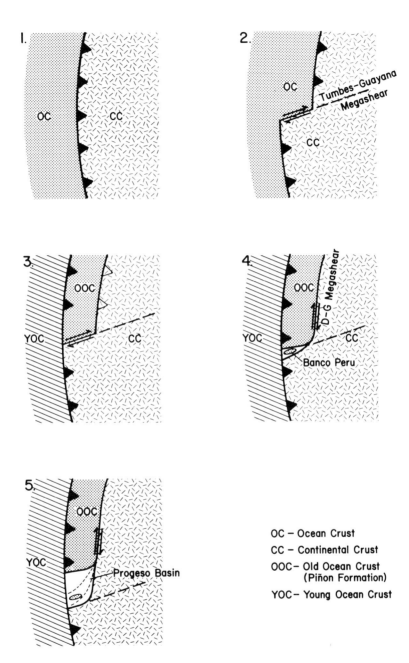

Figure 19. A sequential diagram illustrating a possible explanation for the Progreso Basin, Banco Peru, and subduction of younger oceanic crust under older oceanic crust north of the Progreso Basin. Stage 1 is Jurassic or older; stage 5 is contemporary time, but no time is specified for stages 2 through 4 due to uncertain ages of the rocks and events affecting them. The floor of the Progreso Basin may either be (1) attenuated marginal rocks, (2) a complex of volcanic rocks and sediments, (3) young oceanic crust now subducting, or (4) a combination of all of them.

through 4 are given because the knowledge of ages of rocks and events bearing on them is uncertain. Stage 1, however, would be Jurassic or older, and stage 5 is contemporary time. It shows Banco Peru as an isolated slice of older oceanic crust along a transform (strike-slip) fault activated by the Tumbes-Guayana Megashear. Isolated high-standing blocks are not uncommon in pull-apart basins (Crowell, 1974).

Other hypotheses are that Banco Peru could be a thick section of dolomite on mafic rocks or an intrusion of mafic and ultramafic rocks from melting of the subducting slab. The former would still require an uplifted mafic block. The latter idea, however, is constrained by the fact that the current slab dips 10°. With such a shallow dip, the upper surface would be located 12 km under Banco Peru. It is not conceivable that melting or partial melting could occur at these depths. Theoretical considerations and the trends of volcanoes and trenches call for depths of 100 to 150 km for partial melting to take place (Wyllie, 1971; Marsh, 1976). That would necessitate a dip of 60° or greater on a subducting slab, which is about 30° greater than the steepest dip calculated anywhere along the Peru-Chile Trench subduction zone (Rodriguez and others, 1976). Even if the intrusion were not magmatic, but rather a plastic diapir of ultramafic rock from the mantle, as the Rhonda Complex had been proposed to be (Loomis, 1972), the temperatures should not be high enough for a diapir so close to the trench. Thus, we prefer the explanation shown in Figure 19.

HYDROCARBON ACCUMULATIONS AT CONVERGENT MARGINS

The oil and gas fields of northwest Peru and the Santa Elena Peninsula of Ecuador (Fig. 20) are notably unique among other fields of the world in that they are the only commercial ones located so close to an active convergent margin. Existing productive limits near Talara are as close as 50 km to the axis of the trench, but the true seaward limits of the coastal fields of Peru have not yet been established. The deepest production platform is in 90 m of water about 8 km from the shore. These complexly faulted coastal fields of Peru have attained giant status by having produced more than 941 million barrels of oil by 1976 (Amato, 1977). Those of coastal Ecuador have a cumulative production to 1976 of 110 million barrels. Moreover, there have been several discoveries of oil and gas off the shores of Peru and Ecuador. Bottom-simulating reflectors suggest that there are abundant deep-marine clathrate hydrocarbons.

Exploration for hydrocarbons on active convergent margins currently is not favored by explorationists primarily because they believe that fore-arc basins generally have complex geology as well as reservoirs of poor quality and low geothermal gradients for maturation of organo-rich sediments to hydrocarbons (Yarborough, 1977). Many of the basins lie well below sea level. Off Peru, however, pulses of regional uplift through the Cenozoic that culminated in the Quechua phase exposed the fore-arc basin, allowing easier methods of exploration and drilling compared to those for submerged fore-arc basins. Also, throughout the Cenozoic, continued tensional tectonics provided abundant fault-block traps. Presumably during most of the Cenozoic, oceanographic conditions were similar to present-day ones at this tropical coast on the eastern edge of an ocean. Upwelling favored the production of marine organic matter, some of which was preserved at the oxygen-minimum layer. Proximity to the coast may have accounted for additional, nonmarine, organic detritus. Yet, detrital sediment was not so abundant as to overwhelm by volume the organic sediment. Attenuation of crust under a pull-apart basin may have allowed the flux of subcrustal heat sufficient for hydrocarbon maturation. Space limitations do not allow us to present the details of hydrocarbon generation and entrapment along the Peru-Ecuador margin, much less to review hydrocarbon occurrence along convergent margins in this paper.

CONCLUSIONS

Information about convergent plate margins has accumulated rapidly during the past decade. Many

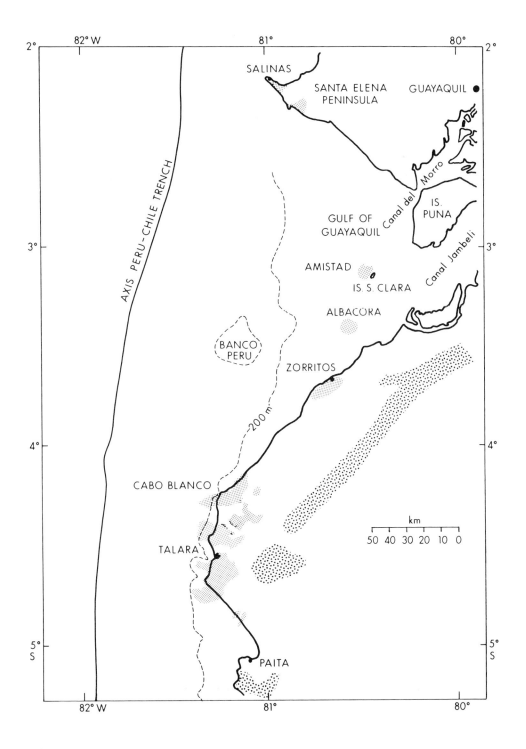

Figure 20. Oil and gas fields of northwestern Peru and southwestern Ecuador. Heavy stippling = massif of metamorphic Paleozoic rocks and later granitics composing the Cordillera Coastal. Fine stippling = oil and gas fields.

features are common from one margin to another, but many others vary markedly. Much important information about subduction remains unknown. We believe that the following conclusions from these South American studies will need to be considered in future syntheses of the tectonics of convergence zone.

1. *Major structural features of Peru and Ecuador can be traced seaward by geophysics, but not small-scale ones.* Fore-arc basins, a pull-apart basin, and alternating horsts and graben exposed along the coast of northwest Peru can be traced by seismic reflection profiles and the pattern of free-air gravity anomalies on to the continental shelf and upper slope. A transform fault, the Dolores-Guayaquil Megashear, can be followed from Ecuador across the continental margin into the trench.

    Neither single-channel nor multi-channel seismic records, however, show the wealth of block faults and olistostrome-like glides known from geologic mapping and subsurface studies of the onshore and offshore oil fields of the fore-arc basin at Talara. To the contrary, the profiles show structures that look like folds.

2. *The zone of plate convergence shows little evidence of compression in the shallow structure.* In a broad traverse from the Nazca lithospheric plate into the South American lithospheric plate, the main indication of crustal shortening is the earthquake evidence of the Nazca slab plunging eastward under the continent. Surficial structural evidence of compressive tectonics is restricted to two belts: (1) a narrow prism immediately landward of the trench, where morphotectonics and seismic reflection records suggest but do not prove imbricate thrust faults and tight folds, and (2) the wide Subandean province of foreland-type folding between the main cordillera and the craton.

    The outer ridge appears to be a simple flexure of the lithosphere before it sinks below the trench; normal faults there, down the outer trench slope, and in the trench itself indicate an extensional regime. Indeed, the apparent decoupling at the trench without the transmitting of appreciable stress from one plate to the other seems to have allowed the degree of subduction of sediment discussed in conclusion 3 below.

    Landward of the small accretionary prism, with its apparent compressional tectonics, all young faulting is extensional: in the upper slope, fore-arc basins, the horsts and graben of the coastal massifs, and the main cordillera. Moreover, although some basin floors have subsided, all of these provinces have generally been warped upward in a series of pulses to the present day.

3. *Subduction off northwest Peru is a very efficient process.* Not only is the accreted prism a small fraction of the total sediment that was available for accretion, but also it is probable that part of the edge of the continental crust has been removed by subduction.

    According to our crude calculations, about four-fifths of the total pelagic, turbidite, and hemipelagic sediment that reaches the Peru Trench is removed by subduction. Different estimates of the thickness and compaction of sediment or of the role of recycling by slumping would not change the estimates by any significant amount. Miscalculations may have been made in (1) the volume of the accreted wedge, or (2) the amount of Nazca-Farallon plate that has been subducted at the present trench, or both. But the multi-channel data, supported by gravity interpretations and the pattern of earthquakes, constrain the former within reasonable limits. Without an obvious post-Cretaceous suture zone in the land geology or other evidence of a change in the trench location, it appears that subduction has continued in the Peru Trench throughout the Cenozoic.

    Igneous and metamorphic rocks are exposed at the coast, and the strike of their structures trends offshore. These parts of the continental crust lie under the landward edge of the fore-arc basins and grabens along the coast. Coastal exposures are thus about 80 km from the trench axis. Interpretations of the style of deformation in reflection profiles and the position of the main gravity-anomaly inflection suggest that the edge of continental crust may be within 20 to 40 km of the trench axis. From any reasonable geometry of a subducting slab and geotherms under them, plutons now at the coast must have been emplaced farther from a trench than they now lie. That consideration, as well as the strike of basement structures, indicates that a substantial width of continental crust is gone. Possibly it was translated as a slice obliquely north or south, but to the south, Peru and Chile present the same problem of continental crust close to the trench, whereas to the north the accreted crustal terrane of coastal Ecuador and Colombia appears to be of oceanic origin. We conclude, therefore, that the lower outer edge of the continental crust has been stoped away by subduction.

4. *Structures at the Gulf of Guayaquil were caused by movements of a small plate.* The tectonic pattern along the western margin of South America is interrupted at the Gulf of Guayaquil. From a triple junction at the Peru Trench, a transform fault—the Dolores-Guayaquil Megashear—strikes northeast under the gulf and thence north-northeast, separating a small plate of coastal Ecuador and Colombia from the South American plate.

Isolation of the small plate of oceanic crust, followed by its northward lateral movement between the Nazca and South American plates, allowed Progreso Basin to form as a pull-apart basin and left Banco Peru as a sliver of former crust.

5. *Coastal northwestern South America is a convergent margin where petroleum has accumulated.* Large fields in a fore-arc basin and new discoveries in a pull-apart basin indicate that conditions indeed exist for the generation and entrapment of hydrocarbons in what generally is considered to be a poor tectonic setting for them.

ACKNOWLEDGMENTS

We are pleased to acknowledge the funding of this work by the National Science Foundation (ID071-04207-A04). We appreciate the interest of the late George Woollard, Vern Kulm, and Donald Hussong and the assistance of Leo Ocola of Peru, Hector Ayon of Ecuador, and Larry Wipperman and William Coulbourn. We appreciate the helpful reviews of this manuscript by Patrick Coleman and Alfred Fischer.

REFERENCES CITED

Amato, F. L., 1977, Petroleum developments in South America, Central America, Mexico, and Caribbean areas in 1976: American Association of Petroleum Geologists Bulletin, v. 60, p. 1578-1635.

Auboin, J., and others, 1973. De quelques problemes géologiques et géomorphologiques de la Cordillere des Andes: Revue de Géographie Physique et de Géologie Dynamique, v. 15, fasc. 1-2, p. 207-216.

Audebaud, E., and others, 1973, Les traits, géologiques essentiels des Andes Centrales (Pérou-Bolivie): Revue de Géographie Physique et de Géologie Dynamique, v. 15, fasc, 1Ç2, p. 73Ç114.

Baldry, R. A., 1938, Slip-planes and breccia zones in the Tertiary rocks of Peru: Quarterly Journal of the Geological Society of London, v. 94, p. 347-358.

Barazangi, M., and Isacks, B. L., 1976, Spatial distribution of earthquakes and subduction of the Nazca plate beneath South America: Geology, v. 4, p. 686-692.

Beck, R. and Lehner, P., 1974, Oceans, new frontier in exploration: American Association of Petroleum Geologists Bulletin, v. 58, p. 376-395.

Bellido, E., 1969, Sinopsis de la geología del Perú: Servicio de Geología y Minería del Perú, Boletín 22, 54 p.

Brown, C. B., 1938, On a theory of gravitational sliding applied to the Tertiary of Ancon, Ecuador: Quarterly Journal of the Geological Society of London, v. 94, p. 359-370.

Bruhn, R. L., and Dalziel, I.W.D., 1977, Destruction of the Early Cretaceous marginal basin in the Andes of Tierra del Fuego, *in* Talwani, M., and Pitman, W. C., III, eds., Island arcs, deep-sea trenches, and back-arc basins: American Geophysical Union, Maurice Ewing Series, v. 1, p. 395-405.

Campbell, C. J., 1974, Ecuadorian Andes, *in* Spencer, A. M., ed., Mesozoic-Cenozoic orogenic belts: Geological Society of London Special Publication 4, p. 725-732.

———1975, Ecuador, *in* Fairbridge, R. W., ed., The encyclopedia of world regional geology, Part 1, Western hemisphere: Stroudsburg, Pa., Dowden, Hutchinson, and Ross, p. 261-270.

Carey, S. W., 1958, The tectonic approach to continental drift, *in* Carey, S. W., convener,

Symposium on the present status of the continental drift hypothesis: Hobart, Tasmania, University of Tasmania, p. 177–355.
Case, J. E., and others, 1973, Trans-Andean geophysical profile, southern Colombia: Geological Society of America Bulletin, v. 84, p. 2895–2905.
———1971, Tectonic investigations in western Columbia and eastern Panama: Geological Society of America Bulletin, v. 83, p. 2685–2712.
Cossio, A., and Jaén, H., 1967, Geología de los cuadrángulos de Puemape, Chocope, Otuzco, Trujillo, Salaverry y Santa: Servicio de Geología y Minería del Perú, Boletín 17, 141 p.
Coulbourn, W. T., and Moberly, R., 1977, Structural evidence for the evolution of fore-arc basins off South America: Canadian Journal of Earth Sciences, v. 14, p. 102–116.
Crowell, J. C., 1974, Origin of late Cenozoic basins in southern California, in Dickinson, W. R., ed., Tectonics and sedimentation: Society of Economic Paleontologists and Mineralogists Special Publication 22, p. 190–204.
De Loczy, L., 1970, Transcurrent faulting in South American tectonic framework: American Association of Petroleum Geologists Bulletin, v. 54, p. 2111–2119.
Dewey, J. F., and others, 1973, Plate tectonics and the evolution of the alpine system: Geological Society of America Bulletin, v. 84, p. 3137–3180.
Dorreen, J. M., 1951, Rubble bedding and graded bedding in Talara Formation of northwestern Peru: American Association of Petroleum Geologists Bulletin, v. 35, p. 1829–1849.
Du Toit, A. L., 1937, Our wandering continents: London, Oliver and Boyd, 366 p.
Faucher, B., and Savoyat, E., 1973, Esquisse géologique des Andes de l'équateur: Revue de Géographie Physique et de Géologie Dynamique, v. 15, p. 115–142.
Feininger, T., 1970, Palestina fault, Colombia: Geological Society of America Bulletin, v. 81, p. 1201–1216.
Fischer, A. G., 1956, Desarrollo geológico del noroeste peruano durante el Mesozoico: Sociedad Geológica del Perú, Boletín, Tomo 30, p. 117–190.
Fyfe, W. S., and McBirney, A. R., 1975, Subduction and the structure of andesite volcanic belts: American Journal of Science, v. 275-A, p. 285–297.
Gansser, A., 1973, Facts and theories on the Andes: Quarterly Journal of the Geological Society of London, v. 129, p. 93–131.
Getts, T. R., 1975, Gravity and tectonics of the Peru-Chile Trench and eastern Nazca plate, 0°–33°S [M.S. thesis]: Honolulu, University of Hawaii, 103 p.
Giletti, B. J., and Day, H. W., 1968, Potassium-argon ages of igneous rocks in Peru: Nature, v. 220, p. 570–572.
Goosens, P. J., 1972, Metallogeny of the Ecuadorian Andes: Economic Geology, v. 67, p. 458–468.
———1973, Características estructurales de la parte noroeste del margen continental de América del Sur relacionadas con la tectónica de placas [abs.]: IUGG Conference on Geodynamics, Lima, Peru.
Goosens, P. J., and Rose, W. I., 1973, Chemical composition and age determination of tholeiitic rocks in basic igneous complex, Ecuador: Geological Society of America Bulletin, v. 84, p. 1043–1052.
———1975, Geochemistry of basalts of the basic igneous complex of northwestern South America and Panama [abs.]: EOS (American Geophysical Union Transactions), v. 50, p. 474.
Goosens, P. J., Rose, W. I., and Flores, D., 1977, Geochemistry of tholeiites of the basic igneous complex of northwestern South America: Geological Society of America Bulletin, v. 88, 1711–1720.
Gough, D. I., 1973, Dynamic uplift of Andean mountains and island arcs: Nature; Physical Science, v. 242, p. 39–41.
͞.m, C. K., and Herrera, L. J., 1963, Role of Subandean fault systems in tectonics of eastern Peru and Ecuador: American Association of Petroleum Geologists, Memoir 2, p. 47–61.
Handschumacher, D. W., 1976, Post-Eocene plate tectonics of the eastern Pacific in Sutton, G. H., and others, eds., The geophysics of the Pacific Ocean basin and its margin: American Geophysical Union Monograph 19, p. 177–202.
Harrington, H. J., 1962, Paleogeographic development of South America: American Association of Petroleum Geologists Bulletin, v. 46, p. 1773–1814.
———1975, South America, in Fairbridge, R. W., ed., The encyclopedia of world regional geology, Part 1, Western hemisphere: Stroudsburg, Pa., Dowden, Hutchinson, and Ross, p. 456–465.
Hayes, D. W., 1966, A geophysical investigation of the Peru-Chile Trench: Marine Geology, v. 4, p. 309–351.
Helwig, J., 1972, Stratigraphy, sedimentation, paleogeography, and paleoclimates of Carboniferous ("Gondwana") and Permian of Bolivia: American Association of Petroleum Geologists Bulletin, v. 56, p. 1008–1033.
Henderson, W. G., 1979, Cretaceous to Eocene volcanic arc activity in the Andes of northern Ecuador: Geological Society of London Journal, v. 136, p. 367–378.
Henderson, W. G., and Evans, C. D. R., 1980, Ecuadorian subduction system: Discussion: American Association of Petroleum Geologists Bulletin, v. 64, p. 280–283.

Herron, E. M., 1972, Sea-floor spreading and the Cenozoic history of the east-central Pacific: Geological Society of America Bulletin, v. 83, p. 1671–1692.

Hey, R. N., 1975, Tectonic evolution of the Cocos-Nazca Rise [Ph.D. thesis]: Princeton University, 169 p.

———1977, Tectonic evolution of the Cocos-Nazca spreading center: Geological Society of America Bulletin, v. 88, p. 1414–1420.

Hosmer, H. L., 1959, Geology and structural development of the Andean system of Peru [Ph.D. thesis]: Ann Arbor, University of Michigan, 291 p.

Howell, D. W., and McDougall, K. A., eds., 1978, Mesozoic paleogeography of the western United States: Pacific Section Society of Economic Mineralogists and Paleontologists, Symposium 2, 573 p.

Hussong, D. M., and others, 1976, Crustal structure of the Peru-Chile Trench: 8°–12°S latitude, in Sutton, G. H., and others, Geophysics of the Pacific Ocean basin and its margin: American Geophysical Union Monograph 19, p. 71–86.

Ingle, J. C., Jr., and Scientific Party, 1975, Site 298, in Karig, D. E., Ingle, J. C., Jr., and others, eds., Initial reports of the Deep Sea Drilling Project, Volume 31: Washington, D.C., U.S. Government Printing Office, p. 317–350.

Irving, E. M., 1975, Structural evolution of the northernmost Andes, Colombia: U.S. Geological Survey Professional Paper 846, 47 p.

James, D. E., 1971, Plate tectonic model for the evolution of the central Andes: Geological Society of America Bulletin, v. 82, p. 3325–3346.

Jenks, W. F., 1956, Peru, in Jenks, W. F., ed., Handbook of South American geology: Geological Society of America Memoir 65, p. 219–247.

Julivert, M., 1973, Les traits structuraux et l'évolution des Andes Colombiennes: Revue de Géographie Physique et de Géologie Dynamique, v. 15, p. 143–156.

Karig, D. E., 1974, Tectonic erosion at trenches: Earth and Planetary Science Letters, v. 21, p. 209–212.

Kennerly, J. B., 1973, Geology of the Loja province, southern Ecuador: Institute of Geological Sciences [U.K.] Report 23, 34 p.

Krause, D. C., 1965, Equatorial shear zone: Canada, Geological Survey, Paper 66-14, p. 400–443.

Kulm, L. D., and Fowler, G. A., 1974, Oregon continental margin structure and stratigraphy, a test of the imbricate thrust model, in Burk, C. A., and Drake, C. L., eds., The geology of continental margins: New York, Springer-Verlag, p. 261–283.

Kulm, L. D., and others, 1973, Tholeiitic basalt ridge in the Peru Trench: Geology, v. 1, p. 11–14.

Kulm, L. D., Schweller, W., and Masias, A., 1977, A preliminary analysis of the subduction process along the Andean continental margin, 6° to 45°S, in Talwani, M., and Pitman, W. C., III, eds., Island arcs, deep sea trenches, and back-arc basins: American Geophysical Union, Maurice Ewing Series, v. 1, p. 285–301.

Kulm, L. D., and others, 1981, Late Cenozoic carbonates on the Peru continental margin: lithostratigraphy, biostratigraphy, and tectonic history, in Kulm, L. D., and others, eds., Nazca plate: Crustal formation and Andean convergence: Geological Society of America Memoir 154 (this volume).

Larson, R. L., and Pitman, W. C., III, 1972, World-wide correlation of Mesozoic magnetic anomalies and its implications: Geological Society of America Bulletin, v. 83, p. 3645–3662.

Lonsdale, P., 1978, Ecuadorian subduction system: American Association of Petroleum Geologists, v. 62, p. 2454–2477.

Loomis, T. P., 1972, Diapiric emplacement of the Rhonda high-temperature ultramafic intrusion, southern Spain: Geological Society of America Bulletin, v. 83, p. 2475–2496.

Malfait, B. I., and Dinkelman, M. G., 1972, Caribbean tectonic and igneous activity and the evolution of the Caribbean plate: Geological Society of America Bulletin, v. 83, p. 251–272.

Mammerickx, J., and others, 1975, Morphology and tectonic evolution of the east-central Pacific: Geological Society of America Bulletin, v. 86, p. 111–118.

Marchant, S., 1961, A photogeological analysis of the structure of the western Guayas province, Ecuador: Quarterly Journal of the Geological Society of London, v. 117, p. 215–232.

Marsh, B. D., 1976, Mechanics of Benioff zone magmatism, in Sutton, G. H., and others, eds., The geophysics of the Pacific Ocean basin and its margin: American Geophysical Union Monograph 19, p. 337–350.

Martinez, M., 1970, Geología de basamento Paleozoico en las Montañas de Amotope y posible origen del petróleo en rocas Paleozoicas del noroeste del Peru: Primer Congreso Latinoamericano de Geología, Tomo 2, p. 105–138.

Masias, J. A., 1976, Morphology, shallow structure, and evolution of the Peruvian continental margin, 6° to 18°S [M.S. thesis]: Corvallis, Oregon State University, 92 p.

Maxwell, J. C., 1959, Turbidite, tectonic and gravity transport, Northern Apennine Mountains, Italy: American Association of Petroleum Geologists Bulletin, v. 43, p. 2701–2719.

Megard, F., 1973, Étude géologique d'une transversale des Andes au-niveau de Pérou Central [Ph.D. thesis]: Académie de Montpellier, Université des Sciences et Techniques du Languedoc, 263 p.

Megard, F., and others, 1971, La chaine hercyniene au Pérou et en Bolivie: Prèmiers Résultats: France, Office de la Recherche Scientifique et Technique Outre-Mer, Cahiers, Serie Geologie, v. 3, p. 5–43.

Moore, J. C., 1975, Selective subduction: Geology, v. 3, p. 530Ç532.

Morris, R. C., and Aleman, A., 1975, Sedimentation and tectonics of middle Cretaceous Copa Sombero Formation in northwest Peru: Sociedad Geológica del Perú, Boletín, Tomo 48, p. 49–64.

Myers, J. S., 1974, Cretaceous stratigraphy and structure, western Andes, between 10œÇ10œ30»: American Association of Petroleum Geologists Bulletin, v. 58, p. 474Ç487.

Ocola, L. C., and Meyer, R. P., 1973, Crustal structure from the Pacific Basin to the Brazilian Shield between 12œ and 30œ south latitude: Geological Society of America Bulletin, v. 84, p. 3387Ç3403.

Paredes, J., 1972, Étude géologique de la Feuille de Jauja [Ph.D. thesis]: Académie de Montpellier, Université des Sciences et Techniques des Languedoc, 79 p.

Paredes, M. P., 1958, Terciario de la Brea y Pariñas y area de Lobitos [M.S. thesis]: Universidad Nacional de San Augustín, Arequipa, 36 p.

Paz, M. H., 1974, Estudio estratigráfico del subsuelo de los yacimentos Zapotal–La Tuna–Coyonitas–Ronchudo Petróleos del Perú, operaciones noroeste, Talara: Abs. III Congreso Peruano de Geología, Sociedad Geológica del Perú, Julio.

Prince, R. A., 1974, Deformation in the Peru-Chile Trench [M.S. thesis]: Corvallis, Oregon State University, 88 p.

Rodriguez, R., Cabré, E. R., and Mercado, A., 1976, Geometry of the Nazca plate and its geodynamic implications, in Sutton, G. H., and others, eds., The geophysics of the Pacific Ocean basin and its margin: American Geophysical Union Monograph 19, p. 87–103.

Ruegg, W., 1967, El margen suroriental de la cuenca Para-Andina de Sechura en el noroeste del Perú: Sociedad Geológica del Perú, Boletín, Tomo 40, p. 99–121.

Rutland, R. W. R., 1971, Andean orogeny and sea-floor spreading: Nature, v. 253, p. 252–255.

Sangree, J. B., and Widmier, J. M., 1978, Seismic stratigraphy and global changes of sea level, Part 9: Seismic interpretation of clastic depositional facies: American Association of Petroleum Geologists Bulletin, v. 62, p. 752-771.

Sauer, W., 1971, Geologie von Ecuador: Berlin, Gebruder Borntrager, 316 p.

Scholl, D. W., von Huene, R., and Ridlon, J. B., 1968, Spreading of the ocean floor: Undeformed sediments in the Peru-Chile Trench: Science, v. 159, p. 869–871.

Schweller, W. J., Kulm, L. D., and Prince, R. A., 1981, Tectonics, structure and sedimentary framework of the Peru-Chile Trench, in Kulm, L. D., and others, eds., Nazca plate: Crustal foramtion and Andean convergence: Geological Society of America Memoir 154 (this volume).

Seely, D. R., Vail, P. R., and Walton, G. G., 1974, Trench slope model, in Burk, C. A., and Drake, C. L., eds., The geology of continental margins: New York, Springer-Verlag, p. 249–260.

Shepherd, G. L., 1979, Shallow crustal structure and marine geology of a convergence zone, northwest Peru and southwest Ecuador [Ph.D. thesis]: Honolulu, University of Hawaii, 201 p.

Shepherd, G. L., and Moberly, R., 1975, Southern extension of the Dolores-Guayaquil Megashear across the continental margin of northwest Peru and the Gulf of Guayaquil [abs.]: EOS (American Geophysical Union Transactions), v. 56, p. 442.

Shepherd, G. L., Wipperman, L. K., and Moberly, R., 1973, Shallow crustal structure of the Peruvian continental margin: Geological Society of America Abstracts with Programs, v. 5, p. 103.

Stauder, W., 1975, Subduction of the Nazca plate under Peru as evidenced by focal mechanisms and by seismicity: Journal of Geophysical Research, v. 80, p. 1053–1064.

Summerhayes, C. P., and others, 1979, Surficial slides and slumps on the continental slope and rise of South West Africa; a reconnaissance study: Marine Geology, v. 31, p. 265–277.

Thalmann, H. E., 1943, Upper Cretaceous limestones near San Juan, Province Chimborazo (western Andes), Ecuador: Geological Society of America Bulletin, v. 54, p. 1827–1828.

——1946, Micropaleontology of Upper Cretaceous and Paleocene in western Ecuador: American Association of Petroleum Geologists Bulletin, v. 30, p. 337–347.

Travis, R. B., 1953, La Brea-Pariñas oil field, northwestern Peru: American Association of Petroleum Geologists Bulletin, v. 37, p. 2093–2118.

van Andel, T. H., Phillips, J. D., and von Herzen, R. P., 1969, Rifting origin for the Vema fracture zone in the North Atlantic: Earth and Planetary Science Letters, v. 5, p. 296–300.

Watts, A. B., and Talwani, M., 1975, Gravity effect of downgoing slabs beneath island arcs: Geological Society of America Bulletin, v. 86, p. 1–4.

Weeks, L. G., 1947, Paleogeography of South America: American Association of Petroleum Geologists Bulletin, v. 31, p. 1194–1241.

Woollard, G. P., and Ocola, L. C., 1973, The tectonic pattern of the Nazca plate: Geofiśica Panamericana, v. 2, p. 125–149.

Wyllie, P. J., 1971, The dynamic earth: New York, John Wiley and Sons, 416 p.

Yarborough, H., 1977, Continental margin types related to plate tectonics and evolution of margins: American Association of Petroleum Geologists, Continuing Education Course Note Series 5, p. A1-A8.

Yeats, R. S., and Hart, S. R., 1976, Introduction and principal results, Leg 34, Deep Sea Drilling Project, *in* Yeats, R. S., Hart, S. R., and others, eds., Initial reports of the Deep Sea Drilling Project, Volume 34: Washington, D.C., U.S. Government Printing Office, 814 p.

Zevallos, O., 1970, Petróleo en rocas del basamento: Primero Congreso Latinoamericano de Geología, Tomo 2, p. 30–62.

Zuñiga, F., and Travis, R. B., 1975, The geology of the coast and continental margin of Peru [abs.]: EOS (American Geophysical Union Transactions), v. 56, p. 911.

MANUSCRIPT RECEIVED BY THE SOCIETY NOVEMBER 12, 1980
MANUSCRIPT ACCEPTED DECEMBER 30, 1980
HAWAII INSTITUTE OF GEOPHYSICS CONTRIBUTION NO. 1102

Printed in U.S.A.

Sedimentary basins of the Peru continental margin: Structure, stratigraphy, and Cenozoic tectonics from 6°S to 16°S latitude

TODD THORNBURG
L. D. KULM
School of Oceanography
Oregon State University
Corvallis, Oregon 97331

ABSTRACT

The morphology and shallow structure of the Peru continental margin has been mapped using bathymetric and seismic reflection profiles from lat 6°S to 16°S. Other geophysical and geologic data are used to constrain interpretations of the margin's deeper structure and to relate the offshore to the onshore Andean geology.

Two prominent structural ridges, subparallel to onshore Andean trends, control the distribution of the offshore Cenozoic sedimentary basins. The Coastal Cordillera, which surfaces north of lat 6°S and south of lat 14°S, can be traced onto the offshore as an Outer Shelf High (OSH); it is evidently cored with Precambrian and Paleozoic metasediments and crystalline rocks. A series of shelf basins is situated between the Coast Range/OSH and the Andean Cordillera: from north to south, these are the Sechura, Salaverry, and East Pisco Basins. A second set of upper-slope basins flanks the Coast Range/OSH to the southwest, limited seaward by an Upper-Slope Ridge (USR) of deformed sediment: from north to south, these are the Trujillo, Lima, and West Pisco Basins. The Yaquina Basin lies within divergent arms of the USR. The shelf and upper-slope basins are set on continental massif. An anastomosing network of elongate ridges and ponded sediments is the surficial expression of the subduction complex, which apparently begins just seaward of the USR.

The effect of the late Paleocene/Eocene Andean orogeny has been extrapolated offshore as a distinct interface of seismic velocity in the Salaverry Basin. Though Cenozoic marine sedimentation in the shelf basins did not begin until after this event, sedimentation in the upper-slope Trujillo Basin may have been more continuous through the early Tertiary. In the Trujillo Basin, the bulk of the nearly 4 km thick sedimentary section is of Paleogene age, while in the adjoining upper-slope Lima Basin to the southeast, the bulk of the nearly 2 km thick sedimentary section is of late Miocene or younger age. Apparently, post-Oligocene tectonism caused uplift, deformation, and a gross reduction of sedimentation in the Trujillo Basin; this event is evidenced by boundaries of differential structural deformation in seismic reflection profiles. In middle to late Miocene time, while orogenic activity affected the inland Andean Cordillera, the upper-slope Lima Basin subsided and began its depositional record. Unconformities in shelf basins apparently reflect the inland tectonism at this time.

The boundary between the Lima and Trujillo Basins, and between the contrasting styles of upper-slope tectonic movement, is near lat 9.5°S, coincident with the present day intersection of the Mendaña Fracture Zone with the continental margin.

A final phase of upper-slope deformation closed the Pliocene. Like earlier tectonic activity, the major break in structural style of this epoch occurs near lat 9.5°S: compressional faulting and folding characterize the younger sediments of the Trujillo Basin, while the Lima Basin appears as a broad, open syncline, distrubed only in its southernmost occurrence.

INTRODUCTION

The Andes of Peru are unique when compared to many of the world's orogenic belts. Generally, Peruvian tectonic movements have been intracratonic: uplift and subsidence, basin formation and destruction have occurred in response to the vertical bobbings of block-faulted continental crust (Myers, 1975). Thrusting is apparently superficial and subordinate (Myers, 1975) save for the eastern foothills of the Subandean foredeep; thick sedimentary turbidite sequences, melange, obducted ophiolites, and suture zones are lacking (Cobbing, 1978); and Precambrian, cratonlike masses are exposed at the coastline (Hosmer, 1959). Scars of collided continents, arcs, and accreted marginal oceanic basins are not found in Peru. Simple oceanic subduction must somehow provide the mechanism for uplifting the Andean crust many kilometres above sea level, though perhaps collision is the final chapter in any orogenic record.

Inspection of a geologic map of Peru (for example, de Almeida and others, 1978) immediately reveals the strong linear nature of the orogenic belts that dominate the western part of the country and trend roughly parallel to the present coastline (Figs. 1, 2B). The geomorphic expressions of these features are, from southwest to northeast, the Coast Range, the coastal shelf and lowlands, and finally, the high Andes, heralded by the Coastal Batholith and folded Mesozoic sediments of the western foothills. The Coast Range is cored with Paleozoic metasediments north of lat 6°S and with Precambrian crystalline rocks south of lat 14°S; this resistant, pre-Mesozoic lineament forms prominent capes at these latitudes. In the intervening area, the sea has transgressed to the Andean foothills, submerging the Coast Range and coastal lowlands as a shallow shelf. The Coast Range, though, can be traced onto the continental margin as a fairly continuous outer shelf structural high (Masias, 1976; Travis and others, 1976; Kulm and others, 1977).

Distinct offsets and punctuations in Andean structural trends occur near these latitudes where the coastal mountains and plains become submerged, suggesting zones of transitional subduction tectonics. At lat 14°S near Pisco, the Abancay Deflection intersects the coast, the extensive Quaternary volcanics of southern Peru terminate, and the aseismic Nazca Ridge abuts the Peru-Chile Trench (Fig. 1). To the north near lat 5°S, the Huancabamba Deflection describes the abrupt juncture of the northwest-trending Peruvian Andes with the northeast-trending Andes that continue into Ecuador (de Almeida and others, 1978). This tectonic "superunit" between lat 6°S is the focus area of the present study.

Our goal is to synthesize a coherent regional description of the structure and stratigraphy of the offshore basins of the Peru Margin. We have established the geographical distribution of these basins and delineated the structural ridges that confine them. Our research shows that the major forearc basins that occupy the shelf and upper slope lie entirely within the continental massif; thus, a comparison of lithologies, events, and deformational styles with onshore geology seems especially pertinent. An understanding of Andean evolution, then is a prerequisite to understanding the patterns of sedimentation and tectonics in the basins of the Peru continental margin.

GEOLOGIC HISTORY

The close of the Paleozoic Era in Peru was marked by strong orogenic activity that increased in

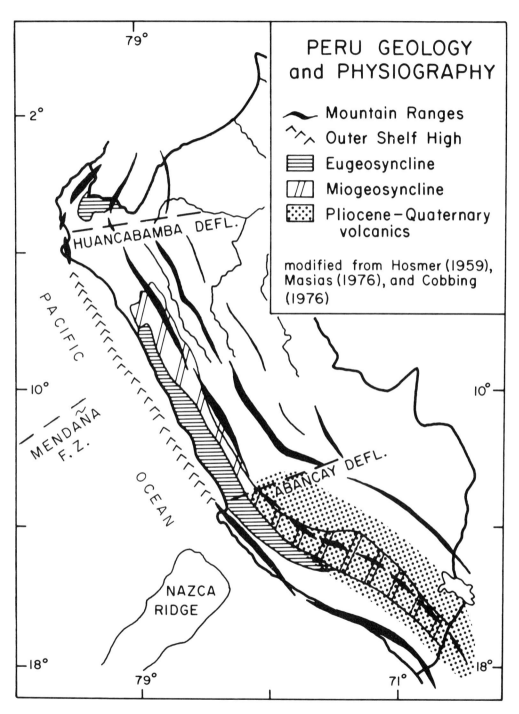

Figure 1. Structural trends, physiography, and geology (including present outcrops of Mesozoic geosynclinal deposits) of the Peru Cordilleran belt and neighboring offshore region.

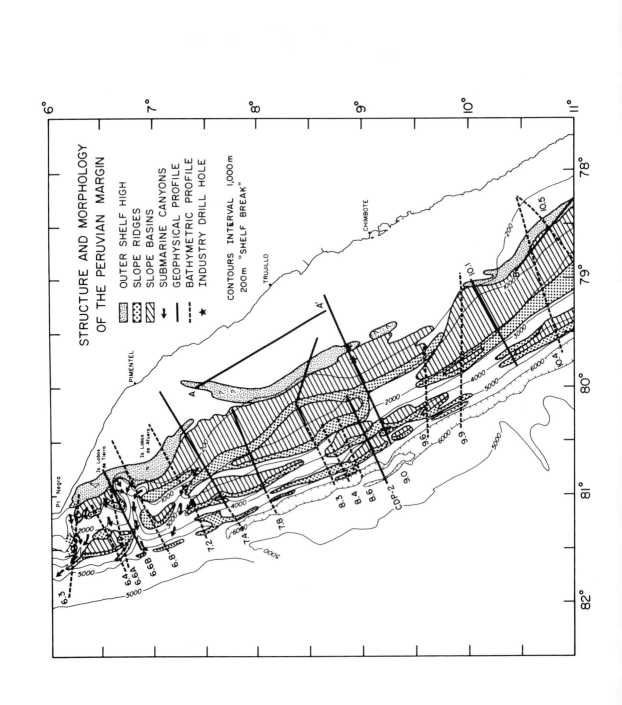

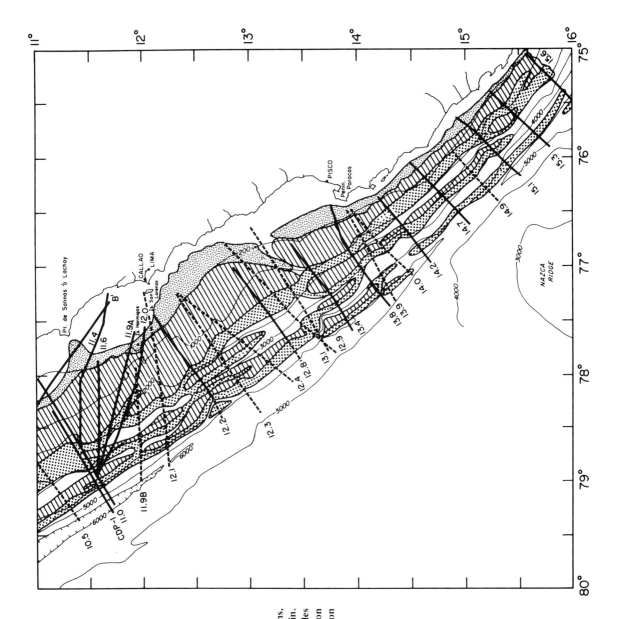

Figure 2A. Structural ridges, sedimentary basins, and submarine canyons of Peru continental margin. Bathymetric and seismic control are plotted; profiles are referred to by latitudinal degree of intersection with the 200 m contour. See Figure 2B for location map.

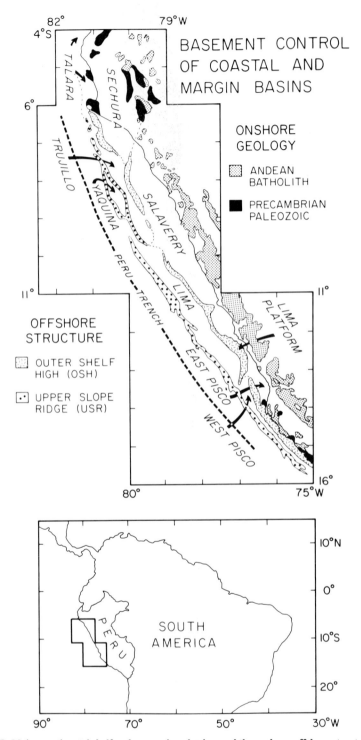

Figure 2B. Major continental shelf and upper-slope basins, and the onshore-offshore structures that control their distribution. The offshore Outer Shelf High (OSH) and Upper-Slope Ridge (USR) are from Figure 2A; onshore geology is from de Almeida and others, 1978. The flecked pattern of the OSH and the stippled pattern of the USR have been standardized for annotation of all geophysical profiles used in this study (Figs. 4A through F, 5 and 6). Location map shown below.

intensity seaward and culminated in a belt of folding, batholithic intrusion, and metamorphism at the edge of the present continental margin (Hosmer, 1959). Today, a record of this event is preserved in the core of the Coast Range and the outer shelf structural high (Figs. 1, 2B). The geologic record of the Early and Middle Triassic is apparently absent throughout the whole of Peru (Jenks, 1956).

Birth of the Andean structural trends occurred in Late Triassic time with the inception of a paired geosynclinal belt: a western eugeosynclinal facies (volcaniclastics, pillow lavas, interbedded shales and tuffs) and an eastern miogeosynclinal facies (limestone, sandstone, shale) (Fig. 1). Tension in the underlying sialic crust resulted in movements along near-vertical fractures and shear zones, which apparently controlled the formation and geometry of these depositional basins (Myers, 1975). The separation of eugeosynclinal and miogeosynclinal facies is distinct north of Lima; to the south, these facies merge and interfinger. Maximum subsidence of the miogeosyncline occurred in Tithonian time (Upper Jurassic), but the thickest accumulations of eugeosynclinal deposits were not until Albian time (Lower Cretaceous) (Cobbing, 1976).

The geosynclinal throughs formed in the shadow of a "continental-island arc" hybrid: volcanism was centered in a western geanticline at the toe or leading edge of the Peruvian craton, coincident with or very near the present Coast Range and outer shelf high of Paleozoic ancestry (James, 1971; Figs. 1, 2B). There is substantial geologic evidence for the existence of a western volcanic source that spewed huge thicknesses of dominantly andesitic volcaniclastics (several kilometres) from Upper Triassic to Upper Cretaceous time, concurrently with basic dikes that fed pillow lavas directly to the eugeosynclinal floor (Cobbing, 1978). Webb (1976) observed that many of the volcaniclastic units thinned to the east, while clast size also diminished eastward. Wilson (1963) documented E-NE paleocurrent directions from Valanginian cross-bedded sands of the Lima area, and eastward-dipping paleoslope features from the middle Albian near Chancay (lat 11.5°S). Myers (1974) noted a facies change from submarine pillows and pyroclastic flows in the west to dominantly air- or water-transported pyroclastics eastward, also during Albian time. The eugeosynclinal volcanics rest directly on the sialic, pre-Mesozoic coastal basement (Hosmer, 1959). The Paleozoic strata of the Coast Range north of lat 6°S are intruded by granitic rocks of pre-Cretaceous (probably Jurassic) age (Travis and others, 1976), and gneiss, schist, and Paleozoic strata of the Coast Range near lat 16°S are cut by quartz diorite intrusives that Stewart and others (1974) have dated at 204 to 157 m.y.; perhaps these intrusions represent the exposed roots of the fossil volcanic belt.

Geosynclinal conditions terminated in an Upper Cretaceous episode of uplift and batholithic intrusion that ushered in the Cenozoic with continental conditions. The Coastal Batholith in Peru is composed of hundreds of plutons that range in composition from gabbro to granite, with tonalite being the most abundant (Cobbing and Pitcher, 1972; Pitcher, 1978; Fig. 2B). Preexisting basement fractures apparently exerted a strong control on the emplacement (Myers, 1975), which spanned the interval between 105 and 60 m.y. (Noble and others, 1979). Passive emplacement to within a shallow 5 km of the surface left sharp, uncomplicated contacts with narrow aureoles of contact metamorphism (Cobbing and Pitcher, 1972).

Deformation of the eugeosynclinal volcanics began as early as the Cenomanian and predated the intrusion of the Coastal Batholith (Hosmer, 1959). Structures were once again controlled by vertical movements of the underlying continental basement blocks: the volcanic strata are locally folded into tight, isoclinal synforms, presumably where they overlie vertical shear zones; elsewhere, deformation is slight, and bedding dips are gentle (Myers, 1975). The miogeosyncline was not deformed until the waning moments of intrusion in early Cenozoic time, after it had received a cover of coarse molassic debris from the rising volcanic belt to the west (Hosmer, 1959). Deformation was more extreme compared to the eugeosynclinal partner, but here, too, thrusting and overturned folds of general NE vergence are supposed to be the superficial response above a décollement surface to vertical, blocky basement activity (Myers, 1975). Interpretation of recent geologic and radiometric data has refined the timing of this tectonic event to within the late Paleocene and/or Eocene (Noble and others, 1979). Tectonic quiescence characterized the Oligocene. Then, major uplift and deformation resumed in middle/late Miocene time (between roughly 15 and 10 m.y. ago), accompanied and followed through the Pliocene by intense volcanism and plutonism (Noble and others, 1974; Farrar and Noble, 1976).

This late Tertiary activity affected the eastern Andean ranges and the Subandean foredeep most directly.

There has been a progressive, though sporadic, eastward migration of magmatism during the evolution of the Andean continental margin: from the Coast Range/outer structural high in the Mesozoic, to the Coastal Batholith in the early Tertiary, to the site of the deformed miogeosyncline in the late Tertiary. James (1971) rationalizes the migration as resulting from: (1) progressive depression of isotherms along the subducting oceanic slab; (2) landward migration of the trench with time; or (3) progressive shallowing of the dip of subduction with time.

MORPHOLOGY AND STRUCTURE OF THE PERU MARGIN

Two persistent structural ridges are traceable along the Peruvian continental margin; these have molded the geometries of the major shelf and upper-slope basins (Figs. 2A, 2B). The pre-Mesozoic lineament that rises above sea level to become the Coast Range near lat 6°S and 14°S is designated the Outer Shelf High (OSH) where it can be traced onto the continental margin (Fig. 2B). Its role as a geanticlinal locus of igneous activity during the development of the Mesozoic geosyncline has been emphasized (see Geologic History). Continued influence as a positive element during the Tertiary has allowed a set of shelf basins to accumulate between the OSH and the onshore Andean Belt. From north to south, these are the Sechura, Salaverry, and East Pisco Basins (Fig. 2B; Travis and others, 1976). To the west, a second set of upper-slope basins is cradled between the OSH and a prominent Upper-Slope Ridge (USR). From north to south, these are the Trujillo, Lima, and West Pisco Basins. The Talara Basin, to the north, lies west of the Paleozoic Coast Range and occupies a correlative upper-slope position. The Yaquina Basin lies entirely within divergent limbs of the USR. The remainder of the Peruvian continental slope is laced with an anastomosing, interweaving net of structural ridges and associated sedimentary basins that have backfilled landward of these ridges. In the north, this ridge and basin terrane is dissected by submarine canyons (Fig. 2A).

These structural/morphological features were correlated between geophysical profiles and geologic sections and delineated in map view (Fig. 2A). The bathymetry of the continental margin, contoured at a scale of 1:1,000,000 (Prince and others, 1980), was used as a base map. Numerous bathymetric profiles (arranged from north to south in Figs. 3A, 3B) and single-channel seismic reflection profiles (Figs. 4A–4F, from north to south) provided a bulk of the raw data, as plotted in Figure 2A. These profiles were collected during the Oregon State University R/V *Yaquina* cruises (1972, 1974) and the Hawaii Institute of Geophysics R/V *Kana Keoki* cruise (1972).

Particular attention was given to mapping the influential OSH and USR. Bathymetry is usually inadequate to establish the presence of the OSH. Very often, the USR controls the first significant gradient increase that can realistically be called a shelf break. As a result, the subsurface meanderings of the OSH are often hidden beneath the flat and featureless morphology of the continental shelf. [Note: regardless of the position of the shelf break, those basins which lie between the OSH and USR are referred to as upper-slope basins; those between the OSH and the Andes are referred to as shelf basins.] In contrast, the basins and ridges of the continental slope (including the USR) are morphologically expressive. Once these features have been ascertained in seismic reflection records, they can be extrapolated onto bathymetric profiles with reasonable confidence (Figs. 3A, 3B).

The OSH can often be recognized from seismic reflection records as a strongly diffracting anticlinal structure or shoaling acoustic basement (Fig. 3A: SP-6.6B; Figs. 4C, 4D, 4E, 4F: SP-10.1, 11.9A, 12.2, 13.8, line B-B'). In other cases, its presence within the subsurface must be inferred to lie between the gently landward-dipping reflectors of the flanking shelf basins and the seaward-dipping reflectors of the upper-slope basins (Figs. 4A, 4B, 4C, 4F: SP-7.4, 9.0, 11.0, 13.4); in some records, the reflectors overlying the OSH are disturbed or convoluted (SP-7.4, 9.0, 11.0). Similarly, the USR is evidenced by strong diffractions, or transparent zones around which reflectors have been domed, deflected, or pinched-out. Gravity and seismic refraction data provide additional documentation of the OSH and USR (Fig. 5). Drill hole information (provided by PetroPeru and Occidental Petroleum), the geologic terrane of offshore islands, and Travis and others' (1976) account of unpublished seismic records further constrain the mapping of the OSH. Descriptions of the major structures are described below, proceeding systematically from north to south along the continental margin and incorporating the various types of supporting evidence where appropriate (see Fig. 2A).

Shelf Basins and the OSH

The Coast Range in northwestern Peru dips below the sea surface at Negra Point to become the Outer Shelf High (Figs. 2A, 2B). The offshore projection includes Paleozoic exposures on the islands of Lobos de Tierra and Lobos de Afuera before complete submergence occurs south of lat 7°S (de Almeida and others, 1978). Travis and others (1976) speak of a N-S trend of subsurface basement highs that separates the Sechura and Salaverry Basins. We postulate this trend to be the southern extension of the eastern arm of the onshore Paleozoic outcrops that circumscribe and delimit the Sechura Basin (Fig. 2B). The structural ridge has been extrapolated from the shoreline near lat 7°S to a juncture with its western brother in a fork near lat 8°S. Seismic refraction along the continental shelf verifies a shoaling of the 5.9 to 6.0 km/sec crustal layer beneath the hypothesized branch of the OSH (line A-A' of Figs. 2A and 5D).

PetroPeru and Occidental Petroleum have released to us information on the industry drill holes near 9°S (Figures 2A, 4B). Both holes bottomed in metamorphic basement: quartz biotite gneiss was penetrated at a depth of 960 m in the landward well (Delfin), and foliated mica schist was penetrated at a depth of 2650 m in the seaward well (Ballena). The landward well is positioned over the western limit of long, continuous, gently landward-dipping reflectors of the Salaverry Basin, and presumably bottomed in the crest of the OSH. The seaward well drilled a section of discontinuous, disturbed reflectors and penetrated the seaward flank of the OSH beneath the cover of an upper-slope basin. Gravity and seismic refraction modeling (Figs. 5A, 5B; Jones, this volume; Hussong and others, 1976) confirm the presence of a high-density, high-velocity ridge beneath the landward well. Limited seismic profiling between lat 9°S and 10°S could not ascertain the OSH and the western boundary of the Salaverry Basin.

Farther south, the OSH is again recognized as a shoaling zone of strong reflectors in SP-10.1, at a subsurface depth of about 1 sec (Fig. 4C). In the region between lat 10°S and 11.5°S, the OSH seems to largely control the morphologic shelf break, which occurs between 200 and 500 m water depth. Near lat 11.5°S, the OSH extends an eastern protuberance to greet the coastal cape, Point de Salina O Lachay, to form the southern boundary of the Salaverry Basin. Features of this area are best seen in the seismic reflection section B-B', which parallels the coastline and traverses the OSH and its eastern extension several times (Fig. 4D). Between lat 11.5°S and 12°S, a small shelf basin may exist. The western boundary of this basin is defined by Is. Hormigas, an offshore island near lat 12°S that is of probable Paleozoic terrane (Kulm and others, 1981b, this volume; Masias, 1976). Alternatively, the transverse ridge at lat 11.5°S may merge southward into the Lima Platform, forming a table of shallow acoustic basement across the hypothesized basin. It is unfortunate that the structure of this region could not be resolved among the confounding bottom multiples on the single-channel reflection records.

Between lat 12°S and 13°S, seismic reflection profiles reveal a gradual shoaling of acoustic basement from the shelf break to the coastline beneath a thin sedimentary cover of only several hundred metres. The diffracting OSH appears to broaden and interfinger with the Mesozoic intrusives and deformed sediments of the Andean foothills, forming a shallow structural platform across the continental shelf. It is referred to as the Lima Platform (Figs. 2A, 2B). Travis and others (1976) describe a similar acoustic platform offshore of Lima that breaks the continuity of the shelf basins; free-air gravity anomalies exhibit positive values over this part of the shelf (Couch and Whitsett, this volume). Seismic profile SP-12.2 (Fig. 4E) shows the tensional breakdown of this feature beneath the shelf break.

Couch and Whitsett's (this volume) free-air gravity map reveals a tongue of positive values extending north into the offshore from the Paracas Peninsula. These positive values are flanked on

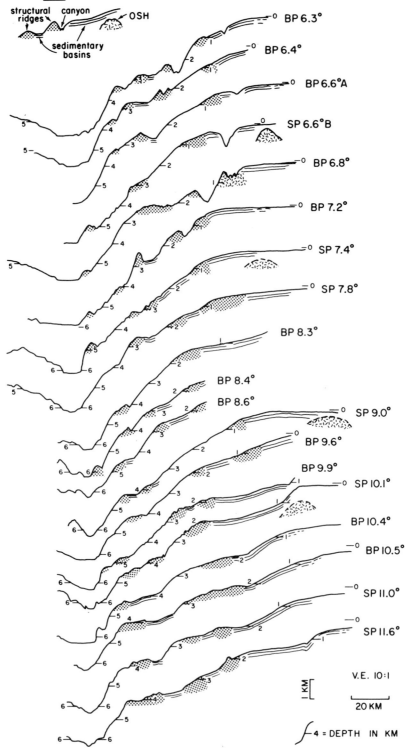

Figure 3A. Bathymetric control (prefix BP) used in Figure 2A. Seismic reflection control (prefix SP; Figs. 4A through F) is also shown here in bathymetric form. Profiles are arranged from north to south. (Note: the double line parallel to the sediment surface used to indicate a sedimentary basin does not imply subsurface structure.)

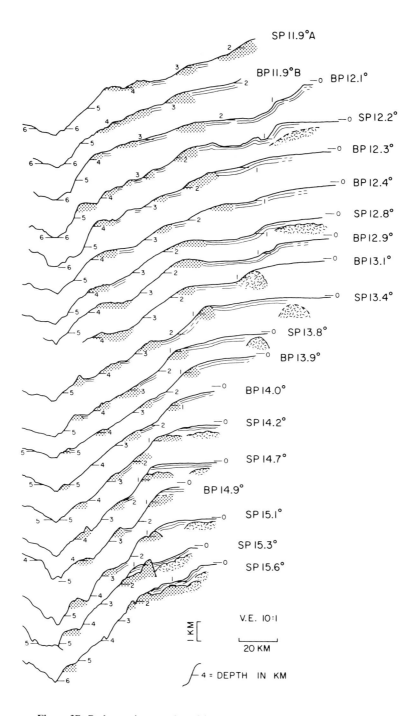

Figure 3B. Bathymetric control used in Figure 2A. See Figure 3A for caption.

Figure 4 (A-F). Seismic reflection line drawings across the Peru continental margin, arranged from north to south. Locations are shown in Figure 2A. Outer Shelf High (OSH) is flecked pattern, Upper-Slope Ridge (USR) is stippled. Seismic unconformites are shown in heavier line. Selected actual profiles are shown beneath their respective line interpretations. Dredge sites of Kulm and others (1981b, this volume) are annotated.

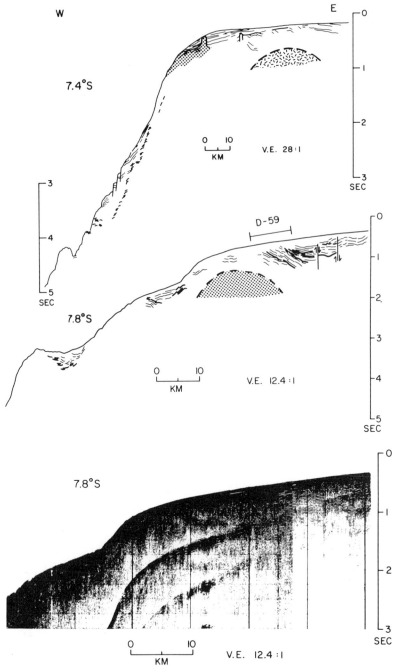

Figure 4A. Seismic reflection line interpretations and actual records across the Trujillo Basin.

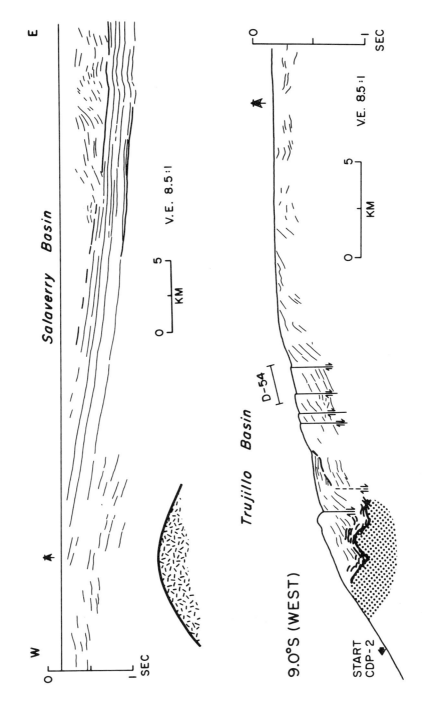

Figure 4B. Seismic reflection line interpretations across the Trujillo and Salaverry Basins. Locations of offshore wells are noted.

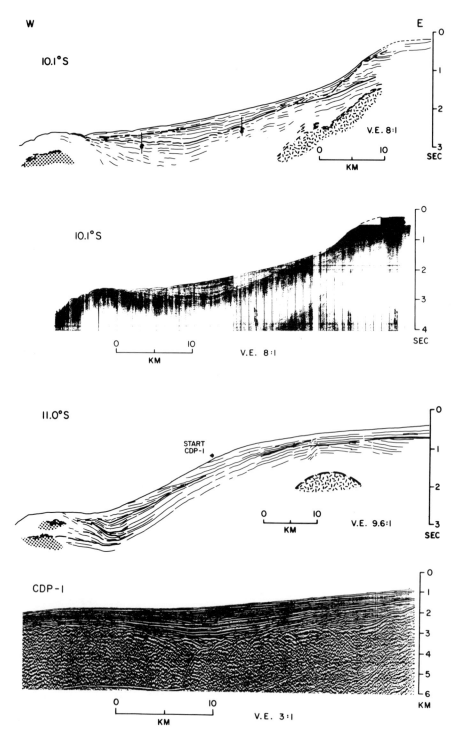

Figure 4C. Seismic reflection line interpretations and actual records across the Lima Basin. Note the good correlation between the single-channel record SP-11.0 and the nearly coincident multichannel record CDP-1 below.

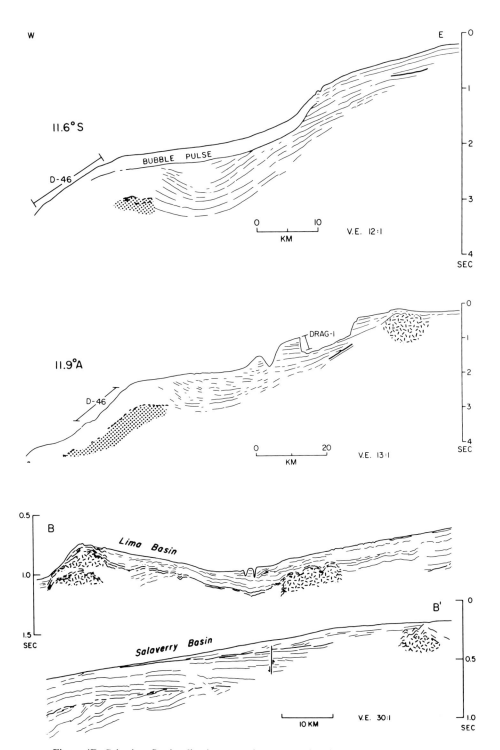

Figure 4D. Seismic reflection line interpretations across the Lima and Salaverry Basins.

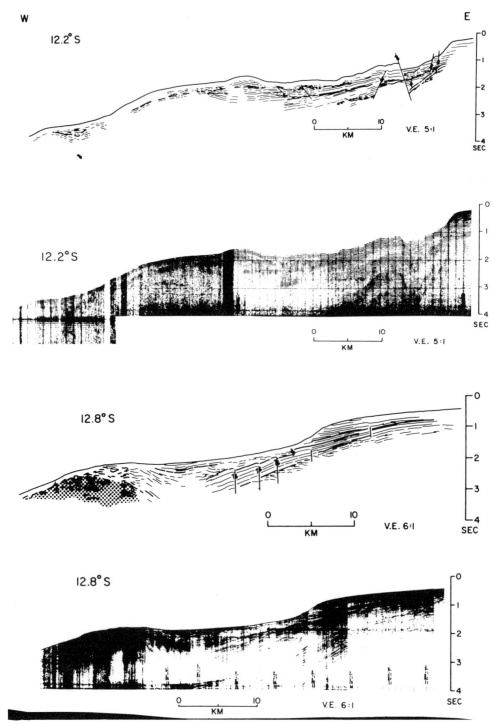

Figure 4E. Seismic reflection line interpretations and actual records across the Lima Basin.

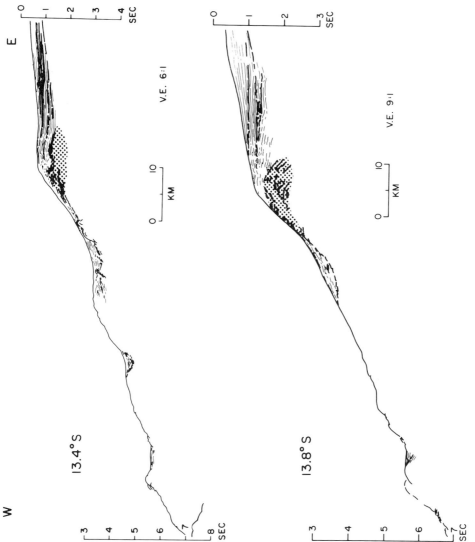

Figure 4F. Seismic reflection line interpretations across the West Pisco Basin. These profiles have been extended shoreward and reproduced in simplified form in Figure 6.

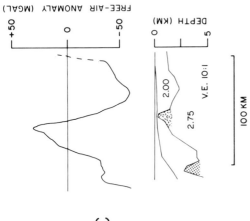
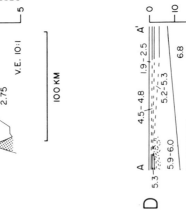

Figure 5. Geophysical modeling (refraction and gravity) of the Peru continental margin. OSH is flecked pattern, USR is stippled. Horizontal scales for 5A, B, and C are identical; vertical exaggeration of 5C is twice that of 5A and B. (A) Gravity profile and resultant crustal density model coincident with seismic reflection profile SP-9.0 across the Trujillo and Salaverry Basins (Figs. 4B, 2A) from Jones (this volume). (B) Seismic refraction crustal velocity model coincident with gravity model 5A and reflection profile SP-9.0 (Figs. 4B, 2A) from Hussong and others (1976). (C) Gravity profile and density model coincident with seismic reflection profile SP-13.4 across the East and West Pisco Basins (Figs. 4F, 2A) from Couch and Whitsett (this volume). (D) Seismic refraction crustal velocity model A-A' (see Fig. 2A for location) across the Salaverry Basin, from Hussong and others (1976).

both sides by closed negative contours. Thus, the OSH is recognized as a ridge of gravity maxima that separates the shelf and upper-slope basins (Fig. 5C). The effect of the OSH on the structure of the basinal sediments is readily seen on SP-13.4 and SP-13.8 (Fig. 6). For this reason, we feel that the Pisco Basin, which includes both the shelf and upper-slope basins according to Travis and others (1976), should be subdivided into a distinct East Pisco Basin, which lies east of the OSH as a shelf basin, and a West Pisco Basin, which lies west of the OSH as an upper-slope basin. In support of this nomenclatural division, the Cenozoic sediments of the opposing basins unconformably onlap their respective sides of the crystalline basement complex in Coast Range outcrops (Travis and others, 1976); the two basins have likely undergone distinct sedimentation histories.

An offshoot of the OSH is postulated to extend S-SW from near lat 13°S, marking the southern boundary of the Lima Basin (Figs. 2A, 2B). This interpretation is somewhat specultive, since the escarpment seen on a single bathymetric profile (Fig. 3B: BP-13.1) is the primary evidence. The interjection of the structural limb seems necessary to account for the rapid change in regional depth between the Lima and West Pisco Basins.

Distribution of Slope Basins, Ridges, and Canyons

Submarine canyons dominate the character of the Peru continental slope north of lat 7.5°S, where they funnel sediment from upper- and middle-slope basins directly into the trench (Fig. 2A). All canyons originate seaward of the OSH; this crystalline ridge seems to provide an effective barrier against further landward encroachment of their erosive heads. However, a major canyon near lat 6.8°S has carved into the OSH to some degree; bathymetric profile BP-6.8°S (Fig. 3A) shows the rugged and dissected appearance of this resistant feature. The slope ridges exert a limited control on

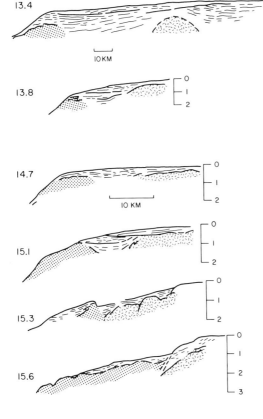

Figure 6. A series of line interpretations of seismic reflection profiles across the West Pisco Basin, from north to south (see Fig. 2A for locations). Note that, proceeding south, the OSH (flecked pattern) and the USR (stippled pattern) shoal, spread laterally, and finally interfinger to form a shallow, contiguous basement beneath a dramatically thinned and restricted section of overlying sedimentary reflectors. This is possibly a result of impingement of the Nazca Ridge against the continental margin. SP-13.4 and 13.8 are of the same horizontal and vertical scale; SP-14.7 through 15.6 are of the same horizontal and vertical scale; thus, note scale change between SP-13.8 and 14.7. Modified from Johnson and Ness (this volume).

canyon development and apparently deflect their downslope progress to some degree; in profile, the canyons are often seen nudged against the landward wall of a prominent ridge (Fig. 3A: BP-6.3, 6.8, 7.2). In many cases, though, the canyons cut directly across the slope ridges, suffering little diversion from a direct gravity-driven path. This is particularly true near the canyon heads. Notably, the extension of the USR near 1,000 m is extremely fragmented in this area (Fig. 2B).

South of lat 7.5°S, canyons are noticeably scarce or absent, and the slope is characterized by extensive upper-slope basins lying between the prominent USR and OSH, of dimensions comparable to the shelf basins. Between lat 7.2°S and 9.6°S, the USR controls the location of the shelf break, which is depressed to 400 to 600 m (Figs. 4A, 4B: SP-7.4, 7.8, 9.0), and forms the western boundary of the Trujillo Basin (Figs. 2A, 2B). The basin is at least 300 km long and approaches 40 km in width. Gravity and refraction modelling clearly indicate the high-density, high-velocity nature of the USR across SP-9.0 (Figs. 5A, 5B; Jones, this volume; Hussong and others, 1976).

A basin 150 km long and up to 35 km wide occurs below the Trujillo Basin. It is centered around the 1,000 m contour (Figs. 2A, 2B) and is named the Yaquina Basin in this study. A 24-channel seismic reflection record across the southern end of this basin (CDP-2 of Kulm and others, 1981a, this volume) shows a 2 km thick sedimentary section containing relatively undisturbed reflectors near the center. These reflectors grade laterally into a highly disturbed and diffracting seismic signature on either side. The disrupted reflectors that surround the undeformed sediments have a domed or anticlinal appearance and are interpreted to be arms of the USR which have diverged to allow the development of the Yaquina Basin between them. Middle-slope basins of more restricted dimensions occur commonly between profiles BP-8.4 and BP-9.0 (Fig. 3A). Below 4,000 m along the entire Peru Margin, steep gradients (up to 1:6) and rough topography restrict basin development to infrequent and isolated ponds on the lower slope.

From lat 9.6°S to 13.5°S, the upper slope has accumulated a striking trough of sediment that Masias (1976) first identified as the Lima Basin (Figs. 3A, 3B: bathymetric profiles 9.9 through 13.1; Figs. 4C, 4D, 4E: seismic profiles 10.1 through 12.8); it reaches 50 km in width. Acoustic basement is quite apparent in a multi-channel reflection line that traces SP-11.0 (Fig. 4C, CDP-1), where a subsurface depth of 1.9 sec is reached at the basin's depocenter. As with the Yaquina Basin to the north, the USR is documented by a zone of structurally disturbed reflectors within the sedimentary sequence. The "submarine canyon" in SP-11.9A is puzzling, since it cannot be traced laterally and is contoured as an extremely localized kink on the base map, despite its impressive dimensions (400 m deep and 15 km wide). The continental slope lying seaward of the northern Lima Basin is extremely rugged from 2,000 m to the trench (Fig. 3A: BP-9.9, 10.1); the entire middle slope of this area seems to lack any appreciable sediment accumulation (Fig. 2A). Downslope of the southern Lima Basin, however, middle-slope sediments are prolific; the basins are elongate and continuous, apparently traceable for hundreds of kilometres within the presently available profiling density (see also Johnson and Ness, this volume).

Between the southern termination of the Lima Basin and the Paracas Peninsula near lat 14°S, the upper and middle slopes remain covered with a generous blanket of basinal sediments. However, where the OSH surfaces in the Coast Range, and where the Nazca Ridge abuts the margin and causes a dramatic shoaling of the trench, sedimentary basins concurrently become shallower and less extensive (Fig. 2A). This progressive abbreviation of sediment cover is best illustrated by a series of reflection profiles across the West Pisco Basin, the southern counterpart of the upper-slope Lima and Trujillo Basins (Fig. 6; modified from Johnson and Ness, this volume). North of lat 14°S, the basin is comparable in width and depth to the neighboring upper-slope basins to the north (Fig. 2A: SP-13.4, 13.8). Ridges of high-density material, the OSH and USR, sharply define the basin's lateral boundaries (Fig. 5C; Couch and Whitsett, this volume). Proceeding south, however, the OSH and USR spread laterally as shallow acoustic platforms, squeezing the interlying sediments into a constricted furrow of only 10 km in width by SP-14.7 and 15.1. In SP-15.3 and 15.6, the expanding structural ridges have pinched together into a continuous, though irregular, acoustic basement. Protruding ridges fragment the thin drape of overlying sediment (generally less than 0.5 sec in thickness) into shallow, restricted ponds.

DISCUSSION

Nature of Basement Structures

The preceding descriptive synthesis has emphasized the structural role of the OSH and the USR antiformal lineanments: that of providing a skeletal framework which molds the geometry and distribution of shelf and upper-slope sedimentary basins along the Peru continental margin (Fig. 2B). The composition of the OSH may be inferred from the lithologies of the Coast Range: Precambrian and Paleozoic crystalline rocks and metasediments cut by Mesozoic intrusives (Travis and others, 1976). This inference is well substantiated by island exposures (Masias, 1976; de Almeida and others, 1978; Kulm and others, 1981b, this volume), drill hole basement lithologies (Kulm and others, 1977 data presented in this paper), and gravity and refraction modeling showing domes of high-density (2.70 to 2.80 g/cm³), high-velocity (5.9 to 6.0 km/sec) material in the subsurface (Fig. 5; Jones, this volume; Couch and Whitsett, this volume; Hussong and others, 1976). In the vicinity of the Lima Platform and the transverse ridge trending offshore from Point de Salinas Ó Lachay (Fig. 2A), this complex is very likely overlain on the east by deformed sediments of the Mesozoic geosyncline and cut in the east by intrusions of the Andean batholith. Together, these lithologies form a shallow, contiguous basement across the continental shelf.

Several lines of evidence indicate that deformed basin sediments may constitute the USR. The USR's velocity structure is a 4.1 km/sec slab within the shallow subsurface, underlain by a puzzling inversion (Fig. 5B). Multichannel seismic reflection profiles across the Yaquina and Lima Basins (CDP-2 of Kulm and others, 1981a, this volume; Fig. 4C: CDP-1) reveal that the disrupted anticlinal reflectors that define the USR lie within a sedimentary sequence which overlies a pronounced acoustic basement, and that this disturbed seismic facies may be laterally correlative to the relatively undisturbed sedimentary reflectors. Also, the submarine canyons north of lat 7.5°S dissect the extension of the USR with seemingly little difficulty, compared to the more resistant OSH (Fig. 2A). However, a striking gravity maximum occurs above the USR near lat 9°S. This maximum is comparable in amplitude to the gravity peak that occurs above the metamorphic OSH at this latitude (Fig. 5A), implying that dense, crystalline material may similarly occupy the core of the USR in some areas. We suspect that the USR is laterally heterogeneous such that sediments of various ages, deformed during various upper-slope orogenies, define its composition.

At lat 9°S, the USR sets at the very edge of a basement block of 5.8 to 6.2 km/sec velocity (Fig. 5B); thus, both the Trujillo and Salaverry Basins appear to lie entirely within the continental massif, while the resticted subduction complex of probable deformed sediments begins just seaward of the USR (Kulm and others, 1981a, this volume). At lat 12°S, refraction and multichannel velocity data demonstrate the presence of continental 6.0 km/sec velocity crust extending seaward to the USR beneath the Lima Basin (Fig. 4C: CDP-1; Fig. 10 of Hussong and others, 1976). Very chaotic sedimentary reflectors overlie a 5.0 km/sec block that occurs trenchward of the continental block and may be fractured or downfaulted continental material.

Structure and Stratigraphy of Forearc Basins: Tectonic Implications

Current concepts of convergent continental margin morphology stress the importance of the subduction complex: an imbricate stack of off-scraped sediment from the descending oceanic plate (Dickinson and Seely, 1979). The subduction complex may develop against the toe of the continental massif; progressive accretion and uplift of this complex then form the seaward structural limit of the ponded forearc basin (Karig and Sharman, 1975; Karig, 1977). If subduction initiates at some distance from the continental massif, a fragment of oceanic crust may become trapped between the subduction complex and the continental massif, forming the floor and seaward wall of the forearc basin (Dickinson and Seely, 1979). Here again, though, the tectonic underplating of thrust packages to the subduction complex is the active mechanism by which the oceanic remnant is uplifted to cause damming of the forearc sediment. In both models, continental massif will form the landward wall and, at most, part of the landward floor of the forearc basin, interfacing with either oceanic crust or the accreted subduction complex beneath the forearc basin in some as yet unknown manner. Variations on subduction complex settings have been applied to numerous continental (and microcontinental) convergent margins about the Pacific, including Alaska (von Huene, 1979), Oregon (Kulm and Fowler, 1974), California (Dickonson and Seely, 1979), Middle America (Karig and others, 1978), northern Chile (Coulbourn and Moberly, 1977), and Sunda (Karig and others, 1979).

Scholl and others (1977) have outlined the main arguments against universally applying the model that tectonically off-scraped material forms the trench-slope break/continental slope of convergent margins. They cite a particular example: the margin of south Chile near lat 41°S is composed entirely of continental massif except for the questionable lowermost slope. This is especially true regarding the basement of the forearc sedimentary basins and is analogous to the situation that we find in Peru. From outcrops and drill holes, there can be no doubt that the OSH is continental in origin. Although the USR is likely composed of deformed sediments, we believe that it, too, rests on the continental block, according to refraction and wide-angle reflection velocities across lat 9° and 12°S (Hussong and others, 1976). Using recent information obtained during the DSDP Japan Trench Transect, Nasu and others (1980) have documented a major forearc basin that lies at greater than 1 km water depth, contains several kilometres of dominantly Neogene sediment, and is floored entirely by seismically determined continental crust.

Like other forearc basins, the shelf and upper-slope basins of Peru will deform in response to differential movements of their bounding structural elements (Seely, 1979), in this case the OSH and USR. Perhaps these more competent (higher-density, higher-velocity) basement strutures serve to transmit stresses resulting from the converging Nazca oceanic and South American continental plates upward into shallower crustal levels, and outward into the less-competent basin sediments. Geophysical investigations, rock dredges, drill hole data, and onshore exposures offer important clues with which to decipher the major tectonic movements that affected the sedimentation histories of the Peru shelf (Sechura, Salaverry, East Pisco) and upper-slope (Trujillo, Lima, West Pisco) basins.

Shelf Basins—Structure. Extensive normal faulting permeates the subaerial outcrops of the coastal basins (Travis and others, 1976), not unlike the blocky tectonics that have historically characterized Cordilleran movements (Myers, 1975; Cobbing, 1976, 1978). Some blocks have been vertically displaced thousands of metres since the Late Cretaceous. Using seismic reflection data, Travis and others (1976) project this structural style onto the continental shelf. Erratic and irregular horst and graben basement features affect the overlying strata in varying degrees. Large-scale gravity-induced slumps and slides have also been interpreted on seismic reflection records. An example is seen on SP-9.0 (East) within the Salaverry Basin (Fig. 4B), where folded and fragmented reflectors unconformably overlie continuous and subparallel reflectors suggestive of soft-sediment deformation above a more competent, intact stratal surface.

Shelf Basins—Stratigraphy. With the waning of the lower Tertiary Andean orogeny, the shelf basins subsided and the Cenozoic record of marine sedimentation began, while continental and volcanic deposition prevailed inland. Up to 3,000 m of strata were laid down in the Sechura Basin in post-Eocene time (Fig. 2B; Travis and others, 1976). A major unconformity in land outcrop sections indicates that the geological record of late Miocene, early and middle Pliocene time is missing (Hosmer, 1959); this may correlate inland with pronounced middle/late Miocene tectonism in the Andean belt (Farrar and Noble, 1976).

Geophysical studies across the northern Salaverry Basin show that the sedimentary sequence is often much thinner than 2 km, as defined by a 1.80 to 2.15 g/cm^3 density and 1.8 to 2.5 km/sec velocity layer (Figs. 5A, 5B, 5D). Below a distinct acoustic interface, several kilometres of 2.65 g/cm^3 density, 4.5 to 5.3 km/sec velocity material is present, which we interpret to be deformed sediments of the Mesozoic geosyncline. The sharp velocity and density discontinuity, then, may represent an unconformity that encompassed latest Cretaceous through much or all of Eocene time. The pronounced angular unconformity within less than 1 sec of penetration in reflection profile B-B' across the southern end of the basin (Figs. 2A, 4D) may correlate with the post–middle Miocene event of the Sechura Basin. Younger reflectors that overlie this unconformity remain subparallel overall, but their individual signatures are jagged and "sawtoothed" in detail, as if bedding interfaces have been disrupted by numerous small-scale offsets. Perhaps a very young (late Neogene) tectonic episode has overprinted the strata in this region.

The East Pisco Basin attains a maximum seismic thickness of 2,500 m (Travis and othrs, 1976); this interpretation is consistent with gravity modeling over the basin (Fig. 5C). According to coastal outcrops, the sedimentary record begins in the middle Eocene, with interruptions or disturbances following the Eocene, Miocene, and Pliocene (Hosmer, 1959).

Upper Slope Basins—Structure. In the single-channel seismic reflection profile SP-7.4 across the northern end of the Trujillo Basin, fragmented bits of convolute reflection are unconformably overlain by a sequence of continuous, subparallel, undisturbed reflectors (Fig. 4A). This pronounced structural break is tentatively dated as post-Oligocene by correlation with a disturbance that greatly reduced Miocene deposition in the neighboring upper-slope Talara Basin to the north (Hosmer, 1959). In SP-9.0 (Fig. 4B) across the southern end of the Trujillo Basin, chaotic and fragmented reflectors are, by analogy of seismic facies, similarly interpreted to predate the Miocene. Micropaleontologic data released to us through PetroPeru confirm that the strata recovered in both drill holes are indeed Paleogene (see Upper Slope Basins—Stratigraphy). The overlying Neogene sequence has apparently accumulated in a small sub-basin extending about 15 km landward of the USR, where deeper penetration of reflective surfaces is recorded. These younger strata are cut by a series of high-angle reverse faults that are upthrown on the seaward side, though reflectors retain a relatively smooth and subparallel expression within the individual fault blocks. Reflectors in the Trujillo Basin across SP-7.8 (Fig. 4A) display a seismic fabric very similar to the young Neogene bundle in SP-9.0. They are faulted in a similar manner and are additionally folded. A maximum principal stress centered within the USR including components of both uplift and compression, could produce the deformational features seen in SP-7.8 and 9.0. Such a stress would be oriented roughly normal to the plane of the subduction interface.

In contrast, the young, undisturbed sequence that overlies the unconformity in SP-7.4 is inferred to be deltalike progradational deposits from Pleistocene low-stands. If the fragmented reflectors that underlie the unconformity are indeed Paleogene, then the bulk of the Neogene section is missing in this part of the basin. Also, if the undisturbed deposits are indeed Pleistocene, an upper limit could be placed on the deformational episode that affected the Neogene sequences in SP-7.8 and 9.0 to the south. The truncated Paleogene strata that overlie the OSH in profiles SP-7.4 and 9.0, and the restricted occurrence of Neogene sediments in the Trujillo Basin, indicate that uplift of the OSH occurred during the post-Oligocene tectonism and perhaps during younger movements as well.

A major structural break occurs near lat 9.5°S, coincidentally where the Mendaña Fracture Zone intersects the margin. Here, the upper-slope Trujillo and Lima Basins are offset, as is the trend of the USR (Fig. 2A). The present study could not determine whether a similar offset occurs in the trend of the OSH. Whereas the Trujillo Basin records a history of tectonic activity (folding, faulting, fragmented reflectors), the northern Lima Basin has suffered only minor warpage, appearing as a broad, open syncline of gently dipping strata (Figs. 4C, 4D: SP-10.1 through 11.9A). The USR appears to have subsided relative to the OSH, evidenced by normal faults of small displacement (SP-10.1) and younger reflectors onlapping strata that have been downwarped seaward around an OSH hinge line (SP-11.0). Paleodepth data from benthic foraminifera of dredged carbonates that were dated by diatoms (dredge locations shown on profile SP-11.9A) substantiate subsidence of 1,100 m since aproximately 3 to 5 m.y. ago at site D-46, and 500 m since approximately 1 m.y. ago at site DRAG-1 (Kulm and others, 1981b, this volume). From the limited age control, we cannot distinguish whether the greater subsidence of the seaward site compared to the landward site has occurred through a greater subsidence rate over a similar time period, or through a similar subsidence rate over a longer time period. Pinched and upturned strata in the immediate vicinity of the USR, however, may be the result of a mild compressional episode (particularly SP-11.0).

In the southern Lima Basin, subsidence is similarly evident. Normal faults cut the sediment surface where the tensional breakdown of the Lima Platform creates the eastern boundary of this upper-slope basin (Fig. 4E: SP-12.2); the locus of youngest sediment accumulation (SP-12.8) has migrated toward the downdropped USR. However, deformational events have complicated the simple syncline that characterizes the reflection profiles of the northern Lima Basin, and tectonism has apparently interrupted the basin's Neogene history of subsidence. Low-angle thrusting appears to have fragmented the seismic depositional sequences of SP-12.2; a series of high-angle reverse faults produces multiple offsets in a highly reflective subsurface unit in SP-12.8. As with the Trujillo Basin, simultaneous uplift and compression of the USR could have produced faults of a near-vertical orientation, upthrown to the seaward side, on SP-12.8; the deformation in SP-12.2 appears to be more strictly compressional. Compression in these upper-slope basins may be the surficial expression of vertical or rotational movements of the underlying basement blocks, similar to the deformational style inland (Cobbing, 1978; Myers, 1975).

The West Pisco Basin exhibits a structural history governed by relative stability. Reflectors are subparallel and gently dipping; unconformities are subtle, and downlap indicates progradation from both landward and seaward sources (Fig. 4F: SP-13.4 and 13.8). Alternatively, rotation of onlap around oscillating landward and seaward hinge lines could produce a similar pattern, mimicking bidirectional progradation. In either case, a vertical seesawing between the OSH and USR must have affected the strata. Because of the ambiguity between progradational downlap versus tectonically tilted onlap on an active margin, we cannot determine whether subaerial exposure and erosion of the USR has ever provided a western source for sediments of the upper-slope basins.

Kulm and others (1981b, this volume) have dredged brecciated dolomicrites of late Miocene to late Pliocene–early Pleistocene age from the Trujillo Basin between lat 8°S and 9°S, and from the Lima Basin between lat 11.5°S and 12°S. The matrix lithology is strongly brecciated along subparallel fracture planes and infilled with a microcrystalline cement. Matrix laminae remain subparallel between breccia blocks, denying a sedimentary origin. The oriented brecciation and the granulated grains in thin section point to a compressive origin for these rocks. In one specimen, the matrix breccia was isotopically dated as late Miocene to late Pliocene, while the cement produced a late Pliocene–early Pleistocene age. One could infer that the brecciation occurred during an episode of compression that climaxed the Tertiary, with cementation of these fractures occurring contemporaneously or immediately following in the earliest Quaternary. Thus, many of the compressive structures observed in Neogene sediments on seismic reflection profiles in the Lima and Trujillo Basins were likely culminated in a pulse that closed the Pliocene. In support of this chronology, Hosmer (1959) discusses Pliocene outcrops in the East Pisco Basin that were folded and faulted prior to deposition of Pleistocene detritus or volcanics. More severe tectonism occurred at this time in the far removed foredeep of the eastern Andean foothills.

In the Lima and West Pisco Basins, nearly flat-lying reflectors truncate against scarps of the present sediment surface (Figs. 4C, 4D, 4E, 4F: SP-11.0, 11.6, 11.9A, 12.2, 12.8, 13.4, and 13.8) and against scarped interfaces within the shallow subsurface (SP-10.1). Recent removal of hundreds of metres of strata is indicated, since these erosional features are not common within the deeper subsurface. In the West Pisco Basin, the truncated sediments occur directly above the USR against a scarp that initiates a significant increase in slope gradient. Slumps, slides, or debris flows may have caused the erosion of these sediments. Such flows can evolve into more stratified turbidite sheet flows (submarine canyons are uncommon seaward of the West Pisco Basin) and finally come to rest in a downslope basin (Underwood and others, 1980). Coulbourn and Moberly (1977) documented similar slumping off the seaward side of an upper-slope structural high near lat 19°S on the southern Peru Margin.

In the Lima Basin, the scarps of stratal truncation are separated by a broad plateau from a significant increase in slope gradient (the morphologic expression of the USR). Chaotic gravity flows seem to be an inadequate mechanism of erosion: the seismic facies of these deposits (chaotic reflectors beneath an irregular, discordant surface; see Sangree and Widmier, 1977) are not recognized on the Lima Basin profiles. The sediment would have to be cleanly and completely removed for a distance of several tens of kilometres across gradients of less than 1:50. Lateral erosion and transport via submarine unidirectional currents may be a more viable mechanism.

Similar surfaces of erosional truncation were documented in the area of the Walvis Shelf in southwest Africa (van Andel and Calvert, 1971), though much thinner stratal sections were involved (less than 100 m) and water depths were much shallower; bottom currents at 200 to 600 m water depth were the postulated agents. Current meter data and geostrophic calculations over the Peru Margin indicate the presence of a poleward-flowing undercurrent whose velocity often reaches 50 cm/sec and infrequently bursts to 70 cm/sec (Huyer, 1980; Brink, and others, 1978). The core of this undercurrent, however, is centered at a water depth of 100 m and dissipates rapidly below 400 m; it is situated over the upper continental slope. The single-channel reflection records exhibit erosional truncation roughly between 500 to 1,000 m water depth (and down to nearly 2 km in SP-11.0). If the margin's present hydrodynamic circulation is an indication of past conditions, then currents powerful enough to cause significant erosion of the sediment surface should not exist below 400 m. The presence of erosionally truncated reflectors at depths much greater than this implies that a considerable amount of subsidence has occurred in the Lima Basin. Based upon benthic foraminifera within dredged rocks of the upper slope, Kulm and others (1981b, this volume) conclude that subsidence of 500 to 1,100 m has occurred in the Lima Basin since the Pliocene–early Pleistocene; reconstruction of this lost elevation would place most of the truncated strata in the water depth range typical of the poleward undercurrent found today. These data supplement the subsidence that is inferred from interpretations of the single-channel reflection records across the Lima Basin. In contrast, foraminiferal studies of the Trujillo Basin show no large-scale vertical movements since possibly late Miocene time.

Upper-Slope Basins—Stratigraphy. The Talara Basin (Fig. 2B) is the only known coastal basin that contains sediments of latest Cretaceous (Campanian-Maestrichtian), Paleocene, and early and middle Eocene age (Travis and others, 1976); negative relief was sustained in the basin in the midst of pervasive Andean tectonism. However, random and intense vertical, blocky movements (displacements as great as 2 to 3 km) rapidly created and destroyed local sub-basins and greatly disrupted stratal continuity. Notably, nearly 90% of Peru's petroleum production is from the lower and middle Eocene strata of this basin. The Coast Range prohibits any of these hydrocarbon-rich, early Tertiary sediments from spilling over into the Sechura Basin.

A seismic refraction profile across the Trujillo Basin near lat 9°S outlines a trough of sedimentary velocity layers approximately 4 km thick (Fig. 5B). The 2.9 to 3.1 km/sec material in the lower part of the upper-slope basin sequence occupies a stratigraphic position correlative to the much higher-velocity 4.8 km/sec material across the crystalline OSH in the flanking Salaverry shelf basin to the east. If the 4.8 km/sec layer of the Salaverry Basin is composed of Mesozoic geosynclinal sediments deformed during the early Tertiary orogenic episode, then the corresponding 2.9 to 3.1 km/sec velocity layer of the Trujillo Basin should represent either geosynclinal sediments that were affected only slightly by the orogeny, or younger Cenozoic sediments. In fact, deposition in this upper-slope basin may have initated and continued, though with sporadic interruptions, in the midst of the early Tertiary tectonism that occurred inland, comparable to the Talara stratigraphy. It is possible that the Talara and Trujillo Basins were a continuous feature during the early Cenozoic.

A gravity profile that traces the refraction line offers further support for this argument (Fig. 5A). The free-air anomaly remains high over the Salaverry Basin, dropping only about 20 mgal from the +75 mgal maximum situated directly above the OSH. A drop of nearly 70 mgal, however, occurs over the Trujillo Basin. This dramatic negative deflection has been modeled by Jones (this volume) as a 2.67 g/cm^3 vertical column of crust extending 13 km into the subsurface, an anomalously low-density structure when compared to the 2.75 to 2.80 g/cm^3 crustal blocks that sandwich it and constitute the bulk of the continental margin. It seems more plausible to consider a more surficial effect when modeling the strong gravity minimum, that is, to thicken and deepen the low-density layers of the Trujillo Basin sediments to better conform to the refraction thickness of 4 km.

By integrating a variety of data sources, the basics of Trujillo stratigraphy may be pieced together in the vicinity of 9°S (Figures 2A, 7). PetroPeru and Occidental Petroleum have provided us with lithologic descriptions of the industry drill holes on the outer continental shelf. The seaward well (Delfin) cut 2,650 m of Paleogene strata. Overlying the metamorphic basement (western flank of the OSH) is reworked basement material, a basal sand, and approximately 300 m of fossiliferous upper Eocene siltstone and claystone. A thin sand unit may mark the Eocene/Oligocene boundary that is unconformable in coastal outcrops (Hosmer, 1959). Approximately 750 m of claystones, siltstones, limestones, and calcareous shales follows stratigraphically. Beginning at a drill hole depth of about 1500 m, very hard, dense, microcrystalline dolomites and dolomitic limestones become commonly interbedded with clays and silts. This lithology may be a diagenetic alteration product of original shelf-slope break deposits (Kulm and others, 1981b, this volume). Graded sands and silts are also evident in this part of the section, consistent with an upper slope environment during middle to late Oligocene time. Finally, a fine to coarse grained sand unit was the youngest lithology cut. The Ballena hole contains a basal sand and an Oligocene section of clays, silts, and sands. Graded beds and dolomitic limestone interbeds are questionably present. A fine to coarse grained sand unit topped the hole, similar to the Delfin well.

As discussed earlier, we interpret the lower 1.5 km of Trujillo Basin sediments to be early Tertiary in age as evidenced by a 2.9 to 3.1 km/s velocity refraction layer that lies basinward and beneath the upper Eocene/Oligocene strata that the industry holes cut. Hosmer's (1959) paleogeographic reconstruction of middle to late Oligocene time shows Pacific drainage of the eastern Andean forelands through an outlet in northwestern Peru. This outlet was closed by tectonism in Miocene time, and freshly eroded Andean detritus was flushed eastward through the newly formed Amazon drainage. Thus, we interpret the graded silt and sand beds that the drill holes recovered in the Oligocene section to be derived from this northern provenance (northwest Peru), with implications that a continuous Talara-Trujillo Basin may have existed at this time. The uppermost coarse sands in the drill holes of the eastern Trujillo Basin may record the early Miocene movements that caused unconformites in coastal outcrops at the close of the Oligocene (Hosmer, 1959). Younger Neogene strata are likely confined to the upper few hundred metres of section in the east, but thicken to approximately 500 m in the western Trujillo Basin, as indicated in seismic reflection profile SP-9.0 by the deeper penetration of relatively undeformed (though offset by faults) reflectors that presumably postdate the early Miocene deformation. The oldest dated rocks dredged from this part of the basin

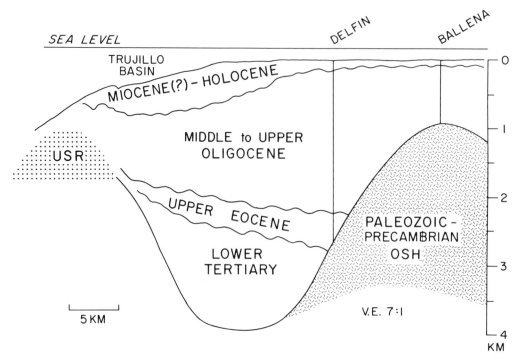

Figure 7. The general stratigraphy (depositional sequences) of the Trujillo Basin near lat 9° S. Wavy lines denote possible unconformities or hiatuses inferred from seismic refraction, reflection, drill hole data (industry wells plotted), and correlation with onshore stratigraphy. See text for discussion.

(D-54 of SP-9.0; Kulm and others, 1981b, this volume) give isotopic and paleontologic ages that are not refined closer than an inclusive late Miocene through late Pliocene bracket.

The filling of the Trujillo Basin, then, was nearly complete by the close of the Oligocene, at which time pronounced uplift must have occurred, and Neogene strata are limited to a small pond in the western part of the basin (Fig. 7). The OSH apparently acted as a positive, isolating element throughout much of the basin's evolution, though subsidence of this feature during middle and late Oligocene time allowed nearly a kilometre of sediment to accumulate at its crest. The Oligocene strata that overlie the OSH have been truncated by post-Oligocene uplift of this feature. If Neogene deposition ever extended farther eastward, it has also been stripped by subsequent uplift of the OSH. If the USR is a deformed and indurated lateral correlative of the Tertiary basinal sediments, then it need not have provided a structural limit to the basin's western boundary until the early Miocene deformation, after which younger sediments onlap the USR (SP-9.0, Fig. 4B).

A multichannel reflection line across the Lima Basin at lat 11.0°S reveals a vivid acoustic basement at a subsurface depth of approximately 2 km (Fig. 4C: CDP-1). Comparison with a refraction profile along lat 12°S (Hussong and others, 1976) indicates that crystalline continental massif may floor these sediments and cause the sharp velocity contrast, since a 2.4 km/sec versus a 6.0 km/sec interface occurs near this depth. Kulm and others (1981b, this volume) have dredged late Miocene to Holocene rocks from this part of the Lima Basin and conclude (from projections of sample locations onto reflection profiles) that deposits of this age account for at least the upper 1.3 km of the stratigraphic section. It follows, then, that Paleogene strata are at most restricted to the several hundred metres of unsampled basal section or may be completely absent. The Lima section provides a sharp contrast to the Trujillo Basin statigraphy, where nearly 4 km of pre-Miocene strata constitute the bulk of the sediment pile. Thus, early Miocene tectonism apparently uplifted the predominantly Paleogene Trujillo Basin, causing a near-fatal reduction of sediment accumulation followed by late Miocene subsidence of the Lima Basin and inception of the predominantly Neogene accumulation. It is noteworthy that the Mendaña Fracture Zone presently marks the boundary and offset between these two upper-slope basins (lat 9.5°S) with their contrasting structures and strata, perhaps signifying its influence as a lithospheric tear that has caused segmented subduction since Miocene time.

Little is known concerning the stratigraphy of the West Pisco Basin. From coastal outcrops and seismic profiling, Travis and others (1976) project 3,000 m of upper Eocene and younger strata into the offshore sequence. In the northern, undisturbed part of the basin, gravity modeling (Fig. 5C) agrees with a similar basinal thickness.

Forearc Basin Tectonics. Available geologic and geophysical evidence indicates that the major Peru forearc basins of the shelf and upper slope are floored entirely by continental crust, while sinuous basins of more stunted dimensions occupy the restricted subduction complex trenchward (Hussong and others, 1976; Kulm and others, 1981a, this volume; Kulm and others, 1981). The USR and upper-slope basins, being at the very toe of the continental massif, may be more susceptible to Nazca–South America plate interactions and more responsive to changes in these interactions over time, compared to the OSH and shelf basins that occupy a more stable interior position. We speculate that in the specific case of Peru upper-slope basins, periods of uplift and compression represent pulses of intensified plate convergence, while periods of subsidence and tension represent quiescent plate activity. This statement cannot be defended without additional drill hole control, but it seems more applicable in Peru than in forearc basins that lie in part, or in full, on subduction complexes, since these features can simultaneously exhibit uplift and subsidence on different sides of the trench-slope break and can be created or destroyed in tens of millions of years (Karig and Sharman, 1975; Seely and others, 1974).

Dating unconformities and deformational episodes in upper-slope basins, then, may give a more direct indication of plate movement than dating inland magmatic/tectonic activity, in which a time lag of several million years could exist. In fact, historical evidence would indicate that antithetic tectonic regimes may exist simultaneously in the upper-slope region versus the inland cordillera. For example, the western volcanic geanticline occupied the toe of the Cretaceous continent while geosynclinal conditions persisted inland (see Geologic History for references). During the early Tertiary uplift and orogeny, which terminated the geosyncline, deposition occurred in the upper-slope Talara and possibly Trujillo Basins. Post-Oligocene uplift and deformation of the Talara and Trujillo Basins did not manifest itself inland until the late Miocene, concurrent with Lima Basin subsidence. Unconformities and deformational episodes in shelf basins seem to predominantly reflect tectonism in the magmatic arc, though they are caught in an intermediary position.

CONCLUSIONS

1. Structural basement ridges subparallel to the onshore Andean trends control the distribution of forearc sedimentary basins. In general, a series of shelf basins (Sechura, Salaverry, East Pisco) is confined between the western Andes and the Coast Range/Outer Shelf High (OSH); a series of upper-slope basins (Trujillo, Lima, West Pisco) is cradled between the Coast Range/OSH and a prominent Upper Slope Ridge (USR).

2. The OSH is composed of Precambrian and Paleozoic crystalline rocks and metasediments cut by Mesozoic intrusives. Deformed Mesozoic sedimentary rocks probably onlap the eastern flank of the OSH and may also constitute effective basement for the overlying Tertiary sediments of the shelf basins. The USR is evidently composed of deformed sediments that appear to be lateral correlatives of the relatively undisturbed sediments that lie within the upper-slope basins. Both shelf and upper-slope basins are floored by the continental massif; the subduction complex apparently lies seaward of the USR.

3. A major change in the structure of the upper-slope basins occurs near lat 9.5°S. North of this latitude, seismic reflectors are fragmented, folded, and cut by high-angle reverse faults, indicating a basinal response to compression and uplift of the USR. South of this latitude, reflectors are smooth and continuous and form a broad syncline indicative of mild, quiet subsidence, though deformational structures overprint this pattern farther south.

4. A major reorganiztion of the structure and sedimentation of the Peru continental margin apparently occurred in early Neogene time. North of lat 9.5°S, in the upper-slope Trujillo Basin, several kilometres of sediments that had accumulated in Paleogene time (a section nearly 4 km thick at lat 9°S) were uplifted and deformed immediately following the Oligocene, and subsequent sedimentation is of greatly reduced thickness. South of lat 9.5°S, in the upper-slope Lima Basin, the bulk of the stratigraphic section (2 km thick near lat 11.5°S) seems to be of late Miocene age or younger; the middle/late Miocene initiated subsidence and sediment accumulation on this part of the upper slope, coincident inland with a period of intense Andean orogeny. Pronounced unconformities reflect these Miocene movements in the shelf basins. A final episode of upper-slope deformation closed the Pliocene and affected the shelf basins to a lesser degree.

ACKNOWLEDGMENTS

This research was sponsored by the Office of International Decade of Ocean Exploration, National Science Foundation, under grants GX 28675, IDOE 71-04208, and OCE 76-05903.

REFERENCES

Brink, K. H., Allen, J. S., and Smith, R. L., 1978, A study of low-frequency fluctuations near the Peru coast: Journal of Physical Oceanography, v. 8, p. 1025–1041.

Cobbing, E. J., 1976, The geosynclinal pair at the continental margin of Peru: Tectonophysics, v. 36, p. 157–165.

——1978, The Andean geosyncline in Peru, and its distinction from Alpine geosynclines: Geological Society of London Journal, v. 135, p. 207–218.

Cobbing, E. J., and Pitcher, W. S., 1972, The coastal batholith of central Peru: Geological Society of London Journal, v. 128, p. 421–460.

Couch, R., and Whitsett, R. M., 1981, Structures of the Nazca Ridge and the continental shelf and slope of southern Peru, in Kulm, L. D., and others, eds., Nazca plate: Crustal formation and Andean convergence: Geological Society of America Memoir 154 (this volume).

Coulbourn, W. T., and Moberly, R., 1977, Structural evidence of the evolution of forearc basins: Canadian Journal of Earth Science, v. 14, p. 102–116.

de Almeida, F.F.M., and others, coordinators, 1978, Tectonic map of South America: Commission for the Geologic Map of the World, Ministry of Mines and Energy, National Department of Mineral Production, Brazil, scale 1:5,000,000.

Dickinson, W. R., and Seely, D. R., 1979, Structure and stratigraphy of forearc regions: American Association of Petroleum Geologists Bulletin, v. 63, p. 2–31.

Farrar, E., and Noble, D. C., 1976, Timing of late Tertiary deformation in the Andes of Peru: Geological Society of America Bulletin, v. 87, p. 1247–1250.

Hosmer, H. L., 1959, Geology and structural development of the Andean system of Peru [Ph.D. thesis]: Ann Arbor, University of Michigan, 281 p.

Hussong, D. M., and others, 1976, Crustal structure of the Peru-Chile Trench: 8°S–12°S latitude, in Sutton, G. H., and others, eds., The geophysics of the Pacific Ocean Basin and its margin, a volume in honor of George P. Woollard: Washington, D.C., American Geophysical Union, Geophysical Monograph 19, p. 71–86.

Huyer, A., 1980, The offshore structure and subsurface expression of sea level variations off Peru, 1976–1977: Journal of Physical Oceanography, v. 10, p. 1755–1768.

James, D. E., 1971, Plate tectonic model for the evolution of the central Andes: Geological Society of America Bulletin, v. 82, p. 3325–3346.

Jenks, W. J., 1956, Handbook of South American geology; an explanation of the geologic map of South America: Geological Society of America Memoir 65, p. 217–247.

Johnson, S. H., and Ness, G. E., 1981, Shallow structures of the Peruvian Margin, 12°S–18°S, in Kulm, L. D., and others, eds., Nazca plate: Crustal formation and Andean convergence: Geological Society of America Memoir 154 (this volume).

Jones, P.R., 1981, Crustal structures of the Peru continental margin and adjacent Nazca plate, 9°S latitude, in Kulm, L. D., and others, eds., Nazca plate: Crustal formation and Andean convergence: Geological Society of America Memoir 154 (this volume).

Karig, D. E., 1977, Growth patterns on the Upper Trench Slope, in Talwani, M., and Pitman, W. E., III, eds., Island arcs, deep sea trenches, and back-arc basins: Washington, D.C., American Geophysical Union, Maurice Ewing Series 1, p. 175–186.

Karig, D. E., and Sharman, G. F., 1975, Subduction and accretion in trenches: Geological Society of America Bulletin, v. 86, p. 377–389.

Karig, D. E., and others, 1978, Late Cretaceous subduction and continental margin truncation along the northern Middle America Trench: Geological Society of America Bulletin, v. 89, p. 265–276.

——1979, Structure and Cenozoic evolution of the Sunda Arc in the Central Sumatra region, in Watkins, J. S., and others, eds, Geological and geophysical investigations of continental margins: Tulsa, American Association of Petroleum Geologists Memoir 29, p. 223–237.

Kulm, L. D., and Fowler, G. A., 1974, Oregon continental margin structure and stratigraphy: A test of the imbricate thrust model, in Burk, C. A., and Drake, C. L., eds., The geology of continental margins: New York, Springer-Verlag, p. 261–283.

Kulm, L. D., Schweller, W. J., and Masias, A., 1977, A preliminary analysis of the subduction processes along the Andean continental margin, 6° to 45°S, in Talwani, M., and Pitman, W. C., III, eds., Island arcs, deep sea trenches, and back-arc basins: Washington, D. C., American Geophysical Union, Maurice Ewing Series 1, p. 285–301.

Kulm, L. D., and others, 1981, Cenozoic structure, stratigraphy, and tectonics of a the central Peru forearc, in Leggett, J. K., ed., Trench and forearc sedimentation and tectonics in modern and ancient subduction zones: London, Geological Society (in press).

——1981a, Crustal structure and tectonics of the central Peru continental margin and trench, in Kulm, L. D., and others, eds., Nazca plate: Crustal formation and Andean convergence: Geological

Society of America Memoir 154 (this volume).
—— 1981b, Late Cenozoic carbonates on the Peru continental margin: Lithostratigraphy, biostratigraphy, and tectonic history, *in* Kulm, L. D., and others, eds., Nazca plate: Crustal formation and Andean convergence: Geological Society of America Memoir 154 (this volume).

Masias, J. A., 1976, Morphology, shallow structure, and evolution of the Peruvian continental margin, 6° to 18°S [M.S. thesis]: Corvallis, Oregon State University, 92 p.

Myers, J. S., 1974, Cretaceous stratigraphy and structure, western Andes of Peru between latitudes 10°–10°30′: American Association of Petroleum Geologists Bulletin, v. 58, p. 474–487.

—— 1975, Vertical crustal movements of the Andes in Peru: Nature, v. 254, p. 672–674.

Nasu, N., and others, 1980, Interpretation of multichannel seismic reflection data, Legs 56 and 57, Japan Trench transect, Deep Sea Drilling Project, *in* Scientific party, Initial reports of the Deep Sea Drilling Project, 56, 57, Part 1: Washington, D. C., U.S. Government Printing Office, p. 489–503.

Noble, D. C., and others, 1974, Episodic Cenozoic volcanism and tectonism in the Andes of Peru: Earth and Planetary Science Letters, v. 21, p. 213–220.

—— 1979, Early Tertiary "Incaic" tectonism, uplift, and volcanic activity, Andes of central Peru: Geological Society of America Bulletin, v. 90, p. 903–907.

Pitcher, W. S., 1978, The anatomy of a batholith: Geological Society of London Journal, v. 135, p. 157–182.

Prince, R., and others, 1980, Bathymetry of the Peru-Chile trench and continental margin: Geological Society of America Map & Chart Series, MC-34, scale 1:1,000,000.

Sangree, J. B., and Widmier, J. M., 1977, Seismic stratigraphy and global changes of sea level, Part 9: Seismic interpretation of clastic depositional facies, *in* Payton, C. E., ed., Seismic stratigraphy—applications to hydrocarbon exploration: Tulsa, American Association of Petroleum Geologists Memoir 26, p. 165–184.

Scholl, D. W., Marlow, M. S., and Cooper, A. K., 1977, Sediment subduction and offscraping at Pacific margins, *in* Talwani, M., and Pitman, W. D., III, eds., Island arcs, deep sea trenches, and back-arc basins: Washington, D.C., American Geophysical Union, Maurice Ewing Series 1, p. 199–210.

Seely, D. R., 1979, The evolution of structural highs bordering major forearc basins, *in* Watkins, J. S., and others, eds., Geological and geophysical investigations of continental margins: Tulsa, American Association of Petroleum Geologists Memoir 29, p. 245–260.

Seely, D. R., Vail, P. R., and Walton, G. G., 1974, Trench slope model, *in* Burk, C. A., and Drake, C. L., eds., The geology of continental margins: New York, Springer-Verlag, p. 249–260.

Stewart, J. W., Everden, J. E., and Snelling, N. J., 1974, Age determinations from Andean Peru: A reconnaissance survey: Geological Society of America Bulletin, v. 85, p. 1107–1116.

Travis, R. B., Gonzales, G., and Pardo, A., 1976, Hydrocarbon potential of coastal basins of Peru, *in* Halbouty, M. T., and others, eds., Circum-Pacific energy and mineral resources: Tulsa, American Association of Petroleum Geologists Memoir 25, p. 331–338.

Underwood, M., Bachman, S. B., and Schweller, W. J., 1980, Sedimentary processes and facies associations within trench and trench-slope settings, *in* Field, M. E., and others, eds., Quaternary depositional environments on the Pacific continental margin: Society of Economic Paleontologists and Mineralogists, Pacific Section, p. 211–229.

van Andel, T. H., and Calvert, S. E., 1971, Evolution of sediment wedge, Walvis Shelf, southwest Africa: Journal of Geology, v. 79, p. 585–602.

von Huene, R., 1979, Structure of the outer convergent margin off Kodiak Island, Alaska, from multichannel seismic records, *in* Watkins, J. S., and others, eds., Geological and geophysical investigations of continental margins: Tulsa, American Association of Petroleum Geologists Memoir 29, p. 261–272.

Wilson, J. J., 1963, Cretaceous stratigraphy of the central Andes of Peru: American Association of Petroleum Geologists Bulletin, v. 47, p. 1–34.

MANUSCRIPT RECEIVED BY THE SOCIETY NOVEMBER 12, 1980
MANUSCRIPT ACCEPTED DECEMBER 30, 1980

Geological Society of America
Memoir 154
1981

Crustal structures of the Peru continental margin and adjacent Nazca plate, 9°S latitude

PAUL R. JONES III*
School of Oceanography
Oregon State University
Corvallis, Oregon 97331

ABSTRACT

Seismic refraction, reflection, and gravity data obtained across the Peru continental margin and Nazca plate at lat 9°S permit a detailed determination of crustal structure. The western portion of the continental shelf basement consists of a faulted outer continental shelf high of Paleozoic or older rocks. It is divided into a deeper western section of velocity 5.0 km/sec. The combined structure forms a basin of depth 2.5 to 3.0 km which contains Tertiary sediments of velocity 1.6 to 3.0 km/sec. The 3 km thick, 4.55 to 5.15 km/sec basement of the eastern shelf shoals shoreward. Together, this basement and the eastern section of the outer continental shelf high form a synclinal basin overlain by Tertiary sediment which have a maximum thickness of 1.8 km and a velocity range of 1.7 to 2.55 km/sec. The gravity model shows a large block of 3.0 g/cm^3 lower crustal material emplaced within the upper crustal region beneath the eastern portion of the continental shelf.

Refraction data indicate a continental slope basement of velocity 5.0 km/sec overlying a slope core material with an interface velocity of 5.6 km/sec. The sedimentary layers of the slope consist of an uppermost layer of slumped sediment with an assumed velocity of 1.7 to 2 km/sec that overlie an acoustic basement of 2.25 to 3.6 km/sec. The high velocities (and densities) of the slope basement suggest the presence of oceanic crustal material overlain by indurated oceanic and continental sediments. This slope mélange may have formed during the initiation of subduction from imbricate thrusting of upper layers of oceanic crust.

A ridgelike structure within the trench advances the seismic arrival times of deeper refractions and supports the suggestion that it is trust-faulted oceanic crust which has been uplifted relative to the trench floor. The model of the descending Nazca plate consists of a 4 km thick upper layer of velocity 5.55 km/sec and a thinner (2.5 km) but faster (7.5 km/sec) lower layer that overlies a Moho of velocity 8.2 km/sec. The gravity model indicates that the plate has a dip of 5° beneath the continental slope and shelf. West of the trench, the lower crustal layers rise upward, which may represent upward flexure of the oceanic plate due to compressive forces resulting from the subduction process.

The upper crustal layers of the 120 km long oceanic plate portion consist of a thin, 1.7 km/sec

---

*Present Address: Union Science & Technology Division, Union Oil Company of California, Brea, California 92621.

sedimentary layer overlying a 5.0 to 5.2 km/sec upper layer. Immediately beneath these layers, a 5.6 to 5.7 km/sec lower layer becomes more shallow to the east within 60 km of the trench, and a deeper 6.0 to 6.3 km/sec layer thickens to the east. The lower crustal model consists of a 7.4 to 7.5 km/sec high velocity layer that varies in thickness from 2.5 km to 4.0 km. The 8.2 km/sec Moho interface varies not more than ±0.5 km from a modeled depth of 10.5 km.

INTRODUCTION

Active continental margins surround most of the Pacific Ocean Basin. One of these is the little known Peru-Chile Trench. The crustal velocities and structures of the Peru continental margin and trench and of part of the Nazca plate displayed by Hussong and others (1976) suggest that the rapidly moving Nazca oceanic crust (10 cm/yr; Minster and others, 1974) is thin and dense relative to other oceanic basins. Structure and velocity distributions in the lower toe of the continental slope suggest uplifted imbricate thrust sheets containing oceanic sediments and rock. It was also found that slope velocities and structures change laterally and are interpreted as highly disrupted, downfaulted continental rocks. The basement rocks of the continental shelf are less faulted and are covered with more than a kilometre of smoothly stratified sediments.

While Hussong and others (1976) reported a preliminary interpretation of HIG refraction data along Line 18-19, the primary purpose of the present study is to determine a more detailed crustal and subcrustal cross section based on this seismic refraction and reflection data from both OSU and HIG data files and on gravity data from HIG. The emphasis is on the lateral changes in the structures seen at intermediate depths (1 to 3 km). The present study makes extensive use of secondary seismic arrivals in the refraction data, well log velocities and depths, near-surface sediment structures, and CDP velocity data to obtain an integrated model of the active Peruvian continental margin at lat 9°S. These data are used to substantiate a hypothesis for the formation of the Peru continental margin.

Methods and Instrumentation

Data Collection. A 360 km long seismic refraction, reflection, and gravity profile located across the continental shelf and slope, trench, and part of the Nazca plate (Fig. 1) was acquired in March 1972 by Oregon State University and the Hawaii Institute of Geophysics. The seismic refraction profile is designated Line 18-19, which lies between 8°26'S lat, 79°06'W long and 9°52'S lat, 81°58'W long, respectively. Other data gathered along the line include a single-channel reflection profile to the trench axis from near shore, a multichannel CDP reflection profile located 25 km to the north (Kulm and others, 1981a, this volume), and a gravity profile over the refraction line (Fig. 1).

The 11-station seismic refraction profile shown in Figure 1 is an adaptation of the land seismic method of shooting overlapping refraction lines. Line 18-19 used more than 400 explosive charges over a profile length that exceeded 360 km. The marine seismic refraction method for this experiment used single receivers and moving shop points (see Hussong and others, 1976). Standard military sonobuoys of the type AN/SSQ-41A were seismic detectors for Line 18-19. Sonic information detected by 4 hydrophones deployed 18 m below the surface was transmitted from the sonobuoy to the ship by a frequency-modulated transmitter in the frequency band of 162 to 174 MHz. The sonic response of the sonobuoy increases at about 5 db/octave in the frequency range from 1 to 1,000 Hz. The transmitted signal was received on a modified police band receiver, amplified, band-pass filtered, and recorded at 50 mm/sec on an oscillographic camera. The recorded traces included high-frequency, low-frequency, and unfiltered sonobuoy signals, as well as the clock channel, and the signal from the streamer that was used to detect the shot break.

The trackline map in Figure 1 shows the location of gravity measurements used for modeling the crustal section. Gravity measurements were obtained by Lacoste and Romberg surface ship gravity meters on board the R/V *Yaquina* (OSU) and R/V *Kana Keoki* (HIG) during the acquisition of

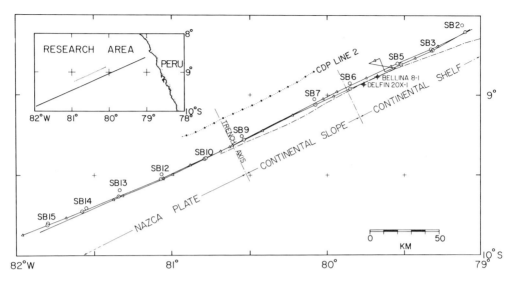

Figure 1. Location of seismic and gravity profiles at lat 9°S. Solid lines represent Line 18-19. The dashed and dotted lines are the 1974 OSU and CDP-2 tracklines. Triangles symbolize satellite navigation fixes; circles with dots and open circles are OSU sonobuoy deployment and HIG midship sonobuoy positions, respectively. Crossed circles are exploratory well locations.

seismic refraction Line 18-19. Due to more extensive coverage, the gravity record obtained by the R/V *Kana Keoki* was used in the crustal modeling. The land gravity data was obtained from the Defense Mapping Agency Aerospace Center in St. Louis, Missouri.

Seismic Ray Tracing Methods. The forward modeling technique of seismic ray tracing in laterally inhomogenous media has been approached by several authors. The methods of Yacoub and others (1968), Jacob (1970), Sorrells and others (1971), and Shah (1973) use velocity models comprised of constant-velocity geological units of arbitrary shape in either two or three dimensions. The methods of Yacoub and others (1968) and Jacob (1970) are poorly suited for seismic refraction exploration, because provision is not made for the critically refracted ray travelling along an interface (head wave). In a step toward further complexity, Gebrande (1976) traces rays through models with two-dimensional elements where velocity gradients are permitted. Velocity-gradient modeling usually requires drill hole control or very close shot spacing to warrant its use.

The ray-tracing technique developed for the analysis of Line 18-19 seismic refraction and reflection data was used on a Data General NOVA minicomputer. Computational speed becomes very important when tracing ray paths in complex geological structures. For this reason, a method similar to that of Sorrells and others (1971) and Shah (1973) was chosen, because it is based on vector operations that are computationally faster than similar methods based on numerical integration (Jacob, 1970) or on transcendental functions (Yacoub and others, 1968; Gebrande, 1976).

A series of interactive computer programs was developed for this study to compute and plot seismic reflection and refraction arrival times and to visually display ray paths traced through a given and possibly complex model (see Jones, 1979 for computer program). This computer technique allows the interpreter to develop complex geological structure to match observed seismic refraction arrivals with those computed from a tentative model, to use other velocity and depth information in order to compensate for near-surface structures that affect the determination of velocities and depths to deep structures, and to model hypothetical cases in order to give a better understanding to the interpretation of seismic reflection and refraction data. The interactive approach to ray trace modeling allows the user to refine a model from the top downward while remaining within the constraints imposed by seismic refraction, reflection, and sonic log data.

RESULTS

Seismic Refraction Modeling

Continental Shelf. The seismic refraction data interpretation and ray trace model for the continental shelf are shown in Figures 2 and 3, which are the eastern and western sections, respectively. The interpretation of both sections suggests a sedimentary basin overlying a hard-rock basement. Substructure was modeled as 3 sedimentary layers overlying a 2-layer rock basement. Extensive modeling of the G_1 arrivals between SB2 and SB5 provided a velocity of 1.7 km/sec that also satisfied the modeling of the upper sediment layer in the seismic reflection profile (western section shown in Fig. 4). This sediment layer can be seen in the seismic reflection profile (Fig. 4). Refracted arrivals from a 1.9 km/sec layer are clearly seen in profiles extending to the west (Figs. 2,3). The shape of the upper surface to the 1.9 km/sec layer is slightly curved on the basis of arrivals G_2 observed from sonobuoys 2 and 5. The 2.3 to 2.55 km/sec layer was modeled from arrivals G_3 detected at short ranges from sonobuoys 2, 3, and 5. At greater distances, the attenuation of seismic energy by the sediments (Hamilton, 1974) probably accounts for the small number of G_3 arrivals observed. An angular unconformity observed at 63 km on a seismic reflection profile (Fig. 4) was an additional constraint to the shape of the eastern sedimentary basin near sonobuoy 5. Layer interfaces identified with the upper surfaces of the 1.7, 1.9, and 2.3 km/sec layers were modeled to produce a time section closely matching the observed profile (Fig. 4).

Layer 4 is divided into a faster western structure (5.7 to 5.9 km/sec) related to the outer continental shelf high and abutting a slower (4.55 to 5.15 km/sec) eastern structure (Fig. 2). The velocity below both structures increases to the west from 6.6 to 7.2 km/sec (G_5 arrivals). The arrivals associated with layer L4-5 are detected only in an eastern travel time branch (G_{4-5}) of sonobuoy 5 (Fig. 2). A large number of models were investigated by ray trace modeling, and it was determined that the 5.7 km/sec, wedge-shaped structure shown gave the best fit to the arrivals.

The model of the western section of the continental shelf (Fig. 3) consists of a multilayered sedimentary basin overlying a two-layer rock basement. Exploratory oil wells Bellena 8-1 and Delfin 20X-1 provided velocity and depth to basement control for the western section. Due to the proprietary nature of the well information, these sonic logs cannot be shown in detail.

Modeling of the sedimentary layers between sonobuoys 5 and 6 was limited mostly to arrivals detected at short ranges. Correct location of layer interfaces between these sonobuoys was achieved by correlation of sonic (velocity) logs of the two exploratory wells (Bellena 8-1 and Delfin 20X-1). Some near-surface indications of the 3.0 km/sec structure between 105 and 115 km can be seen on the tracing of the reflection profiles shown in Figure 5.

The upper surface of the basement structure of the western section is based upon a well-defined set of first arrivals (G_5, 5.0 to 5.65 km/sec), seen in Figure 3. The depth to basement at exploratory well Bellena 8-1 was 0.98 km below sea level. Sonic logging at this well provided an average sediment velocity of 2.7 km/sec immediately overlying a quartz biotite gneiss basement with a velocity of 5.65 km/sec. The depth to basement at exploratory well Delfin 20X-1 was 2.65 km below sea level. Sonic logging indicated a basement velocity of 4.8 km/sec in a highly slickensided and fractured, dark gray phyllite. With this information, the sediment-basement interface was modeled using first arrivals detected by sonobuoys 5, 6, and 7. An improved fit to the observed arrivals was achieved when a basement velocity of 5.0 km/sec was used west of the Delfin well.

A western extension of the 7.2 km/sec interface observed east of sonobuoy 5 was modeled for reflected arrivals (Fig. 3). Due to a limited number of arrivals caused by several explosive misfires, the presence of the 7.2 km/sec interface west of sonobuoy 5 was not well established. Modeling indicated that this interface (using an assumed velocity of 6 km/sec) does not exist farther to the west.

Continental Slope. Figure 6 shows the seismic refraction data interpretation and ray trace model for the continental slope. Two reversed refraction profiles, each 30 km long, were originally intended for this area. Instead, a 56 km long profile was obtained between sonobuoys 7 and 9, because the middle sonobuoy (number 8, not shown) malfunctioned during the shooting. A combination of the longer

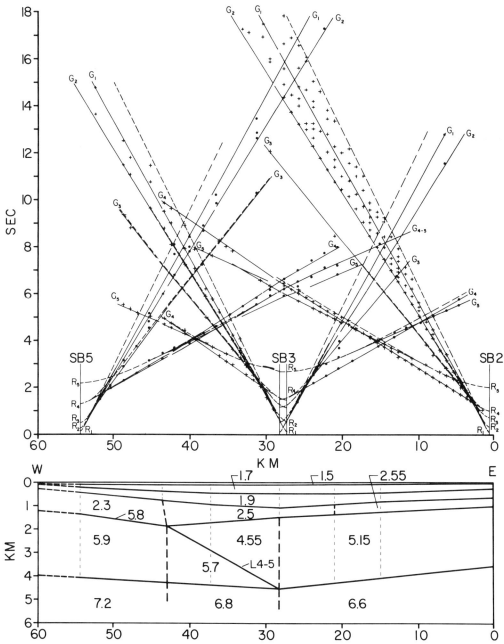

Figure 2. Seismic refraction data interpretation and ray trace model for the eastern continental shelf. R represents a single reflection arrival, and G represents a refraction arrival. The subscripts 1, 2, 3, ... number the sub-bottom layers. Velocities are in km/sec, where assumed values are presented in parentheses. Vertical dashed lines are cell walls of computer model. Vertical exaggeration is 3:1.

profile, frequent explosive charge misfires, and severe topography made interpretation of the data difficult.

The 1.7 km/sec layer seen in Figure 6 was not observed in the refraction data (G_1), but a sedimentary layer overlying an acoustic basement was observed for this area in the 1974 air-gun profile (Figure 5) and in a multichannel profile nearby (that is, CDP line 2; Kulm and others, 1981a, this volume). The distance from sea floor to acoustic basement was modeled by computer program RAYGUN (see Jones, 1979), and the results were incorporated in the ray trace model of the continental slope. A poorly observed 2.25 km/sec layer (interval velocity verified from CDP data) was used to model the velocity medium (G_2 arrivals) between the acoustic basement and a 3.6 km/sec refracting horizon. Between 130 and 145 km in Figure 6 is a series of arrivals located between those arriving for the 3.6 km/sec (G_3) and 5.0 km/sec G_4) refractors. The arrivals may correspond to a refracting layer located in the upper slope. This interface was not modeled because of insufficient data from the reverse line. Between sonobuoys 7 and 9, the majority of refracted first arrivals (G_5) are from a 5.6 km/sec basal refractor that appears to define the core of the continental slope.

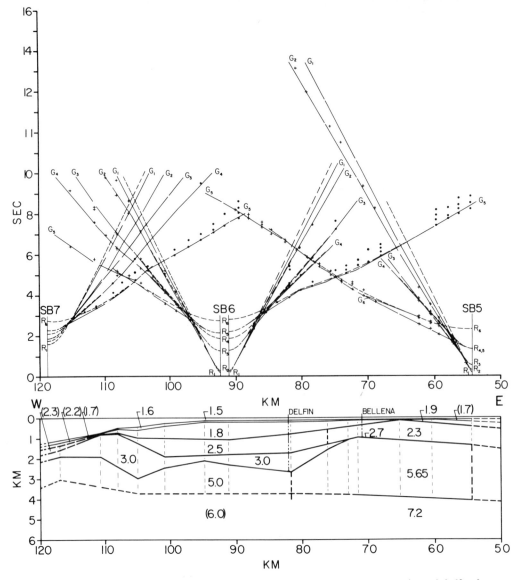

Figure 3. Seismic refraction data interpretation and ray trace model for the western continental shelf and upper slope. See Figure 2 for explanation of symbols. Delfin and Bellena are exploratory wells drilled to basement.

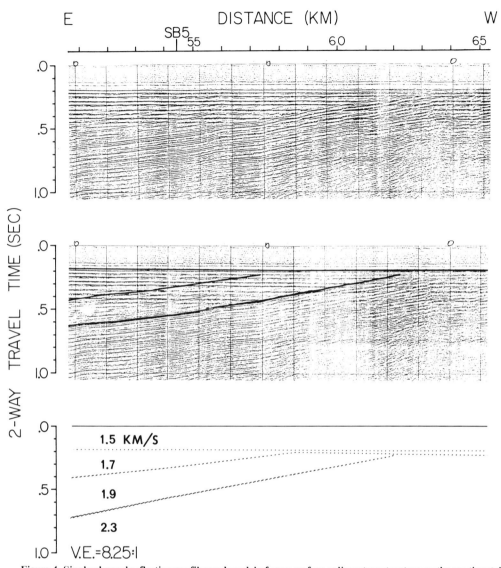

Figure 4. Single-channel reflection profiles and model of near-surface sedimentary structure on the continental shelf near sonobuoy 5. Seismic profiles illustrate angular unconformities near 57 and 63 km. The lower diagram was simulated by program RAYGUN to match the seismic profile.

Hyperbolas for reflections R6 and R7 (Fig. 6) from the major upper layers of the descending Nazca plate were modeled in the data. A velocity of 5.65 km/sec was assumed for the top layer in order to produce a reflecting surface at L6. The surface at L7 was extrapolated from a similar interface located in the Nazca plate model (Figs. 7 and 9). The Moho interface reflections were not modeled due to their extreme depth. The absence of reflected energy at hyperbolas R6 and R7 will be addressed in the discussion.

Trench Area. Figure 7 shows the interpretation of the seismic refraction data and ray trace model for the Peru-Chile Trench and part of the Nazca plate. The velocity-depth model divides into an eastern section (sono-buoys 2 and 10), related to the tectonics of the trench, and into a western section (sonobuoys 10 and 12), related to the Nazca plate.

The eastern area in Figure 7 represents a model based upon an assumed and partially observed upper section (velocities 1.7 to 3.6 km/sec) overlying an observed crustal plate section (velocities 5.55 to 8.2 km/sec). The dimensions and shape of the sedimentary basin (1.7 to 1.9 km/sec layers) located within the trench are modeled from a 1972 single-channel reflection profile (Fig. 20 of Prince, 1974) located at a trench crossing 5 km to the south of Line 18-19. Prince and Kulm (1975) reported the trench fill in this area to be composed of turbidites. Velocities of 1.7 to 1.9 km/sec from Hamilton and others (1974) were used to model the turbidite fill. Due to several misfires of explosives at short distances, it was not possible to analyze the sonobuoy 9 data to the west for the velocity and structure above the 5.55 km/sec layer. A velocity of 2.0 km/sec assumed for the first layer and velocities of 2.25 and 3.6 km/sec were extrapolated from values observed on the slope. The velocities associated with the ridge in the trench (centered around 192 km in Figure 7) are based on the CDP Line 2 interval velocities calculated for this structure. The effect of the ridge structure on deeper layer arrivals is clearly seen on sonobuoys 9 and 10 in Figure 7, where arrival time advancements of up to 0.3 sec are produced. An overall deficiency of well-defined arrivals for the 1.7 to 3.6 km/sec layers indicates a poorly defined upper structure for the descending plate and slope base.

Between sonobuoys 9 and 10, the majority of refracted first arrivals (G_4) are from the 5.55 km/sec upper surface of the descending Nazca plate (Fig. 7). The arrivals labeled G_5 are from a 7.3 km/sec interface located within the plate. Moho arrivals labeled G_6 are also detected in the eastern section.

The western section in Figure 7 represents the eastern Nazca plate prior to subduction. As discussed earlier, refracted arrivals (G_1) of the upper sediment layer are usually not detected in deep-ocean seismic refraction records. An assumed velocity of 1.7 km/sec was used to model the sediment layer and is assumed to extend from the sea floor to the first basement layer.

Although not reproduced here, reduced travel time plots are used to expand the time scale and to separate arrival times observed for the western section (Fig. 7). The ridge structure made it possible to ray trace the 5.2, 5.7, and 6.0 km/sec layers observed between sonobuoys 10 and 12. The shallowing of the upper surfaces of the 7.5 km/sec layer and 8.2 km/sec Moho interface may be related to the tectonics of the discending plate.

Nazca Plate Near the Trench. Figure 8 displays the data and model of the Nazca plate near the trench. The velocity and structure of this 120 km long crustal section are based on observed arrivals on sonobuoys 12, 13, 14, and 15. The sub-bottom model shown in Figure 8 consists of 5 layers overlying an upper mantle of uniform velocity (8.2 km/sec). The thickness of the sediment layer (0.15 to 0.21 km) was computed by the method given earlier. An assumed velocity of 1.7 km/sec was used. The thickness agrees with a sediment isopac map of the Nazca plate (Erlandson and others, this volume) and with the 155 m of sediment overlying a basalt basement at DSDP Site 320 (Yeats and others, 1976).

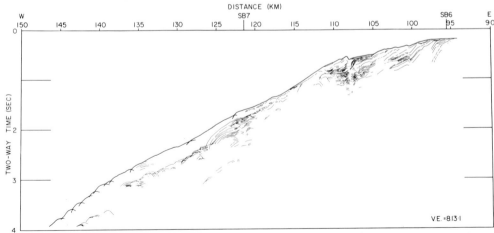

Figure 5. Line drawing interpretation of single-channel refletion profile across the upper and middle continental slope at lat 9°S.

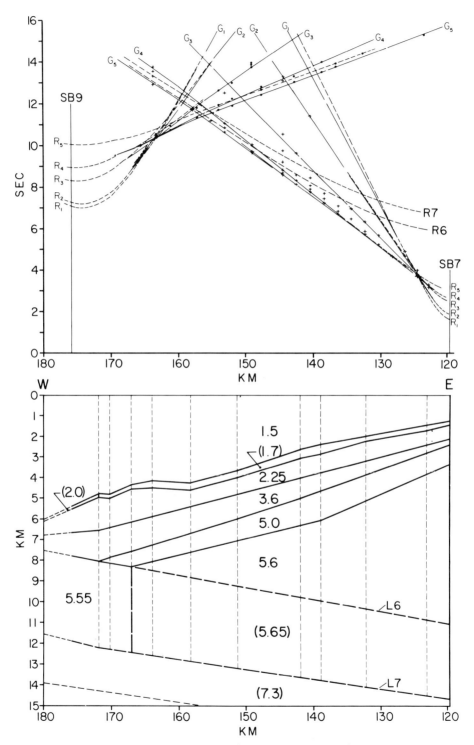

Figure 6. Seismic refraction data interpretation and ray trace model for the middle and lower continental slope. See Figure 2 for explanation of symbols. Layers L6 and L7 are assumed.

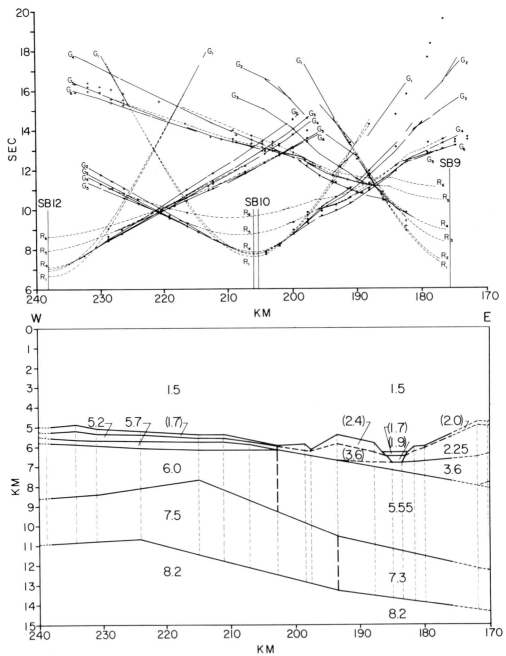

Figure 7. Seismic refraction data interpretation and ray trace model for the trench and part of the Nazca plate. See Figure 2 for explanation of symbols.

Due to the small variation in apparent velocities and data scatter, and perhaps due also to the homogeneous nature of the material, a much simpler model evolved on the plate than on the shelf and slope. A model with relatively constant layer velocities satisfied observed variations in the layer interfaces. The model in Figure 8 shows a uniform thickness in the 5.0 and 5.6 to 5.7 km/sec layers west

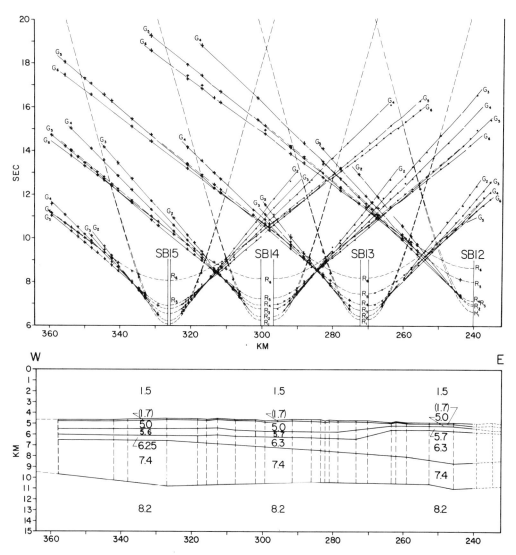

Figure 8. Seismic refraction data interpretation and ray trace model for the Nazca plate. See Figure 2 for explanation of symbols.

of 275 km. Eastward of this location, the 5.0 km/sec layer thins considerabley, and the 5.7 km/sec layer lies closer to the sea floor. The thickness of the 6.34 km/sec layer tapers from 3 km to 0.5 km from east to west. The opposite tapering occurs for the 7.4 km/sec layer (2.3 km to 3.5 km from east to west) so that the overall crustal thickness is relatively uniform (approximately 5.8 km, not including the water layer). The depth of the Moho interface varies not more than ±0.5 km from 10.5 km for the 120 km long model.

Gravity Modeling

Crustal sections are computed across the Andean margin and onto the adjacent Nazca plate following the line integral method of Talwani (Talwani and others, 1959), as adapted by Gemperle (1970, 1975). The crustal and subcrustal model extends to a depth of 70 km and, to avoid edge effects,

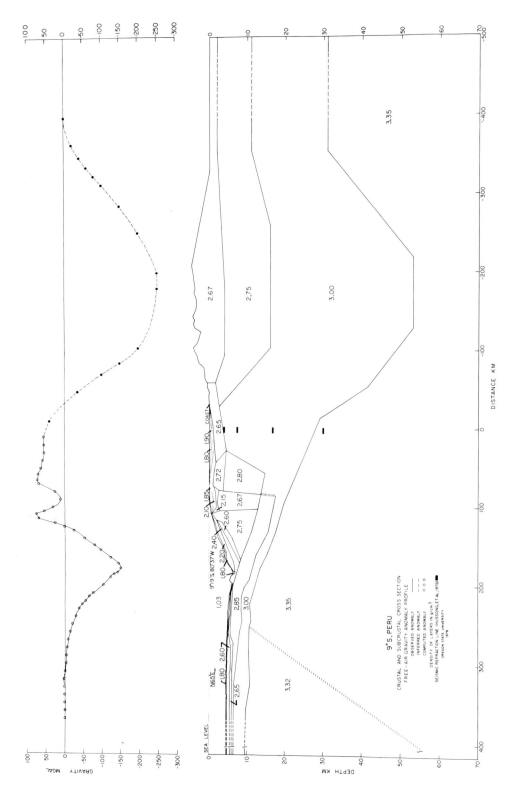

Figure 9. Crustal and subcrustal cross section along Line 18-19 at lat 9° S (see Fig. 1). The vertical exaggeration is 5:1.

extends a large distance to each side of the central area that contains the structure of interest (Fig. 9). The mass column gravity value of 9223.6 mgal for the model is calculated from a mass column given by Barday (1974) plus an additional 20 km section added to the bottom of the mass column.

The velocity-depth model determined above by ray tracing was used to estimate densities. For the continental margin west of 0 km, the velocities in Figures 2, 3, 6, 7, and 8 were used to constrain the initial density values obtained from the Ludwig, Nafe, and Drake (1970) curve. With the few exceptions, noted later, the layer boundaries shown in Figure 9 were not moved during the gravity modeling.

Except for the topography and some surface geology information, the layer boundaries of the gravity model are not constrained east of 0 km (that is, Andean Continent). Land gravity control is based upon 2 coastal gravity values obtained from the Defense Mapping Agency and upon values obtained from a Bouguer gravity anomaly map of South America (Technical Paper No. 73-2, DMAAC).

Seismic refraction Line 20-21 (Hussong and others, 1976), which is located 7 km north of Line 18-19 and is parallel to the coast of Peru, forms the initial boundary control for the subcrustal layers located near 10 km (Fig. 9). Aside from this line, there is no refraction control close to the crustal section along Line 18-19. For this reason, densities and approximate layer thicknesses were obtained from crustal and subcrustal cross sections off southern Peru (Whitsett, 1976). the depth to Moho under the Andes has been found to decrease from 70 km under the Western Cordillera and western Altiplano region at lat 15°S (James, 1971) to 45 km under the Cordillera Central at lat 1°N (Case and others, 1971). Based on this information, the depth to Moho was modeled at 53 km, which is in general agreement with the maximum depth-to-Moho trends of Whitsett (1976).

Continental Shelf. The model of the upper crustal region under the continental shelf combines seismic refraction, gravity, and geological information. The eastern sedimentary basin of the continental shelf is represented by a 3-layer sequence of increasing density from 1.8 to 2.1 g/cm^3 (Fig. 9, 0-60 km). The underlying block of 2.65 g/cm^3 was extended onshore where Cretaceous pillow lavas, cherts, and pyroclastics are mapped on the Geological Map of the Western Cordillera of Northern Peru (Anonymous, 1973). The upward extension of the large block of 3.0 g/cm^3 lower crustal material to the bottom of the 2.65 g/cm^3 layer was necessary in order to obtain the positive gravity anomaly of 50 to 70 mgal observed in this area. The upper surface of the 3.0 g/cm^3 block lies 3.5 km higher than a similar interface detected by Line 20-21 (Hussong and others, 1976). Also, the Moho location in the model is 2 km above the depth located by Line 20-21 (Fig. 9). The difference may be due to errors in the determination of the deeper layers of this line, because the refracted arrivals did not reverse well (Hussong and others, 1976), or because of dip parallel to the margin.

The western edge of the gravity anomaly at 75 km (Fig. 9) represents the western limit of the outer continental high modeled from the seismic refraction data (Fig. 3). The outer continental shelf high is modeled with an upper block of 2.72 g/cm^3 and a lower block of 2.8 g/cm^3 that extends to the subducted plate boundary. Located between a 2.75 g/cm^3 and the outer continental shelf high is a block with a density of 2.67 g/cm^3. Both seismic refraction and exploratory well data indicate the upper surface of this block to be faulted and probably downdropped relative to the surrounding blocks. The surface velocity of 5.0 km/sec for this block]Fag. 3) indicates a density range of 2.45 to 2.75 g/cm^3 (Ludwig and others, 1970). After extensive modeling, a density gf 2.67 g/cm^3 was assigned to the total block to match the observed U-shaped anomaly in Figure 9 at 90 km.

Continental Slope. Modeling of the observed gravity anomaly of the continental slope required few changes in the seismic refraction model (Figs. 2, 3). Modifications were made near 105 km of the seismic model in order to extend the upper boundary of the 2.75 g/cm^3 block closer to the surface. This helped to produce the 75 mgal peak observed at this location. At the slope base, the partially observed layer boundary locations from the seismic refraction modeling (Fig. 3) were used to model the layers of the 1.8, 2.2, 2.4, and 2.6 g/cm^3 densities.

Trench and Oceanic Plate. Near the trench, the descending Nazca plate was seismically modeled as a 2-layer structure consisting of a 2.85 g/cm^3 (5.55 km/sec) layer overlying a thinner 3.0 g/cm^3 (7.5 km/sec) layer (Fig. 9). Eastward of 230 km, it was necessary to shift slightly some of the plate interfaces. This is not significant, since no seismic refraction control exists landward of the slope base. The gravity model suggests that the slope of the descending plate is about 5° down to a depth of 30 km.

West of 230 km, the density layering sequence of 1.8, 2.6, 2.65, 2.85 and 3.0 g/cm^3 represents the oceanic plate. A density of 3.35 g/cm^3 was used for the upper mantle. It was necessary to change the upper mantle from 3.35 to 3.32 g/cm^3 to produce a zero mgal anomaly at western locations far removed from the trench. Without the density transition in the upper mantle, it would have been necessary to invoke large lateral density changes in the lower crust.

GEOLOGICAL INTERPRETATION OF MODELS

Continental Shelf

Precambrian or early Paleozoic rocks are believed to form the core of the outer continental shelf high and partially form the basement of the continental shelf basins near refraction Line 18-19 (Masias, 1976; Kulm and others, 1977; Kulm and others, 1981a, this volume). Early Paleozoic rocks (schist and phyllites) are exposed in the Amotape Mountains of northwestern Peru and in the coastal ranges of southern Peru (Cobbing and Pitcher, 1972; Masias, 1976). Furthermore, Paleozoic rocks crop out on the offshore islands of Lobos de Tierra (lat 6.5°S, long 81.1°W) and Lobos de Afuera (lat 6.9°S, long 80.8°W), Hormigas Island (lat 11.9°S, long 77.8°W) (Masias, 1976; Kulm and others, 1981b, this volume). Unpublished reports from the Bellena 8-1 and Delfin 20X-1 wells (Fig. 1) indicate that basement rocks are composed of quartz biotite gneiss and dark gray phyllite, respectively.

Geological and geophysical evidence suggest that the outer continental shelf high was and perhaps still is a tectonically active structure. Fractured and highly slickensided basement rock, which indicates fault movement between the 5.0 and 5.65 km/sec basement structures (Fig. 3) are reported in the Delfin well logs. Miocene sediments conformably overlie Oligocene sediments that nonconformably overlie the metamorphic basement rocks. This suggests that a major episode of faulting and erosion occurred during or before the Oligocene Epoch. The seismic refraction velocities and layer interfaces for the sediments of this area are based on well velocity correlations and on poorly observed arrival times. Relative subsidence of the central portion of the outer shelf sedimentary basin between 70 and 110 km (Fig. 3) is speculative.

The 2.67 g/cm^3 basement rock, located under the outer sedimentary basin of the continental shelf and required by the gravity model, is considerably less dense than the basement rocks located to the east and west (Figs. 9, 10). The 2.72 to 2.80 g/cm^3 eastern basement is of continental origin and might represent the leading edge of the continental block prior to plate collision (Dietz and Holden, 1974). The source material for the 2.67 g/cm^3 block may have originated from a continental rise prism located at the base of the continental block (Dietz and Holden, 1974), which may subsequently have been trapped and pushed up against the continental block by the subducting plate. If this theory is correct, the event may have occurred as long ago as 15 m.y. (Middle Jurassic), when subduction along the Peru-Chile Trench initiated the major onset of volcanism and orogenic activity on the west coast of South America (Cobbing and Pitcher, 1972).

The sedimentary basin located landward of the outer continental shelf high (Figs. 2, 9, 10) is part of the Salaverry Basin (Masias, 1976). The seismic reflection profiles presented by Masias (1976) and Kulm and others (1981a, this volume) suggest uplift of the seaward edge of the basin based on the landward (eastward) migration of the axis of deposition. The seismic refraction results of Line 18-19 for the sedimentary layers in the Salaverry Basin only weakly support the idea of a landward migration of the axis of deposition, but they do suggest uplift of the seaward edge (for example, Fig. 4). A hiatus of greater than 200 m.y. probably occurs between the metamorphic basement and overlying sediments. Apparently, the outer continental shelf high was nearer to the sea surface in the past and was thus subjected to erosion. Later subsidence allowed the Cenozoic sediments to accumulate, with uplift occurring later in the Cenozoic. The near-surface structures and a large volume of the dolomicrites dredged from the outer shelf are folded and truncated (Jones, 1979; Kulm and others

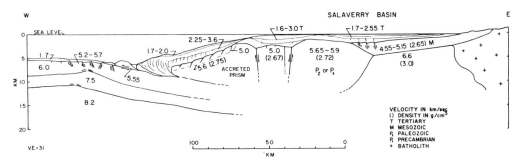

Figure 10. Geophysical and geological model of the continental margin and oceanic plate at lat 9° S.

1981b, this volume). Contour currents moving parallel to the slope edge also may have inhibited deposition in this area.

Figure 9 shows the basement of the eastern portion of the continental shelf to be 2.65 g/cm^3 crustal material. Travis and others (1976) show that sedimentary and volcanic rocks of Mesozoic age overlie Paleozoic strata in northern Peru and suggest that the Mesozoic rocks extend onto the continental shelf. Also, marine deposits of the Late Cretaceous and Tertiary Periods are confined to a narrow coastal belt onshore and are presumed to lie in basins landward of the outer continental shelf high located offshore. Hence, the 2.65 g/cm^3 (4.55 to 5.15 km/sec) basement material probably represents mesozoic rocks, while the overlying 1.0 to 2.1 g/cm^3 (1.7 to 2.55 km/sec) materials represent sediments of Tertiary age (Fig. 10). The lateral velocity change from 5.15 to 4.55 km/sec in the Mesozoic basement may be due to faulting or juxtaposition of different material.

Continental Slope

From velocity analyses of CDP lines obtained at lat 9°S and lat 12°S (Kulm and others 1981a, this volume; Hussong and others, 1976), it appears that the major tectonic disruption of the continental slope basement interface occurs near the base of the slope. This observation agrees with the velocity-depth model for Line 18-19 where no distinct arrivals were noted for a model with a layered slope base (Figs. 6, 7).

A well-defined basement underlies the continental slope landward from the trench. The refraction data indicate that a continental slope basement of velocity 5.0 km/sec overlies a slope core material with an interface velocity of 5.6 km/sec. Line CDP-1 across the continental slope at lat 12°S confirms that a continuous slope basement interface of velocity 5 to 6 km/sec parallels the slope bottom (Hussong and others, 1976). Line CDP-2, located 25 km to the north of Line 18-19 (Fig. 1), also confirms this observation, where as the velocity analysis of Kulm and others (1981a, this volume) suggest basement velocities of 5.0 to 5.3 km/sec. The true velocities may be lower, because the data were obtained while shooting upslope, and no corrections were made for dipping interfaces.

The seismic depth section presented by Kulm and others (1981a, this volume) indicates a poorly defined structure deep within the slope. This material is believed to be formed in part from accreted deposits derived from the offscraped sediments and upper crustal rocks of the descending Nazca plate (Coulbourn and Moberly, 1977; Kulm and others, 1977, 1981a, this volume). If this is true, the model of the continental slope basement based on seismic refraction data of Line 18-19 (Fig. 10) represents only the upper surface of the mélange of accreted deposits; the velocity within the mélange is not well known. The gravity requires a block of density 2.76 g/cm³ materials representing the crust of the Nazca plate and accounting for sediment dewatering and compaction. One must therefore conclude that a large portion of the upper crustal rock of the descending plate (2.6 to 2.85 g/cm³) must be included with the sediments in the mélange.

Tectonically connected with the accretionary process is the formation of sedimentary basins on the continental shelf and slope (Seely and others, 1974; Moore and Karig, 1976; Coulbourn and Moberly,

1977). In addition to the well-developed sedimentary basin on the continental shelf, Line CDP-2 (Fig. 1) clearly shows a 2 km deep sedimentary basin on the upper slope between 120 and 144 km (Kulm and others, 1981a, this volume). The gravity data for Line 18-19 does show a slight downward curvature in the free-air anomaly measured along the upper slope (Fig. 9), indicating a mass deficiency possibly associated with the basin. Gravity modeling of this anomaly indicates that small flexures in the slope basement and the descending plate can account for the observed anomaly. The refraction model (Fig. 6) depicts a continuous, rather than isolated, sedimentary basin for the continental slope. Due to the longer distance between sonobuoys and the fewer number of shot points, the model of the slope tends to integrate the overlying sedimentary velocities and structures, and individual details are lost.

Several speculative interpretations can be made about the slope layers. The uppermost layer of 1.7 to 2.0 km/sec material (Fig. 6) reveals very little structural layering in the single-channel seismic reflection profile in Figure 5. This is confirmed by calcareous breccias dredged from the outermost shelf and upper slope in this region (Kulm and others, 1981b, this volume). Other dredged sedimentary deposits include dolomicrites and glauconitic micrites. The interface of 2.25 km/sec is real, but the velocity is somewhat artificially derived (see model description for Fig. 6). Together, the 2.25 and 3.6 km/sec layers might represent either the terrigenous indurated sediments related to the accretionary process (Hussong and others, 1975), or the carbonates reported by Kulm and others (1981b, this volume), or both. The slope basement is characterized by a highly diffracting interface suggestive of block faulting (Kulm and others, 1981a, this volume). The 5.0 km/sec interface on the slope in Figure 6 may be associated with the disrupted basement, while the 5.6 km/sec interface marks a deeper and thus more uniform region within the accretionary prism (Hussong and others, 1976). Measured velocities in the dolomicrites ranged between 4.5–6.5 km/sec, which suggests that these dense, high-velocity rocks could represent either one or both of these layers.

The location of the Nazca plate under the continental slope was not detected by the refraction data along Line 18-19 (Fig. 6). A possible explanation is that a velocity inversion occurs between the overlying slope material and the upper surface of the plate such that critical refraction cannot occur. A further search for deep reflections (R6 and R7 in Fig. 6) from the plate interface also failed to locate the plate. The CDP depth section (Kulm and others, 1981a, this volume) clearly shows a deep reflector associated with the upper surface (layer 2) of the descending plate. The numerous diffractions in this nonmigrated depth section indicate that the upper surface of the Nazca plate is highly faulted under the slope. Attenuation of seismic energy by crustal materials within the slope cannot account for the absence of distinct reflected energy observed in the seismograms for Line 18-19, because the 2,600 cu in. air-gun source for the CDP-2 data is equivalent to 2.5 pounds of 60% dynamite (Kramer and others, 1968), whereas 3- to 200-pound charges were used in the refraction work. The effect of a highly faulted surface on widely spaced explosive sources would be to scatter the reflecting energy such that reflections received at a point receiver (sonobuoy) are nondistinct. The CDP method uses closely spaced shots (shot spacing was 30 m for the record in Fig. 11) and a 24-channel streamer 1.6 km long. The effectiveness of phase correlation in the CDP method for receiving scattered reflected energy can be easily seen.

Trench Area and Nazca Plate

Prince and Kulm (1975) and Schweller and others (this volume) described intensive deformation in the trench region. Due to the structurally complex nature of the trench area, a simplified model evolved to generate the seismic refraction travel time curves for Figure 7.

Tensional stress along the line of flexure of the descending Nazca plate has been cited as the cause of the normal faulting observed near the trench (Prince, 1974; Prince and Kulm, 1975; Schweller and Kulm, 1978). These authors present seismic reflection records depicting sediment-buried, block-faulted areas seaward of the Peru-Chile Trench which correlate with the 5.2 km/sec layer west at 205 km in Figure 7. The modeling of block faulting was not justified, because the 2 km or greater shot point spacing cannot resolve the randomly sized blocks of 1 to 10 km in length and vertical offsets of 0.2 km or less. The small scatter in the observed arrival times indicates the presence of the broken structure that is not well detected by the seismic refraction data.

A noteworthy feature of the Peru Trench is the prominent ridgelike structure located at 192 km in Figure 7. Schweller and others (this volume) and Prince and Kulm (1975) believe that the ridge represents a portion of thrust-faulted oceanic crust uplifted relative to the floor of the trench. Based on the models presented by Prince and Kulm (1975), the upper interface of the 5.55 km/sec layer under the ridge (Fig. 7) could be interpreted as the location of a thrust fault rather than the top of the Nazca plate. However, due to the complexity of the data, the 5.55 km/sec interface may be considered a simplification of the upper layers of the model west of 203 km. The velocity of the ridge at 192 km is based on a similar feature observed in CDP-2 to the north, where velocity analyses indicated a 2.4 to 3.6 km/sec ridge overlying a 5.2 km/sec interface. Since material recovered from the ridge is basalt (Scheidegger and others, 1978), a velocity of approximately 5 km/sec might be expected for the ridge. The basalt of the ridge may be highly fractured and therefore have a lower interval velocity.

The layer interfaces that represent the descending plate are modeled as plane layers with very little change of dip and no faulting (Figs. 7,8), which, in reality, would be an oversimplification considering the structure of the upper surface (Prince and Kulm, 1975). In modeling the trench area, it was assumed that the major cause of travel time variations are due to the topography and upper-layer structures and not to major structural changes in the deeper layers.

A shallowing of the lower crustal layers is noted near 220 km (Fig. 7), and a similar shallowing of less amplitude is modeled for the gravity model (Fig. 9). The shallowing may represent upward flexure, which is possibly related to a combination of plate bending and compressional forces due to crustal underthrusting. The surficial expression of normal block faulting (Prince and Kulm, 1975; Schweller and Kulm, 1978) is probably directly related to the plate bending observed at depth. Thrust faulting within the plate, as observed by Hussong and others (1975) at lat 12°S is not observed in the crustal section along Line 18-19.

Large-scale crustal thinning seaward of trenches is sometimes noted by a modest gravity high located near the thinned crust. Upward flexure of the oceanic plate as it bends to descend into the trench has been used to explain crustal thinning (Hanks, 1971; Watts and Talwani, 1975; Grow, 1973). The observed gravity of Line 18-19 does not show a well-defined gravity high seaward of the trench (Fig. 9). In this region, from west to east, the crustal structure develops a thickening of the 2.85 g/cm^3 layer and a thinning of the 3.0 g/cm^3 layer, while the upper mantle density changes from 3.32 to 3.35 g/cm^3. The net lateral changes in layer thickness and mantle density tend to compensate each other in the model, with the result that no gravity high is seen either in the observed or modeled gravity. A lateral density change in the upper mantle near subduction zones also has been suggested by other researchers (Hales, 1969; Hussong and others, 1973, 1975). A slightly different version of the Pisco (lat 14°S) crustal and subcrustal cross section (first modeled by Whitsett, 1976) also required a lateral density change in the upper mantle when modeled with the mass column used in this study (R. Couch, pers. commun., 1978).

The irregular layering of the Nazca plate may be due to the Mendena Fracture Zone, which intersects Line 18-19 between 260 km and 310 km in Figure 19. The changes in layer thickness in the crustal section may be related to an age difference of more than 15 m.y. (Herron, 1972) between the younger western and older eastern sections at this intersection.

SUMMARY AND CONCLUSIONS

The basement of the continental shelf is structurally complex and can be divided into eastern and western portions (Figs. 2,3,10). The western portion consists of a faulted outer continental shelf high of Paleozoic or older rocks (Fig. 10). A deeper block to the west has a velocity of 5.0 km/sec and consists of fractured and slickensided phyllite in its upper surface. Basement velocities comparable to this were seen by Fisher and Raitt (1962) on the outer continental shelf 250 km to the south. A shallower but denser block abuts this block to the east and has a velocity of 5.65 to 5.90 km/sec. Quartz biotite gneiss has been obtained from the upper surface of this block, which is believed to be the older of the two blocks (Fig. 10). The combined structure forms a basin 2.5 to 3.0 km thick that contains Tertiary sediments with a velocity of 1.6 to 3.0 km/sec. A hiatus of at least 200 m.y. between basement and overlying sediments suggests that the area was subject to erosion by bottom currents.

Materials 3 km thick with a velocity of 4.55 to 5.15 km/sec shallow to the east beneath sediments covering the eastern portion of the continental shelf. Similar velocities and thicknesses were seen by Hussong and others (1976) 7 km north of Line 18-19. The eastern basement may consist of pillow lavas, cherts, and pyroclastics of Mesozoic age that are confined to the narrow coastal belt onshore. Together, this basement and the eastern section of the outer continental shelf high form the synclinal Salaverry Basin, which contains Tertiary sediments in its upper portion with a maximum thickness of 1.8 km and velocities that range from 1.7 to 2.55 km/sec. Underlying the Mesozoic basement is rock of unknown age that has a velocity of 6.6 km/sec and a density, based on gravity modeling, of 3.0 g/cm^3. The high density of the rock and its location on the eastern continental shelf suggest either crustal rupture and imbricate upthrust of oceanic crust or intrusion at depth under the continental shelf. A similar model was obtained by Whitsett (1976) at Pisco, located 350 km to the south. The similarity suggests that the margins of these areas may have undergone similar deformation at depth.

A well-defined basement underlies the continental slope shoreward from the trench. The refraction data indicate a continental slope basement of velocity 5.0 km/sec overlying a slope core material with an interface velocity of 5.6 km/sec (Fig. 10). The deeper material probably represents the upper surface of a mélange of accreted deposits; however, the velocity within the mélange is not well known. Other researchers (Fisher and Raitt, 1962; Hussong and others, 1976) report similar velocities for the upper basement of the continental slopes off the coast of Peru and Chile. Together, the gravity model, which requires a density of 2.75 g/cm^3 to represent the mélange, and the seismic velocities imply that the slope mélange consists of a larger proportion of oeceanic basalt and metabasalt than oceanic sediments. This could result if the slope mélange formed during the onset of subduction before large volumes of sediments would have been scraped off the descending plate. Once formed, the mélange acts as a trap and forces the subduction of the majority of sediments that enter the trench.

Lack of close data points on the slope resulted in weakly determined sediment velocities and loss of structural details. The sedimentary layers overlying the slope basement consist of an uppermost layer of slumped sediments (1.7 to 2 km/sec) that reveal little structural layering of reflectors. These sediments overlie an acoustic basement of 2.25 to 3.6 km/sec (Fig. 10). This basement probably represents a small volume of consolidated and indurated oceanic sediments that managed to accrete above the slope mélange wedge in the past. Seismic ray trace models show that the slope base is devoid of well-defined layers. This is consistent with the proposed models of accretion by Prince and Kulm (1975) and Kulm and others (1977; 1981a, this volume).

A notable example of the application of seismic ray trace methods occurs in the interpretation of significantly altered arrival times due to the presence of a ridge in the trench. The model of the ridge agrees with the suggestion of Prince and Kulm (1975) that the ridge represents a portion of trust-faulted oceanic crust that has been uplifted relative to the trench floor. Beneath the trench, the descending lithospheric plate is modeled by a 4 km thick upper layer of velocity 5.55 km/sec, which overlies a thinner (2.5 km) but considerably higher-velocity 7.5 km/sec layer. The underlying Moho shows a velocity of 8.2 km/sec and dips at an angle of 5° under the continental margin.

A seismic model with relatively constant velocities satisfies observed variations in the layer interfaces for the Nazca plate seaward from the trench. Upper crustal layers of the modeled plate consist of a thin 1.7 km/sec sedimentary layer overlying a 5.0 to 5.2 km/sec upper layer and a 5.6 to 5.7 km/sec lower layer that shoal to the east within 60 km of the trench, while a deeper, 6.0 to 6.3 km/sec layer thickens to the east. The lower crustal model consists of a 7.4 to 7.5 km/sec layer that varies in thickness from 2.5 to 4.0 km. This high-velocity layer is a predominant feature of the Nazca plate in this region (Hussong and others, 1976). The depth to an 8.2 km/sec Moho interface varies not more than ±0.5 km from 10.5 km for the 120 km long model of the Nazca plate.

Refraction data indicate crustal thickening beneath the trench that is also noted along the Peru-Chile Margin by others (Fisher and Raitt, 1962; Ocola and Meyer, 1973; Hussong and others, 1976).

West of the trench, the higher-velocity crustal layers shallow, and this may represent upward flexure of the oceanic plate. In addition, development of normal faults can be observed in the upper crustal layer just seaward of the trench (Prince, 1974; Prince and Kulm, 1975; Schweller and Kulm, 1978). The combination of crustal thickening, upward flexure, normal faulting, and the ridge in the trench strongly suggest that compressional stresses are present where the plate enters the subduction zone (Fig. 10).

ACKNOWLEDGMENTS

This research was supervised by Dr. Stephen H. Johnson and I am grateful for his time and suggestions in supporting this study. In addition, I would like to thank Dr. Richard Couch for his expert knowledge on the development of gravity models.

This study was conducted for the Nazca Plate Project and was supported through the National Science Foundation, International Decade of Ocean Exploration (Grant OCE76-05903-A02).

REFERENCES CITED

Barday, R., 1974, Structure of the Panama Basin from marine gravity data [M.S. thesis]: Corvallis, Oregon State University, 99 p.

Case, J. E., and others, 1971, Tectonic investigations in western Colombia and eastern Panama: Geological Society of America Bulletin, .v. 82, p. 2685–2712.

Cobbing, E. J., and Pitcher, W. S., 1972, Plate tectonics and the Peruvian Andes: Nature, v. 240, p. 51–53.

Coulbourn, W. T., and Moberly, R., 1977, Structural evidence of the evolution of fore-arc basins off South America: Canadian Journal of Earth Sciences, v. 14, p. 102–116.

Dietz, R. S., and Holden, J. C., 1974, Collapsing continental rises: Actualistic concept of geosynclines—a review, in Dott, R. H., Jr., and Shaver, R. H., eds., Modern and ancient geosynclinal sedimentation: Tulsa, Society of Economic Paleontologists and Mineralogists Special Publication No. 19, p. 14–25.

Erlandson, D., Hussong, D., and Campbell, J., 1981, Sediment and associated structure on the northern Nazca plate, in Kulm, L. D., and others, eds., Nazca plate: Crustal formation and Andean convergence: Geological Society of America Memoir 154 (this volume).

Fisher, R. L., and Raitt, R. W., 1962, Topography and structure of the Peru-Chile Trench: Deep Sea Research, v. 9, p. 423–443.

Gebrande, H., 1976, A seismic-ray tracing method for two-dimensional inhomogeneous media, in Giese, P., and others, eds., Explosion seismology in Central Europe: New York, Springer-Verlag, 429 p.

Gemperle, M., 1970, Two-dimensional gravity cross section computer program TALWANI: Geophysical Data Reduction Technical Report, School of Oceanography, Oregon State Universtiy, Corvallis, 10 p.

——1975, Two-dimensional gravity cross section computer program—source blocks and field points generalized, GRAV2DLD: Supplement to Geophysical Data Reduction Technical Report, School of Oceanography, Oregon State University, Corvallis, 4 p.

Goebel, V., 1974, Modeling of the Peru-Chile Trench from wide-angle reflection profiles [M.S. thesis]: Corvallis, Oregon State University, 73 p.

Grow, J. A., 1973, Crustal and upper mantle structure of the central Aleutian arc: Geological Society of America Bulletin, v. 84, p. 2169–2192.

Hales, A. L., 1969, Gravitational sliding and continental drift: Earth and Planetary Science Letters, v. 6, p. 31–34.

Hamilton, E. L., 1974, Geoacoustic models of the sea floor, in Hampton, L., ed., Physics of sound in marine sediments: New York, Plenum Press, 567 p.

Hamilton, E. L., and others, 1974, Sediment velocities from sonobuoys: Bay of Bengal, Bering Sea, Japan Sea and North Pacific: Journal of Geophysical Research, v. 79, p. 2653–2668.

Hanks, T. C., 1971, The Kuril-trench–Hokkaido rise

system: Large shallow earthquakes and simple models of deformation: Geophysical Journal of the Royal Astronomy Society, v. 23, p. 173-185.
Herron, E. M., 1972, Sea-floor spreading and the Cenozoic history of the East Central Pacific: Geological Society of America Bulletin, v. 83, p. 1671-1692.
Hussong, D. M., and others, 1973, Crustal and upper mantle structure of the Nazca plate [abs.]: International Association of Seismology and Physics of the Earth's Interior, XVII, Lima Program, 221 p.
Hussong, D. M., and others, 1976, Crustal structure of the Peru-Chile Trench: 8°S-12°S latitude, in Sutton, G. H., and others, eds., The geophysics of the Pacific Ocean Basin and its margin: Washington, D.C., Geophysics Monograph, American Geophysical Union, v. 19, p. 71-86.
Hussong, D. M., Wipperman, L. K., and Odegard, M. E., 1975, Compressional faulting of the oceanic crust prior to subduction in the Peru-Chile Trench: Geology, v. 3, p. 601-604.
Jacob, K. H., 1970, Three-dimensional seismic ray tracing in a laterally heterogeneous spherical earth: Journal of Geophysical Research, v. 75, p. 6675-6689.
James, D. E., 1971, Plate tectonic model for the evolution of the Central Andes: Geological Society of America Bulletin, v. 82, p. 3325-3346.
Jones, P., 1979, Seismic ray trace techniques applied to the determination of crustal structures across the Peru continental margin and Nazca plate at 9°S latitude [Ph.D. thesis]: Corvallis, Oregon State University, 156 p.
Kramer, F. S., Peters, R. A., and Walter, W. C., 1968, Seismic energy sources 1968 handbook: Pasadena, California, United Geophysical Corporation, 57 p.
Kulm, L., and others, 1981a, Crustal structure and tectonics of the central Peru continenetal margin and trench, in Kulm, L. D., and others, eds., Nazca plate: Crustal formation and Andean convergence: Geological Society of America Memoir 154 (this volume).
Kulm, L., and others, 1981b, Late Cenozoic carbonates on the Peru continental margin: Lithostratigraphy, biostratigraphy, and tectonic history, in Kulm, L. D., and others, eds., Nazca plate: Crustal formation and Andean convergence: Geological Society of America Memoir 154 (this volume).
Kulm, L. D., Schweller, W. J., and Masias, A., 1977, A preliminary analysis of the subduction processes along the Andean continental margin, 6° to 45°S, in Talwani, M., and Pitman, W. C., III, eds., Island arcs, deep sea trenches and back-arc basins, Maurice Ewing Series 1: Washington, D.C., American Geophysical Union, p. 285-301.
Ludwig, W. J., Nafe, J. E., and Drake, C. L., 1970, Seismic refractions, in Maxwell, A., ed., The sea, Vol. 4, Part I: New York, Wiley, p. 53-84.
Masias, A., 1976, Morphology, shallow structure and evolution of the Peruvian continental margin, 6° to 18°S [M.S. thesis]: Corvallis, Oregon State University, 77 p.
Minster, J. B., and others, 1974, Numerical modeling of instantaneous plate tectonics: Geophysical Journal of the Royal Astronomy Society, v. 36, p. 541-576.
Moore, J. C., and Karig, D. E., Sedimentology, structural geology and tectonics of the Shikoku subduction zone, southwestern Japan: Geological Society of America Bulletin, v. 87, p. 1259-1268.
Ocola, L. C., and Meyer, R. P., 1973, Crustal structure from the Pacific Basin to the Brazilian shield between 12°and 30° south latitude: Geological Society of America Bulletin, v. 84, p. 3387-3404.
Prince, R. A., 1974, Deformation in the Peru Trench, 6° to 10°S [M.S. thesis]: Corvallis, Oregon State University, 91 p.
Prince, R. A., and Kulm, L. D., 1975, Crustal rupture and the initiation of imbricate thrusting in the Peru-Chile Trench: Geological Society of America Bulletin, v. 86, p. 1639-1653.
Scheidegger, K. F., and others, 1978, Fractionation and mantle heterogeneity in basalts from the Peru-Chile Trench: Earth and Planetary Science Letters, v. 37, p. 409-420.
Schweller, W., and Kulm, L. D., 1978, Extensional rupture of oceanic crust in the Chile Trench: Marine Geology, v. 28, p. 271-291.
Schweller, W., Kulm, L., and Prince, R., 1980, Tectonics, structure, and sedimentary framework of the Peru-Chile Trench, in Kulm, L. D., and others, eds., Nazca plate: Crustal formation and Andean convergence: Geological Society of America Memoir 154 (this volume).
Seely, D. R., Vail, P. R., and Walton, G. G., 1974, Trench slope model, in Drake, C. L., and Burk, C. A., eds., The geology of continental margins: New York, Springer-Verlag, p. 249-260.
Shah, P. M., 1973, Ray tracing in three dimensions: Geophysics, v. 38, p. 600-604.
Sorrells, G. G., Crowley, J. B., and Veith, K. F., 1971, Methods for computing ray paths in complex geological structures: Seismological Society of America Bulletin, v. 61, p. 27-53.
Talwani, M., Worzel, J. L., and Landisman, M., 1959, Rapid gravity computations for two-dimensional bodies with application to the Mendocino submarine fracture zone: Journal of Geophysical Research, v. 64, p. 49-59.
Travis, B. R., Gonzales, G., and Pardo, A., 1976, Hydrocarbon potential of coastal basins of Peru, in Halbouty, M. T., and others, eds., Circum-Pacific energy and mineral resources, Memoir 25:

Tulsa, Oklahoma, American Association of Petroleum Geologists, 608 p.

Watts, A. B., and Talwani, M., 1975, Gravity effect of down-going lithosphere slabs beneath island arcs: Geological Society of America Bulletin, v. 86, p. 1-4.

Whitsett, R. M., 1976, Gravity measurements and their structural implications for the continental margin of southern Peru [Ph.D. thesis]: Corvallis, Oregon State University, 82 p.

Yacoub, N. K., Scott, J. H., and McKeown, F. A., 1968, Computer technique for tracing seismic rays in two-dimensional geological models: U.S. Geological Survey Open-File Report, 65 p.

Yeats, R. S., and others, 1976, Initial reports of the deep sea drilling project, Vol. 34: Washington, D.C., National Science Foundation, 814 p.

MANUSCRIPT RECEIVED BY THE SOCIETY NOVEMBER 12, 1980
MANUSCRIPT ACCEPTED DECEMBER 30, 1980

Printed in U.S.A.

Crustal structure and tectonics of the central Peru continental margin and trench

L. D. KULM
R. A. PRINCE[1]
W. FRENCH[2]
S. JOHNSON[3]
A. MASIAS[4]
School of Oceanography
Oregon State University
Corvallis, Oregon 97331

ABSTRACT

The crustal structure and tectonic framework of the central Peru Margin, between lat 7° and 10°S were interpreted using mainly a 102 km long, multichannel seismic section and existing geologic and geophysical data in the region. Thrust faulting occurs in upper layer 2 basalts as the Nazca plate descends beneath the continental slope. This produces basaltic ridges (slabs) within the trench axis and at least 26 km landward beneath the overriding continental plate. Broken low-frequency reflectors within this diffracting subduction complex suggest that ophiolitic slivers of basalt are being incorporated into portions of it, forming a sediment-basalt melange.

Three prominent forearc basins, Salaverry, Trujillo, and Yaquina Basins, occupy the central margin from east to west, respectively. Drill holes penetrated Tertiary sediments on the outer shelf and the nearby eastern flank of the Trujillo Basin and bottomed in a metamorphic arc massif. The massif is correlated with seismic refraction velocities greater than 5.7 km/sec and a density of 2.72 to 2.80 g/cm^3 which underlie the continental shelf. Our interpretation of the seaward limit of the massif is uncertain and depends upon the geophysical cirteria used. Each of three forearc models position the arc massif at about 26, 61, or 115 km landward of the trench, with the subduction complex occupying the region between the trench and the massif. The massif-subduction complex interface should be located through drilling to test the proposed models.

The intramassif basins, Salaverry and Trujillo, have subsided during Tertiary time to allow the accumulation of 2 to 4 km of marine sediment. The Trujullo Basin apparently has not experienced much vertical movement since the late Miocene, based upon microfossil paleodepth indicators found in dolomicrites and glauconitic micrites dredged from the basin. However, abundant brecciated dolomicrites in the same dredges and disturbed strata in reflection records from both the Yaquina and

---

[1]Department of Geology, Box 1846, Brown University, Providence RI 02912
[2]Amoco Production Research Co., Box 50879, New Orleans LA 70150
[3]Amoco Production Company, Research Center, P.O. Box 591, Tulsa OK 74102
[4]Petroleos del Peru, Apartado 3126, Lima, Peru

Trujillo Basins suggest deformation of the basins during the Pliocene-Pleistocene by a compressional regime.

INTRODUCTION

Recent geological-geophysical studies of the Peru-Chile Trench off central Peru (Fig. 1) have revealed regions of extensive crustal rupture in the vicinity of the trench axis (Kulm and others, 1973; Prince and others, 1974; Prince and Kulm, 1975; Masias, 1976; Hussong and others, 1976; Scheidegger and others, 1978; Prince and Schweller, 1978; Schweller and others, this volume). Using single-channel reflection profiles and geologic samples, Prince and Kulm (1975) presented a thrust model for the central Peru Trench which attempted to explain the development of these crustal rupture zones. The acquisition of a 24-channel seismic reflection profile over one of these rupture zones and the adjacent forearc region allows us to test this model more comprehensively. This deep crustal information suggests that the tectonic framework of the trench may influence the basement structure beneath the forearc and consequently the evolution of the subduction complex and associated sedimentary basins of the forearc.

The specific objectives of this study are as follows: (1) to determine the nature of faulting within the oceanic crust near the trench axis and beneath the forearc; (2) to define the subduction complex and arc massif and locate the boundary between these features; and (3) to determine the structure and interrelation of the forearc basins that rest upon the complex and massif.

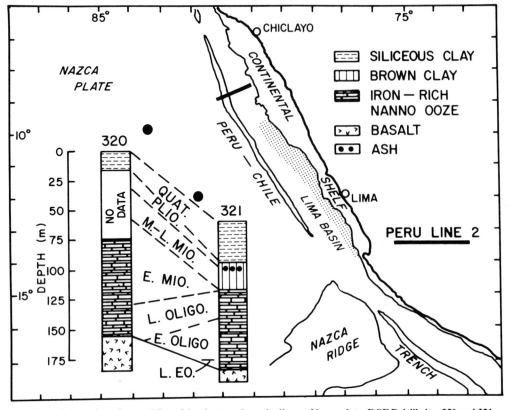

Figure 1. Location of central Peru Margin, trench, and adjacent Nazca plate. DSDP drill sites 320 and 321 shown along eastern edge of plate (data from Yeats, Hart, and others, 1976). Note location of multichannel seismic line 2. Lima Basin shown in dotted pattern.

METHODS

A 101.56 km long, 24-channel seismic reflection line was obtained commercially by Seiscom-Delta for the Nazca Plate Project (Fig. 2). The sound source consisted of two 300 cu. in. and two 1,000 cu. in. air guns fired at a 15-second rate. The length of the active steamer was 5,244 ft (about 1,600 m); band pass filters were set at 8-62 Hz on the binary gain amplifiers. Navigation was done by satellite.

The multichannel data were processed to stacked time section by Seiscom-Delta. These data were further processed by Exxon (Esso Production Research Company) to a nonmigrated depth section using the velocities, with some corrections, obtained by Seiscom-Delta.

Single-channel reflection data were obtained with air guns, and the sediment samples were obtained with free-fall and piston cores as described by Prince and Kulm (1975). Dredge samples from the basalt ridges were obtained on the R/V *Melville F. Drake* cruise during 1975 (see Scheidegger and others, 1978), and dredge samples of sedimentary rocks from the continental slope were obtained on the R/V *Wecoma* cruise during 1977 (see Kulm and others, this volume).

CRUSTAL STRUCTURE AND STRATIGRAPHY

Oceanic Crust

Tholeiitic basalts were recovered from DSDP sites 320 and 321 (Fig. 1) and originated at the fossil Galapagos spreading center (Yeats, Hart, and others, 1976). They are unusually fresh (H_2O content average 0.7%) and are dated at 27 m.y. (site 320) and 38 m.y. (site 321) using the new K^{39}/Sr^{40} technique (Hogan and Dymond, 1976). A classical deep-marine sequence of carbonates was initially deposited on the basaltic crust, with brown clays accumulating later when the flank of the spreading ridge subsided below the calcium carbonate compensation depth (CCD)(Fig. 1). Siliceous hemipelagic clays were then deposited at the drill sites as the Nazca plate converged with the South American continent. Turbidites, which are characteristic of the trench axis, are absent at the drill sites and elsewhere on the plate, which indicates that the Peru-Chile Trench has been a barrier to coarse clastic deposition by bottom-seeking currents on the plate in the late Cenozoic (Yeats, Hart, and others, 1976; Rosato and others, 1975).

Hussong and others (1976), using seismic refraction data, initially defined the oceanic crust of the Nazca plate as a three-layer crust with an average compressional velocity of 5.6 km/sec for layer 2, 6.3 km/sec for layer 3, and 7.3 km/sec for a subcrustal layer lying between layer 2 and the upper mantle (8.2 km/sec). Jones (this volume) reinterpreted these data using a ray tracing technique and subdivided layer 2 into an upper 5.0 km/sec layer and lower 5.6 to 5.7 km/sec layer. Velocities measured in rocks from sites 320 and 321 range from 5.3 to 6.1 km/sec in the upper few tens of metres of layer 2 basalts (Salisbury and Christensen, 1976). Broken zones of pillow basalt fragments and jointed basalt blocks were encountered at both drill sites, eventually terminating the drilling. Apparently, extensive fracturing without later cementation in the upper part of the layer may explain, in part, the somewhat lower velocities obtained for that part of layer 2 in the seismic refraction section. From these data, it appears that a fractured, relatively fresh and unaltered basaltic crust is presently entering the Peru-Chile Trench.

Trench

Prominent ridges were discovered in the Peru-Chile Trench between lat 6°–10°S (Kulm and others, 1973) and 23°–31°S (Schweller and Kulm, 1978). These ridges generally divide the trench into a turbidite-filled inner deep basin and an outer, shallower basin that may contain either turbidites of trench origin or hemipelagic sediments of the oceanic plate (Prince and other, 1974; Schweller and others, this volume). Although the largest ridge in the Peru Trench is 60 km long and is a continuous feature, its morphology changes rather markedly over distances of less than 10 km (Fig. 3). Shoulders

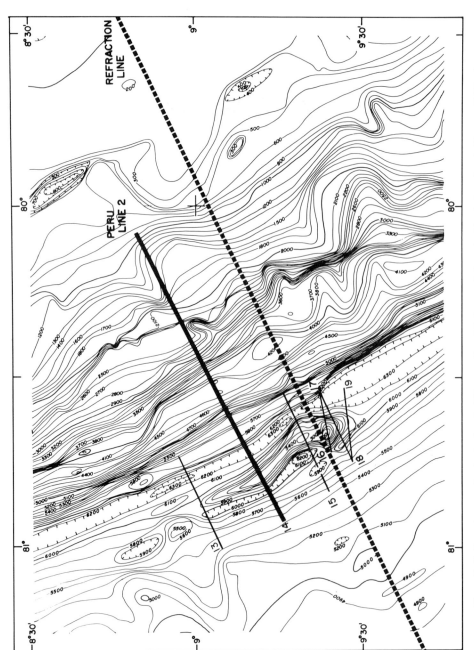

Figure 2. Bathymetry map (metres) of central Peru continental slope and elongate ridge in trench axis. Heavy solid line is position of multichannel reflection line, and dashed line is the seismic refraction line of Jones (this volume). Numbered thin lines are locations of single-channel seismic profiles.

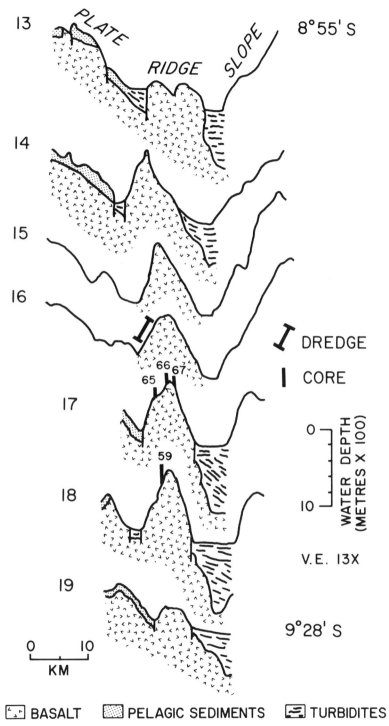

Figure 3. Structure and morphology of basaltic ridge and surrounding trench axis turbidites. Cross sections are schematic drawings of single-channel reflection records (see Fig. 2 for location). Core and dredge samples are shown.

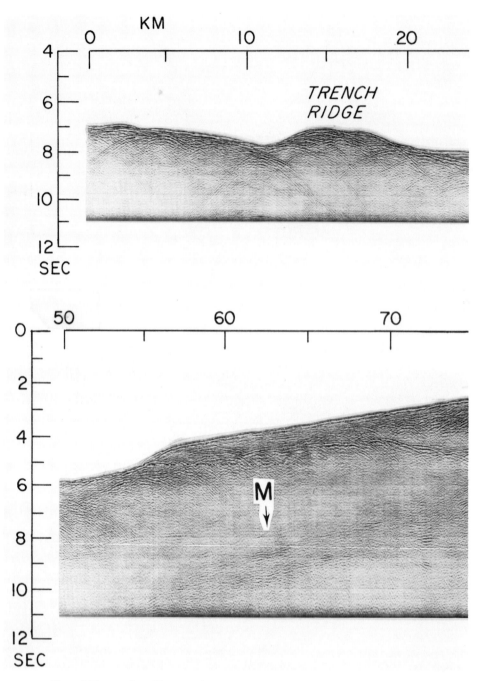

Figure 4. Time section of Peru multichannel seismic reflection line 2 (see Fig. 2 for location). Initial processing of data produced this 1200% stacked time section. Section starts on oceanic plate (km 0 to

and rounded peaks are common, with low-relief hills, either single or double, generally occurring at the termini of the ridge. A subduction-related tectonic origin is indicated by these characteristics and by the fact that turbidites from the trench floor occur on top of the ridge.

The 24-fold seismic reflection profile made across the plate-ridge-trench system exhibits about a 200

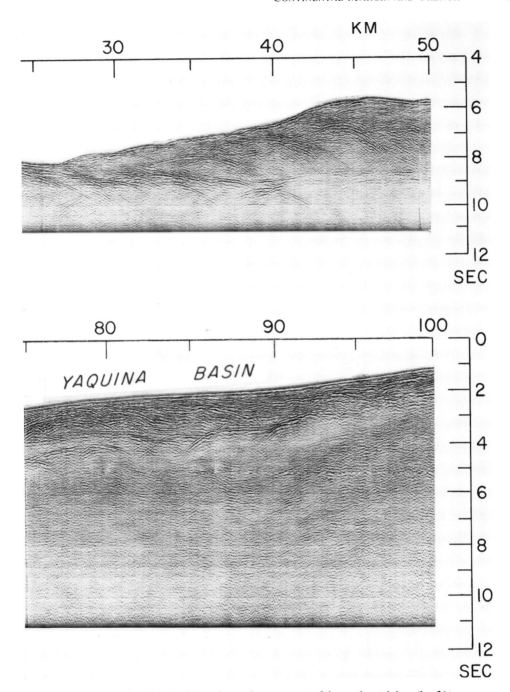

km 10), crosses trench (km 10 to km 26), and extends across most of the continental slope (km 26 to km 100). Letter "M" is water bottom multiple. Vertical exaggeration 2:1.

m thick covering of sediment on the Nazca plate (km 3 to 10). This covering underlies a small, flat-lying, V-shaped basin (km 10 to 12) (Figs. 4,5). A set of diffractions occurs at the intersection of the plate and the steep western scarp of the basaltic ridge. In both the time section (Fig. 4) and the nonmigrated depth section (Fig. 5), it appears that the top of the basaltic layer dives beneath the basin

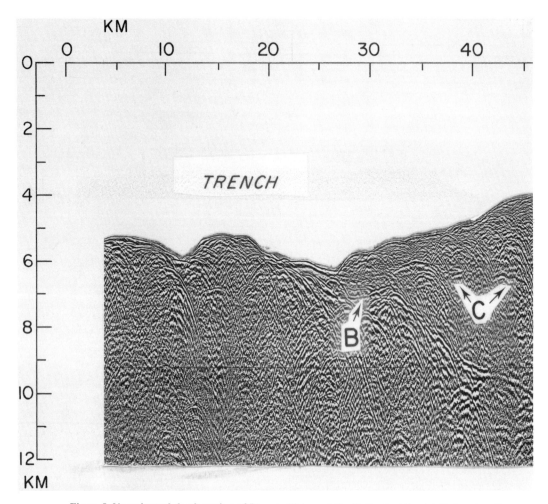

Figure 5. Nonmigrated depth section of Peru multichannel line 2. See text for detailed explanation

and possibly beneath the ridge. An enlarged view of this portion of the time section is shown in Figure 6A.

To partially resolve this structure, several points along the two reflection limbs, dipping in opposite directions under the outer trench basin, were time-migrated (Hagedoorn, 1954). Although both reflection limbs tail into diffractions, a comparison with theoretical diffraction curves verified that the migrated points represent reflection points, not diffractions. The results are shown in Figure 6B. The time picks marked by open triangles and circles migrate to the actual spatial location represented by solid triangles and circles, respectively. Clearly, the top of the basaltic layer extends landward beneath the outer basin and apparently beneath a portion of the ridge. The ridge scarp is also defined and terminates upon intersection with the landward-dipping basaltic layer.

Basalts were dredged from this western scarp (profile 16 of Fig. 3; Fig. 7), indicating exposure of the igneous oceanic crust. The dredge recovered abundant pillow basalts whose megascopic, petrographic, and chemical characteristics are similar to the upper layer 2 tholeiitic basalts drilled at sites 320 and 321 (Fig. 1, Scheidegger and others, 1978). The large amount of chemically unaltered and freshly fractured basaltic material and the general lack of manganese oxide coatings on the rocks indicate relatively recent reexposure to seawater by faulting (Prince and Schweller, 1978).

Because of the low frequencies of the air gun signal, it is difficult to pick a reflector on top of the ridge that represents the boundary between the basalt and the overlying sediment (Figs. 5,7). However,

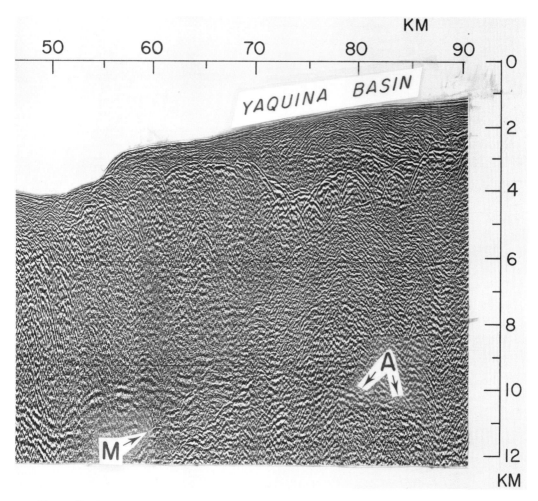

of letters. Vertical exaggeration 3:1.

high-frequency (3.5 kHz) records on the highest elevation of the ridge show a very thin sediment cover with probable rock outcrops. Piston and free-fall cores collected along profiles 17 and 18 (Figs. 3,7) recovered sand turbidites (cores 65-67) and silt turbidites (core 59) with displaced shallow-water benthic foraminifera (Johanna Resig, pers. commun. 1976). Core 67 penetrated the late Quaternary turbidites and recovered fresh tholeiitic basalt (Kulm and others, 1973). These terrigenous turbidites were originally deposited in the trench axis (Rosato, 1974; Rosato and others, 1975). Radiocarbon dating of the youngest turbidite (core 67) indicates that the basaltic crust and overlying sediment have been uplifted 700 to 910 m in the past 3,155 yr (Fig. 7; Prince and Kulm, 1975). Because this ridge system is continuous, the same sequence of events undoubtedly occurred in the vicinity of the multifold line (position of profile 14, Fig. 3).

The multichannel line, as one moves across the ridge, (km 18 to 26 of Figs. 4,5; Fig. 7), shows a thin covering of sediment on the landward (eastern) flank of the basalt ridge. This sediment thickens to form the deeper inner turbidite basin in the trench axis. The sediment layers dip gently landward near the trench axis, but the layering becomes less coherent and generally steeper higher on the ridge flank (Figs. 3, 4). Postdepositional tilting and internal deformation of formerly horizontal trench axis turbidites have resulted in an offset of 350 m between the landward and seaward portions of this inner turbidite basin. These tilted sediments may have originally been continuous with the uplifted turbidites cored on the crest of the ridge. The basaltic basement (layer 2) appears to be faulted into

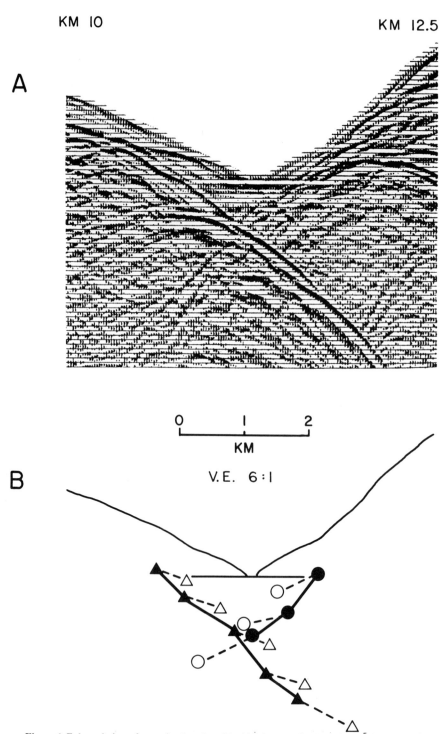

Figure 6. Enlarged view of oceanic plate–basaltic ridge intersection between km 10 and km 12.5 in multichannel record (see Fig. 4). (A) Nonmigrated time section. (B) Time migrated basement reflectors (see text for explanation).

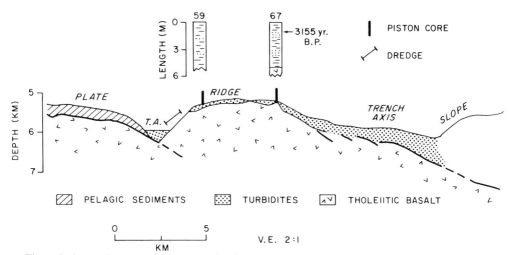

Figure 7. Composite geologic cross section of basalt ridge in trench axis. Geologic data projected onto multichannel depth section from adjacent profiles (Fig. 3). Cross section is the multichannel depth section (see Fig. 2 for location on ridge).

slabs beneath the inner trench sediments with predominantly landward-dipping surfaces (Figs. 5,7). However, there is no apparent continuity of these faults into the overlying sediment.

Forearc

The multichannel record shows a thick diffracting section beneath the continental margin. This section is sandwiched between the strong reflecting acoustic basement of the downbending basalt slab (layer 2) and a strong, shallower reflecting horizon that marks the base of a sedimentary basin (Figs. 4,5). The diffracting mass is believed to be the subduction complex that is overlain by a structurally disturbed forearc basin. The Peruvian forearc basins and their relationship to the various basement structures were studied by Thornburg and Kulm (this volume).

Basaltic Slab. A prominent acoustic basement can be traced continuously 26 km landward from the inner trench basin beneath the inner trench wall (for example, km 26-52), with the exception of a major offset at km 32 (Figs. 4,5). The first segment of the slab has an apparent dip of 10° in a landward direction, and the second segment a dip of 15° before the slab appears to be offset again at km 52. Low-frequency reflectors extend another 18 km landward from this offset beneath the continental margin to the point where they are partially masked by the water-bottom multiple (designated M, Figs. 4,5). Based upon the basalt exposures discovered within the trench axis and the acoustic continuity of the basement reflector, the basaltic crust of the oceanic plate extends landward from at least km 26 to km 52. A number of broken, low-frequency reflectors occur above this portion of the basaltic slab between km 33 and km 36 at a depth of 6.1 to 6.5 km (Fig. 5). These reflectors may be either small broken basaltic slabs or blocks of highly indurated sediments. We favor the former interpretation because the basaltic crust is clearly broken in this region (see later discussion on detached basaltic slabs).

We see the last bit of deep crustal information between km 52 and km 87 (Fig. 5). Intermittent low-frequency reflectors occur between km 52 to km 78. Between km 78 and km 87, a somewhat more continuous, low-frequency, deep reflector (designated A; Fig. 5) occurs at subsurface depth of 10 to 11 km; this feature extends landward for an additional 17 km and has an apparent dip into the continent. A substantial change in the velocity gradient would be required to reverse the dip of this reflector. Although this event is partially obscured by the water-bottom multiple, we believe that it may represent valid information. The nature of these reflectors is speculative and will be treated in the Discussion section.

Forearc Basins. According to Thornburg and Kulm (this volume), the central Peru Margin (lat 7° to 10°S) is separated into five major basins: Sechura, Salaverry, Trujillo, Yaquina, and Lima (Fig. 8). The multichannel profile crosses the southern edge of the Yaquina Basin. A strong and broken reflector (pseudo-acoustic basement) delimits the interface between the basin deposits and the diffracting zone below (km 58 to km 100, Figs. 4,5). Chaotic reflectors within the basin indicate that the basin is disturbed except for that more acoustically coherent portion of the basin located between km 85 to km 100, which is defined as the Yaquina Basin by Thornburg and Kulm (this volume). The disrupted part of the basin, because of its diffracting character in single-channel records (the basement cannot be recognized in these records), is referred to as the Upper-Slope Ridge (USR) by these authors (Fig. 8).

A velocity analysis of the multichannel data in the Yaquina Basin, including the disrupted material, indicates that a 2.0 km thick sedimentary section is present (Fig. 9A, km 58 to km 100). The sediments range from 1.5 to 3.0 km/sec. Seismic refraction data 25 km south of the multichannel line show that sediments with velocities ranging up to 3.6 km/sec are also about 2.0 km thick in the vicinity of the Yaquina Basin (Fig. 9B). Immediately landward in the Trujillo Basin, deposits are about 3.0 km thick and have a similar range in velocity (Fig. 9B). An industry drill hole in the Trujillo Basin (Delfin; Fig. 10) confirms these velocities as well as the depth to metamorphic basement as documented by Jones (this volume). An earlier study of the same refraction data by Hussong and others (1976) produced a thicker section (4.0 versus 3.0 km/sec) in the Trujillo Basin and a higher velocity basement (6.2 versus[1]

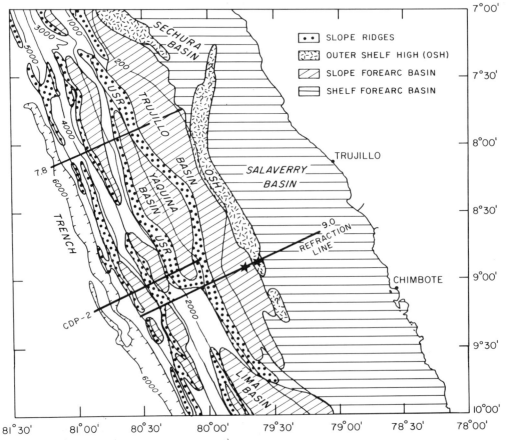

Figure 8. Forearc basins and basement control of basins of the central Peru Margin (modified from Thornburg and Kulm, this volume). Outer shelf high (OSH) consists of metamorphic terrane and slope ridge probably of deformed sediments. Contours in metres. Solid lines are seismic lines, and stars are drill holes.

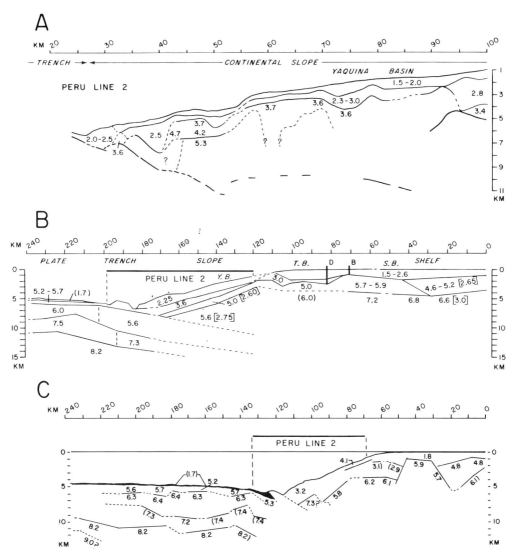

Figure 9. Seismic velocities of the central Peru Margin, trench, and oceanic plate: (A) Velocity analysis of multichannel line 2 (Figs. 4,5). Compressional velocities (km/sec) from Seiscom-Delta computation. Velocities greater than 3.0 km/sec were obtained from analysis of selected velocity spectra and should be considered as approximations only. (B) Seismic refraction and gravity data across the central Peru Margin at lat 9°S (modified from Jones, this volume). Velocities in km/sec and densities [ ] in g/cm³. Basin symbols: YB—Yaquina, TB—Trujillo, SB—Salaverry. Drill holes, D—Delfin and B—Ballena shown as vertical heavy lines. (C) Seismic refraction data, same line as (B) above (modified from Hussong and others, 1976).

5.0) than computed by Jones (Figs. 9B, 9C). These differences in interpretation probably result from the structural ambiguities of these complicated forearc basins (that is, deviations from plane layer theory).

Single-channel reflection records show that the sediments in the adjacent Trujillo Basin also are disturbed by extensive faulting (Figs. 8, 10). Huge blocks of brecciated dolomicrite were dredged from tilted strata (Fig. 10A) and fault blocks (Fig. 10B, long 9°W), as described by Kulm and others (this volume). These intensely sheared rocks exhibit a high degree of grain granulation, typically found in

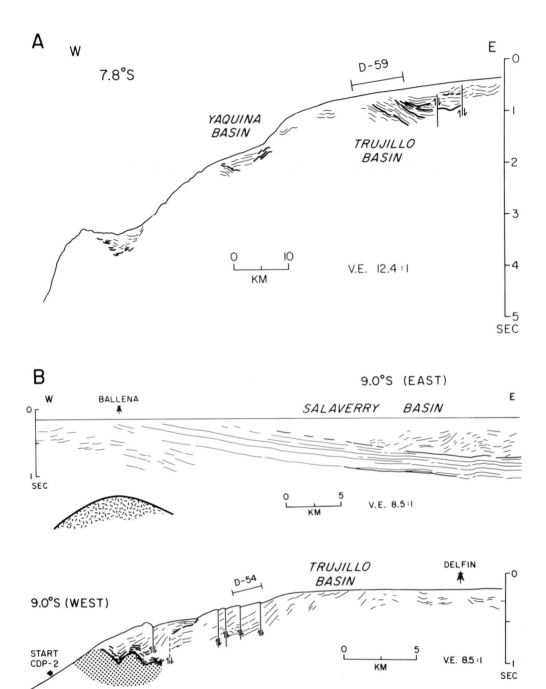

Figure 10. Line drawings of single-channel reflection records of the Yaquina, Trujillo, and Salaverry Basins (A,B) (see Fig. 8 for location). Note locations of dredged D-54 (B) and D-59 (A). Line 9°S is a continuous profile. Angled line pattern beneath drill hole (9°S, east) is Outer Shelf High, and dotted pattern (9°S, west) is deformed slope ridge. Note location of Delfin and Ballena wells (towers).

zones of compression. Measured compressional velocities range from 4.2 to 6.5 km/sec for all carbonates dredged in the basin.

Landward of the metamorphic Outer Shelf High (OSH) (Fig. 8), extensive normal faulting (irregular horst and graben basement structures) characterizes the arc massif over the shelf (Travis and others, 1976). Cenozoic subsidence of the Sechura and Salaverry Basins (Fig. 8) on this block-faulted massif is indicated by deposition of approximately 1.0 to 3.0 km of marine clastic sediments (shale, siltstone, and sandstone), with minor interbedded carbonate, phosphate, and glauconite deposits. Large-scale gravity slumping and sliding and shale flowage occur in the younger strata, as seen in seismic reflection sections (e.g., Fig. 10B, lat 9°S [east]) in the Salaverry Basin. Immediately seaward, the upper-slope Trujillo Basin (Fig. 8) contains more than 3 km of post-Eocene shale, siltstone, marlstone, minor thin limestone beds and, in the basal part, late Eocene sandstone (Travis and others, 1976). This section was initially called the Salaverry Basin section by Travis and others (1976) but now, based upon additional structural data, is called the Trujillo Basin section by Thornburg and Kulm (this volume). The Neogene brecciated dolomicrite, dolomicrite, and glauconitic micrite recovered from the fault blocks in the Trujillo Basin (Figs. 10A, 10B; D-54 and D-59) have paleodepth indicators (benthic foraminifera) typical of the present water depths (Kulm and others, this volume). However, this basin must have subsided in pre-Neogene time to allow 4.0 km of sediment to accumulate; subsequent uplift and truncation of Neogene strata are seen in the reflection records (Thornburg and Kulm, this volume; Figs. 10A, 10B).

Subduction Complex. The subduction complex is defined in geophysical terms as the diffracting zone that is sandwiched between the forearc basin (Yaquina Basin) and the subducting basaltic slab discussed previously (Figs. 4,5). The only coherent reflectors (designated B) are seen in the leading edge, or toe, of the complex (Fig. 5). Sediments appear to be ponded in the structural depression formed by the offset in the basaltic slab. Several broken low-frequency reflectors (designated C) occur at a rather shallow depth within the diffracting mass; they may represent slivers of basalt from subducting slab (see Discussion).

The actual depth of the basaltic slab from km 52 to km 87 is rather subjective, since the depth section was constructed by increasing the velocities downward beneath the Yaquina Basin basement to produce a uniform velocity gradient with depth. As modeled in the depth section, the subduction complex is 6 to 8 km thick beneath the basin.

Geologic evidence for the existence of a subduction complex (accretionary prism) is lacking at lat 9°S in the vicinity of the multichannel line. However, continental accretion occurs further south at lat 15°S on the lower continental slope opposite the Nazca Ridge (Fig. 1; Kulm and other, 1974; Rosato and Kulm, this volume). Here, pelagic sediments from the ridge and turbidites from the trench are accreted to the slope.

DISCUSSION

Deformation of the Basaltic Slab

The 24-channel seismic profile traces the basaltic slab from the seaward side of the trench landward beneath the continental slope. Using these new data and all other available data from the Peru-Chile Trench, Schweller and others (this volume) present a new model to explain the rupture of the basaltic slab and subsequent rapid uplift observed within the trench axis. This model is expanded in the present study to show the possible origin of the broken basaltic slab beneath the forearc and the potential influence of the slab on the development of the outer forearc.

As the oceanic plate encounters resistance from the overriding continental block, both the orientation of principal stresses and the character of faulting undergo a complete transition. Extensional faulting within the oceanic plate on the seaward trench slope gives way to compressional faulting beneath the continental margin. Direct structural evidence of thrusting is seldom seen, because in most regions, compressional faulting is confined to the acoustically obscure region beneath the continental slope. However, we have concluded from previous work that in some areas, the compressional stresses from plate collision can be transmitted seaward within the oceanic plate, resulting in reverse or thrust faulting within the trench axis seaward of the edge of the continental

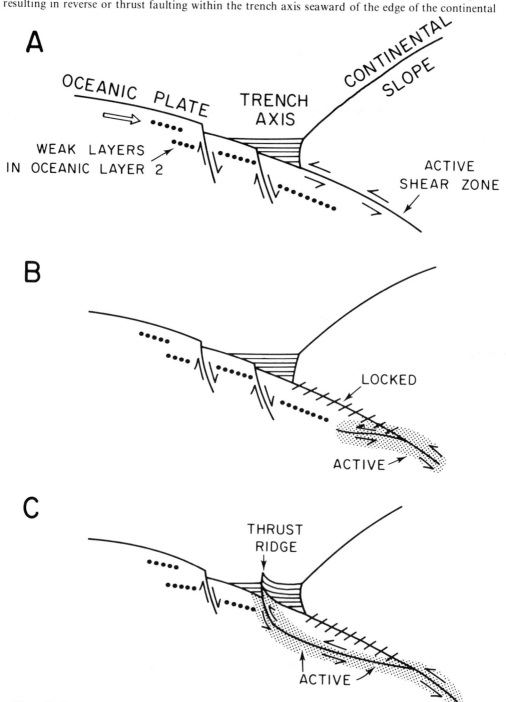

Figure 11. Evolution of basaltic ridge (slab) within the trench axis (after Schweller and others, this volume). See text for explanation.

margin (Prince and Kulm, 1975; Schweller and others, this volume). The axial ridge shown in the single-channel and multichannel profiles (Figs. 3, 7) was apparently created by such a process.

In the Schweller and others model (Fig. 11), a section of the main thrust plane beneath the forearc temporarily locks, so that compressional stresses begin to build in the oceanic plate (Figs. 11A, 11B). When rupture occurs, a low-angle thrust plane splits off the main thrust surface and propagates laterally seaward through oceanic layer 2 basalts (Fig. 11B). This propagation may be aided by the presence of weak horizontal zones of basalt rubble or sediment interlayers with the more competent layer 2 basalt flows in the upper few hundred metres of layer 2 (Aumento and others, 1977). At some point within the trench, the thrust plane bends sharply upward to reach the ocean floor (Fig. 11C). Continued thrusting motion from plate convergence causes uplift along the reversed fault at the seaward edge of the thrust sheet, forming a ridge within the trench axis. High-angle (up to 65°) reverse faults are calculated based upon rates of uplift as high as 22 cm/yr (Prince and Schweller, 1978). Because the angle of the fault is steeper at its leading edge than along the rest of the thrust, the seaward edge of the thrust sheet will be uplifted faster than the landward portion (Schweller and others, this volume). This will produce the landward tilting of the ridge and the overlying trench axis turbidites observed in seismic reflection records (Figs. 3, 4).

Expanding upon the model by Schweller and others, we suggest that the basaltic thrust complex is either decoupled from the oceanic plate in any of several configurations and accreted to the preexisting inner wall of the trench, or it remains coupled to the plate and is carried forward beneath the continental slope (Fig. 12). Decoupling of the already structurally disturbed basaltic mass may be accomplished by the friction created by the overriding continental plate (that is, the subduction complex). The basalt may be planed off as slivers of various dimensions and thicknesses depending upon the internal structure and stratigraphy of the main thrust ridge (Fig. 12, A-C). As noted previously, layer 2A basalts are initially faulted along weak zones during the formation of the trench ridge (Fig. 11). The ridge may contain additional weak zones of basaltic rubble and oceanic plate sediments interbedded with the pillow basalts along which partial decoupling may occur. An imbricated thrust stack of upper layer 2 rocks is an attractive origin for the thrust ridge, but it cannot be confirmed by structure or stratigraphy.

Partially decoupled basalt slivers may range from outcrop-sized blocks imbedded in the terrigenous clastic deposits of the trench and slope to rather large slices of oceanic crust that are several kilometres in areal extend and that resemble ophiolites (Fig. 12C) (see next section). Some of the strong low-frequency reflectors within the diffracting subduction complex (designated C, Fig. 5) may represent the basaltic slivers described here.

If the entire 60 km long, 900 m high ridge is decoupled along the newly activated thrust plane (Figs. 11B, 11C), it may be accreted to the leading edge of the subduction complex by imbricate stacking, creating a sediment-basalt melange (Figs. 12B, 12C). A feature of this magnitude would create a structural barrier parallel to the forearc, behind which sediments from the continental margin may become ponded. The accreted ridge may produce a trench-slope break composed of crystalline rather than sedimentary rocks.

On the other hand, if the ridge remains substantially coupled to the oceanic plate, it will be carried forward beneath the forearc, disrupting the sediments in the overlying subduction complex (Figs. 12D-G). In this model, large irregularities in the basaltic slab beneath the forearc may be former basaltic ridges inheritied from the adjacent trench. The large offset shown in the subducting slab beneath central Peru (km 52, Fig. 5) may be the counterpart of the present ridge in the trench axis. Alternatively, the slab offset beneath the subduction complex may represent a thrust feature (ridge) formed contemporaneously with the trench-thrust feature. In either case, it is unlikely that sediment loading by the relatively thin wedge of the subduction complex caused the slab to break in this fashion.

Basement Thrust Zones Versus Emplacement of Ophiolites

Ophiolotes found in continental terrane are interpreted as pieces of oceanic crust that are transferred to the continental block by some tectonic mechanism (Coleman, 1977). Thick, rather

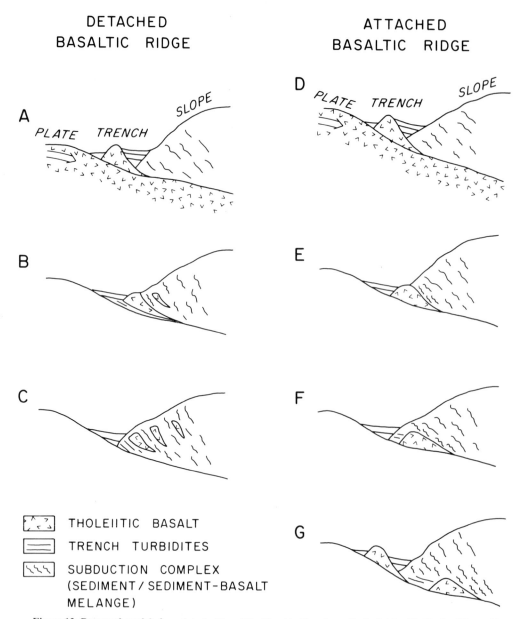

Figure 12. Proposed models for a detached basaltic ridge (A-C) and an attached ridge (D-G). See Figure 11 for initial formation of the ridge in trench axis.

complete ophiolite sequences usually invoke unusual plate-plate interactions for their emplacement, such as the overriding of the elevated crust of a spreading oceanic ridge onto the continental plate (Christensen and Salisbury, 1975). Less spectucular but just as puzzling is the frequent emplacement of ophiolite slivers into sedimentary deposits of continental derivation. This requires the transfer of oceanic basalt slivers from the descending plate to the overriding continental block, with subsequent elevation above sea level. We propose that the elevated basaltic ridges found in the Peru-Chile Trench (Figs. 3,7) are an example of the modern counterparts of ophiolitic slivers commonly found in ancient subduction complexes on land (Coleman, 1977). We further propose that some slivers are emplaced by decoupling, as shown in Figure 12 and described in the previous section. If this is true, then what are the stratigraphic relationships during the early stages of formation of ophiolitic slivers?

Looking at the central Peru Trench region, we see that fresh pillow basalts from the upper 500 m of layer 2 apparently comprise the bulk of the thrust ridge (Fig. 7, Scheidegger and others, 1978). While the oceanic plate displays the classical sequence of carbonates and successor brown clay and hemipelagic deposits on top of the basalt (Fig. 1), core 67 (Fig. 7) demonstrates that the trench turbidites have come in direct contact with a portion of the faulted crust of the ridge. This, together with the thin sediment cover (few tens of metres) over the 5 km wide top of the ridge, indicates that the bulk of the oceanic sedimentary section is missing. The virtual absence of the pelagic deposits requires a complicated structural setting or unusual stratigraphic circumstances. Starting at the base of the sedimentary section, the carbonate deposits may be largely dissolved by seawater, since the trench is below the calcium carbonate compensation depth (Kulm and others, 1974). Extensive faulting of the naturally porous upper layer 2 crust near the trench would enhance the removal of the carbonate facies, which is also commonly missing in ancient deposits. Most of the carbonate dissolution would have to take place within the relatively short time period of approximately 200,000 yr, since the oceanic plate deposits are passing through the 20 km wide trench at the rate of 100 km/m.y. However, the overlying brown clay and hemipelagic deposits, or about one-half of the section (Fig. 1), should still be intact if they are not structurally disturbed.

Previous studies (Prince and Kulm, 1975; Schweller and others, this volume) and this study strongly suggest that the trench ridge is a massive thrust feature elevated high above the trench floor. The oceanic deposits and turbidites from the trench may have been incorporated into the ridge in such a complex fashion that they cannot be deciphered in the multichannel record. Alternatively, these deposits may have been tectonically displaced in some unexplained manner as the ridge grew to its present dimensions. We do not have enough data to speculate further on the facies relationships.

As the basalt ridge comes in contact with the wedge-shaped complex in the continental slope, portions of the already faulted layer 2 crust may be decoupled and, together with the trench turbidites and any oceanic plate deposits, become incorporated into the terrigenous deposits of the subduction complex as slivers of pillow basalt (Fig. 12, A-C). If decoupling of these oceanic slivers occurs as soon as the overriding plate makes contact with the subducting ridge, imbricate stacking and the accompanying uplift could carry this igneous-sediment melange to higher levels within the subduction complex, where the effect of metamorphism is minimal. If decoupling occurs at a later stage in the convergence history, the oceanic slivers are more likely to become metamorphosed as the depth of burial increases along the inclined plate.

Geologic Cross Section of Forearc

Using the geologic and geophysical data available for the central Peru Margin, it is possible to construct a composite cross section across the submerged portion of the Andean forearc in this region. From industry drill holes at lat 9°S (Fig. 8), we know that the arc massif extends seaward to at least the edge of the present continental shelf and that it consists of foliated metamorphic rocks of Paleozoic or Precambrian age (Travis and others, 1976; Kulm and others, 1977). The Amotape Mountains of northeast Peru and the coastal mountains of south central Peru, as well as small islands along the edge of the shelf to the north and to the south of the drill holes, consist of similar metamorphic terrane (Kulm and others, 1977; this volume). Because the actual seaward limit of the massif cannot be defined in geologic terms without additional drill hole data, we will attempt to identify its western boundary using geophysical data.

From the seismic refraction data, we see that the OSH metamorphic basement has the same velocity (5.7 to 5.9 km/sec) in the seismic sections (Figs. 9B, 9C) of both Jones (this volume) and Hussong and others (1976). However, the two sections exhibit different velocities for basement material immediately seaward of the OSH on the upper continental slope. Jones calculated a velocity of 5.0 km/sec and a density of 2.67 g/cm^3 for this material, and Hussong and others, with no gravity data, calculated a velocity of 6.1 to 6.2 km/sec for the same material. The remainder of the middle and lower continental slope shows similar refraction results, especially the 5.6 and 5.8 km/sec velocities for the deepest material, respectively.

To honor both sets of refraction data, we propose three forearc models for the central Peru Margin which place the seaward limit of the arc massif in three different positions depending upon the geophysical criteria selected (Fig. 13). In the first model (Fig. 13A), the massif is located landward 26 km from the trench. Relatively low-velocity (1.7 to 3.6 km/sec) and low-density (1.8 to 2.60 g/cm^3) material, which is probably deformed clastic sediment from the trench and slope, overlies most of the continuous descending basalt slab that comprises the subduction complex. Prominent low-frequency reflectors within the complex (Fig. 5) may be basaltic slivers from the slab (Fig. 12A). Because we cannot rule out the possibility that the deep, broken reflectors from km 52 to km 87 represent the interface between the basaltic slab and an overriding continental metamorphic block (Figs. 5, 13A), or that velocities of 5.6 to 5.8 km/sec and a density of 2.75 g/cm^3 are also representative of a metamorphic block (Figs. 9B,9C), we place the arc massif relatively close to the trench. Virtually all sedimentary forearc basins (Yaquina, Trujillo, and Salaverry) are intramassif basins since they rest upon the arc massif. The Salaverry Basin is subdivided into two basins based upon the high contrast in velocities in seismic section (Jones, this volume; Figs. 9B,9C). The higher-velocity (4.6 to 5.2 km/sec)

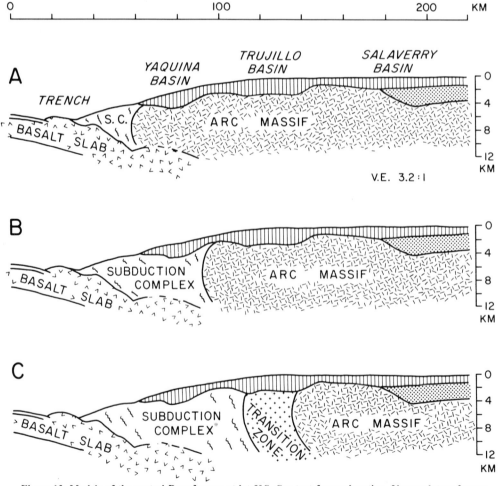

Figure 13. Models of the central Peru forearc at lat 9°S. See text for explanation. Nomenclature for arc elements (subduction complex and arc massif) from Seely (1979).

material may be older Mesozoic lithologies (see Thornburg and Kulm, this volume, for a detailed description of the geology of the Peruvian coast and continental margin). Post–late Eocene sandstone and shale, with minor limestone, characterize the Trujillo Basin (Travis and others, 1976; Thornburg and Kulm, this volume) and probably the low-velocity (1.5 to 2.6 km/sec) material in the adjacent Salaverry Basin. Both basins must have experienced substantial subsidence during Cenozoic time to allow the deposition of from 2.0 to 4.0 km of sediment.

In the second model (Fig. 13B), the arc massif is located 61 km landward of the trench beneath the upper continental slope. In this model, we assume that the deep, discontinuous reflectors from km 52 to km 87 are actually slivers of upper layer 2 basalt from the descending slab (Fig. 12, A-C) and that the overlying diffracting mass is a subduction complex consisting of highly consolidated trench and slope sediment and sediment-basalt melange capable of producing seismic velocities in the 5.6 to 5.8 km/sec range (Figs. 9B,9C) and densities of 2.75 g/cm^3 (Fig. 9B). We further assume that the high, 6.1 to 6.2 km/sec refraction velocites seaward of the OSH are valid (Hussong and others, 1976; Fig. 9C), are more typical of crystalline metamorphic strata than deformed sediments of the subduction complex, and represent the arc massif. The Yaquina Basin would be an accretionary basin since it would rest upon the rising complex (Fig. 13B).

In the third model (Fig. 13C), the arc massif is located near the edge of the continental shelf or near the seawardmost drill site (Fig. 10; lat 9°S [west]). In this model, we designate the low-density (2.67 g/cm^3), 5.0 km/sec material beneath the Trujillo Basin (Jones, this volume, Fig. 9B) a transition zone lying between the massif to the east and the subduction complex to the west. Thornburg and Kulm (this volume) prefer a gravity model whereby the minimum gravity anomaly (70 mgal) over the Trujillo Basin is more of a surficial effect related to 4.0 km thick sedimentary deposits of Hussong and others (1976) with a higher density block (that is, 2.72 to 2.80 g/cm^3) located directly beneath the basin. The massif and subduction complex could interface somewhere within this transition zone. Late Miocene to early Pleistocene dolomicrite and glauconitic micrite were dredged from the Trujillo Basin (Fig. 10, D-54, D-59) and contain benthic foraminifera indicative of approximately the present water depths, which infers relative stability of the basin during most of Neogene time (Kulm and others, this volume). If a subduction complex does underlie the western flank of the Trujillo Basin, it has not been uplifted, nor (based on microfossil data) has it undergone more than 250 m of subsidence during this time. On the other hand, the large volume of highly brecciated dolomicrite recovered with the nondeformed carbonates in the dredges, and the folding and block faulting observed in seismic reflection records along the western flank of the Trujillo Basin, indicate significant internal deformation of this basin. Radiometric dating of the dolomicrite breccia and enclosing cement suggest that the deformation occurred during late Pliocene to early Pleistocene time (Kulm and others, this volume). A similar deformational history is proposed for the Yaquina Basin and the surrounding Upper-Slope High because of the diffracting and faulted character of these deposits in the multichannel record (Figs. 4,5,8). Thornburg and Kulm (this volume) discuss the geologic history of these forearc basins in more detail, and the reader is referred to their paper for additional information on the relationship of the central Peru Margin basins.

All three forearc models are feasible for the central Peru Margin (Figs. 13, 13C). If the deep, broken reflectors landward of the continuous subducting slab represent the large density contrast commonly observed between sediment and basalt rather than between crystalline metamorphic and igneous rocks, then the second model is preferred (Fig. 13B). If the brecciated dolomicrites are indicative of the typical compressive tectonic regime (Kulm and others, this volume) associated with a deforming subduction complex rather than the tensional tectonic regime associated with the arc massif, then the third model is preferred (Fig. 13C). Additional drilling is necessary to test these various forearc models and unravel the geologic history of the region.

CONCLUSIONS

The central Peru Margin and trench from lat 7° to 10°S were studied using a 102 km long, 24-fold seismic reflection profile made across the trench and the adjacent continental slope. These multichannel data were integrated with existing single-channel reflection and seismic refraction data and geologic data to determine the structural and tectonic framework of this portion of the Andean forearc.

A 60 km long and 900 m high ridge (Figs. 3,7) is forming in the trench axis as the result of thrust faulting of upper layer 2 tholeiitic basalts on the descending Nazca plate. The basaltic crust is traced 26 km continuously beneath the continental slope, with another ridgelike offset in this segment, and apparently is traced another 35 km discontinuously as a series of short, broken reflectors (Figs. 4,5). A model utilizing a detached basaltic slab may explain how ophiolitic slivers of upper layer 2 basalts become incorporated into the overriding subduction complex (Figs. 12A,B,C). The stratigraphic implications of initial formation of the sliver are discussed. A highly diffracting mass (subduction complex) overlies these deep reflectors and floors a structurally disrupted forearc basin, Yaquina Basin, on the middle to upper continental slope. The 2 km thick deposits in this basin connect with another upper-slope basin, the Trujillo Basin, which contains a thicker section (up to 4.0 km) of post–late Eocene terrigenous sediments. A 95 km wide shelf basin, Salaverry Basin, lies between a metamorphic OSH which separates the Trujillo and Salaverry Basins, and the Peruvian coastline (Fig. 8).

Because of the limitations and uncertainties in both the geologic and geophysical data, three models are constructed for the central Peru forearc (Fig. 13). The seaward position of the metamorphic terrane of the arc massif varies from 26 to 115 km of the trench axis, depending largely upon the geophysical criteria used to identify the massif. In the first model (Fig. 13A), all three forearc basins are intramassif basins. In the second model, the Yaquina Basin rests upon a subduction complex and is therefore an accretionary basin (Fig. 13B). In the third model, the western flank of the Trujillo Basin rests upon a transitional zone of unknown origin. The actual location of the arc massif–subduction complex interface will determine which model is the correct one.

The Salaverry and Trujillo Basins require 2 to 4 km of subsidence in pre-Neogene time to allow the accumulation of marine sediment. Paleodepth indicators in the carbonates dredged from the west flank of the Trujillo Basin suggest little vertical movement of this region since late Miocene time. On the other hand, brecciated dolomicrites in the same dredges, as well as extensive faulting of the Trujillo and Yaquina Basin strata, suggest deformation by compression during late Pliocene to early Pleistocene time. The central Peru Margin has a complicated tectonic history in time and space which can be best deciphered by drilling.

ACKNOWLEDGMENTS

The authors wish to thank the students and staff of the School of Oceanography, Oregon State University, and the crew of the R/V *Yaquina* for their help with the data collection. Special thanks go to Esso Production Research Company, Houston, Texas, for their assistance in processing the 24-channel seismic profile to a depth section. Peter Vail's kind assistance in this task is greatly appreciated. Financial support for this project was provided by the Office of International Decade of Ocean Exploration, National Science Foundation, under grants GX 28675, IDOE 71-04208, and OCE 76-05903.

REFERENCES CITED

Aumento, F., and others, 1977, Initial reports of the Deep Sea Drilling Project, Volume 37: Washington, D.C., U.S. Government Printing Office, 1008 p.

Christensen, N. I., and Salisbury, M. N., 1975, Structure and constitution of the lower oceanic crust: Reviews of Geophysics and Space Physics, v. 13, p. 57-86.

Coleman, R. G., 1977, Ophiolites: Ancient oceanic lithosphere?: New York, Springer-Verlag, 229 p.

Hagedoorn, J. G., 1954, A process of seismic reflection interpretation: Geophysical Prospecting, v. 2, p. 85-127.

Hogan, L., and Dymond, J., 1976, K-Ar and ^{40}Ar-^{39}Ar dating of site 319 and 321 basalts, in Yeats, R. S., and others, eds., Initial reports of the Deep Sea Drilling Project, Volume 34: Washington, D.C., U.S. Government Printing Office, p. 439-442.

Hussong, D. M., and others, 1976, Crustal structure of the Peru-Chile Trench: 8°S-12°S latitude, in Sutton, G. H., and others, eds., The geophysics of the Pacific Ocean Basin and its margin: American Geophysical Union Geophysical Monograph 19, p. 71-86.

Jones, P. R., III, 1981, Crustal structures of the Peru continental margin and adjacent Nazca plate, 9°S latitude, in Kulm, L. D., and others, eds., Nazca plate: Crustal formation and Andean convergence: Geological Society of America Memoir 154 (this volume).

Kulm, L. D., and others, 1973, Tholeiitic basalt ridge in the Peru Trench: Geology, v. 1, p. 11-14.

Kulm, L. D., and others, 1974, Transfer of Nazca Ridge pelagic sediments to the Peru continental margin: Geological Society of America Bulletin, v. 85, p. 769-780.

Kulm, L. D., Schweller, W. J., and Masias, A., 1977, A preliminary analysis of the geotectonic processes of the Andean continental margin, 6° to 45°S, in Talwani, M., and Pitman, W. C., III, eds., Problems in the evolution of island arcs, deep sea trenches, and back-arc basins: American Geophysical Union, Maurice Ewing Symposium, Series 1, p. 285-301.

Kulm, L. D., and others, 1981, Late Cenozoic carbonates on the Peru continental margin: Lithostratigraphy, biostratigraphy, and tectonic history, in Kulm, L. D., and others, eds., Nazca plate: Crustal formation and Andean convergence: Geological Society of America Memoir 154 (this volume).

Masias, J. A., 1976, Morphology, shallow structure, and evolution of the Peruvian continental margin, 6° to 18°S [M.S. thesis]: Corvallis, Oregon State University, 92 p.

Prince, R. A., and others, 1974, Uplifted turbidite basins on the seaward wall of the Peru Trench: Geology, v. 2, p. 607-611.

Prince, R. A., and Kulm, L. D., 1975, Crustal rupture and the initiation of imbricate thrusting in the Peru-Chile Trench: Geological Society of America Bulletin, v. 86, p. 1639-1653.

Prince, R. A., and Schweller, W. J., 1978, Dates, rates and angles of faulting in the Peru-Chile Trench: Nature, v. 271, p. 743-745.

Rosato, V. J., 1974, Peruvian deep-sea sediments: Evidence for continental accretion [M.S. thesis]: Corvallis, Oregon State University, 93 p.

Rosato, V. J., Kulm, L. D., and Derks, P. S., 1975, Surface sediments of the Nazca plate: Pacific Science, v. 29, p. 117-130.

Rosato, V. J., and Kulm, L. D., 1981, Clay mineralogy of the Peru continental margin and adjacent Nazca plate: Implications for provenance, sea level changes, and continental accretion, in Kulm, L. D., and others, eds., Nazca plate: Crustal formation and Andean convergence: Geological Society of America Memoir 154 (this volume).

Salisbury, M. H., and Christensen, N. I., 1976, Sonic velocities and densities of basalts from the Nazca plate, DSDP leg 34, in Yeats, R. S., Hart, S. R., and others, eds., Initial reports of the Deep Sea Drilling Project, Volume 34: Washington, D.C., U.S. Government Printing Office, p. 543-546.

Scheidegger, K. F., and others, 1978, Fractionation and mantle heterogeneity in basalts from the Peru-Chile Trench: Earth and Planetary Science Letters, v. 37, p. 409-420.

Schweller, W. J., Kulm, L. D., and Prince, R. A., 1981, Tectonics, structure, and sedimentary framework of the Peru-Chile Trench, in Kulm, L. D., and others, eds, Nazca plate: Crustal formation and Andean convergence: Geological Society of America Memoir 154 (this volume).

Schweller, W. J., and Kulm, L. D., 1978, Extensional rupture of oceanic crust in the Chile Trench: Marine Geology, v. 28, p. 271-291.

Seely, D. R., 1979, The evolution of structural highs bordering major forearc basins, in Watkins, J. S., and others, eds., Geological and geophysical investigation of continental margins: American Association of Petroleum Geologists Memoir 29, p. 245-260.

Thornburg, T. M., and Kulm, L. D., 1981, Sedimentary basins of the Peru continental margin: Structure, stratigraphy, and Cenozoic tectonics from 6°S to 16°S latitutde, in Kulm, L. D., and others, eds., Nazca plate: Crustal formation and Andean convergence: Geological Society of America Memoir 154 (this volume).

Travis, R. B., Gonzales, G., and Pardo, A., 1976, Hydrocarbon potential of coastal basins of Peru, *in* Halbouty, M. T., and others, eds., Circum-Pacific energy and mineral resources: American Association of Petroleum Geologists Memoir 25, p. 331–338.

Yeats, R. S., Hart, S. R., and others, 1976, Initial reports of the Deep Sea Drilling Project, Volume 34: Washington, D.C., U.S. Government Printing Office, 803 p.

MANUSCRIPT RECEIVED BY THE SOCIETY NOVEMBER 12, 1980
MANUSCRIPT ACCEPTED DECEMBER 30, 1980

Printed in U.S.A.

Geological Society of America
Memoir 154
1981

Late Cenozoic carbonates on the Peru continental margin: Lithostratigraphy, biostratigraphy, and tectonic history

LaVerne D. Kulm
School of Oceanography
Oregon State University
Corvallis, Oregon 97331

Todd M. Thornburg
School of Oceanography
Oregon State University
Corvallis, Oregon 97331

Hans Schrader
School of Oceanography
Oregon State University
Corvallis, Oregon 97331

Antonio Masias
Petroleos del Peru
Lima, Peru

Johanna M. Resig
Hawaii Institute of Geophysics
University of Hawaii
Honolulu, Hawaii 96822

Leonard Johnson
Office of Naval Research
Arlington, Virginia 22217

ABSTRACT

Dredge samples collected from the continental slope off northern Chile and central to northern Peru elucidate the late Cenozoic geologic history of forearc. Late Miocene siltstone, claystone, and sandstone were recovered in water depths ranging from 2,779 to 8,132 m from the steep slope off Chile. Manganese slabs and crusts on these rocks and the virtual absence of Quaternary sediment cover suggest that this slope is a region of nondeposition.

In marked contrast, fine-grained carbonate rocks were recovered from rock outcrops from 3.5°S to 12°S latitude off Peru in water depths ranging from 202 to 3,749 m. They are dated from late Miocene to middle Pleistocene by floral, faunal, and radiometric techniques. These carbonates are classified as calcareous siltstone, micrite, glauconitic micrite, dolomicrite, and brecciated dolomicrite. The latter two lithologies predominate in all dredges and commonly contain more than 80% dolomite with minor calcite by weight percent. These carbonates are overlain by late Pleistocene and Holocene deposits that consist largely of terrigenous mud.

Microfossil data show that the nearly pure carbonate material was deposited in open marine conditions similar to those found in the upwelling regime on the Peru Margin today. Benthic microfossils indicate that the carbonates in the Lima Basin at 12°S were originally deposited in water depths of 150-500 m, but they were dredged from minimum water depths ranging from 1,639 m to 837 m. We conclude that the seward flank of the basin has subsided 1,100 m since the Pliocene and the landward flank 500 m during the late Pleistocene. The carbonate deposits are estimated to be the dominant lithology in the upper 1.3 km of a 2 km thick sedimentary section from a multichannel seismic section nearby.

Stable isotopic C^{13} and O^{18} values (PDB standard) suggest that the Peru carbonates are organically derived. The δO^{18} values are well constrained between +5.00 and +7.08°/00, while the δC^{13} values range widely between +19.63 and −13.47 °/00. We propose that oxidation of the organic carbon within the upwelling sediments is the most likely carbonate source for the dolomites. Progressive fractionation of C^{13} during the formation of CO_2 within a relatively closed system may be involved to produce the wide range of δC^{13} values, which far exceed the δC^{13} values of shallow-water, deep-water, and evaporitic carbonates. The Peru carbonates apparently formed by direct precipitation as pore space cement or by diagenetic alteration of the existing terrigenous sediments.

In middle Miocene time, the Amazon River drainage shifted from the Pacific to the Atlantic Ocean, greatly reducing the terrigenous sediment input to the Peru margin. Dolomitic strata appear to be prolific following this event, although wholly terrigenous muds began to accumulate on the slope between 0.93 to 0.44 m.y. ago.

The source of magnesium in the dolomicrites is speculative. Magnesium may be diagenetically released to interstitial waters from the tests of benthic foraminifera containing Mg-calcite. Alternatively, magnesium may be derived from the marine waters of restricted marginal seas created by a combination of postdepositional uplift of structural features on the upper slope and the late Miocene and Pliocene eustatic lowerings of sea level. The Pliocene and late Pleistocene subsidence of the slope structures returned the Lima Basin to open marine conditions, and the late Pleistocene terrigenous muds began to accumulate. The apparent lack of dolomitization in these organic-rich muds is difficult to explain without invoking these tectonic processes.

The brecciated dolomicrites and faulted structures in seismic reflection records indicate a complicated deformation history, especially at 8° and 9°S latitude. Dating of the breccia blocks and carbonate cement show that the deformation of some late Miocene to late Pliocene carbonates occurred in late Pliocene to early Pleistocene time. High grain densities and low porosities combined to produce high compressional velocities (4.5 to 6.6 km/sec), especially in the brecciated rocks.

INTRODUCTION

While a substantial amount of geophysical data has been collected on the Chile and Peru continental margins, there is very little information on the lithologies that form the foundation of the margin. Scholl and others (1970), Bandy and Rodolfo (1964), and Fisher and Raitt (1962) described late Cenozoic terrigenous rocks from the northern Chile continental slope (Fig. 1). A set of drill holes on the south-central Chile continental shelf recovered Late Cretaceous to Pliocene terrigenous deposits overlying a metamorphic basement (Mordojovich, 1974), and two drill holes on the outer shelf in the Salaverry Basin off central Peru (Fig. 2) encountered post-Eocene terrigenous deposits overlying a metamorphic basement (Kulm and others, 1977; Travis and others, 1976; Masias, 1976).

In an attempt to better interpret the geophysical data, we utilized an R/V *Wecoma* cruise in 1977 to dredge the continental slope off northern Chile (23°S) and the outer continental shelf and continental slope off central and northern Peru from 12°S to 3°S (Fig. 1). While expecting to find a variety of terrigenous sedimentary rocks, we actually recovered late Cenozoic, fine-grained carbonates and phosphorites at seven different locations spread over a distance of about 950 km along the margin off Peru. This is a most unusual suite of lithologies because terrigenous sedimentation has dominated the Peruvian shelf and slope in late Pleistocene and Holocene time (Krissek and others, 1980; Rosato and Kulm, this volume). The only major carbonate lithofacies known in the coastal region of central Peru are of Mesozoic age (Travis and others, 1976; geologic maps of Peru).

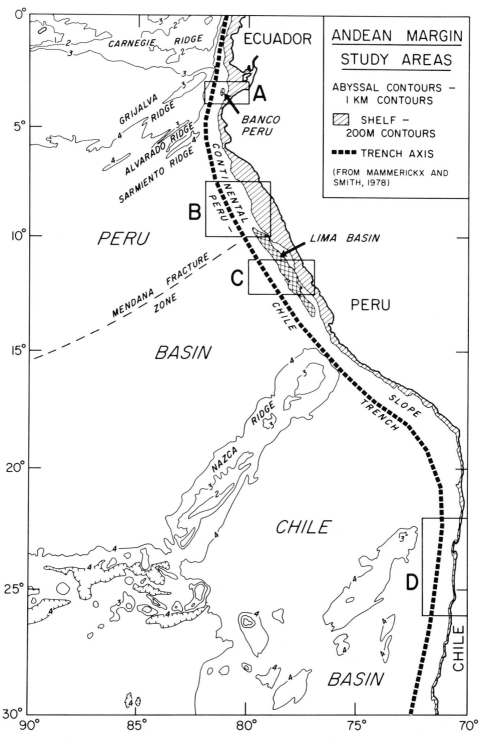

Figure 1. Location map of study areas (A-D) on the Peru-Chile continental margin. Locations of Lima Basin and Banco Peru shown by arrows.

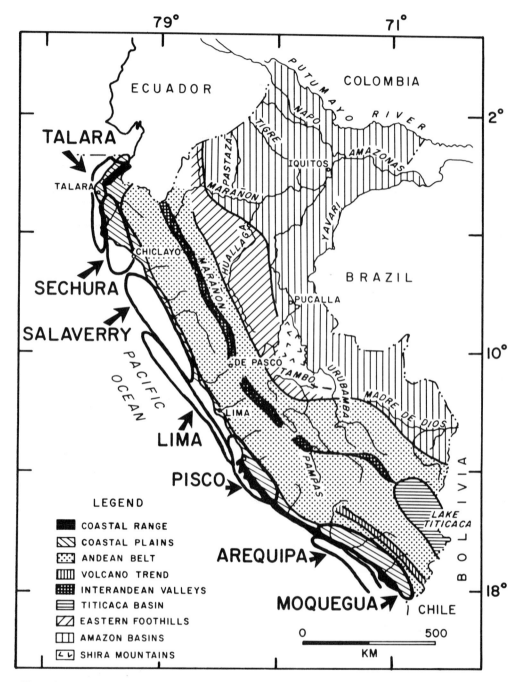

Figure 2. Distribution of sedimentary basins and physiographic provinces of Peru. Lima Basin is positioned on the upper continental slope (see Fig. 1); the other basins are located on the continental shelf and adjacent coastal region. Various features modified from Travis and others (1976).

The objectives of this study are to determine the lithostratigraphy, biostratigraphy, paleoenvironments, and physical properties of these carbonates. This information is crucial to our understanding of the late Cenozoic depositional environments and tectonic history of the outer continental margin. Finally, carbon and oxygen isotopic analyses allow us to make a preliminary interpretation of the origin of the fine-grained carbonate material found in these late Cenozoic rocks.

METHODS

Dredge hauls were taken with a box-type dredge. It was lowered to the bottom, and the vessel laid cable on the bottom as it sailed across the rock outcrop. The dredge was then pulled toward the vessel along the bottom, collecting the samples. The dredge start and stop positions were recorded on a graphic tension device.

Samples for microfossil analyses were taken from the inside of dredged rocks to minimize contamination by younger elements. For diatom analysis (H. Schrader), pea-size subsamples were broken into smaller sizes and boiled for 20 minutes in a 1:1 solution of peroxide and HCl (concentrated) to remove acid solubles and disintegrate samples. Residues were washed 10 times at 90 minutes settling times in 50 ml plastic bottles with demineralized water and were mounted on cover slips using Aroclor 4465 as mounting medium. The total microscopical slide was scanned at 400X magnification (oil immersion planapochromatic objective 40X), and occurrences of diatoms, preservation, and abundance of species were established using a 25% stepwise classification (Schrader and Fenner, 1976). Detailed taxonomy was done at 1000X (planapochromatic objective 100X). Several individuals of uncertain taxonomic position were documented on Polaroid film and are included in this paper (Appendix I, microfiche).

Samples for foraminiferal analysis (J. Resig) were disintegrated by soaking or boiling in water or through displacement of kerosene in the interstices of the dry rock by water. The most strongly consolidated rocks could not be disintegrated and were studied in rock slabs. The loose sediment was processed further on a sieve with openings of 63 μm, and the entire sand fraction was examined for foraminiferal species. Numbers of specimens were estimated in gradational series: A = abundant, many specimens per field of view; C = common, several specimens per field of view; X = frequent to rare, one to several specimens per sample.

Calcite and dolomite weight pecentages were calculated stoichiometrically using Ca and Mg mole percents as determined from atomic absorption and C mole percents from Leco analysis. This method affords an internal check on the accuracy of laboratory methods and on correlation between the two elemental determination techniques, since the summed mole percents of Ca and Mg should equal the mole percents of C in the carbonate fraction. In the majority of instances, the stoichimetry agreed in this manner within an error of 10%, very often within 5%. In a few cases, the C mole percentages significantly exceeded those of Mg and Ca; these were rerun to determine the effect of organic carbon on the Leco analysis. Values of 3 to >5% organic C were obtained on these weakly consolidated rocks. As a check, organic C was also determined for two dense and highly indurated samples that showed no tendency toward excess Leco carbon, yielding values of 1% or less organic C.

In all other cases (including the mole percents recalculated after organic C was determined), significant errors were consistently in the excess of Mg and Ca. This was attributed to leaching of cations from the clay residue upon application of HCl to the sample in preparation of the atomic absorption dilution. Where errors were in excess of 10%, revised dolomite/calcite pecentages were calculated by reducing the summed Ca and Mg to the Leco C mole percentage, fractioning the excess in the ratio of 40% Ca, 60% Mg, which is very roughly representative of the mineralogy of a typical marine clay. The net result was to decrease the weight percent of total carbonate by an amount proportional to the error, and to slightly depress the percent dolomite in the total carbonate fraction.

Petrographic analysis was conducted by the standard point counting technique to determine the percent of various constituents. Harold Enlos of the Department of Geology, Oregon State University, performed this analysis.

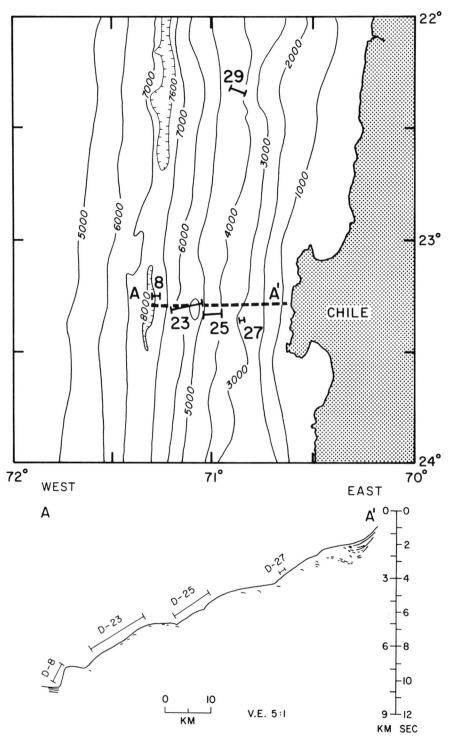

Figure 3. Top: Location of numbered dredge stations off northern Chile (see Area D, Fig. 1). Bathymetry in metres (after Prince and others, 1980). Bottom: Line drawing of seismic reflection record AA' of continental slope near dredge stations.

COLLECTION AND LOCATION OF SAMPLES

During the *Wecoma* W7706 cruise, four dredge stations (23, 25, 27, 29) were occupied on the middle portion of the northern Chile continental slope in water depths ranging from 6,543 to 2,779 m (Fig. 3). One additional dredge sample (8) was collected earlier on the FDRAKE cruise (1975) of the *Melville* on a bench at the base on the continental slope between 8,132 and 7,705 m (Fig. 3). The slope is extremely steep off northern Chile and displays only a small sedimentary section on the uppermost slope in the seismic reflection record (Fig. 3 [AA']).

On the same *Wecoma* cruise, seven dredge stations (Wecoma 46, 48, 54, 55, 59, 75, 80) were successfully occupied on the central and northern Peru outermost continental shelf and upper slope (Figs. 4, 5, 6). An earlier cruise on the *USN Barlett* (1973) dredged the wall of a submarine canyon (DRAG-1) on the landward flank of the Lima Basin between 1,201 and 837 m near Lima, Peru (Figs. 4, 7 [BB']). *Wecoma* dredge 46 was collected from the seaward flank of the Lima Basin on the scarp of a structural bench between 2,263 to 1,639 m (Figs. 4, 7 [BB']). Only one rock specimen was obtained in deep water (3,749 m) at dredge site 48 (Fig. 4).

Farther north (9°S), dredge 54 (Fig. 5) was taken on the scarp of one of the fault blocks on the uppermost continental slope at a depth of 250 to 202 m, as shown in Figure 7 (CC'). In the same area, dredge 55 recovered small rock fragments and mud at 4,479 m. About 110 km to the north (8°S), dredge 59 sampled a rock outcrop on the upper slope at a water depth of 430 m (Figs. 5, 7 [DD']). The structurally well-defined Lima Basin apparently terminates just south of dredge site 54 (Thornburg and Kulm, this volume) (Fig. 1).

Two dredge hauls (75, 80) were taken off Talara, Peru, on the Banco Peru (Fig. 6). This acoustically opaque submarine bank is situated on the outer edge of the continental shelf and is flanked to the east by the Progreso Basin (Figs. 2, 6). Dredge 75 recovered rocks at 292 m from the steep western scarp of Banco Peru, while dredge 80 sampled the flank of a large hill protruding from the deeper part of the bank at 360 m.

Several samples were collected from outcrops on Hormigas Island, located on the outer edge of the continental shelf of Lima, Peru (Fig. 4), by one of the authors, Antonio Masias.

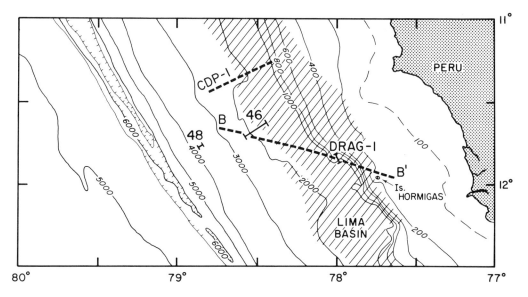

Figure 4. Dredge stations in Lima Basin near 12°S (see Area C, Fig. 1). Bathymetry in metres (after Prince and others, 1980). Seismic profiles BB' and CDP-1 shown as dashed lines (see Figs. 7 and 12, respectively).

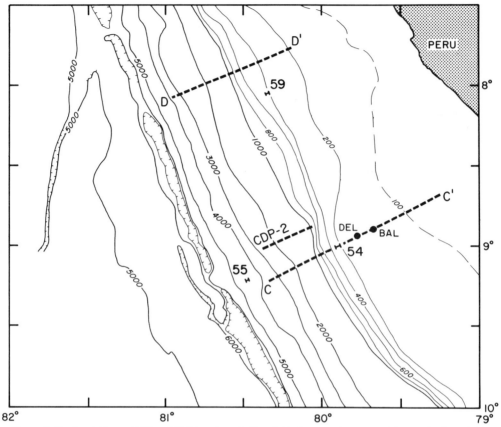

Figure 5. Dredge stations at 8°S and 9°S (see Area B, Fig. 1). Bathymetry in metres (Prince and others, 1980). Seismic profiles CC' and DD' shown as dashed lines (see Fig. 7). Position of Delfin and Ballena wells shown on profile CC'.

LITHOSTRATIGRAPHY

Northern Chile Margin

Lithologies recovered from the northern Chile continental slope (Fig. 3) consist chiefly of sedimentary rocks with occasional crystalline rock fragments. Siltstones and claystones predominate in all dredges except number 27, which contained only one crystalline rock specimen. Several friable, medium-grained sandstone samples were recovered in dredge 25, and one indurated sandstone was found at dredge 8.

Petrographic analyses of the angular to subangular sands (Table 1) and their low matrix content indicate (using the classification of Williams and others [1954]) that they are arkosic arenites. The inclusion of glauconite pellets and the size grading in these deep-water sandstones suggest derivation from shallower marine environments by turbidity currents. The mineralogy and lithic fragments (Table 1) show that the sands were derived from a variety of source rocks, including metamorphic, volcanic, and plutonic rocks. The abundance of biotite and green hornblende along with lithic fragments consisting of feldspar, biotite, hornblende, and muscovite, as well as phyllite, suggest a predominant metamorphic source. In addition, sanidine, tuff fragments, and basaltic hornblende suggest a substantial volcanic source rock, especially siliceous volcanics. Minor plutonic source rocks are indicated by microcline and large quartz grains with fluid inclusions.

The light brown and gray siltstone and claystone are homogeneous with no bedding or size grading visible. They also are barren of foraminifera. However, most of the samples are small and may represent a limited sample of the outcrop. Thin-section analysis indicates that these fine-grained lithologies consist chiefly of quartz, feldspar, mica, heavy minerals, and clay minerals. The provenance of these fine-grained lithologies could not be determined from thin sections.

The crytalline rocks are frequently rounded pebbles of quartzite, basalt, and gneiss derived from metamorphic and volcanic source rocks.

Manganese crusts and manganese covering the siltstones were recovered in dredges 23, 25, and 29. Their abundance indicates that a manganese pavement may exist over a large portion of the continental slope off northern Chile.

Central and Northern Peru Margin

Carbonate and Related Sedimentary Lithologies. The bulk of the consolidated lithologies dredged from the Peru margin are fine-grained carbonates, with phosphorite being present at one station (59).

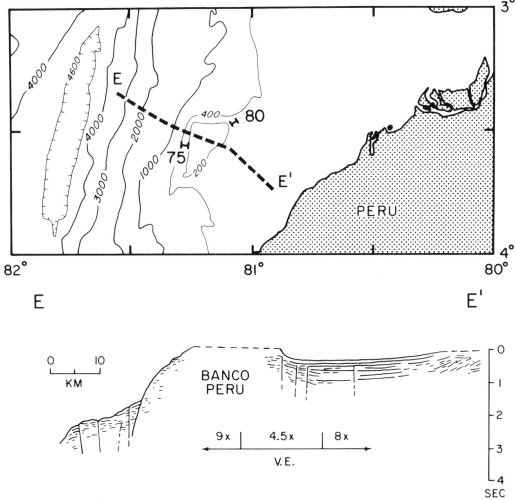

Figure 6. Top: Location of dredge stations on a submarine bank, Banco Peru (see Fig. 1.). Bathymetry in metres (after Prince and others, 1980). Bottom: Line drawing of seismic record EE' over Banco Peru and vicinity (modified from Shepherd, 1979).

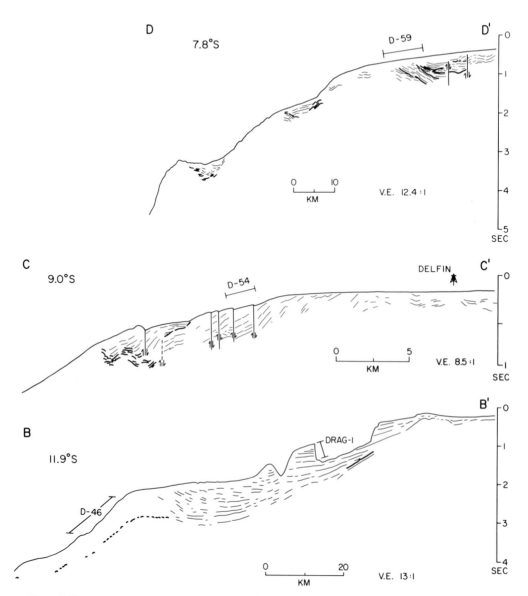

Figure 7. Line drawings of seismic reflection records BB′, CC′, and DD′. See Figure 4 for location of BB′ and Figure 5 for the latter two profiles.

With the exception of the phosphorites forming on the continental margin today (Veeh and others, 1973), there are no modern counterparts for these carbonate rocks.

Chemical analysis (Table 2, microfiche) and petrographic analysis (Table 3) indicate that the light to dark gray carbonates are dolomicrite, brecciated dolomicrite, glauconitic micrite, and micrite. The chemical data show that their total carbonate content (calcite plus dolomite) ranges from 60 to 90% by weight for the 46 samples analyzed (Table 2). The dolomite content ranges from 5% to 99%, with the exception of several low-dolomite, glauconitic micrites and one micrite sample. However, the dolomite content generally remains above 70% weight. X-ray analysis also confirms that dolomite is present. In contrast, the gray to black calcareous siltstones contain less than 25% total carbonate, which is largely calcite (Table 2). Petrographic analysis of 64 thin sections show the fine-grained

TABLE 1. PETROGRAPHIC ANALYSES OF SANDSTONE AND METAMORPHIC ROCKS*

Constituents	Rock type/Sample number		
	Arkosic Arenite	Quartzo-Feldspathic Schist	Quartzite
	25-12	HI-1,2	HI-3
Quartz	21	38	95
Plagioclase	33	32	
Biotite	17	26	T
Muscovite	T		4
Garnet			1
Apatite	1	T	
Opaques	2	4	T
Green hornblende	6		
Untwinned feldspar	5		
Sanidine	4		
Microcline	T		
Basaltic hornblende	T		
Zircon	T		
Glauconite	4		
Collophane	1		
Lithics	6		

Note: T = trace.

*Numbers in percent as determined by point counting.

character and high carbonate content of the Peru rocks. Detailed petrographic analyses of representative lithologies are given in Table 3.

The microfossil content of the carbonate is quite variable, with some samples being barren and others being quite fossiliferous in the same dredge haul. Foraminifera are the most ubiquitous fossils followed by diatoms and radiolarians. The calcareous siltstones are the most fossiliferous lithologies, although numerous dissolution impressions in the highly indurated rocks show that they were also rather fossiliferous.

Evidence of extreme structural deformation is seen most notably in the carbonates of dredges 54 and 59 (Figs. 5, 8C) and, to a lesser extent, in the carbonate from the other dredge sites. These deposits have been broken into small, roughly equidimensional blocks with enclosing carbonate cement.

Dolomicrites. Dolomicrites were found in all dredge hauls collected from the outermost continental

TABLE 3. PETROGRAPHIC ANALYSES OF REPRESENTATIVE CARBONATES*

Constituents	Rock type/Sample number								
	Dolomicrite					Brecciated Dolomicrite		Glauconite Micrite	Phosphorite
	DRAG-1	54-7	59-17	75-1	80-9	46-1	54-5	54-8	59-15
Micrite	71	77	77	71	97	27	28	42	24
Recrystallized micrite	22	10	19	27		2	5	5	20
Fossils	5	11	3	T		3	3	2	
Bone		1						1	2
Collophane	<1	T					T	4	54
Glauconite				T	T			38	
Quartz and plagioclase	1			T	2			1	T
Pyrite								7	
Opaques	1	1	1	2	1	T	T		
Granulated carbonate						14	31		
Coarse replacing carbonate						54	33		

Note: T = trace.

*Numbers in percent as determined by point counting.

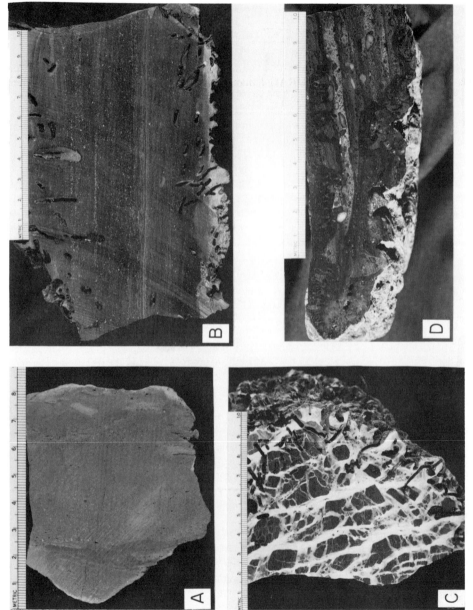

Figure 8. Photograph of rock slices: A.—homogenous dolomicrite; B.—laminated dolomicrite with foraminifera in coarse layers; C.—brecciated dolomicrite; and D.—phosphorite pebbles set in a conglomeratic mass.

shelf and continental slope (Table 2; Figs. 4, 5, 6). The majority of the dolomicrites contain 80% to 99% total carbonate that is mostly dolomite with subordinate calcite (Table 2). These light to dark gray rocks exhibit varying degrees of induration.

In thin section, the dolomicrites (DR-1, 54-7, 75-1, and 80-9) contain about 75% dark micrite and 10–27% recrystallized micrite (Table 3). The dark color of the micrite is probably due to very fine organic material and/or clay minerals. A few organic carbon analyses indicate values of 1% or less. Small amounts of terrigenous debris, glauconite, and collophane are present.

Microfossils (largely foraminifera and fossil fragments) and fish bones may constitute as much as 11% of the rock. They are usually aligned in rows as if paralleling bedding planes (Fig. 8B). These fossils are often completely filled with carbonate, indicating replacement of original skeletal material. Siliceous fossils are more common in DRAG-1 than in the other micrite studies. In some barren samples, especially 59-9, there were impressions of diatoms indicating dissolution after lithification.

Micrite/Calcareous Siltstone. A single carbonate specimen (46-16) is classified as micrite because it contains at least 50% total carbonate that is dominantly calcite with only minor amounts of dolomite (Table 2). This sample is weakly to moderately indurated. The most weakly indurated lithologies are the calcareous siltstones. They contain less than 25% total carbonate, which is almost exclusively calcite. Interestingly, their organic content is quite high (3–5%) as noted in DRAG rocks (DR-3, 5, and 11) (Table 2). Only one sample, the dark black siltstone (46-2), contains a measureable amount of dolomite. It also has a 3% organic carbon content. Petrographic study of the silt fraction of the siltstones indicate that it consists of quartz, feldspar, and heavy minerals.

Glauconitic Micrite. The glauconitic micrite is concentrated in dredge 54 (Fig. 7), although glauconite is present in minor amounts in a large portion of the Peru carbonates. The micrites contain varying quantities of glauconite, which influences their total carbonate content. Chemical data show that calcite is the dominant carbonate mineral, with dolomite being present in subordinate amounts (Table 2).

As shown in the thin section analysis of sample 54-8, this glauconitic micrite contains 38% rounded to angular, green to yellow glauconite (Table 3). The grains are up to 2 mm in diameter and are often surrounded by coarse carbonate that occasionally replaces the glauconite. Fossil and mineral grains are the pellet nuclei. Sparry calcite occurs in veinlets through the glauconite grains or as replacement masses in glauconite. Pyrite is a common inclusion in the glauconite. Collophane pellets, oolites, and bone fragments constitute more than 4% of sample 54-8. Other glauconitic micrites contain abundant coarse sand and pebble-size debris that is probably phosphatic material. Terrigenous minerals are generally minor constituents in the glauconitic micrites.

None of the glauconitic micrites exhibit primary structure, which would indicate downslope transport or sediment reworking by currents.

Brecciated Dolomicrite. Broken breccialike dolomicrite blocks set in a carbonate cement (DR-14, 46-1, 54-4, 59-28) were collected at all dredge sites except 75 and 80 on the margin (Table 2, Fig. 8C). These blocks exhibit similar compositional characteristics as the nondeformed dolomicrites described previously but are disrupted by shear planes, often observed to be oriented in a subparallel manner in at least one direction. They generally contain more than 95% total carbonate, with dolomite being more than 80% of this value.

In thin section, the micrite and recrystallized micrite in the blocks are a dark color probably due to the included organic matter and/or clay minerals. Calcareous fossils are commonly filled with sparry calcite. Siliceous fossils (radiolarians and diatoms) are also present in samples 46-1 and 54-4, as noted in the diatom analysis (Table 5). Collohane and glauconite pellets, as well as quartz and feldspar, are very sparse in the micrite blocks, even at dredge station 54, which also recovered the glauconitic micrites.

The cement of the brecciated dolomicrites can be grouped into two units: granulated carbonate and coarse granular grains, and rhombohedra of carbonate. The granulated carbonate consist of clay-sized carbonate grains in masses and streaks that fill spaces between broken blocks of original rock. This granulated mass is due to mechanical disruption of the original rock. The coarse-replacing carbonate is concentrated along planes of weakness developed by fracturing and granulation. Clear rhombs or clear irregular grains invade the otherwise unbroken dark micrite as if the replacing solutions permeated the original rock. Dolomite appears to be the dominant mineral in the coarse-replacing material, as shown in the chemical analyses (Table 2—microfiche). In the only case in which both matrix and cement could be dated (sample 54-5), the cement proved to be younger, indicating its postdepositional or diagenetic character (Table 2).

Phosphorites. Phosphorites and phosphatic debris are common in dredge 59 (Fig. 7) and probably comprise the bulk of pebbly conglomerates recovered at this site (Fig. 8D). In thin section, more than 50% of sample 59-15 is collophane, which occurs in several petrographic forms including bone fragments, oolites, pellets, and composite grains or masses (Table 3). The oolites are occasionally replaced by cementing carbonate and exhibit a bone fragment, microfossil, or opaque mass as a nucleus. Glauconite and terrigenous grains are very sparse. Calcite serves largely as a cement binding the collophane debris into a solid rock mass.

However, clean rhombohedra are associated with the micritic cement, indicating that at least some of the carbonate may be dolomite. Micrite, usually dark colored, surrounds the collophane clasts as concentric bands and is associated with the microfossils. Recrystallized micrite has dark borders and occurs in irregular patches.

Metamorphic Lithologies. Industry dril holes at 9°S (Figs. 5, 7) indicate that metamorphic strata—biotite gneiss and phyllite—form the acoustic basement of the outer continental shelf in some areas (Kulm and others, 1977). These strata are correlated along an anticlinal structural trend with similar rocks of Paleozoic age in the coastal mountains of northern Peru and of Precambrian age in the coastal mountains of southern Peru (Fig. 2). Hormigas Island, which is located off Lima, Peru, on this outer shelf structural high (Fig. 4), was sampled and found to contain lithologies similar to the pre-Mesozoic metamorphics. Apparently, these older crystalline rocks form a fairly continuous anticlinal structure that separates the shelf and upper-slope basins along the Peru continental margin and causes the shallow acoustic basement that rises toward the island in seismic reflection profiles (see profile B-B' in Figs. 4 and 7; Thornburg and Kulm, this volume).

Two different metamorphic lithologies—quartzite and quartzofeldspathic schist—were collected on Hormigas Island by Antonio Masias. Table 1 shows the simple mineralogy of the quartzite (H1-3), which has elongate quartz grains up to 5 mm long. Veinlets or lenses of a mixture of muscovite, garnet, rare biotite, opaque minerals, and an occasional apatite grain parallel the preferred orientation direction of the quartz grains.

The quartzofeldspathic schist (H1-1,2) is characterized by a poor schistose texture that is imparted both by parallel elongate strands of biotite and opaques and by a tendency of the quartz and feldspar grains to elongate and subparallel. The mineralogy consists chiefly of quartz, plagioclase (An 50), and biotite (Table 1). Both the texture and mineralogy suggest that the rock is a quartzofelspathic schist or perhaps a granulite. The high anorthosite content and the absence of low-rank minerals suggest a fairly high grade of metamorphic rock.

PHYSICAL PROPERTIES OR PERU MARGIN CARBONATES

Nine carbonate samples from three different dredge sites were subjected to porosity, permeability, grain density, and compressional velocity analyses by the Phillips Petroleum Company. After the first three measurements were completed on the dry rock sample, the sample was completely saturated with simulated sea water, and the velocity measurements were made at zero confining pressure.

The dense character of these carbonates is shown by their low porosity and permeability, their high grain densities, and their high compressional (V_p) velocities (Table 4). The highest porosities are associated with the glauconitic micrites (54-8), the phosphorite (59-15), and the dolomicrites that have abundant siliceous microfossils (46-28), which include the broken dolomicrite blocks in the brecciated dolomicrites (54-5). Grain densities, ranging from 2.66–2.80 gm/ml, are quite high, reflecting the high dolomite content of these rocks (Table 2).

The overall high compressional velocities measured in the Peru carbonates reflect the other physical

TABLE 4. PHYSICAL PROPERTIES OF PERU CARBONATES

Sample identification	Porosity (% bulk volume)	Specific gas permeability (millidarcys) horizontal	vertical	Grain density (g/ml)	Length (cm)	Compressional velocity* (m/sec)	(ft/sec)
7706-54-5A†	4.31	<0.007		2.76	3.450	6,272	20,577
-54-5B†	6.62		<0.007	2.77	3.420	6,218	20,400
-46-28A	8.01	0.021		2.66	3.145	6,048	19,842
-46-28B	3.24		0.005	2.77	2.400	6,000	19,685
Drag-1	8.00	0.007		2.76	3.430	5,532	18,150
7706-59-15A	4.24	0.007		2.80	3.335	5,467	17,936
-59-15B	6.02		0.29	2.79 §	3.310	5,092	16,706
-59-17A	0.11	<0.007		2.74	3.380	6,377	20,922
-59-17B	0.44		<0.005	2.76	2.254	6,261	20,541
-59-32A	0.22	<0.007		2.72	3.260	6,151	20,181
-59-32B	0.36		<0.007	2.72	3.005	6,133	20,121
-54-7A	0.90	<0.007		2.74	3.180	5,390	17,683
-54-7B	0.81		<0.007	2.73	3.385	5,460	17,913
-54-8A	13.2	0.15		2.76 §	3.270	4,139	13,580
-54-8B	6.97		0.005	2.68	2.165	4,706	15,440
-46-1A	2.01	<0.007		2.79	3.420	6,615	21,702
-46-1B	1.72		<0.007	2.79	3.335	6,539	21,453
-59-28A	1.95	<0.007		2.78	3.245	6,490	21,292
-59-28B	5.47		<0.005	2.74	1.895	6,113	20,055

Note: Analyses by Phillips Petroleum Company.

*Compressional velocity measurements made under zero net confining pressure.

†'B' samples were drilled vertical to 'A' samples, and vertical to depositional plane where evidence of such appeared possible.

§Partly filled vugs.

properties of these rocks (Table 4). The lowest velocities (avg 4.4 km/sec) were measured in the glauconitic micrite (54-8), the intermediate velocities (avg. about 5.5 km/sec) in the phosphorite (59-15) and two dolomicrites (54-7 and DRAG-1) from two different dredge sites, and the highest velocities in the remaining dolomicrite and brecciated dolomicrite (avg 6.0–6.5 km/sec) from three dredge sites. It is interesting that the brecciated dolomicrites exhibit some of the highest velocities (e.g., 46-1 and 54-5; Table 4). It was determined from the petrographic study that the composition of dolomicrite blocks in the sheared matrix is similar to the undeformed dolomicrites collected in the same dredge (e.g., 54-5 and 54-7, respectively; Table 3). It would appear that the granulation and replacement process that follows the rock deformation may in fact produce a more dense rock by filling the new void spaces and fractures with coarse carbonate which, together with the granulated carbonate, constitutes more than 50% of the rock (Table 2). Since a paired matrix and cement sample from a single brecciated specimen yielded nearly identical C^{13}/O^{18} values, it seems likely that the original dolomicrite was recrystallized in situ to produce the granulated cement within the shear zones during deformation (see section on Origin of Upper Slope Carbonates and Fig. 14).

BIOSTRATIGRAPHY

Diatom Assemblages

Biostratigraphy. The bases for diatom biostratigraphy as used in this paper are recent reports by Burckle (1977, 1978) and Barron (1980) on ranges, data, and first and last "evolutionary" (?) occurrences of species in the circum Pacific (Figure 9). These compilations were made on the basis of assigning each diatom-bearing sample to an interval of the paleomagnetic stratigraphic scale and thus to an absolute time interval. The objective is to establish the ages rather than to detail the taxonomy.

A listing of samples, their diatom content, and stratigraphic position is given in Appendix 1. The complete floral content of the samples is given in Table 5. Only the important species are illustrated in Plates 1 and 2 at the end of this chapter and in Appendix I (microfiche).

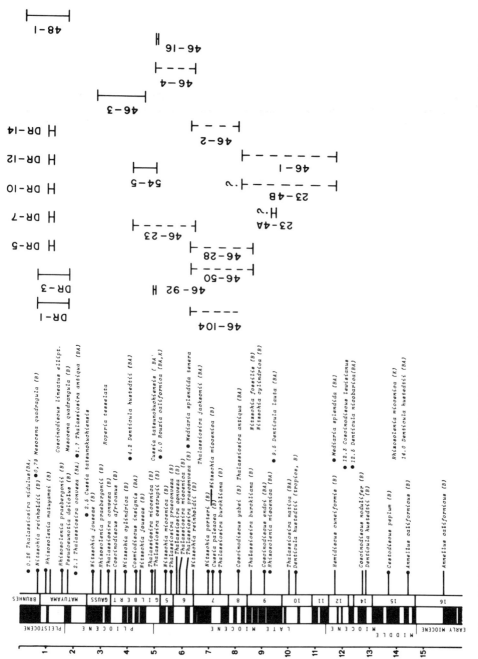

Figure 9. Diatom datum levels as established by Burckle (B), Barron (BA), and Koizumi (K) in Barron. Large dots indicate last occurrences, bars indicate first occurrences. Numbers prior to species names indicate absolute dates in million years of diatom datum levels. Ages of numbered dredge samples given to right of datum levels.

Ecology. All diatom-bearing samples indicate open marine, deep-water (greater than 150 m) sedimentation, similar to conditions found today within the upper continental slope and lens off Peru (Schuette and Schrader, 1979; Krissek and others, 1980).

The admixture of marine benthic species is negligible, and no freshwater diatoms were observed. The environment producing these excellently preserved (in most cases) diatom-rich sediments is comparable to today's coastal upwelling situation off Peru.

Age of Lithologies

Forty-one rock samples on the Peru and Chile margins were examined for diatoms indicative of geological age. Approximately one-half of the samples were barren of microfossils. The remainder of the samples contained enough biostratigraphically important species to date the lithologies. Sixteen samples also were subjected to radiometric dating.

Peru Margin. Siliceous microfossils show that the carbonates on the Peruvian continental slope range in age from late Miocene to middle Pleistocene (see Diatom Assemblages; Table 5, Fig. 9). Additional carbonate samples studied for diatoms by Mobil Research and Development Corporation produced similar ages (Table 2, microfiche). While calcareous microfossils (foraminifera) do not permit specific age determinations, they do suggest that the carbonates are Miocene and post-Miocene in age (see Paleobathymetry of Peru Carbonates).

Mobil Research and Development Corporation made radiometric age determinations on a number of the Peruvian rocks dated by microfossils. Strontium isotopic dating (Sr^{87}/Sr^{86}) indicates that the carbonates have age ranges from middle Miocene to Recent (Table 2). A K/Ar age analysis of the glauconite gives a Pliocene age of 3.6 m.y. Both the microfossils and the radiometric dating show that the carbonates are of late Cenozoic age.

Chile Margin. The terrigenous siltstones collected on the northern Chile margin at dredge station 23 are dated by diatoms as probable late Miocene (Fig. 9). The poorly preserved flora in the sandstones at station 25 may be late Miocene in age or younger. Siltstones from dredge station 29 were barren of microfossils.

PALEOBATHYMETRY OF PERU CARBONATES

Benthic foraminifera indicate that carbonates dredged at stations DRAG, 46, 54, and 59 on the middle and upper continental slope off central Peru (Figs. 4, 5) were originally deposited at outer shelf to upper slope depths. In situ glauconite substantiates these paleodepths. Subsidence of the continental margin is demonstrated in an area at 12°S off Lima, Peru, as described below. If eustatic sea level changes are responsible for this apparent subsidence, they would have to be on the order of from 500 to 1,100 m. Sediment loading of less than 2 km of material on metamorphic continental crust should not produce significant subsidence.

DRAG Station

Pleistocene faunas were contained in the rocks dredged from the wall of the submarine canyon on the landward side of the Lima Basin at water depths of 837 to 1,201 m (Fig. 4; Table 6). These rocks, although in various stages of lithification, contain comparable *Bolivina* assemblages, the differences in species content being readily attributable to lithification and diagenesis. Many highly indurated samples from the dredge could not be disaggregated; only a few of the more common forms were recognized in these samples. *Bolivina seminuda humilis*, a key species linking the assemblages, was identified tentatively in most of the indurated rocks by its test size and shape and chamber configuration in section.

These benthic assemblages of the DRAG rocks compare closely with modern assemblages from the outermost continental shelf–uppermost slope in depths of 100 to 300 m (Resig, this volume). No

TABLE 5. LIST OF SAMPLES CONTAINING DIATOMS WITH INDICATION OF PRESERVATION AND ABUNDANCE

Sample	1	2	3	4	5	6	7	8	9	10	11	12	13	14	15	16	17	18	19	20	21	22	23	24	25	26	27	28	29	30	31	32	33	34	35	36	37	38	39	40	41	42	43	44	45	Taxon
																													R		T	R		T	T												*Nitzschia cylindrica*
																										T		R	T		T	T	R												*fossilis*
																												T	T		T	T	R												*jouseae*
																							T				R		T	R	T	R			T										*marina*
																			T	F		R		F	R	R	R	F	R			R													*porteri*
T				R		T	T	T			T		F							R	F	R	T		R																				*reinholdii*
R	T	R	R	R	R	T	R	R		T	R										R	R	R	R																					*seriata*
	R		R		R		T			R				R	R		R			R	R	T																							*Rhizosolenia alata*
																R				R			R																						*barboi*
R	R	R	R	R	R	R	R					R			R																														*bergonii*
												F				R																													*calcar-avis*
													R																			R													*hebetata forma hiemalis*
R	R	R	R	R																																									*matuyamai*
						T																																							*miocenica*
R				R			R			R	R	R	R	R	F	R	R		R																										*styliformis*
																F	R		R	R																									spp.
R	F		R	R										T						T																									*Roperia tesselata*
													F						T			T		R																					*Rouxia californica*
															R		T																												species
R		R				R	R				F	F	R		R		R			R	R	R	R	F																					*Paralia sulcata*
															R						F																								*Pseudodimerogramma* spp.
R															R									R																					*Pseudoeunotia doliolus*
																R			R																										*Pseudopyxilla americana*
																			R	R	R																								*Stephanopyxis palmeriana*
R	R	R	R	R	F		C	T	F	R				F	R		T	R			R	R																							*turris*
																					F	F																							*Synedra miocenica*
													R								F	F		R	T																				"indica" similar
																								C																					*Thalassionema lineata*
R	T	T	R	R	R	R	F	F		F	F	R	R		F		F	F	C	R	F	F	F	F		C	R																		*nitzschioides*
R	R	R	R	R	R	R	F	R		R		F			R	F	R	F		F	R	R	R	F			F																		*Thalassiothrix "longissima"*
												R			R																														*monospina*
F	F	F	R	R													F	R	F	F			F	F																					*Thalassiosira excentrica*
															R											R	R	R																	*leptopus*
	R	R	R																								R																		*lineata*
																R							R																						*jacksonii*
R	R	R	R	R	R	T	T																F	R																					"oestrupii"
							R	T																																					"nordenskioeldii"
R	R	R	R	R	R	T																	R																						*plicata*
	R			R				R																R																					*symmetrica*
		R		R								F	R	R	R	R			R																										species
				R	R																				R																				*Triceratium alternans*
			R		R											R			R																										*cinnamomeum*
			R	R																																									*Mesocena circulus*
F	A	C	C	F	F	F	A	P	R	R	C	A	A	A	M	A	A	A	A	A	A	A	A	A																					Abundance of diatoms
M	G	G	G	M	G	M	M	F	P	G	M	G	G	G	M	G	G	G	G	G	G	G	M	M																					Preservation of diatoms

DRAG 1
DRAG 3
DRAG 5
DRAG 7
DRAG 10
DRAG 12
DRAG 14
7706-23-4A
4B
25-1
25-3
46-1
46-2
46-3
46-4
46-16
46-23
46-28
46-50
46-92
46-104
48-1
54-5

Species (columns 46–87):

- Actinoptychus splendens
- Actinoptychus undulatus
- Actinocyclus curvatulus
- Actinocyclus divisus
- Actinocyclus ehrenbergii
- Actinocyclus ellipticus
- Asterolampra marylandica
- Asteromphalus arachne
- Asteromphalus spp.
- Biddulphia "aurita" (incl. cornigerum)
- "Cladogramma dubium"
- Chaetoceros spores
- Corethron spp.
- Coscinodiscus africanus
- Coscinodiscus endoi
- Coscinodiscus "lineatus"
- Coscinodiscus lineatus forma elliptica
- Coscinodiscus marginatus
- Coscinodiscus nodulifer
- Coscinodiscus radiatus
- Coscinodiscus "symbolophorus"
- Coscinodiscus tabularis var. egregius
- Coscinodiscus temperei
- Coscinodiscus yabei
- Coscinodiscus (plicatus)
- Cussia tatsunokuchiensis
- Cussia praepaleacea/lancettula
- Cyclotella "striata"
- Cymatosira biharensis
- Delphineis species (coarse)
- Delphineis species (fine)
- Delphineis species (rostrate)
- Denticula hustedtii
- Denticula punctata
- Ditylum brightwellii
- Endictya oceanica
- Eucampia zoodiacus
- Goniothecium "odontella"
- Hemidiscus cuneiformis
- Lithodesmium undulatum
- Mediaria splendida forma tenera
- Mesocena quadrangula

Samples:

DRAG 1, DRAG 3, DRAG 5, DRAG 7, DRAG 10, DRAG 12, DRAG 14, 7706-23-4A, 4B, 25-1, 25-3, 46-1, 46-2, 46-3, 46-4, 46-16, 46-23, 46-28, 46-50, 46-92, 46-104, 48-1, 54-5

Note: Preservation indicated by G = good, M = moderate, P = poor. Abundance indicated by A = abundant, F = few, C = common, R = rare, T = trace.

TABLE 6. FORAMINIFERAL BIOSTRATIGRAPHY AND PALEOBATHYMETRY OF DRAG (DR) DREDGE SAMPLES

FORAMINIFERA A = abundant C = common X = frequent to rare		LITHIFICATION SAMPLES	calcareous siltstone, micrite, dolomicrite		
			DRAG-3,5,11-16	DRAG-2,12	DRAG-1,4,6,7,8,9,10,13,14,15
BENTHIC	Bolivina costata d'Orbigny		X X X X	X ?	
	B. interjuncta Cushman		X		
	B. seminuda humilis Cushman and McCulloch		A A A A	A A	? ? ? C X ? C ? ?
	Bulimina affinis d'Orb. - B. ovula d'Orb.		X	X	
	B. sp. (Resig, Peru Margin)		X X		
	Cancris carmenensis Natland		X		foram traces
	C. inflatus (d'Orbigny)		X X X X		
	Cassidulina cushmani RE & KC Stewart		X X	X	
	C. pulchella d'Orbigny		X C C C	X X	
	Epistominella subperuviana (Cushman)		C X C C	X X	?
	Frondicularia sp.		X		
	Gyroidina rothwelli Natland		X		
	Nonionella auris (d'Orbigny)		X X X X	X X	?
	Siphonodosaria sp.		X		
	Suggrunda sp.		X X X X		
PLANKTONIC	Globigerina bulloides d'Orbigny		X X X X		
	Globigerinita uvula (Ehrenberg)		X X		
	Globigerinoides ruber (d'Orbigny)			X	
	Globorotalia cultrata d'Orbigny			X	
	G. inflata (d'Orbigny)		X X		
	Globorotaloides hexagona (Natland)		X		
	Neogloboquadrina dutertrei (d'Orbigny)		X X X X		
	N. pachyderma (Ehrenberg)		X		
	Orbulina universa d'Orbigny		X		
	Turborotalita quinqueloba (Natland)		C X X C		
BATHYMETRY	PRESENT WATER DEPTH (m)		837 - 1201 (U. Mid. Bathyal)		
	PALEODEPTH (m)		100 - 300 (Outermost Shelf / U. Upper Bathyal)		
STRATIGRAPHY	AGE - FORAMS		Post - Miocene		
	- DIATOMS		Pleistocene		

admixture with assemblages from downslope occurs in these rocks, precluding their deposition in deeper water. Because the dredged rocks were recovered at water depths from 837 to 1,201 m, and because they contain benthic faunas typical of 100 to 300 m water depths, we infer that the region has subsided at least 500 m during the Pleistocene.

Species of the *Bolivina* assemblage of DRAG rocks occur today under oxygen minimum conditions along the Peruvian margin. Floods of diatom frustules, as well as frequent fish scales in these rocks, indicate high productivity of the surface waters. The planktonic foraminifera, dominated by *Turborotalita quinqueloba* with *Neogloboquadrina dutertrei* and *Globigerina bulloides* are indicative of the temperate Peru Current regime.

Station 46

On the seaward side of the Lima Basin, late Miocene to Pliocene rocks were dredged from 1,639 to 2,263 m on the continental slope (Fig. 4). Two lithologies were recovered, each with discrete foraminiferal species (Table 7). Neither of the assemblages are similar in species content to modern faunas of the Peruvian margin.

TABLE 7. FORAMINIFERAL BIOSTRATIGRAPHY AND PALEOBATHYMETRY OF DREDGE 46 SAMPLES

FORAMINIFERA A = abundant C = common X = frequent to rare	LITHOLOGY / SAMPLES	Micrite 7706-46-16	dolomicrites (with foram. layers), breccia, dolomicrite 7706-46-1, -4, -6, -23, -55, -82, -104, -105
Bolivina salinasensis Kleinpell		A	
B. sp.		C	
Bulimina delreyensis Cush. and Gallaher		C	
Buliminella subfusiformis Cushman		A	
Florilus pizarrense (Berry)		X	
Fursenkoina californiensis (Cush.) var. grandis (Cush. and Kleinpell)		C	
F. sp.		X	
Globigerina bulloides d'Orbigny		X	
Gyroidina sp.		X	
Valvulineria californica Cush. var. obesa Cush.		X	
B. spp. indet.			X X X X C C A (forams unidentifiable)
Bulimina sp. (B. delreyensis, smooth vax?)			X X X X C C A
Buliminella sp.			X
Cassidulina sp.			X
Ellipsoglandulina fragilis Bramlette			X X X X X X A
Valvulineria sp. (1.2 mm)			X X X X X
BATHYMETRY	PRESENT WATER DEPTH (m)		1639 – 2263
	PALEODEPTH (m)	50–500	150 – 500
STRATIGRAPHY	AGE – FORAMS	Mid to Late Mio.	late Mio. (or early Plio.?)
	– DIATOMS		Late Mio. to Plio.

A diatomaceous, foraminiferal siltstone (46-16) contains a predominantly benthic fauna (Fig. 10) in which the species *Bolivina salinasensis*, *Bulimina delreyensis*, *Fursenkoina californiensis grandis*, and *Valvulineria californica obesa* indicate a mid to late Miocene age, equivalent to part of the provincial Mohnian stage of California (~6-13 m.y. B.P., Berggren and Van Couvering, 1974). The *Bulimina delreyensis* suite of specimens displays variation in the strength of the costae. A few specimens compare well with the holotype while many specimens have only a few discontinuous costae or lack costae entirely. Specimens of *B. delreyensis* from the Lower Mohnian Santa Ana Formation of California (Smith, 1960) have costae intermediate in development between the holotype and the majority of the Station 46 specimens (R. Cifelli, personal communication); therefore, it is likely that the forms with reduced costae are phenotypes. As before, the predominance of *Bolivina* spp. in the assemblage suggests that oxygen minimum conditions existed on the seafloor and the only planktonic species, *Globigerina bulloides*, probably inhabited the temperate waters of the Peru Current.

Some of the species present are considered to be indicative of upper bathyal (150-500 m) facies in the California Miocene (Ingle, 1973): *Buliminella subfusiformis*, *Fursenkoina californiensis grandis*, and *Valvulineria californica obesa*. On the other hand, the genus *Florilus*, which is fairly well represented in the assemblage, is indicative of the continental shelf. In fact, the type of *F. pizarrense* was taken from a water depth of 16 m off northern Peru (Berry, 1928). Thus, the siltstone was probably deposited at outer shelf to upper bathyal depths, or at upper bathyal depths with influx of shelf specimens. If a depth of 500 m is taken as the maximum water depth represented by the assemblage, the siltstone (46-16) has subsided at least 1,140 m since deposition.

Most of the rocks of the dredge were calcareous siltstone, micrite, and dolomicrite. Benthic foraminifera were abundant, sometimes aligned in layers according to size (Figs. 8B, 11). Specimens

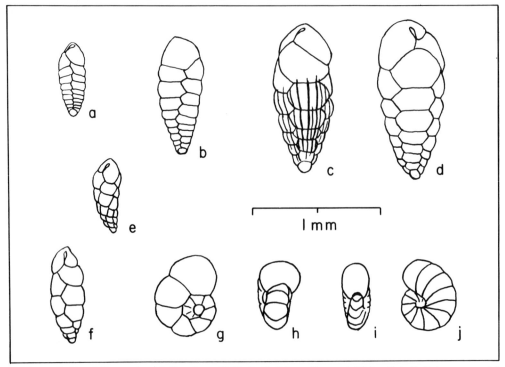

Figure 10. Late Miocene foraminifera from 46-16. A, *Bolivina salinasensis* Kleinpell; b, *Bolivina* sp; c and d, *Bulimina delreyensis* Cushman and Gallaher; e, *Buliminella subfusiformis* Cushman; f, *Fursenkoina californiensis grandis* Cushman and Kleinpell; g and h, *Valvulineria californica obesa* Cushman; i and j, *Florilus pizarrense* Berry. Line drawings from photography, all at the same magnification.

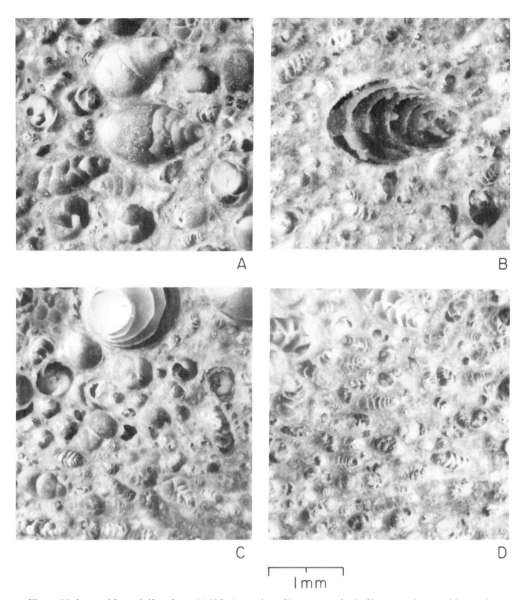

Figure 11. Layered foraminifera from 46-105. A, portion of layer comprised of large specimens with prominent *Ellipsoglandulina fragilis* Bramlette; b, section of *E. fragilis;* c, boundary between large specimens and concentration of small specimens, principally *Bolivina*; d, *Bolivina* layer. Note dense packing of foraminifera and general axial alignment of tests.

were identified in surface relief and in section. The most conspicuous benthic foraminifer of these rocks is the large ovate form, *Ellipsoglandulina fragilis* (Figs. 11A,B). *E. fragilis* was described by Bramlette (Woodring and Bramlette, 1950) from the upper member of the Monterey Formation in the Santa Monica Mountains of California, where it was found in undeformed state only in calcareous concretions. It has been reported in diatomaceous mudstone with *Bolivina* aff. *seminuda forminata*, a species that is part of the modern lower upper bathyal biofacies on the Peru Margin (Resig, this volume). The large *Bulimina* that occurs in the micrites appears to be identical with smooth

B. delreyensis from sample 46-16. The size and configuration of some of the abundant specimens of *Bolivina* in the micrites suggest a species similar to, if not identical, with *B. seminuda humilis*.

Pliocene deposits of bathyal origin exposed in Panama (Coryell and Mossman, 1942) and on an island in the Gulf of California (Natland, 1950) contain species presently living on the Peruvian margin, suggesting that the transition to modern benthic faunas occurred in the early Pliocene. The benthic foraminifera of the Peruvian micrites and dolomicrites are most closely affined to Miocene assemblages and are probably late Miocene in age (corresponding to an Upper Mohnian–Lower Delmontian age of the Upper Monterey Formation) or possibly early Pliocene, before the advent of the modern species assemblages on the Peru Margin.

An upper bathyal origin of the deposit is suggested by the abundance of *Bolivina* similar to *B. seminuda humilis* and by faunal associations of *E. fragilis* in California. The high concentration of tests in layers, according to size and axial alignment, indicates that postmortem redistribution of the assemblage in these layers has occurred. However, more homogeneous-appearing limestone adjoining the layers contains lesser numbers of the same forms, which may have accumulated under normal conditions. This suggests that the tests in the foraminiferal layers were not displaced out of their biofacies. Therefore, subsidence of about 1,100 m is indicated for these rocks.

Stations 54 and 59

The glauconitic micrite, dolomicrite, and brecciated dolomicrite of Stations 54 and 59 are extremely hard, and the foraminifera could not be dislodged from their matrices. Therefore, identifications were made from section and from a few surface reliefs. Most of the components could not be identified.

Samples from the two stations contain different foraminiferal assemblages, although both have in common a preponderance of species of *Bolivina*, and both contain large species of *Bulimina* (Table 8). The micrites of Station 54 are no older than Miocene if identification of *Globigerina bulloides* is correct. Weak evidence indicates a post-Miocene age for those rocks: *Gyroidina multilocula*, a form described from Pliocene rocks of Panama (Coryell and Mossman, 1942) was identified, in relief. A loosely coiled species of *Angulogerina*, similar to *A. albatrossi* Cushman, described from off the west coast of Mexico, is a conspicuous faunal component, along with large specimens of *Globobulimina*. These faunal elements suggest an upper bathyal assemblage.

The carbonate from Station 59 contains large specimens of *Valvulineria* that might be related to *V. californica* or to forms of similar morphology, such as those in the Eocene Chira shale of Peru (Cushman and Stone, 1947). Ingle (1973) considers the *V. californica* group to be indicative of upper bathyal biofacies. The rocks of dredge Station 59 have borings filled with sediment containing Recent foraminifera that occur at the present water depth of the deposit. This assemblage is also incorporated in a phosphorite (?) veneer that was formed on the exposed rocks. None of these Recent species were found in the carbonates.

It appears from the limited amount of information derived from the carbonates of both stations that these carbonate rocks were not moved appreciably (uplifted or subsided) from their environment of origin.

DISCUSSION

Upper Slope Basin Stratigraphic Sections

Lima Basin. The youngest carbonates, early to middle Pleistocene, are located at 12°S in the upper part of the stratigraphic section of the Lima Basin (Figs. 1, 4). Dredge DRAG traversed about 300 to 400 m of section exposed in a submarine canyon (Fig. 7[BB']). This dredge is projected onto the eastern end of a multichannel seismic record (CDP-1) made across the Lima Basin north of the canyon (Figs. 4, 12B). We can only assume that the carbonates were collected from the entire stratigraphic section exposed in the canyon wall and are representative of the section (Fig. 12A).

TABLE 8. FORAMINIFERAL BIOSTRATIGRAPHY AND PALEOBATHYMETRY
OF DREDGES 54 AND 59 SAMPLES

FORAMINIFERA A = abundant C = common X = frequent to rare		LITHOLOGY SAMPLES	dolomicrite, breccia. dolom. 7706-54-5,-7,-10,-11	dolo- micrite 7706-59-6,-9,-12,-13	authigenic phosphorite veneer 7706-59-3,-6,-12,-13
LIMESTONE SPECIES	Angulogerina sp. (loose series, 0.9mm)		X X X		
	Bolivina spp. indet.		X A X X		
	Bulimina sp. (1.0mm)		X X		
	Cancris sp.		X X		
	Globigerina bulloides? d'Orb.		X X		
	Globobulimina sp. (0.9-1.5mm)		X X X X		
	Gyroidina multilocula Coryell and Mossman		X X		
	rotaliform spp.		X X X X		
	Bolivina spp. indet.			X X X X	
	Bulimina aff. affinis d'Orb.			X X X	
	Bulimina sp. (1.0mm)			X X X X	
	Valvulineria sp. (0.9mm)			X X X X	
PHOSPHORITE SPP.	Angulogerina carinata Cush.				X X X
	Bolivina plicata d'Orb.				X X
	B. sp. (0.60-0.75mm)				X X X X
	Cassidulina auka Boltov. and Theyer				X X C
	Epistominella subperuviana (Cush.)				X A A
	Globigerinoides trilobus (Reuss)				X
	Neogloboquadrina dutertrei (d'Orb.)				X
	Uvigerina striata d'Orb.				C X X
BATHYMETRY	PRESENT WATER DEPTH (m)		←— 202-250 —→		←— 430 —→
	PALEODEPTH (m)		←— 150-500 —→		←— 430 —→
STRATIGRAPHY	AGE - FORAMS		←— Post-Mio.? —→	Mio or older	←— Recent —→
	- DIATOMS		Plio.?		

From other studies in the area (Krissek and others, 1980; Rosato and Kulm, this volume), it is evident that the youngest carbonates are overlain by late Pleistocene to Holocene terrigenous sediments. With the exception of the recent formation of phosphorite deposits (Veeh and others, 1973) and concentrations of winnowed foraminiferal tests along the outer edge of the shelf, all late Pleistocene and Holocene sediment on the Peruvian margin is terrigenous and derived from the adjacent Andean continent. From high-resolution 3.5 kHz records, air gun records, and velocity analysis of CDP-1, we estimate that from 50 to 200 m of terrigenous muds form a blanket over the Lima Basin (Figs. 12A,B).

The oldest carbonates at 12°S were apparently collected from the middle and upper part of the lower portion of the stratigraphic section in the Lima Basin (Fig. 12B). Late Miocene to Pliocene lithologies from dredge 46 were recovered from the scarp of a large structural bench on the seaward flank of the Lima Basin (Figs. 4, 7 [BB']). Truncated reflectors inferred from the single-channel record are confirmed in the multichannel record across the bench (Fig. 12B). If superposition is valid in this basin, as the parallel undisturbed reflectors indicate, then the late Miocene carbonates probably

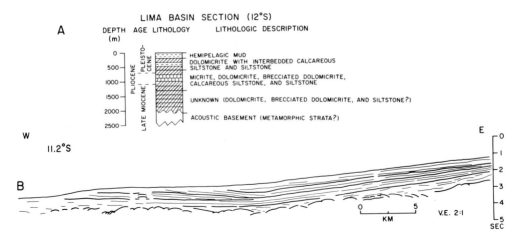

Figure 12. A.—Proposed stratigraphic section for Lima Basin at 12°S. B.—Line drawing of 24-fold seismic reflection record CDP-1. See Figure 4 for location.

represent a lower part and the Pliocene carbonates an intermediate part of the stratigraphic section. From lateral correlation of reflectors, however, it does not appear that we sampled the lowest and oldest part of the section. From this type of uncontrolled dredge data, we cannot calculate the thickness of the sediment record for a given period of geologic time. However, the total Lima Basin section is calculated to be 2 km thick at the center in the depth section of profile CDP-1 (Thornburg and Kulm, this volume). We probably sampled 1.0–1.5 km of the uppermost stratigraphy in the Lima Basin (Fig. 12).

The multifold record shows that the center of the Lima Basin is relatively undisturbed except where the acoustic definition of the structural elements along the seaward flank of the basin deteriorates into a diffracting zone (Fig. 12B). While the brecciated dolomicrites occur in the late Miocene to middle Pleistocene strata at 12°S, they represent less than 5% of the volume of the material dredged at sites DRAG and 46, which is consistent with the relatively undisturbed structure of the basin. It should be pointed out that the acoustic basement of the Lima Basin could consist of metamorphic strata similar to that sampled on Hormigas Island, since the basement rocks have velocities of 5.7–6.0 km/sec (Hussong and others, 1976).

Trujillo Basin. The industry drill holes on the outer edge of the continental shelf at 9°S latitude contain the only other Cenozoic stratigraphic sequence available on the central Peru Margin. These drill holes penetrated sediments in the upper-slope Trujillo Basin, a northern correlative of the Lima Basin. While sketchy lithologic information is published on these holes (Travis and others, 1976), Thornburg and Kulm (this volume) have reconstructed, using various gelogical and geophysical data, the stratigraphic sequence believed to be present at 9°S (Fig. 13). The seismic reflection records show that truncated strata overlie the Paleozoic Outer Shelf High (Fig. 7, CC′), suggesting a lower Miocene period of uplift similar to that in coastal sections. Miocene and younger strata are quite thin in the vicinity of the drill holes and apparently unconformably overlie older strata. We postulate that these younger Neogene units, which contain the carbonates, thicken to several hundred metres in the vicinity of dredge Stations 54 and 59 (Fig. 7, CC′ and DD′; Fig. 13).

Carbonate Depositional Environment

Several different lines of evidence elucidate the depositional environment in which the late Cenozoic carbonates were deposited. Our most reliable data come from the fauna and flora included within the carbonates. The diatom and foraminiferal assemblages both show conclusively that the fine-grained carbonate material and silts and clays were deposited in open marine conditions. This does *not* imply that the carbonate materials were originally formed under these conditions, but rather, that they were ultimately deposited under these conditions.

The most complete paleoenvironmental data set is obtained from dredges DRAG and 46. Marine waters, similar to those occurring in the upwelling regime off Peru today, are documented by the similarity of the diatom assemblages contained in the carbonate rocks to those assemblages found in the Holocene and late Pleistocene sediments beneath the upwelling zone off Peru (see Ecology in Biostratigraphy). The planktonic foraminifera also are indicative of the temperate Peru Current regime off Peru. Furthermore, there are no freshwater or brackish water diatoms or shallow water benthic foraminifera in the carbonates to suggest that the depositional site was ever at or above sea level. The water depth at the depositional site ranged from 50 to 500 m, according to the benthic foraminifera assemblages, and was at least 150 m deep according to the absence of benthic diatom assemblages. Benthic foraminiferal species of the *Bolivina* assemblage in the carbonates are identical to the assemblages found in the modern deposits underlying the oxygen minimum zone (Resig, this volume). Modern terrigenous deposition in the oxygen minimum zone occurs in water depths between 140 and 640 m in a mud lens on the Peruvian continental slope (Krissek and others, 1980), which is the site of high organic carbon accumulation (Müller and Suess, 1979). All data indicate that a similar marine environment existed during the deposition of the early to middle Pleistocene calcareous siltstones.

Less data are available for the remaining dredge sites 54, 59, 75, and 80 to the north, but floral and faunal assemblages indicate that the carbonates were deposited in open marine conditions and at water depths typical of the continental shelf and upper continental slope environments.

Several factors suggest that the carbonates are abundant: (1) the relatively large numbers (seven) of pure carbonate outcrops sampled; (2) the wide geographic distribution of carbonates more than 950

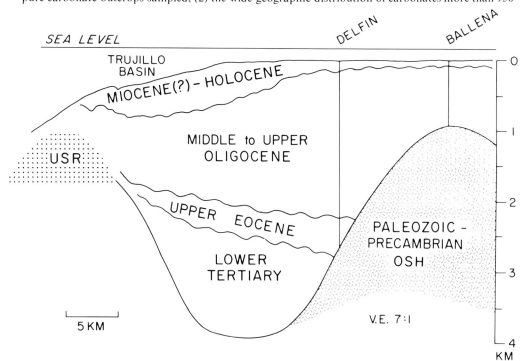

Figure 13. The general stratigraphy (ages of depositional sequences) of the Trujillo Basin near 9° S latitude. Wavy lines denote unconformities or hiatuses. Structural basement ridges form the basin's landward and seaward boundaries: the Outer Shelf High (OSH) and the Upper Slope Ridge (USR), respectively. Drill holes Delfin and Ballena are plotted. Note thickening of Neogene strata to several hundred metres in the western part of the basin. From Thornburg and Kulm, this volume.

km off the Peru slope; (3) the wide range in water depth (205 m to 3,749 m); and (4) the 10 m.y. age span (late Miocene to middle Pleistocene) for all carbonates collected. However, terrigenous deposits, including calcareous siltstone, are probably interbedded with these carbonates; it is more logical to postulate that they represent the majority rather than the minority of lithologies of late Cenozoic age. A word of caution should be noted here. It is possible that the dredge could be sampling the most resistant (carbonate) rock outcrops, thus giving a false indication of the actual volume of carbonate.

It is obvious from our reconstruction of the stratigraphic column of the Lima Basin (Figure 12A) and the previous work on the late Pleistocene and Holocene deposits off Peru that there was a dramatic temporal change in the type of sediments accumulating on the Peruvian continental slope. Fine-grained carbonates appear to dominate the section from late Miocene to middle Pleistocene time, while hemipelagic muds dominate in late Pleistocene to Holocene time. We note that the youngest calcareous siltstone and dolomicrite are dated at 0.93–0.09 m.y. by diatom datums (Table 2), which are related to the paleomagnetic time scale (Fig. 9). Radiolarian datums, tied to the same scale, were established in the terrigenous sediments in long piston cores collected on the middle and lower continental slope off Peru (Rosato and Kulm, this volume). They show that terrigenous sedimentation has dominated the Peruvian margin since at least 0.44 m.y., which places the change from one sediment type to another between 0.44 and 0.93 m.y. and during the most recent series of sea level lowerings in the late Pleistocene.

Several lines of evidence suggest that the supply of terrigenous sedimentation available to the Peru Margin has apparently varied during the past 15 m.y. During the Paleogene, the Amazon River flowed into the Pacific Ocean through an extension of the Gulf of Guayaquil (Jenks, 1956; Hosmer, 1959; Gibbs, 1967), presumably carrying a large volume of terrigenous sediment from the Andes to the continental margin. With the uplift of the Andes in middle Miocene time and subsequent volcanism (Noble, 1974), the Amazon River shifted its flow to the Atlantic Ocean (Jenks, 1956). Carbonate reefs on the Brazilian continental margin were drowned by terrigenous sediments during the middle Miocene (15–20 m.y. ago), indicating the approximate time when the Amazon drainage shifted from the Pacific to the Atlantic (Eric Bosc, pers. commun., 1979). This event must have drastically reduced the amount of sediment supplied to the Peruvian margin until the new drainage system became established along the western flank of the Andes. We propose that during middle and late Pleistocene time, the Andean drainages began to deliver more sediment to the upper slope. This process was enhanced by the increased erosion, downcutting of streams, and shelf bypassing that are typical of eustatic sea level lowerings. Krissek and others (1980) have traced the mineralogy of the modern hemipelagic deposits on the continental slope to their fluvial sources in the various coastal drainage basins of Peru. The slope deposits generally reflect the mineralogy of the adjacent drainage basin on the continent. While many of the Peruvian coastal rivers have an extremely low discharge (Zuta and Guillen, 1970) because of the rather arid condition along the coastal region, enough silt and clay are being discharged to produce a mud lens on the upper slope that overlies the carbonates of the Lima Basin.

Origin of Upper Slope Carbonates

Several distinct and largely exclusive processes might explain the origin of these dense, fine-grained, highly dolomitic carbonate rocks that were recovered predominantly from Peru's upper slope (between 202 and 3,749 m water depth) between 3.5 and 12°S latitude. One common process is that carbonate dust has been wafted in suspension, or carried in more concentrated turbidity currents, away from a reef complex (Scholle, 1977). In the past, reef sources could have been situated either upslope near the shelf break or perched on the structural high that defines the seaward limit of Peru's upper-slope basins. With the exception of microfossils (diatoms, radiolarians) and thin foraminiferal laminae, the complete absence of any recognizable reef debris in megascopic or microscopic examination, or even dissolution ghosts of fossil shell fragments in thin section, makes this explanation undesirable. Pelagic carbonate nannofossil ooze, which has been accreted and uplifted to upper-slope depths on the continental margin, seems an unlikely source of carbonate, since the upper-slope basins are floored entirely by continental massif (Thornburg and Kulm, this volume) and the benthic foraminifera indicate a depositional environment in upper bathyal water depths. Evaporitic dolomites are also excluded on the basis of the upper bathyal forams that are found in these rocks. The environment of formation of the Peru carbonates, then, seems to elude the more classical interpretations. Isotopic analyses were performed on several of the specimens to better understand the origin of these unique rocks.

Carbon and Oxygen Isotopic Analysis: Organic Carbonate Source. Stable isotope C^{13} and O^{18} values (relative to PDB standard) were determined for eight of our dredge samples that are representative of the geographical ranges and lithological variations encountered. Samples were also chosen for their consistency and redundancy in complementary microfossil and radioisotopic age determinations. Our values, plotted on Figure 14, lie well outside the fields of shallow-water skeletal carbonate, deep-water limestones, and evaporitic dolomites given by Milliman (1974), which supports our previous arguments against these modes of origin. The δO^{18} values are well constrained between $+5.00$ and $+7.08^0/_{00}$, yet the δC^{13} values show extreme variability and range from $+19.63$ to $-13.47^0/_{00}$. The C^{13} variability seems independent of age, sample location, weight percent dolomite or carbonate, and, for the most part, lithology. However, the wide range of δC^{13} values exhibits some correlation with degree of induration: soft, black, calcareous siltstone (46-2) had the highest value ($+19.63$), and paired cement and matrix samples from a highly indurated, brecciated dolomicrite (54-5-M, 54-5-C) had the lowest values (-12.86, -13.47, respectively). Compressional velocity information on several of the intermedite, massive dolomicrite lithologies (Table 4) confirms this trend of increasing degree of induration with decreasing δC^{13} values, that is, the compressional velocity of DR-1 $<$ 46-28 $<$ 59-17.

Our stable isotope values plot in a pattern that is strikingly similar to the field of C^{13}/O^{18} determinations which Deuser (1970) obtained from magnesium-poor dolomites (analogous to the Mg/Ca ratios of our own Peru dolomites) that were similarly dredged from upper-slope water depths (57 to 1,000 m) off northeastern United States. His δO^{18} values are closely grouped at $+5.8 \pm 0.7^0/_{00}$, but the δC^{13} values span the exhaustive range from -64 to $+21$ $^0/_{00}$. Deuser (1970) postulated an organic origin for the carbonate in these dolomites (Fig. 14). Specifically, chemically or microbiologically oxidized methane (a process that is not well understood) was proposed to account for the samples with excessively light C^{13} values, as well as the extreme scatter of the stable carbon isotope ratios. Oxidation of the organic matter, and subsequent carbonate formation, presumably occurred diagenetically at some depth below the sediment-water interface, and at some time after deposition.

Detrital organic matter may initially contain δC^{13} values as light as -25 $^0/_{00}$ (Degens, 1969). Preferential incorporation of light carbon during methane formation can further push the C^{13} depletion to extreme negative ratios (Erwin Suess, pers. commun., 1980). Such methane, when oxidized to CO_2 and precipitated as carbonate, presumably accounts for Deuser's anomalously light δC^{13} values. If a finite organic reservoir in a relatively closed system provides the carbonate source (that is, carbonate formation follows burial and some amount of isolation from the water column), then decomposition of the organic matter will initially produce strongly depleted carbonate (via methane and CO_2), but leave a heavier organic residue. As the organic reservoir is progressively exhausted, the residual fraction becomes progressively enriched in C^{13}, and the CO_2 and carbonate, which form from the heavier organic residue, will themselves be heavier. The carbonates that utilize the organically produced CO_2, then, reflect the fractionation process in their extreme range of δC^{13} values (Deuser, 1970).

We conclude that oxidized organic carbon provided the carbonate source for the ubiquitous upper-slope Peru dolomites. They formed by direct precipitation as pore space cement, by diagenetic alteration of existing terrigenous sediment, or by some combination of these processes. The rocks of lightest carbon value are the most indurated, because these rocks presumably formed during the initial stages of diagenesis, when the carbonate-producing organic reservoir (that is, buried sediments rich in organic carbon) was young and abundant, and cementation was more thorough and complete. The organic-rich (reaching $>$20 weight percent organic carbon) upper-slope mud lens sediments that

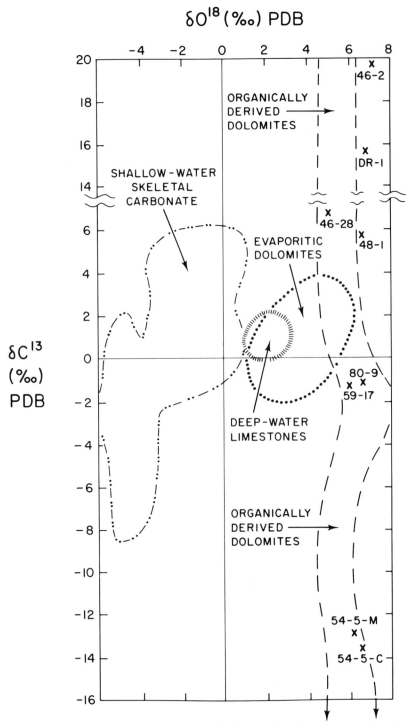

Figure 14. Distribution of carbon and oxygen stable isotope values in various types of marine carbonates. Values of the Peru upper slope carbonates of this study are plotted as "X" and are identified by sample number. Modified from Figures 19 (p. 33) and 93 (p. 307) of Milliman (1974) and from Deuser (1970).

underlie the oxygen minimum zone of the Peru Margin (Krissek and others, 1980) could provide an extensive carbonate source for the dolomites. Microecological studies verify that upwelling conditions characterized the initial sedimentary environment (see Ecology in Biostratigraphy). We believe that the oxygen is not biogenic, since the δO^{18} values bear no relationship to the δC^{13} values is likely hydrogenous in origin.

Since the lowest δC^{13} values are not excessive, it is possible that the organic carbon could be oxidized directly without the involvement of methane. In any case, some sort of pronounced fractionation process must be involved to account for the scatter and excessively heavy δC^{13} values that were encountered, since dilution by deep-water, shallow-water, or evaporitic carbonates cannot produce values heavier than +6 $^0/_{00}$ (Fig. 14). We are not aware of any studies in the marine environment that conclusively indicate a fractionation of C^{13} during direct oxidation of organic carbon; at present, we cannot support or disprove this alternative.

An organic origin has been proposed for authigenic carbonates (precipitative or diagenetic) in various other marine environments. Methane from buried marsh deposits was the presumed carbonate source for aragonite, calcite, and high-magnesium calcite cements in Quaternary littoral sandstones from the outer continental shelf off northeastern United States (Hathaway and Degens, 1969; Allen and others, 1969) and from barrier bars off the Mississippi delta (Roberts and Whelan, 1975); excessively light δC^{13} values provided the primary evidence for methane oxidation. Suess (1979) proposed that microbial decomposition of sedimentary organic matter in an anoxic basin of the Baltic Sea (the Landsort Deep) allowed authigenic Mn-carbonate to precipitate. Since his δC^{13} values of -12 to -14 $^0/_{00}$ fall well within the range of organic detritus, direct oxidation of organic matter apparently proceeded without methane formation. Unlike Deuser's (1970) excessively light δC^{13} values (to 64 $^0/_{00}$), Suess' values compare well with our lowest C^{13} ratios; however, only Deuser's research has documented a *scatter* or *variability* of δC^{13} carbonate values that is similar to our own.

Recent DSDP drilling on the continental margin off northern Honshu, Japan (legs 56 and 57), frequently recovered layers, nodules, and diagenetic cements of various carbonate mineralogies in Neogene sedimentary sections (Arthur and others, 1980). Anoxic conditions and the relatively high organic content of the sediments suggest that the carbonates may be, in part, organically derived. We suspect that stable oxygen and carbon isotope studies would be helpful in understanding the genesis of these carbonates.

Sediment Supply Versus Dolomitization. Because dolomitization of the Peru rocks is a pervasive precipitative/diagenetic pore space process that dominates the lithology of the terrigenous precursor sediment, starved margin conditions are not specifically necessary for the formation of these carbonates. In particular, in the offshore wells near 9°S (Figs. 5, 13), a presumed Oligocene terrigenous sequence of silty clay and graded sands contains occasional thin dolomicrite interbeds (Thornburg and Kulm, this volume). We cannot be certain, though, that these Paleogene carbonates formed under circumstances similar to the dredged Neogene carbonates. However, we do know that a reduction of terrigenous input occurred in middle Miocene time when the Amazon River drainage shifted from the Pacific to the Atlantic Ocean, and that dolomitic strata are apparently prolific following this event (see Carbonate Depositional Environment). Perhaps, for climatological or oceanographic reasons, the organic matter–generating upwelling conditions were not fully initiated along the continental margin until after the dramatic change in terrigenous drainage.

Source of Magnesium. The source of magnesium in the Peru carbonates remains speculative. Milliman (1974) suggests that Mg may be diagenetically liberated to interstitial waters from tests of marine organisms containing Mg-calcite. The Peru rocks often contain abundant benthic foraminifera, which are known to consist of Mg-calcite (Chave, 1962), as well as dissolution ghosts of these fossils.

Deuser (1970), whose carbonates produced stable isotope values similar to our own (see Carbon and Oxygen Isotopic Analysis), postulated that diagenetic alteration of terrigenous sediment to form dolomite occurred in an environment distinct from the original depositional setting. During Pleistocene sea level lowerings, specifically, Mg was concentrated in shallow, saline lagoons behind offshore bars located near the shelf edge.

Similarly, several eustatic low-stands could have been involved in the diagenesis of the Peru carbonates. Late Miocene and late Pliocene lowerings (6.6, 3.8, and 2.8 m.y.; Vail and others, 1977) may have been important during the Neogene history of these rocks. The Upper-Slope Ridge (USR), which forms the seaward structural boundary of the upper-slope Lima and Trujillo Basins (Fig. 13; Thornburg and Kulm, this volume), could have served as a circulation-restricting sill. Although its present morphology is expressed as a prominent bathymetric gradient increase on the continental slope, the USR was likely a more positive feature in the past, before sediment ponding filled up to the relief of its landward flank. Although seismic reflection interpretations and benthic microfossils of the Lima Basin indicate that the USR has subsided relative to the more stable structural ridge which bounds the basin landward (Thornburg and Kulm, this volume), pulses of uplift and compression, particularly a post-Pliocene episode, have deformed strata in the vicinity of the USR in both the Lima and Trujillo Basins and may have reactivated the structure as a positive element. If eustatic sea level lowerings occurred when the USR was of sufficient relief, restricted marginal seas may have characterized the upper-slope basins. The resulting increased salinity and water temperature (as reflected in the high δO^{18} values of the Peru carbonates) may have been conducive to dolomite formation. We emphasize that while the microfossils included in the Peru carbonates are characteristic of 150 to 500 m water depths and open marine conditions, the postdepositional conditions of diagenetic dolomitization and any tectonic movements different from those inferred by the carbonates present position on the sea floor would not be recorded in the microfossil assemblages.

The timing of these uplift and subsidence events may be crucial to the dolomitization process. For example, we have shown that both the landward and seaward flanks of the Lima Basin were formerly at outer-shelf to upper-slope water depths. A relatively small amount of postdepositional uplift (about 100 to 200 m) could have created the restricted marginal basins conducive to dolomitization. The USR beneath the landward flank of the Lima Basin (Fig. 12) did not begin its subsidence until at least early to middle Pliocene time (5.0 to 2.6 m.y.), which would have given ample time for the late Miocene and Pliocene sea level lowering to influence the dolomitization process. Dolomicrite continued to form on the landward flank of the Lima Basin until the late Pleistocene (0.93 to 0.98 m.y. ago), when rapid subsidence of the region commenced removing the sea floor from the influence of the most recent Pleistocene sea level lowerings. Dolomitization terminated between 0.93 to 0.44 m.y. ago, when the more than 50 m of terrigenous mud began to accumulate on the subsiding landward flank of the basin.

On the other hand, terrigenous muds are virtually absent on the upper slope in the vicinity of the Trujillo Basin (Fig. 13; Krissek and others, 1980), and the block-faulted dolomicrites have not experienced any appreciable subsidence since late Miocene time.

In summary, the source of magnesium may have been the abundant benthic foraminifera (containing Mg-calcite), originally deposited in the organic-rich muds, and/or the favorable shallow water conditions produced by postdepositional uplift and structural adjustments.

Tectonics of the Peruvian Margin

The carbonate rock data show that the central Peru Margin has been tectonically active during late Cenozoic time. This is consistent with the recent deformation associated with frequent and large-magnitude earthquakes occurring along the convergence boundary between the Nazca Plate and the South American continent (Stauder, 1975; Barazangi and Isacks, 1976).

The Neogene strata of the Trujillo Basin are quite reflective, but are folded, reverse faulted, and truncated at the sea floor, unlike the generally undisturbed strata of the Lima Basin (see Upper-Slope Basin Stratigraphic Sections, seismic records D-D', C-C', B-B', and CDP-1 of Figs. 7 and 12). The deformation seems indicative of uplift and compression. Perhaps the Trujillo deformational episode is reflected in the much larger percentage of brecciated dolomicrites dredged from sites 54 and 59 versus dredges 46 and DRAG from the Lima Basin (Table 2).

The most recent tectonic movements are documented by the benthic faunas in the carbonates in the Lima Basin. Subsidence of at least 500 m is indicated on the upper slope at dredge site DRAG (Fig. 4). The youngest carbonate deposits at the site are dated at 0.93–0.98 m.y., which means that most of the subsidence occurred during the late Pleistocene and probably is continuing today. Using an age of 1.0 m.y. for these lithologies, we calculate a minimum subsidence rate of 500 m/m.y. for the landward flank of the Lima Basin. At least 1,100 m of subsidence is recorded in the late Miocene to Pliocene carbonates recovered on the seaward flank of the Lima Basin at dredge site 46 (Fig. 4). The most recent subsidence is dated as Pliocene (2.6–5.0 m.y.), based upon the youngest lithologies with shallow-water fauna (Tables 2, 7). Using an age of 4.0 m.y. for these rocks, we calculate a minimum subsidence rate of 275 m/m.y. for the seaward flank of the basin. Paleobathymetric and age data from both dredge sites suggest that the entire Lima Basin at 12°S has undergone continued, large-scale subsidence since perhaps late Miocene or early Pliocene time. Hussong and others (1976) postulated subsidence of this part of the Peru Margin from geophysical data. This is the second occurrence of large-scale subsidence of a convergent margin, with the Japan transect being the first documented occurrence based upon IPOD drill hole data (von Huene and others, 1978). Just how extensive subsidence is along the Peru Margin is yet to be determined from future studies of the lithologies that comprise the outer margin.

CONCLUSIONS

A variety of sedimentary rocks were recovered from 11 dredge stations on the continental margin off northern Chile and central to northern Peru. Late Miocene siltstone, claystone, and sandstone characterize the steep slope off northern Chile in water depths ranging from 2,779 to 8,132 m. The virtual absence of modern sediment, the lack of sediment cover in seismic reflection records, and the occurrence of manganese crusts and slabs indicate that this is a region of nondeposition.

Farther north off central to northern Peru, late Miocene to middle Pleistocene (11.0 to 0.96 m.y.) fine-grained carbonates were dredged from rock outcrops on the outermost continental shelf and the middle to upper continental slope. These carbonates consist of calcareous siltstone, micrite, glauconitic micrite, dolomicrite, and brecciated dolomicrite. Phosphorite is present in the upper-slope lithologies. The bulk of the dredged lithologies are dolomicrite; the dolomite content of this lithology usually exceeds 80% of the rock by weight percent, with calcite and other minerals comprising the remainder of the rock.

A reconstruction of the stratigraphic section in the Lima Basin at 12°S latitude shows that the oldest deposits recovered in the basin are late Miocene to Pliocene dolomicrite with occasional brecciated dolomicrite. Early to middle Pleistocene dolomicrite and calcareous siltstone overlie these older deposits. Terrigenous sedimentation increases in this part of the section and becomes wholly terrigenous in late Pleistocene time (0.93 to 0.44 m.y. ago). This historical change from carbonate to terrigenous deposition is attributed largely to the amount of sediment input to the Peruvian margin from coastal drainages with time. The Amazon River drainage shifted from the Pacific to the Atlantic Ocean in middle Miocene time, markedly reducing the terrigenous input to the margin. With the advent of Pleistocene glaciation, the coastal drainages began to supply a larger volume of sediment.

The carbonates were deposited in normal marine conditions in water depths of 150 to 500 m (similar to the upwelling regime prevalent today), according to the microfossils contained within the rocks. Progressive subsidence of these shallow-water regions is documented by benthic foraminifera at 12°S in the Lima Basin, where carbonates dredged from water depths of 1,639 to 2,200 m and 837 to 1,201 m have subsided 1,100 and 500 m since Pliocene to late Pleistocene time, respectively.

Further north at 9°S and 8°S and at 3.5°S (Banco Peru), late Miocene to Pliocene and possibly Pleistocene dolomicrite and glauconitic micrite crop out on the outer shelf and uppermost slope. Deposits at 9°S and 8°S show no evidence of significant vertical movement, based upon limited microfossil data. However, the abundant amounts of brecciated dolomicrite in these rocks and the warped and block-faulted terrain in the vicinity of these rock outcrops indicate that large-scale deformation of the margin has occurred in this area.

Stable isotopic C^{13} and O^{18} values suggests that the Peru carbonates are organically derived. The δO^{18} values cover a narrow range from +5.00 to +7.08 $^0/_{00}$, while δC^{13} values range widely between +19.63 to −13.47 $^0/_{00}$. On the basis of these data and the paleoenvironmental conditions indicated by the microfossils, we postulate that the source of carbonate for the dolomites is the oxidation of the organic carbon present within the upwelling sediments. These carbonates apparently formed by the direct precipitation as pore space cement or by diagenetic alteration of the existing terrigenous sediments. The sources of magnesium remain speculative. Magnesium may have been diagenetically liberated to interstitial waters, as determined from tests of the abundant benthic foraminifera containing Mg-calcite. Alternatively, magnesium may have been derived from the marine waters of restricted marginal seas created by the reactivation (uplift) of structural features on the flanks of the basins following deposition and by the eustatic lowering of sea level during late Miocene and Pliocene time. Rapid subsidence of these positive features allowed terrigenous muds to accumulate, particularly in the Lima Basin, and may have terminated the dolomitization process in the late Pleistocene.

The ubiquitous dolomicrites and brecciated dolomicrites are characterized by high grain densities, low porosity, and correspondingly high compressional velocities ranging from 4.5 to 6.5 km/sec. The acoustic basement of the outer continental shelf may consist of the metamorphic strata (schist, phyllite, or quartzite) sampled in this study on Hormigas Island at 12°S and in drilled holes at 9°S on the outer continental shelf. Seismic refraction velocities are compatible with metamorphic or crystalline basement material.

ACKNOWLEDGEMENTS

The authors wish to thank the students and staff of the School of Oceanography, Oregon State University, and the captain and crew of the R/V *WECOMA* for their help with the data collection. Special thanks go to S.G. Ostroff and J. R. Rider for the strontium isotope age determinations and diatom ages, respectively, at the Mobil Exploration and Producing Services, Inc. in Dallas, Texas. We greatly appreciate the assistance of Norman Mundorf for obtaining the data on the physical properties of the carbonates at the Phillips Petroleum Company in Bartlesville, Oklahoma. We thank Erwin Suess for advice on the chemical and isotopic analysis of the carbonates. This research was sponsored by the office of International Decade of Ocean Exploration, National Science Foundation, under grants GX 28675, IDOE 71-04208, OCE 76-05903, and NSF grant OCE 77-20624 in the Submarine Geology and Geophysics Branch. Hawaii Institute of Geophysics Contribution No. 1138.

REFERENCES CITED

Allen, R. C., and others, 1969, Aragonite-cemented sandstone from outer continental shelf off Delaware Bay: Submarine lithification mechanism yields product resembling beachrock: Journal of Sedimentary Petrology, v. 39, p. 136–149.

Arthur, M. A., von Huene, R., and Adelseck, C. G., 1980, Sedimentary evolution of the Japan fore-arc region off northern Honshu, legs 56 and 57, Deep-Sea Drilling Project, *in* Scientific party, Initial reports of the Deep-Sea Drilling Project, 56, 57, Part I: Washington, U.S. Government Printing Office, p. 521–568.

Bandy, O. L., and Rodolfo, K. S., 1964, Distribution of foraminifera and sediments, Peru-Chile Trench area: Deep Sea Research, v. 11, p. 817–837.

Barazangi, M., and Isacks, B. L., 1976, Spatial distribution of earthquakes and subduction of the Nazca plate beneath South America: Geology, v. 4, p. 686–692.

Barron, J. A., 1980, Lower Miocene to Quaternary diatom biostratigraphy of DSDP leg 57, off northeastern Japan, *in* Scientific party, Initial reports of the Deep-Sea Drilling Project, 56, 57, Part 2: Washington, D.C., U.S. Government Printing Office, p. 642–686.

Berggren, W. A., and Van Couvering, J., 1974, The late Neogene: biostratigraphy, geochronology and paleoclimatology of the last 15 million years in marine and continental sequences: Palaeogeography, Palaeoclimatology, Palaeoecology, v. 16, p. 1–216.

Berry, E. W., 1928, A new *Nonion* from Peru: Journal of Paleontology, v. 1, p. 269.

Burckle, L. H., 1977, Pliocene and Pleistocene diatom datum levels from the equatorial Pacific: Quaternary Geology, v. 7, p. 330–540.

——1978, Early Miocene to Pliocene diatom datum levels for the equatorial Pacific: Proceedings of the

Geological Society of Indonesia, IGCP Project 114 Meetings, p. 25–44.

Chave, K. E., 1962, Factors influencing the mineralogy of carbonate sediments: Limnology and Oceanography, v. 7, p. 218–223.

Coryell, H. N., and Mossman, R. W., 1942, Foraminifera from the Charco Azul Formation Pliocene, of Panama: Journal of Paleontology, v. 16, p. 233–246.

Cushman, J. A., and Stone, B., 1947, An Eocene foraminiferal fauna from the Chira Shale of Peru: Cushman Foundation for Foraminiferal Research, Special Publication 20, p. 1–27.

Degens, E. T., 1969, Biochemistry of stable carbon isotopes, in Eglinton, G., and Murphy, M.T.J., eds., Organic geochemistry; methods and results: New York, Springer-Verlag, p. 304-329.

Deuser, W. G., 1970, Extreme $^{13}C/^{12}C$ variations in Quaternary dolomites from the continental shelf: Earth and Planetary Science Letters, v. 8, p. 118–124.

Fisher, R. L., and Raitt, R. W., 1962, Topography and structure of the Peru-Chile Trench: Deep-Sea Research, v. 9, p. 423-443.

Gibbs, R. J., 1967, The geochemistry of the Amazon River system: Part I. Factors that control salinity and composition and concentration of the suspended solids: Geological Society of America Bulletin, v. 78, p. 1203–1232.

Hathaway, J. C., and Degens, E. T., 1969, Methane derived marine carbonates of Pleistocene age: Science, v. 165, p. 690–692.

Hosmer, H. L., 1959, Geology and structural development of the Andean system of Peru [Ph.D. thesis]: Ann Arbor, University of Michigan, 281 p.

Hussong, D. M., and others, 1976, Crustal structure of the Peru-Chile Trench: 8°S–12°S latitude, in Sutton, G. H., and others, eds., The geophysics of the Pacific Ocean basin and its margin, a volume in honor of George P. Woollard: Washington, D.C., American Geophysical Union, Geophysical Monograph 19, p. 71–86.

Ingle, J. C., Jr., 1973, Biostratigraphy and paleoecology of early Miocene through early Pleistocene benthonic and planktonic foraminifera, San Joaquin Hills–Newport Bay–Dana Point area, Orange County, California: Society of Economic Paleontologists and Mineralogists Guidebook Field Trip No. 1, Annual Meeting, Los Angeles, p. 18–38.

Jenks, W. J., 1956, Handbook of South American geology; an explanation of the geologic map of South America: Geological Society of America Memoir 65, 378 p.

Krissek, L. A., Scheidegger, K. F., and Kulm, L. D., 1980, Surface sediments of the Peru-Chile continental margin and the Nazca plate: textural and geochemical characteristics and their controls: Geological Society of America Bulletin, v. 91, p. 321–331.

Kulm, L. D., Schweller, W. J., and Masias, A., 1977, A preliminary analysis of the subduction processes along the Andean continental margin, 6° to 45°S, in Talwani, M., and Pitman, W. C., III, eds., Island arcs, deep sea trenches and back-arc basins, Maurice Ewing Series 1: Washington, D. C., American Geophysical Union, p. 285–301.

Masias, J. A., 1976, Morphology, shallow structure, and evolution of the Peruvian continental margin, 6° to 18°S [Master's thesis]: Corvallis, Oregon State University, 92 p.

Milliman, J. D., 1974, Marine carbonates, Part 1: New York, Springer-Verlag, 375 p.

Mordojovich, C., 1974, Geology of a part of the Pacific margin of Chile, in Burk, C., and Drake, C., eds., The geology of continental margins: New York, Springer-Verlag, p. 591–598.

Müller, P. J., and Suess, E., 1979, Productivity, sedimentation rate, and sedimentary organic matter in the oceans—organic preservation: Deep Sea Research, v. 26A, p. 1347–1362.

Natland, M. L., 1950, 1940 E.W. Scripps cruise to the Gulf of California. Part IV, Report on the Pleistocene and Pliocene foraminifera: Geological Society of America Memoir 43, p. 1–55.

Noble, D. C., and others, 1974, Episodic Cenozoic volcanism and tectonism in the Andes of Peru: Earth and Planetary Science Letters, v. 21, p. 213–220.

Prince, R., and others, 1980, Bathymetry of the Peru-Chile Trench and continental margin: Geological Society of America Map and Chart Series, MC–34, scale 1:1,000,000.

Resig, J., 1981, Biogeography of benthic foraminifera of the northern Nazca plate and adjacent continental margin, in Kulm, L. D., and others, eds., Nazca plate: Crustal formation and Andean convergence: Geological Society of America Memoir 154 (this volume).

Roberts, H. H., and Whelan, T., III, 1975, Methane-derived carbonate cements in barrier and beach sands of a subtropical delta complex: Geochimica et Cosmochimica Acta, v. 39, p. 1085–1089.

Rosato, V., and Kulm, L. D., 1981, clay mineralogy of the Peru continental margin and adjacent Nazca plate: implications for provenance, sea level changes, and continental accretion, in Kulm, L. D., and others, eds., Nazca plate: Crustal formation and Andean convergence: Geological Society of America Memoir 154 (this volume).

Scholl, D. W., and others, 1970, Peru-Chile Trench sediments and sea-floor spreading: Geological Society of America Bulletin, v. 81, p. 1339–1360.

Scholle, P. A., 1977, Deposition, diagenesis, and

hydrocarbon potential of "deeper-water" limestones: American Association of Petroleum Geologists Continuing Education Course Note Series #7, 25 p.

Schrader, H. J., and Fenner, J., 1976, Norwegian Sea Cenozoic diatom biostratigraphy and taxonomy, *in* Talwani, M., and others, eds., Initial reports of Deep Sea Drilling Project, volume 38: Washington, D.C., U.S. Government Printing Office, p. 921–1099.

Schuette, G., and Schrader, H., 1979, Diatom taphocoenoses in the coastal upwelling area off western South America: Nova Hedwigia, v. 64, p. 359–378.

Shepherd, G. L., 1979, Shallow crustal structure and marine geology of a convergence zone, northwest Peru and southwest Ecuador [Ph.D. thesis]: Honolulu, University of Hawaii, 201 p.

Smith, P. B., 1960, Foraminifera of the Monterrey Shale and Puente Formation, Santa Ana Mountains and San Juan Capistrano area, California: U.S. Geological Survey Professional Paper 294-M, p. 363–495.

Stauder, W., 1975, Subduction of the Nazca plate under Peru as evidence of focal mechanisms and by seismicity: Journal of Geophysical Research, v. 90, p. 1053–1064.

Suess, E., 1979, Mineral phases formed in anoxic sediments by microbial decomposition of organic matter: Geochimica et Cosmochimica Acta, v. 43, p. 339–352.

Thornburg, T., and Kulm, L. D., 1981, Sedimentary basins of the Peru continental margin: Structure, stratigraphy, and Cenozoic tectonics from 6°S to 16°S latitude, *in* Kulm, L. D., and others, eds., Nazca plate: Crustal formation and Andean convergence: Geological Society of America Memoir 154 (this volume).

Travis, R. B., Gonzales, G., and Pardo, A., 1976, Hydrocarbon potential of coastal basins of Peru, *in* Halbouty, M. T., and others, eds., Circum-Pacific energy and mineral resources: Tulsa, American Association of Petroleum Geologists Memoir 25, p. 331–338.

Vail, P. R., Mitchum, R. M., and Thompson, S., III, 1977, Seismic stratigraphy and global changes of sea level, part 4: Global cycles of relative changes of sea level, *in* Payton, C. E., ed., Seismic stratigraphy—applications to hydrocarbon exploration: Tulsa, American Association of Petroleum Geologists Memoir 26, p. 83–97.

Veeh, H. H., Burnett, W. C., and Soutar, A., 1973, Contemporary phosphorite on the continental margin off Peru: Science, v. 181, p. 845–847.

von Huene, R., and others, 1978, Japan Trench transected: Geotimes, v. 23, p. 16–20.

Williams, H., Turner, F. J., and Gilbert, C. M., 1954, Petrography: San Francisco, W. H. Freeman and Co., 406 p.

Woodring, W. P., and Bramlette, M. N., 1950, Geology and paleontology of the Santa Maria district, California: U.S. Geological Survey Professional Paper 222, p. 1-185.

Zuta, S., and Guillen, O., 1970, Oceanografia de las aguas costeras del Peru [Oceanography of the coastal waters of Peru]: Instituto del Mar del Peru Boletin, v. 2, p. 161-323.

MANUSCRIPT RECEIVED BY THE SOCIETY NOVEMBER 12, 1980
MANUSCRIPT ACCEPTED DECEMBER 30, 1980
CONTRIBUTION 1103, HAWAII INSTITUTE OF GEOPHYSICS

PLATE 1

1. *Thalassiosira* sp., 7706-46-4, 1000X
2. *Thalassiosira jacksonii* (?), 7706-40-104, 1000X
3. *Delphineis* sp., 7706-23-4B, 1000X
4. *Coscinodiscus endoi*, 7706-23-43, 1000X
5. *Thalassiosira jacksonii* (?), 7706-46-16, 1000X
6. *Thalassiosira* sp., 7706-46-92, 1000X
7. *Pseudodimerogramma* sp., 7706-46-92, 1000X
8. *Rhizosolenia* sp., 7706-46-4, 400X
9. *Cussia praepaleacea/lancettula*, 7706-46-1, 1000X
10. *Cussia praepaleacea/lancettula*, 7706-46-104, 1000X
11. *Cussia lancettula*, 7706-46-16, 1000X
12. *Cussia tatsunokuchiensis*, 7706-46-16, 1000X
13. *Denticula hustedtii*, 7706-46-28, 1000X
14. *Nitzschia* sp., 7706-23-4B, 1000X
15. *Rouxia californica*, 7706-23-4A, 1000X
16. *Delphineis* sp., 7706-23-4B, 1000X
17. *Nitzschia proteri* FRENG., 7706-46-104, 1000X
18. *Delphineis* sp., 7706-23-4B, 1000X
19. *Thalassionema nitzschioides* var., 7706-46-50, 1000X

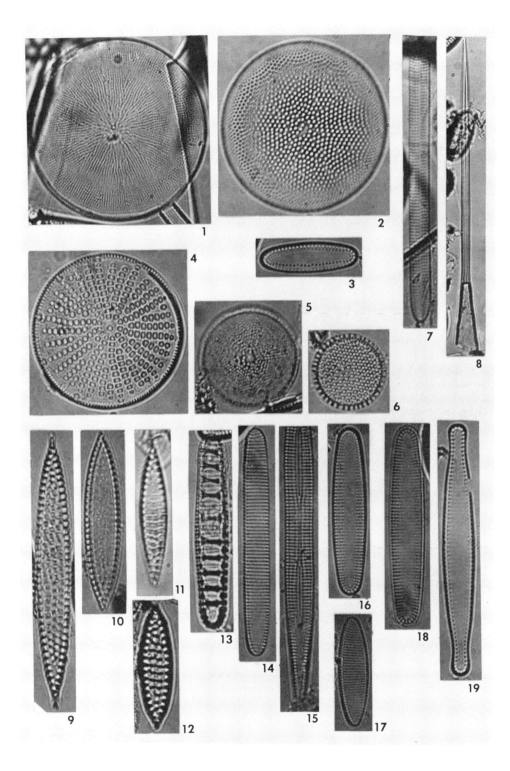

PLATE 2

1. *Thalassiosira* sp., Drag 5, 1000X
2. *Thalassiosira plicata*, Drag 5, 1000X
3. *Thalassiosira* sp., Drag 12, 1000X
4. *"Coscinodiscus" lineatus* forma *ellipta*, Drag 7, 630X
5. *Nitzschia seriata*, Drag 5, 1000X
6. *Mesocena quadrangula*, Drag 5, 1000X
7-8. *Rhizosolenia matuyamai*, Drag 12, 1000X
9. *Delphineis* sp., Drag 10, 1000X
10-11. *Nitzschia* sp., Drag 12, 1000X

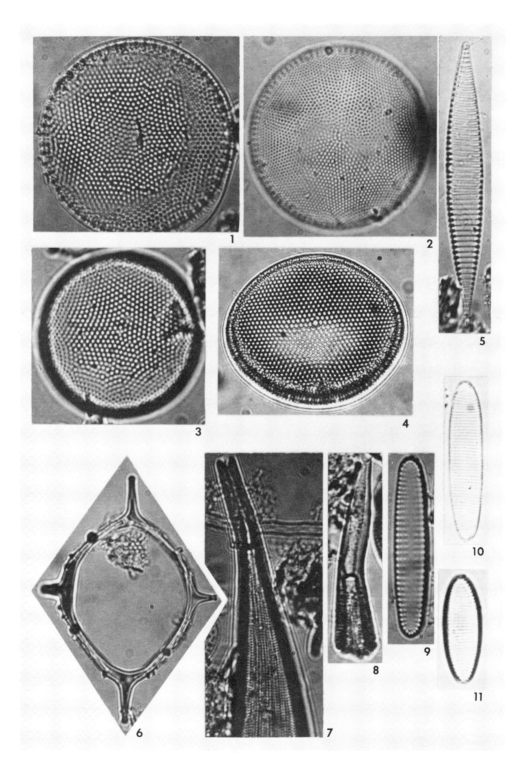

Vertical movement and tectonic erosion of the continental wall of the Peru-Chile Trench near 11°30′S latitude

DONALD M. HUSSONG
Hawaii Institute of Geophysics
University of Hawaii
2525 Correa Road
Honolulu, Hawaii 96822

LARRY K. WIPPERMAN
Mobil Exploration and Producing Services, Inc.
ESC-FO
P.O. Box 900
Dallas, Texas 75221

ABSTRACT

After reevaluation of crustal seismic velocity structure based on refraction data, we are able to present a multichannel seismic reflection profile across the Peru-Chile Trench as a cross section with undistorted reflector depths. These data support previous suggestions that this portion of the western edge of the continent of South America is undergoing tectonic erosion by the subducting oceanic Nazca lithospheric plate. The reflection seismic data also reveal numerous unconformities in the bedded sediments of the upper continental slope. These unconformities have wide lateral extent and are taken to be evidence of periodic vertical tectonic motion, including intermittent episodes of uplift, during the long-term subsidence of the continental wall of the trench. The configuration of deeper crustal reflectors leads us to suggest that the top of the subducting oceanic plate is underthrusting itself, perhaps in response to the compression caused by restraightening of the subducting slab beneath the continental plate. The crustal thickening that would accompany this deformation of the subducted oceanic plate may contribute to the vertical tectonic events and tectonic erosion of the overriding continental plate.

INTRODUCTION

A major task of the Nazca Plate Project was to investigate the tectonics of the convergence of the oceanic Nazca lithospheric plate and the continental South American plate. In particular, we sought to determine what effect the subducted oceanic plate has had on the overriding continental plate. What portion of the oceanic sediments are accreted into the frontal toe of the continental plate, relative to those sediments that are subducted with the rest of the oceanic plate? How has deformation of the downgoing plate affected the overriding plate? How much vertical tectonic motion (subsidence or uplift) has occurred along the continental wall of the trench? Is the edge of the continent being tectonically eroded and consumed with the subducted oceanic plate, or is the western edge of South America being built and expanded by parts of the Nazca plate that are incorporated into the leading edge of the continent rather than subducted?

Although the first two seasons (1972 and 1973) of the Nazca Plate Project had collected voluminous conventional geophysical and geological data off the coasts of southern Ecuador, Peru, and northern Chile, it became apparent that these data were not yielding adequate information about the structure of the continental wall of the trench to provide answers to the questions posed above. Refraction seismic data (for example, Hussong and others, 1976) collected as part of the project provided information on deep structure but lacked resolution. Single-channel reflection seismic profiles did not provide adequate penetration in the complex trench wall. To overcome these limitations, in the fall of 1973 we took advantage of the existence of a commercial oil exploration vessel in the region and hired Seiscom-Delta, Inc., a geophysical contractor, to acquire and process three 24-channel, digital reflection profiles. These profiles extend from the edge of the continental shelf, down the continental slope, across the trench axis, and onto the oceanic plate off Peru (Fig. 1). Each profile is about 100 km long.

The locations of the three profiles were selected to traverse what appeared at that time to be regions with different tectonic characteristics. Based on our profiling data and the seismicity described by Kelleher (1972), Line 1 crossed a region where the sea floor on both sides of the trench is less rugged and therefore less deformed by subduction tectonic activity, and where seismicity is high. A sonobuoy seismic refraction profile (Fig. 2) also crossed the trench at about this latitude. This paper expands the preliminary interpretation of Line 1 described by Hussong and others (1976).

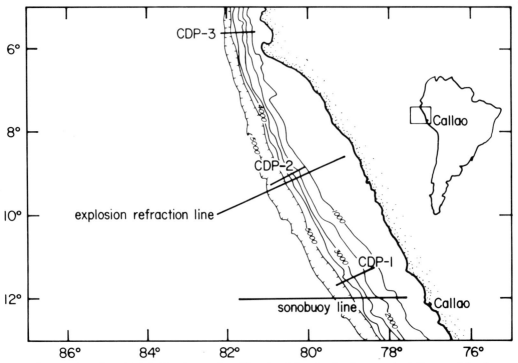

Figure 1. Location of CDP Line 1, as well as Lines 2 and 3 and refraction seismic stations used for the interpretations in this paper. Contour interval, 1 km.

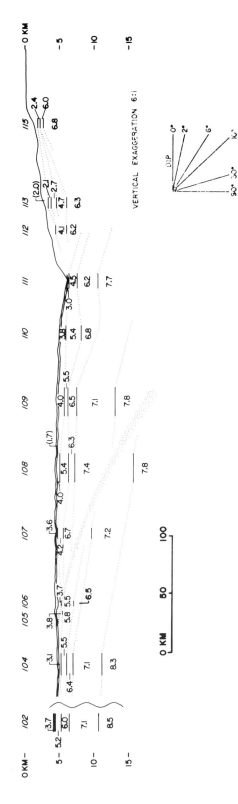

Figure 2. Crustal sections from sonobuoy refraction stations along lat 12°S (from Hussong and others, 1975). Velocities are in km/sec.

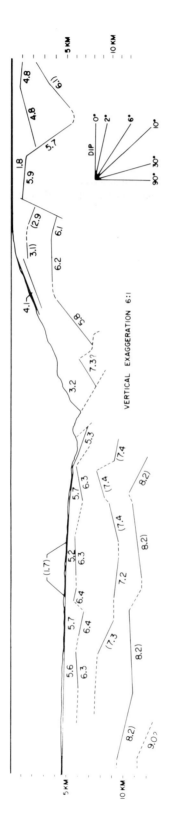

Figure 3. Refraction profile near CDP-2 (from Hussong and others, 1976). Velocities are in km/sec.

Line 2 is in a contrasting region where the continental shelf is exceptionally wide, where the bathymetric relief suggests that the oceanic plate is highly faulted (in response to the bending of the lithosphere prior to subduction), and where trench axis and continental trench-wall sediments are deformed. A large basaltic ridge, which Kulm and others (1973) interpret as a product of presubduction faulting of the oceanic plate, occurs in the trench axis here. In spite of these indications of active tectonics, Kelleher (1972) noted a lack of shallow seismicity in the region of Line 2. Considerable explosion refraction data (Fig. 3) had also been obtained near Line 2. The CDP profile is described by Kulm and others (1981a, this volume).

The third profile was placed offshore of the Huancabamba Deflection (described by Ham and Herrera, 1963) near lat 5°S where the trend of the trench axis and onshore geology changes from roughly NW-SE to N-S. As in the Line 2 area, the bathymetric relief of the oceanic plate suggests that the plate is highly deformed, yet the region is relatively aseismic. By lat 5°S, the thickness of sediment on the oceanic plate has increased to more than five times greater than further south due to the proximity of the equatorial region of high biological productivity (Erlandson and others, this volume). The sediment supply is further augmented in the trench axis due to the influx of terrigenous material from the Rio Guayas system emptying into the Gulf of Guayaquil. The interpretation of Line 3 is described by Shepherd (1979) and Shepherd and Moberly (this volume).

DATA ACQUISITION AND PROCESSING

The seismic reflection data were acquired by the survey vessel *American Delta II*. The sound source was four 2,000 psi air guns (two 300 cu in. and two 1,000 cu in. guns) fired simultaneously at a tow depth of 10 m. The receiving hydrophone array consisted of a 420 m inactive cable and a 24-channel, 1,600 m long, active section. Each channel produced one data trace that was the sum of the 30 hydrophones within that channel. The receiving array was towed at a depth of approximately 15 m. The field data were collected with 24-fold overlap using one shot point every 70 m.

The shipboard digital seismic recording equipment consisted of a Texas Instruments DFS III with dual tape drives and accessory analog recorders and seismic cameras. The signal from the hydrophone array was filtered 8-64 Hz (18 db/octave roll-off), amplified (with a four-second delay) 54 db, digitized at a sampling rate of 4 msec multiplexed, and recorded on tape. The record length for Line 1 was set at 13 sec to permit the recording of all potential reflections down to Moho depths beneath the trench axis.

Ship position and speed were controlled by satellite navigation and doppler sonar. The position accuracy for the system was ±0.6 km, because in water deeper than 200 m the doppler sonar has poor resolution.

The digital field tapes were processed at the Seiscom-Delta, Inc., processing center in Houston, Texas, using procedures that were standard at that time. The processing included spike deconvolution (with a 120 msec operator), CDP gathering, velocity analysis at less than 5 km intervals, and 12-fold stacking. Although considerable interaction between Seiscom-Delta personnel and the authors during this process permitted some iterative determination of filters and stacking velocities, the eventual velocity determinations from wide-angle reflections were inadequate for reliable determination of geologic structure along the profile. Normal moveout (NMO) errors of up to one quarter of the dominant period of the arrival are acceptable for good stacking (Brown, 1969). Thus, for the initial processing of Line 1, where water depths are as much as 6,000 m, the NMO was so little for the hydrophone array used that almost any sediment or rock seismic velocity structure in the oceanic crust provided NMO corrections that were adequate for stacking. A useful "rule of thumb" that is pertinent here is that in order to obtain reliable velocity structure using conventional techniques applied to NMO on wide-angle reflection data, the interpreter must have both good reflectors and a distance from shot to the furthest receiver that is three times as great as the depth to the reflector of interest. Although this problem was not severe in the shallower water close to the continental shelf, it became critical with data from water depths greater than 4 km.

Eventually, in order to determine the depth and geometry of various reflectors with reasonable confidence, it was necessary to reinterpret velocity-depth functions for all of Line 1 using additional data. The revised velocity determinations were done at the Hawaii Institute of Geophysics by completely repicking the Seiscom-Delta semblance plots using nearby seismic refraction stations (Hussong and others, 1975; Hussong and others, 1976) as constraints.

It was observed that the shallow-water sediment seismic velocities in Line 1 were reasonable and consistent within structural units. However, the NMO-derived stacking velocities of the deeper layers, particularly in the oceanic crust, were as low as one-half the velocities calculated from the refraction stations. This occurred because, when faced with very small NMO from the deep reflectors (and thus large uncertainties in the semblance plots), the seismic processors at Seiscom-Delta generally yielded to oil field experience and looked for an rms velocity-depth curve that yielded sediment velocities.

When we repicked the semblance plots, we applied a different prejudice to our approach. The refraction profiles near Line 1 were converted to rms velocity-depth curves and plotted on transparent drafting film. From the Seiscom-Delta semblance plots, a new, generally higher velocity, rms curve was then picked using the refraction-rms curve as a guide. In most cases, this method was successful, indicating that the correct velocity information was present in the spectrum but was obscure and not picked at Seiscom-Delta. In a few cases, we were unable to determine the higher velocity structure, but this occurred because of reflector quality. In those cases, the refraction rms curves were modified slightly for changes in water depth and were used for the depth section generation. Also, the velocity spectra were cut off at too low a velocity in a few cases, so the revised rms curves had to be extrapolated beyond the velocity cutoff. When the iterations and corrections were complete, a refraction-controlled rms profile was constructed to check for uniformity between adjacent rms curves. The new refraction-corrected rms curves were keypunched for computation of velocities and layer thickness, which were then used to hand plot a depth section. Several iterations of the rms correction process were needed to converge on a consistent interval velocity-depth section solution. The seismic time section was converted to depth and plotted for us (Fig. 4) by Exxon Production Research Co., Houston, with the use of the refraction-constrained velocities. A schematic depth section summarizing the major velocity units and interval velocities used to construct the seismic depth section of Figure 4b is shown in Figure 5.

DISCUSSION

The major features of the time section, Figure 4a, of Line 1 have been discussed by Hussong and others (1976). This paper will concentrate on the interpreted depth section and on possible evidence for vertical tectonics in the region.

The primary difference between the travel time and depth seismic sections (Fig. 4b) is the depression of the reflection segment labeled 13 between km 30 and km 60. This reflection is described by Hussong and other (1976) as the top of the oceanic crust that is being subducted beneath the Peruvian continental margin. The abrupt increase in the interval seismic velocity of the overlying rocks from 3.5 km/sec west of km 30 to about 5 km/sec east of km 30 (see Fig. 5) decreases the travel time to reflector B, thereby "pulling up" the reflector east of km 30 on the time section.

The interval velocity of 3.5 km/sec between the trench axis and km 30 is reasonably constrained by the velocity spectra and has an estimated uncertainty of only a few tenths of a km/sec. The higher-velocity material east of km 30 has an estimated uncertainty of less than ±1 km/sec. Therefore, the determinations of these interval velocities ranged between 4 and 6 km/sec, with the average velocity being 5 km/sec. The uncertainty in the interval velocities east of km 30 yields an uncertainty in depth to reflector B of about 1.5 km. In fact, an interval velocity of 4 km/sec was inadvertently used at km 38 (Fig. 4b), causing the sharp bend in reflector B at that point. Using an interval velocity of 5 km/sec would increase the reflector depth near km 38 by about 0.8 km. Note that the V-shaped kink in reflector B at km 20 is also caused by an error in the interval velocity used above the reflector during processing.

The distortion of reflector B in the time section results primarily from velocity contrast in the overlying rocks. For the most part, this distortion is successfully corrected by the depth conversion process using the refraction-constrained velocity-depth functions. Thus, on the depth section (Fig. 4b), reflector B has approximately the same average landward dip beneath the continental trench slope, and for about 10 km seaward of the trench axis, as it does beneath the prism of low seismic velocity (3.5 km/sec and less) material in the toe of the landward slope.

The offsets west of km 30 on reflector B at km 35 and km 47 suggest faulting of at least the top of the oceanic crust. Hussong and others (1976) note that no reasonable velocity distribution in the overlying structure would remove the offsets and suggest that the subducting oceanic crust is underthrusting itself. However, the seismic data alone are consistent with either thrust or normal faulting of the oceanic crust. Normal faulting is commonly inferred on the oceanic plate just seaward of the trench axis and appears to be caused by extension of the top of the oceanic lithosphere as it bends into the subduction zone. If the two offsets of reflector B are normal faults, the large, downdropped blocks are both on the seaward side of the fault, a relationship that is rarely observed seaward of the trench on the Nazca plate. Thrust faults have also occasionally been interpreted as existing on the oceanic plate very near or in the trench axis (Prince and Kulm, 1975; Schweller and others, this volume). The change from extension of the upper oceanic crust on the flexure seaward of the trench axis to compression beneath the continental margin may be caused by (1) locking of the upper oceanic crust against the continental margin (Schweller and others, this volume), (2) flattening of the oceanic plate beneath the continental margin resulting, perhaps, from isostatic loading (discussed in more detail below), or (3) both of these mechanisms.

We favor the thrust fault interpretation for the reflector B offsets, because the direction of faulting is consistent with the previously described thrusting, and because a thrust fault provides a mechanism for relative uplift of the middle slope region with respect to the upper and lower slopes (discussed below).

The seismic structure of the lower slope region of the continental margin was discussed by Hussong and others (1976), who interpreted two bands of reflection hyperbolae as imbricate thrust faults. Several other such bands are apparent on both the time and depth sections and are shown by black lines on the interpreted section (Fig. 6). We also interpret these bands as imbricate thrust planes and suggest that the material with seismic velocity of 3.5 km/sec and less is sedimentary and may be a small accretionary prism of offscraped oceanic plate and trench axis sediments.

In the middle and upper slope regions, the seismic sections show a 1 to 1.5 km thickness of low seismic velocity sediments that are highly stratified in the upper slope and become progressively more faulted and disturbed downslope. These sediments overlie a prominently block-faulted reflector, which, on the basis of well-determined refraction velocities of about 6 km/sec (Hussong and others, 1976), we interpret as continental crystalline basement. The block-faulted nature of the crystalline basement is inferred from short reflector segments with accompanying diffractions. Many of the blocks appear to be rotated so that the original surface dips landward. We observe no evidence for reverse, or thrust, faulting in the seismic data of the middle and upper slopes.

The seismic section extends into the Lima Basin east of km 80. The basin is generally characterized by parallel groups of reflectors bounded by major unconformities. Minor faulting occurs near the eastern end of the seismic profile. West of km 85, the sediments become increasingly faulted until stratigraphic continuity is lost. A simplified line drawing of the time section (Fig. 4a) in the region of the Lima Basin is shown in Figure 7. The unconformities, labeled 1 to 4 on Figure 7, are defined by the truncation of underlying reflectors and by onlap of reflectors against the unconformity surface. Straight reflector segments exhibiting diffractions at the base of the stratified sediments are interpreted as basement fault blocks. Reflectors in the sediments between basement and unconformity 4 are quite irregular. These sediments could be initial progradational sediments (channels, small deltas, etc.) deposited after a marine transgression, or they could be more conformable strata that are extensively faulted. Several coherent reflector segments are shown by the thin lines in Figure 7. The marker horizons in this interval cannot be correlated across the major faults and are shown to illustrate the discordance across unconformity 4. The block faulting of the basement was complete by the time of unconformity 4. Unconformity 3 is observed west of km 97 and truncates the strata above

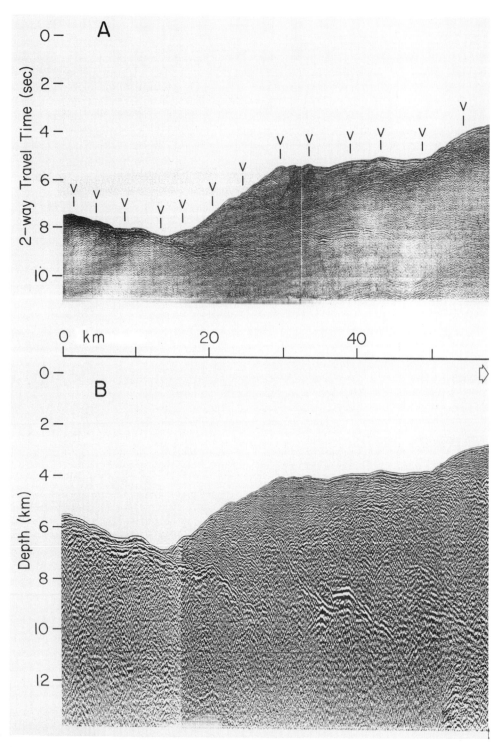

Figure 4. (a) Twelve hundred percent stacked, common-depth-point, seismic reflection line CDP-1 across the Peru-Chile Trench near lat 11° 40′S. Data acquired and processed for this project by Seiscom-Delta, Inc. Data filtered 8-64 Hz; acquired with a 1,600 m, 24-channel cable using two 300 and two 100 cu in. air guns. The small Vs

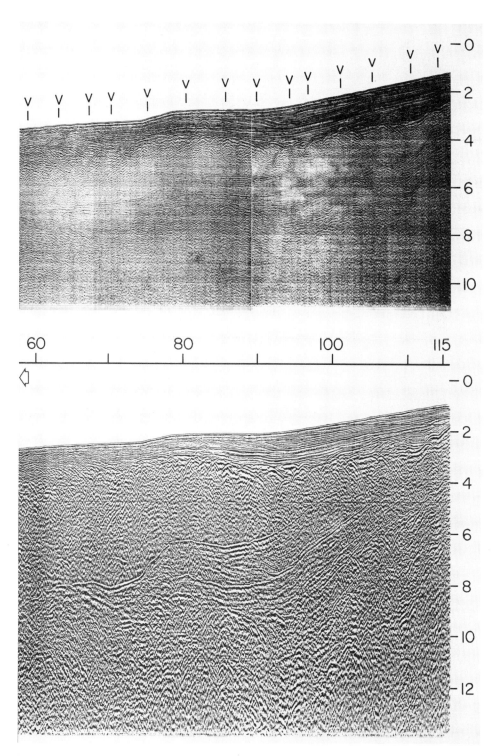

at the top of the section mark the locations of stacking velocity and interval velocity analysis. The section is plotted as two-way travel time. (b) Same as 4(a), but the seismic data have been converted to depth and plotted by Exxon Production Research Company.

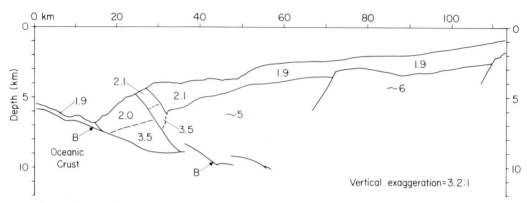

Figure 5. Schematic depth section summarizing the interval velocities used to compute the depth section in Figure 4b. Velocities are in km/sec. The reflector labeled B is interpreted to be the top of the oceanic crust.

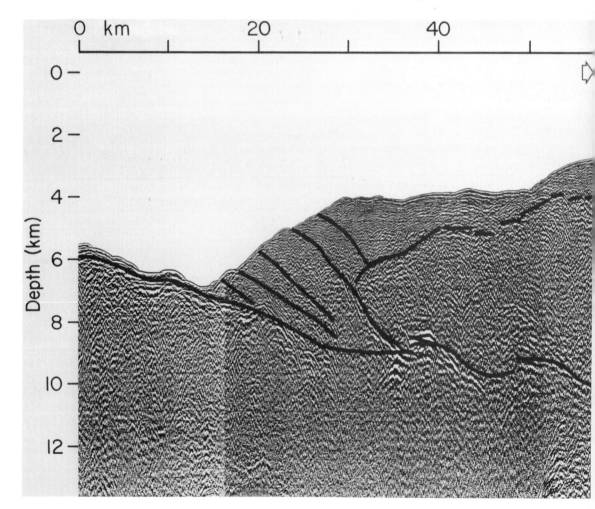

Figure 6. Figure 4b repeated, with prominent reflectors and interpreted imbricate thrusts in the accretionary prism highlighted.

marker horizon A. The strata between unconformity 3 and unconformity 2, as shown by marker horizon B, onlap unconformity 3 and are nearly conformable with unconformity 2. Reflectors between unconformity 2, and those just above marker horizon C, onlap unconformity 2. The strata above unconformity 2 appear to be truncated by the sea floor on the middle slope. The sediments above unconformity 1 onlap the unconformity and appear to be truncated by the sea floor.

The trench slope is blanketed by approximately 100 m of terrigenous and hemipelagic mud based on high-resolution, single channel reflection and 3.5 kHz profiles and piston cores. The CDP seismic data do not resolve this surface layer on the sloping sea floor. Thus, although the deeper strata appear to crop out on the CDP profile, they are actually buried.

Above unconformity 4 and east of km 85, faulting of the sediment column is minor. The three longest faults are shown in Figure 7 to form a graben approximately 9 km wide. The throw of these faults is small and increases with depth, suggesting that faulting has generally been contemporaneous with deposition. While the two west-dipping faults appear to extend very nearly to the surface, the east-dipping fault apparently has not been active since the time of unconformity 1. The faulting is probably due to continued small movements of some of the basement fault blocks.

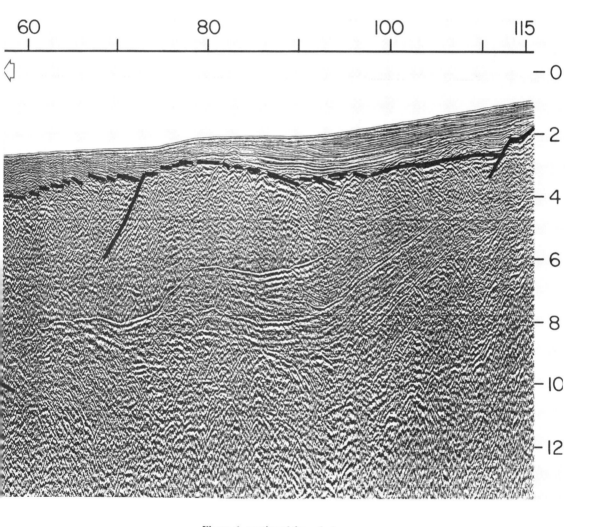

Figure 6. continued from facing page.

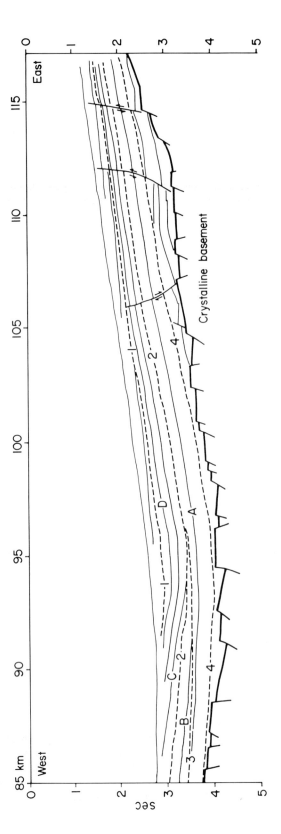

Figure 7. Interpreted line drawing of part of CDP-1 in the Lima Basin. The vertical scale is seconds of two-way travel time. The block-faulted nature of crystalline basement is shown by thick lines bounded by normal faults. The thick dashed lines labeled 1 to 4 represent major unconformities identified on the basis of truncated reflectors. The thin lines (some of them labeled A to D) are marker horizons that illustrate the attitude of reflecting horizons.

Major unconformities on continental shelves and slopes have been attributed to changes in sea level by Vail and others (1977). The unconformities in the Lima Basin are presently at depths of 1 to 3 km. Even accounting for sediment loading, this can be interpreted to suggest that sea level has fluctuated as much as 2 km. Moreover, recent dredging by Kulm and others (1981b, this volume) in the vicinity of the Lima Basin recovered carbonates that range in age from late Miocene to early to middle Pleistocene and are probably representative of much of the Lima Basin stratigraphic column. Kulm and others (1981b, this volume) interpret microfossil assemblages in the carbonates as indicating that the Lima Basin has subsided 1,100 m since late Miocene time, with at least 500 m of subsidence occurring during the Pleistocene. Water depth during the carbonate deposition was as shallow as 50 to 150 m. Thus, the Lima Basin strata were shallow enough to be eroded during low stands of sea level.

While geological and geophysical data are consistent with overall subsidence of the Peru continental margin near lat 12°S, the landward dip of sediment onlapping unconformities 2 and 3 between km 85 and km 94 (Fig. 6) suggests relative uplift (or a substantially lower rate of subsidence) of the midslope region. The onlapping strata are likely to have originally been deposited either horizontally or with a small seaward dip. Coulbourn and Moberly (1977) interpreted single-channel reflection profiles across upper-slope sediment basins on the continental margin of northern Chile as showing relative uplift of the midslope region.

The very slight distortion of the Lima Basin strata east of km 85 indicates that, in spite of the block-faulted nature of the underlying crystalline basement, the upper slope has subsided as a unit. The faults near the eastern end of the seismic reflection profile may indicate differential subsidence, perhaps caused by movement on the preexisting basement faults.

West of km 85 on CDP-1, subsidence has not been as uniform. The high degree of deformation of the overlying sediments suggests that the individual basement blocks are moving separately. The deformation of the top of the crystalline rock and the overlying sediments becomes increasingly severe to the west, toward the trench axis.

TECTONIC IMPLICATIONS

The reinterpretation of the seismic velocity structure of the Peru-Chile Trench near lat 11°30′S and the subsequent presentation of the reflection profile as a true depth section have strengthened the conclusions made by Hussong and others (1976). That paper argues that:

(1) The sediment wedge in the lower trench wall (velocities less than 4 km/sec) is much too small to have served as a trap for more than a small fraction of the volume of oceanic sediment entering the Peru-Chile Trench in Cenozoic time. Most of this sediment must therefore be subducted.

(2) The upper surface of the subducting oceanic plate becomes increasingly deformed immediately after subduction, producing the deep reflector offsets interpreted as thrust faulting 20 to 40 km west of the trench axis.

(3) Although there may be episodes of uplift, on the long term, the western continental margin of Peru that now composes the inner trench wall is subsiding and is being tectonically eroded and subducted along with the oceanic plate material.

Thus, Hussong and others (1976) provided offshore geophysical evidence to support the implication, based on magma trends (James, 1971), continental tectonics (Rutland, 1971), the truncation of onshore geology (Isaacson, 1975), Andean subsidence (Myers, 1975), and the proximity of Paleozoic and older continental cratonic rocks to the trench axis (Cobbing and Pitcher, 1972; Cobbing and others, 1977), of subduction erosion of the western edge of South America.

Our improved seismic velocity modeling and consideration of the unconformities and faulting in the upper kilometre of sediments at the eastern edge of the CDP-1 profile further suggest that the faulting on top of the subducting slab may be significantly contributing to the tectonics of the overriding slab.

We can see from the unconformities in the layered sediments in the Lima Basin, on the eastern one-third of profile CDP-1, that the continental margin has been subjected to both uplift and subsidence. Such major unconformities with wide lateral extent are often caused by subaerial or possibly wave-base erosion. A similar unconformity in the trench wall drilled off Japan during DSDP Leg 57 was caused by subaerial erosion (von Huene, Nasu, and others, 1978). Kulm and others (1981b, this volume) dredged carbonate samples from more than 1 km of water in the Lima Basin. These carbonates were deposited at depths of less than 150 m. Between the two most prominent unconformities in our records, sediment beds tilt and pinch out landward, suggesting a landward migration of the center of deposition, as modeled by Coulbourn and Moberly (1977). Thus, we apparently see several cycles of vertical motion recorded by the Lima Basin sediment. The unconformities mark periods of uplift, probably to above sea level, while the intervening sediments were often deposited in shallow depths and subsided as much as 3 km.

The surface sediments (velocities around 1.9 km/sec) exhibit an increasingly chaotic structure from km 75 to km 50 in CDP-1, but they are probably the same sedimentary sequence that occurs east of km 75. The difference in character can be attributed to disruption of the bedding by faulting in the underlying continental rocks west of km 75 that has disturbed the overlying sediments to a degree that the layering is seismically undetectable. The faulting, which becomes increasingly severe closer to the trench, also causes a decrease in the seismic velocity of the basement crystalline rocks from ~6 km/sec to ~5 km/sec.

A possible mechanism for the faulting and uplift is illustrated by the sketch in Figure 8. The initial bending of the oceanic plate causes extension of the upper oceanic lithosphere. The resultant horst and graben topography is stable when it reaches the trench axis. At this point, however, the oceanic plate straightens again, throwing the upper surface of the plate into compression and causing thrust faults. The straightening of the lithospheric plate and resultant thrust faulting can be observed in our depth section of CDP Line 1.

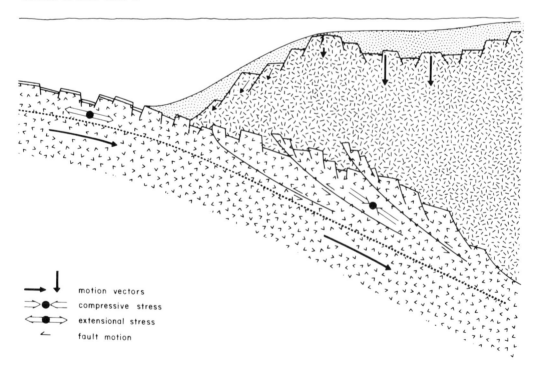

Figure 8. Exaggerated sketch illustrating a possible cause of vertical movement of the continental wall of the trench. Oceanic crystalline crust (checked pattern), continental crystalline crust (line pattern), and continental margin and oceanic plate sediment (dot pattern) are indicated on the diagram. The dashed line in the subducting slab is a zone of neutral stress. The underthrusting periodically thickens portions of the top of the oceanic slab, causing episodic vertical movements of the overriding plate. See legend for explanation of features.

The upper oceanic crustal thrusting effectively thickens areas of the upper crustal layer, producing localized uplift of the overriding plate. Alteration of the subduction geometry and surface irregularities on the oceanic plate (seamounts, etc.) will affect this pattern, resulting in episodic periods of subsidence and uplift. Isostatic loading of the subducting slab and friction caused by the shear zone contact between the slabs might also contribute to the oceanic crust underthrusting itself.

CONCLUSIONS

Over the long term, the landward wall of the Peru-Chile Trench is subsiding because tectonic erosion and subduction of the continental rocks predominate over any accretion of oceanic sediment or crust to the continent. The inner-wall rocks become more fractured and disrupted the closer they are to the trench axis, and tensional faulting and subsidence progressively downdrop blocks of continental material into the trench.

Superimposed on this long-term subsidence are episodic periods of uplift and subsidence that cause the major unconformities and irregular bedding in the less-disturbed sedimentary section at relatively shallow depths on the landward trench wall. Thickening of the top of the oceanic plate, caused by periodic thrust faulting when the top of the subducting plate is compressed by restraightening after initial seaward bending of the trench, may contribute significantly to the vertical movement of the overriding edge of the continental plate.

ACKNOWLEDGMENTS

The authors are indebted to David W. Scholl and Joel S. Watkins for their helpful comments on this paper. This work was supported by the Office of the International Decade of Ocean Exploration, National Science Foundation Grant GY28674.

REFERENCES CITED

Brown, R.J.S., 1969, Normal-moveout and velocity relations for flat and dipping beds and for long offsets: Geophysics, v. 34, p. 180–195.

Cobbing, E. J., Ozard, J. M., and Snelling, N. J., 1977, Reconnaissance geochronology of the crystalline basement rocks of the Coast Cordillera of southern Peru: Geological Society of America Bulletin, v. 88, p. 241–246.

Cobbing, E. J., and Pitcher, W. S., 1972, Plate tectonics and the Peruvian Andes: Nature, v. 246, p. 51–53.

Coulbourn, W. T., and Moberly, R., 1977, Structural evidences of the evolution of fore-arc basins off South America: Canadian Journal of Earth Sciences, v. 14, p. 102–116.

Erlandson, D. L., Hussong, D. M., and Campbell, J. F., 1981, Sediment and associated structure on the northern Nazca plate, in Kulm, L D., and others, eds., Nazca plate: Crustal formation and Andean convergence: Geological Society of America Memoir 154 (this volume).

Ham, C. K., and Herrera, L. J., 1963, Role of Subandean fault system in tectonics of eastern Peru and Ecuador: American Association of Petroleum Geologists, Memoir 2, p. 47–61.

Hussong, D. M., Odegard, M. E., and Wipperman, L. K., 1975, Compressional faulting of the ocean crust prior to subduction in the Peru-Chile Trench: Geology, v. 3, p. 601–604.

Hussong, D. M., and others, 1976, Crustal structure of the Peru-Chile Trench: 8°S–12°S latitude, in Sutton, G. H., and others, eds., The geophysics of the Pacific Ocean Basin and its margin: A volume in honor of G. P. Woollard: American Geophysical Union Geophysical Monograph, v. 19, p. 71–85.

Isaacson, P. E., 1975, Evidence of a western extracontinental land source during the Devonian Period in the central Andes: Geological Society of America Bulletin, v. 82, p. 39–46.

James, D. E., 1971, Plate tectonic model for the evolution of the central Andes: Geological Society of America Bulletin, v. 82, p. 3325–3346.

Kelleher, J. A., 1972, Rupture zones of large South

American earthquakes and some predictions: Journal of Geophysical Research, v. 77, p. 2087-2103.

Kulm, L. D., and others, 1973, Initial reports of the Deep Sea Drilling Project, Vol. 18: Washington, D.C., U.S. Government Printing Office, 930 p.

Kulm, L. D., and others, 1981a, Crustal structure and tectonics of the central Peru continental margin and trench, *in* Kulm, L. D., and others, eds., Nazca plate: Crustal formation and Andean convergence: Geological Society of America Memoir 154 (this volume).

Kulm, L. D., and others, 1981b, Late Cenozoic carbonates on the Peru continental margin: lithostratigraphy, biostratigraphy, and tectonic history, *in* Kulm, L. D., and others, eds., Nazca plate: Crustal formation and Andean convergence: Geological Society of America Memoir 154 (this volume).

Myers, J. S., 1975, Vertical crustal movements of the Andes in Peru: Nature, v. 254, p. 672-674.

Prince, R. A., and Kulm, L. D., 1975, Crustal rupture and the initiation of imbricate thrusting in the Peru-Chile Trench: Geological Society of America Bulletin, v. 86, p. 1639-1653.

Rutland, R.W.R., 1971, Andean orogeny and ocean-floor spreading: Nature, v. 233, p. 252-255.

Schweller, W. J., Kulm, L. D., and Prince, R. A., 1981, Tectonics, structure, and sedimentary framework of the Peru-Chile Trench, *in* Kulm, L. D., and others, eds., Nazca plate: Crustal formation and Andean convergence: Geological Society of America Memoir 154 (this volume).

Shepherd, G., 1979, Shallow crustal structure and marine geology of a convergence zone, northwest Peru and southwest Ecuador [Ph.D. thesis]: Honolulu, University of Hawaii, 201 p.

Shepherd, G., and Moberly, R., 1981, Coastal structure of the continental margin, northwest Peru and southwest Ecuador, *in* Kulm, L. D., and others, eds., Nazca plate: Crustal formation and Andean convergence: Geological Society of America Memoir 154 (this volume).

Vail, P. R., and others, 1977, Seismic stratigraphy and global changes of sea level, *in* Payton, C. E., ed., Seismic stratigraphy—applications to hydrocarbon exploration: Tulsa, American Association of Petroleum Geologists, p. 49-212.

von Huene, R., Nasu, N., and others, 1978, Japan trench transected on Leg 57: Geotimes, v. 23, p. 16-21.

MANUSCRIPT RECEIVED BY THE SOCIETY NOVEMBER 12, 1980
MANUSCRIPT ACCEPTED DECEMBER 30, 1980
HAWAII INSTITUTE OF GEOPHYSICS CONTRIBUTION 1104

Geological Society of America
Memoir 154
1981

Shallow structures of the Peru Margin 12°S - 18°S

S. H. Johnson*

Gordon Ness

School of Oceanography
Oregon State University
Corvallis, Oregon 97331

ABSTRACT

Twenty-six single-channel seismic reflection profiles across the continental margin of Peru south of Lima are used to describe the shallow structures of this portion of the Nazca–South America convergence. The aseismic Nazca Ridge intersects the trench in the center of the study region, and the axis of the trench plunges away on either side of this intersection, more steeply to the south than to the north. The lower continental slope generally has a steep wall, backed by an unfilled bench, except opposite the ridge. Large upper-slope basins are found in the northern and southern portions of the survey area but not opposite the ridge where the slope and shelf are the narrowest. These few relationships are the only ones found in this study that may possibly reflect the interaction of the ridge with the continental margin, and even these may be noncausal. In any event, the limited extent of possible Nazca Ridge effects on the margin is noteworthy, at least for those effects evident in the morphology and shallow structure.

Structural highs that bound the seaward side of the two upper-slope basins are thought to result from a combination of imbricate thrusting and internal flow of material within the accretionary wedge. We argue that the continental margin of southern Peru exhibits characteristics of both processes, based on the character of lower-slope benches and the existence of upper-slope basins.

INTRODUCTION

Shallow structures of convergent continental margins are important indicators of past tectonic movement and give clues to uplift, erosion, deposition, and faulting due to the convergence of oceanic and continental plates. The South American trench system is the major example of oceanic lithosphere underthrusting continental lithosphere. A well-defined Benioff zone marks the subducting slab (Barazangi and Isacks, 1976). The oceanic-continental boundary is identified by a well-developed trench. Subduction-related volcanism is characteristic of the Andean mountain system.

---

*Present Address: Amoco Prod. Res., P.O. Box 591, Tulsa, OK 74102

Comparisons and contrasts are often made between the Peru-Chile and the western Pacific island arc types of subduction systems (Dickinson and Seely, 1979). Until recently, the amount of marine data available from South America has been more limited than from the western Pacific. Geophysical studies of this margin include Fisher and Raitt (1962), Hayes (1966), Scholl and others (1970), Kulm and others (1974), Prince and others (1974), Prince and Kulm (1975), Hussong and others (1976), Coulbourn and Moberly (1977), and Lonsdale (1978). The features of this large and important subduction zone were investigated in reconnaissance scale during the Nazca Plate Project, initiated under the auspices of the International Decade of Ocean Exploration. From 1972 to 1975, large amounts of geological and geophysical data were acquired along the Peru-Chile Trench and adjacent continental margin. In this paper, we concentrate on a small portion of the data set. We present bathymetric and single-channel air gun profiles across the southern Peru Margin and suggest tectonic models to explain the observed features.

Reflection surveys conducted by the R/V *Yaquina* during cruises YALOC 71 and YALOC 73 provide data for the otherwise sparsely studied area between lat 12°S and 18°S. The annotated tracklines shown in Figure 1A, 1B, and 1C provide a schematic view of the structural features of the southern Peru Margin, including the identification of two upper-slope basins and one shelf basin. Seismic reflection lines are too widely spaced (~45 km) to permit the correlation between lines of anything larger than major sub-bottom features of the slope. Bathymetric profiles across the margin are compared in Figure 2.

Dickinson and Seely (1979) summarize common features of arc-trench systems. Where possible, we employ their terminology for structural features while noting that the Peruvian margin lacks certain features of their classification, most notably a volcanic arc between lat 3°S and 16°S. The most characteristic features of this margin, extensive upper-slope basins, are so high on the slope that they may lie landward of the subduction complex. If they do, the generic implications of the term "accretionary basin" proposed by Dickinson and Seely are inapplicable. We call the seaward boundary of an upper-slope basin an "upper-slope break" in preference to an "upper-slope discontinuity" (Karig and Sharman, 1975).

Data

The single-channel reflection records that form the basis of this study were obtained using air guns with capacities of 10 to 300 cu. in. A recording bandwidth of 50-300 Hz provided high resolution at the expense of penetration, although nonmultiple reflections were seen to 3 sec in some upper-slope basins. Records taken during YALOC 71 were made using an 8 sec pulse repetition rate, while the YALOC 73 records were made at a 4 sec rate. Only two records from the 1971 cruise (Fig. 3C) are presented here, although structural information from all available records are transposed onto Figures 1A, 1B, and 1C. Interpretive line drawings of 1973 margin crossings are presented in Figures 3A, 3B, and 3D and show the major features of the margin at an average vertical exaggeration of 6:1. Selected original records reproduced in Figures 4 through 7 show additional details. Figures 1A, 1B, and 1C present sub-bottom characteristics transferred to tracklines drawn over recently compiled bathymetric maps of the area (Prince and others, 1980). These figures allow a comparison of structural and bathymetric features.

Shallow Structures of the Trench

Outer-rise oceanic crust of Eocene age (Herron, 1972; Handschumacher, 1976) is subducting in the direction N 88°E based on the NAZCA-SOAM pole of rotation of Minster and Jordan (1978). This is oblique to the trench, which has an average orientation of 135° in the study region. The trace of the trench changes from convex to concave seaward in the bight at lat 17°S but does not appear to be noticeably affected by the presence of the aseismic Nazca Ridge at lat 14°S in spite of the substantial (1,500m) topographic expression of the ridge. A deep trough parallels the southeastern side of the ridge. Because of these complications, no consistent outer trench rise exists opposite the study area,

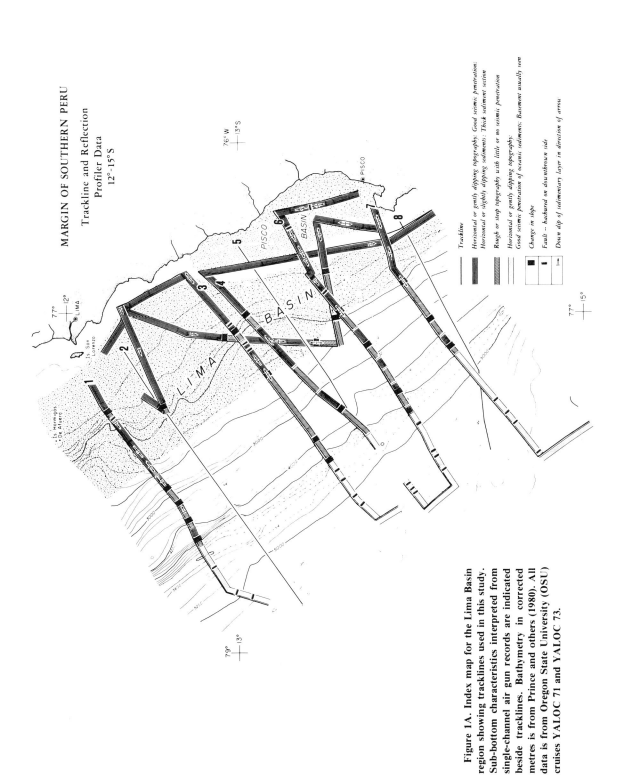

Figure 1A. Index map for the Lima Basin region showing tracklines used in this study. Sub-bottom characteristics interpreted from single-channel air gun records are indicated beside tracklines. Bathymetry in corrected metres is from Prince and others (1980). All data is from Oregon State University (OSU) cruises YALOC 71 and YALOC 73.

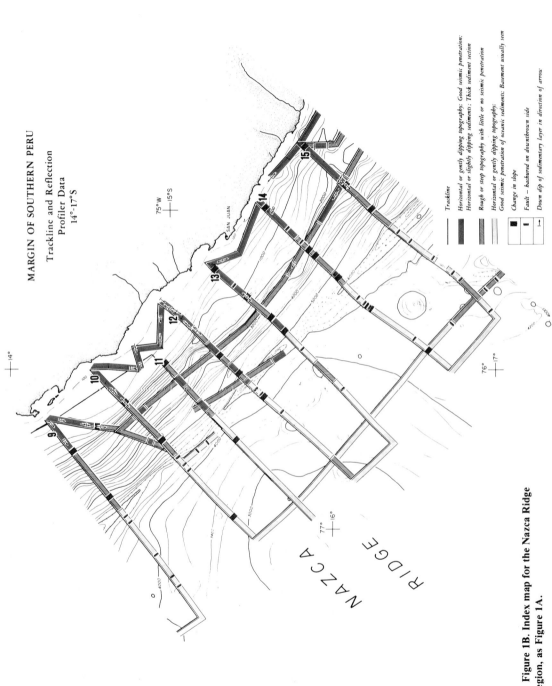

Figure 1B. Index map for the Nazca Ridge region, as Figure 1A.

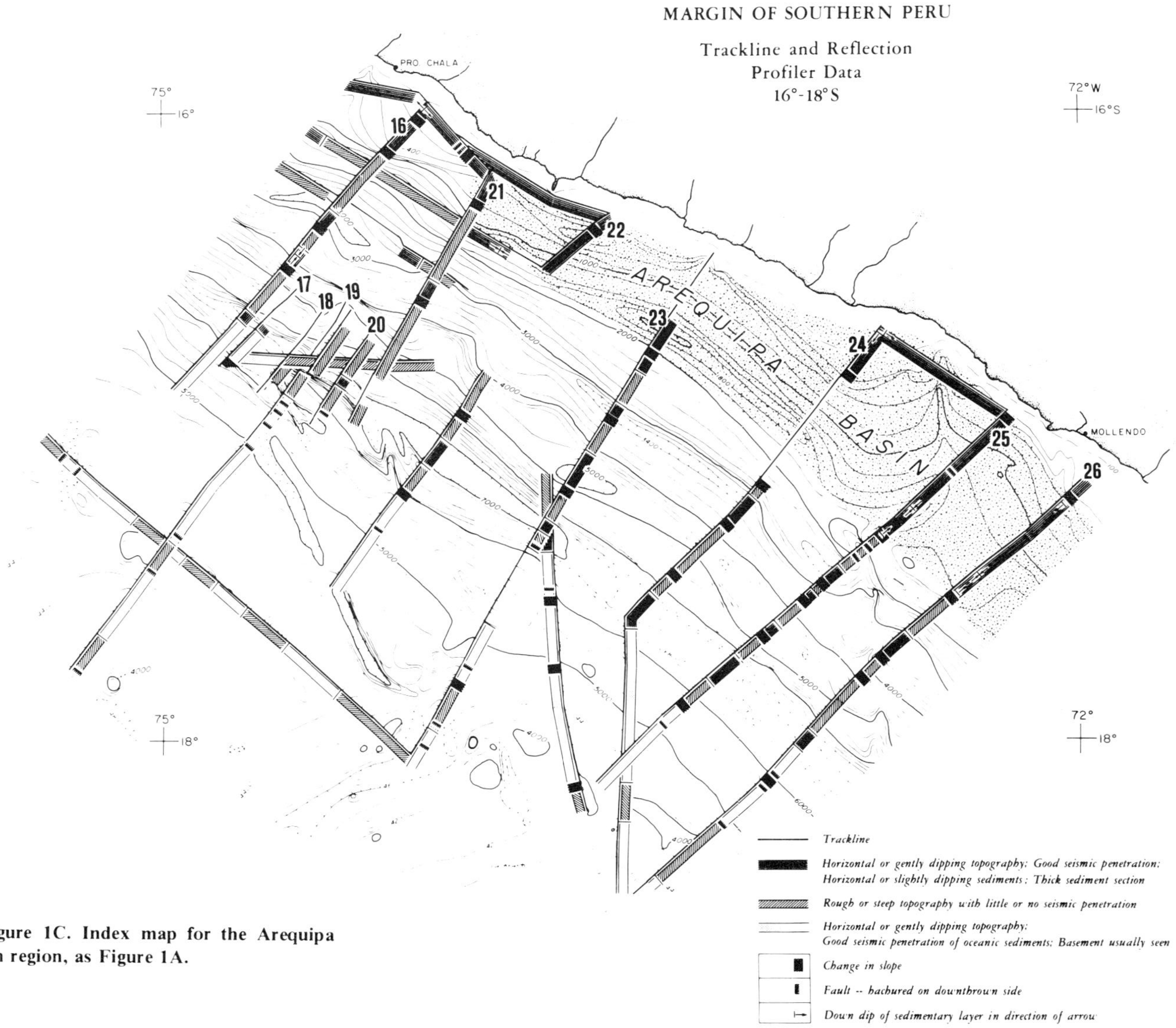

Figure 1C. Index map for the Arequipa Basin region, as Figure 1A.

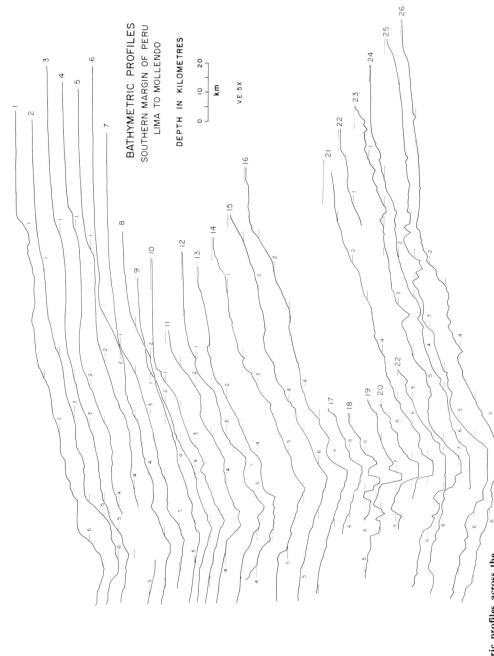

Figure 2. Bathymetric profiles across the southern Peruvian margin.

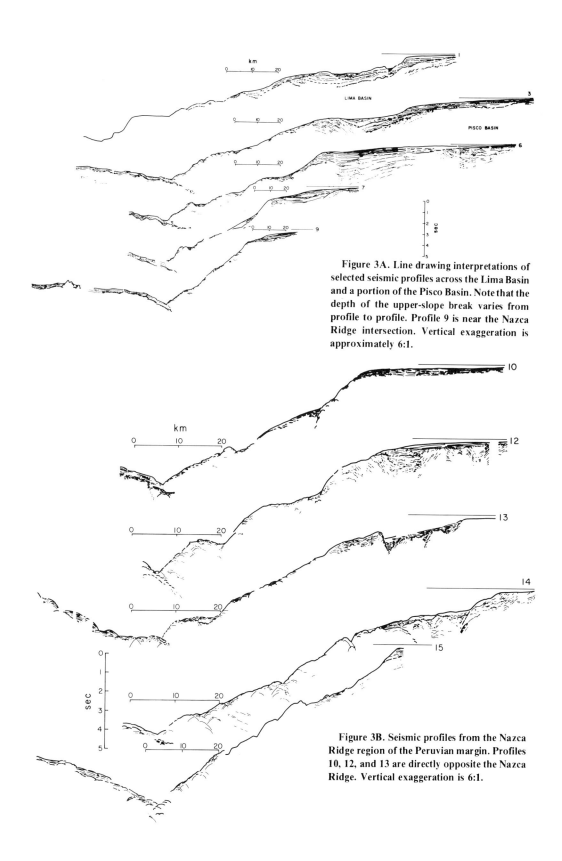

Figure 3A. Line drawing interpretations of selected seismic profiles across the Lima Basin and a portion of the Pisco Basin. Note that the depth of the upper-slope break varies from profile to profile. Profile 9 is near the Nazca Ridge intersection. Vertical exaggeration is approximately 6:1.

Figure 3B. Seismic profiles from the Nazca Ridge region of the Peruvian margin. Profiles 10, 12, and 13 are directly opposite the Nazca Ridge. Vertical exaggeration is 6:1.

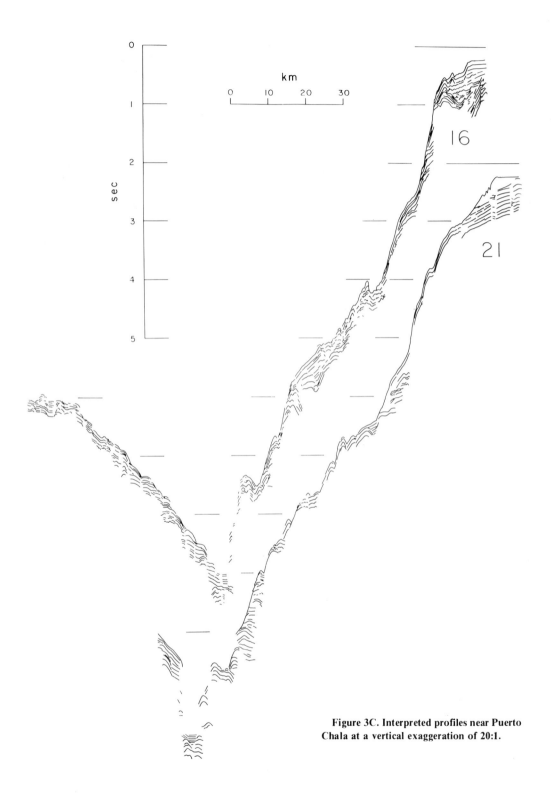

Figure 3C. Interpreted profiles near Puerto Chala at a vertical exaggeration of 20:1.

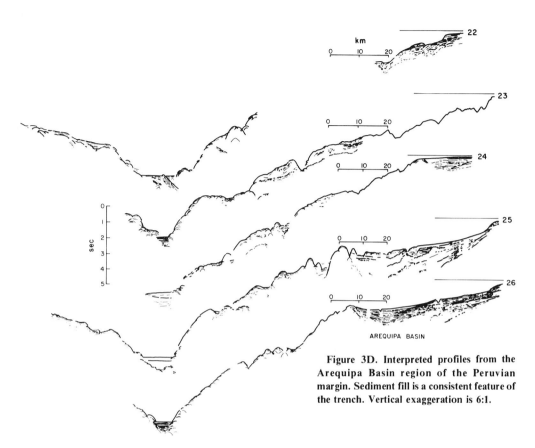

Figure 3D. Interpreted profiles from the Arequipa Basin region of the Peruvian margin. Sediment fill is a consistent feature of the trench. Vertical exaggeration is 6:1.

although the rise is a prominent feature of the Peru Trench north of lat 12°S (Hussong and others, 1976) and of the Chile Trench south of lat 18°S (Coulbourn and Moberly, 1976). The oceanic crust slopes at approximately 3° (with minor normal faulting) into the trench from a distance of about 60 km seaward of the trench axis. In this region, the style of this sloping surface varies greatly, as shown on bathymetric profiles in Figure 2. Here it is seen that the gradient is constant opposite upper-slope basins (profiles 1-10 and 23-26), convex-up opposite the Nazca Ridge (profiles 11-16), and complexly faulted southeast of the Nazca Ridge (profiles 17-22). A 200 m thick cover of pelagic and hemipelagic sediment mantles oceanic basement except over the Nazca Ridge. Here the thickness of calcareous ooze approaches 400 m, because the bottom lies shallower than the calcium carbonate compensation depth (Kulm and others 1974).

Trench Axis

Within the study area, the trench axis deepens away from the Nazca Ridge intersection at a depth of 5 km, to depths of 7 km. The trench axis north of profile 15 is remarkable in that it is nearly devoid of turbidites, probably due to the extremely arid climate and lack of major rivers in the Nazca region of Peru. Profiles 1-15 (Figs. 3A, 3B, and 7) show generally undeformed sediments gently plunging into a V-shaped trench. Oceanic basement can be followed landward beneath lower-slope deposits for distances of 5 km (Figs. 3A, 3B, and 7). In general, layered reflectors terminate at the base of the slope, although in at least one case (profile 7, Fig. 7) there is a suggestion that undeformed oceanic sediments may underlie slope deposits. Profile 13 (Fig. 7) shows what may be an example of imbricate thrust faulting of oceanic basement and is the only such example found in this survey. Blocky bathymetry is seen in profiles 17-21 (Fig. 2), where the trace of the trench changes from convex to concave seaward.

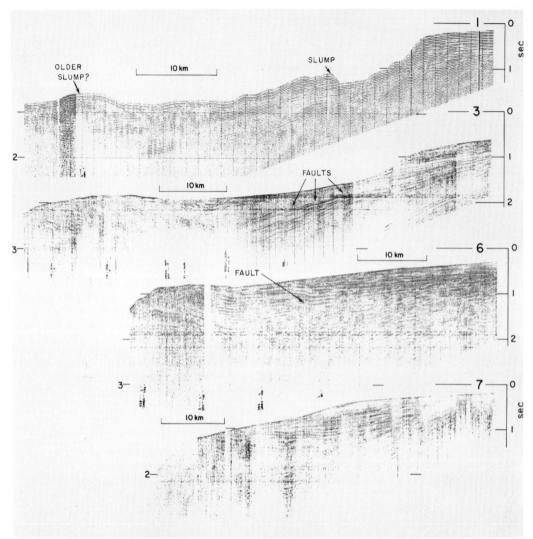

Figure 4. Reproduction of reflection records across the Lima Basin. Reverse motion on vertical faults can be seen on profiles 3 and 6. A large slump feature is present on profile 1.

To the south, turbidites fill the axis in profiles 21-26 (Figs. 3C, 3D, and 7). These sediments show no sign of distortion, although minor tilting is occasionally observed. Profile 22 (Figs. 3D and 7) shows steeply dipping reflectors underlying horizontal trench fill. If this feature is interpreted as a fault block in the trench similar to fault blocks observed elsewhere (Scholl and others, 1970; Prince and Kulm, 1975), then there has been no differential movement with respect to basement since its emplacement.

Lower Slope

Perhaps due to great water depth, even the low vertical exaggeration of most profiles presented here (contrast Figs. 3A, 3B, and 3C) is not sufficient to allow the resolution of details within the seismically chaotic pile of sediments that form the inner wall of the lower slope (here generally deeper than 4 km). Changes of gradient on the slope are often associated with small sediment ponds. In this region, these benches are common and often significant features (Kulm and others, 1977), with the lower one

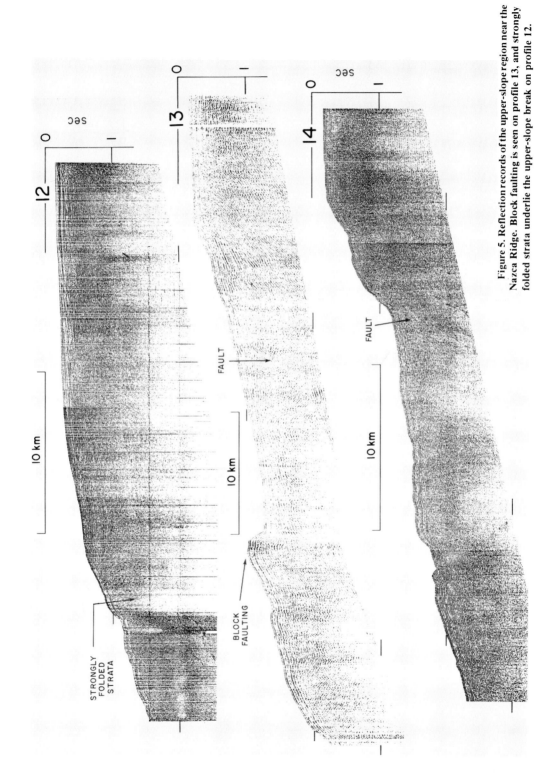

Figure 5. Reflection records of the upper-slope region near the Nazca Ridge. Block faulting is seen on profile 13, and strongly folded strata underlie the upper-slope break on profile 12.

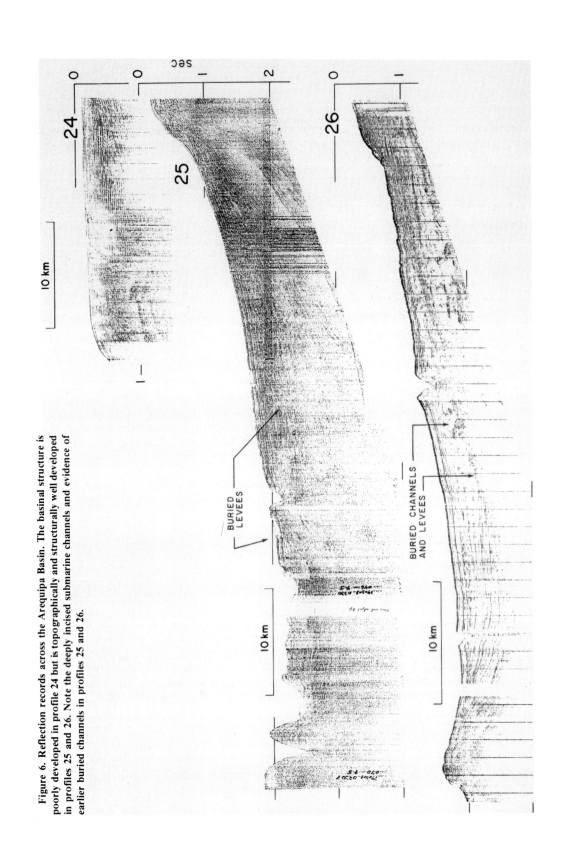

Figure 6. Reflection records across the Arequipa Basin. The basinal structure is poorly developed in profile 24 but is topographically and structurally well developed in profiles 25 and 26. Note the deeply incised submarine channels and evidence of earlier buried channels in profiles 25 and 26.

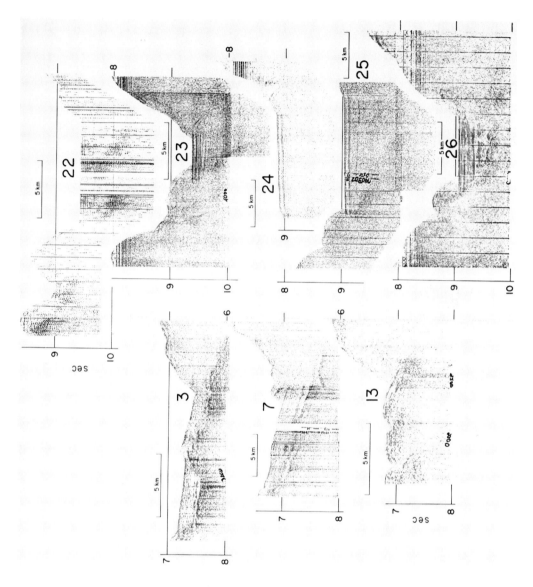

Figure 7. Examples of trench fill and the lowermost slope. Only hemipelagic sediments are evident in profiles 3, 7, and 13; turbidite fill is seen in the remaining profiles. Possible imbricate thrusting of oceanic crust is seen in profile 13.

usually standing 1,000–1,500 m above the trench (Figs. 3A, 3B, 3C, and 3D). Additional benches often occur farther upslope but without the significant ponding of sediments often observed in the western and northern Pacific (Moore and Karig, 1976). The inner trench wall, deeper than about 4 km, steepens on either side of the Nazca Ridge. This is one of the few phenomena that seem to correlate with the presence of the Nazca Ridge.

Middle Slope

The middle slope (2 to 4 km depths) undergoes a marked change in character between Lima and Mollendo. Opposite the Nazca Ridge (profiles 9-15, Fig. 2), the middle slope has a gradient of 1:12 versus 1:16 elsewhere. Mid-slope benches are common on the reflection profiles but are not easily correlated between the profiles. Between lat 12°S and 15°S (Figs. 1A and 3A), mid-slope benches are often filled with undeformed sediments, whereas those between lat 14°S and 17°S (Figs. 1B and 3B) are not. The middle slope between lat 16°S and 18°S (Figs. 1C, 3C, and 3D) has greater bathymetric relief than elsewhere on this margin. The rough, hummocky topography, especially evident on profile 25, and the blocky forms seen on profile 23 are noteworthy. On these profiles, sediment fill is absent in spite of deeply incised benches, and the sub-bottom reflectors present suggest the deformation of shallow material. Gravity modeling along profile 25 (Couch and Whitsett, this volume) indicates that an abnormally thick section (3 km) of low-density (2.20 gm/cc) material comprises the middle slope. Lateral variations between profiles on this region are greater than those just to the north, as can be seen from the bathymetry in Figure 2 and on the reflection profiles in Figure 3D.

Upper Slope

The upper slope off southern Peru does not fit conventional classification schemes (Karig and Sharmon, 1975; Dickinson and Seely, 1979), because the depth of the slope break is highly variable and the slope break does not always coincide with a structural high. For convenience, we define the upper slope in this area as lying between 100 and 2,000 m. Along the upper slope are two regions where the slope gradient is low. These are the locations of two subsurface basins, the Lima Basin to the north, of which only the southern portion is seen on these profiles, and the Arequipa Basin to the south (Masias, 1976). In the intervening region and opposite the Nazca Ridge, basin development is minimal, and subsurface structures are quite different from the basins.

The Lima and Arequipa Basins have many common features. Both underlie water depths between 100 and 1,500 m, have broad synclinal axes, and occasionally exhibit minor faulting. Most profiles show recent out-building of shelf sediments with some instances of slumping. Basin strata appear to have been deposited on horizontal or gently sloping surfaces with occasional angular unconformities, suggestive of periodic episodes of rapid changes in elevation. In general, the outer structural high is composed of a broadly upwarped sedimentary section that may be traced to depths of at least 2 km, the limit of penetration on these profiles.

The Lima Basin (Fig. 1A) extends into the study area from the north. It narrows starting at lat 12.5°S and disappears at lat 14°S, just north of the Nazca Ridge intersection. Sub-bottom structures change rapidly between profiles where the basin axis crosses bathymetry contours at lat 13.5°S (compare profiles 3 and 6, Fig. 4). Seaward convergence of reflectors toward the outer structural high is not characteristic of the Lima Basin except in the uppermost sediments.

The Arequipa Basin lies along the southernmost coastline, from lat 16° to 18°S, beneath a seaward-sloping bottom. Coulbourn and Moberly (1977) describe the southern portion of this basin and point out the seaward convergence of reflectors against the outer structural high, also seen here in Figures 3D and 6. Figure 1C shows that west of long 73°W the basin narrows markedly and loses identity at about long 74°W. Profiles 25 and 26 (Fig. 6) show leveed channels incised up to 160 m into the bottom. Buried channel structures indicate the long influence of turbidity currents on slope sedimentation in this basin. Episodic vertical movements are suggested by angular unconformities. Thinning of layers toward the outer structural high is evidence for either the gradual uplift of the slope break or the downwarping of the basin axis between episodic tilting or uplift events. Basement reflectors are not seen within the limits of seismic penetration, and gravity models indicate a thick, low-density (2.20 gm/cc) section underlying the offshore margin. Pre-Cambrian metamorphic rocks occur immediately onshore (Mapa Geologico Generalizado del Peru, 1969) so that basement either ends near the coast or dips very steeply seaward.

The upper continental slope opposite the Nazca Ridge contrasts strongly with adjacent parts of the continental slope. The upper slope is steeper, and the sub-bottom is more poorly reflective on profiles 9 through 12 (Figs. 3A and 3B). Profiles farther along the margin to the southeast show a more gradual slope gradient similar to the basins, but, in contrast, subsurface reflectors suggest a hard basement with evidence of block faulting.

A third basin is seen on the shelf north of lat 14°S. This is believed to be the seaward extension of the Pisco Basin known on land. Reflectors are seen to depths of 2 sec of two-way reflection time on profile 6, but trackline coverage is sparse, and shallow-water multiples obscure most deep reflections. Distinct magnetic and gravity anomalies and anticlinal structures separate the Pisco Basin from the adjacent Lima Basin.

DISCUSSION

Trench Sediments and Slope Basins

The relative lack of sediments in the trench axis contrasts with the substantial amount of sediment seen in the upper-slope basins of the study area. A reasonable explanation should include the aridity of the adjacent coastal area, the role of the upper-slope basins as sediment traps, and the influence of the Nazca Ridge on the depth of the trench axis.

Although Hayes (1966, 1974) described the trench between lat 8°S and 32°S as a sediment-free province, this is now known to be not strictly true. Prince and Kulm (1975) find nearly continuous fill of the trench axis from lat 6.5°S to as far as lat 10°S. Coulbourn and Moberly (1977) detect ponded sediments in the trench at lat 19.5°S. Records shown in Figures 3 and 7, and sediment mapped in Figures 1A-1C, show the trench to be almost devoid of turbidites from lat 13°S to 17°S. However, from lat 17°S to 18°S, where the trench trends ESE, sediments with horizontal layering are detected on all records. These have also been mapped by Kulm and others (1977).

Schweller and Kulm (1978) comment that sediment supply and trench structure are the two most important factors that regulate the amount of sediment in the trench. The narrow arid coastal plain of southern Peru severely limits the sediment supply available to the offshore. The trench axis shallows to 4,900 m over the Nazca Ridge at lat 15°S and deepens to 7,900 m at lat 19°S, isolating this segment of the trench from axially flowing turbidity currents.

Within this study area, slope structure is probably equally important. A fairly consistent feature of the lower slope is a well-defined bench that typically stands 1,500 m above and extends 20 km landward from the trench axis. Similar benches exist upslope. The uppermost benches are often partly filled by a seaward-sloping veneer of presumably terrigenous sediments up to 300 m thick. On the upper slopes lie the structural highs that bound the upper-slope basins. These basins appear to be formed of terrigenous sediments deposited landward of the upper-slope structural high. The structural high probably rose from a considerable depth so that sediments which lap onto it, and presently dip landward, were originally deposited on the slope dipping seaward. Striking examples of this process are illustrated by Coulbourn and Moberly (1977) for the Iquique and Arica Basins off northern Chile.

We suggest that the major portion of terrestrial sediments that do invade the marine environment is trapped principally within the upper-slope basins and, if not there, in the middle- and lower-slope benches. The small amount of sediments observed in the trench probably travels down-slope through submarine channels such as those observed on profiles 25 and 26 (Figs. 3D and 6) or along the trench axis from the intersection of the Nazca Ridge with the lower slope.

Slope Formation

Currently popular models for convergent margins assume either accretion or consumption of existing material at the outer continental margin, or admit some intermediate condition where material balance is constant. The off-scraping model of Karig (1974) and Karig and Sharman (1975) proposes that pelagic sediments, trench fill, and perhaps oceanic crust accrete to the lower trench slope and that in time these are forced upward between imbricate thrusts toward a "trench-slope break." Seismic reflection profiles supporting this concept have been presented by Beck and Lehner (1974), Seely and others (1974), and others. Hamilton (1977) proposed a modification that emphasizes the possible additional effect of gravity sliding in the development of imbricate thrusts. By contrast, in the subduction model of Scholl and Marlow (1974) and Scholl and others (1977), oceanic deposits are displaced downward together with the oceanic crust beneath the active margin; little, if any material is scraped off at the trench. Kulm and others (1977) extend the consumption model to include removal of lower-slope and even continental crystalline material and their subduction along with oceanic sediments and crust.

Kulm and others (1977) divide the South American subduction zone into three provinces based on criteria that they established to distinguish between these models. Two of the provinces (including the area of this study) are recognized as accretionary (for at least several million years) and one as consuming. Hussong and others (1976) argue, partly on the basis of the sediment volume arriving at the Peru Margin over the past 150 m.y., that the convergence zone has been a consuming boundary throughout its subduction history.

Accretionary models proposed for convergent margins characterize the lower slope as consisting of imbricate thrust sheets, chaotic piles of sediment, or a sediment-basalt melange. Seismic reflection data obtained during the course of this study cannot be used to define either the structure or composition of the lower slope off southern Peru. Imbricate thrust sheets are not observed in the reflection records, although there is evidence for oceanic basement underlying lower-slope material on selected profiles (see profiles 3, 7, and 13 on Fig. 7). Imbricate thrusting of the Peru Margin has been invoked by several authors (Hussong and others, 1976; Kulm and others, 1977; Coulbourn and Moberly, 1977), but the seismic evidence is nondefinitive. The lower slope can be described at best as seismically chaotic, a situation that could be generated by any combination of off-scraping or comsumption models. Uplift of the lower slope is suggested by the presence of pelagic sediments at only one location off southern Peru (Kulm and others, 1974).

Uplift of material over long periods of time to form the upper-slope structural high seems to be a requirement for any convergent model applied to southern Peru. The upper-slope basins of Peru and adjacent northern Chile seem to be best explained by upward movement of the outer margin, or, alternatively, by depression of the shelf and slope landward of the upper-slope structural high. The imbricate thrust model of Seely and others (1974) generates upward movement by continuous underplating of the margin with thrust sheets. Physical model studies by Cowan and Silling (1978) suggest that material might be carried beneath the subduction complex, as suggested by Scholl and Marlow (1974) and Scholl and others (1974), and also that it exerts an upward flow from beneath the complex. Alternating cycles of accretion and consumption over periods of tens of millions of years may also be responsible for vertical movement within the slope.

The large lower-slope bench is another feature of the Peru Margin that must be explained by a tectonic model. This could represent a series of thrust sheets in the imbricate thrust model, a rotating lower lobe in the internal flow model, or a large slump feature associated with a region of high seismicity.

The data presented here do not strongly or exclusively support any particular tectonic model. Instead, features found in several models may be used to explain the characteristics of the southern Peru Margin. These are listed below with reference to Figure 8, which is based on profiles 12 (no upper-slope basin) and 25 (containing an upper-slope basin).

1. Vertical motion is produced by the addition of trench material to the bottom side of the slope wedge either by imbricate thrusting or by internal flow.

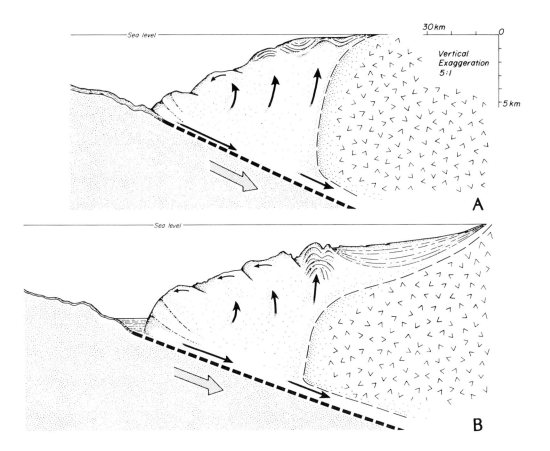

Figure 8. Characteristics of the imbricate thrust model of Seely and others (1974) and the internal flow model of Cowan and Silling (1978) as applied to the development of the southern Peruvian margin. The lowermost lobe accretes as oceanic crust and sediments underthrust it. Most of the material moves beneath the accretionary wedge, where it then either moves upward by some combination of imbricate thrusting and internal flow or is finally subducted beneath the continental crystalline rocks. Near the upper suface of the wedge, gravity processes cause downslope movement of material. (A) The model applied to profile 12, where the edge of the crystalline rocks is bluntly shaped. No upper-slope basin is formed. (B) The model applied to profile 25. Terrestrial sediment moving trenchward is trapped behind the rising accretionary wedge, and an upper-slope basin develops. The structural high bounding the seaward margin of the basin locates the edge of the crystalline rocks.

2. The upper-slope structural high marks the edge of the accretionary prism. An upper-slope basin would form where the crystalline complex was below sea level or was faulted and pulled down by subduction forces.

3. The lowermost bench is probably formed by imbricate thrusting, but simultaneous internal flow may disrupt individual thrust sheets and contribute to the lobate form of the bench.

4. Gravity sliding along the upper surface of the accretionary wedge is a consequence of dewatering of underlying material, the steepness of the slope, and the high seismic activity in the trench region.

5. On the average, roughly half of the oceanic sediments entering the trench must be subducted beneath the crystalline complex. This may occur as the margin alternates between being accretionary and consuming. It may also occur as a continuous process.

CONCLUSIONS

Using 26 single-channel reflection profiles across the southern Peru Margin, we have investigated shallow structures formed where oceanic crust is subducted beneath the edge of the continental crystalline rocks of South America. The Lima and Arequipa upper-slope basins are prominent elongate features of regional size that lie landward of a tectonically formed upper-slope break.

No slope basins are found opposite the Nazca Ridge, but there may be no causal relationship in this since the ridge seems to have very little effect upon other features of the margin. Indeed, the curvature of the mapped trace of the trench is unaffected by the ridge, and the structures and morphology of the slope opposite the ridge are not radically different than elsewhere along this margin. Frankly, we find this lack of relationship surprising. Perhaps the scale of the relationship is larger, or evidence for the relationship lies in deeper structures.

The slope is characterized by large-scale lobate structures having vertical dimensions of 1,000 to 1,500 m and forming a series of benches parallel to the trench. Sediment ponding on the lower slope is rare, probably because of both restricted sediment input due to arid conditions on land and the sediment-trapping effect of benches and basins on the upper slope.

In the study area, turbidites occur in the trench south of lat 17°S. This material probably bypasses slope benches via submarine canyons and channels that are seen on reflection records. Active and buried leveed channels are seen in the Arequipa Basin—evidence that turbidity current flow is an important mechanism in the reworking of basinal deposits in this area.

The data do not exclusively support any particular tectonic model. Instead, we invoke characteristics of imbricate thrusting and internal flow to account for features observed in the trench and upper slope. The proximity of a continental crystalline block to a subduction zone is a situation unlike those observed in the western Pacific, and the broad structural styles of the two areas are different. We feel, however, that internal flow may be an important general process in any subduction zone, including oceanic-oceanic convergences, and that the imbricate underthrusting mechanism, when considered alone, is inadequate to explain the observed phenomena.

REFERENCES

Barazangi, M., and Isacks, B. L., 1976, Spacial distribution of earthquakes and subduction of the Nazca plate beneath South America: Geology, v. 4, p. 686–692.

Beck, R. H., and Lehner, P., 1974, Oceans, new frontier in exploration: American Association of Petroleum Geologists Bulletin, v. 58, p. 376–395.

Couch, R., and Whitsett, R. M., 1981, Structures of the Nazca Ridge and the continental shelf and slope of southern Peru, in Kulm, L. D., and others, eds., Nazca plate: Crustal formation and Andean convergence: Geological Society of America Memoir 154 (this volume).

Coulbourn, W. T., and Moberly, R., 1977, Structural evidence of the evolution of fore-arc basins off South America: Canadian Journal of Earth Science, v. 14, p. 102–116.

Cowan, D. S., and Silling, R. M., 1978, A dynamic scaled model of accretion at trenches and its implications for the tectonic evolution of subduction complexes: Journal of Geophysical Research, v. 83, p. 5389–5396.

Dickinson, W. R., and Seely, D. R., 1979, Structure and stratigraphy of fore-arc regions: American Association of Petroleum Geologists Bulletin, v. 63, p. 2–31.

Fisher, R. L., and Raitt, R. W., 1962, Topography and structure of the Peru-Chile Trench: Deep-Sea Research, v. 9, p. 423–443.

Hamilton, W., 1977, Subduction in the Indonesian region, in Talwani, M., and Pitman, W. C., III, eds., Island arcs, deep sea trenches and back-arc basins: Washington, D.C., American Geophysical Union, Maurice Ewing Series 1, p. 15–31.

Handschumacher, D. W., 1976, Post-Eocene plate tectonics of the eastern Pacific, in Sutton, G. H., and others, eds., The geophysics of the Pacific Ocean Basin and its margin: Washington, D.C., American Geophysical Union Monograph 19, p. 177–202.

Hayes, D. E., 1966, A geophysical investigation of the Peru-Chile Trench: Marine Geology, v. 4, p. 309–351.

——1974, Continental margin of western South America, in Burk, C. A., and Drake, C. L., eds., The geology of continental margins: New York, Springer-Verlag, p. 581–590.

Herron, E. M., 1972, Sea-floor spreading and the Cenozoic history of the east-central Pacific: Geological Society of America Bulletin, v. 83, p. 1671–1692.

Hussong, D. M., and others, 1976, Crustal structure of the Peru-Chile Trench: 8°–12°S latitude, in Sutton, G. H., and others, eds., The geophysics of the Pacific Ocean Basin and its margin: Washington, D.C., American Geophysical Union Monograph 19, p. 71–85.

Karig, D. E., 1974, Evolution of arc systems in the wetern Pacific: Annual Review of Earth and Planetary Science, v. 2, p. 51–75.

Karig, D. E., and Sharman, G. F., 1975, Subduction and accretion in trenches: Geological Society of America Bulletin, v. 86, p. 377–389.

Kulm, L. D., and others, 1974, Transfer of Nazca Ridge pelagic sediments to the Peru continental margin: Geological Society of America Bulletin, v. 85, p. 769–780.

Kulm, L. D., Schweller, W. J., and Masias, A., 1977, A preliminary analysis of the subduction processes along the Andean continental margin, 6° to 45°S, in Talwani, M., and Pitman, W. C., III, eds., Island arcs, deep sea trenches and back-arc basins: Washington, D.C., American Geophysical Union, Maurice Ewing Series 1, p. 285–301.

Lonsdale, P., 1978, Ecuadorian subduction system: American Association of Petroleum Geologists Bulletin, v. 62, p. 2454–2477.

Mapa Geologico Generalizado del Peru, 1969, Republica del Peru, Servicio de Geologia y Mineria, scale 1:2,500,000.

Masias, J. A., 1976, Morphology, shallow structure, and evolution of the Peruvian continental margin, 6° to 18°S [M.S. Thesis]: Corvallis, Oregon State University, 92 p.

Minster, J. B., and Jordan, T. H., 1978, Present-day plate motions: Journal of Geophysical Research, v. 83, p. 5331–5354.

Moore, G. F., and Karig, D. E., 1976, Development of sedimentary basins on the lower trench slope: Geology, v. 4, p. 693–697.

Prince, R. A., and Kulm, L. D., 1975, Crustal rupture and the initiation of imbricate thrusting in the Peru-Chile Trench: Geological Society of America Bulletin, v. 86, p. 1639–1653.

Prince, R. A., and others, 1974, Uplifted turbidite basins on the seaward wall of the Peru Trench: Geology, v. 2, p. 607–611.

——1980, Bathymetry of the Peru-Chile Trench and continental margin: Geological Society of America Map Series, MC-34, scale 1:1,000,000.

Scholl, D. W., and Marlow, M. S., 1974, Global tectonics and the sediments of modern and ancient trenches—some different interpretations, in Kahle, C. F., ed., Plate tectonics, assessments and reassessments: American Association of Petroleum Geologists Memoir 23, p. 255–272.

Scholl, D. W., and others, 1970, Peru-Chile Trench, sediments and sea-floor spreading: Geological Society of America Bulletin, v. 81, p. 1339–1360.

Scholl, D. W., Marlow, M. S., and Cooper, A. K.,

1977, Sediment subduction and offscraping at Pacific margins, *in* Talwani, M., and Pitman, W. C., III, eds., Island arcs, deep-sea trenches and back-arc basins: Washington, D.C., American Geophysical Union, Maurice Ewing Series 1, p. 199–210.

Schweller, W. J., and Kulm, L. D., 1978, Extensional rupture of oceanic crust in the Chile Trench: Marine Geology, v. 28, p. 271–291.

Seely, D. R., Vail, P. R., and Walton, G. G., 1974, Trench slope model, *in* Burk, C. A., and Drake, C. L., eds., The geology of continental margins: New York, Springer-Verlag, p. 249–260.

MANUSCRIPT RECEIVED BY THE SOCIETY NOVEMBER 12, 1980
MANUSCRIPT ACCEPTED DECEMBER 30, 1980

Clay mineralogy of the Peru continental margin and adjacent Nazca plate: Implications for provenance, sea level changes, and continental accretion

Victor J. Rosato*
School of Oceanography
Oregon State University
Corvallis, Oregon 97331

L. D. Kulm
School of Oceanography
Oregon State University
Corvallis, Oregon 97331

ABSTRACT

The clay mineralogy and organic carbon contents of 52 surface and 53 subsurface samples were determined for Quaternary sediments of the Peru continental margin and adjacent Nazca plate. By using Q-mode factor analysis, three factors (oceanic, continental A, and continental B) can explain 99% of the variation in clay mineral composition and organic carbon content in surface sediment. Northeast Nazca plate surface sediment is characterized by an oceanic factor that is dominated by smectite with subordinate illite. Continental margin surface sediment is characterized by the following clay mineral assemblages: (1) continental factor A, which consists of smectite-chlorite mixed-layer clays and the detrital clay minerals illite, kaolinite, and chloride, or (2) continental factor B, which is composed of illite with subordinate kaolinite and chlorite. Upper continental margin (<2,000 m) surface sediment is dominated by either of these continental assemblages, while lower continental margin sediment is characterized by the oceanic or continental-A assemblage.

All subsurface data were compared to the surface factor model by constructing a pseudo-factor matrix to examine the temporal changes in clay mineral assemblages. In the late Quaternary (≤0.44 m.y.b.p.), the boundary between surface sediment with a dominant oceanic factor and sediment with a continental factor (A or B) lies on the continental slope. Earlier in the Quaternary (0.44 to 2.0 m.y.b.p.), this boundary was about 100 km seaward of its present position, a shift that we attribute to regression of the shoreline as a result of glaciation.

By comparing the subsurface factor loadings to the surface factor loadings, displaced sediment can

---

*Present address: Union Oil Company of California, P.O. Box 6176, Ventura, CA 93006

be recognized. Four cores taken on the seaward side of the present Peru Trench axis contained terrigenous turbidites that have been uplifted since deposition. Two cores on the lower continental slope opposite the Nazca Ridge contained pelagic sediment (clay and calcareous ooze) accreted from the Nazca plate. However, most of the 28 cores recovered from the landward wall of the trench were terrigenous turbidites and hemipelagic muds, not accreted Nazca plate sediment.

INTRODUCTION

Since 1971, scientists at Oregon State University have studied the geology, geophysics, and geochemistry of the Peru continental margin and Nazca lithospheric plate. We have examined one aspect of the sedimentary picture, the clay mineralogy and organic carbon content and its relationship to structural settings, in the area between long 72° and 88°W and lat 5° and 18°S (Fig. 1). This work complements recent studies to the west on the mineralogy of metalliferous sediment on the East Pacific Rise (EPR) and Bauer Deep by Heath and Dymond (1977), and to the north on the clay mineralogy of the Eastern Equatorial Pacific by Heath and others (1974).

Previous studies have listed the main differences between typical Nazca plate, Peru-Chile Trench, and continental margin sediments (Rosato and others, 1975; Krissek and others, 1980). Turbidites are restricted to the trench and continental margin except for recently discovered uplifted basins on the seaward side of the Peru Trench (Prince and others, 1974; Schweller and Kulm, 1978). Deep-Sea Drilling Project (DSDP) data for sites 320 and 321 of Leg 34 show that silica-rich silty clay, brown clay, and clacareous ooze are the typical facies of the Nazca plate (Kulm and others, 1976). Earlier work also indicated that there would be differences in the mineralogy of fine-grained sediments (Griffin and others, 1968).

Seely and others (1974) have developed a trench-slope model for continental accretion that predicts that the completely folded and/or thrusted continental slope may be comprised of deformed trench-floor turbidites, abyssal plain sediment, and continental margin sediments. Previous Nazca plate studies off central Peru and northern Chile also suggest that some deposits on the continental slope may be accreted sediments (Masias, 1976; Prince and Kulm, 1975; Coulbourn and Moberly, 1977). Clearly, the ability to determine the original area of deposition of the sediment found on the landward wall of the Peru-Chile Trench would help to substantiate an accretion model.

The objectives of this study are to (1) determine the clay mineralogy and organic carbon content of surface sediment on the Peru continental margin and northeastern Nazca plate, and (2) determine whether sediments found on the acoustically complex landward wall of the Peru Trench are indigenous or accreted deep-sea sediments.

METHODS

Some seismic reflection profiles were made using 40 in^3 air guns and a Teledyne hydrophone streamer coupled to a Huckaby amplifier as the receiver. Filter frequencies were between 30 and 160 Hz. Other profiles were taken with a 300 in^3 air gun as a seismic source, a standard single-channel recording system, and filter frequencies of 40 to 100 Hz.

Clay mineral preparation for 52 surface and 53 subsurface samples consisted of chemical treatments to remove carbonates (acetic acid buffered to pH 5), amorphous Fe-Mn hydroxides (Na-citrate, dithionite method of Mehra and Jackson, 1960), and opal (0.2 M Na_2CO_3). The <2 μm fraction of each sample was separated by settling and saturated with Mg. Three oriented aggregates on porous, Ag-backed X-ray slides were made of each sample. Two slides were sprayed with glycerol and used for semiquantitative calculations of clay mineral abundances (Biscaye, 1965), and one slide was treated with ethylene glycol to determine the degree of expansion of the smectite group of clay minerals. All samples were analyzed by X-ray diffraction. Additional tests such as K-saturation and heating (Weaver, 1956; Hayes, 1973), were run on the surface sample from core 5, which contained a typical

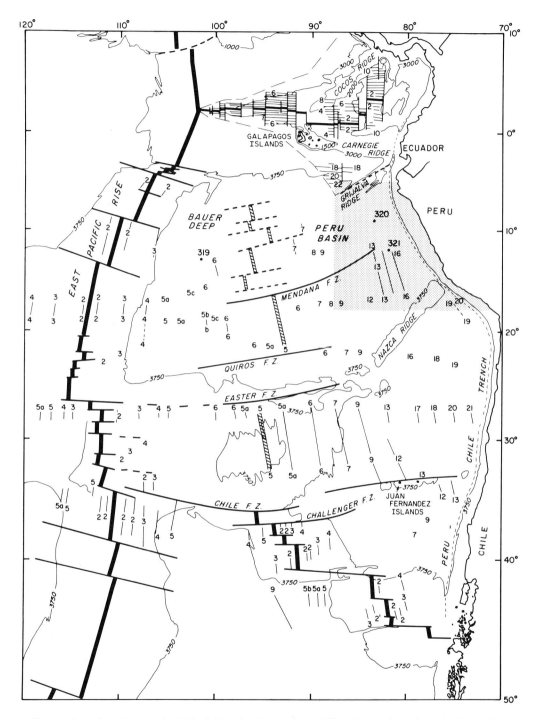

Figure 1. Location of area studies (strippled area) on Nazca plate and Peruvian continental margin. Magnetic anomalies, ridge crests, and fracture zones after Herron (1972).

mixed-layer clay mineral identified as a smectite-chlorite intergrade. The mixed-layer clays were characterized by a broad band on diffractograms between 14 and 20 Å after Mg saturation and glycerol solvation. They lacked resolvable higher-order basal reflections and displayed a resistance to collapse after K-saturation and heating to 300°C. The 10 Å peak intensity relative to the 7 Å peak intensities did not increase after K-saturation, which suggests that vermiculite interlayers are not present. In some of the samples, ethylene glycol expanded the structure to only 16.3 Å, which is less than expected for normal smectites. These properties suggest structural similarities to the smectite-chlorite clay minerals commonly found in soils (Jackson, 1963) and recently synthesized in the laboratory (Carstea, 1967). Oinuma and others (1959) and Powers (1957) have described a similar mixed-layer clay in some Pacific Ocean sediment.

Calcite and dolomite were determined by X-ray diffraction using a ThO_2 spike, which provides two precisely known peaks on either side of the calcite 101 and dolomite 104 peaks. The percentage of Mg in the calcite was estimated from the charts of Chave (1952).

Organic carbon and calcium carbonate abundances were determined with a LECO induction furnace on duplicate samples from the same intervals from which samples were taken for mineralogy.

The radiolarian stratigraphy used here is based on Hays (1970) and Berggren (1972). As discussed in Kulm and others (1974), the ages are based on the extinctions of four radiolarian species:

Species	Extinction	Age
Stylatractus universus (Hays)	0.44 m.y.b.p.	Late Quaternary
Lamprocyclas heteroporous	2.0 m.y.b.p.	Early Quaternary
Stichocorys peregrina (Riedel)	2.7 m.y.b.p.	Late Pliocene
Stichocerys delmontensis (Campbell and Clark)	6.0 m.y.b.p.	Late Miocene

Figure 2 gives the generalized bathymetry and the locations of the cores described. Piston cores, free-fall cores, and multiple-gravity cores were used. Core locations were determined by satellite navigation and bathymetry. Additional sediment samples were provided by Lamont-Doherty Geological Observatory.

SURFACE DISTRIBUTION OF CLAY MINERALS AND ORGANIC CARBON

Distributions of clay minerals on the northeast Nazca plate and Peru continental margin are difficult to explain unequivocally without more data, but regional differeneces in the concentrations are noted. These regional differences allow us to differentiate marine and continentally derived deposits and thus to detect pelagic sediments that may have been accreted to the lower continental slope off Peru.

Smectite and Mixed-Layer Clays

We note several interesting trends in the distribution of smectite and mixed-layer clays (Fig. 3A). There is a strong gradient in smectite concentrations from the continental margin to the Nazca plate between lat 6° and 14°S. The trend may extend farther south, as shown by the absence of smectite and mixed-layer clays in Zen's (1959) upper continental margin cores 31 and 42 (Figs. 2, 3A). The dearth of smectite along the continental margin is unexpected, because soils formed on volcanic terrains in arid climates usually are dominated by smectite and illite (Donoso, 1959; Grim, 1968; Besoain, 1969). Smectite-poor continental margin sediments also are unusual, because volcanic ash is a common accessory component of margin sediments in this area (Zen, 1959; Rosato and others, 1975). Smectite completely dominates the $<2\,\mu$ fraction of partially altered ash layers from three Nazca plate cores (4, 12, 7), which suggests that marine deposits of Andean volcanic ash have weathered to smectites. Continentally derived smectite, then, since it is the finest grained of the clay minerals, is either diluted by other clay minerals weathered from the coastal area of Peru, or winnowed out of margin sediment, or kept in suspension by strong coastal currents.

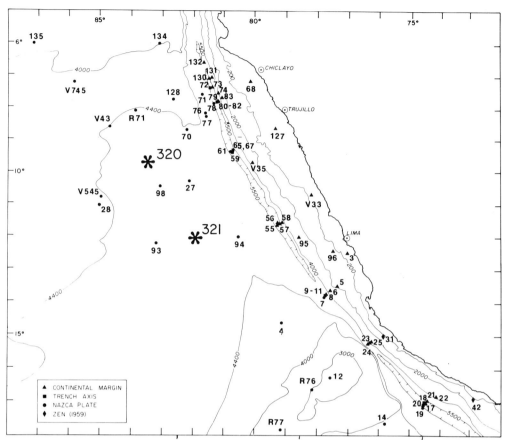

Figure 2. Locations of cores on Peru Margin and Nazca plate. Note position of DSDP drill sites 320 and 321 on Nazca plate.

Another interesting feature of the smectite and mixed-layer clay distributions is the smectite-rich area near the equator. Concentrations are greater than 50%, in agreement with the results of others (Griffin and others, 1968; Heath and others, 1974). This area is close to the Bauer Deep (Fig. 1), where iron-rich smectites are forming (Kendrick, 1974; Eklund, 1974; Yeats, Hart, and others, 1976); to smectite sources in Ecuador (Heath and others, 1974) and northwestern Peru (S. B. Weed, 1974, written commun.); and to areas rich in volcanic debris, such as the Galapagos pedestal and the Carnegie Ridge (Heath and others, 1974).

The concentration of smectite in the Peru Basin is fairly uniform. DSDP sites 320 and 321 (Fig. 2) show that smectite also dominates the older Cenozoic deposits on the Nazca plate. However, the smectite has decreased somewhat in concentration in the younger deposits as the Nazca plate has converged with the South American continent (Kulm and others, 1976).

A prominent lobe of smectite-poor sediment extends seaward from the present trench axis into basins on the eastern edge of the Nazca plate at lat 7°–8°S (Fig. 3A). A northeastern source for this sediment is indicated. Since surface currents flow in the direction opposite to the trend of the lobe, and since there is a topographic low between cores 76 and the continental margin, midwater transport is implied. An alternate explanation is that the sediment may be the fine-grained tail of a turbidite, thus representing sediment displaced from the continental margin prior to recent uplift.

The distribution of smectite south of lat 14°S is different than to the north. Because core 31 and 42 contain no smectite, the concentration gradient between continental margin and Nazca plate sediment

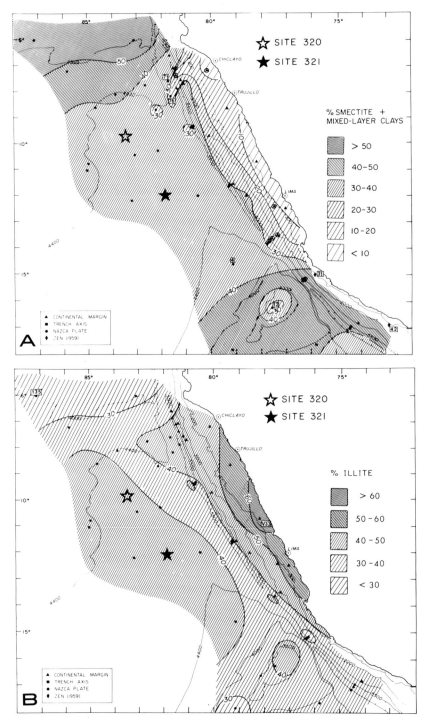

Figure 3. (A) Surface distribution pattern of smectites and mixed-layer clay. Circled symbols (⊛) indicate sediments with mixed-layer clays. Cores discussed in text are indicated by numbers. (B) Surface distribution pattern of illite.

must be shifted markedly toward the shore. Both Nazca plate and lower continental margin sediment contain roughly equivalent amounts of smectite (40-50%) in this area, except one surface sample (12-1) on the northeastern end of the Nazca Ridge, where the smectite is diluted by eolian illite.

Mixed-layer clays are restricted to two areas landward of the Peru Trench (Fig. 3A). There is no evidence to indicate that the mixed-layer clays are forming by diagenesis in the sediments off Peru, but the patchiness in the distribution of the mixed-layer clays makes the source area unclear.

Illite

Illite concentrations in surface sediment generally decrease with increasing distance from the continent (Fig. 3B). Highest concentrations occur on the shelf at lat 11°S (core V33). Lowest concentrations occur on the Nazca Ridge (core R77) and near the equator on the Nazca plate. Illite content, ranging from 40% to 50%, is fairly constant across the Peru Basin. There is a poorly defined concentration gradient from northeast to southwest on the Nazca Ridge.

In the northern hemisphere, illite is thought to be transported mainly by the wind (Griffin and others, 1968). Large rivers, such as the Mississippi and Amazon, also contribute much illite to the ocean.

The distribution of illite off Peru is indicative of both eolian and fluvial inputs. Dispersal in the water column from local sources is shown by the seaward-decreasing illite concentrations on the continental margin, and atmospheric transport is indicated by the relatively high concentrations on topographic highs. The distribution of illite is similar to the surface distribution of quartz in this area, which has been shown to be largely the result of eolian transport (Molina-Cruz and Price, 1977). Because concentrations of illite are low in the trench sediments between the continent and the Nazca Ridge, and high on the northeastern end of the Nazca Ridge, the illite there is probably eolian. The decrease in illite values to the southwest on the Nazca Ridge would then be due to the increasing distance from the continent. Low concentrations of illite in the trench are probably due to dilution by materials carried by the deeper parts of the water column.

Due to the wide spacing of Peruvian shelf samples, we cannot show that particular rivers are sources of illite, but the high illite value found near lat 11°S suggests that this area of the coast may be a source of illite.

Chlorite

Chlorite concentrations generally decrease with increasing distance from the continent without regard to topography (Fig. 4A). Highest concentrations (31%) occur on the continental shelf near lat 13°S, while the lowest concentrations (<10%) occur on the Nazca plate. There are two southward-trending lobes of chlorite-rich surface samples at lat 7° and 13°S. In general, the distribution of chlorite is the inverse of the smectite distribution.

The metamorphic terrains associated with the Upper Cretaceous to lower Tertiary batholiths of coastal Peru are possible sources of the chlorite. Most of these rocks are exposed in the coastal region between lat 8° and 16°S.

Kaolinite

Kaolinite concentrations in surface sediments are quite similar to those of chlorite. Tongues of kaolinite-rich sediment extend southward from near Lima and westward from Chiclayo, which indicates possible sources in those areas (Fig. 4B). Kaolinite generally decreases seaward; however, concentrations on the northeastern end of the Nazca Ridge are slightly lower than to the north and south.

Because kaolinite values are low on the Nazca Ridge, dispersal of kaolinite by winds is not indicated. The lobes of kaolinite-rich sediment extending seaward of Lima and Chiclayo suggest dispersal in the water to the south and west, respectively.

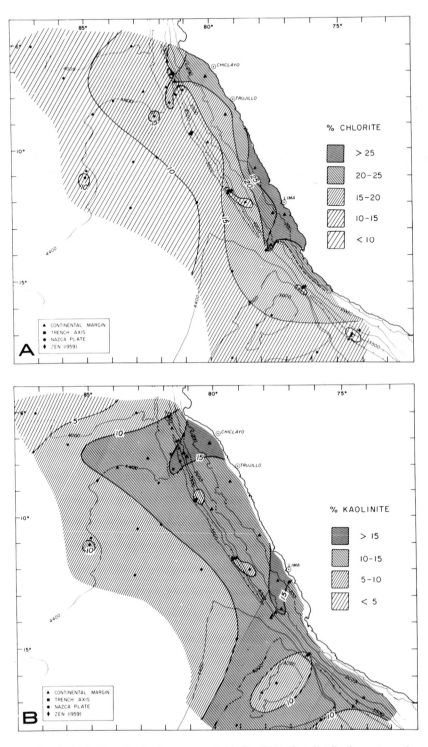

Figure 4. (A) Surface distribution pattern of chlorite. (B) Surface distribution pattern of kaolinite.

Organic Carbon

The distribution of organic carbon in surface sediments is shown in Figure 5. Organic carbon values generally decrease away from the continent. Isopleths of organic carbon content broaden as they approach the equator. There is also a seaward projection of moderately high values (2-3%) over the northeastern terminus of the Nazca Ridge. The highest organic carbon values (7.5%) occur on the upper continental slope and shelf at about lat 13°S, south of Lima. Anomalously low organic carbon contents (<1.5%) are found on the continental shelf in cores 68 and 127.

The primary factor controlling the distribution of organic carbon in surface sediments off Peru is productivity (Rosato, and others 1975; Schuette and Schrader, 1979; and Suess and Müller, 1980): sediments below highly productive upwelled waters generally have significant amounts of organic carbon. We infer that the anomalously low values on the continental shelf between lat 7° and 8°30'S are due to reworking by strong bottom currents; samples from cores 68 and 127 had much higher percentages of sand than shelf samples to the south (cores V33 and 3). Enhanced preservation of organic matter within carbonate tests may lead to the moderate organic carbon contents of the northeasternmost Nazca Ridge samples. Smith and others (1971) have suggested that this is an area where currents converge, a situation that may lead to increased productivity and increased deposition of organic matter.

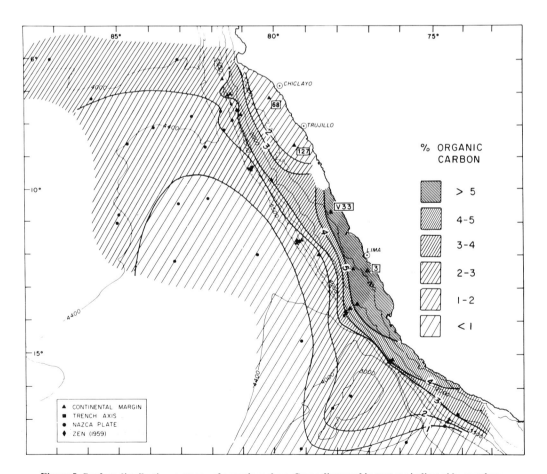

Figure 5. Surface distribution patterns of organic carbon. Cores discussed in text are indicated by numbers.

FACTOR ANALYSIS OF DATA

In order to clarify the interrelationships of the clay mineralogy and organic carbon content, to identify possible source areas, and to create a quantitative basis for identifying displaced fine-grained sediments, the clay mineral and organic carbon contents were subjected to Q-mode factor analysis. This type of analysis as applied to geologic problems is discussed in greater detail in Imbrie and van Andel (1964) and Imbrie and Kipp (1971). The computer program that performs the analysis, CABFAC, is described in Klovan and Imbrie (1971).

Two steps were used in the technique: (1) the surface data were subjected to Q-mode factor anlaysis, and (2) the subsurface data were compared to the surface model by constructing a pseudo-factor matrix. The second step does not construct a new set of factors; rather, it simply multiplies the normalized raw data of the subsurface samples by the varimax factor score matrix from the CABFAC program. As shown below, this allows the recognition of displaced sediment.

The most geologically reasonable factor model that can be constructed from the date is shown in Figure 6. In the model, factors 1, 2, and 3 account for 65%, 25%, and 10% of the variability, respectively. Communalities for all samples were greater than 95.7%, with an average of 99.3%.

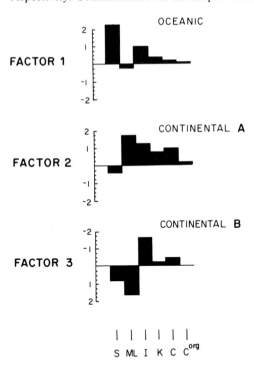

Figure 6. Scaled varimax scores of three factor models. Factor 3 scores are inverted. S = smectite; ML = mixed-layer clay; I = illite; K = kaolinite; C = chlorite; C^{org} = organic carbon.

FACTOR DEFINITION OF SEDIMENT

Factor 1 is dominated by smectite and illite (Fig. 6) and is termed the "oceanic" assemblage because factor loadings tend to increase away from shore (Fig. 7). The areal distribution of factor 1 emphasizes the differences in clay mineral assemblages between upper continental margin and Peru Basin sediments. Generally, sediment west of the axis of the Peru Trench has oceanic factor loadings greater than 0.85. The sample with the highest oceanic factor loading (R77-1) is on the southwestern part of the Nazca Ridge. Lowest values are found on the continental margin opposite the Nazca Ridge. Sediments on the northeastern end of the Nazca Ridge, as well as on the continental margin, have lower oceanic factor loadings because they have higher organic carbon contents and higher proportions of continentally derived clays.

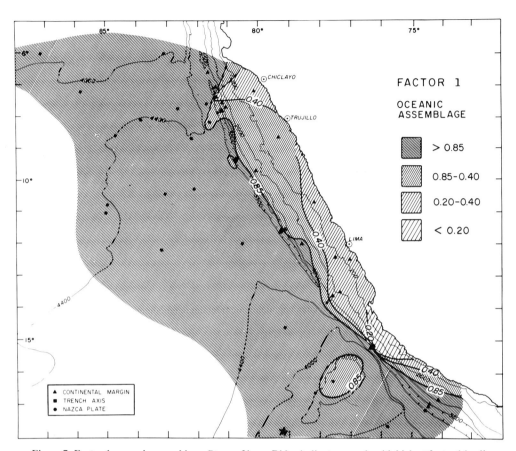

Figure 7. Factor 1, oceanic assemblage. Star on Nazca Ridge indicates sample with highest factor 1 loading (R77-1, Fig. 2).

Factor 2 is termed "continental" assemblage A because it is comprised of detrital clay minerals (illite, kaolinite, and chlorite) and mixed-layer clays. Figure 8 shows that the factor loadings for factor 2 decrease away from shore. Sample 23-1, on the continental slope at lat 15°S, has the highest factor 2 loading. The tongue of high-factor loadings that extends seaward of the trench at lat 7°S, like the surface distribution of chlorite and kaolinite, indicates a source to the northeast. Similarly, the clay minerals in the samples with high factor 2 loadings on the continental margin between lat 13° and 15°S may be derived from the coastal Arequipa Massif, which lies directly opposite the Nazca Ridge. The massif is composed of sillimanite and garnet gneiss, K-feldspar–rich granite, and pegmatite (Cobbing and Pitcher, 1972).

Factor 3 is termed continental assemblage B. It is characterized by an abundance of illite and smaller amounts of kaolinite and chlorite. Sample V33-1 on the continental shelf has the highest factor 3 loading (Fig. 9). Because this vector is represented by negative values, values increase away from shore.

Surface Sediments

The dominant factor in all Nazca plate surface sediments is factor 1 (oceanic assemblage; Figs. 7, 10). Except for core 127, where all three factors are equally important, and core V33, where factors 2 and 3 (continental assemblages A and B) are equally important, the dominant factor in upper continental margin (<2,000 m) sediments is factor 2. Lower continental margin sediments are more

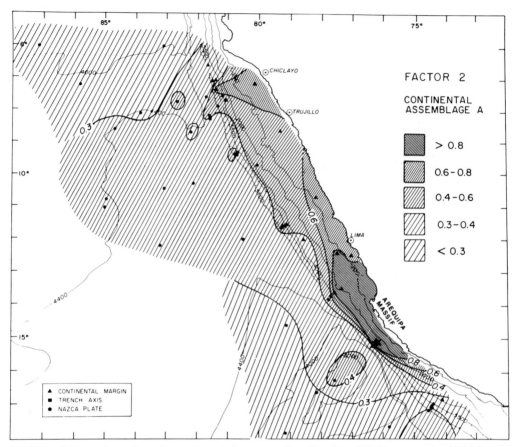

Figure 8. Factor 2, continental assemblage A. Star on southern Peru Margin indicates sample with highest factor 2 loading (23-1, Fig. 2).

variable; they are either dominated by factor 1 or factor 2. Factor 3 does not dominate any of the surface or subsurface sediments.

The boundary between oceanic and continental sediments (as defined by dominance of factors 1 and 2) parallels the continental margin. In most areas, the sampling density is not great enough to precisely define this boundary. However, it lies higher on the continental margin to the north and south of the Nazca Ridge than it does opposite the Nazca Ridge. The boundary is particularly well defined between lat 14° and 15°S, where it lies approximately at the base of the steep continental slope. Between lat 8° and 13°S, the continental slope is not as steep as at lat 15°S, and the oceanic assemblage occurs higher in the slope. The boundary probably occurs on the upper continental slope northeast of core 22 (Fig. 2), because: (1) Zen's (1959) core 42 (Fig. 10A) apparently lacks smectite and mixed-layer clay but contains illite, kaolinite, and chloride; (2) core 42 lies within the region expected to contain from 2% to 4% organic carbon (see Trask, 1961; Rosato and others, 1975); and (3) surface sediment in cores with similar structural settings are dominated by factor 2.

Subsurface Sediments

The dominant factor in the surface sediment is not necessarily the dominant factor in subsurface sediment (Fig. 11). Cores 24, 25, 55, 56, 57, 58, 77, 78, and V35 have sediment at depth that is substantially different from surface sediment in the same core (Fig. 11, open circles). A stratigraphic change in factor dominance implies a change in the source of sediment, a change in dispersal path, or a tectonic juxtaposition of sediment types.

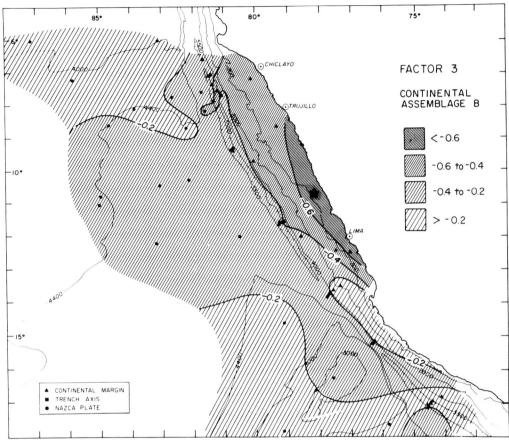

Figure 9. Factor 3, continental assemblage B. Star indicates sample with highest factor 3 loading (V33-1, Fig. 2).

The boundary between sediments with an oceanic factor dominance and those with a continental factor dominance shifted up to 100 km west of the present boundary during the early Quaternary (Figs. 10A, 10B). In cores 24, 74, 77, and 78 (Fig. 11), this shift is due to deep-water deposition of continental-type sediments by turbidity currents (as determined from core textures and sedimentary structures). In cores 55, 56, 57, 58, and V35, where samples were not taken from visibly graded turbidite intervals, factor analysis indicates that hemipelagic sedimentation was dominated by continental assemblage A earlier in the Quaternary. This change occurred fairly recently; for example, within the upper 6 cm of core 57, sediments change from oceanic dominance to continental dominance.

A seaward shift of the boundary between oceanic and continental sediments could have been caused by increased erosion rates and/or lowered sea level (which would cause a westward movement in the shoreline) during Pleistocene glacial stages. The greatest change in the distance between present and past positions of the oceanic-continental sediment boundary generally coincides with the wider shelf from lat 8°30′ to 12°S (Fig. 10). Between lat 13° and 15°S, the continental margin is much steeper than to the north, and little change recurred in the position of the boundary.

Possible Evidence for Continental Accretion

If one uses the factor dominance in surface sediment as an indicator of the present environment of deposition, then displaced subsurface sediments are likely to have a different factor dominance than

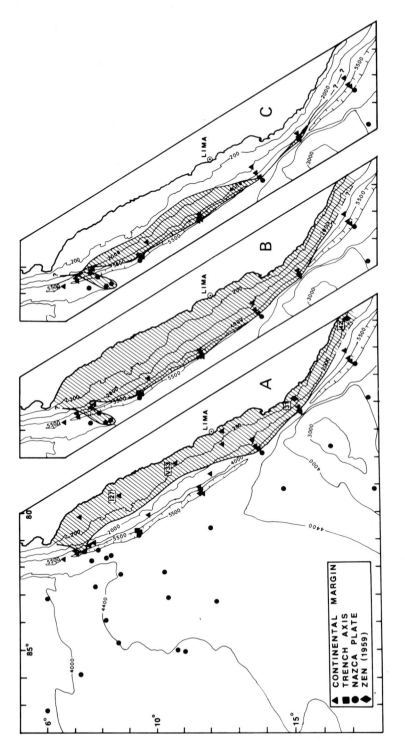

Figure 10. Factor dominance in surface and subsurface samples. (A) Present location of the boundary between oceanic-factor-dominated (no pattern) and continental-factor-dominated (line pattern) surface sediments. (B) Boundary earlier in the Quaternary. (C) Difference between early Quaternary and late Quaternary boundaries.

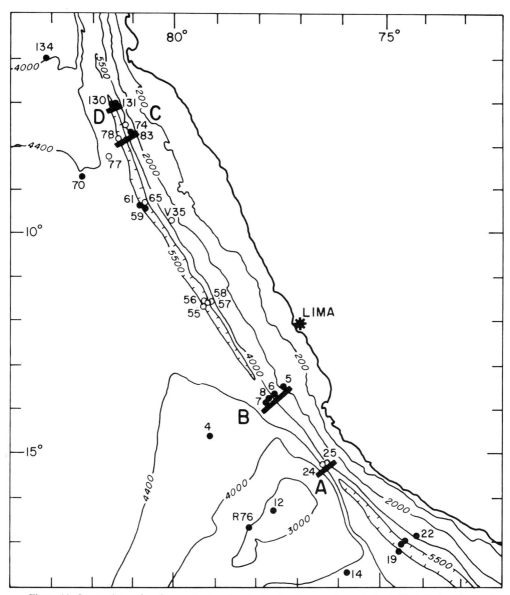

Figure 11. Comparison of surface and subsurface factor dominance. Open circles indicate that surface and subsurface sediments are dominated by different factors. Solid circles indicate no change in factor dominance with depth in core. Heavy solid lines (A, B, C, D) are locations of seismic reflection profiles. Numbers refer to cores discussed in text.

surface sediments from the same core. There are five possible relationships between surface and subsurface sediments (Table 1). Situations 1 and 4 are the most common and do not indicate displacement. This does not mean that displacement has not occurred, only that additional information (such as paleobathymetry from microfossils) is needed.

Situation 3 is common along the Peru continental margin where turbidites and other downslope mechanisms have deposited continental-type sediment in regions normally receiving oceanic-type sediment. An example is core 78 (Fig. 15), which not only demonstrates that turbidity currents

TABLE 1. FACTOR DOMINANCE SITUATIONS

Situation	Surface Factor Dominance	Subsurface Factor Dominance	Physiographic Setting	Interpretation	Examples (Cores)
1	Oceanic	Oceanic	Nazca Plate, Nazca Ridge	Displacement not indicated.	4, R76, 12
2	Oceanic	Oceanic	Continental Slope	(a) Possible continental accretion if accompanied by other data, such as microfossil paleobathymetry or distinctive lithology (calcareous ooze), or (b) Displacement not indicated.	(a) 24 (b) 130, 131
3	Oceanic	Continental	Nazca Plate, Continental Slope	(a) Tectonism, or (b) Change in sedimentation pattern.	(a) 77, 78 (b) V35, 55, 56, 57, 58
4	Continental (A or B)	Continental	Continental Margin	Displacement not indicated.	5, 6
5	Continental (A or B)	Oceanic	Lower Continental Slope	Possible continental accretion.	25

originating on the continental margin have transported sediment westward, but also that uplift has occurred since deposition on the seaward wall of the present Peru Trench axis.

Situations 2 and 5 may indicate that accretion of sediment from the Nazca plate has occurred on the lower continental slope. An example of situation 2 is core 24 (Figs. 13, 16). Core 25 is an example of situation 5. These cores are discussed in the next section.

Figure 12 summarizes our interpretation of displaced sediments using factor analysis. Most of the 28 cores recovered from the landward wall of the trench contained sand-silt terrigenous turbidites and hemipelagic muds. Four cores from the seaward wall of the trench had terrigenous turbidites derived from the continental margin prior to eastward migration of the trench axis. Two cores on the lower trench wall contained oceanic sediment that we infer was accreted from the Nazca plate.

SEISMIC PROFILES

Seismic refection profiles A through D are typical examples of the shallow structure in the trench area. Seismic profile A crosses the margin, trench, and Nazca Ridge (Fig. 13). The mantling carbonate ooze and the underlying acoustic basement on the northeast end of the Nazca Ridge are cut by a series of normal faults. There is a ridge or hill near the narrow trench axis and some ponded sediment on the continental slope at 4,000 m. The lower continental slope, where cores 24 and 25 were recovered, is hummocky with no coherent reflectors.

The tectonic implications and general lithologies of cores 25 and 24 on the lower continental slope, and core 12 on the Nazca Ridge, have been reported in Kulm and others (1974). They demonstrated that the carbonate-rich sections in core 24, which are sandwiched between terrigenous turbidites, came from the Nazca Ridge, the only source of foram-nannofossil ooze in the area. They indicated that these deposits were folded against or thrust beneath the lower slope within the last 400,000 yr. The present study provides additional information about the provenance of the carbonate-poor clay sections in core 24 and supports their conclusion on the origin of the carbonate-rich sections.

From the factor loadings of surface sediment (Figs. 10 and 13), it is clear that the boundary between oceanic-dominated and continental-dominated sediment lies between cores 24 and 25. Nowhere else on the Peru continental margin is this boundary as well defined. The oceanic character of the surface sediment in multiple gravity core 24MG (sample 24-1) is similar to other surface sediment from the

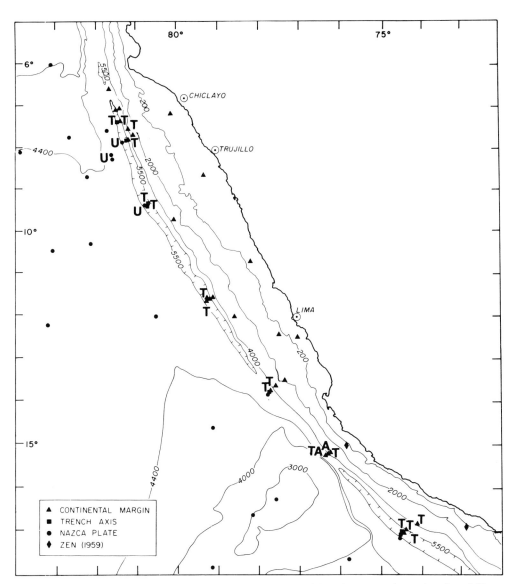

Figure 12. Type of sediments in the trench-margin area. T = turbidites; A = accreted oceanic sediment; U = uplifted turbidites seaward of present trench axis.

lower slope at water depths greater than 2,000 m. The surface sediments in cores 23 and 25 have the same factor loadings as the turbidites from core 24MG (sample 24-2) and the turbidites near the base of the piston core 24P (sample 24-6).

The factor loadings in the carbonate-rich sections of core 24P (samples 24-4, 24-5) are similar to the factor loadings of calcareous ooze in Nazca Ridge cores 12, R76, and R77 (Fig. 2). The carbonate-poor clays in core 24 have virtually the same factor loadings as the calcareous oozes, and they are the same age (Plio-Pleistocene). Since the carbonate sections are accreted, we believe that the clay sections are, too. They may represent material originally deposited on the Nazca Ridge below the calcite compensation depth.

Core 25 was recovered farther up the continental slope along the same profile. The oldest recovered sediment in core 25 is different from the surface sediment in the same core, and the boundary is near

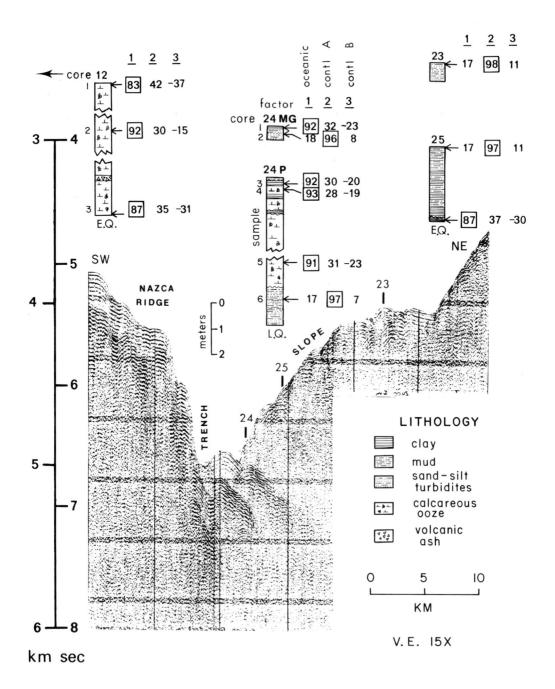

Figure 13. Seismic profile A across end of Nazca Ridge, trench, and continental slope (see Fig. 11 for location). Cores 12 and R76 were recovered on Nazca Ridge about 200 km southwest of core 24 (see Fig. 2). In this figure and in subsequent seismic profiles, the factor loadings have been multiplied by 100 and rounded to two figures. Boxes indicate the dominant factor for each sample. Radiolarian ages are given for the basal sample for each core. L.Q. = late Quaternary (<.4 m.y.); E.Q. = early Quaternary (between 0.4 m.y. and 2.0 m.y.).

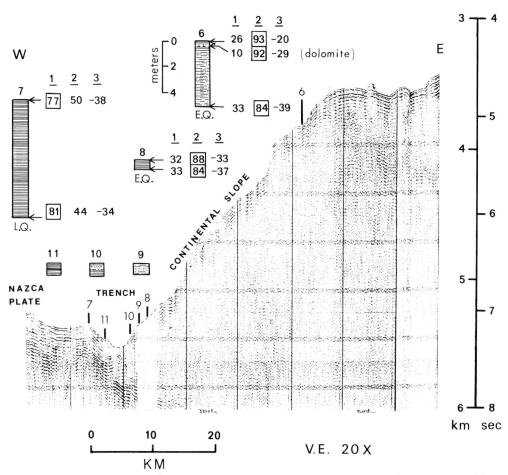

Figure 14. Seismic profile B across trench and continental slope (see Fig. 11 for location). See Figure 13 for explanation of factor data.

286 cm. Olive gray mud occurs above this level, and firm, homogenous, grayish olive clay occurs below (Fig. 13). There are two possible interpretations for the origin of the oldest sediment in core 25: (1) oceanic clay was deposited at this site during the early Quaternary; or (2) the clay section represents accreted oceanic-plate sediment. Because the boundary between oceanic- and continental-factor-dominated sediments shifted westward between lat 8° and 13°S earlier in the Quaternary, it is likely that it also shifted at lat 15°S. If so, early Quaternary sediment at this site would be more continental in character (i.e., have higher continental factor loadings) than late Quaternary sediment. Thus, the sediment at the base of core 25 is probably accreted.

Seismic profile B at lat 14°S (Fig. 14) is unlike profile A. The trench floor is about 15 km wide, the landward wall of the trench dips seaward at about 4°, and the seaward wall of the trench dips toward the continent at about 1°. The acoustic basement of the Nazca plate is covered with 100-250 m of pelagic sediment. Cores 7 and 11 on the seaward side of the trench were composed of homogenous, greenish gray clay. The composition of cores on the landward side of the trench was more variable, ranging from olive gray lutites in core 6 to sand-silt turbidites in cores 9 and 10. Core 6 also contained a 2 cm thick layer of shaly dolomite.

The factor loadings of the sediments were as anticipated, with the trench axis as the approximate dividing line between oceanic and continental sediments. The dolomite, because its carbonate-free fraction is composed chiefly of diatoms (which implies deposition near the coastal upwelling area) and its clay mineral assemblage is continental in character, was probably derived from upslope outcrops. Dolomite has been recovered in several places offshore of Peru, and the subject is discussed in more detail in Kulm and others (this volume).

The trench floor in profile C (Fig. 15) is about 9 km wide and has a partially buried ridge or knoll and 200-500 m of horizontally stratified trench fill. As in other profiles across the trench, the lower continental slope area lacks coherent reflectors. Cores 79, 80, 81, and 82 recovered only short intervals of terrigenous turbidites and hemipelagites. Core 78, on the seaward wall of the trench about 400 m above the trench floor, contained turbidites that were both mineralogically and texturally similar to continental margin sediments. The basal sample of the core had no clear factor dominance, similar to

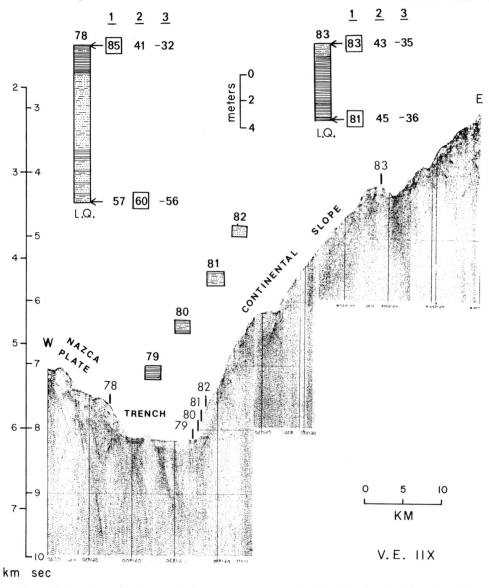

Figure 15. Seismic profile C across trench and continental slope (see Fig. 11 for location). See Figure 13 for explanations.

surface sediment at site 127 on the adjacent continental shelf. We infer that the turbidites in the lower part of core 78 were derived from the continental margin and deposited on the trench floor, and that vertical movements in the late Quaternary-Recent uplifted the site of deposition relative to the present trench axis. Carbon-14 dating of nearby turbidites indicates uplift within the past 5,100 yr (Prince and Kulm, 1975). The relative uplift of the depositional site agrees with the sense of vertical displacement of another basin containing terrigenous turbidites (cores 76 and 77, Fig. 2) on the seaward side of the present trench (Prince, 1974; Prince and others, 1974).

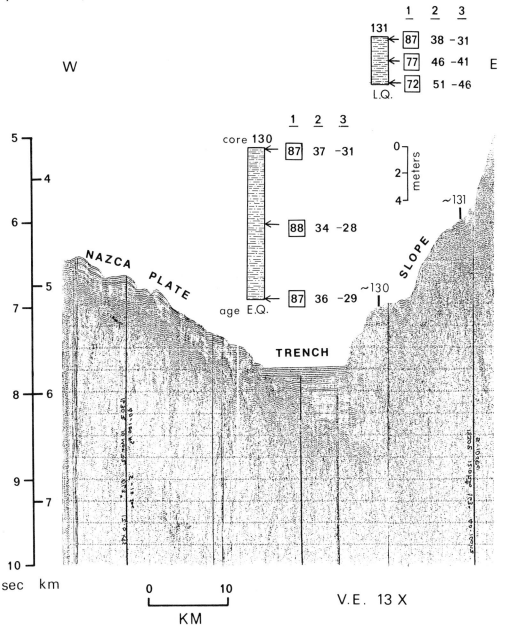

Figure 16. Seismic profile D across trench and continental slope (see Fig. 11 for location). See Figure 13 for explanations.

Core 83 was recovered from 1 150 m high knoll on the middle continental slope. It was composed of mud with an oceanic factor dominance. Since the subsurface sample is similar to the surface sample and the oceanic-continental-factor–dominance boundary is high on the slope rather than near the trench axis, we cannot say that the sediment is displaced.

The northernmost crossing of the Peru Trench is profile D (Fig. 16). The trench floor is about 10 km wide and is filled with up to 600 m of horizontally stratified sediment above the landward-dipping acoustic basement and Nazca plate pelagic sediments. As in previous seismic profiles, the landward wall of the trench lacks coherent reflectors. The homogenous muds and uniform factor loadings in cores 130 and 131 indicate that the oceanic assemblage of clay minerals has been deposited here during the Quaternary.

SUMMARY AND CONCLUSIONS

Sediments on the Peru continental margin and adjacent Nazca oceanic plate display distinctive patterns in the distribution of clay minerals and organic carbon. Spatial and temporal changes in the patterns are illustrated clearly when Q-mode factor analysis is applied to the data. Factor loadings help to define sediment sources, Quaternary changes in the position of the shoreline, and the tectonic displacement of deposits from the Nazca plate to the lower trench slope.

Three factors can explain 99% of the variation in clay mineral composition and organic carbon content in surface sediments on the Peru Margin and adjacent ocean floor (Figs. 7, 8, 9). Nazca plate sediment is dominated by an oceanic factor composed mainly of smectite derived from both authigenic (or diagenetic) and continental sources and minor (inferred) eolian illite. Upper continental margin (<2,000 m) sediments are dominated by continental factor A, consisting of mixed-layer clays, illite, chlorite, and kaolinite. Lower continental margin sediments are dominated by either oceanic or continental A assemblages. Continental factor B, composed mainly of illite and with subordinate chlorite and kaolinite, is not the dominant factor in any area, but it is most important on the continental shelf between lat 9° and 13°S.

The boundary between sediments dominated by an oceanic factor and those with a continental (A or B) factor migrated laterally with time. Earlier in the Quaternary, this boundary was seaward of its present position on the continental slope (Fig. 10), because turbidity currents deposited continental clay mineral assemblages on the seaward side of the present trench axis (cores 61, 76, 77, and 78). In addition, slope sediments were more continental in character during the early Quaternary, possibly because of increased erosion rates and a seaward shift of the shoreline during glacial stages. Turbidites on the seaward side of the trench above the present axis are explained by relative uplift after deposition.

Accreted oceanic sediment on the continental slope may be identified by comparing the surface factor dominance with the subsurface factor dominance and by noting the present position of the factor dominance boundary. Accretion from the Nazca plate is suggested where subsurface sediments are found landward of this boundary. By this criterion, there is one place along the Peru continental slope that contains accreted sediment (Fig. 12). Core 24, which contained calcareous ooze and clay, and core 25, which contained pelagic clay, were recovered from the lower continental slope opposite the Nazca Ridge. These cores are the best documented examples of continental accretion.

Accreted sediment is not the major constituent of near-surface sediments on the landward slope of the Peru Trench. The region is characterized by terrigenous sand-silt turbidites and hemipelagic muds. No deformed trench-floor turbidites were recovered, as postulated for the lower slope along the Middle America Trench by Karig and others (1978). Perhaps more accreted oceanic sediments will be found when longer cores can be recovered from the lower continental slope and when more detailed seismic reflection studies are done.

ACKNOWLEDGEMENTS

We gratefully acknowledge the assistance of the crew of the R/V *Yaquina*, Oregon State graduate students, and technicians in the collection of the data. Dr. S. B. Weed of North Carolina State College supplied unpublished information on the clay mineralogy of Peruvian soils. Roy Capo made available the samples from Lamont-Doherty Geological Observatory, which were from cores collected under ONR Grant N00014-67-A-0108-0004 and NSF Grant GA-35454. This research was funded by NSF Grant GX-28675 under the IDOE Nazca Plate Project. We thank Roger Prince, Chiye Wenkam, William Schweller, Margaret Leinen, and James Hein for their constructive comments. Nick Pisias provided invaluable help with the computer analysis.

REFERENCES CITED

Berggren, W. A., 1972, A Cenozoic time scale—some implications for regional geology and paleobiogeography: Lethaia, v. 5, p. 192–215.

Besoain, E., 1969, Clay mineralogy of volcanic ash soil, *in* Inter-American Institute of Agricultural Science, OAS, Panelon Volcanic Ash Soils in Latin America, July 6-13, 1969, Turrailba, Costa Rica, p.B.1.1–B.1.16.

Biscaye, P. E., 1965, Mineralogy and sedimentation of recent deep-sea clay in the Atlantic Ocean and adjacent seas and ocean: Geological Society of America Bulletin, v. 76, p. 803–832.

Carstea, D. D., 1967, Formation and stability of Al, Fe, and Mg interlayers in montmorillonite and verticulite [Ph.D. thesis]: Corvallis, Oregon State University, 117 p.

Chave, K. E., 1952, A solid-solution between calcite and dolomite: Journal of Geology, v. 60, p. 190–192.

Cobbing, E. J., and Pitcher, W. S., 1972, Plate tectonics and the Peruvian Andes: Nature, v. 240, p. 51–53.

Coulbourn, W. T., and Moberly, R., 1977, Structural evidence of the evolution of fore-arc basins off South America: Canadian Journal of Earth Sciences, v. 14, p. 102–116.

Donoso, L., 1959, Clay fraction of some soils of Chile: Societe Francaise de Mineralogies et de Cristallographic Bulletin, v. 82, p. 361–363.

Eklund, W. A., 1974, A microprobe study of metalliferous sediment components [M.S. thesis]: Corvallis, Oregon State University, 77 p.

Griffin, J. J., Windom, H., and Goldberg, E. D., 1968, The distribution of clay minerals in the world ocean: Deep-Sea Research, v. 15, p. 433–459.

Grim, R. E., 1968, Clay mineralogy: New York, McGraw-Hill Book Co., 596 p.

Hayes, J. B., 1973, Clay petrology of mudstones, leg 18, Deep Sea Drilling Project, *in* Kulm, L. D., von Huene, R., and others, eds., Initial reports of the Deep Sea Drilling Project: Washington, D.C., U.S. Government Printing office, v. 18, p. 903–914.

Hays, J. D., 1970, Stratigraphy and evolutionary trends of Radiolaria in North Pacific deep-sea sediments, *in* Hays, J. D., ed., Geological investigations of the North Pacific: Geological Society of America Memoir 126, p. 185–218.

Heath, G. R., and Dymond, J., 1977, Genesis and transformation of metalliferous sediments from the East Pacific Rise, Bauer Deep and Central Basin, northwest Nazca plate: Geological Society of America Bulletin, v. 88, p. 723–733.

Heath, G. R., and others, 1974, Mineralogy of surface sediments from the Panama Basin, Eastern Equatorial Pacific: Journal of Geology, v. 82, p. 145–160.

Herron, E. M., 1972, Sea floor spreading and the Cenozoic history of the east-central Pacific: Geological Society of America Bulletin, v. 83, p. 1671–1692.

Imbrie, J., and Kipp, N. G., 1971, A new micropaleontological method for quantitative paleoclimatology: Application to a late Pleistocene Caribbean core, *in* Turekian, K. K., ed., The late Cenozoic glacial ages: New York, Yale University Press, p. 71–181.

Imbrie, J., and van Andel, T. H., 1964, Vector analysis of heavy mineral data: Geological Society of America Bulletin, v. 75, p. 1131–1156.

Jackson, M. L., 1963, Interlayering of expansible layer silicates in soils by chemical weathering: Clays and Clay Minerals, v. 11, p. 29–46.

Karig, D. E., and others, 1978, Late Cenozoic subduction and continental margin truncation along the northern Middle America Trench: Geological Society of America Bulletin, v. 89, p. 265–276.

Kendrick, J., 1974, Trace element studies of metalliferous sediments in cores from the East Pacific Rise and Bauer Deep, 10°S [M.S. thesis]: Corvallis, Oregon State University, 117 p.

Klovan, J. E., and Imbrie, J., 1971, An algorithm and Fortran-IV program for large scale Q-mode factor analyses and calculation of factor scores: International Association for Mathematical Geology, Journal, v. 3, p. 61–77.

Krissek, L. A., Scheidegger, K., and Kulm, L. D., 1980, Surface sediments of the Peru-Chile continental margin and the Nazca plate: Geological Society of America Bulletin, v. 91, p. 321–331.

Kulm, L. D., and others, 1974, Transfer of Nazca Ridge sediments to the Peru continental margin: Geological Society of America Bulletin, v. 85, p. 769–780.

Kulm, L. D., and others, 1976, Lithologic evidence for convergence of the Nazca plate with the South American continent, in Yeats, R. S., Hart, S. R., and others, eds., Initial reports of the Deep Sea Drilling Project, Volume 34: Washington, D.C., U.S. Government Printing Office, p. 795–801.

Kulm, L. D., and others, 1981, Late Cenozoic carbonates on the Peru continental margin: Lithostratigraphy, biostratigraphy, and tectonic history, in Kulm, L. D., and others, eds., Nazca plate: Crustal formation and Andean convergence: Geological Society of America Memoir 154 (this volume).

Masias, J. A., 1976, Morphology, shallow structure and evolution of the Peruvian continental margin, 6° to 18°S [M.S. thesis]: Corvallis, Oregon State University, 92 p.

Mehra, O. P., and Jackson, M. L., 1960, Iron oxide removal from soils and clays by a dithiorite-citrate system buffered with sodium bicarbonate, in Swineford, A., ed., Clays and clay minerals, Proceedings, 7th National Conference: London, Pergamon Press, p. 317–327.

Molina-Cruz, A., and Prince, P., 1977, Distribution of opal and quartz on the ocean floor of the subtropical southeastern Pacific: Geology, v. 5, p. 81–84.

Oinuma, K., Kobayashi, K., and Sudo, T., 1959, Clay mineral composition of some recent marine sediments: Journal of Sedimentary Petrology, v. 29, p. 56–63.

Powers, M. C., 1957, Adjustment of land-derived clays to the marine environment: Journal of Sedimentary Petrology, v. 27, p. 355–372.

Prince, R. A., 1974, Deformation in the Peru Trench, 6° to 10°S [M.S. thesis]: Corvallis, Oregon State University, 91 p.

Prince, R. A., and Kulm, L. D., 1975, Crustal rupture and the initiation of imbricate thrusting in the Peru-Chile Trench: Geological Society of America Bulletin, v. 86, p. 1639–1653.

Prince, R. A., and others, 1974, Uplifted turbidite basins on the seaward wall of the Peru Trench: Geology, v. 2, p. 607–611.

Rosato, V. J., Kulm, L. D., and Derks, P. S., 1975, Surface sediments of the Nazca plate: Pacific Science, v. 29, p. 117–130.

Schuette, G., and Schrader, H. J., 1979, Diatom taphocoenoses in the coastal upwelling area off western South America: Nova Hedwigia, v. 64, p. 359–378.

Schweller, W. J., and Kulm, L. D., 1978, Depositional patterns and channelized sedimentation in active Eastern Pacific trenches, in Stanley, D. J., and Kelling, G., eds., Sedimentation in submarine canyons, fans and trenches: Stroudsburg, Pa., Dowden, Hutchinson and Ross, p. 311–324.

Seeley, D. R., Vail, P. R., and Walton, G. G., 1974, Trench slope model, in Burk, C. A., and Drake, C. L., eds., The geology of continental margins: New York, Springer-Verlag, p. 249–260.

Smith, R. L., and others, 1971, The circulation in an upwelling eco-system: the *Pisco* cruise: Investigacion Pesquera, v. 35, p. 9–24.

Suess, E., and Müller, P. J., 1980, Productivity, sedimentation rate and sedimentary organic matter in the oceans. II. Elemental fractionation: Proceedings, Centre Natinal Recherches Scientifiques, Paris (in press).

Trask, P. D., 1961, Sedimentation in a modern geosyncline off the arid coast of Peru and northern Chile, in Proceedings of the 21st International Geological Congress, 1960: Copenhagen, p. 103–118.

Weaver, C. E., 1956, The distribution and identification of mixed-layer clays in sedimentary rocks: American Mineralogist, v. 41, p. 202–221.

Yeats, R. S., Hart, S. R., and others 1976, Initial reports of the Deep Sea Drilling Project, Volume 34: Washington, D.C., U.S. Government Printing Office, 814 p.

Zen, E., 1959, Mineralogy and petrology of marine bottom sediment samples off the coast of Peru and Chile: Journal of Sedimentary Petrology, v. 29, p. 513–539.

MANUSCRIPT RECEIVED BY THE SOCIETY NOVEMBER 12, 1980
MANUSCRIPT ACCEPTED DECEMBER 30, 1980

Structures of the Nazca Ridge and the continental shelf and slope of southern Peru

RICHARD COUCH
School of Oceanography
Oregon State University
Corvallis, Oregon

ROBERT M. WHITSETT
Department of Physics and Computer Science
Pacific Union College
Angwin, California

ABSTRACT

A relatively large density contrast between the Cenozoic and pre-Cenozoic rock that forms the continental margin of southern Peru causes gravity anomalies which outline the topography of the pre-Tertiary rock of the continental shelf of southern Peru. A coastal gravity high of +20 to +50 mgal extends from Mollendo to Lima and is associated with andesites and basalts of Mesozoic age and Precambrian gneisses and granodiorites that crop out along the coast. The gravity anomalies indicate that this coastal structural high extends nearly 100 km out to sea from the Paracas Peninsula southwest of Pisco. Two prominent gravity lows on the margin are those associated with the Pisco Basin on the shelf at lat 13°25′S and the Mollendo Basin on the slope at about lat 17°10′S. Sediment thickness in the Pisco Basin indicated by the gravity anomalies is approximately 2.2 km, and the thickness of the sediments in the Mollendo Basin is approximately 4 km. A series of closed gravity lows occurs on the outer continental shelf and upper slope seaward of the coastal structural high. The lows outline relatively small depositional basins that have sediment thicknesses of approximately 1 km. Between lat 13° and 13°30′S, a marked change occurs in the character of the gravity field of the shelf and slope. The amplitude of the anomaly associated with the coastal structural high decreases abruptly, and a linear negative anomaly of less than -70 mgal, which extends northwestward, indicates a sedimentary basin on the upper continental slope.

The northeastern end of the Nazca Ridge is isostatically compensated by a relatively thick crust whose layers have a slightly lower density than similar layers of the adjacent oceanic crust. Most of the increase in crustal thickness occurs in the basal crustal layer. Approximately 330 km west of southern Peru, the depth to the Mohorovicic discontinuity increases from approximately 10.5 km on the northeast side of the Nazca Ridge to 18 km under the center of the ridge, then rises again to 9.7 km southwest of the ridge. Measurements show that the free-air gravity anomaly along the trench axis has a maximum value 200 km northwest of the point of minimum bathymetric depth of the trench. This requires the rock of a mass column off the northeast end of the Nazca Ridge to be less dense than the rock of a mass column under the axis of the trench for at least 200 km northwest of the Nazca Ridge, and is consistent with subduction of the Nazca Ridge beneath the continental margin of southern Peru.

INTRODUCTION

Wuenschel (1955; cited by Worzel, 1965) discovered the very large negative gravity anomaly associated with the Peru-Chile Trench when he made pendulum measurements of the gravitational attraction of the area in 1947 aboard the submarine *USS Conger*. Surface ship gravity data, obtained by the R/V *Vema* in 1962, 1962 and 1963 and the R/V *Conrad* in 1965 (Hayes, 1966) greatly improved the resolution of the gravity field and at that time made the Peru-Chile Trench one of the best mapped deep-ocean trenches. The data showed that the gravity lows, associated with the trench, extend south of the end of the topographic trench near lat 33°S to the southern tip of South America. This indicated that the trench, as a structural feature, extends 5,500 km from near lat 8°S to near lat 57°S. Although these measurements outlined the large-amplitude gravity anomalies associated with the subduction complex, additional measurements are necessary to resolve the shorter wavelength anomalies caused by marked lateral variations in the geology of the continental slope and shelf that are indicative of the structural evolution of the convergent continental margin.

In 1972 and 1974, personnel of the Geophysics Group, Oregon State University, made gravity measurements aboard the R/V *Yaquina* along the continental margin of Peru between lat 12° and 18°S and over the northeast end of the Nazca Ridge. These measurements constitute a portion of the data gathered along the continental margin of western South America by personnel participating in the Nazca Plate Project during the International Decade of Ocean Exploration. This paper delineates the free-air gravity anomalies off of southern Peru and presents four model cross sections of the continental shelf and slope and a cross section of the northeast end of the Nazca Ridge. Although these models lack sufficient controls to be unique, they are consistent with the measured bathymetry and gravity anomalies and consequently provide constraints on alternative models of postulated geologic structures.

GRAVITY MEASUREMENTS

The trackline map in Figure 1 shows the location of gravity measurements made by Oregon State University personnel during the *Yaquina* Long Cruises of 1971-1972 (YALOC '71, Leg 6) and 1973-1974 (YALOC '73, Legs 6 and 7). Pendulum stations obtained by the *USS Conger* in 1947 (Worzel, 1965), surface-ship gravity measurements made by the *OSS Oceanographer* during 1967 (Sea Gravity Data, Project Operations 476, NOAA), and land gravity measurements compiled by the Instituto Geofisico del Peru and obtained from the Defense Mapping Agency Aerospace Center, St. Louis, Missouri, provide additional data for the study.

Lacoste & Romberg surface-ship gravity meter S-42, aboard the R/V *Yaquina*, yielded measurements at intervals of approximately 1.4 km along the ship's track. Satellite fixes and course and velocity data obtained from the ship's instruments provided navigation parameters for measurement locations and Eötvös corrections. The estimated RMS uncertainty in the measurements based on differences in measurements at trackline crossings is 4 mgal (Whitsett, 1975).

The measurements are referenced to station WH 1068 (Woollard and Rose, 1963) in Callao, Peru, adjusted to the 1967 Potsdam value (International Association of Geodesy, 1971a). Free-air anomalies were computed using the 1971 International Gravity Formula (International Association of Geodesy, 1971b).

Twenty-three crossings of the trench axis occur between lat 12° and 18°S along a segment of the trench that is about 900 km long. The trackline spacing on the shelf is sufficient to resolve features 20 to 30 km or more in horizontal dimension. Over the Nazca Ridge and the abyssal sea floor southwest of

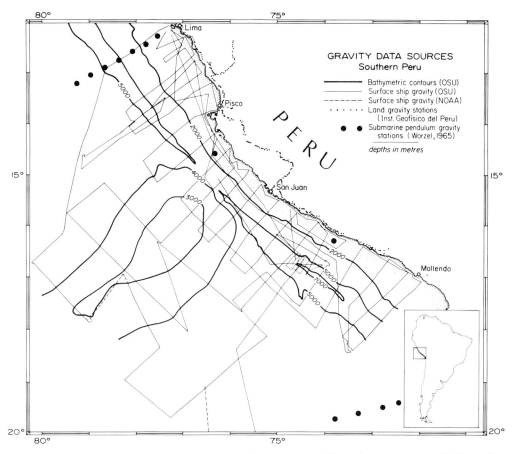

Figure 1. Location map of surface ship, submarine, and terrestrial gravity measurements. Bathymetric contours are in metres.

the trench, tracklines are more widely separated and resolution is limited to features greater than 100 km in horizontal dimension.

FREE-AIR GRAVITY ANOMALIES

Figure 2, a bathymetric map of the continental margin of southern Peru described by Prince and others (1980), outlines the northeast end of the Nazca Ridge and the Peru Trench. The Nazca Ridge is approximately 250 km wide near the trench and rises more than 1,200 m above the surrounding abyssal sea floor. The sea floor is approximately 4,200 m deep northwest of the ridge and 4,400 m deep southeast of the ridge. The Peru Trench, more than 7,400 m deep southeast of the ridge, shoals to approximately 4,800 m between the northeast end of the Nazca Ridge and the continental margin of southern Peru.

Figure 3 shows the free-air gravity anomalies west of southern Peru between lat 12° and 18°S. The anomalies, contoured at 10 mgal intervals, are predominantly negative. The anomalies associated with the Peru-Chile Trench dominate the region and exhibit values of less than -220 mgal. A series of relative gravity lows along the anomaly axis, beginning with the -180 mgal low at lat 15°30′S and extending southeast to adjacent -190, -220, and -210 mgal anomalies, coincide with bathymetric changes along the trench. A significant observation is the absence of anomaly contours coincident with the bathymetric contours that outline the Nazca Ridge. In contrast, the Mendocino Ridge (Dehlinger and others, 1967), the Cocos Ridge and Carnegie Ridges (Barday, 1974), and the Tehuantepec Ridge (Couch and Woodcock, 1981) clearly exhibit pronounced positive anomalies over oceanic ridges that are topographically smaller than the Nazca Ridge. Anomalies greater than +20 mgals over the northeast end of the Nazca Ridge are probably due to the shallow bathymetry, but crustal warping near the subduction zone may also contribute.

Along the axis of the gravity anomaly minima, associated with the Peru-Chile Trench, the highest value is more than -110 mgal near lat 13°45′S, and the lowest value is less than -220 mgal at lat 17°S. The axis of the gravity minima is located 5 to 15 km landward of the bathymetric axis of the trench, north and south of the ridge, and coincides with the axis near the ridge. The most positive anomaly values along the axis of the trench occur near lat 13°45′S, 200 km northwest of a saddle point in the bathymetric axis of the trench near lat 15°15′S. The shallowest point on the trench axis is about 4,800 m below sea level and occurs at the saddle point near lat 15°15′S near the intersection of the Nazca Ridge and the Peru-Chile Trench. The free-air gravity anomaly at this point is about -163 mgal. The ocean depth corresponding to the previously mentioned -110 mgal anomaly is approximately 5,500 m. Hence, between these two points, gravity increases 53 mgal from southeast to northwest, whereas the sea deepens about 600 m. This implies a marked difference in the material and/or the geometry of the subduction complex in the vicinity of the continetal slope along this section of the trench.

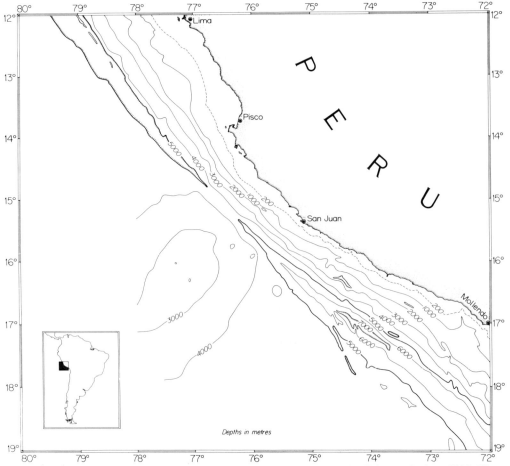

Figure 2. Bathymetric map of the continental margin of southern Peru after Prince and others (1980). The contour interval is 200 m.

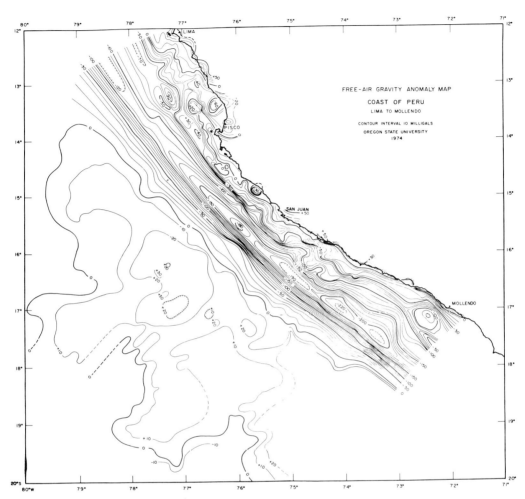

Figure 3. Free-air gravity anomaly map of the continental margin of southern Peru. The dashed lines indicate inferred contours. The estimated RMS uncertainty is 4 mgal.

Two prominent relative gravity lows on the continental margin of southern Peru are those associated with the Pisco Basin (-60 mgal), on the shelf at lat 13°25'S near the coast north of Pisco, Peru, and the Mollendo Basin (-110 mgal), on the continental slope near lat 17°10'S, offshore of Mollendo, Peru. The Mollendo Basin is termed the Arequipa Basin by Masias (1976). A gravity low on the shelf and along the coast near lat 15°S is similar in configuration to the -70 mgal and -90 mgal lows near lat 12°30' and 13°15'S, respectively, on the continental slope and outlines a shelf basin, here named the Caballas Basin (Fig. 14). The basin near lat 15°S produces a gravity anomaly of about -30 mgal in amplitude, but it is not accompanied by the outer shelf high as near Pisco. Assuming a density contrast of 0.75 g/cm³ between the sediments and basement rock, the gravity anomaly suggests a depth to basement of about 1 km for this basin.

The -90 mgal slope basin near lat 13°15'S, which forms the southern terminus of the Lima Basin, differs from the Mollendo Basin in that it is more elongate both gravimetrically and bathymetrically. It is about 20% smaller in area and in gravity anomaly amplitude and has about 50% less bathymetric relief. A bathymetric high just seaward of the basin produces the -50 mgal relative high seaward of the -90 mgal low. The amplitude of the gravity anomaly, associated with the basin, and an assumed density contrast of .75 g/cm³ suggest that about 1 to 1.5 km of sediments fill the basin.

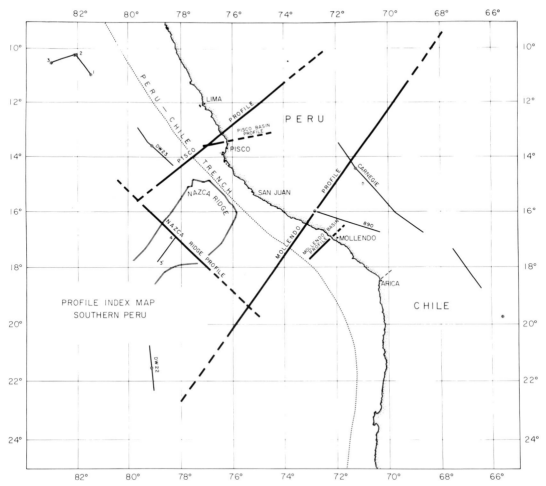

Figure 4. Profile location map. Heavy lines indicate the traces of the crustal cross sections (see also Couch and others, this volume). Light lines indicate the location of seismic refraction lines (Ocola and others, 1971; Shor and others, 1970; Fisher and Raitt, 1962; Woollard, 1960) used to constrain the model cross sections.

A coastal gravity high of +20 to +50 mgl extends from Mollendo to Lima except for two gaps, one between lat 14° and 15°S, where there is a small shelf basin, and the other in the vicinity of the elongate Pisco Basin northwest of Pisco. Outcrops of Precambrian gneisses and granodiorites, in addition to andesites and basalts of Mesozoic age, occur along the coast. The rocks are all relatively dense and are the probable cause of the coastal gravity high. Because of the high density contrast between these rocks and the overlying sediments, calculations based on the observed gravity anomalies yield good estimates of sediment thickness.

Figure 4 shows the location of: (1) a seismic reflection profile and two model crustal cross sections that cross the continental slope and continental shelf through the Pisco Basin; (2) a seismic reflection profile and two model crustal cross sections across the continental margin in the vicinity of the Mollendo Basin; and (3) a model crustal cross section of the Nazca Ridge.

THE PISCO BASIN

The Pisco seismic reflection profile in Figure 5 shows a relatively smooth continental slope, but the sediment layers appear disturbed on nearly the entire slope. The oceanic crust is detectable nearly 10

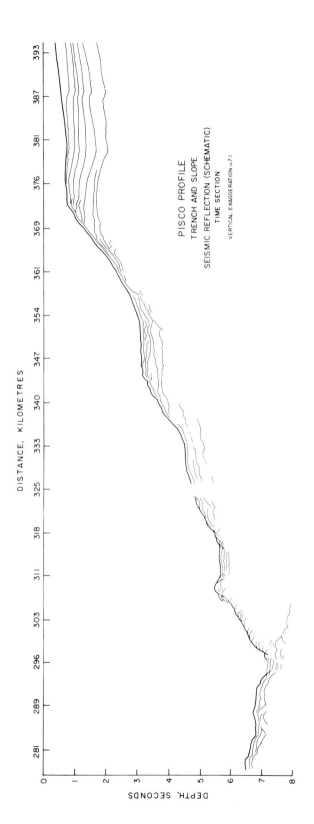

Figure 5. Single-channel seismic reflection profile of the trench and slope along the Pisco Profile located in Figure 4.

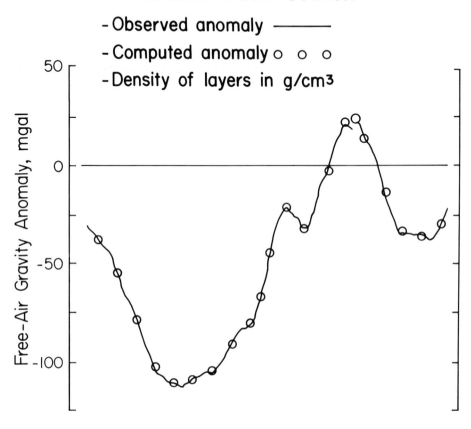
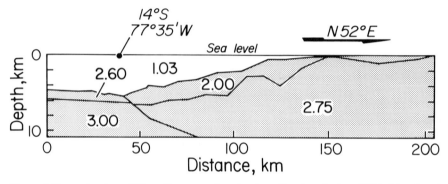

Figure 6. Upper structure of Pisco, Peru, crustal section from Couch and others (1981, this volume).

km landward of the trench beneath the toe of the slope. A small perched basin near km 313 is filled with sedimentary layers that are almost flat lying, and upslope, benches are located near km 333 and km 350. A sharp break between the slope and shelf occurs near km 372, with relatively flat-lying sediments more than 1 km deep occurring between the outer shelf edge and the coast. Upward warping of the sediment layers landward of the shelf break and trucation of the layers at the edge of the break suggest that the outer edge of the continental shelf has been uplifted.

Figure 6 shows a model crustal section along the same profile as the seismic reflection section shown in Figure 5. The model is a portion of the Pisco, Peru, crustal and subcrustal cross section described by Couch and others (1981, this volume). The model ascribes the regional gravity gradient to the dip of the top of the descending oceanic lithosphere beneath the continent, and the short wavelength gravity anomalies to differences in the depth to the interface between the water and sediments and between the sediments and the relatively dense, apparently crystalline rock underlying the sediments. The model suggests that the thickness of the sedimentary rock on the continental slope, whether of marine or terrestrial origin, generally increases from the top of the slope to the trench. The topography of the slope sediments is irregular, and apparently, the underlying crystalline rock is even more irregular, indicating a very complex and extended development of the continental margin. This profile crosses the Pisco Basin north of a minimum gravity anomaly associated with the basin and shows a sediment thickness in the basin at this location of approximately 1 km and a basement very close to the surface near the outer edge of the continental shelf.

Figure 7 shows a model crustal cross section through the deepest part of the Pisco Basin based on the observed gravity anomalies and the assumed densities. The Pisco Basin section is oriented N 79° E and extends from the 2,000 m bathymetric contour at lat 13°35'S to the coast. The block densities in the basin section are identical to the corresponding densities in the Pisco, Peru, section described above. The maximum amplitude of the gravity anomaly over the Pisco Basin is approximately -60 mgl, compared with the -40 mgl on the Pisco section that crosses the basin about 25 km to the north.

The negative regional gravity gradient along the continental margin tends to exaggerate the anomalies associated with basins. Consequently, in the construction of the Pisco Basin and Mollendo Basin sections, described below, the Pisco, Peru crustal and subcrustal cross section described by Couch and others (1981, this volume) establishes the underlying structure for the Pisco Basin section, and the Mollendo, Peru, crustal and subcrustal cross section establishes the underlying structure for the Mollendo Basin section. These larger sections effectively account for the regional gravity gradient, thereby allowing the shorter wavelength anomalies over the shelf and slope to more accurately reflect the upper crustal structure of the continental slope and shelf.

The Pisco Basin section indicates a maximum thickness of the 2.00 g/cm^3 strata of about 2.2 km. Strata in the basin are thought to include Teritiary and Mesozoic deposits and to overlie Precambrian basement rocks of density 2.75 g/cm^3. the model section shows structural highs seaward of the shelf break at km 360 and landward of the break at km 410. Sediments more than 3.5 km thick fill the structural low between the highs and cover the top of the highs to form the outer continental shelf and uppermost continental slope.

Figure 8 shows a seismic reflection profile along part of the Pisco Basin section with the computed contact between the sediments and basement rock of Figure 7 superimposed. The superposition assumes a velocity of 1.8 km/sec in the upper rock layer. The sedimentary layers visible in the reflection profile appear downwarped with respect to a basement high suggested by the gravity model. The layers, however, appear generally conformable with the computed basement, which suggests continuing tectonic deformation of the rocks that form the upper continental slope and outer continental shelf.

THE MOLLENDO BASIN

A seismic reflection profile, obtained across the trench and continental margin of southern Peru approximately 100 km northwest of Mollendo (Fig. 9), reveals a continental slope with very irregular

578   COUCH AND WHITSETT

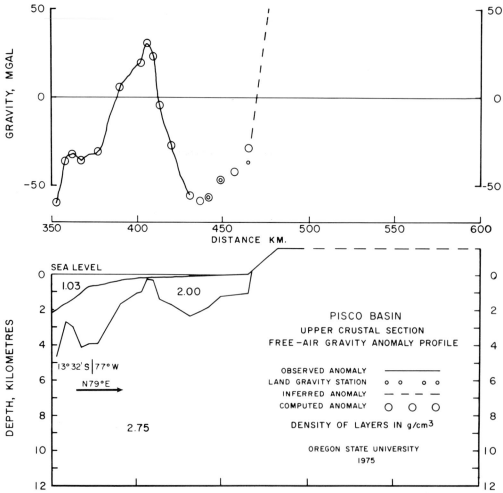

Figure 7. Pisco Basin model cross section. The horizontal scale corresponds to the scale on the Pisco Profile in Figure 5.

topography that extends from the trench nearly to the coastline. The topmost sediments appear disturbed, and the general shape of the slope suggests block faulting on a large scale.

Figure 10 shows the upper part of the Mollendo, Peru, crustal and subcrustal section described by Couch and others (1981, this volume). The section shows an increase in thickness of a block of material with density 2.50 g/cm³ from the coast toward the trench. The block, whose maximum thickness is more than 5 km, overlies rock of density 2.75 g/cm³ and is covered by about 2 km of material of density 2.00 g/cm³. The densities and structure suggest that a thick sequence of sedimentary rocks covers pre-Cenozoic crystalline basement rocks similar to those that crop out along the coast.

Penetration of seismic waves along the Mollendo Basin profile allows the differentiation of upper and lower layers in the model of the basin shown in Figure 11. The interface between the layers of density 1.8 and 2.2 g/cm³ is traceable on the seismic reflection profile in Figure 12 from km 820 to km 920. Wide-angle seismic reflection and refraction measurements (Johnson and others, 1975) constrain the upper basin seaward to about km 873. This upper basin, with its low-density sediment fill, marks that portion of the Mollendo Basin outlined by the gravity low. the computed depth of the interface in Figure 12 assumes a velocity of 1.8 km/sec in the upper layer of sediment. Hence, the estimated sediment thickness in the upper part of the basin, based on gravity measurements and assumed model densities of 1.8 and 2.2 g/cm³, agrees with the thickness indicated by the reflection measurements and an assumed seismic velocity of 1.8 km/sec. In the Mollendo Basin, Cenozoic and Mesozoic sediments are assumed to overlie Precambrian or Mesozoic basement rocks (Masias, 1976). The model crustal section suggests that differential motion in the blocks of basement rock has deformed the overlying sediments. This event was followed by deposition of the sediments (which also show some deformation) in the upper part of the basement. The upper basin layer is about 1.4 km thick, and the deeper sediment layer may be locally as much as 5 km thick.

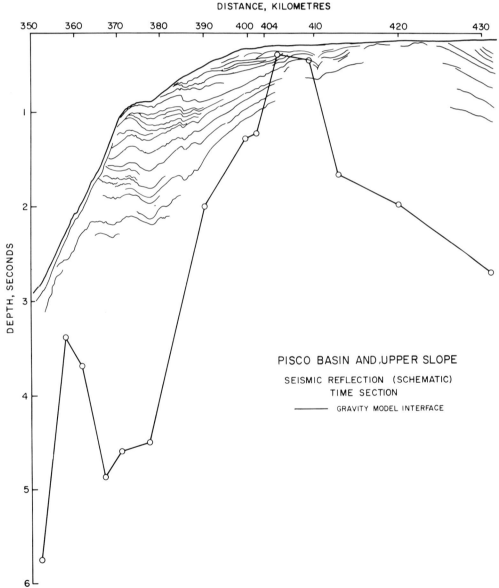

Figure 8. Pisco Basin seismic reflection profile. The horizontal scale and circles are from the Pisco Basin model section in Figure 7. The vertical exaggeration is 28:1 from km 350 to km 404 and 14:1 east of km 404.

THE NAZCA RIDGE

The Nazca Ridge, an aseismic submarine ridge oriented northeast-southwest, adjoins the continental margin of Peru south of Pisco between lat 15° and 16°S. The ridge is about 250 km wide seaward of the trench and rises more than 1,200 m above the sea floor. Surprisinly, however, the gravity anomaly over the ridge is less than +30 mgal at the northeast end and less than +20 mgal within the region mapped in Figure 3.

Figure 4 shows the trace of the Nazca Ridge cross section shown in Figure 13 and the location of seismic refraction lines 1-2, 2-3, and 3'-4 that constrain the model. Gravity mesurements along the trackline (between lat 15°45'S, long 79°30'W, and lat 18°S, long 77°W, in Fig. 1), which traverses the ridge normal to its axis about 270 km southwest of the Peru-Chile Trench, provide the gravity and bathymetric data for the section. Seismic refraction line 3'-4, reported by Hussong (pers. commun., 1975) and oriented approximately along the ridge crest, nearly intersects the section profile. Results from this line indicate northeast-dipping seismic layers. Lines 1-2 and 2-3 are farther from the ridge than DW 23 but are more distant from tectonic disturbances associated with the trench; therefore, these lines were selected to control the west end of the section. Structural control for the southeast end of the Nazca Ridge section is taken from the Mollendo, Peru, section of Couch and others (1981, this volume). Iterative adjustment of the layers until the computed gravity matched the observed gravity anomaly yielded the Nazca Ridge crustal section.

The density contrast between the sea floor and the overlying water, combined with the 1.2 km or more of topographic relief of the ridge, would cause, if uncompensated, gravity anomalies of +75 to +110 mgal over the highest parts of the ridge. Hence, the gravity anomalies indicate a lower density and/or thicker crust under the Nazca Ridge than under the surrounding sea floor. Seismic refraction measurements along line 3'-4 (D. Hussong, pers. commun., 1975) indicate a substantially thicker crust and slightly lower velocities than are found on either side of the ridge. Figure 13 shows hachure lines in the crust on each side of the ridge where model densities were adjusted proportional to the changes in

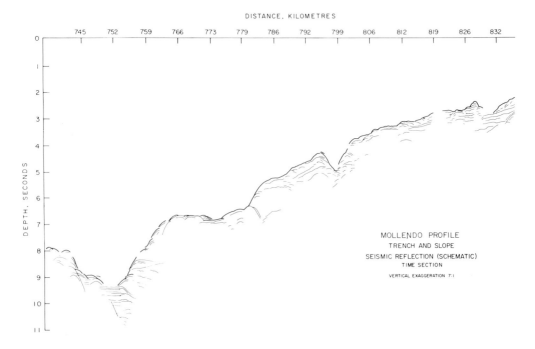

Figure 9. Single-channel seismic reflection profile of the trench and slope along the Mollendo Profile, located in Figure 4.

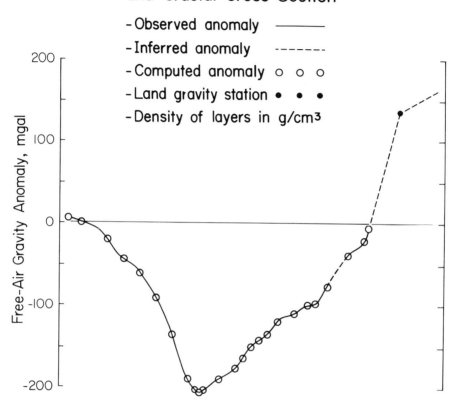
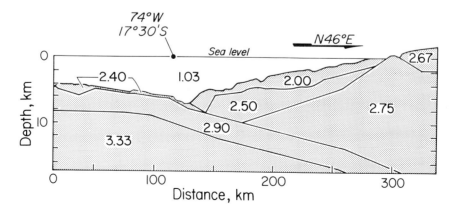

Figure 10. Upper structure of the Mollendo, Peru, crustal section from Couch and others (1981, this volume).

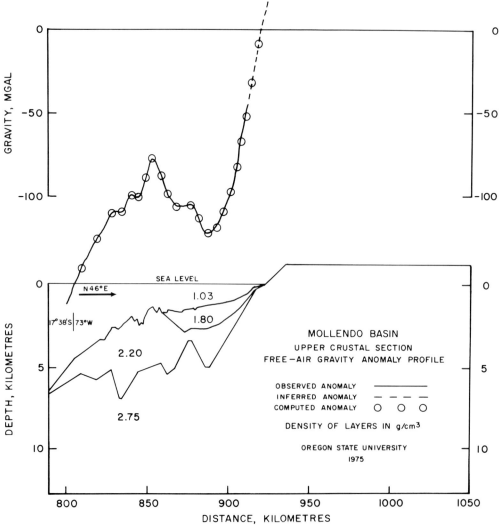

Figure 11. Mollendo Basin cross section. The horizontal scale corresponds approximately with the scale in Figure 9.

seismic velocities. For example, layer densities were reduced 0.03 to 0.04 g/cm³ under the ridge relative to the corresponding densities to the northwest.

The basal crustal layer shows the most pronounced increase in thickness beneath the ridge; however, all layers show an increase in thickness. The depth to the Mohorovicic discontinuity increases from 10.5 km below sea level on the northwest side of the ridge to 18 km under the axis of the ridge, then decreases to 9.7 km southeast of the ridge. Water depths are 100 to 200 m greater on the southeast side of the ridge than on the northwest side of the ridge, and gravity measurements, constrained by seismic refraction measurements (Couch and others, 1981, this volume), indicate that the crust is approximately 0.8 km thicker northwest of the ridge than southeast of the ridge.

Seismic reflection measurements show a sediment layer on the ridge, indicated in the Nazca crustal section as material of density 1.7 g/cm³, that thickens from near zero on each side of the ridge to about 300 m in the middle. This variation in thickness may be related to the dissolution of calcium carbonate,

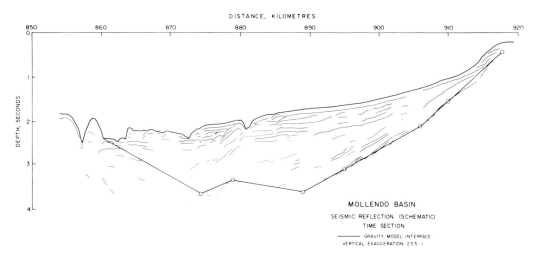

Figure 12. Mollendo Basin seismic reflection profile. The horizontal scale and the circles are from the Mollendo Basin section in Figure 11.

a major constituent of the sediment in this area, at depths greater than 4,000 m (Kulm and others, 1974; Rosato, 1974).

The upper crust in the Nazca crustal cross section shows significant structural variations in the middle and on the southeast flank of the ridge that may be due to large-scale faulting.

DISCUSSION

The continental slope and relatively narrow continental shelf contain numerous, but very small, sediment-filled basins. Figure 14 shows the relative locations of the Pisco, Lima, Caballas, and Mollendo Basins. the largest amplitude gravity anomalies are associated with the Pisco and Mollendo Basins and suggest only 2.2 km and 4.0 km, respectively, of sediment thickness in the basins. The younger sediments in the basins appear slightly deformed and tilted seaward; moreover, the sediments appear to overlie older, highly deformed sediments that form the shelf break and cover the continental slope. The configuration and extent of the sedimentary rocks on the margin suggest that the strata have been oversteepened and in some areas truncated. This indicates an extended and continuing period of deformation for the continental margin and involves vertical displacement of the basement rocks.

A surprising find of the gravity measurements is that the maximum gravity anomaly associated with the axis of the Peru-Chile Trench is displaced approximately 200 km toward the northwest with respect to the minimum bathymetric depth of the trench (Fig. 14). A mass column in the trench axis at the end of the Nazca Ridge has 600 m less water than a similar mass column 200 km along the trench axis northwest of the ridge but also shows more than 50 mgal less gravitational attraction. Less-dense rock beneath the trench and margin at the end of the Nazca Ridge is consistent with subduction of the ridge, which Figure 13 shows to have a relatively thick basal crustal layer. However, this factor does not readily explain the difference that extends along the trench for 200 km northwest of the ridge. Because the motion of the Nazca plate relative to the South American plate is approximately N 81° E (Minster and others, 1974), and because the Nazca Ridge trends N 45° E, the intersection of the ridge and trench must migrate southward. Assuming that the Nazca Ridge was linear and continuous, a convergence rate of 10.3 cm/yr (Minster and others, 1974) indicates that the ridge and trench intersection was 200 km northwest of its present position 3.3 m.y.b.p. Hence, any geologic features between lat 13.5S and the Nazca Ridge and older than 3.3. m.y. are probably not related to the

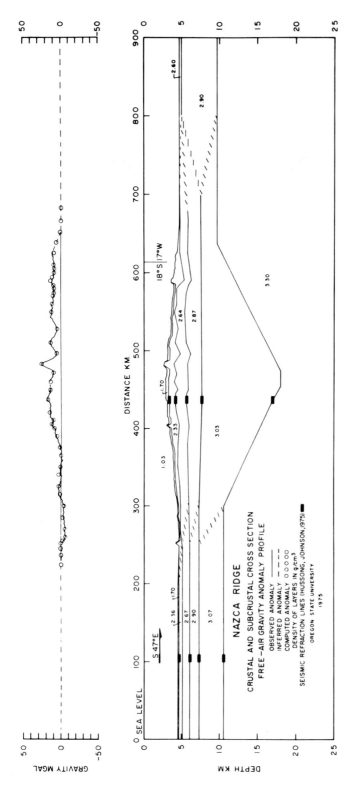

Figure 13. Nazca Ridge crustal cross section. The vertical exaggeration is 10:1.

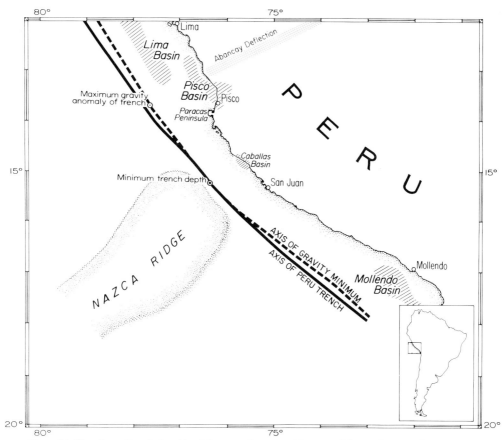

Figure 14. Map shows the relationship of the general structural features of the southern continental margin of southern Peru.

subduction of the ridge. The anomalous gravity anomaly along the trench and slope north of the ridge may reflect other ridge material more recently subducted or a modification of the trench and margin in the subduction zone by subduction of the northeast extension of the Nazca Ridge during the past 3.3 m.y.

The gravity anomaly change observed along the trench near lat 13.5°S, 200 km north of the Nazca Ridge, is approximately on line with the oceanward extension of the Abancay Deflection (Fig. 14), which marks the northern terminus of the late Tertiary and Quaternary volcanism in southern Peru (Masias, 1976; Ministerio de Minas e Energia, 1978), and along which the northern range of the coastal mountains of Peru is set westward with respect to the southern range. The changes in geology that occur in the area designated the Abancay Deflection include features, such as the late Tertiary volcanism, that are older than 3.3 m.y. It is possible that the difference in structure and/or composition of the continental margin and continent west of the Andes may cause a difference in the subduction geometry north and south of the Abancay Deflection. It is therefore postulated that it is this difference which is reflected in the gravity anomalies of the subduction zone.

ACKNOWLEDGEMENTS

We thank Dr. S. Johnson and Messrs. M. Gemperle, G. Connard, G. Ness, and J. Bowers, who provided the expertise and goodwill to make the cruises successful and who assisted in the reduction and analysis of the data. Our thanks also to Captain H. Linse and all of the crew of the R/V *Yaquina* for their services during the survey.

As part of the Nazca Plate Project, this research was supported by the National Science Foundation, International Decade of Ocean Exploration (Grant GX 28675), and by the Office of Naval Research under contract No. 014-67-A-0369-0007, project NR 083-102.

REFERENCES

Barday, R., 1974, Structures of the Panama Basin from marine gravity data [M.S. thesis]: Corvallis, Oregon State University, 99 p.

Couch, R., and Woodcock, S., 1981, Gravity and structure of the continental margin of southwestern Mexico and northwestern Guatemala: Journal of Geophysical Research 86, no. B3, p. 1829-1840.

Couch, R., and others, 1981, Structures of the continental margin of Peru and Chile, *in* Kulm, L. D., and others, eds., Nazca plate: Crustal formation and Andean convergence: Geological Society of America Memoir 154 (this volume).

Dehlinger, P., Couch, R. W., and Gemperle, M., 1967, Gravity and structure of the eastern part of the Mendocino Escarpment: Journal of Geophysical Research, v. 72, no. 3, p. 1233-1247.

Fisher, R. L., and Raitt, R. W., 1962, Topography and structure of the Peru-Chile Trench: Deep-Sea Research, v. 9, p. 423-443.

Hayes, D. E., 1966, A geophysical investigation of the Peru-Chile Trench: Marine Geology, v. 4, p. 309-351.

International Association of Geodesy, 1971a, Geodetic references system, 1967: Bulletin of Geodesy, Special Publication No. 3, 116 p.

International Association of Geodesy, 1971b, The international gravity standardization net 1971 (IGSN 71): Special Publication No. 4, 149 p.

Jenks, W. J., ed., 1956, Handbook of South American geology: An explanation of the geologic map of South America: Geologic Society America Memoir 65, 378 p.

Johnson, S. H., Ness, G. E., and Wrolstad, K. R., 1975, Shallow structures and seismic velocities of the southern Peru Margin: American Geophysical Union Transactions, v. 56, p. 443.

Kulm, L. D., and others, 1974, Transfer of Nazca Ridge pelagic sediments to the Peru continental margin: Geological Society of America Bulletin, v. 85, p. 769-780.

Masias, A., 1976, Morphology, shallow structure, and evolution of the Peruvian continental margin, 6° to 18°W [M.S. thesis]; Corvallis, Oregon State University, 77 p.

Ministerio de Minas e Energia, 1978, Mapa tectonico de America da sul: Departamento Nacional de Producão Mineral, Brasil.

Minster, J. B., and others, 1974, Numerical modeling of instantaneous plate tectonics: Geophysical Journal of the Royal Astronomical Society, v. 36, p. 541-576

Ocola, L. C., Meyer, R. P., and Aldrich, L. T., 1971, Gross crustal structure under the Peru-Bolivia altiplano: Earthquake Notes, v. 42, p. 38-48.

Prince, R. A., and others, 1980, Bathymetry of the Peru-Chile Trench and continental margin: Geological Society of America Map and Chart Series, MC-34, scale 1:1,000,000.

Rosato, V. J., 1974, Peruvian deep-sea sediments: Evidence for continental accretion [M.S. thesis]: Corvallis, Oregon State University, 93 p.

Shor, G. G., Jr., Menard, H. W., and Raitt, R. W., 1970, Structure of the Pacific Basin, *in* A. Maxwell, ed., The Sea, v. 4, Part II: New York, Wiley p. 3-27.

Whitsett, R. M., 1975, Gravity measurements and their structural implications for the continental margin of southern Peru [Ph.D. thesis]: Corvallis, Oregon State University, 82 p.

Woollard, G. P., 1960, Seismic crustal studies during the IGY, 2 continental program (IGY Bulletin No. 34): American Geophysical Union Transactions, v. 41, 351-355.

Woollard, G. P., and Rose, J. C., 1963, International gravity measurements: Menasha, Wisconsin, George Banta Co., p. 121.

Worzel, J. L., 1965, Pendulum gravity measurements at sea, 1936-1959: New York, John Wiley, 422 p.

Wuenschel, P. C., 1955, Gravity measurements and their interpretation in South America between latitudes 15° to 33° S [Ph. D. Thesis]: New York, Columbia University, 210 leaves.

MANUSCRIPT RECEIVED BY THE SOCIETY NOVEMBER 12, 1980
MANUSCRIPT ACCEPTED DECEMBER 30, 1980

> # Tectonics of the Nazca plate and the continental margin of western South America, 18°S to 23°S

WILLIAM T. COULBOURN*
Hawaii Institute of Geophysics
University of Hawaii
Honolulu, Hawaii 96822

ABSTRACT

The morphology and shallow structure of the region bracketed by lat 18°S and 23°S off the South American coast shows that the Nazca plate landward of an outer swell and seaward of the Peru-Chile Trench is broken along an anastomosing network of fault scarps. In contrast, ridges that are discontinuous and variable in strike characterize the continental margin. These structural highs form culminations and depressions along strike and have no magnetic signature. The ridges are uplifted relative to depressions along their landward flanks, and seismic reflectors from basin-filling sediment document a history of rotational movement. Free-air gravity values suggest that the loci of maximum deposition subdivide the three forearc basins of the Arica Bight into subbasins.

Correlation between onshore and offshore tectonic units is tenuous, but the onshore geology suggests that part of the continental crust between the coast and the trench may have been removed. Interpretations of the shallow structure between the coast and the trench, based on single-channel seismic reflection profiles, suggest that the continental margin may either be underlain by granitic fault blocks, or composed of a wedge of deformed and dewatered sediment. The evidence at hand describing the South American subduction zone does not allow an unequivocal choice between the alternatives of subcrustal tectonic erosion of the continental edge or imbricate accretion of a wedge of sediment. Gravity may be the all-important force controlling the form of the hemipelagic cover. Both accretion and tectonic erosion may be acting to some degree with varying quantities of accreted sediment pasted against the disrupted face of the South American plate.

INTRODUCTION

The western margin of South America is an "active", or "Pacific-type" continental margin in that it displays all the characteristics associated with that tectonic setting: trench topography, seismicity,

---

*Present address: Geological Research Division, A-015, Scripps Institution of Oceanography, University of California, San Diego, La Jolla, California 92093.

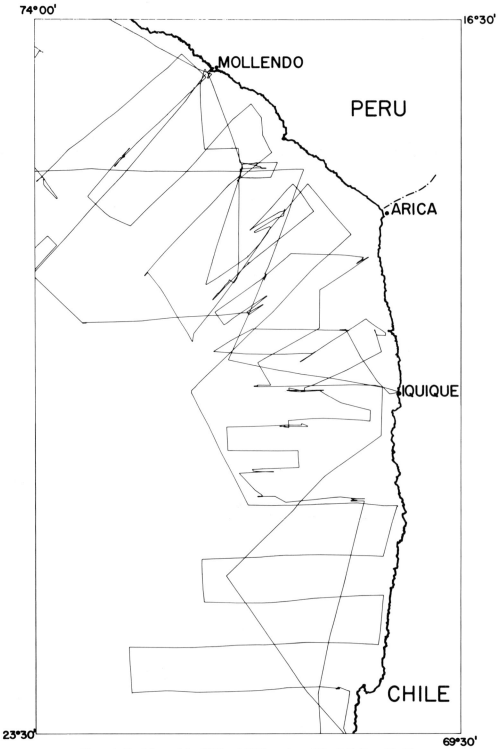

Figure 1. Tracklines of the 1973 and 1974 crusises of the R V *Kana Keoki*.

benches, and macerated sediment and rock (Drake and Burk, 1974; Fisher, 1974). Earthquake foci define segments of the Nazca plate that are being consumed beneath South America (Stauder, 1973; Teisseyre and others, 1974; Carr and others, 1974; Rodriguez and others, 1976; Barazangi and Isacks, 1976). A large gravity minimum parallels the Peru-Chile Trench (Hayes, 1966; Getts, 1975); and implies a mass deficiency interpreted by most investigators to mean that the oceanic plate is being held down. Whether viscous coupling between the asthenosphere and lithosphere allows convection currents to drag the Nazca plate down (after Hess, 1962) or the lithosphere merely slides off welts in the asthenosphere (Jacoby, 1973; Schuiling, 1973), the Nazca and South American plates are converging. How is that interaction expressed in the transition from oceanic to continental structure? Are oceanic sediment and rock transferred to the continental slope? Do near-surface structures constrain the possible geologic structures of the deeper portion of the edge of the continent of South America?

PREVIOUS SURVEYS OF THE CONTINENTAL MARGIN

The Arica Bight segment of the South American continental margin and bordering Nazca plate was surveyed during several marine expeditions. Zeigler and others (1957) summarized work prior to the *Shellback* and *Atlantis* expeditions of 1952 and 1955. Bathymetric profiles from those cruises revealed large offsets seaward of the Peru-Chile Trench and incised canyons along the upper continental margin. Bathymetry from the 1957 and 1958 *Downwind* expeditions showed elevated blocks seaward of the trench and benches or terraces at various crossings of the continental margin (Fisher and Raitt, 1962). During those cruises, seismic refraction lines were shot across the continental margin near Antofagasta. Results from the survey register a velocity of 5.9 km/s for the substrate 1 km below the sediment-water interface. Based on data from the 1962 and 1963 cruises of the R/V *Robert Conrad* and *Vema*, Hayes (1966) mapped the free-air gravity and diagramed crustal sections across the trench. The gravity map outlines the continuous linear trend of the trench and shows a gravity high centered over the Nazca plate at lat 19°S.

From published data, James (1971) proposed a plate tectonic model for the evolution of the Central Andes and compared that scheme with the classic eugeosyncline-miogeosyncline concept. In searching for three distinct sedimentary assemblages of what Dickinson (1971) considered to be a normal cordilleran arc setting, James (1971) found the volcaniclastic wedge of the foredeep to the east of the Andes, but could locate neither the melange or turbidites and ophiolites representing material accreted at a trench, nor the deposits of marine graywacke that typically accumulate in an arc-trench gap. The South American coastal desert is one of the driest regions in the world (Bowman, 1916), and it is not surprising that earlier researchers anticipated a continental margin and trench devoid of sediment. Newell (1949) thought it probable that "sedimentation is very slow in the Krummel Deep since it lies opposite a desert coast drained by only a few small streams." As a part of a study of the climate and relief of the Andes, Garner (1959) suggested that "the Peruvian marine area adjacent to the desert coast is almost certainly starved in a sedimentary sense," and more recently James (1971) conjectured that "there appears to be little sedimentation between the trench and coast." It was surprising, therefore, that the terraces described earlier (Fisher and Raitt, 1962) were discovered during the 1973 and 1974 cruises of the R/V *Kana Keoki* to be large, sediment-filled basins containing the marine graywacke anticipated by previous investigators.

The purpose of this research was to describe and map the features of an area referred to as the Arica Bight (the region offshore of southern Peru and northern Chile), to relate those features to the geologic units observed onshore, and to compare the structures within the Arica Bight to those of other subduction zones.

DATA BASE

During several weeks of 1973 and 1974 the scientific crews of the R/V *Kana Keoki* collected

TECTONIC MAP OF SOUTHERN PERU
AND NORTHERN CHILE (ONSHORE AND OFFSHORE)

LEGEND

SEDIMENTARY ROCKS:

Neogene sediments
  Onshore: Upper Pliocene to Recent continental sediments, primarily alluvial
  Offshore: Undeformed sequences of both pelagic and hemipelagic sediment

Salt flats

Moderately deformed hemipelagic sediments of fore-arc basins

Unresolvable acoustic returns representative of highly deformed sedimentary sequences

Mesozoic and Paleozoic sedimentary rocks
  Chile: Middle Devonian to Upper Cretaceous sedimentary and metasedimentary rocks, primarily submarine, with intercalated volcanics
  Peru: Cretaceous sedimentary rocks

EXTRUSIVE ROCKS:

Quaternary volcanics

Tertiary volcanics

Mesozoic submarine volcanics

INTRUSIVE ROCKS:

Block faulted exposures of acoustic basement, assumed to be granite, age unknown

Jurassic granitoids

Upper Triassic to Upper Paleogene granitoids

Permian to Middle Triassic granitoids

METAMORPHIC ROCKS:

Thick Precambrian detrital section with rare carbonate and volcanogenic sedimentary intercalations, metamorphosed under conditions ranging from greenschist to amphibolite facies

SYMBOLS:

Syncline

Anticline; offshore anticlines are topographic ridges

Steep slopes

Structural highs bordering the seaward margin of fore-arc basins

Steep slopes presumably controlled by normal faulting

Peru-Chile Trench axis

High angle faults

Submarine canyons

UNKNOWN:

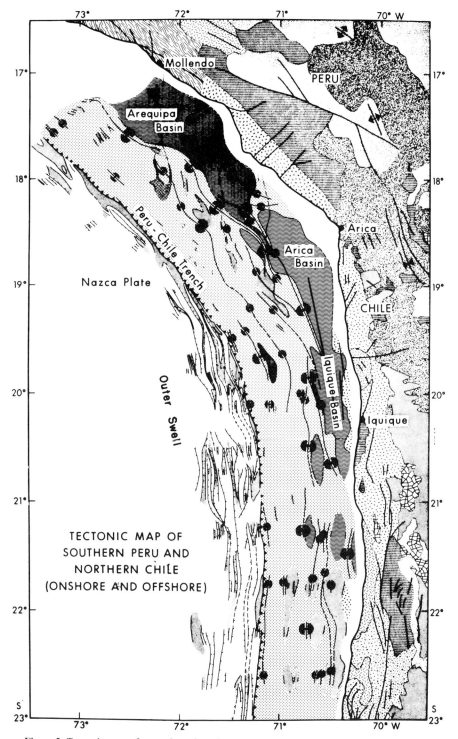

Figure 2. Tectonic map of coastal southern Peru and northern Chile. Legend on facing page.

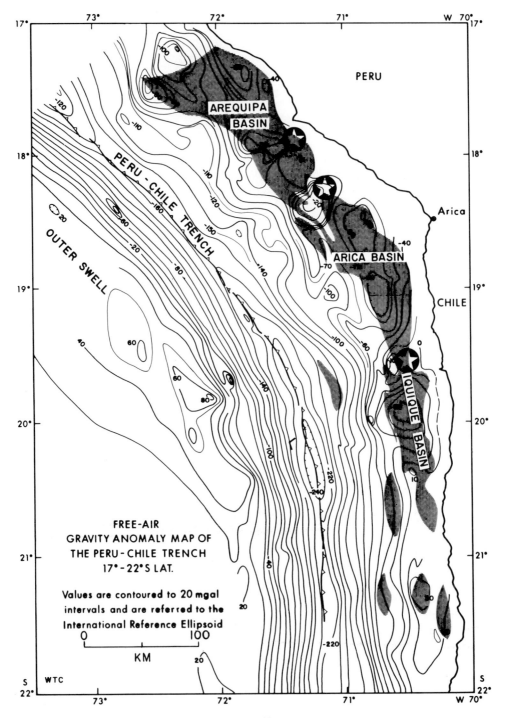

Figure 3. Free-air gravity anomaly map of the Peru-Chile Trench from lat. 17°S to 22°S. The axis of the trench nearly coincides with the gravity minimum. Wavy-line pattern shows the distribution of forearc basins along the continental margin. Localized gravity minima mark depressions in the acoustic basement of the upper continental margin. Stars locate gravity maxima discussed in text.

bathymetric, gravimetric, magnetic, and seismic data along more than 30 crossings of the Peru-Chile Trench in the Arica Bight (Figs. 1 to 5). These data are complemented by 40 sediment samples within the study area. To synthesize the large quantity of seismic reflection data, the profiles were assembled along the trackline grid, and similar features were linked between adjacent profiles. The resulting tectonic map shows that geologic structures can be separated into three categories according to geographic region: (1) the Nazca plate to the Peru-Chile Trench axis, (2) the continental margin to the shore, and (3) the continent landward of the shoreline.

NAZCA PLATE

Wide-beam reflection data along the western edge of the Nazca plate between lat 16°S and 24°S show that the sea floor consists of small-scale, offset blocks (Fig. 6). Latitudinal profiles show that topographic relief caused by horsts and grabens increases eastward from long. 74°W toward the outer swell, a broad arch flanking the trench (Figs. 2, 5; Prince and others, 1980). Sediment accumulations appear to be about 100 m thick above acoustic basement (Fig. 6; Rosato and others, 1975). In general, topographic profiles of trenches are quite similar (see Fig. 19 of Hayes and Ewing, 1970); however, the outer swell at Arica averages several hundred metres, somewhat of an extreme. Outer swells paralleling the Aleutian and Kuril Trenches (Malahoff and Erickson, 1969; Watts and others, 1976) have been attributed to compression of an oceanic plate against an overriding lithospheric plate (Hanks, 1971). Those bathymetric swells are also marked by gravity highs like the ones contoured in Figure 3.

Offsets along the outer slope of the Japan Trench (Ludwig and others, 1966), the Aleutian Trench (Grow, 1973), and the Peru Trench (Prince and Kulm, 1975) are interpreted as extensional faults in agreement with first-motion solutions from similar settings (Isacks and others, 1968; Stauder, 1968). Similarly, the surface of the Nazca plate between the outer swell and the Peru-Chile Trench axis displays a variety of offsets and tilted blocks, also probably extensional, developed along the convex surface of a bending elastic slab (see Hafner, 1951).

Offshore of Peru, as far south as lat 19°S, the plate steps downward toward the trench by means of a series of small offsets. A thin wedge of undeformed sediment fills the trench axis along its southeast-trending segment, except at lat 18°S where it is offset by a ridge (Fig. 2).

Within the Arica Bight a change in the character of faulting and increased relief accompany the change in strike of the trench from southeast to due south. An anastomosing network of fault scarps extends from lat 19°S to 21°S. Profile HH-HH', oriented parallel to the strike of the trench, shows a series of side echos spanning a range of 1,000-m relief (Fig. 7). Discontinuous reflections from the crests of faulted blocks, from salients on the scarp face, and from talus at the base of the scarp demonstrate the irregular and high relief of the area. Dip-slip offsets are as large as 1,000 (Coulbourn and Moberly, 1977). Reflectors suggest antithetic, seaward rotation of the fault blocks. Sediment is redeposited and ponded behind the tilted crests so that fossil assemblages of foraminifera and nanno-fossils are exposed at depths well below the present calcite compensation depth (Coulbourn and Resig, 1979). The lowest block may occupy the central portion of the trench axis, damming undeformed sediment behind it (Fig. 8). The bathyscaph *Archimède* explored a similar setting in the Japan Trench, and visual observations indicated that offsets there are normal faults (Bellaiche, 1967, 1980). Likewise, eight closely spaced holes drilled in the axis of the Middle American Trench encountered nannofossil chalk and basalt at various levels beneath the overburden of turbidites; this situation indicated that normal offsets occur along the downbuckling Cocos plate (von Huene and others, 1980b). The network of seismic reflection profiles suggests that comparable offsets along the subducting Nazca plate trend north to south, oblique to the strike of the Peruvian segment of the trench. That preferred orientation may arise from reactivation of the structural grain inherited during crustal formation at the East Pacific Rise (Coulbourn and Moberly, 1977). Whatever its origin, the effect is to offset the trench axis at the intersections of fault crests and the lowermost continental margin (Fig. 2).

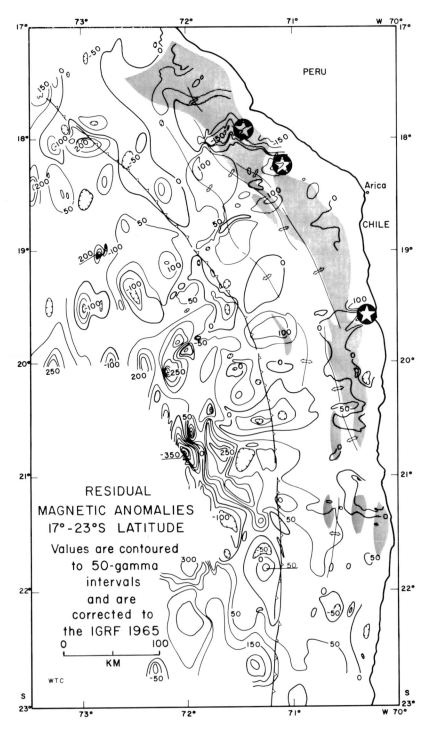

Figure 4. Residual magnetic anomalies for the Arica Bight segment of the Peru-Chile Trench and continental margin. Short-wavelength anomalies of the Nazca plate contrast with long-wavelength anomalies on the continental margin.

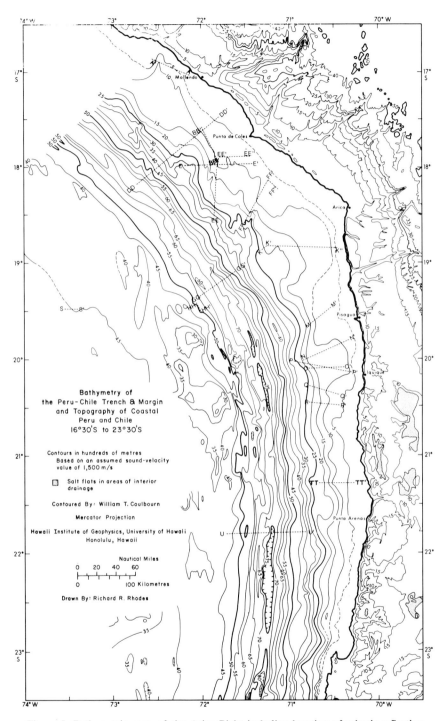

Figure 5. Bathymetric map of the Arica Bight including location of seismic reflection profiles. The data sources include (1) Hawaii Institute of Geophysics surveys; (2) Scripps Institution of Oceanography surveys; and (3) American Geographical Society sheets D-18, E-18, E-19, F-19. For detailed bathymetry, see Prince and others (1980).

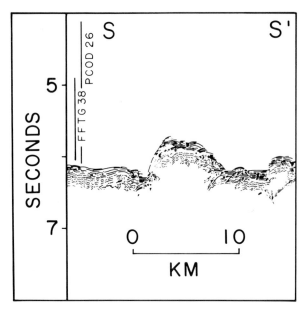

Figure 6. Horsts and grabens typifying the structure of the Nazca plate seaward of the outer swell. Section S-S' is located on Figure 5. Two-way travel time given in seconds for all seismic reflection profiles.

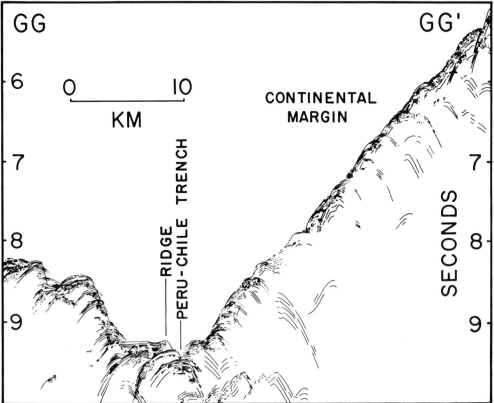

Figure 8. Profile GG-GG' shows three tilted blocks of Nazca plate crust. One of those blocks occupies the trench axis, and its crest protrudes as a low ridge. Turbidites are ponded seaward of the ridge. Acoustically unresolvable returns characterize the continental margin.

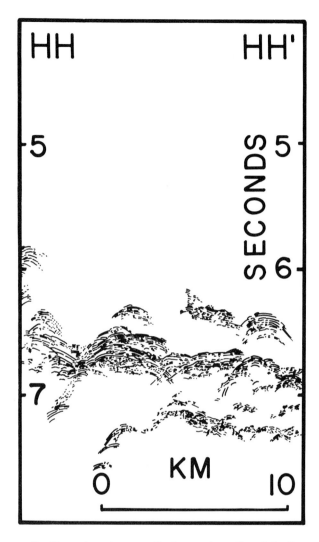

Figure 7. Profile HH-HH' located on the edge of the Nazca plate and oriented parallel to the strike of the trench. The random distribution of side echos, spanning a depth range of more than 1,000 m, demonstrates the high relief of the seaward flank of the trench.

Profiles oriented perpendicular to the strike of the Peru-Chile Trench show axial ridges ponding turbidites seaward and above sediment-free portions of the trench axis (Fig. 8). Although no submarine canyons could be mapped across the deep, landward wall of the trench (Fig. 2), highly stratified reflectors, well-sorted sands, and a sample containing displaced, shallow-water, benthic foraminifera indicate that axial deposits are turbidites derived from the landward slope (Coulbourn, 1980). Where present, horizontal reflectors identify the trench-axis fill (Fig. 9) and abut the toe of the continental margin along a tectonic front (Montecchi, 1976).

Prince and others (1974) described a turbidite basin oriented obliquely to and 700 m above the axis of the Peru Trench. Because the shallow-water benthic foraminifera within that deposit must have come from the continental margin they interpreted the basin as having been uplifted. Prince and Schweller (1978) suggested that many of the scarps within the trench axis formed within the past 500,000 yr; they then calculated rates of uplift for small sections of the axis. To reconcile their bottom samples with structures shown on seismic reflection profiles, they proposed that normal faults offset most of the outer slope, but that those faults are replaced by reverse faults near the trench axis. The seemingly isolated and elevated turbidites of some sections of the outer slope may be produced by mechanisms other than uplift. As shown in Figure 10, the trench axis changes strike at Arica so that

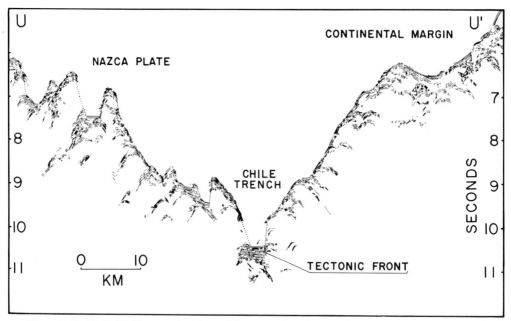

Figure 9. Profile U–U'. Horsts and grabens of the seaward trench wall offshore of Chile. Lowest blocks are tilted toward the trench.

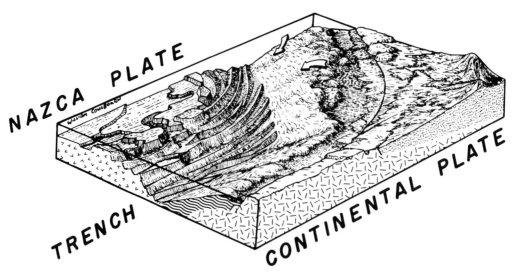

Figure 10. Block diagram of near-surface features of the Arica Bight. Ridges of Nazca plate crust intersect the continental margin of Peru obliquely. Raised portions of those blocks may act as levees that channel turbidity flows from the Peruvian continental margin seaward and above unfilled portions of the trench axis. Linearity of Fault-scarp traces is exaggerated.

blocks of Nazca plate crust intersect the Peru continental margin obliquely. The crests of those tilted fault blocks may act as levees channeling turbidity flows seaward and above isolated, unfilled portions of the trench axis.

Between lat 21°S and Valparaiso, Chile, there is a gradual transition from steplike offsets to horsts and grabens that parallel the outer slope of the trench (Schweller, 1976; Schweller and Kulm, 1975).

Some of those blocks are tilted toward the trench axis (Fig. 9). The trends of fault scarps of the southern section are shown as being straighter than those to the north. The simplified appearance, however, may be due to sparse data coverage in the south. The trench axis is generally V-shaped and sediment-free (Fig. 2).

CONTINENTAL MARGIN

Sharp offsets of the Nazca plate contrast with smooth ridges and terraces of the continental margin. Unresolvable acoustic returns characterize the lower continental slope, so that acoustic basement is exposed along the inner trench slope. Landward-dipping reflectors are, for the most part, limbs of reflection hyperbolas (seismic artifacts rather than geologic structures; Figs. 8 and 9). Subbottom reflectors generally can be resolved only at depths shallower than 4,000 m.

Horizontal reflectors occur in isolated patches and represent undeformed sediment ponded in local depressions. Offshore of Peru, between lat 18°S and 19°S, three of those small ponds can be linked to submarine canyons that continue upslope to larger basins (Fig. 2). Channels flanked by levees cross the forearc basins and trend toward incised channels near the Peruvian coast (Fig. 11). Thus, ponds lower on the continental slope are being filled with sediment that has bypassed catchments farther upslope.

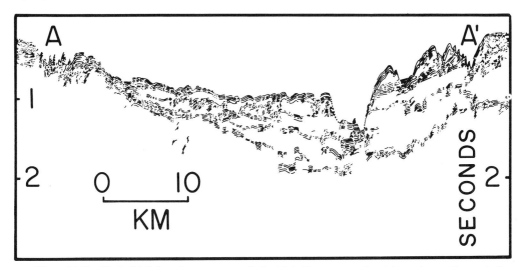

Figure 11. Profile A-A'. Submarine canyons are incised into the upper continental margin near the coast of Peru.

Forearc basins are a second category of sedimentary deposit at the Arica Bight (Fig. 2). At this writing, the structural evolution of similar basins in corresponding tectonic settings is the object of considerable discussion and exploration. For sections of the eastern Sunda and western Banda arcs, Curray and others (1977) and Hamilton (1977) suggested that a prism of tectonized material forms a seaward buttress behind which sediments are ponded over trapped oceanic crust and that oceanic crust underlies most of the arc-trench gap (Dickinson, 1971). These are the residual basins of Dickinson and Seely (1979) in contrast to accretionary basins seaward of the trench-slope break (Moore and Karig, 1976) and seaward of the Aleutians (Atka Basin of Grow, 1973). Dickinson and Seely (1979) termed basins in which "the strata lie unconformably across a structural join between the arc massif on the inner side of the basin and deformed accreted strata of the subduction complex on the outer side of the basin" as constructed basins, and Seely (1979) used the Iquique Basin as an example. At first glance, the three large forearc basins of Arica do seem to fall into that category in which the landward portion

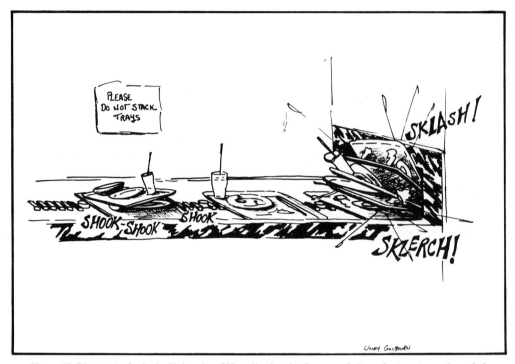

Figure 12. Plate tectonics in the University of Hawaii cafeteria. Conveyor belt carries trays toward restrictive portal. Unaligned trays do not always pass tranquilly through to the dishwasher. Once a tray jams against the wall, successive trays are injected at the bottom of the stack. Intertray paper cups and plates are dewatered and metamorphosed. Accretion and overflow onto floor ceases only when the kitchen worker presses STOP button.

of the basin overlies rocks of the arc-massif, the South American continental block, whereas the seaward portion is ponded behind a deforming subduction complex, a "cafeteria-tray-type" mass of accreted sediment and rock (Fig. 12). The ambiguities of the data at hand, however, leave open the possibility that the forearc basins of the Arica Bight are intramassif basins, completely underlain by rocks of the South American continent. Data from DSDP sites have been interpreted to suggest that the imbricate thrust model of Seely and others (1974) may not apply to the Japanese arc (Kaneps, 1978; von Huene and others, 1980a). Drilling off the Mariana arc (Scientists aboard the *Glomar Challenger*, 1978c) and the Mid-America Trench (von Huene and others, 1980b) provided equivocal results. Discussions within the following sections consider some possible interpretations.

Core samples indicate that turbidites and olive-green to gray hemipelagic mud fill the forearc basins of the continental margin at Arica (Coulbourn, 1980). Those sedimentary layers are recorded on seismic reflection profiles as reflectors that converge seaward toward a structural high (Figs. 13 and 14), and document a history of landward migration of depocenters within each of the large forearc basins of the Arica Bight (Coulbourn and Moberly, 1977; Seely, 1979). This rotational configuration seems to be the typical case for most basins in corresponding tectonic settings, and it has a subaerial analogue at the Arauco Peninsula off central Chile (Kaizuka and others, 1973). There, rotational movement has uplifted and tilted landward the marine terraces on Santa Maria and Mocha Islands relative to a subsiding trough that separates those islands from the mainland. There are exceptions to this sense of motion. For example, single transects of the central Peru Margin show both landward and seaward migration of depocenters (Masias-Echegaray, 1976), and depocenters of the Japan margin are evidently shifting seaward with time (von Huene and others, 1980a).

Undulating ridges are characteristic of the continental margin at Arica and can act as dams to trap sediment (Fig. 2). Ridges are identified as topographic highs or axes along which acoustic basement reaches shallow levels (Figs. 13 and 14). At Arica there is no magnetic or gravimetric pattern that would distinguish the trend of these structural highs. They may coalesce and splay while forming irregular depressions and culminations that expose pasty to semi-indurated, olive-green to gray sediment, poor in microfossils and rich in glauconite (Coulbourn, 1980). With time, sediment ponded by structural highs—that is, the turbidites and hemipelagic sediment represented as subhorizontal reflectors—may be incorporated into the deforming body of the continental margin (Coulbourn and Moberly, 1977; Howell and von Huene, 1978).

In addition to the Arequipa, Arica, and Iquique Basins, five small unnamed basins have been mapped (Fig. 2). The three larger forearc basins differ from the smaller trench-slope basins in size. The former are closer to land and therefore trap more sediment; the later only collect sediment that has bypassed the catchments of the upper slope. Although the three larger basins appear to be connected, they are structurally discrete. Their deeper parts are separate basins and only the youngest sedimentary fill joins one basin to another. Culminations of the acoustic basement at lat 18°20'S and 19°40'S separate depressions over which thick lenses of sediment have accumulated. There is a good correlation between gravity minima and location of large forearc basins. Because there is a good agreement on a large scale, local gravity minima are assumed to identify further subdivisions of the basins where the acoustic basement was too deep to be detected by the single-channel seismic reflection

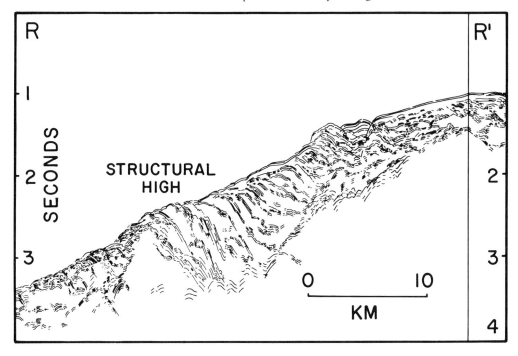

Figure 13. (upper) Profile R-R' across small forearc basin on the continental margin of northern Chile. Reflectors generally converge seaward toward a structural high. (lower) Profile R-R' photographically reduced to a 2:1 vertical exaggeration of topography.

method. The variability in the elevation of acoustic basement, where detected, is demonstrated by profile BB–BB′ (Fig. 15), along the structural high of the Arequipa Basin. If a 2-km/s velocity of sound is assumed for the uppermost sedimentary section, acoustic basement deepens by more than 600 m in a distance of only 10 km. Sequences of profiles across the continental margin, therefore, show widely varying structure (Fig. 14). Sediment is accumulating over an undulatory acoustic basement: the resulting geomorphic patterns resemble those shown in the Landsat images of the Makran of Iran, a similar tectonic setting (Farhoudi and Karig, 1977).

How far seaward can "continental-type" structures be seen on seismic reflection profiles? How great a volume might accreted sediment and rock occupy? A strong reflector forms basement beneath the landward portion of the forearc basins of the Arica Bight (Coulbourn and Moberly, 1977). As off the Sunda Arc (Karig and others, 1979), that reflector extends no farther than the axis of the forearc basins. Within the Arica Bight, the boundary that produces this strong contrast in acoustic impedance has two forms. Along the landward boundary of the Arequipa and Arica Basins at lat 18°10′S, acoustic basement undulates in swells that parallel the strike of the trench axis. In contrast, the acou-

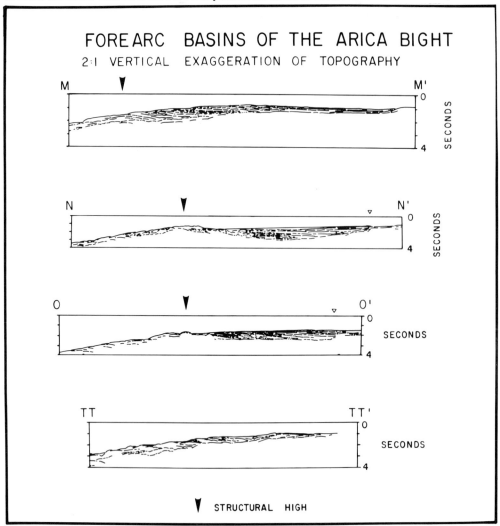

Figure 14. Seismic reflection profiles across the continental margin of Chile. Profiles photographically processed to a 2:1 vertical exaggeration of topography and arranged from north (top) to south (bottom).

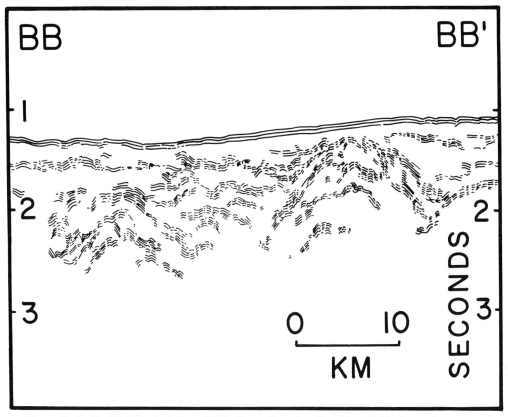

Figure 15. Profile BB-BB' oriented parallel to the coatline of southern Peru. Subbottom reflectors reveal culminations and depressions of acoustic basement beneath the Arequipa Basin.

stic basement of the Iquique Basin rises to the sea floor in a series of stepped, near-vertical offsets, suggestive of block-faulting within a brittle basement. A sediment-rock contact, probably basin-filling turbidites over continental crust, could produce the strong returns observed. The pattern of residual magnetic anomalies for the region suggests that acoustic basement is not oceanic crust (Fig. 4). Long wavelengths and random contours characterize the magnetic residuals mapped along the continental margin in contrast to short wavelengths mapped for the Nazca plate. A similar contrast within magnetic profiles from the Andaman Sea associates broad, low-amplitude, nondefinitive patterns with regions of rapid sedimentation (Curray and others, 1978). With the possible exception of the culminations separating the larger basins of the Arica Bight (located by stars, Fig. 4), short-wavelength anomalies do not coincide with structural highs of the Arica Bight. The cross section that Hamilton (1977) proposed for the subduction system of the Java Trench off southern Sumatra does not, therefore, seem appropriate for the Arica Bight. That configuration places a detached slab of oceanic crust at shallow levels beneath the structural high. A position that shallow should produce short-wavelength residual magnetic anomalies along the structural highs, unless the detached slab at Arica is a composite of tectonically stacked oceanic crust as suggested for the continental margin off Nicoya, Costa Rica (Buffler and Watkins, 1978) and depicted for the Sunda arc (Fig. 7 of Curray and others, 1977). Stacking might cancel the signal by destructive interference of superimposed sources. The geophysical information at hand—stepped offsets in and strong reflections from acoustic basement, and low-amplitude magnetic signal—suggests that continental rocks underlie at least the landward half of the basins off northern Chile. Slices of oceanic crust, if at all present beneath the continental margin within the Arica Bight, must be at considerable depth.

CONTINENTAL TECTONIC UNITS

In order to link the geologic structures depicted by marine geophysical data to those mapped on land, Figure 2 includes a compilation of tectonic units described in several published maps (Blondel, 1964; Dalmayrac and others, 1971; Paredes and Mégard, 1972; Frutos and Ferraris, 1973). Because several sources were used, some tectonic boundaries could not be matched within countries and across the border between Peru and Chile.

Intrusive rocks, representing a wide span of geologic time, crop out along the coast. The oldest among them is the Precambrian coastal batholith of southern Peru. Extrusive rocks cover much of the interior, and Quaternary volcanic rocks are situated landward of their Tertiary equivalents. Salt lakes and unconsolidated Cenozoic sediments cover most of the remaining surface. High-angle faults juxtapose intrusive rocks ranging from Precambrian to Cretaceous age (Wilson and Garcia, 1962); fault trends and lineaments shown in Landsat photographs of southern Peru and northern Chile generally parallel the coastline except at lat 18°10'S and 19°40'S (Rodriguez and others, 1976).

RELATION BETWEEN ONSHORE AND OFFSHORE GEOLOGIC STRUCTURES

Sediment seaward of the trench-slope break and that trapped within the arc-trench gap mask the underlying transition from oceanic crust to continental crust. Landward of the forearc basin, water depths are too shallow for reflection profiling. Relations across the forearc basins can only be inferred by correspondence and extension of geologic trends and structures. Intrusive rocks must compose portions of the continental margin because they front virtually the entire coastline (Fig. 2). A geologic map of the coastal batholith between lat 13°S and 15°30'S indicates that intrusive rocks also border long stretches of the Peruvian coast (Ruegg, 1956). Working in the Ilo quadrangle, near lat 18°S, Navarrez (1964) mapped a promontory of Cretaceous intrusive rocks at Punta Coles, where the ocean surrounds plutonic rocks on three sides. Based on field relations at lat 14°S, Pitcher (1974) stated that the San Nicholas batholith continues seaward from the Paracas Peninsula and forms a portion of the continental shelf. A subrounded granodiorite boulder recovered from an exposure of acoustic basement at rock-dredge station 12 (lat 19°30'S, 380-m water depth) suggests that circumstances similar to those mapped along the Peruvian coast may also exist off Arica. Alternatively, the boulder was transported. It seems appropriate, therefore, to infer that large portions of the margin are underlain by continental rock and further that the acoustic basement landward of the Iquique Basin and at the southern half of the Arica Basin represents the contact of sediment over continental basement. This contention is further supported by the high-angle faulting within the basement (Fig. 14).

The location of extrusive rocks delineates the Quaternary and Tertiary volcanic arcs. Geologic maps show that the magmatic belt narrows from a maximum width of about 400 km in northern Chile to about 100 km at the latitude of Lima, Peru (Blondel, 1964; Paredes and Mégard, 1972). Coats (1962) related volcanism to subduction, a concept that leads to an apparent paradox for this section of the Andes, where the widest segments of the magmatic arc overlie the most steeply dipping sections of the Benioff zone (Stauder, 1973, 1975; Getts, 1975; Whitsett, 1975; Kelleher and McCann, 1976; Mégard and Phillip, 1976). There is some debate regarding the configuration of Nazca plate subduction under Peru, and that section of the oceanic plate may dip landward at 25° to 30°, more steeply than previous extrapolations indicated (Barazangi and Isacks, 1976; Rodriguez and others, 1976; James, 1978).

In northern Chile, extrusive rocks dating from the late Pliocene to Holocene are generally located to the east of their late Miocene and early Pliocene equivalents (Farrar and others, 1970; Frutos and Ferraris, 1973; Stewart and others, 1974). In Peru, Tertiary and Quaternary extrusive rocks mingle in an irregular pattern (Paredes and Mégard, 1972). Bussell and others (1976) demonstrated the superposition of Peruvian ring complexes for a period of 70 m.y. beginning as far back as the Late Cretaceous (95 m.y.). The geologic data show, therefore, a history of continentward migration of igneous activity in northern Chile contrasted with a history of stabilized igneous loci throughout much of Peru. Rutland (1971) attributed the periodic eastward movement of plutonism and volcanism in the Andes to the loss of continental crust along the subduction zone. On the basis of field relations in central Chile, Katz (1971) reached a similar conclusion for the continental margin between lat 35°S and 44°S. It is also possible that the eastward shift of Chilean volcanism is only a response to a decreasing dip of the subducting Nazca plate through time. If the implications drawn by Rutland (1971) and Katz (1971) are applied to Peru, where the volcanic arc has remained stationary, it appears that removal of the continental plate may not be occurring along that section of the subduction zone or merely that the dip of the subducting plate has not changed.

Near-vertical offsets characterize Andean faulting. Geologic cross sections from the Pacific coast to the interior show high-angle faults offsetting basement and overlying sediment in Peru, Chile, and Bolivia (Jenks, 1956; Sonnenberg, 1963; Mégard, 1967; Katz, 1971; Rutland, 1971; Myers, 1975). High-angle faults control the location of ring complexes in central Peru (Bussell and others, 1976), and the eugeosynclinal-miogeosynclinal belts of Peru developed over vertically faulted basement of continental crust (Cobbing, 1976). Crustal extension and differential regional uplift characterize the Quaternary history of northern Chile (Mortimer, 1972; Paskoff, 1977). Overthrusts and transcurrent faults have been mapped in central and northern Peru (Mégard and Phillip, 1976) and along the eastern Andean flank where Devonian and Carboniferous sediments are thrust toward the craton (Dalmayrac and others, 1971). The continental geology shows that to the west of the Andean foreland the convergence of the Nazca and South American plates generally does not produce large-scale compressive structures within the basement rocks of the overriding block. Whatever stresses are imparted by the subducting Nazca lithosphere are manifest principally in vertical oscillations of the crustal blocks of South America (Myers, 1975). How plate convergence of the Andean type could cause a mountainous arc to rise above a subduction zone remains an unsolved problem (Uyeda, 1978). Dewey and Bird (1970) and Le Pichon and others (1976) suggested that thermal processes, a so-called mobile core related to the sinking plate, govern structures cropping out in the mountainous arc of the overriding plate.

At Arica specifically, and along the Peru-Chile Trench in general, the contact between offshore and onshore structures is difficult to locate. As von Huene and others (1979) found for a section across Kodiak Island, the coincidence of a contact or transition with a change of data base — that is, detailed terrestrial field studies in contrast with general marine surveys—prevents a clear understanding of convergent margin processes. Although seemingly buried along the entire length of the Peru-Chile Trench, the contact between onshore and offshore units probably does not bear a fixed relation to distance from the trench axis. Using published geologic field data for the region of Ecuador, Lonsdale (1978) placed the boundary between the South American crust and accreted oeanic crust within the inter-Andean depression, more than 250 km east of the present axis of the trench. Henderson and Evans (1980), however, considered that placement questionable, preferring to locate the boundary at the western margin of the Western Cordillera. Shepherd's (1979) schematic diagram places that boundary as far seaward as the Ecuadorian trench-slope break. Hussong and others (1976) interpreted the seismic velocity structure of the Peru Trench near lat 12°S as evidence that continental crust comprises the subsurface of the middle and upper continental margin. To the south, Scholl and Marlow (1974) interpreted regional geologic studies as evidence that volcanic rocks ranging from Precambrian to Mesozoic age may extend as far seaward as the steep inner wall of the Chile Trench at lat 33°S.

Vertical offsets are visible on the reflection records along the landward border of the Iquique Basin and within the southern portion of the Arica Basin (Fig. 2). As mentioned before, stepped reflectors off northern Chile contrast with the undulatory acoustic basement off southern Peru. Although faults trend perpendicular to the coastline at lat 18°S and 19°40'S and are aligned with culminations bracketing the Arica Basin, offsets cannot be traced offshore on any of the seismic reflection profiles. Culminations seaward of the faults, however, are identified by gravity maxima and by short-wavelength and magnetic patterns (stars in Figs. 3 and 4).

The Andean continental margin seems to be segmented in tectonic response to plate convergence.

Perhaps the leading edge of the South American plate in Chile is crumbling away—or was sheared off laterally as has been suggested for the Middle America Trench (Karig and others, 1978)—whereas the segment off southern Peru may at this time be resisting "attrition," "tectonic-erosion," or "corrosion" by the Nazca plate. Lithofacies data and sediment-isopach maps show a coarsening and thickening of the Devonian rocks of Bolivia toward the northwest (Isaacson, 1975), and Precambrian rocks cropping out in the Peruvian coastal batholith imply that the Peru margin too has been truncated during the geologic past.

In summary, geologic structures generally parallel the coastline both onshore and offshore; the leading edge of the continent is probably beneath the forearc basins but could be placed anywhere between the shoreline and the trench; and indirect evidence implies segmentation of the continental margin. The subaerial geology contains little direct information about the continental margin; therefore, the description of the geology of the Arica Bight must rely on inferences drawn from the available geophysical data and on comparison with structures and rock types of other subduction zones.

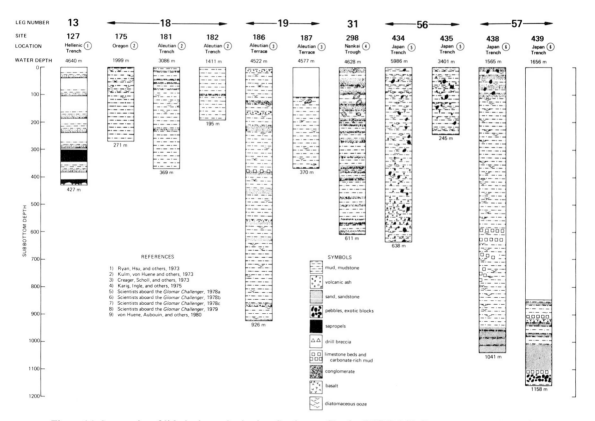

Figure 16. Summaries of lithologies and seismic reflection profiles for DSDP drill sites on convergent margins. Circled numbers identify references cited. No profile was available for Marianas arc drill sites. Site 127: Turbidite and contourites, graded sands and marly oozes, volcanic ash, and sapropels. Site 175: Diatomaceous clayey silt and carbonate-bearing mud, silt, and sand. Site 181: Diatomaceous silty clay, silt, and sand. Site 182: Silty clay and clayey silt. Site 186: Diatomaceous silty clay, volanic ash, sand, and carbonate mud. Site 187: Diatomaceous silty clay. Site 298: Turbidites 0 to 180 proximal, 180 to 611 distal; mud, silt, and sand. Site 434: Diatomaceous mud and mudstone; volcanic ash. Site 435: Diatomaceous mud, volcanic ash, and terrigenous sand. Site 438: Pebbly sand, diatomaceous mud, volcanic ash, limestone, dewatering veinlets. Site 439: Diatomaceous mud and sand; volcanic ash. Site 440: Diatomaceous mud and claystone, volcanic ash and dewatering veinlets. Site 441: Diatomaceous

COMPARISON WITH OTHER SUBDUCTION ZONES

Seismic refraction studies suggest that the Peru-Chile subduction zone differs somewhat from other convergent plate boundaries. Beneath the structural high, compressional wave velocities of 5.9 to 6.2 km/s represent consolidated sediment and rock located 1 to 2 km below the sea floor (Fisher and Raitt, 1962; Hayes, 1966; Hussong and others, 1976). These high velocities exceed those for similar sections across the Aleutian Trench (less than 4.5 km/s; von Huene, 1979), the Japan Trench (5.7 to 5.9 km/s; Uyeda, 1974), and the Sunda arc (less than 5.6 km/s; Curray and others, 1977; 4.6 to 4.9 km/s; Kieckhefer and others, 1980). The concept of an accretionary prism of deformed and dewatered, offscraped sediment and rock as Karig and Sharman (1975) proposed for western Pacific subduction zones may explain low velocities in "accreted prisms" off Indonesia, where great volumes of sediment are entering the subduction zone, but does not seem to accord with the high velocities recorded off South America. Velocities of 6.2 km/s are representative of granitic rock. Gravity contours, moreover, show that minimum values roughly coincide with the trench axis and are not displaced

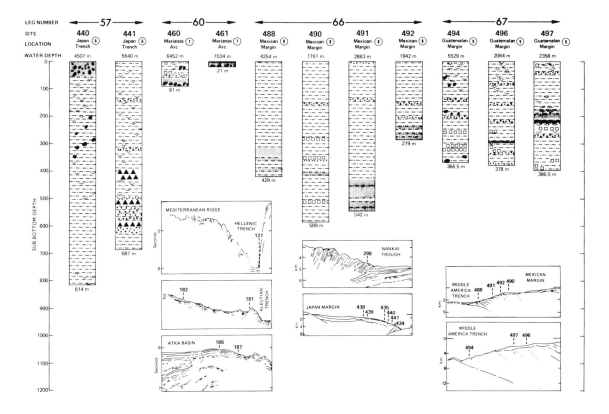

mud and claystone, volcanic ash, and dewatering veinlets. Site 460: Redeposited and mixed sediments, siliceous ooze, calcareous and vitric mud, and pebbles. Site 461: Reworked sediment and cobbles of igneous and metamorphic rock. Site 488: Hemipelagic mud; mudstone and sand. Site 490: Hemipelagic mud and mudstone; gassy frozen sediment. Site 491: Hemipelagic mud, mudstone and sand, and gassy frozen sediment. Site 492: Hemipelagic mud, mudstone and sand, gassy frozen sediment, and slickensided anastomosing fractures. Site 494: Drill breccia, diatomaceous mud, volcanic ash, claystone, and limestone. Site 496: Diatomaceous mud, volcanic ash, carbonate mud, conglomerate, dewatering veinlets, and gassy frozen sediment. Site 497: Diatomaceous mud, volcanic ash, carbonate mud, conglomerate, dewatering veinlets, and gassy frozen sediment. Site 498: Site 498 parallels site 494 and is not shown in this compilation.

landward as would be expected if the continental slope was composed of accreted sediment (Fig. 3 and Hayes, 1966).

Imbricate thrusting (Fig. 12) can explain the characteristic back-tilting of seismic reflectors recorded in transects across the continental margin of Arica (Fig. 13) and similar patterns of reflectors from other convergent margins (Beck and Lehner, 1974; Karig and Sharman, 1975; Moore and Karig, 1976; White and Klitgord, 1976; Curray and others, 1977; Hamilton, 1977). However, the question remains unanswered. Have imbricated sediment and slabs of oceanic crust accreted to form a large prism beyond Andean continental crust (Kulm and others, this volume)? Or most or all of the Nazca oceanic crust and sediment are subducted, leaving behind depositional sequences (Scholl and others, 1977), a thick drape of hemipelagic sediment that veils a basement of subsiding, block-faulted continental crust (Scholl and Marlow, 1974; Hussong and others, 1976).
and others, 1976).

Samples from the International Program of Ocean Drilling (IPOD) of convergent margins are being assessed at the time of this writing. Twenty-three sites have been drilled between the trench axis and the trench-slope break of various convergent margins; a lithologic summary shown in Figure 16 indicates that none have penetrated an imbricated wedge of sediment offscraped from a subducting oceanic plate (Site 498 is not shown in Fig. 16). Most of the sediment retrieved from these wells is olive-green to gray, diatomaceous mud with some intervals rich in carbonate mud, sand, and volcanic ash. Dewatering veinlets are characteristic (Arthur and others, 1980). Samples from the uppermost 200 to 300 metres of those drill holes contain benthic foraminifera displaced beneath their commonly recognized depth ranges (von Huene and others, 1980b). Sites 343, 441, and 494 are particularly important, because they are relatively deep holes drilled into the lowermost slope, yet they did not sample the anticipated suite of accreted sediments. None of the deep-ocean brown clay, biogenic ooze, and chert cored from the westernmost Pacific plate was recovered in the Japan Trench sites, nor were any of the counterparts of the carbonates cored from the Cocos plate recovered in the Guatemala Margin sites (von Huene and others, 1980a, 1980b; Aubouin and others, 1979). The depth of these holes and the absence of offscraped sediment and rock must limit the volume that can be attributed to a hypothetical accretionary prism at those transects. This reduced volume is an unexpected finding because the Guatemala Margin is the "type section" for the trench-slope model as deduced from industry wells and processed multichannel reflection profiles (Seely and others, 1974). In short, DSDP samples from convergent margins drilled to date suggest that subsidence, rather than uplift, is the dominant process in the sedimentary drape. Moore and others (1979) preferred to interpret sandy intervals recovered from Sites 488, 491, and 492 as turbidites uplifted from the trench axis (Fig. 16). Using the presence-absence of calcareous microfossils and extrapolations for the paleo-calcite compensation depth, they computed rates of uplift for these sites. Trace-fossil assemblages at Site 492 imply uplift from more than 4-km depth (Moore and others, 1979). Shipboard paleontological results, however, did not indicate stratigraphic inversions within any of the drill holes, nor could the provenance of the sands be restricted to the trench axis. On the basis of published results for Leg 66, the evidence for accretion and progressive uplift at Sites 488, 491, and 492 seems to be philosophical rather than direct. Karig and others (1975) prefaced their discussion of Site 298 with the statement that "no hole has successfully penetrated the lower slope to reveal how the separation of trench sediments from the downgoing plate occurs." Five years and more than 20 sites later that success remains to be achieved. Drilling results to date do not explain either the uplifted benthic foraminifera off northern California (Silver, 1971) and Sumatra (Moore and others, 1980) or the transferred pelagic sediments on the Peru Margin (Kulm and others, 1974) and offshore of Java (Scripps Institution of Oceanography, unpub. data). Nor do the results demonstrate how subduction complexes may be exposed at Barbados (Speed, 1978) and Nias (Moore and Karig, 1980), which are the culminations of structural highs. Only at 427 subbottom at Site 127 was a sequence encountered (Fig. 16)—horizontally bedded Pliocene ooze 8 m beneath a Cretaceous/Quaternary contact (Ryan and others, 1973; Hsü and Ryan, 1973)—that would seem to embody the features predicted by the trench-slope model (Seely and others, 1974).

The lithologies recovered from DSDP sites on other convergent margins and the relatively high compressional wave velocities computed for the margin at Arica predict that a drape of diatomaceous mud, ash, and sand overlies a basement of subsiding, block-faulted, continental crust within the Arica Bight.

INTERPRETATION AND SPECULATION

If the pattern of deformation recorded by seismic reflection profiles across the forearc basins of the Arica Bight is assumed to be directly related to deeper structures within the margin, two possible end members of a continuum of choices can be drawn (Fig. 17). The accretionary end member shows a thick wedge of deformed and dewatered sediment pasted against the edge of a granitic continental block. A structural high rises as accreted sediment and rock are added to the base of the stack, and the forearc basin straddles the accreted sediment–granite rock contact. Continental rocks are not tectonically eroded from the edge of the South American plate. In contrast, the nonaccretionary end member shows a thin covering of deformed hemipelagic sediment overlying the prow of a faulted continental block. In an independent study, based on the same geophysical data presented in this report, Ferraris (1979) suggested that the "igneous and/or metamorphic basement" constitutes tectonic pillars that form the "Borde de Plataforma," the structural high of this report. Masias-Echegaray (1976) drew comparable profiles for the Peru Margin. To produce the observed geometry of seismic reflectors, the pillar must rise incrementally relative to the "post" or graben that landward underlies the forearc basin. Relatively high-standing pillars would form the culminations of the structural high (Fig. 17). Sediment of the forearc basin accumulates over continental crust only. Most of the pelagic and hemipelagic sediment as well as blocks of granitic rock are subducted. The volume to be filled by an accretionary prism is relatively small. Alternatively, that same space (no. 6 in lower diagram of Fig. 17) could be occupied by a drape of hemipelagic sediment undergoing in-place deformation (Scholl and others, 1977). Assuming that all oceanic-plate sediment is scraped off into an accretionary wedge, Shepherd (1979) computed that 775 km^2 should be represented in a profile crossing the Peru Margin at lat 6°S. The geologic and geophysical data at hand allow room for only 241 km^2. Scholl and Marlow (1974) figured that scrapings along the Chile Margin should have built a wedge 25 km thick and 100 km wide since early Mesozoic time. Clearly some sediment must have been subducted along the Peru-Chile Margin. That being the case, it is only one step further to allow most or all of the pelagic sediment to be removed, the uncertainty of a mechanism notwithstanding.

If the pattern of deformation recorded by seismic reflection profiles at Arica is assumed to be not directly related to deeper structures, the near-surface deformation can be attributed to gravitational creep, irrespective of the structures deep within the continental margin. Elliott (1976) conjectured that imbricate thrusts of the arc-trench gap are driven not by the downgoing slab, but by regional surface slope produced by uplift along the magmatic arc, a point debated by Seely (1977). Elliott's (1976) scheme would account for the juxtaposition of uplifted, continental blocks extended along high-angle faults and an imbricate wedge along the continental slope. The geologic evolution of the northern Apennines has been described by Elter and Trevisan (1973, their Figs. 12 and 15) as "a tectonic pile of mutually overthrust Ligurian units"; this may be a paleogeologic analogue for gravity-produced imbricate wedges of the acoustically impervious portion of some modern subduction zones. Jacobson and others (1979) and Biju-Duval and others (1979) are among the most recent to suggest that olistostromes and imbrication may "act in concert" to produce the diffuse reflectors observed along the landward slopes of trenches. Buffler and others (1979) have extended the possible role of gravity in their discussion of back-tilted reflectors and ridges of the southwestern Gulf of Mexico. Massive slumps, identified by unresolvable acoustic returns and by slope angle, have obliterated the expected subhorizontal reflectors marking portions of the Peru Trench off Ecuador (Fig. 23 of Shepherd, 1979) and segments of the Japan Trench (Fig. 10 of von Huene and others, 1980a). The basins of the Peru-Chile study area may be chance accumulations of turbidites on an unstable drape of hemipelagic sediment. Many profiles across the margin show a series of minor depressions, any one of which may serve as a catchment (Fig. 18). Depressions closest to the source will trap the most sediment, and the added weight of basin-fill may be sufficient to drive the hemipelagic drape downslope and to rotate

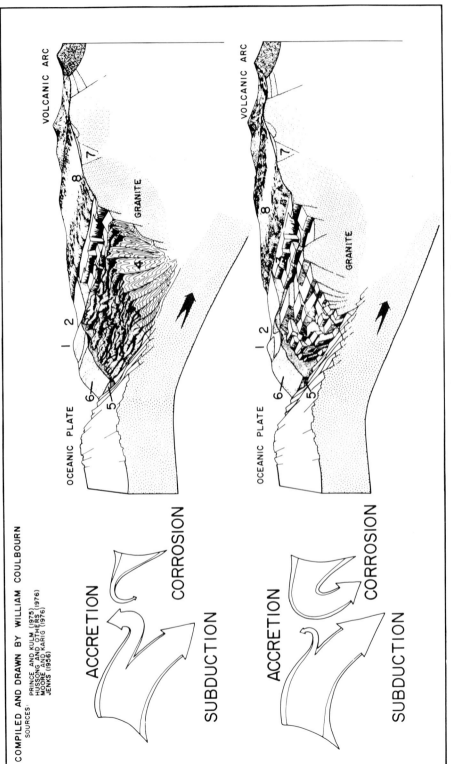

Figure 17. Accretionary and erosional end members for convergent continental margins. Arrows indicate relative amounts of material subducted, accreted, or corroded (that is, tectonically eroded). Key: 1, structural high; 2, forearc basin; 3, block-faulted continental rock; 4, accretionary prism; 5, trench-axis turbidites; 6, deformed and dewatered hemipelagic sediments of the continental margin; 7, Mesozoic sediment; 8, coastal batholith.

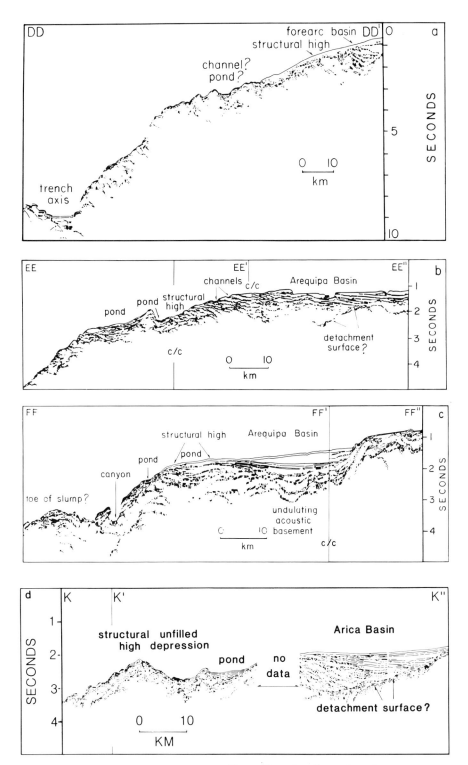

Figure 18 continuation and caption on next page.

Figure 18 continued.

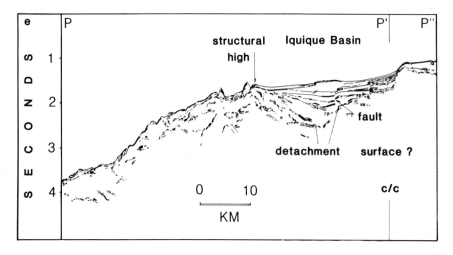

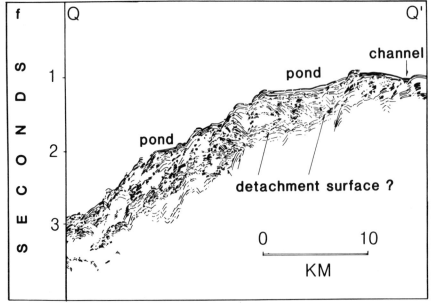

Figure 18. Seismic reflection profiles across the upper continental margin within the Arica Bight. (a) Profile DD-DD′. Transect across the trench and continental margin shows undulating acoustic basement. Basin-filling sediment overflows landward catchment. (b) Profile EE-EE″. Landward-tilted reflectors seaward of the Arequipa Basin. Small sediment ponds form between structural highs. The c/c indicates change course along trackline. (c) Profile FF-FF″. Undulating reflectors beneath the Arequipa Basin and beneath sediment ponds seaward of the basin. Turbidity currents may be channeled in a depression landward of a slump. (d) Profile K-K″. Landward-tilted reflectors of the Arica Basin overlie a well-defined acoustic basement. Data gap is probably a structural high buried by overflowing sediment. Depression seaward of the sediment pond is cut off from source of basin-filling turbidites. (e) Profile P-P″. Structural high seaward of the Iquique Basin is a composite of several ridges. Basin-filling sediment is offset along a high-angle fault. Acoustic bsement beneath the basin may be a detachment surface, "lithification front," or sediment-rock contact. (f) Profile Q-Q′. Irregular landward-dipping reflectors south of the Iquique Basin. Small sediment ponds are formed behind protrusions of deformed, more-consolidated glauconitic mud. Strong reflectors represent a seaward-dipping acoustic basement.

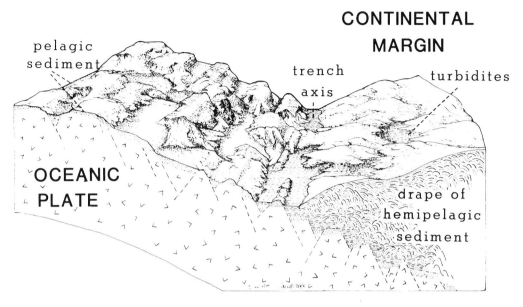

Figure 19. Block diagram suggesting the possible role of gravity in forming the Peru-Chile continental margin. Landward-dipping reflectors formed by the downslope creep of the hemipelagic drape. That sediment covers turbidites accumulating in the trench axis. Those turbidites in turn blanket the subducting, gridlike surface of the Nazca plate. For the sake of clarity, topography is exaggerated in this "Escheresque" rendering.

deeper reflectors landward as the structural high is driven seaward. Acoustic basement would be a detachment surface, perhaps a "lithification front" (Arthur and others, 1980), rather than a depositional contact (Fig. 18). Once initiated, the massive downslope creep could proceed in a wavelike advance of small sediment ponds between wave crests; large upslope waves would advance over and drive forward smaller ponds located seaward and downslope (Fig. 19). A steady-state creep of the slope apron would bury both turbidites and underlying pelagic sedimentary sections arriving on the gridlike surface of the subducting oceanic plate. Compressional deformation arising from plate convergence occurs at great depth and far landward of the trench axis, perhaps in the fashion envisioned by Cowan and Sillen (1978).

What concept is most representative of convergent margins in general? Is a unifying concept attainable, or are trenches truly "disorganized" (Mascle and Le Quellec, 1980)? How assuredly can any of the hypothetical cross sections in print be applied to ancient subduction complexes? The eventual acceptance of a scheme defining convergent-margin tectonics will have far-reaching implications for concepts regarding the evolution of continental crust; however, structure along convergent margins is probably too variable to be described by one "model." Certainly, the variability of structure along the continental margin of the Arica Bight presages the larger-scale variability to be expected among other subduction zones.

ACKNOWLEDGMENTS

The financial support of International Decade of Ocean Exploration Grant ID071-04207-A04 to the Hawaii Institute of Geophysics is gratefully acknowledged. Ralph Moberly, Michael Garcia, and Darrel Cowan offered criticisms and suggestions that considerably improved the manuscript. Part of this research is contained within my doctoral dissertation at the University of Hawaii (Coulbourn,

1977). This work would not have been possible without the enthusiasm of Captain Gary Miller and the crew of the R/V *Kana Keoki* or the support and cooperation of the research staff of the Hawaii Institute of Geophysics.

REFERENCES CITED

Arthur, M. A., Carson, B., and von Huene, R., 1980, Initial tectonic deformation of hemipelagic sediment at the leading edge of the Japan convergent margin, *in* Scientific Party, Initial reports of the Deep Sea Drilling Project Volumes 56 and 57, Part I: Washington, D.C., U.S. Government Printing Office, p. 569–614.

Barazangi, M., and Isacks, B., 1976, Spatial distribution of earthquakes and subduction of the Nazca plate beneath South America: Geology, v. 4, p. 686–692.

Beck, R. H., and Lehner, P., 1974, Oceans, new frontier in exploration: American Association of Petroleum Geologists Bulletin, v. 58, p. 376–395.

Bellaiche, G., 1967, Résultats d'une étude géologique de la fosse du Japon effectuée en bathyscaphe "Archimède": Paris, Comptes Rendus Hebdomadaires des Séances de l'Academie des Sciences, v. 265, p. 1160–1163.

——1980, Sedimentation and structure of the Izu-Ogasawara (Bonin) Trench off Tokyo: New lights on the results of a diving campaign with the bathyscaph "Archimède": Earth and Planetary Science Letters, v. 47, p. 124–130.

Biju-Duval, B., Letouzey, J., and Montadert, L., 1979, Variety of margins and deep basins of the Mediterranean, *in* Watkins, J. S., and others, eds., Geological and geophysical investigations of continental margins: American Association of Petroleum Geologists Memoir 29, p. 293–317.

Blondel, F., 1964, Carte géologique de l'Amerique de Sud: Conselho Nacional de Pesquisas, Brazil.

Bowman, I., 1916, The Andes of southern Peru, first edition: New York, Henry Holt and Co., 336 p.

Buffler, R. T., and Watkins, J. S., 1978, Geologic structure of the continental margin off the Nicoya Peninsula, Costa Rica: Geological Society of America Abstracts with Programs, v. 10, p. 374.

Buffler, R. T., and others, 1979, Anatomy of Mexican ridges, southwestern Gulf of Mexico, *in* Watkins, J. S., and others, eds., Geological and geophysical investigations of continental margins: American Association of Petroleum Geologists Memoir 29, p. 319–327.

Bussell, M. A., Pitcher, W. S., and Wilson, P. A., 1976, Ring complexes of the Peruvian coastal batholith: A long-standing subvolcanic regime: Canadian Journal of Earth Sciences, v. 13, p. 1020–1030.

Carr, M. J., Stoiber, R. E., and Drake, C. L., 1974, The segmented nature of some continental margins, *in* Burk, C. A., and Drake, C. L., eds., The geology of continental margins: New York, Springer-Verlag, p. 105–114.

Coats, R. R., 1962, Magma type and crustal structure in the Aleutian arc, *in* Macdonald, G. A., and Kuno, H., eds., Crust of the Pacific basin: American Geophysical Union Geophysical Monograph 6, p. 92–109.

Cobbing, E. J., 1976, The geosynclinal pair at the continental margin of Peru: Tectonophysics, v. 36, p. 157–165.

Coulbourn, W. T., 1977, Tectonics and sediments of the Peru-Chile Trench and continental margin at the Arica Bight [Ph.D. dissert.]: Honolulu, University of Hawaii, 243 p.

——1980, Relationship between the distribution of foraminifera and geologic structures of the Arica Bight, South America: Journal of Paleontology, v. 54, p. 696–718.

Coulbourn, W. T., and Moberly, R., 1977, Structural evidence of the evolution of fore-arc basins off South America: Canadian Journal of Earth Sciences, v. 14, p. 102–116.

Coulbourn, W. T., and Resig, J. M., 1979, Middle Eocene pelagic microfossils from the Nazca plate: Geological Society of America Bulletin, Part I, v. 90, p. 643–650.

Cowan, D. S., and Sillen, R. M., 1978, A dynamic, scaled model of accretion at trenches and its implications for the tectonic evolution of subduction complexes: Journal of Geophysical Research, v. 83, p. 5389–5396.

Creager, J. S., and others, 1973, Initial reports of the Deep Sea Drilling Project Volume 19: Washington, D.C., U.S. Government Printing Office, p. 217–284.

Curray, J. R., and others, 1977, Seismic refraction and reflection studies of crustal structure of the eastern Sunda and western Banda arcs: Journal of Geophysical Research, v. 82, p. 2479–2489.

——1978, Tectonics of the Andaman Sea and Burma, *in* Watkins, J. S., and others, eds., Geological and geophysical investigations of continental margins: American Association of Petroleum Geologists Memoir 29, p. 189–208.

Dalmayrac, B., and others, 1971, Carte géologique des terrains Paleozoiques et Précambriens du Pérou et de Bolivie: Cahiers, Office de la Recherche

Scientifique et Technique Outre-Mer, Serie Geologie, v. 3, no. 1.
Dewey, J. F., and Bird, J. M., 1970, Mountain belts and the new global tectonics: Journal of Geophysical Research, v. 75, p. 2615–2647.
Dickinson, W. R., 1971, Clastic sedimentary sequences deposited in shelf, slope, and trough settings between magmatic arcs and associated trenches: Pacific Geology, v. 3, p. 15–30.
Dickinson, W. R., and Seely, D. R., 1979, Structure and stratigraphy of forearc regions: American Association of Petroleum Geologists Bulletin, v. 63, p. 2–31.
Drake, C. L., and Burk, C. A., 1974, Geological significance of continental margins, in Burk, C. A., and Drake, C. L., eds., The geology of continental margins: New York, Springer-Verlag, p. 3–10.
Elliott, D., 1976, The motion of thrust sheets: Journal of Geophysical Research, v. 81, p. 949–963.
Elter, P., and Trevisan, L., 1973, Olistostromes in the tectonic evolution of the northern Apennines, in de Jong, K. E., and Scholten, R., eds., Gravity and tectonics: New York, John Wiley and Sons, p. 175–188.
Farhoudi, G., and Karig, D. E., 1977, Makran of Iran and Pakistan as an active arc system: Geology, v. 5, p. 664–668.
Farrar, E., and others, 1970, K-Ar evidence for the post-Paleozoic migration of granitic intrusion foci in the Andes of northern Chile: Earth and Planetary Science Letters, v. 10, p. 60–66.
Ferraris, F. B., 1979, Antecedentes geologicos del borde continental del norte de Chile: Arica, Chile, Segundo Congresso Geologica Chileno, August 1979, p. A1–A3.
Fisher, R. L., 1974, Pacific-type continental margins, in Burk, C. A., and Drake, C. L., eds., The geology of continental margins: New York, Springer-Verlag, p. 25–41.
Fisher, R. L., and Raitt, R. W., 1962, Topographical structure of the Peru-Chile Trench: Deep-Sea Research, v. 9, p. 423–443.
Frutos, J., and Ferraris, F., 1973 Mapa tectónico de Chile, II Congreso Latino-americano de Geología, November, 1973: Santiago, Chile, Instituto de Investigaciónes Geológicas, 2 maps, 53 p.
Garner, H. F., 1959, Stratigraphic-sedimentary significance of contemporary climate and relief in four regions of the Andes mountains: Geological Society of America Bulletin, v. 70, p. 1327–1368.
Getts, T. R., 1975, Gravity and tectonics of the Peru-Chile and eastern Nazca plate, 0°–33°30′S [M.S. thesis]: Honolulu, University of Hawaii, 103 p.
Grow, J. A., 1973, Crustal and upper mantle structure of the Central Aleutian arc: Geological Society of American Bulletin, v. 84, p. 2169–2192.
Hafner, W., 1951, Stress distributions and faulting: Geological Society of American Bulletin, v. 62, p. 373–398.
Hamilton, W., 1977, Subduction in the Indonesian region, in Talwani, M., and Pitman, W. C., III, eds., Island arcs, deep sea trenches, and back-arc basins: American Geophysical Union Maurice Ewing Series, v. 1, p. 15–31.
Hanks, T. C., 1971, The Kuril-Hokkaido trench system: Large shallow earthquakes and simple models of deformation: Royal Astronomical Society Geophysical Journal, v. 23, p. 173–189.
Hayes, D. E., 1966, A geophysical investigation of the Peru-Chile trench: Marine Geology, v. 4, p. 309–351.
Hayes, D. E., and Ewing, M., 1970, Pacific boundary structure, in Maxwell, A. E., ed., The sea, Volume 4, Part II: New York, Wiley Interscience, p. 29–72.
Henderson, W. G., and Evans, C.D.R., 1980, Ecuadorian Subduction system: Discussion: American Association of Petroleum Geologists Bulletin, v. 64, p. 280–283.
Hess, H. H., 1962, History of the ocean basins, in Engel, A.E.J., and others, eds., Petrologic studies: A volume in honor of A. F. Buddington: Boulder, Colorado, Geological Society of America, p. 599–620.
Howell, D. G., and von Huene, R., 1978, Trench-slope basins of Kodiak Island, a modern analog for some Late Cretaceous rocks of central California: Geological Society of America Abstracts with Programs, v. 10, p. 109.
Hussong, D. M., and others, 1976, Crustal structure of the Peru-Chile trench: 18°–12°S latitude, in Sutton, G. H., and others, eds., The geophysics of the Pacific Ocean Basin and its margin, a volume in honor of George P. Woollard: American Geophsyical Union Geophysical Monograph 19, p. 71–86.
Hsü, K. J., and Ryan, W.B.F., 1973, Summary of the evidence for extensional and compressional tectonics in the Mediterranean, in Ryan, W.B.F., Hsü, K. J., and others, eds., Initial reports of the Deep Sea Drilling Project, Volume 13: Washington, D.C., U.S. Government Printing Office, p. 1011–1019.
Isaacson, P. E., 1975, Evidence for a western extracontinental land source during the Devonian in the Central Andes: Geological Society of America Bulletin, v. 86, p. 39–46.
Isacks, B., Oliver, J., and Sykes, L. R., 1968, Seismology and the new global tectonics: Journal of Geophysical Research, v. 73, p. 5855–5899.
Jacobson, R. S., and others, 1979, Seismic refraction and reflection studies in the Timor-Aru trough system and Australian continental shelf, in Watkins, J. S., and others, eds., Geological and geophysical investigations of continental margins:

American Association of Petroleum Geologists Memoir 29, p. 209–222.
Jacoby, W. R., 1973, Gravitational instability and plate tectonics, in de Jong, K. A., and Scholten, R., eds., Gravity and tectonics: New York, John Wiley and Sons, p. 17–33.
James D. E., 1971, Plate tectonic model for the evolution of the central Andes: Geological Society of America Bulletin, v. 82, p. 3325–3346.
—— 1978, Subduction of the Nazca plate beneath central Peru: Geology, v. 6, p. 174–178.
Jenks, W. F., 1956, Peru, in Jenks, W. F., ed., Handbook of South America geology: Geological Society of America Memoir 65, p. 219–247.
Kaizuka, S., and others, 1973, Quaternary tectonic and Recent seismic crustal movements in the Arauco Peninsula and its environs, central Chile: Geographic Reports of the Tokyo Metropolitan University, no. 8, p. 1–49.
Kaneps, A., 1978, In the deep sea: Geotimes, v. 28, p. 21–22.
Karig, D. E., and Sharman, G. F., 1975, Subduction and accretion in trenches: Geological Society of America Bulletin, v. 86, p. 377–389.
Karig, D. E., Ingle, J. C., Jr., and others, 1975, Site 298, in Karig, D. E., Ingle, J. C., Jr., and others, eds., Initial reports of the Deep Sea Drilling Project, Volume 31: Washington, D.C., U.S. Government Printing Office, p. 317–350.
Karig, D. E., and others, 1978, Late Cenozoic subduction and continental margin truncation along the northern Middle America Trench: Geological Society of America Bulletin, v. 89, p. 265–276.
—— 1979, Structure and Cenozoic evolution of the Sunda arc in the central Sumatra region, in Watkins, J. S., and others, eds., Geophysical investigations of continental slopes and rises: American Association of Petroleum Geologists Memoir 29, p. 223–237.
Katz, H. R., 1971, Continental margin in Chile—Is tectonic style compressional or extensional?: American Association of Petroleum Geologists Bulletin, v. 55, p. 1753–1758.
Kelleher, J., and McCann, W., 1976, Buoyant zones, great earthquakes and unstable boundaries of subduction: Journal of Geophysical Research, v. 81, p. 4885–4896.
Kieckhefer, R. M., and others, 1980, Seismic refraction studies of the Sunda trench and forearc basin: Journal of Geophysical Research, v. 85, p. 863–889.
Kulm, L. D., von Huene, R., and others, 1973, Initial reports of the Deep Sea Drilling Project, Volume 18: Washington, D.C., U.S. Government Printing Office, p. 162–180 and 449–508.
Kulm, L. D., and others, 1974, Transfer of Nazca Ridge pelagic sediments to the Peru continental margin: Geological Society of American Bulletin, v. 85, p. 769–780.
—— 1981, Crustal structure and tectonics of the central Peru continental margin and trench, in Kulm, L. D., and others, eds., Nazca plate: Crustal formation and Andean convergence: Geological Society of America Memoir 154 (this volume).
Le Pichon, X. L., Francheteau, J., and Bonnin, J., 1976, Plate tectonics: New York, Elsevier, 311 p.
Lonsdale, P., 1978, Ecuadorian subduction system: American Association of Petroleum Geologists Bulletin, v. 62, p. 2454–2477.
Ludwig, W. J., and others, 1966, Sediments and structure of the Japan Trench: Journal of Geophysical Research, v. 71, p. 2121–2137.
Malahoff, A., and Erickson, B. H., 1969, Gravity anomalies over the Aleutian Trench: EOS (American Geophysical Union Transactions), v. 50, p. 552–555.
Mascle, J., and Le Quellec, P., 1980, Matapan trench (Ionian Sea): Example of trench disorganization?: Geology, v. 8, p. 77–81.
Masias-Echegaray, J. A., 1976, Morphology, shallow structure and evolution of the Peruvian continental margin, 6° to 18°S [M.S. thesis]: Corvallis, Oregon State University, 92 p.
Mégard, F., 1967, Commentaire d'une coupe schematique á travers les Andes centrales du Pérou: Revue de Géographie Physique et de Géologie Dynamique, v. 9, p. 335–346.
Mégard, F., and Phillip, H., 1976, Plio-Quaternary tectonomagmatic zonation and plate tectonics in the central Andes: Earth and Planetary Science Letters, v. 33, p. 231–238.
Montecchi, P. A., 1976, Some shallow tectonic consequences of subduction and their meaning to the hydrocarbon explorationist, in Halbouty, M. T., and others, eds., Circum-Pacific energy and mineral resources: American Association of Petroleum Geologists Memoir 25, p. 189–202.
Moore, G. F., and Karig, D. E., 1976, Development of sedimentary basins on the lower trench slope: Geology, v. 4, p. 693–697.
—— 1980, Structural geology of Nias Island, Indonesia: Implications for subduction zone tectonics: American Journal of Science, v. 280, p. 193–223.
Moore, G. F., and others, 1980, Sedimentology and paleobathymetry of Neogene trench-slope deposits, Nias Island, Indonesia: Journal of Geology, v. 88, p. 161–180.
Moore, J. C., and others, 1979, Progressive accretion in the Middle America Trench, southern Mexico: Nature, v. 281, p. 638–642.
Mortimer, C., 1972, The evolution of the continental margin of northern Chile: International Geological Congress, 24th, Montreal, Section 8, p.

48–52.

Myers, J. S., 1975, Vertical crustal movements of the Andes in Peru: Nature, v. 254, p. 672–674.

Navarrez, S., 1964, Geologia de los cuadrangulos de Illo y Locumba: Lima, Peru, Comision Carta Geologica Nacional, Boletin No. 7, Ministerio de Fomento Y.O.P., 75 p.

Newell, N. D., 1949, Geology of the Lake Titicaca region, Peru and Bolivia: Geological Society of America Memoir 36, 111 p.

Paredes, J., and Mégard, F., 1972, Carte structurale schematique des Andes de Pérou: Laboratoire de Geologie Structurale de Montpellier (privately distributed).

Paskoff, R. P., 1977, Quaternary of Chile: The state of research: Quaternary Research, v. 8, p. 2–31.

Pitcher, W. S., 1974, The Mesozoic and Cenozoic batholiths of Peru: Pacific Geology, v. 8, p. 51–62.

Prince, R. A., and Kulm, L. D., 1975, Crustal rupture and the initiation of imbricate thrusting in the Peru-Chile Trench: Geological Society of America Bulletin, v. 86, p. 1639–1653.

Prince, R. A., and Schweller, W. J., 1978, Dates, rates and angles of faulting in the Peru-Chile Trench: Nature, v. 271, p. 743–745.

Prince, R. A., and others, 1974, Significance of uplifted turbidite basins on the seaward wall of the Peru Trench: Geology, v. 2, p. 607–611.

—— 1980, Bathymetry of the Peru-Chile continental margin and trench, Part I: Geological Society of America Map and Chart Series MC-34, scale 1:1,000,000.

Rodriguez, R. E., Cabre, P.S.J., and Mercado, A., 1976, Geometry of the Nazca plate and its geodynamic implcations, in Sutton, G. H., and others, eds., The geophysics of the Pacific Ocean basin and its margin, a volume in honor of George P. Woollard: American Geophysical Union Geophysical Monograph 19, p. 87–103.

Rosato, V. J., Kulm, L. D., and Derks, P. S., 1975, Surface sediments of the Nazca plate: Pacific Science, v. 29, p. 117–130.

Ruegg, W., 1956, Geologie zwischen canete—San Juan 13°00′–15°24′ Sudperu: Geologischen Rundschau, v. 45, p. 775–858.

Rutland, R.W.R., 1971, Andean orogeny and ocean floor spreading: Nature, v. 233, p. 252–255.

Ryan, W.B.F., Hsu, K. J., and others, 1973, Sites 127 and 128, in Ryan, W.B.F., Hsu, K. J., and others, eds., Initial reports of the Deep Sea Drilling Project, Volume 13: Washington, D.C., U.S. Government Printing Office, p. 243–322.

Scholl, D. W., and Marlow, M. S., 1974, Global tectonics and sediments of modern and ancient trenches: Some different interpretations, in Kahle, C. F., ed., Plate tectonics assessments and reassessments: American Association of Petroleum Geologists Memoir 23, p. 255–272.

Scholl, D. W., Marlow, M. S., and Cooper, A. K., 1977, Sediment subduction and off-scraping at Pacific margins, in Talwani, M., and Pitman, W. C., III, eds., Island arcs, deep sea trenches, and back-arc basins: American Geophysical Union Maurice Ewing Series, v. 1, p. 199–210.

Schuiling, R. D., 1973, Active role of continents in tectonic evolution—Geothermal models, in de Jong, K. A., and Scholten, R., eds., Gravity and tectonics: New York, John Wiley & Sons, p. 35–47.

Schweller, W. J., 1976, Chile Trench: Extensional rupture of oceanic crust and the influence of tectonics on sediment distribution [M.S. thesis]: Corvallis, Oregon State University, 90 p.

Schweller, W. J., and Kulm, L. D., 1975, Crustal rupture of the Nazca plate within the Chile Trench [abs.]: Pacific Science Congress, 13th, p. 413.

Scientists aboard the *Glomar Challenger*, 1978a, Transects begun: Geotimes, v. 23, March, p. 22–26.

Scientists aboard the *Glomar Challenger*, 1978b, Japan Trench transected: Geotimes, v. 23, April, p. 16–21.

Scientists aboard the *Glomar Challenger*, 1978c, Leg 60 ends in Guam: Geotimes, v. 23, October, p. 19–22.

Scientists aboard the *Glomar Challenger*, 1979, Middle America Trench drilled: Geotimes, v. 24, September, p. 20–22.

Seely, D. R., 1977, The significance of landward vergence and oblique structural trends on trench inner slopes, in Talwani, M., and Pitman, W. C., III, eds., Island arcs, deep sea trenches, and back-arc basins: American Geophysical Union Maurice Ewing Series, v. 1, p. 187–198.

—— 1979, The evolution of structural highs bordering major fore-arc basins, in Watkins, J. S., and others, eds., Geophysical investigations of continental slopes and rises: American Association of Petroleum Geologists Memoir 29, p. 245–260.

Seely, D. R., Vail, P. R., and Walton, G. G., 1974, Trench-slope model, in Burk, C. A., and Drake, C. L., eds., The geology of continental margins: New York, Springer-Verlag, p. 249–260.

Shepherd, G. L., 1979, Shallow crustal structure and marine geology of a convergence zone, northwest Peru and southwest Ecuador [Ph.D. thesis]: Honolulu, University of Hawaii, 201 p.

Silver, E. A., 1971, Transitional tectonics and late Cenozoic structure of the continental margin off northernmost California: Geological Society of America Bulletin, v. 82, p. 1–22.

Sonnenberg, F. P., 1963, Bolivia and the Andes, in Childs, O. E., and Beebe, B. W., eds., Backbone of the Americas: American Association of Petroleum Geologists Memoir 2, p. 36–46.

Speed, R. C., 1978, Barbados: Structural analysis of Early Tertiary rocks: Geological Society of America Abstracts with Programs, v. 10, p. 496.

Stauder, W., 1968, Tensional character of earthquake foci beneath the Aleutian trench with relation to sea-floor spreading: Journal of Geophysical Research, v. 73, p. 7693–7701.

—— 1973, Mechanism and spatial distribution of Chilean earthquakes with relation to subduction of the oceanic plate: Journal of Geophysical Research, v. 78, p. 5033–5061.

—— 1975, Subduction of the Nazca plate under Peru as evidenced by focal mechanism and by seismicity: Journal of Geophysical Research, v. 80, p. 1053–1064.

Stewart, J. W., Evernden, J. F., and Snelling, N. J., 1974, Age determinations from Andean Peru: A reconnaissance survey: Geological Society of America Bulletin, v. 85, p. 1107–1116.

Teisseyre, R., and others, 1974, Focus distribution in South American deep-earthquake regions and their relation to geodynamic development: Physics of Earth and Planetary Interiors, v. 9, p. 290–305.

Uyeda, S., 1974, Northwest Pacific trench margins, *in* Burk, C. A., and Drake, C. L., eds., The geology of continental margins: New York, Springer-Verlag, p. 473–491.

—— 1978, Active margins actively considered, a white paper: Geotimes, v. February 23, p. 27–29.

von Huene, R., 1979, Structure of the outer convergent margin off Kodiak island, Alaska, from multichannel seismic records, *in* Watkins, J. S., and others, eds., Geological and geophysical investigations of continental margins: American Association of Petroleum Geologists Memoir 29, p. 261–272.

von Huene, R., and others, 1979, Cross-section, Alaska Peninsula–Kodiak Island–Aleutian Trench: Summary: Geological Society of America Bulletin, Part I, v. 90, p. 427–430.

—— 1980a, Summary, Japan Trench transect, *in* Scientific Party, Initial reports of the Deep Sea Drilling project, Volumes 56 and 57, Part I: Washington, D.C., U.S. Government Printing Office, p. 473–488.

von Huene, R., Aubouin, J., and others, 1980b, Leg 67: The Deep Sea Drilling Project Mid-America Trench transect off Guatemala: Geological Society of America Bulletin, Part I, v. 91, p. 421–432.

Watts, A. B., Talwani, M., and Cochran, J. R., 1976, Gravity field of the northwest Pacific Ocean basin and its margin, *in* Sutton, G. H., and others, eds., The geophysics of the Pacific Oean basin and its margin, a volume in honor of George P. Woollard: American Geophysical Union Geophysical Monograph 19, p. 17–34.

White, R. S., and Klitgord, K., 1976, Sediment deformation and plate tectonics in the Gulf of Oman: Earth and Planetary Science Letters, v. 32, p. 199–209.

Whitsett, R. M., 1975, Gravity measurements and their structural implications for the continental margin of southern Peru [Ph.D. dissertation]: Corvallis, Oregon State University, 82 p.

Wilson, J., and Garcia, W., 1962, Geologia de los cuadrangulos de Pachia y Palca: Lima, Peru, Comision Carta Geologica Nacional, Volume II, Number 4, Ministerio de Fomento Y.O.P., 81 p.

Zeigler, J. M., Athearn, W. D., and Small, H., 1957, Profiles across the Peru-Chile trench: Deep-Sea Research, v. 4, p. 238–249.

MANUSCRIPT RECEIVED BY THE SOCIETY NOVEMBER 12, 1980
MANUSCRIPT ACCEPTED DECEMBER 30, 1980
HAWAII INSTITUTE OF GEOPHYSICS CONTRIBUTION 1105

Geological Society of America
Memoir 154
1981

Biogeography of benthic foraminifera of the northern Nazca plate and adjacent continental margin

JOHANNA M. RESIG
*Hawaii Institute of Geophysics and
Department of Geology and Geophysics
University of Hawaii
Honolulu, Hawaii 96822*

ABSTRACT

Frequency distributions were determined for more than 250 species of benthic foraminifera from 121 core-top samples taken on the Nazca plate, in the Peru-Chile Trench, and on the Peru shelf and slope. According to their species content, the benthic faunas may be regarded as continental margin, plate-bathyal, or trench and plate-abyssal faunas.

About two-thirds of the species from the Peru Margin live in depths shallower than middle bathyal, for which there is no comparable environment on the plate. The displacement of these shallow-dwelling species downslope serves to differentiate the lower-bathyal continental-slope assemblages from those of the plate. Five margin assemblages were recognized on the basis of the shallowest occurrences of species and modes of their percentage distributions. The shallowest assemblages—outer shelf and upper bathyal—are characterized by species of *Bolivina* living under low-O_2 conditions. Many of the species of these assemblages extend northward only as far as Central America. In contrast, assemblages in upper and lower middle-bathyal and lower-bathyal depths are more widespread along the west coast of the Americas; therefore, wider comparison of their distribution and wider potential use of them as paleobathymetric indices are permitted. One species, *Bolivina costata*, was found in all samples from the margin and in some samples from the eastern extremity of the plate; this suggests its potential use as a paleoecologic indicator of plate-boundary conditions.

The predominantly calcareous plate-bathyal populations occupy the sea floor between 2,600- and 4,100-m water depth. About one-fourth of the species of this group occur also in each of the other two groups. Q-mode factor analysis indicates the presence of three principal and two subordinate assemblages. As in other oceans, the deepest calcareous assemblage is dominated by *Nuttallides umbonifera*, which occupies a zone associated with the Antarctic Bottom Water Mass in which there is intensive solution of calcium carbonate. This zone of dissolution occurs between about 3,700 and 4,100 m in the study area. The *Epistominella exigua* Assemblage dominates the shallower parts of the plate between 2,600 and 3,700 m, except for the sea floor beneath the equatorial region of high productivity, where the *Gyroidina turgida* Assemblage predominates.

The trench and plate-abyssal faunas consist entirely of agglutinated species that occur deeper than

619

4,100-m water depth, the regional CCD (carbonate compensation depth). Half of the species are restricted to the abyss. Q-mode factor analysis indicates one principal and two subordinate assemblages. The *Reophax dentaliniformis* Assemblage occupies the Peru-Chile Trench, which is the principal path for the movement of Antarctic Bottom Water to the north. The distribution of the assemblage is also coincident with a high-productivity belt associated with the Peru Current. Two distinctive agglutinated species, *Ammomarginulina* sp. and *Saccammina tubulata*, are restricted to the trench.

Water-mass and substrate preferences of the benthic species may be used to interpret the paleoenvironments of sedimentary sequences in deep-sea cores as well as the origins of sedimentary rocks forming the continental margin. Because this is the first documentation of modern benthic foraminifera in the area, all pertinent species are illustrated.

INTRODUCTION

Foraminifera living on the sea floor are potential indicators of provenance that can be used in reconstructions of the depositional and tectonic history of sedimentary rocks. The northern Nazca plate is a relatively compact area in which to study the distribution of benthic foraminifera from a rise crest where new sea floor is generated to a trench where the sediment load might be subducted beneath or accreted to the continental margin. Major physiographic features of the area include geophysically active and inactive rises, an abyssal plain with positive and negative topographic irregularities, a trench, and a continental slope and shelf. Variability in productivity of surface waters further modifies the substrate and increases the array of environments that govern the distribution of benthic foraminiferal species.

The surface of the Nazca plate lies at lower-bathyal to abyssal depths where, until recently, it was thought that little differentiation occurred within the benthic foraminiferal community. Streeter (1973), however, laid the foundation for future research on deep-sea faunas by showing that the proportions of representation of species of North Atlantic populations change in response to water mass. Coincidently, the recovery by the Deep Sea Drilling Project of ancient bathyal and abyssal sediments emphasized the need for intensive work on the modern counterparts of the fossil faunas, in order to develop paleoecologic indices for interpretation of oceanic plate motion and evolution, as well as for interpretation of processes resulting in oceanic sediment incorporation into the continents. The data presented here are intended to facilitate such interpretations.

STUDY AREA

The portion of the Nazca plate and continental margin studied is bounded by the Galapagos Rift-Carnegie Rise to the north, the continent of South America from the equator to lat 22°S to the east, the Sala Y Gomez Ridge to the south, and the East Pacific Rise spreading center to the west (Fig. 1). The Central Basin and the Bauer Deep are between the East Pacific Rise and the subdued topography of an extinct spreading center, the Galapagos Rise, which separates those depressions from the Peru Basin to the east. The Peru-Chile Trench, marking the subduction zone, is deeper than 7,000 m in the southern portion of the area and is intersected at lat 15°S by the Nazca Ridge, which trends northeast across the Nazca plate.

A generalized pattern of bottom-sediment distribution, compiled by Rosato and others (1975), shows sediment textures in the area consisting of calcareous oozes on the rises and brown clay on the abyssal plain; these sediments grade into greenish-gray fine clastics and sand-sized particles, including pyroclastics, in the trench and on the continental margin (Fig. 2). Organic C content of the continental-margin sediments is as high as 5.6% (Bandy and Rodolfo, 1964); organic C content on the Carnegie Rise is on the order of 1%, more than twice as high as on the East Pacific Rise (Lisitzin, 1972).

From the concentrations of various elements and their interelement associations, Heath and

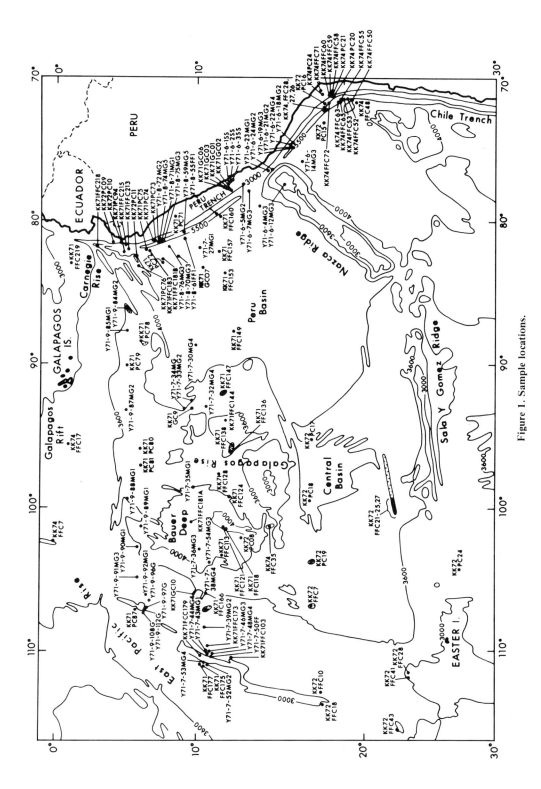

Figure 1. Sample locations.

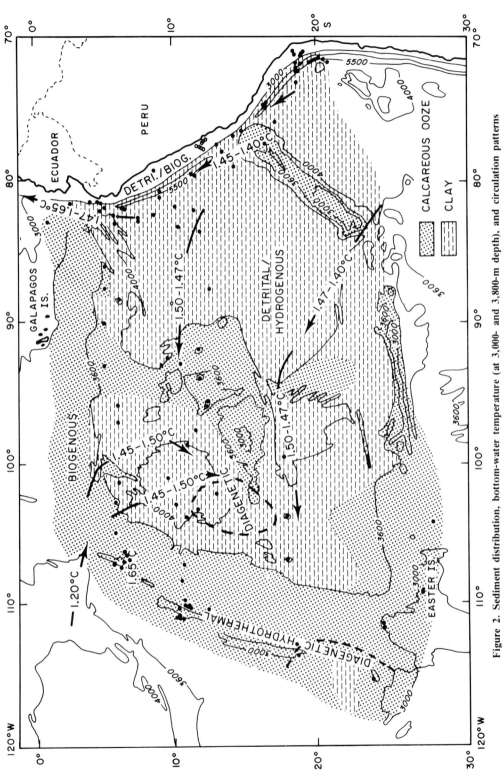

Figure 2. Sediment distribution, bottom-water temperature (at 3,000- and 3,800-m depth), and circulation patterns (arrows). Sediment data from Rosato and others (1975) and Heath and Dymond (1977); oceanographic data from Lonsdale (1976).

Dymond (1977) derived four modes of origin of the noncalcareous sedimentary components: (1) Hydrothermal processes result in the formation of metalliferous sediments, high in Mn, Fe, and Cu, on the East Pacific Rise from lat 10°S to 25°S; some of these sediments are transported by bottom currents into the Bauer Deep; (2) Biogenically derived sediments enriched in Si accumulate on the East Pacific Rise north of lat 10°S and extend eastward toward the Continent; (3) Diagenetic production of smectite characterizes the sediments of the southern Bauer Deep and the East Pacific Rise south of about lat 18°S; (4) Detrital-hydrogenous sediment high in Al and Ni increases progressively eastward across the plate.

Although the detrital component is conspicuous, sediment-accumulation rates are low over the plate, which is shielded from the massive amounts of coarse continental detritus trapped by the trench system (Rosato and others, 1975). The widespread distribution of Mn nodules attests to the slow accumulation of sediments.

Potential temperatures and circulation of the bottom water over the Nazca plate (Lonsdale, 1976) are shown in Figure 2. The bottom water, which originates in the Circumpolar Current, increases in temperature in the course of its passage northward over various sills between the East Pacific Rise and South America. Some bottom water passes eastward over the East Pacific Rise at transform faults to fill the Bauer Deep and redistribute metalliferous sediments formed on the rise. The main path of bottom-water flow to the north is the Peru-Chile Trench, where a deep sill at the nose of the Nazca Ridge permits bottom water to pass unmodified to the north along the trace of the trench axis (Lonsdale, 1976).

Characteristics of water masses and currents impinging on the continental margin are given by Wyrtki (1967) and are presented later. Shallow coastal waters are dominated to a few hundred metres depth by the northward-flowing Peru Current. Temperatures in the current are about 3 °C cooler than at comparable depths 500 km offshore (Reid, 1965; Wooster, 1970). Upwelling associated with the current along the Peruvian coast and along the projection of the current seaward over the Carnegie Rise results in primary productivity of as much as 750 mg C/(m^3 · day) in the waters over those areas, as compared to a maximum of 150 mg C/(m^3 · day) over the abyssal plain (Lisitzin, 1972). The high organic C in the bottom sediments thus corresponds with the areas of high productivity. The depletion of O_2 through the partial oxidation of organic matter in the water column has provided an O_2-minimum layer ($O_2 < 1$ ml/l) that intersects the continental margin between 50 and 450 m at lat 27°S (Reid, 1965) and between 100 and 900 m at lat 15°S (Wooster, 1970). This O_2 depletion excludes certain benthic invertebrates from the outer continental shelf–upper slope and thus promotes accumulation of unspent organic matter there.

PREVIOUS WORK

Benthic foraminifera from the Peru-Chile continental shelf are known from relatively few, geographically disjunct studies, mostly taxonomic. On the basis of the studies of d'Orbigny (1839), Cushman and Kellett (1929), Cushman and McCulloch (1939, 1940, 1942, 1948, 1950), Lalicker and McCulloch (1940), and Boltovskoy and Theyer (1970), various provincial units were recognized by Boltovskoy (1976). One of these, the Peruvian subprovince from about lat 4°S to 15°S, encompasses the shelf samples of the present study. The southern boundary corresponds with a zone of "permanent" upwelling where the Nazca Ridge intersects the trench (Wooster, 1970). Species common in the subprovince include *Bolivina costata, B. punctata, Cancris inflatus, Discorbis corus, Eponides repandus, Nonion pizarrense,* and *Rotalia rosea.* Phleger and Soutar (1973) noted the effect of the O_2-minimum layer in producing populations dominated by *Bolivina* on the continental shelf off Callao, Peru.

Faunas recovered by trawling and coring from the edge of the continental shelf, the slope, and the trench off Peru and Chile were reported by Bandy and Rodolfo (1964). Depth zonations of large species obtained from the trawls as well as small species from core tops were recognized by using upper limits of occurrence to avert the effects of downslope displacement of tests. Khusid (1971) and Saidova

(1971) further discussed the coastal area and trench in terms of generic dominance and biomass. Patterns of abundance and distribution of foraminifera on higher taxonomic levels were later presented for the entire study area (Saidova, 1976).

Studies of the distribution of species of Holocene benthic foraminifera on the northern Nazca plate are few. Coulbourn (1977, 1980) reported the species content of cores from the continental margin and trench between lat 18° and 22°S and interpreted the sedimentary environments and processes in the subduction zone. Resig and others (1980) reported a siliceous foraminifer from the trench and plate margin. Keller and Ingle (1976) and Ingle and others (1980) related benthic foraminiferal distribution to water-mass properties in the Peru-Chile Trench south of the study area, from about lat 33° to 39°S, and Walch (1978) studied living and dead benthic foraminifera in a series of box cores taken westward from the crest of the East Pacific Rise immediately north of the study area.

Publications most useful in identifying the abyssal species are those of Brady (1884) and subsequent notations on his work by Barker (1960), Parker (1964), and Echols (1971). Additional publications particularly useful in identifying the continental-margin species are those of Uchio (1960), Bandy (1961), and Smith (1964).

MATERIAL STUDIED

Samples of 140 piston, free-fall, and gravity core tops taken on Oregon State University's R/V *Yaquina* in 1971 and the University of Hawaii's R/V *Kana Keoki* in 1971, 1972, and 1974 were used in this study (App. 1—microfiche). Of these samples, 13 were from the continental shelf and slope, 64 from the rises and bathyal regions of the plate (<4,100 m), and 63 from the trench and abyssal regions of the plate (>4,100 m). As 19 of the abyssal samples were barren of foraminifera, the foraminiferal assemblages were determined from the remaining 121 samples.

The samples consisted of about 1-cm-thick quarter wedges of surface material from the 5.5-cm-diam cores. Some of the samples representing various sediment types and locations were dried and weighed, so that the absolute abundance of benthic specimens per gram could be calculated (App. 2—microfiche). The remaining samples were wet-sieved (0.63-μm mesh opening) directly without drying, as it was noted from several experiments that some of the delicate agglutinated species from the deeper parts of the plate and trench in particular were destroyed through the drying and rewetting process. In many samples, all of the specimens were identified and counted. When more than a few hundred specimens were contained in the sample, a modified Otto microsplitter was used to obtain aliquots of the sand-sized fraction containing between 200 and 300 benthic foraminifera, from which the percentage representation of the various species was calculated. The entire sample was examined for additional species to improve the continuity of distribution determined for some species and to better record diversity. Any additional species encountered in this manner were designated present at less than 1% of the assemblage. These tabulations are presented in Appendix 3—microfiche.

None of the core-top samples had been placed in preservative, so that specimens living in situ could not be determined on the basis of protoplasmic stain. Although Walch (1978) reported standing crops of 1.5 to 19.3 individuals per square centimetre along a traverse across the East Pacific Rise, the sites lay beneath the equatorial zone of high productivity that fostered the bottom population. Outside such productive areas, deep-sea benthic foraminifera are not particularly abundant, and it is unlikely that staining a small bottom sample would yield a true picture of the modern population. Samples from the shallower regions of the plate were considered modern if they contained brightly pigmented *Globigerina rubescens*, and general distributional patterns were relied upon to determine allochthonous species. In the process of initial examination, nine samples that tapped fossil outcrops or contained a significant quantity of reworked fossils were eliminated from use in this study.

Species identifications were confirmed, where possible, through my comparison of Nazca specimens with types in the U.S. National Museum. A taxonomic list presented at the end of this paper (App. 4—microfiche) contains pertinent remarks.

Because of a substantial difference in species compositon of the three major sampling environments—continental margin, plate bathyal, and trench and plate abyssal—the species-percentage data set was partitioned accordingly, and each of the three groups was subjected to Q-mode factor analysis by using the FORTRAN IV computer program of Davis (1973). In addition to factor loadings, a measure of the correlation of each sample with each factor, the program computes factor scores, a measure of the relative contribution of each species to each of the factors. Thus, the negative score of a species on a factor may result from decreased frequency as well as absence from samples loading high on the factor. The use of this statistical routine facilitated the analysis of a large data set and permitted comparison of the Nazca deep-water assemblages with those defined through factor analysis in other oceanic areas.

Factor analysis proved useful in describing the data sets except for those from the continental margin where sample coverage was insufficient. For that area, factor analysis was abandoned in favor of standard graphic means for depicting assemblage trends. The Fisher α index of diversity was determined for the margin assemblages from graphs given by Murray (1973). This index considers the rarer species but is fairly consistent for different size samples. Its extensive use by Murray (1973) to define various environments invites comparison with the Peru Margin data.

CALCITE-SOLUTION LEVELS

Berger (1968) called attention to a bathymetric interval between what he termed the lysocline and the CCD in which planktonic foraminifera are selectively dissolved; a residue of the more robust species is left. An effect observed in this interval is an increase in the population density of benthic foraminifera (Parker and Berger, 1971) which are generally more resistant to solution than the planktonics. It is likely that this interval also affects the facility of some benthics to secrete calcium carbonate and thus controls to some extent the composition of the abyssal fauna.

The frequency distribution of planktonics was not calculated in this study, but the preservation of assemblages was recognized visually on the basis of fragmentation and species content. According to the ranking of their species in Berger's (1968, 1970) scale of species resistance, planktonic assemblages in samples from the plate and rises showed slight to moderate dissolution between 3,300 and 3,700 m and strong dissolution between 3,850 and 4,100 m. Below 4,100 m there were generally no planktonic foraminifera and only agglutinated benthic foraminifera. This depth is regarded as the regional CCD. The CCD mapped by Berger and Winterer (1974) on the basis of previous carbonate analysis data is somewhat shallower. A lysoclinal gradient between 3,300 and 3,700 m on the plate, becoming deeper toward the East Pacific Rise, corresponds with the areal distribution map of Berger (1970) and Parker and Berger (1971) based on samples showing evidence of partial solution of planktonic assemblages. Broecker and Broecker (1974) recognized a lysocline at 3,950 m on the western flank of the East Pacific Rise at lat 17°S according to the carbonate content of the sediment, which Broecker and Takahashi (1978) related to a decrease in the carbonate ion content of the bottom water. Because water-mass properties differ slightly on the two flanks of the East Pacific Rise (Warren, 1970; Lonsdale, 1976), this latter, sedimentary "lysocline" may correspond to the level of intense solution of planktonic tests that sets in between 3,700 and 3,850 m in the study area.

No new information was derived from this study regarding the intersection of the lysocline and the CCD with the continental margin. Bandy and Rodolfo (1964) measured a marked drop in carbonate content of the sediment to an average of 3% below about 3,300 m on the east side of the trench. Berger (1970) reported a rise in both the level of the lysocline and the CCD from the plate toward the continental intersection.

FORAMINIFERAL DISTRIBUTION

Margin Assemblages

The 13 continental-margin samples were distributed between lat 4°S and 19°S and depths of 82 to

2,286 m. These environments were below the effects of the thermocline (40 to 50 m, Wyrtki, 1964) but were under the influence of an extensive O_2-minimum layer between 50 and 450 m (Reid, 1965) or deeper (Wyrtki, 1967) and various subsurface water masses (Wyrtki, 1967). Benthic foraminifera occurred in concentrations of about 220/g in the O_2-minimum layer and 125/g at middle-bathyal depths. Species inhabiting the margin are illustrated in Plates 1 through 4. Their general bathymetric distribution, relative to structure of the water column, is shown in Figure 3. Through inspection of the data table (App. 3-2), species with similar depth ranges may be grouped into assemblages as follows, with species preceded by an asterisk being significant in more than one assemblage:

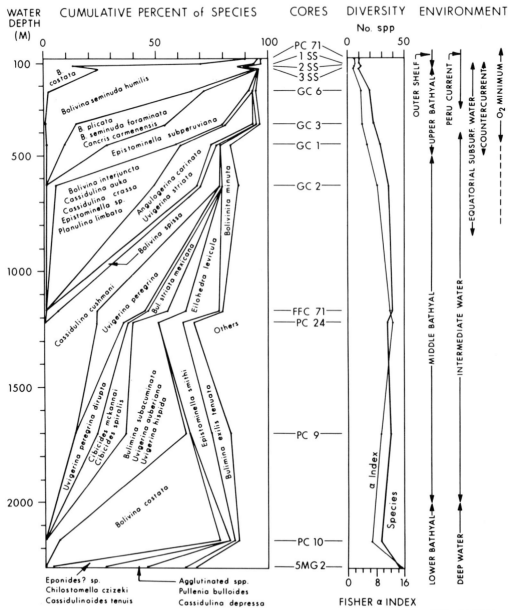

Figure 3. Bathymetric distribution of margin species. Water-mass data from Reid (1975) and Wyrtki (1964, 1967).

Outer-Shelf Assemblage (82 to 150 m). $T = 14$ to $16°C$ (high); $S = 34.8^0/_{00}$ (high); $O_2 = <1$ ml/l (minimum); Fisher α index = 1.7 to 2.3. Species: *Bolivina costata,* 13% to 91%; **B. seminuda humilis,* 5% to 74%.

Remarks. Fisher α indices of <4 are associated with stressed environments such as those where variations in thermal structure are prominent (Murray, 1973). Environments where low-O_2 conditions provide a stress are often inhabited by large populations of *Bolivina* off western North America (for example, Smith, 1963; Harman, 1964; Phleger and Soutar, 1973). Phleger and Soutar (1973) reported 95% dominance of *Bolivina* cf. *pacifica* in the O_2-minimum layer at 180-m depth off Callao, Peru, where the standing crop was 60/cm² of surface sediment.

Bolivina costata is characteristic of the Peru Margin (Cushman and McCulloch, 1942; Boltovskoy, 1976), but has been reported northward to the Costa Rican shelf (Cushman and McCulloch, 1942) and slope (Bandy and Arnal, 1957), where it is rare. Boltovskoy and Gualancañay (1975) did not find it on the intervening Ecuadorian shelf and upper slope. Off Peru, it has been reported from as shallow as 12 m to a depth of about 200 m (Cushman and McCulloch, 1942). The present study confirms a shelf environment for the species, but it was recovered from deeper water also. It made up 72% of the foraminifera in a sand sample from 2,165 m at lat 5°S (KK72-PC10). Shepherd (1979) described a submarine canyon at that location; hence, downslope transport of shelf sediments to that site is suggested. Its frequency of 1% to 3% in other slope samples may also be attributed to reworking; however, the species was also recovered from the eastern margin of the Nazca plate, where no shallow source of sediments was evident. Clearly, further study involving the use of protoplasmic stain is required to resolve the depth range of this species.

Bolivina seminuda humilis was described from about 320 m off Mexico, its northernmost occurrence, and it had been recovered by its authors on the Peru continental shelf and upper slope (Cushman and McCulloch, 1942). Off El Salvador, it is well represented at 360 m (Bandy and Arnal, 1957), between 435 and 800 m (Smith, 1963) within the O_2-minimum layer, and also at 1,700 m, where Smith considered it in situ. In the study area it ranges from the shallowest sample at 82 m to about 500 m, although traces (1% or less) occur lower on the slope, presumably owing to downslope transport. It is the dominant species from about 100 to 300 m—somewhat shallower than off El Salvador.

Upper Bathyal Assemblage (151 to 500 m). $T = 7$ to $13°C$; $S = 34.6^0/_{00}$; $O_2 = <1$ ml/l (minimum); Fisher α index = 3.9 to 5.4. Species: **Bolivina seminuda humilis,* 8% to 64%; *Epistominella subperuviana,* 20% to 37%; *Bolivina plicata,* 5% to 14%; **Uvigerina striata,* 7% to 13%; *Epistominella obesa,* 6% to 7%; *Bolivina seminuda foraminata,* 6%; **Angulogerina carinata,* 2% to 5%; *Cancris carmenensis,* 2% to 4%.

Remarks. In addition to *Bolivina seminuda humilis, Epistominella obesa* has its northernmost occurrence off Mexico (Bandy, 1961). *Cancris carmenensis,* described from fossil deposits in the Gulf of California (Natland, 1950), has been reported living between 435 and 450 m off El Salvador (Smith, 1964). *Uvigerina striata* has not been reported north of Peru, possibly because of misidentification in that region. The four remaining species listed above extend northward to the California margin.

Bandy and Arnal (1957) reported *Bolivina seminuda* vars. and *Epistominella bradyana* (= *subperuviana*) as representative upper-bathyal species off the west coast of Central America, along with *Bolivina acuminata,* which occurred only rarely in the Peruvian samples. *Epistominella obesa* is also indicative of the upper-bathyal depths off El Salvador (Bandy and Arnal, 1957; Smith, 1964) and Peru. *Bolivina plicata* may extend into middle-bathyal depths off Central America (Bandy and Arnal, 1957).

Upper Middle-Bathyal Assemblage (501 to ~1,000 m). $T = 4$ to $7°C$; $S = 34.5^0/_{00}$ (minimum); $O_2 = 1$ to 2 ml/l; Fisher α index = 8.25. Species: **Angulogerina carinata,* 17%; **Uvigerina striata,* 2%; *Bolivina interjuncta,* 4%; *Cassidulina auka,* 15%; *C. crassa,* 8%; *Epistominella* sp., 9%; *Planulina limbata,* 8%; **Cassidulina cushmani,* 8%.

Remarks. This assemblage was present in only one of the samples studied, taken from 632-m water depth. The species association resembles that in upper-bathyal depths, but large *Cassidulina* species are prominent.

Lower Middle-Bathyal Assemblage (from ~1,001 to 2,000 m). $T = 2.5$ to $4°C$; $S = 34.5^0/_{00}$

(minimum); O_2 = 2 to 3 ml/l; Fisher α index = 9.5 to 12.8 Species: *Cassidulina cushmani 13% to 23%; Uvigerina peregrina and var. dirupta, 1% to 22%; U. auberiana, 3% to 11%; Cibicides mckannai, 3% to 9%; C. spiralis, 1% to 2%; Bulimina striata mexicana, 2% to 4%; *B. subcuminata, 3% to 4%; *Uvigerina hispida, 1% to 13%; *Eilohedra levicula, 1% to 8%; *Epistominella smithi, 3% to 14%.

Remarks. Members of this diverse fauna have been reported in these depths elsewhere along the west coast of the Americas (for example, Bandy, 1953; Smith, 1964), but some forms have been found beyond their ranges as presented here (for example, Uchio, 1960). A re-evaluation of all distribution studies with a view toward eliminating noncontemporaneous and allochthonous samples, stabilizing taxonomy, and correlating species ranges with properties of the environment, should explain the bathymetric distribution of this group. A trend in the ornamentation of Uvigerina similar to that presented by Bandy (1953) was noted. Uvigerina peregrina was totally costate at about 1,200 m, whereas the variant showing eruption of the costae into spines on the later chambers was found at about 1,700 m.

Lower-Bathyal Assemblage (2,001 to 2,286 m). T = 2.5 °C; S = 34.6^0/$_{00}$; O_2 = 3 ml/l; Fisher α index = 6.8 and 15.4. Species: *Bulimina subacuminata, 3%; *Uvigerina hispida, 2%; *Eilohedra levicula, 9%; *Epistominella smithi, 3%; Cassidulinoides tenuis, 3% and 4%; Chilostomella czizeki, 1% and 7%; Pullenia bulloides, 4%; Eponides? sp. 1% and 14%.

Remarks. The presence of several new faunal elements below 2,000 m corresponds with the top of the Pacific Deep Water Mass. Agglutinated foraminifera first become prominent in these depths. A sample in which the dominant species was Bolivina costata is considered to be of mixed origin, mostly allochthonous, because of the faunal associations and ecology of this species at shallower depths.

Plate-Bathyal Assemblages

Dominantly calcareous species (Pls. 5 through 8) inhabit the lower-bathyal realm of the plate between 2,600- and 4,100-m water depth. They are present in concentrations of 55 to 1,052 specimens per gram of sediment. Q-mode factor analysis of samples from the rise crests to the abyssal plain yielded five factors that explain 80% of the variance in the data set (Table 1). In the case of each factor, a single species whose frequency distribution represents the distribution of the factor is designated nominal species for the associated assemblages (Table 2).

The *Epistominella exigua* Assemblage (factor 1), which accounts for 26% of the variance, is distributed on the rises from 2,600 m down to a water depth of 3,700 m (Fig. 4). *E. exigua*, which makes up 11% to 27% of the foraminifera in samples loading 0.5 or greater on this factor, also occurs in other assemblages but with lesser representation. *Nuttallides umbonifera* is negatively related to the factor. The assemblage is characterized by well-known cosmopolitan, moderately deep-dwelling species, as well as morphologically unusual species such as *Favocassidulina favus* and *Sigmoilina edwardsi* that are never abundant but occur fairly consistently.

The *Nuttallides umbonifera* Assemblage (factor 2), accounting for 25% of the variance, occurs principally between 3,700-m and 4,100-m water depth and is conspicuous for its shallow incursion onto the East Pacific Rise on the southern margin of the study area (Fig. 5). *N. umbonifera* makes up 7% to 35% of the foraminifera in samples loading 0.5 or greater on factor 2. *E. exigua* and *Cassidulina subglobosa* also score high, and *Eponides tumidulus*, *Favocassidulina* sp., *Melonis pompilioides*, and *Uvigerina senticosa* are associated with this more than any other assemblage.

The *Gyroidina turgida* Assemblage (factor 3), accounting for 20% of the variance, occurs along an east-west belt across the northern margin of the study area (Fig. 6). *G. turgida* makes up 12% to 30% of the foraminifera in samples loading 0.5 or greater on this factor. *E. exigua* and *Bulimina translucens* score negatively in relation to this factor; their percentage representation is relatively low in many of the samples compared to the their preferred habitation area.

The *Bulimina translucens* Assemblage (factor 4), accounting for 6% of the variance, occurs on both flanks of the East Pacific Rise north of lat 15°S between 3,200- and 3,600-m water depth (Fig. 7). *B. translucens* makes up 14% to 26% of the foraminifera in samples loading 0.5 or greater on this factor.

TABLE 1. VARIMAX FACTOR LOADINGS
FOR PLATE-BATHYAL ASSEMBLAGES

Samples	F1	F2	F3	F4	F5	Communality (%)
KK71-FFC103	0.5781	0.6153	-0.3931	-0.2002	0.0054	91
KK71-FFC118	0.3588	0.8740	-0.2895	-0.0468	0.0216	97
KK71-FFC121	0.2851	0.8527	-0.2940	-0.0741	0.0419	89
KK71-FFC124	0.3148	0.8842	-0.2944	-0.0498	0.0031	95
KK71-FFC128	0.2424	0.8435	-0.3658	-0.0755	0.0120	92
KK71-FFC136	0.1875	0.6938	-0.2090	0.0075	0.0030	56
KK71-FFC144	0.2621	0.8387	-0.1611	-0.0735	0.0890	83
KK71-FFC147	0.0669	0.9083	-0.1614	-0.0447	0.0214	87
KK71-FFC149	0.3862	0.8940	-0.1206	-0.0384	-0.0001	95
KK71-FFC173	0.5652	0.2268	-0.1084	-0.7420	0.0290	93
KK71-FFC175	0.7058	0.2785	-0.1771	-0.5444	0.0365	90
KK71-FFC177	0.8501	0.3710	-0.1617	-0.1696	0.0359	92
KK71-FFC179	0.8944	0.1782	-0.3151	-0.0850	0.0096	93
KK71-GC9	0.2816	0.7043	-0.5410	-0.1377	0.0353	88
KK71-GC10	0.6967	0.3000	-0.4601	-0.3234	0.0459	89
KK71-PC72	0.0568	0.0697	-0.1712	-0.0454	0.9391	92
KK71-PC76	0.0671	0.0151	-0.7301	-0.0528	0.3535	66
KK71-PC78	0.1298	0.3765	-0.7296	-0.1282	0.0198	71
KK71-PC79	0.2017	0.5947	-0.6480	-0.1841	0.0334	85
KK71-PC80	0.2970	0.5418	-0.6863	-0.2054	0.0281	90
KK71-PC81	0.5286	0.3733	-0.6857	-0.1751	0.0401	93
KK71-PC87	0.6767	0.2286	-0.4195	-0.3357	0.0596	81
KK72-FFC7	0.4789	0.7332	-0.3306	-0.0554	-0.0092	88
KK72-FFC10	0.5793	0.5566	-0.3311	-0.1073	0.0690	78
KK72-FFC18	0.5196	0.3550	-0.4567	-0.0347	0.0928	62
KK72-FFC28	0.5828	0.6657	0.0141	-0.0184	0.0014	79
KK72-FFC41	0.4471	0.8005	0.0275	-0.2021	0.0146	88
KK72-FFC43	0.4504	0.7196	-0.0500	-0.0819	0.0504	73
KK72-PC11	0.0598	0.0282	-0.0096	0.0081	0.9713	94
KK72-PC18	0.4692	0.8105	-0.2827	-0.0264	-0.0147	96
KK72-PC19	0.4632	0.7197	-0.4403	-0.0516	0.0086	92
KK72-PC24	0.2944	0.5166	-0.2160	-0.1659	-0.0005	43
KK74-FFC35	0.7825	0.4670	-0.0804	-0.0612	0.0014	84
Y71-6-12MG3	0.5530	0.1215	-0.6878	-0.0268	0.2342	84
Y71-7-32MG4	0.6286	0.4428	-0.5881	-0.1586	0.0235	97
Y71-7-33MG2	0.2258	0.5577	-0.7028	-0.1121	0.0097	86
Y71-7-34MG4	0.5806	0.3736	-0.6786	-0.1552	0.0145	97
Y71-7-35MG1	0.5489	0.4164	-0.6768	-0.0676	0.0063	94
Y71-7-38MG4	0.3462	0.4251	-0.7959	-0.0920	0.0067	95
Y71-7-39MG2	0.7070	0.2969	-0.5136	-0.2530	0.0308	91
Y71-7-43MG1	0.6933	0.2419	-0.5093	-0.1950	0.0382	84
Y71-7-44MG4	0.9201	0.2742	-0.1592	-0.1066	0.0302	96
Y71-7-46MG3	0.8199	0.2838	-0.4081	-0.1163	0.0185	93
Y71-7-48MG4	0.8378	0.3590	-0.3081	-0.1729	0.0227	93
Y71-7-50FF	0.7607	0.3508	-0.2402	-0.1526	0.0289	78
Y71-7-53MG4	0.6470	0.2898	-0.3375	-0.1112	0.0369	63
Y71-9-85MG1	0.0975	0.8049	-0.4009	-0.0995	0.0409	84
Y71-9-87MG2	0.5044	0.5696	-0.6055	-0.0772	0.0358	95
Y71-9-88MG1	0.5292	0.2732	-0.7493	-0.0956	0.0291	92
Y71-9-89MG1	0.5525	0.5902	-0.4838	-0.1408	0.0397	90
Y71-9-90MG1	0.4553	0.2215	-0.7338	-0.3457	0.0202	91
Y71-9-91MG3	0.2879	0.1989	-0.8793	-0.0623	0.0166	89
Y71-9-92MG1	0.6956	0.2690	-0.3964	-0.3263	0.0580	83
Y71-9-96G	0.4748	0.1532	-0.4147	-0.7303	0.0149	96
Y71-9-97G	0.8738	0.2126	-0.3028	-0.2193	0.0151	94
Y71-9-108G	0.3314	0.0442	-0.2332	-0.8780	0.0164	93
Y71-9-112G	0.3023	0.0014	-0.6707	-0.5313	-0.0082	82
Average other 5 stations	0.1159	0.1640	-0.1580	-0.0746	0.0444	8
% Variance	26	25	20	6	3	80

Also loading high on factor 4 is the unusual morphotype, *Seabrookia earlandi*. *Bolivina seminuda* and *Quinqueloculina venusta* are associated with this assemblage.

The *Bolivina costata* Assemblage (factor 5), which explains 3% of the variance, is restricted to four samples on the eastern margin of the plate over a wide bathymetric range (Fig. 8). *B. costata* makes up 6% to 63% of the foraminifera in these samples. Of the remaining species important to these samples, only *Bulimina affinis* and *Stainforthia complanata* are not also associated with other assemblages.

Trench and Plate-Abyssal Assemblages

Benthic assemblages consisting totally or predominantly of agglutinate species (Pls. 9 and 10) are

TABLE 2. SPECIES ASSOCIATIONS IN PLATE-BATHYAL SAMPLES

Factor 1 species	Score	Factor 2 species	Score
Epistominella exigua	25.97	Nuttallides umbonifera	32.37
Nuttallides umbonifera	-6.08	Epistominella exigua	10.17
Glomospira charoides	3.33	Cassidulina subglobosa	10.03
Pullenia bulloides	3.33	Melonis affinis	4.05
Fissurina spp.	2.81	Oridorsalis umbonatus	3.83
Cassidulina sp.	2.51	Pullenia bulloides	3.38
Cassidulina laevigata	2.35	Uvigerina senticosa	1.78
Gyroidina turgida	2.32	Glomospira charoides	-1.70
Pullenia sp.	2.07	Cibicidoides wuellerstorfi	1.62
Gyroidina lamarckiana	1.97	Eponides tumidulus	1.61
Sigmoilina edwardsi	1.75	Favocassidulina sp.	1.61
Melonis affinis	-1.47	Laticarinina pauperata	1.33
Uvigerina senticosa	-1.46	Melonis pompilioides	1.25
Favocassidulina favus	1.33		
Cassidulina subglobosa	-1.23		
Cassidulina subglobosa ornata	1.11		
Rhizammina algaeformis	1.06		
Parafissurina spp.	1.02		

Factor 3 species	Score	Factor 4 species	Score
Gyroidina turgida	-28.36	Bulimina translucens	-29.60
Epistominella exigua	4.94	Seabrookia earlandi	-3.80
Bulimina translucens	2.47	Gyroidina lamarckiana	-3.08
Melonis affinis	-2.25	Pullenia sp.	-2.61
Gyroidina lamarckiana	-2.18	Pullenia bulloides	-2.15
Cassidulina subglobosa	-2.10	Nuttallides umbonifera	-2.00
Uvigerina senticosa	-1.43	Bolivina seminuda	-1.81
Siphouvigerina pseudoampullacea	1.22	Gyroidina turgida	-1.63
Nuttallides umbonifera	1.19	Quinqueloculina venusta	-1.16
Fissurina spp.	1.11	Cassidulina depressa	-1.08
		Cassidulina subglobosa	1.00

Factor 5 species	Score
Bolivina costata	49.87
Melonis affinis	8.48
Pullenia bulloides	6.43
Bulimina affinis	5.31
Cibicidoides wuellerstorfi	4.03
Stainforthia complanata	2.97
Fissurina spp.	1.78
Gyroidina lamarckiana	1.71
Epistominella exigua	1.44
Uvigerina senticosa	1.44

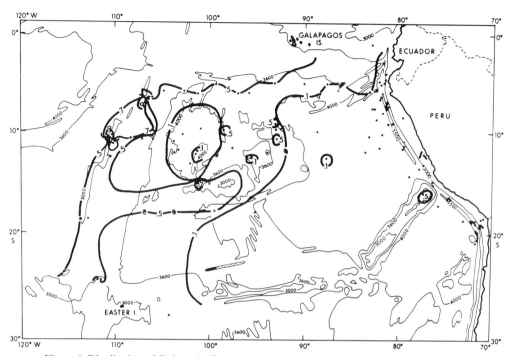

Figure 4. Distribution of *Epistominella exigua* Assemblage; contours represent factor loadings.

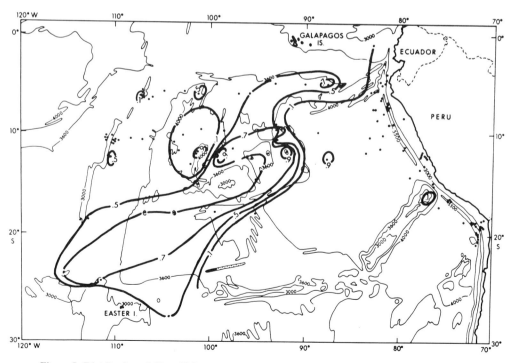

Figure 5. Distribution of *Nuttallides umbonifera* Assemblage; contours represent factor loadings.

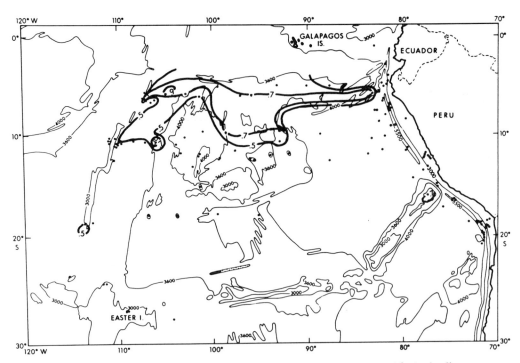

Figure 6. Distribution of *Gyroidina turgida* Assemblage; contours represent factor loadings.

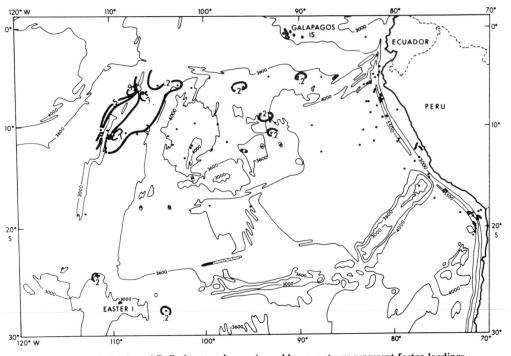

Figure 7. Distribution of *Bulimina translucens* Assemblage; contours represent factor loadings.

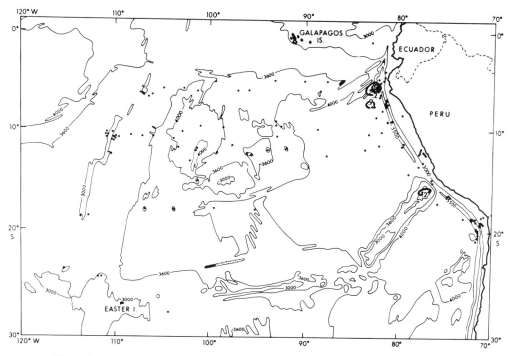

Figure 8. Distribution of *Bolivina costata* Assemblage; contours represent factor loadings.

generally confined to depths greater than 4,100 m in the Peru-Chile Trench and on the adjoining Nazca plate and its depressions, principally the Bauer Deep. Concentrations of tests vary between 0.5 to 5 per gram of sediment. Q-mode factor analysis of the samples with agglutinate assemblages yielded three factors, which accounted for 54% of the variance in the sample set (Table 3). The relatively low percentage of data explained through the factor analysis is a reflection of a lack of species dominance in the diverse populations on parts of the plate. This observation was substantiated by recomputing the data set using three factors and only those species accounting for 3% or more of the populations. The three factors in this recomputation explained only 61% of the data set; therefore, the use of the original computations was justified. The species scoring high on the three factors are listed in Table 4.

Samples loading high on factor 1, accounting for 27% of the variance, are from the trench (Fig. 9) and contain a species association dominated by *Reophax dentaliniformis*. *R. dentaliniformis* makes up 14% to 100% of the foraminifera in samples loading 0.5 or greater on this factor. As the second species of the group, *Spiroplectammina filiformis* occurs in only 76% of the samples loading high on factor 1, it is clear that the relative abundance of *R. dentaliniformis* is the index to factor 1. *Ammomarginulina* sp., occurring in 35% of the samples loading high on factor 1, is limited to the trench.

Samples loading high on factor 2, accounting for 15% of the variance, are distributed in the trench and on the eastern edge of the plate principally to the south of the Nazca Ridge (Fig. 10). *Placopsilinella aurantiaca*, which makes up 8% to 45% of the foraminifera in samples loading 0.5 or greater on factor 2, scored highest, followed by *Glomospira gordialis* and *G. charoides*. *Saccammina tubulata* as well as *Ammomarginulina* sp. are limited to the trench area. The other species characterizing this assemblage are more widely distributed on the plate and occur also in calcareous sediment on the rises, but in relatively low percentages.

Samples loading high on factor 3 occur in the Bauer Deep, in a depression on the plate, and in a few localities marginal to the trench (Fig. 11). Both the number and diversity of foraminifera in the samples were low, and, since these samples occurred adjacent to samples barren of foraminifera,

TABLE 3. VARIMAX FACTOR LOADINGS FOR TRENCH AND PLATE-ABYSSAL ASSEMBLAGES

Samples	F1	F2	F3	Communality (%)
KK71-FFC113	0.0381	-0.1659	-0.9621	95
KK71-FFC153	-0.0213	-0.1367	0.0161	2
KK71-FFC157	0.0032	-0.5035	-0.1461	27
KK71-FFC160	0.7010	-0.1396	0.0029	36
KK71-FFC181B	0.3946	-0.1873	-0.0786	21
KK71-FFC213	0.5231	-0.0233	0.0740	27
KK71-FFC215	0.0381	-0.1659	-0.9621	95
KK71-PC73	0.0295	-0.0019	0.0019	0
KK71-GC7	-0.0047	-0.2782	0.0644	8
KK72-FFC21	0.0349	-0.1427	-0.8860	81
KK72-FFC25	0.0017	0.0126	-0.0569	0
KK72-PC15	0.0350	-0.1420	-0.7809	63
KK74-FFC26	0.6074	-0.0136	0.0348	37
KK74-FFC27	-0.0069	-0.9765	0.0658	96
KK74-FFC28	0.0102	-0.9676	-0.1601	97
KK74-FFC48	0.0014	-0.9043	-0.1466	83
KK74-FFC50	0.0065	-0.6211	-0.5564	69
KK74-FFC53	0.1931	-0.8959	-0.1522	87
KK74-FFC63	-0.0115	-0.2001	-0.4101	21
KK74-FFC65	0.0048	-0.9215	-0.2804	93
KK74-FFC72	0.9622	-0.0039	0.0408	92
KK74-PC21	-0.0103	-0.0101	-0.0035	0
Y71-6-4MG3	0.4317	-0.2358	-0.1782	27
Y71-6-7MG3	0.6543	0.0250	-0.0477	42
Y71-6-14MG3	0.3470	-0.3835	-0.0800	27
Y71-6-18MG2	0.7288	-0.0704	0.0686	53
Y71-6-19MG3	0.1112	-0.8284	-0.1602	73
Y71-6-21MG2	0.8314	0.0057	0.0244	69
Y71-6-22MG4	0.8580	-0.0045	-0.0382	74
Y71-6-23MG1	0.9609	-0.0053	0.0374	92
Y71-6-24MG2	0.6021	-0.1238	-0.1132	38
Y71-7-27MG1	0.4023	-0.2365	0.0550	22
Y71-7-30MG4	0.1382	-0.0849	-0.0982	4
Y71-7-36MG3	0.0065	-0.2712	-0.4537	27
Y71-7-54MG3	0.0362	-0.1403	-0.9233	87
Y71-8-55FF1	0.9665	-0.0530	-0.0380	94
Y71-8-59MG5	0.5928	-0.3274	-0.2948	55
Y71-8-61FF1	0.2565	-0.1207	-0.3447	20
Y71-8-70MG3	0.2949	-0.0893	-0.1883	13
Y71-8-71MG1	0.8395	-0.0866	-0.0136	71
Y71-8-72MG2	0.9159	-0.0358	-0.1021	86
Y71-8-74MG5	0.9811	-0.0081	-0.0006	96
Y71-8-75MG3	0.8360	-0.0363	-0.0788	71
Y71-8-76MG3	0.9141	-0.0293	-0.1504	85
% Variance	27	15	12	54

TABLE 4. SPECIES ASSOCIATIONS IN TRENCH AND PLATE-ABYSSAL SAMPLES

Factor 1 species	Score	Factor 2 species	Score
Reophax dentaliniformis	56.23	Placopsilinella aurantiaca	-39.94
Spiroplectammina filiformis	9.23	Glomospira gordialis	-8.85
Trochammina malovensis	4.59	Glomospira charoides	-8.04
Recurvoides contortus	4.24	Saccammina tubulata	-6.10
Ammomarginulina sp.	3.63	Trochammina soldanii	-5.03
Reophax subdentaliniformis	3.22	Cystammina galeata	-2.45
Textularia wiesneri	2.28	Ammomarginulina sp.	-2.14
Trochammina glabra	1.55		
Reophax spiculifer	1.38		
Trochammina sp. 2	1.16		
Karreriella parkerae	1.06		

Factor 3 species	Score
Recurvoides contortus	-80.76
Reophax nodulosus	-10.21
Placopsilinella aurantiaca	5.79
Cyclammina pusilla	-4.08
Trochammina spp.	-3.56
Rhizammina algaeformis	-3.44
Glomospira charoides	3.07

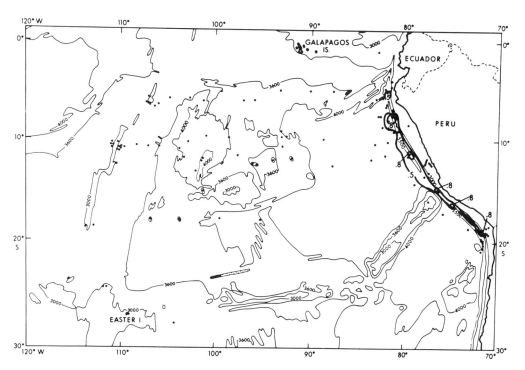

Figure 9. Distribution of *Reophax dentaliniformis* Assemblage; contours represent factor loadings.

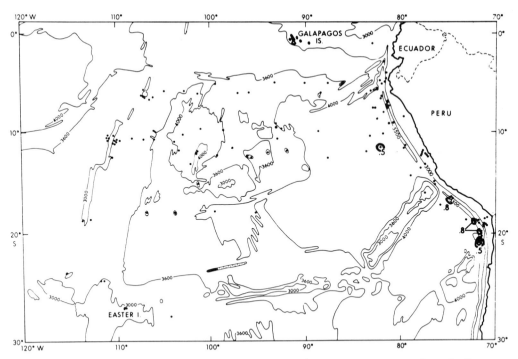

Figure 10. Distribution of *Placopsilinella aurantiaca* Assemblage; contours represent factor loadings.

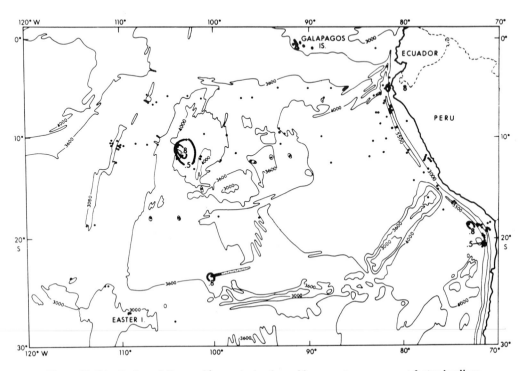

Figure 11. Distribution of *Recurvoides contortus* Assemblage; contours represent factor loadings.

hostile or corrosive environments were suggested. The species scoring highest on factor 3 is *Recurvoides contortus*, and *Reophax nodulosus* is second; both are robust forms.

Summary of Biofacies

Benthic foraminiferal biofacies along an east to west transect across the study area near lat 12°S (Fig. 12) reveal a succession of assemblages that have been defined through factor analysis and visual inspection of data.

The *Epistominella exigua* and *Gyroidina turgida* Assemblages occupy the East Pacific Rise and the lower rise flank, respectively. The *Recurvoides contortus* Assemblage occupies the floor of the Bauer Deep. The *Nuttallides umbonifera* Assemblage occurs on the Galapagos Rise, an elevation in the Bauer Deep to the west, and as a transported deposit from an elevation in the Peru Basin to the east. The Peru Basin is inhabited for the most part by a diversity of agglutinated species in which no species is dominant. These groups of foraminifera fall into no particular assemblages in the factor analysis. At the eastern edge of the Nazca plate the *Placopsilinella aurantiaca* Assemblage occurs, adjacent to the *Reophax dentaliniformis* Assemblage in the Peru Trench.

Benthic foraminifera of the Peru Margin, grouped visually into bathymetric assemblages, are exemplified by the upper-bathyal assemblage in the transect. This assemblage is characterized by low species diversity and high dominance of a few species of *Bolivina*. The assemblage affects other margin deposits through downslope transport of its members.

DISCUSSION OF RESULTS

Certain properties of the physical environment show values that are directly related to the distribution of the various assemblages of benthic foraminifera. It is likely that these properties are agents affecting distribution, but the extent of their influence in relation to less obvious factors is not known.

Temperature is perhaps the most important single agent because of the constraints it places on physiological mechanisms controlling life functions of a species (Valentine, 1973, p. 115-125). Temperature controls the general depth zonation of species into shelf, bathyal, and abyssal assemblages. Other agents may then modify the distribution patterns induced by temperature.

High productivity of the surface water, enhanced by upwelling in the Peru-Chile Current regime and in the equatorial belt, contributes organic matter to the substrate and consequently food to benthic foraminifera. Several assemblages appear to be distributed according to this parameter:

1. Intersection of the oceanographic O_2-minimum layer (which is broad as a result of high productivity) with the shelf and upper slope of the Peru Margin effectively limits the population to the *Bolivina* spp. Assemblage that dominates there.

2. The *Reophax dentaliniformis* Assemblage of the trench follows along the track of the principal movement of bottom water northward along the trench axis. Although it is confined by the 1.48 °C isotherm in the trench, it does not follow that isotherm over the plate where surface productivity is low.

3. The *Gyroidina turgida* Assemblage is associated with the westward deflection of the Peru Current off northern Peru and the zone of equatorial upwelling. In addition to relatively high organic C content, the substrate in the area occupied by this assemblage has a high siliceous component of planktonic origin (Heath and Dymond, 1977).

4. The *Bulimina translucens* Assemblage flanks the East Pacific Rise crest within the equatorial zone of high productivity.

The lysocline at about 3,700 m marks a drop in the level of the carbonate ion concentration at the top of the bottom water mass, which may directly affect the ability of some calcareous species to secrete test material. The intersection of this level with the sea floor approximates the boundary between the factor 1 and factor 2 plate-bathyal assemblages over most of the study area:

1. The *Epistominella exigua* Assemblage is best developed on the rises, from about 2,500-m water

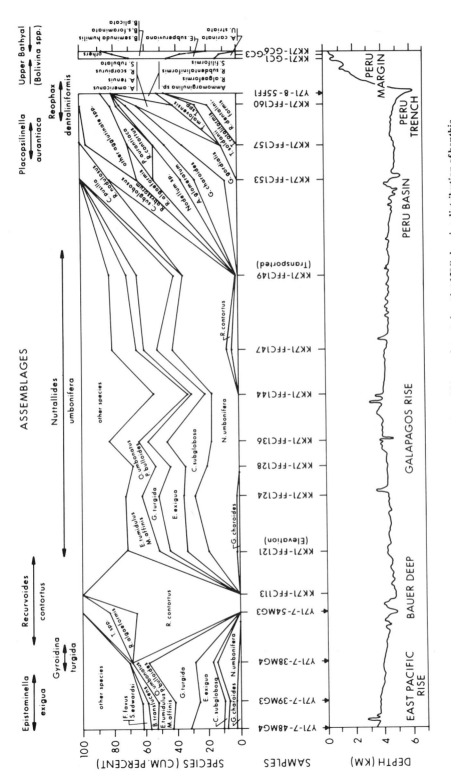

Figure 12. East to west profile across continental margin and Nazca plate at about lat 12°S showing distribution of benthic foraminiferal biofacies. Arrows by sample numbers indicate samples projected to profile.

depth to somewhat in excess of 3,500 m, where the substrate is carbonate ooze and the temperature is on the order of 1.55 to 1.65 °C. Although slight dissolution was noted in the planktonic components of the samples containing this assemblage below about 3,300 m, strong dissolution leading to the destruction of the tests of weaker genera and general fragmentation of planktonic tests was not evident. Several sample loading between 0.45 and 0.6 on this assemblage were taken from water depths of 3,850 to 4,600 m (for example, Y71-7-34MG4, Y71-7-35MG1). These samples are presumed to be displaced from shallower water.

2. In its typical development, the *Nuttallides umbonifera* Assemblage occupies the sea floor between the lysocline (3,700 m) and the CCD (4,100 m), where bottom temperatures are between about 1.55 and 1.50 °C. South of about lat 18°S, incursions of the species *N. umbonifera* onto the East Pacific Rise and Sala Y Gomez Ridge influence the sample loadings to reflect a dominance of the *N. umbonifera* Assemblage there also. The sediment is well-preserved calcareous ooze, accumulating under temperatures of 1.65 to 1.60 °C, an environment contrasting to the principal habitat of the assemblage. A high incidence of smectite, a diagenetic product, in the bottom sediments of this area (Heath and Dymond, 1977) indicates slow sediment accumulation and thus some unusual condition that may have influenced the benthic fauna.

Post-mortem modification has influenced the composition of at least two of the assemblages:

1. The *Bolivina costata* Assemblage found in two samples on the eastern edge of the plate below the CCD contains *Bulimina affinis* and a mixture of species characteristic of other environments. This evidence suggests that the assemblage was displaced from nearby elevations, but a source of *B. costata* as shallow as the outer-shelf to upper-bathyal environment it occupies on the Peru Margin is not apparent.

2. The *Recurvoides contortus* Assemblage, which consists of stoutly constructed species, occurs in widely separated samples all in proximity to sediment that is barren of microfossils. Some of the samples contain only *R. contortus*. This assemblage is best explained as residual, formed through the selective action of corrosive bottom water.

Paleoecologic interpretation within a region relies on the relationships established between present-day environmental parameters and assemblages. A second important consideration is the interregional applicability of these relationships and interpretations.

The lower-bathyal and abyssal species are widespread in the World Ocean, and the distribution of their assemblage associations has been determined by factor analysis in the North Atlantic (Streeter, 1973; Schnitker, 1974), the South Atlantic (Lohmann, 1978), the southeastern Indian Ocean (Corliss, 1976, 1979), and the equatorial western Pacific (Culp, 1977). These data may be directly compared to the assemblage distribution on the lower-bathyal Nazca plate.

In contrast to the study area, the Atlantic Ocean is essentially a carbonate basin, and calcareous assemblages are present to depths of at least 6,000 m (Streeter, 1973). In the western North Atlantic, the lysocline lies at 4,700 ± 200 m from lat 20°N to 40°N and rises to 4,300 ± 200 m south of lat 20°N (Broecker and Takahashi, 1978), where the Atlantic Antarctic Bottom Water is less dilute. Both Streeter (1973) and Schnitker (1974) recognized *Epistominella exigua* and *Nuttallides umbonifera* Assemblages, which they associated with the Arctic and Antarctic Bottom Water Masses, respectively. The *N. umbonifera* Assemblage occurs deeper than 4,000 or 5,000 m in water temperatures of less than 2.0 °C (Streeter) or, more precisely, 1.5 °C (Schnitker). The *E. exigua* Assemblage (or *Cibicides wuellerstorfi–Epistominella exigua* Assemblage of Streeter) occurred between about 1,800 m and 4,000 or 5,000 m, in water temperatures of 2.2 to 3.0 °C (Streeter) or 1.9 °C (Schnitker). These various relationships indicate that the elements common to the distribution of these assemblages in the North Atlantic and on the Nazca plate are (1) the association of the top of the *N. umbonifera* Assemblage with the 1.5 °C isotherm and the lysocline and (2) the definition and succession of these two lower-bathyal to abyssal assemblages. Recently, Bremer and others (1979) reported *N. umbonifera* consistently related to low carbonate in eastern equatorial and South Atlantic basins. Lohmann (1978) recognized *N. umbonifera* as an indicator of Antarctic Bottom Water below 4 km in the western South Atlantic but did not recover *E. exigua* in the overlying water mass that was North Atlantic Deep Water.

In the Indian Ocean, Corliss (1976) defined two assemblages: one, in which *Epistominella exigua* was a dominant species, was associated with the Indian Bottom Water Mass, and the other, in which *Nuttallides umbonifera* was a dominant component, was found in colder Antarctic Bottom Water with temperatures between –0.2 and 0.4 °C. No well-defined relationships of the assemblages to abyssal bathymetry was revealed in the published data. The assemblage containing *N. umbonifera* dominated the high latitudes—a situation reflecting the origins of the Antarctic Bottom Water Mass—and the extention northward of its distribution was considered indicative of the path of flow of those waters. From the low sedimentation rates reported in the area of the northward flow, Corliss postulated that dissolution activity there may have influenced the benthic fauna.

In the western equatorial Pacific, Culp (1977) defined a deep and bottom-water factor below 2,500 m characterized by *Epistominella exigua* and several other species, of which *Melonis pompilioides* exhibited the highest factor loading but was exceeded by *E. exigua* in percentage frequency. From a conventional percentage frequency diagram of species distribution presented by Culp, it is clear that an additional, deeper assemblage can be differentiated on the basis of the percentage frequency of *Nuttallides umbonifera*, which becomes the dominant form below the lysocline at 3,500 m, approximating the top of the bottom water mass. Inclusion of this further subdivision permits the recognition of an *E. exigua* Assemblage from 2,500- to 3,500-m water depth in temperatures of 1.4 to 1.9 °C and a *N. umbonifera* Assemblage below 3,500 m, between the lysocline and the CCD.

Relationships of the other lower-bathyal and abyssal assemblages defined in this study have yet to be determined in other areas. For the present, it suffices that the two deepest carbonate assemblages are the same in many parts of the World Ocean: the *N. umbonifera* Assemblage, associated with Antarctic Bottom Water and temperatures of 1.5 °C or lower, occurs between the lysocline and the CCD; the *E. exigua* Assemblage, immediately shallower, occurs in temperatures reported from 1.60 to 3.0 °C. Bathymetry associated with these biofacies varies with the distribution of water parcels and therefore must be interpreted with discretion.

Areal distribution of select species from the various continental-margin assemblages has been mentioned. Data concerning both the latitudinal and bathymetric extent of these species are relevant to paleoecologic interpretation. Many of the outer-shelf to upper-bathyal species of the Peru Margin extend northward only as far as Central America. This group includes such important faunal constituents as *Bolivina costata, B. seminuda humilis, Cancris carmenensis, Epistominella obesa,* and *Valvulineria inflata*. Particularly at upper-bathyal depths, other species such as *Epistominella subperuviana* (= *E. bradyana*) and *Bolivina spissa* occur off California; thus the area of these species' potential usefulness for paleoecologic interpretation is extended along the west coast of the Americas.

Although the bathymetric succession of species is similar along the western American margin, depth limits of individual species may vary regionally (Smith, 1964). At the present state of the art, paleobathymetric interpretation is most accurate when local environmental relationships are used wherever possible. An association of more widespread application, which has paleoecologic implication that transcends bathymetry, is that of assemblages dominated by *Bolivina* spp. (*B. costata, B. seminuda humilis, B. pacifica*) with the O_2-minimum layer.

CONCLUSIONS

Foraminifera of the Nazca plate and its margins show distinctive faunal associations related to temperature, bathymetry, and other environmental properties that cut across the geographic subdivisions: continental margin, plate-bathyal, and trench and plate-abyssal environments. These assemblages are defined by factor analysis and are characterized by high percentage frequencies of their nominal species.

Two agglutinated assemblages, dominated by *Reophax dentaliniformis* and *Placopsilinella aurantiaca*, are bound to the trench by temperature and productivity of the surface water. Two distinctive agglutinated species associated with these assemblages, *Saccammina tubulata* and

Ammomarginulina sp., are completely restricted to the trench. Predominantly carbonate assemblages of the rises and plate form two major assemblages defined by high percentages of *Epistominella exigua* and *Nuttallides umbonifea*. *N. umbonifera* is found in the top of the Antarctic Bottom Water Mass and the lysocline it creates. *Favocassidulina* sp. is restricted to this assemblage. *E. exigua* inhabits the overlying water mass, or Pacific Deep Water in the study area. A third plate-bathyal assemblage defined by high percentages of the small species, *Gyroidina turgida*, appears to benefit from high production in the surface waters and the resulting accumulation of an organic food source in the bottom sediments, as does the *Bulimina translucens* Assemblage, distributed on the East Pacific Rise in the northern part of the area. Some of these plate carbonate species (for example, *B. translucens*, *Seabrookia earlandi*, *G. turgida*) have never been reported on the continental margin and hence may be true oceanic species. Fossil deposits that originated on the rises and plate or a stratigraphic sequence of progressively deeper biofacies in accordance with the evolution of oceanic crust may be defined by these criteria, in the manner attempted by Resig (1976). In this respect, it is fortuitous that *E. exigua* and *N. umbonifera* range back to at least late Eocene time (Resig, 1976).

The species *Bolivina costata* identifies a factor distributed on the eastern margin of the plate and is present on the Nazca Ridge and in abyssal deposits that were probably displaced from the Sarmiento Ridge and small-scale plate-margin elevations. *B. costata* has a strong affinity with the continental margin, where it is abundant under outer-shelf O_2-minimum conditions. Its concurrence with typical plate faunas might signal the proximity of the plate deposits to the continental margin, in fossil deposits.

The biofacies of outer-shelf, upper-bathyal, upper and lower middle-bathyal, and lower-bathyal depths are recognized on the Peru Margin. *Bolivina* spp. are dominant on the outer shelf and at upper-bathyal depths in the O_2-minimum layer. Shallow-dwelling species are sometimes displaced downslope. The lower-bathyal environment of the margin lacks the full complement of species present in lower-bathyal depths on the plate, which along with displaced species, makes it distinct. Fossil deposits originating on the continental margin such as the upper-bathyal Charco Azul Formation of Panama (Coryell and Mossman, 1942) may be recognized by these criteria.

The determination of provenance for sedimentary rocks composing the continental margin is critical to the interpretation of tectonic processes associated with convergent plate boundaries. The role of subduction of oceanic sediments versus the role of accretion might be resolved by these means.

ACKNOWLEDGMENTS

This research was sponsored by National Science Foundation Grant IDO71-04207-A04. Several individuals aided the assimilation of data, and their help is gratefully acknowledged. Photomicrographs of the specimens were taken by Karen Margolis on a Cambridge S4-10 scanning electron microscope, which was partially funded through National Science Foundation Grant GD-34207. W. Coulbourn provided the computer programs for the species tabulation and Q-mode factor analysis. H. Fortenberry at Oregon State University and N. Penrose at the University of Hawaii were technical assistants for parts of the project. The final version of the manuscript benefited from a review by William Coulbourn, Scripps Institution of Oceanography.

Much of this work was stimulated by L. D. Kulm, Oregon State University, who has continually pointed out situations along the juncture between the Nazca plate and South America where interpretation of foraminiferal paleoecology could contribute toward the determination of tectonic processes and activity.

REFERENCES CITED

Bandy, O. L., 1953, Ecology and paleoecology of some California foraminifera. Part I. The frequency distribution of Recent foraminifera off California: Journal of Paleontology, v. 27, p. 161–182.

———1961, Distribution of foraminifera, radiolaria and diatoms in sediments of the Gulf of California: Micropaleontology, v. 7, p. 1–26.

Bandy, O. L., and Arnal, R., 1957, Distribution of Recent foraminifera off west coast of Central America: American Association of Petroleum Geologists Bulletin, v. 41, p. 2037–2053.

Bandy, O. L., and Rodolfo, K., 1964, Distribution of foraminifera and sediments, Peru-Chile Trench area: Deep-Sea Research, v. 11, p. 817–837.

Barker, R. W., 1960, Taxonomic notes on the species figured by H. B. Brady in his report on the foraminifera dredged by H.M.S. *Challenger* during the years 1873–1876: Society of Economic Paleontologists and Mineralogists Special Publication 9, 238 p.

Berger, W. H., 1968, Planktonic foraminifera: Selective solution and paleoclimatic interpretation: Deep-Sea Research, v. 15, p. 31–43.

———1970, Planktonic foraminifera: Selective solution and the lysocline: Marine Geology, v. 8, p. 111–138.

Berger, W. H., and Winterer, E. L., 1974, Plate stratigraphy and the fluctuating carbonate line, *in* Hsü, K., and Jenkyns, H., eds., Pelagic sediments on land and under the sea: International Association of Sedimentology Special Publication 1, p. 11–48.

Boltovskoy, E., 1976, Distribution of Recent foraminifera of the South American region, *in* Hedley, R. H., and Adams, C. G., eds., Foraminifera, Volume 2: New York, Academic Press, p. 171–236.

Boltovskoy, E., and Gualancañay, E., 1975, Foraminiferos bentónicos actuales de Ecuador. 1. Provincia Esmeraldas: Guayaquil, Ecuador, Instituto Oceanográfico de la Armada, p. 1–56, Pl. 1–10.

Boltovskoy, E., and Theyer, F., 1970, Foraminiferos Recientes de Chile Central: Revista del Museo Argentino de Ciencias Naturales, Hidrobiología, v. 2, p. 279–378, maps 1–2, Pl. 1–6.

Brady, H. B., 1884, Report on the foraminifera dredged by H.M.S. *Challenger*, during the years 1873–1876: Report of the Scientific Results of the Voyage of H.M.S. *Challenger* 1873–1876, v. 9 (Zoology), 814 p.

Bremer, M. L., Edwards, A. S., and Thunell, R. C., 1979, Benthonic foraminiferal biofacies in silled oceanic basins of the eastern Atlantic: Evidence for primary control of their distribution by degree of carbonate saturation: Geological Society of America Abstracts with Programs, v. 11, p. 393–394.

Broecker, W. S., and Broecker, S., 1974, Carbonate dissolution on the western flank of the East Pacific Rise: Society of Economic Paleontologists and Mineralogists Special Publication 20, p. 44–57.

Broecker, W. S., and Takahashi, T., 1978, The relationship between lysocline depth and *in situ* carbonate ion concentration: Deep-Sea Research, v. 25, p. 65–95.

Corliss, B. H., 1976, Recent deep-sea benthic foraminiferal distributions in the southeast Indian Ocean: Antarctic Journal, v. 11, p. 165–167.

———1979, Taxonomy of Recent deep-sea benthonic foraminifera from the southeast Indian Ocean: Micropaleontology, v. 25, p. 1–19.

Coryell, H. N., and Mossman, R. W., 1942, Foraminifera from the Charco Azul Formation, Pliocene of Panama: Journal of Paleontology, v. 16, p. 233–246.

Coulbourn, W., 1977, Tectonics and sediments of the Peru-Chile Trench and continental margin at the Arica Bight [Ph.D. thesis]: Honolulu, University of Hawaii, 243 p.

———1980, Relationship between the distribution of foraminifera and geologic structures of the Arica Bight, South America: Journal of Paleontology, v. 54, p. 696–718.

Culp, S. K., 1977, Recent benthic foraminifera of the Ontong-Java Plateau [M.S. thesis]: Honolulu, University of Hawaii, 67 p.

Cushman, J., and Kellett, B., 1929, Recent foraminifera from the west coast of South America: U.S. National Museum Proceedings, v. 75, art. 25, p. 1–16.

Cushman, J., and McCulloch, I., 1939, A report on some arenaceous foraminifera: Allan Hancock Pacific Expedition, University of Southern California Press, v. 6, no. 1, p. 1–113.

———1940, Some Nonionidae in the collections of the Allan Hancock Foundation: Allan Hancock Pacific Expedition, University of Southern California Press, v. 6, no. 3, p. 145–178.

———1942, Some Virgulininae in the collections of the Allan Hancock Foundation: Allan Hancock Pacific Expedition, University of Southern California Press, v. 6, no. 4, p. 179–230.

———1948, The species of *Bulimina* and related genera in the collections of the Allan Hancock Foundation: Allan Hancock Pacific Expedition, University of Southern California Press, v. 6, no. 5, p. 231–294.

———1950, Some Lagenidae in the collections of the Allan Hancock Foundation: Allan Hancock Pacific Expedition, University of Southern

California Press, v. 6, no. 6, p. 295-362.
Davis, J., 1973, Statistics and data analysis in geology: New York, John Wiley and Sons, Inc., 550 p.
Echols, R. J., 1971, Distribution of foraminifera in sediments of the Scotia Sea area, Antarctic waters: American Geophysical Union Antarctic Research Series, v. 15, p. 93-168.
Harman, R. A., 1964, Distribution of foraminifera in the Santa Barbara Basin, California: Micropaleontology, v. 10, p. 81-96.
Heath, G. R., and Dymond, J., 1977, Genesis and transformation of metalliferous sediments from the East Pacific Rise, Bauer Deep, and Central Basin, northwest Nazca plate: Geological Society of America Bulletin, v. 88, p. 723-733.
Ingle, J., Keller, G., and Kolpack, R., 1980, Benthic foraminiferal biofacies, sediments and water masses of the southern Peru-Chile Trench area, southeastern Pacific Ocean: Micropaleontology, v. 26, p. 113-150.
Keller, G., and Ingle, J., 1976, Recent benthic foraminiferal biofacies and water mass properties, southern Peru-Chile Trench area: Geological Society of America Abstracts with Programs, v. 8, p. 951.
Khusid, T., 1971, Distribution of foraminiferal taxocoenosis and genocoenosis on the South American borderland in the Pacific Ocean: Okeanologia, v. 11, p. 266-268.
Lalicker, C. G., and McCulloch, I., 1940, Some Textulariidae of the Pacific Ocean: Allan Hancock Pacific Expedition, University of Southern California Press, v. 6, no. 2, p. 115-143.
Lisitzin, A., 1972, Sedimentation in the world ocean: Society of Economic Paleontologists and Mineralogists Special Publication 17, 218 p.
Lohmann, G. P., 1978, Abyssal benthic foraminifera as hydrographic indicators in the western South Atlantic Ocean: Journal of Foraminiferal Research, v. 8, p. 6-34.
Lonsdale, P., 1976, Abyssal circulation of the southeastern Pacific and some geological implications: Journal of Geophysical Research, v. 81, p. 1163-1176.
Murray, J. W., 1973, Distribution and ecology of living benthic foraminiferids: New York, Crane Russak and Company, 274 p.
Natland, M. L., 1950, 1940 *E.W. Scripps* cruise to the Gulf of California. Part IV—Report on the Pleistocene and Pliocene foraminifera: Geological Society of America Memoir 43, p. 1-55.
Orbigny, A. d', 1839, Foraminifères, Voyage dans l'Amérique Meridionale: Paris, Pitois-Levrault et C^c, v. 5, 86 p.
Parker, F. L., 1964, Foraminifera from the experimental Mohole drilling near Guadalupe Island, Mexico: Journal of Paleontology, v. 38, p. 617-636.
Parker, F. L., and Berger, W. H., 1971, Faunal and solution patterns of planktonic foraminifera in surface sediments of the South Pacific: Deep-Sea Research, v. 18, p. 73-107.
Phleger, F. B., and Soutar, A., 1973, Production of benthic foraminifera in three east Pacific oxygen minima: Micropaleontology, v. 19, p. 110-115.
Reid, J. L., 1965, Intermediate waters of the Pacific Ocean: Baltimore, Johns Hopkins Press, 85 p.
Resig, J. M., 1976, Benthic foraminiferal stratigraphy, eastern margin, Nazca plate, *in* Yeats, R. S., Hart, S. R., and others, eds., Initial Reports of the Deep Sea Drilling Project, Volume 34: Washington, D.C., U.S. Government Printing Office, p. 743-759.
Resig, J. M., and others, 1980, An extant opaline foraminifer: Test ultrastructure, mineralogy and taxonomy, *in* Sliter, W. V., ed., O. L. Bandy memorial volume: Cushman Foundation for Foraminiferal Research Special Publication 19, p. 205-214.
Rosato, V. J., Kulm, L. D., and Derks, P. S., 1975, Surface sediments of the Nazca plate: Pacific Science, v. 29, p. 117-130.
Saidova, Kh., 1971, On foraminifera distribution near the Pacific coast of South America: Okeanologia, v. 11, p. 256-265.
——1976, Benthic foraminifera of the world ocean: Academy of Sciences of the USSR, Institute of Oceanology, Nauka, 160 p.
Schnitker, D., 1974, West Atlantic abyssal circulation during the past 120,000 years: Nature, v. 248, p. 385-387.
Shepherd, G., 1979, Shallow crustal structure and marine geology of a convergence zone, northwest Peru and southwest Ecuador [Ph.D. dissert.]: Honolulu, University of Hawaii, 201 p.
Smith, P. B., 1963, Quantitative and qualitative analysis of the Family Bolivinitidae: U.S. Geological Survey Professional Paper 429-A, 35 p.
——1964, Ecology of benthonic species. Recent foraminifera off Central America: U.S. Geological Survey Professional Paper 429-B, 55 p.
Streeter, S., 1973, Bottom water and benthonic foraminifera in the North Atlantic—glacial and interglacial contrasts: Quaternary Research, v. 3, p. 131-141.
Uchio, T., 1960, Ecology of living benthonic foraminifera from the San Diego, California area: Cushman Foundation for Foraminiferal Research Special Publication 5, 72 p.
Valentine, J. W., 1973, Evolutionary paleoecology of the marine biosphere: Englewood Cliffs, New Jersey, Prentice-Hall, Inc., 511 p.
Walch, C., 1978, Recent abyssal benthic foraminifera from the eastern equatorial Pacific [M.S. thesis]:

Los Angeles, University of Southern California, 79 p.
Warren, B. A., 1970, General circulation of the South Pacific, *in* Wooster, W. S., ed., Scientific explorations of the South Pacific: Washington, D.C., National Academy of Sciences, p. 33–49.
Wooster, W. S., 1970, Eastern boundary currents in the South Pacific, *in* Wooster, W. S., ed., Scientific explorations of the South Pacific: Washington, D.C., National Academy of Sciences, p. 60–68.
Wyrtki, K., 1964, The thermal structure of the eastern Pacific Ocean: Deutsche Hydrographische Zeitschrift, Ergänzugsheft Reihe A (8°), Nr. 6, 84 p.
—— 1967, Circulation and water masses in the eastern Equatorial Pacific Ocean: International Journal of Oceanology and Limnology, v. 1, p. 117–147.

MANUSCRIPT RECEIVED BY THE SOCIETY NOVEMBER 12, 1980
MANUSCRIPT ACCEPTED DECEMBER 30, 1980
CONTRIBUTION 1106, HAWAII INSTITUTE OF GEOPHYSICS

PLATE SECTION

In the following plate explanations, the species identification is followed by the maximum dimension (in millimeters) of the figured specimen, the sample number, and R if the species is restricted to the indicated assemblage unit in the study area. All photographs were taken on a Cambridge S4-10 scanning electron microscope at varying magnifications.

PLATE 1. MARGIN SPECIES

Figures 1 to 12

1. *Bolivina costata* d'Orbigny. 0.28 mm, Y71-6-1SS.
2. *Bolivina interjuncta* Cushman. 1.10 mm, KK71-GC2, R.
3, 4. *Bolivina plicata* d'Orbigny. 3, microspheric, 0.70 mm; 4, megalospheric, 0.54 mm; both KK71-GC1, R.
5. *Bolivina seminuda* Cushman var. *foraminata* R. E. and K. C. Stewart. 0.50 mm, KK71-GC3, R.
6. *Bolivina seminuda* Cushman var. *humilis* Cushman and McCulloch. 0.45 mm, KK71-GC3, R.
7. *Bolivina spissa* Cushman. 0.63 mm, KK71-GC2, R.
8. *Bolivina subadvena* Cushman var. *sulphurensis* Cushman and Adams. 0.70 mm, KK71-GC2, R.
9. *Bolivinita minuta* (Natland). 0.20 mm, KK71-GC2, R.
10. *Bulimina affinis* d'Orbigny. 0.58 mm, KK72-PC9.
11. *Bulimina exilis* Brady var. *tenuata* (Cushman). 0.50 mm, KK72-PC10, R.
12. *Bulimina striata* d'Orbigny var. *mexicana* Cushman. 0.65 mm, KK72-PC9, R.

PLATE 2. MARGIN SPECIES

Figures 1 to 11

1. *Angulogerina carinata* Cushman. 0.83 mm, KK71-GC2, R.
2. *Uvigerina auberiana* d'Orbigny. 0.25 mm, KK72-PC9, R.
3. *Uvigerina bifurcata* d'Orbigny. 0.78 mm, KK74-PC24, R.
4. *Uvigerina hispida* Schwager. 0.65 mm, KK72-PC9, R.
5. *Uvigerina peregrina* Cushman. 0.60 mm, KK71-GC2, R.
6. *Uvigerina peregrina* Cushman var. *dirupta* Todd. 0.68 mm, KK72-PC9, R.
7. *Uvigerina striata* d'Orbigny. 0.78 mm, KK71-GC1, R.
8. *Cancris inflatus* (d'Orbigny). Oblique dorsal, 0.48 mm, KK71-GC1, R.
9–11. *Cancris carmenensis* Natland. 9, dorsal surface ×1,070; 10, edge, 0.75 mm; 11, dorsal, 0.83 mm; both KK71-GC1, R.

PLATE 3. MARGIN SPECIES

Figures 1 to 12

1, 2. *Epistominella obesa* Bandy and Arnal. 1, edge, 0.30 mm; 2, dorsal, 0.43 mm; both KK71-GC6, R.
3, 6. *Epistominella* sp. 3, edge, 0.33 mm; 6, ventral, 0.43 mm; both KK71-GC2, R.
4, 5. *Epistominella subperuviana* (Cushman). 4, dorsal, 0.25 mm; 5, ventral, 0.25 mm, final chambers broken off; both KK71-GC6, R.
7, 8. *Epistominella smithi* (R. E. and K. C. Stewart). 7, ventral, 0.28 mm; 8, dorsal, 0.33 mm; both KK72-PC9, R.
9, 12. *Eilohedra levicula* (Resig). 9 dorsal, 0.13 mm; 10, ventral, 0.10 mm; both KK72-PC9, R.
10, 11. *Cibicides spiralis* Natland. 10, ventral, 0.30 mm, KK74-FFC71, broken final chambers reveal test filling; 11, dorsal, 0.35 mm, KK72-PC9, R.

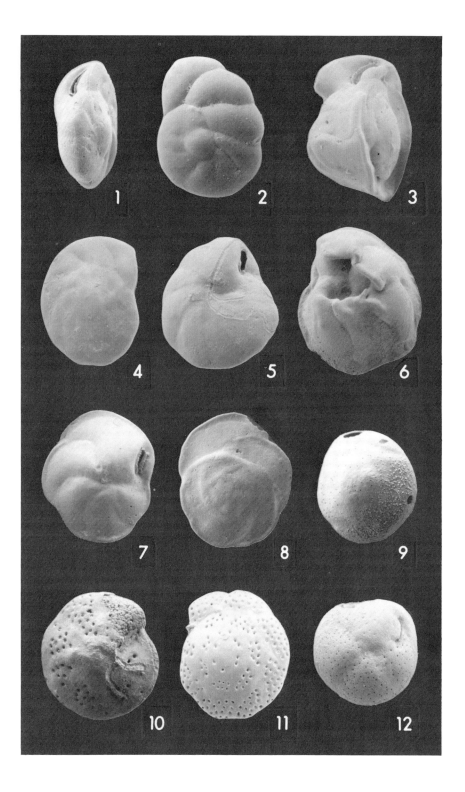

PLATE 4. MARGIN SPECIES

Figures 1 to 15

1–3. *Planulina limbata* Natland. 1, ventral, 0.75 mm; 2, edge, 0.75 mm; 3, dorsal, 1.10 mm; all KK71-GC2, R.
4, 6, 7. *Cibicides mckannai* Galloway and Wissler. 4, edge, 0.38 mm; 6, dorsal, 0.38 mm; 7, ventral, 0.40 mm; all KK72-PC9, R.
5. *Pullenia subcarinata* (d'Orbigny). 0.68 mm, KK71-GC2, R.
8, 9. *Melonis affinis* (Reuss). 8, side, 0.35 mm; 9, edge, 0.38 mm; both KK72-PC9.
10. *Cassidulina cushmani* R. E. and K. C. Stewart. 0.25 mm, KK74-FFC71.
11, 15. *Cassidulina auka* Boltovskoy and Theyer. 11, side, 0.58 mm; 15, edge, 0.55 mm; both KK71-GC2, R.
12, 13. *Cassidulina crassa* d'Orbigny. 12, side, 0.65 mm; 13, edge, 0.58 mm; both KK71-GC2, R.
14. *Cassidulina subglobosa* Brady var. *quadrata* Cushman and Hughes. Edge, 0.45 mm, KK71-GC2, R.

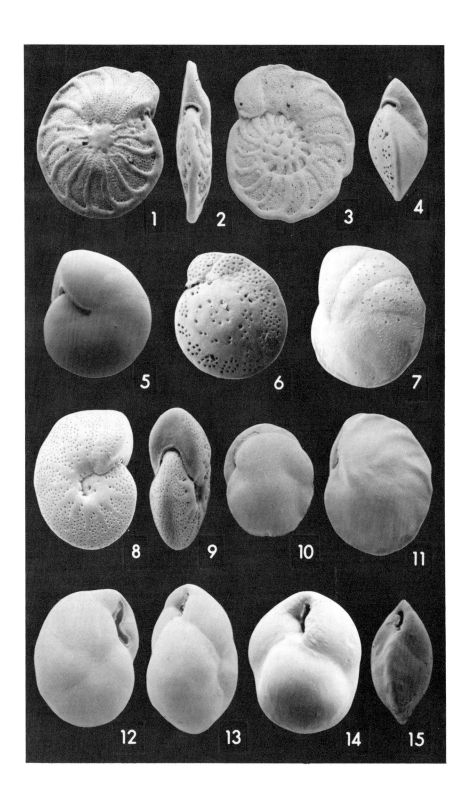

PLATE 5. PLATE-BATHYAL SPECIES

Figures 1 to 17

1. *Rhizammina algaeformis* Brady. 1.38 mm, Y71-9-90MG1. The test composition is mostly tests and fragments of planktonic foraminifera.
2. *Reophax dentaliniformis* Brady. 1.28 mm, Y71-9-87MG2.
3. *Siphotextularia* sp. 0.63 mm, KK71-FFC-121, R.
4. *Trochammina globigeriniformis* (Parker and Jones). 0.40 mm, Y71-9-90MG1.
5. *Ophthalmidium pusillum* (Earland). 0.60 mm, Y71-7-32MG4, R.
6–8. *Quinqueloculina venusta* Karrer. 6, juvenile, 0.40 mm; 7, adult, 0.73 mm; 8, detail of wall ×750; both Y71-9-90MG1, R.
9. *Pyrgo murrhyna* (Schwager). 0.40 mm, Y71-7-46MG3.
10. *Sigmoilina edwardsi* (Schlumberger). 0.50 mm, KK71-PC78, R.
11. *Sigmoilina tenuis* (Czjzek). 0.80 mm, KK72-FFC10.
12. *Seabrookia earlandi* Wright. 0.15 mm, Y71-738MG4, R. The delicate test was broken during SEM preparation.
13. *Nodosaria calomorpha* Reuss. 0.53 mm, Y71-9-112G, R.
14. *Bolivina seminuda* Cushman. 0.48 mm, Y71-9-90MG1.
15. *Bulimina translucens* Parker. 0.28 mm, Y71-9-90MG1, R.
16. *Stainforthia complanata* (Egger). 0.40 mm, Y71-6-12MG3.
17. *Francesita torta* (Cushman). 0.40 mm, KK71-PC78, R.

PLATE 6. PLATE-BATHYAL SPECIES

Figures 1 to 15

1. *Uvigerina asperula* Czjzek. 0.63 mm, Y71-9-88MG1.
2. *Uvigerina bradyana* Fornasini. 0.58 mm, Y71-6-12MG3, R.
3, 4. *Uvigerina senticosa* Cushman. 3, 0.75 mm; 4, 0.58 mm; both KK71-PC78.
5. *Siphouvigerina pseudoampullacea* (Asano). 0.63 mm, Y71-9-96G.
6, 7. *Epistominella exigua* (Brady). 6, ventral, 0.33 mm; 7, dorsal, 0.38 mm; both Y71-6-12MG3.
8. *Laticarinina pauperata* (Parker and Jones). 0.68 mm, Y71-7-44MG4.
9, 10. *Valvulineria rugosa* (d'Orbigny) var. *minuta* (Schubert). 9, dorsal, 0.43 mm, KK71-FFC121; 10, ventral, 0.45 mm, Y71-7-50FF, R.
11, 12. *Heronallenia gemmata* Earland. 11, dorsal, 0.28 mm, KK72-FFC18; 12, ventral, 0.43 mm, Y71-7-39MG2, R.
13–15. *Nuttallides umbonifera* (Cushman). 13, dorsal, 0.45 mm; 14, edge, 0.35 mm; 15, ventral, 0.43 mm; all Y71-9-90MG1, R.

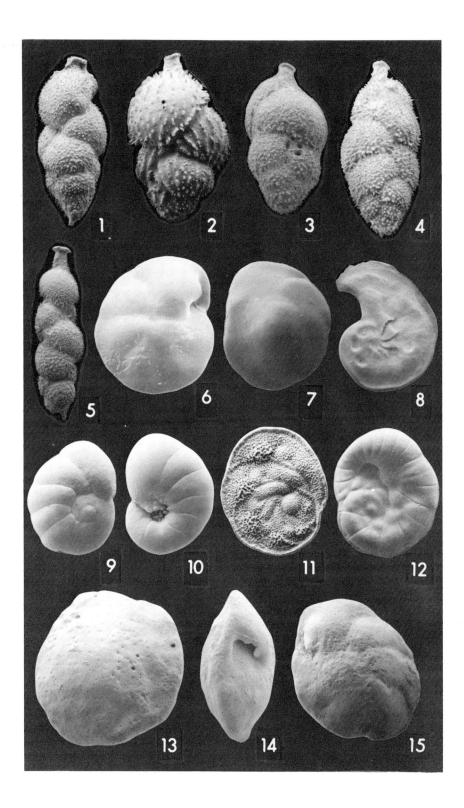

PLATE 7. PLATE-BATHYAL SPECIES

Figures 1 to 16

1–3. *Eponides tumidulus* (Brady). 1, dorsal, circular periphery, 0.23 mm, Y71-9-90MG1; 2, ventral, 0.23 mm, Y71-9-92MG1; 3, dorsal, lobate periphery, 0.25 mm, Y71-9-90 MG1, R.
4. *Coryphostoma subdepressa* (Brady). 0.70 mm, Y71-9-91MG3.
5. *Cassidulina depressa* Asano and Nakamura. 0.18 mm, Y71-6-12MG3.
6. *Cassidulina laevigata* d'Orbigny. 0.28 mm, Y71-9-90MG1.
7. *Cassidulina subglobosa* Brady. Edge, 0.63 mm, KK71-FFC128.
8. *Favocassidulina favus* (Brady). 0.33 mm, Y71-7-44MG4, R.
9. *Favocassidulina* sp. 0.40 mm, KK72-FFC7, R.
10. *Cassidulina* sp. 0.25 mm, Y71-7-44MG4, R.
11. *Astrononion schwageri* (Cushman). 0.28 mm, Y71-7-38MG4, R.
12. *Melonis pompilioides* (Fichtel and Moll). Edge, 0.38 mm, Y71-6-12MG3.
13. *Pullenia bulloides* (d'Orbigny). Edge, 0.33 mm, Y71-6-12MG3.
14. *Pullenia quinqueloba* (Reuss). 0.40 mm, KK71-FFC144.
15, 16. *Pullenia* sp. 15, edge, 0.25 mm; 16, side, 0.28 mm; both Y71-9-90MG1, R.

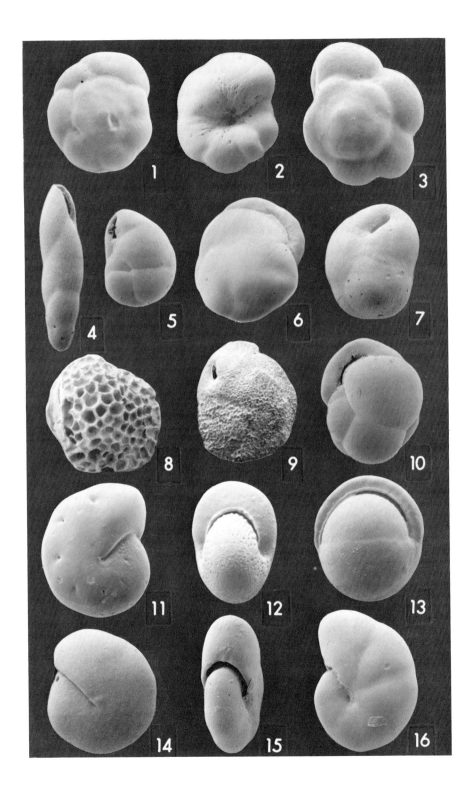

PLATE 8. PLATE-BATHYAL SPECIES

Figures 1 to 18

1, 2. *Gyroidina lamarckiana* (d'Orbigny). 1, dorsal, 0.55 mm, KK72-FFC7; 2, ventral, 0.40 mm, Y71-6-12MG3.
3, 4. *Gyroidina* aff. *zealandica* Finlay. 3, edge, 0.48 mm; 4, dorsal, 0.50 mm; both Y71-9-90MG1, R.
5. *Gyroidina neosoldanii* Brotzen. 1.08 mm, Y71-9-88MG1, R.
6, 7. *Gyroidina turgida* (Phleger and Parker). 6, ventral, 0.15 mm; 7 dorsal, 0.15 mm; both Y71-6-12MG3, R.
8. *Oridorsalis umbonatus* (Reuss). 0.33 mm, KK71-PC78.
9, 10. *Cibicidoides bradyi* (Trauth). 9, dorsal, 0.48 mm, KK71-PC78; 10, ventral 0.30 mm, KK71-FFC128.
11, 12. *Anomalina globulosa* Chapman and Parr. 11, dorsal, 0.48 mm, Y71-7-50FF; 12, edge, 0.48 mm, Y71-9-92MG1.
13, 14. *Cibicidoides cicatricosus* (Schwager). 13, dorsal, 0.75 mm, Y71-7-44Mg4; 14, ventral, 0.83 mm, Y71-9-91MG3, R.
15, 18. *Cibicidoides mundulus* (Brady, Parker and Jones). 15, dorsal, 0.75 mm, Y71-7-33MG2; 18, ventral, 0.68 mm, Y71-7-32MG4.
16, 17. *Cibicidoides wuellerstorfi* (Schwager). 16, ventral, 0.53 mm, Y71-9-97G; 17, dorsal, 0.47 mm, Y71-9-90MG1.

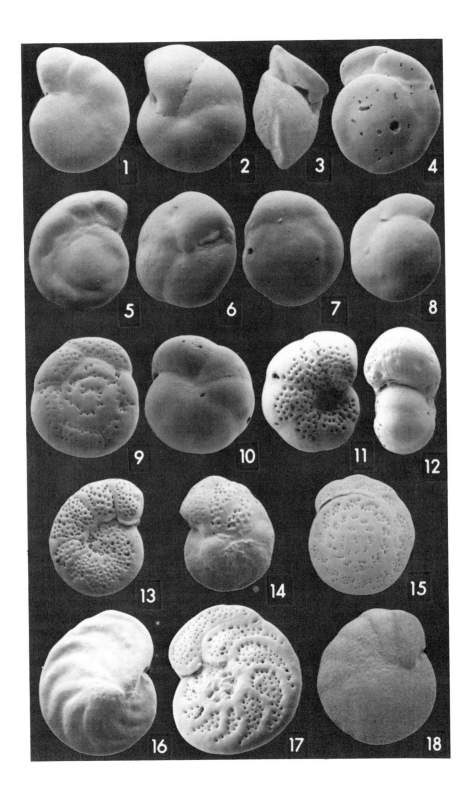

PLATE 9. TRENCH AND PLATE-ABYSSAL SPECIES

Figures 1 to 16

1. *Nodellum membranaceum* (Brady). 1.20 mm, KK71-FFC181B.
2. *Nodellum* sp. 0.50 mm, Y71-7-36-MG3.
3. *Rhizammina algaeformis* Brady. 4.25 mm, Y71-7-27MG1. Test constructed of radiolarians, diatoms, and siliceous fragments, with one conspicuous specimen of *Reophax difflugiformis* Brady.
4. *Bathysiphon filiformis* M. Sars. 1.75 mm, Y71-6-4MG3.
5. *Saccammina tubulata* Rhumbler. 0.33 mm, R; with attached *Placopsilinella aurantiaca* Earland; KK74-FFC53.
6. *Ammodiscus catinus* Hoeglund. 0.25 mm, KK74-FFC26, R.
7. *Ammodiscus tenuis* Brady. 0.95 mm, KK74-FFC50.
8. *Glomospira charoides* (Jones and Parker). 0.20 mm, KK71-FFC181B.
9. *Reophax aduncus* Brady. 0.75 mm, Y71-6-4MG3.
10. *Reophax guttifer* Brady. 0.58 mm, KK71-FFC181B.
11. *Reophax dentaliniformis* Brady. 1.00 mm, Y71-6-4MG3. This large form, which shows constriction in the upper part of the chambers, where specimens often are broken, is typical of the development of the species in the trench environment.
12. *Glomospira gordialis* (Jones and Parker). 0.25 mm, KK71-FFC181B.
13. *Reophax subdentaliniformis* Parr. 0.63 mm, Y71-8-61FF1.
14. *Reophax scorpiurus* Montfort. 0.78 mm, Y71-6-4MG3. The small size and consistent morphology are common in noncalcareous assemblages.
15. *Ammobaculites americanus* Cushman. 0.58 mm, Y71-6-24MG2.
16. *Ammobaculites agglutinans* (d'Orbigny). 0.58 mm, KK71-FFC181B.

PLATE 10. TRENCH AND PLATE-ABYSSAL SPECIES

Figures 1 to 14

1. *Ammobaculites filiformis* Earland. 0.38 mm, KK71-FFC181B.
2. *Ammomarginulina* sp. 0.48 mm, Y71-8-75MG3, R.
3. *Spiroplectammina filiformis* Earland. 0.25 mm, Y71-8-75MG3.
4. *Textularia wiesneri* Earland. 0.33 mm, Y71-6-21MG2, R.
5. *Spirolocammina tenuis* Earland. 0.45 mm, Y71-6-14MG3, R.
6. *Cribrostomoides ringens* (Brady). 0.45 mm, Y71-7-30MG4.
7. *Cribrostomoides subglobosus* (Sars). 0.58 mm, KK71-FFC157.
8. *Cyclammina trullissata* (Brady). 0.68 mm, Y71-6-19MG3.
9, 10. *Cribrostomoides wiesneri* (Parr). 9, detail of wall masonry ×3,975; 10, side, 0.33 mm, Y71-6-4MG3.
11. *Adercotryma glomeratum* (Brady). 0.25 mm, KK71-FFC181B.
12. *Recurvoides contortus* Earland. 0.53 mm, KK71-FFC181B.
13. *Cystammina galeata* (Brady). 0.20 mm, Y71-6-4MG3.
14. *Cystammina pauciloculata* (Brady). 0.30 mm, Y71-8-70MG3.

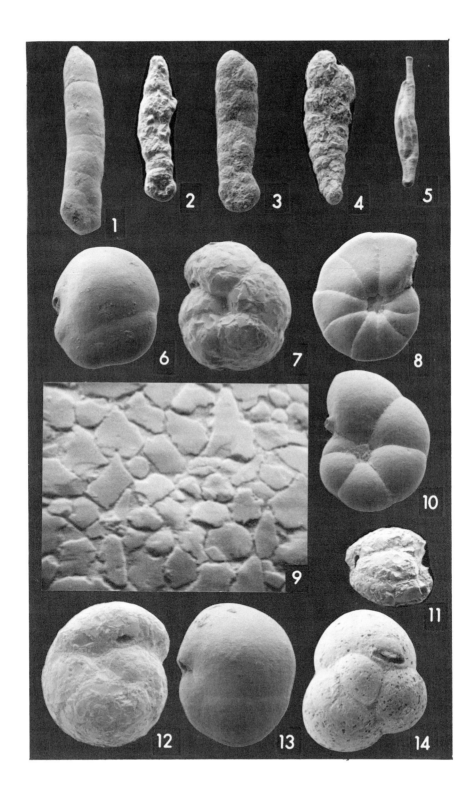

Geological Society of America
Memoir 154
1981

Estimation of depth to magnetic source using maximum entropy power spectra, with application to the Peru-Chile Trench

RICHARD J. BLAKELY
U.S. Geological Survey
345 Middlefield Road
Menlo Park, California 94025

SIAMAK HASSANZADEH
Free University of Iran
Tehran, Iran

ABSTRACT

Estimations of the depth to magnetic sources using the power spectrum of magnetic anomalies generally require long magnetic profiles. The method developed here uses the maximum entropy power spectrum (MEPS) to calculate depth to source on short windows of magnetic data; resolution is thereby improved. The method operates by dividing a profile into overlapping windows, calculating a maximum entropy power spectrum for each window, linearizing the spectra, and calculating with least squares the various depth estimates. The assumptions of the method are that the source is two dimensional and that the intensity of magnetization includes random noise; knowledge of the direction of magnetization is not required. The method is applied to synthetic data and to observed marine anomalies over the Peru-Chile Trench. The analyses indicate a continuous magnetic basement extending from the eastern margin of the Nazca plate and into the subduction zone. The computed basement depths agree with acoustic basement seaward of the trench axis, but deepen as the plate approaches the inner trench wall. This apparent increase in the computed depths may result from the deterioration of magnetization in the upper part of the ocean crust, possibly caused by compressional disruption of the basaltic layer. Landward of the trench axis, the depth estimates indicate possible thrusting of the oceanic material into the lower slope of the continental margin.

INTRODUCTION

Methods of estimating the depth to the source of magnetic anomalies can be divided into two types. The first type assumes that the anomaly is caused by an isolated body with uniform magnetization and uses the shape of the anomaly to calculate the depth (for example, Peters, 1949). A more recent approach uses the statistical properties of a set of anomalies to estimate the depth to an ensemble of sources. Treitel and others (1971), for example, calculated the logarithm of the power spectrum of a magnetic profile. If certain assumptions are satisfied by the source, the logarithm of the spectrum will be a straight line with a slope porportional to the depth to the top of the source.

Those using statistical methods, however, are faced with a dilemma. To determine a good spectral estimate, and hence an accurate depth estimate, requires a long series of data. Using long portions of a profile for each depth estimate reduces the resolution of short-wavelength changes in the source depth. In recent years, a technique for calculating power spectra from short samples has received much attention in the geophysical literature. This method is the maximum entropy power spectrum (MEPS) developed by Burg (1967 and 1968, unpub. ms.; 1972). We have extended the work of Treitel and others (1971) by using the MEPS, so that good depth estimates can be obtained with shorter segments of data; as a result the resolving power of the method is increased.

The technique was tested with computer-generated data and applied to actual data from the Nazca plate. Depth estimates in this area show that the magnetic layer deepens seaward of the trench axis, shallows at the axis, and then deepens again beneath the continental shelf and slope. These observations are explained by the disruption of oceanic crust prior to its consumption by the trench and by faulting of magnetic material into the slope and shelf subsequent to subduction.

METHOD

Consider a magnetized, horizontal layer (Fig. 1). The top of the layer is at depth z and its thickness is t. The x axis is in the direction of the profile and the z axis is positive down. The source is assumed to be two dimensional, that is, the magnetization $m(x)$ varies only in the x direction. The theory discussed here, however, could be extended to magnetizations that vary in both the x and y directions and then be applied to gridded data rather than profiles (Spector and Grant, 1970).

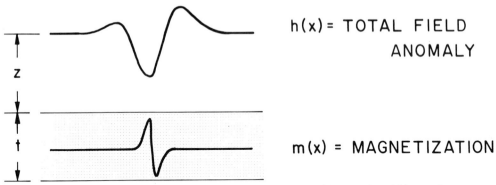

Figure 1. Definitions of model parameters. Depth to the top of the layer is z; thickness of layer is t.

The total field anomaly produced by the layer is given by

$$h(x) = \int_{-\infty}^{\infty} m(u)g(x - u)du ,$$

where $g(x)$ is the Green's function for a magnetized layer. If

$$\int_{-\infty}^{\infty} |m(x)| dx \qquad (1)$$

is finite, then the Fourier transform of $h(x)$ is given by

$$H(k) = M(k)G(k) , \qquad (2)$$

where k is wavenumber and where

$$G(k) = \theta(k)[\exp(-|k|z)][1 - \exp(-|k|t)] , \qquad (3)$$

$$\theta(k) = -2\pi(P - iQ \operatorname{sgn} k) ,$$

$$P = \hat{m}_x\hat{h}_x - \hat{m}_z\hat{h}_z ,$$

and

$$Q = \hat{m}_x\hat{h}_z + \hat{m}_z\hat{h}_x$$

(Blakely and Cox, 1972). In this equation, $\hat{m}_x$ and $\hat{m}_z$ are x and z components of unit vectors in the direction of magnetization; $\hat{h}_x$ and $\hat{h}_z$ are x and z components of the unit vector parallel to the regional field in which the anomaly is measured. It is assumed that these directions do not vary over the length of the profile.

Now we assume that the magnetization of the layer is a random function of x. Under this circumstance, the integral in equation 1 is not finite, and the Fourier transform of $m(x)$ in equation 2 does not strictly exist. However, the Wiener theorem of autocorrelation (Lee, 1960) states that $m(x)$ does posess a power density spectrum given by the Fourier transform of the autocorrelation of $m(x)$. Applying this theorem to equation 2 yields

$$\phi_h(k) = \phi_m(k)|G(k)|^2 , \qquad (4)$$

where $\phi_h(k)$ and $\phi_m(k)$ are the power density spectra of $h(x)$ and $m(x)$, respectively, and where

$$|G(k)|^2 = |\theta(k)|^2[\exp(-2|k|z)][1 - \exp(-|k|t)]^2 \qquad (5)$$

and

$$|\theta(k)|^2 = 4\pi^2(P^2 + Q^2)$$
$$= 4\pi^2(\hat{m}_x^2 + \hat{m}_z^2)(\hat{h}_x^2 + \hat{h}_z^2) .$$

Note that $\theta(k)^2$ is a constant with magnitude $\leq 4\pi^2$, so that the shape of $\phi_h(k)$ does not depend on directions of magnetization or regional field.

Now assume that $m(x)$ is uncorrelated so that

$$E[m(x + \Delta x) \cdot m(x)] = 0, \; \Delta x \neq 0 . \qquad (6)$$

This assumption implies that the average magnetization of any two slices through the layer are independent of each other. Although it is true that magnetic properties possess a high degree of randomness from place to place, the assumption of total randomness is not realistic. For example, given that one slice of crust is normally magnetized, there is a greater probability that the magnetization of a neighboring slice will also be normal rather than reversed. We will see, however, on the basis of empirical testing of the method, that the restriction of uncorrelated magnetization can be relaxed somewhat.

If $J_m(x)$ is uncorrelated as in equation 6, then the magnetization spectrum is a constant:

$$\phi_m(k) = \sigma^2 . \qquad (7)$$

Combining equations 4, 5 and 7 and applying the logarithm to both sides yield

$$\log \phi_h(k) = A - 2|k|z + 2\log[1 - \exp(-|k|t)] , \qquad (8)$$

where

$$A = \log(\sigma^2|\theta(k)|^2) .$$

Equation 8 shows that, as the thickness of the layer increases, the logarithm of the spectrum of the anomaly approaches a straight line with negative slope equal to twice the depth. In previous work by Spector and Grant (1970) and Treitel and others (1971), the thickness of the layer was assumed infinite. We prefer to specify the thickness of the layer, calculate the last term of equation 8, and subtract it from the log power, so that

$$\log \phi'_h(k) = \log \phi_h(k) - 2\log[1 - \exp(-|k|t)]$$

$$= A - 2|k|z . \qquad (9)$$

For typical layer thicknesses, this logarithmic term is not large; we will consider it further in a later section. This adjusted spectrum provides an estimate of z, in principle.

Rather than using traditional methods to calculate $\phi_h(k)$, we have used the maximum entropy power spectrum (MEPS) developed by Burg (1967 and 1968, unpub. ms.; 1972). The method can be summarized as follows. A prediction error filter for a time series is a filter that passes only random noise, so that all "predictable" information about the data is contained in the filter. The inverse of the spectrum of the filter, therefore, is a spectrum of the predictable part of the data and has a resolution superior to traditional spectral estimates. Such a filter can be calculated by least squares from the autocorrelation of the segment of data (Peacock and Treitel, 1969), but this method assumes unrealistically that the data are zero beyond the available segment. Burg (1967 and 1968, unpub. ms.; 1972) suggested (1) that a two-term prediction error filter be calculated by least squares directly from the data segment alone and (2) that higher-order terms be calculated recursively from lower-order terms. His first suggestion eliminates the end-effect problem of autocorrelation, and the second ensures that the prediction error filter will be minimum phase and thus have an inverse spectrum. Finally, the spectrum of the data is given by the reciprocal of the spectrum of the prediction error filter. Lacoss (1971) and Chen and Stegen (1974) have provided mathematical reviews of the MEPS.

Previously, the MEPS has been used to determine sinusoidal components of time series with periods on the same order as the length of the data (Ulrych, 1972; Denham, 1975). Such components appear in the spectrum as very narrow peaks. Our application is different. We expect, according to equation 8, that the spectrum of a typical profile segment will be smooth rather than composed of narrow peaks. Figure 2 shows a short segment of synthetic data and illustrates that the MEPS technique provides a more accurate representation of its spectrum than does the Fourier transform method.

Equation 8 has been programmed in PL-1 computer language to estimate a series of depths automatically from a long profile. The steps in the algorithm are as follows: (1) The profile is broken into segments of length WINDOW, which may overlap. (2) For each segment, the maximum entropy power spectrum $\phi_h(k)$ is computed. (3) The logarithm of each spectrum is calculated, and the effect of layer thickness subtracted, thus linearizing the spectrum. The resulting function is $\log \phi'_h(k)$ in equation 9. (4) Using least squares, a straight line is fit through part of each spectrum. Only a specified band of the spectrum is used for this computation because the highest and lowest wavenumber parts of a spectrum are unreliable. For example, the theoretical power in an anomaly caused by a magnetic layer is negligible at very high wavenumbers. The shape of the spectrum in this region is influenced by noise. (5) The slope of the straight line is used in equation 9 to determine depth.

TESTING

Figures 3 through 7 show examples of tests of this method. In each figure, the geometry of the source is shown by the block diagram in the middle of the figure. This block is given the magnetization $m(x)$ shown at the bottom, and the resulting total field anomaly is at the top of each figure. The depth-to-source method was applied to these profiles to try to reconstruct the depth to the model, and estimated depths are shown by the solid dots superimposed on the block model.

Effect of Profile-Segment Length

Figures 3, 4, and 5 illustrate the influence of profile segment length on calculated depths. For this test, there is a single 2-km offset in the "basement" located midway through the profile. The length of the prediction error filter is 12. Each depth estimate in Figure 3 was calculated from 40-point segments of the profile (WINDOW = 40). The estimated depths are satisfactory with a maximum error of less

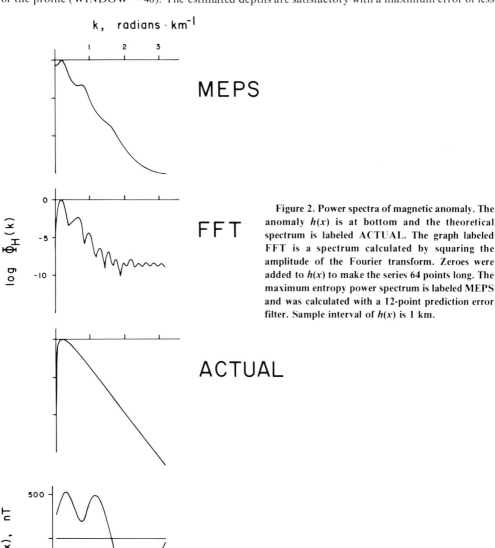

Figure 2. Power spectra of magnetic anomaly. The anomaly $h(x)$ is at bottom and the theoretical spectrum is labeled ACTUAL. The graph labeled FFT is a spectrum calculated by squaring the amplitude of the Fourier transform. Zeroes were added to $h(x)$ to make the series 64 points long. The maximum entropy power spectrum is labeled MEPS and was calculated with a 12-point prediction error filter. Sample interval of $h(x)$ is 1 km.

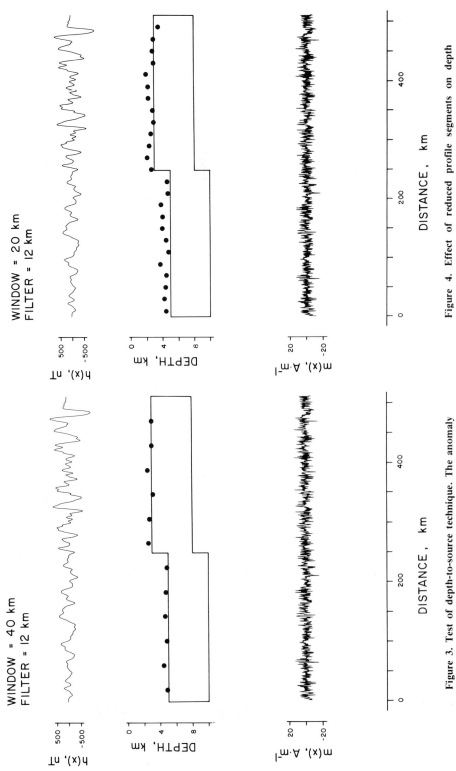

Figure 3. Test of depth-to-source technique. The anomaly $h(x)$ is produced by a source with dimensions shown by block model and with magnetization shown as $m(x)$. Results of depth analysis are shown by the dots. WINDOW is the length of anomaly used for each depth calculation; FILTER is the length of the prediction error filter. Sample interval is 1 km.

Figure 4. Effect of reduced profile segments on depth calculation. See Figure 3 for description.

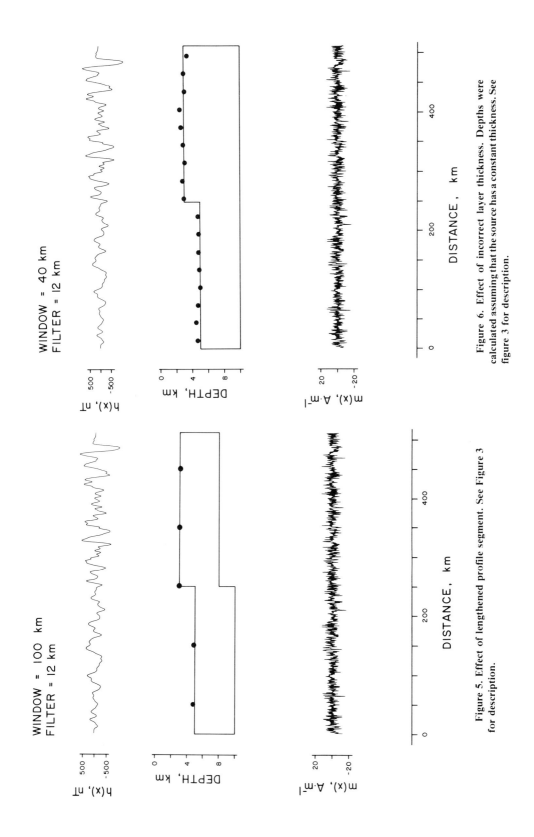

Figure 5. Effect of lengthened profile segment. See Figure 3 for description.

Figure 6. Effect of incorrect layer thickness. Depths were calculated assuming that the source has a constant thickness. See figure 3 for description.

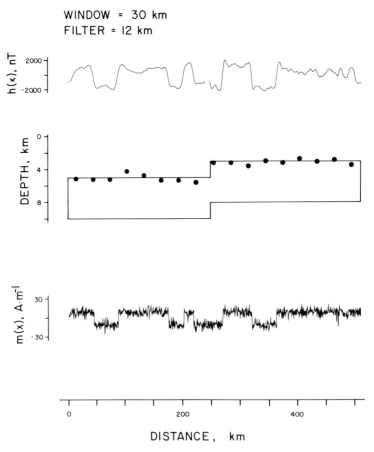

Figure 7. Effect of coherent magnetization. Magnetization $m(x)$ has both a coherent and random component. See Figure 3 for description.

than 0.5 km. Error in depth estimates are small enough that the 2-km offset in basement is easily detected.

The amount of information contained in a series of data is related to the length of data, so it is expected that the estimation of depth will deteriorate for shorter segments of the profile. In Figure 4, WINDOW is halved to 20, and, as expected, the depth estimates deteriorate. It would be difficult to discern the 2-km offset in basement, for example. The estimated depths are not radically incorrect, however. This is an encouraging result because the length of $h(x)$ used for each spectrum in Figure 4 (20 km) is narrower than many of the anomalies of $h(x)$.

Increasing the length of segments improves the estimates of depth. Figure 5 shows the results of increasing WINDOW to 100 km. The estimates are accurate, but the purpose of the method, to search for short-wavelength basement relief, has been defeated.

Effect of Layer Thickness

Figure 6 shows the effect of specifying an incorrect thickness for the layer. The actual thickness of the model changes from 5 km on the left to 7 km on the right. In estimating the depths shown by the dots in Figure 6, the erroneous assumption was made that thickness was a uniform 5 km throughout the block. The results show that, for this test case, thickness is not an essential parameter.

To investigate the influence of thickness, an estimate of the deviation of $\phi'_H(k)$ resulting from an incorrect thickness is required. From equation 9, the power spectrum adjusted for true thickness t is a straight line. Suppose than an erroneous thickness $t + \Delta t$ is assumed in calculating log $\phi'_H(k)$ from log $\phi_H(k)$. The deviation from a straight line will be

$$d(k) = 2\left\{\log\frac{1 - \exp(-|k|t)}{1 - \exp[-|k|(t + \Delta t)]}\right\}. \quad (10)$$

This deviation approaches infinity as k approaches 0, but decreases rapidly as k increases. For the example in Figure 6, the true thickness is 7 km and the assumed thickness is 5 km. The range of log $\phi'_H(k)$ used to estimate the depths in Figure 6 was $0.11 < k < 2.49$ radians/km, so according to equation 10,

$$0.48 < |d(k)| < 7.78 \times 10^{-6}.$$

Because this entire range of log $\phi'_H(k)$ is used to fit a straight line, the error produced by the incorrect thickness in Figure 6 is small.

Effect of Filter Length

The shape of a spectrum calculated from a segment of data using the maximum entropy method depends on the number of terms of the prediction error filter. A prediction error filter with only a few terms produces a smooth spectrum, whereas longer filters cause the spectrum to be more "noisy" (for example, Denham, 1975). Akaike (1969, 1970) has developed a criterion for optimum filter length, but our results are not strongly dependent on this parameter. Although the length of filter influences the details of a spectrum, the *general* properties of a spectrum are not severely affected unless filter lengths are very large or very small. Because depth estimation depends on the slope of a line through a *band* of the spectrum, depth determinations are not critically dependent on filter lengths. On the basis of empirical experiments, satisfactory results are obtained if the number of terms in the filter is between 0.25 and 0.75 of the number of points in the data segment.

Effect of Correlated Magnetization

Uncorrelated magnetization is required for equation 9 to be correct, an assumption not likely to be satisfied by real sources. A more realistic magnetization is used in Figure 7. This magnetization includes two components: a coherent magnetization of ± 10 A/m and a zero-mean, random component with variance 2.5 A^2/m^2. The coherent magnetization is constant for long distances but includes 10 reversals at various locations. The estimates of depth are not severely affected by the addition of the coherent magnetization, but it must be noted that some random component of magnetization is essential to our method. As the magnitude of the random magnetization is decreased relative to the coherent component, the depth estimates deteriorate.

The distinction between coherent and random magnetization is largely one of scale. The random magnetizations used in each of the tests are digital series with sample intervals of 1 km. This implies that at scales of less than 1 km, the magnetization is constant (that is, coherent). Accurate depth estimates are still obtained from random magnetizations, even if the sample interval is greatly reduced. However, amplitudes of anomalies due to the random component are sharply attenuated if the scale of randomness is reduced. Coherent sources then dominate the power of the anomalies and thus produce incorrect estimates of depth. Therefore, the "random component" of magnetization must be uncorrelated over distances roughly equivalent to the sample interval of the profile.

ANALYSIS OF DATA

The data analyzed in this paper are from an area west of the Chilean coastline between lat 25° and 31°S and between long 70° and 75°W. Figure 8 shows the location of the tracklines. These magnetic data were collected by three research vessels: the *Yaquina* of Oregon State University, the *Oceanographer* of the National Oceanic and Atmospheric Administration (NOAA), and the *Melville* of Scripps Institution of Oceanography.

The data are over the Peru-Chile Trench, a major zone of convergence between the oceanic Nazca plate and the South American continent. The rate of convergence of the Nazca plate relative to South America is estimated to be 10 cm/yr (Minster and others, 1974). Recent field studies as a part of the Nazca Plate Project have provided a detailed description of this margin (Kulm and others, 1973; Prince and others, 1974; Prince and Kulm, 1975; Hussong and others, 1975, 1976). The analysis undertaken here is an attempt to investigate the nature of the subduction process as told by the magnetic properties of the oceanic crust. The magnetic patterns in the eastern Pacific Ocean basin date the crust adjacent to the trench between lat 23° and 31°S as approximately 50 m.y. old (Herron, 1972; Hayes, 1974).

In Figure 9, our data are shown plotted along the trackline. The amplitudes of the anomalies rapidly attentuate seaward of the trench axis, approximately coincident with the initial downbending of the oceanic plate at the outer edge of the trench. Studies for other areas have indicated that this attenuation in amplitude cannot be explained solely by the increases in depth of the descending oceanic crust beneath the continental margin (Hayes, 1974).

Depth to magnetic basement was determined from seven of the longest profiles, labeled A through G in Figures 8 and 9, and the results are shown in Figure 10. The values of the log-power spectrum at zero wavenumber and at large wavenumbers were excluded from the calculation. The thickness assigned to the oceanic magnetic layer was 1 km. It is impossible to show whether the magnetization of ocean crust in this area satisfies the assumptions of our model, and the results shown in Figure 10 should be considered with some skepticism. Ideally we would like to show seismic reflection data to help interpret these results. However, only single-channel data exist in this area, and these show only one acoustic horizon slightly below bathymetric depths (L. Kulm, 1978, personal commun.).

The results generally indicate that seaward of the trench axis the magnetic basement lies near the observed topography (Fig. 10). This agrees with seismic refraction studies on the Nazca plate which suggest that the thickness of the sedimentary layer is less than 200 m (Hussong and others, 1976). As the plate approaches the subduction zone, the magnetic basement appears to deepen. Seismic reflection studies of this region, however, do not indicate substantial thickening of the sedimentary layer as the plate approaches the continent, so that the upper surface of basaltic layer 2 (that is, the magnetic layer) is not actually deepening (Schweller and Kulm, 1978). Apparently the spectral depth estimates near the seaward trench-slope are deeper than the actual basement. This would result if the actual thickness of the magnetic layer is larger than the assumed value of 1 km, but the increase in thickness required to raise the estimates to the top of layer 2 results in the magnetic layer extending well below the Curie-point isotherm. Alternatively, the shallow basement that Schweller and Kulm (1978) have described near the trench-slope may have lost part of its magnetization because of tectonic disturbances so that more important magnetic sources are deeper. Since a thickening layer cannot be distinguished from a deepening source by this technique, we cannot decide on the best answer. In fact, the apparent deepening may be a combination of both effects.

The thickening of the oceanic crust would require the presence of horizontal compressional stresses. The presence of a compressional stress regime was suggested by Mendiguren (1971), who determined a dip-slip model with an east-west horizontal pressure axis for a shallow shock in the middle of the Nazca plate. Also, recent studies of the Nazca plate show compressional faulting of the oceanic crust prior to subduction in the Peru-Chile Trench (Hussong and others, 1975), and wide-angle seismic reflection profiling of the Peru-Chile Trench suggests the thickening of the oceanic crust as it plunges beneath the continental slope (Goebel, 1974).

Compressional or tensional stresses may be a cause of demagnetization of the upper part of the

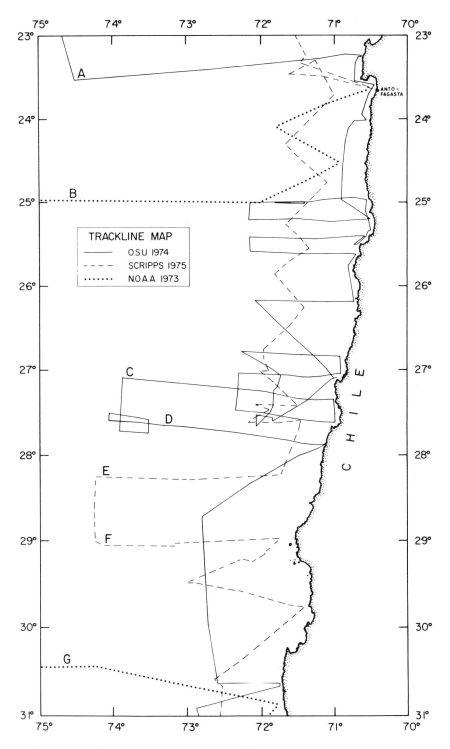

Figure 8. Trackline map of data over the Peru-Chile Trench. Labels A through G refer to analyzed profiles.

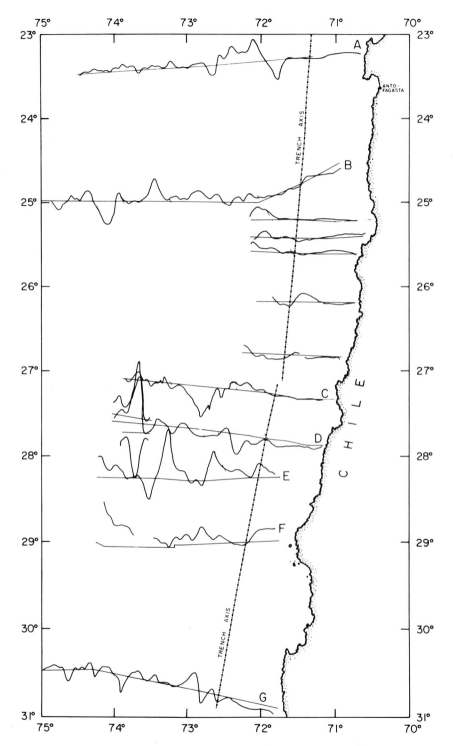

Figure 9. Magnetic anomalies plotted along tracklines. Labels A through G refer to profiles analyzed in this study.

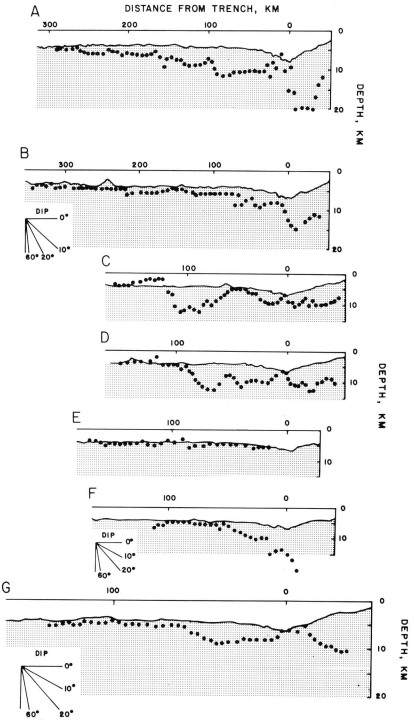

Figure 10. Results of depth determinations on profiles A through G. Horizontal axis is distance from trench axis. Top of patterned area represents bathymetric depth, and heavy dots are estimates of depth to magnetic source. Vertical exaggeration is 3 to 1.

magnetic layer. Faulting, for example, would tend to randomize the magnetization to smaller scales so that deeper, more coherent sources influence the results. There is abundant evidence of extensional faulting of the upper oceanic crust along this section of the Peru-Chile Trench (Schweller and Kulm, 1978), and attenuation of the magnetic anomalies in other trenches is attributed to the demagnetization of the crust by tectonic processes (Grow and Mudie, 1973).

The depth estimates for the regions underlying the continental slope (that is, east of the trench axis) are generally scattered and less reliable since the amplitudes of the anomalies are lower. Where estimates are obtained, they are within the toe of the continental margin. These shallow sources are probably due to rocks with an oceanic origin, because continental rocks in this area have very low magnetic susceptibilities and intensities (Mordojovich, 1974). These shallow sources imply uplift of oceanic material into the continental slope, as suggested for regions farther north by Prince and Kulm (1975) and Hussong and others (1975). Also, geophysical investigations of the continental shelf off western Guatemala suggest the presence of high-density material in the lower slope that may be a melange of sediments and oceanic basement (Woodcock, 1975).

Overall, the magnetic basement appears to be a continuous feature with no major discontinuities except for the mid-plate irregularities observed in profile C, at approximately 100 km from the trench, and in profile D, at approximately 70 km from the trench. It is unlikely that these discontinuities reflect real changes in depth. They are probably caused by the magnetic layer's deviation from the assumptions of our model.

CONCLUSIONS

The method described here provides accurate estimates of depth to magnetic sources when applied to computer-generated anomalies. Of course, these anomalies are forced to meet the requisite assumptions (for example, a source with random magnetization). When applying such techniques to real data, it is essential to beware of results that appear reasonable but are influenced by data that do not satisfy the assumptions.

With this caution in mind, the new method was applied to data from the Peru-Chile Trench. These results show a magnetic layer approaching the trench with a depth equal to layer 2. As the crust nears the seaward slope, the estimated depths increase, apparently because of disruption of the upper part of the magnetic layer. The depths rise again at the trench axis and in the shelf and slope, perhaps because of uplift of deeper oceanic sources.

ACKNOWLEDGMENTS

Preliminary research on the method of depth determination and most of the application to the Peru-Chile Trench was done while we were at Oregon State University and was sponsored by the National Science Foundation, International Decade of Ocean Exploration (Grants GX-28675, IDOE 71-04208 A07, and OCE76-05903 A02).

REFERENCES CITED

Akaike, H., 1969, Power spectrum estimation through auto-regressive model fitting: Annals of Institute of Statistical Mathematics, v. 21, p. 407–419.

——1970, A fundamental relation between predictor identification and power spectrum estimation: Annals of Institute of Statistical Mathematics, v. 22, p. 219–233.

Blakely, R. J., and Cox, A., 1972, Identification of short polarity events by transforming marine magnetic profiles to the pole: Journal of Geophysical Research, v. 77, p. 4339–4349.

Burg, J. P., 1972, The relationships between maximum entropy spectra and maximum likelihood spectra: Geophysics, v. 37, p. 375–376.

Chen, W. Y., and Stegen, G. R., 1974, Experiments with maximum entropy power spectra of sinusoids: Journal of Geophysical Research, v. 79, p. 3019–3022.

Denham, C. R., 1975, Spectral analysis of paleomagnetic time series: Journal of Geophysical Research, v. 80, p. 1897–1901.

Goebel, V., 1974, Modeling of the Peru-Chile Trench from wide angle reflection profiles [M.S. thesis]: Corvallis, Oregon State University, 88 p.

Grow, J. A., and Mudie, J. D., 1973, Attenuation of Pacific magnetic anomalies in the Aleutian Trench [abs.]: EOS (American Geophysical Union Transactions), v. 54, p. 330.

Hayes, D. E., 1974, Continental margin of western South America, in Burk, C. A., and Drake, C. L., eds., The geology of continental margins: New York, Springer-Verlag, p. 581–590.

Herron, E. M., 1972, Sea-floor spreading and the Cenozoic history of the east-central Pacific: Geological Society of America Bulletin, v. 83, p. 1671–1692.

Hussong, D. M., Odegard, M. E., and Wipperman, L. K., 1975, Compressional faulting of the oceanic crust prior to subduction in the Peru-Chile Trench: Geology v. 3, p. 601–604.

Hussong, D. M., and others, 1976, Crustal structure of the Peru-Chile Trench; 8°–12°S latitude, in Sutton, G. H., and others, eds., The geophysics of the Pacific Ocean basin and its margin: American Geophysical Union Geophysical Monograph 19, p. 71–85.

Kulm, L. D., and others, 1973, Tholeiitic basalt ridge in the Peru Trench: Geology, v. 1, p. 11–14.

Lacoss, R. T., 1971, Data adaptive spectral analysis methods: Geophysics, v. 36, p. 661–675.

Lee, Y. W., 1960, Statistical theory of communication: New York, John Wiley and Sons, 509 p.

Mendiguren, J. A., 1971, Focal mechanism of a shock in the middle of the Nazca plate: Journal of Geophysical Research, v. 76, p. 3861–3879.

Minster, J. B., and others, 1974, Numerical modeling of instantaneous plate tectonics: Geophysics Journal, v. 36, p. 541–576.

Mordojovich, K. C., 1974, Geology of the part of the Pacific margin of Chile, in Burk, C. A., and Drake, C. L., eds., The geology of continental margins: New York, Springer-Verlag, p. 591–597.

Peacock, K. L., and Treitel, S., 1969, Predictive deconvolution: Theory and practice: Geophysics, v. 34, p. 155–169.

Peters, L. J., 1949, The direct approach to magnetic interpretation and its practical application: Geophysics, v. 14, p. 290–320.

Prince, R. A., and Kulm, L. D., 1975, Crustal rupture and the initiation of imbricate thrusting in the Peru Trench: Geological Society of America Bulletin, v. 86, p. 1639–1653.

Prince, R. A., and others, 1974, Significance of uplifted turbidite basins on the seaward wall of the Peru Trench: Geology, v. 2, p. 607–611.

Schweller, W. J., and Kulm, L. D., 1978, Extensional rupture of oceanic crust in the Chile Trench: Marine Geology, v. 28, p. 271–291.

Spector, A., and Grant, F. S., 1970, Statistical models for interpreting aeromagnetic data: Geophysics, v. 35, p. 293–302.

Treitel, S., and others, 1971, The spectral determination of depths to buried magnetic basement rocks: Geophysics Journal, v. 24, p. 415–429.

Ulrych, T., 1972, Maximum entropy power spectrum of long period geomagnetic reversals: Nature, v. 235, p. 218–219.

Woodcock, S. F., 1975, Crustal structure of the Tehuantapec ridge and adjacent continental margins of southwestern Mexico and western Guatemala [M.S. thesis]: Corvallis, Oregon State University, 52.

MANUSCRIPT RECEIVED BY THE SOCIETY NOVEMBER 12, 1980
MANUSCRIPT ACCEPTED DECEMBER 30, 1980

Printed in U.S.A.

An active spreading center collides with a subduction zone: A geophysical survey of the Chile Margin triple junction

E. M. HERRON*
S. C. CANDE
B. R. HALL
*Lamont-Doherty Geological Observatory
of Columbia University
Palisades, New York 10964*

ABSTRACT

The triple junction between the Nazca, Antarctic, and South American crustal plates has been mapped at lat 46.4°S, long 75.7°W. The Chile Ridge, an active spreading center opening at a half-rate of 28 mm/yr, can be traced into the axis of the Chile Trench; at the triple junction itself, the axial rift valley abuts the inner wall of the trench. The spreading process does not appear to be modified by proximity to a subduction zone until the actual point of collision: magnetic anomalies formed by sea-floor spreading can be traced to the inner wall of the trench; the depth of the axis of the Chile Ridge does not change with distance from the trench; the free-air gravity minimum associated with the trench decreases in amplitude owing to the elevated bathymetry at the triple junction, but this gravity anomaly continues across the junction; the thick wedge of trench-floor sediments is displaced only at the triple junction itself; and evidence for uplift of the continent also appears to be restricted to the immediate vicinity of the junction. We estimate that the present trench-ridge-trench triple junction has existed for about 300,000 yr. Before the collision of this segment of the Chile Ridge axis with the trench, the triple junction was of the trench-transform-trench type.

INTRODUCTION

The events associated with the collision of an active spreading center and a subduction zone are of considerable interest to geologists. Past collisions have been recognized along several margins of the Pacific basin for example, in Japan (Uyeda and Miyashiro, 1974), Alaska (Pitman and Hayes; 1968; Grow and Atwater, 1970; DeLong and others, 1978, 1979), North America (Atwater, 1970), and South America (Herron and others, 1977). These studies and others have investigated the possible geologic consequences of ridge-crest subduction (for example, Bird and Dewey, 1970; Dewey, 1976; DeLong and Fox, 1977; Menard, 1978; DeLong and others, 1979). At present there are only a few places where

*Order of the first two authors chosen by lot.

a collision is taking place between an active ridge and a trench. Two of them, the collision of the Juan de Fuca Ridge with North America and the collision of the East Pacific Rise with Central America, have been studied extensively. In both places, the interaction between the plates being subducted (former pieces of the Farallon plate) and the overriding plate has been complex (Menard, 1978). In this paper, we examine the collision of the Chile Ridge with the Chile Trench.

The intersection of the Chile Ridge with the Chile Trench occurs at the triple junction between the Nazca, South American, and Antarctic plates (referred to here as the Chile Margin triple junction). The Chile Ridge, which is the boundary between the Antarctic and Nazca plates, terminates in a second triple junction to the northwest at the intersection of the Pacific-Antarctic Ridge and the East Pacific Rise. These plates and their respective boundaries are shown schematically in the inset in Figure 1. On the basis of a study of worldwide plate motions, Minster and Jordan (1978) estimated that in the vicinity of the Chile Margin triple junction the Nazca plate is converging on South America at a rate of 90 mm/yr in a N79°E direction and Antarctica is converging on South America at a rate of 21 mm/yr in a S88°E direction.

The Chile Ridge was not recognized as a major plate boundary in the early papers on plate tectonics (for example, Isacks and others, 1968; Le Pichon, 1968). Seismic activity along a rather broad zone extending northwest from southern South America to the axis of the East Pacific Rise was known, but because the area was remote and other data were sparse, the structure of the area in which these seismic events occurred was essentially unknown. The first analyses of the magnetic anomalies associated with the Chile Ridge reflect this lack of data. Although the Chile Ridge trends mainly northwest from the Chile Trench to the axis of the East Pacific Rise, Morgan and others (1969), using aeromagnetic data and following Heezen's bathymetric chart (1968), assumed that individual ridge segments trended slightly east of due north. Herron and Hayes (1969), using shipboard data collected by the USNS *Eltanin*, proposed that the ridge segments trended north-northwest. More recent studies by Mammerickx and others (1974), Klitgord and others (1973), Molnar and others (1975), Handschumacher (1976), and Weissel and others (1977) show that the present strike of Chile Ridge segments is close to 350°. The work presented here for the region of the Chile Margin triple junction substantiates a ridge trend of about 350° (Fig. 1).

The intersection of the Chile Ridge and the Peru-Chile Trench has not been the specific subject of any previous studies. Although the large number of ship tracks shown in Figure 2 crossing the region of the Chile Margin triple junction would seem to provide good control, the quantity of reliable data (that is, satellite-navigated) is small. Consequently, in 1975, we carried out a marine geophysical survey of the Chile Margin triple junction on *Conrad* cruises 1802 and 1803. This project attempted to study the following problems: (1) What is the nature and history of the triple junction between the Nazca, Antarctic, and South American plates? (2) Are any submarine or continental geologic features in the vicinity of the triple junction directly related to the presence of the triple junction?

SURVEY OF THE CHILE MARGIN TRIPLE JUNCTION

Bathymetric, gravimetric, and magnetic profiles from these cruises illustrate the principal features in the area of the triple junction (Figs. 2 and 3): the crest of the actively spreading Chile Ridge, the major fracture zones, and the axis of the Peru-Chile Trench. Magnetic anomalies are identified according to the system of Heirtzler and others (1968). These data are presented and discussed in more detail in the following sections.

Bathymetry

In Figure 4 we present a contoured bathymetric map of the triple junction area based mainly on the data from C1802 and C1803 (C = *Conrad*; see caption for Fig. 2) but including many other available data. Additional data consist of a few satellite-navigated tracks (C1503, C2107, EL25, GL35, CATO) and a large number of older tracks positioned only by celestial navigation (for example, *Eltanin* cruises

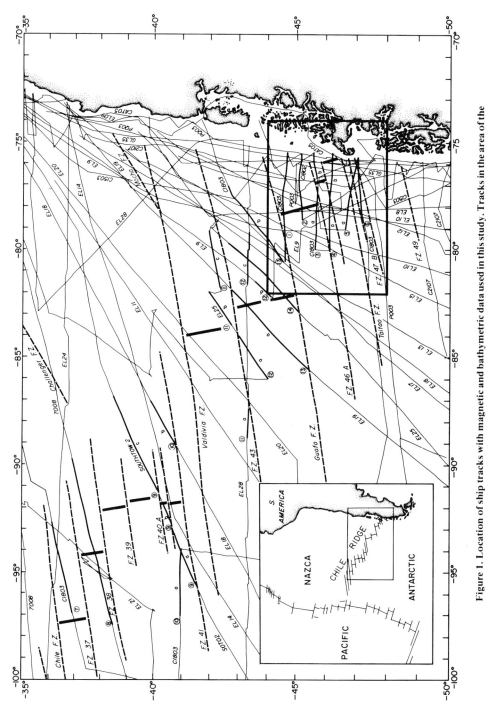

Figure 1. Location of ship tracks with magnetic and bathymetric data used in this study. Tracks in the area of the Chile Margin triple junction (box) are shown in detail in Figure 2. Identifying numbers 7 through 14 refer to the data shown in Figure 10.

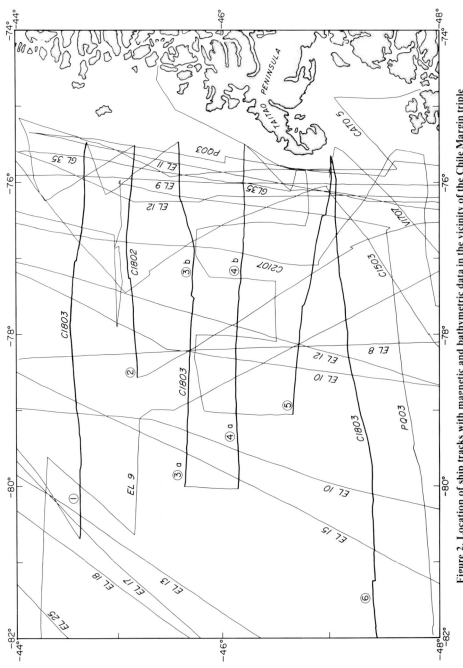

Figure 2. Location of ship tracks with magnetic and bathymetric data in the vicinity of the Chile Margin triple junction. Cruise identifications are as follows: C1803 = R/V *Conrad* cruise 18, leg 3; V1707 = R/V *Vema*; EL = USNS *Eltanin*; GL = *Glomar Challenger*; PQ = Piqueros cruise of R/V *Thomas Washington*; CATO = R/V *Melville*. Circled numbers indicate location of data displayed in the following figures. Position of circled numbers 3b and 4b indicate beginning of respective lines.

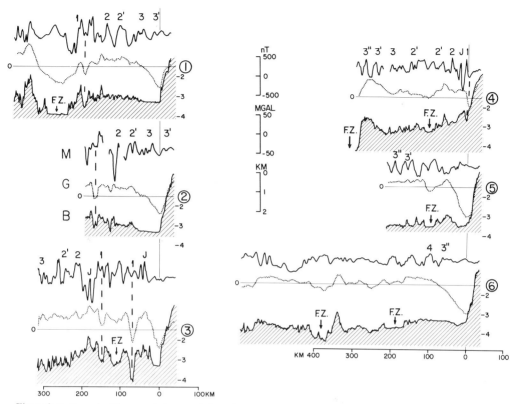

Figure 3. Bathymetry (B, corrected metres), free-air gravity (G), and total-intensity magnetic anomaly (M) data in the area of the triple junction (locations shown in Fig. 2) projected perpendicular to the margin. Magnetic anomaly identification follows nomenclature of Heirtzler and others (1968). Dashed vertical lines mark position of axis of rift valley.

8 through 20, V1707, PQ03). When crossover errors were encountered, the older tracks were adjusted to agree with the satellite-navigated data.

An orthogonal system of rift valleys and fracture zones is the most prominent feature on the bathymetric map. This system intersects the Peru-Chile Trench at lat 46°S. The large fracture zone at the northern edge of the map is the Guafo Fracture Zone, which offsets the rift valley by 300 km in the left-lateral sense. Near lat 46°S there are two additional fracture zones which offset the rift valley by 70 km and 60 km, respectively. We refer to these unnamed fractures zones as Fracture Zone 46A and Fracture Zone 46B on the basis of the approximate latitude at which they intersect the ridge crest. We use this same procedure for naming other fracture zones throughout the paper. Additional buried fracture zones to the south can be identified from the magnetic and seismic-profiler data. These buried fracture zones are discussed in the section on magnetic data.

A rift valley characterizes the actively spreading zone of the Chile Ridge in the three ridge segments shown in the bathymetric map (indicated by arrows near patterned area in Fig. 4). On the southernmost ridge segment the rift valley is located at the trench axis adjacent to the inner slope of the trench (see profile 4, Fig. 6b). The southern end of this segment of the rift valley passes beneath the inner wall of the trench at lat 46.4°S, long 75.7°W, which is the location of the triple junction. Between Fracture Zone 46A and the triple junction there is a gradual narrowing of the inner slope from north to south. The slope of the inner wall in this region is approximately 15°. Immediately south of the triple junction, there is a seaward projection of the inner slope above the location where the Taitao Fracture Zone passes beneath the inner slope. South of the intersection of the Taitao

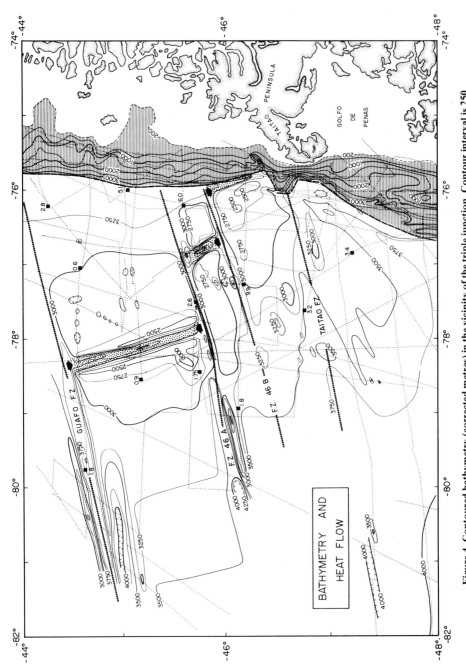

Figure 4. Contoured bathymetry (corrected metres) in the vicinity of the triple junction. Contour interval is 250 m; 200-m isobath based on bathymetric map of Mammerickx and others (1974). Heat-flow measurements are given in HFU (1 HFU = 41.8 mW/m^2). Arrows by patterned areas indicate axes of rift valleys. The area between the base of the inner slope and the 200-m isobath is ruled. TMP = Tres Montes Peninsula.

Fracture Zone with the trench, adjacent to the Tres Montes Peninsula, the inner slope has its narrowest width along the entire margin. The distance from the trench axis to land is less than 20 km. Between lat 47° and 48°S, adjacent to the Golfo de Peñas, the inner slope gradually widens until, south of lat 48°S, it is nearly twice as wide as it is in the region north of the triple junction.

Heat Flow

The heat-flow data presented with the bathymetric map in Figure 4 reflect the presence of the Chile Ridge spreading center and its intersection with the Peru-Chile Trench. Relatively high heat-flow values characterize the entire area of Figure 4 but are consistently found at stations located on the abyssal plain of the trench where values up to 6 HFU were recorded. Measurements on the Chile Ridge, away from the trench, vary between 0.6 and 9.9 HFU; most of the values are close to 2 HFU. The high values observed near the trench reflect the proximity of the ridge axis and suggest that the sediments have acted as a thermal blanket on the oceanic crust by restricting hydrothermal circulation through cracks and thus allowing measurement of the actual heat flow conducted through the crust. Although the data are insufficient to permit a detailed study of the heat-flow pattern in the vicinity of the triple junction, the data do in general support current ideas on the role of circulating water in modifying the observed flow of heat from the sea floor. The association of the high heat-flow values in the thickly sedimented areas on young crust adjacent to the trench supports the ideas of Lister (1972) and Anderson and others (1977), who proposed that hydrothermal circulation near unsedimented ridge axes is the primary cause for the observed scatter of heat-flow data at mid-ocean ridge stations. Only when crust is sufficiently buried by sediment does this circulation stop and the true heat flow from the lithosphere become observable.

Gravity Data and Seismic Reflection Profiles

In Figure 5 we show a contoured free-air gravity anomaly map for the triple-junction area. The predominant features in the map are the large negative gravity anomalies associated with the Peru-Chile Trench axis north and south of the triple junction. The gravity minimum south of the triple junction is considerably broader and more negative than the minimum north of the triple junction, perhaps a reflection of the greater sediment fill to the south. Prominent gravity lows also mark the location of the major fracture zones and the rift valleys associated with the Chile Ridge spreading center.

Representative single-channel seismic-profiler records across the Chile Margin north of, at, and south of the triple junction (lines 2, 4, and 6, respectively) are shown in Figure 6. These records show the major sedimentary features near the triple junction: (1) a dramatic increase in sediment fill in the trench from north of the triple junction to south of it, (2) the thin sediment cover where the ridge crest intersects the trench, and (3) disturbed sediments in the trench south of the triple junction. The profile south of the triple junction (line 6) which is distinguished by the presence of a broad V-shaped basin instead of the expected flat trench floor leads us to suggest that the topography here is controlled by either recent faulting or strong bottom currents.

Magnetic Data

In Figure 7 we show magnetic anomaly profiles in the vicinity of the triple junction projected perpendicular to the ridge crest. Since the original track lines 3, 4, and 5 crossed fracture zones, the two lowermost projected profiles are made up from sections of different tracks. We modeled the profiles using the polarity-reversal time scale of LaBrecque and others (1977). Although there are some local variations in the magnetic anomaly pattern, we found that all data can be accounted for by the same model and by assuming symmetrical spreading.

As a result of the model study, we found that the spreading has slowed considerably since formation of the youngest crust of anomaly 3' (5.12 m.y. B.P.) before which the half-spreading rate was 56

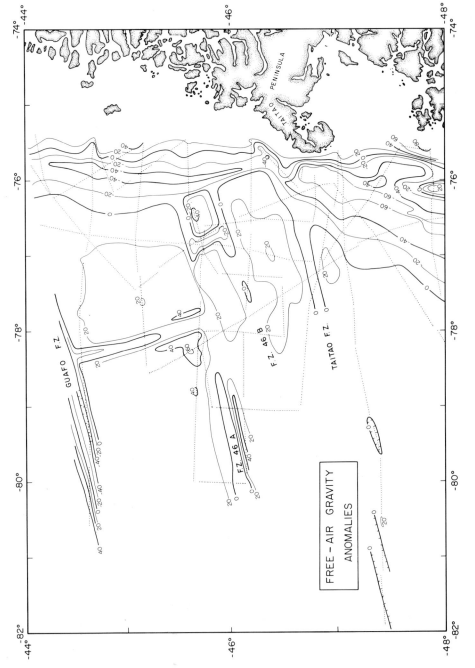

Figure 5. Contoured free-air gravity anomalies in vicinity of triple junction. Contour interval is 20 mgal.

mm/yr. During anomaly-3 time (between 3.5 and 5.12 m.y. B.P.), the half-rate was 40 mm/yr, followed by a half-rate of 30 mm/yr until the beginning of anomaly-2 time (1.84 m.y. B.P.) with a brief interval of faster spreading (37 mm/yr) continuing until the start of the central anomaly at 0.7 m.y. B.P. The half-spreading rate averaged for the central anomaly has been only 28 mm/yr. Because the most recent slowdown may not have been right at the Brunhes/Matuyama boundary (0.7 m.y. B.P.), it is possible that the present spreading rate is even slower than 28 mm/yr. Although in Figure 7 we modeled the spreading-rate changes as occurring in discrete steps at discrete times, owing to uncertainties in the time scale and in the data, the spreading-rate changes may have occurred over a period of time or may even have been continuous.

From the magnetic data (and seismic-profiler records) we can locate the major fracture zones south of the triple junction (see Fig. 8). Two fracture zones, the Taitao Fracture Zone and Fracture Zone 47A, intersect the trench just south of the triple junction. On the basis of the tentative identification of anomaly 4 along profile 6, these two fracture zones together represent a left-lateral offset of 130 km at the ridge crest. If it is assumed that there were no irregularities in spreading, the age of the crust at the trench axis in this area should be about 4.5 m.y. (anomaly 3). Additional fracture zones intersect the trench at lat 47°S (Fracture Zone 47B) and lat 49°S (Fracture Zone 49). The age of the crust at the trench axis south of these fracture zones is 9 m.y. (anomaly 5) and 17 m.y. (anomaly 5D), respectively.

In Figure 9 we present a summary of the sea-floor spreading pattern over the entire Chile Ridge. Figure 9 includes the results of Herron (1972), Klitgord and others (1973), Molnar and others (1975), Handschumacher (1976), and Weissel and others (1977) as well as the results of this study. The locations of the fracture zones in Figure 9 are based both on the magnetic data and the bathymetry of Mammerickx and Smith (1980). In Figure 10 we compared the spreading-rate model based on the data near the triple junction (Fig. 7) to composite projected profiles at various locations along the Chile Ridge. We believe that this model fits these data quite well. The main misalignments in Figure 10 appear to be due to intervals of asymmetric sea-floor spreading along some segments of the Chile Ridge. This is particularly noticeable for the segments between Fracture Zones 39 and 40A.

DISCUSSION

Sea-Floor Spreading near the Trench Axis

The most surprising observation associated with the Chile Margin triple junction is that the presence of the active subduction zone at the Peru-Chile Trench appears to have had very little effect on the sea-floor spreading process near the trench axis. This is shown in a comparison (Fig. 11) of the bathymetry, gravity, and magnetic data across the seaward side of the spreading center at the axis of the trench (line 4b) and across the spreading center at a distance of 200 km from the trench (line 2). It is apparent from Figure 11 that (1) the depth of the crest of the outer wall of the rift valleys is the same to within 50 m (about 2,500 m), (2) the half-width of the rift valleys (from spreading center to the outer wall) is the same to within 1 km (8 km), (3) the spreading rates, as measured by the magnetic anomalies, are nearly identical, and (4) the relative shape of the free-air gravity profiles is nearly the same (although the regional level is 30 mgal lower at the trench). The quality of the individual magnetic anomalies is quite good right up to the trench and triple junction; this indicates that crustal accretion on the seaward side of the spreading center (Antarctic plate) adjacent to the inner wall of the trench is continuing in spite of its proximity to the trench.

The magnetic data also seem to rule out the possibility that the Nazca plate is breaking up into smaller fragments near the triple junction. The uniformity of the magnetic anomaly patterns both in the area of the triple junction (Fig. 8) and north of the Guafo Fracture Zone (Fig. 9) over the past 20 m.y. implies that the Nazca plate is behaving as a single unit. This is contrary to the situation described by Menard (1978) for the Farallon plate as portions of the Pacific-Farallon Ridge were subducted along the western margin of North and Central America. Menard (1978) found that small portions of the Farallon plate separated from the main plate section and rotated, presumably to allow the ridge

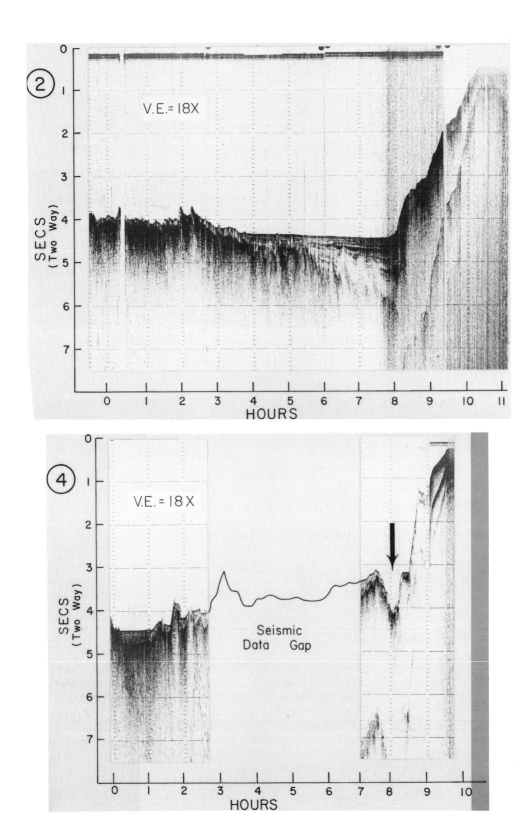

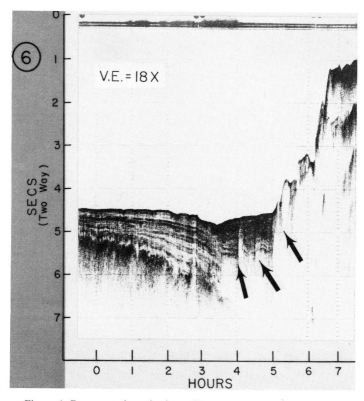

Figure 6. Representative seismic-profile records across the trench in the vicinity of the triple junction. On line 4, the axis of the rift valley is indicated by arrow. On line 6, arrows show location of disturbed sediments.

crest to become more parallel to the trench. He called this process "pivoting subduction." A diagnostic feature of this phenomenon is the existence of a pivot near the ridge-trench intersection with a variation in spreading rates away from the pivot along the ridge and a consequent fanning of the magnetic anomaly pattern. However, no fanning of anomalies is observed along the Chile Ridge. The geometry of the plate interaction may explain why "pivoting subduction" is not observed off Chile. The angle between the spreading center on the Chile Ridge and the Peru-Chile Trench is only about 15° and perhaps is too small to require pivoting. Also, the offsets along the Chile Ridge displace the ridge crest away from the trench in the opposite sense to examples cited by Menard (1978).

The Accretionary Prism

The collision of the ridge crest with the trench leads to the almost complete destruction of the accretionary prism. The accretionary prism observed landward of the rift valley just north of the triple junction (line 4, Fig. 6) is much narrower than the accretionary prism observed farther north (line 2, Fig. 6). Either the sediments which make up the accretionary prism are carried down with the subducting plate and/or they are eroded away and redeposited on the trench floor north and south of the triple junction. After the ridge-trench collision, the accretionary prism slowly rebuilds, as indicated in the general widening of the inner wall of the trench shown in Figure 4. The building of a broad accretionary prism south of the triple junction is attributable to several factors including (1) a high rate of sediment influx, (2) the slow rate of Antarctic–South American convergence, and (3) the fact that the relative age of the crust being subducted is gradually increasing rather than decreasing as it is north of the triple junction.

Interaction with the South American Plate

Previous studies of the margin of southern Chile have shown that about 20 m.y. B.P. this triple junction was located near the western end of the Strait of Magellan at lat 52°S (Herron and others, 1977; Weissel and others, 1977). Structures generated by the passage of the triple junction have not been identified in the continental geology, and the magnetic data between lat 52° and 48°S are still too sparse to allow for a detailed discussion of the history of the triple junction during the past 20 m.y. However, we can give a preliminary estimate of the timing of more recent events near the triple junction if we assume that the estimated convergence rate for the Antarctic and South American plates of 21 mm/yr (Minster and Jordan, 1978) is roughly correct for the past 10 m.y.

If we assume a convergence rate of 21 mm/yr for the Antarctic–South American plates, the present ridge that extends north from the triple junction would have arrived at the trench about 0.3 m.y. B.P. Before that time the triple junction would have been a trench-transform-trench triple junction. The small section of ridge between Fracture Zone 47A and Fracture Zone 47B would have collided with the trench about 2.5 m.y. B.P.; the large section of ridge between Fracture Zone 47B and Fracture Zone 49 would have collided about 6 m.y. B.P. We can also predict that the triple junction should evolve into a trench-transform-trench triple junction in about 0.3 m.y. when Fracture Zone 46B will intersect the

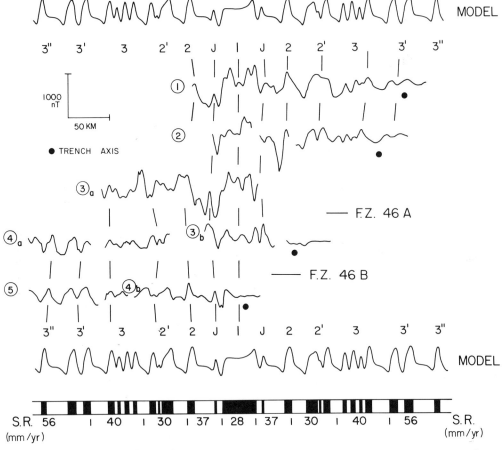

Figure 7. Projected magnetic anomaly profiles aligned with respect to the axial anomaly and compared to a model profile. The half-spreading rates (S.R.) for the model are in millimetres per year. Other model parameters: top of layer = 3 km, bottom of layer = 3.5 km, magnetization = 10 A/m, azimuth = 80°, $I_0 = -44°$, $D_0 = 17°$, $I_r = -63°$, $D_r = 0°$.

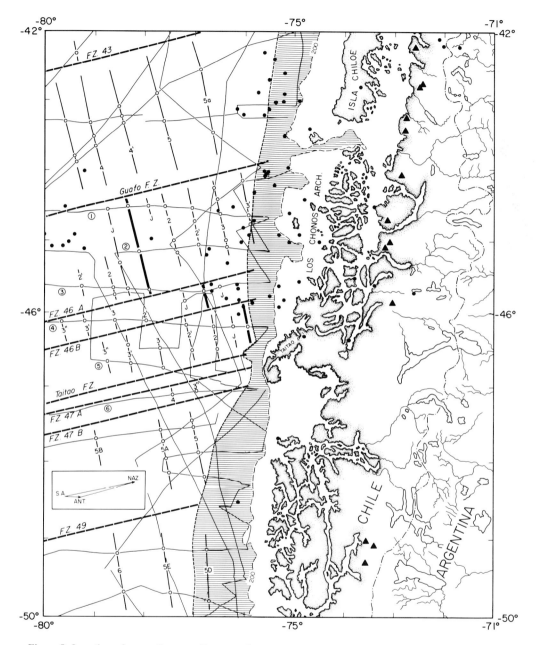

Figure 8. Location of magnetic anomalies, spreading centers, and fracture zones near the Chile Margin triple junction. Filled triangles show location of recent and active volcanoes (from Stern and others, 1976). Filled circles are locations of preliminarily determined earthquake epicenters between 1961 and 1977. Open circles are locations of magnetic anomalies along the trackline. Inner slope is ruled as in Figure 4. Relative-motion vectors for vicinity of triple junction are from Minster and Jordan (1978).

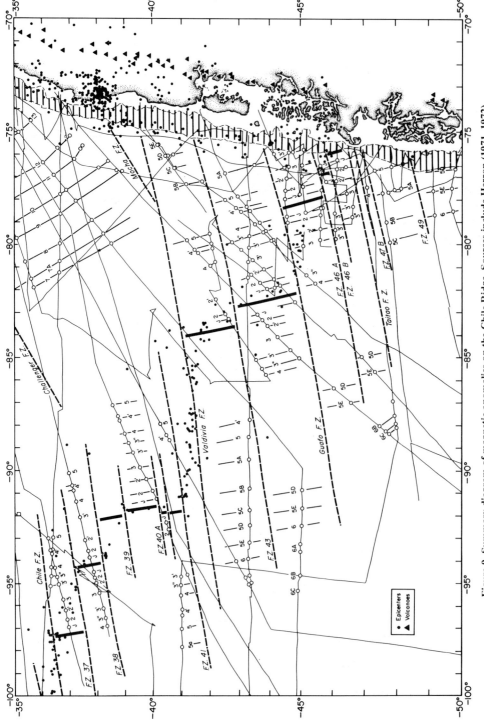

Figure 9. Summary diagram of magnetic anomalies on the Chile Ridge. Sources include Herron (1971, 1972), Klitgord and others (1973), Molnar and others (1975), Handschumacher (1976), Weissel and others (1977), and this study.

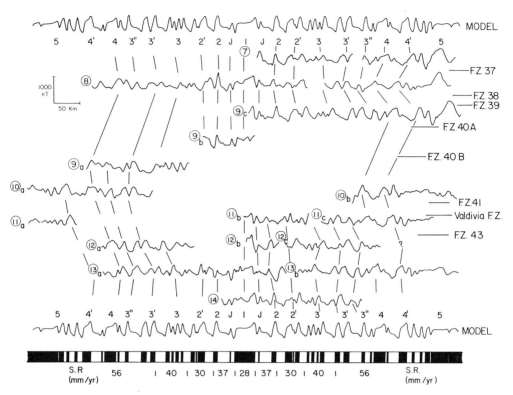

Figure 10. Projected magnetic anomaly profiles between the Guafo Fracture Zone and Chile Fracture Zone aligned with respect to the axial anomaly. For location of tracks, see Figure 1. Model parameters same as Figure 7.

trench. Furthermore, the triple junction should maintain that configuration for about 0.8 m.y., at which time the next section of ridge crest will reach the trench.

Topography

Between lat 42°S and Cape Horn, the Taitao Peninsula opposite the Chile Margin triple junction is the only projection of the mainland toward the Pacific Ocean. Both north and south of this peninsula the margin is composed of islands and fjords. On the southwestern end of the Taitao Peninsula, slightly deformed marine sediments of Tertiary age have been observed by Dalziel and others (1975), evidence for relatively recent uplift of the area. These sediments contain large amounts of coarse andesitic to mafic volcanic detritus that may have been derived from the east (Dalziel and others, 1975). DeLong and Fox (1977) proposed that the approach of an active spreading center to a subduction zone causes gradual shoaling of the continental margin before the actual collision occurs. In their model, the extent of the uplifted region is controlled by the geometry of the ridge system, the spreading rate across the ridge, and the convergence rate across the subduction zone. At the Chile Margin triple junction, the only evidence in support of this model seems to be the Taitao Peninsula. We do not feel that the applicability of this model can be properly tested without additional data which date changes in the elevation of the margin north of the triple junction.

Farther north, the landward extension of the Guafo Fracture Zone coincides with the shelf edge between Isla Chiloe and the Los Chonos Archipelago (see Fig. 8). These topographic features may be controlled by the differences in age and physical properties of the crust in the downgoing Nazca plate on either side of the Guafo Fracture Zone.

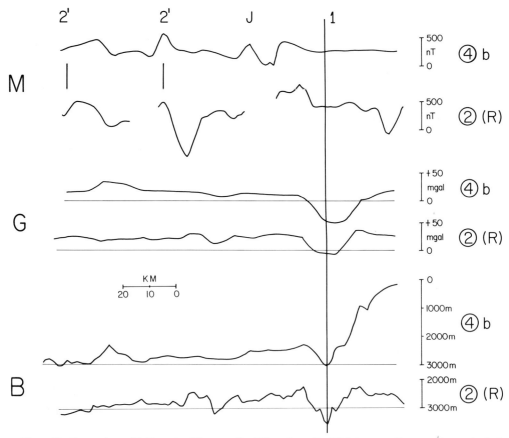

Figure 11. Comparison of bathymetry (B), magnetics (M), and gravity (G) data across the spreading center just north of the triple junction (line 4) and between Fracture Zone 46A and the Guafo Fracture Zone (line 2). Line 2 has been reversed (R). Profiles are projected perpendicular to spreading axis. Heavy vertical line marks axis of rift valley.

Seismicity

A number of workers have studied the seismicity along the continental margin of Chile. Kausel and Lomnitz (1968) observed that south of the triple junction, activity is practically absent. Forsyth (1975) pointed out that the slow convergence rate south of the triple junction was inadequate to explain the distribution of seismicity. He attributed the lack of seismic activity south of lat 46°S to a softening of the subducting slab due to an elevated thermal regime. Stauder (1973) studied the mechanisms and spatial distribution of Chilean earthquakes north of lat 46°S. More recently, Barazangi and Isacks (1976), studying the seismicity of the western margin of South America north of lat 45°S, relocated earthquakes and argued for segmentation of the descending Nazca plate into discrete sections. They indicated that one slab segment extends from lat 33°S to 45°S, just north of the triple junction. Within this area, the depth of the deepest earthquake decreases southward from 170 km to less than 70 km.

Earthquake activity drops off about 60 km north of the triple junction (Fig. 8). There are virtually no earthquakes along the margin adjacent to the actively spreading segment of ridge crest just north of the triple junction, although the convergence rate here is still 90 mm/yr. The lack of seismic activity along this segment of the subduction zone may be related to the elevated thermal regime associated with the ridge crest, as was proposed by Forsyth (1975) for the area south of the triple junction.

Volcanicity

The chain of recent volcanoes in Chile south of lat 33°S has been studied by numerous authors (for example, Pichler and Zeil, 1970; Vergara, 1970; Vergara and Gonzales-Ferran, 1972; Vergara and Munizaga, 1974; Vergara and Drake, 1976; Drake, 1976). These studies tend to concentrate only on the more northern volcanoes. However, as pointed out by Fuenzalida (1974) and Stern and others (1976), the chain continues to lat 46°S, the latitutde of the triple junction. Volcanic activity resumes at about lat 49°S, (see Figs. 8 and 10). The gap in seismicity between lat 46°S and 49°S corresponds to the region between the Taitao Fracture Zone and Fracture Zone 49. As we showed above, this corresponds to the region in which segments of the ridge crest were subducted between 6 and 2.5 m.y. ago. Volcanic activity reappears landward of the zone where the ridge crest was subducted roughly 10 to 15 m.y. B.P. The observed gap in volcanic activity appears to be of slightly shorter duration than the 20- to 30-m.y. gap predicted on theoretical grounds by DeLong and others (1979).

SUMMARY

1. The triple junction between the Nazca, Antarctic, and South American plates is now a trench-ridge-trench triple junction at lat 46.4°S, long 75.7°W.
2. Just north of the triple junction an active spreading center, characterized by a rift valley, abuts the inner wall of the trench.
3. Spreading along the entire length of the Chile Ridge has slowed from a half-rate of 56 mm/yr before 5.12 m.y. B.P. to 28 mm/yr, averaged over the past 700,000 yr. This slowdown has been irregular: spreading slowed to 30 mm/yr 3.5 m.y. ago and then increased to 37 mm/yr between 1.84 and 0.7 m.y. B.P.
4. The sea-floor spreading process along the seaward side of the ridge crest adjacent to the inner wall of the trench is the same as that along the Chile Ridge axis away from the trench.
5. There is no evidence for "pivoting subduction" of the Nazca plate or spreading-rate variation along the Chile Ridge.
6. The only obvious long-term submarine topographic indication of ridge-crest subduction is a doubling in the width of the inner wall of the trench from north to south of the triple junction. This is probably due to the decrease in convergence rates from roughly 90 to 21 mm/yr as estimated by Minster and Jordan (1978). Near the triple junction the inner wall of the trench is extemely narrow.
7. On land, the seaward projection of the Taitao Peninsula reflects the proximity of the triple junction and a recently subducted (2.5 m.y. B.P.) section of ridge crest. The inlet between Isla Chiloe and the Los Chonos Archipelago at lat 44°S lies along the landward extension of the Guafo Fracture Zone and may be controlled by the difference in the age of the crust on either side of the fracture zone.
8. Contemporary seismic activity effectively stops about 60 km north of the triple junction.
9. A gap in recent volcanicity extends from landward of the triple junction to landward of the fracture zone at lat 49°S, corresponding to the section where the ridge crest has been subducted within the past 6 m.y.

ACKNOWLEDGMENTS

This research was supported by National Science Foundation Grant OCE 76 01811. We thank R. N. Anderson and A. B. Watts for their careful and constructive review of this paper.

REFERENCES CITED

Anderson, R. N., Langseth, M. G., and Sclater, J. G., 1977, The mechanism of heat transfer through the floor of the Indian Ocean: Journal of Geophysical Research, v. 82, p. 3391–3409.

Atwater, T., 1970, Implications of plate tectonics for the Cenozoic tectonic evolution of western North America: Geological Society of America Bulletin, v. 81, p. 3513–3536.

Barazangi, M., and Isacks, B. L., 1976, Spatial distribution of earthquakes and subduction of the Nazca plate beneath South America: Geology, v. 4, p. 686–692.

Bird, J. M., and Dewey, J. F., 1970, Lithosphere plate–continental margin tectonics and the evolution of the Appalachian orogen: Geological Society of America Bulletin, v. 81, p. 1031–1060.

Dalziel, I.W.D., deWitt, M. J., and Ridley, W. I., 1975, Structure and petrology of the Scotia arc and the Patagonian Andes: R/V HERO cruise 75-4: Antarctic Journal of the United States, v. 10, p. 307–310.

DeLong, S. E., and Fox, P. J., 1977, Geological consequences of ridge subduction, *in* Talwani, M., and Pitman, W. C., III, eds., Island arcs, deep sea trenches and back-arc basins: American Geophysical Union Maurice Ewing Series, v. 1, p. 221–228.

DeLong, S. E., Fox, P. J., and McDowell, F. W., 1978, Subduction of the Kula Ridge at the Aleutian Trench: Geological Society of America Bulletin, v. 89, p. 83–95.

DeLong, S. E., Schwarz, W. M., and Anderson, R. N., 1979, Thermal effects of ridge subduction: Earth and Planetary Science Letters, v. 44, p. 239–246.

Dewey, J. F., 1976, Ophiolite obduction: Tectonophysics, v. 31, p. 93–120.

Drake, R. E., 1976, Chronology of Cenozoic igneous and tectonic events in the central Chilean Andes—Latitudes 35°30′ to 36°S: Journal of Volcanology and Geothermal Research, v. 1, p. 265–284.

Forsyth, D. W., 1975, Fault plane solutions and tectonics of the South Atlantic and Scotia Sea: Journal of Geophysical Research, v. 80, p. 1429–1443.

Fuenzalida, R., 1974, The Hudson volcano, *in* Proceedings, Symposium on Andean and Antarctic Volcanology Problems: International Association of Volcanology and Chemistry of the Earth's Interior, Santiago, p. 1–10.

Grow, J. A., and Atwater, T., 1970, Mid-Tertiary tectonic transition in the Aleutian arc: Geological Society of America Bulletin, v. 81, p. 3715–3722.

Handschumacher, D. W., 1976, Post-Eocene plate tectonics of the eastern Pacific, *in* Sutton, G. D., and others, eds., The geophysics of the Pacific Ocean basin and its margin: American Geophysical Union Geophysical Monograph 19, p. 117–202.

Heezen, B. C., 1968, Physiographic chart of the Atlantic Ocean: National Geographic Magazine, v. 133, no. 6.

Heirtzler, J. K., and others, 1968, Marine magnetic anomalies, geomagnetic field reversals, and motions of the ocean floor and continents: Journal of Geophysical Research, v. 73, p. 2119–2136.

Herron, E. M., 1971, Crustal plates and sea-floor spreading in the southeastern Pacific: Antarctic Oceanography I, Antarctic Research Series, v. 15, p. 229–237.

———1972, Sea-floor spreading and the Cenozoic history of the east-central Pacific: Geological Society of America Bulletin, v. 83, p. 1671–1692.

Herron, E. M., and Hayes, D. E., 1969, A geophysical study of the Chile Ridge: Earth and Planetary Science Letters, v. 6, p. 77–83.

Herron, E. M., and others, 1977, Post Miocene tectonics of the margin of southern Chile, *in* Talwani, M., and Pitman, W. C., III, eds., Island arcs, deep sea trenches and back-arc basins: American Geophysical Union Maurice Ewing Series, v. 1, p. 273–283.

Isacks, B., Oliver, J., and Sykes, L., 1968, Seismology and the new global tectonics: Journal of Geophysical Research, v. 73, p. 5855–5900.

Kausel, E., and Lomnitz, C., 1968, Tectonics of Chile: Mexico City, Universidad National Autónoma de Mexico, Instituto de Geofísica, Pan-American Symposium on the Upper Mantle, v. 2, p. 47–67.

Klitgord, K. D., and others, 1973, Fast sea-floor spreading on the Chile Ridge: Earth and Planetary Science Letters, v. 20, p. 93–99.

LaBrecque, J. L., Kent, D. V., and Cande, S. C., 1977, Revised magnetic polarity time scale for Late Cretaceous and Cenozoic time: Geology, v. 5, p. 330–335.

Le Pichon, X., 1968, Sea-floor spreading and continental drift: Journal of Geophysical Research, v. 73, p. 3661–3698.

Lister, C.R.B., 1972, On the thermal balance of a midocean ridge: Royal Astronomical Society Geophysical Journal, v. 26, p. 465–509.

Mammerickx, J., and Smith, S. M., 1980, Bathymetry of the Southeast Pacific: Geological Society of America Map and Chart Series, MC-26, scale 1:6,442,194.

Mammerickx, J., and others, 1974, Bathymetry of the South Pacific: Scripps Institution of Oceanography IMR Technical Report 53A, Chart 20, 1st edition.

Menard, H. W., 1978, Fragmentation of the Farallon plate by pivoting subduction: Journal of Geology, v. 86, p. 99–110.

Minster, J. B., and Jordan, T. H., 1978, Present-day plate motions: Journal of Geophysical Research, v. 83, p. 5331–5354.

Molnar, P., and others, 1975, Magnetic anomalies, bathymetry, and the tectonic evolution of the South Pacific since the Late Cretaceous: Royal Astronomical Society Geophysical Journal, v. 40, p. 383–420.

Morgan, W. J., Vogt, P. R., and Falls, D. F., 1969, Magnetic anomalies and sea-floor spreading on the Chile Rise: Nature, v. 222, p. 137.

Pichler, H., and Zeil, W., 1970, Chilean "andesites"—Crustal or mantle derivation?, in Symposium on the results of upper mantle investigations with emphasis on Latin America: Buenos Aires, International Upper Mantle Project, Science Report no. 37, v. 2, p. 361–370.

Pitman, W. C., III, and Hayes, D. E., 1968, Sea-floor spreading in the Gulf of Alaska: Journal of Geophysical Research, v. 73, p. 6571–6580.

Stauder, W., 1973, Mechanism and spatial distribution on Chilean earthquakes with relation to subduction of the oceanic plate: Journal of Geophysical Research, v. 78, p. 5033–5061.

Stern, C., Skewes, M. A., and Duran, M., 1976, Volcanismo orogénico en Chile austral: Primer Congreso Geológico Chileno, p. F195–F212.

Uyeda, S., and Miyashiro, A., 1974, Plate tectonics and the Japanese Islands: A synthesis: Geological Society of America Bulletin, v. 85, p. 1159–1170.

Vergara, M., 1970, Note on the zonation of the upper Cenozoic volcanism of the Andean area of central-south Chile and Argentina, in Symposium on the results of upper mantle investigations with emphasis on Latin America: Buenos Aires, International Upper Mantle Project, Science Report no. 37, v. 2, p. 381–397.

Vergara, M., and Munizaga, F., 1974, Age and evolution of the upper Cenozoic andesitic volcanism in central-south Chile: Geological Society of America Bulletin, v. 85, p. 603–606.

Vergara, M., and Drake, R., 1976, Evidencias de periodicidad en el volcanismo Cenozoico de los Andes centrales: Primer Congresso Geológico Chileno, p. F153–F161.

Vergara, M., and Gonzalez-Ferran, O., 1972, Structural and petrological characteristics of the late Cenozoic volcanism from Chilean Andean region and west Antarctica: Krystalinikum, v. 9, p. 157–184.

Weissel, J. K., Hayes, D. E., and Herron, E. M., 1977, Plate tectonic synthesis: The displacements between Australia, New Zealand and Antarctica since the Late Cretaceous: Marine Geology, v. 25, p. 231–277.

MANUSCRIPT RECEIVED BY THE SOCIETY NOVEMBER 12, 1980
MANUSCRIPT ACCEPTED DECEMBER 30, 1980
L-DGO CONTRIBUTION NO. 2950

Geological Society of America
Memoir 154
1981

Structures of the continental margin of Peru and Chile

RICHARD COUCH
School of Oceanography
Oregon State University
Corvallis, Oregon 97330

ROBERT WHITSETT
Department of Physics and Computer Science
Pacific Union College
Angwin, California 94508

BRUCE HUEHN
Amoco Production Company
Security Life Building
Denver, Colorado 80202

LUIS BRICENO-GUARUPE
Universidad Nacional de Colombia
Departamento de Geociencias
Ciudad Universitaria, Bogota, Colombia, S.A.

ABSTRACT

Model crustal cross sections contrained by gravity and seismic refraction data indicate that pre-Cenozoic continental rocks extend to within 50 km of the trench off Mollendo and 20 km off Pisco in southern Peru and to within 30 km of the trench off northern and central Chile. The oceanic plate dips beneath the continental margin at approximately 5.1° west of Pisco, 3.8° west of Mollendo, 3.8° off northern Chile, and 4.1° off central Chile. The dip increases in the vicinity of the coast and coastal mountains in all sections. The oceanic crust decreases in thickness near the outer gravity high seaward of the trench. Magnetic anomaly 12 is identified beneath the continental slope of central Chile, and anomaly 24 is beneath the margin of northern Chile.

The presence of the marine magnetic anomalies beneath the margin, the shallow dip of the oceanic plate, and the continuity of the model structures suggest that the subduction process is relatively smooth in central and northern Chile and off Mollendo south of the Nazca Ridge. However, the computed model of the continental margin near Pisco north of the ridge indicates structural complications in the lower continental crust and/or uppermost mantle between the coast and the west flank of the Andes. The complications are likely associated with ongoing crustal deformation in the vicinity of the zone of contact between the subducting and overriding plates.

Free-air gravity anomalies and bathymetry suggest that the continental margin of Chile is segmented at intervals of approximately 280 km. However, only small structural differences are noted along the margin. Based on sediment thickness on the Nazca plate, plate-closure rates, and the interpretation of continental-margin low-density rocks as post-Paleozoic sediments, calculations indicate that the sediment now on the margin is less than 30% of the sediments transported to the margin during post-Paleozoic time. The decrease in volume of the oceanic sediments caused by compaction and dewatering (indicated by the difference in densities of the sediments) is at least partially offset by the addition of terrigenous sediments. These observations indicate that not all of the oceanic and terrigenous sediments have been accreted to the margin and that consumption of the material must occur, at least on occasion, during the subduction of the Nazca plate.

INTRODUCTION

Convergence of the Nazca and South American plates causes major and continuing tectonism along the western continental margin of South America. Over time, the tectonism has produced the dominant morphologic features along the margin, the steep topographic gradient between the Peru-Chile Trench and the Andes Mountains that often exceeds a 12-km change in elevation over a horizontal distance of 250 km. Although the rate of convergence between the plates changes only gradually with latitude and the ages of the two plates along the join vary only a relatively small amount, marked along-trench variations in trench depth, trench curvature, slope gradient, coastal geology, elevation of the Andes Mountains, volcanism, and depths to earthquake hypocenters indicate the subduction process is indeed variable and complex along western South America. This paper presents six crustal cross sections of the continental margin of western South America, based on gravity and seismic refraction measurements. Figure 1 shows the location of the six model sections. they are referenced to a common mass column, thus, they allow a comparison along the margin of the volume of low-density material on the continental shelf and slope, the dip of the oceanic crust beneath the margin, and the large-scale structure of the margin.

CRUSTAL AND SUBCRUSTAL STRUCTURE

In general, model cross sections based on gravity measurements outline two-dimensional mass distributions for the crust and upper mantle that will produce the gravity anomalie observed along the section normal to known or postulated structures of interest. In addition to the bathymetric or topographic, seismic, and magnetic data that help constrain the model cross sections, the amplitude and wavelength of the gravity anomalies together with bounds on model density values (prescribed by geology) limit the source depths of the anomalies. In the subduction zone of South America, the interpretation of earthquake hypocenters as indicative of a descending lithospheric plate suggests that variations in mass distribution extend to depths of several hundred kilometres; however, the amplitude and wavelength of the gravity anomalies caused by these mass distributions are, at present, not resolvable. The gravity data do not "require" variations in mass at those depths. Hence, an attempt was made in the model cross sections presented in this paper to limit their complexity to that which the data require.

Figure 2 shows the location of the Pisco, and Mollendo cross sections; they are oriented approximately normal to the continental margin of southern Peru and approximately parallel to and on opposite sides of the Nazca Ridge. Figure 2 also shows the location of seven seismic refraction lines. The Scripps Expedition Downwind, 1957–1958, yielded two-ship refraction lines DW22 (Shor and others, 1970) and DW23 (Fisher and Raitt, 1962). the Nazca Plate Project, conducted jointly by the Hawaii Institute of Geophysics and Oregon State University, yielded two-ship refraction lines 1–2, 2–3 and 3'–4 (D. M. Hussong and S. H. Johnson, 1975, personal commun.). Land refraction data include a reversed line from the Carnegie Institute Peru-Bolivia Altiplano Refraction of 1968 (Ocola and others,

1971) and the unreversed line 890 (McConnell and McTaggart-Cowan, 1963) from Seismic Crustal Studies during the International Geophysical Year (Woollard, 1960). Seismic reflection measurements obtained concurrently with the gravity measurements provide information on water depths and sediment thicknesses in the abyssal ocean and near-surface structure on the continental shelf. Land gravity data compiled by the Instituto de Geofísica del Perú (data tape obtained from the Defense Mapping Agency Aerospace Center, St. Louis, Missouri) provide gravity and elevation information ashore. The empirical relations between seismic velocity and density (Ludwig and others, 1970) and the mapped surface geology guided the selection of model densities. The crustal sections

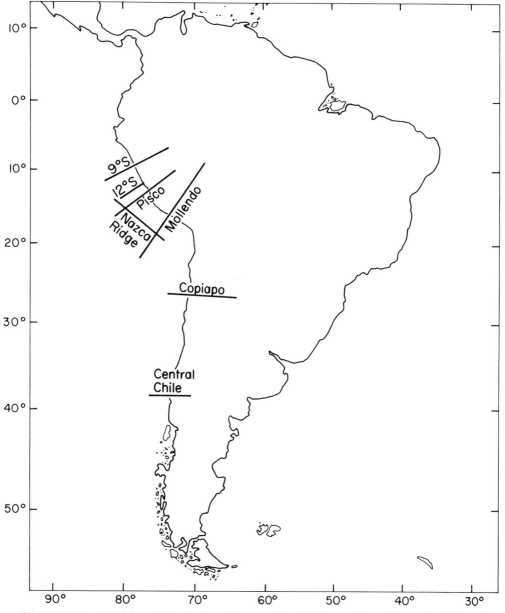

Figure 1. Index map of six crustal and subcrustal cross sections that transect the western continental margin of South America.

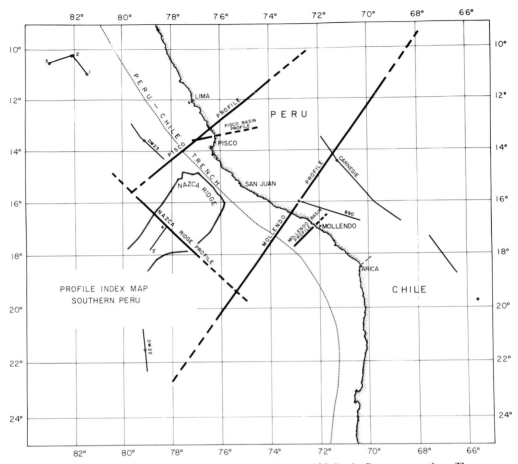

Figure 2. Index map showing the location of the Pisco, Peru, and Mollendo, Peru, cross sections. The narrow lines show the location of seismic refraction lines that constrain the model sections. The Nazca Ridge, Pisco Basin, and Mollendo Basin sections are shown in Couch and Whitsett (this volume).

assume two-dimensional structure, a standard mass column of 70 km and 9,255 mgal that corresponds to a zero free-air gravity anomaly (Barday, 1974; Whitsett, 1975), and no lateral variations in density below 70-km depth. Iterative adjustment of layer boundaries, constrained by water depth, land elevation, abyssal sediment thickness and seismic refraction data, made until computed gravity agreed with the observed free-air gravity anomalies, yielded the Pisco and Mollendo, Peru, crustal and subcrustal cross sections. Couch and Whitsett (this volume) describe the Nazca Ridge crustal and subcrustal cross section that is consistent with and that intersects the westernmost ends of the Pisco and Mollendo cross sections.

Pisco Cross Section

Figure 2 shows the trace of the Pisco cross section and the location of seismic refraction line DW23 used to constrain the model section. The dashed lines show extensions of the sections outside the area of gravity control. The area includes the northwest flank of the Andes east of Huancayo, Peru, and the area of intersection of the Pisco and Nazca Ridge sections. Couch and Whitsett (this volume) show the gravity and seismic reflection lines and describe the free-air gravity anomalies, and Johnson and Ness (this volume) decribe the bathymetry and shallow structure in the vicinity of the Pisco section.

The Pisco and Nazca Ridge sections have equivalent mass columns where they intersect; however,

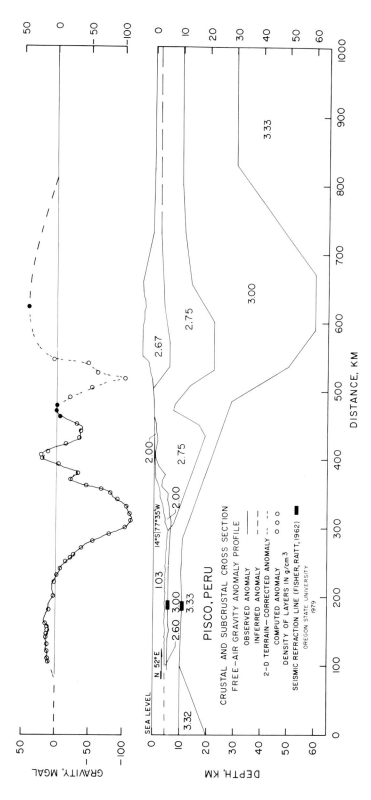

Figure 3. Pisco, Peru, crustal and subcrustal cross section. The vertical exaggeration is 4:1.

the two upper-crustal layers and the two lower-crustal layers of the Nazca Ridge section described by Whitsett, 1975) have been combined to yield single upper- and lower-crustal layers, respectively, in the Pisco section.

No land refraction data are available for the Pisco section; hence, densities and approximate layer thicknesses were extrapolated from the Carnegie data (Fig. 2) with a reduction of the crustal root beneath the Andes from 67 to 60 km, as suggested by the analysis of Rayleigh and Love waves by James (1971). Land gravity stations were occupied along a road that extends from the coast northeastward to approximately long. 76°W in the Andes. Because these stations are located in a canyon, simple terrain corrections were made to the data to more accurately reflect the anomalies over the west flank of the Andes (Whitsett, 1975).

Figure 3 shows the Pisco, Peru, crustal and subcrustal cross section. The section shows a two-layer oceanic crust, approximately 5 km thick, apparently dipping beneath the continental slope and shelf. The continental slope and shelf are formed of two blocks with a relatively high density contrast between them. The upper block is interpreted as sedimentary rock of $2.00\text{-g}/cm^3$ average density overlying a block of crystalline rock of $2.75\text{-g}/cm^3$ average density. This implies that crystalline rock, similar to the rock that crops out just east of the coast, extends seaward to the toe of the continental slope.

The interpretation of a block of $3.0\text{-g}/cm^3$ lower-crustal material intruding the $2.75\text{-g}/cm^3$ material under the coast (center of Fig. 3 at coordinate 500 km) results from the following combination of geologic and gravity constraints: (1) The $2.75\text{-g}/cm^3$ rock, at shallow depths beneath the continental shelf, extends seaward from the coast and is observed on seismic reflection records south of the Pisco section (Couch and Whitsett, this volume). (2) Shoaling of the $2.75\text{-g}/cm^3$ basement rock under the shelf (at coordinates 370 and 400 km, Fig. 3) requires the $3.00\text{-g}/cm^3$ material to be deeper under the shelf to fit the observed gravity. (3) Because the $2.75\text{-g}/cm^3$ rocks extend to the surface east of the coast and lie at a relatively shallow depth west of the coast (at coordinate 400 km, Fig. 3), the free-air anomaly requires a high-density block ($3.0 \text{ g}/cm^3$ in this model) at a relatively shallow depth beneath the coastal gravity high (at coordinate 475 km, Fig. 3). The amplitude of the high-density structure would be less if $3.3\text{-g}/cm^3$ mantle or similar rock were included in the upbowed portion of the model. However, the anomaly wavelength and the rock densities limit the depth of the anomaly source to lower-crustal depths.

The approximately 60-km depth of the Moho (which agrees with the results of James, 1971) and the structure under the Andes are constrained by only one gravity measurement in the Andes, the topography, reasonable crustal densities, and the standard mass column.

Mollendo Cross Section

The Mollendo, Peru, crustal and subcrustal cross section, shown in Figure 4, transects the continental margin of southern Peru along the Mollendo profile shown in Figure 2. Refraction control from DW22 is projected normal to the profile onto its southwest extension, refraction line 890 is projected orthogonally onto the profile from its midpoint, and the Carnegie line intersects the profile on the Altiplano near coordinate 1,185 km in Figure 4.

Couch and Whitsett (this volume) show the location of the ship track along which the marine gravity measurements and the terrestrial stations in Figure 4 were made; they describe the free-air gravity anomalies near the Mollendo crustal section. Land gravity data were taken along roads that cross the trace of the section near the coast, on the southwest side of the Andes at coordinate 965 km, and on the Altiplano between Lake Titicaca and Cuzco near coordinate 1,200 km. Gravity data were also taken along a road near coordinate 1,090 km and were projected onto the section at that coordinate. Two isolated stations in the Amazon Basin are close to the profile and were projected onto it at coordinates 1,320 and 1,560 km.

The Mollendo crustal section in Figure 4 shows oceanic crust 5 km thick, excluding water, approximately 500 km west of the coast; this crust thins to about 3 km near the Peru-Chile Trench. The Moho dips beneath the continental margin at a shallow angle from a depth of 8.5 km just seaward of

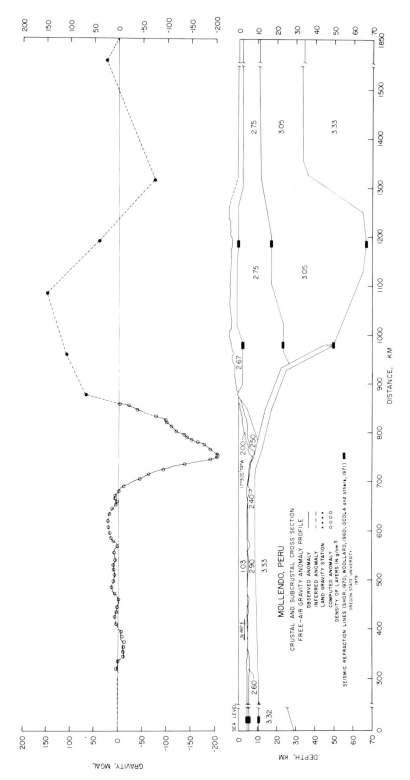

Figure 4. Mollendo, Peru, crustal and subcrustal cross section. the vertical exaggeration is 4:1.

the trench to about 18 km near the coast. Seismic refraction measurements show that the Moho is nearly 60 km below sea level about 100 km inland and that it dips steeply beneath the west flank of the Andes and reaches its greatest depth of more than 70 km at coordinate 1,190 km. A negative anomaly extends along the east flank of the Andes from at least lat 10° to 25°S (Grow and Bowin, 1975). The gravity minimum of −75 mgal on the lowlands near the Andes is probably due to the thick, relatively low density crust beneath the Andes.

Free-Air Gravity Anomalies on the Continental Margin of Northern Chile

Figure 5 shows the location of surface-ship, submarine and terrestrial gravity measurements on the continental margin of northern Chile. The data include measurements made by Oregon State University in 1972 and 1974 and by the National Oceanic and Atmospheric Administration in 1973, submarine pendulum stations reported by Worzel (1965), and marine and terrestrial data (to 1976) obtained from the Defense Mapping and Aerospace Center, St. Louis, Missouri. The root-mean-square uncertainty in measurements, estimated from a small number of differences at trackline crossings, is approximately 5 mgal.

Figure 6 shows a free-air gravity anomaly map of the continental margin of northern Chile between lat 23° and 28°30′S. The large negative gravity anomaly associated with the trench dominates the region and exhibits the maximum gradient over the seaward wall of the trench. Contiguous gravity minima of less than −200, −240, and −250 mgal occur along and slightly landward of the bathymetric axis of the trench. A series of gravity highs with values over +70 mgal occur along the coast. The gravity minima along the trench and the coastal gravity highs indicate large structural changes and strongly suggest that the continental margin is segmented along its length. Segment boundaries— approximately normal to the trench and coastline—are evident near lat 25° and 27°30′S; another is suggested near lat 22°30′S. The segments are about 280 km long.

Figure 7 shows a bathymetric map of the continental margin of northern Chile between lat 25°30′ and 28°30′S. South of lat 27°30′S, trench depths are less than 6,200; north of lat 27°30′S, trench depths exceed 7,400. Ridges and troughs occur on the seaward wall of the trench north of lat 27°30′S but are not apparent in Figure 7 south of that latitude. Hence, the morphology of the trench changes coincidentally with the postulated segment boundary at lat 27°30′S.

Figure 7 shows the trace of the Copiapó crustal and subcrustal cross section of northern Chile. The section, oriented S88°E, intersects the coast near lat 26°47′S and extends from the abyssal sea floor west of the trench to the Altiplano east of the Andes. Figure 8 shows the Copiapó cross section. The section assumes the same mass column and was constructed in the same manner as the Pisco and Mollendo sections. No seismic refraction data were available for the section; however, the results reported by Fisher and Raitt (1962) for the margin near Antofagasta, Chile, provided a guide to the selection of layer thicknesses and densities at sea. Rock types indicated by geologic maps of Chile (for example, Zeil, 1964; Kausel and Lomnitz, 1969) suggested the densities of the near-surface coastal rocks. Bouguer gravity anomalies on the continent of South America were obtained on tapes from the Defense Mapping Agency Aerospace Center in St. Louis, Missouri.

The Copiapó cross section indicates that the thickness of oceanic crustal rock west of northern Chile is less than 4 km, and the bathymetry suggests that the oceanic crust is fractured seaward of the trench. A thick section of relatively low density rock (2.50 g/cm^3) forms the continental slope, and the rock of the continental block extends westward to within about 50 km of the trench. The limited gravity data suggest a crustal thickness of approximately 50 km beneath the Andes of northern Chile.

Figure 8 shows two interpretations of the observed marine magnetic anomalies. In the left center of the figure, the magnetic anomalies are interpreted as remanent sea-floor anomalies in the 2.60- and 2.90-g/cm^3 crustal blocks; as such, anomalies 23A and 24 are identified beneath the continental slope almost to the coast. In the right center of Figure 8, the magnetic anomalies are attributed to the induced magnetization of the continental slope and shelf rocks; this interpretation requires magnetization of material contiguous with the oceanic crust similar to the other interpretation.

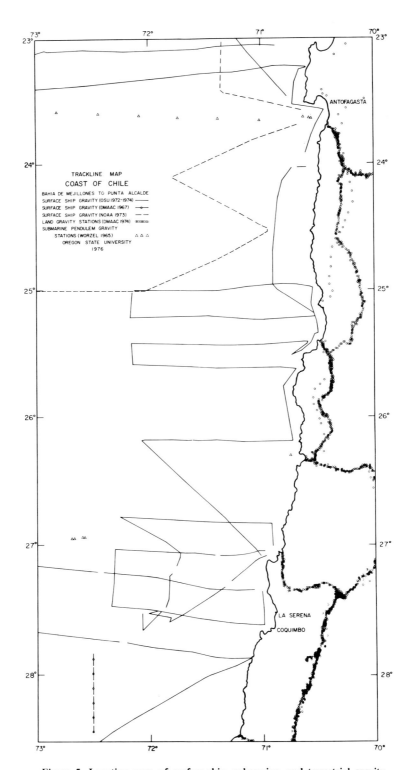

Figure 5. Location map of surface-ship, submarine, and terrestrial gravity measurements on the continental margin of northern Chile.

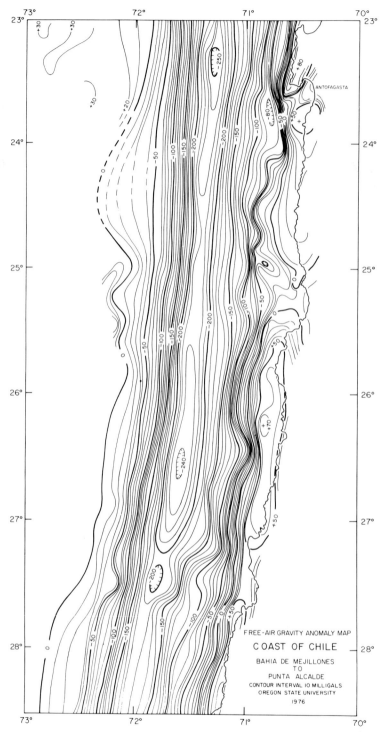

Figure 6. Free-air gravity anomaly map of the continental margin of northern Chile. The contour interval is 10 mgal, and the estimated root-mean-square uncertainty is 5 mgal.

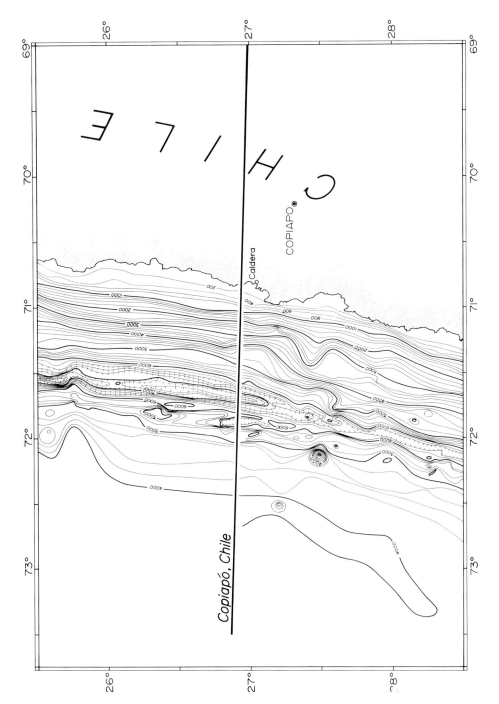

Figure 7. Bathymetric map of the Peru-Chile Trench between lat 25° 30′ and 28° 30′S and location map of the Copiapó, Chile, cross section shown in Figure 8. The contours are corrected metres at 200-m intervals (after Prince and others, 1980).

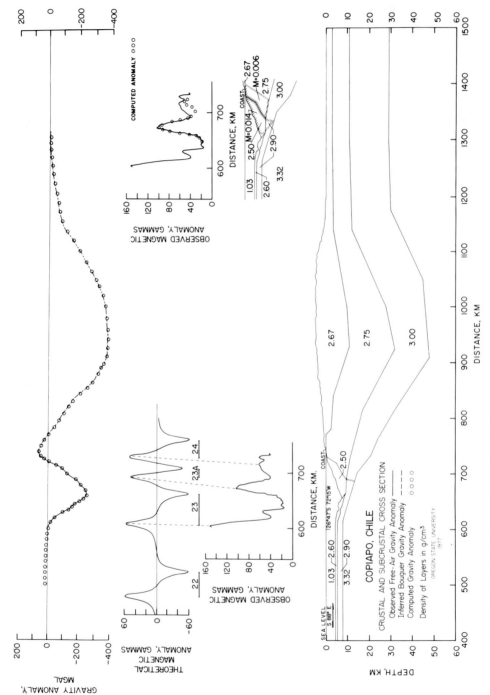

Figure 8. Copiapó, Chile, crustal and subcrustal cross section and two possible interpretations of the observed magnetic anomalies. The vertical exaggeration is 4:1.

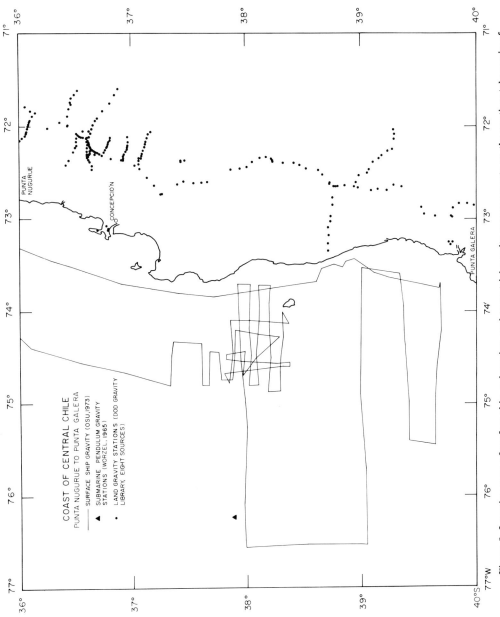

Figure 9. Location map of surface-ship, submarine, and terrestrial gravity measurements on the continental margin of central Chile.

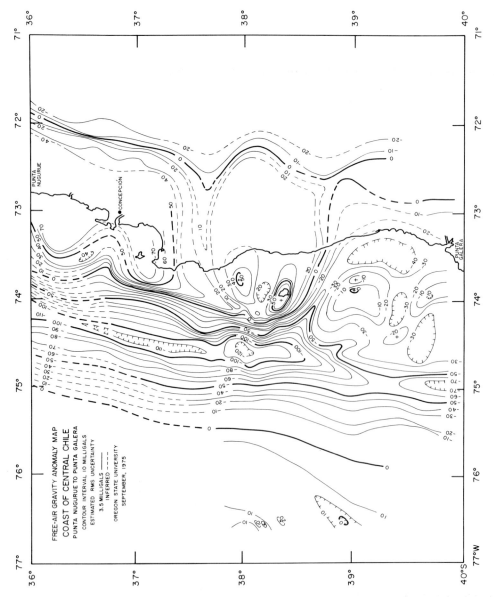

Figure 10. Free-air gravity anomaly map of the continental margin of central Chile. The contour interval is 10 mgal, and the estimated root-mean-square uncertainty is 5 mgal.

Although both satisfy the data, the interpretation that assumes tht remanent magnetization causes the observed anomalies is thought to be the more appropriate.

Cross Sections of the Continental Margin of Central Chile

Figure 9 shows surface-ship gravity tracklines and submarine stations west of central Chile and terrestrial stations east of the coast between Punta Nugurue and Punta Galera, Chile. Figure 10 shows a free-air gravity anomaly map of the continental margin of central Chile between lat 36° and 40°S based on the above measurements. Differences at line crossings yield an estimated root-mean-square uncertainty of 3.5 mgal for the measurements.

Negative anamalies of less than −110 mgal associated with the Peru-Chile Trench dominate the free-air gravity anomaly map of central Chile. Free-air anomalies on the continental shelf are low in amplitude, closed, and generally negative south of approximately lat 38.5°S, whereas to the north the anomalies are larger and predominantly positive. The character of the anomalies associated with the trench and slope also change markedly near lat 38.5°S. The gravity anomalies suggest a segment boundary across the continental margin of central Chile near lat 38.5°S and possibly another approximately 280 kn to the north near lat 36°S.

Line A–A' in Figure 11 shows the N90°E transect of the central Chile crustal and subcrustal cross section across the continental margin of Chile, south of Concepción, near lat 38°S. The Peru-Chile Trench at the location of the central Chile section is approximately 4,700 m deep and contains a relatively large thickness of flat-lying sediments. South of lat 38.5°S the trench is narrower and less deep, and the continental slope is less steep.

Figure 12 shows the central Chile cross section. It is referenced to the same mass column as the sections described above and is constrained by marine gravity measurements west of the coast and a few terrestrial stations east of the coast obtained on tape from the Defense Mapping Agency Aerospace Center. The section shows oceanic crust approximately 5 km thick in the abyssal ocean west of central Chile. The crust appears to thin just seaward of the trench and then dip at a shallow angle beneath the continental slope. The crust beneath the Andes of central Chile, as indicated by the section, is about 40 km thick. Relatively low density rock (about 2.30 g/cm^3) forms the continental slope and shelf and extends landward of the coast. Short-wavelength gravity anomalies on the continental shelf and slope are attributed to irregularities in the surface of the continental rocks (2.75 g/cm^3) beneath the less-dense slope and shelf deposits. Except near the toe of the slope, the shelf and slope deposits are less than 4 km thick.

The remanent marine magnetic anomalies west of the coast compare extremely well with theoretical anomalies computed for the latitutde of the section; this indicates that anomaly 12 exists beneath the continental slope of central Chile. The amplitudes and character of the observed remanent magnetization suggest that the dip of the underthrusting oceanic crust is shallow, consistent with the structure indicated by the model cross section (Fig. 12), and that the remanent magnetization is little affected by the underthrusting process along the continental margin.

Cross Sections of Northern Peru

Figure 13 shows a cross section of the continental margin of Peru at lat 12°S. The section, prepared by V. Goebel (1974, unpub.), is constrained by marine gravity measurements and referenced to the same mass column as the other sections described above. In addition, sonobuoy refraction measurements (heavy lines in Fig. 13) constrain the depths to the layers and provide a guide to the rock densities via the empirical relations (Ludwig and others, 1970) between seismic velocities and densities.

The cross section of the Peru-Chile Trench at lat 12°S includes a basal layer of oceanic crust with a density of 3.22 g/cm^3 that was detected by the sonobuoy refraction measurements. An offset in the basal layer just seaward of the trench (at coordinate 150 km, Fig. 13) suggests faulting of the oceanic crust just before it is subducted under the continental slope. Several layers of relatively low density sedimentary rock form the toe of the continental slope between the trench and what appears to be

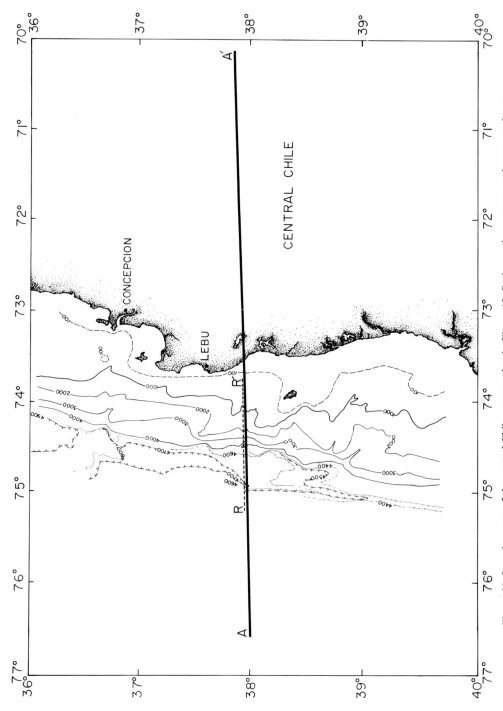

Figure 11. Location map of the central Chile cross section in Figure 12. Bathymetric contours are in corrected meters.

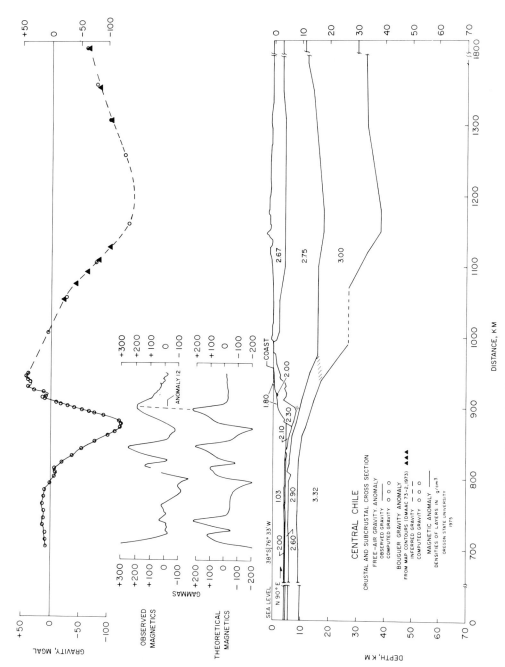

Figure 12. Central Chile crustal and subcrustal cross section. The vertical exaggeration is 4:1.

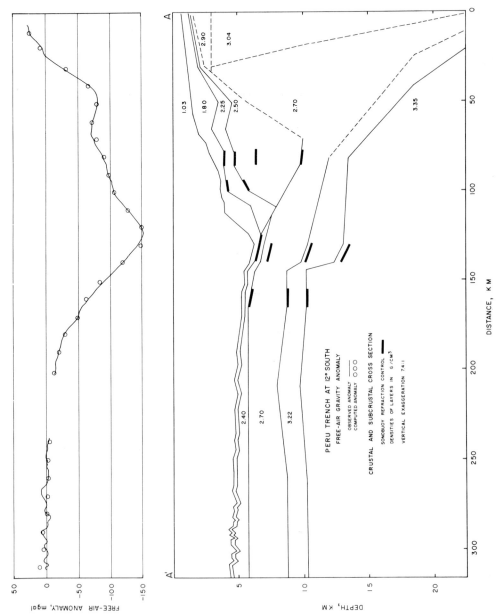

Figure 13. Crustal and subcrustal cross section of the Peru-Chile Trench at lat 12°S. The vertical exaggeration is 7.4:1. (From V. Goebel, 1974, unpub.)

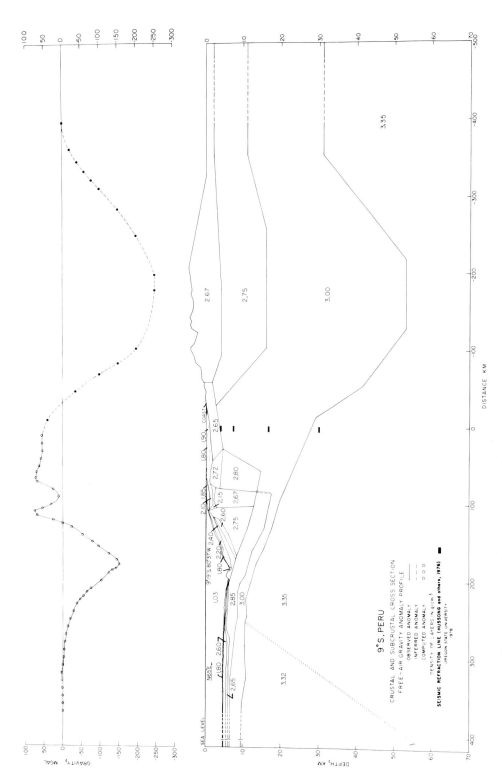

Figure 14. Crustal and subcrustal cross section of the continental margin of Peru at lat 9° S. The vertical exaggeration is 5:1. (From Jones, this volume.)

high-density rock of the continental block. This section, like the Pisco section, requires two blocks of high-density material—2.9 g/cm³ on top of 3.04 g/cm³—in the upper crust to account for the observed gravity anomalies near and landward of the coast (extreme right, Fig. 13).

Figure 14 shows a crustal and subcrustal section through the continental margin of Peru at lat 9°S. The section (computed and described by Jones, this volume) is well constrained by seismic refraction measurements and is referenced to the same mass column as the previously described crustal sections. This section also shows 3.00-g/cm³ rock at shallow depths in the vicinity of and seaward of the coast. Hussong and others (1976) showed seismic refration velocities of 6.0 km/s at depths of about 10 km in the vicinity of the high-density block in the lat 9°S section.

CHARACTERISTICS OF THE CONTINENTAL MARGIN OF SOUTH AMERICA

North of the intersection of the Nazca Ridge and the continental margin of Peru, gravity measurements between the coast and the Andes indicate that relatively dense rocks exist in the crust of the continent. The Pisco crustal section (Fig. 3) attributes the increase in mass west of the Andes as an upbowing of lower-crustal 3.00-g/cm³ material. Alternatively, the dense shallow structure could include part of the upper mantle emplaced by fracturing and thrusting of the subducting lithosphere. Another explanation might include diapiric upwelling of lower-crustal rock during formation of a batholith.

The structure indicated by the shallow, high-density rocks shown in the sections at approximately lat 9°, 12°, and 14°S (Figs. 14, 13, and 3, respectively) apparently widens and shoals toward the north. The structure seems to be centered near the coast at lat 14°S and is predominantly offshore at lat 9°S. The general trend of the structure is approximately 10° to 15° counterclockwise from the trend of the coast and the mapped geologic structures that parallel the coast. Masias (1976) and Kulm and Schweller (1977) described the occurrence of Precambrian to Paleozoic metamorphic strata (for example, gneiss, schist, phyllites, and quartzites) along the coast and on the continental shelf of Peru. The position of these metamorphic rocks (as indicated by samples from islands and from bore holes; Masias, 1976) indicates that they may be juxtaposed to, or overlie and/or form part of, the west flank of the high-density structure. Although the velocities and densities deduced from the seismic refraction and gravity measurements are compatible with schistose crustal rock, they are more likely indicative of a gabbroic or other mafic rock complex that forms a major part of the crystalline continental block of western Peru.

Because of the relatively high density contrast between the crystalline rocks of the continental block and the rocks that form the continental slope and upper continental shelf, models based on gravity measurements provide a good estimate of the volume of sedimentary rock on the continental margin. Table 1 lists the area of sediment in each of the six cross sections described in this paper. The material included in the cross-section estimates is all of density less than 2.6 g/cm³ and likely includes unconsolidated oceanic and terrestrial sediments, well-indurated and dewatered sediments, and probably some low-grade metamorphic rocks. Table 1 shows a surprisingly constant cross-sectional area of low-density along the continental margin. The range is 255 to 480 km² with the average near 345 km². Mordojovich (1974) has indicated that deposits in the wells on the continental shelf off Chile from lat 35° to 39°S are as old as Late Cretaceous, and Masias (1976) has indicated that wells off central Peru contain early and late Cenozoic deposits; hence, the sediments on the continental margin are at least as old as Cretaceous. If we assume that the Nazca plate near the trench is covered with approximately 100 m of sediment, that the convergence rate is approximately 10 cm/yr, and that a 20% reduction in cross-sectional area occurs owing to induration and dewatering, then we would expect about 1,200 km² of sediments to have been accreted to the continent in the past 150 m.y. Table 1 suggests that only 30% of this amount is actually present; furthermore, some of this must derive from terrestrial sources. These observations indicate tht not all of the oceanic sediments on the Nazca plate or those shed from the continent are accreted to the continental margin; consumption of the material must occur, at least on occasion, throughout geologic time.

TABLE 1. CROSS-SECTIONAL AREAS OF SEDIMENTS ON THE CONTINENTAL MARGINS OF PERU AND CHILE

Cross section	Lat (°S)	Sediment cross section (km^2)	Dip of margin thrust (°)
9°S Peru	9.2	380	4.4
Peru Trench at 12°S	12.0	385	4.4-6.6
Pisco, Peru	13.5	255	5.1
Mollendo, Peru	17.0	480	3.8
Copiapó, Chile	26.7	260	3.8
Central Chile	38.0	305	4.1
Avg.		345	4.6

Except for the Pisco section, all sections show that the largest volume of sediments on the continental margin is located at the toe of the continental slope between the trench axis and the rocks of the continental block. Precambrian and Paleozoic rocks of the continental block crop out along the coast of western South America, and their surface dips more steeply seaward than the actual sea-floor topography. This suggests either that the continent and hence the trench is migrating seaward with time or that rocks of the continental block are being consumed or re-emplaced in the subduction process.

The six sections presented here show a thin crust in the vicinity of the trench, and most sections suggest the crust is bowed up and fractured seaward of the trench. In his interpretation of gravity measurements over the Peru-Chile Trench, Hayes (1966) also indicated that the oceanic crust is thinnest beneath and just seaward of the trench. Because the decrease in water depth due to lithospheric flexure seaward of the trench is less than that expected from the indications of the gravity measurements, we postulate that the crust thins seaward of the trench (as shown in Figs. 3, 4, 8, 12, 13, and 14) and/or an increase in density occurs in the uppermost mantle beneath the gravity high.

Table 1 lists the average dip of the top of the oceanic crust beneath the continental margin as indicated by the six cross sections. The dip of the oceanic plate beneath the continental margin is approximately 5.1° west of Pisco, 3.8° west of Mollendo, 3.8° off northern Chile, and 4.1° off central Chile. Although the differences are small, the dip of the thrust plane is greater north of the Nazca Ridge than south of the ridge, and the dip increases in the vicinity of the coast and coastal mountains in all sections.

Figure 15 shows the Pisco and Mollendo cross sections with a vertical-to-horizontal scale of 1:1. Earthquake hypocenters that occurred between 1931 and 1973 within 100 km of the plane of the cross section are projected onto the section (hypocenter data obtained on magnetic tape from the National Earthquake Information Service, U.S. Geological Survey, U.S. Department of the Interior for the years 1900-1973). Figure 15 shows that most of the earthquakes occurred in the upper mantle between the trench and the western flank of the Andes; very few occurred along the contact between the oceanic and continental blocks, as defined by the model sections. Stauder (1975), using very similar hypocenter projections, noted that the dip of the Benioff zone north of the Nazca Ridge is approximately 20°, whereas south of the ridge the dip is approximately 30°; in both areas the dip decreases eastward beneath the Andes and the Altiplano. Stauder (1975) concluded that the subduction process is different on either side of the ridge, and our results show a distinctly different continental-margin

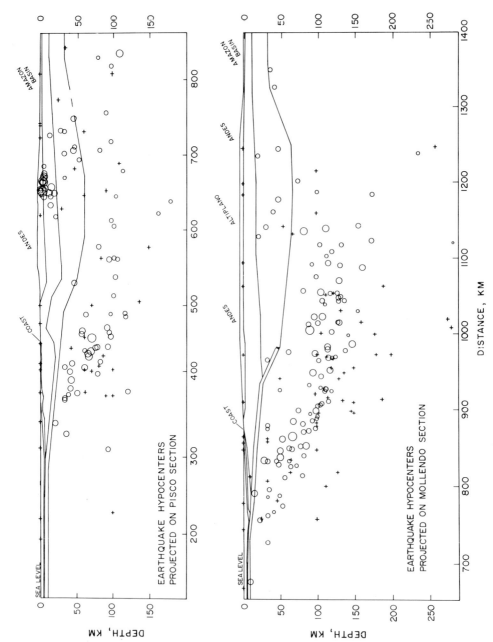

Figure 15. Pisco and Mollendo crustal and subcrustal cross sections with superimposed earthquake hypocenters. The vertical exaggeration of the sections is 1:1.

structure on opposite sides of the Nazca Ridge. Moreover, the gravity and bathymetric measurements suggest that the continental margin of Chile is segmented at intervals of approximately 280 km; however, only small differences are noted in the structure along the continental margin south of the Nazca Ridge.

ACKNOWLEDGMENTS

We thank L. D. Kulm, S. Johnson, R. Blakely, M. Gemperle, G. Connard, G. Ness, and J. Bowers for successful cruises off South America. Our thanks also to Captain H. Linse and the crew of the R/V *Yaquina* for their efforts during the surveys. We also thank V. Goebel for permission to use his unpublished cross section (Fig. 13).

As part of the Nazca Plate Project, this research was supported by the National Science Foundation, International Decade of Ocean Exploration (Grant GX 28675) and by the Office of Naval Research under contract No. 014-67-A-0369-0007, Project NR 083-102.

REFERENCES CITED

Barday, R., 1974, Structure of the Panama Basin from marine gravity data [M.S. thesis]: Corvallis, Oregon State University, 99 p.

Couch, R. W., and Whitsett, R. M., 1981, Structures of the Nazca Ridge and continental shelf and slope of southern Peru, *in* Kulm, L. D., and others, eds., Nazca plate: Crustal formation and Andean convergence: Geological Society of America Memoir 154 (this volume).

Fisher, R. L., and Raitt, R. W., 1962, Topography and structure of the Peru-Chile Trench: Deep-Sea Research, v. 9, p. 423–443.

Grow, J. A., and Bowin, C. O., 1975, Evidence for high-density crust and mantle beneath the Chile Trench due to the descending lithosphere: Journal of Geophysical Research, v. 80, p. 1449–1458.

Hayes, D. E., 1966, A geophysical investigation of the Peru-Chile Trench: Marine Geology, v. 4, p. 309–351.

Hussong, D. M., and others, 1976, Crustal structure of the Peru-Chile Trench; 8° to 12°S lat, *in* Sutton, G. H., and others, eds., The geophysics of the Pacific ocean basin and its margin: American Geophysical Union Geophysical Monograph 19, p. 177–202.

James, D. E., 1971, Andean crustal and upper mantle structure: Journal of Geophysical Research, v. 76, p. 3246–3271.

Johnson, S. H., and Ness, G. E., 1981, Shallow structures of the Peruvian margin 12°–18°S, *in* Kulm, L. D., and others, eds., Nazca plate: Crustal formation and Andean convergence: Geological Society of America Memoir 154 (this volume).

Jones, P., 1981, Crustal structures of the Peru continental margin and adjacent Nazca plate, 9°S lat., *in* Kulm, L. D., and others, eds., Nazca plate: Crustal formation and Andean convergence: Geological Society of America Memoir 154 (this volume).

Kausel, E., and Lomnitz, C., 1969, Tectonics of Chile, *in* Petrolofia, Volume 2, Pan American Symposium on the upper mantle: Mexico Instituto Geofisica, p. 47–68.

Kulm, L. D., and Schweller, W. J., 1977, A preliminary analysis of the subduction processes along the Andean continental margin, 6° to 45°S, *in* Talwani, M., and Pitman, W. C., III, eds., Island arcs, deep-sea trenches and back-arc basins: American Geophysical Union Maurice Ewing Series, v. 1, p. 285–301.

Ludwig, W. J., Nafe, J. E., and Drake, C. L., 1970, Seismic refraction, *in* Maxwell, A., ed., The sea, Volume 4, Part I: New York, John Wiley, p. 53–84.

Masias, J. A., 1976, Morphology, shallow structure, and evolution of the Peruvian continental margin, 6° to 18° S [M.S. thesis]: Corvallis, Oregon State University, 92 p.

McConnell, R. K., and McTaggart–Cowan, G. H., 1963, Crustal seismic refraction profiles (a compilation): University of Toronto Institute of Earth Sciences, Scientific Report No. E.

Mordojovich, C., 1974, Geology of a part of the Pacific margin of Chile, *in* Burk, C. A., and Drake, C. L., eds., The geology of the continental margins: New York, Springer-Verlag, p. 591–598.

Ocola, L. D., Meyer, R. P., and Aldrich, L. T., 1971, Gross crustal structure under Peru-Bolivia Altiplano: Earthquake Notes, v. 42, p. 33–48.

Prince, R. A., and others, 1980, Bathymetry of the Peru-Chile Trench and continental margin: Geological Society of America Map and Chart Series, MC-34, Scale 1:1,000,000.

Shor, G. G., Jr., Menard, H. W., and Raitt, R. W., 1970, Structure of the Pacific basin, *in* Maxwell, A., ed., The sea, Volume 4, Part II: New York, John Wiley p. 3–27.

Stauder, W., 1975, Subduction of the Nazca plate under Peru as evidenced by focal mechanisms and by seismicity: Journal of Geophysical Research, v. 80, p. 1053–1064.

Whitsett, R. M., 1975, Gravity measurements and their structural implications for the continental margin of southern Peru [M.S. thesis]: Corvallis, Oregon State University, 82 p.

Woollard, G. P., 1960, Seismic crustal studies during the IGY. 2, Continental program (IGY Bulletin No. 34): American Geophysical Union Transactions, v. 41, p. 351–355.

Worzel, J. L., 1965, Pendulum gravity measurements at sea, 1936–1959: New York, John Wiley, 422 p.

Zeil, W., 1964, Geologie von Chile: Gebruder Borntraefer, Berlin N. Kolassee.

MANUSCRIPT RECEIVED BY THE SOCIETY NOVEMBER 12, 1980
MANUSCRIPT ACCEPTED DECEMBER 30, 1980

Printed in U.S.A.

ANDEAN CONVERGENCE ZONE

… Geological Society of America
Memoir 154
1981

Volcanic gaps and the consumption of aseismic ridges in South America

AMOS NUR
ZVI BEN-AVRAHAM
Department of Geophysics
Stanford University
Stanford, California 94305

ABSTRACT

The gaps of volcanic activity and the associated shallow-dipping seismicity in South America can be explained by the consumption of the thick-rooted, buoyant, aseismic Nazca and Juan Fernandez Ridges and perhaps also the Cocos Ridge. The ridges erase the trench where they collide with the overriding continent. The point of collision migrates north or south along the plate boundary, depending on the orientation of the ridge relative to the direction of plate motion. This migration leaves behind a zone in which subduction is temporarily stopped; lack of subduction leads to the cessation of volcanism, perhaps owing to lack of water needed for partial melting.

Although the present aseismic ridges probably consist of basaltic cumulates, there is some indication that earlier-consumed parts of these ridges (or different, previously consumed ridges) contained continental fragments that are now embedded in the western coast of South America.

INTRODUCTION

One of the most important aspects of the theory of plate tectonics is the recognition that seismicity, plate motion, and volcanism are intimately related—particularly in subduction domains. The nature of these relationships is understood, however, only in a most general sense. For example, although most regions in which oceanic crust is being subducted also possess long chains of volcanoes situated parallel to the trench, some regions do not. Furthermore, seismic activity associated with the downgoing slab is often observed to depths of 600 km or more, although sometimes it is absent. Thus, there are deviations from the general patterns, including seismic gaps or otherwise anomalous seismicity, lack of volcanism, and the lack of trench. It is through the analysis of these exceptions to the common patterns that insight may be gained into the more basic nature of the underlying tectonic processes.

The subduction region in the western margin of South America displays a remarkable variety of combinations of seismicity, volcanism, and morphology. The oceanic Nazca plate is being consumed at the Peru-Chile Trench from Columbia in the north to southern Chile, a process which is accompanied by volcanism on land, seismic activity along Benioff zones to 600-km depth, and uplifted continental crust. However, clear anomalies are apparent: where the Nazca Ridge—towering more than 1,500 m above the sea floor—collides with South America, the trench is greatly diminished in depth, and volcanism stops. Furthermore, a 1,500-km-long volcanic gap (that is, an area that lacks present day volcanism despite lying along a trench) exists north of this point. In addition, the dip of the seismic plane is anomalously shallow in this area. A similar volcanic gap is present farther south, just north of where the Juan Fernandez Ridge collides with the Chile coast. A third gap extends south of the point (about lat 10°N) at which the Cocos Ridge meets the coast of Panama.

Barazangi and Isacks (1976) suggested that the seismicity and volcanic activity patterns are related in that volcanism is controlled by the dip of the subducting slab and the amount of asthenospheric material between the slab and the overriding South American lithosphere. Their model did not provide a reason for the latitudinal variation in the dip. We will show that the oceanic ridges may well be responsible. Noble and McKee (1977) have further proposed that the flattened slab is associated with the breakup of the normally subducting slab; the old, detached, normally dipping slab continued to sink, whereas the new edge of the oceanic plate moved eastward at a low angle. They further suggested that the rupture of the slab was somehow related to a compressive episode in mid-Miocene time. Pilger (1977, 1981) related the volcanic gaps to the actual consumption of the Nazca and Juan Fernandez Ridges and suggested that underthrusting of the aseismic ridges could result in low-angle subduction. We will show in some detail in this paper that the consumption of the Nazca Ridge could indeed be the underlying cause for the plate breakup.

In this paper we attempt to explain the volcanic gaps and the shallow dip of the seismic plane. We will show how certain observations can be related to the consumption of oceanic crust that is anomalous because it contains either embedded continental fragments or thick basaltic ridges.

TECTONIC ELEMENTS IN WESTERN CENTRAL AND SOUTH AMERICA

Several investigators have shown that it is convenient to divide the length of western South America into large segments on the basis of their geologic features. In this paper we follow the detailed analysis by Barazangi and Isacks (1976), who have clearly shown the relation between volcanism, seismicity, and crustal structure. We will consider the presence or absence of several tectonically important elements: trench, volcanic activity, deep seismicity, steep Benioff zone, and old continental rocks. Figure 1 shows some of these elements for the South and Central American segments, and Table 1 summarizes the elements for each segment. Considering recent volcanism alone, we notice the following segments: From lat 2°S to 15°S and from lat 28°S to 33°S there are two long gaps of volcanic activity. From lat 15°S to 28°S and lat 33°S to 44°S there are chains of volcanoes which are active or have recently been active. An additional segment to the north, from lat 2°S to 5°N, also exhibits recent volcanic activity, followed by an inactive zone in Panama, south of the area where the Cocos Ridge collides with Central America. Although the gaps are associated with a trench—a clear indication for ongoing subduction—the inclination of the seismic activity zones under the continent is very shallow, about 10°. The shallow seismicity in the Nazca gap is complemented by a cluster of very deep earthquakes at about 600-km depth. No deep seismicity is observed in the Juan Fernandez gap. Today's volcanic gaps, however, have experienced volcanic activity in the geological past (for example, Noble and others, 1974, 1979). Hence, any explanation for the presence of these gaps must also account for their transient nature.

An important part of the consumption zones are the aseismic ridges—the Nazca and Juan Fernandez Ridges in the south and the Carnegie and Cocos Ridges to the north (Fig. 1). The nature of these ridges, like other aseismic ridges, is not well established. Some, such as the Mascarene and Ontong-Java Plateaus, may be submerged continental fragments (for example, Nur and Ben-Avraham, 1978); they have thick crusts and a velocity structure that is very typical of continents. Others may be basaltic piles—perhaps similar in origin to Iceland—with thick accumulations of lava flows. Others may be extinct volcanic arcs (Karig, 1972). Whatever their nature, most ridges must have deep roots, since they are in isostatic equilibrium. Ridges which rise 2 to 4 km above the adjacent ocean

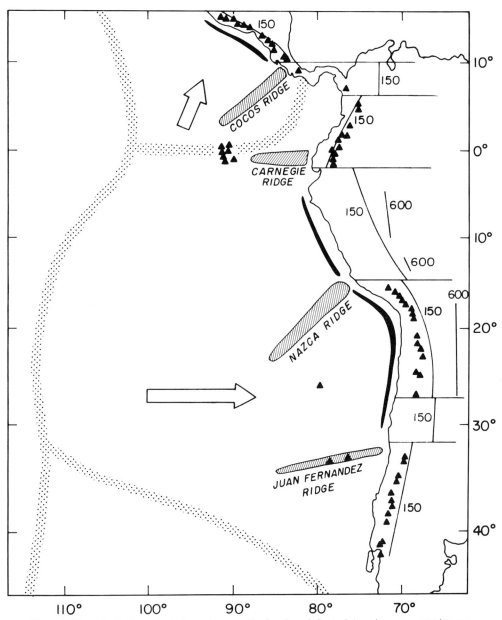

Figure 1. Tectonically important elements along the South and Central American consumption zone, including trench segments (heavy lines), active volcanoes (triangles), and depths to Benioff zone (numbers, in kilometres) as shown by seismic activity. (See also Table 1.) Also shown are aseismic ridges (lined pattern) west of the trench.

floor may have roots as much as 20 or 30 km deep, depending on the density contrasts between the ridge and the surrounding normal oceanic lithosphere. Such roots have in fact been observed for several aseismic ridges and plateaus: the Shatsky Rise (Den and others, 1969), the Ontong-Java Plateau (Furumoto and others, 1976), Manihiki Plateau (Sutton and others, 1971), Bowers Ridge (Ludwig and others, 1971), and the Seychelles Islands (Laughton and others, 1970), among others.

Because the roots of these ridges and other aseismic ridges are lighter than the surrounding oceanic

TABLE 1. TECTONIC ELEMENTS ALONG THE COCOS AND NAZCA PLATE MARGINS

Latitude	Trench	Volcanics	Deep seismicity	Mountain plateau
15°–10°N	Yes	Yes	No	Narrow
10°–05°N	No	No	No	No
05°N–02°S	No	Yes	No	No
02°–15°S	Yes	No	Yes	Narrow
15°–28°S	Yes	Yes	Yes	Wide
28°–33°S	Yes	No	No	Narrow
33°–45°S	No	Yes	No	No

lithosphere, they do not sink as does normal oceanic lithosphere at subduction plate boundaries. Vogt and others (1976) have shown numerous examples of such fragments which have erased the trenches into which they extend: the Benham Rise, the Marcus Ridge, and the Louisville Ridge in the western Pacific and Barbados Ridge in the Atlantic. Timor represents a similar situation in the Indian Ocean. Chung and Kanamori (1978) have shown that several of these fragments such as the Louisville Ridge are associated with a marked decrease of seismic activity, particularly deep seismic activity, along the associated Benioff zone.

The same features are also observed to varying degrees for the three aseismic ridges considered here: the Juan Fernandez, Nazca, and Cocos Ridges. All three ridges tend to obscure the trench segments where they approach the plate boundary in South America and Panama. Furthermore, the Cocos and Nazca Ridges have been shown to have relatively deep, low-density roots (Whitsett, 1976), which in all likelihood are responsible for the isostatic compensation of the ridges. Both the Juan Fernandez and Nazca Ridges are probably also associated with the diminution of deep seismicity, including large seismic gaps below the depth of 150 km. Most dramatically, all three ridges are associated with the abrupt changes in volcanic activity along the margin of the continental side of the subduction zone: north of where the Juan Fernandez and Nazca Ridges meet the continent, the typically continuous chain of subduction-associated active volcanoes becomes discontinuous for hundreds of kilometres, as shown in Figure 1. The edge of the Cocos Ridge in contrast is associated with a volcanic gap to the south.

In the following section we show that these volcanic gaps may be simply related to the consumption of the aseismic ridges and the inability of their light roots to sink into the athenosphere.

A TECTONIC MODEL

We suggest that the main features of the large tectonic segments in South and Central America are directly related to the collision and consumption of the ridge that is being carried by the ocenaic plate, as proposed by Vogt and others (1976). In particular Vogt and others (1976) reviewed a dozen or so cases where ridges approach consumption zones, disrupt the continuity of the trench, and diminish the subduction process. This has been associated with the relatively low density of the ridges. Barazangi and Isacks (1976, 1979), Isacks and Barazangi (1977), and W. Y. Chung and H. Kanamori (1980, personal communs.) have also suggested that the shallow dip of seismic zones and the lack of deep seismicity may be related to the buoyancy of these ridges. Here we wish to extend these relations to an additional feature—the activity of volcanoes in the consumption zone.

In Figure 2 we show schematically the geometric configuration of the Juan Fernandez and Nazca Ridges, the direction of motion of the Nazca plate, and the trench at the consuming South American plate boundary. If a ridge is oblique to the direction of plate motion, the point at which the ridge meets the continent moves along the boundary. Assuming that the original ridges were roughly straight, similar to the still-remaining portions, we can relate the displacement of the ridge D, the distance swept by the portion consumed L, plate velocity v, the direction of plate motion ϕ relative to the ridge direction, and the time since the ridge first collided with the continent Δt by using the equations

$$L = D \tan \phi \tag{1a}$$

and

$$D = v \Delta t . \tag{1b}$$

The velocity w at which the oblique ridge sweeps the plate boundary is thus

$$w = \frac{L}{\Delta t} = v \tan \phi . \tag{2}$$

For simplicity we have ignored here the finite width of the ridge. For the Nazca Ridge, $v \approx 10$ cm/yr, $\phi \approx 45°$, tan $\phi = 1$, and hence $w \approx 10$ cm/yr. Accordingly, the Nazca Ridge sweeps the consumption zone from north to south at a rate of 10 cm/yr, or 100 km/m.y. The 1,500-km gap would

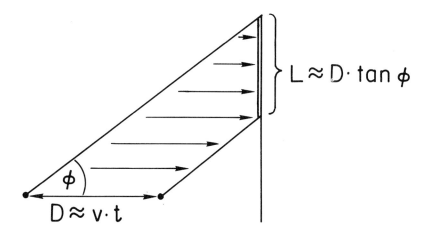

Figure 2. Geometric configuration of the Nazca and Juan Fernandez Ridges relative to the directions of motion of the Nazca plate.

thus require about 15 m.y. to develop. The Juan Fernandez Ridge, with the same plate velocity $v \approx 10$ cm/yr but with $\phi \approx 15°$ and tan $\phi = 0.3$, has a sweep velocity of only $w \approx 3$ cm/yr or 30 km/m.y.

What kind of geologic evidence might reflect the oblique consumption of the ridges? The most promising features are the volcanic gaps, with their associated low-dipping seismic zones. The Nazca and Juan Fernandez Ridges meet the trench at the southern end of the associated prominent volcanic gaps, and volcanoes are active to the south of both ridges. The northern gap (north of the Nazca Ridge) is approximately 1,500 km long, whereas the southern gap (north of the Juan Fernandez Ridge) is only 500 km long. During the past several million years these ridges were carried with the oceanic Nazca plate. Presumably, the easternmost portions of the ridges have already been consumed either by subduction or by collision with the South America continent. In Figure 3 we sketch schematically a possible history of a ridge as a function of geologic time.

Like Pilger (1977), we suggest that it is the oblique collision of the ridges which is responsible for producing the volcanic gaps. During subduction of normal oceanic lithosphere typically dipping 30° to 45°, volcanic activity is continuously maintained. When a buoyant fragment arrives at the trench, the fragment cannot be subducted very easily. The cessation or diminution of subduction could then be the cause for the cessation of volcanism—perhaps due to the absence or great reduction in water available for melting associated with the downgoing slab. This is supported by the observed geometric relationship shown in Figure 1: the volcanic gap associated with the Nazca Ridge is about 1,500 km long, whereas the Juan Fernandez gap is 500 km long—a ratio which is quite close to the corresponding ratio of tan $45°$/tan $15° \approx 3$.

Both Figures 3 and 4 show in a cartoon fashion the process we envision: An elongated ridge-like fragment, oriented obliquely to the direction of plate motion, approaches the trench. Because of the buoyancy of the ridge's low-density rocks, the trench is erased. This is followed by disruption of the downgoing slab, perhaps in the form of normal faulting (for example, Eisler and Kanamori, 1979). By this process the low-density material is incorporated into the overriding plate—in the form of embedded fragments or crustal undercoating, as suggested, for example, by Dewey (1972)—depending perhaps on the density of the ridge material. The diminished subduction process reduces the intensity of melting at depth, which is reflected in the cessation of volcanic activity. Because of the oblique orientation of the ridge relative to the movement of the oceanic plate, this sequence of events migrates along the plate boundary, as shown in Figures 2 and 3. After the ridge segment becomes well embedded or consumed, subduction of oceanic crust begins anew. A new trench forms and a new seismic slab is available for subduction seaward of the old one. This sequence of events may lead to an apparent flattening of the seismic active zone, as observed by Hasegawa and Sacks (1979) and Kelleher and McCann (1977), for example. When the new subduction process develops sufficiently, volcanic activity is renewed and perhaps offset seaward relative to the older volcanic zone.

Unlike the two southern ridges, the volcanic gap associated with the Cocos Ridge is to the south of the ridge. This relationship is clearly consistent with the model presented: because the direction of movement of the Cocos plate is more northerly than the orientation of the Cocos Ridge, the Cocos Ridge sweeps the collision zone from south to north. Although this is in agreement with the model, the volcanic gap here may be related also to the presence of a spreading ridge (see Dickinson and Snyder, 1979). Whether this produces a major effect on volcanism is not clear, but other spreading ridges such as the Chile Rise are known to be associated with regions of low volcanic activity.

The evidence for the past existence and consumption of aseismic ridges can be found in the history of volcanic activity—revealed by gaps of activity in the volcanic chain, migration of gaps with time, and renewal of volcanic activity. As we have shown already, the existing volcanic gaps are in good agreement with this model. However, further study and analysis of the patterns of activity and migration in late Tertiary and early Quaternary time (for example, Noble and others, 1974; Thorpe and Francis, 1979; Drake, 1976) are clearly still needed.

Aside from the volcanic gaps and their temporal histories, we also expect a different kind of record to be found in the geology of the region in which the ridge has been consumed, perhaps in the form of allochthonous fragments of the ridge embedded in the autochthonous crust. In the following section we discuss possible evidence for such fragments in conjunction with the Nazca Ridge.

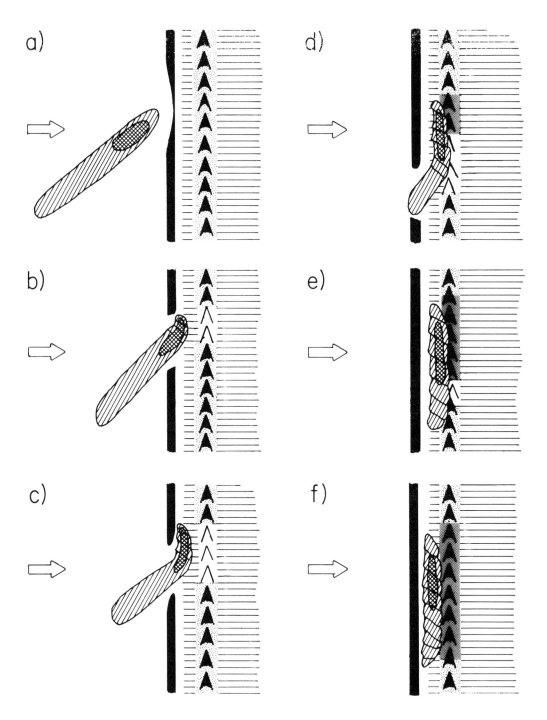

Figure 3. Cartoon depicting the consumption of aseismic ridges in plan view. (a) Ridge is carried by oceanic plate to plate boundary. (b) Early consumption—trench is filled, and volcanic gap begins to develop. (c) Volcanic gap grows, as more of the ridge is plastered onto the overriding plate. (d) Volcanic gap grows and volcanic activity is resumed at the northern end of the gap. (e) Volcanic activity is resumed and volcanic gap is diminished. (f) Volcanic activity is continuous across the previous gap.

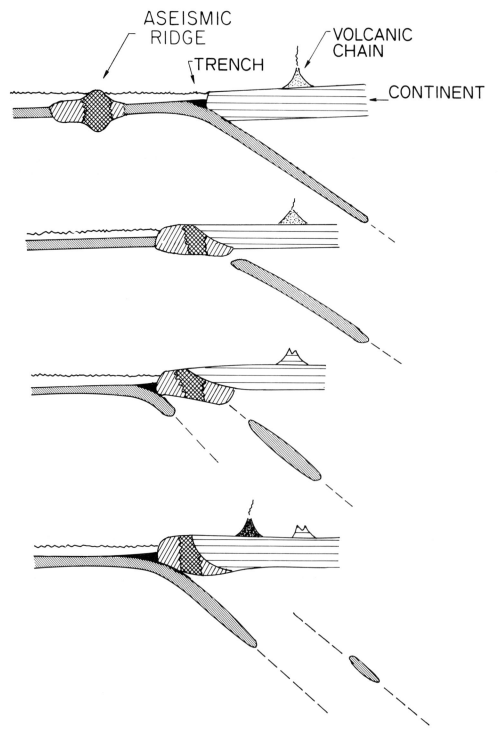

Figure 4. Cartoon depicting the stages of the consumption process in cross section, including initial collision, cessation of volcanic activity, detachment of the downgoing slab, development of new slab, and renewal of volcanic activity.

EMBEDDED FRAGMENTS

It has long been suggested that continental fragments may be embedded in the western coast of Peru, as shown in Figure 5 (Kulm and others, 1977; James, 1971; Helwig, 1973). The ages of some of these fragments have been determined to be about 500 to 800 m.y. (Dalmayrac and others, 1977). Geophysical evidence further suggests westward extension of these continental rocks toward the trench. It has been suggested that these rocks have been exposed by a poorly defined tectonic erosion process and belong to cratonic South America. James (1971), on the other hand, suggested that the presence of these very old continental basement rocks between the trench and the orogenic belt can be explained by an arc which has drifted in from the west. Dalziel (1979) and Bruhn and Dalziel (1977) suggested that such an arc first originated in South America, next migrated westward, and then returned to be consumed. We proposed instead that the "arc" might have been part of a pre-Triassic central Pacific continental mass which we called Pacifica (Nur and Ben-Avraham, 1978).

The existence of an arc and its incorporation into Peru's coast fit well with the consumption of the Nazca Ridge. A speculative possibility is illustrated in Figure 6, where we compare the morphology of the Nazca Ridge with that of the Mascarene Plateau in the Indian Ocean. Both have elongated and submerged portions about 1,000 to 2,000 km long and 300 to 500 km wide. At least the northwest part of the Mascarene Plateau is continental, including the Seychelles Islands and their surroundings (Laughton and others, 1970). Conceivable, today's Nazca Ridge may also be the tail of a larger fragment, parts of which might also have been continental. Thus in addition to the effects of the ridge consumption on seismicity and volcanism, it could also explain the puzzling presence along the coast of continental crustal fragments—perhaps they were carried in from the Pacific Ocean.

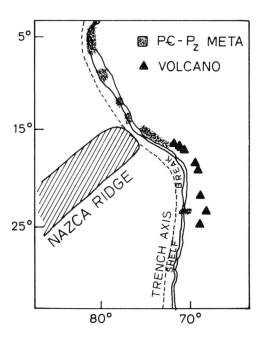

Figure 5. The Peru coast and Nazca Ridge. Active volcanos and Precambrian-Paleozoic metamorphic rocks outcrops are shown (modified from Kulm and others, 1977).

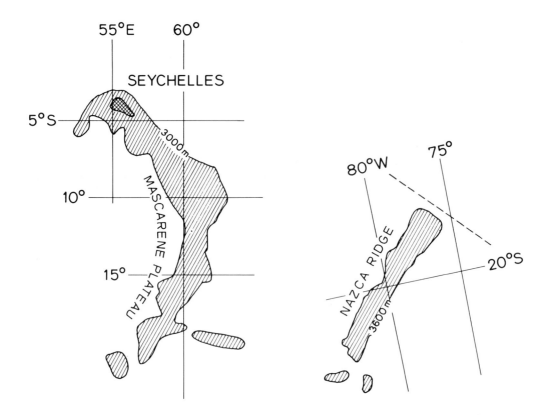

Figure 6. Comparison between bathymetric outlines of the Nazca Ridge and the Mascarene Plateau. The plateau in the Indian Ocean, consists at least in part of a continental fragment, as indicated by the crustal structure under the Seychelles Islands. It is possible that the Nazca Ridge was also associated with a continental fragment, now embedded in the coast of Peru. The cutoff of the Nazca Ridge is represented by the dashed line.

CONCLUSIONS

We suggest that the presence of aseismic ridges within the consumption regions of the Nazca and Cocos oceanic plates is responsible for the volcanic gaps and apparent shallow seismicity in these gaps in Central and South America. The buoyancy of the ridge material breaks the continuity of the subducted plate, which may lead to reduced water supply required for melting of magma. The same effects should occur at other plate boundaries whenever light, buoyant material is consumed.

Some of the aseismic ridges, such as the Nazca Ridge, may have also involved continental fragments which have been embedded in the overriding plate, such as in the Pasific coast of Peru.

ACKNOWLEDGMENT

This research was supported in part by Grant EAR 78-12972 from the Geophysics Program, Earth Science Division, National Science Foundation.

REFERENCES CITED

Barazangi, M., and Isacks, B., 1976, Spatial distribution of earthquakes and subduction of the Nazca plate beneath South America: Geology, v. 4, p. 686–692.

—— 1979, Subduction of the Nazca plate beneath Peru: Evidence from spatial distribution of earthquakes: Royal Astronomical Society Geophysical Journal v. 57, p. 537–555.

Bruhn, R. L., and Dalziel, I.W.D., 1977, Destruction of the Early Cretaceous marginal basin in the Andes of Tierra del Fuego, in Talwani, M., and Pitman, W. C., III, eds., Island arcs, deep sea trenches and back-arc basins: American Geophysical Union Maurice Ewing Series, v. 1, p. 395–405.

Chung, W. Y., and Kanamori, H., 1978, Tectonic anomalies in aseismic ridge subduction in the New Hebrides arc: Tectonophysics, v. 50, p. 29–40.

Dalmayrac, B., Lancelot, J. R., and Leyreloup, A., 1977, Two-billion-year granulites in the late Precambrian metamorphic basement along the southern Peruvian coast: Science, v. 198, p. 49–51.

Dalziel, I.W.D., 1979, The role of microplates in the evolution of the continental margins of the southeast Pacific: Geological Society of America Abstracts with Programs, v. 11, p. 409.

Den, N., and others, 1969, Seismic-refraction measurements in the northwest Pacific basin: Journal of Geophysical Research, v. 74, p. 1421–1434.

Dewey, J. F., 1972, Plate tectonics: Scientific American, v. 226, p. 56–68.

Dickinson, W. R., and Snyder, W. S., 1979, Geometry of triple junctions related to the San Andreas transform: Journal of Geophysical Research, v. 84, p. 561–572.

Drake, R. E., 1976, The chronology of Cenozoic igneous and tectonic events in the Central Chilean Andes, in Ferran, O. G., ed., Proceedings, Symposium on Andean and Antarctic volcanology problems: Rome, International Association on Volcanology and Chemistry of the Earth's Interior, Special Series, p. 670–697.

Eisler, H. K., and Kanamori, H., 1979, A large normal fault earthquake at the junction of Louisville ridge and the Tonga trench and its geophysical implication: EOS (American Geophysical Union Transactions), v. 60, p. 878.

Furumoto, A. S., and others, 1976, Seismic studies on the Ontong-Java Plateau, 1970: Tectonophysics, v. 34, p. 71–90.

Hasegawa, A., and Sacks, I. S., 1979, Subduction of the Nazca plate beneath Peru as determined from seismic observations: EOS (American Geophysical Union Transactions), v. 60, p. 876.

Helwig, J., 1973, Plate tectonic model for the evolution of the Central Andes: Discussion: Geological Society of America Bulletin, v. 84, p. 1493–1496.

Isacks, B. L., and Barazangi, M., 1977, Geometry of Benioff zones: Lateral segmentation and downwards bending of the subducted lithosphere, in Talwani, M., and Pitman, W. C., III, eds., Island arcs, deep sea trenches and back-arc basins: American Geophysical Union Maurice Ewing Series, v. 1, p. 99–114.

James, D. E., 1971, Plate tectonic model for the evolution of the Central Andes: Geological Society of America Bulletin, v. 82, p. 3326–3346.

Karig, D. E., 1972, Remnant arcs: Geological Society of America Bulletin, v. 83, p. 1057–1068.

Kelleher, J., and McCann, W., 1977, Bathymetric highs and the development of convergent plate boundaries, in Talwani, M., and Pitman, W. C., III, eds., Island arcs, deep sea trenches and back-arc basins: American Geophysical Union Maurice Ewing Series, v. 1, p. 115–122.

Kulm, L. D., Schweller, W. J., and Masias, A., 1977, A preliminary analysis of the subduction processes along the Andean continental margin, 6° to 45°S, in Talwani, M., and Pitman, W. C., III, eds., Island arcs, deep sea trenches and back-arc basins: American Geophysical Union Maurice Ewing Series, v. 1, p. 285–301.

Laughton, A. S., Matthews, D. H., and Fisher, R. L., 1970, The structure of the Indian Ocean, in Maxwell, A. E., ed., The sea, Volume 4, Part II: New York, Wiley-Interscience, p. 543–586.

Ludwig, W. J., and others, 1971, Structure of Bowers Ridge, Bering Sea: Journal of Geophysical Research, v. 76, p. 6350–6366.

Noble, D. C., and McKee, E. H., 1977, Spatial distribution of earthquakes and subduction of the Nazca plate beneath South America: Comment: Geology, v. 5, p. 576–578.

Noble, D. C., and others, 1974, Episodic Cenozoic volcanism and tectonism in the Andes of Peru: Earth and Planetary Science Letters, v. 21, p. 213–220.

Noble, D. C., McKee, E. H., and Mégard, F., 1979, Early Tertiary "Incaic" tectonism, uplift, and volcanic activity, Andes of Central Peru: Geological Society of America Bulletin, Part I, v. 90, p. 903–907.

Nur, A., and Ben-Avraham, Z., 1978, Speculations on mountain building and the lost Pacific continent: Journal of Physics of the Earth, v. 26, p. 521–537.

Pilger, R. H., Jr., 1977, Plate reconstructions, aseismic ridges and low-angle subduction beneath the Andes [abs.]: EOS (American Geophysical Union Transactions), v. 58, p. 1232.

—— 1981, Plate reconstructions, aseismic ridges and

low-angle subduction beneath the Andes: Geological Society of America Bulletin, v. 92 (in press).

Sutton, G. H., Maynard, G. L., and Hussong, D. M., 1971, Widespread occurrence of a high-velocity basal layer in the Pacific crust found with repetitive sources and sonobuoys (with discussion), *in* Structure and physical properties of the Earth's crust: American Geophysical Union Geophysical Monograph 14, p. 193–209.

Thorpe, R. S., and Francis, P. W., 1979, Variations in Andean andesite composition and their petrogenetic significance: Tectonophysics, v. 57, p. 53–70.

Vogt, P. R., and others, 1976, Subduction of aseismic ridges: Effects on shape, seismicity, and other characteristics of consuming plate boundaries: Geological Society of America Special Paper 172, p. 59.

Whitsett, R. M., 1976, Gravity measurements and their structural implications for the continental margin of southern Peru [Ph.D. thesis]: Corvallis, Oregon State University, 82 p.

MANUSCRIPT RECEIVED BY THE SOCIETY NOVEMBER 12, 1980
MANUSCRIPT ACCEPTED DECEMBER 30, 1980

Geological Society of America
Memoir 154
1981

Geological and geophysical variations along the western margin of Chile near lat 33° to 36°S and their relation to Nazca plate subduction

ALLEN LOWRIE*
U.S. Naval Ocean Research and Development Activity
Code 360
Bay St. Louis, Mississippi 39529

RICHARD HEY*
Hawaii Institute of Geophysics
University of Hawaii
Honolulu, Hawaii 96822

ABSTRACT

There are several types of geologic and geophysical variations along the western margin of Chile from lat 33° to 36°S. The elevation of the Andean mountain peaks averages about 6 km north of lat 33°S, but 300 km farther south, the average elevation of the Andean peaks has decreased to 3 km. The width of the Andes also decreases abruptly from 425 to 260 km in this region. A similar elevation change is noted for the Coastal Range mountain peaks whose average elevation decreases abruptly from about 1.8 to 0.8 km between lat 33° and 34°S. The Central Valley, which lies between the Andes and the Coastal Range, decreases about 300 m in elevation near lat 34°S but maintains a constant elevation to the south.

Volcanism is discontinuous. North of lat 27.5°S, the Chilean volcanoes have been active during Quaternary time. Between lat 27.5° and 33.5°S Quaternary volcanism is absent, but reappears at lat 33.5°S to form the southern Chilean volcanic belt.

Other variations include a major gravitational change near lat 34°S—which implies a change in crustal thickness along the center of the Andean Cordillera—and the termination of the main porphyry copper belt of northern Chile near lat 34°S.

The occurrence of these changes and discontinuities between lat 33° and 34°S suggests that they result from some type of interaction between the Nazca and South American plates.

INTRODUCTION

Many types of geologic and geophysical variations occur along or transverse to consuming plate margins (see Vogt and others, 1976; Carr and others, 1973, 1974; Sillitoe, 1974; and Isacks and

---
*Present Addresses: (Lowrie) U.S. Naval Oceanographic Office, Code 7221, NSTL Station, Mississippi 39529; (Hey) Scripps Institution of Oceanography, University of California, San Diego, La Jolla, California 92093.

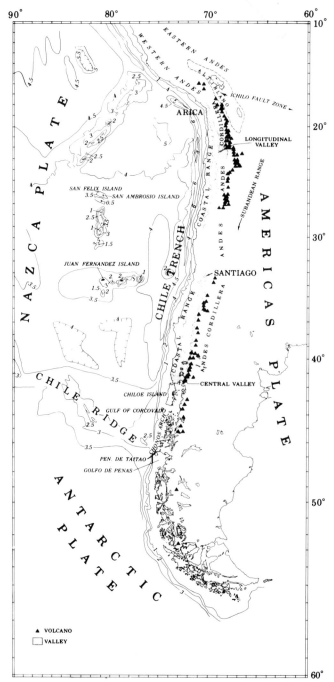

Figure 1. Map of study area showing major physiographic features, geomorphic regions, and place names mentioned in the text. Active and dormant volcanoes are shown as filled triangles (see Casertano, 1963; Pichler and Zeil, 1969; and Plafker and Savage, 1970). Bathymetry simplified from Fisher and Raitt (1962), Scholl and others (1970), and "World Maps" (from Defense Mapping Agency Topographic Command (1973).

Barazangi, 1977; for recent summaries with numerous references). We will document in detail several such variations along the Peru-Chile Trench that were described earlier by Lowrie and Hey (1976). These occur along the western margin of Chile near lat 33°S, the boundary between the physiographic provinces of Norte Chico and Zona Central (Munoz, 1956; Sillitoe, 1974; Koster, 1975; see Figs. 1, 2, 3, and 4). We will also discuss the various Nazca–South America plate interactions that may have produced these variations.

ABRUPT CHANGES AND DISCONTINUITIES

Figure 2 shows the maximum elevation of the Andes Mountains from lat 25° to 50°S plotted against latitude at 0.5° intervals. North of lat 33°S, the maximum elevations average about 6 km; however, between lat 33° and 34°S, the peak elevations begin to decrease. Within a distance of 300 km to the south (lat 36°S), the average maximum elevation has dropped to 3 km. The average width of the Andes also decreases from about 425 to 260 km between lat 33° and 35°S (Fig. 3). A similar change is noted in the Coastal Range where the average maximum elevation decreases from about 1.8 to 0.8 km between lat 33° and 34°S (Fig. 4). The northern limit of the widespread Precambrian rocks and Paleozoic intrusives occurs at about lat 33°S in the Coast Range (this is boundary 13 of Sillitoe, 1974). A structural feature, known as the Longitudinal Valley (see Koster, 1975) to the north and the Central Valley to the south, lies between the Andes and the Coast Range (Figs. 1 and 4; Segerstrom, 1964). Whereas the northern limit of the Central Valley occurs near lat 33°S, the valley actually drops in elevation about 300 m near lat 34°S.

The Chilean volcanic belt extends from about lat 16° to 45°S with a marked absence of Quaternary volcanoes between lat 27.5° and 33.5°S (Casertano, 1963; Ruiz and others, 1965). The position of the Central Valley correlates with the southern group of Quaternary volcanoes (Fig. 1): their northern limit is lat 33.5°S, near the marked change in elevation of the Central Valley floor.

On the coast the marine terraces, representing former shorelines, are younger and lower in elevation south of lat 35°S than they are to the north (Fig. 4; Fuenzalida and others, 1965). Also, the amount of sediment fill in the Peru-Chile Trench increases dramatically south of lat 33°S because of a structural offset in oceanic layer 2 in the trench axis (Fig. 4; Hayes, 1966; Ewing and others, 1969; Scholl and others, 1970; von Huene, 1974; Schweller and others, this volume, their Fig. 4).

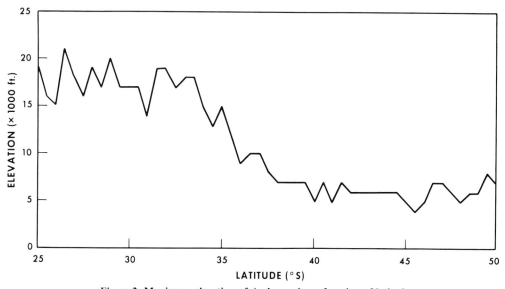

Figure 2. Maximum elevation of Andes peaks as function of latitude.

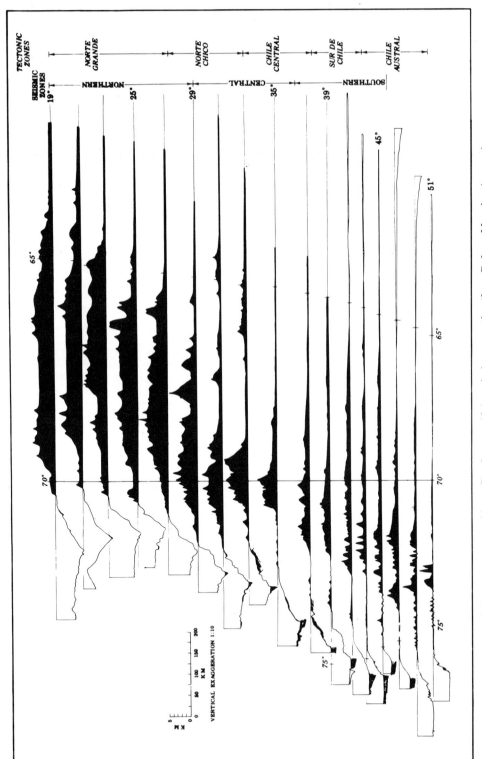

Figure 3. East-west topographic profiles drawn at 10:1 vertical exaggeration (from Defense Mapping Agency Aerospace Center, 1977b). Offshore bathymetry and sediment thickness from Fisher and Raitt (1962), Hayes (1966), Scholl and others (1970), and D. W. Scholl and R. von Huene (1965, unpub. data).

The Benioff zone is longer and its dip shallower north of lat 37°S than it is to the south (Fig. 9 of Stauder, 1973). South of lat 34°S there are few intermediate-focus earthquakes, and the dip of the Benioff zone is 20° to 30° (Fig. 4). North of lat 34°S, the intermediate-focus earthquakes increase in frequency, and the dip is 15° to 25° (Fig. 4; Barazangi and Isacks, 1976; Stauder, 1973).

A large free-air gravity anomaly between lat 33° and 34°S extends eastward from the Coast Range through the Andean Cordillera to the Chaco-Pampas Basin. The maximum gradient exceeds 150 mgal (Defense Mapping Agency Aerospace Center, 1977a) in a north-south direction between these latitudes (G. P. Woollard, 1975, personal commun.; Lomnitz, 1962). The gravity data suggest that the crustal thickness along the center of the Andean Cordillera changes in a north-south direction.

Finally, the known belt of porphyry copper deposits that occurs along the western margin of South America terminates near lat 34°S (Fig. 5; Ruiz and Ericksen, 1962; Sillitoe, 1974, his boundary 15).

We speculate that the present variations along the western margin of Chile were caused by the reactivation of pre-existing zones of weakness, which are related to the evolution of the subduction zone. Subduction began about 110 m.y. ago as a result of the initiation of sea-floor spreading in the South Atlantic Ocean. The subduction zone presumably evolved in the manner described by Isacks and others (1968), Hamilton (1969), Isacks and Molnar (1971), Dickinson (1973), and Sykes (1972, 1978).

Sykes (1978) has documented intraplate reactivation through the most recent tectonism of pre-existing zones of weakness in areas of continental fragmentation. Similarly, some continental margins are segmented in areas of plate convergence (Carr and others, 1974). By conceptual extension, pre-existing weakened zones in these margins are reactivated by the young tectonism. In the case of Chile, the youngest tectonism is related to the present phase of subduction, as shown by the granitic intrusions [for example, Guest, 1969; Clark and others, 1967; Hamilton, 1969 (correlation of batholiths and andesitic volcanism); Farrar and others, 1970; and Dickinson, 1973 (emplacement within a plate-convergence province)].

DISCUSSION

The variations described along the western margin of Chile are characterized by their abrupt boundaries, large magnitudes, and diversity. Such characteristics argue for a plate tectonic origin rather than a geomorphic origin related to differential erosion. It is difficult to explain these variations in the present-day plate tectonic setting since the Nazca plate is underthrusting the South American plate from at least lat 3°S to 46°S, and the instantaneous rate of subduction along the Peru-Chile Trench must be nearly uniform if the Nazca and South American plates behave as single rigid plates, a theory that is supported by plate-motion models (Minster and Jordan, 1978; Hey and others, 1977). Today, there is no major first-order plate tectonic discontinuity (plate boundary) in the vicinity of lat 33°S, the location of the variations described above. In contrast, there is a first-order plate tectonic discontinuity at lat 46°S where spreading occurs at the Nazca–South American–Antarctic triple junction. North of this boundary the Nazca–South American plate convergence rate is about 88 mm/yr; to the south the Antarctic–South American convergence rate is only about 21 mm/yr (Minster and Jordan, 1978). Other types of major variations are described by Sillitoe (1974) at lat 46°S.

On the basis of these considerations, we propose some second-order plate tectonic explanations which may have produced the variations observed along the Chile Margin in the vicinity of lat 33°S. We will also propose an additional speculative explanation involving a first-order plate tectonic discontinuity, the "broken-plate" model of Hey and Lowrie (1976).

Subduction Angle Versus Torn Slab

Evidence from focal mechanisms and hypocenter distributions suggests that the downgoing Nazca plate is broken into a series of tongues that are absorbed independently. Such evidence also suggests

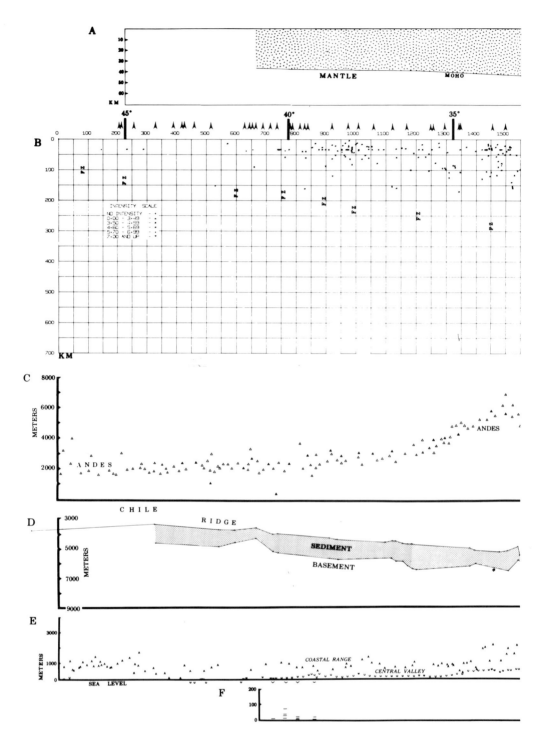

Figure 4. Sections parallel to coast of Chile. (A) Crustal section inferred from gravity, taken from Lomnitz (1962, Fig. 6). (B) All earthquakes of the Chilean seismic zone from 1964 to 1972 and extending from lat 18°30′ to 47°S. Vertical and horizontal lines are at 50-km spacing. Individual earthquakes are denoted by a corresponding intensity symbol. The latitude marks at 5° intervals refer to all profiles. Vertical arrows above the profile denote active or dormant volcanoes (Plafker and Savage, 1970; Pichler and Zeil, 1969; Casertano, 1963). (C) Heights of volcanoes (see above for sources) and mountains along the Andes (from Defense Mapping Agency Aerospace

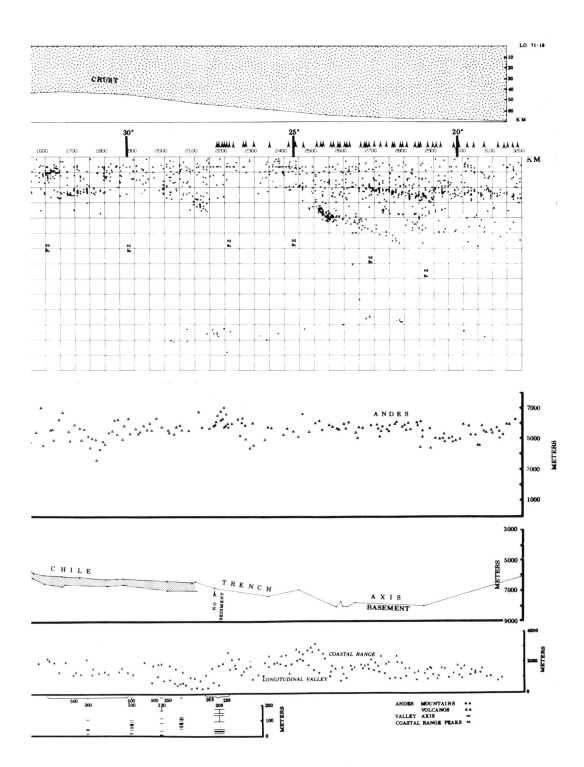

Center, 1977b). (D) Depths to the floor of the Peru-Chile Trench and amount of sediment fill (see Fig. 1) measured in travel-time units and not corrected for velocity in sediments, from Fisher and Raitt (1962), Hayes (1966), Scholl and othrs (1970), von Huene (1974), and D. W. Scholl and R. von Huene (unpub. data). (E) Composite of two profiles of the elevations of the Coastal Range peaks and average elevations along central valleys (from Defense Mapping Agency Aerospace Center, 1977b). (F) Composite of former sea levels, taken from Fuenzalida and others (1965).

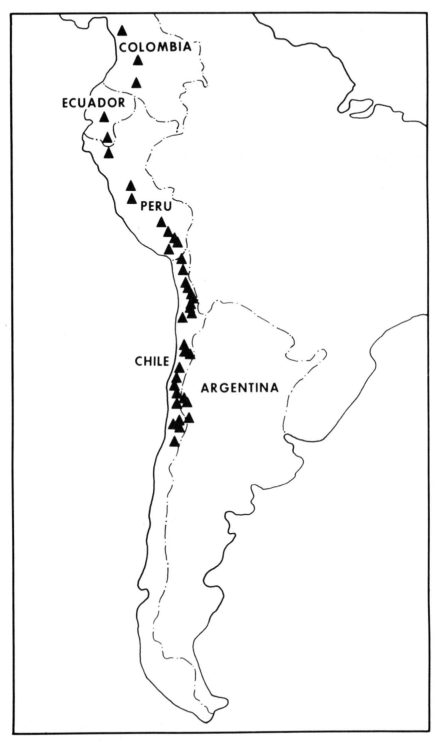

Figure 5. Distribution of known porphyry copper deposits along the western margin of South America (Ruiz and Ericksen, 1962).

that, from lat 27° to 34°S, the subducted slab moves horizontally eastward under South America at depths of about 120 km (Stauder, 1973; Barazangi and Isacks, 1976; Isacks and Barazangi, 1977). Barazangi and Isacks have noted that areas with flat (horizontal) Benioff zones correlate with an absence of Quaternary volcanism in both Chile (Stauder, 1973) and Peru (Isacks and Molnar, 1971; Sykes, 1972; Stauder, 1975). Sykes (1972) suggested that the absence of volcanism is due to high compressive stress in the continental plate of Peru; this situation might result from the strong coupling between the flat subducted slab and the overriding continental block described by Isacks and Barazangi (1977). The lack of volcanism might also result from the absence of a wedge of asthenosphere between the upper and lower plates, as suggested by Barazangi and Isacks (1976). Barazangi and Isacks note that this could explain the low attenuation of seismic waves in the uppermost mantle in areas of horizontal subduction (see also Molnar and Oliver, 1969; James, 1971; Barazangi and others, 1975).

Finally, D'Angelo and Le Bert (1969) suggested that great thrust faults dipping eastward and cropping out landward of the Coastal Range could have sealed the conduits of magma ascension.

Hanus and Vanek (1978) noted that the presence of an aseismic gap between shallow- and deep-focus earthquakes along the western margin of South America was necessary for (but not sufficient) to explain overlying andesitic volcanism. They proposed that the aseismic gap indicated partial melting of the downgoing plate and that this melting is the source of the andesitic magma. They also noted that depths of earthquake foci substantially decrease where major transverse structures on the oceanic plate intersect the trench.

A tear fault in the slab has been used to explain variations in continental geology along the Peru-Chile Trench and other subduction zones (Mogi, 1969; Katsumata and Sykes, 1969; Stoiber and Carr, 1973; Carr and others, 1974; Sillitoe, 1974; Swift and Carr, 1974; Carr, 1976). However, discontinuities in seismicity at lat 15°, 25°, and 27° to 28°S along the Peru-Chile continental margin (Stauder, 1973; Barazangi and Isacks, 1976) do not correlate with geologic variations of a magnitude correspnding to those noted at lat 33° to 36°S.

Interference by Aseismic Ridges

In various subduction zones around the world, Vogt and others (1976) and Kelleher and McCann (1976) documented transverse variations similar to those discussed here. They proposed that the variations resulted from the interference of aseismic ridges and seamount chains with the subduction process (Vogt, 1973). They specifically discussed some of the same variations observed along Chile and indicated that these resulted from the interaction of the Juan Fernandez islands and seamounts with the subduction zone. Whether this ridge marks a hot-spot trace (Vogt and others, 1976) or a mantle hot line (Bonatti and Harrison, 1976; Bonatti and others, 1977), it would be less dense than the surrounding crust and could conceivably cause interference with subduction, leading to transverse variations. Vogt and others (1976) noted many transverse(?) variatons near the intersection of the Nazca aseismic ridge with the trench (see also Zentilli, 1974, and Bonatti and others, 1977). Bonatti and others (1977) suggested that these variations could be caused by the extension of their proposed Easter "mantle hot line" across the trench and beneath South America or, more likely, by the subduction of the anomalous lithosphere produced along the line. A similar model could be developed for the Juan Fernandez island chain.

Threshold Effect

The intersection of the Challenger Fracture Zone with the Peru-Chile Trench near lat 33°S (Fig. 1) implies that lithospheric plates of different ages are subducted north and south of this boundary (see Frankel and McCann, 1979, for discussion of a similar situation found along the South Sandwich arc). Magnetic anomaly correlations of Morgan and others (1969) indicate that lithosphere of anomaly-21 age (50 m.y. old, according to the LaBrecque and others, 1977, time scale) is subducted north of the Challenger Fracture Zone and anomaly-17 age lithosphere (40 m.y. old) is subducted south of the fracture zone. Because the lithosphere thicknesses will be very similar (Sclater and others, 1971; Parker and Oldenburg, 1973), they could not produce the observed variations unless some threshold effect is involved (a similar argument holds for subduction velocity, which is nearly uniform along the trench). Molnar and Atwater (1978) proposed that sea floor older than about 50 m.y. may be dense enough to sink under its own weight, whereas subduction of younger lithosphere may require an additional force. An empirically derived formula for lithosphere thickness is 9.3 times the square root of the sea-floor age in millions of years (Parker and Oldenburg, 1973); this yields thicknesses for 40- and 50-m.y.-old lithosphere of 59.5 and 66 km, respectively. If a density of 3.55 g/cm^3 is accepted, the gravitational body force acting on a downgoing slab would be about 20 and 22×10^6 g/cm^3, respectively. Thus, differences in lithosphere thickness and gravitational forces on 40- and 50-m.y.-old slabs on either side of the Challenger Fracture Zone will be approximately 10%—not a marked difference.

Changes in the dehydration characteristics of the lithosphere being subducted may offer an explanation of the observed variations. Fox and DeLong (1976), Anderson and others (1978), and Anderson and Langseth (1977) proposed that layer 2A (the uppermost oceanic crustal layer) is in a "healing" process from the time of formation until it is about 40 m.y. old, that is, crustal fractures are filled in with sediments and minerals such as calcite and zeolite. Supporting data from Houtz and Ewing (1976) indicate that the velocity of layer 2A increases with age until the sea floor is about 40 m.y. old when the velocities of layers 2A and 2B become identical. The older healed crust would be relatively impermeable; thus, water released by dehydration may be unable to escape immediately from the slab and may instead be carried down some distance by subduction. Younger crust would be more permeable, and the water released by dehydration would be freed from the slab almost immediately (in the initial and shallower stages of subduction); relatively less magmatism would result (Anderson and others, 1978; DeLong and Fox, 1977).

With regard to the three parameters related to the subducted lithosphere, we conclude that lithosphere thickness and changes in gravitational force show only minor changes between lat 33° and 36°S along the western margin of Chile and that changes in dehydration characteristics cannot be quantified at this time.

Differential Erosion

Differential erosion along the length of Chile has been discussed by Scholl and others (1970). Accepting their minimal denudation rates (their Table 1) and using a 2.7-g/cm^3 crustal density and a 3.3-g/cm^3 mantle density (Press and Siever, 1974), we find that the total denudation for the past 20 m.y., corrected for isostatic adjustment, would be 20, 72, and 400 m for the areas between lat 23 and 26°S, 29 and 33°S, and 36 and 42°S, respectively. The resultant 400-m decrease in elevation calculated for the area between lat 36° and 42°S would be approximately an order of magnitude less than the observed average maximum elevation in the area between lat 32° and 36°S. Thus, erosion cannot totally explain the observed Andean morphologic changes.

Broken Plate

Another speculative explanation for the observed variations along the Chile Margin is our broken-plate model (Hey and Lowrie, 1976). This model was developed to explain apparent differences in Pacific-Nazca separation rates north and south of the Pacific-Nazca-Antarctic triple junction between 10 and 20 m.y. B.P. North of the junction, two spreading centers were active during this time according to Herron (1972): the East Pacific Rise, which was spreading at 100 mm/yr, and the now-fossil Galapagos Rise (lat 28°S), which was spreading at 180 mm/yr (total spreading rates). Herron's rate for the Galapagos Rise is corroborated by the present separation of anomaly 7 on the Pacific and Nazca plates (Handschumacher, 1976), which implies an average separation rate of 180 mm/yr during the past 30 m.y. This is the same rate that has held for the past 10 m.y. (Herron, 1972); therefore, the total separation rate between 10 and 20 m.y. B.P. was probably 180 mm/yr whether or not the fossil Galapagos Rise was active at that time just north of the triple junction. South of the junction the total opening rate was the sum of the opening rates on the Pacific-Antarctica and Nazca-Antarctica (Chile

Ridge) rises. The Pacific-Antarctica Rise was spreading at 50 mm/yr at lat 45°S (Molnar and others, 1975). The Chile Ridge was opening at about 60 to 80 mm/yr near lat 40°S (derived from Morgan and others, 1969; Klitgord and others, 1974; Molnar and others, 1975; and Handschumacher, 1976). The total separation rate between the Pacific plate and the subducting plate was 110 to 130 mm/yr.

If these published rates are accepted, it seems likely that this total separation-rate difference is too great to result from changes in velocity due to different distances from the profiles to the triple junction. If we are justified in using a scalar approximation for this area, as spreading on all rises is essentially east-west, then total separation rates were different north and south of some boundary between lat 28° and 45°S. This suggests that either the Pacific or Nazca plate was fractured across its width between 10 and 20 m.y. B.P. One simple plate configuration, which apparently satisfies the published correlations and spreading rates, is that the Nazca plate was broken into two plates along the Challenger Fracture Zone (Fig. 1) during this time interval (10 to 20 m.y. B.P.), with subduction north of this boundary 50 to 70 mm/yr faster than to the south. These plates were welded together later into the single Nazca plate that exists today and apparently has existed for the past 10 m.y. Carrying our speculation one step further, we suggest that this hypothetical plate boundary intersected South America at a point located near the geologic and geophysical variations discussed above.

A definitive test of our model could be made by determining the patterns of the relative positions of magnetic anomalies 5 and 6 north and south of the triple junction and by making detailed plate reconstructions at the times of anomalies 5, 6, and 7. In addition, our model predicts that the Challenger Fracture Zone should extend continuously from the trench to at least the nearest 10-m.y.-old isochron (anomaly 5).

A possible explanation of why the plate should break is suggested by the presence of detached seismic slabs along the Peru-Chile Trench (for example, Barazangi and Isacks, 1976). The presence of detached slabs implies that trench pull is an important part of the driving mechanism. Therefore, subduction of part of a plate might slow or pause for a few million years following detachment of part of the subducting slab; this situation could lead to a broken plate and segmented and discontinuous subduction.

Similar broken-plate explanations have been proposed by Menard (1978), who suggested that the Farallon plate occasionally fractured across its width to allow subduction to become perpendicular to existing trends. Hey (1977) noted that the splitting of the Farallon plate into the Cocos and Nazca plates at about 25 m.y. B.P. allowed subduction to become perpendicular to both the Mid-America and Peru-Chile Trenches, and he suggested that this might help explain why the Farallon plate fractured at a particular location.

Plate Tectonics and Copper

The southern boundary of known porphyry copper prospects in Chile is located at about lat 34°S (Fig. 5). This boundary or discontinuity could have resulted from a subduction rate 50 mm/yr faster on the north side of our hypothesized plate boundary than south of it. The difference in subduction rates would result in 50 km more lithosphere subducted per unit length of trench per million years north of the boundary than south.

CONCLUSIONS

There are many geologic and geophysical variations near lat 33° to 34°S along the western margin of Chile. There are large and abrupt decreases in elevation of both the Andes Mountains and Coastal Range and also a marked decrease in the width of the Andes near lat 33°S. This is the northern limit of the southern Chilean volcanic belt and the Central Valley of Chile. It is the southern boundary of known porphyry copper deposits. There is a major gravitational change here as well.

Many explanations for these variations have been proposed, including changes in the dip of the downgoing slab, tear faults in the slab, interference of the Juan Fernandez aseismic ridge with subduction in the Peru-Chile Trench, threshold effects related to increased plate density with age and/or the healing of layer 2A, and a broken-plate model (proposed by us) in which the Nazca plate fractured across its width, with different subduction rates north and south of the Challenger Fracture Zone. These hypotheses are not necessarily mutually exclusive, and all or some of them may well explain the variations described in this paper. For example, if the Nazca plate broke between 10 and 20 m.y. B.P. and the rate of subduction was different across the resulting fracture zone, there must have been a corresponding tear in the downgoing slab that might well have resulted in changes in the dip of the slab. Volcanism may have occurred along the break in the oceanic plate, and thus formed the Juan Fernandez Ridge.

ACKNOWLEDGMENTS

We thank G. L. Johnson, W. Ruddiman, P. R. Vogt, W. J. Morgan, L. R. Sykes, J. B. Coleman, and E. M. Herron for their helpful comments and for critically reading the manuscript at various stages. D. Scholl and R. von Huene kindly provided unpublished line drawings of all their Peru-Chile Trench seismic profiles; and we benefited from access to an unpublished manuscript by S. E. DeLong, J. F. Casey, and D. R. Spydell. Thanks and appreciation are extended to the officers and men of the USN ships, Kane and Bartlett, aboard which most of this work was conceived and developed. B. Grosvenor, C. Fruik, S. Edwards, B. Trott, G. Kent, and S. Tamura helped with the manuscript. R. Hey was supported by National Science Foundation Grants OCE 76-21966 and OCE 78-19816, the Office of Naval Research, and the International Decade of Ocean Exploration, Nazca Plate Project.

REFERENCES CITED

Anderson, R. N., and Langseth, M. G., 1977, The mechanisms of heat transfer through the floor of the ocean: Journal of Geophysical Research, v. 82, p. 3391–3409.

Anderson, R. N., DeLong, S. E., and Schwarz, W. M., 1978, Thermal model for subduction with dehydration in the downgoing slab: Journal of Geology, v. 86, p. 731–739.

Barazangi, M., and Isacks, B. L., 1976, Spatial distribution of earthquakes and subduction of the Nazca plate beneath South America: Geology, v. 4, p. 686–692.

Barazangi, M., Pennington, W., and Isacks, B., 1975, Global study of seismic wave attenuation in the upper mantle behind island arcs using P_p waves: Journal of Geophysical Research, v. 80, p. 1079–1092.

Bonatti, E., and Harrison, C.G.A., 1976, Hot lines in the earth's mantle: Nature, v. 263, p. 402–404.

Bonatti, E., and others, 1977, Easter volcanic chain (Southeast Pacific): A mantle hot line: Journal of Geophysical Research, v. 82, p. 2457–2478.

Carr, M. J., 1976, Underthrusting and Quaternary faulting in northern Central America: Geological Society of America Bulletin, v. 87, p. 825–829.

Carr, M. J., Stoiber, R. E., and Drake, C. L., 1973, Discontinuities in the deep seismic zones under the Japanese arcs: Geological Society of America Bulletin, v. 84, p. 2917–2929.

—— 1974, The segmented nature of some continental margins, in Burk, C. A., and Drake, C. L., eds., The geology of continental margins: New York, Springer-Verlag, Inc., p. 105–114.

Casertano, L., 1963, General characteristics of active Andean volcanoes and a summary of their activities during recent centuries: Seismological Society of America Bulletin, v. 53, p. 1415–1433.

Clark, A. H., and others, 1967, Implications of the isotopic ages of ignimbrite flows, southern Atacama Desert, Chile: Nature, v. 215, p. 723–724.

D'Angelo, E. P., and Le Bert, L. A., 1969, Relación entre estructura y volcanismo cuaternario andino en Chile: Pan-America Symposium on the Upper Mantle, Group 2, Upper Mantle Petrology and Tectonics, p. 39–46.

Defense Mapping Agency Aerospace Center, 1977a, Free-air gravity anomaly map of south America, first edition: St. Louis, scale 1:10, 000,000, Mercator projection.

—— 1977b, Operational Navigation Charts S-21, Q-26, P-26: St. Louis, scale 1:1,000,000.

Defense Mapping Agency Topographic Command, 1973, World maps: Washington, D.C., ser. 11-44, scale 1:22,000,000, Mercator projection.

DeLong, S. E., and Fox, P. J., 1977, Geological consequences of ridge subduction, in Talwani, M., and Pitman, W. C., III, eds., Island arcs, deep sea trenches, and back-arc basins: American Geophysical Union Maurice Ewing Series, v. 1, p.

Dickinson, W. R., 1973, Width of modern arc-trench gaps proportional to past duration of igneous activity in associated magmatic arc: Journal of Geophysical Research, v. 78, p. 3376–3389.

Ewing, M., Houtz, R., and Ewing, J., 1969, South Pacific sediment distribution: Journal of Geophysical Research, v. 74, p. 2477–2511.

Farrar, E., and others, 1970, K-Ar evidence for the post-Paleozoic migration of granitic foci in the Andes of northern Chile: Earth and Planetary Science Letters, v. 10, p. 60–66.

Fisher, R. L., and Raitt, R. W., 1962, Topography and structure of the Peru-Chile trench: Deep-Sea Research, v. 9, p. 423–443.

Fox, P. J., and DeLong, S. E., 1976, Healing of oceanic layer 2A: Geophysical and geological consequences: EOS (American Geophysical Union Transactions), v. 57, p. 329.

Frankel, A., and McCann, W., 1979, Moderate and large earthquakes in the South Sandwich arch: Indicators of tectonic variations along subduction zone: Journal of Geophysical Research, v. 84, p. 5571–5577.

Fuenzalida, H., and others, 1965, High stands of Quaternary sea level along the Chilean coast: Geological Society of America Special Paper 84, p. 473–496.

Guest, J. E., 1969, Upper Tertiary ignimbrites in the Andean cordillera of part of the Antofagasta province, northern Chile: Geological Society of America Bulletin, v. 80, p. 337–362.

Hamilton, W., 1969, The volcanic central Andes—A modern model for the Cretaceous batholiths and tectonics of western North American, in McBirney, A. R., ed., Proceedings of the andesite conference, Oregon, July 1968, International Upper Mantle Project, Science Report Number 16: Oregon, Department of Mineral Industries, Bulletin no. 65, p. 175–184.

Handschumacher, D. W., 1976, Post-Eocene plate tectonics of the eastern Pacific, in Sutton, G. H., and others, eds., The geophysics of the Pacific Ocean basin and its margins (Woollard volume): American Geophysical Union Geophysical Monograph 19, p. 177–202.

Hanus, V., and Vanek, J., 1978, Morphology of the Andean Wadati-Benioff zone, andesitic volcanism, and tectonic features of the Nazca plate: Tectonophysics, v. 44, p. 65–77.

Hayes, D. W., 1966, A geophysical investigation of the Peru-Chile trench: Marine Geology, v. 4, p. 309–351.

Herron, E. M., 1972, Sea-floor spreading and the Cenozoic history of the east-central Pacific: Geological Society of America Bulletin, v. 83, p. 1671–1691.

Hey, R. N., 1977, Tectonic evolution of the Cocos-Nazca spreading center: Geological Society of America Bulletin, v. 83, p. 1404–1420.

Hey, R. N., and Lowrie, A., 1976, Speculative plate tectonic explanation for the geological and geophysical discontinuities in Chile near 33°-36°S: EOS (American Geophysical Union Transactions), v. 57, p. 1004.

Hey, R. N., Johnson, G. L., and Lowrie, A., 1977, Recent tectonic evolution of the Galapagos area and plate motions in the East Pacific: Geological Society of America Bulletin, v. 88, p. 1385–1403.

Houtz, R., and Ewing, J., 1976, Upper crustal structure as a function of plate age: Journal of Geophysical Research, v. 81, p. 2490–2498.

Isacks, B. L., and Barazangi, M., 1977, Geometry of Benioff zones: Lateral segmentation and downwards bending of the subducted lithosphere, in Talwani, M., and Pitman, W. C., III, eds., Island arcs, deep sea trenches, and back-arc basins: American Geophysical Union Maurice Ewing Series, v. 1, p. 99–114.

Isacks, B. L., and Molnar, P., 1971, Distribution of stresses in the descending lithosphere from a global survey of focal mechanism solutions of mantle earthquakes: Reviews of Geophysics and Space Physics, v. 9, p. 103–174.

Isacks, B. L., Oliver, J., and Sykes, L. R., 1968, Seismology and the new global tectonics: Journal of Geophysical Research, v. 73, p. 5855–5900.

James, D. E., 1971, Plate tectonic model for the evolution of the Central Andes: Geological Society of America Bulletin, v. 82, p. 3325–3346.

Katsumata, M., and Sykes, L. R., 1969, Seismicity and tectonics of the western Pacific: Izu-Mariana-Caroline and Ryukyu-Taiwan regions: JOurnal of Geophysical Research, v. 74, p. 5923–5948.

Kelleher, J., and McCann, W., 1976, Buoyant zones, great earthquakes, and unstable boundaries of subduction: Journal of Geophysical Research, v. 81, p. 4885–4896.

Klitgord, J. D., and others, 1974, Fast sea-floor spreading on the Chile ridge: Earth and Planetary Science Letters, v. 20, p. 93–99.

Koster, R., 1975, Chile, in Fairbridge, R. W., ed., Encyclopedia of world regional geology, Part I, Western Hemisphere (including Antarctica and Australia): Stroudsburg, Pennsylvania, Dowden, Hutchinson, and Ross, Inc., p. 237–243.

LaBrecque, J. S., Kent, D. V., and Cande, S. C., 1977, Revised magnetic polarity time scale for the Late Cretaceous and Cenozoic: Geology, v. 5, p. 330–335.

Lomnitz, C., 1962, On Andean structure: Journal of Geophysical Research, v. 76, p. 351–363.

Lowrie, A., and Hey, R. N., 1976, Geological and geophysical discontinuities in Chile near 33°-35°S:

EOS (American Geophysical Union Transactions), v. 57, p. 332.
Menard, H. W., 1978, Fragmentation of the Farallon plate by pivoting subduction: Journal of Geology, v. 86, p. 99–110.
Minster, J. B., and Jordan, T. H., 1978, Present-day plate motions: Journal of Geophysical Research, v. 83, p. 5331–5354.
Mogi, K., 1969, Some features of recent seismic activity in and near Japan (2): Activity before and after great earthquakes: Earthquake Research Institute Bulletin, v. 47, p. 395–417.
Molnar, P., and Atwater, T., 1978, Interarc spreading and Cordilleran tectonics as alternates related to the age of subducted oceanic lithosphere: Earth and Planetary Science Letters, v. 41, p. 330–340.
Molnar, P., and Oliver, J., 1969, Lateral variations of attenuation in the upper mantle and discontinuities in the lithosphere: Journal of Geophysical Research, v. 74, p. 2648–2682.
Molnar, P., and others, 1975, Magnetic anomalies, bathymetry and the tectonic evolution of the South Pacific since the Late Cretaceous: Royal Astronomical Society Geophysical Journal, v. 40, p. 383–520.
Morgan, W. J., Vogt, R. P., and Fall, D. F., 1969, Magnetic anomalies and sea-floor spreading on the Chile rise: Nature, v. 222, p. 137–142.
Munoz, C., 1956, Geology of Chile, in Handbook of South America geology: Geological Society of America Memoir 65, p. 189–214.
Parker, R. L., and Oldenburg, D. W., 1973, Thermal model of ocean ridges: Nature, v. 242, p. 137–139.
Pichler, H., and Zeil, W., 1969, Andesites of the Chilean Andes, in McBirney, A. R., ed., Proceedings of the andesite conference, Oregon, July 1968, International Upper Mantle Project, Science Report Number 16: Oregon, Department of Mineral Industries, Bulletin no. 65, p. 165–174.
Plafker, G., and Savage, J. C., 1970, Mechanism of the Chilean earthquakes of May 21 and 22, 1960: Geological Society of America Bulletin, v. 81, p. 1001–1030.
Press, F., and Siever, R., 1974, Earth: San Francisco, W. H. Freeman and Company, 945 p.
Ruiz, F. C. and others, 1965, Geología y yacimientos metalíferos de Chile: Santiago, Instituto de Investigaciones Geológicas de Chile, Boletín, 385 p.
Ruiz, F. C., and Ericksen, G. E., 1962, Metallogenic provinces of Chile, S. A.: Economic Geology, v. 57, p. 91–106.
Sclater, J. G., Anderson, R. N., and Bell, M. L., 1971, Elevation of ridges and evolution of the central eastern Pacific: Journal of Geophysical Research, v. 76, p. 7888–7915.
Scholl, D. W., and others, 1970, Elevation of ridges and evolution of the central eastern Pacific: Geological Society of America Bulletin, v. 81, p. 1339–1360.
Schweller, W., Kulm, L. D., and Prince, R., 1981, Tectonics, structure, and sedimentary framework of the Peru-Chile Trench, in Kulm, L. D., and others, eds., Nazca plate: Crustal formation and Andean convergence: Geological Society of America Memoir 154 (this volume).
Segerstrom, K., 1964, Quaternary history of Chile: Brief outline: Geological Society of America Bulletin, v. 75, p. 157–170.
Sillitoe, R. H., 1974, Tectonic segmentation of the Andes: Implications for magmatism and metallogeny: Nature, v. 250, p. 542–545.
Stauder, W., 1973, Mechanism and spatial distribution of Chilean earthquakes with relation to subduction of the oceanic plate: Journal of Geophysical Research, v. 78, p. 5033–5061.
——— 1975, Subduction of the Nazca plate under Peru as evidenced by focal mechanisms and by seismicity: Journal of Geophysical Research, v. 80, p. 1053–1064.
Stoiber, R. W., and Carr, M. J., 1973, Quaternary volcanic and tectonic segmentation of Central America: Bulletin Volcanologique, v. 37, p. 304–325.
Swift, S., and Carr, M., 1974, The segmented nature of the Chilean seismic zone: Physics of the Earth and Planetary Interiors, v. 9, p. 183–191.
Sykes, L. R., 1972, Seismicity as a guide to global tectonics and earthquake prediction: Tectonophysics, v. 13, p. 393–414.
——— 1978, Intraplate seismicity, reactivation of preexisting zones of weakness, alkaline magmatism, and other tectonism postdating continental fragmentation: Reviews of Geophysics and Space Physics, v. 16, p. 621–688.
Vogt, P. R., 1973, Subduction and aseismic ridges: Nature, v. 241, p. 189–191.
Vogt, P. R., and others, 1976, Subduction of aseismic oceanic ridges: Effects on shape, seismicity, and other characteristics of consuming plate boundaries: Geological Society of America Special Paper 172, 59 p.
von Huene, R., 1974, Modern trench sediments, in Burk, C. A., and Drake, C. L., eds., The geology of continental margins: New York, Springer-Verlag, p. 207–211.
Zentilli, M., 1974, Geological evolution and metallogenetic relationships in the Andes of northern Chile between 26° and 29° South [Ph.D. thesis]: Kingston, Ontario, Queen's University, 446 p.

MANUSCRIPT RECEIVED BY THE SOCIETY NOVEMBER 12, 1980
MANUSCRIPT ACCEPTED DECEMBER 30, 1980

Printed in U.S.A.

Chile Margin near lat 38°S: Evidence for a genetic relationship between continental and marine geologic features or a case of curious coincidences?

E. M. HERRON

Lamont-Doherty Geological Observatory
of Columbia University
Palisades, New York 10964

INTRODUCTION

In a companion paper presented in this volume, Herron and others have described the geophysical characteristics of the Chile Margin triple junction at lat 46°S (see their Fig. 1). At this triple junction, an active spreading center, the Chile Ridge, has been traced into an active subduction zone, the Chile Trench. The data that were presented in the comparison paper are primarily marine observations, and at the present time, few observations on the continent have been cited to genetically link the marine and continental tectonic observations in the area of the triple junction—for example, the oceanward projection of the continent at the Taitao Peninsula and the suggested extension of the Guafo Fracture Zone landward of the trench axis onto the shelf. In this brief note, I present some observations made in the area 900 km north of the triple junction where I suspect that recent geologic features observed on the continent may have been generated in response to the *local* subduction processes. The data may turn out to be simply a series of curious coincidences, and additional work is required to fully assess the meaning of these observations.

Along the margin of central Chile, the Chile Trench is now subducting oceanic crust which was generated at two spreading centers (Fig. 1). North of lat 38°S, the trench is subducting crust formed at the Pacific-Farallon spreading center at least 35 m.y. ago. Magnetic lineations associated with this spreading system trend 330° near the continental margin. These lineations have been discussed previously in papers by Herron (1972) and Handschumacher (1976). Fracture zones associated with this spreading center trend 060°, and the southernmost fracture zone in this system is the Mocha Fracture Zone (Handschumacher, 1976), as shown in Figure 1. Between lat 40°S and the Chile Margin triple junction at lat 46°S, the Chile Trench is subducting crust that was generated within the past 20 m.y. at the Chile Ridge (the Antarctic-Nazca spreading center). Magnetic lineations associated with the Chile Ridge trend 350° and fracture zones trend 080°. Along the continental margin the northern limit of the crust generated at the Chile Ridge axis is defined by the Valdivia Fracture Zone (Fig. 1).

Between the Valdivia and Mocha Fracture Zones and the Chile Trench, a triangular area is defined.

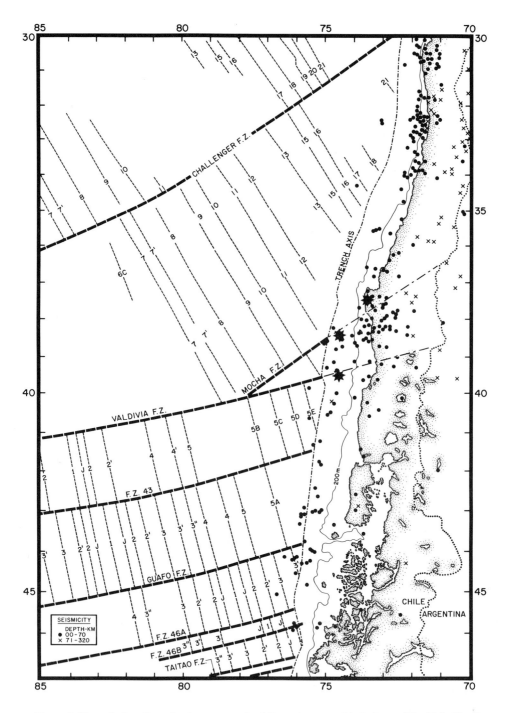

Figure 1. Magnetic lineations, fracture zones, seismicity, and geomorphic features of the Chile Margin between lat 30° and 47°S. The seismicity data are from Barazangi and Isacks (1976), and the magnetic lineation data primarily are from Herron and others (this volume). Locations for the major events of the 1960 earthquake sequence are indicated by the large stars. Possible extensions of the Mocha and Valdivia Fracture Zones onto the continental margin are suggested by the light dashed lines.

A magnetic lineation has been identified tentatively as 7' associated with the Pacific-Farallon set on two *Yaquina* 735 tracks that cross the area. The curious coincidence or the evidence of a genetic relationship between subduction and continental geologic processes is located in the vicinity of this triangular area which delineates the boundary between oceanic crust of different age and of different structural grain.

DISTRIBUTION OF EARTHQUAKES AND PATTERN OF DEFORMATION ASSOCIATED WITH MAJOR EVENTS

On May 21 and 22, 1960, the margin of central Chile was jolted by a series of extremely large earthquakes, the largest of which exceeded magnitude 8. Locations for the main events are poorly determined, but Duda (1963) and Plafker and Savage (1970) placed the main events at lat 37.5°S, long 73.5°W; lat 38.5°S, long 74.5°W; and lat 39.5°S, long 74.5°W. The latter two events occurred within 1 min of each other and may reflect a complex multiple rupture on a fault (Plafker and Savage, 1970). As shown in Figure 1, these locations, if accurate, place the main events on or very close to the Mocha and Valdivia Fracture Zones between the axis of the Chile Trench and the continent. The 1960 earthquake sequence caused the margin of Chile from lat 37° to 46°S to rupture (Plafker and Savage, 1970). Only the section of the Chile Margin that lies south of the Mocha Fracture Zone was involved in the deformation pattern. The coincidence of the extent of the rupture zone with the young oceanic crust in the subduction zone suggests that the physical properties of this relatively young and warm oceanic crust controlled the deformation associated with this earthquake sequence.

More recently, Barazangi and Isacks (1976) studied the pattern of seismicity along the margin of South America north of lat 45°S. Their relocated positions for earthquakes in the region between lat 30° and 45°S are presented in Figure 1. Barazangi and Isacks identified several discrete segments of the subduction zone along the South American margin, but treated the area between lat 20° and 45°S as forming a single subduction segment. They did note that south of lat 38°S, earthquake foci rarely occur at depths greater than 70 km. Minster and others (1974) and Minster and Jordan (1978) showed that the Nazca plate is converging on the South American continent at a rate of 80 to 100 mm/yr in the area between lat 30° and 46°S. The absence of deep earthquakes in the region between lat 38° and 46°S cannot be explained by citing a change in the convergence rate across the margin. As Herron and others (this volume) indicated, there is no evidence to support the break-up of the southern portions of the Nazca plate as it approaches South America.

In addition to the absence of deep-focus events south of lat 38°S, the position of the seismically active pattern of the epicenters appears to change abruptly at lat 38°S. Referring to the relocated events shown in Figure 1, it is apparent that north of lat 38°S, seismicity is confined to the area landward of the shelf break (the 200-m contour). South of lat 38°S, the seismicity shifts seaward and is most frequent in the region of the trench floor and inner wall of the trench.

FREE-AIR GRAVITY FIELD ALONG THE MARGIN OF CENTRAL CHILE

The free-air gravity field along the margin between lat 34° and 42°S is shown in Figure 2. The data along the margin near lat 38°S are presented to show the high degree of correlation between the changes in the amplitude of the free-air gravity minimum associated with the trench and the Mocha Fracture Zone. The amplitude of the free-air minimum decreases abruptly by 50 mgal (from −100 to −50) on crossing the Mocha Fracture Zone from north to south, and in contrast to other local gravity anomalies along the trench axis at lat 34° and 36°S, the amplitude of the minimum associated with the trench does not increase again south of the Mocha Fracture Zone beyond −50 mgal. This change in the gravity field cannot be due only to the gradually shoaling bathymetry. The depth of the trench axis decreases continuously from 5,000 m at lat 34°S to about 3,800 m at lat 42°S, according to the most recent bathymetric chart of Mammerickx and Smith (1978). The change in the gravity field at the

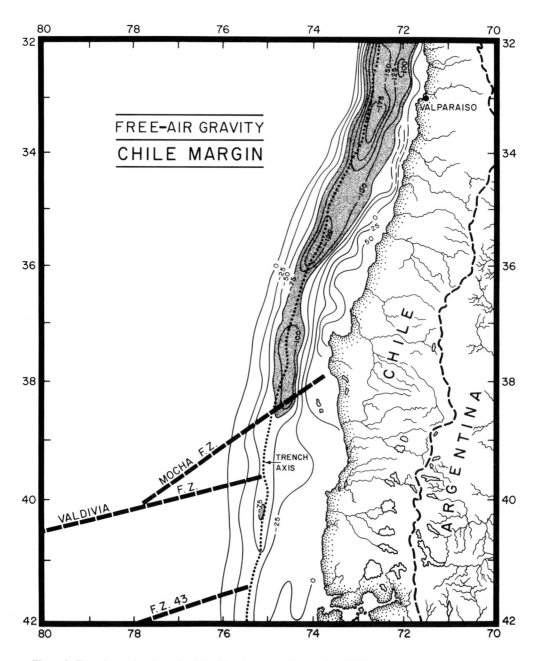

Figure 2. Free-air gravity along the Chile Margin in the vicinity of lat 38°S. Note the abrupt and permanent decrease in the amplitude of the minimum free-air anomaly (dense stippled pattern) associated with the Chile Trench that occurs at the Mocha Fracture Zone. Contours given in milligals.

GEOMORPHIC AND STRUCTURAL GEOLOGY OBSERVATIONS ON THE CHILE MARGIN

The continental edge in the area bounded by the Mocha and Valdivia Fracture Zones is controlled by a north-trending fault (Mordojovich, 1974). This fault was active during the 1960 earthquake. North-trending faults are also postulated for the entire shelf in this region, and the western margin of Mocha Island has been identified as a fault scarp (Mordojovich, 1974). North of the Mocha Fracture Zone and south of the Valdivia Fracture Zone there is no evidence to suggest that the margin of Chile is controlled by active faults. If one extends the trends of the Mocha and Valdivia Fracture Zones landward to the continental divide (coincident in this area with the political boundary between Argentina and Chile shown in Fig. 1), there appears to be a direct correlation between the extension of the fault traces and right-lateral offsets of the continental divide. South of the extended fault trace for the Valdivia Fracture Zone, the Central Valley of Chile is marked by a chain of lakes which continues southward to lat 42°S where the Central Valley is submerged in the Gulf of Corcovado.

SUMMARY AND QUESTIONS FOR FURTHER STUDY

The observations outlined above may constitute only a case of "missing the forest for the trees." However, the association of a rather large and significant change in the age of the crust being subducted, from 35 m.y. on the north to less than 20 m.y. on the south, combined with the change in the structural "grain" of the crust in the two regions, makes me very suspicious that in this area, the geologic situation is conducive to the formation of geologic features that can directly correlate with the *local* subduction processes.

J. Bodine and S. C. Cande (in prep.) are developing a model to explain the marine and continental data along the Chile Margin between lat 30° and 48°S through the combined use of the gravity data and data on the thermal properties of subducting oceanic crust of different ages. The model developed by DeLong and Fox (1977) which predicts regional uplifting of the continental margin as a ridge crest approaches the subduction zone (see paper by Herron and others, this volume) does not appear to be applicable to the Chile Margin. I suspect that the conditions which are conducive to the formation of correlatable marine and continental features are so restricted that such correlations persist only for a geologic instant and hence are rarely observable. The marine and continental features discussed in this note are cited as evidence that at the present time, directly correlatable structural features can be identified on the margin of central Chile.

ACKNOWLEDGMENTS

This research was supported by National Science Foundation Grant OCE 76 01811. I thank S. C. Cande, A. Watts, and R. N. Anderson for their helpful critique of this note.

REFERENCES CITED

Barazangi, M., and Isacks, B. L., 1976, Spatial distribution of earthquakes and subduction of the Nazca plate beneath South America: Geology, v. 4, p. 686–692.

DeLong, S. E., and Fox, P. J., 1977, Geological consequences of ridge subduction, *in* Talwani, M., and Pitman, W. C., III, eds., Island arcs, deep sea trenches and back-arc basins: American Geo-

physical Union, Maurice Ewing Series, v. 1, p. 221–228.
Duda, S. J., 1963, Strain release in the circum-Pacific belt, Chile: Journal of Geophysical Research, v. 68, p. 5531–5544.
Handschumacher, D. W., 1976, Post-Eocene plate tectonics of the eastern Pacific, *in* Sutton, G. D., and others, eds., The geophysics of the Pacific Ocean basin and its margin: American Geophysical Union Geophysical Monograph 19, p. 177–202.
Herron, E. M., 1972, Sea-floor spreading and the Cenozoic history of the east-central Pacific: Geological Society of America Bulletin, v. 38, p. 1671–1672.
Herron, E. M., Cande, S. C., and Hall, B. R., 1981, An active spreading center collides with a subduction zone: A geophysical survey of the Chile Margin triple junction, *in* Kulm, L. D., and others, eds., Nazca plate: Crustal formation and Andean convergence: Geological Society of America Memoir 154 (this volume).
Mammerickx, J., and Smith, S. M., 1978, Bathymetry of the southeast Pacific: Geological Society of America Map and Chart Series, MC-26, scale 1:6,442,194.

Minster, J. B., and Jordan, T. H., 1978, Present day plate motions: Journal of Geophysical Research, v. 83, p. 5331–5354.
Minster, J. B., and others, 1974, Numerical modeling of instantaneous plate tectonics: Royal Astronomical Society Geophysical Journal, v. 36, p. 541–576.
Mordojovich, C., 1974, Geology of a part of the Pacific margin of Chile, *in* Burk, C. A., and Drake, C. L., eds., The geology of continental margins: New York, Springer-Verlag, p. 591–598.
Plafker, G., and Savage, J. C., 1970, Mechanism of the Chilean earthquakes of May 21 and 22, 1960: Geological Society of America Bulletin, v. 81, p. 1001–1030.

MANUSCRIPT RECEIVED BY THE SOCIETY NOVEMBER 12, 1980

MANUSCRIPT ACCEPTED DECEMBER 30, 1980

L-DGO CONTRIBUTION NO. 2951

Geological Society of America
Memoir 154
1981

Convergence and mineralization—Is there a relation?

C. WAYNE BURNHAM
Department of Geosciences
Pennsylvania State University
University Park, Pennsylvania 16802

ABSTRACT

Mineralization in regions of plate convergence is dominantly of the hydrothermal type, either directly associated with high-level (subvolcanic) intrusive bodies or with manifestations of explosive volcanism. In both associations, a strong case can be made for a close connection between mineralization and the processes that operate during generation, emplacement, and solidification of hydrous felsic magmas, whether such magmas are generated from appropriate source rocks in a subduction zone or in the lower continental crust.

The thesis espoused here is that magmas of appropriate compositions, including abnormally high metal, S, and H_2O contents, are normal products of partially melting nonporous mafic amphibolites of typical oceanic tholeiite composition in the upper parts of a subducting plate. Furthermore, the depth at which these magmas are generated is fully consistent with the depth to the top of the seismic zone under many volcanic arcs. There is no need, therefore, to call upon abnormally metalliferous source rocks or other exotic genetic schemes to account for the association of mineralization with plate convergence. Hydrothermal mineralization is linked to convergence, as to other tectonic regimes, through hydrous magmas.

INTRODUCTION

An attractively simple model of ore genesis in the Peruvian and Chilean Andes involves remobilization of certain metals from metalliferous sediments on the Nazca plate during subduction. A more extreme version of this model involves derivation of the ore-forming fluids by metamorphic dehydration of subducted oceanic crust and thus implies a direct relation between mineralization and plate convergence. For reasons discussed below, this latter version does not appear to be tenable. Furthermore, abnormally metalliferous source rocks do not appear to be required for the formation of the ore deposits that are typical of the Andes and other regions of plate convergence.

Mineralization in regions of plate convergence is typically of the hydrothermal type and is associated, spatially and temporally, with calc-alkalic intrusive or extrusive igneous activity. In intrusive environments, mineralization is commonly of the porphyry type, and a strong case can be made for the hydrothermal ore-forming fluids having been derived directly from crystallizing hydrous magmas. In many massive sulfide and vein-type deposits in extrusive rocks and volcanogenic sediments, on the other hand, an equally strong case can be made for the hydrothermal ore-forming fluids having been derived largely from local groundwaters of meteoric or seawater origin. Despite the nonmagmatic origin of these fluids, however, hydrous magmas play an important role in the formation of the deposits, in addition to providing sources of thermal energy to drive convectively circulating groundwater systems. Explosive volcanism, which creates extensive fracture systems and other geologic conditions conducive to large-scale convective circulation of fluids, may be linked directly to the large increase in volume that necessarily attends crystallization in shallow-seated bodies of hydrous magma (Burnham, 1972, 1979).

Thus, there appears to be a relation—direct in some cases and indirect in others—between hydrothermal mineralization and hydrous magmas of appropriate compositions, wherever such magmas have been emplaced at sufficiently shallow depths to cause extensive fracturing upon solidification. The question posed in the title of this communication might, therefore, be rephrased: "Convergence, hydrous magmas, and mineralization—is there a relation?" Even if this question can be answered affirmatively, however, it does not imply that hydrous, ore-bearing magmas of requisite composition are unique to regions of plate convergence. Hydrous magmas of requisite composition can be generated in other tectonic regimes from appropriate source rocks.

CONVERGENCE AND HYDROUS MAGMA GENERATION

Metamorphism of Subducted Oceanic Crust

In most models for the consumption of oceanic crust and lithosphere in regions of plate convergence, oceanic lithosphere, carrying with it basaltic crust and minor intercalated sediments, descends in the vicinity of oceanic trenches at angles ranging from about 20° to 50° or more (Allen and others, 1975). In the model of Turcotte and Schubert (1973), average dips of Benioff zones and widths of arc-trench gaps were combined with an assumed initial melting temperature of 1,000 °C to obtain an estimate of the geothermal gradient in slip zones of about 11 °C/km. This estimated gradient, while perhaps realistic for slip zones, does not take into consideration the possibility that seismic activity, which defines Benioff zones, may be concentrated in the cooler inner parts of descending plates. The dips of Benioff zones, therefore, may be steeper than the dips of slip zones. Also, an assumed initial melting temperature of 1,000 °C, although again perhaps realistic for partially hydrated oceanic crust (mafic amphibolite), appears to be applicable only for geothermal gradients of about 13 °C/km (Fig. 1). According to the experimental results of Allen and others (1972) and the thermodynamic calculations of Burnham (1979), melting of partially hydrated (nonporous) mafic amphibolite would begin along a geothermal gradient of 11 °C/km at a temperature of about 860 °C and a depth of 75 to 80 km. These melting relations will be discussed further after briefly examining the mode of occurrence of H_2O and its behavior during premelting metamorphism.

Before descent of oceanic crust, H_2O occurs as water in interstitial pores of the sediments, as well as in fractures and vesicles of the basalts, and as hydroxyl in clays of the sediments and chloritic minerals or serpentine of the basalts. Upon descent, probably much of the water-rich sediments accumulates as a thickened wedge that forms the inner wall of the trench; hence the average total H_2O content of the uppermost few kilometres of the descending plate probably does not normally exceed 2 or 3 wt %. As temperatures increase by conduction and frictional heating near the slip zone, the H_2O-rich upper part of the plate must undergo metamorphism. The initial metamorphic reactions that occur in basaltic rocks as temperature and pressure increase are predominantly hydration reactions that yield greenschist-facies, and locally blueschist-facies, mineral assemblages. These reactions can consume far more H_2O (up to 13 wt % in chlorite and serpentine) than is generally available; furthermore, they release sufficient heat (exothermic) to raise the temperature as much as 2 °C for every percent of rock hydrated. Thus, it is probable that essentially all of the initial pore water in sea-floor basalts becomes

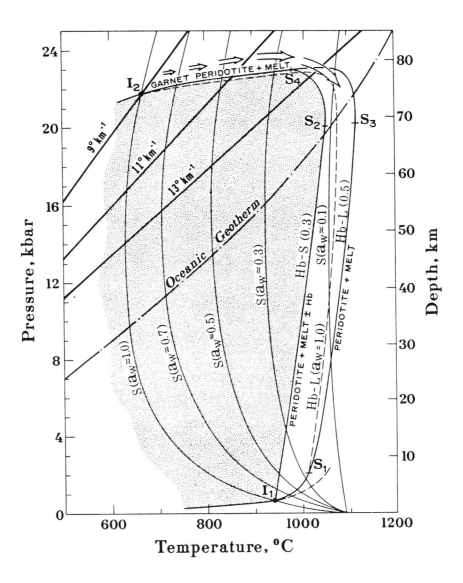

Figure 1. Pressure-temperature projection of melting relations for an amphibolite of olivine tholeiite composition. Nonporous amphibolite is stable (no melt formed) throughout the P-T range covered by the stippled pattern, but undergoes partial melting along the pseudo-univariant boundary I_1-S_2-S_4-I_2. Hornblende is not stable in this bulk composition at temperatures and pressures above the pseudo-univariant boundary I_1-S_1-S_3-S_4-I_2. Arrows indicate the postulated direction of migration of hydrous partial melts formed in subduction zones where the average geothermal gradient is between 9 and 13 °C/km, as discussed in the text; the oceanic geotherm is from Ringwood and others (1964). Melting relations are based on the experimental results of Allen and others (1972), Boyd (1959), Eggler and Burnham (1973), Hamilton and others (1964), Helz (1973), Hill and Boettcher (1970), Holloway (1973), Holloway and Burnham (1972), and Yoder and Tilley (1962), as well as thermodynamic calculations by Burnham (1979). The curves labelled $S(a_w \approx 1.0)$ to $S(a_w \approx 0.1)$ are solidi (beginning of melting) for constant activities of H_2O (a_w) in the range of 1.0 to 0.1, respectively. The pseudo-univariant Hb-S (0.3) boundary between I_1 and S_2 represents the beginning of melting of nonporous amphibolite, which occurs at a constant mole fraction of H_2O (X_w^m) in the melt of 0.3 (~3.0 wt % H_2O). Also, the Hb-L (0.5) boundary between S_1 and S_4 represents the maximum thermal stability of hornblende (Hb), which occurs at a constant $X_w^m \approx 0.5$ (~6.4 wt % H_2O).

bound in hydrous minerals at this stage and that virtually none of it is released from the descending oceanic plate at depths shallower than about 40 km.

Upon further descent of the plate and heating to temperatures of 550 to 600 °C, the upper part passes into the amphibolite facies of metamorphism. The more highly hydrated minerals of the lower-grade facies react to produce correspondingly larger amounts of less-hydrated amphibole, which generally contains 2.0 to 3.0 wt % H_2O. Unless the bulk H_2O content of the uppermost few kilometres of descending plate exceeds this amount, however, significant quantities of free H_2O are not released from the system, because basalts can be converted almost entirely into compositionally equivalent amphibole (Yoder and Tilley, 1962). Furthermore, any free H_2O that might escape into the overlying, generally hotter (Toksöz and others, 1971) peridotitic lithosphere would be immediately absorbed in amphibolitization and serpentinization reactions. Therefore, the metamorphic dehydration hypothesis for the derivation of hydrothermal ore-forming fluids in regions of plate convergence does not appear to be tenable.

Partial Melting of Subducted Oceanic Crust

In light of the above discussion, it appears probable that at pressures greater than about 10 kb in a subduction zone, amphibolitized oceanic crust (olivine tholeiite) will have essentially zero porosity and all H_2O will be bound as hydroxyl, mainly in hornblende. Under these conditions, as shown in Figure 1, a non-porous mafic amphibolite is stable to depths of about 75 to 80 km ($P \approx$ 22 to 23 kb), whether the average geothermal gradient in the slip zone is 11 °C/km (Turcotte and Schubert, 1973), as low as 9 °C/km, or as high as 13 °C/km. At greater depths (pressures) within this range of geothermal gradients, hornblende undergoes incongruent melting, the solid residue from which is garnet peridotite, the prevailing rock type in the suboceanic lithosphere at these depths. Under higher geothermal gradients, hornblende also melts incongruently, but at correspondingly shallower depths along "curve" S_4-S_2-I_1 in Figure 1. (Owing to compositional variations in hornblende solid solutions, phase boundaries I_2-S_4-S_2-I_1 and I_2-S_4-S_3-S_1-I_1 are not univariant, as implied. Instead, reaction occurs over limited ranges of pressure and temperature for a given bulk composition. For a more detailed discussion of these melting relations, see Burnham, 1979).

Melts produced along the boundary I_2-S_4-S_2-I_1 range in composition from rhyodacitic to quartz dioritic, depending mainly on the degree of melting. The degree of melting, in turn, depends on the bulk H_2O content of the original amphibolite and, for the boundary segment between I_2 and S_2, the temperature at which the ambient geotherm intersects the boundary (beginning of melting). Along the beginning-of-melting "curve" between S_2 and I_1, on the other hand, the degree of melting is independent of temperature, because the H_2O content of the melts is fixed at about 3.0 wt % by the hornblende-melt equilibrium.

The dependency of the degree of melting on bulk H_2O content and temperature between I_2 and S_2 in Figure 1 stems from the fact that a specific minimum activity of H_2O is required for melt to coexist with other incongruent melting products of mafic amphibolite at a given pressure and temperature. These minimum activities of H_2O for stable existence of melt in an amphibolite of olivine tholeiite composition have been calculated by Burnham (1979) and are shown as isoactivity (a_w) contours in Figure 1. As indicated in the figure, the minimum activity of H_2O for stable existence of melt at I_2 on the 9 °C/km geotherm is essentially 1.0 (H_2O-saturated solidus) which, at this temperature (660 °C) and pressure (21.7 kb), requires an H_2O content of the melt of about 27 wt %. Thus, in nonporous mafic amphibolite with a bulk H_2O content of 1.3 wt % (Nockolds's, 1954, average hornblendite), for example, only about 5% melt is formed. At the intersection of the 13 °C/km geotherm and the melting "curve" between S_4 and S_2 (1,015 °C and 22.3 kb), on the other hand, $a_w \approx 0.17$ and the H_2O content of the melt is approximately 4.0 wt %; hence the same amphibolite will yield 33% melt. Similarly, this same amphibolite will yield approximately 13% melt that contains 9.8 wt % H_2O at the intersection of the boundary "curve" and the 11 °C/km geotherm of Turcotte and Schubert (1973). Of course, if the initial H_2O content of the amphibolite is less than 1.3 wt %, the amounts of melt formed in these examples would be proportionately less; however, the H_2O contents of the melts are fixed at the same values by the conditions of hornblende-melt equilibrium (P, T, a_w).

It is evident from the foregoing discussions that fluid-absent partial melting of nonporous mafic amphibolite yields H_2O-rich melts (3 to 27 wt % H_2O), whether melting occurs at 1 or 23 kb. Also, within a geophysically reasonable range of average geothermal gradients in subduction (slip) zones (9 to 13 °C/km), hydrous melts containing from 4 to 27 wt % H_2O are produced over a very narrow depth (pressure) interval (Fig. 1). It is this feature of the melting relations of a mafic amphibolite that is thought (Burnham, 1979) to be responsible for the remarkable alignment of volcanic centers equidistant from trench axes in regions of plate convergence. The connection between this nearly isobaric generation of hydrous magmas and volcanism, however, is not as simple or direct as this statement might imply.

Hydrous magmas generated along the nearly isobaric boundary I_2-S_4 are incapable of upward intrusion into the overlying peridotitic lithosphere by virtue of their high H_2O contents, low temperatures, and the fact that they formed in equilibrium with a peridotitic residue. Any reduction in pressure, as must accompany upward movement, would result in complete reaction of the hydrous melt with mantle peridotite to reform hornblende—a process akin to "underplating". [It is of interest to note that this lithospheric "underplating" is not restricted to the high-pressure regime. It should occur wherever an H_2O-bearing melt encounters peridotitic or gabbroic rocks within the P-T stability field of nonporous amphibolite ($P < I_2$-S_4, $T < S_4$-I_1 in Fig. 1).]

In order for the hydrous melts produced on the high-pressure side of I_2-S_4 to escape this fate (that is, the descending plate) and rise into the overlying lithosphere, they must be heated to sufficiently high temperatures that the increased degree of melting will reduce their H_2O content to about 3 wt % or less. This condition is met, as indicated in Figure 1, where temperatures exceed about 1,025 to 1,050 °C, but such temperatures at depths of 75 to 80 km exist only at some distance away from the descending plate, in the overlying lithosphere (Toksöz and others, 1971). The actual lateral distance from the slip zone at which these temperatures prevail depends primarily on the rate of subduction. The faster the rate, the greater lateral distance the melts must migrate before they attain sufficiently high temperatures, low H_2O contents, and hence large mass to prevent recrystallization of hornblende (except near the margins) as they ascend to shallower and cooler depths. Thus, variations in the widths of volcanic arc–trench gaps, from one region of plate convergence to another, may reflect different combinations of rates and angles of subduction. Narrow gaps presumably reflect slow rates and steep angles of subduction, whereas very wide gaps, as in South America and Japan, may reflect fast rates and shallow angles. [A subduction rate of 8 cm/yr at an angle of 45° (Toksöz and others, 1971, Fig. 7) yields an arc-trench gap of about 175 km, whereas a 1 cm/yr rate at the same angle (Fig. 10) yields a gap of about 80 km (negligible lateral migration of melt).] Cardwell and Isacks (1978) and Isacks and Barazangi (1977), on the other hand, have presented evidence that the volcanoes in the Indonesian and many other arc-trench systems are located about 100 km above the top of the seismic zone. Hence, if the top of the seismic zone is roughly coincident with the slip zone, then a 100-km depth beneath the volcanoes would suggest a lateral migration of melts of only about 25 km or less before ascent to the surface. In any case, the geophysical evidence supports the hypothesis that the melting relations of hornblende in rocks of basaltic composition are the principal control on the location of extrusive and intrusive igneous activity in regions of plate convergence, as well as on the compositions of the magmas generated.

In their lateral migration into the hotter lithosphere, the H_2O-rich melts thus produced must dissolve a certain amount of the lower-melting fraction of the lithosphere, depending on the temperature at which they separate from the descending plate. A melt that separates from the plate near I_2 in Figure 1, as an extreme example, will dissolve nearly nine times its initial mass before it can circumvent the "hornblende barrier." Obviously, such a magma will have lost most of its original identity and have taken on many of the minor-element and isotopic characteristics of the overlying lithosphere. On the other hand, a melt that separates at 1,000 °C and is further heated to 1,150 °C in passing through the overlying lithosphere will increase in mass only about 50% and, hence, may retain much of its original character. Such a melt, incidentally, contains about 1.5 wt % H_2O and is tholeiitic in composition.

Behavior of Metal Sulfides During Partial Melting

The experimental results of Haughton and others (1974) indicate that the solubility of S in anhydrous basaltic melts at constant temperature and fugacities of O_2 and S_2 is directly dependent on the FeO content of the melt. For a Hawaiian basalt melt at 1,200 °C with an O_2 fugacity of the QFM buffer and an S_2 fugacity of about 1 bar, they calculated a solubility of 0.1 wt % S, which is in good agreement with that measured in quenched glasses by Moore and Fabbi (1971). They also estimated, in agreement with earlier workers, that the solubility decreased by a factor of about five for each 100 °C decrease in temperature between 1,200 and 1,000 °C, which would imply a solubility of only about 40 ppm S at 1,000 °C. Furthermore, in more felsic melts with lower Fe contents, solubilities of S presumably would be even less. On the other hand, experiments conducted in my laboratory on Fe-poor granitic melts at 850 °C—also with an O_2 fugacity of the QFM buffer and an S_2 fugacity of about 1 bar, and in equilibrium with an aqueous phase initially 1.0 m in NaHS—yielded a content of 0.17 wt % S at 2.0-kb H_2O pressure (~6.3 wt % H_2O in the melt).

This extremely large apparent discrepancy is not a discrepancy at all, but another illustration of the profound effects of H_2O on equilibrium relations in magmatic systems. The solution reaction of H_2O with silicate melts, as discussed by Burnham (1979 and elsewhere) forms hydroxyl $(OH)^-$ complexes in the melt, which in turn react with sulfides (principally pyrrhotite) to form bisulfide $(SH)^-$ complexes in the melt. For the solution of stoichiometric FeS, the reaction may be written: $2FeS(s) + SiO_2(m) + 2(OH)^-(m) \rightleftarrows Fe_2SiO_4(m) + 2(SH)^-(m)$, where (s) and (m) refer to the solid and melt phases, respectively. From this equilibrium reaction it is evident that increasing the H_2O content of the melt, which increases the activity of $(OH)^-(m)$, shifts equilibrium toward increased solubilities of sulfide minerals. At the same time, however, increasing the H_2O content also increases the fugacity of H_2S, through the equilibrium reaction $(SH)^-(m) + (OH)^-(m) \rightleftarrows H_2S(v) + O^{2-}(m)$, where (v) refers to the vapor phase, and this promotes precipitation of sulfide minerals (limits sulfide solubilities).

The H_2O content of the melt at which the latter equilibrium limits the solubility of pyrrhotite, for a given temperature and pressure, is the subject of current experimental investigation. The data available suggest that it is in the order of 4 wt % H_2O at 2 kb and somewhat higher at higher pressures. The corresponding maximum solubility of pyrrhotite at 2 kb is about 0.5 to 0.6 wt % (0.2 wt % S), and it probably increases significantly at very high pressures. Thus, it is reasonable to assume that, in melts formed from amphibolitized oceanic basalt in subduction zones at temperatures between 900 °C (~10 wt % H_2O) and 1,050 °C (~3 wt % H_2O), the solubility of pyrrhotite will be at least 0.5 wt %. At lower temperatures, the solubility of pyrrhotite may be somewhat less, owing to the higher H_2O contents of the melts.

According to the summary data of Engel and others (1965), the average FeS and Cu contents of oceanic tholeiites are 0.1 wt % and 77 ppm, respectively. Upon amphibilitization of these basalts during subduction, essentially all of this Cu may be expected to combine with the "FeS" to form a pyrrhotite solid solution in which the Cu content is about 7.7 wt %, and upon partial melting of the amphibolite, this cupriferous pyrrhotite will dissolve in accordance with the solubility relations discussed above. Thus, under conditions where partial melt constitutes 20% or less of the total and its H_2O content is in the range 3 to 10 wt %, the Cu content of the melt will be about 385 ppm. Such high Cu contents are rarely encountered in unaltered igneous rocks associated with porphyry Cu or other types of hydrothermal ore deposits. Therefore, it is unnecessary to call upon abnormally Cu-rich rocks, such as the metalliferous sediments of the Nazca plate, for the sources of ore-bearing magmas.

This same argument is equally applicable to other metals and other tectonic regimes. Mafic amphibolites appear to be adequate source rocks for most metal sulfide-rich hydrous magmas, whether in a subduction zone or in the lower continental crust. Similarly, metasedimentary rocks in the lower continental crust appear to be appropriate sources of hydrous granitic magmas rich in Sn and other less-chalcophile elements.

THE LINK BETWEEN CONVERGENCE AND MINERALIZATION

It has been argued above that subduction (convergence) of amphibolitized oceanic basalts results in generation of hydrous magmas whose compositions, at the time of ascent toward the surface, are appropriate to yield the common igneous rocks of volcanic arcs. It was argued further that, by virtue of their high H_2O contents, the magmas thus generated have a high S-carrying capacity (high solubilities of metal sulfides); hence they are effective in extracting metals from their source rocks. Also by virtue of their high H_2O content at the time of ascent (~3 wt % H_2O), some of these magmas are endowed with the capacity for extensive hydrothermal activity, as well as for explosive eruption, upon reaching the near-surface environment. The critical link between convergence and mineralization, then, is the hydrous magmas generated by partial melting of appropriate source rocks (amphibolitized oceanic crust) in subduction zones.

It should be emphasized, again, that this relationship between hydrothermal mineralization and H_2O-rich felsic magmas is not restricted to convergent tectonic regimes. It may exist wherever partially hydrated rocks undergo appropriate degrees of partial melting and the resulting magmas are emplaced in the near-surface (uppermost 6 to 8 km) environment. The critical factor in the relationship is the H_2O content of the magma, both at its source and at the site of emplacement. Felsic magmas that contain less than about 2 wt % H_2O are relatively ineffective in concentrating metals and S from their source rocks and in producing extensive hydrothermal activity or explosive volcanism near the surface. On the other hand, magmas containing more than about 3 wt % H_2O in the upper mantle or 6 wt % in the lower continental crust are incapable of reaching the near-surface environment without crystallizing completely. Perhaps this narrow range of optimum H_2O content is the main reason that major hydrothermal ore deposits are such anomalous features of the Earth's crust.

REFERENCES CITED

Allen, J. C., and others, 1972, The role of water in the mantle of the earth: The stability of amphiboles and micas: Inernational Geological Congress, 24th, Montreal, Section 2, p. 231–240.

Allen, J. C., Boettcher, A. L., and Marland, G., 1975, Amphiboles in andesite and basalt: I. Stability as a function of P-T-f_{O_2}: American Mineralogist v. 60, p. 1069–1085.

Boyd, F. R., 1959, Hydrothermal investigations of amphiboles, in Abelson, P., ed., Researches in geochemistry: New York, John Wiley, p. 377–396.

Burnham, C. W., 1972, The energy of explosive volcanic eruptions: Earth and Mineral Sciences, v. 41, p. 69–70.

———1979, Magmas and hydrothermal fluids, in Barnes, H. L., ed., Geochemistry of hydrothermal ore deposits, 2nd edition: New York, John Wiley & Sons, Inc., p. 71–136.

Cardwell, R. K., and Isacks, B. L., 1978, Geometry of the subducted lithosphere beneath the Banda Sea in eastern Indonesia from seismicity and fault plane solutions: Journal of Geophysical Research, v. 83, p. 2825–2838.

Eggler, D. H., and Burnham, C. W., 1973, Crystallization and fractionation trends in the system andesite-H_2O-CO_2-O_2 at pressure to 10 kb: Geological Society of America Bulletin, v. 84, p. 2517–2532.

Engel, A.E.J., Engel, C. G., and Havens, R. G., 1965, Chemical characteristics of oceanic basalts and the upper mantle: Geological Society of America Bulletin, v. 76, p. 719–734.

Hamilton, D. L., Burnham, C. W., and Osborn, E. F., 1964, The solubility of water and effects of oxygen fugacity and water content on crystallization in mafic magmas: Journal of Petrology, v. 5, p. 21–39.

Haughton, D. R., Roeder, P. L., and Skinner, B. J., 1974, Solubility of sulfur in mafic magmas: Economic Geology, v. 69, p. 451–467.

Helz, R. T., 1973, Phase relations of basalts in their melting range at $P_{H_2O} = 5$ kb as a function of oxygen fugacity: Part I. Mafic phases: Journal of Petrology, v. 14, p. 249–302.

Hill, R.E.T., and Boettcher, A. L., 1970, Water in the earth's mantle: Melting curves of basalt-water and basalt–water–carbon dioxide: Science, v. 167, p. 980–981.

Holloway, J. R., 1973, The system pargasite-H_2O-CO_2: A model for melting of a hydrous mineral with a mixed-volatile fluid—I. Experimental results to 8 kbar: Geochimica et Cosmochimica Acta, v. 37, p. 651–666.

Holloway, J. R., and Burnham, C., 1972, Melting relations of basalt with equilibrium-water pressure less than total pressure: Journal of Petrology, v. 13, p. 1–12.

Isacks, B. L., and Barazangi, M., 1977, Geometry of Benioff zones: Lateral segmentation and downwards bending of the subducted lithosphere, in Talwani, M., and Pitman, W. C., III, eds., Island arcs, deep sea trenches, and back-arc basins: American Geophysical Union Maurice Ewing Series, v. 1, p. 99–114.

Moore, J. G., and Fabbi, B. P., 1971, An esitmate of the juvenile sulfur content of basalt: Contributions to Mineralogy and Petrology, v. 33, p. 118–127.

Nockolds, S. R., 1954, Average chemical compositions of some igneous rocks: Geological Society of America Bulletin, v. 65, p. 1007–1032.

Ringwood, A. E., MacGregor, I. D., and Boyd, F. R., 1964, Petrologic composition of the upper mantle: Carnegie Institution of Washington Year Book 63, p. 147–152.

Toksöz, M. N., Minear, T. W., and Julian, B. R., 1971, Temperature field and geophysical effects of a downgoing slab: Journal of Geophysical Research, v. 76, p. 1113–1138.

Turcotte, D. L., and Schubert, G., 1973, Frictional heating on the descending lithosphere: Journal of Geophysical Research, v. 78, p. 5876–5886.

Yoder, H. S., Jr., and Tilley, C. E., 1962, Origin of basalt magmas: An experimental study of natural and synthetic rock systems: Journal of Petrology, v. 3, p. 342–532.

MANUSCRIPT RECEIVED BY THE SOCIETY NOVEMBER 12, 1980
MANUSCRIPT ACCEPTED DECEMBER 30, 1980

Role of subducted continental material in the genesis of calc-alkaline volcanics of the central Andes

DAVID E. JAMES
Department of Terrestrial Magnetism
Carnegie Institution of Washington
5241 Broad Branch Road, N.W.
Washington, D.C. 20015

ABSTRACT

Late Cenozoic calc-alkaline lavas of the central Andean continental volcanic arc are characterized by high $^{87}Sr/^{86}Sr$ ratios (0.705 to 0.708) and high K, Rb, Sr, Ba, LREE, and Ni concentrations relative to island-arc volcanics of comparable SiO_2 content. Conflicting published interpretations of these "anomalous" compositions include (1) sialic contamination of a mantle-derived magma of low $^{87}Sr/^{86}Sr$ and (2) melting of enriched continental lithospheric mantle. The $\delta^{18}O$ values for these volcanics, however, lie almost entirely in the range +7.0 to +7.5$^0/_{00}$, slightly higher than accepted mantle values of +5.5 to +6.5$^0/_{00}$, but too low to be compatible with the large degree of crustal contamination required to alter the $^{87}Sr/^{86}Sr$ ratios of a mantle-derived magma. Moreover, $\delta^{18}O$ values tend to decrease slightly with increasing $^{87}Sr/^{86}Sr$. Fractional crystallization as a major factor is precluded by the strong covariance of trace-element concentrations with Sr-isotope ratios.

Melting of sialic material in the subduction zone can account for observed $^{87}Sr/^{86}Sr$ ratios, $\delta^{18}O$ values, and trace-element abundances in the lavas of southern Peru. The data fit a petrogenetic model similar to that proposed by Nicholls and Ringwood in which magma derived from a subducted slab of oceanic crust and continental material rises into and reacts with overlying mantle to yield calc-alkaline magma. Sr-isotope ratios and trace-element abundances of magmas generated by this process are dominated by the slab-derived material, whereas O-isotope ratios and major-element abundances are dominated by the volumetrically more important mantle material.

O-isotope data on the Peruvian volcanics suggest that about one part slab melt equilibrates with four parts mantle material. Sr-isotope ratios and Sr abundances indicate that approximately 60% to 90% of the slab melt is derived from sialic material and the remainder from quartz eclogite. Predicted magmatic compositions match observed compositions when partial melting of the slab is assumed to be 10% to 20% and partial melting of modified mantle is assumed to be 7% to 15%.

INTRODUCTION

Late Cenozoic calc-alkaline lavas of the Central Andean Volcanic arc have been erupted from a chain of stratovolcanoes that extend from northern Chile into southern Peru. The volcanics are unusual in that they are characterized by high $^{87}Sr/^{86}Sr$ low $^{206}Pb/^{204}Pb$ (Tilton, 1979) low $^{143}Nd/^{144}Nd$ (Tilton, 1979; De Paolo and Wasserburg, 1977) and generally high K, Rb, Sr, Ba, LREE (light rare-earth elements), and Ni concentrations relative to island-arc volcanics of comparable SiO_2 content and depth to underlying Benioff zone. Conflicting interpretations of these "anomalous" isotope ratios and trace-element abundances include (1) sialic contamination of a mantle-derived magma of low $^{87}Sr/^{86}Sr$ (see Klerkx and others, 1977; Briqueu and Lancelot, 1979) and (2) melting of enriched continental lithospheric mantle (James and others, 1976). An alternative model involving partial melting of subducted oceanic crust with subsequent rise of magma into overlying mantle which reacts to yield andesitic magma (Thorpe and others, 1976) appears unable to account for the high $^{87}Sr/^{86}Sr$ ratios observed.

The extent to which crustal material is involved in magma genesis or incorporated into a magma after it has formed may be estimated from the O-isotope ratios of the rock—a result of most crustal material being richer in ^{18}O than mantle material. Thus, the O-isotope data alone show that the hypothesis of James and others (1976) cannot be wholly correct, for Andean volcanics exhibit slight ^{18}O enrichment relative to volcanics derived from the mantle. The nature and degree of inferred sialic contamination may be investigated through O- and Sr-isotope and trace-element variations observed in the lavas of southern Peru.

High $^{87}Sr/^{86}Sr$ ratios and $\delta^{18}O$ values of the Central Andean andesites may be due either to melting of sialic material at the source or to crustal contamination. The case against crustal contamination of a parental magma with a "normal" Sr-isotope ratio (that is, $^{87}Sr/^{86}Sr < 0.704$) has been reviewed previously by James and others (1976). The crux of the argument lies in the high Sr concentrations (500 to 900 ppm) of the eruptives. Those high Sr concentrations require that for a parent magma of low $^{87}Sr/^{86}Sr$, either very large amounts of contaminant were assimilated or the contaminant has a high $^{87}Sr/^{86}Sr$ ratio (probably >0.73); however, plots of other trace elements against $^{87}Sr/^{86}Sr$ ratios (see Fig. 7) reveal systematic variations that would require implausibly high trace-element concentrations in any contaminant with $^{87}Sr/^{86}Sr$ higher than about 0.715. Since the ancient crustal rocks which underlie the Central Andes rarely contain more Sr than a few hundred parts per million, even a contaminant with $^{87}Sr/^{86}Sr$ ratio as high as 0.715 implies bulk assimilation of at least one part contaminant to one part parent magma. The $\delta^{18}O$ values reported in this paper preclude such large-scale assimilation unless the contaminant has a $\delta^{18}O$ value significantly lower than that assumed (see Taylor, 1968) for crustal material.

Selective partitioning into the parent magma of trace elements derived from crustal contaminants is improbable because plagioclase, hornblende, mica, and K-feldspar are common crustal phases. Plagioclase, with a high partition coefficient (C_{solid}/C_{melt}) for Sr (Arth, 1976), is on the andesite liquidus so that crustal plagioclase will tend to be a residual phase that retains Sr. Thus the amount of contaminating Sr is not likely to be proportionately much greater than the bulk quantity of contaminant added to the melt and possibly may be even less. Similar arguments can be made for other trace elements. K is a stoichiometric component in both biotite and K-feldspar and has a partition coefficient of about 1 in hornblende. The partition coefficient for Ba is very large in both biotite and K-feldspar; the partition coefficient for Rb is large (2 to 3) in biotite and 0.3 to 0.4 in both hornblende and K-feldspar; and the partition coefficient for Sr is large in both plagioclase and K-feldspar (Arth, 1976). The effect of high partition coefficients is to preclude significant trace-element enrichment during partial melting of crustal materials at crustal depths.

Despite compelling arguments against crustal contamination as a cause of high $^{87}Sr/^{86}Sr$ ratios, the $\delta^{18}O$ values require some sialic participation in the magmatic process. If the contamination does not occur in the crust, it may occur in the subduction zone. The model considered here for the origin of Andean andesite is a variation on that proposed by Nicholls and Ringwood (1973) in which subducted oceanic crust undergoes partial melting; the melt rises into and reacts with the overlying mantle to produce melts of calc-alkaline composition (Nicholls, 1975). Subsequent workers have embellished the model by considering melting of oceanic and continental sediments as well as oceanic basaltic crust (for example, Kay and others, 1978; Magaritz and others, 1978; Kay, 1977; Whitford and others, 1977; Thorpe and others, 1976). For some island arcs (for example, Aleutians, Banda arc) a significant component of subducted continental sedimentary material appears necessary to explain isotope ratios and trace-element abundances of the lavas (Armstrong, 1968, 1971; Kay and others, 1978; Whitford and others, 1977; Magaritz and others, 1978).

One obvious difference between purely oceanic island arcs and those of the continental margin is the availability of mature sialic crustal material for subduction. Trenches bordering continental margins will receive large amounts of continentally derived detritus which may be swept down the subduction zone. That material, largely graywacke in character, will include volcanic and plutonic detritus of the volcanic arc as well as an admixture of mature sialic material of relatively high $^{87}Sr/^{86}Sr$ ratio (>0.71). The bulk composition of the graywacke will thus have higher $^{87}Sr/^{86}Sr$, K, Rb, Ba, Sr, and possibly $\delta^{18}O$ than subducted oceanic crust. Although ocean-floor basalt is likely to be involved to some extent in Andean subduction-zone melting, observed $^{87}Sr/^{86}Sr$ ratios of altered basalt (Hart, 1971; Dasch and others, 1973) are not high enough to account for ratios measured in the Andean lavas. Thus, any magma generated in the subduction zone and involving melting (at least in part) of continentally derived graywacke will display important compositional differences, notably in Sr-isotope ratios and contents of some trace elements, from magmas generated by subduction-zone processes involving only subducted oceanic crust and sediments.

If graywacke or other sialic material eroded during plate motion from the leading edge of the continent is subducted to 150-km depth and subsequently involved in magma genesis, such material will influence isotope ratios and trace-element abundances in ways fundamentally different from sialic material that contaminates mantle-derived magma as it rises through the crust. Moreover, it is important to recognize that the processes described here are *not* equivalent to magma mixing. The distinction may be shown by considering a mechanism whereby slab-derived magma (in this case, of continental rather than oceanic basalt parentage) rises into and equilibrates with overlying mantle material to yield calc-alkaline magma. In such a model, the following processes are important in determining the isotope ratios and trace-element abundances of the magma:

1. Trace elements are highly concentrated in the slab-derived melt. Major crustal phases that tend to hold trace elements (plagioclase, hornblende, and biotite) are unstable at depths below about 100 km. Sanidine may be stable but would probably be the first phase to melt (Stern and Wyllie, 1973). Thus, major residual phases in material of graywacke composition would likely be clinopyroxene, garnet, quartz (coesite), and possible kyanite (Stern and Wyllie, 1978), all of which have relatively low partition coefficients for K, Rb, Sr, Ba and LREE. Partial melting products of metamorphosed graywacke (that is, granulite) should, therefore, be enriched in K, Rb, Sr, Ba, LREE, H_2O and SiO_2. Similar enrichment will occur during melting of subducted oceanic crust, but trace-element abundances for comparable degrees of partial melting will generally be lower, though this depends on the mineral assemblages involved.

2. Mixing and equilibration of a slab-derived melt are with bulk mantle material, not with a mantle-derived melt. Trace element concentrations in the unmodified mantle are so low that the trace-element abundances and, in particular, the $^{87}Sr/^{86}Sr$ ratios of the modified mantle will be dominated by the abundances and ratios of the slab-derived melt even if the weight proportion of that melt is small relative to the mantle material with which it equilibrates. Thus, relatively small weight proportions of sialic contaminant will have a gretly disporportionate effect on the Sr-isotope ratios. For example, if a slab-derived melt with 500 ppm Sr and $^{87}Sr/^{86}Sr$ of 0.708 equilibrates with mantle material with 25 ppm Sr and $^{87}Sr/^{86}Sr$ of 0.703 in the weight proportion one part melt to four parts mantle material, the modified mantle will have an $^{87}Sr/^{86}Sr$ ratio of 0.7072. Regardless of the way in which the modified mantle undergoes subsequent melting, the melting product will have that $^{87}Sr/^{86}Sr$ ratio.

3. O-isotope abundances will be approximately a simple linear function of the bulk proportion of slab-derived melt to total mantle material with which the melt equilibrates and of the $\delta^{18}O$ of the two materials. If graywacke retains a $\delta^{18}O$ of about $+12^0/_{00}$ at depth (Magaritz and Taylor, 1976) and

mantle material is $+6^0/_{00}$, the measured $\delta^{18}O$ of $+7.2^0/_{00}$ in the lavas requires about one part slab to four parts mantle material. Thus, although the trace-element abundances and Sr-isotope ratios of partial melting products of the modified mantle are dominated by the slab-derived material, the O-isotope value and bulk (major-element) chemical composition of the melt are dominated by material of mantle composition.

4. Partial melting of mantle material which has been modified by rising slab-derived melt will result in trace-element abundances which are dependent primarily on the phase proportions, melting relations, and partition coefficients of the modified mantle. Mantle material, even highly modified by the influx of slab-derived melt, will strongly partition trace elements into the magma, so that the degree of enrichment of trace elements in the modified mantle will in general be reflected in high trace-element concentrations of the derivative magma.

In this paper, I attempt to show that the Andean data may plausibly be interpreted in terms of the model described above. That interpretation, if correct, explains the differences between oceanic and

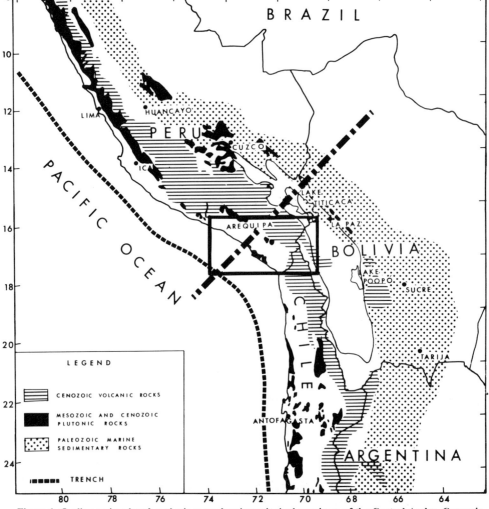

Figure 1. Outline regional and geologic map showing principal provinces of the Central Andes. Cenozoic volcanic rocks crop out principally along the Western Cordillera, Paleozoic metasedimentary rocks along the Eastern Cordillera. The outline of the Altiplano is shown between the two ranges. Dashed line perpendicular to the arc is the line of cross section shown in Figure 2. Area of study is shown by heavy-lined rectangle.

continental volcanic arcs and provides evidence that crustal material eroded from the continents is recycled through subduction-zone magmatism.

TECTONIC SETTING AND EVIDENCE FOR SUBDUCTION OF CONTINENTAL SEDIMENTS

The Central Andean orogen is composed of three principal provinces (Fig. 1): the Western Cordillera composed of the calc-alkaline volcanic arc which is situated 150 to 175 km above the Benioff zone; the Eastern Cordillera consisting of crumpled Paleozoic marine sedimentary rocks that have been intruded by magmas of Mesozoic and Cenozoic age; and the Altiplano, a vast intermontane plateau separating the Western and Eastern Cordilleras that stands at an elevation of about 4 km and may be underlain by as much as 30 km of continental and marine sedimentary rocks (James, 1971a).

The area of study lies entirely within the Western Cordillera of southern Peru (see Fig. 1). The volcanics studied are of Pliocene to Quaternary age and constitute the major extrusive units exposed in the region. Depths to the Benioff zone beneath the Western Cordillera are shown in plan view in Figure 3 and in cross section in Figure 2. In the area of Quaternary volcanism in southern Peru and northern Chile, the well-defined Benioff zone dips at an angle of about 30°. The crust beneath the Western Cordillera is extraordinarily thick; it attains about 75 km at its greatest thickness, some 40 km thicker than "normal" continental crust at sea level (James, 1971b). Despite the great crustal thickness, it can be seen in Figure 2 that the base of the crust beneath the volcanic arc is many tens of kilometres above the subduction plane, a fact that raises serious difficulties for the hypothesis proposed by Klerkx and others (1977) in which sialic material is supposedly added to the base of the crust along the subduction plane. Moreover, rigid continental lithospheric mantle extends nearly to the subduction zone in the region of southern Peru (Sacks, 1977); therefore, it is rheologically implausible that solid sialic material could rise through the mantle to underplate the crust.

Although there is latitudinal variation in the morphology and structure of the trench (Kulm and Schweller, 1977), it is clear that voluminous detritus, shed from the deeply eroded orogenic belt, must have been consumed in the trench. The immense volumes of detritus that have been derived from the

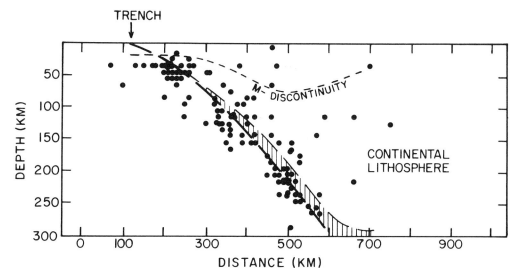

Figure 2. Schematic cross section through southern Peru showing seismicity and crustal structure. Line of the cross section is perpendicular to the trench and corresponds to the dashed line shown in Figure 1. Vertical hatching signifies zone of high attenuation sandwiched between the upper boundary of the downgoing plate and the overlying continental lithosphere. Closed circles are earthquake hypocenters.

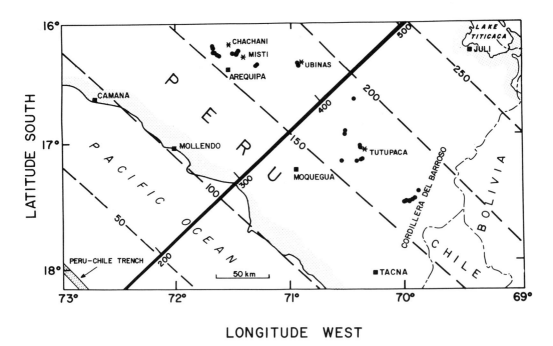

Figure 3. Map of southern Peru showing sample localities. Kilometer markings along cross-section line (solid black) are keyed to those of Figure 2. Contours indicating depth to Benioff zone are shown as dashed lines.

unroofing of the Andean orogen may in part be found in the thick sedimentary sequences blanketing the Altiplano to the east, but sedimentary wedges of comparable thickness are not found to the west of the volcanic arc. Gilluly (1971) noted that offshore the Pacific coast of the United States, sediments are less than one-sixth as voluminous as offshore the Atlantic Coast, despite the roughly comparable areas tributary to the two regions, and concluded that large volumes of continentally derived sediments have been swept down the subduction zone. The coast of southern Peru is an analogous situation in which it appears likely that continentally derived sediments have been continuously subducted.

Local Geology and Sample Locations

Sample localities in southern Peru are shown in Figure 3. The geology of the region has been described in detail elsewhere (Lefèvre, 1973; James and others, 1976; James, 1971a), but a short summary is necessary for the discussion that follows. The rock-stratigraphic terminology follows James and others (1976). According to that classification, the volcanics of southern Peru may be divided into the Barroso volcanics and the Arequipa volcanics. In addition, this study includes the Tutupaca volcanics, Ubinas volcanics, and the Misti volcanics (see Fig. 3). The Barroso volcanics are Pliocene to Pleistocene in age (Wilson and Garcia, 1962), whereas the other volcanics are probably Quaternary. The Tutupaca samples came from within the region of the Barroso volcanics but are from a younger volcanic center. The Ubinas and Misti samples are from young, still active stratovolcanoes.

In general, the more northerly samples (Arequipa, Misti, and Ubinas) are slightly fresher petrographically than those from locations farther south. This is consistent with data on water content, which show the Arequipa volcanics to have 0.25% to 0.5% H_2O and the Barroso volcanics to have 0.4% to 0.8% H_2O. None of the samples studied show any appreciable petrographic evidence of either high- or low-temperature alteration. Phenocrysts, almost entirely plagioclase, are clear and unaltered, although there are minor differences from sample to sample.

ANALYTICAL TECHNIQUES

Rock samples were prepared by removing all cracks, veins, and weathered surfaces and then crushed, split several times and finally pulverized in a chrome steel shatter box. Plagioclase mineral separates were obtained by sieving a disk-milled fraction, washing in water and acetone, and then removing all groundmass and mafic minerals using a Franz magnetic separator. Separates so obtained are nearly pure plagioclase.

Rb, Sr, Ba, and Ni were determined by X-ray fluorescence. Estimated accuracy is ±3% for Rb and Sr, ±15% for Ba, and ±10% for Ni. Rare-earth element (REE) analyses (reported previously by Whitford, 1977) were by isotope dilution.

Sr-isotope ratios were measured to an analytical precision of better than ±0.0001 at the 95% confidence level. All ratios are normalized to $^{86}Sr/^{88}Sr = 0.1194$ and an Eimer and Amend standard $SrCO_3$ value of 0.70800. O-isotope analyses were performed using the method of BrF_5 dissolution and collection as described in detail by Clayton and Mayeda (1963). Mass-spectrometer precision at the 1σ level is about $\pm 0.02^0/_{00}$ and estimated overall accuracy is $\pm 0.1^0/_{00}$. Corrections were made for machine and interference effects (Craig, 1957).

O-isotope ratios are expressed relative to standard mean ocean water (SMOW) by the relationship

$$\delta^{18}O(^0/_{00}) = \left[\frac{(^{18}O/^{16}O)_{obs.} - (^{18}O/^{16}O)_{SMOW}}{(^{18}O/^{16}O)_{SMOW}}\right] \times 1000$$

Measured $\delta^{18}O$ values are normalilzed to an assumed California Institute of Technology rose quartz standard value of $+8.45^0/_{00}$ (M. Magaritz, 1977, oral commun.). For that assumed standard, NBS no. 28 (National Bureau of Standards) white quartz has a $\delta^{18}O$ of $9.55^0/_{00}$.

Weathering

Most whole-rock $\delta^{18}O$ values are higher than those of the plagioclase phenocryst separates (see Magaritz and others, 1978). This result was unexpected because the rocks are young volcanics exhibiting little petrographic evidence of weathering. Evidence of late-stage deuteric or hydrothermal alteration such as veining or reaction rims around phenocrysts is virtually absent. Rock samples with H_2O contents as low as 0.25% show high $\delta^{18}O$ relative to their plagioclase phenocrysts.

Use of plagioclase phenocryst separates is motivated by two primary considerations: (1) plagioclase, which appears in large quantity in most of the volcanics, is the only volumetrically significant phenocryst; and (2) feldspars tend to be closest in $\delta^{18}O$ to the melt itself (Taylor, 1968, p. 26). Extrapolated results on diffusion of O in feldspars (Gilette and others, 1978) show very low diffusion coefficients ($\sim 10^{-28}$ cm^2/s) at surface temperatures and thus preclude significant low-temperature exchange of O between feldspar phenocrysts in late Cenozoic lavas and meteoric water. That conclusion is supported by measurements which show no variation in $\delta^{18}O$ between phenocrysts of different grain size. The possibility that phenocrysts and matrix were not in isotopic equilibrium at the time of extrusion is unlikely in view of the fact that $^{87}Sr/^{86}Sr$ ratios are similar for both. It thus appears that the $\delta^{18}O$ discrepancy between whole-rock and phenocrysts is the result of whole-rock— but not plagioclase phenocryst—exchange with meteoric water.

ANALYTICAL RESULTS

Analytical data for Sr- and O-isotope ratios and for K, Rb, Sr, Ba, Cs, Ni, and REE are summarized graphically in Figures 4 through 9 and in Table 1. The data are grouped by regional geochemical affinity into Barroso, Tutupaca, Ubinas, Arequipa, and Misti volcanics. A regional trend of southeast to northwest character is evident in both trace-element abundances and Sr-isotope ratios of the

TABLE 1. END-MEMBER COMPOSITIONS OF AREQUIPA AND BARROSO VOLCANICS AND AVERAGE COMPOSITIONS OF SUNDA ARC ANDESITE* AND FRANCISCAN METAGRAYWACKE†

	K (o/o)	Rb (ppm)	Sr (ppm)	Ba (ppm)	K/Rb	K/Ba	Rb/Sr	K/Sr	$\frac{K_2O}{(K_2O+Na_2O)}$	Ce (ppm)	Yb (ppm)	$(Ce/Yb)_N$	Ni (ppm)	MgO (o/o)	SiO_2 (o/o)	$^{87}Sr/^{86}Sr$	$\delta^{18}O$ (o/oo)
Barroso volcanics	1.6 2.7	48 108	730 530	580 970	335 250	27.6 27.8	0.07 0.20	21.9 50.9	0.36 0.45	51 63	1.6 1.3	7.6 11.6	22 33	3.7 2.6	57 63	0.7055 0.7068	+7.3
Arequipa volcanics	1.6 2.6	36 79	850 640	620 1100	440 330	25.8 23.6	0.04 0.12	18.8 40.6	0.33 0.43	42 75	1.3 1.3	7.8 14	50 20	3.7 3.0	57 59	0.7071 0.7080	+7.16
End-point avg. Barroso Arequipa	2.1 2.1	78 58	630 745	775 860	290 360	27.1 24.4	0.12 0.07	33.3 24.4	0.41 0.38	57 59	1.4 1.3	9.6 10.9	28 35	3.2 3.4	60 58	0.7062 0.7076	+7.3 +7.16
Sunda Arc Andesite	1.5	62	360	420	248	36	0.17	41.7	0.34	40.9	2.1	4.64	8	2.9	59.0		
Franciscan Metagraywacke	1.3	60	200	400	217	32.5	0.3	65	0.31	80	2.8	6.8		2.6	68	0.706-0.710	+12.0

*From Whitford, 1975a.

†From Bailey and others, 1964; Peterman and others, 1967; and Magaritz and Taylor, 1976.

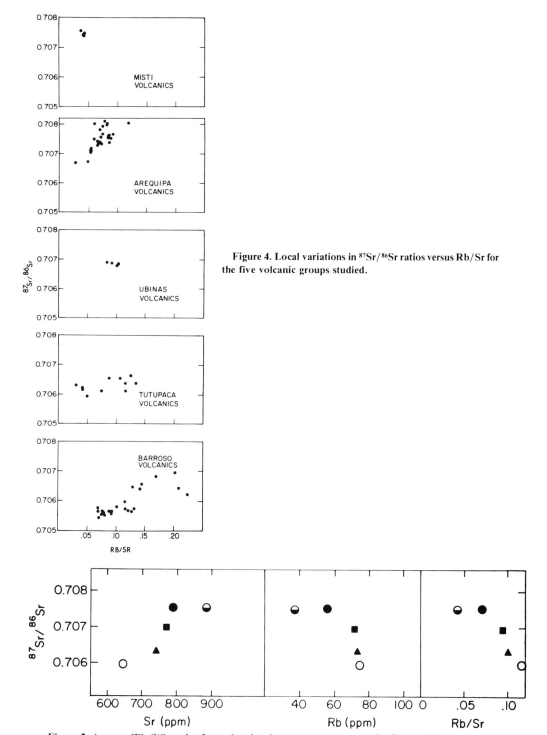

Figure 4. Local variations in $^{87}Sr/^{86}Sr$ ratios versus Rb/Sr for the five volcanic groups studied.

Figure 5. Average $^{87}Sr/^{86}Sr$ ratios for each volcanic group versus average Sr, Rb, and Rb/Sr. Key: closed circle = Arequipa; semi-closed circle = Misti; closed square = Ubinas; closed triangle = Tutupaca; open circle = Barroso.

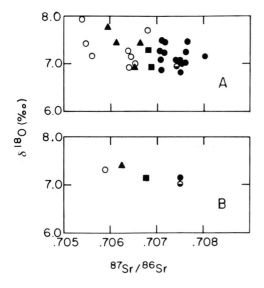

Figure 6. $\delta^{18}O$ versus $^{87}Sr/^{86}Sr$ for individual plagioclase separates (A) and averaged by group (B). Symbols same as in Figure 5.

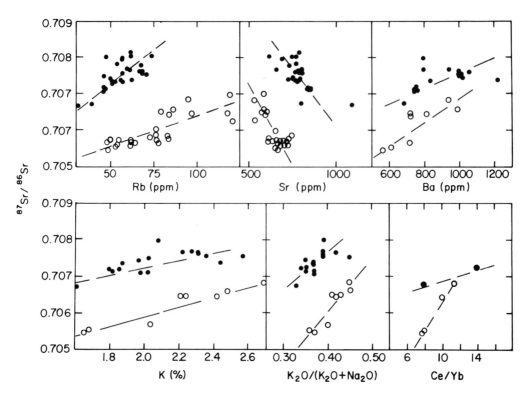

Figure 7. K, Rb, Sr, Ba, and $K_2O/(K_2O + Na_2O)$ versus $^{87}Sr/^{86}Sr$ within Arequipa and Barroso groups. Symbols are as in Figure 5.

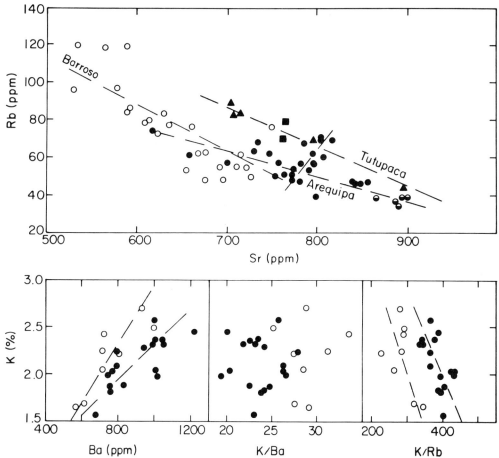

Figure 8. Rb versus Sr and K versus Ba, K/Ba, and K/Rb for Barroso and Arequipa volcanics. Symbols are as in Figure 5.

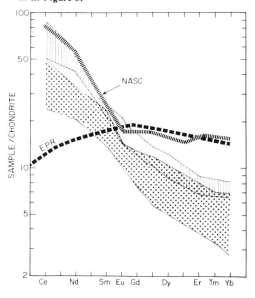

Figure 9. Chondrite-normalized REE patterns for Arequipa and Barroso volcanics (vertical hatching) (from Whitford, 1977) compared to data from northern Chile (stippled pattern) (Thorpe and others, 1976). Data from Nazca plate basalts (EPR) (Schilling and Bonatti, 1975) and the North American shale composite (NASC) (given by Nance and Taylor, 1976) are shown for comparison.

various volcanic groups (Fig. 5). Within each volcanic group, however, distinct local geochemical trends are superimposed on the regional variation. An important exception to both regional and local variation is the O-isotope ratios of the lavas. Within the scatter of the data, $\delta^{18}O$ values exhibit neither regional nor local variations, although there may be a small (statistically unproved) regional decrease in $\delta^{18}O$ with increasing $^{87}Sr/^{86}Sr$.

A summary compilation of regional and local geochemical variation is given in the form of end-member compositions in Table 2. Only the local variations for Arequipa and Barroso volcanics are included in Table 2 because those are the only volcanic groups for which a full data set is available.

Isotope Data

Sr-isotope ratios for the five volcanic series studies are plotted against Rb/Sr ratios in Figure 4. The Barroso, Tutupaca, and Arequipa volcanics define statistically significant pseudoisochrons of ~100 to

TABLE 2. COMPOSITIONAL SUMMARY OF ROCK TYPES USED IN PETROGENETIC MODELLING CALCULATIONS

	Franciscan Metagraywacke	Tonalite*	Andean intrusive (avg)	Average ocean floor basalt†	Mantle§	Modified mantle
Major elements (wt %)						
SiO_2	68.1	59.1	59.7	49.9	45.2	
TiO_2	0.24	0.79	0.68	1.4	0.71	
Al_2O_3	13.7	18.2	17.1	16.0	3.5	
Fe_2O_3	1.3	2.3			0.46	
FeO	2.4	3.6	7.52#	9.3#	8.0	
MnO	0.07	0.11	0.15	0.17	0.14	
MgO	2.6	2.5	2.6	8.7	37.5	
CaO	2.6	5.9	6.1	11.3	3.1	
Na_2O	3.4	3.8	3.5	2.75	0.6	
K_2O	1.6	2.2	2.5	0.27	0.13	
H_2O	1.4	0.86	n.d.			
Mode (%)						
Qtz	n.d.	13	n.d.			
K-spar		4.5				
Plag		59				
Bt		12.5				
Hb		9				
Opx					17	18
Cpx				47.5	12	55
Oliv					57	5
Gar				47.5	14	14
Phlog						8
Accessories		2		5		
Trace elements (ppm)						
Sr	200	n.d	445	110	20	100-140
Rb	60		94	1.8	2.5	35-50
Ba	400		526	13	4	140-200
Ni	n.d.		11.6	120	2000	
Ce	n.d.		n.d.	10.3 (est)	1.7	30-40
Yb	n.d.		n.d.	2.9	0.8	0.75
$^{87}Sr^{86}Sr$	0.706-0.710		0.7066	0.7040	0.7030	0.7055-0.7068
$\delta^{18}O$	+12.0 o/oo		n.d.	+12.0 o/oo	+6.0 o/oo	7.2 o/oo

Note: n.d. = data not available.

*Stern and others, 1975.

†Gill, 1974.

§Ringwood, 1975.

#Total Fe as FeO.

~400 m.y. Along the arc, average $^{87}Sr/^{86}Sr$ ratios *increase* from southeast to northwest and Rb/Sr ratios *decrease* (Sr increases and Rb decreases); this pattern produces a negative regional pseudoisochron (Fig. 5).

Measured $\delta^{18}O$ of the plagioclase separates lie mostly between $+7.0^0/_{00}$ and $+7.50^0/_{00}$. Individual $\delta^{18}O$ values are plotted against whole-rock $^{87}Sr/^{86}Sr$ in Figure 6A, and average $\delta^{18}O$ is plotted versus average $^{87}Sr/^{86}Sr$ in Figure 6B for the five volcanic groups studied. The $\delta^{18}O$ values show no significant correlation with $^{87}Sr/^{86}Sr$, although lower $\delta^{18}O$ values are apparently associated with higher $^{87}Sr/^{86}Sr$ ratios. This result is opposite to what would be anticipated if the high $\delta^{18}O$ ratios of the lavas were due to crustal contamination of a mantle-derived magma. All of the $\delta^{18}O$ values measures are slightly higher then $\delta^{18}O$ associated with volcanics of purely oceanic-mantle origin ($+6.0^0/_{00} \pm 0.5^0/_{00}$, Taylor, 1968) but do overlap $\delta^{18}O$ values measured for calc-alkaline volcanics of island arcs, where $\delta^{18}O$ may be as high as $+8.0^0/_{00}$ (Taylor, 1968).

Trace-Element and REE Geochemistry

In most instances, trace-element concentrations in the lavas of southern Peru show a strong correlation with $^{87}Sr/^{86}Sr$. That correlation precludes simple differentiation processes (or those of partial melting) as the major cause of the trace element variations observed. Because Sr-isotope ratios do not change during differentiation, any variation in isotope ratios and, by inference, any trace-element variation that correlates strongly with isotope ratios must be due either to source inhomogeneity or to contamination.

Elemental variations in K, Rb, Sr, Ba, $K_2O/(K_2O + Na_2O) = k$, and Ce/Yb within the Barroso and Arequipa volcanics are plotted versus $^{87}Sr/^{86}Sr$ in Figure 7. All of these quantities show some correlation with the $^{87}Sr/^{86}Sr$ ratio, and some show remarkably consistent covariance. Of the quantities plotted in Figure 7, only Sr decreases with increasing $^{87}Sr/^{86}Sr$ ratio within a given volcanic group. K, Rb, Ba, k, and Ce/Yb increase with increasing $^{87}Sr/^{86}Sr$.

Trace-element variations within the Barroso and Arequipa volcanics are shown as Rb versus Sr, Ba versus K, K/Rb versus K, and K/Ba versus K in Figure 8. In general, K, Rb, and Ba behave coherently, whereas Sr behaves in the opposite sense. Regionally averaged Rb and Sr concentrations and Rb/Sr ratios correlate with $^{87}Sr/^{86}Sr$ in an opposite sense to that observed in local variations. It is significant that the highest Sr-isotope ratios are measured in rocks with very high Sr concentrations (~800 to 900 ppm), a critical factor in evaluating crustal contamination as a cause of the high $^{87}Sr/^{86}Sr$ ratios observed.

Relative to island-arc volcanics (see Table 1), all of the Andean volcanics are greatly enriched in Sr and Ba and somewhat enriched in K and possibly Rb. K/Rb ratios in the Arequipa volcanics are significantly higher than in the Barroso volcanics and are about the same as those of Sunda arc averages (Whitford, 1975), whereas K/Ba ratios in both Arequipa and Barroso volcanics are significantly lower, a reflection of relatively high Ba abundances. Rb/Sr ratios of the Andean rocks are also somewhat lower than are those of comparable calc-alkaline volcanics of oceanic island arcs.

REE concentrations in four Barroso volcanics and two Arequipa volcanics were measured by Whitford (1977; Fig. 9). Measured concentrations show a pronounced LREE-enrichment pattern even relative to other Central Andean volcanics such as those from northern Chile studied by Thorpe and others (1976; see Fig. 9). Slight negative Eu anomalies are evident in samples PE 144 and 146, but no significant anomalies occur in other samples.

Ni analyses and a few Cs determinations have also been done on the rocks. Ni concentrations are significantly higher (20 to 50 ppm) than those measured in comparable calc-alkaline rocks of island arcs [7 to 10 ppm (Gill, 1974; Whitford, 1975)] but are typical of or even slightly lower than Ni concentrations measured elsewhere in the Central Andes (Thorpe and others, 1976). The Cs content (0.5 to 1.1 ppm) of those Peru samples for which Cs was determined is significantly lower than that measured in island-arc calc-alkaline rocks (2 to 4 ppm, Whitford, 1975).

Trace-element data are summarized as end-member compositions in Table 1. Local variations are best seen within the Barroso volcanics and to a lesser extent within the Arequipa volcanics.

Considering only the Barroso volcanics, the local geochemical trends referred to *increasing* $^{87}Sr/^{86}Sr$ are (1) increasing K, Rb, Ba, Ce, Ni(?), Rb/Sr, k, and Ce/Yb, (2) decreasing Sr, K/Rb, and Yb, and (3) roughly constant K/Ba. Similar variations characterize the Arequipa volcanics except that the Ni variation may be reversed and there is no discernible decrease in Yb with increasing $^{87}Sr/^{86}Sr$ ratio.

Average regional trace-element concentrations vary with *increasing* $^{87}Sr/^{86}Sr$ in the following fashion: (1) increasing Sr, Ba, k, Ni, and K/Rb, (2) decreasing R, Rb/Sr, and K/Rb, and (3) constant (within geological scatter) K, Ce, Yb, MgO, SiO_2, and Ce/Yb.

Regional correlations of Sr, Rb, and K/Rb with $^{87}Sr/^{86}Sr$ ratios are in the opposite sense to those observed within the local volcanic groups. Ba, k, and possibly Ni behave the same both regionally and locally. Not enough data on REE are available to judge whether they behave in the same or opposite sense regionally and locally.

MAGMA GENESIS

The data described above may be cast in terms of the Nicholls and Ringwood model described earlier in this paper. Isotope ratios during all phases of magma genesis may be determined by the usual mixing equations. Calculated trace-element concentrations, however, depend on a number of variables, including (1) the concentrations of trace elements in the original solid, (2) the proportions in which crystalline phases are present in the source material, (3) the degree to which those crystalline phases participate in the melting process, (4) the overall degree of partial melting, (5) the type of melting (batch or fractional), and (6) the partition coefficients of the various elements for the phases present.

None of these variables is known, but the major uncertainties lie in the mineralogy and melting relations of both the slab and the modified mantle. In addition, there is convincing evidence that partition coefficients, poorly determined in general, vary not only with temperature and pressure but also with composition of the melt. As an added complication, K is a stoichiometric component in certain minor phases. When one or more of those phases is present, K cannot be treated as a trace element. In such instances, the concentrations of K (and probably Rb) will be controlled largely by the degree to which the K-bearing phases enter the melt. The partition coefficients I have used are those compiled from the literature by Arth (1976) for various minerals in equilibrium with different melt compositions. Although the processes that lead to the generation of calc-alkaline magma can in principle be modeled quantitatively, the practical limitations of the model allow at best for a partly quantitative, partly semiquantitative treatment of trace-element and isotope variations. Melting calculations involve both slab and modified mantle.

Slab Melting

The composition of subducted material beneath the Central Andes may be inferred by comparison with continental-margin ophiolite complexes such as the Franciscan melange of California. Studies by Bailey and others (1964) showed that 80% of Franciscan rocks are graywacke or metagraywacke, 20% are basalts and greenstones, and cherts and other deep-sea sediments occur in trace amounts.

The composition of the Franciscan formation is summarized in Tables 1 and 2. The major-element composition given in Table 2 is for average metagraywacke (Bailey and others, 1964). Rb, Sr, and $^{87}Sr/^{86}Sr$ data for Franciscan graywacke are from Peterman and others (1967). The $\delta^{18}O$ values, ranging from $+11^0/_{00}$ to $+13^0/_{00}$ have been reported by Magaritz and Taylor (1976) for both Franciscan graywackes and greenstones. REE and Ba concentrations are here assumed to be the average of Australian post-Archean sediments (chiefly graywacke) (Nance and Taylor, 1976).

Graywacke of the Andean region should consist predominantly of volcano-plutonic debris derived from Late Cretaceous–early Tertiary igneous rocks. For comparison, therefore, average compositions of Andean batholithic rocks of the Arequipa region are given in Table 3. That composition is roughly the same as the composition of the tonalite used by Stern and others (1975) for

TABLE 3. MINERALOGY AND ASSUMED MELTING PROPORTIONS FOR GRANULITE AND MODIFIED MANTLE*

	Granulite (Metagraywacke)	Granulite (Tonalite + 5% H_2O)	Modified mantle
Subsolidus mineral phase proportions			
ol	..	..	0.05
opx	..	..	0.18
cpx	0.40-0.35	0.43	0.55
gar	0.15-0.20	0.13	0.14
phlogopite	..	..	0.08
sanidine	0.10	..	..
qtz (ct)	0.35	0.27	..
kyanite	..	0.17	..
Phase melting proportions (0-20% melt)			
ol	..	..	0.01
opx	..	..	0.04
cpx	0.25-0.20	0.29	0.45-0.57
gar	0.00-0.10	0.00	0.20
phlogopite	..	..	0.30-0.18
sanidine	0.30-0.40	..	..
qtz (ct)	0.45-0.30	0.59	..
kyanite	..	0.12	..

*Granulite (Tonalite + 5% H_2O) composition from Stern and Wyllie, 1978.

experimental studies under hydrous conditions at high temperature and pressure (see Table 2). Trace-element abundances and Sr-isotope ratios of the tonalite studied by Stern and others (1975) are assumed to be the same as for the average intrusive of the Andean batholith near Arequipa.

Average composition of subducted oceanic basalt is taken from Gill (1974, with references to the sources of the analyses) and summarized in Table 3.

At subduction-zone depths of 150 to 175 km, a rock of graywacke composition would probably be granulite, and oceanic basalt would be quartz eclogite. In general, experimental data on rocks of graywacke (tonalite) composition are available only to pressures of about 30 to 35 kb, whereas magma generation occurs at pressures of 50 to 60 kb. It appears, however, that most of the major phase changes occur by 30 kb (Wyllie, 1977). More serious than the lack of data at higher pressures is the uncertainty over the amount of H_2O present in the slab. The stability of important minor phases such as sanidine and phlogopite is governed largely by the amount of H_2O present. Studies show that at 30 to 35 kb with 5 wt % H_2O, sanidine is entirely dissolved in the vapor phase (Stern and Wyllie, 1978). The mineral assemblage at subsolidus temperatures for 5 wt % H_2O is clinopyroxene, quartz (coesite), garnet, kyanite, and vapor. The relative weight proportions of those minerals present for different degrees of partial melting have been measured by Stern and Wyllie (1978), and the relative proportions in which the various phases enter the melt can be calculated. The initial mineral proportions in the granulite and the appropriate "eutectic" phase proportions (that is, the proportion in which the different phases enter the melt) for up to 20% partial melting are given in Table 4.

It seems likely that 5 wt % H_2O is substantially more than will be present in the subduction zone at 150 km. Even in metagraywacke of the Franciscan formation, *total* H_2O is typically 3%, and most of that is contained in hydrous phases (Bailey and others, 1964). One would anticipate that at subduction-zone depths, free vapor would exist only in very small quantities so that a potassic phase, probably K-feldspar (sanidine), could be present in significant amounts, and kyanite would be absent. This supposition is supported by anhydrous experiments on diorite (60% SiO_2, 1.3% K) by Green (1970) which reveal that significant amounts of K(?)-feldspar, here assumed to be sanidine, and no measurable amount of kyanite ($\leq 5\%$) are present at pressures of 30 to 35 kb and temperatures of about 900 °C. The anhydrous mineral assemblage at those P-T conditions is clinopyroxene $\gg$ quartz (coesite) $>$ garnet $>$ K-feldspar. Sanidine may be expected to persist as long as the volume of the aqueous phase is small. For the melting model considered here, I have assumed that sanidine is present

TABLE 4. PARTITION COEFFICIENTS FOR MELTING OF GRANULITE AND MODIFIED MANTLE*

Trace element	Granulite (Dacite-Rhyolite Melt)					Modified Mantle (Basalt-Andesite Melt)				
	cpx	gar	phlog	san	qtz	oliv	opx	cpx	gar	phlog
K	0.037	0.020	..	..	0.00	0.00680	0.0140	0.0110	0.0150	..
Rb	0.032	0.0035	(0.940)	(0.340)	0.00	0.00980	0.0220	0.0150	0.0420	(3.00)
Sr	0.516	0.015	0.672	3.87	0.00	0.0140	0.0170	0.120	0.0120	0.81
Ba	0.131	0.017	6.00	6.12	0.00	0.00990	0.0130	0.0130	0.0230	1.09
Ce	0.50	0.350	0.23	0.044	0.00	0.00690	0.0240	0.070	0.0280	0.03
Yb	1.58	39.90	0.17	0.012	0.00	0.0140	0.0340	0.160	11.50	0.04

*Arth, 1976.

in the granulite in sufficient quantity (~10%) to hold most of the K. If sanidine is present, it will affect the concentrations of other trace elements for which it has a high partition coefficient, notably Rb, Ba, and Sr. For comparatively small degrees of partial melting (<20%), retention of trace elements in sanidine may be important.

Mineralogic compositions and assumed melting proportions for granulite are given in Table 3. Partition coefficients as compiled by Arth (1976) are given in Table 4. Concentrations of trace elements for varying degrees of partial melting are shown in Figure 10, based on initial trace-element abundances given in Table 2. Shown also in Figure 10 are average trace-element concentrations calculated by Gill (1974) for melting of oceanic crust using slightly different partition coefficients. It is important to note that relatively small degrees of partial melting of eclogite (~10%) will produce Sr concentrations much greater than those of granulite partial melts for similar or greater degrees of

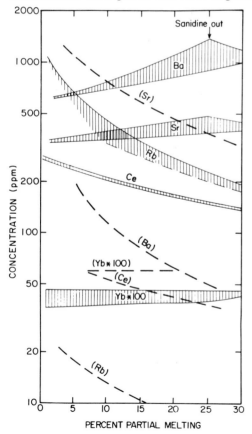

Figure 10. Trace-element concentrations versus degree of partial melting for granulite and oceanic crust (eclogite), the latter from Gill (1974). Partition coefficients used by Gill are not identical to those used for the granulite calcualtions. Dashed lines with element symbols in parentheses are for eclogite melting. Vertical hatching denotes range of elemental abundances for different granulite mineralogies.

partial melting, whereas K, Rb, Ba, and LREE will be much less abundant in eclogite partial melt than in granulite partial melt for *all* degrees of melting (see Fig. 10). This observation is crucial with regard to local geochemical variations where Sr concentrations are negatively correlated with $^{87}Sr/^{86}Sr$ ratios whereas K, Rb, Ba, and LREE concentrations are positively correlated.

At any given temperature, the degree of melting of granulite should be greater than that of quartz eclogite. Although the temperature of the solidus at 30 kb under hydrous conditions is nearly the same for basalt as it is for tonalite (Stern and Wyllie, 1973), granulite will melt more quickly with rising temperature than will eclogite owing to the greater abundance of early melting phases such as quartz and sanidine. From 10% to 20% melting of tonalite with 5% water occurs over the temperature range of about 750 to 800 °C at 30 kb (Stern and Wylie, 1978). In that range of melting, approximately 60% of the partial melt consists of quartz (see Table 3). If eclogite consists of 50% clinopyroxene, 40% garnet, and 10% quartz and if the minerals melt at about the same rate as in granulite, only about half as much melt will be produced from eclogite over the same temperature interval.

Modified Mantle Melting

Few experimental results have been published on reactions that may occur between wet silicic melt from the slab and peridotitic mantle. From results that are available (for example, Modreski and Boettcher, 1973; Bravo and O'Hara, 1975; Boettcher and others, 1975) it appears evident that the reactions will involve the breakdown of olivine with formation of pyroxene (probably mostly clinopyroxene) and phlogopite.

Results of Boettcher and others (1975) suggest that phlogopite is a very refractory hydrous mineral, stable at high temperatures and pressures. Whether phlogopite is a residual phase is important in determining the concentration of K and Rb in the calc-alkaline melt. If phlogopite is a residual phase, it will control both K and Rb concentrations in the magma, with K entering the melt in proportion to the amount of phlogopite melted. Rb would be controlled by the K-Rb partition coefficient for phlogopite, here taken to be ~3 (Beswick, 1973). Thus, magma generated from modified mantle in which phlogopite is a significant residual phase will have a higher K/Rb ratio than the source material.

Local Geochemical Variations

O-isotope data show that the proportion of slab-derived melt to mantle material with which it equilibrates remains nearly constant at about 1:4. The $^{87}Sr/^{86}Sr$ ratios within the individual volcanic groups, however, commonly vary over a wide interval. It is evident, therefore, that the slab-derived melt varies either in $^{87}Sr/^{86}Sr$ ratio, Sr concentration, or both. One obvious means of achieving that variability is for the slab-derived melt to consist of differing proportions of granulite partial melt and eclogite partial melt. Thus, larger proportions of eclogite partial melt relative to granulite partial melt will produce modified mantle material of lower $^{87}Sr/^{86}Sr$ ratio.

Figure 11 shows theoretical mixing curves between graywacke partial melts and mantle material plotted as functions of $\delta^{18}O$ and $^{87}Sr/^{86}Sr$. In general, the Sr concentrations of the slab melt will be 10 to 40 times those of the peridotite mantle, whereas O concentrations will be roughly comparable. The resultant mixing curves are hyperbolas, strongly convex downward, that reflect the disparate Sr concentrations of the end members. The observed $^{87}Sr/^{86}Sr$ variation can be modeled over a relatively small range of Sr concentrations by assuming differing proportions of the granulite and eclogite end members.

Although a quantitative assessment of the degree of partial melting of slab and modified mantle material is hampered by serious uncertainties, it is instructive to consider at least one hypothetical case which can account for some of the geochemical variations observed. For that case, it is assumed that within the slab the temperature is such that granulite and eclogite are 20% and 10% partially melted, respectively. For those degrees of melting, Sr concentrations in eclogite melts are significantly higher than those in granulite melts (Fig. 10). Under these hypothetical conditions, observed Sr-isotope ratios of the Barroso volcanics require that for the highest $^{87}Sr/^{86}Sr$ ratios the slab-derived melt would

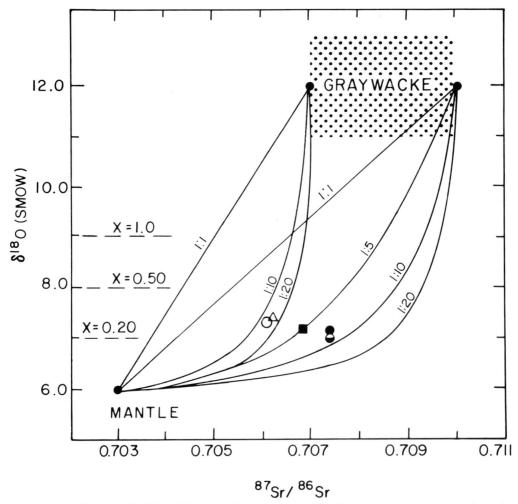

Figure 11. $\delta^{18}O$ versus $^{87}Sr/^{86}Sr$ mixing curves between partial melt with graywacke-type isotope ratios and material with mantle-type isotope ratios. Proportions shown with each curve denote trace-element concentration in mantle to trace-element concentration in slab-derived melt (for example, 1:20 signifies 1 ppm Sr in mantle to 200 ppm Sr in melt). X denotes bulk weight proportion of slab-derived melt to 1 part mantle material. O concentrations assumed equal in the two materials. Approximate range of graywacke compositions shown by dotted pattern.

consist of about 60% granulite melting product and 40% eclogite melting product. In all cases, magmas from granulite and eclogite are assumed *not* to mix with one another before separating from their respective source materials.

Corresponding degrees of partial melting of modified mantle can be estimated by comparing observed Sr concentrations with those calculated from partial melting of modified mantle (Fig. 12). For the Barroso volcanics, the degree of partial melting that satisfies observed Sr concentrations is 13% to 15%. The range of values shown by hatching in Figure 12 defines the interval between sanidine-bearing granulite and kyanite-bearing granulite with 5 wt % H_2O.

Trace-element model abundances can be calculated over the 13% to 15% melting interval obtained above and then compared to concentrations observed in the lavas. Some general features include the following: (1) Calculated Ba concentrations are close to observed concentrations. (2) K concentrations require that significant amounts of phlogopite remain as a residual phase. Calculated

K/Rb ratios vary in the correct sense and proportion to observed K/Rb ratios but tend consistently to be too high (calculated Rb too low). (3) Calculated LREE (Ce) abundances are consistently much higher than those observed, and calculated HREE (Yb) abundances are much lower. These discrepancies are difficult to explain but could be due in part to incorrect partition coefficients, especially for Yb (Shimizu, 1975), or to incorrect starting composition in the granulite.

The high observed Ni concentrations (relative to island-arc volcanics) may be due to lesser amounts of olivine in the modified mantle resulting from reaction with a realtively more silicic melt than one derived solely from eclogite. Smaller proportions of olivine will result in a corresponding decrease in the bulk Ni partition coefficient. Low Cs contents in the lavas may be due to residual phlogopite in the modified mantle, since phlogopite has a high partition coefficient for Cs.

Regional Geochemical Variations

The increase in Sr and Ba concentrations and the increase in K/Rb ratios in the lavas trending southeast to northwest along the arc suggest that the overall degree of partial melting of the modified mantle may be decreasing to the northwest. Independent evidence for that conclusion comes from the general decrease in Quaternary volcanism from southeast to northwest, with Quaternary volcanism ceasing entirely just north of Arequipa.

If the modified mantle source region of the Arequipa volcanics has the same trace-element composition as the source region of the Barroso volcanics, partial melting of 7% to 10% produces melts with Sr concentrations similar to those observed in the Arequipa volcanics (650 to 850 ppm; Fig. 12). When phlogopite is a residual phase, smaller degrees of melting should result in increasing K/Rb ratios, consistent with the observed trend. Also consistent with observed variations are the increases in Ba and LREE predicted for lower degrees of partial melting.

Higher $^{87}Sr/^{86}Sr$ ratios in the Arequipa lavas cannot be due to varying degrees of partial melting but may be due either to relatively larger proportions of granulite partial melt or, more probably, to subducted continental sedimentary material with a slightly higher $^{87}Sr/^{86}Sr$ ratio.

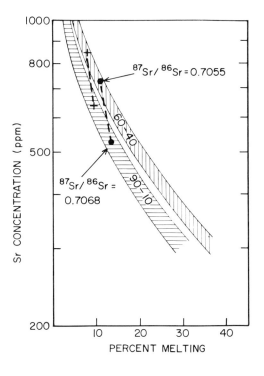

Figure 12. Sr concentrations in partial melting products of modified mantle for different degrees of granulite and eclogite participation. Range of compositions shown by hatched areas are for different mineralogies and mineralogic melting proportions in granulite. Proportionality 90–10 (60–40) signifies that the slab-derived melt is made up of 90% (60%) partial melt from granulite and 10% (40%) partial melt for eclogite. In all cases, granulite is presumed to undergo 20% partial melting. The closed circles indicate where the end-member Sr concentrations of the Barroso volcanics intersect the partial melting curves. End-member $^{87}Sr/^{86}Sr$ ratios are shown for each point of intersection. Crosses show the Sr concentration–melting curve intersection for Arequipa volcanics. Weight proportion of slab melt to mantle is 1:4.

CONCLUSIONS

Unusual isotope ratios and trace-element abundances of Central Andean andesitic lavas may plausibly be explained as due to subduction-zone melting of sialic material and small amounts of oceanic crust. Observed trace-element and isotopic variations are modeled by a petrogenetic scheme in which mantle material, modified by partial melts rising from granulite and quartz eclogite of the descending slab, produces calc-alkaline magma of distinctive isotopic character.

O-isotope measurements indicate that subduction-zone contamination of the overlying mantle involves about one part slab-derived melt to four parts mantle material. Very high concentrations of trace elements in slab-derived melt and very low concentrations in mantle material account for the high trace-element abundances and Sr-isotope ratios observed in the lavas, whereas the volumetric dominance of mantle material accounts for the nearly "normal" major-element composition and O-isotope ratios observed.

Involvement of sialic material in subduction-zone petrogenesis signifies that continental material is being recycled back into the curst in the Central Andes. That recycling of "aged" sialic material may provide an explanation for the systematic increase through time of $^{87}Sr/^{86}Sr$ ratios of Andean volcanics (McNutt and others, 1975; James and others, 1976). That is, as the Andean arc evolves and older sialic material is evaluated and eroded in the process of Cordilleran development, increasing volumes of old radiogenic detritus are shed off the continent and into the trench where they are subsequently subducted, melted, and their trace elements returned to the continent.

ACKNOWLEDGMENTS

I thank T. Hoering and D. Rumble of the Geophysical Laboratory of the Carnegie Institution of Washington for use of their O extraction line and mass spectrometer and M. Magaritz for assistance in some of the earlier O-isotope analyses. Special thanks go to M. Feigenson, now at Princeton University, for sample preparation and many of the Rb, Sr, and $^{87}Sr/^{86}Sr$ analyses. Some of the major- and trace-element analyses were done on XRF equipment at the University of Montreal formerly maintained by B. Gunn. Discussions with D. Whitford, A. Hofmann, and M. Magaritz helped me to clarify many of the arguments. Computer programs for calculating trace-element abundances during partial melting were provided by W. White. Sample collection was done with C. Brooks (University of Montreal) and A. Cuyubamba (McMaster University) and with the active cooperation of the Geological Survey of Peru, Ing. E. Bellido, Director.

REFERENCES CITED

Armstrong, R. L., 1968, A model for the evolution of strontium and lead isotopes in a dynamic earth: Reviews of Geophysics, v. 6, p. 175–199.

——1971, Isotopic and chemical constraints on models of magma genesis in volcanic arcs: Earth and Planetary Science Letters, v. 12, p. 137–142.

Arth, J. G., 1976, Behavior of trace elements during magmatic processes—A summary of theoretical models and their applications: U.S. Geological Survey Journal of Research, v. 4, no. 1, p. 41–47.

Bailey, E. H., Irwin, W. P., and Jones, D. L., 1964, Franciscan and related rocks, and their significance in the geology of western California: California Division of Mines and Geology Bulletin, v. 183, p. 177.

Beswick, A. E., 1973, An experimental study of alkali metal distributions in feldspars and micas: Geochimica et Cosmochimica Acta, v. 37, p. 183–208.

Boettcher, A. L., Mysen, B. O., and Modreski, P. J., 1975, Melting in the mantle: Phase relationships in natural and synthetic peridotite-H_2O-CO_2 systems at high pressures, *in* Ahrens, L. H., and others, eds., Physics and chemistry of the earth: Oxford, Pergamon, p. 855–868.

Bravo, M. S., and O'Hara, M. J., 1975, Partial melting of phlogopite-bearing synthetic spinel- and garnet-lherzolites, *in* Ahrens, L. H., and others, eds., Physics and chemistry of the earth: Oxford, Pergamon, p. 845–854.

Briqueu, L., and Lancelot, J. R., 1979, Rb-Sr systematics and crustal contamination models for calc-alkaline igneous rocks: Earth and Planetary Science Letters, v. 43, p. 385–396.

Clayton, R. N., and Mayeda, T., 1963, The use of bromine pentafluoride in the extraction of oxygen from oxides and silicates for isotopic analysis: Geochimica et Cosmochimica Acta, v. 27, p. 43–52.

Craig, H., 1957, Isotopic standards for carbon and oxygen and correction factors for mass-spectrometric analysis of carbon dioxide: Geochimica et Cosmochimica Acta, v. 12, p. 133.

Dasch, E. J., Hedge, C. E., and Dymond, J., 1973, Effect of sea-water interaction on strontium isotope composition of deep-sea basalts: Earth and Planetary Science Letters, v. 19, p. 177–183.

De Paolo, D. J., and Wasserburg, G. J., 1977, The sources of island arcs as indicated by Nd-Sr isotopic studies: Geophysical Research Letters, v. 4, p. 465–468.

Giletti, B. J., Semet, M. P., and Yund, R. A., 1978, Studies in diffusion—III. Oxygen in feldspars: An ion microprobe determination: Geochimica et Cosmochimica Acta, v. 42, p. 45–57.

Gill, J. B., 1974, Role of underthrust oceanic crust in the genesis of a Fijian calc-alkaline suite: Contributions to Mineralogy and Petrology, v. 43, p. 29–45.

Gilluly, J., 1971, Plate tectonics and magmatic evolution: Geological Society of America Bulletin, v. 82, p. 2383–2396.

Green, T. H., 1970, High pressure experimental studies on the mineralogical constitution of the lower crust: Physics of the Earth and Planetary Interiors, v. 3, p. 441–450.

Hart, S. R., 1971, Dredge basalts: Some geochemical aspects: EOS [American Geophysical Union Transactions], v. 52, p. 376.

James, D. E., 1971a, Plate tectonic model for the evolution of the central Andes: Geological Society of America Bulletin, v. 82, p. 3325–3346.

—— 1971b, Andean crustal and upper mantle structure: Journal of Geophysical Research, v. 76, p. 3246–3271.

James, D. E., Brooks, C., and Cuyubamba, A., 1976, Andean Cenozoic volcanism: Magma genesis in the light of strontium isotopic composition and trace-element geochemistry: Geological Society of America Bulletin, v. 87, p. 592–600.

Kay, R. W., 1977, Geochemical constraints on the origin of Aleutian magmas, in Talwani, M., and Pitman, W. C., III, eds., Island arcs, deep sea trenches and back-arc basins: American Geophysical Union Maurice Ewing Series, v. 1, p. 229–242.

Kay, R. W., Sun, S.-S., and Lee-Hu, C.-N., 1978, Pb and Sr isotopes in volcanic rocks from the Aleutian Islands and Pribilof Islands, Alaska: Geochimica et Cosmochimica Acta, v. 42, p. 263–273.

Klerkx, J., and others, 1977, Strontium isotopic composition and trace element data bearing on the origin of Cenozoic volcanic rocks of the central and southern Andes: Journal of Volcanology and Geothermal Research, v. 2, p. 49–71.

Kulm, L. D., and Schweller, W. J., 1977, A preliminary analysis of the subduction process along the Andean continental margin, 6° to 45°S, in Talwani, M., and Pitman, W. C., III, eds., Island arcs, deep sea trenches and back-arc basins: American Geophysical Union Maurice Ewing Series, v. 1, p. 285–301.

Lefèvre, C., 1973, Les caractères magmatiques du volcanisme plio-quaternaire des Andes dans le Sud de Pérou: Contributions to Mineralogy and Petrology, v. 41, p. 259–272.

Magaritz, M., and Taylor, H. P., Jr., 1976, Oxygen, hydrogen and carbon isotope studies of the Franciscan formation, Co st Ranges, California: Geochimica et Cosmochimica Acta, v. 40, p. 215–234.

Magaritz, M., Whitford, D. J., and James, D. E., 1978, Oxygen isotopes and the origin of high $^{87}Sr/^{86}Sr$ andesites: Earth and Planetary Science Letters, v. 40, p. 220–230.

McNutt, R. H., and others, 1975, Initial $^{87}Sr/^{86}Sr$ ratios of plutonic and volcanic rocks of the central Andes between latitudes 23° and 29° south: Earth and Planetary Science Letters, v. 27, p. 305–313.

Modreski, P. J., and Boettcher, A. L., 1973, Phase relationships of phlogopite in the system K_2O-MgO-CaO-Al_2O_3-SiO_2-H_2O to 35 kilobars; a better model for micas in the interior of the earth: American Journal of Science, v. 273, p. 385–414.

Nance, W. B., and Taylor, S. R., 1976, Rare earth element patterns and crustal evolution—I. Australian post-Archean sedimentary rocks: Geochimica et Cosmochimica Acta, v. 40, p. 1539–1551.

Nicholls, I. A., 1975, The origin of magmas at convergent plate boundaries: Australian Society of Exploration Geophysicists Bulletin, v. 6, p. 75–76.

Nicholls, I. A., and Ringwood, A. E., 1973, Effect of water on olivine stability in tholeiites and the production of silica-saturated magmas in the island arc environment: Journal of Geology, v. 81, p. 285–300.

Peterman, Z. E., and others, 1967, Sr^{87}/Sr^{86} ratios in some eugeosynclinal sedimentary rocks and their bearing on the origin of granitic magma in orogenic belts: Earth and Planetary Science Letters, v. 2, p. 433–439.

Ringwood, A. E., 1975, Composition and petrology of

the earth's mantle: New York, McGraw-Hill, 618 p.

Sacks, I. S., 1977, Interrelationships between volcanism, seismicity, and anelasticity in western South America: Tectonophysics, v. 37, p. 131–139.

Schilling, J.-G., and Bonatti, E., 1975, East Pacific ridge (2°S–19°S) versus Nazca intraplate volcanism: rare-earth evidence: Earth and Planetary Science Letters, v. 25, p. 93–102.

Shimizu, N., 1975, Rare earth elements in garnet and clinopyroxenes from garnet lherzolite nodules in kimberlites: Earth and Planetary Science Letters, v. 25, p. 26–32.

Stern, C. R., and Wyllie, P. J., 1973, Melting relations of basalt-andesite-rhyolite-H_2O and a pelagic red clay: Contributions to Mineralogy and Petrology, v. 42, p. 313–323.

—— 1978, Phase compositions through crystallization intervals in basalt-andesite-H_2O at 30 kilobars with implications for subduction zone magmas: American Mineralogist, v. 63, p. 641–663.

Stern, C. R., Huang, W. L., and Wyllie, P. J., 1975, Basalt-andesite-rhyolite-H_2O crystallization intervals with excess H_2O and H_2O undersaturated liquidus surfaces to 35 kilobars, with implications for magma genesis: Earth and Planetary Science Letters, v. 28, p. 189–196.

Taylor, H. P., 1968, The oxygen isotope geochemistry of igneous rocks: Contributions to Mineralogy and Petrology, v. 19, p. 1–71.

Thorpe, R. S., Potts, P. J., and Francis, P. W., 1976, Rare earth data and petrogenesis of andesite from the north Chilean Andes: Contributions to Mineralogy and Petrology, v. 54, p. 65–78.

Tilton, G. R., 1979, Isotopic studies of Cenozoic Andean calc-alkaline rocks: Carnegie Institution of Washington Year Book 78, p. 298–304.

Whitford, D. J., 1975, Geochemistry and petrology of volcanic rocks from the Sunda arc, Indonesia [Ph.D. thesis]: Canberra, Australia, Australian National University.

—— 1977, Rare earth element abundances in late-Cenozoic lavas from the Peruvian Andes: A preliminary report: Carnegie Institution of Washington Year Book 76, p. 840–844.

Whitford, D. J., Compston, W., and Nicholls, I. A., 1977, Geochemistry of late Cenozoic lavas from eastern Indonesia: Role of subducted sediments in petrogenesis: Geology, v. 5, p. 571–575.

Wilson, J., and García, W., 1962, Geología de los cuadrángulos de Pachía y Palca: Boletín de la Comisión de la carta geológica nacional del Perú, v. 2, no. 4.

Wyllie, P. J., 1977, Crustal anatexis: An experimental review: Tectonophysics, v. 43, p. 41–71.

MANUSCRIPT SUBMITTED AUGUST 18, 1978
MANUSCRIPT RECEIVED BY THE SOCIETY NOVEMBER 12, 1980
MANUSCRIPT ACCEPTED DECEMBER 30, 1980

Geological Society of America
Memoir 154
1981

Isotopic composition of Pb in Central Andean ore deposits

GEORGE R. TILTON
ROBERT J. POLLAK
Department of Geological Sciences
University of California
Santa Barbara, California 93106

ALAN H. CLARK
RONALD C. R. ROBERTSON
Department of Geological Sciences
Queen's University
Kingston, Ontario, K7L 3N6

ABSTRACT

Pb-isotope ratios are reported for 21 specimens of sulfides from 17 hydrothermal ore deposits, representative of several richly mineralized areas of the ensialic Central Andes. The deposits range in age from Middle or Late Triassic to mid-Miocene—embracing much of the time span of Andean metallogenesis—and include members of the copper deposit–dominated *Cordillera Occidental* and of the northern domains of the tin belt of the Bolivian *Cordillera Oriental*.

Pb from five deposits of Paleocene-Eocene age in northern Chile (lat 26° to 28°S) yields positive correlations of $^{207}Pb/^{204}Pb$ and $^{208}Pb/^{204}Pb$ with $^{206}Pb/^{204}Pb$. The slope of the regression line in the $^{207}Pb/^{204}Pb$ versus $^{206}Pb/^{204}Pb$ diagram is too great to correspond to a meaningful Pb-Pb age; instead, this slope is ascribed to mixing between two dominant sources of Pb. There is no relationship between the Pb-isotope ratios in ore and the age or rock type of the local, pre-Andean, basement terrane. In isotope correlation diagrams, published Pb data from coeval igneous rocks of the area generally plot in the same field as the Pb from ores, but show greater scatter. These observations suggest that the ore Pb and rock Pb are cogenetic and that ore Pb has not been affected by interaction between hydrothermal fluids and the underlying basement rocks.

In isotope correlation diagrams, both ore Pb and rock Pb plot distinctly above the regression lines for Pb from oceanic volcanic rocks; hence, models in which such volcanic rocks are the sole source of Pb must be eliminated. Pb from pelagic sediments of the Pacific Ocean and Nazca plate are systematically displaced below the ore regression line and therefore cannot account for the ore Pb. Mixtures of sediment and volcanic rock Pb likewise cannot produce the Pb-isotope ratios of the ore. The ore Pb is markedly enriched in radiogenic ^{207}Pb relative to comparable assemblages in most other circum-Pacific segments, which seems to require significant contributions of Pb from old sialic crust. The crustal component may contain Pb from subducted pelagic sediments, but an additional component enriched in ^{207}Pb is required to account completely for all of the ore Pb data. The Pb-isotope relations for the northern Chile ores are ascribed to mixing of a mantle component, which could be derived from subducted oceanic basalt, and a more radiogenic Pb derived from upper or intermediate continental crust. In the case of the igneous rocks, parameters such as the bulk chemical composition require that the crustal Pb component was introduced by selective contamination rather than by bulk mixing of the parent materials. Pb is therefore inferred to have behaved as a relatively mobile element during magma genesis.

Parallel experiments are reported from the northern Bolivian tin belt, where tin and tungsten mineralization is associated, in part, with granites displaying strong evidence of interaction with metapelitic crustal rocks. Two subprovinces are distinguished on the basis of ore Pb isotopes, as well as age. The Oligocene-Miocene ores in both the tin belt and the contiguous Altiplano have Pb-isotope ratios that plot in isotope correlation diagrams approximately along the same regression lines defined by the Chilean ores. The Triassic-Cretaceous ores, however, define regression lines that are displaced toward lower ^{206}Pb/^{204}Pb. In the ^{207}Pb/^{204}Pb versus ^{206}Pb/^{204}Pb diagram, the separation of the regression lines corresponds closely to the age difference. No rock Pb data are available from the tin belt. The Bolivian trends are explained in the same manner as those of the Chilean ores, that is, by mixing of a mantle and a crustal Pb component. Although metallogenetically distinctive, the tin belt is not delimited by a specific compositional range of Pb-isotope ratios. In particular, Pb-rich ores associated with mafic to intermediate igneous rocks in the tin belt have Pb-isotope ratios that are similar to those of the approximately coeval Sn-bearing deposits. Pb-isotope data obtained directly on the cassiterite in the tin ores may help to determine whether the Sn and Pb are derived from separate sources.

INTRODUCTION

Clarification of the controls on the nature, intensity, and boundaries of metallogenetic domains in the post-Paleozoic Central Andean orogen (lat 5° to 35°S) will require an understanding of both the source regions of the ore metals and the processes of shallow-crustal metal concentration.

Whereas considerable progress has been made in the reconstruction of metal-concentrating hydrothermal systems in individual ore deposits (for example, El Salvador, Chile; Gustafson and Hunt, 1975), there are few known constraints on ore-metal source regions. It is generally accepted that ore deposition in the Andean cordilleras has been a product of tectonic and magmatic events associated with the prolonged subduction of eastern Pacific oceanic lithosphere beneath the South American continental margin, but there is little agreement regarding the critical metallogenetic determinants (Ericksen, 1975; Sillitoe, 1976; Clark and others, 1976; Noble, 1976). Most petrochemical research to date has focused on the igneous rocks of the region rather than on the ores, and the majority of such studies have concerned only the youngest volcanism (Neogene-Holocene), which, except very locally, is not demonstrably associated with metallic mineralization.

McNutt and others (1979) presented Pb-isotope data for 12 volcanic and plutonic rocks from an intensely mineralized transect of the Andean orogen in northern Chile and contiguous northwestern Argentina (lat 26° to 29°S). On the basis of isotopic and other geochemical data (McNutt and others, 1975; Clark and others, 1976; Dostal and others, 1977), McNutt and others (1979) concluded that the rocks, which range in age from early Mesozoic to Pleistocene, originated through a two-stage magma-generation process. Pb- and Sr-isotope ratios were considered to have been primarily controlled by partial melting of subducted oceanic lithosphere, and the immediate source of the magmas was thought to have been continental-mantle peridotite, enriched in large-ion lithophile elements (LILE) through interaction with melts or aqueous fluids rising from the subduction zone. McNutt and others (1979) also noted that the rock Pb-isotope ratios are compatible with a source in the sialic crust, as well as in layer 1 of the subducted oceanic crust. However, they considered wholesale crustal anatexis an improbable mechanism for magma generation, in view of the intermediate average composition of the Andean igneous suite, the minor-element contents (including rare-earth elements), and the generally

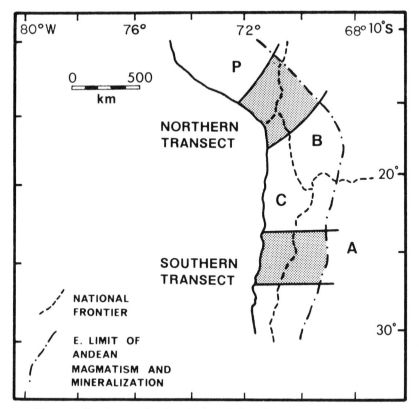

Figure 1. Sketch map showing the boundaries of the southern and northern transects across the Central Andes (P, Peru; B, Bolivia; C, Chile; and A, Argentina).

low initial Sr-isotope ratios (McNutt and others, 1975). Since no ore Pb data were determined, only tentative conclusions could be reached concerning the genesis of the associated mineralization, and these conclusions involved the assumption that rock Pb and ore Pb are cogenetic.

The present paper reports Pb-isotope data for 21 ore samples from 17 metallic ore deposits in two Andean transects (Fig. 1), including the study area of McNutt and others (1979), herein termed the "southern transect" (Clark and others, 1976). The remaining samples are from a second transect across northern Bolivia and southernmost Peru: the "northern transect." The studied specimens span an age range from Triassic to late Miocene or early Pliocene and thus embrace almost the entire duration of Andean metallogenesis. The origin of the ore deposits will be considered on the basis of the Pb data, together with other geochemical observations made since the preparation of the discussion by McNutt and others (1979).

GEOLOGIC SETTING

Petrographic and Metallogenetic Subprovinces

With respect to the nature of the igneous rocks and associated ores, it is probably valid to distinguish two major domains in the Central Andes of southern Peru, northern Chile, western Bolivia, and northwestern Argentina. Of these, the Main arc—comprising the now-active volcanic belt and its volcano-plutonic antecedents of Mesozoic and Cenozoic age—underlies much of the *Cordillera Occidental*, or *Principal*, and its Pacific slopes. The more restricted Eastern arc defines the continentward limit of extensive magmatism in southern Peru and Bolivia (see Sillitoe, 1976) and forms the core of the *Cordillera Oriental* of Bolivia and eastern Peru.

Throughout northern Chile, the Main arc has displayed a remarkably systematic eastward migration of the major focus of igneous activity, and of hydrothermal mineralization, beginning in the Early Jurassic and continuing into the early Miocene (Farrar and others, 1970; Clark and others, 1976; McBride, 1977). The predominantly calc-alkaline rocks are largely intermediate (tonalitic to andesitic) in composition, but display an overall eastward enrichment in K_2O and in incompatible elements (Lefèvre, 1973; Dostal and others, 1977; Thorpe and Francis, 1979); this trend results in a geographic and temporal transition from suites characteristic of island arcs, in the vicinity of the present coastline, through continental-margin calc-alkaline assemblages, to rocks of shoshonitic affinity farthest inland (according to the criteria of Jakes and White, 1972). These markedly potassic inland rocks were emplaced over enormous areas of the Eastern Andes during a brief Neogene (Early Miocene–Early Pliocene) episode. This event has been ascribed (Clark and others, 1976) to an approximately 10- to 15-m.y.-long period of subduction somewhat shallower than had previously prevailed, broadly coincident with the reorganization of plate boundaries in the eastern Pacific (Herron, 1972; Handschumacher, 1976), and an inferred concomitant increase in convergence rate.

Important mineralization has occurred in the Central Andes since the Middle Triassic, and, moreover, the nature of the metallogenesis has changed significantly since that time (Clark, 1976). For example, extensive Cu mineralization has taken place from the Jurassic to the Pliocene, and the major Cu porphyry deposits range in age at least from Paleocene to early Pliocene (Quirt and others, 1971). Conversely, emplacement of hydrothermal, polymetallic Pb-Zn-Ag (-Cu) ores was markedly concentrated during one or more brief Miocene metallogenetic episodes (Noble, 1977).

Our sampling of large and small base- and precious-metal ore deposits in northern Chile, northwestern Argentina, southern Peru, and the Bolivian Altiplano focuses on the Tertiary metallogenetic episodes and includes hydrothermal deposits of Paleocene, Eocene, and Miocene age.

Although several authors (Petersen, 1970; Sillitoe, 1972; Ericksen, 1975) have emphasized the existence of a broad, longitudinal zonation of the ore metals in the Central Andes, such a pattern is lacking in the southern transect (Clark and others, 1976; McNutt and others, 1979); in other transects, the zonation is replaced by more complex distributions (for example, between lat 14° and 16°S; Bellido and de Montreuil, 1972).

Of the major ore metals, Sn alone is almost entirely restricted to the Eastern arc, particularly in Bolivia, but also in contiguous southern Peru and northern Argentina. Radiometric dating (Clark and Farrar, 1973; Evernden and others, 1977; Grant and others, 1979; S. L. McBride, R.C.R. Robertson, A. H. Clark, and E. Farrar, work in progress) has delimited two brief metallogenetic episodes of Sn and polymetallic mineralization in this region, of Middle to Late Triassic and Late Oligocene to early Miocene age, the latter being subdivided into two distinct phases. Both episodes, which largely affected separate segments of the Bolivian *Cordillera Oriental,* are represented in our sampling.

The locations of the sampled ore deposits and prospects are given in Table 1, together with the Pb-isotope analytical data.

Southern Transect

The southern transect (Fig. 2), from the Atacama Province of Chile to La Rioja and Catamarca Provinces of Argentina, reveals an unusually comprehensive history of Andean magmatism, extending from the Late Triassic to the Pleistocene. A previous paper (Clark and others, 1976) gives an overview of the geology and metallogeny of this region.

On the basis of stratigraphic relationships and extensive K-Ar dating (Quirt, 1972; McBride, 1972), the volcano-plutonic complexes apparently were emplaced during a succession of episodes at 191 to 176, 156 to 137, 128 to 117, 107 to 87, 67 to 59, 45 to 40, 25 to 15, and, mainly in Argentina, 11 to 5 m.y. B.P. Thereafter, voluminous andesitic-dacitic volcanism occurred through the Pliocene and early Pleistocene to construct the *Cordillera Principal,* which rises along the Chile-Argentina border. Areally extensive rhyodacitic ignimbrite flows were erupted at intervals from the Late Cretaceous to

TABLE 1. ISOTOPIC COMPOSITION OF LEAD IN ANDEAN SULFIDE ORES

Locality (mine)	Lat - Long	Sample no.	Age* (m.y.)	Mineral†	$\frac{^{206}Pb}{^{204}Pb}$	$\frac{^{207}Pb}{^{204}Pb}$	$\frac{^{208}Pb}{^{204}Pb}$	Pb (ppm)
			Southern Transect					
Chile								
Unnamed (Queb. Buenos Aires)	27°21';70°08'	RL-248	60-63	C	18.661	15.664	38.657	49
Faro (Queb. Carrillos)	27°30';70°07'	RL-336	60-63	C	18.609	15.646	38.650	8
Descubridora (Queb. Puquios)	27°10';69°52'	RL-338	60-63	C	18.540	15.620	38.461	48
Unnamed (Queb. S. Andrés)	27°08';69°44'	RL-347	60-63	C	18.772	15.697	38.985	295
El Salvador	26°15'18";69°34'24"	ES-35	46	Py	18.511	15.593	38.444	0.55
El Salvador	26°15'18";69°34'24"	Z-534B	46	Py	18.529	15.620	38.523	70
Argentina								
Capillitas	27°21';66°24'	JC-103	7-9	G	18.825	15.664	38.884	..
Capillitas	27°21';66°24'	JC-104	7-9	G	18.835	15.668	38.920	..
			Northern Transect					
Peru								
Toquepala	17°13'17";70°39'06"	SP-31	57	C	18.519	15.629	38.533	11
Toquepala	17°13'17";70°39'06"	SP-160	57	C	18.716	15.664	38.617	..
Bolivia (Cordillera Oriental--Tin Belt)								
Monolito	15°47'24";68°39'00"	CR-342	83	G	18.407	15.640	38.506	..
Mercedes	15°52'00";68°36'18"	CR-50A	210	C	18.801	15.807	39.201	20
Don Carlos	16°20'42";68°06'54"	CR-22	211	C	18.434	15.666	38.623	550
Chojlla	16°24'06";67°46'24"	CR-65B	203	C	18.519	15.689	38.641	10
Urania	16°41'18";67°47'18"	CR-103B	26-28	Ph	18.598	15.686	38.757	11
Urania	16°41'18";67°47'18"	CR-103B	26-28	C	18.734	15.694	38.938	220
Viloco	16°51'42";67°30'54"	CR-256	26	C	18.755	15.671	38.958	8
Argentina	16°58'54";67°19'48"	CR-115	26	C	18.765	15.701	39.047	1,900
Pacuni	16°59'42";67°19'24"	CR-297	26	G	18.692	15.646	38.874	..
Bolivia (Altiplano)								
Maltilde	15°45'18";68°58'12"	CR-130	?	G	18.347	15.627	38.458	..
Tarpa (Quimsa Chata)	16°37'48";68°40'24"	CR-389	12	G	18.644	15.657	38.911	..

*Analyses were made predominantly on the basis of K-Ar ages of ores or mean ages of associated intrusions or volcano-plutonic complexes (after Quirt, 1972; Caelles and others, 1971; Gustafson and Hunt, 1975; McBride, 1977, and unpub. data; Evernden and others, 1977).

†C = chalcopyrite; Ph = pyrrhotite; Py = pyrite; G = galena.

the Pliocene, in part coevally with the development of the pluton-cored stratovolcanoes, but predominantly during the brief intervals separating the later events.

Each of the main volcano-plutonic episodes generated a broadly longitudinal magmatic subprovince, with a maximum width from east to west of about 40 km, although it is uncertain whether the subprovinces are continuous along the length of the southern transect. The mean rate of eastward migration of the main Andean magmatic arc increased from 0.6 mm/yr in the Mesozoic to 1.1 mm/yr in the Paleogene. The Miocene episode of volcanism and epizonal plutonism reached inland more than 500 km from the present Peru-Chile Trench to affect an extremely wide area of the Chilean cordillera and northwestern Argentina. Since the early Pliocene, the magmatic arc has contracted areally and is now superimposed on the western part of the Neogene subprovince.

The Andean orogenic assemblages constitute a comparatively thin veneer on a largely authochthonus, predominantly crystalline basement terrane, the *exposed* parts of which range in age from Cambrian, or latest Proterozoic, to Early Triassic. In the western part of the transect, K-Ar (Quirt, 1972; McBride and others, 1976) and Rb-Sr (Halpern, 1978) dating has shown that the basement granitoid plutons are predominantly of Permian age. Farther east, in the Argentinian Sierras Pampeanas, Late Ordovician–Silurian and Carboniferous episodes of batholith emplacement have been delimited (McBride and others, 1976), and latest Precambrian intrusions and metasedimentary rocks are probably also present (Halpern, 1978).

The westward decrease in the age of the Paleozoic orogenic belts probably reflects the progressive accretion of the continental crust about the Precambrian nucleus (Caelles, 1979; McBride and others, 1976) and results in a west-to-east increase in the age discordance between the Andean rocks (and ores) and their foundations. Preliminary plate tectonic reconstructions (Caelles, 1979) for the basement suggest that the Paleozoic orogens were ensimatic; certainly, there is no direct evidence for the existence of early Precambrian crust beneath this Andean segment (Zentilli, 1974).

Scattered K-Ar (Quirt, 1972; McBride, 1972) and Rb-Sr (Gustafson and Hunt, 1975) dates on ore deposits in the transect suggest a close temporal association of hydrothermal and magmatic activities since the Jurassic. Thus, the above-mentioned magmatic subprovinces are cospatial with metallogenetic subprovinces, or domains.

This Andean transect is intensely mineralized. The distribution of mineral deposits in the region has been summarized by Ruiz and Ericksen (1962) and by Angelelli (1950); Zentilli (1974) and Haynes (1975) have provided metallogenetic syntheses on the basis of plate tectonic relationships (see Clark and others 1976; McNutt and others, 1979). Mineralization is very widespread, and the majority of Andean igneous centers have associated base- and precious-metal deposits, with the exception of the Early Jurassic quartz dioritic plutons of the immediate coastal region and the undissected stratovolcanoes of the *Cordillera Principal*. Cu mineralization almost everywhere dominates, but Fe and Ag are locally more important. The Miocene metallogenetic episode, which affected large areas of

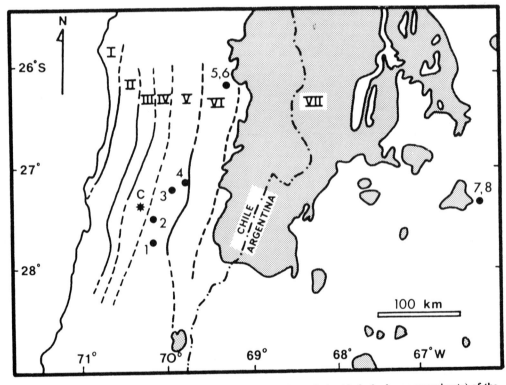

Figure 2. Sample locations in the southern transect. The boundaries (dashed where approximate) of the major volcano-plutonic subprovinces are indicated, largely after Quirt (1972), Zentilli (1974), and Haynes (1975). The subprovinces are I, Early Jurassic; II, Late Jurassic; III, Early Cretaceous; IV, middle to Late Cretaceous; V, Paleocene; VI, late Eocene–early Oligocene; VII, middle Miocene–early Pliocene (the Neogene "break-out episode"). The active volcanic arc lies along the Chile-Argentina frontier north of lat 27° 20′S. A Late Oliogocene-Early Miocene arc, not shown in the diagram, lies close to the western boundary of field VII. Sample numbers 1 to 8 correspond to specimen numbers RL-248, RL-336, RL-338, RL-347, ES-35, Z-534, JC-103, and JC-104; respectively, in Table 1. C = the city of Copiapó, Chile.

Argentina, did not yield the diversity of metals common in more northerly transects of the Central Andes; Cu is abundant, Pb and Zn mineralization is sparse, and Sn deposits of Andean age are lacking.

Sampled Ore Deposits. Of the six deposits examined in the present study, four (represented by samples RL-248, RL-336, RL-338, and RL-347, Table 1) are small deposits lying within the Paleocene (67 to 59 m.y. B.P.) metallogenetic domain of the Copiapó mining camp. In this district of the Andean Precordillera, andesitic, dacitic, and rhyolitic flows of the Paleocene continental Hornitos Formation (Segerstrom and Parker, 1959) are crosscut by several extensive felsic subvolcanic domes and by elongated granodioritic and monzogranitic plutons. The immediate host rocks of these small mineralized systems range from Triassic grits, constituting the local base of the Andean supracrustal succession (as at RL-347), to Paleocene volcanics and volcaniclastic sediments of the Hornitos Formation. Sample RL-248 is from a Cu-Ag–bearing breccia body at the margin of a trachytic stock, and RL-336 was collected from Mina Faro, a small vein working situated close to the contact of a granodioritic pluton.

Far more extensive mineralization is represented by samples ES-35 and Z-534B, both from the El Salvador mine. This Cu porphyry deposit, with original reserves of 300 million tons of 1.6% Cu ore, has been described by Gustafson and Hunt (1975), who demonstrated with unusual clarity the time and space relations of stockwork mineralization and the evolution of the Eocene Indio Muerto intermediate to silicic volcanic center. Subvolcanic emplacement of granodioritic-rhyolitic porphyries took place 46 m.y. ago into approximately 50-m.y.-old rhyolitic domes. Mineralization began at this time, but culminated at 41 m.y. B.P., during intrusion of a granodioritic porphyry complex.

Much of the economic-grade mineralization at El Salvador has been affected by supergene sulfide enrichment, but the two samples analyzed in this study are of unaltered hyopgene assemblages. Specimen ES-35 is of stockwork pyrite-chalcopyrite mineralization from the Inca adit level (2,400 m above sea level). Quartz-anhydrite-sulfide veins here cut probable Late Cretaceous andesites displaying intense K-silicate alteration (biotite–alkali feldspar–anhydrite). The veins are probably representative of the early-formed "A" quartz-vein type defined by Gustafson and Hunt (1975). In contrast, sample Z-534B is from a very coarse grained quartz-pyrite-enargite-sphalerite-chalcopyrite vein, with strongly sericitized wall rocks. This sample (unlocated, but from the deep core zone) is characteristic of the late-stage "D" veins of the deposit. Fluid inclusion study of these specimens supports the assignment of the samples to the above vein types. Quartz in ES-35 contains several inclusions similar to the type I of Gustafson and Hunt (1975), which display large daughter crystals of halite and sylvite and remain unhomogenized at 550 °C. Z-534B contains relatively abundant two-phase inclusions with homogenization temperatures of 270 to 295 °C.

The two samples from El Salvador were selected to look for any variation in Pb-isotope ratio from the early to the late stages of development of the mineralized center.

The remaining two samples (JC-103, JC-104) from the southern transect are from Capillitas, once the most productive Cu mine in northwestern Argentina. Situated at the eastern limit of the southern transect, the mining district lies within the Sierras Pampeanas structural province and is underlain predominantly by Paleozoic metasedimentary rocks and granitic batholiths. At Capillitas, several steeply inclined polymetallic veins are associated with a subvolcanic rhyolite-dacite center, emplaced into Late Ordovician–Silurian granites (McBride and others, 1976). The age of mineralization is undetermined, but this center is a satellite of the large, partially eroded, Farallón Negro stratovolcano, which contains several Cu-Mo-Au porphyry and vein deposits. K-Ar dating (Caelles and others, 1971; McBride, 1972) has delimited a 5-m.y. interval (10.6 to 6 m.y. B.P.) in which volcanism, intravolcanic intrusion, and hydrothermal mineralization occurred in the Farallón Negro and nearby Cerro Rico centers. The Capillitas veins, therefore, are considered to be of Late Miocene or, perhaps, Early Pliocene age. The Farallón Negro volcanics have shoshonitic compositions (Caelles, 1979) and were erupted through the Paleozoic crystalline basement of the Andean foreland during the Neogene expansion of the magmatic arc.

The Capillitas veins, with pyritic and sericitic envelopes, comprise early quartz-pyrite zones, the economically important quartz-enargite-tennantite-chalcopyrite-bornite-chalcocite assemblage, and the late galena-sphalerite-rhodochrosite assemblage (Angelelli and others, 1974). The two galenas analyzed in this study are from the late assemblage.

Northern Transect

Southernmost Peru. In broad terms, the geologic evolution of the *Cordillera Occidental* of southern Peru was similar to that of northern Chile across the southern transect. Magmatic activity and associated mineralization began by the Early Jurassic in the coastal zone (James and others, 1975; McBride, 1977); throughout the Mesozoic and Paleogene, the activity migrated inland at essentially the same rate as that defined in more detail in the southern transect. Subsequently, Neogene volcanism and high-level plutonism extended inland at least as far as the Lake Titicaca area. This episode, probably in part of Late Oligocene-Miocene age, corresponds closely to that recognized farther south and generated a wide spectrum of calc-alkaline compositions, with shoshonitic suites farther inland (Lefèvre, 1973).

In two respects, however, this area differs from that to the south. In northern Chile, the locus of magmatism was consistently displaced away from the continental margin at approximately 30-m.y. intervals; the northern transect, however, shows considerable superposition of igneous activity of widely different ages. Thus, the *Cordillera de la Costa* is underlain by probable Early Jurassic marine volcanics, intruded by granitoid plutons of both Late Jurassic and mid-Cretaceous age (McBride, 1977). Similarly, 100 km inland, Paleocene-Eocene volcanic and plutonic rocks are partially overlain by Neogene-Quaternary volcanics.

Of greater potential significance from the standpoint of the Pb-isotope ratios of Andean ores is the widespread occurrence (Fig. 3) of early Precambrian rocks in the Andean basement of this transect. At least one of the late Precambrian-Paleozoic episodes of orogeny and granitoid plutonism recognized farther south is represented in coastal southern Peru, but this is itself superimposed on a 2,000-m.y.-

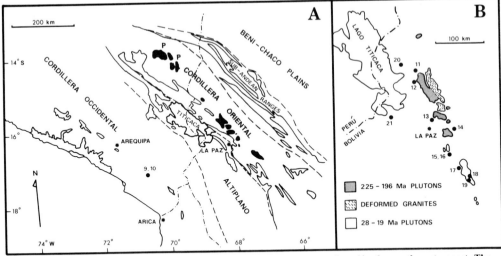

Figure 3. Sample locations and major tectonic-physiographic subprovinces in the northern transect. The exposures of pre-Andean (that is, pre-middle Triassic) basement rocks are shown in Figure 3A (fine shading), largely after Megard and others (1971). Basement rocks in southern Peru (the Arequipa massif) range in age from 2,000 m.y. to middle Paleozoic (Shackleton and others, 1979), whereas the exposed basement of the Cordillera Oriental in northern Bolivia is Paleozoic (Ordovician to Permian). Black areas in Figure 3A are Phanerozoic granitoid plutons of the Cordillera Oriental (those labeled P are Permian). The ages (Ma = millions of years) of granitoid plutons (after Evernden and others, 1977, and McBride, 1977) in the northern Bolivian tin belt are shown in Figure 3B. Sample numbers 9 to 21 correspond to specimen numbers SP-31, SP-160, CR-342, CR-50A, CR-22, CR-65B, CR-103B (two analyses), CR-256, CR-115, CR-297, CR-130, and CR-389, respectively, in Table 1.

old metamorphic complex (Shackleton and others, 1979), which includes a considerable proportion of granulite-facies rocks. The continuity of this ancient basement assemblage with the partly coeval Brazilian Shield has been suggested by preliminary paleomagnetic studies (Shackleton and others, 1979).

The two analyzed sulfide samples (SP-31, SP-160; Table 1; Fig. 3) are from the large Toquepala Cu-Mo mine, Moquegua Department, Peru. Stockwork mineralization in this complex porphyry deposit is associated with a small, intensely-altered, porphyritic dacite stock and with a major tourmaline-cemented hydrothermal breccia pipe (Richard and Courtright, 1958). The age of mineralization has not been directly determined, but the nearby, broadly comparable Quellaveco and Cuajone porphyry deposits are of Early Eocene age (Estrada, 1975; McBride, 1977), while pre-ore grandodiorite has yielded a K-Ar age of 58.7 m.y. (McBride, 1977). James and others (1974) determined a Rb-Sr age of 69 m.y. for the pre-ore Toquepala volcanics.

Of the analyzed specimens, one (SP-160) is of hypogene chalcopyrite from chalcopyrite-pyrite-quartz-anhydrite veinlets intersected in diamond-drill core 358 m below the (1976) base of the open pit. The associated hydrothermal alteration is of the phyllic facies (sericite-quartz-pyrite), but textural relationships suggest that this has been superimposed on an earlier potassic silicate assemblage. This sample was taken from well below the zone of intense supergene sulfide enrichment exposed in the open pit. On the other hand, sample SP-31, from the 3,220-m level of the pit, is of chalcopyrite from a pyritic veinlet, forming part of a stockwork in intensely sericitized granodiorite, situated close to the base of a zone of supergene sulpide enrichment. However, the analyzed sulfides show no evidence of replacement by chalcocite.

Altiplano. The extensive, approximately 3,800-m-high, *Altiplano* of western Bolivia and southern Peru overlies the thickest section of sialic crust in the Central Andes (James, 1971), the M-discontinuity lying at a depth of greater than 70 km. The essentially planar topography is a Neogene-Quaternary erosional and depositional composite, but this is underlain by a continental clastic succession of unusual thickness for the Central Andes (in excess of 10,000 m). Sedimentation to the east of the progressively uplifted and migrating Main arc of the *Cordillera Occidental* apparently began in the Late Cretaceous (Ahlfeld, 1972; Evernden and others, 1977). Episodic volcanism and epizonal plutonism extended from the Late Cretaceous to the Holocene (Evernden and others, 1977), but was concentrated in the Miocene. The volcanic association is dominated in many areas by felsic ash-flow tuffs, in part coeval with ignimbritic formations of the Pacific slope of the *Cordillera Occidental* (Tosdal and others, 1981).

The nature of the pre-Cretaceous basement of the Bolivian *Altiplano* is imperfectly known. However, Carboniferous-Permian continental sedimentary and volcanic rocks, which cover wide areas of southeastern Peru, probably do not extend southward far beyond the Lake Titicaca region. Lehmann (1978) described a drillhole through the *Altiplano* sediments some 50 km south of the lake, which demonstrated that here the Cretaceous strata lie directly on a basement of late Precambrian deformed granitoids, which yielded a Rb-Sr whole-rock model age of about 1,050 m.y.; the basement unconformity lies at an altitude of 1,184 m. Evernden and others (1977) reported a late Precambrian (527 m.y. B.P.) K-Ar date for weathered red granite, again underlying Cretaceous strata in the northwestern *Altiplano*. Elsewhere in the northern *Altiplano*, Tertiary clastics contain cobbles of late Precambrian gneisses (Evernden and others, 1977), presumed to have been derived from the crystalline basement of the *Cordillera Oriental*. Thus, there is no evidence, as yet, for the existence of ancient basement beneath the Altiplano.

Apart from several moderately important red-bed deposits, most metallic mineralization in the Altiplano is associated with subvolcanic stocks (Ahlfeld, 1967) of Miocene age (Everden and others, 1977; McBride, 1977) and, at least in part, of shoshonitic character. The ores are polymetallic, and include Pb-Zn-Ag and Cu-Au veins; in comparison with northwestern Argentina (that is, in the eastern part of the southern transect), the relative importance of Pb-Zn and Cu mineralization appears to be reversed. However, Sn minerals are extremely rare in both regions; only toward the eastern margin of the *Altiplano* do small amounts of Sn and W mineralization herald the approach to the *Cordillera Oriental* tin belt (Ahlfield, 1967).

The two deposits selected for analysis from the Bolivian Altiplano polymetallic province are from the vicinity of Lake Titicaca. The Quimsa Chata camp, located about 12 km south of Tiahuanacu, comprises several systems of barite-bearing veins and a small breccia pipe, worked in the past mainly for Pb and Ag (as argentian tetrahedrite), but also for sphalerite and chalcopyrite (Ahlfeld and Schneider-Scherbina, 1964). The analyzed sample (CR-389) was taken from a Pb-rich ore zone in the small abandoned Mina Tarpa working (Pb-Zn-(?)Ag), situated south of Hacienda Achaca. Mineralization is related spatially to a Miocene (12.3 m.y. B.P.; Evernden and others, 1977) stock of porphyritic dacite and to satellitic dikes cutting red sandstones of early Neogene age (Corocoro Formation).

The much larger Matilde vein deposit, the source of the analyzed galena in CR-130, is the most important Zn producer in Bolivia; the mine lies 12 km northeast of the town of Ancoraimes, to the east of Lake Titicaca. Two major veins, the Matilde and Maravillas, have been traced for considerable distances both vertically and horizontally. The geologic relationships of the mineralization are unclear (Ahlfeld and Schneider-Scherbina, 1964). The host rocks are sandstones and shales, of Devonian or, possibly, Early Carboniferous (Tournaisian: Laubacher, 1978) age, but no igneous rocks crop out in the mine area. Thus, a genetic relationship with the predominantly dacitic subvolcanic stocks (Rivas, 1968) exposed along the eastern shore of Lake Titicaca cannot be demonstrated.

The veins in the Matilde mine comprise crustified siderite, with abundant sphalerite, lesser galena, and minor pyrite and chalcopyrite. The analyzed sample was taken from the 205-m level of the Maravillas vein.

Cordillera Oriental—Tin Belt. The general geology of the northern part of the Bolivian tin belt has been described by, among others, Turneaure and Welker (1947), Ahlfeld (1967), and Turneaure (1971). Martinez and Tomasi (1978) presented a revised tectonic map of the region. There are few detailed descriptions of individual ore deposits in the area, but Ahlfeld and Schneider-Scherbina (1964) provided summaries of the geology of the majority of the significant deposits, and Kelly and Turneaure (1970) carried out reconnaissance mineralogical and fluid inclusion studies. This northerly segment of the tin belt is coextensive with the *Cordilleras Muñecas, Real, Quimsa Cruz,* and *Santa Vera Cruz,* the last three of which constitute an essentially linear mountain chain with many glaciated peaks rising above 5,500 m.

Mineralization, predominantly of vein type, occurs within several granitoid stocks and small batholiths and, more extensively, in their metamorphic aureoles. Lower Paleozoic clastic sedimentary rocks formed the host rocks; they are now largely metamorphosed to greenschist facies. The sediments appear to have been deposited in a relatively narrow, but deep, northwest-trending trough, which delimited the eastern margin of the Paleozoic orogens (Mégard and others, 1971). This "Hercynian" sedimentary trough has generally been assumed to be ensialic, but the nature of its basement is unresolved. If the 500- to 1,000 m.y.-old metagranitic rocks described by Evernden and others (1977) and Lehmann (1978) are representative of the basement beneath the *Altiplano* (see above), the *Cordillera Oriental* could overlie the boundary between this Proterozoic terrane and the little-known 2,000-m.y.-old Guaporé craton of the Brazilian Shield (Cordani and others, 1972). The lower Paleozoic sedimentary rocks of the region may have been derived through erosion of both older and younger Precambrian metamorphic and igneous rocks.

Continuing petrologic studies by one of us (Robertson) have demonstrated that the granitoid massifs of this part of the tin belt are composite intrusions, comprising granodioritic and granitic (sensu stricto) units. Several of the stocks in the region are peraluminous and contain cordierite or garnet megacrysts, and may thus represent "S-type" granitoids (products of anatexis of continental crust; Chappell and White, 1974). The overall compositional range of plutonic rocks in the *Cordillera Oriental* differs from that in the *Cordillera Occidental* of southern Peru and northern Chile, where quartz diorites and granodiorites predominate and true granites are of limited occurrence.

South of lat 17°20'S, the average elevation of the *Cordillera Oriental* decreases as its width broadens from 50 to more than 150 km. In central and southern Bolivia, the magmatic belt comprises numerous subvolcanic igneous centers which host the world's largest bedrock Sn deposits, as well as several very important Ag and base-metal vein systems. In contrast, the Sn production of the northern domains of the tin belt has been less extensive, and Ag is present in only minor amounts. This area, however, is a major source of W.

Most of the ore deposits of the plutonic domain are characteristically clustered wolframite- or cassiterite-bearing quartz veins, containing abundant base-metal sulfides. Locally, other metals or metalloids, such as Ni (Don Carlos) and Bi (Mercedes), are abundant. Pegmatitic deposits containing Sn and/or Li are developed in the *Cordillera Real*, but have not supported a large production (Turneaure and Welker, 1947).

Earlier geochronologic studies of the plutonic rocks of the northern tin belt (Evernden and others, 1977; Clark and Farrar, 1973) suggested that, despite the striking similarities of all of the granitic stocks and ore deposits of the region, intrusion and mineralization took place during two widely separated intervals, in the early Mesozoic and mid-Tertiary. More detailed studies (McBride, 1977; S. L. McBride, R.C.R. Robertson, A. H. Clark, and E. Farrar, work in progress) have confirmed these relationships and have delimited two plutonic and metallogenetic episodes at 225 to 196 m.y. B.P. (Middle to Late Triassic) and 28 to 16 m.y. B.P. (Late Oligocene to Early Miocene). The products of these episodes do not appear to overlap areally, the younger domain lying to the south of the earlier. There is no good evidence for granitoid magmatism or W-Sn mineralization during the long hiatus between the two episodes, although in the Late Cretaceous (from 83 to 79 m.y. B.P.; S. L. McBride; R.C.R. Robertson, A. H. Clark, and E. Farrar, work in progress) the *Cordillera Oriental* experienced a pulse of basaltic-andesitic minor intrusion and volcanism, with which several small Pb-Ag vein deposits appear to have been related. These include the Monolito deposit discussed below. McBride and others (1977) interpreted the intermediate (Jurassic-Paleogene) K-Ar ages shown by several deformed granitic bodies in the *Cordillera Real* as resulting from a brief mid-Tertiary episode of compressional tectonism. Earlier work (Mégard and others 1971; Bard and others, 1974) had assigned these intrusions to an early Hercynian event in the pre-Andean bsement.

Sampled Ore Deposits. In the present discussion, the sampled ore deposits will be briefly described in order from north to south; thus, the Mesozoic mineralization will be dealt with initially, to be followed by that of Tertiary age.

The first deposit to be discussed is the most problematic in terms of its geologic relationships. Sample CR-342 was taken from the small abandoned Monolito Pb mine, located near the town of Sorata. This vein deposit is close to the contact between a small intrusive and extrusive andesitic complex and its host rocks of Devonian shale and Cretaceous sandstone (Rosenblum, 1968). A K-Ar whole-rock date for a fresh andesite from the Sorata area (82.7 ± 1.0 m.y. B.P.; S. L. McBride, R.C.R. Robertson, A. H. Clark, and E. Farrar, work in progress) is in permissive agreement with the Late Cretaceous age assigned on stratigraphic grounds to mafic volcanics of the Puca Group by Russo and Rodrigo (1965). The age relationships make it very unlikely that the Monolito deposit is related to the nearby Illampu granitoid batholith of Triassic age.

The three analyzed ore samples (from Mercedes, Don Carlos, and Chojlla mines; see Table 1) of inferred or proved Triassic age are for vein deposits which display a close spatial association with clearly Middle to Late Triassic granitoid intrusions; the veins lie within metamorphosed Lower Paleozoic clastics. The samples were selected to embrace the widest compositional spectrum represented in the mineralization of this subprovince.

The Mercedes del Illampu deposit, located in the Millipaya mining district at the northwestern margin of the Illampu batholith (Sorata), is rich in Cu, W, and Bi. The vein material comprises quartz, abundant chalcopyrite (the main ore mineral in sample CR-50A), pyrite, and arsenopyrite, with lesser wolframite, bismuthinite, and sphalerite. The deposit was mined for W (Ahlfeld and Schneider-Scherbina, 1964).

The small Don Carlos vein deposit (sample CR-22) displays an unusual variety of minerals: chalcopyrite, cassiterite, stannite, galena, sphalerite, quartz and ankerite, and a complex nickel arsenide assemblage. The vein was emplaced close to the contact of the small, intensely greisened Chacaltaya stock, a satellite of the large Huayña Potosí pluton. Late Triassic K-Ar ages (210 to 213 m.y.) have been determined for hydrothermal muscovites in the stock and an associated Sn vein deposit. No mafic rocks are exposed in the vicinity of the Ni-bearing Don Carlos deposit.

The last deposit sampled in the Triassic subprovince is the large Chojlla W-Sn vein system, probably the major single source of W in the tin belt. The mine is situated at relatively low altitude (2,000 m) on the east slope of the *Cordillera Real.* Ahlfeld and Schneider-Scherbina (1964) and Michel and Reutter (1977) described the geology of the deposit. A subparallel series of greisen-bordered quartz veins, containing wolframite, cassiterite, and base-metal sulfides, is hosted by Ordovician metaclastics immediately overlying a greisened aplogranite cupola, itself presumably an apophysis of the nearby Taquesi granitoid pluton. K-Ar ages of hydrothermal muscovites from the stock and mineralized veins range from 196 to 203 m.y. (S. L. McBride, R.C.R. Robertson, A. H. Clark, and E. Farrar, work in progress). Evernden and others (1977) earlier determined ages as young as 183 m.y. for muscovite from the mine. We tentatively conclude that the mineralized veins were emplaced in the Late Triassic, but that they were affected by later tectonic activity, perhaps in the mid-Tertiary (McBride and other, 1977), which resulted in some resetting of the mineral ages. However, there is no clear field evidence for the suggestion of Schneider-Scherbina (1962) that the deposit comprises an early (that is, Triassic) episode of W mineralization and a later (probably Tertiary) Sn-rich stage of deposition. The analyzed sample from Chojlla (CR-65B) is of chalcopyrite, associated with wolframite and cassiterite, from the no. 7 central vein on the -60-m level of the mine.

The remaining four specimens are from important Sn-W vein deposits in the southern segment of the plutonic subprovince of the tin belt, in which the age of the agranitoid intrusions is clearly Late Oligocene (about 28 m.y.) to Early Miocene (about 19 m.y.; S. L. McBride, R.C.R. Robertson, A. H. Clark, and E. Farrar, work in progress). Although only a single ore deposit has been specifically dated, we assume that mineralization shortly followed intrusion of the nearby, or host, stocks.

The northernmost igneous body in this younger subprovince is the Illimani plutonic-volcanic cener, which underlies the 6,460-m-high Cerro Illimani. The complex comprises both coarse granodiorites and extrusive and subvolcanic dacites; it was emplaced in the Late Oligocene (28 to 27 m.y. B.P.). The most important mine in the district was the Urania, which exploited a vein in Silurian (?) phyllites and quartzites adjacent to a small apophysis of the Illimani stock. The complex vein was worked for wolframite and scheelite and is also rich in chalcopyrite, pyrrhotite, and pyrite, with minor arsenopyrite and sphalerite. Coexisting chalcopyrite and pyrrhotite were separately analyzed (sample CR-103B).

Lying to the south-southeast of the Illimani center, the more extensive Quimsa Cruz (Tres Cruces) granitoid batholith is associated with widespread and locally intense Sn and W mineralization, including several of the most important Sn-producing mines in the plutonic segment of the tin belt. This composite intrusion comprises earlier granodiorites (24.2 to 25.0 m.y. B.P.; S. L. McBride, R.C.R. Robertson, A. H. Clark, and E. Farrar, work in progress) and younger two-mica granites (23.6 to 23.9 m.y. B.P.). We have analyzed sulfide minerals from three major mines in this region. Of these, the famous Viloco (Araca) Sn-W vein system (Ahlfeld and Schneider-Scherbina, 1964; Turneaure and Welker, 1947) straddles the southwest contact of the Quimsa Cruz granite. Ores of Mo, W, and Sn were apparently deposited successively, and at increasing distances from the intrusive contact. The analyzed chalcopyrite (sample CR-256) was taken from the Nueva vein in the most westerly section of the mine, San Antonio. Here, on the 4,280-m level, the sulfide-rich quartz vein cuts chloritized Lower Devonian shales.

The two remaining specimens are from the highly productive Caracoles mining camp, located at the southeastern extremity of the batholith. Sample CR-115 is chalcopyrite from the 133-m level of the Argentina Sn mine. The Gloria vein here cuts granite and is rich in tourmaline, quartz, cassiterite, sphalerite, bismuthinite, chalcopyrite, and arsenopyrite. The nearby Pacuni mine (sample CR-297, galena) comprises both wolframite- and cassiterite-rich veins, again in granite. The anlayzed sample (CR-297) is a galena from the 5,064-m level of the P-6-North vein, a Sn-rich body with minor Pb and Zn sulfides.

Our sampling does not extend southward into the most productive, "subvolcanic," segments of the tin belt, where mineralization was in part coeval with that in the southern part of the plutonic domain (Grant and others, 1979; S. L. McBride, R.C.R. Robertson, A. H. Clark, and E. Farrar, work in progress) but also extended into the late Early Miocene.

ANALYTICAL METHODS

The chemical and mass spectrometric methods closely followed those described in Meijer (1976) and are discussed in detail in Pollak (1977). Chalcopyrite samples, with admixed pyrite, and, in one specimen, pyrrhotite, were hand-picked from the crushed ore. Analyses were carried out on 6 to 10-mg fragments, generally from single grains or grain aggregates. The samples were dissolved in $8N$ HNO_3 in teflon containers, heated to dryness, redissolved in $6N$ HCl, and re-evaporated to dryness. The residues were dissolved in one drop of $3M$ HCl, and then 1 ml of $1N$ HBr was added. The solutions were loaded onto 1-ml ion exchange columns containing Dowex 1, ×8, 100 to 200 mesh. The Pb was purified by washing the columns with $1N$ HBr and $3N$ HCl, and finally by removal with $6N$ HCl. After evaporation, the purified Pb was loaded for mass spectrometric analysis by the silica gel–phosphoric acid method.

Galena samples were dissolved in $3N$ HCl and processed on identical columns, but $1N$ HBr was not added. All blanks for chemical processing were in the range of 0.2 to 0.5 ng.

The ratios given in Table 1 are based on two or more independent determinations, including separate chemical processing. On the basis of replicate analyses of Pb standard NBS 981 (Catanzaro and others, 1968), a correction of 0.15% per mass unit was applied to the measured ratios to convert them to absolute values. The isotopic ratios are believed to be accurate to within $\pm 0.06\%$ (1δ) for ^{206}Pb, $\pm 0.09\%$ fr $^{207}Pb/^{204}Pb$, and $\pm 0.12\%$ for $^{208}Pb/^{204}Pb$. The isotopic analyses were typically made on 0.5 μg of Pb.

RESULTS

The Pb-isotope ratios for the ore samples are listed in Table 1 together with the Pb concentrations, when known. The estimated ages of the mineral deposits are based predominantly on the K-Ar geochronologic studies of Quirt (1972) and McBride (1972, 1977) and, in some cases, on general stratigraphic relationships.

In Figures 4 and 5, the $^{207}Pb/^{204}Pb$ and $^{206}Pb/^{204}Pb$ ratios of the ores are compared to those of several selected lithologic assemblages which represent potential Pb sources, on the assumption that subduction beneath the Pacific margin of the South American continent has persisted at least since Late Triassic time (Clark and others, 1976). These include ocean-ridge basalts and other basaltic rocks from ocean basins, taken mainly from the data summarized in Church and Tatsumoto (1976); metalliferous sediments from the Nazca plate (Dymond and others, 1973); and pelagic sediments from the Pacific Ocean and Nazca plate (Chow and Patterson, 1962; Meijer, 1976; Reynolds and Dasch, 1971). The data of Chow and Patterson have been converted to absolute ratios by normalizing their ratios for the California Institute of Technology reference sample to those given by Catanzaro (1967). In Figure 4 the new ore Pb data are compared to the rock Pb ratios reported by McNutt and others (1979) for the southern transect. The present-day ratios are plotted; the rock ages range from Pliocene to Jurassic. The initial $^{206}Pb/^{204}Pb$ ratios are shown, where known, for the Cretaceous and Jurassic rocks. Since Th concentrations were not determined, the initial $^{208}Pb/^{204}Pb$ ratios are unknown.

DISCUSSION OF METALLOGENETIC AND PETROGENETIC MODELS

Our ore Pb data are of a reconnaissance nature and may not be representative of the post-Paleozoic Andean metallogenesis. In particular, our coverage has emphasized Cenozoic ores at the expense of earlier mineralization. Although all of the Pb data share certain characteristics, it is probably premature, on both analytical and geologic grounds, to ascribe all the isotopic ratios to a single population and, hence, to a unified genetic model. For this reason, we will examine the possible significance of the data on a geographic basis, although we thereby run the risk of separating essentially cogenetic samples. In the following discussion, we will take into account the known, and

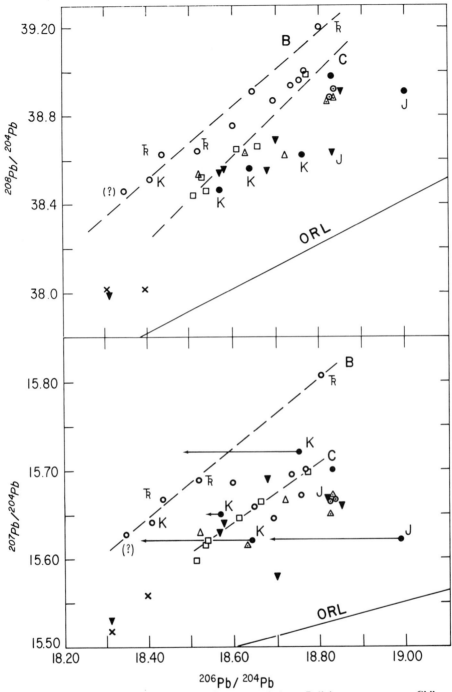

Figure 4. Pb-isotope correlation diagram. Ores: open circles = Bolivia; open squares = Chile; open triangles = Peru (this study); open triangles with a dot = Peru (Doe and Zartman, 1979); open circles with a dot = Argentina. X = Metalliferous sediments from the East Pacific Rise (Dymond and others, 1975). Igneous rocks of southern transect from McNutt and others (1979): solid triangles = volcanic rocks; solid circles = plutonic rocks. Regression lines defined as follows: B, for Bolivian Triassic-Cretaceous ores; C, for Chilean ores; ORL, for oceanic volcanic rocks. Arrows give in situ decay corrections for Jurassic and Cretaceous rocks where known.

probable, make-up of the broad geologic environments in which the ores occur, including the basement rocks, as well as present knowledge of the compositions and probable genesis of the associated Andean igneous rocks. Our discussion will focus on the relationships between $^{207}Pb/^{204}Pb$ and $^{206}Pb/^{204}Pb$, but similar trends are shown by the $^{208}Pb/^{204}Pb$ ratios.

Southern Transect

Northern Chile. The most significant aspect of the northern Chilean ore Pb ratios is that they define linear trends in the correlation diagram of Figure 4. The slope of line C in the lower part of Figure 4 corresponds to a Pb-Pb age of 3.7 b.y., a value quite certainly too old to have any time significance. The correlation is undoubtedly a reflection of the mixing of Pb from two sources, one unusually rich in ^{207}Pb. The nature of these components will be discussed below. It is noteworthy that the components appear to have maintained their identity over the time span of 50 m.y. represented by the samples. It is also possibly significant that the more massive El Salvador deposit contains less radiogenic Pb (that is, lower isotope ratios with respect to ^{204}Pb) than do most of the remaining ores, which come from smaller deposits.

This is the only region of the Central Andes in which comparative Pb-isotope data are available for local and essentially coeval rocks (McNutt and others, 1979) and mineral deposits (Fig. 4). The igneous rocks, which include extrusive and plutonic members, occupy a wider composition field than do the ores, even when the apparently aberrant, ^{206}Pb-deficient sample No. Z-815 of McNutt and others (1979) is omitted. The spread appears to reflect mainly the wider age range of the rock samples. If the data for the six Cenozoic (Paleocene-Eocene) Chilean ore samples are compared to those for the essentially coeval (Late Cretaceous–Miocene) rocks in Figure 4, the two groups are very similar. In subsequent discussions we will therefore assume that the rock Pb and ore Pb derive from common sources and that any petrogenetic model derived for one type applies equally to the other.

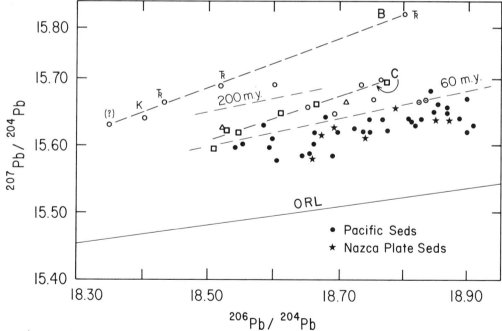

Figure 5. Pb-isotope ratios of ores compared with ratios from pelagic sediments and manganese nodules of the Pacific Ocean, taken from Chow and Patterson (1962), Meijer (1976), Reynolds and Dasch (1971). Ore symbols and line designations have the same meanings as in Figure 4. The numbered dashed lines give the estimated mean isotope ratios for Pb in pelagic sediments 60 and 200 m.y. ago. See text for explanation.

The ores are compared with Pb from Pacific pelagic sediments in Figure 5. There is little overlap in Pb-isotope ratios between ore and present-day sediment. However, since the ores are up to 45 to 60 m.y. old, any contributions from pelagic sediments would probably have involved sediments that were subducted about 55 to 70 m.y. ago. By using the evolution curves of Stacey and Kramers (1975) for average terrestrial Pb, we estimate that $^{206}Pb/^{204}Pb$ ratios would then have been lower on the average by approximately 0.1 compared to present ratios, whereas the $^{207}Pb/^{204}Pb$ ratios would have been essentially unchanged. One of the dashed lines in Figure 5 depicts average pelagic sediment Pb about 60 m.y. ago according to this estimate. Even with this modification the overlap between ores and sediments is poor; even if pelagic sediments have contributed to the Pb in the ores, we conclude that still other sources are required for the more radiogenic component and that complete mixing must have occurred between the various sources.

Since the igneous rocks and ores appear to contain similar Pb ratios, it is of interest to examine the Sr-isotope data for igneous rocks from the southern transect (McNutt and others, 1975), which include all of the samples studied for Pb (McNutt and others, 1979). McNutt and others (1975) noted that the initial $^{87}Sr/^{86}Sr$ ratios of the samples of Cretaceous to Neogene age increase with decreasing age; the ratios also increase from west to east since the rocks are aligned in north-trending belts with age increasing to the west. Those authors proposed a model involving mixtures of Sr from a subducted oceanic lithospheric slab and the overlying mantle peridotite to account for the secular trend in the Sr ratios. The higher $^{87}Sr/^{86}Sr$ ratios in the younger rocks were ascribed to greater amounts of Sr from a phlogopite phase in the mantle peridotite. McNutt and others (1975) did not completely rule out pelagic sediments or basal crustal melting as alternative sources of the radiogenic Sr. In any case, the Sr data require some kind of mixing between a minimum of two sources, the question being whether all of the Sr derives from the mantle, or whether a crustal component is present. We will further evaluate this question in the following discussions.

Figures 4 and 5 show that the Pb ratios in the ores cannot be accounted for by Pb from the subducted ocean crust. The ore Pb ratios clearly plot above the regression line for oceanic volcanic rocks (ORL in Fig. 4). It is also difficult to account for the ore Pb ratios by mixtures of Pb from volcanic rocks and pelagic sediment since the sediment has $^{206}Pb/^{204}Pb$ ratios that are generally too high to satisfy the requirements for the radiogenic component. In particular there is as yet no evidence for a sediment having a $^{207}Pb/^{204}Pb$ ratio of 15.70 and a $^{206}Pb/^{204}Pb$ ratio of 18.77, the minimum ratios required to account for the San Andrés ore Pb (RL-347). We conclude that we must examine other mixing models.

The most obvious source of Pb having high $^{207}Pb/^{204}Pb$ ratios is geologically old continental crust, either in upper-crustal sialic rocks (for example, Rosholt and others, 1973) or lower-crustal granulitic rocks (Montgomery and Hurley, 1978; Barreiro and Tilton, 1980; Tilton and Barreiro, 1980). The granulitic Pb is distinguished from upper-crustal Pb by having high $^{207}Pb/^{204}Pb$ ratios relative to $^{206}Pb/^{204}Pb$ ratios. In addition the $^{208}Pb/^{206}Pb$ ratios in granulites correspond to high Th/U ratios (Montgomery and Hurley, 1978; Barreiro and Tilton, 1980; Tilton and Barreiro, 1980); the $^{208}Pb/^{204}Pb$ and $^{206}Pb/^{204}Pb$ ratios may even be negatively correlated (Barreiro and Tilton, 1980 Tilton and Barreiro, 1980). In contrast, upper-crustal rocks on the average yield a positive correlation between $^{208}Pb/^{204}Pb$ and $^{206}Pb/^{204}Pb$ ratios that corresponds to a normal Th/U of about 3.8.

Precambrian rocks are not known in the study area; the oldest rocks appear to be of Early Paleozoic age (Caelles, 1979). It is possible, however, that the sedimentary rocks in the area contain detritus from the Precambrian craton to the north and east of the transect, which could have contributed to the Pb in the rocks and ores, either by direct contamination or by subduction mechanisms. We will therefore first consider mixing models in which crustal rocks are the source of the high $^{207}Pb/^{204}Pb$ component; these models emphasize igneous rock data. The strongest evidence favoring a crustal source perhaps comes from Sr-isotope data, based on the relatively high initial $^{87}Sr/^{86}Sr$ ratios shown by Neogene andesites and dacites between lat 15° and 29°S. Klerkx and others (1977) emphasized the existence of a correlation between present crustal thickness and initial Sr ratios for young volcanic rocks of the Andes as a whole: low ratios (approximately 0.704) are shown by volcanics in Ecuador and central Chile, where crustal thicknesses probably do not exceed 40 km, while in southern Peru and northern

Chile, higher ratios are found in areas where the depth of the M-discontinuity is greater than 50 km (James, 1971). Klerkx and others (1977) proposed that the magmas with high $^{87}Sr/^{86}Sr$ originate by "mixing between a subcrustal magma of basaltic affinity and upper crustal rocks."

Briqueu and Lancelot (1979) presented additional Sr evidence for crustal contamination of andesitic lavas from southern Peru and northern Chile. First, they showed that in a $^{87}Sr/^{86}Sr$ versus $1/^{86}Sr$ diagram the least-differentiated members of the Arequipa and Barroso volcanics define a line of negative slope, consistent with two-component mixing. Significantly, the mixing line intersects the field of other Ecuadoran andesites that have uniformly low $^{87}Sr/^{86}Sr$ ratios of 0.7045 and occur where the crust is only about 40 km thick (Francis and others, 1977). The Ecuadoran lavas presumably define a mantle component. To account for the bulk chemical composition of the rocks, Briqueu and Lancelot (1979) emphasized that any Sr contamination from the crustal component must be selective. James (1978) has in effect made the same observation. The remaining Arequipa and Barroso volcanics, and the San Pedro and San Pablo lavas from northern Chile, plot in such a way as to require open-system behavior during lava formation, that is, addition of crustal Sr during fractional crystallization of the magma. Such a process seems likely to approach the complexities that must exist in natural settings. The Sr data of McNutt and others (1975) plot in a manner similar to those for the northern Chile lavas in a $^{87}Sr/^{86}Sr$ versus $1/^{86}Sr$ diagram. These data accord with and strengthen the observations of Klerkx and others (1977).

McNutt and others (1979) proposed that Pb in the igneous rocks plotted in Figure 4 was derived by a two-stage mechanism in which the immediate source was continental-mantle peridotite enriched in Pb from subducted pelagic sediment and basalt. These authors also noted that the Pb ratios are compatible with a source in the sialic crust by selective transfer of Pb to the invading magmas. In either case, a substantial component of Pb from sialic crustal rocks is incorporated into the lavas. It seems significant that the 20 to 60 m.y.-old rock samples of McNutt and others (1979) plot approximately along the mixing line defined by the Chilean ores in Figure 4.

Finally, we note that the Chilean ore Pb and rock Pb plot in nearly the same field in Figure 4 as Pb from the Oligocene ore of the western Bolivian tin belt, where the igneous rocks associated with the ores bear strong evidence of crustal anatexis. We shall discuss this point more fully below.

We emphasize that Precambrian crustal rocks could have supplied the radiogenic Pb required to account for the Chilean data in Figure 4 for both igneous and ore Pb; also, there is good evidence in at least some cases for crustal mixing processes from Sr-, Pb-, and O-isotope data.

Basically, there are four ways in which crustal Pb might be incorporated into magmas and ores:

1. rock-meteoric water interaction in the shallow Andean crust (this mechanism may have had particular importance for the ore deposits).

2. Contamination, probably often selective, of mantle-derived melts rising through the crust.

3. Direct thickening of the continental-margin crust as a result of the underthrusting of slabs of upper-crustal igneous and sedimentary rocks detached from the leading edge of the continental plate, followed by partial melting of the accreted material.

4. Subduction and partial melting of clastic sediments (graywackes) formed through erosion of supracrustal and basement rocks.

From the standpoint of Pb isotopes, all four mechanisms are equivalent since they result in addition of Pb from crustal materials to the ore solutions and magmas. Other data, which we do not propose to discuss here, are required to choose between them. We emphasize again that mechanisms (2) and (3) generally will require selective contamination through solution transport of Pb in place of mechanical mixing in order to account for the bulk chemistry of the rocks in relation to their Pb isotopes. James (1978) has shown that mechanical mixing of graywacke sediments with mantle-derived components might account for many of the chemical and isotopic data for the Arequipa and Barroso lavas, although it is not known whether such a model could satisfactorily explain the Pb data (Tilton, 1979).

The second Pb component in the crustal mixing models is undoubtedly of mantle origin. This seems required because calc-alkaline lavas also occur in oceanic settings where continental crustal materials have not been available to add Pb to the magmas (Sinha and Hart, 1972; Meijer, 1976). In the present case the mantle end member might be given by the El Salvador ore, the largest of the analyzed Chilean deposits. Moreover, the Pb ratios from El Salvador agree closely with ratios from the Disputada Mine ore at lat 33°S (Tilton, 1979), where $^{87}Sr/^{86}Sr$ ratios are uniformly about 0.704 (R. E. Drake, 1978, personal commun.). We also note that the metalliferous sediments plot along the mixing trends for the Chilean ores in Figure 4. The Pb data therefore allow the mantle component to be represented by metalliferous sediments. We see no way to choose between these two possible mantle sources at present.

Alternatively, we consider the possibility that the Chilean ore Pb and rock Pb are derived solely from mantle sources. Again at least two components are required to account for the range of Pb-isotope ratios in the ores. As noted above, McNutt and others (1975) favored a mantle-disequilibrium partial-melting model to account for the Sr-isotope trends in the plutonic and volcanic rocks from the southern transect. This was subsequently modified (McNutt and others, 1979) to allow for an origin from layers 1 and 2 of subducted oceanic lithosphere. For a mantle origin, one component with $^{206}Pb/^{204}Pb \geq 18.77$, $^{207}Pb/^{204}Pb \geq 15.70$, and $^{208}Pb/^{204}Pb \geq 38.99$ is required, that is, Pb at least as radiogenic as that in the San Andrés ore (RL-347). The ratios must be such as to plot along the mixing line defined by the ores. There is at present no evidence for such Pb from either oceanic or continental rocks that are of probable mantle origin. Since these are mainly basaltic rocks, for which we currently have limited information in continental terranes, it is possible that, as data accumulate, such highly radiogenic Pb might be found.

If the magma-generation process does involve only mantle materials, the mechanism must explain why the highly radiogenic Sr is found in southern Peru and northern Chile, but not in Ecuador and central Chile. In particular, the tectonic settings in northern and central Chile seem very similar.

Although it is impossible to rule out a mantle source for the radiogenic Pb component of the northern Chile ores and rocks, present evidence strongly favors a sialic crustal origin. It is less clear whether the addition mechanism is a direct process, such as wall-rock interaction, or an indirect one involving selective contamination from subducted pelagic and/or continental sediments. In principle, the dilemma will be resolved by determining the extent to which the component rich in LILE (large-ion lithophile elements) and radiogenic Pb and Sr is controlled by crustal thickness and age of basement complex in various localities.

Argentina. Pb data are available for the Capillitas ore and a neighboring latite-andesite (sample JC 27 of McNutt and others, 1979). These samples are from the shoshonitic province of the transect, as discussed above. Little can be stated on the basis of the present limited sampling. Pb of the Capillitas ore plots at the right-hand extremity of the ore field in Figure 4. This could be a reflection either of the younger age compared to the other ores or perhaps of the petrologic province. On the other hand, Pb from the latite plots well within the field of the Chilean rocks and ores. On the basis of present data, speculation on the origin of the Pb seems premature. More detailed studies of Pb from rocks of the shoshonite province of southeastern Peru are underway in our laboratories.

Northern Transect

Many of the arguments advanced in the previous section, and particularly those regarding the origin of the igneous rocks, are germane to the discussion of at least the *Cordillera Occidental* region of the northern transect that crosses southernmost Peru and probably to the discussion of the Bolivian *Altiplano* as well. Thus, the petrogenetic data and their interpretation will be reviewed only briefly. The most significant difference is that the northern transect is located within the Precambrian craton.

Southernmost Peru. The petrochemical relationships among igneous rocks of this area have been studied extensively by D. E. James and his colleagues. James (1978), revising his earlier hypothesis for the origin of the Quaternary andesitic rocks of the area, has developed a model compatible with their major- and minor-element chemistry and with their Sr- and O-isotope ratios. The 1978 hypothesis favors generation of the magmas in the mantle wedge overlying the subduction zone, through partial melting of LILE-enriched periodotite. The high initial Sr-isotope ratios (to 0.708) and relatively high $\delta^{18}O$ values are interpreted as reflecting the involvement of considerable volumes of subducted graywackes in the original phase of magma generation, rather than contamination of the magmas during their passage through the sialic crust. In contrast, Zentilli and Dostal (1977) tentatively recognized direct interaction with the crust, on the basis of the correlation between U contents of volcanic rocks and present crustal thicknesses.

More recent Pb-isotope work on the young lavas (Arequipa and Barroso volcanics) studied by James (Tilton, 1979) and neighboring granulitic rocks of the Precambrian basement complex (Barreiro and Tilton, 1980; Tilton and Barreiro, 1980) indicates interaction between magmas and Precambrian lower crust. Tilton (1979) favored a mechanism rather like that of Zentilli and Dostal (1977) to account for the Pb data.

The differing compositions of the two sulfide samples from the Toquepala porphyry deposit are of interest. Of these, the one with more radiogenic Pb (chalcopyrite, SP-160) is that from a deep drill hole, well beneath the effects of supergene sulfide enrichment; the one with less radiogenic Pb (chalcopyrite, SP-30), although showing no microscopic evidence of chalcocite replacement, may have exchanged Pb with meteoric fluids during weathering in the Miocene (Tosdal, 1978). It may be significant that the two ore samples plot approximately along the mixing trend defined by the Chilean ores. Thus, although Pb-isotope ratios from the Quaternary igneous rocks from southern Peru differ distinctly from those of the southern transect, the Early Eocene Toquepala ores do not differ greatly from the southern transect rocks and ores on the basis of the limited sampling in Peru.

The Toquepala ore Pb values display positive correlations between $^{208}Pb/^{204}Pb$ and $^{206}Pb/^{204}Pb$ ratios. This contrasts with the granulite mixing trends for the Arequipa and Barroso volcanics, for which these ratios are negatively correlated (Tilton, 1979). The Toquepala ore Pb values therefore share with the rocks and ores of the southern transect the imprint of *upper-crustal* involvement, direct or indirect.

The contrast in Pb-isotope ratios in the ore and neighboring volcanic rocks is perhaps surprising. The overall distribution of ancient granulites in the thick (approximately 70 km; James, 1971) crust of the transect is unknown, but such rocks constitute no more than 50% of the exposed basement assemblages grouped as the Arequipa massif (for example, Shackleton and others, 1979). Therefore, although granulitic metamorphic conditions very probably prevail in the present deep crust, large volumes of Andean magma may not have come into contact with 2,000-m.y.-old granulites but, rather, with lower-grade Proterozoic or Paleozoic crystalline rocks comparable to the basement rocks of the southern transect.

The present extreme thickness of the crust in southern Peru might, in itself, lead to unusually extensive interaction between crust and rising magmas. Geomorphologic studies (Tosdal, 1978) of the Pacific slope of the *Cordillera Occidental* in this region reveal that episodic Andean uplift—presumably in reaction to progressive thickening of the sialic crust through magmatic activity—has taken place at least since the Late Oligocene. Conditions for crustal contamination may, therefore, have been less favorable before Pliocene time. Alternatively, it is possible that, if James's (1978) petrogenetic model is valid, the pre-Neogene magmas (and ores) may not display the effects of granulite Pb because subduction of substantial volumes of the granulites may not have occurred before the late Tertiary; the granulites are exposed extensively along the present coast of southern Peru, but whether they extend to the southwest across the narrow continental shelf is unknown. In any case, as in Chile, involvement of Pb from crustal sources is favored to account fully for the range of Pb ratios observed at Toquepala.

Northern Bolivia *(Altiplano* and *Cordillera Oriental).* In northern Bolivia, as noted above, our more comprehensive sampling of Triassic, probably Cretaceous, and Oligocene-Miocene ores in the tin belt and its surroundings delimits two compositionally distinct Pb-isotope subprovinces, which transgress the *Altiplano–Cordillera Oriental* boundary. For this reason, we will discuss the implications of the Pb data according to the compositional clans rather than on the basis of tectonic setting or mineralization type. No rock Pb data are available for this region.

The older ore deposits from the northern segment of the tin belt are represented by the three vein deposits of Triassic age associated with the Illampu (Mercedes), Huayña Potosí (Don Carlos), and Taquesi (Chojlla) granitic plutons; by the Monolito Pb-Ag veins of probable Late Cretaceous age; and by the Matilde Zn-Pb veins of uncertain age. Although the latter two Pb-rich deposits are less enriched in ^{207}Pb than the three Sn-bearing deposits, all five samples define linear arrays in Pb-isotope correlation diagrams (Fig. 4). We infer that the Pb in these contrasted types of mineralization was derived from identical or similar sources. Thus, it would appear that the presence of Sn in the ores is not correlated with a specific set of Pb-isotope ratios and that the nature of the associated igneous rocks (if any) is not a critical determinant.

As in the case of the northern Chile ores, we believe the Pb ratios in the ores cannot be entirely accounted for by partial melting of subducted oceanic volcanic rocks; by Pb from pelagic sediments, even when corrections are made for the age of the required sediments (see Fig. 5); or by mixtures of volcanic and sediment Pb. Instead, the high ^{207}Pb/^{204}Pb ratios of the ore Pb probably indicate the involvement of old sialic crust in the evolution of the various ore types. In particular, the very high ^{207}Pb/^{204}Pb ratio of 15.80 for the Mercedes deposit is strong evidence for such a component. As at Toquepala, the ^{208}Pb/^{204}Pb and ^{206}Pb/^{204}Pb ratios are positively correlated, which again argues for upper-crustal mixing, rather than granulite mixing as observed in the Arequipa and Barroso volcanics of southern Peru.

The contrast in isotopic ratios between the older (Triassic) and younger (Late Oligocene–Early Miocene) Sn-W ores of the tin belt can be ascribed mainly to difference in age. Thus, the mean terrestrial Pb evolution curves of Stacey and Kramers (1975) predict a difference of about 0.28 in the ^{206}Pb/^{204}Pb ratios between 200- and 25-m.y.-old rocks, in close agreement with the observed difference.

Significantly, the Quimsa Chata Pb-Zn deposit and the younger Sn ores of the *Cordilleras Illimani* and Quimsa Cruz have Pb-isotope ratios which plot, with somewhat greater scatter, along the mixing trend defined by the Cenozoic Cu-dominant deposits of northern Chile and southern Peru. Thus, the marked metallogenetic specialization of the tin belt is *not* reflected in a unique set of Pb-isotope ratios.

As in the southern transect, there is little support in this area for a significant derivation of ore Pb from the *immediate* country rocks of the mineralized veins. All sampled deposits in the Triassic domain have lower Paleozoic metasedimentary host rocks, but the Oligocene-Miocene ores of the tin belt lie both within and outside of the granitoid plutons, and there are no apparent differences in ore Pb ratios in the two settings. Thus, we conclude that, if sialic crust contributed Pb to the mineralization, this occurred at a crustal level below the penetration of the hydrothermal systems.

The upper-crustal "fingerprint" shown by the Triassic Sn-bearing deposits is in conformity with long-accepted theories regarding the origin of Sn deposits, and, broadly, with the nature of the associated granitoid plutonic rocks. There are few published data for the northern Bolivian intrusions, but continuing petrologic studies by one of us (Robertson) indicate that, although displaying considerable compositional variation, the plutonic rocks of the *Cordillera Oriental* are readily distinguished, as a suite, from those of the more westerly parts of the Central Andes. Thus, granites (sensu stricto—Streckeisen, 1967) are widespread; these are associated with granodiorites, but quartz diorites (tonalites) and diorites are rare or absent in most intrusive centers.

Whereas the Triassic and Oligocene-Miocene rocks of the northern tin belt do not conform in all respects to the "S-type" granitoid clan of Chappell and White (1974) and White and Chappell (1977), they display many features, in addition to their overall compositional range, which are strongly suggestive of an origin through anatexis of metapelitic crustal rocks or, at least, of extensive involvement with such rocks. The granitoids of the tin belt are peraluminous: coarse, euhedral, cordierite occurs as an accessory mineral in several granitic plutons, both in northern Bolivia and in southern Peru, and cannot be ascribed to incorporation of metasedimentary rocks in the zone of emplacement. This aluminosilicate is observed in both Triassic and Tertiary plutons, but garnet, again apparently of deep-seated origin, has been described only from the early Miocene granodioritic Kari Kari batholith (located approximately at lat 19°48′S) in the central part of the tin belt (Wolf, 1973). Most intrusive units in the northern tin belt, including both granites and granodiorites, contain abundant biotite, and hornblende is of very restricted occurrence. Some granites contain apparently primary muscovite. The plutonic rocks have low oxidation ratios [that is, molecular ($2Fe_2O_3 \times 100$)/ ($2Fe_2O_3 + FeO$) < 60], while the low total contents of Fe-Ti oxide minerals, among which ilmenite predominates, are reflected in low magnetic susceptibilities (<150×10^{-3} S.I. units). Thus, the tin belt plutons correspond to the "ilmenite-series" granitoids of Kanaya and Ishihara (1973) and Ishihara (1977), who have emphasized the strong correlation between Sn mineralization and such magmas, which are inferred to have been reduced through interaction with C-bearing sedimentary rocks. Continuing studies on the O-isotope ratios of the granitoid rocks (F. Longstaffe, unpub. data) reveal numerous whole-rock values for $\delta^{18}O$ in excess of $+10^0/_{00}$. R. H. McNutt (unpub data) has determined that the initial Sr-isotope ratios of both Triassic and Oligocene-Miocene rocks are close to 0.708, a relatively high value when compared to the majority of ratios found for plutonic rocks of the *Cordillera Occidental* (McNutt and others, 1975).

These compositional and petrographic features are in strong contrast to those of the *Cordillera Occidental* and require considerable interaction between felsic melts and reduced, metasedimentary, sialic crust. Although at least a partial source of the magmas in the mantle cannot be ruled out, most of their compositional parameters would be satisfied by an origin through partial melting of clastic sediments.

Lack of information on the deeper structure of the *Cordillera Oriental* of northern Bolivia inhibits identification of the crustal source we infer for the ore Pb. However, as noted above, Proterozoic metamorphic and granitoid rocks underlie the northern *Altiplano* at shallow depths and, although the stratigraphic thickness of lower Paleozoic sedimentary rocks in the *Cordillera Oriental* is considerable, the schematic structural profiles of Mégard and others (1971) suggest that the sub-Paleozoic unconformity lies at no more than 10 km. The nature of the Precambrian rocks inferred to lie beneath the Paleozoic basin is unknown, but Mégard and others (1971) have shown that low-grade metasedimentary rocks predominate in the Precambrian basement of the Central Andes.

In summary, we propose that a substantial proportion of the ore Pb in both plutonic subprovinces of the tin belt had an origin in upper, or intermediate, sialic crust. The broad congruence in the two age domains of ore Pb ratios, on the one hand, and the composition of the associated granitoid rocks, on the other, lead us to favor a model in which the ores derived their Pb and other metals from the felsic magmas, and, therefore, had an ultimate source, or partial source, in metapelitic crustal rocks undergoing anatexis. Such an origin is unlikely for the mafic rocks associated with the Monolito Pb-Ag veins, but, again, the Pb in that deposit almost certainly had a similar crustal origin.

Despite the apparent absence of mafic igneous rocks with ages similar to those of the granitic plutons, the essentially linear trends delineated by the Triassic and Oligocene-Miocene ore Pb ratios implies the involvement of a second, weakly radiogenic, Pb source. Because crustal anatexis, particularly in the Triassic subprovince, probably required the addition of heat from the mantle, we favor a location for this second source either in the mantle wedge overlying the subduction zone, or in the metabasaltic crust of the subducted slab itself. This conjunction of crustal and subcrustal sources was, as previously argued, probably also instrumental in generating the isotopic ratios of ores and igneous rocks in southern Peru and in the *Altiplano,* although the involvement of metapelitic materials is not in evidence outside of the tin belt.

CONCLUSIONS

One of the most remarkable features of the Pb data is the similarity in isotope patterns from each of the study areas in spite of their greatly differing tectonic settings. Thus both the northern Chile and Bolivian ores contain Pb that was apparently derived by mixing between upper- or intermediate-crustal and mantle sources. The Chilean Pb occurs outside the known limit of rocks of the Precambrian craton, while the Bolivian ores are within the craton. The mixing model holds for Bolivian ores of Triassic-Cretaceous and Oligocene-Miocene age even though the crust should have evolved and thickened considerably over the time interval between the two ore sequences.

Whereas, in the Bolivian tin belt, the petrography and chemistry of the granitic rocks reveal the direct involvement of metapelitic crustal rocks in the magma-generation systems and therefore imply that the radiogenic nature of the ore Pb was there attained through direct interaction within the sialic crust, the mechanism responsible for the enrichment in radiogenic Pb of the ores and rocks of the

Cordillera Occidental is more problematic. For the igneous rocks it is difficult to discriminate between the subduction and partial melting of continent-derived sialic sediments or graywackes, on the one hand, and direct crustal "contamination" of the rising calc-alkaline magmas, on the other, when only the data available for the southern transect are considered. The latter process requires that extensive introduction of crustal Pb into the magmas has taken place, but that, in all or most other respects, the melts were little affected by their passage through the crust. Thus, crustal contamination has the aspect of a *deus ex machina*, one not clearly required by the overall petrochemistry of the Andean igneous suite. However, such a mechanism would satisfy the lack of correlation between the essentially consistent Pb ratios and the broad areal and temporal compositional evolution of the rocks throughout the Andean orogeny. The similarity in the Pb compositions of Cenozoic ores in the tin belt and in the *Cordillera Occidental* may, in itself, constitute support for the occurrence of crustal contamination in the latter region.

The new ore Pb data reported herein have ambiguous implications concerning the parameters controlling the boundaries of metallogenetic subprovinces in the Central Andes. Although our sampling of *Cordillera Occidental* ores has focused on Cu deposits at the expense, for example, of ores rich in Pb, Zn, Ag, Au, and Fe, all, or most, epigenetic hydrothermal mineralization in the western part of the orogen probably has similar Pb-isotope ratios, revealing significant contributions from both the sialic crust and the subjacent mantle. Thus, the chalcophile ore metals other than Pb may also have a dual source in the sialic crust and underlying mantle or subducted plate.

It is possible, however, that one or all of these metals are predominantly or entirely of subcrustal origin, that is, they have been less "mobile" than Pb in the vicinity of magma diapirs rising through the sialic crust. In that case, the locations and intensities of concentrations of the metals in the orogen may reflect temporal and/or spatial irregularities in their abundance and extraction from the mantle wedge overlying the subduction zone. The absence of a correlation between the trace abundances of Cu, Zn, and so forth in the igneous rocks of the *Cordillera Occidental* and the bulk rock compositions (for example, K_2O indices, LILE contents) suggests, however, that no matter what the ultimate source of the metals may have been, the degree of their concentration as mineable deposits is primarily dependent on the nature of epizonal magma–hydrothermal fluid interactions. This implies that, even if the source regions for the Pb in the Andean ores were firmly established, this would not necessarily provide constraints of significance to an understanding of metallogenetic subprovince relationships or to mineral exploration in the *Cordillera Occidental*.

When the northern Bolivian tin belt is considered, the Pb-isotope data are again not readily amenable to practical application. Whereas the Triassic W-Sn deposits have Pb ratios compatible with broadly accepted theories of crustal metal derivation, the Pb-rich ores in the same area display essentially identical Pb-isotope compositions, despite their association in some cases with mafic rocks of probable mantle origin. Moreover, the ores of the more important Oligocene-Miocene Sn-producing subprovince are not distinguishable on the basis of their Pb-isotope ratios from those of the Cu subprovince of the *Cordillera Occidental* or from Pb-rich deposits on the *Altiplano*. The data are, however, in permissive agreement with an origin for the Sn in the sialic crust.

ACKNOWLEDGMENTS

Field studies, constituting part of the Central Andean Metallogenetic Project, were supported through National Research Council of Canada (now NSERC) grants to Clark, and were carried out with the generous logistic assistance of the Instituto de Investigaciones Geológicas de Chile, the Dirección Nacional de Geología y Minería, Argentina, the Servicio Geológico de Bolivia, and Southern Peru Copper Corporation. We are particularly indebted to Srs. Carlos Ruiz F. (I.I.G.), Dr. J.C.M. Turner (DGM), Srs. Salomón Rivas and José Torrez (GEOBOL), and to Sr. Armando Plazolles (SPCC), for their generous cooperation in establishing the field programs in Chile, Argentina, Bolivia, and Peru, respectively. Numerous mine staffs kindly assisted us in the collection of specimens. Samples were also collected by Drs. R. B. Lortie, M. Zentilli, and J. C. Caelles, who are thanked for their collaboration. We thank Ms. Lyndy Davis for careful typing of a difficult manuscript.

Pb-isotope analysis at the University of California at Santa Barbara was financed by grants from the National Science Foundation to Tilton.

REFERENCES CITED

Ahlfeld, F., 1967, Metallogenetic epochs and provinces of Bolivia: Mineralium Deposita, v. 2, p. 291–311.

—— 1972, Geología de Bolivia: La Paz, Los Amigos del Libro, 190 p.

Ahlfeld, F., and Schneider-Scherbina, A., 1964, Los yacimientos minerales y de hidrocarburos de Bolivia: Servicio de Geología de Bolivia, Boletín, v. 5, 388 p.

Angelelli, V., 1950, Recursos minerales de la República Argentina, Parte I: Yacimientos metalíferos: Buenos Aires, Museo Argentino de Ciencias Naturales, Instituto Nacional de Investigación de las Ciencias Naturales, Revista, Geología, v. II, 535 p.

Angelelli, V., Schlamuk, I., and Cagnoni, J., 1974, La rodocrosita del yacimiento cuprífero de Capillitas, Dpto. Andagala, Prov. de Catamarca, República Argentina: Asociacion Geológica Argentina, Revista, v. 29, p. 105–127.

Bard, J. P., and others, 1974, Relations entre tectonique, métamorphisme et mise en place d'un granite eohercynien à deux micas dans la Cordillère Réal de Bolivie (massif de Zongo-Yani): Office de la Recherche Scientifique et Technique Outre-Mer, Cahiers, Série Géologie, v. 6, p. 3–18.

Barreiro, B. A., and Tilton, G. R., 1980, U-Th-Pb studies on Quaternary volcanics and Precambrian basement rocks from southern Peru: Geological Society of America, Abstracts with Programs, v. 12, p. 95.

Bellido, E., and de Montreuil, L., 1972, Aspectos generales de la metalogenía del Perú: Lima, Peru, Servicio de Geología y Minería, Boletín, v. 1, 149 p.

Briqueu, L., and Lancelot, J. R., 1979, Rb-Sr systematics and crustal contamination models for calc-alkaline igneous rocks: Earth and Planetary Science Letters, v. 43, p. 385–396.

Caelles, J. C., 1979, The geological evolution of the Sierras Pampeanas massif, La Rioja and Catamarca provinces, Argintina [Ph.D. thesis]: Kingston, Queen's University, 514 p.

Caelles, J. C., and others, 1971, Potassium-argon ages of porphyry copper deposits and associated rocks in the Farallón Negro–Capillitas district, Catamarca, Argentina: Economic Geology, v. 66, p. 961–964.

Catanzaro, E. J., 1967, Absolute isotopic abundance ratios of three common lead reference samples: Earth and Planetary Science Letters, v. 3, p. 343–346.

Catanzaro, E. J., and others, 1968, Absolute isotopic abundance ratios of common, equal-atom and radiogenic lead isotopic standards: Journal of Research of the National Bureau of Standards, v. 72A, p. 261–267.

Chappell, B. W., and White, A.J.R., 1974, Two contrasting granite types: Pacific Geology, v. 8, p. 173–174.

Chow, T. J., and Patterson, C. C., 1962, The occurrence and significance of lead isotopes in pelagic sediments: Geochimica et Cosmochimica Acta, v. 26, p. 263–308.

Church, S. E., and Tatsumoto, M., 1976, Lead isotope relations in oceanic ridge basalts from the Juan de Fuca-Gorda Ridge area, N.W. Pacific Ocean: Contributions to Mineralogy and Petrology, v. 53, p. 253–279.

Clark, A. H., 1976, Metallogenic epochs of the Central Andes: A peliminary assessment [abs.]: Economic Geology, v. 72, p. 726–727.

Clark, A. H., and Farrar, E., 1973, The Bolivian tin province: Notes on the available geochronological data: Economic Geology, v. 68, p. 102–106.

Clark, A. H., and others, 1976, Longitudinal variations in the metallogenetic evolution of the Central Andes: A progress report: Geological Association of Canada Special Paper 14, p. 23–58.

Cordani, U. G., Amaral, G., and Kawashita, K., 1973, The Precambrian evolution of South America: Geologische Rundschau, v. 62, p. 309–317.

Doe, B. R., and Zartman, R. E., 1979, Plumbotectonics: The Phanerozoic, in Barnes, H. L., ed., Geochemistry of hydrothermal ore deposits: New York, John Wiley, p. 22–70.

Dostal, J., and others, 1977, Geochemistry of volcanic rocks of the Andes between 26° and 28° South: Contributions to Mineralogy and Petrology, v. 63, p. 113–128.

Dymond, J., and others, 1973, Origin of metalliferous sediments from the Pacific Ocean: Geological Society of America Bulletin, v. 84, p. 3355–3372.

Ericksen, G. E., 1975, Metallogenetic provinces of the southeastern Pacific: U.S. Geological Survey,

Project Report, Circum-Pacific Investigation, CP-1 Open-File Report, 52 p.

Estrada, F., 1975, Geología de Quellavaco: Boletín de la Sociedad Geológica del Perú, v. 46, p. 65–86.

Evernden, J. E., Kriz, S. J., and Cherroni, C., 1977, Potassium-argon ages of some Bolivian rocks: Economic Geology, v. 72, p. 1042–1061.

Farrar, E., and others, 1970, K-Ar evidence for the post-Paleozoic migration of granitic intrusive foci in the Andes of northern Chile: Earth and Planetary Science Letters, v. 10, p. 60–66.

Francis, P. W., Moorbath, S., and Thorpe, R. S., 1977, Strontium isotope data for recent andesites in Ecuador and North Chile: Earth and Planetary Science Letters, v. 37, p. 197–202.

Grant, J. N., and others, 1979, K-Ar ages of igneous rocks and mineralizatin in part of the Bolivian tin belt: Economic Geology, v. 74, p. 838–851.

Gustafson, L. B., and Hunt, J. P., 1975, The porphyry copper deposit at El Salvador, Chile: Economic Geology, v. 70, p. 857–912.

Halpern, M., 1978, Geological significance of Rb-Sr isotopic data of northern Chile crystalline rocks of the Andean orogen between latitudes 23° and 27° South: Geological Society of America Bulletin, v. 89, p. 522–532.

Handschumacher, D. W., 1976, Post-Eocene plate tectonics of the Eastern Pacific, in Woollard, G. P., and others, eds., The geophysics of the Pacific Ocean and its margin: American Geophysical Union Geophysical Monograph 19, p. 177–202.

Haynes, S. J., 1975, Granitoid petrochemistry, metallogeny, and lithospheric plate tectonics, Atacama, Chile [Ph.D. thesis]: Kingston, Queen's University, 330 p.

Herron, E. M., 1972, Sea-floor spreading and the Cenozoic history of the east-central Pacific: Geological Society of America Bulletin, v. 83, p. 1671–1692.

Ishihara, S., 1977, The magnetite-series and ilmenite-series granitic rocks: Mining Geology, v. 27, p. 293–305.

Jakes, D., and White, A.J.R., 1972, Major and trace element abundances in volcanic rocks of orogenic areas: Geological Society of America Bulletin, v. 83, p. 29–40.

James, D. E., 1971, Plate tectonic model for the evolution of the central Andes: Geological Society of America Bulletin, v. 82, p. 3325–3346.

—— 1978, On the origin of the calc-alkaline volcanics of the Central Andes: A revised interpretation: Carnegie Institution of Washington Year Book 77, p. 562–590.

James, D. E., Brooks, C., and Cuyubamba, A., 1974, Strontium isotopic composition and K, Rb, Sr geochemisty of Mesozoic volcanic rocks of the Central Andes: Carnegie Institution of Washington Year Book 73, p. 970–983.

—— 1975, Early evolution of the Central Andean volcanic arc: Carnegie Institution of Washington Year Book 74, p. 247–250.

Kanaya, H., and Ishihara, S., 1973, Regional variations of magnetic susceptibility of the granitic rocks in Japan: Journal of the Japanese Association of Mineralogists, Petrologists, and Economic Geologists, v. 68, p. 211–224.

Kelly, W. C., and Turneaure, F. S., 1970, Mineralogy, paragenesis and geothermometry of the tin and tungsten deposits of the Eastern Andes, Bolivia: Economic Geology, v. 65, p. 609–680.

Klerkx, J., Deutsch, S., Pichler, H., and Zeil, W., 1977, Strontium isotopic composition and trace element data bearing on the origin of Cenozoic volcanic rocks of the Central and Southern Andes: Journal of Volcanology and Geothermal Research, v. 2, p. 49–71.

Laubacher, G., 1978, Estudio geológico de la region norte del Lago Titicaca: Lima, Peru, Instituto Geológico y Minería, Boletín, v. 5, 120 p.

Lefèvre, C., 1973, Les caractères magmatiques du volcanisme plio-quarternaire des Andes dans le sud du Pérou: Contributions to Mineralogy and Petrology, v. 41, p. 259–272.

Lehmann, B., 1978, A Precambrian core sample from the Altiplano, Bolivia: Geologische Rundschau, v. 67, p. 270–278.

Martinez, C., and Tomasi, P., 1978, Carte structurale des Andes septentrionales de Bolivie (1:1,000,000): Office de la Recherche Scientifique et Technique Outre-Mer, Cahiers, Not. Explicative No. 77, 48 p.

McBride, S. L., 1972, A potassium-argon age investigation of igneous and metamorphic rocks from Catamarca and La Rioja Provinces, Argentina [M.Sc. thesis]: Kingston, Queen's University, 101 p.

—— 1977, A K-Ar study of the Cordillera Real, Bolivia, and its regional setting [Ph.D. thesis]: Kingston, Queen's University, 231 p.

McBride, S. L., 1972, A potassium-argon age investigation of igneous and metamorphic rocks latitudes 25°–30°S: Earth and Planetary Science Letters, v. 29, p. 373–383.

McNutt, R. H., and others, 1975, Initial $^{87}Sr/^{86}Sr$ ratios of plutonic and volcanic rocks of the Central Andes between latitudes 26° and 29°S: Earth and Planetary Science Letters, v. 27, p. 305–313.

McNutt, R. H., Clark, A. H., and Zentilli, M., 1979, Lead isotopic compositions of Andean igneous rocks, latitudes 26° to 29°S: Petrologic and metallogenic implications: Economic Geology, v. 74, p. 827–837.

Mégard, F., Dalmayrac, B., Laubacher, G., Marocco, R., Martinez, C., Paredes, J., and Tomasi, P.,

1971, La Châine hercynienne au Pérou et en Bolivie: Premiers résultats: Office de la Recherche Scientifique et Technique Outre-Mer, Cahiers, Série Géologie, v. 3, p. 5–44.

Meijer, A., 1976, Pb and Sr isotopic data bearing on the origin of the volcanic rocks from the Mariana island-arc system: Geological Society of America Bulletin, v. 87, p. 1358–1369.

Michel, H., and Reutter, K.-J., 1977, Die W-Sn-Lagerstätte Chojlla, Cordillera Real, Bolivien. Teil 1: Nebengestein and Tektonik: Mineralium Deposita, v. 12, p. 247–262.

Montgomery, C. W., and Hurley, P. M., 1978, Total-rock U-Pb and Rb-Sr systematics in the Imataca Series, Guayana Shield, Venezuela: Earth and Planetary Science Letters, v. 39, p. 281–290.

Noble, D. C., 1977, A summary of radiometric age determinations on rocks from Peru: Isochron/West, v. 19, p. 7–18.

Noble, J. A., 1976, Metallogenic provinces of the Cordillera of western North and South America: Mineralium Deposita, v. 11, p. 219–233.

Petersen, U., 1970, Metallogenic provinces in South America: Geologische Rundschau, v. 59, p. 834–897.

Pollak, R. J., 1977, Reconnaissance study of the isotopic composition of lead in some Andean ores [M.A. thesis]: Santa Barbara, University of California, 53 p.

Quirt, G. S., 1972, A potassium-argon geochronological investigation of the Andean mobile belt of north-central Chile [Ph.D. thesis]: Kingston, Queen's University, 240 p.

Quirt, G. S., and others, 1971, Potassium-argon ages of porphyry copper deposits in northern and central Chile: Geological Society of America Abstracts with Programs, v. 3, p. 676–677.

Reynolds, P. H., and Dasch, E. J., 1971, Lead isotopes in marine manganese nodules and the ore-lead growth curve: Journal of Geophysical Research, v. 76, p. 5124–5129.

Richard, K., and Courtright, J. H., 1958, Geology of Toquepala, Peru: Mining and Engineering v. 10, p. 262–266.

Rivas, S. W., 1968, Geología de la parte norte del Lago Titicaca: Servicio de Geología de Bolivia, Boletín 2, 88 p.

Rosenblum, S., 1968, Estudio geológico y mineralógico del Distrito de Sorata: Servicio de Geología de Bolivia, unpub. report 10-25.

Rosholt, J. N., Zartman, R. E., and Nkomo, I. T., 1973, Lead isotope systematics and uranium depletion in the Granite Mountains, Wyoming: Geological Society of America Bulletin, v. 84, p. 989–1002.

Ruiz, C., and Ericksen, G. E., 1962, Metallogenic provinces of Chile: Economic Geology, v. 57, p. 91–106.

Russo, A., and Rodrigo, G., 1965, Estratigrafía y paleogeografía del Grupo Puca en Bolivia: Boletín del Instituto Boliviana del Petróleo, v. 5, p. 5–51.

Schneider-Scherbina, A., 1962, Über metallogenetische Epochen Boliviens und den hybriden Charakter des sogenannten Zinn-Silber Formation: Geologische Jahresbericht, v. 81, p. 157–170.

Segerstrom, K., and Parker, R. L., 1959, Cuadrángulo Cerrillos, Província de Atacama: Chile, Instituto de Investigaciones Geológicas, Carta I, no. 1.

Shackleton, R. M., and others, 1979, Structure, metamorphism and geochronology of the Arequipa Massif of coastal Peru: Geological Society of London Journal, v. 136, p. 195–214.

Sillitoe, R. H., 1972, Relation of metal provinces in western America to subduction of oceanic lithosphere: Geological Society of America Bulletin, v. 83, p. 813–818.

——1976, Andean mineralization: A model for the metallogeny of convergent plate margins: Geological Association of Canada Special Paper, v. 14, p. 59–100.

Sinha, A. K., and Hart, S. R., 1972, A geochemical test of the subduction hypothesis for the generation of island arc magmas: Carnegie Institution of Washington Year Book 71, p. 309–312.

Stacey, J. S., and Kramers, J. D., 1975, Approximation of terrestrial lead isotopes evolution by a two-stage model: Earth and Planetary Science Letters, v. 26, p. 207–221.

Streckeisen, A., 1967, Classification and nomenclature of igneous rocks: Neues Jahrbuch fuer Mineralogie, Abhandlungen, v. 107, p. 144–240.

Thorpe, R. S., and Francis, P.W., 1979, Variations in Andean andesite compositions and their petrogenetic significance: Tectonophysics, v. 57, p. 53–70.

Tilton, G. R., 1979, Isotopic studies of Cenozoic Andean calc-alkaline rocks: Carnegie Institution of Washington Year Book 78, p. 298–304.

Tilton, G. R., and Barreiro, B. A., 1980, Origin of lead in Andean calc-alkaline lavas, southern Peru: Science, v. 210, p. 1245–1247.

Tosdal, R. M., 1978, The timing of the geomorphic and tectonic evolution of the southernmost Peruvian Andes [M.S. thesis]: Kingston, Queen's University, 138 p.

Tosdal, R. M., Farrar, E., and Clark, A. H., 1981, K-Ar geochronology of the Late Cenozoic volcanic rocks of the Cordillera Occidental, southernmost Peru: Journal of Volcanology and Geothermal Research (in press), v.

Turneaure, F. S., 1971, The Bolivian tin-silver province: Economic Geology, v. 66, p. 215–225.

Turneaure, F. S., and Welker, K. K., 1947, The ore deposits of the Eastern Andes of Bolivia: The

Cordillera Real: Economic Geology, v. 42, p. 595–625.

White, A.J.R., and Chappell, B. R., 1977, Ultrametamorphism and granitoid genesis: Tectonophysics, v. 43, p. 7–22.

Wolf, M., 1973, Zum Magmatismus der Cordillera de Potosi in Bolivien: Freiberger Forschungshefte, v. 275, 174 p.

Zentilli, M., 1974, Geological evolution and metallogenetic relationships in the Andes of Northern Chile between 26° and 29° South [Ph.D. thesis]: Kingston, Queen's University, 446 p.

Zentilli, M., and Dostal, J., 1977, Uranium in volcanic rocks from the Central Andes: Journal of Volcanology and Geothermal Research, v. 2, p. 251–258.

MANUSCRIPT RECEIVED BY THE SOCIETY NOVEMBER 12, 1980
MANUSCRIPT ACCEPTED DECEMBER 30, 1980

Printed in U.S.A.

Epilogue: Geostill reconsidered

Cyrus W. Field
E. Julius Dasch
*Department of Geology
Oregon State University
Corvallis, Oregon 97331*

It is held that the whole Cordilleran metallization is based on the gradual eastward movement of deep-seated magmas beginning with the original impulse from the Pacific. That impulse is no doubt caused by the reaction between the North American continent and the deep sima layers of the Pacific, but this question had better be left to the geophysicists. The theory held predicates the gradual movement of molten magma, or at least a gradual eastward progress of liquifaction, in a layer perhaps 60 kilometers below the surface.

Waldemar Lindgren, 1933

ABSTRACT

The best evidence for a direct relation between plate convergence and ore genesis is the close spatial occurrence of many ore deposits and active or relict convergent plate boundaries. Speculations abound on the source of the metals and the role of metalliferous sediment in the formation of these mineral deposits, but both remain intriguing and enigmatic problems.

INTRODUCTION

A central theme of the Nazca Plate Project has been an understanding and evaluation of the economic significance of metalliferous sediments found on the plate. In earlier planning it was hoped to evaluate also the possible relation of these sediments to the major ore deposit regime of the Andes. The Geostill concept was informally proposed in 1973 by Edward M. Davin, Program Manager of the International Decade of Ocean Exploration, to emphasize possible interrelationships between geologic processes of the ocean basins and those of the adjoining continents as implied by plate tectonic theory. The concept suggests that magmatism and metallogeny of the mid-ocean ridge systems and those of the continental cordilleras are genetically linked as a consequence of subduction and partial melting of oceanic crust.

In 1975, members of the Nazca group convened a National Science Foundation workshop—the Geostill Conference (Dasch, 1976)—to explore the possible relation of subduction and ore genesis, with the Nazca plate–South America block transition as a focus. This international conference, held at Salishan Lodge, Oregon, was attended by about 60 scientists with a wide variety of academic and industrial specialities. Partly as an outgrowth of the Nazca Plate Project and the Geostill Conference, members of Oregon State University, the National Oceanic and Atmospheric Administration, and the U.S. Geological Survey convened a meeting in 1980 of about 85 scientists—the Cascadia Conference—to explore the much more extensive geologic aspects of plate convergence, focusing largely on the ocean-land transition of the Pacific Northwest. Although the specific issue of ore deposits was not addressed, the conference (also held at Salishan Lodge) dealt extensively with magmatic processes associated with convergence and thus provided an update on parts of the Geostill concept.

As discussed by Woollard and Kulm (this volume), the Nazca Plate Project has undergone major shifts in purpose and direction. Because of these changes, continental aspects of the study, especially including studies of Andean igneous rocks and ore deposits, were greatly curtailed.

Nevertheless, as an addendum to the preceding group of papers on Andean convergence, we review very briefly our present understanding of the relation of ocean-land convergence, magmatism, and ore deposits.

GEOSTILL CONCEPT

The Geostill hypothesis suggests that oceanic crust with its veneer of metalliferous sediment is subducted, undergoes at least partial melting, and generates calc-alkalic magmas with contained metals and fluids that ascend and ultimately form the andesitic host rocks and associated hydrothermal ore deposits that are typical of cordilleran terranes. The subparallel juxtaposition of ocean trenches and nearby tectonically active cordilleras is well established, as is the spatial and temporal coincidence of magmatism and metallization in those cordilleras. Before plate tectonic theory, several perceptive investigators such as Lindgren (1915, 1933), Buddington (1933), Turneaure (1955), Lacy (1957), Moore (1959), and Gilluly (1965), among others, had previously noted systematic regional variations in the distribution, age, and chemistry of plutonic and volcanic rocks and/or of their associated ores and metals, within the cordilleras and with respect to the continental margins of western North and South America. The most prophetic statement relative to magmas and mineral deposits, and to the impending revolution in earth sciences, clearly was that of Lindgren (introductory quote; 1933).

It remained for Sillitoe (1972a, 1972b) to formulate the modern framework for cordilleran metallogeny. This model, which represents the Geostill concept in the boradest sense, integrates geologic, geochemical, and geophysical information for the ocean basins and continents into a unified hypothesis of magmatism and metallogeny. Moreover, Sillitoe (1972a, 1972b) explicitly identified metalliferous sediments and underlying basalts of the oceanic crust as the potential source of metals in the hydrothermal ore deposits associated with igneous rocks of the adjacent continental cordilleras.

For the present discussion of the Nazca plate and the Andean convergent zone, relevant geographic and geologic features are illustrated in Figure 1. Also shown are the ages (in millions of years) of oceanic crust as derived from the distribution of magnetic anomalies published by Herron (1972) and of the radiometrically dated porphyry-type deposits as reported by Hollister (1974), Stillitoe (1976, 1977), Oyarzun and Frutos (1980), and references cited therein.

Metalliferous sediments, the postulated source of metals in hydrothermal deposits of the Andes, are a geographically and stratigraphically appreciable component (3% to 30%) of the sedimentary veneer that overlies ocean-floor basalts of the Nazca plate. In contrast to more typical marine sediments, they are characterized by anomalous concentrations of transition metals, including Fe and Mn, as well as other elements (Bostrom and Peterson, 1966; Dymond and others, 1973, 1976; Bostrom and others, 1976; Field and others, 1976; Heath and Dymond, 1977; Dymond, this volume). Absolute concentrations of the metals are typically as follows: Fe, 5% to 20%, Mn, 1% to 5%; Co, 50 to 300 ppm; Cu, 100 to 1,200 ppm; Pb, 25 to 100 ppm; Mo, 5 to 200 ppm; Ni, 50 to 1,200 ppm; Ag, 1 to 4 ppm; and Zn, 50 to 500 ppm. For a further discussion of the economic potential of these sediments, see Field and others (this volume).

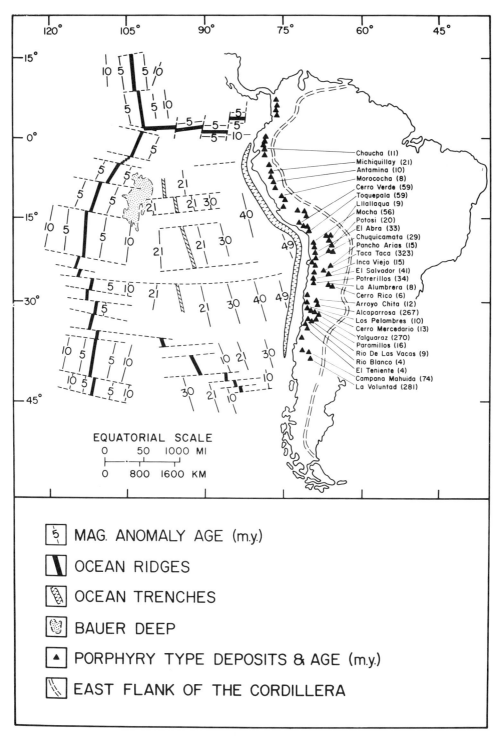

Figure 1. Nazca plate and Andean convergent zone. See text for sources of data.

ORE DEPOSITS OF THE ANDES

The Andean Cordilleras have long represented the epitome of mineral riches, regardless of one's professional viewpoint, be it conquistador, economic geologist, or international banker. The richer ores first were derived from small high-grade vein and replacement deposits of base and precious metals. However, the large low-grade porphyry-type deposits have become the dominant factor in the mining industry of the South American Cordilleras over the past 75 years. These deposits are of hydrothermal origin, as are the fissure and replacement deposits with which they may be associated, and are mined for Cu and recoverable amounts of Mo, Ag, and Au as by-products. For 1979, we estimate the total value of metallic mineral exports from Chile and Peru to equal or exceed $3.5 billion. Approximately 80% of this value was derived from the production of just 10 porphyry-type deposits now in operation. The potential for additional mines and the likelihood of their development are assured. For this discussion, Andean metallogeny as portrayed in Figure 1 is represented exclusively by the distribution of porphyry-type mines and prospects. The majority of investigators attribute the origin of porphyry-type deposits to magmatic hydrothermal processes. Support for this interpretation is the intimate spatial association with plutonic host rocks; contemporaneity of mineralization with the plutons and nearby volcanic country rocks; textural and paragenetic relations between mineral components of the host rocks and ores; and data from coexisting mineral assemblages, fluid inclusions, and isotopic geochemistry that collectively imply a high-temperature magmatic affiliation between host rocks and ores. Moreover, these deposits commonly occupy the deep central core or larger mining districts that exhibit well-defined concentric patterns of hydrothermally imposed chemical and mineralogic zonations, as documented by Lowell and Guilbert (1970), Rose (1970), and Gustafson and Hunt (1975). The deposits are localized in composite intrusions, stock to batholith in size, and normally in direct association with later quartz- and feldspar-bearing porphyritic phases. Comagmatic volcanic rocks may be in close proximity, and Sillitoe (1973) has suggested that the upper parts of these hydrothermally mineralized plutonic-volcanic complexes were marked by stratovolcanoes. Thus, the porphyry-type deposits are not simply illustrative of cordilleran metalogeny, they also serve to define in time and space the calc-alkaline magmatic arcs that are characteristic of these tectonically active regions.

THE PROBLEM

Magmatic processes leading to the formation of metalliferous sediments of the Nazca plate and of hydrothermal mineral deposits of the Andean Cordilleras can be amply documented, but the roles of subduction and partial melting of oceanic crust to form the igneous rocks and ores of the continental crust, as postulated by the Geostill concept, cannot at present be resolved. Although the rate and geometry of subduction may be reasonably established in some places, the question of *what* is subducted (as well as how much) is uncertain. In a few localities it appears that the sedimentary wedge [as well as a part of the oceanic crust (ophiolites)] may be accreted to continental margins quantitatively; in other places, for example, along parts of the western South America boundary, the most reasonable interpretation of the geophysical and sediment-budget information suggests that significant amounts of sediment have been subducted in the recent geologic past. Clearly, well-integrated geophysical information (seismic, gravity, magnetotelluric) is needed along selected profiles, both normal and longitudinal to the convergent margin. Suggestive of plate convergence and compatible with subduction are the apparently systematic longitudinal variations of diminishing age, increasing H_2O content, and increasing initial Sr-isotope ratios for plutonic and volcanic rocks from west to east across the cordillera (Clark and others, 1976). These temporal and compositional variations are accompanied by apparent longitudinal changes in the metallic constituents of associated hydrothermal deposits (from Fe, through Cu-Au-Mo and Cu-Pb-Zn-Ag, to Sn-W-Ag-Bi) as described by Sillitoe (1972a, 1976) and previous investigators. Armstrong and Hein (1973) have found that processes involving the subduction and recycling of crustal materials are consistent with a model for the evolution of Earth's crust based on computer simulation of evolving Pb- and Sr-isotope ratios with geologic time. This theoretical conclusion is supported by geophysical evidence for the subduction-related depletion or consumption of sediment from parts of the western margin of South America (Hussong and others, 1976; Kulm and others, 1977; Hussong and Wipperman, this volume). Despite this theoretical support, the majority of earth scientists continue to regard the proximity and subparallelism of volcanoes, plutons, and mineral deposits to active or relict convergent plate margins as the best evidence for subduction and for the possible genetic relations between oceanic and continental crust.

Recent summaries of Andean metallogeny by Hollister (1974), Noble (1976), Sillitoe (1976), Petersen (1979), and Oyarzun and Frutos (1980) have been undertaken with a focus toward plate tectonic theory as proposed by Sillitoe (1972a, 1972b). Although there remains a critical shortage of factual data because of the enormous size of this terrane, the elegantly detailed and systematic investigations at both local and regional scales by Gustafson and Hunt (1975) and Clark and others (1976) have clearly established the standard for future work. Most regional overviews have been directed primarily to the location and age of specific types of mineral deposits, their contained metals, and associated host rocks. This approach is likely to generate oversimplifications, if not errors, particularly in frontier areas such as the Andes, because of new exploration discoveries, revisions of prior geologic interpretations, and changes in economic, political, and technological factors. The distribution of porphyry-type deposits illustrated in Figure 1 documents a regionally extensive and temporally long-lasting (323 to 3 m.y. B.P.) succession of common magmatic-metallogenetic events throughout most of the South American cordillera. Although a few groups of near-similar radiometric ages suggest possibly discrete episodes of hydrothermal metallization, which are in accord with distinct pulses of magmatism reported by other investigators, the available data do not provide compelling evidence of clear-cut systematic regional variations or of locally coherent epochs, except over relatively short distances. The Bolivian Sn-bearing porphyries at Llallagua, Potosi, and four nearby occurrences (see Fig. 1) described by Sillitoe (1977) constitute a geographically distinct geochemical anomaly (see Lehmann and Pichler, 1980) because all other deposits shown are of the Cu-Mo affinity. Nonetheless, their location fits the general Cordilleran trend of the porphyry-type deposits, and the ages of these and other hydrothermal deposits of Sn (about 6 to 200 m.y.) in Bolivia, as reported by Evernden and others (1977), are temporally consistent with the overall range of metallogenetic events. With the exceptions of the Bolivian tin belt that extends into south-central Peru and of Early Cretaceous contact metasomatic deposits of Fe in the coastal provinces of northern Chile and southern Peru, hydrothermal deposits of other metals show less-pronounced geographic coherence. For example, W is widely distributed throughout the Central Andes from at least southern Bolivia to northern Peru. Moreover, its occurrence is not geologically unique because W may be found in association with polymetallic ores of the regionally extensive fissure and replacement deposits as well as with the porphyry-type deposits. In addition, the level of erosion may be an important factor in obscuring more systematic trends for the distribution of metals and the ages of related magmatism and mineralization. Because hydrothermal systems commonly exhibit a three-dimensional zonation of metals and minerals on a district-wide scale that extends upward and outward from a deep central core of porphyry-type mineralization, the level of erosional exposure may exert a profound influence on the constituents that compose a particular orebody; other intangibles such as economic constraints and geologic uncertainties may also affect current understanding.

Further work—particularly integrated geophysical, geologic, and geochemical investigations—must be directed to the most promising transects and profiles of the Andean convergent zone before an internally consistent model can be developed and applied to the search for mineral deposits. We regard the following considerations as most relevant to such a model: (1) Differences that may relate to the presence of accretionary wedges versus "missing" or depleted sections of sediment, or to the recycling of continental detritus at the plate boundary. (2) Variations that may relate to the abundance, proportions of lithic components, and chemistry of the sedimentary veneer on the adjoining oceanic plate. (3) Effects imposed by the tectonic regime such as fracture segmentation of the oceanic plate, rates of plate convergence, and the geometry, seismicity, and depth of the Benioff zone (see Barazangi and Isacks, 1976). (4) Calculations of mass balance for metals and other elements to determine if concentrations in rocks and ores of the cordillera are consistent with those of potential sources in crust and/or mantle. (5) Differences that may relate to variations in the thickness, age, rock type, and chemistry of continental crust and to possible contributions from pre-existing sources of metal in the upper continental crust (Petersen, 1979). (6) Construction of isochronous and time-transgressive geologic and plate tectonic sections to determine, in part, whether igneous rocks and ores were derived from single and continuously evolving or separate sources of magma.

With fairness to the skeptics, we conclude by noting that results of broadly integrated multi-disciplinary investigations may diminish rather than enhance, some facets of plate tectonic theory that are currently regarded with favor. For example, Noble (1976) has rejected the subduction hypothesis to account for metallogeny in the western cordillera of North and South America. He attributed metallogeny to processes operative on a primitive, heterogeneous source in the upper mantle, as deduced from discontinuities in the crustal occurrences of metals and from spatial and temporal distributions of igneous rocks, epigenetic ores, and structural features. In addition, Robyn and others (1980) have discounted the roles of both subduction and oceanic crust in the formation of late Tertiary calc-alkaline igneous rocks and associated porphyry-type deposits in the Cascade Range of Washington. However, it must be noted that the authors of papers in this volume concerning Andean magmatism and metallogeny are generally receptive of subduction processes at convergent plate boundaries. Although they present some evidence for components derived from continental crust, none of the data including Pb isotopes of igneous rocks and ores (Tilton and others, this volume) and trace elements and O and Sr isotopes of late Cenozoic calc-alkaline volcanics (James, this volume) dictate a significant contribution from marine sediments. Moreover, on the basis of experimental data for hydrous silicate melts, Burnham (this volume) specifically dismisses the necessity of an unusually metal-rich component, such as metalliferous sediments, to account for the sulfide mineral deposits formed by ore-bearing magmas.

REFERENCES CITED

Armstrong, R. L., and Hein, S. M, 1973, Computer simulation of Pb and Sr isotope evolution of the Earth's crust and upper mantle: Geochimica et Cosmochimica Acta, v. 37, p. 1–18.

Barazangi, M. and Isacks, B. L., 1976, Spatial distribution of earthquakes and subduction of the Nazca plate beneath South America: Geology, v. 4, p. 686–692.

Bostrom, K., and Peterson, M.N.A., 1966, Precipitates from hydrothermal exhalations on the East Pacific Rise: Economic Geology, v. 61, p. 1258–1265.

Bostrom, K., and others, 1976, Geochemistry and origin of East Pacific sediments sampled during DSDP Leg 34, in Yeats, R. S., Hart, S. R., and others, eds., Initial reports of the Deep Sea Drilling Project, Volume 34: Washington, D.C., U.S. Government Printing Office, p. 559–574.

Buddington, A. F., 1933, Correlation of kinds of igneous rocks with kinds of mineralization, in Ore deposits of the western states (Lindgren volume): New York, American Institute of Mining and Metallurgical Engineers, p. 350–385.

Burnham, C. W., 1981, Convergence and mineralization: Is there a relation?, in Kulm, L. D., and others, eds., Nazca plate: Crustal formation and Andean convergence: Geological Society of America Memoir 154 (this volume).

Clark, A. H., and others, 1976, Longitudinal variations in the metallogenic evolution of the Central Andes—A progress report, in Strong, D. F. ed., Metallogeny and plate tectonics: Geological Association of Canada Special Paper 14, p. 23–58.

Dasch, E. J., 1976, Nazca plate studied: Geotimes, v. 21, no. 4, p. 24–25.

Dymond, J., 1981, The geochemistry of Nazca plate surface sediments: An evaluation of hydrothermal, biogenic, detrital, and hydrogenous sources, in Kulm, L. D., and others, eds., Nazca plate: Crustal formation and Andean convergence: Geological Society of America Memoir 154 (this volume).

Dymond, J., and others, 1973, Origin of metalliferous sediments from the Pacific Ocean: Geological Society of America Bulletin, v. 84, p. 3355–3372.

Dymond, J., Corliss, J. B., and Stillinger, R., 1976, Chemical composition and metal accumulation rates of metalliferous sediments from sites 319, 320, and 321, in Yeats, R. S., Hart, S. R., and others, eds., Initial reports of the Deep Sea Drilling Project, Volume 34: Washington, D.C., U.S. Government Printing Office, p. 575–588.

Evernden, J. F., Kriz, S. J., and Cherroni, M., 1977, Potassium-argon ages of some Bolivian rocks: Economic Geology, v. 72, p. 1042–1061.

Field, C. W., and others, 1976, Metallogenesis in southeast Pacific ocean—Nazca Plate Project, in Halbouty, M. T., and others, eds., Circum-Pacific energy and mineral resources: American Association of Petroleum Geologists Memoir 25, p. 539–550.

Field, C. W., Wetherell, D. G., and Dasch, E. J., 1981, Economic appraisal of Nazca plate metalliferous sediments, in Kulm, L. D., and others, eds., Nazca plate: Crustal formation and Andean convergence: Geological Society of America Memoir 154 (this volume).

Gilluly, J., 1965, Volcanism, tectonism, and plutonism in the western United States: Geological Society of America Special Paper 80, 69 p.

Gustafson, L. B., and Hunt, J. P., 1975, The porphyry copper deposit at El Salvador, Chile: Economic Geology, v. 70, p. 857–912.

Heath, G. R., and Dymond, J., 1977, Genesis and transformation of metalliferous sediments from the East Pacific Rise, Bauer Deep, and Central Basin, northwest Nazca plate: Geological Society of America Bulletin, v. 88, p. 723–733.

Herron, E. M., 1972, Sea-floor spreading and the Cenozoic history of the east-central Pacific: Geological Society of America Bulletin, v. 83, p. 1671–1692.

Hollister, V. F., 1974, Regional characteristics of porphyry copper deposits of South America: Society of Mining Engineers, American Institute of Mining, Metallurgical and Petroleum Engineers, Transactions, v. 255, p. 45–53.

Hussong, D. M., and Wipperman, L. K., 1981, Vertical movement and tectonic erosion of the continental wall of the Peru-Chile Trench near 11°30′S, in Kulm, L. D., and others, eds., Nazca plate: Crustal formation and Andean convergence: Geological Society of America Memoir 154 (this volume).

Hussong, D. M., and others, 1976, Crustal structure of the Peru-Chile Trench—8°S–12°S latitude, in Sutton, G. H., and others, eds., The geophysics of the Pacific ocean basin and its margin: American Geophysical Union Monograph 19, p. 71–86.

James, D. E., 1981, Role of subducted continental material in the genesis of calc-alkaline volcanics of the central Andes, in Kulm, L. D., and others, eds., Nazca plate: Crustal formation and Andean convergence: Geological Society of America Memoir 154 (this volume).

Kulm, L.D., Schweller, W. J., and Masias, A., 1977, A preliminary analysis of the subduction processes along the Andean continental margin, 6° to 45° S, in Talwani, M., and Pitman, W. C., III, eds., Island arcs, deep sea trenches and back-arc basins: American Geophysical Union Maurice Ewing Series, v. 1, p. 285–301.

Lacy, W. C., 1957, Differentiation of igneous rocks and ore deposition in Peru: American Institute of Mining, Metallurgical and Petroleum Engineers Transactions, v. 208, p. 559–562.

Lehmann, B., and Pichler, H., 1980, Tin distribution in mid-Andean volcanic rocks: Mineralium Deposita, v. 15, p. 35–59.

Lindgren, W., 1915, The igneous geology of the Cordillera and its problems, in Problems of American geology: New Haven, Yale University Press, p. 234–286.

———1933, Differentiation and ore deposition, Cordilleran region of the United States, in Ore deposits of the western states (Lindgren volume): New York, American Institute of Mining and Metallurgical Engineers, p. 152–180.

Lowell, J. D., and Guilbert, J. M., 1970, Lateral and vertical alteration-mineralization zoning in porphyry ore deposits: Economic Geology, v. 65, p. 373–408.

Moore, J. G., 1959, The quartz diorite boundary line in the western United States: Journal of Geology, v. 67, p. 198–210.

Noble, J. A., 1976, Metallogenic provinces of the cordillera of western North and South America: Mineralium Deposita, v. 11, p. 219–233.

Oyarzun, M. J., and Frutos, J., 1980, Metallogenesis and porphyry deposits of the Andes, in Colloque C 1, Ressources Minerales: Bureau de Recherches Géologiques and Minières, Memoire 106, p. 50–62.

Petersen, U., 1979, Metallogenesis in South America—Progress and problems: Episodes, v. 1979, p. 3–11.

Robyn, T. L, Henage, L. F., and Hollister, V., 1980, Possible nonsubduction associated porphyry ore deposits, Pacific Northwest: Geological Society of America Abstracts with Programs, v. 12, p. 150 (manuscript to be published in symposium volume, Mineral deposits of the Pacific Northwest, 1981, as U.S. Geological Survey Open-File Report 81-355, p. 153–165).

Rose, A. W., 1970, Zonal relations of wallrock alteration and sulfide distribution at porphyry copper deposits: Economic Geology, v. 65, p. 920–936.

Sillitoe, R. H., 1972a, Relation of metal provinces in western America to subduction of oceanic lithosphere: Geological Society of America Bulletin, v. 83, p. 813–818.

———1972b, A plate tectonic model for the origin of porphyry copper deposits: Economic Geology, v. 67, p. 184–197.

———1973, The tops and bottoms of porphyry copper deposits: Economic Geology, v. 68, p. 799–815.

—— 1976, Andean mineralization—A model for the metallogeny of convergent plate margins, *in* Strong, D. F., ed., Metallogeny and plate tectonics: Geological Association of Canada Special Paper 14, p. 59–100.

—— 1977, Permo-Carboniferous, Upper Cretaceous, and Miocene porphyry copper-type mineralization in the Argentinian Andes: Economic Geology, v. 72, p. 99–109.

Tilton, G. R., and others, 1981, Isotopic compositions of Pb in Central Andean ore deposits, *in* Kulm, L. D., and others, eds., Nazca plate: Crustal formation and Andean convergence: Geological Society of America Memoir 154 (this volume).

Turneaure, F. S., 1955, Metellogenetic provinces and epochs: Economic Geology Fiftieth Anniversary Volume, Part I, p. 38–98.

Woollard, G., and Kulm, L. D., 1981, History of project: Organization of monograph, brief description of the region, *in* Kulm, L. D., and others, eds., Nazca plate: Crustal formation and Andean convergence: Geological Society of America Memoir 154 (this volume).

MANUSCRIPT RECEIVED BY THE SOCIETY NOVEMBER 12, 1980
MANUSCRIPT ACCEPTED DECEMBER 30, 1980

Typeset by CGE Services, Inc., Denver, Colorado 80231
Printed in the U.S.A. by Malloy Lithographing, Inc., Ann Arbor, Michigan 48106

WITHDRAWN